INTERNATIONAL UNION OF CRYSTALLOGRAPHY
TEXTS ON CRYSTALLOGRAPHY

IUCr BOOK SERIES COMMITTEE

E. N. Baker, *New Zealand*
J. Bernstein, *Israel*
G. R. Desiraju, *India*
A. M. Glazer, *UK*
J. R. Helliwell, *UK*
P. Paufler, *Germany*
H. Schenk (*Chairman*), *The Netherlands*

IUCr Monographs on Crystallography

1. *Accurate molecular structures*
   A. Domenicano and I. Hargittai, *editors*
2. *P. P. Ewald and his dynamical theory of X-ray diffraction*
   D. W. J. Cruickshank, H. J. Juretschke, and J. Kato, *editors*
3. *Electron diffraction techniques, Volume 1*
   J. M. Cowley, *editor*
4. *Electron diffraction techniques Volume 2*
   J. M. Cowley, *editor*
5. *The Rietveld method*
   R. A. Young, *editor*
6. *Introduction to crystallographic statistics*
   U. Shmueli and G. H. Weiss
7. *Crystallographic instrumentation*
   L. A. Aslanov, G. V. Fetisov, and G. A. K. Howard
8. *Direct phasing in crystallography*
   C. Giacovazzo
9. *The weak hydrogen bond*
   G. R. Desiraju and T. Steiner
10. *Defect and microstructure analysis by diffraction*
    R. L. Snyder, J. Fiala, and H. J. Bunge
11. *Dynamical theory of X-ray diffraction*
    A. Authier
12. *The chemical bond in inorganic chemistry*
    I. D. Brown
13. *Structure determination from powder diffraction data*
    W. I. F David, K. Shankland, L. B. McCusker, and Ch. Baerlocher, *editors*
14. *Polymorphism in molecular crystals*
    J. Bernstein

OXFORD
UNIVERSITY PRESS

Great Clarendon Street, Oxford OX2 6DP

Oxford University Press is a department of the University of Oxford.
It furthers the University's objective of excellence in research, scholarship,
and education by publishing worldwide in

Oxford New York

Auckland Cape Town Dar es Salaam Hong Kong Karachi
Kuala Lumpur Madrid Melbourne Mexico City Nairobi
New Delhi Shanghai Taipei Toronto

With offices in

Argentina Austria Brazil Chile Czech Republic France Greece
Guatemala Hungary Italy Japan Poland Portugal Singapore
South Korea Switzerland Thailand Turkey Ukraine Vietnam

Oxford is a registered trade mark of Oxford University Press
in the UK and in certain other countries

Published in the United States
by Oxford University Press Inc., New York

© Wai-Kee Li, Gong-Du Zhou and Thomas C.W. Mak

The moral rights of the authors have been asserted
Database right Oxford University Press (maker)

First published 2008

All rights reserved. No part of this publication may be reproduced,
stored in a retrieval system, or transmitted, in any form or by any means,
without the prior permission in writing of Oxford University Press,
or as expressly permitted by law, or under terms agreed with the appropriate
reprographics rights organization. Enquiries concerning reproduction
outside the scope of the above should be sent to the Rights Department,
Oxford University Press, at the address above

You must not circulate this book in any other binding or cover
and you must impose this same condition on any acquirer

British Library Cataloguing in Publication Data
Data available

Library of Congress Cataloging in Publication Data
Data available

Typeset by Newgen Imaging Systems (P) Ltd, Chennai, India
Printed in Great Britain
on acid-free paper by
Antony Rowe Ltd, Chippenham, Wiltshire

ISBN 978–0–19–921694–9
ISBN 978–0–19–921695–6

1 3 5 7 9 10 8 6 4 2

Preface

The original edition of this book, written in Chinese for students in mainland China, was published in 2001 jointly by Chinese University Press and Peking University Press. The second edition was published by Peking University Press in 2006. During the preparation of the present English edition, we took the opportunity to correct some errors and to include updated material based on the recent literature.

The book is derived from lecture notes used in various courses taught by the three authors at The Chinese University of Hong Kong and Peking University. The course titles include Chemical Bonding, Structural Chemistry, Structure and Properties of Matter, Advanced Inorganic Chemistry, Quantum Chemistry, Group Theory, and Chemical Crystallography. In total, the authors have accumulated over 100 man-years of teaching at the two universities. The book is designed as a text for senior undergraduates and beginning postgraduate students who need a deeper yet friendly exposure to the bonding and structure of chemical compounds.

Structural chemistry is a branch of science that attempts to achieve a comprehensive understanding of the physical and chemical properties of various compounds from a microscopic viewpoint. In building up the theoretical framework, two main lines of development—electronic and spatial—are followed. In this book, both aspects and the interplay between them are stressed. It is hoped that our presentation will provide students with sufficient background and factual knowledge so that they can comprehend the exciting recent advances in chemical research and be motivated to pursue careers in universities and research institutes.

This book is composed of three Parts. Part I, consisting of the first five chapters, reviews the basic theories of chemical bonding, beginning with a brief introduction to quantum mechanics, which is followed by successive chapters on atomic structure, bonding in molecules, and bonding in solids. Inclusion of the concluding chapter on computational chemistry reflects its increasing importance as an accessible and valuable tool in fundamental research.

Part II of the book, again consisting of five chapters, discusses the symmetry concept and its importance in structural chemistry. Chapter 6 introduces students to symmetry point groups and the rudiments of group theory without delving into intricate mathematical details. Chapter 7 covers group theory's most common chemical applications, including molecular orbital and hybridization theories, molecular vibrations, and selection rules. Chapter 8 utilizes the symmetry concept to discuss the bonding in coordination complexes. The final two chapters address the formal description of symmetry in the crystalline state and the structures of basic inorganic crystals and some technologically important materials.

Part III constitutes about half of the book. It offers a succinct description of the structural chemistry of the elements in the Periodic Table. Specifically, the main-group elements (including noble gases) are covered in the first seven chapters, while the last three deal with the rare-earth elements, transition-metal clusters, and supramolecular systems, respectively. In all these chapters, selected examples illustrating interesting aspects of structure and bonding, generalizations of structural trends, and highlights from the recent literature are discussed in the light of the theoretical principles presented in Parts I and II.

In writing the first two Parts, we deliberately avoided the use of rigorous mathematics in treating various theoretical topics. Instead, newly introduced concepts are illustrated with examples based on real chemical compounds or practical applications. Furthermore, in our selective compilation of material for presentation in Part III, we strive to make use of the most up-to-date crystallographic data to expound current research trends in structural inorganic chemistry.

On the ground of hands-on experience, we freely make use of our own research results as examples in the presentation of relevant topics throughout the book. Certainly there is no implication whatsoever that they are particularly important or preferable to alternative choices.

We faced a dilemma in choosing a fitting title for the book and eventually settled on the present one. The adjective "inorganic" is used in a broad sense as the book covers compounds of representative elements (including carbon) in the Periodic Table, organometallics, metal–metal bonded systems, coordination polymers, host–guest compounds and supramolecular assemblies. Our endeavor attempts to convey the message that inorganic synthesis is inherently less organized than organic synthesis, and serendipitous discoveries are being made from time to time. Hopefully, discussion of bonding and structure on the basis of X-ray structural data will help to promote a better understanding of modern chemical crystallography among the general scientific community.

Many people have contributed to the completion of this book. Our past and present colleagues at The Chinese University of Hong Kong and Peking University have helped us in various ways during our teaching careers. Additionally, generations of students have left their imprint in the lecture notes on which this book is based. Their inquisitive feedback and suggestions for improvement have proved to be invaluable. Of course, we are solely responsible for deficiencies and errors and would most appreciate receiving comments and criticisms from prospective readers.

The publication of the original Chinese edition was financed by a special grant from Chinese University Press, to which we are greatly indebted. We dedicate this book to our mentors: S.M. Blinder, You-Qi Tang, James Trotter and the late Hson-Mou Chang. Last but not least, we express our gratitude to

Preface

our wives, Oi-Ching Chong Li, Zhi-Fen Liu, and Gloria Sau-Hing Mak, for their sacrifice, encouragement and unflinching support.

Wai-Kee Li
The Chinese University of Hong Kong

Gong-Du Zhou
Peking University

Thomas Chung Wai Mak
The Chinese University of Hong Kong

Contents

List of Contributors		xxi
Part I	**Fundamentals of Bonding Theory**	1

1 Introduction to Quantum Theory   3

1.1	Dual nature of light and matter	4
1.2	Uncertainty principle and probability concept	5
1.3	Electronic wavefunction and probability density function	6
1.4	Electronic wave equation: the Schrödinger equation	10
1.5	Simple applications of the Schrödinger equation	13
	1.5.1 Particle in a one-dimensional box	13
	1.5.2 Particle in a three-dimensional box	17
	1.5.3 Particle in a ring	21
	1.5.4 Particle in a triangle	23
References		28

2 The Electronic Structure of Atoms   29

2.1	The Hydrogen Atom	29
	2.1.1 Schrödinger equation for the hydrogen atom	29
	2.1.2 Angular functions of the hydrogen atom	31
	2.1.3 Radial functions and total wavefunctions of the hydrogen atom	34
	2.1.4 Relative sizes of hydrogenic orbitals and the probability criterion	38
	2.1.5 Energy levels of hydrogenic orbitals; summary	42
2.2	The helium atom and the Pauli exclusion principle	42
	2.2.1 The helium atom: ground state	43
	2.2.2 Determinantal wavefunction and the Pauli Exclusion Principle	48
	2.2.3 The helium atom: the $1s^1 2s^1$ configuration	51
2.3	Many-electron atoms: electronic configuration and spectroscopic terms	54
	2.3.1 Many-electron atoms	54
	2.3.2 Ground electronic configuration for many-electron atoms	55
	2.3.3 Spectroscopic terms arising from a given electronic configuration	56
	2.3.4 Hund's rules on spectroscopic terms	60
	2.3.5 j–j Coupling	62
2.4	Atomic properties	64
	2.4.1 Ionization energy and electron affinity	64

		2.4.2	Electronegativity: the spectroscopic scale	67
		2.4.3	Relativistic effects on the properties of the elements	71
	References			76

3 Covalent Bonding in Molecules — 77

3.1 The hydrogen molecular ion: bonding and antibonding molecular orbitals — 77
    3.1.1 The variational method — 77
    3.1.2 The hydrogen molecular ion: energy consideration — 79
    3.1.3 The hydrogen molecular ion: wavefunctions — 82
    3.1.4 Essentials of molecular orbital theory — 84

3.2 The hydrogen molecule: molecular orbital and valence bond treatments — 85
    3.2.1 Molecular orbital theory for H_2 — 85
    3.2.2 Valence bond treatment of H_2 — 86
    3.2.3 Equivalence of the molecular orbital and valence bond models — 89

3.3 Diatomic molecules — 91
    3.3.1 Homonuclear diatomic molecules — 92
    3.3.2 Heteronuclear diatomic molecules — 96

3.4 Linear triatomic molecules and sp^n hybridization schemes — 99
    3.4.1 Beryllium hydride, BeH_2 — 99
    3.4.2 Hybridization scheme for linear triatomic molecules — 100
    3.4.3 Carbon dioxide, CO_2 — 101
    3.4.4 The sp^n ($n = 1-3$) hybrid orbitals — 104
    3.4.5 Covalent radii — 109

3.5 Hückel molecular orbital theory for conjugated polyenes — 110
    3.5.1 Hückel molecular orbital theory and its application to ethylene and butadiene — 110
    3.5.2 Predicting the course of a reaction by considering the symmetry of the wavefunction — 113

References — 116

4 Chemical Bonding in Condensed Phases — 118

4.1 Chemical classification of solids — 118
4.2 Ionic bond — 121
    4.2.1 Ionic size: crystal radii of ions — 121
    4.2.2 Lattice energies of ionic compounds — 124
    4.2.3 Ionic liquids — 126
4.3 Metallic bonding and band theory — 128
    4.3.1 Chemical approach based on molecular orbital theory — 128
    4.3.2 Semiconductors — 130
    4.3.3 Variation of structure types of 4d and 5d transition metals — 131
    4.3.4 Metallic radii — 132
    4.3.5 Melting and boiling points and standard enthalpies of atomization of the metallic elements — 133

4.4	Van der Waals interactions		134
	4.4.1 Physical origins of van der Waals interactions		135
	4.4.2 Intermolecular potentials and van der Waals radii		138
References			139

5 Computational Chemistry — 140

- 5.1 Introduction — 140
- 5.2 Semi-empirical and *ab initio* methods — 141
- 5.3 Basis sets — 142
  - 5.3.1 Minimal basis set — 142
  - 5.3.2 Double zeta and split valence basis sets — 143
  - 5.3.3 Polarization functions and diffuse functions — 143
- 5.4 Electron correlation — 144
  - 5.4.1 Configuration interaction — 145
  - 5.4.2 Perturbation methods — 146
  - 5.4.3 Coupled-cluster and quadratic configuration interaction methods — 146
- 5.5 Density functional theory — 147
- 5.6 Performance of theoretical methods — 148
- 5.7 Composite methods — 151
- 5.8 Illustrative examples — 152
  - 5.8.1 A stable argon compound: HArF — 152
  - 5.8.2 An all-metal aromatic species: Al_4^{2-} — 154
  - 5.8.3 A novel pentanitrogen cation: N_5^+ — 156
  - 5.8.4 Linear triatomics with noble gas–metal bonds — 158
- 5.9 Software packages — 162
- References — 163

Part II Symmetry in Chemistry — 165

6 Symmetry and Elements of Group Theory — 167

- 6.1 Symmetry elements and symmetry operations — 167
  - 6.1.1 Proper rotation axis C_n — 167
  - 6.1.2 Symmetry plane σ — 168
  - 6.1.3 Inversion center i — 169
  - 6.1.4 Improper rotation axis S_n — 169
  - 6.1.5 Identity element E — 170
- 6.2 Molecular point groups — 170
  - 6.2.1 Classification of point groups — 170
  - 6.2.2 Identifying point groups — 178
  - 6.2.3 Dipole moment and optical activity — 179
- 6.3 Character tables — 180
- 6.4 The direct product and its use — 185
  - 6.4.1 The direct product — 185
  - 6.4.2 Identifying non-zero integrals and selection rules in spectroscopy — 187
  - 6.4.3 Molecular term symbols — 189
- References — 193
- Appendix 6.1 Character tables of point groups — 195

7 Application of Group Theory to Molecular Systems — 213

- 7.1 Molecular orbital theory — 213
  - 7.1.1 AH_n ($n = 2$–6) molecules — 214
  - 7.1.2 Hückel theory for cyclic conjugated polyenes — 221
  - 7.1.3 Cyclic systems involving d orbitals — 227
  - 7.1.4 Linear combinations of ligand orbitals for AL_4 molecules with T_d symmetry — 228
- 7.2 Construction of hybrid orbitals — 232
  - 7.2.1 Hybridization schemes — 232
  - 7.2.2 Relationship between the coefficient matrices for the hybrid and molecular orbital wavefunctions — 233
  - 7.2.3 Hybrids with d-orbital participation — 234
- 7.3 Molecular vibrations — 236
  - 7.3.1 The symmetries and activities of the normal modes — 236
  - 7.3.2 Some illustrative examples — 239
  - 7.3.3 CO stretch in metal carbonyl complexes — 246
  - 7.3.4 Linear molecules — 252
  - 7.3.5 Benzene and related molecules — 254
- References — 259

8 Bonding in Coordination Compounds — 261

- 8.1 Crystal field theory: d-orbital splitting in octahedral and tetrahedral complexes — 261
- 8.2 Spectrochemical series, high spin and low spin complexes — 263
- 8.3 Jahn–Teller distortion and other crystal fields — 265
- 8.4 Octahedral crystal field splitting of spectroscopic terms — 267
- 8.5 Energy level diagrams for octahedral complexes — 268
  - 8.5.1 Orgel diagrams — 268
  - 8.5.2 Intensities and band widths of d–d spectral lines — 271
  - 8.5.3 Tanabe–Sugano diagrams — 274
  - 8.5.4 Electronic spectra of selected metal complexes — 274
- 8.6 Correlation of weak and strong field approximations — 279
- 8.7 Spin–orbit interaction in complexes: the double group — 280
- 8.8 Molecular orbital theory for octahedral complexes — 282
  - 8.8.1 σ bonding in octahedral complexes — 283
  - 8.8.2 Octahedral complexes with π bonding — 285
  - 8.8.3 The eighteen-electron rule — 288
- 8.9 Electronic spectra of square planar complexes — 289
  - 8.9.1 Energy level scheme for square-planar complexes — 289
  - 8.9.2 Electronic spectra of square-planar halides and cyanides — 291
- 8.10 Vibronic interaction in transition metal complexes — 294
- 8.11 The 4f orbitals and their crystal field splitting patterns — 295
  - 8.11.1 The shapes of the 4f orbitals — 295
  - 8.11.2 Crystal field splitting patterns of the 4f orbitals — 297
- References — 298

9 Symmetry in Crystals — 300

- 9.1 The crystal as a geometrical entity — 300
  - 9.1.1 Interfacial angles — 300
  - 9.1.2 Miller indices — 301
  - 9.1.3 Thirty-two crystal classes (crystallographic point groups) — 301
  - 9.1.4 Stereographic projection — 303
  - 9.1.5 Acentric crystalline materials — 304
- 9.2 The crystal as a lattice — 307
  - 9.2.1 The lattice concept — 307
  - 9.2.2 Unit cell — 307
  - 9.2.3 Fourteen Bravais lattices — 309
  - 9.2.4 Seven crystal systems — 309
  - 9.2.5 Unit cell transformation — 310
- 9.3 Space groups — 312
  - 9.3.1 Screw axes and glide planes — 312
  - 9.3.2 Graphic symbols for symmetry elements — 313
  - 9.3.3 Hermann–Mauguin space-group symbols — 316
  - 9.3.4 International Tables for Crystallography — 316
  - 9.3.5 Coordinates of equipoints — 317
  - 9.3.6 Space group diagrams — 320
  - 9.3.7 Information on some commonly occurring space groups — 323
  - 9.3.8 Using the International Tables — 323
- 9.4 Determination of space groups — 323
  - 9.4.1 Friedel's law — 323
  - 9.4.2 Laue classes — 325
  - 9.4.3 Deduction of lattice centering and translational symmetry elements from systemic absences — 328
- 9.5 Selected space groups and examples of crystal structures — 333
  - 9.5.1 Molecular symmetry and site symmetry — 333
  - 9.5.2 Symmetry deductions: assignment of atoms and groups to equivalent positions — 333
  - 9.5.3 Racemic crystal and conglomerate — 338
  - 9.5.4 Occurrence of space groups in crystals — 338
- 9.6 Application of space group symmetry in crystal structure determination — 339
  - 9.6.1 Triclinic and monoclinic space groups — 340
  - 9.6.2 Orthorhombic space groups — 343
  - 9.6.3 Tetragonal space groups — 345
  - 9.6.4 Trigonal and rhombohedral space groups — 347
  - 9.6.5 Hexagonal space groups — 350
  - 9.6.6 Cubic space groups — 353
- References — 362

10 Basic Inorganic Crystal Structures and Materials — 364

- 10.1 Cubic closest packing and related structures — 364
  - 10.1.1 Cubic closest packing (ccp) — 364

	10.1.2 Structure of NaCl and related compounds	366
	10.1.3 Structure of CaF$_2$ and related compounds	370
	10.1.4 Structure of cubic zinc sulfide	371
	10.1.5 Structure of spinel and related compounds	373
10.2	Hexagonal closest packing and related structures	375
	10.2.1 Hexagonal closest packing (hcp)	375
	10.2.2 Structure of hexagonal zinc sulfide	376
	10.2.3 Structure of NiAs and related compounds	376
	10.2.4 Structure of CdI$_2$ and related compounds	377
	10.2.5 Structure of α-Al$_2$O$_3$	379
	10.2.6 Structure of rutile	380
10.3	Body-centered cubic packing and related structures	381
	10.3.1 Body-centered cubic packing (bcp)	381
	10.3.2 Structure and properties of α-AgI	383
	10.3.3 Structure of CsCl and related compounds	384
10.4	Perovskite and related compounds	385
	10.4.1 Structure of perovskite	385
	10.4.2 Crystal structure of BaTiO$_3$	388
	10.4.3 Superconductors of perovskite structure type	389
	10.4.4 ReO$_3$ and related compounds	390
10.5	Hard magnetic materials	391
	10.5.1 Survey of magnetic materials	391
	10.5.2 Structure of SmCo$_5$ and Sm$_2$Co$_{17}$	393
	10.5.3 Structure of Nd$_2$Fe$_{14}$B	393
References		395

Part III Structural Chemistry of Selected Elements 397

11 Structural Chemistry of Hydrogen 399

11.1	The bonding types of hydrogen	399
11.2	Hydrogen bond	403
	11.2.1 Nature and geometry of the hydrogen bond	403
	11.2.2 The strength of hydrogen bonds	405
	11.2.3 Symmetrical hydrogen bond	406
	11.2.4 Hydrogen bonds in organometallic compounds	408
	11.2.5 The universality and importance of hydrogen bonds	409
11.3	Non-conventional hydrogen bonds	411
	11.3.1 X–H$\cdots\pi$ hydrogen bond	411
	11.3.2 Transition metal hydrogen bond X–H$\cdots$M	412
	11.3.3 Dihydrogen bond X–H$\cdots$H–E	413
	11.3.4 Inverse hydrogen bond	415
11.4	Hydride complexes	416
	11.4.1 Covalent metal hydride complexes	417
	11.4.2 Interstitial and high-coordinate hydride complexes	419
11.5	Molecular hydrogen (H$_2$) coordination compounds and σ-bond complexes	422
	11.5.1 Structure and bonding of H$_2$ coordination compounds	422
	11.5.2 X–H σ-bond coordination metal complexes	424

		11.5.3 Agostic bond	425
		11.5.4 Structure and bonding of σ complexes	428
	References		430

12 Structural Chemistry of Alkali and Alkaline-Earth Metals — 432

- 12.1 Survey of the alkali metals — 432
- 12.2 Structure and bonding in inorganic alkali metal compounds — 433
  - 12.2.1 Alkali metal oxides — 433
  - 12.2.2 Lithium nitride — 435
  - 12.2.3 Inorganic alkali metal complexes — 436
- 12.3 Structure and bonding in organic alkali metal compounds — 442
  - 12.3.1 Methyllithium and related compounds — 442
  - 12.3.2 π-Complexes of lithium — 443
  - 12.3.3 π-Complexes of sodium and potassium — 445
- 12.4 Alkalides and electrides — 446
  - 12.4.1 Alkalides — 446
  - 12.4.2 Electrides — 447
- 12.5 Survey of the alkaline-earth metals — 449
- 12.6 Structure of compounds of alkaline-earth metals — 450
  - 12.6.1 Group 2 metal complexes — 450
  - 12.6.2 Group 2 metal nitrides — 451
  - 12.6.3 Group 2 low-valent oxides and nitrides — 452
- 12.7 Organometallic compounds of group 2 elements — 454
  - 12.7.1 Polymeric chains — 454
  - 12.7.2 Grignard reagents — 454
  - 12.7.3 Alkaline-earth metallocenes — 455
- 12.8 Alkali and alkaline-earth metal complexes with inverse crown structures — 456
- References — 458

13 Structural Chemistry of Group 13 Elements — 460

- 13.1 Survey of the group 13 elements — 460
- 13.2 Elemental Boron — 461
- 13.3 Borides — 464
  - 13.3.1 Metal borides — 464
  - 13.3.2 Non-metal borides — 467
- 13.4 Boranes and carboranes — 470
  - 13.4.1 Molecular structure and bonding — 470
  - 13.4.2 Bond valence in molecular skeletons — 472
  - 13.4.3 Wade's rules — 473
  - 13.4.4 Chemical bonding in *closo*-boranes — 475
  - 13.4.5 Chemical bonding in *nido*- and *arachno*-boranes — 477
  - 13.4.6 Electron-counting scheme for macropolyhedral boranes: *mno* rule — 479
  - 13.4.7 Electronic structure of β-rhombohedral boron — 481
  - 13.4.8 Persubstituted derivatives of icosahedral borane $B_{12}H_{12}^{2-}$ — 482
  - 13.4.9 Boranes and carboranes as ligands — 483

13.4.10 Carborane skeletons beyond the icosahedron	485
13.5 Boric acid and borates	486
13.5.1 Boric acid	486
13.5.2 Structure of borates	487
13.6 Organometallic compounds of group 13 elements	490
13.6.1 Compounds with bridged structure	490
13.6.2 Compounds with π bonding	491
13.6.3 Compounds containing M–M bonds	492
13.6.4 Linear catenation in heavier group 13 elements	494
13.7 Structure of naked anionic metalloid clusters	494
13.7.1 Structure of $Ga_{84}[N(SiMe_3)_2]_{20}^{4-}$	495
13.7.2 Structure of NaTl	495
13.7.3 Naked Tl_n^{m-} anion clusters	496
References	498

14 Structural Chemistry of Group 14 Elements   500

14.1 Allotropic modifications of carbon	500
14.1.1 Diamond	500
14.1.2 Graphite	501
14.1.3 Fullerenes	502
14.1.4 Amorphous carbon	506
14.1.5 Carbon nanotubes	507
14.2 Compounds of carbon	509
14.2.1 Aliphatic compounds	509
14.2.2 Aromatic compounds	510
14.2.3 Fullerenic compounds	511
14.3 Bonding in carbon compounds	517
14.3.1 Types of bonds formed by the carbon atom	517
14.3.2 Coordination numbers of carbon	520
14.3.3 Bond lengths of C–C and C–X bonds	520
14.3.4 Factors influencing bond lengths	520
14.3.5 Abnormal carbon–carbon single bonds	524
14.3.6 Complexes containing a naked carbon atom	527
14.3.7 Complexes containing naked dicarbon ligands	529
14.4 Structural chemistry of silicon	533
14.4.1 Comparison of silicon and carbon	533
14.4.2 Metal silicides	534
14.4.3 Stereochemistry of silicon	535
14.4.4 Silicates	540
14.5 Structures of halides and oxides of heavier group 14 elements	544
14.5.1 Subvalent halides	544
14.5.2 Oxides of Ge, Sn, and Pb	546
14.6 Polyatomic anions of Ge, Sn, and Pb	547
14.7 Organometallic compounds of heavier group 14 elements	549
14.7.1 Cyclopentadienyl complexes	549
14.7.2 Sila- and germa-aromatic compounds	550

14.7.3	Cluster complexes of Ge, Sn, and Pb	551
14.7.4	Metalloid clusters of Sn	553
14.7.5	Donor–acceptor complexes of Ge, Sn and Pb	554
References		557

15 Structural Chemistry of Group 15 Elements 561

15.1 The N_2 molecule, all-nitrogen ions and dinitrogen complexes 561
    15.1.1 The N_2 molecule 561
    15.1.2 Nitrogen ions and catenation of nitrogen 561
    15.1.3 Dinitrogen complexes 564
15.2 Compounds of nitrogen 569
    15.2.1 Molecular nitrogen oxides 569
    15.2.2 Oxo-acids and oxo-ions of nitrogen 575
    15.2.3 Nitrogen hydrides 578
15.3 Structure and bonding of elemental phosphorus and P_n groups 579
    15.3.1 Elemental phosphorus 579
    15.3.2 Polyphosphide anions 581
    15.3.3 Structure of P_n groups in transition-metal complexes 581
    15.3.4 Bond valence in P_n species 584
15.4 Bonding type and coordination geometry of phosphorus 586
    15.4.1 Potential bonding types of phosphorus 586
    15.4.2 Coordination geometries of phosphorus 587
15.5 Structure and bonding in phosphorus–nitrogen and phosphorus–carbon compounds 590
    15.5.1 Types of P–N bonds 590
    15.5.2 Phosphazanes 591
    15.5.3 Phosphazenes 593
    15.5.4 Bonding types in phosphorus–carbon compounds 596
    15.5.5 π-Coordination complexes of phosphorus–carbon compounds 600
15.6 Structural chemistry of As, Sb, and Bi 602
    15.6.1 Stereochemistry of As, Sb, and Bi 602
    15.6.2 Metal–metal bonds and clusters 605
    15.6.3 Intermolecular interactions in organoantimony and organobismuth compounds 607
References 608

16 Structural Chemistry of Group 16 Elements 610

16.1 Dioxygen and ozone 610
    16.1.1 Structure and properties of dioxygen 610
    16.1.2 Crystalline phases of solid oxygen 612
    16.1.3 Dioxygen-related species and hydrogen peroxide 613
    16.1.4 Ozone 614
16.2 Oxygen and dioxygen metal complexes 616
    16.2.1 Coordination modes of oxygen in metal–oxo complexes 616
    16.2.2 Ligation modes of dioxygen in metal complexes 616
    16.2.3 Biological dioxygen carriers 618

16.3 Structure of water and ices — 619
    16.3.1 Water in the gas phase — 620
    16.3.2 Water in the solid phase: ices — 620
    16.3.3 Structural model of liquid water — 623
    16.3.4 Protonated water species, H_3O^+ and $H_5O_2^+$ — 627
16.4 Allotropes of sulfur and polyatomic sulfur species — 627
    16.4.1 Allotropes of sulfur — 626
    16.4.2 Polyatomic sulfur ions — 630
16.5 Sulfide anions as ligands in metal complexes — 631
    16.5.1 Monosulfide S^{2-} — 631
    16.5.2 Disulfide S_2^{2-} — 632
    16.5.3 Polysulfides S_n^{2-} — 632
16.6 Oxides and oxoacids of sulfur — 634
    16.6.1 Oxides of sulfur — 634
    16.6.2 Oxoacids of sulfur — 637
16.7 Sulfur–nitrogen compounds — 641
    16.7.1 Tetrasulfur tetranitride, S_4N_4 — 641
    16.7.2 S_2N_2 and $(SN)_x$ — 642
    16.7.3 Cyclic sulfur–nitrogen compounds — 643
16.8 Structural chemistry of selenium and tellurium — 644
    16.8.1 Allotropes of selenium and tellurium — 644
    16.8.2 Polyatomic cations and anions of selenium and tellurium — 644
    16.8.3 Stereochemistry of selenium and tellurium — 649
References — 652

17 Structural Chemistry of Group 17 and Group 18 Elements — 654

17.1 Elemental halogens — 654
    17.1.1 Crystal structures of the elemental halogens — 654
    17.1.2 Homopolyatomic halogen anions — 654
    17.1.3 Homopolyatomic halogen cations — 656
17.2 Interhalogen compounds and ions — 657
    17.2.1 Neutral interhalogen compounds — 657
    17.2.2 Interhalogen ions — 659
17.3 Charge-transfer complexes of halogens — 660
17.4 Halogen oxides and oxo compounds — 662
    17.4.1 Binary halogen oxides — 662
    17.4.2 Ternary halogen oxides — 664
    17.4.3 Halogen oxoacids and anions — 666
    17.4.4 Structural features of polycoordinate iodine compounds — 668
17.5 Structural chemistry of noble gas compounds — 670
    17.5.1 General survey — 670
    17.5.2 Stereochemistry of xenon — 671
    17.5.3 Chemical bonding in xenon fluorides — 672
    17.5.4 Structures of some inorganic xenon compounds — 674
    17.5.5 Structures of some organoxenon compounds — 677

	17.5.6 Gold–xenon complexes	678
	17.5.7 Krypton compounds	679
References		680

18 Structural Chemistry of Rare-Earth Elements — 682

18.1 Chemistry of rare-earth metals — 682
    18.1.1 Trends in metallic and ionic radii: lanthanide contraction — 682
    18.1.2 Crystal structures of the rare-earth metals — 683
    18.1.3 Oxidation states — 684
    18.1.4 Term symbols and electronic spectroscopy — 685
    18.1.5 Magnetic properties — 687
18.2 Structure of oxides and halides of rare-earth elements — 688
    18.2.1 Oxides — 688
    18.2.2 Halides — 689
18.3 Coordination geometry of rare-earth cations — 690
18.4 Organometallic compounds of rare-earth elements — 694
    18.4.1 Cyclopentadienyl rare-earth complexes — 694
    18.4.2 Biscyclopentadienyl complexes — 696
    18.4.3 Benzene and cyclooctatetraenyl rare-earth complexes — 697
    18.4.4 Rare-earth complexes with other organic ligands — 697
18.5 Reduction chemistry in oxidation state +2 — 699
    18.5.1 Samarium(II) iodide — 699
    18.5.2 Decamethylsamarocene — 700
References — 701

19 Metal–Metal Bonds and Transition-Metal Clusters — 703

19.1 Bond valence and bond number of transition-metal clusters — 703
19.2 Dinuclear complexes containing metal–metal bonds — 705
    19.2.1 Dinuclear transition-metal complexes conforming to the 18-electron rule — 707
    19.2.2 Quadruple bonds — 708
    19.2.3 Bond valence of metal–metal bond — 711
    19.2.4 Quintuple bonding in a dimetal complex — 712
19.3 Clusters with three or four transition-metal atoms — 713
    19.3.1 Trinuclear clusters — 713
    19.3.2 Tetranuclear clusters — 714
19.4 Clusters with more than four transition-metal atoms — 715
    19.4.1 Pentanuclear clusters — 715
    19.4.2 Hexanuclear clusters — 715
    19.4.3 Clusters with seven or more transition-metal atoms — 717
    19.4.4 Anionic carbonyl clusters with interstitial main-group atoms — 718
19.5 Iso-bond valence and iso-structural series — 719
19.6 Selected topics in metal–metal interactions — 721
    19.6.1 Aurophilicity — 721
    19.6.2 Argentophilicity and mixed metal complexes — 724

	19.6.3 Metal string molecules	724
	19.6.4 Metal-based infinite chains and networks	729
References		731

20 Supramolecular Structural Chemistry 733

- 20.1 Introduction 733
  - 20.1.1 Intermolecular interactions 733
  - 20.1.2 Molecular recognition 734
  - 20.1.3 Self-assembly 735
  - 20.1.4 Crystal engineering 735
  - 20.1.5 Supramolecular synthon 737
- 20.2 Hydrogen-bond directed assembly 738
  - 20.2.1 Supramolecular architectures based on the carboxylic acid dimer synthon 740
  - 20.2.2 Graph-set encoding of hydrogen-bonding pattern 742
  - 20.2.3 Supramolecular construction based on complementary hydrogen bonding between heterocycles 744
  - 20.2.4 Hydrogen-bonded networks exhibiting the supramolecular rosette pattern 744
- 20.3 Supramolecular chemistry of the coordination bond 752
  - 20.3.1 Principal types of supermolecules 752
  - 20.3.2 Some examples of inorganic supermolecules 753
  - 20.3.3 Synthetic strategies for inorganic supermolecules and coordination polymers 757
  - 20.3.4 Molecular polygons and tubes 760
  - 20.3.5 Molecular polyhedra 762
- 20.4 Selected examples in crystal engineering 768
  - 20.4.1 Diamondoid networks 768
  - 20.4.2 Interlocked structures constructed from cucurbituril 772
  - 20.4.3 Inorganic crystal engineering using hydrogen bonds 776
  - 20.4.4 Generation and stabilization of unstable inorganic/organic anions in urea/thiourea complexes 780
  - 20.4.5 Supramolecular assembly of silver(I) polyhedra with embedded acetylenediide dianion 784
  - 20.4.6 Supramolecular assembly with the silver(I)-ethynide synthon 792
  - 20.4.7 Self-assembly of nanocapsules with pyrogallol[4]arene macrocycles 797
  - 20.4.8 Reticular design and synthesis of porous metal–organic frameworks 799
  - 20.4.9 One-pot synthesis of nanocontainer molecule 804
  - 20.4.10 Filled carbon nanotubes 804
- References 808

Index 811

List of Contributors

Wai-Kee Li obtained his B.S. degree from University of Illinois in 1964 and his Ph.D. degree from University of Michigan in 1968. He joined the Chinese University of Hong Kong in July 1968 to follow an academic career that spanned a period of thirty-eight years. He retired as Professor of Chemistry in August 2006 and was subsequently conferred the title of Emeritus Professor of Chemistry. Over the years he taught a variety of courses in physical and inorganic chemistry. His research interests in theoretical and computational chemistry have led to over 180 papers in international journals.

Gong-Du Zhou graduated from Xichuan University in 1953 and completed his postgraduate studies at Peking University in 1957. He then joined the Chemistry Department of Peking University and taught Structural Chemistry, Structure and Properties of Matter, and other courses there till his retirement as a Professor in 1992. He has published more than 100 research papers in X-ray crystallography and structural chemistry, together with over a dozen Chinese chemistry textbooks and reference books.

Thomas Chung Wai Mak obtained his B.Sc. (1960) and Ph.D. (1963) degrees from the University of British Columbia. After working as a NASA Postdoctoral Research Associate at the University of Pittsburgh and an Assistant Professor at the University of Western Ontario, in June 1969 he joined the Chinese University of Hong Kong, where he is now Emeritus Professor of Chemistry and Wei Lun Research Professor. His research interest lies in inorganic synthesis, chemical crystallography, supramolecular assembly and crystal engineering, with over 900 papers in international journals. He was elected as a member of the Chinese Academy of Sciences in 2001.

Fundamentals of Bonding Theory

The theory of chemical bonding plays an important role in the rapidly evolving field of structural inorganic chemistry. It helps us to understand the structure, physical properties and reactivities of different classes of compounds. It is generally recognized that bonding theory acts as a guiding principle in inorganic chemistry research, including the design of synthetic schemes, rationalization of reaction mechanisms, exploration of structure–property relationships, supramolecular assembly and crystal engineering.

There are five chapters in Part I: Introduction to quantum theory, The electronic structure of atoms, Covalent bonding in molecules, Chemical bonding in condensed phases and Computational chemistry. Since most of the contents of these chapters are covered in popular texts for courses in physical chemistry, quantum chemistry and structural chemistry, it can be safely assumed that readers of this book have some acquaintance with such topics. Consequently, many sections may be viewed as convenient summaries and frequently mathematical formulas are given without derivation.

The main purpose of Part I is to review the rudiments of bonding theory, so that the basic principles can be applied to the development of new topics in subsequent chapters.

Introduction to Quantum Theory

In order to appreciate fully the theoretical basis of atomic structure and chemical bonding, we need a basic understanding of the quantum theory. Even though chemistry is an experimental science, theoretical consideration (especially prediction) is now playing a role of increasing importance with the development of powerful computational algorithms. The field of applying quantum theoretical methods to investigate chemical systems is commonly called quantum chemistry.

The key to theoretical chemistry is molecular quantum mechanics, which deals with the transference or transformation of energy on a molecular scale. Although the quantum mechanical principles for understanding the electronic structure of matter has been recognized since 1930, the mathematics involved in their application, i.e. general solution of the Schrödinger equation for a molecular system, was intractable at best in the 50 years or so that followed. But with the steady development of new theoretical and computational methods, as well as the availability of larger and faster computers with reasonable price tags during the past two decades, calculations have sometimes become almost as accurate as experiments, or at least accurate enough to be useful to experimentalists. Additionally, compared to experiments, calculations are often less costly, less time-consuming, and easier to control. As a result, computational results can complement experimental studies in essentially every field of chemistry. For instance, in physical chemistry, chemists can apply quantum chemical methods to calculate the entropy, enthalpy, and other thermochemical functions of various gases, to interpret molecular spectra, to understand the nature of the intermolecular forces, etc. In organic chemistry, calculations can serve as a guide to a chemist who is in the process of synthesizing or designing new compounds; they can also be used to compare the relative stabilities of various molecular species, to study the properties of reaction intermediates and transition states, and to investigate the mechanism of reactions, etc. In analytical chemistry, theory can help chemists to understand the frequencies and intensities of the spectral lines. In inorganic chemistry, chemists can apply the ligand filed theory to study the transition metal complexes. An indication that computational chemistry has been receiving increasing attention in the scientific community was the award of the 1998 Nobel Prize in chemistry to Professor J.A. Pople and Professor W. Kohn for their contributions to quantum chemistry. In Chapter 5, we will briefly describe the kind of questions that may be fruitfully treated by computational chemistry.

In this chapter, we discuss some important concepts of quantum theory. A clear understanding of these concepts will facilitate subsequent discussion of bonding theory.

1.1 Dual nature of light and matter

Around the beginning of the twentieth century, scientists had accepted that light is both a particle and a wave. The wave character of light is manifested in its interference and diffraction experiments. On the other hand, its corpuscular nature can be seen in experiments such as the photoelectric effect and Compton effect. With this background, L. de Broglie in 1924 proposed that, if light is both a particle and a wave, a similar duality also exists for matter. Moreover, by combining Einstein's relationship between energy E and mass m,

$$E = mc^2, \tag{1.1.1}$$

where c is the speed of light, and Planck's quantum condition,

$$E = h\nu, \tag{1.1.2}$$

where h is Planck's constant and ν is the frequency of the radiation, de Broglie was able to arrive at the wavelength λ associated with a photon,

$$\lambda = \frac{c}{\nu} = \frac{hc}{h\nu} = \frac{hc}{mc^2} = \frac{h}{mc} = \frac{h}{p}, \tag{1.1.3}$$

where p is the momentum of the photon. Then de Broglie went on to suggest that a particle with mass m and velocity v is also associated with a wavelength given by

$$\lambda = \frac{h}{mv} = \frac{h}{p}, \tag{1.1.4}$$

where p is now the momentum of the particle.

Before proceeding further, it is instructive to examine what kinds of wavelengths are associated with particles having various masses and velocities, as shown in Table 1.1.1.

By examining the results listed in Table 1.1.1, it is seen that the wavelengths of macroscopic objects will be far too short to be observed. On the other hand, electrons with energies on the order of 100 eV will have wavelengths between 100 and 200 pm, approximately the interatomic distances in crystals. In 1927, C.J. Davisson and L.H. Germer obtained the first electron diffraction pattern of a crystal, thus proving de Broglie's hypothesis experimentally. From then on, scientists recognized that an electron has dual properties: it can behave as both a particle and a wave.

Table 1.1.1. Wavelength of different particles travelling with various velocities ($\lambda = h/p = h/mv$; $h = 6.63 \times 10^{-27}$ erg s $= 6.63 \times 10^{-34}$ J s)

Particles	m / kg	v / m s^{-1}	λ / pm
Electron at 298 K	9.11×10^{-31}	1.16×10^5	6270*
1-volt electron	9.11×10^{-31}	5.93×10^5	1230
100-volt electron	9.11×10^{-31}	5.93×10^6	123†
He atom at 298 K	6.65×10^{-27}	1.36×10^3	73.3
Xe atom at 298 K	2.18×10^{-25}	2.38×10^2	12.8
A 100-kg sprinter running at world-record speed	1.00×10^2	1.00×10^1	6.63×10^{-25}

$$*E = \tfrac{3}{2}kT = \tfrac{1}{2}mv^2; v = \left(\tfrac{3kT}{m}\right)^{\tfrac{1}{2}} = \left(\tfrac{3 \times 1.38 \times 10^{-23} \text{J K}^{-1} \times 298 \text{K}}{9.11 \times 10^{-31} \text{kg}}\right)^{\tfrac{1}{2}} = 1.16 \times 10^5 \text{ m s}^{-1};$$

$$\lambda = \tfrac{h}{mv} = \tfrac{6.63 \times 10^{-34} \text{Js}}{9.11 \times 10^{-31} \text{kg} \times 1.16 \times 10^5 \text{m s}^{-1}} = 6.27 \times 10^{-9}\text{m} = 6270 \text{ pm}.$$

$$†E = 100 \text{ eV} = 100 \times 1.60 \times 10^{-19} \text{ J} = 1.60 \times 10^{-17} \text{J} = \tfrac{1}{2}mv^2,$$

$$v = \left(\tfrac{2E}{m}\right)^{\tfrac{1}{2}} = \left(\tfrac{2 \times 1.60 \times 10^{-17} \text{J}}{9.11 \times 10^{-31} \text{kg}}\right)^{\tfrac{1}{2}} = 5.93 \times 10^6 \text{ m s}^{-1};$$

$$\lambda = \tfrac{h}{mv} = \tfrac{6.63 \times 10^{-34} \text{Js}}{9.11 \times 10^{-31} \text{kg} \times 5.93 \times 10^6 \text{m s}^{-1}} = 1.23 \times 10^{-10} \text{ m} = 123 \text{ pm}.$$

This wavelength is similar to the atomic spacing in crystals.

1.2 Uncertainty principle and probability Concept

Another important development in quantum mechanics is the Uncertainty Principle set forth by W. Heisenberg in 1927. In its simplest terms, this principle says, "The position and momentum of a particle cannot be simultaneously and precisely determined." Quantitatively, the product of the uncertainty in the x component of the momentum vector (Δp_x) and the uncertainty in the x direction of the particle's position (Δx) is on the order of Planck's constant:

$$(\Delta p_x)(\Delta x) \sim \frac{h}{4\pi} = 5.27 \times 10^{-35} \text{J s}. \tag{1.2.1}$$

While h is quite small in the macroscopic world, it is not at all insignificant when the particle under consideration is of subatomic scale. Let us use an actual example to illustrate this point. Suppose the Δx of an electron is 10^{-14} m, or 0.01 pm. Then, with eq. (1.2.1), we get $\Delta p_x = 5.27 \times 10^{-21}$ kg m s^{-1}. This uncertainty in momentum would be quite small in the macroscopic world. However, for subatomic particles such as an electron, with mass of 9.11×10^{-31} kg, such an uncertainty would not be negligible at all. Hence, on the basis of the Uncertainty Principle, we can no longer say that an electron is precisely located at this point with an exactly known velocity. It should be stressed that the uncertainties we are discussing here have nothing to do with the imperfection of the measuring instruments. Rather, they are inherent indeterminacies. If we recall the Bohr theory of the hydrogen atom, we find that both the radius of the orbit and the velocity of the electron can be precisely calculated. Hence the Bohr results violate the Uncertainty Principle.

With the acceptance of uncertainty at the atomic level, we are forced to speak in terms of probability: we say the probability of finding the electron within this volume element is how many percent and it has a probable velocity (or momentum) such and such.

1.3 Electronic wavefunction and probability density function

Since an electron has wave character, we can describe its motion with a wave equation, as we do in classical mechanics for the motions of a water wave or a stretched string or a drum. If the system is one-dimensional, the classical wave equation is

$$\frac{\partial^2 \Phi(x,t)}{\partial x^2} = \frac{1}{v^2} \frac{\partial^2 \Phi(x,t)}{\partial t^2}, \quad (1.3.1)$$

where v is the velocity of the propagation. The wavefunction Φ gives the displacement of the wave at point x and at time t. In three-dimensional space, the wave equation becomes

$$\left(\frac{\partial^2}{\partial x^2} + \frac{\partial^2}{\partial y^2} + \frac{\partial^2}{\partial z^2} \right) \Phi(x,y,z,t) = \nabla^2 \Phi(x,y,z,t) = \frac{1}{v^2} \frac{\partial^2 \Phi(x,y,z,t)}{\partial t^2}. \quad (1.3.2)$$

A typical wavefunction, or a solution of the wave equation, is the familiar sine or cosine function. For example, we can have

$$\Phi(x,t) = A \sin(2\pi/\lambda)(x - vt). \quad (1.3.3)$$

It can be easily verified that $\Phi(x,t)$ satisfies eq. (1.3.1). An important point to keep in mind is that, in classical mechanics, the wavefunction is an amplitude function. As we shall see later, in quantum mechanics, the electronic wavefunction has a different role to play.

Combining the wave nature of matter and the probability concept of the Uncertainty Principle, M. Born proposed that the electronic wavefunction is no longer an amplitude function. Rather, it is a measure of the probability of an event: when the function has a large (absolute) value, the probability for the event is large. An example of such an event is given below.

From the Uncertainty Principle, we no longer speak of the exact position of an electron. Instead, the electron position is defined by a probability density function. If this function is called $\rho(x,y,z)$, then the electron is most likely found in the region where ρ has the greatest value. In fact, $\rho \, d\tau$ is the probability of finding the electron in the volume element $d\tau (\equiv dxdydz)$ surrounding the point (x,y,z). Note that ρ has the unit of volume^{-1}, and $\rho \, d\tau$, being a probability, is dimensionless. If we call the electronic wavefunction ψ, Born asserted that the probability density function ρ is simply the absolute square of ψ:

$$\rho(x,y,z) = |\psi(x,y,z)|^2. \quad (1.3.4)$$

Since ψ can take on imaginary values, we take the absolute square of ψ to make sure that ρ is positive. Hence, when ψ is imaginary,

$$\rho(x,y,z) = |\psi(x,y,z)|^2 = \psi^*\psi, \tag{1.3.5}$$

where ψ^* is the complex conjugate (replacing i in ψ by $-i$) of ψ.

Before proceeding further, let us use some numerical examples to illustrate the determination of the probability of locating an electron in a certain volume element in space. The ground state wavefunction of the hydrogen atom is

$$\psi_{1s} = \left(\pi a_0^3\right)^{-\frac{1}{2}} e^{-\frac{r}{a_0}}, \tag{1.3.6}$$

where r is the nucleus–electron separation and a_0, with the value of 52.9 pm, is the radius of the first Bohr orbit (hence is called Bohr radius). In the following we will use ψ_{1s} to determine the probability P of locating the electron in a volume element $d\tau$ of 1 pm^3 which is $1a_0$ away from the nucleus.

At $r = 1a_0$,

$$\psi_{1s} = \left(\pi a_0^3\right)^{-\frac{1}{2}} e^{-\frac{a_0}{a_0}} = \left[\pi(52.9\,\text{pm})^3\right]^{-\frac{1}{2}} e^{-1} = 5.39 \times 10^{-4}\,\text{pm}^{-\frac{3}{2}},$$

$$\rho = |\psi_{1s}|^2 = \left(5.39 \times 10^{-4}\,\text{pm}^{-\frac{3}{2}}\right)^2 = 2.91 \times 10^{-7}\,\text{pm}^{-3},$$

$$P = |\psi_{1s}|^2\,d\tau = 2.91 \times 10^{-7}\,\text{pm}^{-3} \times 1\,\text{pm}^3 = 2.91 \times 10^{-7}.$$

In addition, we can also calculate the probability of finding the electron in a shell of thickness 1 pm which is $1a_0$ away from the nucleus:

$$d\tau = (\text{surface area}) \times (\text{thickness})$$
$$= (4\pi r^2 dr) = 4\pi(52.9\,\text{pm})^2 \times 1\,\text{pm}$$
$$= 3.52 \times 10^4\,\text{pm}^3,$$

and

$$P = |\psi_{1s}|^2\,d\tau = 2.91 \times 10^{-7}\,\text{pm}^{-3} \times 3.52 \times 10^4\,\text{pm}^3 = 1.02 \times 10^{-2}.$$

In other words, there is about 1% chance of finding the electron in a spherical shell of thickness 1 pm and radius $1a_0$.

In Table 1.3.1, we tabulate $\psi_{1s}, |\psi_{1s}|^2, |\psi_{1s}|^2 d\tau$ (with $d\tau = 1\,\text{pm}^3$), and $4\pi r^2|\psi_{1s}|^2 dr$ (with $dr = 1$ pm), for various r values.

As $|\psi|^2 d\tau$ represents the probability of finding the electron in a certain region in space, and the sum of all probabilities is 1, ψ must satisfy the relation

$$\int |\psi|^2\,d\tau = 1. \tag{1.3.7}$$

Table 1.3.1. Values of ψ_{1s}, $|\psi_{1s}|^2$, $|\psi_{1s}|^2 d\tau$ (with $d\tau = 1$ pm^3), and $4\pi r^2|\psi_{1s}|^2 dr$ (with $dr = 1$ pm) for various nucleus–electron distances

r(pm)	0	26.45 (or $a_0/2$)	52.9 (or a_0)	100	200
ψ_{1s} (pm$^{-\frac{3}{2}}$)	1.47×10^{-3}	8.89×10^{-4}	5.39×10^{-4}	2.21×10^{-4}	3.34×10^{-5}
$|\psi_{1s}|^2$ (pm^{-3})	2.15×10^{-6}	7.91×10^{-7}	2.91×10^{-7}	4.90×10^{-8}	1.12×10^{-9}
$|\psi_{1s}|^2 d\tau$	2.15×10^{-6}	7.91×10^{-7}	2.91×10^{-7}	4.90×10^{-8}	1.12×10^{-9}
$4\pi r^2 |\psi_{1s}|^2 dr$	0	6.95×10^{-3}	1.02×10^{-2}	6.16×10^{-3}	5.62×10^{-4}

When ψ satisfies eq. (1.3.7), the wavefunction is said to be normalized. On the other hand, if $\int |\psi|^2 d\tau = N$, where N is a constant, then $N^{-1/2}\psi$ is a normalized wavefunction and $N^{-1/2}$ is called the normalization constant.

For a given system, there are often many or even an infinite number of acceptable solutions: $\psi_1, \psi_2, \ldots, \psi_i, \psi_j, \ldots$ and these wavefunctions are "orthogonal" to each other, i.e.

$$\int \psi_i^* \psi_j d\tau = \int \psi_j^* \psi_i d\tau = 0. \tag{1.3.8}$$

Combining the normalization condition (eq. (1.3.7)) and the orthogonality condition (eq. (1.3.8)) leads us to the orthonormality relationship among the wavefunctions

$$\int \psi_i^* \psi_j d\tau = \int \psi_j^* \psi_i d\tau$$
$$= \delta_{ij} = \begin{cases} 0 & \text{when } i \neq j \\ 1 & \text{when } i = j \end{cases}, \tag{1.3.9}$$

where δ_{ij} is the Kronecker delta function. Since $|\psi|^2$ plays the role of a probability density function, ψ must be finite, continuous, and single-valued.

The wavefunction plays a central role in quantum mechanics. For atomic systems, the wavefunction describing the electronic distribution is called an atomic orbital; in other words, the aforementioned 1s wavefunction of the ground state of a hydrogen atom is also called the 1s orbital. For molecular systems, the corresponding wavefunctions are likewise called molecular orbitals.

Once we know the explicit functions of the various atomic orbitals (such as 1s, 2s, 2p, 3s, 3p, 3d,…), we can calculate the values of ψ at different points in space and express the wavefunction graphically. In Fig. 1.3.1(a), the graph on the left is a plot of ψ_{1s} against r; the sphere on the right shows that ψ has the same value for a given r, regardless of direction (or θ and φ values). In Fig. 1.3.1(b), the two graphs on the left plot density functions $|\psi_{1s}|^2$ and $4\pi r^2|\psi_{1s}|^2$ against r. The probability density describing the electronic distribution is also referred to as an electron cloud, which may be represented by a figure such as that shown on the right side of Fig. 1.3.1(b). This figure indicates that the 1s orbital has maximum density at the nucleus and the density decreases steadily as the electron gets farther and farther away from the nucleus.

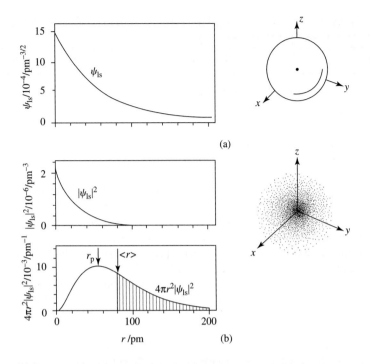

Fig. 1.3.1.
(a) The 1s wavefunction and (b) the 1s probability density functions of the hydrogen atom.

In addition to providing probability density functions, the wavefunction may also be used to calculate the value of a physical observable for that state. In quantum mechanics, a physical observable A has a corresponding mathematical operator $\hat{A}$. When $\hat{A}$ satisfies the relation

$$\hat{A}\psi = a\psi, \qquad (1.3.10)$$

ψ is called an eigenfunction of operator $\hat{A}$, and a is called the eigenvalue of the state described by ψ. In the next section, we shall discuss the Schrödinger equation,

$$\hat{H}\psi = E\psi. \qquad (1.3.11)$$

Here ψ is the eigenfunction of the Hamiltonian operator $\hat{H}$ and the corresponding eigenvalue E is the energy of the system.

If ψ does not satisfy eq. (1.3.10), we can calculate the expectation value (or mean) of A, $<A>$, by the expression

$$<A> = \frac{\int \psi^* \hat{A} \psi \, d\tau}{\int |\psi|^2 \, d\tau}. \qquad (1.3.12)$$

If ψ is a normalized wavefunction, eq. (1.3.12) becomes

$$<A> = \int \psi^* \hat{A} \psi \, d\tau. \qquad (1.3.13)$$

Fundamentals of Bonding Theory

In the ground state of a hydrogen atom, there is no fixed r value for the electron, i.e. there is no eigenvalue for r. On the other hand, we can use eq. (1.3.13) to calculate the average value of r:

$$\begin{aligned}<r> &= \left(\pi a_0^3\right)^{-1} \int_0^{2\pi} d\phi \int_0^{\pi} \sin\theta d\theta \int_0^{\infty} e^{-\frac{r}{a_0}} r e^{-\frac{r}{a_0}} r^2 dr \\ &= \left(\pi a_0^3\right)^{-1} \cdot 4\pi \int_0^{\infty} e^{-\frac{2r}{a_0}} r^3 dr \\ &= 3a_0/2. \end{aligned} \quad (1.3.14)$$

In carrying out the integration in eq. (1.3.14), we make use of $d\tau = r^2 dr \sin\theta\, d\theta\, d\varphi$ and $0 \leq r < \infty, 0 \leq \theta \leq \pi$, and $0 \leq \phi \leq 2\pi$.

In Fig. 1.3.1(b), in the plot of $4\pi r^2|\psi_{1s}|^2$ against r, the mean value $<r>$ is marked on the r axis, separating the graph into two parts. These two parts have unequal areas, the unshaded area being larger than the shaded area. This result implies that it is more likely to find a r value smaller than $<r>$ than one larger than $<r>$. In addition, the r value corresponding to the maximum in the $4\pi r^2|\psi_{1s}|^2$ function is labelled r_p. At $r_p, r = a_0$ and r_p is called the most probable electron distance.

1.4 Electronic wave equation: The Schrödinger equation

In 1926, E. Schrödinger developed his famous wave equation for electrons. The validity of the Schrödinger equation rests solely on the fact that it leads to the right answers for a variety of systems. As in the case of Newton's equations, the Schrödinger equation is a fundamental postulate that cannot be deduced from first principles. Hence what is presented below is merely a heuristic derivation. In this presentation, we can see how the particle character is incorporated into a wave equation.

We start with eq. (1.3.2), the general differential equation for wave motion:

$$\nabla^2 \Phi(x,y,z,t) = \frac{\partial^2 \Phi(x,y,z,t)}{v^2 \partial t^2}. \quad (1.3.2)$$

Note that wavefunction Φ has time t as one of its variables. Since our primary concern is the energy of a system and this energy is independent of time (we are ignoring the process of radiation here), we need an equation that is time independent. The wavefunctions obtained from a time-independent equation are called standing (or stationary) waves. To obtain such an equation, we assume $\Phi(x,y,z,t)$ has the form

$$\Phi(x,y,z,t) = \psi(x,y,z)g(t), \quad (1.4.1)$$

where ψ is a function of space coordinates and g is a function of time t. For standing waves, there are several acceptable g functions and one of them is

$$g(t) = e^{2\pi i \nu t}, \quad (1.4.2)$$

where frequency ν is related to propagation velocity v and wavelength λ by

$$\nu = \frac{v}{\lambda}. \tag{1.4.3}$$

If we substitute

$$\Phi(x, y, z, t) = \psi(x, y, z)e^{2\pi i \nu t} \tag{1.4.4}$$

into eq. (1.3.2), we get

$$e^{2\pi i \nu t}\nabla^2 \psi = \frac{1}{v^2}\psi\frac{\partial^2 e^{2\pi i \nu t}}{\partial t^2}$$
$$= -4\pi^2 \nu^2 v^{-2} e^{2\pi i \nu t}\psi, \tag{1.4.5}$$

or, upon canceling $e^{2\pi i \nu t}$,

$$\nabla^2 \psi = -4\pi^2 \nu^2 v^{-2} \psi. \tag{1.4.6}$$

Now we incorporate the corpuscular character ($\lambda = h/p$) into eq. (1.4.6)

$$v = \nu\lambda = \nu\left(\frac{h}{p}\right) = \frac{h\nu}{p}, \tag{1.4.7}$$

and eq. (1.4.6) becomes

$$\nabla^2 \psi = \left(\frac{-4\pi^2 p^2}{h^2}\right)\psi. \tag{1.4.8}$$

If we rewrite p^2 in terms of kinetic energy T, or total energy E, and potential energy V,

$$p^2 = 2mT = 2m(E - V), \tag{1.4.9}$$

the wave equation now has the form

$$\nabla^2 \psi = \left(\frac{-8m\pi^2}{h^2}\right)(E - V)\psi. \tag{1.4.10}$$

Rearranging eq. (1.4.10) yields,

$$\left[\left(\frac{-h^2}{8\pi^2 m}\right)\nabla^2 + V\right]\psi = E\psi, \tag{1.4.11}$$

or

$$\hat{H}\psi = E\psi, \tag{1.4.12}$$

where the Hamiltonian operator $\hat{H}$ is defined as

$$\hat{H} = \left(\frac{-h^2}{8\pi^2 m}\right)\nabla^2 + V. \tag{1.4.13}$$

In other words, $\hat{H}$ has two parts: kinetic energy operator $(-h^2/8\pi^2 m)\nabla^2$ and potential energy operator V.

To summarize, the quantum mechanical way of studying the electronic structure of an atom or a molecule consists of the following steps:

(1) Write down the Schrödinger equation of the system by filling in the proper potential energy.
(2) Solve the differential equation to obtain the electronic energies E_i and wavefunctions ψ_i, $i = 1, 2, \ldots$.
(3) Use the wavefunctions ψ_i to determine the probability density functions $|\psi_i|^2$ and the expectation values of physical observables.

In the following, we use the ground state wavefunction ψ_{1s} [eq. (1.3.6)] to determine the energy E_{1s} of this state as well as the expectation values of the kinetic energy, $<T>$, and potential energy, $<V>$.

By applying the Hamiltonian operator of the hydrogen atom on ψ_{1s}, we can readily obtain E_{1s}:

$$\hat{H}\psi_{1s} = \left[\left(\frac{-h^2}{8\pi^2 m}\right)\nabla^2 - \left(\frac{e^2}{4\pi\varepsilon_0 r}\right)\right]\psi_{1s}$$
$$= E_{1s}\psi_{1s}, \tag{1.4.14}$$

and

$$E_{1s} = \left(\frac{1}{4\pi\varepsilon_0}\right)\left(\frac{-e^2}{2a_0}\right) = -2.18 \times 10^{-18}\,\text{J} = -13.6\,\text{eV}. \tag{1.4.15}$$

The expectation value of potential energy, $<V>$, can also be computed easily:

$$<V> = \int_0^{2\pi} d\phi \int_0^{\pi} \sin\theta\, d\theta \int_0^{\infty} \left(\pi a_0^3\right)^{-1} e^{-\frac{r}{a_0}} \left(\frac{-e^2}{4\pi\varepsilon_0 r}\right) e^{-\frac{r}{a_0}} r^2 dr$$
$$= \left(\frac{-1}{4\pi\varepsilon_0}\right)\left(\frac{e^2}{a_0}\right). \tag{1.4.16}$$

The quantity $<T>$ is simply the difference between E_{1s} and $<V>$:

$$<T> = E_{1s} - <V> = \left(\frac{1}{4\pi\varepsilon_0}\right)\left(\frac{e^2}{2a_0}\right). \tag{1.4.17}$$

Introduction to Quantum Theory

Comparing eqs. (1.4.16) and (1.4.17), we get

$$<T> = -\frac{<V>}{2}. \tag{1.4.18}$$

which is called the *virial theorem* for atomic and molecular systems.

In the following section, we treat several systems quantum mechanically to illustrate the method introduced here.

1.5 Simple applications of the Schrödinger equation

In the four examples given below, a particle (or electron) is allowed to move freely. The only difference is the shape of the "box," in which the particle travels. As will be seen later, different shapes give rise to different boundary conditions, which in turn lead to different allowed energies (eigenvalues) and wavefunctions (eigenfunctions).

1.5.1 Particle in a one-dimensional box

In this system, the box has only one dimension, with length a. The potential energy is zero inside the box and infinity at the boundary and outside the box. In other words, the electron can move freely inside the box and it is impossible for it to get out of the box. Mathematically

$$V = \begin{cases} 0, & 0 < x < a; \\ \infty, & 0 \geq x \text{ or } x \geq a. \end{cases} \tag{1.5.1}$$

So the Schrödinger equation has the form

$$\left(\frac{-h^2}{8\pi^2 m}\right)\left(\frac{d^2\psi}{dx^2}\right) = E\psi, \tag{1.5.2}$$

or

$$\frac{d^2\psi}{dx^2} = \left(\frac{-8\pi^2 mE}{h^2}\right)\psi = -\alpha^2\psi, \tag{1.5.3}$$

with

$$\alpha^2 = \frac{8\pi^2 mE}{h^2}. \tag{1.5.4}$$

Solutions of eq. (1.5.3) are

$$\psi = A \sin \alpha x + B \cos \alpha x, \tag{1.5.5}$$

where A and B are constants to be determined by the boundary conditions defined by eq. (1.5.1). At $x = 0$ or $x = a$, the potential barrier is infinitely high and the particle cannot be found at or around those points, i.e.

$$\psi(0) = \psi(a) = 0. \tag{1.5.6}$$

With $\psi(0) = 0$, we get

$$B = 0. \tag{1.5.7}$$

Also, $\psi(a) = 0$ yields

$$A \sin \alpha a = 0, \tag{1.5.8}$$

which means either A or $\sin \alpha a$ vanishes. Since the former is not acceptable, we have

$$\sin \alpha a = 0, \tag{1.5.9}$$

or

$$\alpha a = n\pi, \quad n = 1, 2, 3, \ldots. \tag{1.5.10}$$

So, the wavefunctions have the form

$$\psi_n(x) = A \sin\left(\frac{n\pi x}{a}\right), \quad n = 1, 2, 3, \ldots. \tag{1.5.11}$$

The constant A can be determined by normalization:

$$\int_0^a |\psi_n|^2 dx = A^2 \int_0^a \sin^2\left(\frac{n\pi x}{a}\right) dx = 1, \tag{1.5.12}$$

which leads to the following form for the wavefunctions:

$$\psi_n(x) = \left(\frac{2}{a}\right)^{\frac{1}{2}} \sin\left(\frac{n\pi x}{a}\right). \tag{1.5.13}$$

Note that ψ has the unit of length$^{-1/2}$ and ψ^2 has the unit of length^{-1}.

By combining eqs. (1.5.4) and (1.5.10), the energy of the system can also be determined:

$$\alpha = \frac{n\pi}{a} = \left(\frac{8\pi^2 mE}{h^2}\right)^{\frac{1}{2}} \tag{1.5.14}$$

or

$$E_n = \frac{n^2 h^2}{8ma^2}, \quad n = 1, 2, 3, \ldots. \tag{1.5.15}$$

Figure 1.5.1 summarizes the results of this particle-in-a-box problem. From this figure, it is seen that when the electron is in the ground state ($n = 1$), it is most likely found at the center of the box. On the other hand, if the electron is in the first excited state ($n = 2$), it is most likely found around $x = a/4$ or $x = 3a/4$.

Introduction to Quantum Theory

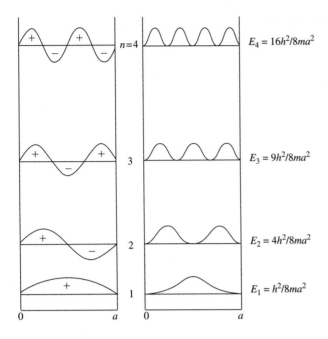

Fig. 1.5.1.
Pictorial representations of E_n, ψ_n (left), and $|\psi_n|^2$ (right) for the particle in a one-dimensional box problem.

In the following, we calculate the expectation values $<p_x>$ and $<p_x^2>$, where p_x is the momentum of the particle (recall this is a one-dimensional system), for the ground state of this system. First,

$$<p_x^2> = 2m <T> = 2mE$$
$$= (2m)\left(\frac{h^2}{8ma^2}\right) = \frac{h^2}{4a^2}. \quad (1.5.16)$$

Next, for $<p_x>$, we need to make use of the fact that the quantum mechanical operator $\hat{p}_x$ for p_x is $(-ih/2\pi)(\partial/\partial x)$. So

$$<p_x> = \left(\frac{2}{a}\right)\int_0^a \sin\left(\frac{\pi x}{a}\right)\left(\frac{-ih}{2\pi}\right)\left(\frac{\partial}{\partial x}\right)\sin\left(\frac{\pi x}{a}\right)dx = 0. \quad (1.5.17)$$

So the mean momentum is zero, as the electron is equally likely to travel to the left or to the right. On the other hand, $<p_x^2>$ is not zero, as the square of a momentum is always positive. From statistics, the uncertainty of momentum, Δp_x, may be expressed in terms of $<p_x>$ and $<p_x^2>$:

$$\Delta p_x = \left[<p_x^2> - <p_x>^2\right]^{\frac{1}{2}}$$
$$= \frac{h}{2a}. \quad (1.5.18)$$

In the following, again for the ground state, we calculate the mean value of position x as well as that of x^2:

$$<x> = \left(\frac{2}{a}\right) \int_0^a \sin\left(\frac{\pi x}{a}\right) x \sin\left(\frac{\pi x}{a}\right) dx$$
$$= \frac{a}{2}. \qquad (1.5.19)$$

This result can be obtained by simply examining the function $|\psi_1|^2$ shown in Fig. 1.5.1. Meanwhile,

$$<x^2> = \left(\frac{2}{a}\right) \int_0^a \sin\left(\frac{\pi x}{a}\right) x^2 \sin\left(\frac{\pi x}{a}\right) dx$$
$$= a^2 \left(\frac{1}{3} - \frac{1}{2\pi^2}\right). \qquad (1.5.20)$$

Now we are ready to determine the uncertainty in x:

$$\Delta x = \left[<x^2> - <x>^2\right]^{\frac{1}{2}}$$
$$= a\left[\left(\frac{1}{12}\right) - \left(\frac{1}{2\pi^2}\right)\right]^{\frac{1}{2}}. \qquad (1.5.21)$$

The product $\Delta x \cdot \Delta p_x$ satisfies the Uncertainty Principle:

$$\Delta x \Delta p_x = a\left[\left(\frac{1}{12}\right) - \left(\frac{1}{2\pi^2}\right)\right]^{\frac{1}{2}} \left(\frac{h}{2a}\right) = 1.14 \left(\frac{h}{4\pi}\right) > \frac{h}{4\pi}. \qquad (1.5.22)$$

There is a less mathematical way to show that the results of this one-dimensional box problem do conform to the Uncertainty Principle. The ground state, or minimum, energy of this system is $h^2/8ma^2$, which has a positive value. On the other hand, when this system is treated classically, the minimum energy would be zero. The residual energy of the (quantum) ground state, or the energy above the classical minimum, is called the zero-point energy. The existence of this energy implies that the kinetic energy, and hence the momentum, of a bound particle cannot be zero. If we take the ground state energy to be $p_x^2/2m$, we get the minimum momentum of the particle to be $\pm h/2a$. The uncertainty in momentum, Δp_x, may then be approximated to be h/a. If we take the uncertainty in position, Δx, to be the length of the box, a, then $\Delta x \Delta p_x$ is (approximately) h, which is in accord with the Uncertainty Principle.

The results of the particle in a one-dimensional box problem can be used to describe the delocalized π electrons in (linear) conjugated polyenes. Such an approximation is called the free-electron model. Take the butadiene molecule $CH_2=CH-CH=CH_2$ as an example. The four π electrons of this system would fill up the ψ_1 and ψ_2 orbitals, giving rise to the $(\psi_1)^2(\psi_2)^2$ configuration. If we excite one electron from the ψ_2 orbital to the ψ_3 orbital, we need an energy of

$$\Delta E = \frac{5h^2}{8ma^2} = \frac{hc}{\lambda}. \qquad (1.5.23)$$

The length of the box, a, may be approximated in the following way. Typical C–C and C=C bond lengths are 154 and 135 pm, respectively. If we allow the π electrons to move a bit beyond the terminal carbon atoms, the length of the box may be rounded off to (4 × 1.40 =) 560 pm. If a is taken to be this value, λ in eq. (1.5.23) can be calculated to be 207.0 nm. Experimentally, butadiene absorbs light at $\lambda = 210.0$ nm. So, even though the model is very crude, the result is fortuitously good.

The free-electron model breaks down readily when it is applied to longer polyenes. For hexatriene, we have $a = 6 \times 1.40 = 840$ pm, $\Delta E = E_4 - E_3$, and $\lambda = 333$ nm. Experimentally this triene absorbs at $\lambda = 250$ nm. For octatetraene, the box length a now becomes 1120 pm, and $\Delta E = E_5 - E_4$ with $\lambda = 460$ nm, compared to the experimental value of 330 pm. Despite this shortcoming, the model does predict that when a conjugated polyene is lengthened, its absorption band wavelength becomes longer as well.

When a polyene reaches a certain length, its absorption wavelength will appear in the visible region, i.e. λ is between 400 and 700 nm. When this occurs, the polyene is colored. One of the better known colored polyenes is β-carotene, which is responsible for orange color of carrots. It has the structure:

This polyene has 11 conjugated π bonds, with $\lambda = 450$ nm. Carotene can be cleaved enzymatically into two units of all-*trans*-vitamin A, which is a polyene with five conjugated π bonds:

The absorption peak of this compound appears at $\lambda = 325$ nm. This molecule plays an important role in the chemistry of vision.

1.5.2 Particle in a three-dimensional box

The potential energy of this system has the form

$$V = \begin{cases} 0, & 0 < x < a \quad \text{and} \quad 0 < y < b \quad \text{and} \quad 0 < z < c, \\ \infty, & 0 \geq x \geq a \quad \text{or} \quad 0 \geq y \geq b \quad \text{or} \quad 0 \geq z \geq c. \end{cases} \quad (1.5.24)$$

So the Schrödinger equation now is

$$\left(\frac{-h^2}{8\pi^2 m}\right)\nabla^2\psi = E\psi. \tag{1.5.25}$$

To solve this equation, we make use of the technique of separation of variables

$$\psi(x,y,z) = X(x)Y(y)Z(z), \tag{1.5.26}$$

where X, Y, and Z are one-variable functions involving variables, x, y, and z, respectively. Substituting eq. (1.5.26) into eq. (1.5.25) leads to

$$\nabla^2\psi = \left[\left(\frac{\partial^2}{\partial x^2}\right) + \left(\frac{\partial^2}{\partial y^2}\right) + \left(\frac{\partial^2}{\partial z^2}\right)\right]XYZ = \left(\frac{-8\pi^2 mE}{h^2}\right)XYZ, \tag{1.5.27}$$

or

$$YZ\left(\frac{\partial^2 X}{\partial x^2}\right) + XZ\left(\frac{\partial^2 Y}{\partial y^2}\right) + XY\left(\frac{\partial^2 Z}{\partial z^2}\right) = \left(\frac{-8\pi^2 mE}{h^2}\right)XYZ. \tag{1.5.28}$$

Dividing eq. (1.5.28) by XYZ yields.

$$\frac{1}{X}\left(\frac{\partial^2 X}{\partial x^2}\right) + \frac{1}{Y}\left(\frac{\partial^2 Y}{\partial y^2}\right) + \frac{1}{Z}\left(\frac{\partial^2 Z}{\partial z^2}\right) = \frac{-8\pi^2 mE}{h^2}. \tag{1.5.29}$$

Now it is obvious that each of the three terms on the left side of eq. (1.5.29) is equal to a constant:

$$\frac{1}{X}\left(\frac{\partial^2 X}{\partial x^2}\right) = -\alpha_x^2, \tag{1.5.30}$$

$$\frac{1}{Y}\left(\frac{\partial^2 Y}{\partial y^2}\right) = -\alpha_y^2, \tag{1.5.31}$$

$$\frac{1}{Z}\left(\frac{\partial^2 Z}{\partial z^2}\right) = -\alpha_z^2, \tag{1.5.32}$$

with the constraint on the constants being

$$\alpha_x^2 + \alpha_y^2 + \alpha_z^2 = \frac{8\pi^2 mE}{h^2}. \tag{1.5.33}$$

In other words, each degree of freedom makes its own contribution to the total energy:

$$\alpha_x^2 = \frac{8\pi^2 mE_x}{h^2}, \quad \alpha_y^2 = \frac{8\pi^2 mE_y}{h^2}, \quad \alpha_z^2 = \frac{8\pi^2 mE_z}{h^2}, \tag{1.5.34}$$

and

$$E = E_x + E_y + E_z. \tag{1.5.35}$$

Each of eqs. (1.5.30) to (1.5.32) is similar to that of the one-dimensional problem, eq. (1.5.2) or (1.5.3). Hence the solutions of eqs. (1.5.30) to (1.5.32) can be readily written

$$X_j(x) = \left(\frac{2}{a}\right)^{\frac{1}{2}} \sin\left(\frac{j\pi x}{a}\right), \quad j = 1, 2, 3, \ldots, \tag{1.5.36}$$

$$Y_k(y) = \left(\frac{2}{b}\right)^{\frac{1}{2}} \sin\left(\frac{k\pi y}{b}\right), \quad k = 1, 2, 3, \ldots, \tag{1.5.37}$$

$$Z_\ell(z) = \left(\frac{2}{c}\right)^{\frac{1}{2}} \sin\left(\frac{\ell\pi z}{c}\right), \quad \ell = 1, 2, 3, \ldots. \tag{1.5.38}$$

The total wavefunction, as given by eq. (1.5.26), then becomes

$$\psi_{j,k,\ell}(x,y,z) = \left(\frac{8}{abc}\right)^{\frac{1}{2}} \sin\left(\frac{j\pi x}{a}\right) \sin\left(\frac{k\pi y}{b}\right) \sin\left(\frac{\ell\pi z}{c}\right),$$

$$j, k, \ell = 1, 2, 3, \ldots. \tag{1.5.39}$$

Now ψ has the unit of volume$^{-1/2}$ and $|\psi|^2$ has the unit of volume^{-1}, as expected for a probability density function of a three-dimensional system.

As in the case of the wavefunctions, the energy of the system is also dependent on three quantum numbers:

$$E_{j,k,\ell} = (E_x)_j + (E_y)_k + (E_z)_\ell$$

$$= \left(\frac{h^2}{8m}\right)\left[\left(\frac{j^2}{a^2}\right) + \left(\frac{k^2}{b^2}\right) + \left(\frac{\ell^2}{c^2}\right)\right], \quad j, k, \ell = 1, 2, 3, \ldots. \tag{1.5.40}$$

When the box is a cube, i.e. $a = b = c$, eq. (1.5.40) becomes

$$E_{j,k,\ell} = \left(\frac{h^2}{8ma^2}\right)(j^2 + k^2 + \ell^2), \quad j, k, \ell = 1, 2, 3, \ldots. \tag{1.5.41}$$

An interesting feature of the energy expression given by eq. (1.5.41) is that different states, with different sets of quantum numbers and different wavefunctions, can have the same energy. When different states have the same energy, they are called degenerate states. For examples,

$$E_{1,1,2} = E_{1,2,1} = E_{2,1,1} = \frac{3h^2}{4ma^2}, \tag{1.5.42}$$

or

$$E_{1,2,3} = E_{2,1,3} = E_{1,3,2} = E_{3,1,2} = E_{3,2,1} = E_{2,3,1} = \frac{7h^2}{4ma^2}. \tag{1.5.43}$$

Now we will apply the particle in a three-dimensional box model to a chemical problem. When sodium vapor is passed over a crystal of NaCl, the crystal exhibits a greenish-yellow color, which is the result of the process

$$\delta\text{Na(g)} + \text{NaCl(c)} \longrightarrow (\text{Na}^+)_{1+\delta}(\text{Cl}^-e_\delta^-)(\text{c}), \quad \delta \ll 1.$$

In this "solid-state reaction," the adsorbed Na atoms ionize on the crystal surface, and the excess electrons diffuse into the interior and occupy vacant anionic sites, while an equal number of chloride ions migrate toward the surface to preserve charge balance. Trapped electrons occupying anionic vacancies are called color centers or F-centers (F stands for *farbe*, meaning color in German). A trapped electron within the lattice of crystalline $(\text{Na}^+)_{1+\delta}(\text{Cl}^-e_\delta^-)$ is depicted in Fig. 1.5.2. The color observed in such experiments can be related with a transition between two energy levels of a (free) particle trapped in a box. Before treating this problem quantitatively, it is of interest to note that the color observed in such experiments depends on the nature of the host crystal, but not on the source of the electron. Thus heating KCl in potassium vapor gives a purple color, and NaCl heated in the same potassium vapor emits a greenish-yellow color.

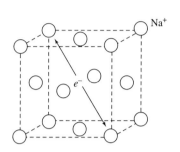

Fig. 1.5.2.
A color center (marked e^-) in a sodium halide crystal. Note that the electronic position is an anionic site. Also, for simplicity, anions are not shown here.

Experimentally, the absorption maxima for NaCl and NaBr crystals have energies (ΔE) 4.32×10^{-19} J (corresponding to $\lambda = 460$ nm) and 3.68×10^{-19} J ($\lambda = 540$ nm), respectively. If we take this ΔE as the energy difference between the two lowest levels of a three-dimensional box, $E_{1,1,1}$ and $E_{1,1,2}$, we can readily calculate the dimension of the box (denoted as ℓ below):

$$\Delta E = \frac{3h^2}{8m\ell^2}.$$

In the above expression we have replaced the dimension of the box a in eq. (1.5.41) by ℓ. Now the ℓ values for NaCl and NaBr can be easily calculated to be 647 and 701 pm, respectively. It is noted that the ℓ value of NaBr "box" is longer than that of NaCl by 54 pm.

The cubic unit-cell dimensions (a) of NaCl and NaBr crystals are 563 and 597 pm, respectively (Table 10.1.4). As shown in Fig.1.5.2, the side (denoted as ℓ' below) of the three-dimensional box for an electron occupying a color center can *at most* be the body diagonal of the unit cell minus twice the cationic radius (the "true" value should be somewhat less):

$$\ell' = (3)^{\frac{1}{2}}a - 2r_{\text{Na}^+}.$$

With r_{Na^+} being 102 pm, we find that the (maximum) ℓ' values for NaCl and NaBr crystals are 771 and 830 pm, respectively. Because of the crudeness of the model, these ℓ' values differ to some extent from the ℓ values calculated earlier. Still, a good qualitative correlation between the two crystals is obtained: the difference between the two ℓ values, 54 pm, is in good accord with the difference between the two ℓ' values, 59 pm.

The topic of color centers in NaCl and related crystals will be discussed more fully in Section 10.1.2.

1.5.3 Particle in a ring

In this system, the electron can move only along the circumference and therefore polar coordinates can be used to advantage, as shown in Fig. 1.5.3:

$$V = \begin{cases} 0, & r = R; \\ \infty & r \neq R. \end{cases}$$

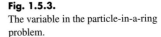

Fig. 1.5.3.
The variable in the particle-in-a-ring problem.

The Schrödinger equation is

$$\left(\frac{-h^2}{8\pi^2 m}\right) \frac{d^2\psi(x)}{dx^2} = E\psi(x). \quad (1.5.44)$$

We can change the variable x to the angular variable ϕ by

$$x = \phi R. \quad (1.5.45)$$

Now eq. (1.5.44) becomes

$$\frac{d^2\psi(\phi)}{d\phi^2} = \left(\frac{-8\pi^2 mER^2}{h^2}\right) \psi(\phi) = -m_\ell^2 \psi(\phi), \quad (1.5.46)$$

with

$$m_\ell^2 = \frac{8\pi^2 mER^2}{h^2}. \quad (1.5.47)$$

Solutions to eq. (1.5.46) are obvious:

$$\psi(\phi) = Ae^{im_\ell \phi}. \quad (1.5.48)$$

Since $\psi(\phi)$ must be single-valued,

$$\psi(\phi + 2\pi) = \psi(\phi) = Ae^{im_\ell \phi} = Ae^{im_\ell(\phi + 2\pi)}. \quad (1.5.49)$$

In other words,

$$e^{im_\ell(2\pi)} = \cos 2m_\ell \pi + i \sin 2m_\ell \pi = 1. \quad (1.5.50)$$

For eq. (1.5.50) to hold,

$$m_\ell = 0, \pm 1, \pm 2, \ldots. \quad (1.5.51)$$

So the wavefunctions have the form

$$\psi_{m_\ell}(\phi) = Ae^{im_\ell \phi}, \quad m_\ell = 0, \pm 1, \pm 2, \ldots. \quad (1.5.52)$$

The constant A can again be determined with the normalization condition:

$$A^2 \int_0^{2\pi} |\psi_{m_\ell}|^2 d\varphi = A^2 \int_0^{2\pi} \psi_{m_\ell}^* \psi_{m_\ell} d\phi = A^2 \int_0^{2\pi} d\phi = 2\pi A^2 = 1, \quad (1.5.53)$$

Fig. 1.5.4.
The eigenvalues and eigenfunctions of the particle-in-a-ring problem.

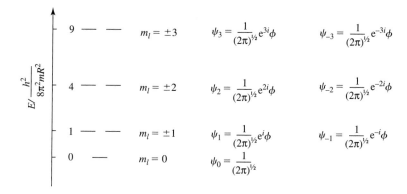

or

$$A = (2\pi)^{-\frac{1}{2}}. \qquad (1.5.54)$$

So the normalized wavefunctions are

$$\psi_{m_\ell} = (2\pi)^{-\frac{1}{2}} e^{im_\ell \phi}, \quad m_\ell = 0, \pm 1, \pm 2, \ldots. \qquad (1.5.55)$$

The allowed energy values for this system can be determined using eq. (1.5.47):

$$E_{m_\ell} = \frac{m_\ell^2 h^2}{8\pi^2 mR^2}, \quad m_\ell = 0, \pm 1, \pm 2, \ldots. \qquad (1.5.56)$$

In other words, only the ground state is non-degenerate, while all the excited states are doubly degenerate. The quantum mechanical results of the particle-in-a-ring problem are summarized in Fig. 1.5.4.

If we apply the free-electron model to the six π electrons of benzene, we see that the ψ_0, ψ_1, and ψ_{-1} orbitals are filled with electrons, while ψ_2 and ψ_{-2} and all the higher levels are vacant. To excite an electron from ψ_1 (or ψ_{-1}) to ψ_2 (or ψ_{-2}), we need an energy of

$$\Delta E = \frac{3h^2}{8\pi^2 mR^2} = hc/\lambda.$$

If we take $R = 140$ pm, we get $\lambda = 212$ nm. Experimentally, benzene absorbs weakly at 268 nm.

Next we apply this simple model to the annulenes. These compounds are monocyclic conjugated polyenes with the general molecular formula $(CH)_n$ with n even. Thus benzene may be considered as [6]-annulene. As n increases, essentially all [n]-annulenes are non-planar. For instance, cyclooctatetraene, $(CH)_8$, has the well-know "tub" structure. However, [18]-annulene is nearly planar, as shown below (bond lengths in picometers are displayed in bold italic font).

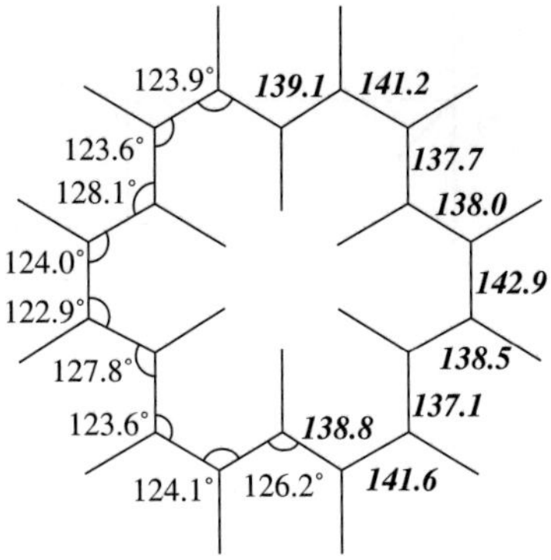

[Figure: [18]-annulene structure with bond angles labeled: 123.9°, 139.1, 141.2, 123.6°, 137.7, 128.1°, 138.0, 124.0°, 142.9, 122.9°, 127.8°, 138.5, 123.6°, 137.1, 138.8, 124.1°, 126.2°, 141.6]

Adding up the 18 bond lengths around the ring, we obtain a value of 2510 pm. If we take this to be the circumference of the molecular ring, the radius R of the ring is approximately 400 pm. Thus the lowest electronic transition of this molecule requires an energy of

$$\Delta E = \frac{9h^2}{8\pi^2 mR^2} = hc/\lambda.$$

Here λ is calculated to be 572 nm. Experimentally, [18]-annulene absorbs at $\lambda = 790$ nm. Clearly the "agreement" is not very good. But this crude model does arrive at an absorption wavelength that has the correct order of magnitude.

In Section 1.5.1, it was mentioned that the energy of the lowest state of a particle confined in a one-dimensional box is not zero and this residual energy is a consequence of the Uncertainty Principle. Yet the ground state energy of the particle-in-a-ring problem is zero. Does this mean the present result is in violation of the Uncertainty Principle? The answer is clearly no, and the reason is as follows. In a one-dimensional box, variable x starts from 0 and ends at a, the length of the box. Hence Δx can at most be a. On the other hand, in a ring, cyclic variable ϕ does not lie within a finite domain. In such a situation, the uncertainty in position cannot be estimated.

1.5.4 Particle in a triangle

To conclude this chapter, we present the quantum mechanical results of the particle-in-a-triangle problem. Before going into details, we first need to note that, if the "box" is a scalene triangle, no analytical solution is known. Indeed, the Schrödinger equation is exactly solvable for only a few triangular systems. In addition, for all these solvable (two-dimensional) cases, the wavefunctions are no longer the simple products of two functions each involving only one variable.

In the following discussion, some language and notation of group theory are used for convenience. The meaning of this language or notation is made clear in Chapter 6.

(1) Isosceles right triangle

The coordinate system chosen for this problem is shown in Fig. 1.5.5.

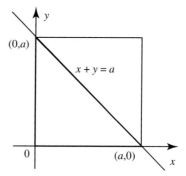

Fig. 1.5.5.
Coordinate system for the problem of a particle in an isosceles right triangle.

The Schrödinger equation in this case is simply the two-dimensional analog of eqs. (1.5.2) and (1.5.25):

$$\left(\frac{-h^2}{8\pi^2 m}\right)\left(\frac{\partial^2 \psi}{\partial x^2} + \frac{\partial^2 \psi}{\partial y^2}\right) = E\psi \qquad (1.5.57)$$

The boundary conditions for this problem are ψ vanishes when $x = 0$, $y = 0$ or $x + y = a$. If we apply the technique of separation of variables introduced in Section 1.5.2 to eq. (1.5.57), we can easily obtain the following wavefunctions and energy values:

$$\psi_{j,k}(x,y) = \left[\left(\frac{2}{a}\right)^{\frac{1}{2}}\sin\left(\frac{j\pi x}{a}\right)\right]\left[\left(\frac{2}{a}\right)^{\frac{1}{2}}\sin\left(\frac{k\pi y}{a}\right)\right], \qquad (1.5.58)$$

$$E_{j,k} = \left(\frac{h^2}{8ma^2}\right)(j^2 + k^2), \quad j,k = 1,2,3,\ldots. \qquad (1.5.59)$$

These are the solutions of the particle-in-a-square problem.

Since the expression for $E_{j,k}$ is symmetrical with respect to the exchange of quantum numbers j and k, i.e. $E_{j,k} = E_{k,j}$. In other words, $\psi_{j,k}$ and $\psi_{k,j}$ are degenerate wavefunctions.

It is clear that wavefunctions $\psi_{j,k}$ satisfy the boundary condition of ψ vanishes when $x = 0$ or $y = 0$, but not the condition of ψ vanishes when $x+y = a$. In order to satisfy the latter condition, we first linearly combine $\psi_{j,k}$ and $\psi_{k,j}$:

$$\psi'_{j,k} = (2)^{-\frac{1}{2}}(\psi_{j,k} + \psi_{k,j})$$

$$= \left[\frac{(2)^{\frac{1}{2}}}{a}\right]\left[\sin\left(\frac{j\pi x}{a}\right)\sin\left(\frac{k\pi y}{a}\right) + \sin\left(\frac{k\pi x}{a}\right)\sin\left(\frac{j\pi y}{a}\right)\right] \qquad (1.5.60)$$

$$\psi''_{j,k} = (2)^{-\frac{1}{2}}(\psi_{j,k} - \psi_{k,j})$$

$$= \left[\frac{(2)^{\frac{1}{2}}}{a}\right]\left[\sin\left(\frac{j\pi x}{a}\right)\sin\left(\frac{k\pi y}{a}\right) - \sin\left(\frac{k\pi x}{a}\right)\sin\left(\frac{j\pi y}{a}\right)\right]. \qquad (1.5.61)$$

When $j = k$, $\psi'_{j,k}$ is simply $\psi'_{j,j}$ (aside from a numerical factor) and $\psi''_{j,j}$ vanishes; hence neither $\psi'_{j,j}$ nor $\psi''_{j,j}$ is an acceptable solution. On the other hand, when $j = k \pm 1, k \pm 3, \ldots$, $\psi'_{j,k}$ vanishes when $x + y = a$; also, when $j = k \pm 2, k \pm 4, \ldots$, $\psi''_{j,k}$ vanishes under the same condition. Thus, $\psi'_{j,k}$ ($j = k \pm 1, k \pm 3, \ldots$) and $\psi''_{j,k}$ ($j = k \pm 2, j \pm 4, \ldots$) are the solutions to the isosceles triangle problem. It is clear that these functions are no longer products of two functions each involving only one variable. In addition, the energy expression now becomes:

$$E_{j,k} = \left(\frac{h^2}{8ma^2}\right)(j^2 + k^2), \quad j, k = 1, 2, 3, \ldots \text{ and } j \neq k. \qquad (1.5.62)$$

Since only one wavefunction can be written for a set of quantum numbers (j, k), i.e. $\psi'_{j,k} = \psi'_{k,j}$ and $\psi''_{j,k} = \psi''_{k,j}$ (aside from a negative sign), the "systematic" degeneracy in the particle-in-a-square problem, i.e. $E_{k,j} = E_{j,k}$ in eq. (1.5.59), is now removed. This is expected as there is a reduction in symmetry from D_{4h} (square box) to C_{2v} (isosceles triangular box). Still, some "accidental" degeneracies remain in the right-triangular case; for example, $E_{1,8} = E_{4,7}$. These degeneracies may be dealt with by techniques of elementary number theory, which is clearly beyond the scope of this book. Additional discussions may be found in the references listed at the end of this chapter.

(2) Equilateral triangle

When the triangle is equilateral, with the coordinates as shown in Fig. 1.5.6, the boundary conditions for eq. (1.5.57) now take the form $\psi = 0$ for $y = 0$, $y = (3)^{1/2}x$ or $y = (3)^{1/2}(a - x)$.

There are a number of ways to solve this problem, and the results obtained by these methods appear to be quite different at first sight. In fact, in some formulations, it takes multiple sets of quantum number to specify a single state. In any event, the mathematics involved in all of these treatments is beyond the scope of this book. Here we merely present the results of one approach and discuss the energies and wavefunctions of this system from the symmetry viewpoint.

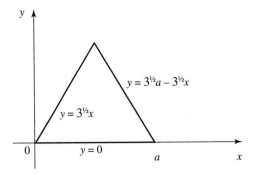

Fig. 1.5.6.
Coordinate system for an equilateral triangle. Equations of the three sides specify the boundary conditions.

Table 1.5.1. Energies and symmetry of the wavefunctions for the first seven levels of a particle in an equilateral triangle.

(p, q)	Energy*	Symmetry
$(1, 0)$	1	A_1
$(1^1/_3, ^1/_3)$	$2^1/_3$	E
$(2, 0)$	4	A_1
$(1^2/_3, ^2/_3)$	$4^1/_3$	E
$(2^1/_3, ^1/_3)$	$6^1/_3$	E
$(2, 1)$	7	A_1, A_2
$(3, 0)$	9	A_1

* In units of ground state energy E_0, $2h^2/3ma^2$.

For this two-dimensional system, it again takes a set of two quantum numbers to specify a state. The energies of this system may be expressed as

$$E_{p,q} = (p^2 + pq + q^2)\left(\frac{2h^2}{3ma^2}\right)$$

$$\text{with } q = 0, \frac{1}{3}, \frac{2}{3}, 1, \ldots, \quad p = q+1, q+2, \ldots. \quad (1.5.63)$$

So the ground state energy E_0 is

$$E_0 = E_{1,0} = \frac{2h^2}{3ma^2}. \quad (1.5.64)$$

The wavefunctions $\psi_{p,q}$ may be classified according to their symmetry properties. If we take the symmetry point group of this system to be C_{3v}, there are three symmetry species in this group: A_1 (symmetric with respect to all operations of this group), A_2 (symmetric with respect to the threefold rotations but antisymmetric with respect to the vertical symmetry planes), and E (a two-dimensional representation).

For a level with energy $E_{p,0}$, the wavefunction $\psi_{p,0}$ has A_1 symmetry and this level is non-degenerate. On the other hand, a level defined by positive integral quantum numbers p and q with energy $E_{p,q}$ is doubly degenerate. One wavefunction of this level has A_1 symmetry and the other one has A_2 symmetry. When p and q are non-integers $(1/3, 2/3, 1^1/_3, \ldots,$ etc.$)$, the doubly degenerate functions form an E set. Table 1.5.1 lists the quantum numbers, energies and wavefunction symmetry for the first seven states for a particle in an equilateral triangle.

The wavefunctions $\psi_{p,q}$ have the form

$$\psi_{p,q}(A_1) = \cos\left[\frac{q(3)^{1/2}\pi x}{A}\right]\sin\left[\frac{(2p+q)\pi y}{A}\right]$$
$$-\cos\left[\frac{p(3)^{1/2}\pi x}{A}\right]\sin\left[\frac{(2q+p)\pi y}{A}\right]$$
$$-\cos\left[\frac{(p+q)(3)^{1/2}\pi x}{A}\right]\sin\left[\frac{(p-q)\pi y}{A}\right] \quad (1.5.65)$$

when $q = 0$,

$$\psi_{p,0}(A_1) = \sin\left[\frac{2p\pi y}{A}\right] - 2\sin\left[\frac{p\pi y}{A}\right]\cos\left[\frac{p(3)^{1/2}\pi x}{A}\right] \qquad (1.5.66)$$

$$\psi_{p,q}(A_2) = \sin\left[\frac{q(3)^{1/2}\pi x}{A}\right]\sin\left[\frac{(2p+q)\pi y}{A}\right]$$
$$- \sin\left[\frac{p(3)^{1/2}\pi x}{A}\right]\sin\left[\frac{(2q+p)\pi y}{A}\right]$$
$$+ \sin\left[\frac{(p+q)(3)^{1/2}\pi x}{A}\right]\sin\left[\frac{(p-q)\pi y}{A}\right]. \qquad (1.5.67)$$

In eqs. (1.5.65) to (1.5.67), A represents the altitude of the triangle. The two wavefunctions that make up the E set may also be expressed by eqs. (1.5.65) and (1.5.67), but quantum numbers p and q now have non-integral values.

The wavefunctions for the first seven states with energies and symmetries summarized in Table 1.5.1 are graphically illustrated in Fig. 1.5.7. It is now clear that all levels with non-integral quantum numbers are doubly degenerate and their respective wavefunctions form an E set. Also, the A_1 wavefunctions are symmetric with respect to the threefold rotations as well as to the symmetry planes, while A_2 wavefunctions are symmetric with respect to the threefold rotations but antisymmetric with respect to the symmetry planes.

(3) The 30°–60°–90° triangle

This is a triangle that is half of an equilateral triangle. From Fig. 1.5.7, it is obvious that all the A_2 functions and one component from each pair of the E functions possess a nodal plane which bisects the equilateral triangle into two 30°–60°–90° triangles. Thus a particle confined to a 30°–60°–90° triangle has energies given by eq. (1.5.67), with the allowed quantum numbers $q = 1/3, 2/3, 1, \ldots$

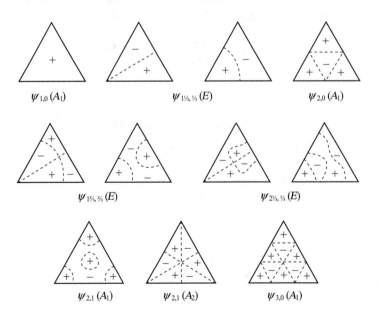

Fig. 1.5.7.
Graphical representations of the wavefunctions for the first seven states of the equilateral triangle.

and $p = q + 1, q + 2, \ldots$. Among the seven states listed in Table 1.5.1, only one component of the E states and the A_2 states are the acceptable solutions of the 30°–60°–90° triangle.

References

1. P. W. Atkins and J. de Paula, *Atkins' Physical Chemistry*, 8th edn., Oxford University Press, Oxford, 2006.
2. R. J. Silbey and R. A. Alberty, *Physical Chemistry*, 3rd edn., Wiley, New York, 2001.
3. I. N. Levine, *Physical Chemistry*, 5th edn., McGraw-Hill, Boston, 2002.
4. R. S. Berry, S. A. Rice, and J. Ross, *Physical Chemistry*, 2nd edn., Oxford University Press, New York, 2000.
5. R. Grinter, *The Quantum in Chemistry: An Experimentalist's View*, Wiley, New York, 2005.
6. S. M. Blinder, *Introduction to Quantum Mechanics*, Elsevier/Academic Press, Amsterdam, 2004.
7. P. W. Atkins and R. S. Friedman, *Molecular Quantum Mechanics*, 3rd ed., Oxford University Press, New York, 1997.
8. V. Magnasco, *Elementary Methods of Molecular Quantum Mechanics*, Elsevier, Amsterdam, 2007.
9. G. C. Schatz and M. A. Ratner, *Quantum Mechanics in Chemistry*, Prentice-Hall, Englewood Cliffs, NJ, 1993.
10. M. A. Ratner and G. C. Schatz, *Introduction to Quantum Mechanics in Chemistry*, Prentice Hall, Upper Saddle River, NJ, 2001.
11. J. Simons and J. Nichols, *Quantum Mechanics in Chemistry*, Oxford University Press, New York, 1997.
12. F. L. Pilar, *Elementary Quantum Chemistry*, 2nd edn., McGraw-Hill, New York, 1990.
13. I. N. Levine, *Quantum Chemistry*, 5th edn., Prentice-Hall, Upper Saddle River, NJ, 2000.
14. J. E. House, *Fundamentals of Quantum Chemistry*, 2nd edn., Elsevier/Academic Press, San Diego, CA, 2004.
15. C. S. Johnson, Jr. and L. G. Pedersen, *Problems in Quantum Chemistry and Physics*, Addison-Wesley, Reading, MA, 1974; repr. Dover, New York, 1987.
16. G. L. Squires, *Problems in Quantum Mechanics: with Solutions*, Cambridge University Press, Cambridge, 1995.
17. W.-K. Li, Degeneracy in the particle-in-a-square problem. *Am. J. Phys.* **50**, 666 (1982).
18. W.-K. Li, A particle in an isosceles right triangle. *J. Chem. Educ.* **61**, 1034 (1984).
19. W.-K. Li and S. M. Blinder, Particle in an equilateral triangle: exact solution for a particle in a box. *J. Chem. Educ.* **64**, 130–2 (1987).

The Electronic Structure of Atoms

2

Now we are ready to apply the method of wave mechanics to study the electronic structure of the atoms. At the beginning of this chapter, we concentrate on the hydrogen atom, which consists of one proton and one electron. After treating the hydrogen atom, we will proceed to the other atoms in the Periodic Table.

2.1 The hydrogen atom

2.1.1 Schrödinger equation for the hydrogen atom

If we place the nucleus of the hydrogen atom at the origin of a set of Cartesian coordinates, the position of the electron would be given by x, y, and z, as shown in Fig. 2.1.1. However, the solution of the Schrödinger equation for this system becomes intractable if it is done in Cartesian coordinates. Instead, this problem is solved using polar spherical coordinates r, θ, and ϕ, which are also shown in Fig. 2.1.1. These two sets of coordinates are related by:

$$\begin{cases} z = r\cos\theta \\ x = r\sin\theta\cos\phi \\ y = r\sin\theta\sin\phi \end{cases} \quad (2.1.1)$$

Some other useful relationships are

$$r^2 = x^2 + y^2 + z^2, \quad (2.1.2)$$

$$\tan\phi = \frac{y}{x}. \quad (2.1.3)$$

Also, these two sets of variables have different ranges:

$$-\infty < x, y, z < \infty; \quad (2.1.4)$$

$$\begin{cases} 0 \leq r < \infty \\ 0 \leq \theta \leq \pi \\ 0 \leq \phi \leq 2\pi. \end{cases} \quad (2.1.5)$$

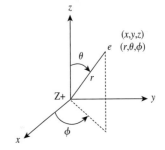

Fig. 2.1.1.
The coordinate system for a hydrogen-like (one-electron) atom; $Z = 1$ for the H atom.

Fundamentals of Bonding Theory

The Laplacian operator ∇^2 has the form

$$\nabla^2 = \frac{\partial^2}{\partial x^2} + \frac{\partial^2}{\partial y^2} + \frac{\partial^2}{\partial z^2}$$

$$= \frac{1}{r^2}\frac{\partial}{\partial r}\left(r^2\frac{\partial}{\partial r}\right) + \frac{1}{r^2\sin\theta}\frac{\partial}{\partial \theta}\left(\sin\theta\frac{\partial}{\partial \theta}\right) + \frac{1}{r^2\sin^2\theta}\frac{\partial^2}{\partial \phi^2}. \quad (2.1.6)$$

The volume element $d\tau$ is given by

$$d\tau = dxdydz = r^2 dr \sin\theta d\theta d\phi. \quad (2.1.7)$$

For the hydrogen atom and other hydrogenic ions, the potential energy V is simply the attraction between the proton and the electron:

$$V = -\frac{Ze^2}{4\pi\varepsilon_0 r}. \quad (2.1.8)$$

So the Schrödinger equation for this system is

$$\left[\left(-\frac{h^2}{8\pi^2 m}\right)\nabla^2 - \left(\frac{Ze^2}{4\pi\varepsilon_0 r}\right)\right]\psi(r,\theta,\phi) = E\psi(r,\theta,\phi), \quad (2.1.9)$$

or

$$\frac{1}{r^2}\frac{\partial}{\partial r}\left(r^2\frac{\partial}{\partial r}\psi\right) + \frac{1}{r^2\sin\theta}\frac{\partial}{\partial \theta}\left(\sin\theta\frac{\partial\psi}{\partial \theta}\right)$$

$$+ \frac{1}{r^2\sin^2\theta}\frac{\partial^2\psi}{\partial \phi^2} + \frac{8\pi^2 m}{h^2}\left(E + \frac{Ze^2}{4\pi\varepsilon_0 r}\right)\psi = 0. \quad (2.1.10)$$

To solve eq. (2.1.10), we first assume that the function ψ (with three variables) is a product of three functions, each of one variable:

$$\psi(r,\theta,\phi) = R(r)\Theta(\theta)\Phi(\phi). \quad (2.1.11)$$

Upon substituting, we can "factorize" eq. (2.1.10) into the following three equations, each involving only one variable:

$$\frac{d^2\Phi}{d\phi^2} = -m_\ell^2 \Phi, \quad (2.1.12)$$

$$\frac{m_\ell^2 \Theta}{\sin^2\theta} - \frac{1}{\sin\theta}\frac{d}{d\theta}\left(\sin\theta\frac{d\Theta}{d\theta}\right) - \beta\Theta = 0, \quad (2.1.13)$$

$$\frac{1}{r^2}\frac{d}{dr}\left(r^2\frac{dR}{dr}\right) - \frac{\beta}{r^2}R + \frac{8\pi^2 m}{h^2}\left(E + \frac{Ze^2}{4\pi\varepsilon_0 r}\right)R = 0. \quad (2.1.14)$$

In eqs. (2.1.12) to (2.1.14), m_ℓ and β are so-called "separation constants" and they will eventually lead to quantum numbers. Also, $R(r)$ is called the radial function, while the product $\Theta(\theta)\Phi(\phi)$, henceforth called $Y(\theta,\phi)$, is the angular function:

$$Y(\theta,\phi) = \Theta(\theta)\Phi(\phi). \quad (2.1.15)$$

2.1.2 Angular functions of the hydrogen atom

After solving eqs. (2.1.12) and (2.1.13), it is found that, in order for functions to be meaningful, the separation constants m_ℓ and β must take the following values:

$$m_\ell = 0, \pm 1, \pm 2, \ldots, \quad (2.1.16)$$

$$\beta = \ell(\ell + 1), \ell = 0, 1, 2, \ldots, \quad (2.1.17)$$

and

$$m_\ell = -\ell, -\ell + 1, \ldots, \ell. \quad (2.1.18)$$

The integers ℓ and m_ℓ are called azimuthal (or angular momentum) and magnetic quantum numbers, respectively.

Moreover, the function $\Phi(\phi)$ depends on m_ℓ and hence is written as $\Phi_{m_\ell}(\phi)$, while $\Theta(\theta)$ depends on both ℓ and m_ℓ and is written as $\Theta_{\ell,m_\ell}(\theta)$. Their product $Y(\theta, \phi)$ thus also depends on both ℓ and m_ℓ:

$$Y_{\ell,m_\ell}(\theta, \phi) = \Theta_{\ell,m_\ell}(\theta) \Phi_{m_\ell}(\phi). \quad (2.1.19)$$

The $Y_{\ell,m_\ell}(\theta, \phi)$ functions are called spherical harmonics. They determine the angular character of the electronic wavefunction and will be of primary consideration in the treatment of directional bonding.

Spherical harmonics, which describes the angular parts of the atomic orbitals, are labeled by their ℓ values according to the scheme

$$\begin{array}{l} \ell = 0, 1, 2, 3, 4, \ldots, \\ \text{Label} = \text{s}, \text{p}, \text{d}, \text{f}, \text{g}, \ldots. \end{array} \quad (2.1.20)$$

Table 2.1.1 lists the first few, and the most often encountered, spherical harmonics.

The spherical harmonics $Y_{\ell,m_\ell}(\theta, \phi)$ form an orthonormal set of functions:

$$\int_0^{2\pi} \int_0^{\pi} Y^*_{\ell,m_\ell} Y_{\ell',m'_\ell} \sin\theta \, d\theta \, d\phi = \int_0^{2\pi} \Phi^*_{m'_\ell} \Phi_{m_\ell} d\phi \cdot \int_0^{\pi} \Theta^*_{\ell',m'_\ell} \Theta_{\ell,m_\ell} \sin\theta \, d\theta$$

$$= \begin{cases} 1, & \text{when } m'_\ell = m_\ell \text{ AND } \ell' = \ell; \\ 0, & \text{when } m'_\ell \neq m_\ell \text{ OR } \ell' \neq \ell. \end{cases} \quad (2.1.21)$$

Note that the quantum number m_ℓ appears in the exponential function $e^{im_\ell \phi}$ in the spherical harmonics. The Y_{ℓ,m_ℓ} functions, being complex, cannot be conveniently drawn in real space. However, we can linearly combine them to make

Table 2.1.1. Explicit expressions for the spherical harmonics with $\ell = 0, 1, 2, 3$

$Y_{0,0} = \frac{1}{(4\pi)^{1/2}}$	$Y_{3,3} = \left(\frac{35}{64\pi}\right)^{1/2} \sin^3\theta e^{3i\phi}$
$Y_{1,1} = \left(\frac{3}{8\pi}\right)^{1/2} \sin\theta e^{i\phi}$	$Y_{3,2} = \left(\frac{105}{32\pi}\right)^{1/2} \sin^2\theta \cos\theta e^{2i\phi}$
$Y_{1,0} = \left(\frac{3}{4\pi}\right)^{1/2} \cos\theta$	$Y_{3,1} = \left(\frac{21}{64\pi}\right)^{1/2} (5\cos^2\theta - 1)\sin\theta e^{i\phi}$
$Y_{1,-1} = \left(\frac{3}{8\pi}\right)^{1/2} \sin\theta e^{-i\phi}$	$Y_{3,0} = \left(\frac{63}{16\pi}\right)^{1/2} \left(\frac{5}{3}\cos^3\theta - \cos\theta\right)$
$Y_{2,2} = \left(\frac{15}{32\pi}\right)^{1/2} \sin^2\theta e^{2i\phi}$	$Y_{3,-1} = \left(\frac{21}{64\pi}\right)^{1/2} (5\cos^2\theta - 1)\sin\theta e^{-i\phi}$
$Y_{2,1} = \left(\frac{15}{8\pi}\right)^{1/2} \sin\theta \cos\theta e^{i\phi}$	$Y_{3,-2} = \left(\frac{105}{32\pi}\right)^{1/2} \sin^2\theta \cos\theta e^{-2i\phi}$
$Y_{2,0} = \left(\frac{5}{16\pi}\right)^{1/2} (3\cos^2\theta - 1)$	$Y_{3,-3} = \left(\frac{35}{64\pi}\right)^{1/2} \sin^3\theta e^{-3i\phi}$
$Y_{2,-1} = \left(\frac{15}{8\pi}\right)^{1/2} \sin\theta \cos\theta e^{-i\phi}$	
$Y_{2,-2} = \left(\frac{15}{32\pi}\right)^{1/2} \sin^2\theta e^{-2i\phi}$	

Table 2.1.2. The real angular functions of the s, p, d and f orbitals

$\ell = 0$ s	$\left(\frac{1}{4\pi}\right)^{1/2}$	$\ell = 3$ f_{z^3}	$\frac{1}{4}\left(\frac{7}{\pi}\right)^{1/2} (5\cos^3\theta - 3\cos\theta)$
$\ell = 1$ p_z	$\left(\frac{3}{4\pi}\right)^{1/2} \cos\theta$	f_{xz^2}	$\frac{1}{8}\left(\frac{42}{\pi}\right)^{1/2} \sin\theta(5\cos^2\theta - 1)\cos\phi$
p_x	$\left(\frac{3}{4\pi}\right)^{1/2} \sin\theta \cos\phi$	f_{yz^2}	$\frac{1}{8}\left(\frac{42}{\pi}\right)^{1/2} \sin\theta(5\cos^2\theta - 1)\sin\phi$
p_y	$\left(\frac{3}{4\pi}\right)^{1/2} \sin\theta \sin\phi$	f_{xyz}	$\frac{1}{4}\left(\frac{105}{\pi}\right)^{1/2} \sin^2\theta \cos\theta \sin 2\phi$
$\ell = 2$ d_{z^2}	$\frac{1}{4}\left(\frac{5}{\pi}\right)^{1/2} (3\cos^2\theta - 1)$	$f_{z(x^2-y^2)}$	$\frac{1}{4}\left(\frac{105}{\pi}\right)^{1/2} \sin^2\theta \cos\theta \cos 2\phi$
d_{xz}	$\frac{1}{4}\left(\frac{15}{\pi}\right)^{1/2} \sin 2\theta \cos\phi$	$f_{x(x^2-3y^2)}$	$\frac{1}{8}\left(\frac{70}{\pi}\right)^{1/2} \sin^3\theta \cos 3\phi$
d_{yz}	$\frac{1}{4}\left(\frac{15}{\pi}\right)^{1/2} \sin 2\theta \sin\phi$	$f_{y(3x^2-y^2)}$	$\frac{1}{8}\left(\frac{70}{\pi}\right)^{1/2} \sin^3\theta \sin 3\phi$
$d_{x^2-y^2}$	$\frac{1}{4}\left(\frac{15}{\pi}\right)^{1/2} \sin^2\theta \cos 2\phi$		
d_{xy}	$\frac{1}{4}\left(\frac{15}{\pi}\right)^{1/2} \sin^2\theta \sin 2\phi$		

the imaginary parts vanish. For example:

$$\frac{1}{(2)^{1/2}}(Y_{1,1} + Y_{1,-1}) = \frac{1}{(2)^{1/2}} \cdot \frac{1}{2} \cdot \left(\frac{3}{2\pi}\right)^{1/2} \sin\theta \left(e^{i\phi} + e^{-i\phi}\right)$$

$$= \frac{1}{4}\left(\frac{3}{\pi}\right)^{1/2} \sin\theta (\cos\phi + i\sin\phi + \cos\phi - i\sin\phi)$$

$$= \frac{1}{2}\left(\frac{3}{\pi}\right)^{1/2} \sin\theta \cos\phi. \qquad (2.1.22)$$

As $\sin\theta \cos\phi$ describes the angular dependence of the x component of r [eq. (2.1.1)], the combination $(1/(2)^{1/2})(Y_{1,1} + Y_{1,-1})$ is hence called

The Electronic Structure of Atoms

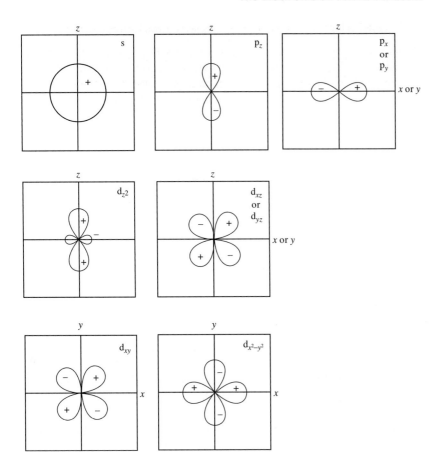

Fig. 2.1.2.
The angular functions of the s, p, and d orbitals.

the p_x orbital. The most often used real angular functions are summarized in Table 2.1.2.

The real angular functions given in Table 2.1.2 can be drawn readily. In Fig. 2.1.2, the shapes of the s, p, and d orbitals are shown, along with the signs of the lobes in each angular function, and the radial function $R(r)$ is assumed to be a constant. We note once again that these figures represent only the shapes of the orbitals. The curves outlining the shapes are not the contour lines of the atomic orbitals shown in some subsequent figures such as Figs. 2.1.4 and 2.1.6.

From Fig. 2.1.2, we can see that if we have an electron occupying an s orbital, we will have equal probability of finding this electron in all possible directions. On the other hand, for a p_x electron, we will most likely find it along the $+x$ or $-x$ axis. For an electron in the d_{xy} orbital, the electron is likely found if we look for it in the xy plane and along the directions of $x = y$ or $x = -y$. For the remaining orbitals, the directions along which the electron is most likely to be found can be determined readily with the help of these figures.

2.1.3 Radial functions and total wavefunctions of the hydrogen atom

The remaining equation to be solved is the radial equation [eq. (2.1.14)], bearing in mind that β is now $\ell(\ell+1)$ [eq. (2.1.17)]. Hence the solution $R(r)$ also depends on ℓ. In addition, the principal quantum number n arises from the solution of this equation. Thus the radial functions depend on both n and ℓ, and we write them as $R_{n,\ell}(r)$. Specifically, n and ℓ take the values

$$n = 1, 2, 3, 4, \ldots, \tag{2.1.23}$$

$$\ell = 0, 1, 2, 3, \ldots, n-1. \tag{2.1.24}$$

Also, as energy E appears only in the radial equation, it might be expected to be dependent on n and ℓ and independent of m_ℓ. As it turns out, it only depends on n.

The explicit forms of the radial functions of hydrogenic orbitals 1s through 4f are listed in Table 2.1.3, where Z is the nuclear charge of the atom and a_0 is the Bohr radius:

$$a_0 = \varepsilon_0 h^2 / \pi m e^2 = 0.529 \times 10^{-10} \text{ m} = 52.9 \text{ pm}. \tag{2.1.25}$$

Note that the radial functions $R_{n,\ell}$ also form an orthonormal set of functions:

$$\int_0^\infty R_{n',\ell} R_{n,\ell} r^2 dr = \begin{cases} 1, & \text{when } n' = n; \\ 0, & \text{when } n' \neq n. \end{cases} \tag{2.1.26}$$

The first six radial functions given in Table 2.1.3 are plotted in Fig. 2.1.3. The squares of these functions, $|R_{n,\ell}(r)|^2$, which is related to the probability density

Table 2.1.3. The radial functions for hydrogenic orbitals with $n = 1$–4

1s	$R_{1,0} =$	$\left(\frac{Z}{a_0}\right)^{\frac{3}{2}} 2 e^{-Zr/a_0}$
2s	$R_{2,0} =$	$\left(\frac{Z}{2a_0}\right)^{\frac{3}{2}} \left(2 - \frac{Zr}{a_0}\right) e^{-Zr/2a_0}$
2p	$R_{2,1} =$	$\left(\frac{Z}{2a_0}\right)^{\frac{3}{2}} \left(\frac{Zr}{a_0 (3)^{1/2}}\right) e^{-Zr/2a_0}$
3s	$R_{3,0} =$	$\left(\frac{Z}{3a_0}\right)^{\frac{3}{2}} \left[2 - \left(\frac{4Zr}{3a_0}\right) + \left(\frac{4}{27}\right)\left(\frac{Zr}{a_0}\right)^2\right] e^{-Zr/3a_0}$
3p	$R_{3,1} =$	$\left(\frac{Z}{3a_0}\right)^{\frac{3}{2}} \left(\frac{2(2)^{1/2}}{9}\right) \left[\left(\frac{2Zr}{a_0}\right) - \frac{1}{3}\left(\frac{Zr}{a_0}\right)^2\right] e^{-Zr/3a_0}$
3d	$R_{3,2} =$	$\left(\frac{Z}{3a_0}\right)^{\frac{3}{2}} \left(\frac{4}{27(10)^{1/2}}\right) \left(\frac{Zr}{a_0}\right)^2 e^{-Zr/3a_0}$
4s	$R_{4,0} =$	$\left(\frac{Z}{4a_0}\right)^{\frac{3}{2}} \left[2 - \left(\frac{3Zr}{2a_0}\right) + \left(\frac{1}{4}\right)\left(\frac{Zr}{a_0}\right)^2 - \left(\frac{1}{96}\right)\left(\frac{Zr}{a_0}\right)^3\right] e^{-Zr/4a_0}$
4p	$R_{4,1} =$	$\left(\frac{Z}{4a_0}\right)^{\frac{3}{2}} \left(\frac{(5)^{1/2}}{2(3)^{1/2}}\right) \left[\left(\frac{Zr}{a_0}\right) - \left(\frac{1}{4}\right)\left(\frac{Zr}{a_0}\right)^2 + \left(\frac{1}{80}\right)\left(\frac{Zr}{a_0}\right)^3\right] e^{-Zr/4a_0}$
4d	$R_{4,2} =$	$\left(\frac{Z}{4a_0}\right)^{\frac{3}{2}} \left(\frac{1}{8(5)^{1/2}}\right) \left[\left(\frac{Zr}{a_0}\right)^2 - \left(\frac{1}{12}\right)\left(\frac{Zr}{a_0}\right)^3\right] e^{-Zr/4a_0}$
4f	$R_{4,3} =$	$\left(\frac{Z}{4a_0}\right)^{\frac{3}{2}} \left(\frac{1}{96(35)^{1/2}}\right) \left(\frac{Zr}{a_0}\right)^3 e^{-Zr/4a_0}$

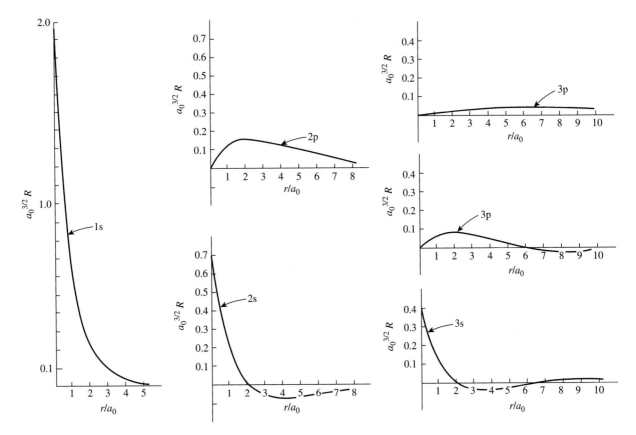

Fig. 2.1.3.
The first six radial functions of the hydrogen atom drawn to the same scale.

function, will be discussed in detail later. At this point, we are in a position to write down the first few total wavefunctions $\psi(r,\theta,\phi)$, which are simply the products of the angular functions (Table 2.1.2) and the radial functions (Table 2.1.3). These total wavefunctions are listed in Table 2.1.4.

The hydrogenic wavefunctions have the general analytical form:

$$\psi_{n,\ell,m_\ell}(r,\theta,\phi) = R_{n,\ell}(r)\Theta_{\ell,m_\ell}(\theta)\Phi_{m_\ell}(\phi), \qquad (2.1.27)$$

with each part normalized to unity. The total wavefunctions of some representative orbitals are plotted in Fig. 2.1.4.

From Fig. 2.1.4, it is seen that, while all s orbitals are spherically symmetric, the 2s orbital is larger than the 1s, and the 3s is even larger. Note that the 2s orbital has one node at $r = 2a_0$ and the 3s orbital has two nodes at $r = 1.91a_0$ and $r = 7.08a_0$ (see below). On the other hand, the $2p_z$ orbital is no longer spherically symmetric. Instead, its absolute value has a maximum at $\theta = 0°$ and $\theta = 180°$ and the function vanishes at $\theta = 90°$. In other words, the xy plane is a nodal plane. As the figure indicates, the function has a positive value for $0° < \theta < 90°$ and it becomes negative for $90° < \theta < 180°$. The wavefunctions of the other orbitals may be interpreted in a similar manner.

Table 2.1.4. The total wavefunctions of atomic orbitals with $n = 1\text{--}4$

$$\psi(1s) = \left(\left(\frac{1}{\pi}\right)^{1/2}\right)\left(\frac{Z}{a_0}\right)^{3/2} e^{-Zr/a_0}$$

$$\psi(2s) = \left(\frac{1}{4}\left(\frac{1}{2\pi}\right)^{1/2}\right)\left(\frac{Z}{a_0}\right)^{3/2}\left(2 - \frac{Zr}{a_0}\right) e^{-Zr/2a_0}$$

$$\psi(2p_x) = \left(\frac{1}{4}\left(\frac{1}{2\pi}\right)^{1/2}\right)\left(\frac{Z}{a_0}\right)^{5/2} r e^{-Zr/2a_0} \sin\theta \cos\phi$$

$$\psi(2p_y) = \left(\frac{1}{4}\left(\frac{1}{2\pi}\right)^{1/2}\right)\left(\frac{Z}{a_0}\right)^{5/2} r e^{-Zr/2a_0} \sin\theta \sin\phi$$

$$\psi(2p_z) = \left(\frac{1}{4}\left(\frac{1}{2\pi}\right)^{1/2}\right)\left(\frac{Z}{a_0}\right)^{5/2} r e^{-Zr/2a_0} \cos\theta$$

$$\psi(3s) = \left(\frac{1}{81}\left(\frac{1}{3\pi}\right)^{1/2}\right)\left(\frac{Z}{a_0}\right)^{3/2}\left[27 - 18\left(\frac{Zr}{a_0}\right) + 2\left(\frac{Zr}{a_0}\right)^2\right] e^{-Zr/3a_0}$$

$$\psi(3p_x) = \left(\frac{1}{81}\left(\frac{2}{\pi}\right)^{1/2}\right)\left(\frac{Z}{a_0}\right)^{5/2}\left[6 - \left(\frac{Zr}{a_0}\right)\right] r e^{-Zr/3a_0} \sin\theta \cos\phi$$

$$\psi(3p_y) = \left(\frac{1}{81}\left(\frac{2}{\pi}\right)^{1/2}\right)\left(\frac{Z}{a_0}\right)^{5/2}\left[6 - \left(\frac{Zr}{a_0}\right)\right] r e^{-Zr/3a_0} \sin\theta \sin\phi$$

$$\psi(3p_z) = \left(\frac{1}{81}\left(\frac{2}{\pi}\right)^{1/2}\right)\left(\frac{Z}{a_0}\right)^{5/2}\left[6 - \left(\frac{Zr}{a_0}\right)\right] r e^{-Zr/3a_0} \cos\theta$$

$$\psi(3d_{xy}) = \left(\frac{1}{81}\left(\frac{1}{2\pi}\right)^{1/2}\right)\left(\frac{Z}{a_0}\right)^{7/2} r^2 e^{-Zr/3a_0} \sin^2\theta \sin 2\phi$$

$$\psi(3d_{x^2-y^2}) = \left(\frac{1}{81}\left(\frac{1}{2\pi}\right)^{1/2}\right)\left(\frac{Z}{a_0}\right)^{7/2} r^2 e^{-Zr/3a_0} \sin^2\theta \cos 2\phi$$

$$\psi(3d_{yz}) = \left(\frac{1}{81}\left(\frac{2}{\pi}\right)^{1/2}\right)\left(\frac{Z}{a_0}\right)^{7/2} r^2 e^{-Zr/3a_0} \sin\theta \cos\theta \sin\phi$$

$$\psi(3d_{z^2}) = \left(\frac{1}{81}\left(\frac{1}{6\pi}\right)^{1/2}\right)\left(\frac{Z}{a_0}\right)^{7/2} r^2 e^{-Zr/3a_0} (3\cos^2\theta - 1)$$

$$\psi(3d_{xz}) = \left(\frac{1}{81}\left(\frac{2}{\pi}\right)^{1/2}\right)\left(\frac{Z}{a_0}\right)^{7/2} r^2 e^{-Zr/3a_0} \sin\theta \cos\theta \cos\phi$$

$$\psi(4s) = \left(\frac{1}{16}\left(\frac{1}{\pi}\right)^{1/2}\right)\left(\frac{Z}{a_0}\right)^{3/2}\left[2 - \frac{3}{2}\left(\frac{Zr}{a_0}\right) + \frac{1}{4}\left(\frac{Zr}{a_0}\right)^2 - \frac{1}{96}\left(\frac{Zr}{a_0}\right)^3\right] e^{-Zr/4a_0}$$

$$\psi(4p_x) = \left(\frac{1}{32}\left(\frac{5}{\pi}\right)^{1/2}\right)\left(\frac{Z}{a_0}\right)^{5/2}\left[1 - \frac{1}{4}\left(\frac{Zr}{a_0}\right) + \frac{1}{80}\left(\frac{Zr}{a_0}\right)^2\right] r e^{-Zr/4a_0} \sin\theta \cos\phi$$

$$\psi(4p_y) = \left(\frac{1}{32}\left(\frac{5}{\pi}\right)^{1/2}\right)\left(\frac{Z}{a_0}\right)^{5/2}\left[1 - \frac{1}{4}\left(\frac{Zr}{a_0}\right) + \frac{1}{80}\left(\frac{Zr}{a_0}\right)^2\right] r e^{-Zr/4a_0} \sin\theta \sin\phi$$

$$\psi(4p_z) = \left(\frac{1}{32}\left(\frac{5}{\pi}\right)^{1/2}\right)\left(\frac{Z}{a_0}\right)^{5/2}\left[1 - \frac{1}{4}\left(\frac{Zr}{a_0}\right) + \frac{1}{80}\left(\frac{Zr}{a_0}\right)^2\right] r e^{-Zr/4a_0} \cos\theta$$

$$\psi(4d_{xy}) = \left(\frac{1}{256}\left(\frac{3}{\pi}\right)^{1/2}\right)\left(\frac{Z}{a_0}\right)^{7/2}\left[1 - \frac{1}{12}\left(\frac{Zr}{a_0}\right)\right] r^2 e^{-Zr/4a_0} \sin^2\theta \sin 2\phi$$

$$\psi(4d_{x^2-y^2}) = \left(\frac{1}{256}\left(\frac{3}{\pi}\right)^{1/2}\right)\left(\frac{Z}{a_0}\right)^{7/2}\left[1 - \frac{1}{12}\left(\frac{Zr}{a_0}\right)\right] r^2 e^{-Zr/4a_0} \sin^2\theta \cos 2\phi$$

$$\psi(4d_{yz}) = \left(\frac{1}{128}\left(\frac{3}{\pi}\right)^{1/2}\right)\left(\frac{Z}{a_0}\right)^{7/2}\left[1 - \frac{1}{12}\left(\frac{Zr}{a_0}\right)\right] r^2 e^{-Zr/4a_0} \sin\theta \cos\theta \sin\phi$$

$$\psi(4d_{xz}) = \left(\frac{1}{128}\left(\frac{3}{\pi}\right)^{1/2}\right)\left(\frac{Z}{a_0}\right)^{7/2}\left[1 - \frac{1}{12}\left(\frac{Zr}{a_0}\right)\right] r^2 e^{-Zr/4a_0} \sin\theta \cos\theta \cos\phi$$

$$\psi(4d_{z^2}) = \left(\frac{1}{256}\left(\frac{1}{\pi}\right)^{1/2}\right)\left(\frac{Z}{a_0}\right)^{7/2}\left[1 - \frac{1}{12}\left(\frac{Zr}{a_0}\right)\right] r^2 e^{-Zr/4a_0} (3\cos^2\theta - 1)$$

$$\psi(4f_{z^3}) = \left(\frac{1}{3072}\left(\frac{1}{5\pi}\right)^{1/2}\right)\left(\frac{Z}{a_0}\right)^{9/2} r^3 e^{-Zr/4a_0} (5\cos^3\theta - 3\cos\theta)$$

$$\psi(4f_{xz^2}) = \left(\frac{1}{6144}\left(\frac{6}{5\pi}\right)^{1/2}\right)\left(\frac{Z}{a_0}\right)^{9/2} r^3 e^{-Zr/4a_0} \sin\theta (5\cos^2\theta - 1) \cos\phi$$

$$\psi(4f_{yz^2}) = \left(\frac{1}{6144}\left(\frac{6}{5\pi}\right)^{1/2}\right)\left(\frac{Z}{a_0}\right)^{9/2} r^3 e^{-Zr/4a_0} \sin\theta (5\cos^2\theta - 1) \sin\phi$$

$$\psi(4f_{xyz}) = \left(\frac{1}{3072}\left(\frac{3}{\pi}\right)^{1/2}\right)\left(\frac{Z}{a_0}\right)^{9/2} r^3 e^{-Zr/4a_0} \sin^2\theta \cos\theta \sin 2\phi$$

$$\psi(4f_{z(x^2-y^2)}) = \left(\frac{1}{3072}\left(\frac{3}{\pi}\right)^{1/2}\right)\left(\frac{Z}{a_0}\right)^{9/2} r^3 e^{-Zr/4a_0} \sin^2\theta \cos\theta \cos 2\phi$$

$$\psi(4f_{x(x^2-3y^2)}) = \left(\frac{1}{6144}\left(\frac{2}{\pi}\right)^{1/2}\right)\left(\frac{Z}{a_0}\right)^{9/2} r^3 e^{-Zr/4a_0} \sin^3\theta \cos 3\phi$$

$$\psi(4f_{y(3x^2-y^2)}) = \left(\frac{1}{6144}\left(\frac{2}{\pi}\right)^{1/2}\right)\left(\frac{Z}{a_0}\right)^{9/2} r^3 e^{-Zr/4a_0} \sin^3\theta \sin 3\phi$$

The Electronic Structure of Atoms

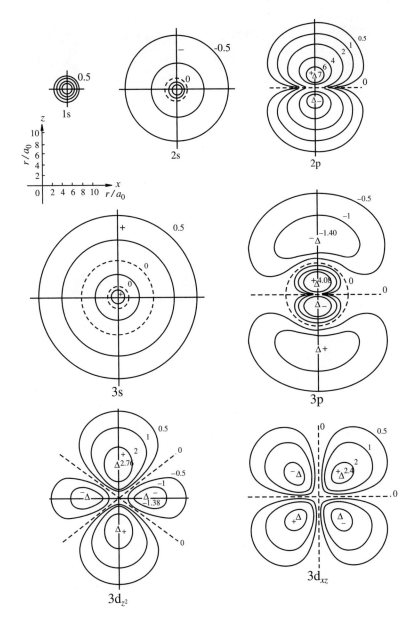

Fig. 2.1.4.
The total wavefunctions of some hydrogenic orbitals. The + and − symbols give the sign of the function in the region indicated, while the dashed lines show the positions of the nodes. The values of ψ contours shown have been multiplied by a factor of 100.

In Fig. 2.1.4, the outer lines represent the value of $\psi = 0.5 \times 10^{-2}$ or -0.5×10^{-2}, in which $\psi_{1s}, r = 2.4a_0$; $\psi_{2s}, r = 10.2a_0$; $\psi_{3s}, r = 15.8a_0$. These r values and other r values in p and d orbitals can be used to indicate the relative sizes of the orbitals. The radii of node spheres which are shown by broken lines are as follows: $\psi_{2s}, r = 2a_0$; $\psi_{3s}, r = 1.91a_0$ and $7.08a_0$; $\psi_{3p_z}, r = 6a_0$.

The following points are noted from the plots in Figs. 2.1.3 and 2.1.4:

(1) The orbitals increase in size as the principal quantum number n increases.
(2) Only s orbitals have finite density at the nucleus.

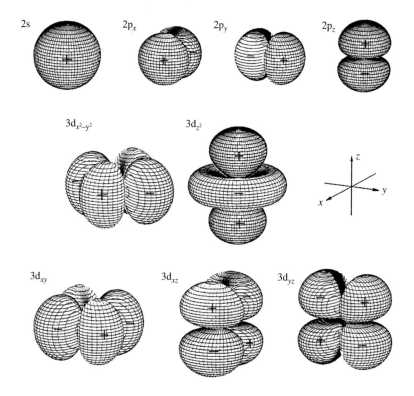

Fig. 2.1.5.
The three-dimensional shapes of the 2s, 2p, and 3d orbitals.

(3) The number of nodes for functions $R_{n,\ell}$ is $n - \ell - 1$.
(4) Among orbitals with the same n, those with smaller ℓ have greater density close to the nucleus, but their principal maximum is further out.

To summarize at this point, it is reiterated that wavefunction $\psi(r, \theta, \phi)$ is a function of r, θ, and ϕ. When the point is moved to another location in space, the value of the wavefunction is changed accordingly. The three-dimensional shape of each orbital can be represented by a contour surface, on which every point has the same value of ψ. The three-dimensional shapes of nine hydrogenic orbitals (2s, 2p, and 3d) are displayed in Fig. 2.1.5. In these orbitals, the nodal surfaces are located at the intersections where ψ changes its sign. For instance, for the $2p_x$ orbital, the yz plane is a nodal plane. For the $3d_{xy}$ orbital, the xz and yz plane are the nodal planes.

2.1.4 Relative sizes of hydrogenic orbitals and the probability criterion

In the Bohr model of the hydrogen atom, the radius of an orbit in terms of unit a_0 is given by n^2, and the sizes of the K, L, M and N shells are in the ratio 1:4:9:16.

For hydrogenic orbitals, the average distance of the electron from the nucleus, $\bar{r}$, can be evaluated exactly from the equation

$$\bar{r} = \int_0^\infty r D_{n,\ell} dr = n^2 [3/2 - \{\ell(\ell+1)\}/2n^2]. \tag{2.1.28}$$

Table 2.1.5. Average distances ($\bar{r}$), most probable distances (r_{max}), and associated probabilities (P) of hydrogenic orbitals

Atomic orbital	$\bar{r}$	Size ratio	P	r_{max}	P
1s	1.5	1	0.58	1	0.32
2s	6	4	0.54	5.24	0.39
2p	5	3.33	0.56	4	0.37
3s	13.5	9	0.49	13.07	0.44
3p	12.5	8.33	0.49	12	0.44
3d	10.5	7	0.55	9	0.39
4s	24	16	0.44	24.62	0.48
4p	23	15.33	0.44	23.58	0.48
4d	21	14	0.45	21.21	0.47
4f	18	12	0.46	16	0.41

The $\bar{r}$ values for orbitals 1s through 4f and their size ratios, calculated relative to the 1s value as a standard, are given in the second and third columns of Table 2.1.5.

For any given orbital, the probability of finding the electron in the volume element $d\tau = r^2\, dr \sin\theta\, d\theta d\phi$ [eq. (2.1.7)] in the vicinity of the point (r, θ, ϕ) is

$$P = |\psi_{n,\ell,m_\ell}(r,\theta,\phi)|^2 r^2 dr \sin\theta d\theta d\phi. \tag{2.1.29}$$

If we integrate this expression over all possible values of θ and ϕ, we get

$$\int_0^\pi |\Theta_{\ell,m_\ell}|^2 \sin\theta d\theta \cdot \int_0^{2\pi} |\Phi_{m_\ell}|^2 d\phi \cdot |R_{n,\ell}|^2 r^2 dr$$
$$= r^2 R_{n,\ell}^2 dr \equiv D_{n,\ell} dr, \tag{2.1.30}$$

which represents the probability of finding the electron between distances r and $r + dr$ from the nucleus regardless of direction. The $D_{n,\ell}$ function in eq. (2.1.30) is called the radial probability distribution function. It is incorrect to write $D_{n,\ell}$ as $4\pi r^2 R_{n,\ell}^2$, which has the extraneous factor 4π, or as $4\pi r^2 \psi^2$ which holds only for the spherically symmetric s orbitals.

Figure 2.1.6 shows plots of the $D_{n,\ell}$ functions against r for the first few hydrogenic orbitals. The positions of the maxima and minima in each curve are as follows:

1s: maximum at $1a_0$;
2s: maxima at 0.76 and $5.24a_0$; minimum at $2a_0$;
2p: maximum at $4a_0$;
3s: maxima at 0.73, 4.20 and $13.07a_0$; minima at 1.91 and $7.08a_0$;
3p: maxima at 3 and $12a_0$; minimum at $6a_0$;
3d: maximum at $9a_0$.

Take the 1s electron as an example. In the Bohr theory, the electron moves in a fixed orbit of radius $1a_0$. On the other hand, in the wave mechanical treatment, the electron can in principle be found at any distance from the nucleus, and the most probable nucleus-electron separation is $1a_0$. Here we can see

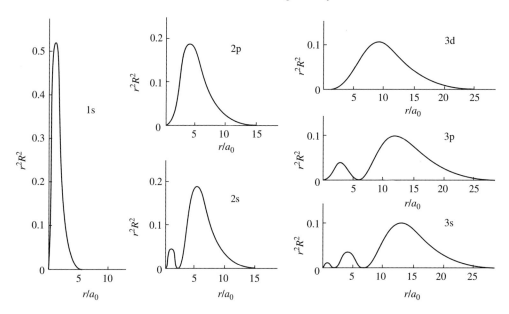

Fig. 2.1.6.
The radial probability function $D_{n,\ell}$ or $|rR_{n,\ell}|^2$, drawn to the same scale, of the first six hydrogenic orbitals.

the philosophical difference between these two models. In addition, while we should look for a $2p_x$ or $3p_x$ electron along the $+x$ or $-x$ direction, Fig. 2.1.6 shows that the $2p_x$ electron is most likely found at a point $4a_0$ from the nucleus, whereas the corresponding distance for a $3p_x$ electron is $12a_0$.

For hydrogenic orbitals with $n - \ell > 1$, the most probable distance $r_{\max}$ may be defined as the distance of the "principal maximum," namely the outermost and most prominent of the maxima of the $D_{n,\ell}$ function from the nucleus. The $r_{\max}$ values of the hydrogenic orbitals are listed in the fifth column of Table 2.1.5. It is seen that the size ratios for the 1s, 2p, 3d, and 4f orbitals, which satisfy the relation $\ell = n - 1$, are exactly the same as those of the Bohr orbits, which vary as n^2. Furthermore, for orbitals in the same shell, $r_{\max}$ decreases as ℓ increases, as in the case of $\bar{r}$. The data in Table 2.1.5 show that orbital size ratios based on $\bar{r}$ are smaller than those based on $r_{\max}$.

A plausible estimate of the spatial extension of a hydrogenic orbital is the radius of a spherical boundary surface within which there is a high probability of finding the electron. In order to develop this criterion on a quantitative basis, it is useful to define a cumulative probability function $P_{n\ell}(\rho)$ which gives the probability of finding the electron at a distance less than or equal to ρ from the nucleus:

$$P_{n\ell}(\rho) = \int_0^\rho D_{n,\ell} dr. \qquad (2.1.31)$$

The function $P_{n\ell}(\rho)$ can be expressed in closed form, and the expressions for 1s through 4f orbitals are listed in Table 2.1.6. They can be used to give some measure of correlation between the various criteria for orbital size. For

Table 2.1.6. Cumulative probability functions for hydrogenic orbitals

Atomic orbital	$P_{n\ell}(\rho)$
1s	$1 - e^{-2\rho}(1 + 2\rho + 2\rho^2)$
2s	$1 - e^{-\rho}(1 + \rho + \rho^2/2 + \rho^4/8)$
2p	$1 - e^{-\rho}(1 + \rho + \rho^2/2 + \rho^3/6 + \rho^4/24)$
3s	$1 - e^{-2\rho/3}(1 + 2\rho/3 + 2\rho^2/9 + 4\rho^4/81 - 8\rho^5/729 + 8\rho^6/6561)$
3p	$1 - e^{-2\rho/3}(1 + 2\rho/3 + 2\rho^2/9 + 4\rho^3/81 + 2\rho^4/243 - 4\rho^5/2187 + 4\rho^6/6561)$
3d	$1 - e^{-2\rho/3}(1 + 2\rho/3 + 2\rho^2/9 + 4\rho^3/81 + 2\rho^4/243 + 4\rho^5/3645 + 4\rho^6/32805)$
4s	$1 - e^{-\rho/2}(1 + \rho/2 + \rho^2/8 + 3\rho^4/128 - \rho^5/128 + 13\rho^6/9216 - \rho^7/9216 + \rho^8/294912)$
4p	$1 - e^{-\rho/2}(1 + \rho/2 + \rho^2/8 + \rho^3/8 + \rho^4/384 - \rho^5/960 + 7\rho^6/15360 - \rho^7/20480 + \rho^8/491520)$
4d	$1 - e^{-\rho/2}(1 + \rho/2 + \rho^2/8 + \rho^3/48 + \rho^4/384 + \rho^5/3840 + \rho^6/46080 - \rho^7/184320 + \rho^8/1474560)$
4f	$1 - e^{-\rho/2}(1 + \rho/2 + \rho^2/8 + \rho^3/48 + \rho^4/384 + \rho^5/3840 + \rho^6/46080 + \rho^7/645120 + \rho^8/10321920)$

Table 2.1.7. Relative sizes of hydrogenic orbitals based on different probability criteria

Atomic orbital	$P = 0.50$		$P = 0.90$		$P = 0.95$		$P = 0.99$	
	ρ	Ratio	ρ	Ratio	ρ	Ratio	ρ	Ratio
1s	1.34	1	2.66	1	3.15	1	4.20	1
2s	5.80	4.38	9.13	3.43	10.28	3.27	12.73	3.03
2p	4.67	3.49	7.99	3.00	9.15	2.91	11.61	2.76
3s	13.62	10.19	19.44	7.31	21.39	6.80	25.46	6.06
3p	12.57	9.40	18.39	6.91	20.34	6.46	24.41	5.81
3d	10.01	7.48	15.80	5.94	17.76	5.64	21.86	5.20
4s	24.90	18.63	33.62	12.64	36.47	11.59	42.35	10.08
4p	23.87	17.86	32.59	12.25	35.45	11.26	41.32	9.83
4d	21.60	16.15	30.31	11.39	33.17	10.54	39.06	9.29
4f	17.34	12.97	25.99	9.77	28.87	9.17	34.81	8.28

any specified distance ρ, the probability of finding the electron within the range $0 < r \leq \rho$ can be obtained by direct substitution into these expressions.

The cumulative probability functions listed in Table 2.1.6 can be used to give some measure of correlation between the various criteria for orbital size. Once a particular value for P is chosen, it can be equated to each of the expresssions in turn and the resulting transcendental equation solved numerically. The ρ values thus obtained for $P = 0.50, 0.90, 0.95$, and 0.99 are tabulated in Table 2.1.7. The significant quantites are the size ratios, with the 1s ρ value as standard, which provide a rational scale of relative orbital size based on any adopted probability criterion. As the prescribed P value approaches unity, the size ratios gradually decrease in magnitude and orbitals in the same shell tend to converge to a similar size.

The probabilty criterion $P = 0.95$ (or 0.99) yields the most compact scale of orbital size, according to which the size of the first four qunatum shells are approximately in the ratio of 1:3:6:10. In the fourth and last columns of Tables 2.1.5, the probabilities corresponding to $\bar{r}$ and r_{max} are listed, respectively. Scrutiny of the data in Tables 2.1.5 and 2.1.7 show that adoption of $\bar{r}$ as an indication of orbital size corresponds fairly closely to a 50% probability criterion. The use of r_{max} may be roughly matched with a lower criterion of 40%, but the correlation is not as satisfactory.

2.1.5 Energy levels of hydrogenic orbitals; summary

As mentioned previously, energy E appears only in the radial equation [eq. (2.1.14)], which does not involve m_ℓ. Hence we readily note that s orbitals are non-degenerate, p orbitals are triply degenerate, d orbitals have fivefold degeneracy, and f orbitals have sevenfold degeneracy. As it turns out, *for the hydrogen atom only*, E is also independent of ℓ:

$$E_n = \frac{-me^4 Z^2}{8\varepsilon_0^2 h^2 n^2} = \frac{-Z^2 e^2}{8\pi \varepsilon_0 a_0 n^2} = -13.6 \left(\frac{Z^2}{n^2}\right) \text{ eV}. \qquad (2.1.32)$$

To summarize, each hydrogenic orbital is described by three quantum numbers and each quantum number is responsible for some properties:

(1) Principal quantum number n, $n = 1, 2, \ldots$: solely responsible for the energy of the orbital and largely accountable for the orbital size;
(2) Azimuthal quantum number ℓ, $\ell = 0, 1, 2, \ldots, n-1$: mainly accounting for the shape of the orbital; and
(3) Magnetic quantum number m_ℓ, $m_\ell = -\ell, -\ell+1, \ldots, \ell$: describes the orientation of the orbital.

Comparing the results of the Bohr and the wave mechanical models, we find that:

(a) Both theories lead to the same energy expression.
(b) It can be shown that wavefunctions may be used to determine other atomic properties such as the intensity of the spectral lines, etc.
(c) Quantum numbers appear naturally in solving of the Schrödinger equation (even though this has not been explicitly shown here). But they are inserted arbitrarily in the Bohr model.
(d) Electrons in the Bohr theory occupy planet-like orbits. In the wave mechanical model they occupy delocalized orbitals. Experimental evidences support the Schrödinger picture.

2.2 The helium atom and the Pauli Exclusion Principle

Before we embark on this topic, we first introduce the "atomic units," which are defined in Table 2.2.1.

In the quantum mechanical treatment we have studied so far, we always try to obtain the wavefunction ψ and energy E by solving the Schrödinger equation:

$$\hat{H}\psi = E\psi. \qquad (2.2.1)$$

However, it often happens that an exact solution of the Schrödinger equation is impossible. Instead we usually rely on approximation methods. If we have ψ'

Table 2.2.1. Atomic units (a.u.)

Length	1 a.u. = a_0 = 5.29177 × 10^{-11} m (Bohr radius)
Mass	1 a.u. = m_e = 9.109382 × 10^{-31} kg (rest mass of electron)
Charge	1 a.u. = $|e|$ = 1.6021764 × 10^{-19} C (charge of electron)
Energy	1 a.u. = $\frac{e^2}{4\pi\varepsilon_0 a_0}$ = 27.2114 eV (potential energy between two electrons when they are 1 a_0 apart)
Angular momentum	1 a.u. = $\frac{h}{2\pi}$ ($\equiv \hbar$) = 1.05457 × 10^{-34} J s
Scale factor	$4\pi\varepsilon_0 = 1$

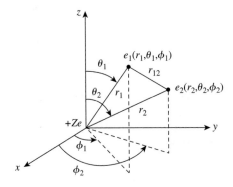

Fig. 2.2.1.
The coordinate system for the helium atom.

as an approximation solution of eq. (2.2.1), its energy is simply

$$E = \frac{\int \psi'^* \hat{H} \psi' d\tau}{\int \psi'^* \psi' d\tau}. \qquad (2.2.2)$$

2.2.1 The helium atom: ground state

Now we turn to the helium atom, whose coordinate system is shown in Fig. 2.2.1. The potential energy of this system (in a.u.) is

$$V = -\frac{Z}{r_1} - \frac{Z}{r_2} + \frac{1}{r_{12}}, \qquad (2.2.3)$$

and the Schrödinger equation takes the form

$$\left(-\frac{1}{2}\nabla_1^2 - \frac{1}{2}\nabla_2^2 - \frac{Z}{r_1} - \frac{Z}{r_2} + \frac{1}{r_{12}}\right)\psi = E\psi, \qquad (2.2.4)$$

where $\psi = \psi(r_1, \theta_1, \phi_1, r_2, \theta_2, \phi_2) = \psi(1, 2)$. Note that operators ∇_1^2 and ∇_2^2 affect the coordinates of electrons 1 and 2, respectively. Equation (2.2.4) cannot be solved exactly and, in the following treatment, we will attempt to solve it by approximation. As a first (and very drastic) approximation, we ignore the electronic repulsion term $\frac{1}{r_{12}}$ in eq. (2.2.3) or eq. (2.2.4). Now eq. (2.2.4) becomes

$$\left[-\frac{1}{2}\nabla_1^2 - \frac{Z}{r_1}\right]\psi(1,2) + \left[-\frac{1}{2}\nabla_2^2 - \frac{Z}{r_2}\right]\psi(1,2) = E\psi(1,2). \qquad (2.2.5)$$

If we separate the variables in $\psi(1,2)$,

$$\psi(1,2) = \psi_1(1)\psi_2(2), \tag{2.2.6}$$

and substitute this version of $\psi(1,2)$ into eq. (2.2.5), we have

$$\psi_2(2)\left[-\frac{1}{2}\nabla_1^2 - \frac{Z}{r_1}\right]\psi_1(1) + \psi_1(1)\left[-\frac{1}{2}\nabla_2^2 - \frac{Z}{r_2}\right]\psi_2(2) = E\psi_1(1)\psi_2(2). \tag{2.2.7}$$

Dividing eq. (2.2.7) by $\psi_1(1)\psi_2(2)$ leads to

$$\frac{\left[-\frac{1}{2}\nabla_1^2 - \frac{Z}{r_1}\right]\psi_1(1)}{\psi_1(1)} + \frac{\left[-\frac{1}{2}\nabla_2^2 - \frac{Z}{r_2}\right]\psi_2(2)}{\psi_2(2)} = E. \tag{2.2.8}$$

It is obvious that both terms on the left side of eq. (2.2.8) must each be equal to a constant, corresponding to E_a and E_b, respectively, and

$$E_a + E_b = E. \tag{2.2.9}$$

So the simplified Schrödinger equation eq. (2.2.7) can be separated into two (familiar) equations, each involving only the coordinates of one electron:

$$\left[-\frac{1}{2}\nabla_1^2 - \frac{Z}{r_1}\right]\psi_1(1) = E_a\psi_1(1); \tag{2.2.10}$$

$$\left[-\frac{1}{2}\nabla_2^2 - \frac{Z}{r_2}\right]\psi_2(2) = E_b\psi_2(2). \tag{2.2.11}$$

We have already solved eqs. (2.2.10) and (2.2.11) in the treatment of the hydrogen atom; the only difference is that now $Z = 2$. If we take the ground state wavefunction and energy (in a.u.),

$$\psi_1(1) = \psi_{1s}(1) = \frac{1}{(\pi)^{1/2}}Z^{\frac{3}{2}}e^{-Zr_1}; \tag{2.2.12}$$

$$\psi_2(2) = \psi_{1s}(2) = \frac{1}{(\pi)^{1/2}}Z^{\frac{3}{2}}e^{-Zr_2}, \tag{2.2.13}$$

and

$$\psi(1,2) = \left(\frac{1}{(\pi)^{1/2}}Z^{\frac{3}{2}}e^{-Zr_1}\right)\left(\frac{1}{(\pi)^{1/2}}Z^{\frac{3}{2}}e^{-Zr_2}\right); \tag{2.2.14}$$

$$E_a = E_b = E_{1s} = -\frac{1}{2}Z^2 \text{ a.u.,} \tag{2.2.15}$$

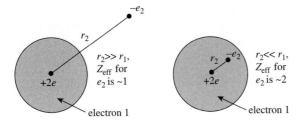

Fig. 2.2.2.
Screening of the nucleus by one of the electrons in the helium atom.

and

$$E = -Z^2 \text{ a.u.} = -108.8 \text{ eV}. \tag{2.2.16}$$

The experimental energy is -79.0 eV.

To improve on this result, we need to understand what causes the error in the solution given by eq. (2.2.14), aside from the fact that electronic repulsion has been ignored. From $\psi_1(1)$, $\psi_2(2)$, and $\psi(1,2)$ given by eqs. (2.2.12) to (2.2.14), we can see that, in our approximation, each electron "sees" a nuclear charge of $+2$. As illustrated in Fig. 2.2.2, when electron 2 is completely outside electron 1, the charge it "sees" is about $+1$, as the $+2$ nuclear charge is shielded by the -1 electron cloud. On the other hand, if electron 2 is deep inside the electron cloud of electron 1, the charge it "sees" is close to $+2$. Hence a more realistic picture would be that each electron should "see" an "effective nuclear charge" (Z_{eff}) between 1 and 2:

$$1 < Z_{\text{eff}} < 2. \tag{2.2.17}$$

If that is the case, $\psi_1(1)$ and $\psi_2(2)$ in eqs. (2.2.12) and (2.2.13) now become

$$\psi_1(1) = \frac{1}{(\pi)^{1/2}} Z_{\text{eff}}^{\frac{3}{2}} e^{-Z_{\text{eff}} r_1}, \tag{2.2.18}$$

$$\psi_2(2) = \frac{1}{(\pi)^{1/2}} Z_{\text{eff}}^{\frac{3}{2}} e^{-Z_{\text{eff}} r_2}, \tag{2.2.19}$$

Again,

$$\psi(1,2) = \psi_1(1)\psi_2(2). \tag{2.2.6}$$

To determine the energy of the trial wavefunction given by eq. (2.2.18), (2.2.19), and (2.2.6), we make use of eq. (2.2.2):

$$E = E(Z_{\text{eff}})$$
$$= \frac{\int\int \psi(1,2)\left[-\frac{1}{2}\nabla_1^2 - \left(\frac{Z}{r_1}\right) - \frac{1}{2}\nabla_2^2 - \left(\frac{Z}{r_2}\right) + \left(\frac{1}{r_{12}}\right)\right]\psi(1,2)d\tau_1 d\tau_2}{\int\int |\psi(1,2)|^2 \, d\tau_1 d\tau_2}. \tag{2.2.20}$$

It should be clear that E is dependent on Z_{eff}. To calculate E is not very difficult, except for the integral involving $1/r_{12}$. Still, we are going to skip the mathematical details here and will only be concerned with the results:

$$E = \left(Z_{\text{eff}}^2 - 2ZZ_{\text{eff}} + \frac{5Z_{\text{eff}}}{8}\right) \text{a.u.} \tag{2.2.21}$$

To determine the optimal (or best) value for Z_{eff}, we vary Z_{eff} until E is at the minimum:

$$\frac{dE}{dZ_{\text{eff}}} = 2Z_{\text{eff}} - 2Z + \frac{5}{8} = 0, \tag{2.2.22}$$

or

$$Z_{\text{eff}} = Z - \left(\frac{5}{16}\right). \tag{2.2.23}$$

For the specific case of helium, $Z = 2$, we get

$$Z_{\text{eff}} = \frac{27}{16} = 1.6875. \tag{2.2.24}$$

This value of Z_{eff} is within the range we wrote down in eq. (2.2.17). Substituting eq. (2.2.23) into eq. (2.2.21) leads to

$$E_{\min} = -Z_{\text{eff}}^2 \text{ a.u.} = -\left[Z - \left(\frac{5}{16}\right)\right]^2 \text{ a.u.} \tag{2.2.25}$$

For helium, with $Z = 2$, we have

$$E_{\min} = -(1.6875)^2 \text{ a.u.} = -2.85 \text{ a.u.} = -77.5 \text{ eV}, \tag{2.2.26}$$

recalling that experimentally helium has the energy of -79.0 eV. So, with such a simple treatment, a very reasonable result is obtained. However, to close the remaining gap of 1.5 eV, considerable effort is required (see below).

What we have applied to helium is called the variational method, which usually consists of the following steps:

(1) Set up a trial function with one or more parameters $\psi(\alpha_1, \alpha_2, \ldots)$;
(2) Calculate $E(\alpha_1, \alpha_2, \ldots) = \frac{\int \psi^* \hat{H} \psi \, d\tau}{\int \psi^* \psi \, d\tau}$.
(3) Obtain the optimal values of $\alpha_1, \alpha_2, \ldots$ and hence the minimum energy by setting $\partial E/\partial \alpha_1 = \partial E/\partial \alpha_2 = \cdots = 0$.

In 1959, C. L. Pekeris, by optimizing a 1,078-term wavefunction for helium, obtained an energy essentially identical to the experimental value.

Before concluding this section, we will take a closer look at the relatively simple trial wavefunctions for helium, i.e., those functions with one or two adjustable parameters. While neglecting the mathematical details of the calculations, we will pay particular attention to the physical meaning of the parameters. In so doing we will acquire a better understanding on the design of trial wavefunctions.

First we rewrite the trial wavefunction for helium given by eqs. (2.2.18), (2.2.19), and (2.2.6) as

$$\psi_a = e^{-\alpha(r_1+r_2)}.$$

As mentioned previously, parameter α may be viewed as the effective nuclear charge "felt" by either one of the two electrons. Such an interaction is commonly called the screening effect. Furthermore, as described by this wavefunction, the two electrons move independently of each other, i.e., angular correlation is ignored. Electron correlation may be taken as the tendency of the electrons to avoid each other. For helium, angular correlation describes the two electrons' inclination to be on opposite sides of the nucleus. On the other hand, radial correlation, or screening effect, is the tendency for one electron to be closer to the nucleus, while the other one is farther away. A one-parameter trial function that does take angular correlation into account is

$$\psi_b = e^{-Z(r_1+r_2)}(1 + cr_{12}).$$

As will be shown later, ψ_b is actually a better trial function than ψ_a. Thus it is surprising, and a pity, that ψ_b has not been included in most of the text books.

Proceeding to trial functions with two parameters, we can see that a function that includes both the screening effect and angular correlation is

$$\psi_c = e^{-\alpha(r_1+r_2)}(1 + cr_{12}).$$

Clearly this function is an improvement on both ψ_a and ψ_b.

Recall ψ_a assumes that both electrons in helium experience the same effective nuclear charge α. While this may be so in an average sense, such an approximation fails to take into account that, at a given instant, the two electrons are not likely to be equidistant from the nucleus and hence the effective nuclear charges they feel should not be the same. Taking this into consideration, C. Eckart proposed the following trial function in 1930:

$$\psi_d = e^{-\alpha r_1}e^{-\beta r_2} + e^{-\beta r_1}e^{-\alpha r_2}.$$

It is clear in ψ_d one electron is allowed to move closer to the nucleus and the other is allowed to be farther away. Once again, angular correlation is not explicitly included in this function. It is noted that this function has two terms because the two electrons are indistinguishable from each other. If we only take the first term, electron 1 is labeled by effective nuclear charge α, while the other electron is labeled by β. This is not allowed in quantum mechanics. The indistinguishability of electrons will also be discussed in the next section.

The results for the aforementioned four trial wavefunctions are summarized in Table 2.2.2.

Examining the tabulated results, it is clear that, between the one-parameter trial functions, ψ_b is "better" than ψ_a. Physically, this implies that angular

Table 2.2.2. Summary of four trial wavefunctions of helium atom in its ground state

Trial wavefunction	Optimized parameter value(s)	IE(calc) (eV)*
$\psi_a = e^{-\alpha(r_1+r_2)}$	$\alpha = 1.6875$ (exact)	23.1
$\psi_b = e^{-Z(r_1+r_2)}(1+cr_{12})$	$c = 0.5572$	23.9
$\psi_c = e^{-\alpha(r_1+r_2)}(1+cr_{12})$	$\alpha = 1.8497; c = 0.3658$	24.2
$\psi_d = e^{-\alpha r_1}e^{-\beta r_2} + e^{-\beta r_1}e^{-\alpha r_2}$	$\alpha = 1.1885; \beta = 2.1832$	23.8

* IE, ionization energy. The experimental IE of He is 24.6 eV.

correlation is a more important factor than the screening effect. Indeed, ψ_b is even slightly better than the two-parameter function ψ_d. This once again underscores the importance of angular correlation. In ψ_d, we can see that one optimized parameter is close to 2, the nuclear charge, while the other is close to 1. That means one electron is close to the nucleus and the other is effectively screened by the inner electron. Finally, since ψ_c takes both screening effect and angular correlation into consideration, it is naturally the best trial function among the four considered here.

2.2.2 Determinantal wavefunction and the Pauli Exclusion Principle

In our treatment of helium, each electron occupies a 1s-like orbital, and hence the wavefunctions given in eqs. (2.2.18), (2.2.19), and (2.2.6) may be written as

$$\psi_{\text{He}}(1,2) = 1s(1)1s(2). \tag{2.2.27}$$

However, we cannot handle atoms with more than two electrons in the same manner. For instance, for lithium, the wavefunction

$$\psi_{\text{Li}}(1,2,3) = 1s(1)1s(2)1s(3) \tag{2.2.28}$$

is not acceptable at all, since it denotes an electronic configuration $(1s^3)$ that violates the Pauli Exclusion Principle. This principle states that no two electrons can have the same set of quantum numbers.

For atomic systems, it is often said that each electron is defined by four quantum numbers: n, ℓ, m_ℓ, and m_s. Actually, there is a fifth quantum number, s, which has the value of $1/2$ for all electrons. Quantum number m_s can be either $1/2$ or $-1/2$, corresponding to spin function α (spin up) and β (spin down), respectively. Spin functions α and β form an orthonormal set,

$$\int \alpha\alpha \, d\gamma = \int \beta\beta \, d\gamma = 1, \tag{2.2.29}$$

and

$$\int \alpha\beta \, d\gamma = 0, \tag{2.2.30}$$

where γ is called a spin variable. If we add spin function to the spatial wavefunction given by eq. (2.2.27) to yield the total wavefunction for helium,

$$\psi_p(1,2) = 1s\alpha(1)1s\beta(2). \tag{2.2.31}$$

However, function ψ_p implies that electrons are distinguishable: electron 1 has its spin up and electron 2 spin down. Similarly, we should also consider the following in the total wavefunction for helium:

$$\psi_q(1,2) = 1s\alpha(2)1s\beta(1). \tag{2.2.32}$$

There are two ways of combining ψ_p and ψ_q in order to make the two electrons indistinguishable:

(1) $\dfrac{1}{(2)^{1/2}}[\psi_p(1,2) + \psi_q(1,2)] = \dfrac{1}{(2)^{1/2}}[1s\alpha(1)1s\beta(2) + 1s\alpha(2)1s\beta(1)],$

$$\tag{2.2.33}$$

which is symmetric with respect to the exchange of the two electrons;

(2) $\dfrac{1}{(2)^{1/2}}[\psi_p(1,2) - \psi_q(1,2)] = \dfrac{1}{(2)^{1/2}}[1s\alpha(1)1s\beta(2) - 1s\alpha(2)1s\beta(1)],$

$$\tag{2.2.34}$$

which is antisymmetric with respect to the exchange of the two electrons. Note that in expressions (2.2.33) and (2.2.34), the factor $1/(2)^{1/2}$ is a normalization constant.

It is found that electronic wavefunctions must be antisymmetric with respect to the exchange of the coordinates of any two electrons. Hence, only the function given by expression (2.2.34) is acceptable for the helium atom:

$$\psi(1,2) = \dfrac{1}{(2)^{1/2}}[1s\alpha(1)1s\beta(2) - 1s\alpha(2)1s\beta(1)]$$

$$= 1s(1)1s(2)\left\{\dfrac{1}{(2)^{1/2}}[\alpha(1)\beta(2) - \beta(1)\alpha(2)]\right\}$$

$$= \text{(symmetric spatial part)} \times \text{(antisymmetric spin part)}. \tag{2.2.35}$$

In our previous treatment of helium, we only used the spatial function $1s(1)1s(2)$ to calculate the energy, while ignoring the spin part. This is acceptable because energy is independent of spin.

In eq. (2.2.35) we see that the helium wavefunction can be factorized into spatial (or orbital) and spin parts. It should be noted that such a factorization is possible only for two-electron systems. Yet another way of writing $\psi(1,2)$ of He as given in eq. (2.2.35) is

$$\psi(1,2) = \dfrac{1}{(2)^{1/2}}[1s\alpha(1)1s\beta(2) - 1s\alpha(2)1s\beta(1)]$$

$$= \dfrac{1}{(2)^{1/2}}\begin{vmatrix} 1s\alpha(1) & 1s\beta(1) \\ 1s\alpha(2) & 1s\beta(2) \end{vmatrix}. \tag{2.2.36}$$

This is called a Slater determinant, in honor of physicist J. C. Slater. Since a determinant changes its sign upon the exchange of any two rows or columns, any wavefunction written in determinantal form must be antisymmetric with respect to electron exchange. If we take the determinantal function in eq. (2.2.36) and exchange its two rows:

$$\psi(2,1) = \frac{1}{(2)^{1/2}} \begin{vmatrix} 1s\alpha(2) & 1s\beta(2) \\ 1s\alpha(1) & 1s\beta(1) \end{vmatrix}$$

$$= \frac{1}{(2)^{1/2}}[1s\alpha(2)1s\beta(1) - 1s\alpha(1)1s\beta(2)] = -\psi(1,2). \quad (2.2.37)$$

Also, such an antisymmetric function automatically satisfies the Pauli Exclusion Principle. For instance, if we write a function for helium having both electrons in the 1s orbital with spin up,

$$\frac{1}{(2)^{1/2}} \begin{vmatrix} 1s\alpha(1) & 1s\alpha(1) \\ 1s\alpha(2) & 1s\alpha(2) \end{vmatrix} = \frac{1}{(2)^{1/2}}[1s\alpha(1)1s\alpha(2) - 1s\alpha(1)1s\alpha(2)] = 0. \quad (2.2.38)$$

Such a function vanishes because any determinant with two identical rows or columns vanishes. In other words, any system having both electrons in the 1s orbital with α spin cannot exist. Now we see there is an alternative way of saying the Pauli Exclusion Principle: A wavefunction for a system with two or more electrons must be antisymmetric with respect to the interchange of labels of any two electrons.

Returning to the lithium atom, it is now clear the 1s orbital cannot accommodate all three electrons. Rather, two electrons are in the 1s orbital with opposite spins, while the remaining electron is in the 2s orbital with either α or β spin. Hence two determinantal wavefunctions can be written for the ground state of lithium:

$$\psi(1,2,3) = \frac{1}{(6)^{1/2}} \begin{vmatrix} 1s\alpha(1) & 1s\beta(1) & 2s\alpha(1) \\ 1s\alpha(2) & 1s\beta(2) & 2s\alpha(2) \\ 1s\alpha(3) & 1s\beta(3) & 2s\alpha(3) \end{vmatrix}, \quad (2.2.39)$$

and

$$\psi(1,2,3) = \frac{1}{(6)^{1/2}} \begin{vmatrix} 1s\alpha(1) & 1s\beta(1) & 2s\beta(1) \\ 1s\alpha(2) & 1s\beta(2) & 2s\beta(2) \\ 1s\alpha(3) & 1s\beta(3) & 2s\beta(3) \end{vmatrix}. \quad (2.2.40)$$

Since there are two determinantal functions for lithium, its ground state is a spin doublet. On the other hand, only one determinantal function can be written for helium [eq. (2.2.36)], and its ground state is a spin singlet.

Another notation describing the ground state of helium and lithium is $1s^2$ and $1s^2 2s^1$, respectively. From this notation, we can tell quickly which orbitals accommodate the electrons and how the electron spins are related to each other. Such an assignment for the electrons is called the electronic configuration of the atom.

2.2.3 The helium atom: the $1s^12s^1$ configuration

When one of the electrons in helium is excited from the 1s to the 2s orbital, the configuration $1s^12s^1$ is obtained. Taking the indistinguishibility of electrons into account, two spatial wavefunctions can be written:

symmetric spatial part: $\quad \dfrac{1}{(2)^{1/2}}[1s(1)2s(2) + 1s(2)2s(1)],$ (2.2.41)

antisymmetric spatial part: $\quad \dfrac{1}{(2)^{1/2}}[1s(1)2s(2) - 1s(2)2s(1)].$ (2.2.42)

On the other hand, four functions can be written for the spin part:

symmetric spin part: $\quad \alpha(1)\alpha(2),$ (2.2.43)

symmetric spin part: $\quad \beta(1)\beta(2),$ (2.2.44)

symmetric spin part: $\quad \dfrac{1}{(2)^{1/2}}[\alpha(1)\beta(2) + \alpha(2)\beta(1)],$ (2.2.45)

antisymmetric spin part: $\quad \dfrac{1}{(2)^{1/2}}[\alpha(1)\beta(2) - \alpha(2)\beta(1)].$ (2.2.46)

Since the total (spatial × spin) wavefunction must be antisymmetric, four acceptable functions can be written from expressions (2.2.41) to (2.2.46):

$$\psi_1(1,2) = \frac{1}{(2)^{1/2}}[1s(1)2s(2) + 1s(2)2s(1)] \cdot \frac{1}{(2)^{1/2}}[\alpha(1)\beta(2) - \alpha(2)\beta(1)]$$

$$= \frac{1}{(2)^{1/2}}\left[\frac{1}{(2)^{1/2}}\begin{vmatrix} 1s\alpha(1) & 2s\beta(1) \\ 1s\alpha(2) & 2s\beta(2) \end{vmatrix} - \frac{1}{(2)^{1/2}}\begin{vmatrix} 1s\beta(1) & 2s\alpha(1) \\ 1s\beta(2) & 2s\alpha(2) \end{vmatrix}\right]. \quad (2.2.47)$$

$$\psi_2(1,2) = \frac{1}{(2)^{1/2}}[1s(1)2s(2) - 1s(2)2s(1)] \cdot \alpha(1)\alpha(2)$$

$$= \frac{1}{(2)^{1/2}}\begin{vmatrix} 1s\alpha(1) & 2s\alpha(1) \\ 1s\alpha(2) & 2s\alpha(2) \end{vmatrix}. \quad (2.2.48)$$

$$\psi_3(1,2) = \frac{1}{(2)^{1/2}}[1s(1)2s(2) - 1s(2)2s(1)] \cdot \beta(1)\beta(2)$$

$$= \frac{1}{(2)^{1/2}}\begin{vmatrix} 1s\beta(1) & 2s\beta(1) \\ 1s\beta(2) & 2s\beta(2) \end{vmatrix}. \quad (2.2.49)$$

$$\psi_4(1,2) = \frac{1}{(2)^{1/2}}[1s(1)2s(2) - 1s(2)2s(1)] \cdot \frac{1}{(2)^{1/2}}[\alpha(1)\beta(2) + \alpha(2)\beta(1)]$$

$$= \frac{1}{(2)^{1/2}}\left[\frac{1}{(2)^{1/2}}\begin{vmatrix} 1s\alpha(1) & 2s\beta(1) \\ 1s\alpha(2) & 2s\beta(2) \end{vmatrix} + \frac{1}{(2)^{1/2}}\begin{vmatrix} 1s\beta(1) & 2s\alpha(1) \\ 1s\beta(2) & 2s\alpha(2) \end{vmatrix}\right]. \quad (2.2.50)$$

Fundamentals of Bonding Theory

As $E = \int \psi^* \hat{H} \psi d\tau$ and $\hat{H}$ is spin-independent, it is clear that ψ_2, ψ_3, and ψ_4 have the same energy and form a spin triplet, while ψ_1 has different energy value and is called a spin singlet. In the following, we denote the orbital (or spatial) part of the singlet [eq. (2.2.41)] and triplet [eq. (2.2.42)] functions by $^1\psi$ and $^3\psi$, respectively:

$$^1\psi(1,2) = \frac{1}{(2)^{1/2}}[1s(1)2s(2) + 1s(2)2s(1)], \qquad (2.2.51)$$

$$^3\psi(1,2) = \frac{1}{(2)^{1/2}}[1s(1)2s(2) - 1s(2)2s(1)]. \qquad (2.2.52)$$

Now we evaluate the energies ^1E and ^3E for the singlet and triplet states, respectively. Recall, for helium, in atomic units

$$\hat{H} = \left(-\frac{1}{2}\nabla_1^2 - \frac{Z}{r_1}\right) + \left(-\frac{1}{2}\nabla_2^2 - \frac{Z}{r_2}\right) + \frac{1}{r_{12}}. \qquad (2.2.53)$$

It can be shown that

$$^1E = \int\int {}^1\psi^* \hat{H} \, {}^1\psi \, d\tau_1 d\tau_2$$
$$= E_{1s} + E_{2s} + J + K, \qquad (2.2.54)$$

where J (coulomb integral) and K (exchange integral) have the form

$$J = \int\int 1s(1)2s(2)\left(\frac{1}{r_{12}}\right)1s(1)2s(2) d\tau_1 d\tau_2 = \frac{34}{81} \text{a.u.} = 0.420 \text{ a.u.}, \qquad (2.2.55)$$

$$K = \int\int 1s(1)2s(2)\left(\frac{1}{r_{12}}\right)1s(2)2s(1) d\tau_1 d\tau_2 = \frac{32}{729} \text{a.u} = 0.044 \text{ a.u.} \qquad (2.2.56)$$

In deriving eq. (2.2.54), we have skipped some mathematical steps. Recall from the treatment of hydrogen atom:

$$E_n = -\frac{Z^2}{2n^2} \text{ a.u.} \qquad (2.2.57)$$

Hence, for helium, with $Z = 2$,

$$E_{1s} = -2 \text{ a.u.}, \qquad (2.2.58)$$

$$E_{2s} = -1/2 \text{ a.u.}, \qquad (2.2.59)$$

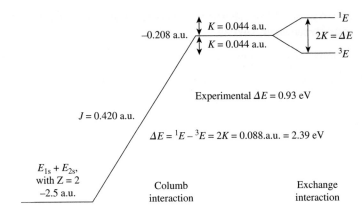

Fig. 2.2.3.
Determination of the energies of the singlet and triplet states arising from the $1s^12s^1$ configuration of the helium atom.

and

$$^1E = (-2.500 + 0.420 + 0.044) \text{ a.u.} = -2.036 \text{ a.u.} \quad (2.2.60)$$

In a similar manner, it can be shown that

$$^3E = \left(-2\tfrac{1}{2} + J - K\right) \text{ a.u.}$$
$$= (-2.500 + 0.420 - 0.044) \text{ a.u.} = -2.124 \text{ a.u.} < {}^1E. \quad (2.2.61)$$

The way we have arrived at the energies of the singlet and triplet states arising from the $1s^12s^1$ configuration is illustrated pictorially in Fig. 2.2.3. From the above discussion and the example of the $1s^12s^1$ configuration for the helium atom, it is seen that more than one state can arise from a given configuration. Furthermore, the energy difference for these states is due to the electronic repulsion term.

To conclude this section, we note the following:

(1) For the ground configuration $1s^2$ of helium, only one wavefunction can be written [eq. (2.2.36)]. Hence the ground state of helium is a singlet.
(2) For the ground configuration $1s^22s^1$ of lithium, two wavefunctions can be written [eqs. (2.2.39) and (2.2.40)]. Hence the ground state of lithium is a doublet.
(3) For the excited configuration $1s^12s^1$ of helium, two states can be derived. One is a singlet and the other is a triplet, with the latter having the lower energy.

As a final remark, from (3), it can be seen that, it is not proper to speak of an electronic transition by mentioning only the configurations involved, such as $1s^2 \rightarrow 1s^12s^1$. This is because, in this particular case, it does not specify the final state of the transition. Hence, to cite an electronic transition, we need to specify both the initial and final electronic states. Only stating the initial and final configurations is not precise enough. We will have more discussion on this later.

2.3 Many-electron atoms: electronic configuration and spectroscopic terms

2.3.1 Many-electron atoms

For an atom with n electrons, the Schrödinger equation has the form (in a.u.)

$$\left(-\frac{1}{2}\sum_i \nabla_i^2 - \sum_i \frac{Z}{r_i} + \sum_{i<j} \frac{1}{r_{ij}}\right)\psi(1,2,\ldots,n) = E\psi(1,2,\ldots,n). \tag{2.3.1}$$

This equation cannot be solved exactly. The most often used approximation model to solve this equation is called the self-consistent field (SCF) method, first introduced by D. R. Hartree and V. A. Fock. The physical picture of this method is very similar to our treatment of helium: each electron "sees" an effective nuclear charge contributed by the nuclear charge and the remaining electrons.

Figure 2.3.1 shows the radial distribution functions of sodium ($1s^22s^22p^63s^1$). The shaded portion shows the electronic distribution of Na$^+$ ($1s^22s^22p^6$), which indicates that the K and L shells of Na$^+$ are two concentric rings close to the nucleus; such a picture resembles the orbits in the Bohr theory. Also shown in Fig. 2.3.1 are the 3s, 3p, and 3d radial distribution functions. As a general rule, s orbitals are least screened (it "sees" the largest effective nuclear charge), or most penetrating. As ℓ increases, the orbital penetrates less into the region close to the nucleus so that the effective nuclear charge experienced by the electron decreases. As a result, for atoms with more than one electron, orbitals having the same n but different ℓ values are not degenerate. Recall, for the hydrogen atom, orbitals with the same n but different ℓ values have the same energy [eq. (2.1.28)],

$$E_n = -\frac{Z^2}{2n^2} \text{ a.u.,} \tag{2.3.2}$$

but such degeneracies no longer hold true for atoms other than hydrogen.

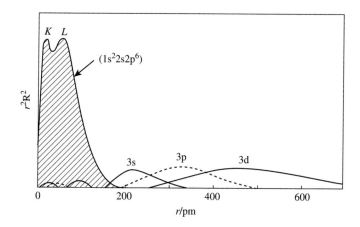

Fig. 2.3.1.
The radial distribution functions of 3s, 3p, and 3d orbitals together with that of the sodium core ($1s^22s^22p^6$).

In an atom with more than one electron, each electron (or orbital) is characterized by the following quantum numbers:

n: It largely determines the energy associated with the orbital and its size.

ℓ: It indicates the shape of the orbital; also, the electronic orbital angular momentum has a magnitude of $[(\ell(\ell+1)]^{1/2}(h/2\pi)$, or $[\ell(\ell+1)]^{1/2}$ a.u.

m_ℓ: It indicates the orientation of the orbital; also, the z component of the orbital angular momentum of the electron is $m_\ell(h/2\pi)$ or m_ℓ a.u.

s: It has a value of 1/2 for all electrons; also the electronic spin angular momentum has a magnitude of $[s(s+1)]^{1/2}$ a.u. or $^1\!/\!_2(3)^{1/2}$ a.u.

m_s: It has a value of either 1/2 or $-1/2$; the z component of the spin angular momentum of an electron is m_s a.u.

Note the ℓ plus m_ℓ and s plus m_s make two equivalent pairs of angular momentum quantum numbers. The former pair is for the orbital motion, and the latter for the spin motion.

2.3.2 Ground electronic configuration for many-electron atoms

By applying the Pauli Exclusion Principle, we can put each successive electron into the available lowest-energy orbital to yield the following electronic configurations for the first 30 elements in the periodic table (Table 2.3.1).

In the last column of Table 2.3.1, [Ar] denotes that the first 18 electrons of these elements have the Ar configuration. For Sc ($Z = 21$), we have $1s^22s^22p^63s^23p^6$ for the first 18 electrons. For electron no. 19, $E(3d) < E(4s)$, and hence the electron enters into the 3d orbital. For electron no. 20, $E(4s) < E(3d)$, so it enters into the 4s orbital. For electron no. 21, $E(4s) < E(3d)$, and thus it enters into the 4s orbital. Finally, we have the configuration $[Ar]3d^14s^2$ for Sc, as well as $[Ar]3d^14s^1$ and $[Ar]3d^1$ for Sc^+ and Sc^{2+}, respectively.

It is now clear that the energies of the atomic orbitals do not follow an immutable sequence. Rather, the energy ordering depends on both the nuclear charge and electronic configuration under consideration. For example, for electron no. 19 in both K ($Z = 19$) and Ca ($Z = 20$), we have $E(4s) < E(3d)$. For electron no. 19 in Sc ($Z = 21$), it becomes $E(3d) < E(4s)$. Furthermore, for electrons no. 20 and no. 21 of Sc, the ordering is reversed.

Table 2.3.1. Electronic configurations for the first 30 elements

H	$1s^1$	Na	$1s^22s^22p^63s^1$	Sc	$[Ar]3d^14s^2$
He	$1s^2$	Mg	$1s^22s^22p^63s^2$	Ti	$[Ar]3d^24s^2$
Li	$1s^22s^1$	Al	$1s^22s^22p^63s^23p^1$	V	$[Ar]3d^34s^2$
Be	$1s^22s^2$	Si	$1s^22s^22p^63s^23p^2$	Cr	$[Ar]3d^54s^1$
B	$1s^22s^22p^1$	P	$1s^22s^22p^63s^23p^3$	Mn	$[Ar]3d^54s^2$
C	$1s^22s^22p^2$	S	$1s^22s^22p^63s^23p^4$	Fe	$[Ar]3d^64s^2$
N	$1s^22s^22p^3$	Cl	$1s^22s^22p^63s^23p^5$	Co	$[Ar]3d^74s^2$
O	$1s^22s^22p^4$	Ar	$1s^22s^22p^63s^23p^6$	Ni	$[Ar]3d^84s^2$
F	$1s^22s^22p^5$	K	$1s^22s^22p^63s^23p^64s^1$	Cu	$[Ar]3d^{10}4s^1$
Ne	$1s^22s^22p^6$	Ca	$1s^22s^22p^63s^23p^64s^2$	Zn	$[Ar]3d^{10}4s^2$

Note that there are two anomalies in the first transition series: $[Ar]3d^54s^1$ (instead of $[Ar]3d^44s^2$) for Cr and $[Ar]3d^{10}4s^1$ (instead of $[Ar]3d^94s^2$) for Cu. These two configurations arise from the extra stability of a half-filled or completely filled subshell. Such stability comes from the spherically symmetric electron density around the nucleus for these configurations. Take the simpler case of p^3 as an example. The angular portion of the density function is proportional to

$$|p_x|^2 + |p_y|^2 + |p_z|^2$$
$$= \text{constant}[(\sin\theta\cos\phi)^2 + (\sin\theta\sin\phi)^2 + (\cos\theta)^2]$$
$$= \text{constant}. \qquad (2.3.3)$$

Similar spherical symmetry also occurs for configurations d^5 and f^7.

2.3.3 Spectroscopic terms arising from a given electronic configuration

We are now in position to derive the electronic states arising from a given electronic configuration. These states have many names: spectroscopic terms (or states), term symbols, and Russell–Saunders terms, in honor of spectroscopists H. N. Russell and F. A. Saunders. Hence, the scheme we use to derive these states is called Russell–Saunders coupling. It is also simply referred to as L–S coupling.

Each electronic state is defined by three angular momenta: total orbital angular momentum vector **L**, total spin angular momentum vector **S**, and total angular momentum vector **J**. These vectors have the following definitions:

$$\mathbf{L} = \sum_i \boldsymbol{\ell}_i, \qquad (2.3.4)$$

$$\mathbf{S} = \sum_i \mathbf{s}_i, \qquad (2.3.5)$$

and

$$\mathbf{J} = \mathbf{L} + \mathbf{S}. \qquad (2.3.6)$$

In eqs. (2.3.4) and (2.3.5), $\boldsymbol{\ell}_i$ and $\mathbf{s}_i$ are the orbital angular momentum vector and spin angular momentum vector of an individual electron, respectively. For a state defined by quantum numbers L, S, and J, the magnitudes of the total orbital angular momentum, total spin angular momentum, and total angular momentum vectors of the system are $[L(L+1)]^{1/2}$, $[S(S+1)]^{1/2}$, and $[J(J+1)]^{1/2}$ a.u., respectively.

Referring to eq. (2.3.6), in quantum mechanics, when we add two vectors **A** and **B** to yield a resultant **C**, C can take on the values of $A+B, A+B-1,\ldots,|A-B|$ only, where A, B, and C are the quantum numbers for the vectors **A**, **B**, and **C**, respectively. Hence,

$$J = L+S, L+S-1, \ldots, |L-S|. \qquad (2.3.7)$$

Furthermore, each state is labeled by its L value according to:

L value	0	1	2	3	4	5	6...
State	S	P	D	F	G	H	I ...

A term symbol is written in the short-hand notation

$$^{2S+1}L_J.$$

So if we have a state $^4G_{3\frac{1}{2}}$, we can see immediately this state has $L = 4$, $S = 1^1/2$, and $J = 3^1/2$. From eq. (2.3.7), we see that, for $L = 4$ and $S = 1^1/2$, J can be $5^1/2$, $4^1/2$, $3^1/2$, and $2^1/2$. So $3^1/2$ is one of the allowed values. The superscript $2S+1$ in the term symbol is called the spin multiplicity of the state. When a state has spin multiplicity $1, 2, 3, 4, 5, 6, \ldots$, the state is called a singlet, doublet, triplet, quartet, quintet, sextet,$\ldots$, respectively. Usually the number of allowed J values is the same as the state's spin multiplicity; an example is the aforementioned ^4G. However, there are exceptions. For example, for the state ^2S, we have $L = 0$ and $S = 1/2$, and $J = 1/2$ only.

To derive the electronic states, we sometimes also need the z components of angular momenta $\mathbf{L}$ and $\mathbf{S}$, called L_z and S_z, respectively. The magnitudes of L_z and S_z are M_L and M_S a.u., respectively, where

$$M_L = \sum_i (m_l)_i = L, L-1, \ldots, -L; \quad (2.3.8)$$

$$M_S = \sum_i (m_s)_i = S, S-1, \ldots, -S. \quad (2.3.9)$$

Similarly, the total angular momentum vector $\mathbf{J}$ also has its z component J_z, with its corresponding magnitude being M_J a.u., where

$$M_J = M_L + M_S = J, J-1, \ldots, -J. \quad (2.3.10)$$

Note that the sums in eqs. (2.3.8) to (2.3.10) are numerical sums, not vector sums.

When we only have one electron in a configuration, e.g., $1s^1$ for H, we get

$$L = \ell_1 = 0,$$

which indicates an S state. Also,

$$S = s_1 = 1/2, \quad \text{or} \quad 2S+1 = 2.$$

So we have only one term, ^2S. The allowed J value is $1/2$. Hence $^2S_{1/2}$ is the only state arising from the s^1 configuration. Similarly, it is easy to show the following:

Configuration	State(s)
s^1	$^2S_{1/2}$
p^1	$^2P_{1\frac{1}{2}}, {}^2P_{1/2}$
d^1	$^2D_{1\frac{1}{2}}, {}^2D_{2\frac{1}{2}}$
f^1	$^2F_{2\frac{1}{2}}, {}^2F_{3\frac{1}{2}}$

Now we proceed to the configuration s^2 of, say, helium. Here these electrons are called equivalent electrons, since they have the same n value and also the same ℓ value. For such systems, we need to bear the Exclusion Principle in mind. Now the two electrons can only be accommodated in the following manner:

$$1s \uparrow\downarrow \qquad M_L = \sum_i (m_\ell)_i \qquad M_S = \sum_i (m_s)_i$$
$$n = 1 \quad \ell = 0, \; m_\ell = 0, \quad M_L = 0 + 0 = 0 \quad M_S = 1/2 + (-1/2) = 0.$$

So the only (M_L, M_S) combination is $(0,0)$. Such a combination is also called a microstate. Here the only allowed value of $M_L = 0$; the same is true for M_S. From eqs. (2.3.8) and (2.3.9), the only allowed value for L is 0; the same also holds for S. So the only state from the s^2 configuration is $^1S_0$. Indeed, $^1S_0$ (with $S = 0$ and $L = 0$) is the only allowed state for all filled configurations:

Configuration	State
s^2	$^1S_0$
p^6	$^1S_0$
d^{10}	$^1S_0$
f^{14}	$^1S_0$

For the excited configuration $1s^1 2s^1$ of helium, we have the following:

$n=1, \ell=0, m_\ell=0$	$n=2, \ell=0, m_\ell=0$	M_L	M_S
↑	↑	0	1
↑	↓	0	0
↓	↑	0	0
↓	↓	0	−1

These four microstates lead to an (M_L, M_S) distribution shown below:

$$\begin{array}{c|ccc} M_L = 0 & 1 & 2 & 1 \\ \hline & -1 & 0 & 1 \end{array} \; M_S$$

For such a simple distribution, it is obvious that only two terms can arise from it: ^1S [which requires one microstate: $(M_L = 0, M_S = 0)$] and ^3S [which requires three microstates: $(0,1)$, $(0,0)$, and $(0,-1)$]. Previously we have already concluded that the $1s^1 2s^1$ configuration of helium has two states, one singlet and one triplet, with the triplet state having the lower energy. When we put in the proper J values, the states arising from configuration $ns^1 n's^1$ are $^1S_0$ and $^3S_1$.

The Electronic Structure of Atoms

An example often used in texts is the $1s^2 2s^2 2p^2$ configuration of carbon. Since all filled subshells lead to $L = 0$ and $S = 0$, we only need to be concerned with p^2 here. There are 15 microstates for this configuration:

$m_\ell = 1$	$m_\ell = 0$	$m_\ell = -1$	M_L	M_S
↑↓			2	0
	↑↓		0	0
		↑↓	−2	0
↑	↑		1	1
↓	↓		1	−1
↑	↓		1	0
↓	↑		1	0
	↑	↑	−1	1
	↓	↓	−1	−1
	↑	↓	−1	0
	↓	↑	−1	0
↑		↑	0	1
↓		↓	0	−1
↑		↓	0	0
↓		↑	0	0

These 15 microstates have the distribution shown below:

M_L			
2		1	
1	1	2	1
0	1	3	1
−1	1	2	1
−2		1	
	−1	0	1 M_S

This large distribution is made up of the following three smaller distributions:

^1S:

M_L	
0	1
	0 M_S

^3P:

M_L			
1	1	1	1
0	1	1	1
−1	1	1	1
	−1	0	1 M_S

^1D:

M_L	
2	1
1	1
0	1
−1	1
−2	1
	0 M_S

So three terms arise from configuration p^2: ^1D, ^3P, and ^1S. If we include the allowed J values, we then have $^1D_2$, $^3P_2$, $^3P_1$, $^3P_0$, and $^1S_0$. Levels such as $^3P_2$, $^3P_1$, and $^3P_0$ which differ only in their J values are called multiplets. For light elements, the levels of a multiplet will have slightly different energies, while levels from different terms will have larger energy differences.

By writing out the microstates of a given configuration, it is easy to see that configurations ℓ^n and $\ell^{4\ell+2-n}$ have the same states. This is because the number

Table 2.3.2. Spectroscopic terms arising from configurations s^n, p^n, and d^n

Configuration	LS terms	Configuration	LS terms
p^1, p^5	^2P	d^2, d^8	^1S, ^1D, ^1G, ^3P, ^3F
p^2, p^4	^1S, ^1D, ^3P	d^3, d^7	$^2D(2)$, ^2P, ^2F, ^2G, ^2H, ^4P, ^4F
p^3	^2P, ^2D, ^4S	d^4, d^6	$^1S(2)$, $^1D(2)$, ^1F, $^1G(2)$, ^1I, $^3P(2)$, ^3D, $^3F(2)$, ^3G, ^3H, ^5D
d^1, d^9	^2D	d^5	^2S, ^2P, $^2D(3)$, $^2F(2)$, $^2G(2)$, ^2H, ^2I, ^4P, ^4D, ^4F, ^4G, ^6S

of ways of assigning electrons to orbitals is equal to the number of ways of assigning vacancies to the same orbitals. Hence carbon ($1s^22s^22p^2$) and oxygen ($1s^22s^22p^4$) have the same states. Table 2.3.2 lists the term symbols (L and S values only) arising from a given equivalent electronic configuration. We note that configurations with only one electron in a subshell and configurations with filled subshells have already been dealt with.

Note that the last term given for each configuration is the ground term (see discussions on Hund's rule below). The number in brackets indicates the number of times that term occurs, e.g., there are two different ^2D terms for the d^3 configuration.

2.3.4 Hund's rules on spectroscopic terms

When there are many terms arising from a given configuration, it would be of interest to know the relative order of their energies, or at least which is the ground term. To accomplish this, we make use of Hund's rule, in honor of physicist F. Hund:

(1) Of the terms arising from a given configuration, the lowest in energy is the one having the highest spin multiplicity.
(2) Of two or more terms with the same multiplicity, the lowest in energy is the one with the largest L value.
(3) For a configuration less than half-filled, the lowest state is the one with the smallest J value (the multiplet is then called "normal"); if the configuration is more than half-filled, the lowest state is the one with the largest J value (the multiplet is "inverted").

Note that, strictly speaking, Hund's rules should be applied only to equivalent electronic configurations and for the determination of only the ground state. By applying these rules, we can see that both carbon ($2p^2$) and oxygen ($2p^4$) have ^3P as the ground term. However, the ground state of carbon is $^3P_0$, while that for oxygen is $^3P_2$. A complete energy ordering for the state of carbon is given in Fig. 2.3.2. Also, a state $^{2S+1}L_J$ has a $(2J+1)$–fold degeneracy, as $M_J = J, J-1, \ldots, -J$. This degeneracy will be manifested when we apply an external magnetic field to a sample and then take its spectrum (Zeeman effect).

Before leaving this topic, we once again note that electronic transitions occur between states and may or may not involve a change in configuration. Take the states in Fig. 2.3.2 as an example. Transition from ground state $^3P_0$ to excited state $^1D_2$ ($^3P_0 \rightarrow {}^1D_2$) [note that this is not an allowed transition in the first

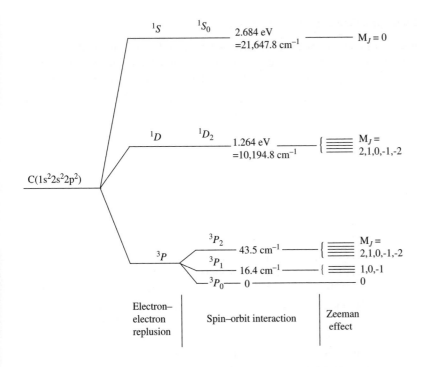

Fig. 2.3.2.
Energies of the spectroscopic state for the ground state configuration of carbon (not to scale!).

place!] does not involve a change in configuration, as both the initial and final states of the electronic transition arise from configuration $2p^2$.

To conclude this section, we illustrate how to obtain the ground term of a given configuration without writing out all the microstates. Again we take p^2 as an example. When we place there two electrons into the three p orbitals we bear in mind that the ground term requires the maximum S (first priority) and L values. To achieve this, we place the electron in the following manner:

m_ℓ	1	0	−1
	↑	↑	

$M_L = 1 \quad M_S = 1$

Since M_L runs from L to $-L$ and M_S runs from S to $-S$, the maximum L and maximum S values are both 1 in this case. Hence, the ground term is ^3P.

Two more examples are given below:

(1) p^3

m_ℓ	1	0	−1
	↑	↑	↑

$M_L = 0 \quad M_S = 1^1/_2$

We thus have $L = 0$ and $S = 1^1/_2$ and the ground term is ^4S.

(2) d^7

m_ℓ	2	1	0	−1	−2
	↑↓	↑↓	↑	↑	↑

$M_L = 3 \quad M_S = 1^1/_2$

Hence we have $L = 3$ and $S = 1^1/_2$ and the ground term is ^4F.

2.3.5 j–j Coupling

In the L–S coupling scheme we have just discussed, it is assumed that the electronic repulsion is much larger than the spin–orbit interaction. This assumption certainly holds for lighter elements, as we have seen in the case of carbon atom (Fig. 2.3.2). However, this assumption becomes less and less valid as we go down the Periodic Table. The breakdown of this coupling scheme is clearly seen from the spectral data given in Table 2.3.3.

Table 2.3.3. Spectral data (in cm^{-1}) for the electronic states arising from configurations np^2, $n = 2$–6

	$^3P_0$	$^3P_1$	$^3P_2$	$^1D_2$	$^1S_0$
C	0.0	16.4	43.5	10193.7	21648.4
Si	0.0	77.2	223.31	6298.8	15394.2
Ge	0.0	557.1	1409.9	7125.3	16367.1
Sn	0.0	1691.8	3427.7	8613.0	17162.6
Pb	0.0	7819.4	10650.5	21457.9	29466.8

In the extreme case where the spin–orbit interaction is much larger than electronic repulsion, total orbital angular momentum L and total spin angular momentum S are no longer "good" quantum numbers. Instead, states are defined by total angular momentum J, which is the vector sum of all the total angular momenta j values of the individual electrons:

$$\mathbf{J} = \sum_i \mathbf{j}_i, \qquad (2.3.11)$$

where

$$\mathbf{j}_i = \boldsymbol{\ell}_i + \mathbf{s}_i. \qquad (2.3.12)$$

We can illustrate this scheme with a simple example. Consider the electronic configuration p^1s^1. For the p electron, we have $\mathbf{j}_1 = \boldsymbol{\ell}_1 + \mathbf{s}_1$, which means that j_1 can be either $1\frac{1}{2}$ and $1/2$. On the other hand, for the s electron, we have $j_2 = 1/2$. When $j_1 = 1\frac{1}{2}$ and $j_2 = 1/2$ are coupled, $\mathbf{J} = \mathbf{j}_1 + \mathbf{j}_2$ and quantum number J can be 1 or 2. When $j_1 = 1/2$ and $j_2 = 1/2$ are coupled, J can be 0 or 1. So for the p^1s^1 configuration, there are four states: $J = 0, 1, 1, 2$. These states are "correlated" to the Russell–Saunders terms in the manner shown in Fig. 2.3.3.

To conclude this section, we take up a slightly more complex example: the case of two equivalent p electrons. In the j–j coupling scheme, the Pauli's principle becomes "no two electrons can have the same set of quantum numbers (n, ℓ, j, m_j)." For configuration np^2, $\ell_1 = \ell_2 = 1$, $s_1 = s_2 = 1/2, j_1, j_2 = 1\frac{1}{2}$ or $1/2$. Once again, as in the case of L–S coupling, there are now 15 microstates, as listed in Table 2.3.4.

Note that in Table 2.3.4 labels 1 and 2 are for convenience only. It is important to bear in mind that electrons are indistinguishable. The five electronic states derived using this scheme may be correlated to those derived with the L–S coupling scheme in Fig. 2.3.2. This correlation diagram is shown in Fig. 2.3.4.

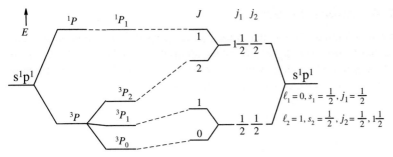

Fig. 2.3.3.
Correlation diagram for electronic states arising from configuration s^1p^1 employing L–S and j–j coupling schemes.

Table 2.3.4. The microstates for configuration np^2 employing the j–j coupling scheme

j_1	j_2	$(m_j)_1$	$(m_j)_2$	M_J	J
1/2	1/2	1/2	−1/2	0	0
1/2	1½	1/2	1½	2	
		1/2	1/2	1	
		1/2	−1/2	0	
		1/2	−1½	−1	2, 1
		−1/2	1½	1	
		−1/2	1/2	0	
		−1/2	−1/2	−1	
		−1/2	−1½	−2	
1½	1½	1½	1/2	2	
		1½	−1/2	1	
		1½	−1½	0	2, 0
		1/2	−1/2	0	
		1/2	−1/2	−1	
		−1/2	−1½	−2	

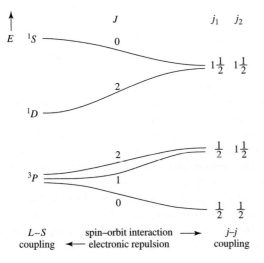

Fig. 2.3.4.
Correlation diagram for electronic states arising from configuration np^2 employing L–S and j–j coupling schemes.

2.4 Atomic properties

In this section we discuss the various atomic properties that are the manifestation of the electronic configurations of the atoms discussed in the previous sections. These properties include ionization energy, electron affinity, electronegativity, etc. Other properties such as atomic and ionic radii will be discussed in subsequent chapters, as these properties are related to the interaction between atoms in a molecule. Toward the end of this section, we will also discuss the influence of relativistic effects on the properties of elements.

2.4.1 Ionization energy and electron affinity

(1) Ionization energy

The first ionization energy I_1 of an atom is defined as the energy required to remove the outermost electron from a gaseous atom in its ground state. This energy may be expressed as the enthalpy change of the process

$$A(g) \rightarrow A^+(g) + e^-(g), \quad \Delta H = I_1. \tag{2.4.1}$$

The second ionization energy I_2 of A is then the energy required to remove an electron from A^+:

$$A^+(g) \rightarrow A^{2+}(g) + e^-(g), \quad \Delta H = I_2. \tag{2.4.2}$$

The third, fourth, and subsequent ionization energies may be defined in a similar manner. Table 2.4.1 lists the I_1 and I_2 values of the elements. The variations of I_1 and I_2 against atomic number Z are shown in Fig. 2.4.1.

Examining the overall trend of I_1 shown in Fig. 2.4.1, it is seen that I_1 decreases as Z increases. On the other hand, within the same period, I_1 increases with Z. Also, the noble gases occupy the maximum positions, as it is difficult to remove an electron from a filled configuration (ns^2np^6). Similarly, but to a less extent, Zn, Cd, and Hg, with configuration $(n-1)d^{10}ns^2$, occupy less prominent maxima. Also, N, P, and As, with half-filled subshells, are also in less prominent maximum positions as well.

In the I_1 curve shown in Fig. 2.4.1, the minimum positions are taken up by the alkali metals. These elements have general configuration ns^1 and this lone electron is relatively easy to remove. On the other hand, in the I_2 curve, the alkali metals now occupy the maximum positions, as the singly charged cations of the alkali metals have the noble gas electronic configuration. Additionally, the I_2 values of the alkali metals are very large, in the range $2.2–7.3 \times 10^3$ kJ mol^{-1}. The chemical properties of these elements may be nicely explained by their ionization energy data. Similarly, elements B, Al, Ga, In, and Tl, with only one lone electron in the p orbital, also have relatively low I_1 values.

The two valence electrons of alkali earth metals (Be, Mg, Ca, Sr, and Ba) are comparatively easy to ionize, as the data in Fig. 2.4.1 show. These elements occupy the minimum positions in the I_2 curve. Hence the chemistry of these elements is dominated by their M^{2+} cations. The only exception is Be, which, with the largest $(I_1 + I_2)$ value in the family, forms mainly covalent compounds.

Table 2.4.1. Ionization energies (MJ mol^{-1}) I_1 and I_2 and the electron affinities (kJ mol^{-1}) Y of the elements

Z	Element	I_1	I_2	Y	Z	Element	I_1	I_2	Y
1	H	1.3120		72.77	47	Ag	0.7310	2.074	125.7
2	He	2.3723	5.2504	—	48	Cd	0.8677	1.6314	—
3	Li	0.5203	7.2981	59.8	49	In	0.5583	1.8206	29
4	Be	0.8995	1.7571	—	50	Sn	0.7086	1.4118	121
5	B	0.8006	2.4270	27	51	Sb	0.8316	1.595	101
6	C	1.0864	2.3526	122.3	52	Te	0.8693	1.79	190.1
7	N	1.4023	2.8561	−7	53	I	1.0084	1.8459	295.3
8	O	1.3140	3.3882	141.0	54	Xe	1.1704	2.046	—
9	F	1.6810	3.3742	327.9	55	Cs	0.3757	2.23	45.49
10	Ne	2.0807	3.9523	—	56	Ba	0.5029	0.9653	—
11	Na	0.4958	4.5624	52.7	57	La	0.5381	1.067	50
12	Mg	0.7377	1.4507	—	58	Ce	0.528	1.047	50
13	Al	0.5776	1.8167	44	59	Pr	0.523	1.018	50
14	Si	0.7865	1.5771	133.6	60	Nd	0.530	1.034	50
15	P	1.0118	1.9032	71.7	61	Pm	0.536	1.052	50
16	S	0.9996	2.251	200.4	62	Sm	0.543	1.068	50
17	Cl	1.2511	2.297	348.8	63	Eu	0.547	1.085	50
18	Ar	1.5205	2.6658	—	64	Gd	0.592	1.17	50
19	K	0.4189	3.0514	48.36	65	Tb	0.564	1.112	50
20	Ca	0.5898	1.1454	—	66	Dy	0.572	1.126	50
21	Sc	0.631	1.235	—	67	Ho	0.581	1.139	50
22	Ti	0.658	1.310	20	68	Er	0.589	1.151	50
23	V	0.650	1.414	50	69	Tm	0.5967	1.163	50
24	Cr	0.6528	1.496	64	70	Yb	0.6034	1.175	50
25	Mn	0.7174	1.5091	—	71	Lu	0.5235	1.34	50
26	Fe	0.7594	1.561	24	72	Hf	0.654	1.44	—
27	Co	0.758	1.646	70	73	Ta	0.761		60
28	Ni	0.7367	1.7530	111	74	W	0.770		60
29	Cu	0.7455	1.9579	118.3	75	Re	0.760		15
30	Zn	0.9064	1.7333	—	76	Os	0.84		110
31	Ga	0.5788	1.979	29	77	Ir	0.88		160
32	Ge	0.7622	1.5372	120	78	Pt	0.87	1.7911	205.3
33	As	0.944	1.7978	77	79	Au	0.8901	1.98	222.7
34	Se	0.9409	2.045	194.9	80	Hg	1.0070	1.8097	—
35	Br	1.1399	2.10	324.6	81	Tl	0.5893	1.9710	30
36	Kr	1.3507	2.3503	—	82	Pb	0.7155	1.4504	110
37	Rb	0.4030	2.633	46.89	83	Bi	0.7033	1.610	110
38	Sr	0.5495	1.0643	—	84	Po	0.812		180
39	Y	0.616	1.181	0.0	85	At	—		270
40	Zr	0.660	1.267	50	86	Rn	1.0370		—
41	Nb	0.664	1.382	100	87	Fr	—		
42	Mo	0.6850	1.558	100	88	Ra	0.5094	0.9791	
43	Tc	0.702	1.472	70	89	Ac	0.49	1.17	
44	Ru	0.711	1.617	110	90	Th	0.59	1.11	
45	Rh	0.720	1.744	120	91	Pa	0.57		
46	Pd	0.805	1.875	60	92	U	0.59		

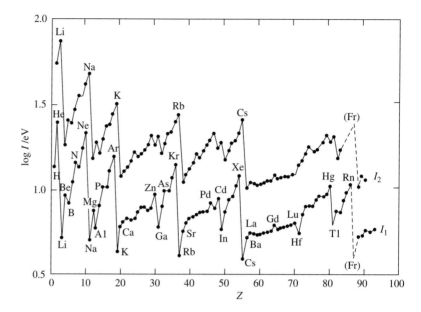

Fig. 2.4.1.
The variation of ionization energies, I_1 and I_2 against the atomic number Z.

Before leaving this topic, it should be noted that ionization energy is only one of the factors that influence the chemistry of the elements. For instance, even though Ca^+ is easier to form than Ca^{2+}, we have never observed the existence of Ca^+. This is because the lattice energy of CaX_2 is much larger than that of CaX, enough to compensate the cost of the second ionization energy, which comes to about 10^3 kJ mol^{-1}. In solution, the hydration energy of Ca^{2+} is much larger than that of Ca^+, thus favoring the existence of Ca^{2+} over Ca^+. In short, the cation of calcium exists exclusively in the form Ca^{2+}, either in solution or in the crystalline state.

(2) Electron affinity

The electron affinity Y of an atom is defined as the energy released when the gaseous atom in its ground state captures an electron to form an anion:

$$A(g) + e^-(g) \rightarrow A^-(g), \quad -\Delta H = Y. \tag{2.4.3}$$

In other words, the electron affinity of A is simply the ionization energy of A^-. The electron affinities of the elements are also listed in Table 2.4.1. As the singly charged anions of a number of elements including Zn, Cd, Hg, and the noble gases, are unstable, the electron affinities of these elements are missing from the table. Also, there is one element, nitrogen, which has a negative electron affinity. This indicates that N^- is an unstable species and it loses an electron spontaneously.

In general, electron affinities have smaller values than ionization energies. Chlorine has the largest electron affinity, 348.8 kJ mol^{-1}, which is still smaller than the smallest ionization energy, 375.7 kJ mol^{-1}, of Cs.

Among the electron affinities listed in Table 2.4.1, the halogens have the largest values. This is because these atoms will attain the noble gas configuration upon capturing an electron. The members of the oxygen family, neighbors of the halogens, also have fairly large electron affinities. Another element that has a relatively large electron affinity is gold, which, as in the case of halogens, also forms salts such as CsAu. We shall have a more detailed discussion on this element in the section on relativistic effects. Finally, all the lathanides have similar electron affinities, about 50 kJ mol^{-1}. This is the value listed for all lathanides in Table 2.4.1.

There is no element that has a positive second electron affinity; i.e., the electron affinity of any anion A$^-$ is always negative. This is understandable as it is difficult to bring together two negatively charged particles. Thus the electron affinities of O$^-$ and S$^-$ are -744 and -456 kJ mol^{-1}, respectively. Even though O^{2-} and S^{2-} are unstable in gas phase, they can exist in solution or in the solid state. The stability of these divalent anions arises from lattice energy, solvation energy, etc. These ions may also hydrolyze to form more stable species, e.g., O^{2-} + H$_2$O → 2OH$^-$.

2.4.2 Electronegativity: the spectroscopic scale

Electronegativity is a qualitative concept with a long history and wide application. As early as 1835, Berzelius considered chemical bonding as electrostatic interaction. Also, he took metals and their oxides as electropositive substances and nonmetals and their oxides as electronegative ones. Almost one hundred years later, in 1932, Pauling defined electronegativity as "the power of atom in a molecule to attract electrons to itself." Quantitatively he proposed a scale, hereafter denoted as χ_P, in the following manner: the difference between the electronegativities of elements A and B, $|\chi_A - \chi_B|$, is proportional to $\Delta^{1/2}$, where Δ is the difference between the bond energy of AB and the mean of the bond energies of A$_2$ and B$_2$. In 1934 to 1935, Mullikan proposed the χ_M scale that the power of an atom to attract electrons should be the average of its ionization energy and electron affinity. Later in 1958, Allred and Rochow proposed their χ_{AR} scale. They assumed that the electronegativity of an atom is proportional to the effective nuclear charge of its valence electron and inversely proportional to the square of its atomic radius.

More recently, in 1989, Allen developed his spectroscopic scale, χ_S, where he equated the electronegativity of an atom to its configuration energy (CE)

$$\text{CE} = \frac{n\varepsilon_s + m\varepsilon_p}{n+m}, \qquad (2.4.4)$$

where n and m are the number of s and p valence electrons, respectively, and ε_s and ε_p are the corresponding one-electron energies. These energies are the multiplet-averaged total energy differences between the ground state neutral and singly charged cation. As accurate atomic energy level data are readily available, the CE of an atom can be easily calculated. Table 2.4.2 lists the Pauling and spectroscopic electronegativities of the elements.

Before leaving this topic, we discuss in more detail the determination of CE and χ_S values of an element. Table 2.4.3 lists the ε_s and ε_p, and CE values of the first eight elements in the Periodic Table. For the first four elements, where

Table 2.4.2. Electronegativities of the elements*

H 2.20 2.300							He — 4.160
Li 0.98 0.912	Be 1.57 1.576	B 2.04 2.051	C 2.55 2.544	N 3.04 3.066	O 3.44 3.610	F 3.98 4.193	Ne — 4.787
Na 0.93 0.869	Mg 1.31 1.293	Al 1.61 1.613	Si 1.90 1.916	P 2.19 2.253	S 2.58 2.589	Cl 3.16 2.869	Ar — 3.242
K 0.82 0.734	Ca 1.00 1.034	Ga 1.81 1.756	Ge 2.01 1.994	As 2.18 2.211	Se 2.55 2.424	Br 2.96 2.685	Kr 3.34 2.966
Rb 0.82 0.706	Sr 0.95 0.963	In 1.78 1.656	Sn 1.96 1.824	Sb 2.05 1.984	Te 2.10 2.158	I 2.66 2.359	Xe 2.95 2.582
Cs 0.79 0.659	Ba 0.89 0.881	Tl 2.04 1.789	Pb 2.33 1.854	Bi 2.02 2.01	Po 2.0 2.19	At 2.2 2.39	Rn — 2.60

Sc 1.36 1.19	Ti 1.54 1.38	V 1.63 1.53	Cr 1.66 1.65	Mn 1.55 1.75	Fe 1.83 1.80	Co 1.88 1.84	Ni 1.91 1.88	Cu 1.90 1.85	Zn 1.65 1.59
Y 1.22 1.12	Zr 1.33 1.32	Nb — 1.41	Mo 2.16 1.47	Tc — 1.51	Ru — 1.54	Rh 2.28 1.56	Pd 2.20 1.59	Ag 1.93 1.87	Cd 1.69 1.52
Lu — 1.09	Hf — 1.16	Ta — 1.34	W 2.36 1.47	Re — 1.60	Os — 1.65	Ir 2.20 1.68	Pt 2.28 1.72	Au 2.54 1.92	Hg 2.00 1.76

* χ_P values are given above the χ_S results.

Table 2.4.3. The ε_s, ε_p, and CE values of the first eight elements

Element	Electronic configuration	ε_p (eV)	ε_s (eV)	CE/eV	χ_S
H	$1s^1$	—	13.60	13.60	2.300
He	$1s^2$	—	24.59	24.59	4.160
Li	$1s^2 2s^1$	—	5.39	5.39	0.912
Be	$1s^2 2s^2$	—	9.32	9.32	1.576
B	$1s^2 2s^2 2p^1$	8.29	14.04	12.13	2.051
C	$1s^2 2s^2 2p^2$	10.66	19.42	15.05	2.544
N	$1s^2 2s^2 2p^3$	13.17	25.55	18.13	3.066
O	$1s^2 2s^2 2p^4$	15.84	32.36	21.36	3.610

only S states are involved, the CE value is simply I_1 of the element concerned. For the remaining four elements, let us take O as an example.

For O, with configuration $2s^2 2p^4$ and its states ^3P, ^1D, and ^1S, we can determine its multiplet-averaged energy, using the available spectroscopic data. For

O^+, with configuration $2s^2 2p^3$ and its states ^4S, ^2D, and ^2P, we can also determine this cation's multiplet-averaged energy. Then ε_p is simply the difference between these two average energies. To determine ε_s, we consider configuration $2s^1 2p^4$ and its states ^4P, ^2D, ^2P, and ^2S and calculate the average energy of this configuration. Then ε_s is simply the energy difference between configurations $2s^2 2p^4$ and $2s^1 2p^4$. Upon calculating the CE of an element, we can convert it to χ_S by multiplying the factor (2.30/13.60). As we can see from this procedure, χ_S is related to the average one-electron valence shell ionization energy of an atom, which, we would intuitively think, should be related to its power to attract electrons in a molecule.

Figures 2.4.2 and 2.4.3 show the variations of electronegativities of representative elements and transition metals, respectively, across a given period. As we can see from these two figures, χ_S increases with Z in a period. The noble gases have the largest χ_S values for elements of the same period, as they have very strong tendency to retain the electrons. Among all elements, Ne is the most electronegative, and F, He, and O are the next ones. Noble gas Xe is less electronegative than F and O. Hence xenon fluorides and oxides are capable of existence. Also, Xe and C have comparable electronegativities, and Xe–C bond can be formed under suitable conditions. One such example is $[F_5C_6XeNCMe]^+[(C_6F_5)_2BF_2]^-$MeCN, and the structure of the cation of this compound is shown in Fig. 2.4.4.

From the electronegativities shown in Figs. 2.4.2 and 2.4.3, it is seen that metals are less electronegative, with $\chi_S < 2$, nonmetals are more electronegative, with $\chi_S > 2$. Around $\chi_S \sim 2$, we have metalloid elements such as B, Si Ge, Sb, and Bi. Most of these elements have semi-conducting properties. For elements of the same group, χ_S decreases as we go down the Periodic Table. However, because of relativistic effects, for transition metals from group 7 to

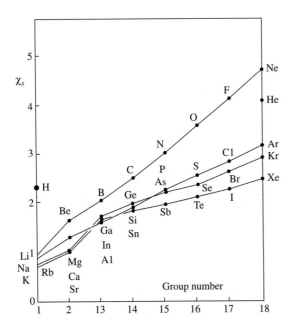

Fig. 2.4.2.
Variation of electronegativities of representative elements across a given period of the periodic table. Elements of period 6 have electronegativities very similar to those of elements of period 5. For clarity, electronegativities of the elements of period 6 are not shown.

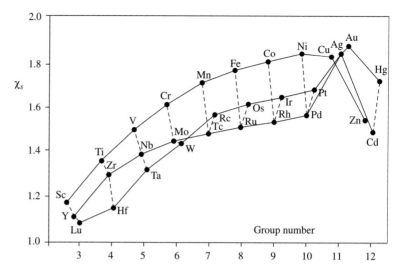

Fig. 2.4.3.
Variation of electronegativities of transition metals across a given period of the periodic table.

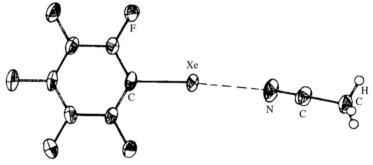

Fig. 2.4.4.
Structure of cation $[F_5C_6XeNCMe]^+$: C–Xe = 209.2 pm, Xe–N = 268.1 pm, C–Xe–N = 174.5°.

group 12, the elements of the sixth period are more electronegative than the corresponding elements of the fifth period.

In 1963, N. Barlett discovered that PtF_6 can oxidize O_2 to form salt-like $[O_2]^+[PtF_6]^-$. As the I_1 values of Xe (1.17×10^3 kJ mol^{-1}) and O_2 (1.18×10^3 kJ mol^{-1}) are similar, he reacted Xe with PtF_6 to form the first noble gas compound $Xe^+[PtF_6]^-$. Soon afterwards, XeF_2 and XeF_4 were synthesized and a whole new area of chemistry was discovered. Now many compounds with bonds such as Xe–F, Xe–O, Xe–N, and Xe–C have been synthesized; an example is shown in Fig. 2.4.4. Even though Kr has a slightly higher I_1 value than Xe, compounds such as KrF, $[KrF]^+[Sb_2F_{11}]^-$, and $CrOF_4 \cdot KrF_2$ (structure shown below) have already been made.

Since Ar has an I_1 value lower than that of F, it is not unreasonable to expect that argon fluorides will be synthesized soon. Recently HArF was detected in a photodissociation study of HF in an Ar environment (see Section 5.8.1). With infrared spectroscopic studies of species such as $H^{40}ArF$, $D^{40}ArF$, and $H^{36}ArF$, it was determined that the bond lengths of H–Ar and F–Ar are 133 and 197 pm, respectively. As of now, He and Ne are the only two elements that do not form any stable compounds.

2.4.3 Relativistic effects on the properties of the elements

In recent years, relativistic effects on the chemical properties of atoms have received considerable attention. In the theory of relativity, when an electron is traveling with high velocity v, its mass m is related to its rest mass m_o in the following way,

$$m = m_o \Big/ \left[1 - \left(\frac{v}{c}\right)^2\right]^{1/2}, \tag{2.4.5}$$

where c is the speed of light. In atomic units,

$$c = 137 \text{ a.u.}$$

The average radial velocity $<v_{\text{rad}}>$ of a 1s electron in an atom is approximately Z a.u., where Z is the atomic number. Hence, for an electron in Au, with $Z = 79$, we have

$$\frac{v_{\text{rad}}}{c} \sim \frac{79}{137} = 0.58. \tag{2.4.6}$$

Substituting this ratio into eq. (2.4.5), we get

$$m = \frac{m_o}{[1 - (0.58)^2]^{1/2}} = 1.23 m_o. \tag{2.4.7}$$

In other words, the mass of the electron has increased to $1.23 m_o$, which influences significantly the radial distribution of the electron. Recall from Bohr's atomic theory that the radius of the nth orbit is given by

$$r_n = \frac{n^2 h^2 \varepsilon_0}{\pi m e^2}.$$

Hence the ratio between the relativistic 1s radius, $<r_{1s}>_R$, to its non-relativistic counterpart, $<r_{1s}>_{NR}$ is simply

$$<r_{1s}>_R / <r_{1s}>_{NR} = (1.23 m_o)^{-1} / m_o^{-1} = 0.81.$$

That is, relativistic effects have effectively contracted the 1s orbital by 19%.

The contraction of the 1s orbital in heavy atoms also results in a contraction of the s orbitals in higher quantum shells of that atom, which in turn leads to

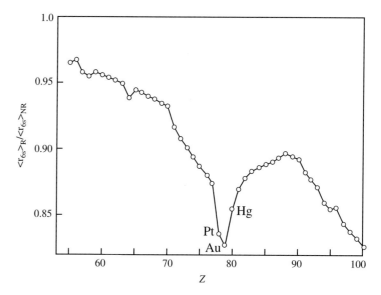

Fig. 2.4.5.
Calculated orbital contraction ratio $\langle r_{6s}\rangle_R / \langle r_{6s}\rangle_{NR}$ as a function of atomic number Z.

lowering of the energies of these orbitals. This is known as the direct relativistic effect, which accounts for orbital contraction and stabilization in s orbitals and, to a lesser extent, in p orbitals.

In contrast, the valence d and f orbitals in heavy atoms are expanded and destabilized by the relativistic effects. This is because the contraction of the s orbitals increases the shielding effect, which gives rise to a smaller effective nuclear charge for the d and f electrons. This is known as the indirect relativistic orbital expansion and destabilization. In addition, if a filled d or f subshell lies just inside a valence orbital, that orbital will experience a larger effective nuclear charge which will lead to orbital contraction and stabilization. This is because the d and f orbitals have been expanded and their shielding effect accordingly lowered.

For heavy atoms with different Z values and various numbers of d and f electrons, the two aforementioned relativistic effects will lead to different degrees of 6s orbital contraction. Figure 2.4.5 shows the ratio $\langle r_{6s}\rangle_R / \langle r_{6s}\rangle_{NR}$ as a function of Z.

As we can see from Fig. 2.4.5, three of the elements that exhibit the largest relativistic effects are Au, Pt, and Hg. In these atoms, the (mostly) filled 4f and 5d orbitals lie just inside the 6s valence orbital. In the following paragraphs, we discuss some examples where relativistic effects are manifested in the atomic structure and properties of the elements.

(1) Ground electronic configuration

In the fifth period of the Periodic Table, we find transition metal elements with configurations $4d^n 5s^1$ and $4d^n 5s^0$. For the corresponding elements in the sixth period, the configurations become $5d^{n-1} 6s^2$ and $5d^{n-1} 6s^1$, respectively. This change is due to the aforementioned stabilization of the 6s orbital caused by the relativistic effects. Table 2.4.4 lists the configuration of the elements concerned.

The Electronic Structure of Atoms

Table 2.4.4. The ground electronic configuration of some elements of the fifth and sixth periods

Period	Group					
	5	6	7	8	9	10
5th	Nb d^4s^1	Mo d^5s^1	Tc d^5s^2	Ru d^7s^1	Rh d^8s^1	Pd $d^{10}s^0$
6th	Ta d^3s^2	W d^4s^2	Re d^5s^2	Os d^6s^2	Ir d^7s^2	Pt d^9s^1

(2) Lanthanide contraction

The properties of the elements of the sixth period are influenced by lanthanide contraction: a gradual decrease of the atomic radius with increasing atomic number from La to Lu. The elements of groups 5 to 11 for the fifth and sixth periods have comparable structural parameters. For instance, Nb and Ta, as well as the pair Mo and W, have very similar ionic radii, when they have the same oxidation number. As a result, it is very difficult to separate Nb and Ta, and it is also not easy to separate Mo and W. Similarly, Ag and Au have nearly the same atomic radius, 144 pm. Recent studies of the coordination compounds of Ag(I) and Au(I) indicate that the covalent radius of Au is even shorter than that of Ag by about 8 pm. In elementary textbooks the phenomenon of lanthanide contraction is attributed to incomplete shielding of the nucleus by the diffuse 4f inner subshell. Recent theoretical calculations conclude that lanthanide contraction is the result of both the shielding effect of the 4f electrons and relativistic effects, with the latter making about 30% contribution.

(3) Effect of the $6s^2$ inert pair

In the group 13 triad gallium/indium/thallium, the experimental energy separations of the $(ns^2np^1)^2P$ and $(ns^1np^2)^2P$ states are 4.7, 4.3, and 5.7 eV, respectively. The unusually large energy separation in Tl is another manifestation of the relativistic stabilization of the 6s orbital. Also, whereas the ionic radius of Tl^+ (150 pm) is larger than that of In^+ (140 pm), the average of the second and third ionization energies, $1/2(I_2+I_3)$, of Tl (4.848×10^3 kJ mol^{-1}) is larger than that of In (4.524×10^3 kJ mol^{-1}). In other words, even through Tl^+ is larger than In^+, the 6s electrons of Tl^+ are more difficult to detach. Hence Tl is predominantly monovalent as well as trivalent, whereas Ga and In are both trivalent. The same kind of reasoning is valid for Pb and Bi, which tend to favor the lower oxidation states Pb^{2+} and Bi^{3+}, respectively, with the $6s^2$ electron pair remaining intact.

(4) Comparison between gold and mercury

Elements Au and Hg differ by only one electron, with the following configurations:

$$_{79}\text{Au}: [\text{Xe}]4f^{14}5d^{10}6s^1;$$

$$_{80}\text{Hg}: [\text{Xe}]4f^{14}5d^{10}6s^2.$$

Because of the contraction and stabilization of the 6s orbital, the outermost, or valence, shell of Au is formed by both the 5d and 6s orbitals. Indeed, electronically, Au is halogen-like, with one electron missing from the pseudo noble gas (closed subshell) configuration. Hence, similar to the existence of halogen X_2 molecule, gold also forms the covalent Au_2 molecule. In addition, gold also forms ionic compounds such as RbAu and CsAu, in which the Au^- anion has the pseudo noble gas electronic configuration.

Mercury, with filled 5d and 6s subshells, also has the pseudo noble gas configuration. In the gas phase, Hg exists as a monatomic "molecule." From Fig. 2.4.1, Hg has an I_1 value similar to those of noble gases; it occupies a maximum position in the I_1 curve. The interaction between mercury atoms is of the van der Waals type. Hence, Hg and Au have some notably different properties:

	Hg	Au
Density (g cm^{-3})	13.53	19.32
Melting point (°C)	−39	1064
ΔH_{fusion} (kJ mol^{-1})	2.30	12.8
Conductivity (kS m^{-1})	10.4	426

Finally, it should be mentioned that mercury forms the cation Hg_2^{2+} (as in Hg_2Cl_2), which is isoelectronic to Au_2.

(5) Melting points of the metals

The variation of the melting points of the transition metals, as well as those of the alkali metals and alkali earth metals of the same period, are displayed in Fig. 2.4.6. It is seen that the uppermost curve (that for the elements of the sixth period) starts from Cs, increasing steadily and reaching a maximum at W. Beyond W, the curve starts to decrease and reach the minimum at Hg at the end. It is believed that this trend is the result of the relativistic effects.

In view of the previously discussed contraction and stabilization of the 6s orbital, the valence shell of the elements of the sixth period consists of one s and five d orbitals. In the metallic state, the six valence orbitals of each atom overlap effectively with the six valence orbitals of the other atoms. For a metal with N atoms, with a total of $6N$ valence orbitals, $3N$ of the molecular orbitals are bonding. The energies of these bonding orbitals are very close to each other, forming a bonding band in the process. The remaining $3N$ molecular orbitals are antibonding. These antibonding molecular orbitals are also with energies very close to each other and they form an antibonding energy band. It should be noted that the overlap between the valence orbital is very effective, due to the high symmetry of the coordination, which is essentially the closest packed situation. As a result, there are only bonding and antibonding energy bands, and there are no nonbonding energy bands.

As we go across the period, we feed electrons to the molecular orbitals, starting from the bonding orbitals. As the bonding energy band starts to fill, the metallic bond becomes stronger and stronger. At W, with six electrons per atom, all the bonding molecular orbitals are filled, which leads to the strongest metallic bond and highest melting point. Once we go beyond W, the antibonding

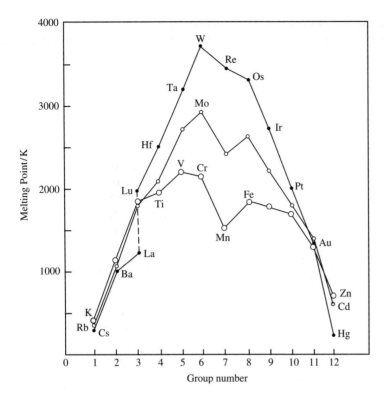

Fig. 2.4.6.
Melting points of the elements.

molecular orbitals start to fill, weakening the metallic bond in the process. Hence the melting point starts to decrease once we get to Re. At the end, at Hg, with 12 valence electrons per atom, all the bonding and antibonding molecular orbitals are fully occupied. There is no net bonding effect, as in the case of noble gases. Hence Hg has the lowest melting point. A similar situation exists for the elements of the two previous periods. However, the variation is less dramatic. This is because the relativistic effects are most prominent for the elements of the sixth period.

The variation of other physical properties such as hardness and conductivity of the metals may be analyzed in a similar manner.

(6) Noble gas compounds of gold

Relativity plays an important role in stabilizing elusive species such as $XeAuXe^+$ (a structural analog of the $ClAuCl^+$ anion), which have been observed experimentally by mass spectroscopy. A theoretical study of Pyykkö has shown that about half of the bonding energy in $AuXe^+$ arises from relativistic effects. A nice example that illustrates the importance of relativistic effects in the stabilization of gold compounds is $[AuXe_4^{2+}][Sb_2F_{11}^-]_2$, the first bulk compound with covalent bonding between a noble gas and a noble metal. Crystals of this compound remain stable up to $-40°C$. The dark red tetraxenonogold(II) cation $[AuXe_4]^{2+}$ has a square-planar configuration with an average Au–Xe bond distance of 274 pm. In $[AuXe_4]^{2+}$, xenon functions as a σ donor toward Au^{2+}, and calculation shows that a positive charge of about $+0.4$ resides on each ligand atom. The Xe–Au bond in $[AuXe_4]^{2+}$ may be contrasted with the Xe–F bond in XeF_4, in which a charge transfer of electrons occurs from xenon to fluorine.

References

1. M. Karplus and R. N. Porter, *Atoms and Molecules: An Introduction for Students of Physical Chemistry*, Benjamin, New York, 1970.
2. J. E. Huheey, E. A. Keiter and R. L. Keiter, *Inorganic Chemistry: Principles of Structure and Reactivity*, 4th edn., HarperCollins, New York, 1993.
3. C. E. Housecroft and A. G. Sharpe, *Inorganic Chemistry*, 2nd edn., Prentice-Hall, London, 2004.
4. D. M. P. Mingos, *Essential Trends in Inorganic Chemistry*, Oxford University Press, Oxford, 1998.
5. P. W. Atkins and J. de Paula, *Atkins' Physical Chemistry*, 8th edn., Oxford University Press, Oxford, 2006.
6. R. S. Berry, S. A. Rice and J. Ross, *Physical Chemistry*, 2nd edn., Oxford University Press, New York, 2000.
7. F. L. Pilar, *Elementary Quantum Chemistry*, 2nd edn., McGraw-Hill, New York, 1990.
8. I. N. Levine, *Quantum Chemistry*, 5th edn., Prentice Hall, Upper Saddle River, 2000.
9. D. D. Fitts, *Principles of Quantum Mechanics as Allied to Chemistry and Chemical Physics*, Cambridge University Press, Cambridge, 1999.
10. E. U. Condon and H. Odabaşi, *Atomic Structure*, Cambridge University Press, Cambridge, 1980.
11. J. Emsley, *The Elements*, 3rd edn., Clarendon Press, Oxford, 1998.
12. D. H. Rouvray and R. B. King (eds.), *The Periodic Table: Into the 21st Century*, Research Studies Press Ltd., Baldock, Hertfordshire, 2004.
13. K. Balasubramanian (ed.), *Relativistic Effects in Chemistry*, Part A and Part B, Wiley, New York, 1997.
14. J. B. Mann, T. L. Meek and L. C. Allen, Configuration energies of the main group elements. *J. Am. Chem. Soc.* **122**, 2780–3 (2000).
15. J. B. Mann, T. L. Meek, E. T. Knight, J. F. Capitani and L. C. Allen, Configuration energies of the d-block elements. *J. Am. Chem. Soc.* **122**, 5132–7 (2000).
16. P. Pyykkö, Relativistic effects in structural chemistry. *Chem. Rev.* **88**, 563–94 (1988).
17. N. Kaltsoyannis, Relativistic effects in inorganic and organometallic chemistry. *Dalton Trans.* 1–11 (1997).
18. T. C. W. Mak and W.-K. Li, Probability of locating the electron in a hydrogen atom. *J. Chem. Educ.* **77**, 490–1 (2000).
19. T. C. W. Mak and W.-K. Li, Relative sizes of hydrogenic orbitals and the probability criterion. *J. Chem. Educ.* **52**, 90–1 (1975).
20. W.-K. Li, Two-parameter wave functions for the helium sequence. *J. Chem. Educ.* **64**, 128–9 (1987).
21. W.-K. Li, A lesser known one-parameter wave function for the helium sequence and the virial theorem. *J. Chem. Educ.* **65**, 963–4 (1988).
22. S. Seidel and K. Seppelt, Xenon as a complex ligand: the tetraxenongold(II) cation in $AuXe_4^{2+}(Sb_2F_{11}^-)_2$. *Science* **297**, 117–18 (2000).

Covalent Bonding in Molecules

3

After a discussion on atomic structure, we now consider the electronic structure in molecules. It is clear that, whatever theory is chosen to treat this problem, it must be able to answer "Why do molecules form at all?" For instance, when two H atoms collide, the H_2 molecule is formed. However, when two He atoms are brought together, no He_2 is formed.

One way to answer the aforementioned question is "Two atoms form a molecule because the energy of the whole is lower than the sum of the energies of its parts." So the theory we need should bring about a decrease in energy when a molecule is formed by two atoms. As in the case of an atom, the energy of a molecule can also be calculated using quantum mechanical methods. In this chapter, we first use the simplest diatomic molecules H_2^+ and H_2 as examples to illustrate the basic concepts of chemical bonding. Then we will turn to other more complicated molecules.

3.1 The hydrogen molecular ion: bonding and antibonding molecular orbitals

3.1.1 The variational method

To apply the variational method, we first need to have a trial function. When the system under study is a molecule composed of n atoms, the trial function, which is to become a molecular orbital, is a linear combination of atomic orbitals (LCAO):

$$\psi = c_1\phi_1 + c_2\phi_2 + \cdots + c_n\phi_n, \qquad (3.1.1)$$

where $\phi_1, \phi_2, \ldots, \phi_n$ are known atomic functions, and the coefficients $c_1, c_2, \ldots, c_n$ are parameters to be optimized. In the following we use the simplest LCAO to illustrate the variational method:

$$\psi = c_1\phi_1 + c_2\phi_2. \qquad (3.1.2)$$

The energy of this wavefunction is

$$E = \frac{\int (c_1\phi_1 + c_2\phi_2)\hat{H}(c_1\phi_1 + c_2\phi_2)d\tau}{\int (c_1\phi_1 + c_2\phi_2)^2 d\tau}$$

$$= \frac{c_1^2 \int \phi_1 \hat{H} \phi_1 d\tau + 2c_1 c_2 \int \phi_1 \hat{H} \phi_2 d\tau + c_2^2 \int \phi_2 \hat{H} \phi_2 d\tau}{c_1^2 \int \phi_1^2 d\tau + 2c_1 c_2 \int \phi_1 \phi_2 + c_2^2 \int \phi_2^2 d\tau}$$

$$= \frac{c_1^2 H_{11} + 2c_1 c_2 H_{12} + c_2^2 H_{22}}{c_1^2 S_{11} + 2c_1 c_2 S_{12} + c_2^2 S_{22}}, \quad (3.1.3)$$

where we have made use of the Hermitian property of $\hat{H}$

$$\int \phi_1 \hat{H} \phi_2 d\tau = \int \phi_2 \hat{H} \phi_1 d\tau \quad (3.1.4)$$

and also introduced the notation

$$H_{ij} = \int \phi_i \hat{H} \phi_j d\tau \quad (3.1.5)$$

and

$$S_{ij} = \int \phi_i \phi_j d\tau. \quad (3.1.6)$$

To obtain the minimum energy of the system, we set

$$(\partial E/\partial c_1) = 0, \quad (3.1.7)$$

and

$$(\partial E/\partial c_2) = 0. \quad (3.1.8)$$

Applying eqs. (3.1.7) and (3.1.8) to eq. (3.1.3), we obtain

$$\begin{cases} (H_{11} - ES_{11})c_1 + (H_{12} - ES_{12})c_2 = 0 \\ (H_{12} - ES_{12})c_1 + (H_{22} - ES_{22})c_2 = 0. \end{cases} \quad (3.1.9)$$

These equations are called secular equations. Clearly, $c_1 = c_2 = 0$ is a solution of eq. (3.1.9). But it is a meaningless solution, as ψ becomes zero. In order for c_1 and c_2 to be non-vanishing, we need to have

$$\begin{vmatrix} H_{11} - ES_{11} & H_{12} - ES_{12} \\ H_{12} - ES_{12} & H_{22} - ES_{22} \end{vmatrix} = 0. \quad (3.1.10)$$

This is known as a secular determinant, where the only unknown is E. The remaining quantities in the determinant, H_{ij} and S_{ij}, may be evaluated readily according to their definitions given in eqs. (3.1.5) and (3.1.6) and using the known functions ϕ_i and ϕ_j. Upon obtaining E from eq. (3.1.10), we can substitute E into eq. (3.1.9) to solve for c_1 and c_2. Since there are two roots of E from the secular determinant, there are also two corresponding sets of coefficients, which lead to two molecular orbitals.

If our trial function has the form given in eq. (3.1.1), the corresponding secular determinant has the dimensions $n \times n$:

$$\begin{vmatrix} H_{11} - ES_{11} & H_{12} - ES_{12} & \cdots & H_{1n} - ES_{1n} \\ H_{12} - ES_{12} & H_{22} - ES_{22} & \cdots & H_{2n} - ES_{2n} \\ \vdots & \vdots & \vdots & \vdots \\ H_{1n} - ES_{1n} & H_{2n} - ES_{2n} & \cdots & H_{nn} - ES_{nn} \end{vmatrix} = 0. \quad (3.1.11)$$

There are now n roots of E. Substituting each E into the secular equations

$$\begin{cases} (H_{11} - ES_{11})c_1 + (H_{12} - ES_{12})c_2 + \ldots + (H_{1n} - ES_{1n})c_n = 0 \\ (H_{12} - ES_{12})c_1 + (H_{22} - ES_{22})c_2 + \ldots + (H_{2n} - ES_{2n})c_n = 0 \\ \quad \vdots \\ (H_{1n} - ES_{1n})c_1 + (H_{2n} - ES_{2n})c_2 + \ldots + (H_{nn} - ES_{nn})c_n = 0 \end{cases} \quad (3.1.12)$$

we get a set of n coefficients. In other words, there are altogether n sets of coefficients, or n molecular orbitals. To briefly summarize at this point, starting with n atomic orbitals ϕ_i ($i = 1, 2, \ldots, n$) in eq. (3.1.1), we can construct n independent molecular orbitals using the variational method.

3.1.2 The hydrogen molecular ion: energy consideration

Now we consider the simplest molecular system, namely the hydrogen molecular ion H_2^+. The Schrödinger equation for H_2^+, in a.u., is

$$\left[-\frac{1}{2} \nabla^2 - (1/r_a) - (1/r_b) + (1/r_{ab}) \right] \psi = E\psi, \quad (3.1.13)$$

where r_a, r_b, and r_{ab} (treated as a constant for the time being) are defined in Fig. 3.1.1.

When the nuclei are infinitely apart and the electron is on atom a, we have

$$\psi = 1s_a = [1/(\pi)^{1/2}]e^{-r_a} = \psi_a. \quad (3.1.14)$$

Similarly, when the electron resides on atom b instead, we have

$$\psi = 1s_b = [1/(\pi)^{1/2}]e^{-r_b} = \psi_b. \quad (3.1.15)$$

If the two nuclei are allowed to come together, it is reasonable to expect that the "molecular orbital" ψ_{MO} would be some kind of combination of atomic functions (3.1.14) and (3.1.15). This is the basis of the LCAO approximation for describing the chemical bonding in molecular systems. Thus we now have

$$\psi_{MO} = c_1 \psi_a + c_2 \psi_b, \quad (3.1.16)$$

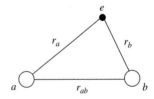

Fig. 3.1.1.
The coordinate system of H_2^+.

Fig. 3.1.2.
The overlap integral
$S_{ab} = \int (1s_a)(1s_b)d\tau$, as represented by the shaded volume.

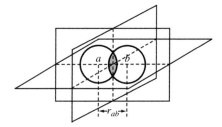

where coefficients c_1 and c_2 are to be determined by the variational method. To do that, we need to solve the secular determinant for E:

$$\begin{vmatrix} H_{aa} - ES_{aa} & H_{ab} - ES_{ab} \\ H_{ab} - ES_{ab} & H_{bb} - ES_{bb} \end{vmatrix} = 0. \qquad (3.1.17)$$

The solution of E involves the evaluation of six integrals: $S_{aa}, S_{bb}, S_{ab}, H_{aa}, H_{bb}$, and H_{ab}. From the normalization relationship, we have

$$S_{aa} = \int |1s_a|^2 d\tau = S_{bb} = \int |1s_b|^2 d\tau = 1. \qquad (3.1.18)$$

In other words, S_{aa} and S_{bb} are simply normalization integrals.

Integral S_{ab} is called the overlap integral. Physically it represents the common volume of the two atomic orbitals. A pictorial representation of S_{ab} is given in Fig. 3.1.2. It is clear that S_{ab} varies with r_{ab}, the internuclear distance: when $r_{ab} = 0$, $S_{ab} = 1$; when $r_{ab} \to \infty$, $S_{ab} \to 0$. Mathematically, it can be shown that, in a.u.,

$$S_{ab} = \int (1s_a)(1s_b)d\tau = \left[1 + r_{ab} + \left(\frac{r_{ab}^2}{3}\right)\right]e^{-r_{ab}}. \qquad (3.1.19)$$

Overlap integral S_{ab} is often used as a measure of the strength of the chemical bond: the bond is strong if S_{ab} is large, while the bond is weak if S_{ab} is small.

Now we deal with the integrals involving Hamiltonian operator $\hat{H}$, in a.u.,

$$H_{aa} = H_{bb} = \int (1s_a)\left[-\frac{1}{2}\nabla^2 - \frac{1}{r_a} - \frac{1}{r_b} + \frac{1}{r_{ab}}\right](1s_a)d\tau$$

$$= E_{1s} + \frac{1}{r_{ab}} + J, \qquad (3.1.20)$$

where

$$J = \int (1s_a)\frac{1}{r_b}(1s_a)d\tau = \int (1s_b)\frac{1}{r_a}(1s_b)d\tau$$

$$= -\frac{1}{r_{ab}} + \left(1 + \frac{1}{r_{ab}}\right)e^{-2r_{ab}}. \qquad (3.1.21)$$

So

$$H_{aa} = H_{bb} = -\frac{1}{2} + \left(1 + \frac{1}{r_{ab}}\right)\exp[-2r_{ab}]. \quad (3.1.22)$$

$$H_{ab} = \int (1s_a)\left[-\frac{1}{2}\nabla^2 - \frac{1}{r_a} - \frac{1}{r_b} + \frac{1}{r_{ab}}\right](1s_b)d\tau$$

$$= S_{ab}\left(E_{1s} + \frac{1}{r_{ab}}\right) + K, \quad (3.1.23)$$

where

$$K = -(1 + r_{ab})e^{-r_{ab}}. \quad (3.1.24)$$

So

$$H_{ab} = \left(\frac{r_{ab}^2}{3} + r_{ab} + 1\right)\exp[-r_{ab}] \cdot \left(\frac{1}{r_{ab}} - \frac{1}{2}\right) - (r_{ab} + 1)e^{-r_{ab}}$$

$$= -\left(\frac{r_{ab}^2}{6} + \frac{7}{6}r_{ab} + \frac{1}{2} - \frac{1}{r_{ab}}\right)e^{-r_{ab}}. \quad (3.1.25)$$

Thus H_{aa}, H_{bb}, and H_{ab} are dependent on r_{ab} as well. Note that the S_{ab} integrals are dimensionless, while the H integrals have the energy unit. Also, all H integrals are negative.

Expanding secular determinant (3.1.17), we have

$$(H_{aa} - E)^2 - (H_{ab} - ES_{ab})^2 = 0, \quad (3.1.26)$$

which leads to roots

$$E_S = \frac{H_{aa} + H_{ab}}{1 + S_{ab}} = -\frac{1}{2} + \frac{1}{r_{ab}} - \frac{1 + r_{ab}(r_{ab} + 1)e^{-r_{ab}} - (r_{ab} + 1)e^{-2r_{ab}}}{r_{ab}\left[1 + e^{-r_{ab}}\left(\frac{r_{ab}^2}{3} + r_{ab} + 1\right)\right]}, \quad (3.1.27)$$

and

$$E_A = \frac{H_{aa} - H_{ab}}{1 - S_{ab}} = -\frac{1}{2} + \frac{1}{r_{ab}} - \frac{1 - r_{ab}(r_{ab} + 1)e^{-r_{ab}} - (r_{ab} + 1)e^{-2r_{ab}}}{r_{ab}\left[1 - e^{-r_{ab}}\left(\frac{r_{ab}^2}{3} + r_{ab} + 1\right)\right]}, \quad (3.1.28)$$

with

$$E_S < E_A.$$

If we minimize E_S by setting $dE_S/dr_{ab} = 0$, we get the optimal bond length

$$r_e = 2.494 \text{ a.u.} = 131.9 \text{ pm}. \tag{3.1.29}$$

Experimentally, the internuclear separation in H_2^+ is 2 a.u. or 105.8 pm. When $r_{ab} = 2.494$ a.u., $S_{ab} = 0.460$ (a rather large value for S_{ab}). If we substitute $r_{ab} = 2.494$ a.u. into eq. (3.1.27), we get

$$E_S = \left[-\frac{1}{2} + \left(\frac{1}{r_{ab}}\right) - 0.466 \right] \text{a.u.} \tag{3.1.30}$$

The electronic dissociation energy (D_e) of H_2^+ is

$$\begin{aligned} D_e &= E(H) - E_S(H_2^+) \\ &= \{-1/2 - [-1/2 + (1/2.494) - 0.466]\} \text{a.u.} \\ &= 0.065 \text{ a.u.} = 1.77 \text{ eV}. \end{aligned} \tag{3.1.31}$$

This is only in fair agreement with the experimental value of 2.79 eV for D_e. An improvement in the calculation can be made if we replace the nuclear charge Z in $1s_a$ and $1s_b$ by the effective nuclear charge Z_{eff} as a parameter. With $Z_{\text{eff}} = 1.239$, we have $D_e = 2.34$ eV and $r_e = 106$ pm.

If we plot the E_S and E_A, as given in eqs. (3.1.27) and (3.1.28), against r_{ab}, we get the curves shown in Fig. 3.1.3. These are called potential energy curves. In the figure, r_e is the equilibrium bond length, and D_e is the electronic dissociation energy. It should be noted that molecular vibration exists even at 0 K, so that it does not require all of D_e to dissociate the molecule. Rather, the dissociation energy D_0, as shown in Fig. 3.1.3, will suffice. The difference between D_e and D_0 is the zero-point vibrational energy (ZPVE) of the molecule.

Returning to the hydrogen molecular ion, we see that the energy of the system, $E(H_2^+)$, is lower than $E(H) + E(H^+)$. In other words, the quantum mechanical results "predict" a stable H_2^+ species.

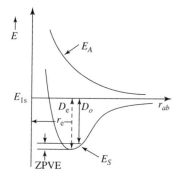

Fig. 3.1.3.
Energies E_S and E_A of H_2^+ as a function of r_{ab}.

3.1.3 The hydrogen molecular ion: wavefunctions

Upon obtaining the energies of the molecular orbitals by solving the secular determinant [eq. (3.1.17)], we are now ready to proceed to determine the coefficients c_1 and c_2 of eq. (3.1.16). To do this we need to solve the secular equations

$$\begin{cases} (H_{aa} - E)c_1 + (H_{ab} - ES_{ab})c_2 = 0 \\ (H_{ab} - ES_{ab})c_1 + (H_{bb} - E)c_2 = 0. \end{cases} \tag{3.1.32}$$

When $E = E_S = (H_{aa} + H_{ab})/(1 + S_{ab})$, we have $c_1 = c_2$, or

$$\psi_S = c_1(\psi_a + \psi_b) \equiv c_1(1s_a + 1s_b). \tag{3.1.33}$$

Upon normalization, ψ_S becomes

$$\psi_S = (2 + 2S_{ab})^{-1/2}(1s_a + 1s_b). \tag{3.1.34}$$

Wavefunction ψ_S is called a bonding orbital, as its energy is lower than that of its constituent atomic orbitals (c.f. Fig. 3.1.3).

When $E = E_A = (H_{aa} - H_{ab})/(1 - S_{ab})$, upon substituting into eq. (3.1.32), we obtain $c_1 = -c_2$, or

$$\psi_A = c_1(\psi_a - \psi_b) = c_1(1s_a - 1s_b). \tag{3.1.35}$$

Upon normalization, ψ_A becomes

$$\psi_A = (2 - 2S_{ab})^{-1/2}(1s_a - 1s_b) \tag{3.1.36}$$

Wavefunction ψ_A is called an antibonding molecular orbital, as its energy is higher than that of its constituent atomic orbitals.

The more detailed energy diagram in Fig. 3.1.3 can be simplified to that shown in Fig. 3.1.4, where we see that atomic orbitals $1s_a$ and $1s_b$ are combined to form two molecular orbitals ψ_S (to be called σ_{1s} later) and ψ_A (to be called σ_{1s}^*). Also, σ_{1s} with energy lower than those of atomic orbitals is a bonding orbital; σ_{1s}^* with energy higher than those of atomic orbitals is an antibonding orbitals. The ground electronic configuration for H_2^+ is thus $(\sigma_{1s})^1$.

An important point regarding the energy level diagram in Fig. 3.1.4 is that, when a bonding and an antibonding molecular orbital are formed by two atomic orbitals, the antibonding effect (the energy difference between 1s and σ_{1s}^*) is greater than the bonding effect (the energy difference between 1s and σ_{1s}). This is not difficult to see if we examine the expressions for E_S and E_A given in eqs. (3.1.27) and (3.1.28). The denominators for E_S and E_A are $1 + S_{ab}$ and $1 - S_{ab}$, respectively. Recall that $0 < S_{ab} < 1$. So $1 + S_{ab} > 1$ and $1 - S_{ab} < 1$ and this leads to the result that the bonding effect is smaller than the antibonding effect. If we ignore the overlap of the atomic orbitals by setting $S_{ab} = 0$, the bonding effect would be the same as the antibonding effect.

Now we examine the bonding orbital σ_{1s} and antibonding orbital σ_{1s}^* as well as their probability density functions $|\sigma_{1s}|^2$ and $|\sigma_{1s}^*|^2$. A schematic representation of σ_{1s} is shown in Fig. 3.1.5(a). In this combination of two 1s orbitals, electron density accumulates in the internuclear region. Also, σ_{1s} has cylindrical symmetry around the internuclear axis.

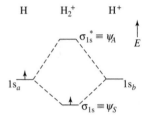

Fig. 3.1.4.
A simplified energy level diagram for H_2^+.

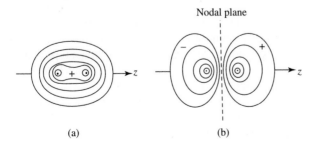

Fig. 3.1.5.
Molecular orbitals of H_2^+: (a) σ_{1s}, (b) σ_{1s}^*.

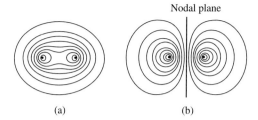

Fig. 3.1.6.
Probability density distribution of H_2^+ plotted along the internuclear axis: (a) $|\sigma_{1s}|^2$, (b) $|\sigma_{1s}^*|^2$.

In Fig. 3.1.5(b), a schematic drawing for σ_{1s}^* is shown. It is seen that this combination of 1s orbitals has no charge accumulation between the nuclei. Indeed, in the nodal plane there is zero probability of finding an electron. As in the case of σ_{1s}, the σ_{1s}^* orbital also has cylindrical symmetry around the molecular axis.

The accumulation of charge density between the nuclei in σ_{1s} is clearly seen when we consider the probability density function $|\sigma_{1s}|^2$:

$$|\sigma_{1s}|^2 = |1s_a + 1s_b|^2$$
$$= |1s_a|^2 + 2|1s_a||1s_b| + |1s_b|^2. \tag{3.1.37}$$

When we plot this function along the internuclear axis, Fig. 3.1.6(a) is obtained. In this figure, the build-up of charge density between the nuclei is obvious.

For the antibonding orbital σ_{1s}^*, we have the probability density function

$$|\sigma_{1s}^*|^2 = |1s_a - 1s_b|^2$$
$$= |1s_a|^2 - 2|1s_a||1s_b| + |1s_b|^2. \tag{3.1.38}$$

If we plot $|\sigma_{1s}^*|^2$ along the internuclear axis, we get Fig. 3.1.6(b). Now there is a clear deficiency of charge density between the nuclei.

3.1.4 Essentials of molecular orbital theory

In sections 3.1.2 and 3.1.3, we use the example of H_2^+ to illustrate the molecular orbital theory. In particular, we note the following:

(1) Since molecular orbitals are linear combinations of atomic orbitals, it follows that n atomic orbitals will generate n molecular orbitals.
(2) If we have n atomic orbitals forming n molecular orbitals, "usually" half of the molecular orbitals are bonding, while the other half are antibonding. However, this condition does not always hold. For instance, if we have three atomic orbitals to form three molecular orbitals, obviously half of the latter cannot be bonding. Also, there is an additional type that we have not yet considered: nonbonding molecular orbitals, which by definition neither gains nor loses stability (or energy).
(3) Bonding molecular orbitals have two characteristics: its energy is lower than those of the constituent atomic orbitals and there is a concentration of charge density between the nuclei.

Table 3.1.1. Molecular orbital theory applied to H_2^+, H_2, He_2^+, and He_2

Molecule	Configuration	Bond energy (kJ mol^{-1})	Bond length (pm)	Remark
H_2^+	σ_{1s}^1	255	106	The bond is formed by one bonding electron.
H_2	σ_{1s}^2	431	74	There are now two bonding electrons. Hence the bond is stronger and shorter than that in H_2^+.
He_2^+	$\sigma_{1s}^2 \sigma_{1s}^{*1}$	251	108	This bond is slightly weaker than that in H_2^+, since antibonding effect is greater than bonding effect.
He_2	$\sigma_{1s}^2 \sigma_{1s}^{*2}$	Repulsive state		The gain of energy by the bonding electrons is more than offset by the antibonding electrons. Hence there is no bond in He_2.

(4) On the other hand, the energy of an antibonding molecular orbital is higher than those of the constituent atomic orbitals. Also, the wavefunction of an antibonding orbital has one or more nodes between the nuclei. Hence there is a deficiency of charge density between the nuclei.

To conclude this section, we apply the energy level diagram in Fig. 3.1.4 to four simple molecules having one to four electrons. The results are summarized in Table 3.1.1.

Examining Table 3.1.1, we see that our simple treatment is able to rationalize, at least semi-quantitatively, the formation of H_2 and the non-existence of He_2.

3.2 The hydrogen molecule: molecular orbital and valence bond treatments

3.2.1 Molecular orbital theory for H_2

In our previous discussions on H_2^+, we saw that the lone electron occupies the σ_{1s} molecular orbital:

$$\sigma_{1s} = (2 + 2S_{ab})^{-1/2}(1s_a + 1s_b). \qquad (3.2.1)$$

The molecular orbital theory was first introduced by F. Hund (of the Hund's rule fame) and R. S. Mullikan, with the latter winning the Nobel Prize in chemistry in 1966. Since σ_{1s} can accommodate two electrons, in molecular orbital theory, the wavefunction for H_2 is

$$\begin{aligned}\psi(1,2) &= \sigma_{1s}(1)\sigma_{1s}(2) \\ &= (2 + 2S_{ab})^{-1/2}[1s_a(1) + 1s_b(1)] \\ &\quad \times (2 + 2S_{ab})^{-1/2}[1s_a(2) + 1s_b(2)]. \end{aligned} \qquad (3.2.2)$$

From this function, we can see that both electrons in H_2 reside in the ellipsoidal σ_{1s} orbital. This situation is similar to that of the helium atom, where both electrons reside in the spherical 1s orbital, with wavefunction 1s(1)1s(2). It is important to appreciate that, in σ_{1s}, *the identity of the atomic orbital is lost*. This is the essential tenet of molecular orbital theory.

When we use eq. (3.2.2) to determine the energy of H_2, we obtain $D_e = 260$ kJ mol^{-1} and $r_e = 85$ pm, while the experimental results are 458 kJ mol^{-1} and 74 pm, respectively. If we vary the nuclear charge, we obtain $Z_{\text{eff}} = 1.197$, $D_e = 337$ kJ mol^{-1}, and $r_e = 73$ pm. So while the quantitative results may only be called fair, we do get a stable H_2 molecule.

3.2.2 Valence bond treatment of H_2

Historically, molecular orbital theory was preceded by an alternative and successful description of the bonding in H_2. In 1927, W. Heitler and F. London proposed the valence bond theory, in which each electron resides in an atomic orbital. In other words, in this model, *the identity of the atomic orbital is preserved*. There are two ways in which the two electrons in H_2 can be accommodated in the pair of 1s atomic orbitals:

(a) Electron 1 in a 1s orbital centered at nucleus a and electron 2 in a 1s orbital centered at nucleus b. Mathematically:

$$\psi_\text{I}(1,2) = 1s_a(1)1s_b(2). \tag{3.2.3}$$

(b) Electron 1 in orbital $1s_b$ and electron 2 in $1s_a$, or,

$$\psi_\text{II}(1,2) = 1s_b(1)1s_a(2). \tag{3.2.4}$$

The valence bond wavefunction for H_2 is simply a linear combination of ψ_I and ψ_II:

$$\begin{aligned}\psi(1,2) &= c_\text{I}\psi_\text{I} + c_\text{II}\psi_\text{II} \\ &= c_\text{I} 1s_a(1)1s_b(2) + c_\text{II} 1s_b(1)1s_a(2),\end{aligned} \tag{3.2.5}$$

where the coefficients c_I and c_II are to be determined by the variational method. To do that, we once again need to solve a secular determinant:

$$\begin{vmatrix} H_\text{I\,I} - ES_\text{I\,I} & H_\text{I\,II} - ES_\text{I\,II} \\ H_\text{I\,II} - ES_\text{I\,II} & H_\text{II\,II} - ES_\text{II\,II} \end{vmatrix} = 0, \tag{3.2.6}$$

Upon solving for E, we substitute E into the following secular equations to obtain coefficient c_I and c_II:

$$\begin{cases} (H_\text{I\,I} - ES_\text{I\,I})c_\text{I} + (H_\text{I\,II} - ES_\text{I\,II})c_\text{II} = 0 \\ (H_\text{I\,II} - ES_\text{I\,II})c_\text{I} + (H_\text{II\,II} - ES_\text{II\,II})c_\text{II} = 0 \end{cases}. \tag{3.2.7}$$

With $H_\text{I\,I} = H_\text{II\,II}$ and $S_\text{I\,I} = S_\text{II\,II}$, the roots of eq. (3.2.6) are

$$E_+ = (H_\text{I\,I} + H_\text{I\,II})/(1 + S_\text{I\,II}), \tag{3.2.8}$$

$$E_- = (H_\text{I\,I} - H_\text{I\,II})/(1 - S_\text{I\,II}). \tag{3.2.9}$$

A diagram illustrating the various distances in H_2 is shown in Fig. 3.2.1. The Hamiltonian operator for this system, in a.u., is

$$\hat{H} = -\frac{1}{2}\nabla_1^2 - \frac{1}{2}\nabla_2^2 - \frac{1}{r_{a1}} - \frac{1}{r_{b1}} - \frac{1}{r_{a2}} - \frac{1}{r_{b2}} + \frac{1}{r_{12}} + \frac{1}{r_{ab}}. \quad (3.2.10)$$

In the following, we evaluate the various H_{ij} and S_{ij} integrals:

$$S_{\text{I I}} = S_{\text{II II}} = \iint 1s_a(1)1s_b(2) \cdot 1s_a(1)1s_b(2) d\tau_1 d\tau_2$$

$$= \int |1s_a(1)|^2 d\tau_1 \cdot \int |1s_b(2)|^2 d\tau_2 = 1. \quad (3.2.11)$$

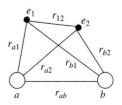

Fig. 3.2.1.
The molecular system of H_2.

$$S_{\text{I II}} = \iint 1s_a(1)1s_b(2) \cdot 1s_b(1)1s_a(2) d\tau_1 d\tau_2$$

$$= \int 1s_a(1)1s_b(1) d\tau_1 \cdot \int 1s_a(2)1s_b(2) d\tau_2$$

$$= S_{ab}^2, \quad (3.2.12)$$

where S_{ab} is simply the overlap integral introduced in the treatment of H_2^+. Meanwhile,

$$H_{\text{I I}} = H_{\text{II II}} = \iint 1s_a(1)1s_b(2)\hat{H} 1s_a(1)1s_b(2) d\tau_1 d\tau_2$$

$$= 2E_{1s} + \frac{1}{r_{ab}} + J_1 - 2J_2, \quad (3.2.13)$$

where J_1 and J_2 are called Coulomb integrals:

$$J_1 = \iint \frac{1}{r_{12}} |1s_a(1)1s_b(2)|^2 d\tau_1 d\tau_2, \quad (3.2.14)$$

$$J_2 = \iint \frac{1}{r_{a2}} |1s_a(1)1s_b(2)|^2 d\tau_1 d\tau_2$$

$$= \iint \frac{1}{r_{b1}} |1s_a(1)1s_b(2)|^2 d\tau_1 d\tau_2. \quad (3.2.15)$$

Finally,

$$H_{\text{I II}} = \iint 1s_a(1)1s_b(2)\hat{H} 1s_b(1)1s_a(2) d\tau_1 d\tau_2$$

$$= 2E_{1s}S_{ab}^2 + \frac{S_{ab}^2}{r_{ab}} + K_1 - 2K_2, \quad (3.2.16)$$

where K_1 and K_2 are called resonance integrals:

$$K_1 = \iint \frac{1}{r_{12}} |1s_a(1)1s_b(2)1s_b(1)1s_a(2)| \, d\tau_1 d\tau_2, \quad (3.2.17)$$

$$K_2 = \iint \frac{1}{r_{a1}} |1s_a(1)1s_b(2)1s_b(1)1s_a(2)| \, d\tau_1 d\tau_2$$

$$= \iint \frac{1}{r_{b1}} |1s_a(1)1s_b(2)1s_b(1)1s_a(2)| \, d\tau_1 d\tau_2. \quad (3.2.18)$$

Substituting H_{ij} and S_{ij} into eqs. (3.2.8) and (3.2.9), we obtain

$$E_+ = 2E_{1s} + \frac{1}{r_{ab}} + \frac{J_1 - 2J_2 + K_1 - 2K_2}{1 + S_{ab}^2} \quad (3.2.19)$$

$$E_- = 2E_{1s} + \frac{1}{r_{ab}} + \frac{J_1 - 2J_2 - K_1 + 2K_2}{1 - S_{ab}^2} \quad (3.2.20)$$

Note the integrals J_1, J_2, K_1, and K_2 are all functions of r_{ab}.

When we plot E_+ and E_- against r_{ab}, we get the energy curves shown in Fig. 3.2.2. So, once again, the valence bond treatment yields a stable H_2 molecule, even though the quantitative results do not match exactly the experimental data.

Upon substituting E_+ into eq. (3.2.7), we get $c_I = c_{II}$. After normalization, the wavefunction becomes

$$\psi_+ = (2 + 2S_{ab}^2)^{-1/2}[1s_a(1)1s_b(2) + 1s_b(1)1s_a(2)]. \quad (3.2.21)$$

Also, with E_-, we get $c_I = -c_{II}$. After normalization, the wavefunction becomes:

$$\psi_- = (2 - 2S_{ab}^2)^{-1/2}[1s_a(1)1s_b(2) - 1s_b(1)1s_a(2)]. \quad (3.2.22)$$

After a simple valence bond treatment of H_2, we now proceed to study the excited states of H_2. Through this discussion, we will recognize that the molecular orbital and valence bond treatments, after modification, can bring about the same quantitative results.

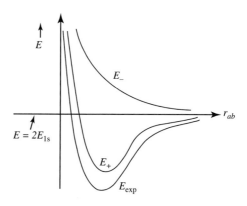

Fig. 3.2.2.
Valence bond energies E_+ and E_- plotted as functions of r_{ab}.

3.2.3 Equivalence of the molecular orbital and valence bond models

As discussed previously, the bonding molecular orbital of H_2 is

$$\sigma_{1s} = (2 + 2S_{ab})^{-1/2}(1s_a + 1s_b), \quad (3.2.23)$$

and the antibonding molecular orbital has the form

$$\sigma_{1s}^* = (2 - 2S_{ab})^{-1/2}(1s_a - 1s_b). \quad (3.2.24)$$

The ground configuration of H_2 is σ_{1s}^2, with the (unnormalized) wavefunction

$$\psi_1(\sigma_{1s}^2) = [1s_a(1) + 1s_b(1)] \times [1s_a(2) + 1s_b(2)]. \quad (3.2.25)$$

Upon adding the spin part, the total wavefunction for the ground state of H_2 is

$$\psi_1(\sigma_{1s}^2) = [1s_a(1) + 1s_b(1)][1s_a(2) + 1s_b(2)][\alpha(1)\beta(2) - \beta(1)\alpha(2)]$$
$$= \begin{vmatrix} \sigma_{1s}\alpha(1) & \sigma_{1s}\beta(1) \\ \sigma_{1s}\alpha(2) & \sigma_{1s}\beta(2) \end{vmatrix}. \quad (3.2.26)$$

Clearly, this is a spin singlet state.

Similarly, the excited configuration σ_{1s}^{*2} has the wavefunction

$$\psi_2(\sigma_{1s}^{*2}) = [1s_a(1) - 1s_b(1)][1s_a(2) - 1s_b(2)][\alpha(1)\beta(2) - \beta(1)\alpha(2)]$$
$$= \begin{vmatrix} \sigma_{1s}^*\alpha(1) & \sigma_{1s}^*\beta(1) \\ \sigma_{1s}^*\alpha(2) & \sigma_{1s}^*\beta(2) \end{vmatrix}. \quad (3.2.27)$$

This is also a spin singlet state. Furthermore, this is a repulsive state; i.e., it is not a bound state.

For the excited configuration $\sigma_{1s}^1 \sigma_{1s}^{*1}$, there are two states, one singlet and one triplet. This situation is similar to that found in the excited configuration $1s^1 2s^1$ of the helium atom. The singlet and triplet wavefunctions for these excited configurations are:

$$\psi_3(\sigma_{1s}^1 \sigma_{1s}^{*1}; S=0) = [\sigma_{1s}(1)\sigma_{1s}^*(2) + \sigma_{1s}(2)\sigma_{1s}^*(1)][\alpha(1)\beta(2) - \beta(1)\alpha(2)] \quad (3.2.28)$$

$$\psi_4(\sigma_{1s}^1 \sigma_{1s}^{*1}; S=1) = [\sigma_{1s}(1)\sigma_{1s}^*(2) - \sigma_{1s}(2)\sigma_{1s}^*(1)] \begin{cases} [\alpha(1)\beta(2) + \beta(1)\alpha(2)] \\ \alpha(1)\alpha(2). \\ \beta(1)\beta(2) \end{cases} \quad (3.2.29)$$

On the other hand, the (un-normalized) ground (ψ_+) and excited (ψ_-) state valence bond wavefunctions are

$$\psi_+ = 1s_a(1)1s_b(2) + 1s_b(1)1s_a(2) \quad (3.2.30)$$
$$\psi_- = 1s_a(1)1s_b(2) - 1s_b(1)1s_a(2). \quad (3.2.31)$$

Expanding the wavefunction in eq. (3.2.25), we get

$$\psi_1(\sigma_{1s}^2) = [1s_a(1)1s_b(2) + 1s_b(1)1s_a(2)] + [1s_a(1)1s_a(2) + 1s_b(1)1s_b(2)]$$
$$= \psi_+ + \psi_i, \qquad (3.2.32)$$

where

$$\psi_i = 1s_a(1)1s_a(2) + 1s_b(1)1s_b(2). \qquad (3.2.33)$$

Wavefunction ψ_i refers to the situation where both electrons are on nucleus a or nucleus b, i.e., ionic structures. Now it is obvious that the valence bond wavefunction ψ_+ considers only covalent structure, while the molecular orbital wavefunction ψ_1 has an equal mixture of covalent and ionic contributions. Similarly, expanding the wavefunction in eq. (3.2.27) yields

$$\psi_2(\sigma_{1s}^{*2}) = -[1s_a(1)1s_b(2) + 1s_b(1)1s_a(2)] + [1s_a(1)1s_a(2) + 1s_b(1)1s_b(2)]$$
$$= -\psi_+ + \psi_i. \qquad (3.2.34)$$

So the wavefunction for the excited configuration σ_{1s}^{*2} is also an equal mixture of covalent and ionic parts, except now that the linear combination coefficients have different signs.

Solving eqs. (3.2.32) and (3.2.34) for ψ_+, we get

$$\psi_+ = \psi_1 - \psi_2. \qquad (3.2.35)$$

So, put another way, the valence bond wavefunctions for the ground state of H_2 has equal contributions from configurations σ_{1s}^2 and σ_{1s}^{*2}, and the combination coefficients have different signs. Such a wavefunction, which employs a mixture of configurations to describe the electronic state of an atom or molecule, is called a configuration interaction (CI) wavefunction.

It is obvious that the deficiency of ψ_+ is that it does not take ionic contributions (ψ_i) into account. Conversely, ψ_1 suffers from too much (50%) contribution from the ionic structures. To get the wavefunction with optimal ionic contribution, we set

$$\psi'_+ = c_+\psi_+ + c_i\psi_i, \qquad (3.2.36)$$

where coefficient c_+ and c_i are to be determined through the following 2×2 secular determinant and the associated secular equations:

$$\begin{vmatrix} H_{++} - ES_{++} & H_{+i} - ES_{+i} \\ H_{+i} - ES_{+i} & H_{ii} - ES_{ii} \end{vmatrix} = 0 \qquad (3.2.37)$$

$$\begin{cases} (H_{++} - ES_{++})c_+ + (H_{+i} - ES_{+i})c_i = 0 \\ (H_{+i} - ES_{+i})c_+ + (H_{ii} - ES_{ii})c_i = 0 \end{cases}. \qquad (3.2.38)$$

The result of this optimization process is $c_i/c_+ = 0.16$. In other words, the optimal wavefunction has ionic and covalent contributions in the ratio of about 1:6. Such a dominance of covalent character is expected for a molecule such as H_2.

Table 3.2.1. Optimized dissociation energies and equilibrium bond lengths from trial wavefunctions for H_2 with adjustable parameters

Trial wavefunction	D_e(kJ mol^{-1})	r_e(pm)
$[1s_a(1) + 1s_b(1)][1s_a(2) + 1s_b(2)]$, eq. (3.2.1); $Z = 1$	260.0	85.2
$[1s_a(1) + 1s_b(1)][1s_a(2) + 1s_b(2)]$, eq. (3.2.1); $Z_{\text{eff}} = 1.197$	336.5	73.0
$1s_a(1)1s_b(2) + 1s_b(1)1s_a(2)$, eq. (3.2.30); $Z = 1$	304.5	86.8
$1s_a(1)1s_b(2) + 1s_b(1)1s_a(2)$, eq. (3.2.30); $Z_{\text{eff}} = 1.166$	364.9	74.6
$c[1s_a(1)1s_b(2) + 1s_b(1)1s_a(2)] + [1s_a(1)1s_a(2) + 1s_b(1)1s_b(2)]$, eqs. (3.2.30), (3.2.33), and (3.2.36); $Z = 1$ and $c = 6.322$	311.6	88.4
$c[1s_a(1)1s_b(2) + 1s_b(1)1s_a(2)] + [1s_a(1)1s_a(2) + 1s_b(1)1s_b(2)]$, eqs. (3.2.30), (3.2.33), and (3.2.36); $Z_{\text{eff}} = 1.194$ and $c = 3.78$	388.4	75.6
$\phi_a(1)\phi_b(2) + \phi_b(1)\phi_a(2)$, $\phi = 1s + \lambda 2p_z$ (a "polarized" atomic function); $Z_{\text{eff}}(1s) = Z_{\text{eff}}(2p) = 1.190, \lambda = 0.105$	389.8	74.9
$c[\phi_a(1)\phi_b(2) + \phi_b(1)\phi_a(2)] + [1s_a(1)1s_a(2) + 1s_b(1)1s_b(2)]$, $\phi = 1s + \lambda 2p_z; Z_{\text{eff}}(1s) = Z_{\text{eff}}(2p) = 1.190, c = 5.7, \lambda = 0.07$	397.7	74.6
Four-parameter function by Hirschfelder and Linnett (1950)	410.1	76.2
Thirteen-term function by James and Coolidge (1933)	455.7	74.1
One hundred-term function by Kołos and Wolniewicz (1968)	458.1	74.1
Experimental	458.1	74.1

Similarly, we can also improve ψ_1 by mixing in an optimal amount of ψ_2:

$$\psi_1' = c_1\psi_1 + c_2\psi_2. \tag{3.2.39}$$

The optimal c_2/c_1 ratio is -0.73. More importantly, the improved valence bond wavefunction ψ_+' and the improved molecular orbital wavefunction ψ_1' are one and the same, thus showing these two approaches can lead to identical quantitative results.

Table 3.2.1 summarizes the results of various approximate wavefunctions for the hydrogen molecule. This list is by no means complete, but it does show that, as the level of sophistication of the trial function increases, the calculated dissociation energy and bond distance approach closer to the experimental values. In 1968, W. Kołos and L. Wolniewicz used a 100-term function to obtain results essentially identical to the experimental data. So the variational treatment of the hydrogen molecule is now a closed topic.

3.3 Diatomic molecules

After the treatments of H_2^+ and H_2, we are ready to take on other diatomic molecules. Before we do that, we first state two criteria governing the formation of molecular orbitals:

(1) For two atomic orbitals ϕ_a and ϕ_b to form a bonding and antibonding molecular orbitals, ϕ_a and ϕ_b must have a non-zero overlap, i.e.,

$$S_{ab} = \int \phi_a \phi_b \, d\tau \neq 0. \tag{3.3.1}$$

Furthermore, for $S_{ab} \neq 0$, ϕ_a and ϕ_b must have the same symmetry. In Fig. 3.3.1, we can see that there is net overlap in the left and middle cases,

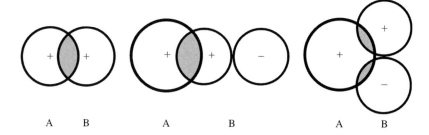

Fig. 3.3.1.
Non-zero overlap in the two cases shown on the left and in the middle, and zero overlap in the case shown on the right.

while S_{ab} vanishes in the case shown on right. The concept of symmetry will be studied more fully in Chapters 6 and 7, when the topic of group theory is taken up.

(2) The second criterion is the energy factor: for ϕ_a and ϕ_b to have significant bonding (and antibonding) effect, they should have similar energies. This is why, in our treatments of H_2 and H_2^+, we are only concerned with the interaction between two 1s orbitals. We ignore the interaction between, say, the 1s orbital on H_a and the 2s orbital on H_b. These two atomic orbitals do have non-zero overlap between them. But these orbitals do not bond (or interact) effectively because they have very different energies.

3.3.1 Homonuclear diatomic molecules

Now we proceed to discuss homonuclear diatomics with 2s and 2p valence orbitals. Starting from these atomic orbitals, we will make (additive and subtractive) combinations of them to form molecular orbitals. In addition, we also need to know the energy ordering of these molecular orbitals.

We first consider the 2s orbitals. As in the cases of H_2^+ and H_2, the combination of $(2s_a + 2s_b)$ leads to a concentration of charge between the nuclei. Hence it is a σ bonding orbital and is called σ_s. On the other hand, the $(2s_a - 2s_b)$ combination has a nodal plane between the nuclei and there is a charge deficiency in this region. Hence this is a σ antibonding orbital, which is designated as σ_s^*.

For the 2p orbitals, there are two types of overlap. We first note that the two p_z orbitals both lie along the internuclear axis (conventionally designated as the z axis), while the p_x and p_y orbitals are perpendicular to this axis. Once again, $2p_{za} + 2p_{zb}$ leads to a bonding orbital called σ_z, while the combination $2p_{za} - 2p_{zb}$ is an antibonding molecular orbital called σ_z^*. It is seen that these two molecular orbitals also have cylindrical symmetry around the internuclear axis. So both of them are σ orbitals.

The overlap between the $2p_{xa}$ and $2p_{xb}$ orbitals occurs in two regions, which have the same size and shape but carry opposite signs: one above the yz plane and the other below it. As there is no longer cylindrical symmetry around the nuclear axis, the molecular orbital is not a σ type. Rather, it is called a π_x orbital, which is characterized by a change in sign across the yz plane (a nodal plane). In an analogous manner, the combination $2p_{xa} - 2p_{xb}$ leads to the antibonding π_x^* orbital, which is composed of four lobes of alternating signs partitioned by two nodal planes. Similarly, there are the corresponding bonding $\pi_y(2p_{ya} + 2p_{yb})$ and antibonding $\pi_y^*(2p_{ya} - 2p_{yb})$ molecular orbitals.

Covalent Bonding in Molecules

Before we proceed to discuss the energy order of these molecular orbitals, it is important to note that the $2p_x$ orbital on atom a has zero overlap with either $2p_y$ or $2p_z$ on atom b; in other words, the pair of orbitals do not have compatible symmetry. Therefore, among the six 2p orbitals on the two atoms, $2p_{xa}$ interacts only with $2p_{xb}$, $2p_{ya}$ with $2p_{yb}$, and $2p_{za}$ with $2p_{zb}$. Furthermore, the π_x and π_y orbitals have the same energy; i.e., they are doubly degenerate. Similarly, the π_x^* and π_y^* orbitals compose another degenerate set at a higher energy.

To summarize briefly at this point: referring to Fig. 3.3.2, we have started with eight atomic orbitals (one 2s and three 2p orbitals on each atom) and have constructed eight molecular orbitals, σ_{2s}, σ_{2s}^*, σ_z, σ_z^*, π_x, π_x^*, π_y, and π_y^*. The relative energies of these molecular orbitals can be determined from experiments such as spectroscopic measurements or from calculations. For homonuclear diatomics, there are two energy level schemes, as shown in Fig. 3.3.2.

In each scheme, the eight molecular orbitals form six energy levels and can accommodate up to 16 electrons. The scheme on the right is applicable to atoms whose 2s–2p energy difference is small, while the scheme on the left is for atoms with a large 2s–2p energy gap. (Recall, from Fig. 3.3.1, s orbitals can overlap with p orbitals to form σ molecular orbitals. Whether this interaction is important depends on the energy difference between the interacting atomic orbitals.)

The experimental 2s–2p energy differences of the elements of the second period are summarized in Table 3.3.1. It can be readily seen that O and F have the largest 2s–2p gap. Hence the energy ordering shown on the left side of Fig. 3.3.2 is applicable to O_2 and F_2. On the other hand, the energy scheme on the right side of Fig. 3.3.2 is applicable for Li_2, Be_2, B_2, C_2, and N_2. Furthermore, the only difference between the two scheme is the relative ordering of the σ_z and

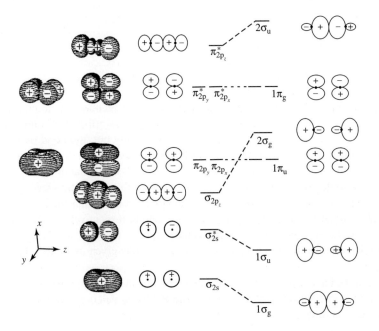

Fig. 3.3.2.
The shapes and energy ordering of the molecular orbitals for homonuclear diatomic molecules. The scheme on the left is applicable to O_2 and F_2, while that on the right is applicable to other diatomics of the same period.

Table 3.3.1. The 2s–2p energy gap of the second-row elements

Element	Li	Be	B	C	N	O	F
$-E_{2s}$(eV)	5.39	9.32	12.9	16.6	20.3	28.5	37.8
$-E_{2p}$(eV)	3.54	6.59	8.3	11.3	14.5	13.6	17.4
$E_{2p} - E_{2s}$(eV)	1.85	2.73	4.6	5.3	5.8	14.9	20.4

degenerate (π_x, π_y) levels. For O_2 and F_2, there is no significant s–p mixing, due to the larger 2s–2p energy gaps of O and F. On the other hand, for the remaining homonuclear diatomics of the same period, through s–p mixing, the molecular orbitals are no longer called σ_s, σ_s^*, σ_z, σ_z^*, etc. Instead they are now called σ_g, σ_u, etc., where g indicates centrosymmetry and u indicates antisymmetry with respect to the molecular center.

Table 3.3.2. Electronic configurations and structural parameters of homonuclear diatomic molecules of the second period

X_2	Electronic configuration	Bond length (pm)	Bonding energy (kJ mol^{-1})	Bond order
Li_2	$1\sigma_g^2$	267.2	110.0	1
Be_2	$1\sigma_g^2 1\sigma_u^2$	—	—	0
B_2	$1\sigma_g^2 1\sigma_u^2 1\pi_u^2$	158.9	274.1	1
C_2	$1\sigma_g^2 1\sigma_u^2 1\pi_u^4$	124.25	602	2
C_2^{2-}	$1\sigma_g^2 1\sigma_u^2 1\pi_u^4 2\sigma_g^2$	120	—	3
N_2	$1\sigma_g^2 1\sigma_u^2 1\pi_u^4 2\sigma_g^2$	109.76	941.69	3
N_2^+	$1\sigma_g^2 1\sigma_u^2 1\pi_u^4 2\sigma_g^1$	111.6	842.15	2½
N_2^{2-}	$\sigma_s^2 \sigma_s^{*2} \sigma_z^2 \pi_x^2 \pi_y^2 \pi_x^{*1} \pi_y^{*1}$	122.4	—	2
O_2	$\sigma_s^2 \sigma_s^{*2} \sigma_z^2 \pi_x^2 \pi_y^2 \pi_x^{*1} \pi_y^{*1}$	120.74	493.54	2
O_2^+	$\sigma_s^2 \sigma_s^{*2} \sigma_z^2 \pi_x^2 \pi_y^2 \pi_x^{*1}$	112.27	626	2½
O_2^-	$\sigma_s^2 \sigma_s^{*2} \sigma_z^2 \pi_x^2 \pi_y^2 \pi_x^{*2} \pi_y^{*1}$	126	392.9	1½
O_2^{2-}	$\sigma_s^2 \sigma_s^{*2} \sigma_z^2 \pi_x^2 \pi_y^2 \pi_x^{*2} \pi_y^{*2}$	149	138	1
F_2	$\sigma_s^2 \sigma_s^{*2} \sigma_z^2 \pi_x^2 \pi_y^2 \pi_x^{*2} \pi_y^{*2}$	141.7	155	1

Table 3.3.2 summarizes the various properties of second-row homonuclear diatomic molecules. In the last column of the table, we list the "bond order" between atoms A and B in the molecule AB. Simply put, the bond order is a number that gives an indication of its strength relative to that of a two-electron single bond. Thus the bond order of H_2^+ (σ_{1s}^1) is 1/2, while that of H_2 (σ_{1s}^2) is 1. For a system with antibonding electrons, we take the simplistic view that one antibonding electron "cancels out" one bonding electron. Thus the bond orders in He_2^+ ($\sigma_{1s}^2 \sigma_{1s}^{*1}$) and He_2 ($\sigma_{1s}^2 \sigma_{1s}^{*2}$) are 1/2 and 0, respectively, and helium is not expected to form a diatomic molecule.

Some interesting points are noted from the results given in Table 3.3.2:

(1) The bond in Li_2 is longer and weaker than that in H_2 (74 pm; 431 kJ mol^{-1}). This is because the bond in Li_2 is formed by 2s valence electrons that lie outside filled 1s atomic orbitals.

(2) The diatomic molecule Be_2 is unstable; the ground state is a repulsive state. As in He_2, the stabilization gained by the bonding electrons is more than offset by the destabilizing antibonding electrons.

(3) The bond in B_2 is stronger than that in Li_2 because B has a smaller atomic radius than Li. Also, B_2 has two unpaired electrons in the $1\pi_u$ orbital and is hence a paramagnetic species.

(4) The bond length and bond energy in C_2 are compatible with the double bond predicted by molecular orbital theory. It is noted that the $2\sigma_g$ orbital is only slightly higher in energy than $1\pi_u$. Indeed, C_2 absorbs light in the visible region at 19,300 cm^{-1}. This corresponds to exciting an electron from the $1\pi_u$ molecular orbital to the $2\sigma_g$ orbital. This is a fairly small excitation energy for a diatomic molecule, since electronic excitations for other diatomics are usually observed in the ultraviolet region.

(5) The bond in N_2 is a triple bond, in agreement with the Lewis structure :N≡N: of this molecule. We now can see that the lone pairs in the Lewis structure correspond to the $1\sigma_g^2$ and $1\sigma_u^2$ electrons; the two pairs of bonding and antibonding electrons result in no net bonding. The three bonds in the Lewis structure correspond to the $1\pi_u^4$ and $2\sigma_g^2$ electrons in molecular orbital theory. The electronic excitation from the $2\sigma_g$ orbital to $1\pi_g$ orbital occurs at around 70,000 cm^{-1} in the vacuum ultraviolet region. The experimental energy ordering of the molecular orbitals for the N_2 molecule is shown in Fig. 3.3.3(a). Upon losing one electron to form N_2^+, there is not much change in bond length, as the electron is from a weakly bonding $2\sigma_g$ orbital. In a recently determined crystal structure of SrN_2, the bond length found for the diazenide anion (N_2^{2-}) is compatible with the theoretical bond order of 2.

(6) Molecular orbital theory predicts that O_2 is paramagnetic, in agreement with experiment. Note that the Lewis structure of O_2 does not indicate that it has two unpaired electrons, even through it does imply the presence of a double bond. In fact, the prediction/confirmation of paramagnetism in O_2 was one of the early successes of molecular orbital theory. Also, the ions O_2^+ (dioxygen cation), O_2^- (superoxide anion), and O_2^{2-} (peroxide anion) have bond orders $2^{1/2}$, $1^{1/2}$, and 1, respectively. The experimental energy levels of the molecular orbital for the O_2 molecule are shown in Fig. 3.3.3(b).

(7) Fluorine F_2, is isoelectronic with O_2^{2-}. Hence they have similar bond lengths as well as similar bond energies. Finally, molecular orbital theory

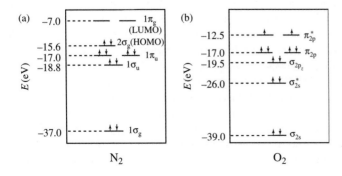

Fig. 3.3.3.
Energy level diagrams for N_2 and O_2.

predicts that the ground electronic configuration of Ne₂ (with two more electrons than F₂) leads to a repulsive state. So far there is no experimental evidence for the existence of Ne₂, in agreement with molecular orbital results.

3.3.2 Heteronuclear diatomic molecules

(1) The hydrogen fluoride molecule

Before discussing heteronuclear diatomic XY, where both X and Y are second-row atoms, we first take HF as an example for detailed molecular orbital treatment. Since the H 1s orbital (with energy -13.6 eV) and F 2p (-17.4 eV) have similar energies, while that of F 2s (-37.8 eV) is much lower, we only need to consider the interaction between H 1s and F 2p orbitals.

If we take the internuclear axis as the z axis, it is clear that the F $2p_z$ and the H 1s orbital overlap. The bonding orbital (called σ_z) is represent by the combination

$$\sigma_z = c_1[1s(H)] + c_2[2p_z(F)]. \tag{3.3.2}$$

The coefficients c_1 and c_2 give the relative contributions of the H 1s and F $2p_z$ orbitals to the σ_z orbital. Unlike the case of homonuclear diatomics, these coefficients are no longer equal, since the interacting atomic orbitals have different energies.

As we know, bonding electrons are "pulled" toward the more electronegative atom, which has the more stable valence orbitals. Therefore, in eq. (3.3.2), c_2 is greater than c_1; i.e., the F $2p_z$ contributes more than H 1s to the bonding molecular orbital σ_z. It is always true that the atomic orbital on the more electronegative atom contributes more to the bonding molecular orbital.

The antibonding orbital (called σ_z^*) between F $2p_z$ and H 1s orbitals has the form

$$\sigma_z^* = c_3[1s(H)] - c_4[2p_z(F)]. \tag{3.3.3}$$

Since most of the $2p_z$ orbital on F is "used up" in the formation of σ_z, it is now clear that, in σ_z^*, $c_3 > c_4$. In other words, the atomic orbital on the more electropositive atom always contributes more to the antibonding molecular orbital.

The F $2p_x$ and F $2p_y$ orbitals are suitable for forming π molecular orbitals. However, in the HF molecule, the hydrogen 1s orbital does not have the proper symmetry to overlap with the F $2p_x$ or F $2p_y$ orbital. Thus, the F $2p_x$ and F $2p_y$ orbitals are nonbonding orbitals. By definition, a nonbonding molecular orbital in a diatomic molecule is simply an atomic orbital on one of the atoms; it gains or loses no stability (or energy) and its charge density is localized at the atom where the aforementioned atomic orbital originates.

The energy level diagram for HF is shown in Fig. 3.3.4. By the dotted lines, we can see that σ_z is formed by the H 1s and F 2p ($2p_z$ in this case) orbitals. Since the energy of σ_z is closer to that of F 2p than that of H 1s, it is clear the F 2p orbital contributes more to the σ_z orbital. On the other hand, the H

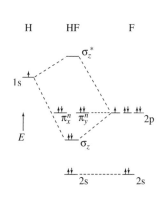

Fig. 3.3.4.
Energy level diagram of HF.

1s contributes more to the σ_z^* orbital. Also, the degenerate nonbonding π_x^n and π_y^n orbitals are formed solely by F $2p_x$ and F $2p_y$ orbitals, respectively, without any energy change.

From the energy level diagram in Fig. 3.3.4, it follows naturally that the ground configuration of HF is $2s^2\sigma_z^2(\pi_x^n = \pi_y^n)^4$. Among the four pairs of electron in this molecule, only the pair in σ_z is shared by H and F, while the other three pairs are localized at F. This bonding picture is in total agreement with the Lewis formula:

$$\text{H}\!:\!\ddot{\underset{\cdot\cdot}{\text{F}}}\!:$$

Since the two electrons in σ_z are not equally shared by H and F, it would be of interest to determine the charge separation in this molecule. This information is furnished by the experimentally determined dipole moment, 6.06×10^{-30} C m, of the molecule. Also, the electrons are closer to F than to H; i.e., the negative end of the dipole points toward F.

The H–F bond length is 91.7 pm. If the lone electron of H is transferred to F, giving rise to the ionic structure H^+F^-, the dipole moment would be

$$(91.7 \times 10^{-12}\text{m}) \times (1.60 \times 10^{-19}\text{C}) = 1.47 \times 10^{-29}\text{C m}.$$

Hence we see that the electron of H is only partially transferred to F. The ionic percentage of the HF may be estimated to be

$$\frac{6.06 \times 10^{-30}\text{C m}}{1.47 \times 10^{-29}\text{C m}} \times 100\% = 41\%.$$

(2) Heteronuclear diatomic molecules of the second-row elements

Now we describe the bonding in a general diatomic molecule XY, where both X and Y are second-row elements and Y is more electronegative than X. The molecular orbital energy level diagram is shown in Fig. 3.3.5. The σ and π bonding and antibonding orbitals are formed in the same manner as for X_2, but the coefficients of the orbitals on Y are larger than those on X for the bonding orbitals, and the converse holds for the antibonding orbitals. In other words, the electrons in bonding orbitals are more likely to be found near the more electronegative atom Y, while the electrons in the antibonding orbitals are more likely to be near the more electropositive atom X. In Fig. 3.3.5, the molecular orbitals no longer carry the subscripts u and g, since there is no center of symmetry in XY.

The bonding properties of several representative diatomics are discussed below.

(a) BN (with eight valence electrons): This paramagnetic molecule has the ground configuration of $1\sigma^2 1\sigma^{*2} 1\pi^3 2\sigma^1$, with two unpaired electrons. For this system, the 2σ orbital energy is higher than that of 1π by about the same energy required to pair two electrons. In any event, the bond order is 2. The bond lengths of C_2 and BN are 124.4 and 128 pm, respectively. The

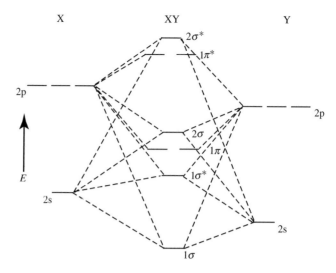

Fig. 3.3.5.
Energy level diagram of a heteronuclear diatomic molecule.

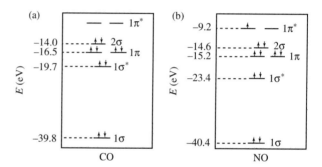

Fig. 3.3.6.
Energy level diagrams of CO and NO.

bond energy of BN is 385 kJ mol^{-1}, suspiciously low compared with 602 kJ mol^{-1} for C$_2$. Clearly more experimental work is required in this case.

(b) BO, CN, and CO$^+$ (with nine valence electrons): The electronic configuration for these molecules is $1\sigma^2 1\sigma^{*2} 1\pi^4 2\sigma^1$, with bond order $2\frac{1}{2}$. They all have bond lengths shorter than BN (or C$_2$), 120 pm for BO, 117 pm for CN, and 112 pm for CO$^+$. Also, they have fairly similar bond energies: 800, 787, and 805 kJ mol^{-1} for BO, CN, and CO$^+$, respectively, all of which are greater than that of C$_2$.

(c) NO$^+$, CO, and CN$^-$ (with ten valence electrons): Here the electronic configuration is $1\sigma^2 1\sigma^{*2} 1\pi^4 2\sigma^2$, with a bond order of 3. They have similar bond lengths: 106, 113, and 114 pm for NO$^+$, CO, and CN$^-$, respectively. The bond energy of carbon monoxide (1070 kJ mol^{-1}) is slightly greater than that of N$_2$ (941 kJ mol^{-1}). The energy level diagram for CO is shown in Fig. 3.3.6(a).

(d) NO (with eleven valence electrons): The ground electronic configuration is $1\sigma^2 2\sigma^{*2} 1\pi^4 2\sigma_z^2 1\pi^{*1}$ and the bond order is $2\frac{1}{2}$. The bond length of NO, 115 pm, is longer than those of both CO and NO$^+$. Its bond dissociation energy, 627.5 kJ mol^{-1}, is considerably less than those of CO and N$_2$. The importance of NO in both chemistry and biochemistry will be discussed

3.4 Linear triatomic molecules and sp^n hybridization schemes

In this section, we first discuss the bonding in two linear triatomic molecules: BeH$_2$ with only σ bonds and CO$_2$ with both σ and π bonds. Then we go on to treat other polyatomic molecules with the hybridization theory. Next we discuss the derivation of a self-consistent set of covalent radii for the atoms. Finally, we study the bonding and reactivity of conjugated polyenes by applying Hückel molecular orbital theory.

3.4.1 Beryllium hydride, BeH$_2$

The molecular orbitals of this molecule are formed by the 2s and 2p orbitals of Be and the 1s orbitals of H$_a$ and H$_b$. Here we take the molecular axis in BeH$_2$ as the z axis, as shown in Fig. 3.4.1.

To form the molecular orbitals for polyatomic molecules AX$_n$, we first carry out linear combinations of the orbitals on X and then match them, taking into account their symmetry characteristics, with the atomic orbitals on the central atom A.

For our simple example of BeH$_2$, the valence orbitals on H$_a$ and H$_b$, 1s$_a$ and 1s$_b$, can form only two linear (and independent) combinations: 1s$_a$ + 1s$_b$ and 1s$_a$ − 1s$_b$. We can see that combination 1s$_a$ + 1s$_b$ matches in symmetry with the Be 2s orbital. Hence they can form both bonding and antibonding molecular orbitals:

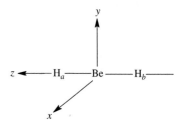

Fig. 3.4.1.
Coordinate system for BeH$_2$.

$$\sigma_s = c_1 2s(Be) + c_2(1s_a + 1s_b), \quad c_2 > c_1 \qquad (3.4.1)$$

$$\sigma_s^* = c_1' 2s(Be) - c_2'(1s_a + 1s_b), \quad c_1' > c_2'. \qquad (3.4.2)$$

The relative magnitudes of coefficients c_1 and c_2, as well as those of c_1' and c_2', are determined by the relative electronegatives of the atoms concerned, which are reflected by the relative energies of the atomic orbitals.

Similarly, the combination $1s_a - 1s_b$ has a net overlap with the $2p_z$ orbital of Be. They form bonding and antibonding molecular orbitals in the following manner:

$$\sigma_z = c_3 2p_z(Be) + c_4(1s_a - 1s_b), \quad c_4 > c_3 \qquad (3.4.3)$$

$$\sigma_z^* = c_3' 2p_z(Be) - c_4'(1s_a - 1s_b), \quad c_3' > c_4'. \qquad (3.4.4)$$

Finally, the $2p_x$ and $2p_y$ orbitals on Be are not symmetry-compatible with the $1s_a$ or $1s_b$ orbitals (or their linear combinations). Hence they are nonbonding orbitals:

$$\begin{cases} \pi_x^n = 2p_x(Be) \\ \pi_y^n = 2p_y(Be) \end{cases}. \qquad (3.4.5)$$

Table 3.4.1. Summary of the formation of the molecular orbitals in BeH$_2$

Orbital on Be	Orbitals on H	Molecular orbitals
2s	$(2)^{-1/2}(1s_a + 1s_b)$	σ_s, σ_s^*
2p$_z$	$(2)^{-1/2}(1s_a - 1s_b)$	σ_z, σ_z^*
$\begin{cases} 2p_x \\ 2p_y \end{cases}$	—	$\begin{cases} \pi_x^n \\ \pi_y^n \end{cases}$

There are totally six molecular orbitals (σ_s, σ_s^*, σ_z, σ_z^*, π_x^n, and π_y^n) formed by the six atomic orbitals (2s and 2p orbitals on Be and 1s orbitals on the hydrogens). Note that the σ molecular orbitals have cylindrical symmetry around the molecular axis, while the nonbonding π orbitals do not. Another important characteristic of these orbitals is that they are "delocalized" in nature. For example, an electron occupying the σ_s orbital has its density spread over all three atoms. Table 3.4.1 summarizes the way the molecular orbitals of BeH$_2$ are formed by the atomic orbitals on Be and H, where the linear combinations of H orbitals are normalized.

The energy level diagram for BeH$_2$, shown in Fig. 3.4.2, is constructed as follows. The 2s and 2p of Be are shown on the left of the diagram, while the 1s orbitals of the hydrogens are shown on the right. Note that the H 1s orbitals are placed lower than either the Be 2p or Be 2s orbitals. This is because Be is more electropositive than H. The molecular orbitals—bonding, antibonding, and nonbonding—are placed in the middle of the diagram. As usual, the bonding orbitals have lower energy than the constituent atomic orbitals, and correspondingly the antibonding orbitals are of higher energy. The nonbonding orbitals have the same energy as their parent atomic orbitals. After constructing the energy level diagram, we place the four valence electrons of BeH$_2$ in the two lowest molecular orbitals, leading to a ground electronic configuration of $\sigma_s^2 \sigma_z^2$. In this description, the two electron-pair bonds are spread over all these atoms. The delocalization of electrons is an important feature of the molecular orbital model.

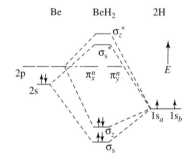

Fig. 3.4.2.
Energy level diagram of BeH$_2$.

Concluding the molecular orbital treatment of BeH$_2$, we can see that the two (filled) bonding molecular orbitals σ_s and σ_z have different shapes and different energies. This is contrary to our intuition for BeH$_2$: we expect the two bonds in BeH$_2$ to be identical (in shape as well as in stability) to each other. In any event, this is the picture provided by the molecular orbital model.

3.4.2 Hybridization scheme for linear triatomic molecules

If we prefer to describe the bonding of a polyatomic molecule using localized two-center, two-electron (2c-2e) bonds, we can turn to the hybridization theory, which is an integral part of the valence bond method. In this model, for AX$_n$ systems, we linearly combine the atomic orbitals on atom A in such a way that the resultant combinations (called hybrid orbitals) point toward the X atoms. For our BeH$_2$ molecule in hand, two equivalent, colinear hybrid orbitals are constructed from the 2s and 2p$_z$ orbitals on Be, which can overlap with the two 1s hydrogen orbitals to form two Be–H single bonds. (The 2p$_x$ and 2p$_y$

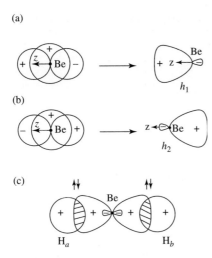

Fig. 3.4.3.
The formation of the two sp hybrid orbitals in BeH$_2$ [(a) and (b)] and the two equivalent bonds in BeH$_2$ (c).

orbitals do not take part in the hybridization scheme, otherwise the resultant hybrid orbitals would not point directly at the hydrogens.) If we combine the 2s and 2p$_z$ orbitals in the following manner:

$$h_1 = (2)^{-1/2}(2s + 2p_z), \quad (3.4.6)$$

$$h_2 = (2)^{-1/2}(2s - 2p_z), \quad (3.4.7)$$

the hybrid orbitals h_1 and h_2 would overlap nicely with the 1s orbitals on H$_a$ and H$_b$, respectively, as shown in Fig. 3.4.3.

The two bonding orbitals in BeH$_2$ have the wavefunctions

$$\psi_1 = c_5 h_1 + c_6 1s_a, \quad c_6 > c_5 \quad (3.4.8)$$

$$\psi_2 = c_5 h_2 + c_6 1s_b, \quad c_6 > c_5. \quad (3.4.9)$$

So now we have two equivalent bonding orbitals ψ_1 and ψ_2 with the same energy. Moreover, ψ_1 and ψ_2 are localized orbitals: ψ_1 is localized between Be and H$_a$ and ψ_2 between Be and H$_b$. They are 2c-2e bonds.

3.4.3 Carbon dioxide, CO$_2$

Carbon dioxide is a linear molecule with both σ and π bonds. The coordinate system chosen for CO$_2$ is shown in Fig. 3.4.4. Once again, the molecular axis is taken to be the z axis. The atomic orbitals taking part in the bonding of this molecule are the 2s and 2p orbitals on C and the 2p orbitals on O. There are a total of ten atomic orbitals and they will form ten molecular orbitals.

The σ orbitals in CO$_2$ are very similar to those in BeH$_2$. The only difference is that the oxygens make use of their 2p$_z$ orbitals instead of the 1s orbitals used

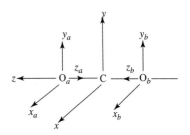

Fig. 3.4.4.
The coordinate system of CO$_2$.

Fig. 3.4.5.
(a) The linear combination $(x_a + x_b)$, which overlaps with the $2p_x$ orbital on C, and (b) linear combination $(x_a - x_b)$, which does not overlap with the $2p_x$ orbital on C. Here x_a and x_b represent $2p_x(a)$ and $2p_x(b)$, respectively.

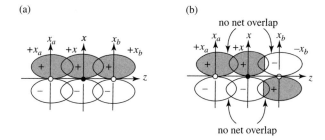

by the hydrogens in BeH$_2$. The σ orbitals thus have the wavefunctions

$$\sigma_s = c_7 2s(C) + c_8[2p_z(a) + 2p_z(b)], \quad c_7 > c_8 \tag{3.4.10}$$

$$\sigma_s^* = c_7' 2s(C) - c_8'[2p_z(a) + 2p_z(b)], \quad c_8' > c_7' \tag{3.4.11}$$

$$\sigma_z = c_9 2p_z(C) + c_{10}[2p_z(a) + 2p_z(b)], \quad c_{10} > c_9 \tag{3.4.12}$$

$$\sigma_z^* = c_9' 2p_z(C) - c_{10}'[2p_z(a) + 2p_z(b)], \quad c_9' > c_{10}'. \tag{3.4.13}$$

The π molecular orbitals are made up of the $2p_x$ and $2p_y$ orbitals of the three atoms. Let's take the $2p_x$ orbitals first. The two $2p_x$ orbitals can be combined in two ways:

$$2p_x(a) + 2p_x(b) \tag{3.4.14}$$

$$2p_x(a) - 2p_x(b). \tag{3.4.15}$$

Combination (3.4.14) overlaps with the C $2p_x$ orbital as shown in Fig. 3.4.5(a). Since, for our linear molecule, the x and y axes are equivalent (and not uniquely defined), we can readily write down the following π bonding and antibonding molecular orbitals:

$$\pi_x = c_{11} 2p_x(C) + c_{12}[2p_x(a) + 2p_x(b)], \quad c_{12} > c_{11} \tag{3.4.16}$$

$$\pi_x^* = c_{11}' 2p_x(C) - c_{12}'[2p_x(a) + 2p_x(b)], \quad c_{11}' > c_{12}' \tag{3.4.17}$$

$$\pi_y = c_{11} 2p_y(C) + c_{12}[2p_y(a) + 2p_y(b)], \quad c_{12} > c_{11} \tag{3.4.18}$$

$$\pi_y^* = c_{11}' 2p_y(C) - c_{12}'[2p_x(a) + 2p_x(b)], \quad c_{11}' > c_{12}' \tag{3.4.19}$$

On the other hand, combination (3.4.15) has zero overlap with the C $2p_x$ orbital [Fig. 3.4.5(b)] and is therefore a nonbonding orbital. Indeed, we have two equivalent nonbonding orbitals:

$$\pi_x^n = 2p_x(a) - 2p_x(b) \tag{3.4.20}$$

$$\pi_y^n = 2p_y(a) - 2p_y(b). \tag{3.4.21}$$

As previously mentioned, ten molecular orbitals are formed. Table 3.4.2 summarizes the formation of the molecular orbitals in CO$_2$, where the linear combinations of O orbitals are normalized. Note that all bonding and antibonding orbitals spread over all three atoms, while the nonbonding orbitals have no participation from C orbitals.

Covalent Bonding in Molecules

Table 3.4.2. Summary of the formation of the molecular orbitals in CO_2

Orbital on C	Orbitals on O*	Molecular orbitals
2s	$(2)^{1/2}(z_a + z_b)$	σ_s, σ_s^*
$2p_z$	$(2)^{1/2}(z_a - z_b)$	σ_z, σ_z^*
$\begin{cases} 2p_x \\ 2p_y \end{cases}$	$\begin{cases} (2)^{1/2}(x_a + x_b) \\ (2)^{1/2}(y_a + y_b) \\ (2)^{1/2}(x_a - x_b) \\ (2)^{1/2}(y_a - y_b) \end{cases}$	$\begin{cases} \pi_x \\ \pi_y \end{cases} \begin{cases} \pi_x^* \\ \pi_y^* \end{cases}$ $\begin{cases} \pi_x^n \\ \pi_y^n \end{cases}$

* Here x_a represents $2p_x(a)$. Similar abbreviations are also used to designate other orbitals on the oxygen atoms.

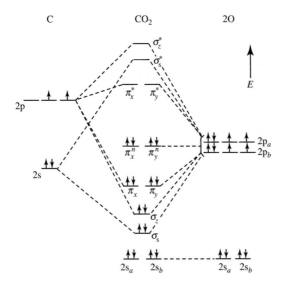

Fig. 3.4.6. Energy level diagram of CO_2.

The energy level diagram for CO_2 is shown in Fig. 3.4.6. Note that the oxygen 2p orbitals are more stable than their corresponding orbitals on carbon. The 16 valence electrons in CO_2 occupy the orbitals as shown in Fig. 3.4.6 and the ground configuration for CO_2 is

$$2s_a^2 2s_b^2 \sigma_s^2 \sigma_z^2 (\pi_x = \pi_y)^4 (\pi_x^n = \pi_y^n)^4.$$

So there are two σ bonds, two π bonds, and four nonbonding electron pairs localized on the oxygen atoms.

In the valence bond or hybridization model for CO_2, we have two resonance (or canonical) structures, as shown in Fig. 3.4.7. In both structures, the two σ bonds are formed by the sp hybrids on carbon with the $2p_z$ orbitals on the oxygens. In the left resonance structure, the π bonds are formed by the $2p_x$ orbitals on C and O_a and the $2p_y$ orbitals on C and O_b. In the other structure, the π bonds are formed by the $2p_x$ orbitals on C and O_b and the $2p_y$ orbitals on C and O_a. The "real" structure is a resonance hybrid of these two extremes. In effect, once again, we get two σ bonds, two π bonds, and four "lone pairs" on the two oxygens. This description is in total agreement with the molecular orbital picture. The only difference is that electron delocalization in CO_2 is

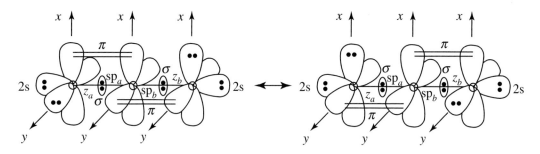

Fig. 3.4.7.
The resonance structures of CO_2.

inherent in the molecular orbital model, whereas description according to the valence-bond hybridization scheme requires the concept of resonance between two canonical structures.

3.4.4 The sp^n ($n = 1$–3) hybrid orbitals

(1) The sp hybridization scheme

Recalling from Section 3.4.1, we use the s orbital and the p_z orbital to form two equivalent hybrid orbitals, one pointing in the $+z$ direction and the other in the $-z$ direction. These two orbitals are called sp hybrids, since they are formed by one s and one p orbital. The wavefunctions of the sp hybrid orbitals are given by eqs. (3.4.6) and (3.4.7). In matrix form the wavefunction are

$$\begin{vmatrix} h_1 \\ h_2 \end{vmatrix} = \begin{vmatrix} 1/(2)^{1/2} & 1/(2)^{1/2} \\ 1/(2)^{1/2} & -1/(2)^{1/2} \end{vmatrix} \begin{vmatrix} s \\ p_z \end{vmatrix} = \begin{vmatrix} a & b \\ c & d \end{vmatrix} \begin{vmatrix} s \\ p_z \end{vmatrix} \quad (3.4.22)$$

We now use this 2×2 coefficient matrix to illustrate the relationships among the coefficients:

(a) Since each atomic orbital is "used up" in the construction of the hybrids, $a^2 + c^2 = 1$ and $b^2 + d^2 = 1$.
(b) Since each hybrids is normalized, $a^2 + b^2 = 1$ and $c^2 + d^2 = 1$.
(c) Since hybrids are orthogonal to each other, $ac + bd = 0$.

(2) The sp^2 hybridization scheme

If we use the s orbital and the p_x and p_y orbitals to form three equivalent orbitals h_1, h_2, and h_3, these orbitals are called sp^2 hybrids. Furthermore, they lie in the xy plane and form $120°$ angles between them. Now the hybrids have the wavefunctions

$$\begin{vmatrix} h_1 \\ h_2 \\ h_3 \end{vmatrix} = \begin{vmatrix} a & b & c \\ d & e & f \\ g & j & k \end{vmatrix} \begin{vmatrix} s \\ p_x \\ p_y \end{vmatrix}. \quad (3.4.23)$$

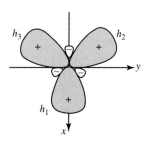

Fig. 3.4.8.
A coordinate system for the sp^2 hybrid orbitals.

If the three hybrids have the orientations as shown in Fig. 3.4.8, we then get the following results for the coefficients. Since the s orbital is (equally) split among the three equivalent hybrids,

$$a^2 = d^2 = g^2 = 1/3, \quad (3.4.24)$$

$$a = d = g = 1/(3)^{1/2}. \quad (3.4.25)$$

(a) Since h_1 lies on the x axis, p_y cannot contribute to h_1,

$$c = 0. \tag{3.4.26}$$

(b) Relation $a^2 + b^2 + c^2 = 1$ leads to

$$b = \left(\frac{2}{3}\right)^{1/2}. \tag{3.4.27}$$

(c) Orbital p_x contributes equally to h_2 and h_3 and $b^2 + e^2 + j^2 = 1$. Hence

$$e = j = -\left(\frac{1}{6}\right)^{1/2}. \tag{3.4.28}$$

Note that both e and j are negative because h_2 and h_3 project on the $-x$ direction.

(d) Orbital p_y contributes equally to h_2 and h_3 (but $f > 0$ and $k < 0$) and $c^2 + f^2 + k^2 = 1$. Hence

$$f = 1/(2)^{1/2}, \tag{3.4.29}$$

$$k = -1/(2)^{1/2}. \tag{3.4.30}$$

Collecting all the coefficients, we have

$$\begin{vmatrix} h_1 \\ h_2 \\ h_3 \end{vmatrix} = \begin{vmatrix} 1/(3)^{1/2} & (2/3)^{1/2} & 0 \\ 1/(3)^{1/2} & -1/(6)^{1/2} & 1/(2)^{1/2} \\ 1/(3)^{1/2} & -1/(6)^{1/2} & -1/(2)^{1/2} \end{vmatrix} \begin{vmatrix} s \\ p_x \\ p_y \end{vmatrix}. \tag{3.4.31}$$

The correctness of these coefficients can be checked in many ways. For example,

$$ad + be + cf = \frac{1}{(3)^{1/2}} \frac{1}{(3)^{1/2}} + \left(\frac{2}{3}\right)^{1/2}\left(-\frac{1}{(6)^{1/2}}\right) + 0\left[-\frac{1}{(2)^{1/2}}\right] = 0. \tag{3.4.32}$$

Also, the angle θ between h_2 and the $+y$ axis can be calculated:

$$\theta = \tan^{-1}[(6)^{-1/2}/(2)^{-1/2}] = \tan^{-1}(3)^{-1/2} = 30°. \tag{3.4.33}$$

To confirm that h_1, h_2, and h_3 are equivalent to each other, we can calculate their hybridization indices and see that they are identical. The hybridization index n of a hybrid orbital is defined as

$$m = \frac{\text{total p orbital population}}{\text{total s orbital population}} = \frac{\sum |\text{p orbital coefficients}|^2}{|\text{s orbital coefficients}|^2}. \tag{3.4.34}$$

Fundamentals of Bonding Theory

For h_1, h_2, and h_3 given in eq. (3.4.31), $n = 2$. Hence they are called sp² hybrids. The two hybrid orbitals in BeH₂ [eq. (3.4.22)] have $n = 1$. So they are sp hybrids.

A molecule whose bonding can be readily rationalized by the sp² hybridization scheme is BF₃. As indicated by the Lewis formulas shown below, there are three canonical structures, each with three σ bonds, one π bond, and eight lone pairs on the F atoms. The three σ bonds are formed by the overlap of the sp² hybrids (formed by 2s, $2p_x$, and $2p_y$) on B with the $2p_z$ orbital on each F atom, while the π bond is formed by the $2p_z$ orbital on B and the $2p_x$ orbital on one of the F atoms. This description is in accord with the experimental FBF bond angles of 120°. Also, it may be concluded that the bond order for the B–F bonds in BF₃ is 1¹/₃.

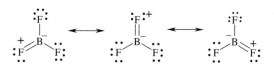

(3) The sp³ hybridization scheme

Now we take up the construction of the sp³ hybrid orbitals. It is well known that the angle between two sp³ hybrids is about 109°. With the aid of Fig. 3.4.9, this angle can now be calculated exactly:

$$\theta = \cos^{-1}\left(-\frac{1}{3}\right) \sim 109°28'16''.$$

If the four hybrids h_1, h_2, h_3, and h_4 points to atoms a, b, c, and d, respectively, as shown in Fig. 3.4.9, it can be readily shown that

$$\begin{vmatrix} h_1 \\ h_2 \\ h_3 \\ h_4 \end{vmatrix} = \begin{vmatrix} 1/2 & 1/2 & 1/2 & 1/2 \\ 1/2 & -1/2 & -1/2 & 1/2 \\ 1/2 & 1/2 & -1/2 & -1/2 \\ 1/2 & -1/2 & 1/2 & -1/2 \end{vmatrix} \begin{vmatrix} s \\ p_x \\ p_y \\ p_z \end{vmatrix} \quad (3.4.35)$$

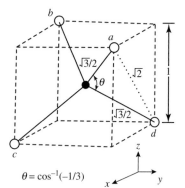

Fig. 3.4.9.
The angle formed by two sp³ hybrids.

If we adopt the coordinates system in Fig. 3.4.10 for the four sp³ hybrids, we can readily get

$$\begin{vmatrix} h_1 \\ h_2 \\ h_3 \\ h_4 \end{vmatrix} = \begin{vmatrix} 1/2 & (3/2)^{1/2} & 0 & 0 \\ 1/2 & -1/[2(3)^{1/2}] & 0 & (2/3)^{1/2} \\ 1/2 & -1/[2(3)^{1/2}] & 1/(2)^{1/2} & -1/(6)^{1/2} \\ 1/2 & -1/[2(3)^{1/2}] & -1/(2)^{1/2} & -1/(6)^{1/2} \end{vmatrix} \begin{vmatrix} s \\ p_x \\ p_y \\ p_z \end{vmatrix}. \quad (3.4.36)$$

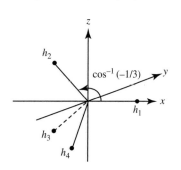

Fig. 3.4.10.
The "original" (1931) Pauling coordinate system for the sp³ hybrids.

It is of interest to note that, in his original construction of the sp³ hybrids, L. Pauling adopted this coordinate system.

It can be easily checked that the hybrids given in either eq. (3.4.35) or eq. (3.4.36) are sp³ hybrids; i.e., their hybridization index (population ratio of p orbitals and s orbital) is 3. The four sp³ hybrids of C are directed toward the corners of a tetrahedron and suitable for forming four localized bonding orbitals with four hydrogen 1s orbitals. Such a bonding picture for CH₄ is familiar to all chemistry students.

Another sp³ hybridized system is the sulfate anion SO_4^{2-}, in which the four oxygen atoms are also arranged in a tetrahedral manner. However, in addition to the four σ bonds, this anion also has two π bonds:

$$\text{[Lewis structure of } SO_4^{2-}\text{]}$$

It is clear that there are six equivalent resonance structures for SO_4^{2-}. Hence the bond order of S–O bonds in SO_4^{2-} is 1½.

(4) Inequivalent hybrid orbitals

We conclude this section by discussing systems where there are two types of hybrid orbitals. One such example is NH₃, where there are three equivalent bond hybrids and one lone pair hybrid. Another example is the water molecule. As shown in Fig. 3.4.11, hybrids h_1 and h_2 are lone pairs orbitals, while hybrids h_3 and h_4 are used to form bonding orbitals with hydrogen 1s orbitals. If we let the coefficient for the s orbital in h_1 be $(a/2)^{1/2}$, the other coefficients for all hybrids can be expressed in terms of a:

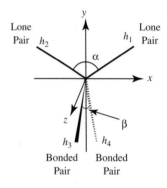

Fig. 3.4.11.
Hybrids h_1 and h_2 are for lone pairs, and h_3 and h_4 are for bonded pairs.

$$\begin{vmatrix} h_1 \\ h_2 \\ h_3 \\ h_4 \end{vmatrix} = \begin{vmatrix} (a/2)^{1/2} & 1/(2)^{1/2} & [(1-a)/2]^{1/2} & 0 \\ (a/2)^{1/2} & -1/(2)^{1/2} & [(1-a)/2]^{1/2} & 0 \\ [(1-a)/2]^{1/2} & 0 & -(a/2)^{1/2} & 1/(2)^{1/2} \\ [(1-a)/2]^{1/2} & 0 & -(a/2)^{1/2} & 1/(2)^{1/2} \end{vmatrix} \begin{vmatrix} s \\ p_x \\ p_y \\ p_z \end{vmatrix}.$$

(3.4.37)

Now we can derive a relationship between angles α (formed by h_1 and h_2) and β (formed by h_3 and h_4):

$$\cot\left(\frac{\alpha}{2}\right) = \frac{[(1-a)/2]^{1/2}}{(1/2)^{1/2}} = (1-a)^{1/2}. \quad (3.4.38)$$

$$\cot\left(\frac{\beta}{2}\right) = \frac{(a/2)^{1/2}}{(1/2)^{1/2}} = a^{1/2}. \quad (3.4.39)$$

Combining eqs. (3.4.38) and (3.4.39),

$$\cot^2\left(\frac{\alpha}{2}\right) + \cot^2\left(\frac{\beta}{2}\right) = 1. \quad (3.4.40)$$

For H$_2$O, $<$ HOH $= \beta = 104.5°$. Now we can calculate the angle formed by the two lone pairs:

$$\alpha = 115.4°. \quad (3.4.41)$$

The angle between lone pairs is larger than that between two bonded pairs, in agreement with the VSEPR (valence-shell electron-pair repulsion) theory.

Now we derive a relationship between the hybridization index of a bonded pair (n_b) with that of a lone pair (n_ℓ):

$$n_\ell = n_1 = n_2 = \frac{[1/2 + (1-a)/2]}{(a/2)}$$

$$= \frac{(2-a)}{a},$$

or

$$a = \frac{2}{(n_\ell + 1)}. \quad (3.4.42)$$

$$n_b = n_3 = n_4 = \frac{[(1+a)/2]}{[(1-a)/2]}$$

$$= \frac{(1+a)}{(1-a)},$$

or

$$a = \frac{(n_b - 1)}{(n_b + 1)}. \quad (3.4.43)$$

Combining eqs. (3.4.42) and (3.4.43) we can obtain a relationship between n_ℓ and n_b:

$$n_b = \frac{(n_\ell + 3)}{(n_\ell - 1)}. \quad (3.4.44)$$

Finally, we relate bond angle β with parameter a:

$$\cos \beta = \cos^2\left(\frac{\beta}{2}\right) - \sin^2\left(\frac{\beta}{2}\right)$$

$$= \left[\frac{a}{2}\bigg/\left(\frac{1}{2}+\frac{a}{2}\right)\right] - \left[\frac{1}{2}\bigg/\left(\frac{1}{2}+\frac{a}{2}\right)\right] = \frac{(a-1)}{(a+1)}. \quad (3.4.45)$$

With eq. (3.4.43), we get

$$\cos \beta = -1/n_b, \text{ or } n_b = -\sec \beta. \quad (3.4.46)$$

For water, $\beta = 104.5°$, $n_b = 3.994$, $n_\ell = 2.336$, $a = 0.600$.

It should be noted that, in this treatment, it is assumed that bonded hybrids h_3 and h_4 point directly at the hydrogens. In other words, the O–H bonds are "straight," not "bent."

3.4.5 Covalent radii

A covalent bond is formed through the overlap of the atomic orbitals on two atoms. All atoms have their structural characteristics; the bond length and bond energy of a covalent bond between two atoms are manifestations of these characteristics. Based on a large quantity of structural data, it is found that there is a certain amount of constancy for the lengths of the single bonds formed between atoms A and B. This kind of constancy also exists for the lengths of the double bonds and triple bonds formed between the same two atoms. From these structural data chemists arrive at the concept of a covalent radius for each element. It is expected that the length of bond A–B is approximately the sum of covalent radii for elements A and B.

The covalent radius of a given element is determined in the following manner. Experimentally, the single bonds formed by the overlap of sp^3 hybrids in C (diamond), Si, Ge, Sn (gray tin) have lengths 154, 235, 245, and 281 pm, respectively. Hence the covalent radii for the single bonds of these elements are 77, 117, 122, and 140 pm, respectively. From the average length of C–O single bonds, the single-bond covalent radius for O can be determined to be 74 pm. Table 3.4.3 lists the covalent radii of the main group elements.

Different authors sometimes give slightly different covalent radii for the same element; this is particularly true for alkali metals and alkali earth metals. Such a variation arises from the following factors:

(a) Atoms A and B have different electronegativities and the bonds they form are polar, having a certain amount of ionic character.
(b) The bond length usually varies with coordination number. It is often that the higher the coordination number, the longer the bonds.
(c) The length of a bond is also influenced by the nature of the neighboring atoms or group.

As pointed out in Sections 14.3.3 and 14.3.4, the length of a C–C single bond ranges from 136 to 164 pm. Hence the data given in Table 3.4.3 are for reference

Table 3.4.3. The covalent radii, in picometers (pm), of main group elements

Single bond	H 37										He —
	Li 134	Be 111			B 88	C 77	N 74	O 74	F 71	Ne —
	Na 154	Mg 136			Al 125	Si 117	P 110	S 102	Cl 99	Ar —
	K 196	Ca 174	Cu 117	Zn 125	Ga 122	Ge 122	As 122	Se 117	Br 114	Kr 110
	Rb —	Sr 192	Ag 133	Cd 141	In 150	Sn 140	Sb 141	Te 137	I 133	Xe 130
	Cs —	Ba 198	Au 125	Hg 144	Tl 155	Pb 154	Bi 152			
	B	C	N	O	P	S				
Double bond	79	67	62	60	100	94				
Triple bond	71	60	55	—	—	—				

only. The bond lengths arising from these covalent radii should not be taken as very accurate structural data.

3.5 Hückel molecular orbital theory for conjugated polyenes

In this section, we first study the π bonding in conjugated polyenes by means of the Hückel molecular orbital theory. Then we will see how the wavefunctions obtained control the course of the reaction for these molecules.

3.5.1 Hückel molecular orbital theory and its application to ethylene and butadiene

A schematic energy level diagram for conjugated polyenes is shown in Fig. 3.5.1. It is obvious that the chemical and physical properties of these compounds are mostly controlled by the orbitals within the dotted lines; i.e., the π and π^* molecular orbitals, as well as the nonbonding orbitals, if there are any. Hence, to study the electronic structure of these systems, as a first approximation, we can ignore the σ and σ^* orbitals and concentrate on the π and π^* orbitals.

In conjugated polyenes, each carbon atom contributes one p orbital and one electron to the π bonding of the system, as illustrated in Fig. 3.5.2. So the π molecular orbitals have the general form

$$\psi = \sum c_i \phi_i$$
$$= c_1 \phi_1 + c_2 \phi_2 + \ldots + c_n \phi_n. \tag{3.5.1}$$

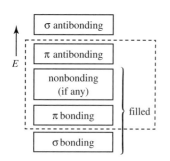

Fig. 3.5.1.
Schematic energy level diagram for conjugated polyenes.

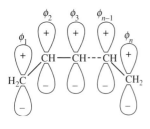

Fig. 3.5.2.
Labeling of the π atomic orbitals in a conjugated polyene.

The values of the coefficient c_i can be determined "variationally"; i.e., they are varied until the total energy reaches a minimum. The energies of the molecular orbitals are obtained by solving for values of E in the secular determinant:

$$\begin{vmatrix} H_{11} - ES_{11} & H_{12} - ES_{12} & \ldots & H_{1n} - ES_{1n} \\ H_{12} - ES_{12} & H_{22} - ES_{22} & \ldots & H_{2n} - ES_{2n} \\ \vdots & \vdots & & \vdots \\ H_{1n} - ES_{1n} & H_{2n} - ES_{2n} & \ldots & H_{nn} - ES_{nn} \end{vmatrix} = 0. \tag{3.5.2}$$

Different methods of approximation means different ways of calculating integrals H_{ij} and S_{ij}. The Hückel approximation, first proposed by German physicist E. Hückel, is very likely to be simplest:

(1) $S_{ij} = \int \phi_i \phi_j d\tau = \begin{cases} 0, & \text{if } i \neq j, \\ 1, & \text{if } i = j. \end{cases}$

That is, there is no overlap between orbitals on different atoms. Note that this drastic approximation violates the basic bonding principle: To form a bond there must an overlap of orbitals.

(2) $H_{ii} = \int \phi_i \hat{H} \phi_i d\tau = \alpha$.

H_{ii} is called the Coulomb integral and is assumed to be the same for each atom. Also, $\alpha < 0$.

(3) $H_{ij} = \int \phi_i H \phi_j d\tau = \beta$ if ϕ_i and ϕ_j are neighboring atomic orbitals.
H_{ij} is called the resonance integral. Also, $\beta < 0$.
(4) $H_{ij} = 0$ if ϕ_i and ϕ_j are not neighboring atomic orbitals.
In other words, the interaction between non-neighboring orbitals is ignored.

After solving for E (n of them in all) in eq. (3.5.2), we can then substitute each E in the following secular equations to determine the values of coefficient c_i:

$$\begin{cases} (H_{11} - ES_{11})c_1 + (H_{12} - ES_{12})c_2 + \ldots + (H_{1n} - ES_{1n})c_n = 0 \\ (H_{12} - ES_{12})c_1 + (H_{22} - ES_{22})c_2 + \ldots + (H_{2n} - ES_{2n})c_n = 0 \\ \vdots \qquad \vdots \qquad \vdots \\ (H_{1n} - ES_{1n})c_1 + (H_{2n} - ES_{2n})c_2 + \ldots + (H_{nn} - ES_{nn})c_n = 0 \end{cases} \quad (3.5.3)$$

Now we apply this approximation to the π system of a few polyenes.

(1) Ethylene

The two atomic orbitals participating in π bonding are shown in Fig. 3.5.3. The secular determinant has the form

$$\begin{vmatrix} \alpha - E & \beta \\ \beta & \alpha - E \end{vmatrix} = \begin{vmatrix} x & 1 \\ 1 & x \end{vmatrix} = 0 \quad (3.5.4)$$

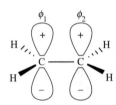

with $x = (\alpha - E)/\beta$. Solving eq. (3.5.4) leads to $x = \pm 1$, or

$$E_1 = \alpha + \beta = E(\pi), \quad (3.5.5)$$
$$E_2 = \alpha - \beta = E(\pi^*). \quad (3.5.6)$$

Fig. 3.5.3.
The atomic orbitals for π bonding in ethylene.

Substituting E_1 and E_2 into the equations

$$\begin{cases} (\alpha - E)c_1 + \beta c_2 = 0 \\ \beta c_1 + (\alpha - E)c_2 = 0 \end{cases} \quad (3.5.7)$$

we can obtain the molecular orbital wavefunctions:
When $E = E_1$, we have $c_1 = c_2$, or

$$\psi_1 = \psi(\pi) = (2)^{-1/2}(\phi_1 + \phi_2). \quad (3.5.8)$$

When $E = E_2$, we have $c_1 = -c_2$, or

$$\psi_2 = \psi(\pi^*) = (2)^{-1/2}(\phi_1 - \phi_2). \quad (3.5.9)$$

The two π electrons in ethylene occupy the $\psi(\pi)$ orbital. Hence the energy for a π bond in the Hückel model is

$$E_\pi = 2\alpha + 2\beta. \quad (3.5.10)$$

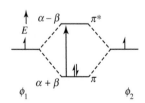

The energy level diagram for the π orbitals in ethylene is very simple and shown in Fig. 3.5.4. Note that in this approximation the bonding effect is equal to the antibonding effect. This arises from ignoring the overlap.

Fig. 3.5.4.
The energy level diagram for the π molecular orbitals of ethylene.

Fundamentals of Bonding Theory

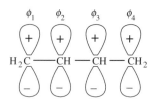

Fig. 3.5.5.
The atomic orbitals for π bonding in butadiene.

(2) Butadiene

Now we proceed to use the same method to treat a higher homolog of ethylene, namely butadiene. The atomic orbitals taking part in the π bonding of this molecule is shown in Fig. 3.5.5. For this system, the secular determinant is

$$\begin{vmatrix} \alpha - E & \beta & 0 & 0 \\ \beta & \alpha - E & \beta & 0 \\ 0 & \beta & \alpha - E & \beta \\ 0 & 0 & \beta & \alpha - E \end{vmatrix} = \begin{vmatrix} x & 1 & 0 & 0 \\ 1 & x & 1 & 0 \\ 0 & 1 & x & 1 \\ 0 & 0 & 1 & x \end{vmatrix} = 0, \quad (3.5.11)$$

where, again, $x = (\alpha - E)/\beta$. Expanding this determinant, we get

$$x^4 - 3x^2 + 1 = 0,$$

or

$$x = [\pm(5)^{1/2} \pm 1]/2 = \pm 1.618, \pm 0.618. \quad (3.5.12)$$

The energies and the wavefunctions are tabulated in Table 3.5.1. The wavefunctions are shown pictorially in Fig. 3.5.6. Note that as the number of nodes in a wavefunction increases, so does the energy associated with the wavefunction.

Table 3.5.1. The Hückel energies and wavefunction of the π molecular orbitals of butadiene

x	Energy	Wavefunction
1.618	$E_4 = \alpha - 1.618\beta$	$\psi_4(\pi^*) = 0.372\phi_1 - 0.602\phi_2 + 0.602\phi_3 - 0.372\phi_4$
0.618	$E_3 = \alpha - 0.618\beta$	$\psi_3(\pi^*) = 0.602\phi_1 - 0.372\phi_2 - 0.372\phi_3 + 0.602\phi_4$
−0.618	$E_2 = \alpha + 0.618\beta$	$\psi_2(\pi) = 0.602\phi_1 + 0.372\phi_2 - 0.372\phi_3 - 0.602\phi_4$
−1.618	$E_1 = \alpha + 1.618\beta$	$\psi_1(\pi) = 0.372\phi_1 + 0.602\phi_2 + 0.602\phi_3 + 0.372\phi_4$

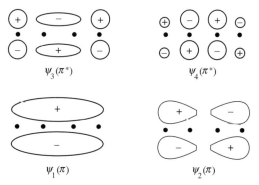

Fig. 3.5.6.
The π and π^* molecular orbitals of butadiene.

Since the first two orbitals (ψ_1 and ψ_2) are filled,

$$E_\pi = 2(\alpha + 1.618\beta) + 2(\alpha + 0.618\beta) = 4\alpha + 4.472\beta. \quad (3.5.13)$$

If the two π bonds in butadiene *were* localized, they would have energy $2(2\alpha + 2\beta) = 4\alpha + 4\beta$. Hence, by allowing the four π electrons to delocalize over the entire molecular skeleton, there is a gain in stability, which is called delocalization energy (DE). For butadiene,

$$\begin{aligned}\text{DE} &= 4\alpha + 4.472\beta - (4\alpha + 4\beta) \\ &= 0.472\beta < 0.\end{aligned} \quad (3.5.14)$$

By combining high-level *ab initio* calculations with high-resolution infrared spectroscopy, the equilibrium bond lengths in *s-trans*-butadiene have been determined to an unprecedented precision of 0.1 pm. The values found for the pair of π-electron delocalized double bonds and the delocalized central single bond are 133.8 and 135.4 pm, respectively. The data provide definitive structural evidence that validates the fundamental concepts of π-electron delocalization, conjugation, and bond alternation in organic chemistry.

3.5.2 Predicting the course of a reaction by considering the symmetry of the wavefunction

The molecular orbitals of butadiene, shown in Fig. 3.5.6, can be used to predict, or at least to rationalize, the course of concerted reactions (those which take place in a single step without involvement of intermediates) it would undergo. For instance, experimentally it is known that different cyclization products are obtained from butadiene by heating and upon light irradiation.

In 1965, the American chemists R. B. Woodward and R. Hoffmann ("conservation of orbital symmetry" or Woodward–Hoffmann rules) and Japanese chemist K. Fukui ("frontier orbital theory") proposed theories to explain these results as well as those for other reactions. (Woodward won the Nobel Prize in Chemistry in 1965 for his synthetic work. In 1981, after the death of Woodward, Hoffmann and Fukui shared the same prize for the theories discussed here.)

These theories assert that the pathway of a chemical reaction accessible to a compound is controlled by its highest occupied molecular orbital (HOMO). For the thermal reaction of butadiene, which is commonly called "ground-state chemistry," the HOMO is ψ_2 and lowest unoccupied molecular orbital (LUMO) is ψ_3 (Fig. 3.5.6). For the photochemical reaction of butadiene, which is known to be "excited-state chemistry," the HOMO is ψ_3 (Fig. 3.5.6).

The thermal and photochemical cyclization of a butadiene bearing different substituents at its terminal carbon atoms is represented by eqs. (3.5.15) and (3.5.16), respectively. Note that for a short conjugated polyene consisting of four or six carbon atoms, conformational interconversion between their transoid (or *s-trans*) and cisoid (or *s-cis*) forms takes place readily.

114　　　　　　　　　　Fundamentals of Bonding Theory

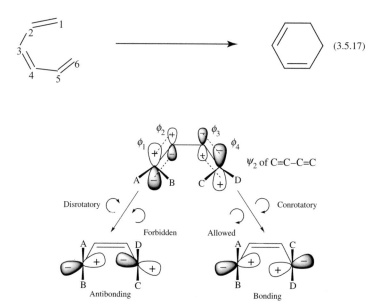

For wavefunction ψ_2, the terminal atomic orbitals ϕ_1 and ϕ_4 have the relative orientations as shown in Fig. 3.5.7. It is evident that a conrotatory process leads to a bonding interaction between ϕ_1 and ϕ_4, while a disrotatory process leads to an antibonding interaction between ϕ_1 and ϕ_4. Clearly the conrotatory process prevails in this case.

Conversely, for wavefunction ψ_3, the terminal atomic orbitals ϕ_1 and ϕ_4 have the relative orientations shown in Fig. 3.5.8. Here a conrotatory pathway yields an antibonding interaction between the terminal atomic orbitals, while a disrotatory step leads to a stabilizing bonding interaction. Hence the disrotatory process wins out in this case.

To conclude this section, we apply this theory to the cyclization of hexatriene:

Fig. 3.5.7.
Thermal cyclization of butadiene.

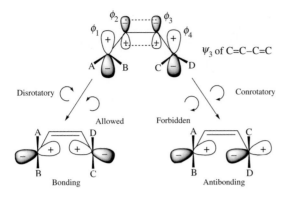

Fig. 3.5.8. Photocyclization of butadiene.

Table 3.5.2. The Hückel energies and wavefunctions of the π molecular orbitals of hexatriene

x	Energy	Wavefunction
1.802	$E_6 = \alpha - 1.802\beta$	$\psi_6 = 0.232\phi_1 - 0.418\phi_2 + 0.521\phi_3 - 0.521\phi_4 + 0.418\phi_5 - 0.232\phi_6$
1.247	$E_5 = \alpha - 1.247\beta$	$\psi_5 = 0.418\phi_1 - 0.521\phi_2 + 0.232\phi_3 + 0.232\phi_4 - 0.521\phi_5 + 0.418\phi_6$
0.445	$E_4 = \alpha - 0.445\beta$	$\psi_4 = 0.521\phi_1 - 0.232\phi_2 - 0.418\phi_3 + 0.418\phi_4 + 0.232\phi_5 - 0.521\phi_6$
-0.445	$E_3 = \alpha + 0.445\beta$	$\psi_3 = 0.521\phi_1 + 0.232\phi_2 - 0.418\phi_3 - 0.418\phi_4 + 0.232\phi_5 + 0.521\phi_6$
-1.247	$E_2 = \alpha + 1.247\beta$	$\psi_2 = 0.418\phi_1 + 0.521\phi_2 + 0.232\phi_3 - 0.232\phi_4 - 0.521\phi_5 - 0.418\phi_6$
-1.802	$E_1 = \alpha + 1.802\beta$	$\psi_1 = 0.232\phi_1 + 0.418\phi_2 + 0.521\phi_3 + 0.521\phi_4 + 0.418\phi_5 + 0.232\phi_6$

The secular determinant of hexatriene is

$$\begin{vmatrix} x & 1 & 0 & 0 & 0 & 0 \\ 1 & x & 1 & 0 & 0 & 0 \\ 0 & 1 & x & 1 & 0 & 0 \\ 0 & 0 & 1 & x & 1 & 0 \\ 0 & 0 & 0 & 1 & x & 1 \\ 0 & 0 & 0 & 0 & 1 & x \end{vmatrix} = 0, \text{ with } x = (\alpha - E)/\beta. \quad (3.5.18)$$

The solutions of this determinant, along with the energies and wavefunctions of the π molecular orbitals, are summarized in Table 3.5.2. While these wavefunctions are not pictorially shown here, we can readily see that the number of nodes of the wavefunctions increases as their energies increase. Indeed, $\psi_1, \psi_2, \ldots, \psi_6$ have 0, 1, ..., 5 nodes, respectively. Also,

$$E_\pi = 2(\alpha + 1.802\beta) + 2(\alpha + 1.247\beta) + 2(\alpha + 0.445\beta)$$
$$= 6\alpha + 6.988\beta. \quad (3.5.19)$$
$$\text{DE} = 6\alpha + 6.988\beta - 3(2\alpha + 2\beta)$$
$$= 0.988\beta < 0. \quad (3.5.20)$$

For the thermal and photochemical cyclization of hexatriene, the controlling HOMO's are ψ_3 and ψ_4, respectively. As shown in Fig. 3.5.9, the allowed

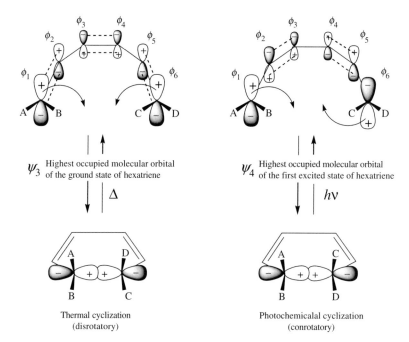

Fig. 3.5.9.
Thermal and photochemical cyclization reactions of hexatriene.

pathway for the thermal reaction is disrotatory. On the other hand, the allowed pathway for the photochemical reaction is conrotatory. These results are just the opposite of those found for the cyclization reactions of butadiene (Figs. 3.5.7 and 3.5.8).

References

1. R. McWeeny, *Coulson's Valence*, 3rd edn., Oxford University Press, Oxford, 1979.
2. J. N. Murrell, S. F. A. Kettle and J. M. Tedder, *The Chemical Bond*, Wiley, Chichester, 1978.
3. E. Cartmell and G. W. A. Fowles, *Valency and Molecular Structure*, 4th edn., Butterworth, London, 1977.
4. R. L. DeKock and H. B. Gray, *Chemical Structure and Bonding*, Benjamin/Cummings, Menlo Park, 1980.
5. M. Karplus and R. N. Porter, *Atoms & Molecules: An Introduction for Students of Physical Chemistry*, Benjamin, New York, 1970.
6. N. W. Alcock, *Bonding and Structure: Structural Principles in Inorganic and Organic Chemistry*, Ellis Horwood, New York, 1990.
7. R. J. Gillespie and I. Hargittai, *The VSEPR Model of Molecular Geometry*, Allyn and Bacon, Boston, 1991.
8. R. J. Gillespie and P. L. A. Popelier, *Chemical Bonding and Molecular Geometry from Lewis to Electron Densities*, Oxford University Press, New York, 2001.
9. B. M. Gimarc, *Molecular Structure and Bonding: The Quantitative Molecular Orbital Approach*, Academic Press, New York, 1979.
10. Y. Jean and F. Volatron (translated by J. Burdett), *An Introduction to Molecular Orbitals*, Oxford University Press, New York, 1993.

11. V. M. S. Gil, *Orbitals in Chemistry: a Modern Guide for Students*, Cambridge University Press, Cambridge, 2000.
12. J. G. Verkade, *A Pictorial Approach to Molecular Bonding and Vibrations,* 2nd edn., Springer, New York, 1997.
13. J. K. Burdett, *Molecular Shapes: Theoretical Models of Inorganic Stereochemistry,* Wiley, New York, 1980.
14. A. Rauk, *Orbital Interaction Theory of Organic Chemistry*, 2nd edn., Wiley, New York, 2001.
15. F. Weinhold and C. Landis, *Valency and Bonding: A Natural Bond Orbital Donor-Acceptor Perspective*, Cambridge University Press, Cambridge, 2005.
16. T. A. Albright, J. K. Burdett and M.-H. Whangbo, *Orbital Interactions in Chemistry*, Wiley, New York, 1985.
17. T. A. Albright and J. K. Burdett, *Problems in Molecular Orbital Theory,* Oxford University Press, New York, 1992.
18. I. Fleming, *Pericyclic Reactions*, Oxford University Press, Oxford, 1999.
19. W.-K. Li, MO and VB wave functions for He_2. *J. Chem. Educ.* **67**, 131 (1990).
20. M. J. S Dewar, H. Kollmar and W.-K. Li, Valence angles and hybridization indices in "sp^3 hybridized" AX_2Y_2 systems. *J. Chem. Educ.* **52**, 305–306 (1975).

4 Chemical Bonding in Condensed Phases

Studies of structural chemistry in the solid state are of particular interest in view of the following justifications:

(a) The majority of the important materials used in modern optical, electronic, and magnetic applications are solids.
(b) Our present understanding of structure-property correlation originates mainly from results obtained in the gaseous and solid states. Based on this foundation, the structure of matter in the liquid phase can be deduced and further investigations then carried out.
(c) Starting from the knowledge of the fixed atomic positions in the solid state of a given compound, new materials can be designed by adding atoms into the lattice or removing atoms from the lattice, and such new materials will have different properties.
(d) The atoms in a solid are bonded together through many types of interactions, the nature of which often determines the properties of the material.

In this chapter, the basic types of chemical bonds existing in condensed phases are discussed. These interactions include ionic bonds, metallic bonds, covalent bonding (band theory), and intermolecular forces. In Chapter 10, the structures of some inorganic crystalline materials will be presented.

4.1 Chemical classification of solids

Solids are conveniently classified in terms of the types of chemical bonds that hold the atoms together, as shown in Table 4.1.1. The majority of solids involve

Table 4.1.1. Classification of simple solids

Class	Example
Ionic	NaCl, MgO, CaF_2, CsCl
Covalent	C (diamond), SiO_2 (silica)
Molecular	Cl_2, S_8, $HgCl_2$, benzene
Metallic	Na, Mg, Fe, Cu, Au

Table 4.1.2. Some examples of complex solids held together by a combination of different bond types

Bond types involved	Example
Ionic, covalent	ZnS, TiO_2
Ionic, covalent, van der Waals	CdI_2
Ionic, metallic	NbO, TiO_x ($0.75 < x < 1.25$)
Ionic, covalent, metallic, van der Waals	ZrCl
Ionic, covalent, metallic	$K_2Pt(CN)_4Br_{0.3} \cdot 3H_2O$, $(SNBr_x)_\infty$ ($0.25 < x < 1.5$)
Covalent, metallic, van der Waals	C (graphite)
Covalent, hydrogen bond	Ice
Ionic, covalent, metallic, van der Waals	TTF:TCNQ, Tl_2RbC_{60}

TTF, tetrathiafulvalene; TCNQ, tetracyanoquinodimethane.

more complex chemical bonding, and some examples are given in Table 4.1.2. Here the term "metallic" indicates electron delocalization over the entire solid.

In ZnS the transfer of two electrons from the metal atom to the non-metal is incomplete, so that the bond between the formal Zn^{2+} and S^{2-} ions has significant covalent character. The same is true for compounds containing metal ions with a high formal charge such as TiO_2.

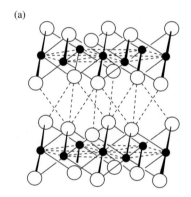
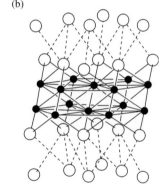

Fig. 4.1.1.
Layer structure in (a) CdI_2 and (b) ZrCl.

In the layer structure of CdI_2, the iodine atoms belonging to adjacent layers are connected by van der Waals forces (Fig. 4.1.1(a)). The metal atoms in TiO and NbO are in a low oxidation state and the excess electrons can form metal–metal bonds extending in one and two dimensions, which are responsible for the metallic character of these oxides. The sheet structure of ZrCl is composed of four homoatomic layers in the sequence $\cdots ClZrZrCl \cdots ClZrZrCl \cdots$, as shown in Fig. 4.1.1(b). Each Zr has six metal neighbors in the same layer at 342 pm, and three metal neighbors in the next layer at 309 pm, compared to an average distance of 320 pm in α-Zr. There is strong metal–metal bonding in the double Zr layers. Each Zr has three Cl neighbors in the adjacent layer at 263 pm, indicating both covalent and ionic bonding. The $Cl \cdots Cl$ inter-sheet distances at $\sim$360 pm are normal van der Waals contacts.

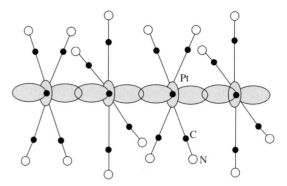

Fig. 4.1.2.
Overlap of $5d_{z^2}$ orbitals along a chain of Pt centers in the crystal structure of $K_2Pt(CN)_4Br_{0.3} \cdot 3H_2O$.

The partially oxidized complex $K_2Pt(CN)_4Br_{0.3} \cdot 3H_2O$ behaves as a one-dimensional conductor, owing to overlap of the platinum $5d_{z^2}$ orbitals atoms along a chain composed of stacking of square-planar $[Pt(CN)_4]^{1.7-}$ units (Fig. 4.1.2).

The 1:1 solid adduct of tetrathiafulvalene (TTF) and tetracyanoquinodimethane (TCNQ) is the first-discovered "molecular metal," which consists of alternate stacks each composed of molecules of the same type (Fig. 4.1.3). A charge transfer of 0.69 electron per molecule from the HOMO (mainly S atom's lone pair in character) of TTF to the LUMO of TCNQ results in two partially filled bands, which account for the electrical conductivity of TTF:TCNQ.

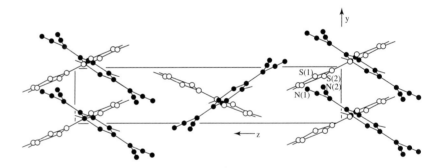

Fig. 4.1.3.
Crystal structure of the "molecular metal" TTF:TCNQ.

A binary compound A_nB_m consists of atoms of two kind of elements, A and B, and its bond type depends on the average value of the electronegativities of the two elements ($\bar{\chi}$, mean of χ_A and χ_B), and on the difference of the electronegativities of the two elements ($\Delta\chi = |\chi_A - \chi_B|$). Figure 4.1.4 shows the distribution of the compounds A_nB_m in a diagram of the "$\bar{\chi}-\Delta\chi$" relationship. This diagram may be divided into four regions which correspond to the four bond types: ionic (I), metallic (II), semi-metallic (III), and covalent (IV). For example, the compound SF_4 has the following parameters: $\chi_F = 4.19$, $\chi_S = 2.59$, $\bar{\chi} = 3.39$, and $\Delta\chi = 1.60$, which place the compound in region IV. Thus the bond type of SF_4 is covalent. In another example, the parameters of ZrCl are $\chi_{Zr} = 1.32$, $\chi_{Cl} = 2.87$, $\bar{\chi} = 2.20$, and $\Delta\chi = 1.55$, which place the compound on the border of regions I and IV. So the bond in ZrCl is expected to exhibit both ionic and covalent characters.

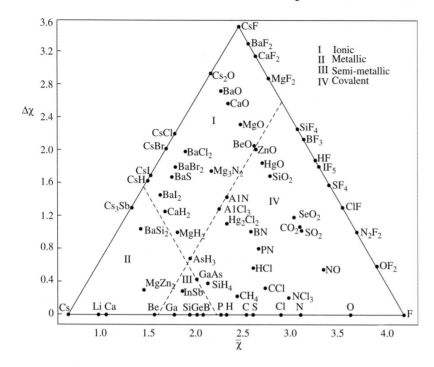

Fig. 4.1.4.
Bond types and electronegativities of some binary compounds.

4.2 Ionic bond

4.2.1 Ionic size: crystal radii of ions

A set of empirical ionic radii can be derived from the direct measurement of internuclear distances in crystal structures. The additivity of ionic radii is substantiated by the near constancy of the differences in internuclear distances Δr between the alkali metal halides, as shown in Table 4.2.1.

By assuming a value for the radius of each single ion (e.g., 140 pm for O^{2-} and 133 pm for F^-), a set of self-consistent ionic radii can be generated. There exist several compilations of ionic radii, and some selected representative values for ions with coordination number 6 are listed in Table 4.2.2.

Some comments on the ionic radii collected in Table 4.2.2 are as follows:

(1) Values of the ionic radii are derived from experimental data, which give the internuclear distances and electron densities, and generally take the distance of contacting neighbor ions to be the sum of the ionic radii of the cation and anion:

Internuclear distance between contacting neighbors = $r_{\text{cation}} + r_{\text{anion}}$.

Thus, a hard sphere model is assumed for the ionic crystal, with ions of opposite charge touching one another in the crystal lattice. Such an approximation means that the assignment of individual radii is somewhat arbitrary. The values listed in Table 4.2.2 are a set of data which assume that the radius of O^{2-} is 140 pm and that of F^- is 133 pm.

Fundamentals of Bonding Theory

Table 4.2.1. Internuclear distance (pm) and their difference in alkali metal halides

Cation	Anion								Avg Δr
	F^-		Cl^-		Br^-		I^-		
Li^+	201		257		275		300		
		28		24		22		23	24
Na^+	229		281		297		323		
		37		33		32		30	33
K^+	266		314		329		353		
		16		13		15		13	14
Rb^+	282		327		344		366		
		18		29		27		29	26
Cs^+	300		356		371		395		

Anion	Cation								Avg Δr
	Li^+		Na^+		K^+		Rb^+		Cs^+	
F^-	201		229		266		282		300	
		56		52		48		45		56	51
Cl^-	257		281		314		327		356	
		18		16		15		17		15	16
Br^-	275		297		329		344		371	
		25		26		24		22		24	24
I^-	300		323		353		366		395	

(2) The valence state of ions listed in Table 4.2.2 are obtained from the oxidation states of the atoms in the compounds. They are only formal values and they do not indicate the number of transferred electrons. In other words, the bonding type between the atoms is not considered. These radii are effective ionic radii and they can be used for rough estimation of the packing of ions in crystals and other calculations.

(3) There is an increase in size with increasing coordination number and for a given coordination number with increasing Z within a periodic group. In general:

Coordination number:	4	6	8	12
Relative radius:	0.94	1.00	1.03	1.12

For cations with high charge the decrease in radii as the coordination number decreases can be quite large. For instance, the radii of Mn^{7+} in six- and four-coordinate geometry are 46 and 25 pm, respectively. Some effective ionic radii of four-coordinate high-valence cations are listed below:

Cation	B^{3+}	C^{4+}	Cr^{6+}	Mn^{7+}	Mo^{6+}	P^{5+}	Pb^{4+}	S^{6+}	Se^{6+}	Si^{4+}	Sn^{4+}	V^{5+}	W^{6+}	Zr^{4+}
Radius (pm)	11	15	26	25	41	17	65	12	28	26	55	35.5	42	59

(4) The ionic radii are generally irregular, slowly decreasing in size with increasing Z for transition metals of the same charge. Also, the high-spin (HS) ions have larger radii than low-spin (LS) ions of the same species and charge.

Table 4.2.2. Ionic radii (in pm)* with coordination number 6

Ion	Radius	Ion	Radius	Ion	Radius	Ion	Radius	Ion	Radius
Ac^{3+}	112	Cr^{2+}(HS)	80	Li^+	76	Pa^{4+}	90	Sm^{3+}	95.8
Ag^+	115	Cr^{3+}	61.5	Lu^{3+}	86.1	Pa^{5+}	78	Sn^{4+}	69.0
Ag^{2+}	94	Cr^{4+}	55	Mg^{2+}	72.0	Pb^{2+}	119	Sr^{2+}	118
Ag^{3+}	75	Cr^{5+}	49	Mn^{2+}(LS)	67	Pb^{4+}	77.5	Ta^{3+}	72
Al^{3+}	53.5	Cr^{6+}	44	Mn^{2+}(HS)	83.0	Pd^{2+}	86	Ta^{4+}	68
Am^{3+}	97.5	Cs^+	167	Mn^{3+}(LS)	58	Pd^{3+}	76	Ta^{5+}	64
Am^{4+}	85	Cu^+	77	Mn^{3+}(HS)	64.5	Pd^{4+}	61.5	Tb^{3+}	92.3
As^{3+}	58	Cu^{2+}	73	Mn^{4+}	53	Pm^{3+}	97	Tb^{4+}	76
As^{5+}	46	Cu^{3+}	54	Mn^{7+}	46	Po^{4+}	94	Tc^{4+}	64.5
At^{7+}	62	Dy^{2+}	107	Mo^{4+}	65.0	Po^{6+}	67	Tc^{5+}	60
Au^+	137	Dy^{3+}	91.2	Mo^{5+}	61	Pr^{3+}	99	Tc^{7+}	56
Au^{3+}	85	Er^{3+}	89.0	Mo^{6+}	59	Pr^{4+}	85	Te^{2-}	221
Au^{5+}	57	Eu^{2+}	117	N^{3+}	16	Pt^{2+}	80	Te^{4+}	97
B^{3+}	27	Eu^{3+}	94.7	N^{5+}	13	Pt^{4+}	62.5	Te^{6+}	56
Ba^{2+}	135	F^-	133	Na^+	102	Pt^{5+}	57	Th^{4+}	94
Be^{2+}	45	F^{7+}	8	Nb^{3+}	72	Pu^{4+}	86	Ti^{2+}	86
Bi^{3+}	103	Fe^{2+}(LS)	61	Nb^{4+}	68	Pu^{5+}	74	Ti^{3+}	67.0
Bi^{5+}	76	Fe^{2+}(HS)	78.0	Nb^{5+}	64	Pu^{6+}	71	Ti^{4+}	60.5
Bk^{3+}	96	Fe^{3+}(LS)	55	Nd^{3+}	98.3	Rb^+	152	Tl^+	150
Bk^{4+}	83	Fe^{3+}(HS)	64.5	Ni^{2+}	69.0	Re^{4+}	63	Tl^{3+}	88.5
Br^-	196	Fe^{4+}	58.5	Ni^{3+}(LS)	56	Re^{5+}	58	Tm^{2+}	103
Br^{7+}	39	Fr^+	180	Ni^{3+}(HS)	60	Re^{6+}	55	Tm^{3+}	88.0
C^{4+}	16	Ga^{3+}	62.0	Ni^{4+}	48	Re^{7+}	53	U^{3+}	102.5
Ca^{2+}	100	Gd^{3+}	93.8	No^{2+}	110	Rh^{3+}	66.5	U^{4+}	89
Cd^{2+}	95	Ge^{2+}	73	Np^{2+}	110	Rh^{4+}	60	U^{5+}	76
Ce^{3+}	101	Ge^{4+}	53	Np^{3+}	101	Rh^{5+}	55	U^{6+}	73
Ce^{4+}	87	Hf^{4+}	71	Np^{4+}	87	Ru^{3+}	68	V^{2+}	79
Cf^{3+}	95	Hg^+	119	Np^{5+}	75	Ru^{4+}	62.0	V^{3+}	64.0
Cf^{4+}	82.1	Hg^{2+}	102	Np^{6+}	72	Ru^{5+}	56.5	V^{4+}	58
Cl^-	181	Ho^{3+}	90.1	Np^{7+}	71	S^{2-}	184	V^{5+}	54
Cl^{7+}	27	I^-	220	O^{2-}	140	S^{4+}	37	W^{4+}	66
Cm^{3+}	97	I^{5+}	95	OH^-	137	S^{6+}	29	W^{5+}	62
Cm^{4+}	85	I^{7+}	53	Os^{4+}	63.0	Sb^{3+}	76	W^{6+}	60
Co^{2+}(LS)	65	In^{3+}	80.0	Os^{5+}	57.5	Sb^{5+}	60	Xe^{8+}	48
Co^{2+}(HS)	74.5	Ir^{3+}	68	Os^{6+}	54.5	Sc^{3+}	74.5	Y^{3+}	90.0
Co^{3+}(LS)	54.5	Ir^{4+}	62.5	Os^{7+}	52.5	Se^{2-}	198	Yb^{2+}	102
Co^{3+}(HS)	61	Ir^{5+}	57	P^{3+}	44	Se^{4+}	50	Yb^{3+}	86.8
Co^{4+}	53	K^+	138	P^{5+}	38	Se^{6+}	42	Zn^{2+}	74.0
Cr^{2+}(LS)	73	La^{3+}	103.2	Pa^{3+}	104	Si^{4+}	40.0	Zr^{4+}	72

* Selected from R. D. Shannon. *Acta Crystallogr.* **A32**, 751–67 (1976). Notations HS and LS refer to high- and low-spin state, respectively.

Because of the diversity of the chemical bonds, the bond type usually varies with the coordination number, especially for the ions with high valence and low coordination number. Figure 4.2.1 shows the relationship between the ionic radii and the number of d electrons in the first series of transition metals.

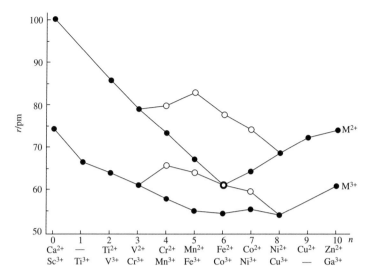

Fig. 4.2.1.
Relationship between the ionic radii and the number of d electrons of the first series of transition metals. Open circles denote high-spin complexes.

4.2.2 Lattice energies of ionic compounds

The lattice energy U of an ionic compound is defined as the energy required to convert one mole of crystalline solid into its component cations and anions in their thermodynamic standard states (non-interacting gaseous ions at standard temperature and pressure). It can be calculated using either the Born–Landé equation

$$U = \frac{N_0 A Z_+ Z_- e^2}{4\pi \epsilon_0 r_e}\left(1 - \frac{1}{m}\right) \quad (4.2.1)$$

or the Born–Mayer equation

$$U = \frac{N_0 A Z_+ Z_- e^2}{4\pi \epsilon_0 r_e}\left(1 - \frac{\rho}{r_e}\right), \quad (4.2.2)$$

where N_0 is Avogardro's number, A is the Madelung constant, Z_+ and Z_- are the charges on the cation and anion, r_e is the equilibrium interionic distance, and ρ is a parameter in the repulsion term $-be^{-r/\rho}$, which is found to be nearly constant for most crystals with a value of 34.5 pm, and m is a parameter in the repulsion term a/r^m for interionic interaction, which is usualy assigned an integral value of 9. Table 4.2.3 gives the Madelung constants A for a number of ionic crystals. The values listed refer to the structural type with unit charges ($Z_+ = Z_- = 1$) at the ion sites.

Kapustinskii noted that if the Madelung constant A is divided by the number of ions per formula unit for a number of crystal structures, nearly the same value is obtained. Furthermore, as both A/n and r_e increase with the coordination number, their ratio A/nr_e is expected to be approximately the same from one structure to another. Therefore, Kapustinskii proposed that the structure of any ionic solid is energetically equivalent to a hypothetical rock-salt structure and its lattice energy can be calculated using the Madelung constant of NaCl and the appropriate ionic radii for (6,6) coordination.

Table 4.2.3. Madelung constants of selected crystals

Formula	Space group	Compound name	Coordination mode	Madelung constant
NaCl	$Fm\bar{3}m$	rock salt	(6,6)	1.7476
CsCl	$Pm\bar{3}m$	cesium chloride	(8,8)	1.7627
β–ZnS	$F\bar{4}3m$	zinc blende (sphalerite)	(4,4)	1.6381
α–ZnS	$P6mc$	wurzite	(4,4)	1.6413
CaF_2	$Fm\bar{3}m$	fluorite	(8,4)	2.5194
Cu_2O	$Pn3m$	cuprite		2.2212
TiO_2	$P4/mnm$	rutile	(6,3)	2.408
TiO_2	$C4/2mc$	anatase		2.400
β–SiO_2	$P6_222$	β-quartz	(4,2)	2.220
α–Al_2O_3	$R\bar{3}c$	corundum	(6,4)	4.172

Substituting $r_e = r_+ + r_-$ (in pm), $\rho = 34.5$ pm, $A = 1.7476$, and values for N_0, e and ε_0 into the Born–Mayer equation gives the Kapustinskii equation:

$$U = \frac{1.214 \times 10^5 Z_+ Z_- n}{r_+ + r_-}\left(1 - \frac{34.5}{r_+ + r_-}\right) \text{ kJ mol}^{-1}. \quad (4.2.3)$$

The lattice energies calculated using this equation are compared with those obtained from the Born–Haber cycle in Table 4.2.4.

Table 4.2.4. Lattice energies (in kJ mol^{-1}) of some alkali metal halides and divalent transition metal chalcogenides

Ionic Compound	Born–Haber	Kapustinskii	Ionic compound	Born–Haber	Kapustinskii
LiF	1009	952	CsF	715	713
LiCl	829	803	CsCl	640	625
LiBr	789	793	CsBr	620	602
LiI	734	713	CsI	587	564
NaF	904	885	MnO	3812	3895
NaCl	769	753	FeO	3920	3987
NaBr	736	734	CoO	3992	4046
NaI	688	674	NiO	4075	4084
KF	801	789	CuO	4033	4044
KCl	698	681	ZnO	3971	4142
KBr	672	675	ZnS(zinc blende)	3619	3322
KI	632	614	ZnS (wurtzite)	3602	3322
RbF	768	760	MnS	3351	3247
RbCl	678	662	MnSe	3305	3083
RbBr	649	626	ZnSe	3610	3150
RbI	613	590			

A useful application of the Kapustinskii equation is the prediction of the existence of previously unknown compounds. From Table 4.2.5, it is seen that all dihalides of the alkali metals with the exception of CsF_2 are unstable with respect to their formation from the elements. However, CsF_2 is unstable with respect to disproportionation: the enthalpy of the reaction $CsF_2 \rightarrow CsF + 1/2 F_2$ is -405 kJ mol^{-1}.

Table 4.2.5. Predicted enthalpies of formation (in kJ mol^{-1}) of some metal dihalides and oxides using the Kapustinskii equation

M	MF$_2$	MCl$_2$	MBr$_2$	MI$_2$	MO
Li	—	4439	4581	4740	4339
Na	1686	2146	2230	2427	2184
K	435	854	975	1117	1017
Rb	163	548	661	—	787
Cs	−126	213	318	473	515
Al	−774	−272	−146	8	−230
Cu	−531	−218	−142	−21	−155
Ag	−205	96	167	282	230

In cases where the lattice energy is known from the Born–Haber cycle, the Kapustinskii equation can be used to derive the ionic radii of complex anions such as SO_4^{2-} and PO_4^{3-}. The values determined in this way are known as thermochemical radii; some values are shown in Table 4.2.6.

Table 4.2.6. Thermochemical radii of polyatomic ions*

Ion	Radius (pm)	Ion	Radius (pm)	Ion	Radius (pm)	Ion	Radius (pm)
$AlCl_4^-$	295	$GeCl_6^{2-}$	328	O_2^-	158	$SnCl_6^{2-}$	349
BCl_4^-	310	GeF_6^{2-}	265	O_2^{2-}	173	$TiBr_6^{2-}$	352
BF_4^-	232	HCO_3^-	156	O_3^-	177	$TiCl_6^{2-}$	331
BH_4^-	193	HF_2^-	172	OH^-	133	TiF_6^{2-}	289
CN^-	191	HS^-	207	$PtCl_4^{2-}$	293	UCl_6^{2-}	337
CNS^-	213	HSe^-	205	$PtCl_6^{2-}$	313	VO_3^-	182
CO_3^{2-}	178	HfF_6^{2-}	271	PtF_6^{2-}	296	VO_4^{3-}	260
ClO_3^-	171	$MnCl_6^{2-}$	322	ReF_6^{2-}	277	WCl_6^{2-}	336
ClO_4^-	240	MnF_6^{2-}	256	RhF_6^{2-}	264	$ZnCl_4^{2-}$	286
CoF_6^{2-}	244	MnO_4^-	229	S_2^-	191	$ZrCl_6^{2-}$	358
CrF_6^{2-}	252	N_3^-	195	SO_4^{2-}	258	ZrF_6^{2-}	273
CrO_4^{2-}	256	NCO^-	203	$SbCl_6^-$	351		
$CuCl_4^{2-}$	321	NO_2^-	192	SeO_3^{2-}	239	Me_4N^+	201
$FeCl_4^-$	358	NO_3^-	179	SeO_4^{2-}	249	NH_4^+	137
$GaCl_4^-$	289	NbO_3^-	170	SiF_6^{2-}	259	PH_4^+	157

* Selected from: H. D. B. Jenkins and K. P. Thakur, *J. Chem. Educ.* **56**, 576–7 (1979).

4.2.3 Ionic liquids

An ionic liquid is a liquid that consists only of ions. However, this term includes an additional special definition to distinguish it from the classical definition of a molten salt. While a molten salt generally refers to the liquid state of a high-melting salt, such as molten sodium chloride, an ionic liquid exists at much lower temperatures (approx. < 100 °C). The most important reported ionic liquids are composed of the following cations and anions:

Cations:

R–N⁺(imidazolium ring)N–R'
imidazolium ion

R = R'= methyl (MMIM)
R = methyl, R'= ethyl (EMIM)
R = methyl, R'= n-butyl (BMIN)

pyridinium ion

ammonium ion: R'''(R')N⁺(R)R''

phosphonium ion: R'''(R')P⁺(R)R''

Anions: $AlCl_4^-$, $Al_2Cl_7^-$, BF_4^-, Cl^-, $AlEtCl_3^-$, PF_6^-, NO_3^-, $CF_3SO_3^-$

The development of ionic liquids has revealed that the ions possess the following structural features:

(1) The ionic radii have large values. The key criterion for the evaluation of an ionic liquid is its melting point. Comparison of the melting points of different chlorides salts, as listed in Table 4.2.7, illustrates the influence of the cation clearly: High melting points are characteristic for alkali metal chlorides, which have small cationic radii, whereas chlorides with suitable large organic cations melt at much lower temperatures. Besides the effect of the cation, an increasing size of anion with the same charge leads to a further decrease in melting point, as shown in Table 4.2.7.
(2) There are weak interactions between the ions in an ionic liquid, but the formation of hydrogen bonds is avoided as far as possible.
(3) In general, the cations and anions of ionic liquids are composed of organic and inorganic components, respectively. The solubility properties of an ionic liquid can be achieved by variation of the alkyl group on the cations and by choice of the counter-anions.

The physical and chemical properties of the ionic liquids that make them interesting as potential solvents and in other applications are listed below.

(1) They are good solvents for a wide range of inorganic and organic materials at low temperature, and usual combinations of reagents can be brought into the same phase. Ionic liquids represent a unique class of new reaction media for transition metal catalysis.
(2) They are often composed of poorly coordinating ions, so they serve as highly polar solvents that do not interfere with the intended chemical reactions.

Table 4.2.7. Melting point of some salts and ionic liquids

Salt/ionic liquid	mp (°C)	Salt/ionic liquid	mp (°C)
NaCl	801	[EMIM]NO₂	55
KCl	772	[EMIM]NO₃	38
[MMIM]Cl	125	[EMIM]AlCl₄	7
[EMIM]Cl	87	[EMIM]BF₄	6
[BMIM]Cl	65	[EMIM]CF₃SO₃	−9
[BMIM]CF₃SO₃	16	[EMIM]CF₃CO₂	−14

(3) They are immiscible with a number of organic solvents and provide a non-aqueous, polar alternative for two-phase systems. Hydrophobic ionic liquids can also be used as immiscible polar phases with water.
(4) The interacting forces between ions in the ionic liquids are the strong electrostatic Coulombic forces. Ionic liquids have no measurable vapor pressure, and hence they may be used in high-vacuum systems to overcome many contaminant problems. Advantage of their non-volatile nature can be taken to conduct product separation by distillation with prevention of uncontrolled evaporation.

4.3 Metallic bonding and band theory

Metallic structure and bonding are characterized by delocalized valence electrons, which are responsible for the high electrical conductivity of metals. This contrasts with ionic and covalent bonding in which the valence electrons are localized on particular atoms or ions and hence are not free to migrate through the solid. The physical data for some solid materials are shown in Table 4.3.1.

The band theory of solids has been developed to account for the electronic properties of materials. Two distinct lines of approach will be described.

4.3.1 Chemical approach based on molecular orbital theory

Table 4.3.1. Conductivities of some metals, semiconductors, and insulators

Category	Material	Conductivity (Ω^{-1} m^{-1})	Band gap (eV)
Metals	Copper	6.0×10^7	0
	Sodium	2.4×10^7	0
	Magnesium	2.2×10^7	0
	Aluminium	3.8×10^7	0
Zero-band gap semiconductor	Graphite	2×10^5	0
Semiconductors	Silicon	4×10^{-4}	1.11
	Germanium	2.2×10^{-4}	0.67
	Gallium arsenide	1.0×10^{-6}	1.42
Insulators	Diamond	1×10^{-14}	5.47
	Polythene	10^{-15}	

A piece of metal may be regarded as an infinite solid where all the atoms are arranged in a close-packed manner. If a large number of sodium atoms are

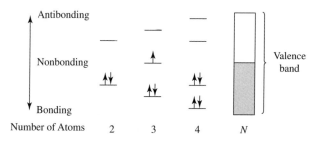

Fig. 4.3.1.
The evolution of a band of molecular orbitals as the number of contributing atoms such as Na increases. The shaded part in the band of N atoms indicates its containing electrons.

brought together, the interaction of the 3s valence orbitals will generate a very closely spaced set of molecular orbitals, as shown in Fig. 4.3.1. It is more appropriate to refer to such molecular orbitals as energy states since each is delocalized over all atoms in the metal, and the entire spectrum of energy states is known as a band.

The energy states in a band are not evenly distributed, and the band structure is better described by a plot of energy E against a density of states function $N(E)$ which represents the number of energy states lying between E and δE, as shown in Fig. 4.3.2.

The number of states within each band is equal to the total number of orbitals contributed by the atoms. For one mole of metal, a s band will consist of N_0 states and a 3p band of $3N_0$ states. These bands will remain distinct if the s–p separation is large, as shown in Fig. 4.3.3(a). However, if the s–p separation is small, the s and p bands overlap extensively and band mixing occurs. This situation, as illustrated in Fig. 4.3.3(b), applies to the alkaline-earth metals and main group metals.

The electronic properties of a solid are closely related to the band structure. The band that contains the electrons of highest energy is called the valence band, and the lowest unoccupied energy levels above them are called the conduction band.

There are four basic types of band structure as shown in Fig. 4.3.3. In (a) the valence band is only partially filled by electrons, and metals such as Na and Cu with a half-filled s band are well described by this diagram. The energy corresponding to the highest occupied state at 0 K is known as the Fermi energy E_F.

The alkaline-earth metals and group 12 metals (Zn, Cd, Hg) have the right number of electrons to completely fill an s band. However, s–p mixing occurs and the resulting combined band structure remains incompletely filled, as shown in Fig. 4.3.3(b). Hence these elements are also good metallic conductors.

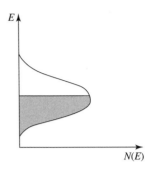

Fig. 4.3.2.
Density of states diagram for the 3s band of sodium.

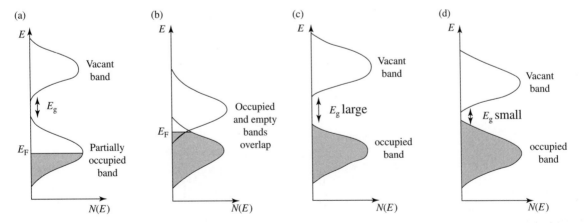

Fig. 4.3.3.
Schematic representation of band structure and the primary density of states situations: (a) metal will no overlapping bands, (b) metal with overlapping bands, (c) an insulator, and (d) a semiconductor.

Insulators have a completely filled valence band separated by a large gap from the empty conduction band, as illustrated in Fig. 4.3.3(c). Diamond is an excellent insulator with a band gap of 5.47 eV. Figure 4.3.3(d) shows a situation in which a fully occupied band is separated from a vacant band by a small band gap. This property characterizes a semiconductor, the subject matter of the next section.

4.3.2 Semiconductors

Intrinsic semiconductors have a band structure similar to that of insulators, except that the gap is smaller, usually in the range 0.5 to 3 eV. A few electrons may have sufficient thermal energy to be promoted to the empty conduction band. Each electron leaving the valence band not only creates a charge carrier in the conduction band but also leaves behind a hole in the valence band. This hole provides a vacancy which can promote the movement of electrons in the valence band.

The presence of an impurity such as an As or a Ga atom in silicon leads to an occupied level in the band gap just below the conduction band or a vacant level just above the valence band, respectively. Such materials are described as extrinsic semiconductors. The n-type semiconductors have extra electrons provided by donor levels, and the p-type semiconductors have extra holes originating from the acceptor levels. Band structures of the different types of semiconductors are shown in Fig. 4.3.4.

In solid-state electronic devices, a p–n junction is made by diffusing a dopant of one type into a semiconductor layer of the other type. Electrons migrate from the n-type region to the p-type region, forming a space charge region where there are no carriers. The unbalanced charge of the ionized impurities causes the bands to bend, as shown in Fig. 4.3.5, until a point is reached where the Fermi levels are equivalent.

A p–n junction can be used to rectify an electric current, that is, to make it much easier to pass in one direction than the other. The depletion of carriers for the junction effectively forms an insulating barrier. If positive potential is applied to the n-type side (a situation known as reverse bias), more carriers are removed and the barrier becomes wider. However, under forward bias the n-type is made more negative relative to the p side, so that the energy barrier is decreased and the carriers may flow through.

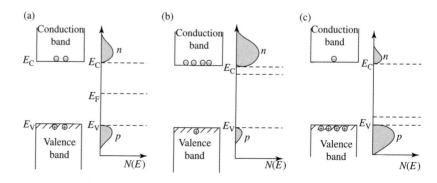

Fig. 4.3.4. Density of states diagrams for (a) an intrinsic semiconductor, (b) an n-type semiconductor, and (c) a p-type semiconductor.

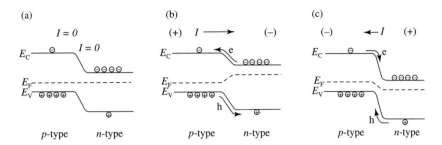

Fig. 4.3.5.
Schematic representation of a *p–n* junction: (a) No electric field, (b) positive electric field to *p*-region, and (c) positive electric field to *n*-region.

4.3.3 Variation of structure types of 4d and 5d transition metals

Calculated relative energies from the density of states curves including the d orbitals have been used to rationalize the changes in metallic structure observed across the 4d and 5d transition series, as summarized in Table 4.3.2. To understand the variation of these structures, we first determine the average electronic energy for a model transition metal with body-centered cubic packing (bcp), hexagonal closest packing (hcp), and cubic closest packing (ccp) structures. Then we plot the energy differences of these structures as a function of the number of d electrons. As shown in Fig. 4.3.6, the solid line indicates the energy of bcp minus the energy of ccp (bcp–ccp), and the broken line indicates the energy of hcp minus the energy of ccp (hcp–ccp).

Table 4.3.2. Crystal structure of 4d and 5d transition metals

4d element and number of d electrons	Structure	ΔE (hcp – ccp)	ΔE (bcp – ccp)	5d element and number of d electrons	Structure	ΔE (hcp – ccp)	ΔE (bcp – ccp)
Y (1)	hcp	−	+	La (1)	hexagonal	−	+
Zr (2)	hcp	−	+	Hf (2)	hcp	−	+
Nb (3)	bcp	+	−	Ta (3)	bcp	+	−
Mo (4)	bcp	+	−	W (4)	bcp	+	−
Tc (5)	hcp	∼0	−	Re (5)	hcp	∼0	−
Ru (6)	hcp	−	+	Os (6)	hcp	−	+
Rh (7)	ccp	−	+	Ir (7)	ccp	−	+
Pd (8)	ccp	∼0	+	Pt (8)	ccp	∼0	+
Ag (9)	ccp	+	−	Au (9)	ccp	+	−
Cd (10)	hcp	0	0	Hg (10)	trigonal	0	0

Studying the plots in Fig. 4.3.6, we can see that for d electron counts of 1 and 2, the preferred structure is hcp. For $n = 3$ or 4, the bcp structure is the most stable. These results are in agreement with the observation as listed in Table 4.3.2. For d electron counts of 5 or more, the hcp and ccp structures have comparable energies. However, the hcp structure is correctly predicted to be more stable for metals with six d electrons, and the ccp for later transition elements. These calculations show how the structures of metallic elements are determined by rather subtle differences in the density of states, which in turn are controlled by the different types of bonding interaction present.

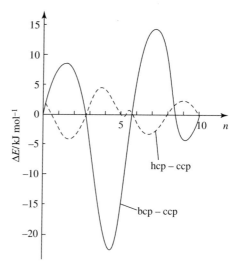

Fig. 4.3.6.
The computed relative energy of bcp – ccp (solid line) and hcp – ccp, (broken line) as a function of the number of d electrons.

4.3.4 Metallic radii

The metallic radius (r_{met}) is defined as one half of the internuclear distance between neighboring metal atoms in a metallic crystal, and is dependent upon the coordination number (CN). The relative metallic radius for different polymorphs of the same metal varies with CN. When CN decreases, r_{met} also decreases. The relative r_{met} values are estimated to be as follows:

Coordination number:	12	8	6	4
Relative radius:	1.00	0.97	0.96	0.88

For example, metallic barium has a bcp lattice with $a = 502.5$ pm (Table 10.3.2). From these results, the metallic radius of Ba atom can be calculated:

$$r_{met} \text{ (CN = 8)} = 217.6 \text{ pm},$$
$$r_{met} \text{ (CN = 12)} = 224.3 \text{ pm}.$$

The values of r_{met} listed in Table 4.3.3 refer to 12-coordinate metal centers. Since not all metals actually adopt structures with equivalent 12-coordinate atoms such as the ccp structure, the r_{met} values are estimated from the experimental values. For the hcp structure, there are two sets of six-coordinate atoms, which are not exactly equivalent to each other. In such circumstances, we usually adopt the average value of the two sets. Gallium possesses a complex structure in which each atom has one neighbor at 244 pm, two at 270 pm, two at 273 pm, and two at 279 pm, and its r_{met} can be estimated as 141 pm.

The metallic radii may be used to define the relative sizes of atoms, which has implications for their structural and chemical properties. For example, going down a given group in the Periodic Table, the metallic radii of the

Table 4.3.3. Metallic radii (in pm) of the elements

Li	Be											Al			
157	112											143			
Na	Mg														
191	160														
K	Ca	Sc	Ti	V	Cr	Mn	Fe	Co	Ni	Cu	Zn	Ga	Ge		
235	197	164	147	135	129	137	126	125	125	128	137	141	137		
Rb	Sr	Y	Zr	Nb	Mo	Tc	Ru	Rh	Pd	Ag	Cd	In	Sn	Sb	
250	215	182	160	147	140	135	134	134	137	144	152	167	158	169	
Cs	Ba	La	Hf	Ta	W	Re	Os	Ir	Pt	Au	Hg	Tl	Pb	Bi	
272	224	188	159	147	141	137	135	136	139	144	155	171	175	182	
Fr	Ra	Ac													
280	247	190													

	Ce	Pr	Nd	Pm	Sm	Eu	Gd	Tb	Dy	Ho	Er	Tm	Yb	Lu
	183	183	182	181	180	204	180	178	177	177	176	175	194	174
	Th	Pa	U	Np	Pu	Am	Cm	Bk	Cf					
	180	164	154	155	159	173	174	170	169					

elements increase, the first ionization energies generally decrease, and the electronegativities also decrease.

4.3.5 Melting and boiling points and standard enthalpies of atomization of the metallic elements

Table 4.3.4 lists the melting and boiling points and enthalpies of atomization ($\Delta H°_{at}$) of the metallic elements. These data show that, in general, these properties are closely correlated. The $\Delta H°_{at}$ values reflect the strength of the metal–metal bonds in their metallic structures. For example, the sixth row elements show an overall increase in $\Delta H°_{at}$ from Cs to W and then there is a marked decrease from W to Hg. Because of the relativistic effects on the sixth row elements (which have been discussed in section 2.4), the energies of 6s and 5d orbitals are very similar. Hence, for a metal containing N atoms, there are $6N$ valence orbitals. The band structure of the metal corresponds to an infinite set of delocalized molecular orbitals which extend throughout the structure. Half of these closely spaced orbitals are bonding ($3N$ orbitals) and half are antibonding ($3N$ orbitals). If the bands of molecular orbitals are filled in the aufbau fashion, the $\Delta H°_{at}$ values should increase from a low initial value for d^0s^1 (metal Cs) to a maximum at d^5s^1 (metal W), and a minimum for the completely filled shell $d^{10}s^2$ (metal Hg), where all the bonding and antibonding orbitals are filled and may be approximated as nonbonding. These features of the band structure of metals are in accord with the data listed in Table 4.3.4 and with the discussion given in section 2.4.

In the fourth row elements, the large exchange energies associated with the $3d^54s^1$ (Cr) and $3d^54s^2$ (Mn) configurations (both with either half-filled or completely filled orbitals) result in non-aufbau configurations for these elements when the metallic valence bands are populated. For these non-aufbau configurations, some electrons are not paired up, leading to magnetic properties and occupation of antibonding levels in the band. Consequently, there is a

Table 4.3.4. Melting points [mp (K), upper], boiling points [bp (K), middle], and standard enthalpies of atomization [ΔH°_{at} (kJ mol^{-1}), bottom] of metallic elements

Li	Be												
454	1551												
1620	3243												
135	309												
Na	**Mg**										**Al**		
371	922										934		
1156	1363										2740		
89	129										294		
K	**Ca**	**Sc**	**Ti**	**V**	**Cr**	**Mn**	**Fe**	**Co**	**Ni**	**Cu**	**Zn**	**Ga**	**Ge**
337	1112	1814	1933	2160	2130	1517	1808	1768	1726	1357	693	303	1211
1047	1757	3104	3560	3650	2945	2235	3023	3143	3005	2840	1180	2676	3103
78	150	305	429	459	349	220	351	382	372	305	115	256	334
Rb	**Sr**	**Y**	**Zr**	**Nb**	**Mo**	**Tc**	**Ru**	**Rh**	**Pd**	**Ag**	**Cd**	**In**	**Sn**	**Sb**
312	1042	1795	2125	2741	2890	2445	2583	2239	1825	1235	594	429	505	904
961	1657	3611	4650	5015	4885	5150	4173	4000	3413	2485	1038	2353	2543	1908
69	139	393	582	697	594	585	568	495	393	255	100	226	290	168
Cs	**Ba**	**La**	**Hf**	**Ta**	**W**	**Re**	**Os**	**Ir**	**Pt**	**Au**	**Hg**	**Tl**	**Pb**	**Bi**	**Po**
302	1002	1194	2503	3269	3680	3453	3327	2683	2045	1338	234	577	601	545	527
952	1910	3730	5470	5698	5930	5900	5300	4403	4100	3080	630	1730	2013	1883	1235
66	151	400	661	753	799	707	628	564	510	324	59	162	179	179	101
Fr	**Ra**	**Ac**													
300	973	1324													
—	—	—													
—	—	—													

	Ce	**Pr**	**Nd**	**Pm**	**Sm**	**Eu**	**Gd**	**Tb**	**Dy**	**Ho**	**Er**	**Tm**	**Yb**	**Lu**
	1072	1204	1294	1441	1350	1095	1586	1629	1685	1747	1802	1818	1097	1936
	3699	3785	3341	~3000	2064	1870	3539	3396	2968	2835	3136	2220	1466	3668
	314	333	284	—	192	176	312	391	293	251	293	247	159	428
Th	**Pa**	**U**	**Np**	**Pu**	**Am**	**Cm**	**Bk**	**Cf**	**Es**	**Fm**	**Md**	**No**	**Lr**	
2023	1845	1408	917	913	1444	1173	1323	1173	1133	1800	1100	1100	1900	
—	—	—	—	—	—	—	—	—	—	—	—	—	—	
—	—	—	—	—	—	—	—	—	—	—	—	—	—	

weakening of the metal–metal bond and Cr and Mn metals have the anomalously low ΔH°_{at} values listed in Table 4.3.4.

4.4 Van der Waals interactions

The collective term "intermolecular forces" refers to the interactions between molecules. These interactions are different from the those involved in covalent, ionic, and metallic bonds. Intermolecular interactions include forces between ions or charged groups, dipoles or induced dipoles, hydrogen bonds, hydrophobic interactions, $\pi \cdots \pi$ overlap, nonbonding repulsions, etc. The energies of these interactions (usually operative within a range of 0.3–0.5 nm) are generally below 10 kJ mol^{-1}, or one to two order of magnitude smaller than that of an ordinary covalent bond. These interactions are briefly described below.

The interaction between charge groups, such as –COO$^-\cdots\,^+$H$_3$N–, gives rise to an energy which is directly proportional to the charges on the groups and inversely proportional to the distance between them. The interactions among dipoles and induced dipoles are proportional to r^{-6}, where r is the separation between the interacting bodies. These interactions are collectively called van der Waals forces and they will be discussed in some detail later in this section. Hydrogen bonding is one of the most important intermolecular interactions and it will be described in some detail in section 11.2.

In protein molecules, hydrophobic side-chain groups such as phenylalanine, leucine, and isoleucine aggregate together, forming a hydrophobic zone in the interior of the molecule. The forces that hold these groups together are the hydrophobic interactions. These interactions include both energetic and entropy effects. It is of interest to note that these hydrophobic groups unite together not because they prefer each other's company, but because all of them are "repelled" by water.

The $\pi \cdots \pi$ overlap interaction is the force that holds together two or more aromatic rings that tend to be parallel to each other in molecular packing. The most well-known example is the interlayer interaction in graphite, where the distance separating layers measures 335 pm.

Finally, nonbonding repulsion exists in all types of groups. It is a short-range interacting force and is on the order of r^{-9} to r^{-12}.

4.4.1 Physical origins of van der Waals interactions

The interactions between molecules are repulsive at short range and attractive at long range. When the intermolecular separation is small, the electron clouds of two adjacent molecules overlap to a significant extent, and the Pauli exclusion principle prohibits some electrons from occupying the overlap region. As the electron density in this region is reduced, the positively charged nuclei of the molecules are incompletely shielded and hence repel each other.

The long-range attractive interaction between molecules, generally known as van der Waals interaction, becomes significant when the overlap of electron clouds is small. There are three possible contributions to this long-range interaction, depending on the nature of the interacting molecules. These physical origins are discussed below.

(1) Electrostatic contribution

A polar molecule such as HCl possesses a permanent dipole moment μ by virtue of the non-uniform electric charge distribution within the neutral molecule. The electrostatic energy $U_{\mu\mu}$ between two interacting dipoles μ_1 and μ_2 is strongly dependent on their relative orientation. If all relative orientations are equally probable and each orientation carries the Boltzmann weighting factor $e^{-U_{\mu\mu}/kT}$, the following expression is obtained:

$$<U_{\mu\mu}>_{el} = -\frac{2\mu_1^2\mu_2^2}{3(4\pi\epsilon_0)^2 kTr^6}. \qquad (4.4.1)$$

This expression shows that the attractive electrostatic interaction, commonly known as the Keesom energy, is inversely proportional to the sixth power of the intermolecular separation, and the temperature dependence arises from orientational averaging with Boltzmann weighting.

Some non-polar molecules such as CO_2 possess an electric quadrupole moment Q which can contribute to the electrostatic energy in a similar manner. The results for dipole–quadrupole and quadrupole–quadrupole interactions are

$$<U_{\mu Q}>_{el} = -\frac{\mu_1^2 Q_2^2 + \mu_2^2 Q_1^2}{(4\pi\epsilon_0)^2 kTr^8} \qquad (4.4.2)$$

$$<U_{QQ}>_{el} = -\frac{14 Q_1^2 Q_2^2}{5(4\pi\epsilon_0)^2 kTr^{10}}. \qquad (4.4.3)$$

(2) Induction contribution

When a polar molecule and a non-polar molecule approach each other, the electric field of the polar molecule distorts the electron charge distribution of the non-polar molecule and produces an induced dipole moment within it. The interaction of the permanent and induced dipoles then results in an attractive force. This induction contribution to the electrostatic energy is always present when two polar molecules interact with each other.

The average induction energy (called Debye energy) between a polar molecule with dipole moment μ and a non-polar molecule with polarizability α is

$$<U_{\mu\alpha}>_{ind} = -\frac{\mu^2 \alpha}{(4\pi\epsilon_0)^2 r^6}. \qquad (4.4.4)$$

Note that the induction energy, unlike the electrostatic energy, is not temperature dependent although both vary as r^{-6}.

For two interacting polar molecules, the induction energy averaged over all orientations is

$$<U_{\mu\mu}>_{ind} = -\frac{\mu_1^2 \alpha_2 + \mu_2^2 \alpha_1}{(4\pi\epsilon_0)^2 r^6}. \qquad (4.4.5)$$

(3) Dispersion contribution

The electrons in a molecule are in constant motion so that at any instant even a non-polar molecule such as H_2 possesses an instantaneous electric dipole which fluctuates continuously in time and orientation. The instantaneous dipole in one molecule induces an instantaneous dipole in a second molecule, and interaction of the two synchronized dipoles produces an attractive dispersion energy (known as London energy). In the case of interaction between two neutral, non-polar molecules the dispersion energy is the only contribution to the long-range energy.

When an instantaneous dipole is treated as a charge of $-q$ oscillating about a charge of $+q$ with frequency ν_0, the dispersion energy for two identical molecules in their ground states is

$$U_{disp} = -\frac{3\alpha^2 h\nu_0}{4(4\pi\epsilon_0)^2 r^6}. \qquad (4.4.6)$$

Chemical Bonding in Condensed Phases

For the most general case of the interaction between two polar molecules, the total long-range intermolecular energy is

$$U = U_{el} + U_{ind} + U_{disp}. \tag{4.4.7}$$

For two neutral polar molecules of the same kind that are free to rotate, the intermolecular energy is

$$U = -\frac{1}{(4\pi\epsilon_0)^2 r^6}\left[\frac{2\mu^4}{3kT} + 2\mu^2\alpha + \frac{3\alpha^2 h\nu_0}{4}\right]$$

$$= -(c_{el} + c_{ind} + c_{disp})r^{-6}. \tag{4.4.8}$$

Table 4.4.1 lists typical values of dipole moments and polarizabilities of some simple molecules and the three coefficients of the r^{-6} term at 300 K. Except for H_2O which is small and highly polar, the dispersion term dominates the long-range energy. The induction term is always the least significant.

Table 4.4.1. Comparison of contributions to long-range intermolecular energy for like pairs of molecules at 300 K

Molecule	$10^{30}\mu$ (C m*)	$10^{30}\alpha(4\pi\varepsilon_0)^{-1}$ (m³)	$10^{79}c_{el}$ (J m²)	$10^{79}c_{ind}$ (J m²)	$10^{79}c_{disp}$ (J m²)
Ar	0	1.63	0	0	50
Xe	0	4.0	0	0	209
CO	0.4	1.95	0.003	0.06	97
HCl	3.4	2.63	17	6	150
NH_3	4.7	2.26	64	9	133
H_2O	6.13	1.48	184	10	61

* Dipole moments are usually given in Debye units. In the rationalized SI system, the unit used is Coulomb meter (C m); 1 Debye = 3.33×10^{-30} C m.

The magnitudes of the three attractive components of the intermolecular energy per mole of gas of some simple molecules are compared in Table 4.4.2, along with the bond energy of the polyatomic species and the enthalpy of sublimation (ΔH_s°). The calculation is based on the assumption that the molecules interact in pairs at a separation δ where the total potential energy $U(\delta) = 0$.

Table 4.4.2. Intermolecular energies, bond energies and enthalpies of sublimation for some simple molecules

Molecule	δ (nm)	Attractive energy (kJ mol⁻¹)			Single-bond energy (kJ mol⁻¹)	ΔH_s° (kJ mol⁻¹)
		U_{el}	U_{ind}	U_{disp}		
Ar	0.33	0	0	−1.2	—	7.6
Xe	0.38	0	0	−1.9	—	16
CO	0.36	-4×10^{-5}	-8×10^{-4}	−1.4	343	6.9
HCl	0.37	−0.2	−0.07	−1.8	431	18
NH_3	0.260	−6.3	−0.9	−13.0	389	29
H_2O	0.265	−16.0	−0.9	−5.3	464	47

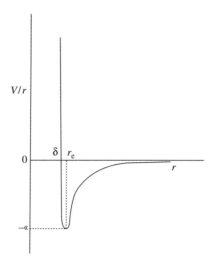

Fig. 4.4.1.
Lennard–Jones 12–6 potentials:
$V(r) = 0$ at $r = \delta$; $V(r) = -\varepsilon$ at $r = r_e = 2^{1/6}\delta$.

It must be emphasized that the above discussion of long-range attractive interactions is very much simplified. A rigorous treatment of the subject is obviously very complicated, or even impossible if the molecules of interest are large and have complex structures.

4.4.2 Intermolecular potentials and van der Waals radii

A widely used energy function V that describes intermolecular interaction as a function of intermolecular separation r is the Lennard–Jones 12–6 potential:

$$V(r) = 4\varepsilon[(\delta/r)^{12} - (\delta/r)^6]. \tag{4.4.9}$$

The resulting potential energy curve is shown in Fig. 4.4.1 in which r_e is the equilibrium distance.

Some values for the parameters ε (usually given as ε/k, where k is Boltzmann's constant) and δ are given in Table 4.4.3. For simple and slightly polar substances, ε and δ^3 are approximately proportional to the critical temperature T_C and the critical volume V_C, respectively.

Crystal structure data show that intermolecular nonbonded distances between pairs of atoms vary over a very narrow range. In the absence of hydrogen

Table 4.4.3. Parameters of the Lennard–Jones 12–6 potential function

Molecule	ε/k (K)	10δ (nm)	T_C (K)	V_C (cm^3 mol^{-1})
Ne	47.0	2.72	44.4	41.7
Ar	141.2	3.336	150.8	74.9
K	191.4	3.575	209.4	91.2
Xe	257.4	3.924	289.7	118
N_2	103.0	3.613	126.2	89.5
O_2	128.8	3.362	154.6	73.4
CO_2	246.1	3.753	304.2	94
CH_4	159.7	3.706	190.5	99

Table 4.4.4. van der Waals radii of atoms (pm)*,†

H 120										He 140
Li 182						C 170	N 155	O 152	F 147	Ne 154
Na 227	Mg 173					Si 210	P 180	S 180	Cl 175	Ar 188
K 275		Ni 163	Cu 143	Zn 139	Ga 187	Ge 219	As 185	Se 190	Br 185	Kr 202
		Pd 163	Ag 172	Cd 162	In 193	Sn 217		Te 206	I 198	Xe 216
	U 186	Pt 175	Au 166	Hg 170	Tl 196	Pb 202				CH$_3$ 200

* For the methyl group, 200 pm; half thickness of phenyl group, 185 pm.
† Taken from A. Bondi, *J. Phys. Chem.* **68**, 441–451 (1964); **70**, 3006–3007 (1966).

bonding and donor–acceptor bonding, the contact between C, N, and O atoms is around 370 pm for a wide variety of compounds. This observation leads to the concept of van der Waals radii (r_{vdw}) for atoms, which can be used to calculate average minimum contacts between atoms belonging to neighboring molecules in a condensed phase. The r_{vdw} values of some common elements are show in Table 4.4.4. The van der Waal radius of an atom is close to the δ parameter of the Lennard–Jones 12–6 potential function, and it can also be correlated with the size of the outermost occupied atomic orbital. For instance, a carbon 2p orbital enclosing 99% of its electron density extends from the nucleus to about 190 pm, as compared to $r_{vdw} = 170$ pm for carbon.

References

1. J. Emsley, *The Elements*, 3rd edn., Clarendon Press, Oxford, 1998.
2. D. M. P. Mingos, *Essential Trends in Inorganic Chemistry*, Oxford University Press, Oxford, 1998.
3. M. Ladd, *Chemical Bonding in Solids and Fluids*, Ellis Horwood, Chichester, 1994.
4. W. E. Dasent, *Inorganic Energetics*, 2nd edn., Cambridge University Press, Cambridge, 1982.
5. D. Pettifor, *Bonding and Structures of Molecules and Solids*, Clarendon Press, Oxford, 1995.
6. P. A. Cox, *The Electronic Structure and Chemistry of Solids*, Oxford University Press, Oxford, 1987.
7. H. Ibach and H. Lüth, *Solid State Physics: An Introduction to Principles of Materials Science*, 2nd edn., Springer, Berlin, 1995.
8. W. Gans and J. C. A. Boeyens (eds.), *Intermolecular Interactions*, Plenum, New York, 1998.
9. J. I. Gersten and F. W. Smith, *The Physics and Chemistry of Materials*, Wiley, New York, 2001.
10. T. Welton, Room-temperature ionic liquids. Solvents for synthesis and catalysis. *Chem. Rev.* **99**, 2071–83 (1999).
11. P. Wasserscheid and W. Keim, Ionic liquids—new 'solutions' for transition metal catalysis. *Angew. Chem. Int. Ed.* **39**, 3772–89 (2000).

5 Computational Chemistry

5.1 Introduction

Traditionally chemistry is an experimental science and, for a long time, mathematics played only a very minor role. For many decades, in order to justify doing computational chemistry, it was almost obligatory to start a talk or preface a book on theoretical chemistry by quoting P. A. M. Dirac (1902–84), one of the greatest physicists of the last century:

The underlying physical laws necessary for the mathematical theory of a large part of physics and the whole of chemistry are thus completely known, and the difficulty is only that the exact application of these laws leads to equations much too complicated to be soluble.

Dirac made this statement in 1929, when he was only 27 years old. He was at the University of Cambridge where he was appointed to the chair once occupied by Sir Isaac Newton. In a way, this statement by Dirac reflects a good news – bad news situation: The good news is that we know how to do it in theory, and the bad news is that we cannot do it in practice!

The physical laws and the mathematical theory Dirac referred to in the aforementioned quote were, of course, the essence of quantum theory, which is briefly described in Chapter 1. As we recall, when we treat a chemical problem computationally, we are usually confronted by the task of solving the Schrödinger equation

$$\hat{H}\psi = E\psi,$$

where $\hat{H}$, the Hamiltonian operator, is a mathematical entity characteristic of the system under study. In other words, in this equation, $\hat{H}$ is a known quantity, while ψ and E are the unknowns to be determined. Once we get E, we know the electronic energy of the system; once we get ψ, we have an idea of how the electrons are distributed in a molecule. From these solutions, useful information regarding the system under study, such as structure, energetics, and reactivity, may be obtained, or at least inferred.

It is important that the energy E of the system is one of the solutions of the Schrödinger equation. As we mentioned in Chapter 3, atoms and molecules are rather "simple-minded" species. Their behavior is entirely dictated by the energy factor. For instance, two atoms will combine to form a molecule if the formation leads to a lowering in energy. Also, a reaction will proceed spontaneously if again there is a lowering in energy. If a reaction does not take place spontaneously, we can still "kick-start" it by supplying energy to it. Hence, energy is the all important property if we wish to understand or predict the behavior of a molecular species.

As early as the 1930s, physicists and chemists knew that solving the Schrödinger equation held the key to understanding the nature of bonding and the reactivity of molecules. However, the task remained to be mathematically daunting for more than half of a century. Indeed, at the beginning, drastic approximations (such as those used in the Hückel molecular orbital theory introduced in Chapter 3) were made in order to make the problem solvable. Consequently, the results obtained in such a manner were at best qualitative. After decades of work on the development of theoretical models as well as the related mathematics, along with the concomitant increase in computing power made possible by faster processors and enormous data storage capacity, chemists were able to employ calculations to solve meaningful chemical problems in the latter part of the last century. At the present stage, in many cases we can obtain computational results almost as accurate as those from experiments, and such findings are clearly useful to experimentalists. As mentioned in Chapter 1, an indication that theoretical calculations in chemistry have been receiving increasing attention in the scientific community was the award of the Nobel Prize to W. Kohn and J. A. Pople for their contributions to quantum chemistry. It is of interest to note that, in his Nobel lecture, Pople made reference to the Dirac statement mentioned earlier.

The title of this chapter is clearly too ambitious: it can easily be the title of a series of monographs. In fact, the aim of this chapter is confined to introducing the kind of results that may be obtained from calculations and in what way calculations can complement experimental research. Hence, the discussion is essentially qualitative, with practically all the theoretical and mathematical details omitted.

5.2 Semi-empirical and *ab initio* methods

For a molecule composed of n atoms, there should be $3n-6$ independent structural parameters (bond lengths, bond angles, and dihedral angles). This number may be reduced somewhat if symmetry constraint is imposed on the system. For instance, for H_2O, there are three structural parameters (two bond lengths and one bond angle). But if C_{2v} symmetry (point groups will be introduced in Chapter 6) is assumed for the molecule, which is entirely reasonable, there will only be two unique parameters (one bond length and one bond angle). The electronic energy of the molecule is then a function of these two parameters; when these parameters take on the values of the equilibrium bond length and bond angle, the electronic energy is at its minimum. There are many cases where symmetry reduces the number of structural parameters drastically. Take benzene as an example. For a 12-atom molecule with no symmetry, there will be 30 parameters. On the other hand, if we assume the six C–H units form a regular hexagon, then there will be only two independent parameters, the C–C and C–H bond lengths. The concepts of symmetry will be discussed in the next chapter.

There are two general approaches to solving the Schrödinger equation of a molecular system: semi-empirical and *ab initio* methods. The semi-empirical methods assume an approximate Hamiltonian operator and the calculations are further simplified by approximating integrals with various experimental data such as ionization energies, electronic spectral transition energies, and bond energies. An example of these methods is the Hückel molecular orbital theory described in Chapter 3. When applied to conjugated polyenes, this method uses a one-electron Hamiltonian and treats the Coulomb integral α and resonance integral β as adjustable parameters.

On the other hand, *ab initio* (meaning "from the beginning" in Latin) methods use a "correct" Hamiltonian operator, which includes kinetic energy of the electrons, attractions between electrons and nuclei, and repulsions between electrons and those between nuclei, to calculate all integrals without making use of any experimental data other than the values of the fundamental constants. An example of these methods is the self-consistent field (SCF) method first introduced by D. R. Hartree and V. Fock in the 1920s. This method was briefly described in Chapter 2, in connection with the atomic structure calculations. Before proceeding further, it should be mentioned that *ab initio* does not mean "exact" or "totally correct." This is because, as we have seen in the SCF treatment, approximations are still made in *ab initio* methods.

Another approach closely related to the *ab initio* methods that has gained increasing prominence in recent years is the density functional theory (DFT). This method bypasses the determination of the wavefunction ψ. Instead, it determines the molecular electronic probability density ρ directly and then calculates the energy of the system from ρ.

Ab initio and DFT calculations are now routinely applied to study molecules of increasing complexity. Sometimes the results of these calculations are valuable in their own right. But more often than not these results serve as a guide or as a complement to experimental work. Clearly computational chemistry is revolutionizing how chemistry is done. In the remaining part of this chapter, the discussion will be devoted to *ab initio* and DFT methods.

An *ab initio* calculation is defined by two "parameters": the (atomic) basis functions (or basis sets) employed and the level of electron correlation adopted. These two topics will be described in some detail (and qualitatively) in the next two sections.

5.3 Basis sets

The basis sets used in *ab initio* calculations are composed of atomic functions. Pople and co-workers have devised a notation for various basis sets, which will be used in the following discussion.

5.3.1 Minimal basis set

A minimal basis set consists of just enough functions required to accommodate all the filled orbitals in an atom. Thus, for hydrogen and helium, there is only one s-type function; for elements lithium to neon, this basis set has 1s, 2s, and

2p functions (making a total of five); and so on. The most popular minimal basis set is called STO-3G, where three Gaussian functions (of the form $r^{n-1}e^{-\alpha r^2}$) are used to represent one Slater-type orbital (STO, which has the general form of $r^{n-1}e^{-\alpha r}$).

The STO-3G basis set does surprisingly well in predicting molecular geometries, even though it has been found that the success is partly due to a fortuitous cancellation of errors. On the other hand, not surprisingly, the energetics results obtained using minimal basis sets are not very good.

5.3.2 Double zeta and split valence basis sets

In order to have a better basis set, we can replace each STO of a minimal basis set by two STOs with different orbital exponent ζ (zeta). This is known as a double zeta basis set. In this basis, there is a linear combination of a "contracted" function (with a larger zeta) and a "diffuse" function (with a smaller zeta) and the coefficients of these combinations are optimized by the SCF procedure. Using H_2O as an example, a double zeta set has two 1s STOs on each H, two 1s STOs, two 2s STOs, two $2p_x$, two $2p_y$ and two $2p_z$ STOs on oxygen, making a total of 14 basis functions.

If only valence orbitals are described by double zeta basis, while the inner shell (or core) orbitals retain their minimal basis character, a split valence basis set is obtained. In the early days of computational chemistry, the 3-21G basis was fairly popular. In this basis set, the core orbitals are described by three Gaussian functions. The valence electrons are also described by three Gaussians: the inner part by two Gaussians and the outer part by one Gaussian. More recently, the popularity of this basis set is overtaken by the 6-31G set, where the core orbitals are a contraction of six Gaussians, the inner part of the valence orbitals is a contraction of three Gaussians, and the outer part is represented by one Gaussian.

Similarly, the basis set can be further improved if three STOs (with three different zetas, of course) are used to describe each orbital in an atom. Such a basis is called a triple zeta set. Correspondingly, the 6-311G set is a triple-split valence basis, with the core orbitals still described by six Gaussians and the valence orbitals split into three functions described by three, one, and one Gaussians, respectively.

5.3.3 Polarization functions and diffuse functions

The aforementioned split valence (or double zeta) basis sets can be further improved if polarization functions are added to the mix. The polarization functions have a higher angular momentum number ℓ; so they correspond to p orbitals for hydrogen and helium and d orbitals for elements lithium to neon, etc. So if we add d orbitals to the split valence 6-31G set of a non-hydrogen element, the basis now becomes 6-31G(d). If we also include p orbitals to the hydrogens of the 6-31G(d) set, it is then called 6-31G(d,p).

Why may the inclusion of polarization functions improve the quality of the basis set? Let us take the hydrogen atom, which has a symmetrical electronic distribution, as an example. When a hydrogen atom is bonded to another atom,

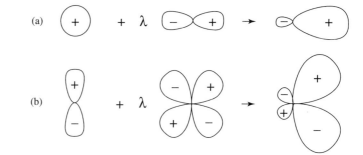

Fig. 5.3.1.
The addition of a polarization functions yields a distorted, but more flexible, orbital.

its electronic distribution will be attracted toward the other nucleus. Such a distortion is similar to mixing a certain amount of p orbital to the hydrogen's 1s orbital, yielding a kind of sp hybrid, as illustrated in Fig. 5.3.1(a). For a p orbital of a carbon atom, mixing in some amount of d orbital will also cause a distortion (or polarization) of its electronic distribution, as shown in Fig. 5.3.1(b). It is not difficult to see that the addition of polarization functions to a basis set improves its flexibility to describe certain bonding situations. For example, the 6-31G(d,p) set is found to describe the bonding in systems with bridging hydrogen atoms particularly well.

The basis sets discussed so far are suitable for systems where electrons are tightly bound. Hence they are not adequate to describe anions where the extra electron is usually quite loosely held. Indeed, even the very large basis sets such as 6-311G(d,p) do not have functions that have significant amplitude far away from the nucleus. To remedy this situation, diffuse functions are added to the basis sets. The additional functions are described as diffuse because they have fairly small (on the order of 10^{-2}) zeta values. If we add diffuse functions to the non-hydrogen atoms of the 6-31G(d,p) set, the basis now becomes 6-31+G(d,p). If we add diffuse functions to the hydrogens of the 6-31+G(d,p) set, it is now called 6-31++G(d,p).

With the addition of polarization functions and/or diffuse functions to the basis sets, the Pople notation can become rather cumbersome. For example, the 6-311++G(3df,2pd) set has a single zeta core and triple zeta valence shell, diffuse functions for all the atoms. Regarding polarized functions, there are three sets of d functions and one set of f functions on the non-hydrogens and two sets of p functions and one set of d orbitals on the hydrogens.

Before leaving this topic, it should be mentioned that, in addition to the (Pople) basis sets discussed so far, there are others as well. The more popular ones include the Dunning–Huzinaga basis sets, correlation consistent basis sets, etc. These functions will not be described here.

5.4 Electron correlation

The SCF method, or the Hartree–Fock (HF) theory, assumes that each electron in a molecule moves in an average potential generated by the nuclei and the remaining electrons. This assumption is flawed in that it ignores electron correlation. What is electron correlation? Simply put, and as introduced in our discussion in the Introduction, it is the tendency for electrons to avoid each other in an atom or in a molecule. Electrons do not like each other and wish to stay as far apart as possible in order to lower the energy of the system. Since the HF theory neglects electron correlation, correlation energy is defined as the difference between the Hartree–Fock energy and the exact energy.

There are various ways to account for the electron correlation effects. In the following we will classify these methods into three groups: configuration interaction, perturbation, and coupled cluster methods.

5.4.1 Configuration interaction

In our treatment on the hydrogen molecule, we mentioned that the ground state wavefunction can be improved if the ground configuration σ_{1s}^2 function is admixed with a certain amount of the excited configuration σ_{1s}^{*2} wavefunction. Such a method is called configuration interaction (CI), as these two configurations interact with each other (to lower the energy of the system). Note that another excited configuration, $\sigma_{1s}^1 \sigma_{1s}^{*1}$, cannot take part in the interaction with the two aforementioned configurations. This is because $\sigma_{1s}^1 \sigma_{1s}^{*1}$ does not have the proper symmetry properties, even though its energy is lower than that of σ_{1s}^{*2}.

Clearly more than one excited configuration can take part in CI. A full CI is the most complete treatment possible within the limitation of the basis set chosen. In most cases, a full CI is prohibitively expensive to carry out and we need to limit the excited states considered. If we consider only those wavefunctions that differ from the HF ground state wavefunction by one single spin orbital, we are said to be doing a configuration interaction singles (CIS) calculation. Similarly, there are those calculations that involve double substitutions (configuration interaction doubles, CID). Sometimes even a full CIS or CID calculation can be very difficult. In that case, we can restrict the unoccupied molecular orbitals taking part in the interaction. If we combine CIS and CID, we will then have configuration interaction singles and doubles (CISD). Still higher levels of CI treatments include CISDT and CISDTQ methods, where T and Q denote triple and quadruple excitations, respectively. At this point, it is convenient to introduce the following notation: a CISD calculation with the 6-31G(d) basis set is simply written as CISD/6-31G(d). Other types of calculations using different kinds of basis functions may be written in a similar manner. Furthermore, the longer notation CISD/6-311++G(d,p)//HF/6-31G(d) signifies an energy calculation at the CISD level with the 6-311++G(d,p) basis set using a molecular geometry optimized at the HF level with the 6-31G(d) basis. In this example, the computationally expensive geometry optimization is done at the lower level of theory of HF/6-31G(d), while a "single-point" calculation at the higher level of theory of CISD/6-311++G(d,p) is carried out in order to obtain a more reliable energetic value.

In the CI methods mentioned so far, only the mixing coefficients of the excited configurations are optimized in the variational calculations. If we optimize both the coefficients of the configurations and those of the basis functions, the method is called MCSCF, which stands for multiconfiguration self-consistent field calculation. One popular MCSCF technique is the complete active-space

SCF (CASSCF) method, which divides all the molecular orbitals into three sets: those doubly occupied orbitals which do not take part in the CI calculations, those vacant orbitals which also do not participate in the CI exercise, and those occupied orbitals and vacant orbitals that form the "active space." The list of configurations that take part in the CI calculation can be generated by considering all possible substitutions of the active electrons among the active orbitals.

Finally, it should be stressed again that CI calculations are variational; i.e., the energy obtained cannot be lower than the exact energy, as stipulated by the variational principle.

5.4.2 Perturbation methods

Perturbation theory is a standard approximation method used in quantum mechanics. Early in 1930s, C. Møller and M. S. Plesset (MP) applied this method to treat the electron correlation problem. They found that, using the HF wavefunction as the zeroth-order wavefunction, the sum of the zeroth-order and first-order energies correspond to the HF energy. So the electron correlation energy is the sum of the second-order, third-order, fourth-order, etc., corrections. Even though these methods have long been available in the literature, the second-order (MP2) energy correction was not routinely calculated until the 1980s. Currently, even third-order (MP3) and fourth-order (MP4) corrections have become very common.

Because of its computational efficiency and good results for molecular properties, notably structural parameters, the MP2 level of theory is one of the most popular methods to include correlation effects on computed molecular properties. The other widely applied method is the density functional method, which will be introduced later.

Unlike the CI methods, which are variational, the MP corrections are perturbational. That is, they can in practice lead to an energy that is lower than the exact energy. In addition, an analysis of the trends in the MP2, MP3, and MP4 energies for many systems indicates that the convergence of perturbation theory is slow or even oscillatory.

5.4.3 Coupled-cluster and quadratic configuration interaction methods

Another way to improve the HF description is the coupled-cluster (CC) approach, where the CC wavefunction ψ_{cc} is written as an exponential of a cluster operator $\hat{T}$ working on the HF wavefunction ψ_0

$$\psi_{cc} = e^{\hat{T}}\psi_0 = (\hat{T}_1 + \hat{T}_2 + \hat{T}_3 + \ldots)\psi_0, \qquad (5.4.1)$$

where $\hat{T}_1$ creates single excitations, $\hat{T}_2$ creates double excitations, and so on. Experience shows that the contribution of $\hat{T}_i$ decreases rapidly after $i = 2$. Hence when a CC calculation ends after $i = 2$, we have the coupled-cluster singles and doubles (CCSD) method. If we go further to include the $\hat{T}_3$ terms, we then have the CCSDT method, where all the triples are also involved in the calculation. Note that the CC methods introduced here are not variational.

Currently the full CCSDT model is far too expensive for routine calculations. To save time, we first carry out a CCSD calculation, which is then followed by a computation of a perturbative estimate of the triple excitations. Such an approximate method is called CCSD(T).

In the 1980s, Pople and co-workers developed the non-variational quadratic configuration interaction (QCI) method, which is intermediate between CC and CI methods. Similar to the CC methods, QCI also has the corresponding QCISD and QCISD(T) options. Both the CCSD(T) and QCISD(T) have been rated as the most reliable among the currently computationally affordable methods.

5.5 Density functional theory

As mentioned earlier, density functional theory (DFT) does not yield the wavefunction directly. Instead it first determines the probability density ρ and calculates the energy of the system in terms of ρ. Why is it called density functional theory and what is a functional anyway? We can define functional by means of an example. The variational integral $E(\varphi) = \int \varphi \hat{H} \varphi d\tau / \int \varphi \varphi d\tau$ is a functional of the trial wavefunction φ and it yields a number (with energy unit) for a given φ. In other words, a functional is a function of a function. So, in the DFT theory, the energy of the system is a functional of electron density ρ, which itself is a function of electronic coordinates.

In the 1960s, Kohn and co-workers showed that for molecules with a non-degenerate ground state, the energy and other electronic properties can be determined in terms of the electronic probability density ρ of the ground state. It was later proved this is also true for systems with degenerate ground states. Soon afterwards, physicists applied the first version of DFT (something called local spin density approximation, LSDA) to investigate the electronic structure of solids and it quickly became a popular method for studying solids. However, chemists were slow to pick up this method, probably because of numerical difficulties. In the early 1980s, after the numerical difficulties were essentially resolved, DFT LSDA calculations were applied to molecules and good results were obtained for molecular species, especially for structural parameters. After about ten years of advancement on this theory, DFT methods were added to the popular software packages such as *Gaussian*, and since then DFT calculations have experienced an explosive growth.

As in the case of *ab initio* calculations, which range from the rather crude HF method to the very sophisticated CCSD(T) or QCISD(T) method, DFT also offers a variety of functionals for users to opt for. More commonly used DFT levels of theory include local exchange and correlation functional SVWN (a synonym for LSDA in the *Gaussian* package) and its variant SVWN5, as well as gradient-corrected functional BLYP and hybrid functionals B3LYP and B3PW91. Among these functionals, B3LYP appears to be the most popular one. In addition, it is noted that DFT methods also make use of the kind of basis sets that have been employed in *ab initio* calculations. So a typical DFT calculation has a notation such as B3LYP/6-31G(d).

One important reason for DFT's ever-increasing popularity is that even the most elementary calculation includes correlation effects to a certain extent but takes roughly the same amount of time as a HF calculation, which does not take correlation into account. Indeed, there are some DFT advocates who believe DFT will replace HF as well as those correlation methods based on the HF wavefunctions (such as MP, CC, and CI), but whether this will come to fruition remains to be seen.

5.6 Performance of theoretical methods

How good are the *ab initio* and DFT methods in providing an adequate treatment of a molecule of interest? Clearly, a proper answer to this and related questions requires at least a major review article, if not a monograph (or maybe even a series of monographs). Still, it is useful to have a clear idea of the applicability and limitation of each chosen method. In this section, we concentrate on the performance of various theoretical methods in yielding the geometric parameters (bond lengths, bond angles, and dihedral angles) as well as the vibrational frequencies of a given molecule. In the next section, we shall discuss various composite methods, i.e., those that approximate the energy of a very high level of theory with a combination of lower level energetic calculations. In addition, the (energetic) performance of these methods will be assessed.

The structural parameters and vibrational frequencies of three selected examples, namely, H_2O, O_2F_2, and B_2H_6, are summarized in Tables 5.6.1 to 5.6.3, respectively. Experimental results are also included for easy comparison. In each table, the structural parameters are optimized at ten theoretical levels, ranging from the fairly routine HF/6-31G(d) to the relatively sophisticated QCISD(T)/6-31G(d). In passing, it is noted that, in the last six correlation methods employed, CISD(FC), CCSD(FC),…, QCISD(T)(FC), "FC" denotes the "frozen core" approximation. In this approximation, only the correlation energy associated with the valence electrons is calculated. In other words, excitations out of the inner shell (core) orbitals of the molecule are not considered. The basis of this approximation is that the most significant chemical changes occur in the valence orbitals and the core orbitals remain essentially intact. On

Table 5.6.1. Structural parameters (in pm and degrees) and vibrational frequencies (in cm^{-1}) of H_2O calculated at various *ab initio* and DFT levels

Level of theory	O–H	H–O–H	ν_1 (A_1)	ν_2 (A_1)	ν_3 (B_2)
HF/6-31G(d)	94.7	105.5	1827	4070	4189
MP2(Full)/6-31G(d)	96.9	104.0	1736	3776	3917
MP2(Full)/6-311+G(d,p)	95.9	103.5	1628	3890	4009
B3LYP/6-31G(d)	96.9	103.6	1713	3727	3849
B3LYP/6-311+G(d,p)	96.2	105.1	1603	3817	3922
CISD(FC)/6-31G(d)	96.6	104.2	1756	3807	3926
CCSD(FC)/6-31G(d)	97.0	104.0	1746	3752	3878
QCISD(FC)/6-31G(d)	97.0	104.0	1746	3749	3875
MP4SDTQ(FC)/6-31G(d)	97.0	103.8	1743	3738	3869
CCSD(T)(FC)/6-31G(d)	97.1	103.8	1742	3725	3854
QCISD(T)(FC)/6-31G(d)	97.1	103.8	1742	3724	3853
Experimental	95.8	104.5	1595	3652	3756

Table 5.6.2. Structural parameters (in pm and degrees) and vibrational frequencies (in cm^{-1}) of O_2F_2 calculated at various *ab initio* and DFT levels

Level of theory	O–O	O–F	F–O–O	F–O–O–F	Vibrational frequencies					
HF/6-31G(d)	131.1	136.7	105.8	84.2	210	556	709	1135	1145	1161
MP2(Full)/6-31G(d)	129.1	149.5	106.9	85.9	210	456	566	657	782	1011
MP2(Full)/6-311+F(d,p)	112.9	185.0	114.7	90.2	112	192	354	496	567	2032
B3LYP/6-31G(d)	126.6	149.7	108.3	86.7	215	481	575	732	787	1125
B3LYP/6-311G+(d,p)	121.6	155.1	109.8	88.4	217	443	528	682	691	1233
CISD(FC)/6-31G(d)	131.8	141.5	105.8	84.8	209	522	647	952	992	1080
CCSD(FC)/6-31G(d)	131.4	147.2	106.2	85.7	202	481	583	761	837	1009
QCISD(FC)/6-31G(d)	127.5	154.1	107.6	86.7	200	475	578	751	823	1006
MP4SDTQ(FC)/6-31G(d)	131.3	147.6	106.4	85.6	196	391	518	599	693	997
CCSD(T)(FC)/6-31G(d)	127.2	154.7	107.8	86.7	190	392	514	616	690	1065
QCISD(T)(FC)/6-31G(d)	128.8	152.8	107.2	86.7	189	386	510	617	688	1074
Experimental	121.7	157.5	109.5	87.5	202	360	466	614	630	1210

the other hand, in one of the methods listed in these tables, MP2(Full), "Full" indicates that all orbitals, both valence and core, are included in the correlation calculation.

When these tables are examined, it is found that, for structural parameters, most theoretical methods yield fair to excellent results when compared to the experimental data. Also, in general (though not always true), the agreement between calculated and experimental values improves gradually as we go to more and more sophisticated methods. On the other hand, it is seen that, for O_2F_2, one level of theory, MP2(Full)/6-311+G(d,p), gives highly unsatisfactory results. In particular, the calculated O–F bond (185.0 pm) is about 28 pm too long, while the O–O bond (112.9 pm) is about 10 pm too short, giving rise to an erroneous bonding description of $F^{\delta+}\ldots{}^{\delta-}O–O^{\delta-}\ldots{}^{\delta+}F$. Since there is no way to tell beforehand which method will lead to unacceptable results for a particular chemical species, it is often advantageous to carry out a series of calculations to make certain that convergence is attained. In Tables 5.6.1 to 5.6.3, the last five levels of theory give essentially the same results for each molecule, which lends credence to validity of the calculations. In summary, the calculated bond lengths can be within ±2 pm of the experimental value, while the corresponding accuracy for bond angles and dihedral angles is about ±2°. For some molecules, their energies may not be very sensitive to the change of some dihedral angles. In such cases, the calculated dihedral angles may be less reliable.

Let us now turn our attention to the calculated vibrational frequencies of H_2O, O_2F_2, and B_2H_6. First of all, it should be mentioned that the calculation of these frequencies is a computationally "expensive" task. As a result, high-level calculations of vibrational frequencies are performed only for relatively small systems. When the calculated frequencies are examined and compared with experimental data, it is found that the former are often larger than the latter. Indeed, after an extensive comparison between calculation and experiment, researchers have arrived at a scaling factor of 0.8929 for the HF/6-31G(d) frequencies. In other words, vibrational frequencies calculated at this level are

Table 5.6.3. Structural parameters (in pm and degrees) and vibrational frequencies (in cm^{-1}) of B_2H_6 calculated at various *ab initio* and DFT levels

Level of theory	B-H$_t$*	B-H$_b$†	H$_t$-B-H$_t$	H$_b$-B-H$_b$	Vibrational frequencies								
HF/6-31G(d)	118.5	131.5	122.1	95.0	409	828	897	900	999	1071	1126	1193	1286
					1303	1834	1933	2067	2301	2735	2753	2828	2843
MP2(Full)/6-31G(d)	118.9	130.9	121.7	96.2	363	842	878	907	970	999	1017	1083	1240
					1248	1817	1970	2112	2277	2693	2707	2793	2805
MP2(Full)/6-311+F(d,p)	118.7	131.5	122.3	95.7	360	826	869	902	955	981	1010	1081	1218
					1230	1766	1932	2039	2214	2643	2659	2741	2755
B3LYP/6-31G(d)	119.1	131.7	121.9	95.6	354	799	851	889	946	977	1000	1054	1206
					1211	1732	1864	2022	2205	2640	2653	2732	2745
B3LYP/6-311G+(d,p)	118.6	131.6	121.9	95.8	358	797	849	894	936	949	992	1022	1190
					1199	1701	1845	1980	2166	2598	2611	2686	2701
CISD(FC)/6-31G(d)	119.1	131.2	121.6	96.0	377	841	872	898	968	1018	1019	1094	1235
					1245	1815	1945	2092	2268	2678	2694	2774	2786
CCSD(FC)/6-31G(d)	119.4	131.3	121.6	96.1	369	836	862	886	958	1001	1003	1073	1220
					1228	1805	1933	2080	2249	2649	2663	2744	2756
QCISD(FC)/6-31G(d)	119.4	131.4	121.6	96.1	369	835	861	885	957	1000	1001	1072	1219
					1227	1801	1929	2076	2246	2646	2661	2742	2753
MP4SDTQ(FC)/6-31G(d)	119.4	131.5	121.7	96.1	362	833	862	885	957	987	999	1063	1218
					1225	1798	1935	2077	2242	2650	2664	2748	2759
CCSD(T)(FC)/6-31G(d)	119.5	131.5	121.6	96.1	364	830	858	880	953	991	997	1062	1214
					1222	1796	1925	2070	2237	2638	2653	2734	2746
QCISD(T)(FC)/6-31G(d)	119.5	131.5	121.6	96.1	363	830	857	880	953	990	997	1062	1214
					1222	1795	1924	2069	2236	2637	2652	2733	2745
Experimental	119.4	132.7	121.7	96.4	368	794	833	850	915	950	973	1012	1177
					1180	1602	1768	1915	2104	2524	2525	2591	2612

*H$_t$, terminal hydrogen
†H$_b$, bridging hydrogen.

on the average about 10% too large. On the other hand, the scaling factor for the MP2(Full)/6-31G(d) frequencies is 0.9661. While this scaling factor is very close to 1, the absolute discrepancy can still be in the range of 100-200 cm^{-1}, as some vibrational frequencies, especially those involving the stretching motions of X–H bonds, have energies of a few thousand wavenumbers.

5.7 Composite methods

It has long been a computational chemist's goal to have a method at his disposal that would yield energetic values that have "chemical accuracy," which translates to an uncertainty of ± 1 kcal mol^{-1}, or about ± 4 kJ mol^{-1}. Currently, high-level methods such as QCISD(T), CCSD(T), or MP6 with very large basis sets may be able to achieve this. However, such calculations are only possible for very small systems. To get around this difficulty, composite methods that would allow calculations on systems with a few non-hydrogen atoms have been proposed.

Since 1990, Pople and co-workers have been developing a series of Gaussian-n (Gn) methods: G1 in 1989, G2 in 1990, G3 in 1998, and G4 in 2007. In addition, a number of variants of the G2 and G3 methods have also been proposed. Since G1 did not last very long, and G4 has not yet been widely used, only G2 and G3 will be discussed here. The G2 method is an approximation for the level of theory of QCISD(T)/6-311+G(3df,2p). In this method, instead of directly doing such an expensive calculation, a series of less time-consuming single-points are carried out and the G2 energy is obtained by applying additivity rules to the single-points and including various corrections such as zero-point vibrational energy (ZPVE) correction and empirical "higher level" correction (HLC). On the other hand, the G3 method is an approximation for QCISD(T)/G3Large, where G3Large is an improved version of the aforementioned 6-311+G(3df,2p) basis set. Once again, for the G3 method, a series of single-points are performed (which are less expensive than the G2 single-points) and various corrections such as ZPVE, HLC, and spin–orbit interaction (newly included for G3) are taken into account.

To test the accuracy of the G2 method, Pople and co-workers used a set of very accurate experimental data consisting of 55 atomization energies, 38 ionization energies, 25 electron affinities, and 7 proton affinities of small molecules. Later, these workers also proposed an extended G2 test set of 148 gas-phase heats of formation. For this extended set of data, the average absolute errors for G2 and G3 are 6.7 and 3.8 kJ mol^{-1}, respectively. Furthermore, it is noted that G3 is actually less expensive than G2, which shows the importance of designing a basis set judiciously. Experience indicates that, for systems of up to 10 non-hydrogen atoms, the expected absolute uncertainty for G2/G3 is about 10 to 15 kJ mol^{-1}.

In addition to the Gaussian-n methods, there is a variety of CBS (complete basis set) methods, including CBS-Q, CBS-q, and CBS-4, which have also enjoyed some popularity among computational chemists. In these methods, special procedures are designed to estimate the complete-basis-set limit energy by extrapolation. Similar to the Gn methods, single-point calculations are also required in the CBS methods. For the aforementioned extended G2 test set of 148 heats of formation, the absolute errors for the CBS-Q, CBS-q, and CBS-4 models are 6.7, 8.8, and 13.0 kJ mol^{-1}, respectively.

Another series of composite computational methods, Weizmann-n (Wn), with $n = 1$–4, have been recently proposed by Martin and co-workers: W1 and W2 in 1999 and W3 and W4 in 2004. These models are particularly accurate for thermochemical calculations and they aim at approximating the CBS limit at the CCSD(T) level of theory. In all Wn methods, the core–valence correlations, spin–orbit couplings, and relativistic effects are explicitly included. Note that in G2, for instance, the single-points are performed with the frozen core (FC) approximation, which was discussed in the previous section. In other words, there is no core–valence effect in the G2 theory. Meanwhile, in G3, the core–valence correlation is calculated at the MP2 level with a valence basis set. In the Wn methods, the core–valence correlation is done at the more advanced CCSD(T) level with a "specially designed" core–valence basis set.

In the W3 and W4 methodologies, connected triple and quadruple correlations are incorporated in order to correct the imperfection of the CCSD(T) wavefunction. As a result, the very sophisticated W3 and W4 methods are only applicable to molecular systems with no more than three non-hydrogen atoms. For the extended G2 test set, the uncertainty for W1 is ± 1.98 kJ mol^{-1}. On the other hand, for the original G2 test set, the absolute errors for the W2 and W3 methods are 1.50 and 0.89 kJ mol^{-1}, respectively. Tested with a very small test data set consisting of 19 atomization energies, the respective absolute errors for the W2, W3, W4a, and W4b methods are 0.96, 0.64, 0.60, and 0.71 kJ mol^{-1}, where W4a and W4b are two variants of the W4 method.

Before presenting some case examples, we note that *ab initio* calculations are usually carried out for individual molecules (or ions or radicals). Hence, strictly speaking, theoretical results should only be compared with gas-phase experimental data. Cautions should be taken when computational results are compared with data obtained from experiments in the solid state, liquid or solution phase.

5.8 Illustrative examples

In the foregoing discussion, the computational methods currently being used are very briefly introduced. Indeed, in most cases, only the language and terminology of these methods are described. Despite this, we can now use a few actual examples found in the literature to see how computations can complement experiments to arrive at meaningful conclusions.

5.8.1 A stable argon compound: HArF

Up to the year of 2000, no compound of helium, neon, or argon was known. However, that did not discourage theoretical chemists from nominating candidate compounds to the synthetic community. In a conference in London in August 2000, there were two poster papers with the following intriguing titles: "Prediction of a stable helium compound: HHeF" by M. W. Wong and "Will

HArF be the first observed argon compound?" by N. Runeberg and co-workers. The formal paper by Wong was published in July 2000, which includes the calculated bond lengths of the linear triatomics HHeF, HNeF, and HArF, as well as the vibrational frequencies of these molecules. Interestingly, one month later, Runeberg and co-workers published an experimental article reporting the detection of HArF. This compound was made by the photolysis of HF in a solid argon matrix and then characterized by infrared spectroscopy. It was found that the measured vibrational frequencies were in good agreement with the calculated ones. One year later, calculations of still higher quality on HArF were reported by the Runeberg group.

Before discussing the computational results in more details, let us first consider the bonding of HArF in elementary and qualitative terms. In hindsight, it is perhaps not very surprising that HArF should be a stable molecule, especially since analogous Kr- and Xe-containing compounds are known to exist. In any event, the bonding of HArF may be understood in terms of a three-center four-electron ($3c$–$4e$) bond, as shown in Fig. 5.8.1. In this figure, it is seen that the 1s orbital of H and the 2p orbitals of Ar and F combine to form three molecular orbitals: σ, σ^n, and σ^*, with bonding, nonbonding, and antibonding characteristics, respectively. Also, the two lower molecular orbitals are filled with electrons, leading to a $3c$–$4e$ bond for this stable molecule.

Let us now consider the calculated results. The calculated bond lengths of HArF are summarized in Table 5.8.1, while the calculated vibrational frequencies, along with the experimental data, are listed in Table 5.8.2.

Examining the results given in these two tables, it is seen that, for this small molecule, very advanced calculations can be carried out. In the tables, all the methods employed have been introduced in the previous sections. For the basis sets, aug-cc-pVnZ stands for augmented correlation consistent polarized valence n zeta, with $n = 2$–5 referring to double, triple, quadruple, and quintuple, respectively. Clearly, these basis functions are specially designed for

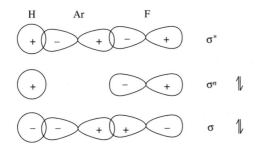

Fig. 5.8.1.
The $3c$–$4e$ bond in HArF.

Table 5.8.1. Calculated structural parameters (in pm) of HArF

Method	Basis set	H–Ar	Ar–F
CCSD	cc-pVTZ	133.4	196.7
CCSD(T)	aug-cc-pVDZ	136.7	202.8
CCSD(T)	aug-cc-pVTZ	133.8	199.2
CCSD(T)	aug-cc-pVQZ	133.4	198.0
CCSD(T)	aug-cc-pV5Z	132.9	196.9

Table 5.8.2. Calculated and experimental vibrational frequencies (in cm^{-1}) of HArF

Method	Basis set	ν(Ar–F)	δ (H–Ar–F)	ν(Ar–H)
MP2	aug-cc-pVDZ	458	706	2220
CCSD	cc-pVTZ	488	767	2.60
CCSD(T)	aug-cc-pVDZ	461	674	1865
CCSD(T)	aug-cc-pVTZ	474	725	2053
CCSD(T)	aug-cc-pVQZ	480	729	2097
CCSD(T)*	aug-cc-pVQZ	463	686	1925
Experimental		435.7	687.0	1969.5

* Corrected for anharmonicity and matrix effects.

correlated calculations. The bond lengths listed in Table 5.8.1 are very consistent with one another and they should be reliable estimates. Meanwhile, as can be seen from the results in Table 5.8.2, the only physical data of the molecule, the vibrational frequencies, are in good accord with the calculated results, thus leading us to believe that this is indeed a linear molecule. Furthermore, according to these calculations, HArF should be stable in the gas phase as well. This awaits experimental confirmation.

As far as charge distribution is concerned, the charges on H, Ar, and F have been estimated to be 0.18, 0.66, and −0.74 a.u., respectively. This indicates a significant charge transfer from the F atom to the ArH moiety. In other words, the bonding of this molecule may be portrayed by the ionic description of HAr$^{\delta+}$F$^{\delta-}$. The calculations of both groups also report the reaction profile of the dissociation reaction HArF → Ar + HF. Both calculations agree that the transition state of this reaction has a bent structure, with H–Ar–F angle of about 105°. The barriers of dissociation calculated by Wong and Runeberg and co-workers are 96 and 117 kJ mol^{-1}, respectively. These results indicate that HArF is trapped in a fairly deep potential well. Similarly, Wong reported that, for HHeF, the barrier of dissociation is about 36 kJ mol^{-1}. While this potential is not as deep as that of HArF, it is still not insignificant. This may provide an incentive for experimentalists to prepare this compound.

To conclude, HArF represents one of those rare examples for which calculations preceded experiment. In addition, it nicely demonstrates that computational results can play an indispensable role in the characterization and understanding of small molecules.

5.8.2 An all-metal aromatic species: Al_4^{2-}

Benzene $(CH)_6$, of course, is the most prototypal aromatic system. When one or more of the CH groups are replaced by other atom(s), a heterocyclic aromatic system is obtained. Well-known examples include pyridine $N(CH)_5$ and pyrimidine $N_2(CH)_4$, while lesser known cases are phosphabenzene $P(CH)_5$ and arsabenzene $As(CH)_5$. There are also systems where the heteroatom is a heavy transition metal; examples include $L_nOs(CH)_5$ and $L_nIr(CH)_5$.

However, it is rare when all the atoms in an aromatic species are metals. One such system was synthesized in 2001 by A. I. Boldyrev and L.-S. Wang and their colleagues. Using a laser vaporization supersonic cluster source and a Cu/Al alloy as the target, these scientists were able to produce the anion $CuAl_4^-$. Similarly, when Al/Li_2CO_3 and Al/Na_2CO_3 were used as targets, anions $LiAl_4^-$ and $NaAl_4^-$ were prepared, respectively. After these anions were synthesized, their vertical detachment energies (VDEs) were measured by photoelectron spectroscopy. As its name implies, VDE is simply the energy required to detach an electron from a bonding orbital of a species and, during the ionization process, the structure of the species remains unchanged. Subsequently, the VDE data were analyzed with the aid of *ab initio* calculations.

Before discussing the computational results, we will attempt to understand the bonding involved in the synthesized anions, using the elementary theories introduced in Chapter 3. First of all, it should be noted that the aromatic system referred to above is not the MAl_4^- anion as a whole. Instead, structurally, the MAl_4^- anion should be considered as consisting of an M^+ cation coordinated to a square-planar Al_4^{2-} unit and it is the dianion that has aromatic character. For Al_4^{2-}, there are 14 valence electrons and the resonance structures can be easily written:

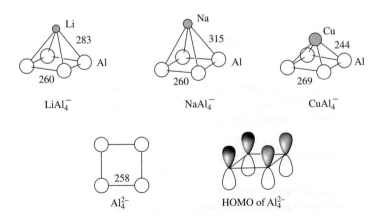

In other words, for this square-planar dianion, there are four σ bonds and one π bond, as well as two lone pairs. Hence each Al–Al linkage has a bond order of about $1\frac{1}{4}$. Additionally, there are two π electrons in this planar ring system, satisfying the $(4n+2)$-rule for aromatic species.

Now let us turn to the computational results. The structures of the square planar Al_4^{2-}, and square pyramidal $LiAl_4^-$, $NaAl_4^-$ and $CuAl_4^-$ are shown in Fig. 5.8.2. It is noted that the structure of $CuAl_4^-$ has been optimized at the MP2/6-311+G(d) level, while the theoretical level for the remaining three species is CCSD(T)/6-311+G(d). When these structures are examined, it is seen that the four-membered ring does not undergo significant structural change upon coordination to a M^+ cation. This is especially true when the metal is lithium

Fig. 5.8.2.
Optimized structures (with bond lengths in pm) of $LiAl_4^-$, $NaAl_4^-$, $CuAl_4^-$, and Al_4^{2-}, and the HOMO of Al_4^{2-}.

Table 5.8.3. Experimental and calculated vertical electron detachment energies (VDEs, in kJ mol^{-1}) for LiAl$_4^-$, NaAl$_4^-$, and CuAl$_4^-$

Species	VDE (experimental)	VDE* (calculated)	Molecular orbital involved
LiAl$_4^-$	207.4 ± 5.8	201.7	$3a_1$
	212.3 ± 5.8	209.4	$1b_1$
	272.1 ± 7.7	259.5	$2a_1$
	298.1 ± 3.9	286.6	$1b_2$
NaAl$_4^-$	196.8 ± 4.8	185.3	$3a_1$
	201.7 ± 4.8	197.8	$1b_1$
	260.5 ± 4.8	243.1	$2a_1$
	285.6 ± 4.8	275.9	$1b_2$
CuAl$_4^-$	223.8 ± 5.8	223.8	$2b_1$
	226.7 ± 5.8	230.6	$4a_1$
	312.6 ± 8.7	323.2	$2b_2$
	370.5 ± 5.8	352.2	$3a_1$

* VDE's are calculated at the theoretical level of OVGF/6-311+G(2df), based on the structures shown in Fig. 5.8.2.

or sodium. When M$^+$ is Cu$^+$, there is a change of 0.11 Å in the Al–Al bond lengths. Such a relative large change may be due to the fact that Cu$^+$ is bigger than Li$^+$ or Na$^+$; it may also be due to the fact that the structure of CuAl$_4^-$ is optimized at a different theoretical level. In any event, the optimized structures of these four species support the argument that Al$_4^{2-}$ is an aromatic species. Also shown in Fig. 5.8.2 is the HOMO of Al$_4^{2-}$, which is very similar to the most stable π molecular orbital of benzene shown in Fig. 7.1.12. The only difference is that benzene is a six-membered ring, while Al$_4^{2-}$ is a four-membered ring. Also, while benzene has two more filled delocalized π molecular orbitals (making a total of three to accommodate its six π electrons), Al$_4^{2-}$ has only one delocalized π molecular orbital for its two π electrons.

The calculated and experimental VDEs of LiAl$_4^-$, NaAl$_4^-$, and CuAl$_4^-$ are summarized in Table 5.8.3. The calculated VDEs have been obtained using the outer valence Green function (OVGF) method with the 6-311+G(2df) basis, based on the structures shown in Fig. 5.8.2. The OVGF model is a relatively new method designed to calculate correlated electron affinities and ionization energies. As seen from the table, the agreement between the calculated results and the experimental data is, on the whole, fairly good. Such an agreement is in support of the proposed square-pyramidal structure for the MAl$_4^-$ anions. The notation for the molecular orbitals shown in the table will be discussed in Chapters 6 and 7.

5.8.3 A novel pentanitrogen cation: N$_5^+$

Even though nitrogen is a very common and abundant element, there are very few stable polynitrogen species, i.e., those compounds containing only nitrogen atoms. Among the few known examples, dinitrogen, N$_2$, isolated in 1772, is the most familiar. Another one is the azide anion, N$_3^-$, which was discovered in 1890. Other species such as N$_3$ and N$_4$, as well as their corresponding cations,

N_3^+ and N_4^+, have been observed only as free gaseous or matrix-isolated ions or radicals. The only polynitrogen species synthesized on a macroscopic scale thus far is the pentanitrogen cation N_5^+ by Christe and co-workers in 1999.

Before discussing N_5^+, maybe we should ask why polynitrogen species are so interesting. It is noted that the bond energy of the N≡N triple bond (in N_2, 942 kJ mol^{-1}) is much larger than those of the N–N single bond (160 kJ mol^{-1}) and the N=N double bond (418 kJ mol^{-1}). In other words, any dissociation of a polynitrogen species to N_2 molecules will be highly exothermic. Hence, polynitrogen compounds are potential high-energy density materials (HEDM), which, by definition, have high ratios between the energy released in a fragmentation reaction and the specific weight. Clearly, HEDM have applications as explosives and in rocket propulsion.

The first synthesis of a N_5^+ compound was accomplished in the following straightforward manner:

$$N_2F^+AsF_6^- + HN_3 \xrightarrow[-78°C]{HF} N_5^+AsF_6^- + HF.$$

In the authors' words, the product "$N_5^+AsF_6^-$ is a highly energetic, strongly oxidizing material that can detonate violently." Upon synthesizing the compound, low-temperature Raman ($-130°C$) and infrared ($-196°C$) spectra were recorded, and high-level *ab initio* calculations were performed in order to determine the structure of the cation and to make assignments of the spectra.

The structure of the N_5^+ cation, optimized at the CCSD(T)/6-311+G(2d) level, is shown in Fig. 5.8.3. This structure is V-shaped, having two short terminal bonds and two longer inner bonds. With such a structure, it is not difficult to visualize a simple bonding description for the cation. In this bonding picture, which is also shown in the same figure, the hybridization scheme for the central N atom is sp^2, while that for its neighboring nitrogen atoms is sp. For the central atom, the *exo* sp^2 hybrid and the p$_z$ orbital are filled by four valence electrons, while the remaining two sp^2 hybrids interact with two terminal N_2 units to complete the bonding scheme. Based on such a description, we can quickly write down the following resonance structures for the N_5^+ cation:

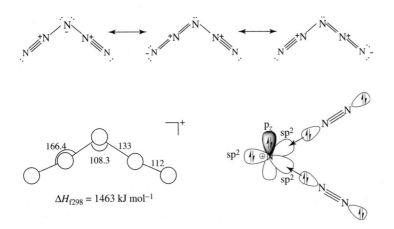

Fig. 5.8.3.
Optimized V-shape structure (bond lengths in pm and bond angles in degrees) of N_5^+, and its bonding scheme. The heat of formation is calculated at the G3 level.

Table 5.8.4. Raman and infrared bands (in cm^{-1}) of N_5^+ (in solid $N_5^+AsF_6^-$) and their assignments based on calculated harmonic frequencies (in cm^{-1}) of gaseous N_5^+

Raman	Infrared	Assignment	Calculated*
2271	2270	$\nu_1(A_1)$	2229
2211	2210	$\nu_7(B_2)$	2175
—	1088	$\nu_8(B_2)$	1032
871	872	$\nu_2(A_1)$	818
672	—	$\nu_3(A_1)$	644
—	—	$\nu_5(A_2)$	475
—	420	$\nu_6(B_1)$	405
—	—	$\nu_9(B_2)$	399
209	—	$\nu_4(A_1)$	181

* At the theoretical level of CCSD(T)/6-311+G(2d). The group-theoretic notation of the vibrational spectra will be made clear in Chapters 6 and 7.

While the three resonance structures may not be of equal importance, it is safe to say that the two terminal N–N bonds have a bond order between 2 and 3, and the remaining two bonds are somewhere between a single bond and a double bond. The optimized bond lengths are in good accord with this description. If, for simplicity, we consider only these three resonance structures and assume they contribute equally, the bond orders of the outer and inner N–N bonds are $2^{2/3}$ and $1^{1/3}$, respectively.

Based on the optimized structure, the harmonic vibrational frequencies of N_5^+ have been calculated at the CCSD(T)/6-311+G(2d) level, and these results are listed in Table 5.8.4, along with the experimental data. Comparing the experimental and calculated results, we can see that there is fairly good agreement between them, bearing in mind that the calculations are done on individual cations and experimental data are measured in the solid state. Such an agreement lends credence to the structure optimized by theoretical methods.

Before leaving this novel cation, it is of interest to point out that in the literature, four more structures (optimized at the MP2/6-31G(d) level) have been considered for the N_5^+ cation, and these are shown in Fig. 5.8.4. However, as can be seen from the G3 heats of formation listed in Figs. 5.8.3 and 5.8.4 for these five isomers, the V-shaped isomer is by far the most stable. The other four isomers are much less stable by about 500 to 1,000 kJ mol^{-1}.

5.8.4 Linear triatomics with noble gas–metal bonds

Since the turn of the present century, a series of linear complexes with the general formula NgMX (Ng = Ar, Kr, Xe; M = Cu, Ag, Au; X = F, Cl, Br) have been prepared and characterized by physical methods. These complexes were prepared by laser ablation of the metal from its solid and letting the resulting plasma react with the appropriate precursor. The complexes formed were then stabilized in a supersonic jet of argon gas. Characterization of these complexes was carried out mainly by microwave spectroscopy.

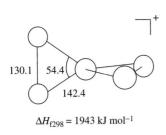

$\Delta H_{f298} = 1943$ kJ mol^{-1}

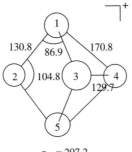

$r_{15} = 207.2$
$r_{23} = r_{24} = 209.4$
$\Delta H_{f298} = 2045$ kJ mol^{-1}

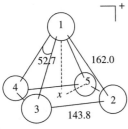

$r_{1x} = 126.1$
$r_{2x} = r_{3x} = r_{4x} = r_{5x} = 101.7$
$\Delta H_{f298} = 2118$ kJ mol^{-1}

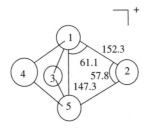

$r_{23} = r_{24} = r_{34} = 231.0$
$\Delta H_{f298} = 2462$ kJ mol^{-1}

Fig. 5.8.4.
Optimized structures (bond lengths in pm and bond angles in degrees) and G3 heats of formation of four other N_5^+ isomers.

The linear structure of NgMX is found to be rigid with small centrifugal distortion constants. (A molecule's centrifugal distortion constant is a measure of its flexibility, and this quantity may be determined from its microwave spectrum.) The Ng–M bond lengths are fairly short, and they range from ∼225 pm for Ar–Cu, ∼245 pm for Ar–Au, ∼260 pm for Ar–Ag, ∼265 pm for Kr–Ag, etc. In addition, based on the centrifugal distortion constants, the Ng–M stretching frequencies can be estimated and they are found to be above 100 cm^{-1}. Summarized in Table 5.8.5 are the experimental and calculated Ng–M bond lengths and stretching frequencies as well as force constants and computed electronic dissociation energies for the Ng–M bonds in NgMX complexes. The calculated results were obtained at the MP2 correlated level. The basis set used for F was 6-311G(d,p) and those for the remaining elements were of comparable quality (such as triple zeta sets). In the following paragraphs we shall discuss these results in some detail from four different, yet interconnected, perspectives.

(1) Ng–M bond lengths

Comparing the experimental and calculated Ng–M bond lengths tabulated in Table 5.8.5, it is seen that the *ab initio* methods adopted do reproduce this quantity very well. In order to obtain a better understanding of the nature of the Ng–M bonds, let us compare these bond lengths with values estimated from

Table 5.8.5. Experimental and calculated (in brackets) Ng–M bond lengths and stretching frequencies v as well as computed electronic dissociation energies D_e for the Ng–M bonds in NgMX complexes

NgMX	Ng–M (pm)	v(Ng–M) (cm^{-1})	v(M–X) (cm^{-1})	D_e(kJ mol^{-1})
ArCuF	222 (219)	224 (228)	621 (674)	44
ArCuCl	227 (224)	197 (190)	418 (456)	33
ArCuBr	230 (226)	170 (164)	313 (350)	—
KrCuF	232 (232)	(185)	—	48
ArAgF	256 (256)	141 (127)	513 (541)	14
ArAgCl	261 (259)	135 (120)	344 (357)	16
KrAgF	259 (260)	125 (113)	513 (544)	17
KrAgCl	264 (263)	117 (105)	344 (352)	15
KrAgBr	266 (269)	106 (89)	247 (255)	17
XeAgF	265 (268)	130	—	43
ArAuF	239 (239)	221 (214)	544 (583)	55
ArAuCl	247 (246)	198 (184)	383 (413)	42
ArAuBr	250 (249)	178 (165)	264 (286)	—
KrAuF	246 (245)	176 (184)	544	58
KrAuCl	252 (251)	161 (163)	383 (409)	44
XeAuF	254 (255)	169 (165)	—	97

standard parameters such as van der Waals radius (r_{vdW}), ionic radius (r_{ion}), and covalent radius (r_{cov}). For instance, we may take the sums r_{cov}(Ng) + r_{cov}(M^I) as a covalent limit and r_{vdW}(Ng) + r_{ion}(M$^+$) as a van der Waals limit.

Among the NgMX complexes, the Ar–Ag bond length is closer to the van der Waals limit than the covalent limit: the Ar–Ag bond lengths in ArAgX range from 256 to 264 pm, while the r_{vdW}(Ar) + r_{ion}(Ag$^+$) sum is 269 pm and the r_{cov}(Ar) + r_{cov}(Ag^I) sum is 226 pm. At the other end of the spectrum, the Xe–Au bond length (254 pm) in XeAuF is very close to the covalent limit: the r_{cov}(Xe) + r_{cov}(Au^I) sum is 257 pm, while the van der Waals limit of r_{vdW}(Xe) + r_{ion}(Au$^+$) is much longer at 295 pm. Summarizing all the results for all the NgMX complexes, it may be concluded that covalent character increases in the orders of Ar < Kr < Xe and Ag < Cu < Au.

(2) Stretching frequencies

The small centrifugal distortion constants (4 to 95 kHz) of the NgMX complexes are consistent with the high rigidity of their linear skeleton. These constants correlate well with the short Ng–M bond lengths, implying strong Ng–M bonding. They also lead to high Ng–M stretching vibrational frequencies, all of which exceed 100 cm^{-1}. The Ng–M and M–X frequencies calculated with the MP2 method agree fairly well with the experimental data, as shown in Table 5.8.5.

When the Ng–M stretching force constants of the NgMX complexes listed in Table 5.8.5 are examined, it is seen that these values are approximately half of that for the bond stretching vibration of KrF$_2$, ranging from 30 N m^{-1} for ArAgBr (recall that the Ar–Ag bond length in ArAgX is near the "van der Waals limit") to 137 N m^{-1} for XeAuF (whose Xe–Au bond length is essentially at the "covalent limit"). In comparison, the corresponding value for Ar–NaCl, which may be considered as a benchmark van der Waals complex, is 0.6 N m^{-1}. Thus in this respect the Ng–M bonding in NgMX is considerably stronger than typical van der Waals interaction.

(3) Electronic dissociation energies

As shown in Table 5.8.5, the calculated Ng–M dissociation energies cover a rather wide range, from 14 kJ mol^{-1} in Ar–AgF to 97 kJ mol^{-1} in Xe–AuF. In comparison, for the aforementioned van der Waals complex Ar–NaCl, the corresponding value for the Ar–Na bond is estimated to be 8 kJ mol^{-1}. Also, the mean Kr–F and Xe–F bond energies in KrF$_2$ and XeF$_2$ are 49 and 134 kJ mol^{-1}, respectively.

It is instructive to compare the Ng–M bond energies with purely electrostatic induction energies such as dipole-induced dipole and charge-induced dipole interactions. Without going into the computational details, we simply note that the dipole-induced dipole energy is ∼35% of the charge-induced dipole interaction. Furthermore, for NgAuF complexes, the charge-induced dipole interaction is ∼10% of the dissociation energy, while the corresponding value for NgCuF and NgAgF complexes is ∼60%. In comparison, for the "typical" van der Waals complex Ar–NaCl, the charge-induced dipole interaction exceeds its dissociation energy. Based on these results, it may be concluded that, in NgCuF and NgAgF complexes, the Ng–M bonds are unlikely to be electrostatic in nature. For NgAuF complexes, particularly for XeAuF, the Xe–Au bond is almost certainly not electrostatic.

(4) Electronic distribution of the Ng–M bond

We can also obtain some idea about the nature of the Ng–M bond by studying the electronic distribution in the complexes. For KrAuF, there is a σ-donation of 0.21 electron from Kr to AuF accompanied by a smaller π-donation. A comparable result is also found for KrAuCl, while the corresponding values for ArAuF and ArAuCl are 0.12–0.14 electron. For NgAgX complexes, the calculated donation from Ng to AgX is significantly less, about 0.06–0.07 electron. For XeAuF, the complex with the largest dissociation energy so far, the corresponding donation is about 0.26 electron. For XeAgF, the donation is ∼0.1 electron less than that in XeAuF. So the "interaction strength" in these complexes once again follows the trends mentioned previously: Ar < Kr < Xe and Ag < Cu < Au.

In summary, the physical properties of the NgMX complexes may vary quantitatively, but these properties remain remarkably similar qualitatively as we go from one complex to the next. For instance, all complexes feature a short, rigid Ng–M bond with stretching frequency above 100 cm^{-1}. Also, the Ng–M dissociation energies is relatively large, and there is noticeable electronic rearrangement upon the formation of the Ng–M bond. Neither the dissociation energies nor the electronic re-distribution upon bond formation can be fully accounted for by electrostatic interaction alone. For XeAuF, the complex with the largest interaction strength, it is fairly certain that the Xe–Au bond is covalent in nature. At the same time, we may also conclude that there is Ng–M chemical bonding, to a greater or lesser degree, in all the remaining NgMX complexes.

Before leaving the NgMX complexes and the compound HArF introduced in Section 5.8.1, it is noted that a fuller discussion on the chemistry of noble gases may be found in Section 17.5, including a subsection on gold–xenon complexes in Section 17.5.6.

With reference to the four examples given in this section, each literature article describing the preparation of the relevant compound(s) also includes a computational section reporting the calculated results, which are used to interpret the experimental data obtained by various physical methods. Clearly, computation has become a very common tool to the experimentalists, not much different from an infrared spectrophotometer or a mass spectrometer. Another point worth noting is that, while high-level calculations can give us a more detailed bonding description, in the examples shown here, we can make use of the simple concepts introduced in Chapters 2 and 3, such as Lewis electron-pair, σ and π bonds, hybridization schemes, resonance theory, ionic radii, covalent radii, and van der Waals radii, to obtain a qualitative, but useful, bonding description of the novel molecular species.

5.9 Software packages

Currently there are numerous software packages that perform the *ab initio* or DFT calculations introduced in this section. Most of these programs are available commercially, but there are a few distributed to the scientific community free of charge. Some popular programs are briefly described below, and the list is by no means exhaustive.

(1) Gaussian (http://www.gaussian.com): This is probably the most popular package around. It was first released in 1970 as *Gaussian 70* and the latest version is *Gaussian 03*, issued in 2003. It does essentially every type of *ab initio*, DFT, and semi-empirical methods and it is easy to use. Its popularity is well earned.

(2) Gamess (http://www.msg.ameslab.gov/GAMESS): This is another popular package and it is available free of charge. It is able to calculate molecular properties at various correlated levels, but it requires a good understanding of the theories.

(3) MOLPRO (http://www.molpro.net): This program is highly optimized for CCSD calculations and it has a special input format which allows the user to define sophisticated computational steps.

(4) SPARTAN (http://www.wavefun.com): This is a very user-friendly program and it is popular among experimental organic and organometallic chemists. Most of the calculations beyond HF level are carried out by a modified Q-Chem (see below) package that comes with SPARTAN.

(5) Q-Chem (http://www.q-chem.com): This commercial package is able to perform energy calculations and geometry optimizations at ground state and excited states at various *ab initio* and DFT levels. It also performs calculations of NMR chemical shifts, solvation effects, etc.

(6) NWChem (http://www.emsl.pnl.gov/docs/nwchem/newchem.html): This program is free of charge for faculty members or staff members of academic institutions.

(7) MQPC (http://www.mqpc.org): This program is also available free of charge and its source code is accessible for modification by users.
(8) MOPAC 2000 (http://www.schrodinger.com/Products/mopac.html): This commercial package is mainly for semi-empirical calculations.
(9) AMPAC (http://www.semichem.com/ampac/index.html): This is another commercial package for semi-empirical calculations.
(10) HyperChem (www.hyper.com): This is an inexpensive PC-based package for molecular modeling and visualization.

References

1. P. W. Atkins and R. S. Friedman, *Molecular Quantum Mechanics*, 3rd edn., Oxford University Press, Oxford, 1997.
2. M. A. Ratner and G. C. Schatz, *Introduction to Quantum Mechanics in Chemistry*, Prentice-Hall, Upper Saddle River, NJ, 2001.
3. D. D. Fitts, *Principles of Quantum Mechanics: As Applied to Chemistry and Chemical Physics*, Cambridge University Press, Cambridge, 1999.
4. J. Simons and J. Nicholls, *Quantum Mechanics in Chemistry*, Oxford University Press, New York, 1997.
5. J. P. Lowe, *Quantum chemistry*, 2nd edn., Academic Press, San Diego, CA, 1993.
6. I. N. Levine, *Quantum Chemistry*, 5th edn., Prentice Hall, Upper Saddle River, NJ, 2000.
7. A. Szabo and N. S. Ostlund, *Modern Quantum Chemistry*, Dover, Mineola, 1996.
8. T. Veszprémi and M. Fehér, *Quantum Chemistry: Fundamentals to Applications*, Kluwer/Plenum, New York, 1999.
9. R. Grinter, *The Quantum in Chemistry: An Experimentalist's View*, Wiley, Chichester, 2005.
10. I. Mayer, *Simple Theorems, Proofs, and Derivations in Quantum Chemistry*, Kluwer/Plenum, New York, 2003.
11. F. Jensen, *Introduction to Computational Chemistry*, Wiley, Chichester, 1999.
12. E. Lewars, *Computational Chemistry: Introduction to the Theory and Applications of Molecular and Quantum Mechanics*, Kluwer Academic, Boston, 2003.
13. A. R. Leach, *Molecular Modelling: Principles and Applications*, 2nd edn., Prentice Hall, Harlow, 2001.
14. W. J. Hehre, L. Radom, P. v. R. Schleyer and J. A. Pople, *Ab initio Molecular Orbital Theory*, Wiley, New York, 1986.
15. R. G. Parr and W. Yang, *Density-Functional Theory of Atoms and Molecules*, Oxford University Press, New York, 1989.
16. J. B. Foresman and A. Frisch, *Exploring Chemistry with Electronic Structure Methods*, 2nd edn., Gaussian Inc., Pittsburgh, 1996.
17. T. M. Klapötke and A. Schulz, *Quantum Chemical Methods in Main-Group Chemistry*, Wiley, Chichester, 1998.
18. J. Cioslowski (ed.), *Quantum-Mechanical Prediction of Thermochemical Data*, Kluwer, Dordrecht, 2001.
19. J. A. Pople, M. Head-Gordon, D. J. Fox, K. Raghavachari and L. A. Curtiss, Gaussian-1 theory: A general procedure for prediction of molecular energy. *J. Chem. Phys.* **90**, 5622–9 (1989).
20. L. A. Curtiss, K. Raghavachari, G. W. Trucks and J. A. Pople, Gaussian-2 theory for molecular energies of first- and second-row compounds. *J. Chem. Phys.* **94**, 7221–30 (1991).

21. L. A. Curtiss, K. Raghavachari, P. C. Redfern, V. Rassolov and J. A. Pople, Gaussian-3 (G3) theory for molecules containing first and second-row atoms. *J. Chem. Phys.* **109**, 7764–76 (1991).
22. J. W. Ochterski, G. A. Petersson and J. A. Montgomery, Jr., A complete basis set model chemistry. V. Extensions to six or more heavy atoms. *J. Chem. Phys.* **104**, 2598–619 (1996).
23. J. A. Montgomery Jr., M. J. Frisch, J. W. Ochterski and G. A. Petersson, A complete basis set model chemistry. VI. Use of density functional geometries and frequencies. *J. Chem. Phys.* **110**, 2822–7 (1999).
24. J. A. Montgomery, Jr., M. J. Frisch, J. W. Ochterski and G. A. Petersson, A complete basis set model chemistry. VII. Use of the minimum population localization method. *J. Chem. Phys.* **112**, 6532-42 (2000).
25. J. M. L. Martin and G. de Oliveira, Towards standard methods for benchmark quality *ab initio* thermochemistry - W1 and W2 theory. *J. Chem. Phys.* **111**, 1843–56 (1999).
26. S. Parthiban and J. M. L. Martin, Assessment of W1 and W2 theories for the computation of electron affinities, ionization potentials, heats of formation, and proton affinities. *J. Chem. Phys.* **114**, 6014–29 (2001).
27. A. D. Boese, M. Oren, O. Atasoylu, J. L. M. Martin, M. Kállay and J. Gauss, W3 theory: Robust computational thermochemistry in the kJ/mol accuracy range. *J. Chem. Phys.* **120**, 4129–42 (2004).
28. M. W. Wong, Prediction of a metastable helium compound: HHeF. *J. Am. Chem. Soc.* **122**, 6289–90 (2000).
29. N. Runeberg, M. Pettersson, L. Khriachtchev, J. Lundell and M. Räsänen, A theoretical study of HArF, a newly observed neutral argon compound. *J. Chem. Phys.* **114**, 836–41 (2001).
30. L. Khriachtchev, M. Pettersson, N. Runeberg and M. Räsänen, A stable argon compound. *Nature* **406**, 874–6 (2000).
31. X. Li, A. E. Kuznetsov, H.-F. Zhang, A. I. Boldyrev and L.-S. Wang, Observation of all-metal aromatic molecules. *Science* **291**, 859–61 (2001).
32. K. O. Christe, W. A. Wilson, J. A. Sheehy and J. A. Boatz, N_5^+: A novel homoleptic polynitrogen ion as a high energy density material. *Angew. Chem. Int. Ed.* **38**, 2004–9 (1999).
33. X. Wang, H.-R. Hu, A. Tian, N. B. Wong, S.-H. Chien and W.-K. Li, An isomeric study of N_5^+, N_5, and N_5^-: A Gaussian-3 investigation. *Chem. Phys. Lett.* **329**, 483–9 (2000).
34. J. M. Thomas, N. R. Walker, S. A. Cooke and M. C. L. Gerry, Microwave spectra and structures of KrAuF, KrAgF, and KrAgBr; ^{83}Kr nuclear quadrupole coupling and the nature of noble gas-noble metal halide bonding, *J. Am. Chem. Soc.* **126**, 1235–46 (2004).
35. S. A. Cooke and M. C. L. Gerry, XeAuF, *J. Am. Chem. Soc.* **126**, 17000–8 (2004).
36. L. A. Curtiss, P. C. Redfern, and K. Raghavachari, Gaussian-4 theory, *J. Chem. Phys.* **126**, 084108 (2007).
37. L. A. Curtiss, P. C. Redfern and K. Raghavachari, Gaussian-4 theory using reduced order perturbation theory, *J. Chem. Phys.* **127**, 124105 (2007).

Symmetry in Chemistry

In the study of the structure and properties of molecules and crystals, the concept of symmetry is of fundamental importance. Symmetry is an abstract concept associated with harmony and balance in nature or in social relationship. Yet in chemistry this ever-evolving concept does have a very practical role to play.

The main advantage of studying the symmetry characteristics of a chemical system is that we can apply symmetry arguments to solve physical problems of chemical interest. Specifically, we utilize a mathematical tool called group theory to simplify the physical problem and to yield solutions of chemical significance. The advantage of this method becomes more obvious when the symmetry of the chemical system increases. Quite often, for highly symmetric molecules or crystals, very complex problems can have simple and elegant solutions. Even for less symmetric systems, symmetry arguments can lead to meaningful results that cannot be easily obtained otherwise.

Before applying group theory, we need to recognize the symmetry properties of a chemical system. For individual molecules, we only need to consider the symmetry of the species itself, but not the symmetry that may exist between the species and its neighbors. In trying to determine the symmetry of a molecule, we need to see whether it has any *symmetry elements*, which are defined as geometrical entities such as an axis, a plane, or a point about which *symmetry operations* can be carried out. A symmetry operation on a molecule may be defined as an exchange of atoms in the molecule about a symmetry element such that the molecule's outward appearance, including orientation and location, remains the same after the exchange. Chapters 6 and 7 discuss molecular symmetry and elementary group theory, as well as chemical applications of group theory. In Chapter 8, the coordination bond is described, again with symmetry and group theory playing an important role. Also, the presentation of group theory in these chapters is informal and includes many illustrations and examples, while essentially all the mathematical derivations are omitted.

In contrast to discrete molecules, crystals have a lattice structure exhibiting three-dimensional periodicity. As a result, we need to consider additional symmetry elements that apply to an infinitely extended object, namely the translations, screw axes, and glide planes. Chapters 9 and 10 introduce the concept and nomenclature of space groups and their application in describing the structures of crystals, as well as a survey of the basic inorganic crystalline materials.

Symmetry and Elements of Group Theory

6

In this chapter, we first discuss the concept of symmetry and the identification of the point group of any given molecule. Then we present the rudiments of group theory, focusing mainly on the character tables of point groups and their use.

6.1 Symmetry elements and symmetry operations

Symmetry is a fundamental concept of paramount importance in art, mathematics, and all areas of natural science. In the context of chemistry, once we know the symmetry characteristics (i.e., point group) of a molecule, it is often possible for us to draw qualitative inferences about its electronic structure, its vibrational spectra, as well as other properties such as dipole moment and optical activity.

To determine the symmetry of a molecule, we first need to identify the *symmetry elements* it may possess and the *symmetry operations* generated by these elements. The twin concepts of symmetry operation and symmetry element are intricately connected and it is easy to confuse one with the other. In the following discussion, we first give definitions and then use examples to illustrate their distinction.

A symmetry operation is an atom-exchange operation (or more precisely, a coordinate transformation) performed on a molecule such that, after the interchange, the equivalent molecular configuration is attained; in other words, the shape and orientation of the molecule are not altered, although the position of some or all of the atoms may be moved to their equivalent sites. On the other hand, a symmetry element is a geometrical entity such as a point, an axis, or a plane, with respect to which the symmetry operations can be carried out. We shall now discuss symmetry elements and symmetry operations of each type in more detail.

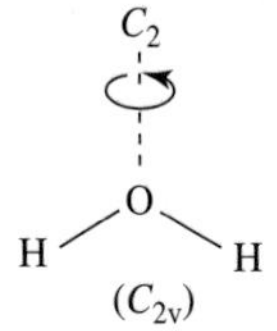

Fig. 6.1.1.
The C_2 axis in H_2O.

6.1.1 Proper rotation axis C_n

This denotes an axis through the molecule about which a rotation of $360°/n$ can be carried out. For example, in H_2O the C_2 axis is a symmetry element (Fig. 6.1.1), which gives rise to the symmetry operation $\boldsymbol{C_2}$. (In this book all symmetry *operations* will be indicated in bold font.) Meanwhile, NH_3 has a symmetry element C_3 (Fig. 6.1.2), but now there are two symmetry operations

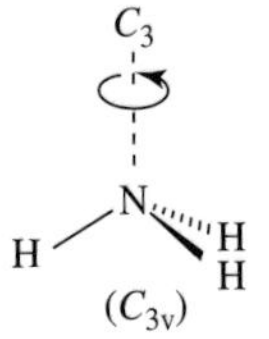

Fig. 6.1.2.
The C_3 axis in NH_3.

Fig. 6.1.3.
Two orientations of CH$_4$ showing the directions of the C_3 and C_2 axes.

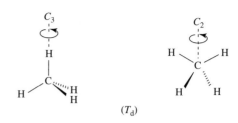

Fig. 6.1.4.
Two views of SF$_6$ showing the C_4 and C_3 axes. In the figure on the right, the C_3 axis is perpendicular to the paper and passes through the S atom. In both figures, a secondary C_2 axis (the same one!) is also shown.

generated by this element: a rotation of 120° and another of 240°. Sometimes it is easy to identify the C_n axis in a molecule (as in the cases of H$_2$O and NH$_3$), and sometimes it is not. To clearly see one particular rotational axis, often it is advantageous to draw the molecule in a certain orientation. Figure 6.1.3 shows CH$_4$ in two orientations; for one it is easy to see the C_3 axis, and for the other the C_2 axis. Similarly, Fig. 6.1.4 shows SF$_6$ in two orientations, one for the C_4 axis and the other for the C_3 axis. Cyclohexane has one C_3 axis and three C_2 axes perpendicular to it; these are clearly shown in Fig. 6.1.5. In this molecule, the C_3 axis is called the principal symmetry axis.

Fig. 6.1.5.
Two views of cyclohexane showing the principal C_3 axis (perpendicular to the paper in the figure on the right) and three secondary C_2 axes. In the figure on the right, the + and − signs label the C atoms lying above and below the mean plane of the molecule, respectively.

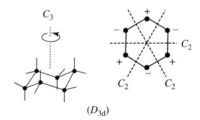

6.1.2 Symmetry plane σ

This denotes a plane through the molecule, about which a reflection operation may be carried out. The symbol σ originates from the German word *Spiegel*, which means a mirror. In H$_2$O (Fig. 6.1.1), when one of the hydrogen atoms is substituted by its isotope D, the C_2 axis no longer exists. However, the molecular plane is still a symmetry plane.

Symmetry (or mirror) planes may be further classified into three types. First, there are the vertical planes σ_v whose common intersection constitutes the rotation axis C_n. For example, it is obvious that there are two σ_v's in H$_2$O (Fig. 6.1.1) and three σ_v's in NH_3 (Fig. 6.1.2). Generally speaking, for a molecule with a rotational axis C_n, the identification of one σ_v necessarily implies the presence of n σ_v's.

Secondly, there is the horizontal plane σ_h which lies perpendicular to the principal axis C_n. Examples are given in Figs. 6.1.6 and 6.1.7. Note that some highly symmetric molecules have more than one σ_h; for instance, octahedral SF_6 has three σ_h's that are mutually orthogonal to one another.

Lastly, in molecules with a principal rotational axis C_n and n secondary C_2 axes perpendicular to it, sometimes there are additionally n vertical planes which bisect the angles formed between the C_2 axes. These vertical planes are called σ_d, or dihedral planes. In cyclohexane, there are three σ_d's, each lying between a pair of C_2 axes (Fig. 6.1.5). It should be noted that sometimes σ_v's and σ_d's cannot be differentiated unambiguously. For example, in BrF_4^- (Fig. 6.1.8), in addition to the principal C_4 axis, there are four C_2 axes perpendicular to C_4, and four vertical planes containing the C_4 axis. By convention, the two C_2 axes passing through the F atoms are called C_2', while those which do not pass through the F atoms are the C_2'' axes. Furthermore, the vertical planes containing the C_2' axes are called σ_v's, while those containing the C_2'' axes are σ_d's. In group theory, it can be shown that the C_2' axes form a *class* of symmetry operations. Other distinct classes are composed of the C_2'' axes, the σ_v's, and the σ_d's.

Fig. 6.1.6.
The molecular plane in *trans*-N_2F_2 is a σ_h.

Fig. 6.1.7.
The plane containing all carbon atoms in all-*trans*-1,5,9-cyclododecatriene is a σ_h.

6.1.3 Inversion center *i*

The inversion operation is carried out by joining a point to the inversion center (or center of symmetry) and extending it an equal distance to arrive at an equivalent point. Molecules which possess an inversion center are termed centrosymmetric. Among the eight examples given so far, SF_6 (Fig. 6.1.4), cyclohexane (Fig. 6.1.5), *trans*-N_2F_2 (Fig. 6.1.6), and BrF_4^- (Fig. 6.1.8) are centrosymmetric systems. Molecules lacking an inversion center are called non-centrosymmetric.

6.1.4 Improper rotation axis S_n

This denotes an axis about which a rotation-reflection (or improper rotation) operation may be carried out. The rotation-reflection operation S_n involves a

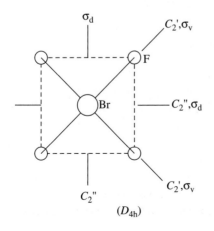

Fig. 6.1.8.
The C_2' and C_2'' symmetry axes in BrF_4^-. By convention, the symmetry planes containing the C_2' axes are designated as σ_v's, while those containing C_2'' are the σ_d's.

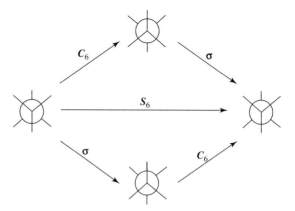

Fig. 6.1.9.
The S_6 operation in ethane.

rotation of the molecule by $360°/n$ about the improper rotation axis S_n, which is followed by a reflection of all atoms through a plane perpendicular to the S_n axis. Note that the rotation-reflection operation S_n can be carried out in either order, i.e., $S_n = C_n\sigma = \sigma C_n$. An example of the S_6 operation in ethane is shown in Fig. 6.1.9. Now it is obvious that the three C_2 axes in CH_4 (Fig. 6.1.3) coincide with the corresponding S_4 axes, the C_4 and C_3 axes in SF_6 (Fig. 6.1.4) are S_4 and S_6 axes, respectively, and the C_3 axis in cyclohexane (Fig. 6.1.5) is also an S_6 axis. Finally, it is pointed out that operation S_1 is the same as σ (e.g., HOD), and the operation S_2 is equivalent to the inversion i (e.g., *trans*-N_2F_2 in Fig. 6.1.6).

6.1.5 Identity element E

The symbol E comes from the German word *Einheit* meaning unity. This element generates an operation E which leaves the molecule unchanged. All molecules possess such an element or operation. The need for this seemingly trivial "do nothing" or "leave it alone" operation arises from the mathematical requirements of group theory, as we shall see in Chapter 7. Note that in some books the symbol I (for identity) is used in place of E.

6.2 Molecular point groups

6.2.1 Classification of point groups

When we say a molecule belongs to a certain point group, it is meant that the molecule possesses a specific, self-consistent set of symmetry elements. The most common point groups are described below with illustrative examples.

(1) Point group C_1

This group has only one symmetry element: identity element E; i.e., the molecule concerned is asymmetric. Examples include methane derivatives with the central carbon atom bonded to four different groups, e.g., CHFClBr.

(2) Point group C_s

This group has only two symmetry elements: E and σ. The aforementioned HOD belongs to this group. Other examples include thionyl halide SOX_2 and secondary amines R_2NH (Fig. 6.2.1).

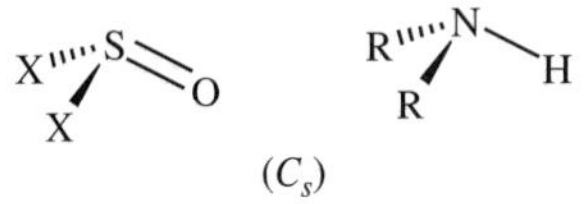

(3) Point group C_i

This group has only two symmetry elements: E and i. There are not many molecules with this kind of symmetry. Two examples are given in Fig. 6.2.2.

(4) Point group C_n

This group has only symmetry elements E and C_n. Examples for the C_n group are shown in Fig. 6.2.3.

Note that the term "dissymmetric" is reserved for describing a molecule not superimposable on its own mirror image. Accordingly, an object with no improper rotation (rotation-reflection) axis must be dissymmetric. All asymmetric objects are dissymmetric, but the converse is not necessarily true. In fact, those molecules that possess one or more rotational axes of any order as the *only* symmetry elements are dissymmetric; for instance, the 1,3,5-triphenylbenzene molecule in the three-leaved propeller configuration (point group C_3) and the $[Co(en)_3]^{3+}$ cation (point group D_3) are dissymmetric species. Note that the term "asymmetric unit" has a special meaning in crystallography, as explained in Section 9.3.6.

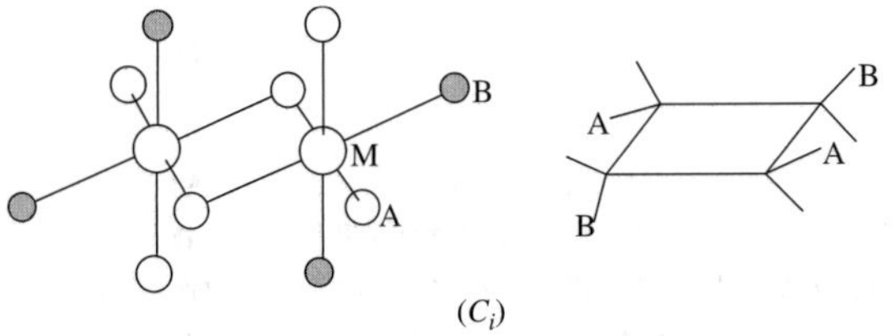

Fig. 6.2.1.
Examples of molecules with C_s symmetry: thionyl halide (SOX_2) and secondary amines R_2NH.

Fig. 6.2.2.
Examples of molecules with C_i symmetry.

(5) Point group C_{nv}

This group has symmetry elements E, a rotational axis C_n, and n σ_v planes. The H_2O (Fig. 6.1.1) and NH_3 (Fig. 6.1.2) molecules have C_{2v} and C_{3v} symmetry, respectively. Other examples are given in Fig. 6.2.4.

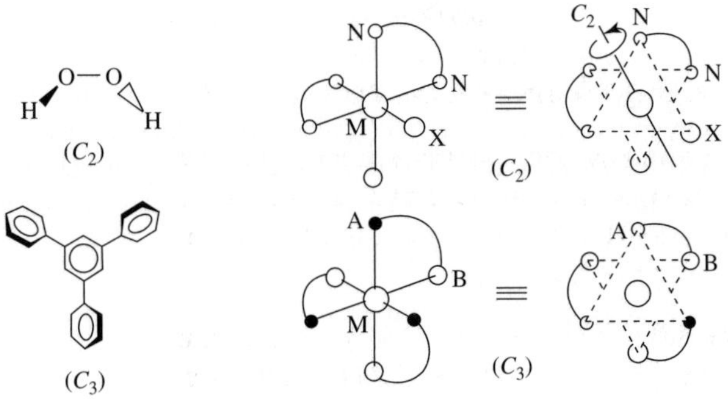

Fig. 6.2.3.
Examples of molecules with C_n symmetry. The fragment N⌒N represents the bidentate chelating ligand ethylenediamine $H_2NCH_2CH_2NH_2$, while A⌒B represents a bidentate chelating ligand with different coordinating sites such as $Me_2NCH_2CH_2NH_2$.

Fig. 6.2.4.
Examples of molecules with C_{nv} symmetry: (a) BrF$_5$, (b) (η^5-C$_5$H$_5$)Ni(NO) with the NO group linearly coordinated to the Ni center, and (c) (η^6-C$_6$H$_6$)Cr(η^6-C$_6$F$_6$).

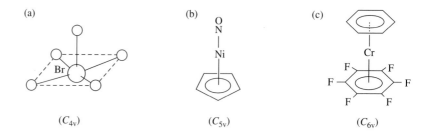

(6) Point group C_{nh}

This group has symmetry element E, a rotational axis C_n, and a horizontal plane σ_h perpendicular to C_n. Note that S_n also exists as a consequence of the elements already present (C_n and σ_h). Also, when n is even, the presence of i is again a necessary consequence. Examples include *trans*-N$_2$F$_2$ (C_{2h}; Fig. 6.1.6), all-*trans*-1,5,9-cyclododecatriene (C_{3h}; Fig. 6.1.7), and boric acid B(OH)$_3$ (C_{3h}; Fig. 13.5.1).

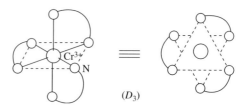

Fig. 6.2.5.
Representative example of tris-chelated metal complexes with D_3 symmetry: [Cr(en)$_3$]$^{3+}$. The bidentate ligand ethylenediamine is represented by N⌒N.

(7) Point group D_n

This group has a principal C_n axis with n secondary C_2 axes perpendicular to it. The symbol D arises from the German *Diedergruppe*, which means dihedral group. Examples of this group are uncommon, as the presence of D_n is usually accompanied by other symmetry elements. An organic molecule of this type is cycloocta-1,5-diene (D_2). The most important as well as interesting inorganic examples are the chiral tris-chelated transition metal complexes such as Mn(acac)$_3$ (acac = acetylacetonate) and tris(ethylenediamine)chromium(III), which is illustrated in Fig. 6.2.5.

(8) Point group D_{nh}

This group has symmetry element E, a principal C_n axis, n secondary C_2 axes perpendicular to C_n, and a σ_h also perpendicular to C_n. The necessary consequences of such combination of elements are a S_n axis coincident with the C_n axis and a set of n σ_v's containing the C_2 axes. Also, when n is even, symmetry center i is necessarily present. The BrF$_4^-$ molecule has point group symmetry D_{4h}, as shown in Fig. 6.1.8. Examples of other molecules belonging to point groups D_{2h}, D_{3h}, D_{5h} and D_{6h} are given in Fig. 6.2.6.

(9) Point group D_{nd}

This group consists of E, a principal C_n axis, n C_2's perpendicular to C_n, and n σ_d's between the C_2 axes. A necessary consequence of such a combination

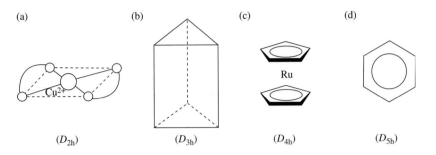

Fig. 6.2.6.
Examples of molecules with D_{nh} symmetry: (a) [Cu(en)$_2$]$^{2+}$, the bidentate chelating ligand ethylenediamine is represented by N⌒N; (b) triprismane C$_6$H$_6$; (c) ruthenocene (η^5-C$_5$H$_5$)$_2$Ru; and (d) benzene C$_6$H$_6$.

is the presence of symmetry element S_{2n}. The cyclohexane molecule shown in Fig. 6.1.5 has D_{3d} symmetry. Two other D_{nd} examples are given in Fig. 6.2.7.

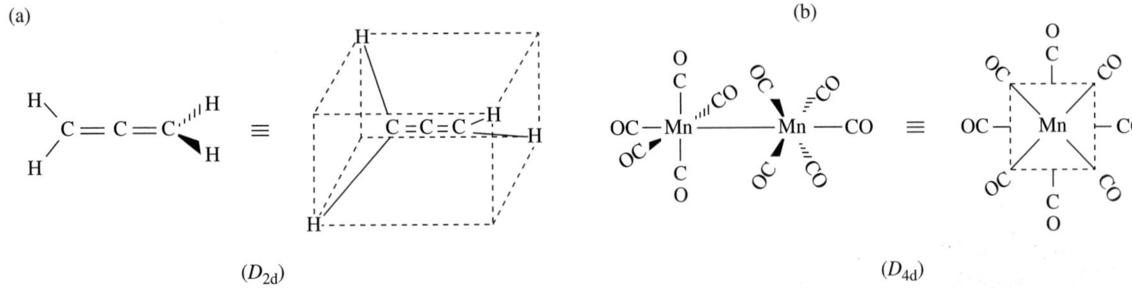

Fig. 6.2.7.
Examples of molecules with D_{nd} symmetry: (a) allene and (b) Mn$_2$(CO)$_{10}$.

A general geometrical relationship holds for any molecule that can be represented as two regular polygons of n vertices (n-gons) separated by a distance along the principal C_n rotation axis. In the fully eclipsed conformation, i.e., when the two polygons exactly overlap when viewed along C_n, the point group is D_{nh}. In the perfectly staggered conformation, the point group is D_{nd}. In the skew (or *gauche*) conformation, the point group is D_n. Common examples illustrating such patterns are ethane and sandwich-type metallocenes (η-C$_n$H$_n$)$_2$M.

(10) Point group S_n

This point group has only two symmetry elements: E and S_n. Since carrying out the S_n operation n times should generate the identity operation E, integer n must be even. Also, $n \geq 4$. Examples of this point group are rare; two are given in Fig. 6.2.8. Molecules belonging to S_n with $n \geq 6$ are quite uncommon. In hexakis(pyridine N-oxide)coblalt(II) perchlorate [Fig. 6.2.8(d)], the pyridine N-oxide ligand is coordinated to the metal center at a Co–O–N bond angle of 119.5(2)°; the dihedral angle between the plane containing these three atoms and the pyridine ring is 72.2°.

(11) Point groups $D_{\infty h}$ and $C_{\infty v}$

Symmetric linear molecules such as H$_2$, CO$_2$, and HC≡CH have $D_{\infty h}$ symmetry, while unsymmetric linear molecules such as CO, HCN, and FCCH belong to the $C_{\infty v}$ point group.

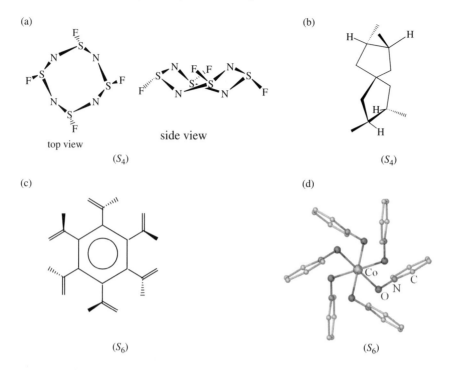

Fig. 6.2.8.
Examples of molecules with S_n symmetry: (a) $N_4(SF)_4$, (b) 2,3,7,8-tetramethylspiro[4.4]nonane, (c) $C_6(CMe=CH_2)_6$, and (d) $[(C_5H_5NO)_6Co]^{2+}$.

(12) High-symmetry point groups T_d, O_h, and I_h

Tetrahedral molecules such as CH_4 (Fig. 6.1.3) have 24 symmetry operations (E, $8C_3$, $3C_2$, $6S_4$ and $6\sigma_d$) and belong to the T_d point group. In comparison, octahedral molecules such as SF_6 (Fig. 6.1.4) have 48 symmetry operations (E, $8C_3$, $6C_2$, $6C_4$, $3C_2 = 3C_4^2$, i, $6S_4$, $8S_6$, $3\sigma_h$, and $6\sigma_d$) and belong to the O_h point group. It would be instructive for students to examine all the operations for these two important high-symmetry point groups with the aid of molecular models.

Among organic molecules, cubane C_8H_8 (O_h) and adamantane $(CH)_4(CH_2)_6$ (T_d) stand out distinctly by virtue of their highly symmetric cage structures. In contrast to the scarcity of organic analogs such as hexamethylenetetramine $N_4(CH_2)_6$ and hexathiaadamantane $(CH)_4S_6$, there exists an impressive diversity of inorganic and organometallic compounds containing cubane-like and adamantane-like molecular skeletons. In current usage, the term cubane-like (cubanoid) generally refers to a M_4Y_4 core composed of two interpenetrating tetrahedra of separate atom types, resulting in a distorted cubic framework of idealized symmetry T_d (Fig. 6.2.9(a)). Similarly, the adamantane-like (adamantoid) M_4Y_6 skeleton consists of a M_4 tetrahedron and a Y_6 octahedron which interpenetrate, giving rise to a fused system of four chair-shaped rings with overall symmetry T_d (Fig. 6.2.9 (b)). If M stands for a metal atom, the compound is often described as tetranuclear or tetrameric. In either structural type additional ligand atoms or groups L and L′ may be symmetrically appended

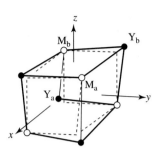 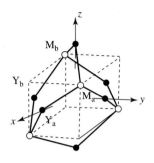

Fig. 6.2.9.
(a) Cubane-like M_4Y_4 core referred to a Cartesian frame. (b) Adamantane-like M_4Y_6 core referred to a Cartesian frame. The reference cube is outlined by broken lines. Atoms of type M and Y are represented by open and solid circles, respectively.

to core atoms M and Y, respectively. The two families of cubane-like and adamantane-like cage compounds may be formulated as $(L_mM)_4(YL'_n)_4$ (with $m = 0, 1, 3$; $n = 0, 1$) and $(L_mM)_4(YL'_n)_6$ (with $m = 0, 1$; $n = 0, 1, 2$), respectively. Not included in these two categories are structurally related compounds such as the oligomeric silasesquioxane $Me_8Si_8O_{12}$ and basic beryllium acetate $Be_4O(CH_3COO)_6$, in which the core atoms are bridged by oxygen atoms.

Table 6.2.1. Relations between bond angles Y–M–Y(α) and M–Y–M(β) in M_4Y_4 and M_4Y_6 cores

	M_4Y_4 core, see Fig. 6.2.9(a)	M_4Y_6 core, see Fig. 6.2.9(b)	
Atom positions expressed in terms of u and v	M_a (uuu) Y_b ($\bar{v}vv$)	M_a (uuu) Y_a ($00\,v$)	
	M_b ($\bar{u}\bar{u}u$) Y_b ($v\bar{v}\bar{v}$)	M_b ($\bar{u}\bar{u}u$) Y_b ($0\bar{v}0$)	
$d\sin(\alpha/2) =$	$v(2)^{\frac{1}{2}}$	$v/(2)^{\frac{1}{2}}$	
$d\sin(\beta/2) =$	$u(2)^{\frac{1}{2}}$	$u(2)^{\frac{1}{2}}$	
$d^2 =$	$3u^2 - 2uv + 3v^2$	$3u^2 - 2uv + v^2$	
$\sin(\alpha/2) =$	$[\sin(\beta/2) + \{6 - 8\sin^2(\beta/2)\}^{\frac{1}{2}}]/3$ (1)	$[\sin(\beta/2) + (2)^{\frac{1}{2}}\cos(\beta/2)]/2$	(3)
$\sin(\beta/2) =$	$[\sin(\alpha/2) + \{6 - 8\sin^2(\alpha/2)\}^{\frac{1}{2}}]/3$ (2)	$[2\sin(\alpha/2) + \{6 - 8\sin^2(\alpha/2)\}^{\frac{1}{2}}]/3$	(4)

Equations (1) to (4) may be recast in the following forms:

$$\cos\alpha = [4 - 7\cos\beta - 4(1 + \cos\beta - 2\cos^2\beta)^{\frac{1}{2}}]/9 \quad (1')$$
$$\cos\beta = [4 - 7\cos\alpha - 4(1 + \cos\alpha - 2\cos^2\alpha)^{\frac{1}{2}}]/9 \quad (2')$$
$$\cos\alpha = (1 - \cos\beta)/4 - \sin\beta/(2)^{\frac{1}{2}} \quad (3')$$
$$\cos\beta = [1 - 4\cos\alpha - 8(1 + \cos\alpha - 2\cos^2\alpha)^{\frac{1}{2}}]/9 \quad (4')$$

In describing the idealized T_d cage structure of either the cubane or adamantane type, it is customary to give the bond length d, the Y–M–Y angle α, and the M–Y–M angle β. While the parameter d readily conveys an appreciation of molecular size, it is not generally appreciated that only one bond angle suffices to define the molecular geometry uniquely. Derivation of the bond-angle relationship between α and β for both structural types are outlined in Table 6.2.1. If an external ligand atom L is bound to core atom M, the bond angle L–M–Y (γ) may be calculated from α using the equation $\sin(\pi - \gamma) = 2\sin(\alpha/2)/(3)^{1/2}$ for local C_{3v} symmetry. For the M_4Y_6 core, if a pair of ligand atoms L' are symmetrically attached to atom Y to attain local C_{2v} symmetry, the bond angle L'–Y–L' (δ) may be calculated from β using the equation $\cot^2(\delta/2) + \cot^2(\beta/2) = 1$.

Table 6.2.2. Application of bond-angle relationships to cubane-like and adamantane-like compounds

Complex	Core	Crystallographic site symmetry	Bong angles (°)		S
			Experimental values	α_c calculated from $\bar{\beta}$	
$[(\eta^5\text{-}C_5H_5)_4Fe_4(CO)_4]$	Fe_4C_4	$1(C_1)$	$\bar{\beta} = 78.8, \sigma_{max} = 0.3$	100.2	2.2
$(Et_4N)_2[(PhCH_2S^*)_4Fe_4S_4]$	Fe_4S_4	$1(C_1)$	$\bar{\beta} = 73.81, \sigma_{max} = 0.06$	104.09	5.7
$[(\eta^5\text{-}C_5H_5)_4Co_4S_4]$	Co_4S_4	$2(C_2)$	$\bar{\beta} = 95.31, \sigma_{max} = 0.06$	84.43	37
$(Me_4N)_2[Cu_4OCl_{10}]$	Cu_4OCl_6	$23\ (T)$	$\bar{\beta} = 81.5, \sigma_{max} = 1.2$	119.1	0
		$\bar{4}(23)$	$\bar{\beta} = 81.3, \sigma_{max} = 0.9$	119.1	1.1
$(Et_2NHCH_2CH_2NHEt_2)_2[Cu_4OCl_{10}]$	Cu_4OCl_6	$2(C_2)$	$\bar{\beta} = 81.5, \sigma_{max} = 0.3$	119.1	18
$K_4[Cu_4OCl_{10}]$	Cu_4OCl_6	$2(C_2)$	$\bar{\beta} = 81.1, \sigma_{max} = 0.3$	119.2	23
$(Et_2NH_2)_4Cu_4OCl_{10}$	Cu_4OCl_6	$1(C_1)$	$\bar{\beta} = 80.4, \sigma_{max} = 0.1$	119.3	103

If either α or β is known for a given structure, eqs. (1)–(4) in Table 6.2.1 can be used to evaluate the remaining angle. Another possible use of these equations is in assessing the extent to which the cores of related compounds deviate from idealized T_d symmetry. For this purpose we assume that there is no systematic variation in the observed M–Y bond distances and define a goodness-of-fit index S by the equation:

$$S = \sum_{i,j} \frac{[(\alpha_i - \alpha_c)^2/n_\alpha + (\beta_j - \bar{\beta})^2/n_\beta]^{\frac{1}{2}}}{\sigma_{max}} \quad (5)$$

where n_α and n_β are, respectively, the numbers of independently measured angles α_i and β_j, $\bar{\beta}$ is the average of β_j, α_c is calculated from $\bar{\beta}$ with equation (1) or (1′) for cubane-like and (3) or (3′) for adamantane-like systems, and σ_{max} is the maximum value of the estimated standard deviations of individual bond angles. In Table 6.2.2, the computed S values of 2.2, 5.7, and 37 for the cubane-like Fe_4C_4, Fe_4S_4, and Co_4S_4 cores, respectively, indicate increasing deviations from idealized T_d symmetry. This is in accord with the assignment of the core symmetries in the literature as T_d, D_{2d}, and lower than D_{2d}, respectively. In the case of the adamantane-like $[Cu_4OCl_{10}]^{4-}$ anion (Cu coordinated by three bridging Cl, an external Cl, and the O atom at the center), which occupies sites of different symmetry in four reported crystal structures, consideration of the relative S values in Table 6.2.2 leads to the conclusion that it conforms to T_d symmetry in its tetramethylammonium salt, and becomes increasingly more distorted in the N,N,N',N',-tetraethylenediammonium, potassium, and diethylammonium salts. The present procedure, when applied to a series of related molecules, yields a set of numbers which allow one to spot a trend and distinguish between clear-cut cases.

The most characteristic feature of point group I_h is the presence of six fivefold rotation axes, ten threefold rotation axes, and 15 twofold rotation axes. A σ plane lies perpendicular to each C_2 axis. Group I_h has 120 symmetry operations (E, $12C_5$, $12C_5^2$, $20C_3$, $15C_2$, i, $12S_{10}$, $12S_{10}^3$, $20S_6$, and 15σ). Well-known cage-like molecules that belong to this point group are icosahedral $B_{12}H_{12}^{2-}$ and

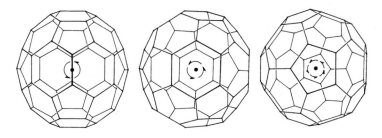

Fig. 6.2.10.
Perspective views of C_{60} along its C_2, C_3 and C_5 axes (I_h symmetry).

dodecahedrane $C_{20}H_{20}$. Fullerene C_{60} is a truncated icosahedron that exhibits I_h symmetry, and the C_2, C_3, and C_5 axes of this cage system are shown in Fig. 6.2.10.

The point groups T_d and O_h, together with their subgroups T, T_h, and O, are known as cubic groups because they are geometrically related to the cube. The point groups I_h and I are known as icosahedral groups. These seven point groups reflect part or all of the symmetry properties of the five Platonic solids: the tetrahedron, the cube, the octahedron, the pentagonal dodecahedron, and the icosahedron. Note that the eight-coordinate dodecahedral complexes such as $[Mo(CN)_8]^{4-}$, $[ZrF_8]^{4-}$, and $[Zr_4(OH)_8(H_2O)_{16}]^{8+}$ are based on the triangulated dodecahedron, which has D_{2d} symmetry.

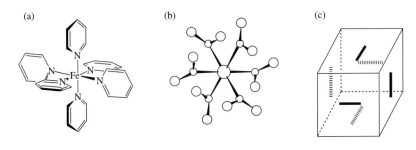

Fig. 6.2.11.
Examples of molecules with T_h symmetry: (a) $[Fe(C_5H_5N)_6]^{2+}$ and (b) $[M(NO_2)_6]^{n-}$ viewed along a C_3 axis. In (c), the three pairs of bars (each bar representing a planar ligand) lying on the faces of a reference cube exhibit the same symmetry.

(13) Rarely occurring point groups O, T, T_h, I, and K_h

Point groups O is derived from O_h by removing all symmetry elements except the rotational axes. A rare example is octamethylcubane, $C_8(CH_3)_8$, in which the CH_3 group is rotated about an exo-polyhedral C–C bond to destroy the σ_d plane; note that this simultaneously eliminates symmetry elements i, σ_h, S_4, and S_6. Similarly, point group T is derived from T_d by leaving out the σ_d planes. Examples of molecules that possess point symmetry T are tetrahedral $Pt(PF_3)_4$ and $Si(SiMe_3)_4$, in which the two sets of substituents attached to atoms forming the corresponding Pt–P or Si–Si single bond take the *gauche* conformation.

Point group T_h has 24 symmetry operations (E, $4C_3$, $4C_3^2$, $3C_2$, i, $4S_4$, $4S_4$, and $3\sigma_h$). Inorganic complexes that belong to this point group are $[Fe(C_5H_5N)_6]^{2+}$ and $[M(NO_2)_6]^{n-}$ (M = Co, n = 3; M = Ni, n = 4), as shown in Figs. 6.2.11(a) and (b), respectively. In each complex, three pairs of monodentate ligands lie on mutually orthogonal mirror planes (Fig. 6.2.11(c)).

Point group I is derived from I_h by deleting the σ planes normal to the 15 C_2 axes. It is inconceivable that any molecule having symmetry I can ever be constructed.

A finite object that exhibits the highest possible symmetry is the sphere, whose point group symbol K_h is derived form the German word *Kugel*. It has an infinite number of rotation axes of any order in every direction, all passing through the center, as well as an infinite number of σ_h planes each perpendicular to a C_∞ axis. Obviously, K_h is merely of theoretical interest as no chemical molecule can possess such symmetry.

6.2.2 Identifying point groups

To help students to determine the symmetry point group of a molecule, various flow charts have been devised. One such flow chart is shown in Table 6.2.3. However, experience indicates that, once we are familiar with the various operations and with visualizing objects from different orientations, we will dispense with this kind of device.

Often it is difficult to recognize the symmetry elements present in a molecule from a perspective drawing. A useful technique is to sketch a projection of the molecule along its principal axis. Plus and negative signs may be used to indicate the relative locations of atoms lying above and below the mean plane, respectively. Cyclohexane has been used as an example in Fig. 6.1.5. The identification of the point groups of some other non-planar cyclic molecules with the aid of projection diagrams is illustrated in Table 6.2.4.

Table 6.2.3. Flow chart for systematic identification of the point group of a molecule

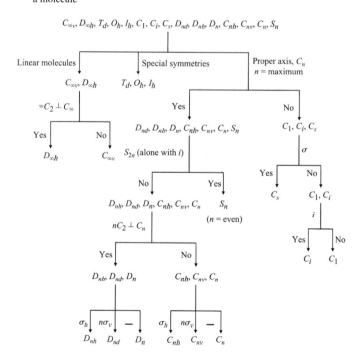

Table 6.2.4. Point group identification of some cyclic molecules

Compound	Perspective diagram	Projection diagram	Point group
$N_4(SF)_4$			S_4
S_8			D_{4d}
C_8H_8			D_{2d}
S_4N_4			D_{2d}

6.2.3 Dipole moment and optical activity

The existence of a dipole moment tells us something about the molecular geometry or symmetry, and vice versa. For example, the existence of a dipole moment for H_2O and NH_3 implies that the former cannot be linear and the latter cannot be trigonal planar. Indeed, it is not difficult to show that molecules that possess dipole moments belong only to point groups C_n, C_s, or C_{nv}.

For a molecule to be optically active, it and its mirror image cannot be superimposed on each other; i.e., they are dissymmetric (see Section 6.1.3). In the language of symmetry operations, this means that an optically active molecule cannot have any S_n element. In particular, since S_1 is σ and S_2 is i, any molecule that possesses a symmetry plane or a center of inversion cannot be optically active.

6.3 Character tables

In the next chapter, we will present various chemical applications of group theory, including molecular orbital and hybridization theories, spectroscopic selection rules, and molecular vibrations. Before proceeding to these topics, we first need to introduce the character tables of symmetry groups. It should be emphasized that the following treatment is in no way mathematically rigorous. Rather, the presentation is example- and application-oriented.

According to group theory, for a molecule belonging to a certain symmetry point group, each electronic or vibrational wavefunction of this molecule must have the symmetry of one of the irreducible representations of the point group. For simplicity, we sometimes replace the mathematical term "irreducible representation" with the simple name "symmetry species." For a given point group, there are only a limited number of symmetry species, and the properties of each symmetry species are contained in the characters of the species. The characters of all the irreducible representations of a point group are summarized in the character table of this group. A typical character table, that of the C_{2v} group, will now be examined in detail.

As shown in Table 6.3.1, we can see that the C_{2v} character table is divided into four parts: Areas I to IV. We now discuss these four areas one by one.

Area I. In this area, we see that the four symmetry operations of the C_{2v} group, E, C_2, $\sigma_v(xz)$, $\sigma'_v(yz)$, constitute four classes of operations; i.e., each class has one operation. According to group theory, for any point group, the number of irreducible representations is equal to the number of (symmetry operation) classes. Here, for C_{2v}, there are four irreducible representations and the characters of the representations are given in this area. For the first representation, called A_1, all the characters are 1, which signifies that any wavefunction with A_1 symmetry is symmetric (with character 1) with respect to all four symmetry operations. (Examples will be given later.) On the other hand, representation A_2 is symmetric with respect to operation E and C_2, but antisymmetric (with character -1) with respect to the two mirror reflections. For the remaining two representations, B_1 and B_2, they are antisymmetric with respect to C_2 and to one of the mirror reflections.

Additionally, it is noted that, mathematically, each irreducible representation is a square matrix and the character of the representation is the sum of the diagonal matrix elements. In the simple example of the C_{2v} character table, all the irreducible representations are one-dimensional; i.e., the characters are simply the lone elements of the matrices. For one-dimensional representations, the character for operation R, $\chi(R)$, is either 1 or -1.

Table 6.3.1. Character table of point group C_{2v}

C_{2v}	E	C_2	$\sigma_v(xz)$	$\sigma'_v(xz)$		
A_1	1	1	1	1	z	x^2, y^2, z^2
A_2	1	1	-1	-1	R_z	xy
B_1	1	-1	1	-1	x, R_y	xz
B_2	1	-1	-1	1	y, R_x	yz
Area II		Area I			Area III	Area IV

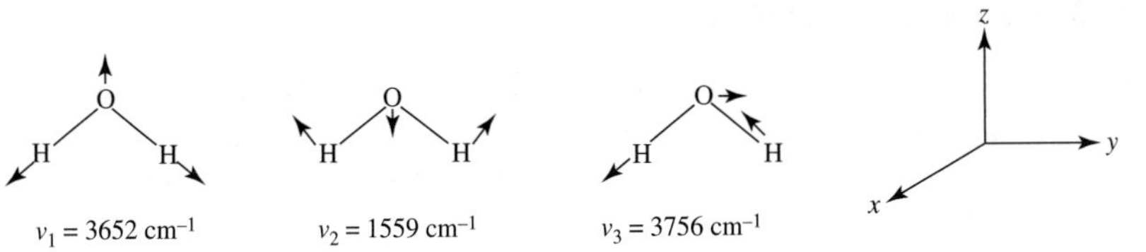

$\nu_1 = 3652$ cm^{-1} $\nu_2 = 1559$ cm^{-1} $\nu_3 = 3756$ cm^{-1}

Fig. 6.3.1.
Normal vibrational modes of the water molecule; ν_1 (symmetric stretch) and ν_2 (bending) have A_1 symmetry, while ν_3 (asymmetric stretch) has B_2 symmetry.

Before leaving the discussion of this area, let us consider a specific chemical example. The water molecule has C_{2v} symmetry, hence its normal vibrational modes have A_1, A_2, B_1, or B_2 symmetry. The three normal modes of H$_2$O are pictorially depicted in Fig. 6.3.1. From these illustrations, it can be readily seen that the atomic motions of the symmetric stretching mode, ν_1, are symmetric with respect to C_2, $\sigma_v(xz)$ and $\sigma'_v(yz)$; thus ν_1 has A_1 symmetry. Similarly, it is obvious that the bending mode, ν_2, also has A_1 symmetry. Finally, the atomic motions of the asymmetric stretching mode, ν_3, is antisymmetric with respect to C_2 and $\sigma_v(xz)$, but symmetric with respect to $\sigma'_v(yz)$. Hence ν_3 has B_2 symmetry. This example demonstrates all vibrational modes of a molecule must have the symmetry of one of the irreducible representations of the point group to which this molecule belongs. As will be shown later, molecular electronic wavefunctions may be also classified in this manner.

Area II. In this area, we have the Mulliken symbols for the representations. The meaning of these symbols carry is summarized in Table 6.3.2.

Referring to the C_{2v} character table and the nomenclature rules in Table 6.3.2, we can see that all symmetry species of the C_{2v} group are one-dimensional; i.e., they are either A or B representations. The first two symmetry species, with $\chi(C_2) = 1$, are clearly A representations, while the remaining two are B representations. To these representations, subscripts 1 or 2 can be easily added, according to the rules laid down in Table 6.3.2. (It is noted that the classification of B_1 and B_2 cannot be unambiguously assigned.) Also, the first representation, with all characters equal to 1, is called the totally symmetric representation (sometimes denoted as Γ_{TS}). Such a representation exists for all groups, even though Γ_{TS} is not called A_1 in all instances.

Before concluding the discussion on the notation of the irreducible representations, we use C_{2v} point group as an example to repeat what we mentioned previously: since this point group has only four symmetry species, A_1, A_2, B_1, and B_2, the electronic or vibrational wavefunctions of all C_{2v} molecules (such as H$_2$O, H$_2$S) must have the symmetry of one of these four representations. In addition, since this group has only one-dimensional representations, we will discuss degenerate representations such as E and T in subsequent examples.

Area III. In this part of the table, there will always be six symbols: x, y, z, R_x, R_y, R_z, which may be taken as the three components of the translational vector (x, y, z) and the three rotations around the x, y, z axes (R_x, R_y, R_z). For the C_{2v} character table, z appears in the row of A_1. This means that the z component of the translational vector has A_1 symmetry. Similarly, the x and y components have B_1 and B_2 symmetry, respectively. The symmetries of rotations R_x, R_y, and R_z can be seen from this table accordingly.

Table 6.3.2. Mulliken symbols for irreducible representations

Dimensionality

A: One-dimensional, i.e., $\chi(E) = 1$; symmetric with respect to C_n, i.e., $\chi(C_n) = 1$
B: One-dimensional, i.e., $\chi(E) = 1$; antisymmetric with respect to C_n, i.e., $\chi(C_n) = -1$
E: Two-dimensional, i.e., $\chi(E) = 2$
T: Three-dimensional, i.e., $\chi(E) = 3$
G: Four-dimensional, i.e., $\chi(E) = 4$
H: Five-dimensional, i.e., $\chi(E) = 5$

Superscript

$'$: Symmetric with respect to σ_h, i.e., $\chi(\sigma_h) = 1$
$''$: Antisymmetric with respect to σ_h, i.e., $\chi(\sigma_h) = -1$

Subscript

(i) g: Symmetric with respect to i, i.e., $\chi(i) = 1$
u : Antisymmetric with respect to i, i.e., $\chi(i) = -1$

(ii) For A and B representations,
1 : Symmetric with respect to C_2 or σ_v, i.e., $\chi(C_2)$ or $\chi(\sigma_v) = 1$
2 : Antisymmetric with respect to C_2 or σ_v, i.e., $\chi(C_2)$ or $\chi(\sigma_v) = -1$

(iii) For E and T representations, numerical subscripts 1, 2, ..., are also employed in a general way which does not always indicate symmetry relative to C_2 or σ_v.

For linear molecules with $C_{\infty v}$ and $D_{\infty h}$ symmetry

Σ^+ One-dimensional, symmetric with respect to σ_v or C_2, i.e., $\chi(E) = 1$, $\chi(\sigma_v)$ or $\chi(C_2) = 1$
Σ^- One-dimensional, antisymmetric with respect to σ_v or C_2, i.e., $\chi(E) = 1$, $\chi(\sigma_v)$ or $\chi(C_2) = -1$
Π Two-dimensional, i.e., $\chi(E) = 2$
Δ Two-dimensional, i.e., $\chi(E) = 2$
Φ Two-dimensional, i.e., $\chi(E) = 2$

Lower case letters ($a, b, e, t, \sigma, \pi, \ldots$) are used for the symmetries of individual orbitals; capital letters ($A, B, E, T, \Sigma, \Pi, \ldots$) for the symmetries of the overall states.

Another way of rationalizing these results is to treat x, y, z as p_x, p_y, p_z orbitals of central atom A in an AH$_n$ molecule. A molecule with C_{2v} symmetry is H$_2$S. A convention to set up the Cartesian coordinates for this molecule is as follows. Take the principle axis (C_2 in this case) as the z axis. Since H$_2$S is planar, we take the x axis to be perpendicular to the molecular plane. Finally, the y axis is taken so as to form a right-handed system. Following this convention, the p_x, p_y, p_z orbitals on sulfur in a H$_2$S molecule are shown in Fig. 6.3.2. When the orientations of these orbitals are examined, it is obvious that the p_z orbital is symmetric with respect to all four operations of the C_{2v} point group: E, C_2, $\sigma_v(xz)$, $\sigma'_v(yz)$. Thus the p_z orbital has A_1 symmetry. On the other hand, the p_x orbital is symmetric with respect to E and $\sigma_v(xz)$, but antisymmetric with respect to C_2 and $\sigma'_v(yz)$; hence p_x has B_1 symmetry.

Now it can be easily deduced that the p_y orbital has B_2 symmetry.

Area IV. This area contains the quadratic terms $x^2, y^2, z^2, xy, yz, xz$ or their linear combinations. Sometimes, instead of x^2 and y^2, we use their combinations $x^2 + y^2$ and $x^2 - y^2$. Following the discussion on Area III for the H$_2$S molecule, one might be inclined to consider these six entities as possible d functions of central atom A in an AH$_2$ molecule, as shown in Fig. 6.3.3. Noting that there exist only five independent d orbitals each possessing two nodes, the $x^2 + y^2$ function can be ruled out as it is devoid of nodal property. Furthermore, the

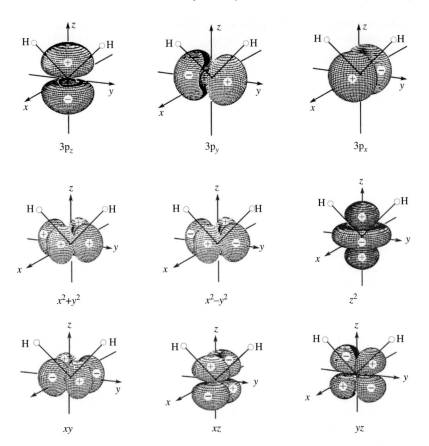

Fig. 6.3.2.
The three p orbitals of sulfur in a H$_2$S molecule, which lies in the yz plane.

Fig. 6.3.3.
The quadratic functions $x^2 + y^2$, $x^2 - y^2$, z^2 (more correctly $3z^2 - 1$), xy, xz, and yz. The $x^2 + y^2$ function cannot represent a d orbital because it does not have a node. The remaining five are taken as pictorial representations of the five d orbitals of central atom A in the AH$_2$ molecule. Note that xy, xz, yz, and $x^2 - y^2$ each has two nodal planes, whereas the nodes of z^2 are the curved surfaces of two co-axial cones sharing a common vertex.

presence of $x^2 - y^2$ implies that we can write down the equivalent terms $y^2 - z^2$ and $z^2 - x^2$; however, these three functions are not linearly independent. If $x^2 - y^2$ is chosen, the sum of $y^2 - z^2$ and $z^2 - x^2$ gives $y^2 - x^2$, which is the same as $x^2 - y^2$ except for a reversal in sign, but the difference of $y^2 - z^2$ and $z^2 - x^2$ gives the new function $2z^2 - x^2 - y^2$, which can be simplified to $3z^2 - 1$ since $x^2 + y^2 + z^2 = 1$ (polar coordinate r with unit length). In practice, it is conventional to denote the $3z^2 - 1$ function simply as z^2. The five d orbitals are therefore labelled z^2, $x^2 - y^2$, xy, yz, and xz.

Examining the six quadratic functions displayed in Fig. 6.3.3, it is clear that $x^2 + y^2$, $x^2 - y^2$, and z^2 are symmetric with respect to all four operations of the C_{2v} point group. Hence they belong to the totally symmetric species A_1, as do x^2 and y^2 since they are merely linear combinations of $x^2 + y^2$ and $x^2 - y^2$. Similarly, the xy function is symmetric with respect to E and C_2 and antisymmetric with respect to the reflection operations, and hence it has A_2 symmetry. The symmetries of the xz and yz functions can be determined easily in an analogous manner.

The character tables of most familiar point groups are listed in Appendix 6.1 at the end of this chapter. It is noted that some collections of character tables, such as those given in Appendix 6.1, also include an "Area V," where the symmetry species of the seven f orbitals (on the central atom) is indicated. Since the f orbitals are treated in Section 8.11 of this book, we will ignore this area in this initial discussion on the character tables.

Before we leave the C_{2v} character table, it is noted that the characters of all groups satisfy the "orthonormal" relation:

$$\frac{1}{h}\sum_R \chi_i(\mathbf{R})\chi_j(\mathbf{R}) = \delta_{ij} = \begin{cases} 1 & \text{when } i = j \\ 0 & \text{when } i \neq j \end{cases}, \quad (6.3.1)$$

where h is the order of the group (the number of operations in the group; for C_{2v}, $h = 4$) and $\chi_i(\mathbf{R})$ is the character of operation $\mathbf{R}$ for the irreducible representation Γ_i, etc. The relation can be illustrated by the following two examples:

When $\Gamma_i = \Gamma_j = A_2$, eq. (6.3.1) becomes

$$\frac{1}{4}[(1)(1) + (1)(1) + (-1)(-1) + (-1)(-1)] = 1. \quad (6.3.2)$$

When $\Gamma_i = A_2$ and $\Gamma_j = B_1$, eq. (6.3.1) now has the form

$$\frac{1}{4}[(1)(1) + (1)(-1) + (-1)(1) + (-1)(-1)] = 0. \quad (6.3.3)$$

After a fairly lengthy discussion on the character table of the C_{2v} group, we will proceed to those for the C_{3v} and C_{4v} point groups. The character tables of these two groups are shown in Tables 6.3.3 and 6.3.4, respectively. When Table 6.3.3 is examined, it is seen that the six operations of the C_{3v} group are divided into three classes: E; C_3, and C_3^{-1}; the three σ_v's. Hence there are three symmetry species for this group. According to group theory, if the dimensions of the irreducible representations of a group are called $\ell_1, \ell_2, \ldots$, then we have

$$\sum_i \ell_i^2 = h, \quad (6.3.4)$$

Table 6.3.3. Character table of point group C_{3v}

C_{3v}	E	$2C_3$	$3\sigma_v$		
A_1	1	1	1	z	$x^2 + y^2, z^2$
A_2	1	1	−1	R_z	
E	2	−1	0	$(x, y)(R_x, R_y)$	$(x^2 − y^2, xy)(xz, yz)$

Table 6.3.4. Character table of point group C_{4v}

C_{4v}	E	$2C_4$	C_2	$2\sigma_v$	$2\sigma_d$		
A_1	1	1	1	1	1	z	$x^2 + y^2, z^2$
A_2	1	1	1	−1	−1	R_z	
B_1	1	−1	1	1	−1		$x^2 − y^2$
B_2	1	−1	1	−1	1		xy
E	2	0	−2	0	0	$(x, y)(R_x, R_y)$	(xz, yz)

where, once again, h is the order of the group. So, for the C_{3v} group, with $h = 6$, $\ell_1 = 1$, $\ell_2 = 1$, and $\ell_3 = 2$.

When examining Table 6.3.3, the first one-dimensional (totally symmetric) representation is again called A_1. The second one-dimensional representation, which is symmetric with respect to the C_3 and C_3^{-1} operations and antisymmetric with respect to the three σ_v operations, is called A_2. From these results, it is now obvious that, for a given representation, operations belonging to the same class have the same character. Also, the A_1 and A_2 representations are "orthogonal" to each other, as stipulated by eq. (6.3.1):

$$\frac{1}{h}\sum_R \chi_{A_1}(R)\chi_{A_2}(R) = \left(\frac{1}{6}\right)[(1)(1) + 2(1)(1) + 3(1)(-1)] = 0. \quad (6.3.5)$$

The last symmetry species of the C_{3v} group is a two-dimensional E representation. If we consider the p orbitals of phosphorus in PF$_3$, a C_{3v} molecule, we can see that the p$_z$ orbital has A_1 symmetry. On the other hand, the p$_x$ and p$_y$ orbitals form an inseparable set with E symmetry. In other words, no single function can have E symmetry and it takes two functions to form an E set. It is straightforward to show that the E representation is orthogonal to both A_1 and A_2. Similarly, the $x^2 - y^2$ and xy functions for an E set and the xz and yz functions constitute yet another E set. In passing, it is pointed that the notation of representation E originates from the German word *entartet*, meaning degenerate or abnormal. It should not be confused with the identity operation E, which, as already mentioned, is from *Einheit* meaning unity.

A similar situation exists for the C_{4v} group, as shown in Table 6.3.4. Now there are five classes of operations and hence five irreducible representations. With $h = 8$, there are four one-dimensional and one two-dimensional representations [cf. eq. (6.3.4)]. For a hypothetical AX$_4$ (or AX$_5$) molecule with C_{4v} symmetry, the p$_z$ orbital on A has A_1 symmetry (note that the C_4 axis is taken as the z axis), while the p$_x$ and p$_y$ functions form an E pair. Regarding the d orbitals on A, the d$_{z^2}$ orbital has A_1 symmetry, the d$_{x^2-y^2}$ orbital has B_1 symmetry, the d$_{xy}$ orbital has B_2 symmetry, while the d$_{xz}$ and d$_{yz}$ orbitals form an E set.

6.4 The direct product and its use

6.4.1 The direct product

In this section we discuss the direct product of two irreducible representations. The concept of direct product may seem abstract initially. But it should become easily acceptable once its applications are discussed in subsequent sections.

When we obtain the direct product Γ_{ij} of two irreducible representations Γ_i and Γ_j, i.e.,

$$\Gamma_i \times \Gamma_j = \Gamma_{ij} \text{ (sometimes written as } \Gamma_i \otimes \Gamma_j\text{)}. \quad (6.4.1)$$

Γ_{ij} has the following properties:

(1) The dimension of Γ_{ij} is the product of the dimensions of Γ_i and Γ_j;

Table 6.4.1. Direct products of the representations of the C_{3v} group

C_{3v}	E	$2C_3$	$3\sigma_v$	
A_1	1	1	1	
A_2	1	1	−1	
E	2	−1	0	
$A_1 \times A_1$	1	1	1	$\equiv A_1$
$A_1 \times A_2$	1	1	−1	$\equiv A_2$
$A_1 \times E$	2	−1	0	$\equiv E$
$A_2 \times A_2$	1	1	1	$\equiv A_1$
$A_2 \times E$	2	−1	0	$\equiv E$
$E \times E$	4	1	0	$\equiv A_1 + A_2 + E^*$

* $n(A_1) = (1/6)[(4)(1) + 2(1)(1) + 3(0)(1)] = 1$;
$n(A_2) = (1/6)[(4)(1) + 2(1)(1) + 3(0)(-1)] = 1$;
$n(E) = (1/6)[(4)(2) + 2(1)(-1) + 3(0)(0)] = 1$.

(2) The character of Γ_{ij} for operation $\mathbf{R}$ is equal to the product of the characters of Γ_i and Γ_j for the same operation:

$$\chi_{ij}(\mathbf{R}) = \chi_i(\mathbf{R})\chi_j(\mathbf{R}). \tag{6.4.2}$$

(3) In general, Γ_{ij} is a reducible representation, i.e., a combination of irreducible representations. The number of times the irreducible representation Γ_k occurs in the direct product Γ_{ij} can be determined by:

$$n_k = \frac{1}{h} \sum_{\mathbf{R}} \chi_{ij}(\mathbf{R})\chi_k(\mathbf{R}). \tag{6.4.3}$$

Now let us apply these formulas to some examples. All the direct products formed by the three irreducible representations of the C_{3v} group are summarized in Table 6.4.1. These results show that the direct product of two irreducible representations is sometimes also an irreducible representation. When it is not, its constitution can be readily determined using eq. (6.4.3).

When eq. (6.4.3) is examined more closely, it is seen that, the division by h, the group order, makes the equation untenable for groups with an infinite order, namely, $C_{\infty v}$ and $D_{\infty h}$. Fortunately, in most cases the decomposition of a reducible representation for such groups can be achieved by inspection. Also, a work-around method is available in the literature (see references 19 and 20 listed at the end of the chapter). In any event, examples will be given later.

Based on the rather limited examples of direct products involving the representations of the C_{3v} group, we can draw the following conclusions (or guidelines) regarding the totally symmetric representation Γ_{TS}:

(1) The direct product of any representation Γ_i with Γ_{TS} is simply Γ_i.
(2) The direct product of any representation Γ_i with itself, i.e., $\Gamma_i \times \Gamma_i$, is or contains Γ_{TS}. In fact, only the direct product of a representation with itself is or contains Γ_{TS}.

These results are useful in identifying non-zero integrals involving in various applications of quantum mechanics to molecular systems. This is the subject to be taken up in the next section.

6.4.2 Identifying non-zero integrals and selection rules in spectroscopy

The importance of direct products becomes more obvious when we consider some of the integrals we encounter in quantum chemistry. Take the simple case of

$$I_1 = \int \psi_i \psi_j d\tau. \tag{6.4.4}$$

An example of I_1 is the overlap integral with ψ_i being an orbital on one atom and ψ_j an orbital on a different atom. Integral I_1 will vanish unless the integrand is invariant under all symmetry operations of the point group to which the molecule belongs. This condition is a generalization of the simple case of

$$I_2 = \int_{-\infty}^{\infty} y dx = \int_{-\infty}^{\infty} f(x) dx. \tag{6.4.5}$$

When $f(x)$ is an odd function, i.e., $f(-x) = -f(x)$, $I_2 = 0$. Integral I_2 vanishes because the integrand is not invariant to the inversion operation.

Returning to integral I_1, when we say that integrand $\psi_i \psi_j$ is invariant to all symmetry operations, it means, in group theory language, the representation for $\psi_i \psi_j$ (denoted Γ_{ij}) is or contains Γ_{TS}. In order for Γ_{ij} to be or contain Γ_{TS}, Γ_i (representation for ψ_i) and Γ_j (representation for ψ_j) must be one and the same. In chemical bonding language, for two atomic orbitals to have non-zero overlap, they must have the same symmetry.

Another integral that appears often in quantum chemistry is the energy interaction integral between orbitals ψ_i and ψ_j

$$I_3 = \int \psi_i \hat{H} \psi_j d\tau, \tag{6.4.6}$$

where $\hat{H}$ is the Hamiltonian operator. Since $\hat{H}$ is the operator for the energy of the molecule, it must be invariant to all symmetry operations; i.e., it has the symmetry of Γ_{TS}. It follows that the symmetry of the integrand of I_3 is simply $\Gamma_i \times \Gamma_{TS} \times \Gamma_j = \Gamma_i \times \Gamma_j$.

For $\Gamma_i \times \Gamma_j$ to be or to contain Γ_{TS}, $\Gamma_i = \Gamma_j$. In non-mathematical language, only orbitals of the same symmetry will interact with each other.

Integrals of the form

$$I_4 = \int \psi_i \hat{A} \psi_j d\tau, \tag{6.4.7}$$

where $\hat{A}$ is a certain quantum mechanical operator, are also common in quantum chemistry. For these integrals to be non-vanishing, $\Gamma_i \times \Gamma_A \times \Gamma_j$ is or contains Γ_{TS}. For such condition to hold, $\Gamma_i \times \Gamma_j$ must be or contain Γ_A. (So I_3 is a special case of I_4.) In the following, we consider a special case of I_4.

In spectroscopy, for an electronic transition between the ith state with wavefunction ψ_i and the jth state with wavefunction ψ_j, its intensity in the x, y, or

z direction (denoted as I_x, I_y, and I_z, respectively) is proportional to the square of an integral similar to I_4:

$$I_x \propto \left| \int \psi_i x \psi_j d\tau \right|^2, \qquad (6.4.8)$$

$$I_y \propto \left| \int \psi_i y \psi_j d\tau \right|^2, \qquad (6.4.9)$$

$$I_z \propto \left| \int \psi_i z \psi_j d\tau \right|^2. \qquad (6.4.10)$$

Stating the conditions for I_x, I_y, and I_z to be non-zero amounts to laying down the selection rule for the electric dipole transitions. Based on the foregoing discussion, it is now obvious that, for a transition between states with wavefunctions ψ_i and ψ_j to be allowed, direct product $\Gamma_i \times \Gamma_j$ must be or contain Γ_x, Γ_y, or Γ_z.

Let us now apply this method to see whether the electron transition between states A_2 and B_2 is allowed for a molecule with C_{2v} symmetry (such as H$_2$S):

$\Gamma_i = A_2, \Gamma_x = B_1, \Gamma_j = B_2 : A_2 \times B_1 \times B_2 = A_1$, allowed in the x direction;

$\Gamma_i = A_2, \Gamma_y = B_2, \Gamma_j = B_2 : A_2 \times B_2 \times B_2 = A_2$, forbidden in the y direction;

$\Gamma_i = A_2, \Gamma_z = A_1, \Gamma_j = B_2 : A_2 \times A_1 \times B_2 = B_1$, forbidden in the z direction.

So the transition $A_2 \leftrightarrow B_2$ is allowed in the x direction, or x-polarized. The polarization of other transitions can be determined in a similar manner. The polarization of the allowed transitions for a C_{2v} molecule is summarized in Fig. 6.4.1.

A well-known selection rule concerning centrosymmetric systems (those with a center of inversion) is the Laporte's rule. For such systems, states are either g (even) or u (odd). Laporte's rule states that only transitions between g and u states are allowed; i.e., transitions between two g states and those between two u states are forbidden. With the foregoing discussion, this rule can now be easily proved. For centrosymmetric molecules, the three components of the dipole moment vector are all u. For g $\leftrightarrow$ g transitions, the overall symmetry

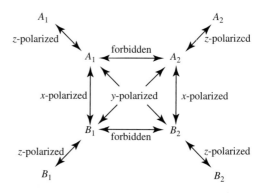

Fig. 6.4.1.
Polarization of all electric dipole transitions for a C_{2v} molecule.

of the transition dipole moment is g × u × g = u (antisymmetric) and hence the transition is not allowed. For u ↔ u transitions, we have u × u × u = u and the transitions are again forbidden. For u ↔ g ones, we get u × u × g = g (symmetric) and they are allowed.

In this section, we have shown that, by determining the symmetry of the integrand, we can identify the non-zero integrals and, in turn, derive the various selection rules in electronic spectroscopy.

6.4.3 Molecular term symbols

In our treatment on the bonding in linear molecules, we called the molecular orbitals of these systems either σ or π orbitals. We now know that σ and π are actually irreducible representations of the $C_{\infty v}$ or $D_{\infty h}$ groups. For nonlinear molecules, names such as σ and π no longer apply. Instead, we should use the symbols of the irreducible representations of the symmetry group to which the molecule belongs. So for the C_{2v} molecules such as H_2S and H_2O, the molecular orbitals are a_1, a_2, b_1, or b_2 orbitals. (Again we note that we use lower case letters for individual orbitals and capital letters for the overall states. This is a convention we also use for atoms: we have s, p, d, ... atomic orbitals and $S, P, D, \ldots$ atomic states.) If the molecular orbitals have the symmetries of the irreducible representations, so do the states, or the molecular term symbols, arising from the electronic configurations. This is the subject to be taken up in this section.

When the electronic configuration has only filled orbitals, there is only one state: $^1\Gamma_{TS}$, where Γ_{TS} is the totally symmetric representation. If the configuration has only one open shell (having symmetry Γ_i) with one electron, we again have only one state: $^2\Gamma_i$. When there are two open shells (having symmetries Γ_i and Γ_j) each with one electron, the symmetry of the state is $\Gamma_i \times \Gamma_j = \Gamma_{ij}$ and the overall states (including spin) are $^3\Gamma_{ij}$ and $^1\Gamma_{ij}$.

Let us now apply this method to a specific example. Consider the ethylene molecule with D_{2h} symmetry. As can be seen from the character table of the D_{2h} point group (Table 6.4.2), this group has eight symmetry species. Hence the molecular orbitals of ethylene must have the symmetry of one of these eight representations. In fact, the ground electronic configuration for ethylene is

$$(1a_g)^2(1b_{1u})^2(2a_g)^2(2b_{1u})^2(1b_{2u})^2(3a_g)^2(1b_{3g})^2(1b_{3u})^2(1b_{2g})^0.$$

Table 6.4.2. Character table of point group D_{2h}

D_{2h}	E	$C_2(z)$	$C_2(y)$	$C_2(x)$	i	$\sigma(xy)$	$\sigma(xz)$	$\sigma(yz)$		
A_g	1	1	1	1	1	1	1	1		x^2, y^2, z^2
B_{1g}	1	1	−1	−1	1	1	−1	−1	R_z	xy
B_{2g}	1	−1	1	−1	1	−1	1	−1	R_y	xz
B_{3g}	1	−1	−1	1	1	−1	−1	1	R_x	yz
A_u	1	1	1	1	−1	−1	−1	−1		
B_{1u}	1	1	−1	−1	−1	−1	1	1	z	
B_{2u}	1	−1	1	−1	−1	1	−1	1	y	
B_{3u}	1	−1	−1	1	−1	1	1	−1	x	

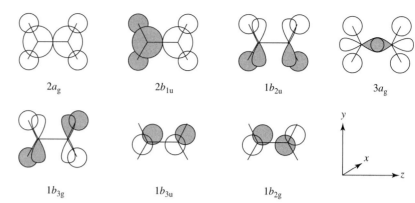

Fig. 6.4.2.
Some molecular orbitals of ethylene (point group D_{2h}). All are completely filled except $1b_{2g}$ in the ground state. The x axis points into the paper.

So the eight pairs of electrons of this molecule occupy *delocalized* molecular orbitals $1a_g$ to $1b_{3u}$, while the first vacant orbital is $1b_{2g}$. Note that the names of these orbitals are simply the symmetry species of the D_{2h} point group. In other words, molecular orbitals are labeled by the irreducible representations of the point group to which the molecule belongs. So for ethylene there are three filled orbitals with A_g symmetry; the one with the lowest energy is called $1a_g$, the next one is $2a_g$, etc. Similarly, there are two orbitals with B_{1u} symmetry and they are called $1b_{1u}$ and $2b_{1u}$. All the molecular orbitals listed above, except the first two, are illustrated pictorially in Fig. 6.4.2. By checking the D_{2h} character table with reference to the chosen coordinate system shown in Fig. 6.4.2, it can be readily confirmed that these orbitals do have the labeled symmetry. In passing, it is noted that the two filled molecular orbitals of ethylene not displayed in Fig. 6.4.2, $1a_g$ and $1b_{1u}$, are simply the sum and difference, respectively, of the two carbon 1s orbitals.

So for the ethylene molecule, with an electronic configuration having only filled orbitals, the ground state is simply ^1A_g, where A_g is the totally symmetric species of the D_{2h} point group. For the ethylene cation, with configuration $\ldots(1b_{3u})^1$, the electronic state is $^2B_{3u}$. For ethylene with the excited configuration $\ldots(1b_{3u})^1(1b_{2g})^1$, the electronic states are $^1B_{1u}$ and $^3B_{1u}$, as $B_{3u} \times B_{2g} = B_{1u}$.

If, in the configuration, there are a number of completely filled orbitals and one open (degenerate) shell with two (equivalent) electrons, there will be singlet as well as triplet states. To determine the spatial symmetry of these states, we make use of the formulas:

$$n = 2: \quad \chi(\boldsymbol{R}, \text{singlet}) = \tfrac{1}{2}[\chi^2(\boldsymbol{R}) + \chi(\boldsymbol{R}^2)], \quad (6.4.11)$$

$$\chi(\boldsymbol{R}, \text{triplet}) = \tfrac{1}{2}[\chi^2(\boldsymbol{R}) - \chi(\boldsymbol{R}^2)]. \quad (6.4.12)$$

In eqs. (6.4.11) and (6.4.12), $\chi^2(\boldsymbol{R})$ is the square of $\chi(\boldsymbol{R})$, while $\chi(\boldsymbol{R}^2)$ is the character of operation $\boldsymbol{R}^2$, which, according to group theory, must also be an operation of the group.

Table 6.4.3. Derivation of the electronic states arising from configuration $(e)^2$ of a molecule with C_{3v} symmetry

C_{3v}		E	$2C_3$	$3\sigma_v$	
A_1		1	1	1	
A_2		1	1	-1	
E		2	-1	0	
R		E	C_3	σ_v	
R^2		E	C_3	E	
$\chi(R)$ in E representation		2	-1	0	
$\chi^2(R)$ in E representation		4	1	0	
$\chi(R^2)$ in E representation		2	-1	2	
$\chi(R, \text{singlet}) = 1/2[\chi^2(R) + \chi(R^2)]$		3	0	1	$\equiv {}^1E + {}^1A_1$
$\chi(R, \text{triplet}) = 1/2[\chi^2(R) - \chi(R^2)]$		1	1	-1	$\equiv {}^3A_2$

For a molecule with C_{3v} symmetry, when there are two electrons in an e orbital; i.e., when we have a configuration of $(e)^2$ (note that it takes four electrons to completely fill the doubly degenerate e orbital), we can use eqs. (6.4.11) and (6.4.12) to derive the electronic states of this configuration, as shown in Table 6.4.3.

According to Hund's rule, among the three electronic states derived, ${}^3A_2$ should have the lowest energy. So for a molecule with C_{3v} symmetry, configurations involving e orbitals will lead to the following electronic states: $(e)^1$, ${}^2E$; $(e)^2$, ${}^3A_2$ (lowest energy), ${}^1E$, ${}^1A_1$; $(e)^3$, ${}^2E$; $(e)^4$, ${}^1A_1$. Note that $(e)^1$ and $(e)^3$, being conjugate configurations, have the same state(s), analogous to the p^1 and p^5 configurations for atomic systems.

Now let us turn to a three-dimensional irreducible representation for the first time. Take the T_{2g} representation of a molecule with O_h symmetry as an example. For configurations $(t_{2g})^1$ and $(t_{2g})^5$, the only electronic state is ${}^2T_{2g}$. (Note that a t_{2g} orbital can accommodate up to six electrons.) For configurations $(t_{2g})^2$ and $(t_{2g})^4$, the states are ${}^3T_{1g}$ (lowest energy), ${}^1T_{2g}$, ${}^1E_g$, and ${}^1A_{1g}$. The detailed derivation of these states is summarized in Table 6.4.4. For $(t_{2g})^6$, a closed-shell configuration, the only term is ${}^1A_{1g}$, a singlet state with totally symmetric spatial wavefunction. For the only remaining configuration, $(t_{2g})^3$, there will be doublet and quartet states. To derive these states, we require the following formulas:

$$n = 3: \quad \chi(R, \text{doublet}) = \frac{1}{3}[\chi^3(R) - \chi(R^3)], \tag{6.4.13}$$

$$\chi(R, \text{quartet}) = \left(\frac{1}{6}\right)[\chi^3(R) - 3\chi(R)\chi(R^2) + 2\chi(R^3)]. \tag{6.4.14}$$

In eqs. (6.4.13) and (6.4.14), $\chi(R^3)$ is the character of operation R^3 which of course is also one of the operations in the group. Applying these equations, we can derive the following states for the $(t_{2g})^3$ configuration: ${}^4A_{2g}$ (lowest energy), ${}^2T_{1g}$, ${}^2T_{2g}$, and ${}^2E_g$, as shown in Table 6.4.4.

Table 6.4.4. Derivation of the electronic states arising from the configurations $(t_{2g})^2$ and $(t_{2g})^3$ of a coordination complex with O_h symmetry

O_h	E	$8C_3$	$6C_2$	$6C_4$	$3C_4^2$	i	$6S_4$	$8S_6$	$3\sigma_h$	$6\sigma_d$	
A_{1g}	1	1	1	1	1	1	1	1	1	1	
A_{2g}	1	1	−1	−1	1	1	−1	1	1	−1	
E_g	2	−1	0	0	2	2	0	−1	2	0	
T_{1g}	3	0	−1	1	−1	3	1	0	−1	−1	
T_{2g}	3	0	1	−1	−1	3	−1	0	−1	1	
A_{1u}	1	1	1	1	1	−1	−1	−1	−1	−1	
A_{2u}	1	1	−1	−1	1	−1	1	−1	−1	1	
E_u	2	−1	0	0	2	−2	0	1	−2	0	
T_{1u}	3	0	−1	1	−1	−3	−1	0	1	1	
T_{2u}	3	0	1	−1	−1	−3	1	0	1	−1	
R	E	C_3	C_2	C_4	C_4^2	i	S_4	S_6	σ_h	σ_d	
R^2	E	C_3	E	C_4^2	E	E	C_4^2	C_3	E	E	
R^3	E	E	C_2	C_4	C_4^2	i	S_4	i	σ_h	σ_d	
$\chi(R)$ in T_{2g}	3	0	1	−1	−1	3	−1	0	−1	1	
$\chi(R^2)$ in T_{2g}	3	0	3	−1	3	3	−1	0	3	3	
$\chi(R^3)$ in T_{2g}	3	3	1	−1	−1	3	−1	3	−1	1	
$\chi(R,$ singlet$)$	6	0	2	0	2	6	0	0	2	2	$\equiv {}^1A_{1g} + {}^1E_g + {}^1T_{2g}$
$\chi(R,$ triplet$)$	3	0	−1	1	−1	3	1	0	−1	−1	$\equiv {}^3T_{1g}$
$\chi(R,$ doublet$)$	8	−1	0	0	0	8	0	−1	0	0	$\equiv {}^2E_g + {}^2T_{1g} + {}^2T_{2g}$
$\chi(R,$ quartet$)$	1	1	−1	−1	1	1	−1	1	1	−1	$\equiv {}^4A_{2g}$

For the sake of completeness, we give below the formulas for open-shell systems with four or five equivalent electrons:

$$n = 4: \quad \chi(R, \text{singlet}) = \left(\frac{1}{12}\right)[\chi^4(R) - 4\chi(R)\chi(R^3) + 3\chi^2(R^2)], \tag{6.4.15}$$

$$\chi(R, \text{triplet}) = \left(\frac{1}{8}\right)[\chi^4(R) - 2\chi^2(R)\chi(R^2) + 2\chi(R^4) - \chi^2(R^2)], \tag{6.4.16}$$

$$\chi(R, \text{quintet}) = \left(\frac{1}{24}\right)[\chi^4(R) - 6\chi^2(R)\chi(R^2) + 8\chi(R)\chi(R^3) - 6\chi(R^4) + 3\chi^2(R^2)]. \tag{6.4.17}$$

$$n = 5: \quad \chi(R, \text{doublet}) = \left(\frac{1}{24}\right)[\chi^5(R) - 2\chi^3(R)\chi(R^2) - 4\chi^2(R)\chi(R^3) + 6\chi(R)\chi(R^4) + 3\chi(R)\chi^2(R^2) - 4\chi(R^2)\chi(R^3)], \tag{6.4.18}$$

$$\chi(R, \text{quartet}) = \left(\frac{1}{30}\right)[\chi^5(R) - 5\chi^3(R)\chi(R^2) + 5\chi^2(R)\chi(R^3) + 5\chi(R^2)\chi(R^3) - 6\chi(R^5)], \tag{6.4.19}$$

Table 6.4.5. The direct product of $\Pi \times \Pi$ or a molecule with $C_{\infty v}$ symmetry

$C_{\infty v}$	E	$2C_\infty^\phi$	...	$\infty \sigma_v$
Σ^+	1	1	...	1
Σ^-	1	1	...	-1
Π	2	$2\cos\phi$	...	0
Δ	2	$2\cos 2\phi$	...	0
Φ	2	$2\cos 3\phi$	...	0
⋮	⋮	⋮		⋮
$\Gamma(\Pi \times \Pi)$	4	$4\cos^2\phi = 2 + 2\cos 2\phi$	...	0

$$\chi(R, \text{sextet}) = \left(\frac{1}{120}\right)[\chi^5(R) - 10\chi^3(R)\chi(R^2) + 20\chi^2(R)\chi(R^3)$$
$$- 30\chi(R)\chi(R^4) + 15\chi(R)\chi^2(R^2)$$
$$- 20\chi(R^2)\chi(R^3) + 24\chi(R^5)]. \qquad (6.4.20)$$

Equations (6.4.15) to (6.4.20) may be of use for systems with I_h symmetry. For such systems, we may be dealing with orbitals with G or H symmetry, which have four- and five-fold degeneracy, respectively.

To conclude this section, we discuss an example involving a linear molecule, for which eq. (6.4.3) is not applicable. Consider configuration $(1\pi)^1(2\pi)^1$ for linear molecule XYZ with $C_{\infty v}$ symmetry. As shown in Table 6.4.5, we can decompose the direct product $\Pi \times \Pi$ into the symmetry species of the $C_{\infty v}$ group, without using eq. (6.4.3) explicitly.

To decompose $\Gamma(\Pi \times \Pi)$, by inspecting $\chi(C_\infty^\phi)$, we can see that $\Gamma(\Pi \times \Pi)$ must contain Δ once. Elimination of Δ from $\Gamma(\Pi \times \Pi)$ yields characters (2, 2, ..., 0), which is simply the sum of Σ^+ and Σ^-. Hence, after considering the spin part, it can be readily seen that the states arising from configuration $(1\pi)^1(2\pi)^1$ are $^3\Delta$, $^1\Delta$, $^3\Sigma^+$, $^1\Sigma^+$, $^3\Sigma^-$, and $^1\Sigma^-$.

In addition to the derivation of selection rules in electronic spectroscopy and electronic states from a given configuration, there are many other fruitful applications of group theory to chemical problems, some of which will be discussed in the next chapter.

References

1. I. Hargittai and M. Hargittai, *Symmetry through the Eyes of a Chemist*, 2nd edn., Plenum Press, New York, 1995.
2. A. Vincent, *Molecular Symmetry and Group Theory*, 2nd edn., Wiley, Chichester, 2001.
3. R. L. Carter, *Molecular Symmetry and Group Theory*, Wiley, New York, 1998.
4. Y. Öhrn, *Elements of Molecular Symmetry*, Wiley, New York, 2000.
5. F. A. Cotton, *Chemical Applications of Group Theory*, 3rd edn., Wiley, New York, 1990.
6. G. Davison, *Group Theory for Chemists*, Macmillan, London, 1991.
7. A. M. Lesk, *Introduction to Symmetry and Group Theory for Chemists*, Kluwer, Dordrecht, 2004.

8. R. L. Flurry, Jr., *Symmetry Groups: Theory and Chemical Applications*, Prentice-Hall, Englewood Cliffs, NJ, 1980.
9. S. F. A. Kettle, *Symmetry and Structure: Readable Group Theory for Chemists*, Wiley, Chichester, 1995.
10. M. F. C. Ladd, *Symmetry and Group Theory in Chemistry*, Ellis Horwood, Chichester, 1998.
11. D. C. Harris and M. D. Bertolucci, *Symmetry and Spectroscopy: An Introduction to Vibrational and Electronic Spectroscopy*, Oxford University Press, New York, 1978.
12. B. D. Douglas and C. A. Hollingsworth, *Symmetry in Bonding and Spectra: An Introduction*, Academic Press, Orlando, 1985.
13. P. R. Bunker and P. Jensen, *Fundamentals of Molecular Symmetry*, Institute of Physics Publishing, Bristol, 2005.
14. P. W. M. Jacobs, *Group Theory with Applications in Chemical Physics*, Cambridge University Press, Cambridge. 2005.
15. B. S. Tsukerblat, *Group Theory in Chemistry and Spectroscopy: A Simple Guide to Advanced Usage,* Academic Press, London, 1994.
16. S. K. Kim, *Group Theoretical Methods and Applications to Molecules and Crystals*, Cambridge University Press, Cambridge, 1999.
17. W.-K. Li, Identification of molecular point groups. *J. Chem. Educ.* **70**, 485–7 (1993).
18. T. C. W. Mak and F.-C. Mok, Inorganic cages related to cubane and adamantane. *J. Cryst. Mol. Struct.* **8**, 183–91 (1978).
19. D. P. Strommen and E. R. Lippincott, Comments on infinite point groups. *J. Chem. Educ.* **49**, 341–2 (1972).
20. W. McInerny, A simple reduction process for the normal vibrational modes occurring in linear molecules. *J. Chem. Educ.* **82**, 140–4 (2005).
21. R. B. Shirts, Correcting two long-standing errors in point group symmetry character tables. *J. Chem. Educ.* **84**, 1882–4 (2007).

Appendix 6.1 Character tables of point groups

1. The non-axial groups

C_1	E
A	1

C_s	E	σ_h		
A'	1	1	x, y, R_z	x^2, y^2, z^2, xy	$xz^2, yz^2, x(x^2-3y^2), y(3x^2-y^2)$
A''	1	−1	z, R_x, R_y	yz, xz	$z^3, xyz, z(x^2-y^2)$

C_i	E	i			
A_g	1	1	R_x, R_y, R_z	$x^2, y^2, z^2, xy, xz, yz$	
A_u	1	−1	x, y, z		all cubic functions

2. The C_n groups

C_2	E	C_2			
A	1	1	z, R_z	x^2, y^2, z^2, xy	$z^2, xyz, z(x^2-y^2)$
B	1	−1	x, y, R_x, R_y	yz, xz	$xz^2, yz^2, x(x^2-3y^2), y(3x^2-y^2)$

C_3	E	C_3	C_3^2			$\epsilon = \exp(2\pi i/3)$
A	1	1	1	z, R_z	x^2+y^2, z^2	$z^3, x(x^2-3y^2), y(3x^2-y^2)$
E	$\begin{Bmatrix}1\\1\end{Bmatrix}$	$\begin{matrix}\epsilon\\\epsilon^*\end{matrix}$	$\begin{matrix}\epsilon^*\\\epsilon\end{matrix}$	$(x, y), (R_x, R_y)$	$(x^2-y^2, xy), (yz, xz)$	$(xz^2, yz^2), [xyz, z(x^2-y^2)]$

C_4	E	C_4	C_2	C_4^3			
A	1	1	1	1	z, R_z	x^2+y^2, z^2	z^3
B	1	−1	1	−1		x^2-y^2, xy	$xyz, z(x^2-y^2)$
E	$\begin{Bmatrix}1\\1\end{Bmatrix}$	$\begin{matrix}i\\-i\end{matrix}$	$\begin{matrix}-1\\-1\end{matrix}$	$\begin{matrix}-i\\i\end{matrix}\Bigr\}$	$(x,y), (R_x, R_y)$	(xz, yz)	$(xz^2, yz^2), [x(x^2-3y^2), y(3x^2-y^2)]$

C_5	E	C_5	C_5^2	C_5^3	C_5^4			$\epsilon = \exp(2\pi i/5)$
A	1	1	1	1	1	z, R_z	x^2+y^2, z^2	z^3
E_1	$\begin{Bmatrix}1\\1\end{Bmatrix}$	$\begin{matrix}\epsilon\\\epsilon^*\end{matrix}$	$\begin{matrix}\epsilon^2\\\epsilon^{2*}\end{matrix}$	$\begin{matrix}\epsilon^{2*}\\\epsilon^2\end{matrix}$	$\begin{matrix}\epsilon^*\\\epsilon\end{matrix}\Bigr\}$	$(x,y), (R_x, R_y)$	(yz, xz)	(xz^2, yz^2)
E_2	$\begin{Bmatrix}1\\1\end{Bmatrix}$	$\begin{matrix}\epsilon^2\\\epsilon^{2*}\end{matrix}$	$\begin{matrix}\epsilon^*\\\epsilon\end{matrix}$	$\begin{matrix}\epsilon\\\epsilon^*\end{matrix}$	$\begin{matrix}\epsilon^{2*}\\\epsilon^2\end{matrix}\Bigr\}$		(x^2-y^2, xy)	$[xyz, z(x^2-y^2)], [x(x^2-3y^2), y(3x^2-y^2)]$

C_6	E	C_6	C_3	C_2	C_3^2	C_6^5			$\epsilon \exp(2\pi i/6)$
A	1	1	1	1	1	1	z, R_z	x^2+y^2, z^2	z^3
B	1	−1	1	−1	1	−1			$x(x^2-3y^2), y(3x^2-y^2)$
E_1	$\begin{Bmatrix}1\\1\end{Bmatrix}$	$\begin{matrix}\epsilon\\\epsilon^*\end{matrix}$	$\begin{matrix}-\epsilon^*\\-\epsilon\end{matrix}$	$\begin{matrix}-1\\-1\end{matrix}$	$\begin{matrix}-\epsilon\\-\epsilon^*\end{matrix}$	$\begin{matrix}\epsilon^*\\\epsilon\end{matrix}\Bigr\}$	$(x,y), (R_x, R_y)$	(xz, yz)	(xz^2, yz^2)
E_2	$\begin{Bmatrix}1\\1\end{Bmatrix}$	$\begin{matrix}-\epsilon^*\\-\epsilon\end{matrix}$	$\begin{matrix}-\epsilon\\-\epsilon^*\end{matrix}$	$\begin{matrix}1\\1\end{matrix}$	$\begin{matrix}-\epsilon^*\\-\epsilon\end{matrix}$	$\begin{matrix}-\epsilon\\-\epsilon^*\end{matrix}\Bigr\}$		(x^2-y^2, xy)	$[xyz, z(x^2-y^2)]$

C_7	E	C_7	C_7^2	C_7^3	C_7^4	C_7^5	C_7^6			$\epsilon = \exp(2\pi i/7)$
A	1	1	1	1	1	1	1	z, R_z	x^2+y^2, z^2	z^3
E_1	$\begin{cases}1\\1\end{cases}$	$\epsilon \\ \epsilon^*$	$\epsilon^2 \\ \epsilon^{2*}$	$\epsilon^3 \\ \epsilon^{3*}$	$\epsilon^{3*} \\ \epsilon^3$	$\epsilon^{2*} \\ \epsilon^2$	$\epsilon^* \\ \epsilon$	$(x, y), (R_x, R_y)$	(xz, yz)	(xz^2, yz^2)
E_2	$\begin{cases}1\\1\end{cases}$	$\epsilon^2 \\ \epsilon^{2*}$	$\epsilon^{3*} \\ \epsilon^3$	$\epsilon^* \\ \epsilon$	$\epsilon \\ \epsilon^*$	$\epsilon^3 \\ \epsilon^{3*}$	$\epsilon^{2*} \\ \epsilon^2$		(x^2-y^2, xy)	$[xyz, z(x^2-y^2)]$
E_3	$\begin{cases}1\\1\end{cases}$	$\epsilon^3 \\ \epsilon^{3*}$	$\epsilon^* \\ \epsilon$	$\epsilon^2 \\ \epsilon^{2*}$	$\epsilon^{2*} \\ \epsilon^2$	$\epsilon \\ \epsilon^*$	$\epsilon^{3*} \\ \epsilon^3$			$[x(x^2-3y^2), y(3x^2-y^2)]$

C_8	E	C_8	C_4	C_2	C_4^3	C_8^3	C_8^5	C_8^7			$\epsilon = \exp(2\pi i/8)$
A	1	1	1	1	1	1	1	1	z, R_z	x^2+y^2, z^2	z^3
B	1	-1	1	1	1	-1	-1	-1			
E_1	$\begin{cases}1\\1\end{cases}$	$\epsilon \\ \epsilon^*$	$i \\ -i$	$-1 \\ -1$	$-i \\ i$	$-\epsilon^* \\ -\epsilon$	$-\epsilon \\ -\epsilon^*$	$\epsilon^* \\ \epsilon$	$(x, y), (R_x, R_y)$	(xz, yz)	(xz^2, yz^2)
E_2	$\begin{cases}1\\1\end{cases}$	$i \\ -i$	$-1 \\ -1$	$1 \\ 1$	$-1 \\ -1$	$-i \\ i$	$i \\ -i$	$-i \\ i$		(x^2-y^2, xy)	$[xyz, z(x^2-y^2)]$
E_3	$\begin{cases}1\\1\end{cases}$	$-\epsilon \\ -\epsilon^*$	$i \\ -i$	$-1 \\ -1$	$-i \\ i$	$\epsilon^* \\ \epsilon$	$\epsilon \\ \epsilon^*$	$-\epsilon^* \\ -\epsilon$			$[x(x^2-3y^2), y(3x^2-y^2)]$

3. The D_n groups

D_2	E	$C_2(z)$	$C_2(y)$	$C_2(x)$			
A	1	1	1	1		x^2, y^2, z^2	xyz
B_1	1	1	-1	-1	z, R_z	xy	$z^3, z(x^2-y^2)$
B_2	1	-1	1	-1	y, R_y	xz	$yz^2, y(3x^2-y^2)$
B_3	1	-1	-1	1	x, R_x	yz	$xz^2, x(x^2-3y^2)$

(x axis coincident with C_2)

D_3	E	$2C_3$	$3C_2$			
A_1	1	1	1		x^2+y^2, z^2	$x(x^2-3y^2)$
A_2	1	1	-1	z, R_z		$z^3, y(3x^2-y^2)$
E	2	-1	0	$(x,y), (R_x, R_y)$	$(x^2-y^2, xy), (xz, yz)$	$(xz^2, yz^2), [xyz, z(x^2-y^2)]$

(x axis coincident with C_2')

D_4	E	$2C_4$	$C_2(=C_4^2)$	$2C_2'$	$2C_2''$			
A_1	1	1	1	1	1		x^2+y^2, z^2	
A_2	1	1	1	-1	-1	z, R_z		z^3
B_1	1	-1	1	1	-1		x^2-y^2	xyz
B_2	1	-1	1	-1	1		xy	$z(x^2-y^2)$
E	2	0	-2	0	0	$(x,y), (R_x, R_y)$	(xz, yz)	$(xz^2, yz^2), [x(x^2-3y^2), y(3x^2-y^2)]$

(x axis coincident with C_2)

D_5	E	$2C_5$	$2C_5^2$	$5C_2$			
A_1	1	1	1	1		x^2+y^2, z^2	
A_2	1	1	1	-1	z, R_z		z^3
E_1	2	$2\cos 72°$	$2\cos 144°$	0	$(x,y), (R_x, R_y)$	(xz, yz)	$(xz^2, yz^2), [x(x^2-3y^2), y(3x^2-y^2)]$
E_2	2	$2\cos 144°$	$2\cos 72°$	0		(x^2-y^2, xy)	$[xyz, z(x^2-y^2)], [x(x^2-y^2)], y(3x^2-y^2)]$

D_6	E	$2C_6$	$2C_3$	C_2	$3C_2'$	$3C_2''$		(x axis coincident with C_2')
								x^2+y^2, z^2	z^3
A_1	1	1	1	1	1	1		
A_2	1	1	1	1	−1	−1	z, R_z	
B_1	1	−1	1	−1	1	−1			$x(x^2-3y^2)$
B_2	1	−1	1	−1	−1	1			$y(3x^2-y^2)$
E_1	2	1	−1	−2	0	0	$(x, y), (R_x, R_y)$	(xz, yz)	(xz^2, yz^2)
E_2	2	−1	−1	2	0	0		(x^2-y^2, xy)	$[xyz, z(x^2-y^2)]$

4. The C_{nv} groups

C_{2v}	E	C_2	$\sigma_v(xz)$	$\sigma_v'(yz)$			
A_1	1	1	1	1	z	x^2, y^2, z^2	$z^3, z(x^2-y^2)$
A_2	1	1	−1	−1	R_z	xy	xyz
B_1	1	−1	1	−1	x, R_y	xz	$xz^2, x(x^2-3y^2)$
B_2	1	−1	−1	1	y, R_x	yz	$yz^2, y(3x^2-y^2)$

C_{3v}	E	$2C_3$	$3\sigma_v$			
A_1	1	1	1	z	x^2+y^2, z^2	$z^3, x(x^2-3y^2)$
A_2	1	1	−1	R_z		$y(3x^2-y^2)$
E	2	−1	0	$(x, y), (R_x, R_y)$	$(x^2-y^2, xy), (xz, yz)$	$(xz^2, yz^2), [xyz, z(x^2-y^2)]$

C_{4v}	E	$2C_4$	C_2	$2\sigma_v$	$2\sigma_d$			
A_1	1	1	1	1	1	z	x^2+y^2, z^2	
A_2	1	1	1	-1	-1	R_z		
B_1	1	-1	1	1	-1		x^2-y^2	
B_2	1	-1	1	-1	1		xy	
E	2	0	-2	0	0	$(x, y), (R_x, R_y)$	(xz, yz)	

C_{5v}	E	$2C_5$	$2C_5^2$	$5\sigma_v$			
A_1	1	1	1	1	z	x^2+y^2, z^2	z^3
A_2	1	1	1	-1	R_z		
E_1	2	$2\cos 72°$	$2\cos 144°$	0	$(x, y), (R_x, R_y)$	(xz, yz)	(xz^2, yz^2)
E_2	2	$2\cos 144°$	$2\cos 72°$	0		(x^2-y^2, xy)	$[xyz, z(x^2-y^2)], [x(x^2-3y^2), y(3x^2-y^2)]$

C_{6v}	E	$2C_6$	$2C_3$	C_2	$3\sigma_v$	$3\sigma_d$			
A_1	1	1	1	1	1	1	z	x^2+y^2, z^2	z^3
A_2	1	1	1	1	-1	-1	R_z		
B_1	1	-1	1	-1	1	-1			$x(x^2-3y^2)$
B_2	1	-1	1	-1	-1	1			$y(3x^2-y^2)$
E_1	2	1	-1	-2	0	0	$(x, y), (R_x, R_y)$	(xz, yz)	(xz^2, yz^2)
E_2	2	-1	-1	2	0	0		(x^2-y^2, xy)	$[xyz, z(x^2-y^2)]$

5. The C_{nh} groups

C_{2h}	E	C_2	i	σ_h		
A_g	1	1	1	1	R_z	x^2, y^2, z^2, xy
B_g	1	−1	1	−1	R_x, R_y	xz, yz
A_u	1	1	−1	−1	z	
B_u	1	−1	−1	1	x, y	

C_{3h}	E	C_3	C_3^2	σ_h	S_3	S_3^5			$\epsilon = \exp(2\pi i/3)$
A'	1	1	1	1	1	1	R_z	x^2+y^2, z^2	
E'	$\begin{Bmatrix}1\\1\end{Bmatrix}$	ϵ ϵ^*	ϵ^* ϵ	1 1	ϵ ϵ^*	ϵ^* ϵ	(x, y)	(x^2-y^2, xy)	(xz^2, yz^2)
A''	1	1	1	−1	−1	−1	z		z^3
E''	$\begin{Bmatrix}1\\1\end{Bmatrix}$	ϵ ϵ^*	ϵ^* ϵ	−1 −1	$-\epsilon$ $-\epsilon^*$	$-\epsilon^*$ $-\epsilon$	(R_x, R_y)	(xz, yz)	$[xyz, z(x^2-y^2)]$

C_{4h}	E	C_4	C_2	C_4^3	i	S_4^3	σ_h	S_4		
A_g	1	1	1	1	1	1	1	1	R_z	x^2+y^2, z^2
B_g	1	−1	1	−1	1	−1	1	−1		x^2-y^2, xy
E_g	$\begin{Bmatrix}1\\1\end{Bmatrix}$	i $-i$	−1 −1	$-i$ i	1 1	i $-i$	−1 −1	$-i$ i	(R_x, R_y)	(xz, yz)
A_u	1	1	1	1	−1	−1	−1	−1	z	z^3
B_u	1	−1	1	−1	−1	1	−1	1		$xyz, z(x^2-y^2)$
E_u	$\begin{Bmatrix}1\\1\end{Bmatrix}$	i $-i$	−1 −1	$-i$ i	−1 −1	$-i$ i	1 1	i $-i$	xy	$(xz^2, yz^2), [x(x^2-3y^2), y(3x^2-y^2)]$

C_{5h}	E	C_5	C_5^2	C_5^3	C_5^4	σ_h	S_5	S_5^7	S_5^3	S_5^9			$\epsilon = \exp(2\pi i/5)$
A'	1	1	1	1	1	1	1	1	1	1	R_z	x^2+y^2, z^2	
E_1'	$\begin{cases}1\\1\end{cases}$	ϵ ϵ^*	ϵ^2 ϵ^{2*}	ϵ^{2*} ϵ^2	ϵ^* ϵ	1 1	ϵ ϵ^*	ϵ^2 ϵ^{2*}	ϵ^{2*} ϵ^2	ϵ^* ϵ	(x, y)		
E_2'	$\begin{cases}1\\1\end{cases}$	ϵ^2 ϵ^{2*}	ϵ^* ϵ	ϵ ϵ^*	ϵ^{2*} ϵ^2	1 1	ϵ^2 ϵ^{2*}	ϵ^* ϵ	ϵ ϵ^*	ϵ^{2*} ϵ^2		(x^2-y^2, xy)	
A''	1	1	1	1	1	-1	-1	-1	-1	-1	z		
E_1''	$\begin{cases}1\\1\end{cases}$	ϵ ϵ^*	ϵ^2 ϵ^{2*}	ϵ^{2*} ϵ^2	ϵ^* ϵ	-1 -1	$-\epsilon$ $-\epsilon^*$	$-\epsilon^2$ $-\epsilon^{2*}$	$-\epsilon^{2*}$ $-\epsilon^2$	$-\epsilon^*$ $-\epsilon$	(R_x, R_y)	(xz, yz)	(xz^2, yz^2)
E_2''	$\begin{cases}1\\1\end{cases}$	ϵ^2 ϵ^{2*}	ϵ^* ϵ	ϵ ϵ^*	ϵ^{2*} ϵ^2	-1 -1	$-\epsilon^2$ $-\epsilon^{2*}$	$-\epsilon^*$ $-\epsilon$	$-\epsilon$ $-\epsilon^*$	$-\epsilon^{2*}$ $-\epsilon^2$			$[x(x^2-3y^2), y(3x^2-y^2)]$ z^3 $[xyz, z(x^2-y^2)]$

C_{6h}	E	C_6	C_3	C_2	C_3^2	C_6^5	i	S_3^5	S_6^5	σ_h	S_6	S_3			$\epsilon = \exp(2\pi i/6)$
A_g	1	1	1	1	1	1	1	1	1	1	1	1	R_z	x^2+y^2, z^2	
B_g	1	-1	1	-1	1	-1	1	-1	1	-1	1	-1			
E_{1g}	$\begin{cases}1\\1\end{cases}$	ϵ ϵ^*	$-\epsilon^*$ $-\epsilon$	-1 -1	$-\epsilon$ $-\epsilon^*$	ϵ^* ϵ	1 1	ϵ ϵ^*	$-\epsilon^*$ $-\epsilon$	-1 -1	$-\epsilon$ $-\epsilon^*$	ϵ^* ϵ	(R_x, R_y)	(xz, yz)	
E_{2g}	$\begin{cases}1\\1\end{cases}$	$-\epsilon^*$ $-\epsilon$	$-\epsilon$ $-\epsilon^*$	1 1	$-\epsilon^*$ $-\epsilon$	$-\epsilon$ $-\epsilon^*$	1 1	$-\epsilon^*$ $-\epsilon$	$-\epsilon$ $-\epsilon^*$	1 1	$-\epsilon^*$ $-\epsilon$	$-\epsilon$ $-\epsilon^*$		(x^2-y^2, xy)	
A_u	1	1	1	1	1	1	-1	-1	-1	-1	-1	-1	z		
B_u	1	-1	1	-1	1	-1	-1	1	-1	1	-1	1			z^3 $x(x^2-3y^2), y(3x^2-y^2)$
E_{1u}	$\begin{cases}1\\1\end{cases}$	ϵ ϵ^*	$-\epsilon^*$ $-\epsilon$	-1 -1	$-\epsilon$ $-\epsilon^*$	ϵ^* ϵ	-1 -1	$-\epsilon$ $-\epsilon^*$	ϵ^* ϵ	1 1	ϵ ϵ^*	$-\epsilon^*$ $-\epsilon$	(x, y)		(xz^2, yz^2)
E_{2u}	$\begin{cases}1\\1\end{cases}$	$-\epsilon^*$ $-\epsilon$	$-\epsilon$ $-\epsilon^*$	1 1	$-\epsilon^*$ $-\epsilon$	$-\epsilon$ $-\epsilon^*$	-1 -1	ϵ^* ϵ	ϵ ϵ^*	-1 -1	ϵ^* ϵ	ϵ ϵ^*			$[xyz, z(x^2-y^2)]$

6. The D_{nh} groups

D_{2h}	E	$C_6(z)$	$C_2(y)$	$C_2(x)$	i	$\sigma(xy)$	$\sigma(xz)$	$\sigma(yz)$		
A_g	1	1	1	1	1	1	1	1		x^2, y^2, z^2
B_{1g}	1	1	-1	-1	1	1	-1	-1	R_z	xy
B_{2g}	1	-1	1	-1	1	-1	1	-1	R_y	xz
B_{3g}	1	-1	-1	1	1	-1	-1	1	R_x	yz
A_u	1	1	1	1	-1	-1	-1	-1		
B_{1u}	1	1	-1	-1	-1	-1	1	1	z	
B_{2u}	1	-1	1	-1	-1	1	-1	1	y	
B_{3u}	1	-1	-1	1	-1	1	1	-1	x	

D_{3h}	E	$2C_3$	$3C_2$	σ_h	$2S_3$	$3\sigma_v$			(x axis coincident with C_2)
A_1'	1	1	1	1	1	1		x^2+y^2, z^2	$x(x^2-3y^2)$
A_2'	1	1	-1	1	1	-1	R_z		$y(3x^2-y^2)$
E'	2	-1	0	2	-1	0	(x,y)	(x^2-y^2, xy)	(xz^2, yz^2)
A_1''	1	1	1	-1	-1	-1			
A_2''	1	1	-1	-1	-1	1	z		z^3
E''	2	-1	0	-2	1	0	(R_x, R_y)	(xz, yz)	$[xyz, z(x^2-y^2)]$

xyz
$z^3, z(x^2-y^2)$
$yz^2, y(3x^2-y^2)$
$xz^2, x(x^2-3y^2)$

D_{4h}	E	$2C_4$	C_2	$2C_2'$	$2C_2''$	i	$2S_4$	σ_h	$2\sigma_v$	$2\sigma_d$			(x axis coincident with C_2')
A_{1g}	1	1	1	1	1	1	1	1	1	1			x^2+y^2, z^2
A_{2g}	1	1	1	−1	−1	1	1	1	−1	−1	R_z		
B_{1g}	1	−1	1	1	−1	1	−1	1	1	−1			x^2-y^2
B_{2g}	1	−1	1	−1	1	1	−1	1	−1	1			xy
E_g	2	0	−2	0	0	2	0	−2	0	0	(R_x, R_y)		(xz, yz)
A_{1u}	1	1	1	1	1	−1	−1	−1	−1	−1			
A_{2u}	1	1	1	−1	−1	−1	−1	−1	1	1	z		
B_{1u}	1	−1	1	1	−1	−1	1	−1	−1	1			z^3
B_{2u}	1	−1	1	−1	1	−1	1	−1	1	−1			xyz
E_u	2	0	−2	0	0	−2	0	2	0	0	(x,y)		$z(x^2-y^2)$ $(xz^2, yz^2), [x(x^2-3y^2), y(3x^2-y^2)]$

D_{5h}	E	$2C_5$	$2C_5^2$	$5C_2'$	σ_h	$2S_5$	$2S_5^3$	$2\sigma_v$			(x axis coincident with C_2')
A_1'	1	1	1	1	1	1	1	1			x^2+y^2, z^2
A_2'	1	1	1	−1	1	1	1	−1	R_z		
E_1'	2	2 cos 72°	2 cos 144°	0	2	2 cos 72°	2 cos 144°	0	(x,y)		(xz^2, yz^2)
E_2'	2	2 cos 144°	2 cos 72°	0	2	2 cos 144°	2 cos 72°	0			(x^2-y^2, xy) $[x(x^2-3y^2), y(3x^2-y^2)]$
A_1''	1	1	1	1	−1	−1	−1	−1			
A_2''	1	1	1	−1	−1	−1	−1	1	z		z^3
E_1''	2	2 cos 72°	2 cos 144°	0	−2	−2 cos 72°	−2 cos 144°	0	(R_x, R_y)		(xz, yz)
E_2''	2	2 cos 144°	2 cos 72°	0	−2	−2 cos 144°	−2 cos 72°	0			$[xyz, z(x^2-y^2)]$

D_{6h}	E	$2C_6$	$2C_3$	C_2	$3C_2'$	$3C_2''$	i	$2S_3$	$2S_6$	σ_h	$3\sigma_d$	$3\sigma_v$			(x axis coincident with C_2')
A_{1g}	1	1	1	1	1	1	1	1	1	1	1	1		x^2+y^2, z^2	
A_{2g}	1	1	1	1	−1	−1	1	1	1	1	−1	−1	R_z		
B_{1g}	1	−1	1	−1	1	−1	1	−1	1	−1	1	−1			
B_{2g}	1	−1	1	−1	−1	1	1	−1	1	−1	−1	1			
E_{1g}	2	1	−1	−2	0	0	2	1	−1	−2	0	0	(R_x, R_y)	(xz, yz)	
E_{2g}	2	−1	−1	2	0	0	2	−1	−1	2	0	0		(x^2-y^2, xy)	
A_{1u}	1	1	1	1	1	1	−1	−1	−1	−1	−1	−1			
A_{2u}	1	1	1	1	−1	−1	−1	−1	−1	−1	1	1	z		z^3
B_{1u}	1	−1	1	−1	1	−1	−1	1	−1	1	−1	1			$x(x^2-3y^2)$
B_{2u}	1	−1	1	−1	−1	1	−1	1	−1	1	1	−1			$y(3x^2-y^2)$
E_{1u}	2	1	−1	−2	0	0	−2	−1	1	2	0	0	(x, y)		(xz^2, yz^2)
E_{2u}	2	−1	−1	2	0	0	−2	1	1	−2	0	0			$[xyz, z(x^2-y^2)]$

D_{8h}	E	$2C_8$	$2C_8^3$	$2C_4$	C_2	$4C_2'$	$4C_2''$	i	$2S_8^3$	$2S_8$	$2S_4$	σ_h	$4\sigma_v$	$4\sigma_d$			(x axis coincident with C_2')
A_{1g}	1	1	1	1	1	1	1	1	1	1	1	1	1	1		x^2+y^2, z^2	
A_{2g}	1	1	1	1	1	−1	−1	1	1	1	1	1	−1	−1	R_z		
B_{1g}	1	−1	−1	1	1	1	−1	1	−1	−1	1	1	1	−1			
B_{2g}	1	−1	−1	1	1	−1	1	1	−1	−1	1	1	−1	1			
E_{1g}	2	$(2)^{1/2}$	$-(2)^{1/2}$	0	−2	0	0	2	$(2)^{1/2}$	$-(2)^{1/2}$	0	−2	0	0	(R_x, R_y)	(xz, yz)	
E_{2g}	2	0	0	−2	2	0	0	2	0	0	−2	2	0	0		(x^2-y^2, xy)	
E_{3g}	2	$-(2)^{1/2}$	$(2)^{1/2}$	0	−2	0	0	2	$-(2)^{1/2}$	$(2)^{1/2}$	0	−2	0	0			
A_{1u}	1	1	1	1	1	1	1	−1	−1	−1	−1	−1	−1	−1			
A_{2u}	1	1	1	1	1	−1	−1	−1	−1	−1	−1	−1	1	1	z		z^3
B_{1u}	1	−1	−1	1	1	1	−1	−1	1	1	−1	−1	−1	1			
B_{2u}	1	−1	−1	1	1	−1	1	−1	1	1	−1	−1	1	−1			
E_{1u}	2	$(2)^{1/2}$	$-(2)^{1/2}$	0	−2	0	0	−2	$-(2)^{1/2}$	$(2)^{1/2}$	0	2	0	0	(x, y)		(xz^2, yz^2)
E_{2u}	2	0	0	−2	2	0	0	−2	0	0	2	−2	0	0			$[xyz, z(x^2-y^2)]$
E_{3u}	2	$-(2)^{1/2}$	$(2)^{1/2}$	0	−2	0	0	−2	$(2)^{1/2}$	$-(2)^{1/2}$	0	2	0	0			$[x(x^2-3y^2), y(3x^2-y^2)]$

7. The D_{nd} groups

D_{2d}	E	$2S_4$	C_2	$2C_2'$	$2\sigma_d$		(x axis coincident with C_2')
A_1	1	1	1	1	1		x^2+y^2, z^2
A_2	1	1	1	−1	−1	R_z	
B_1	1	−1	1	1	−1		x^2-y^2
B_2	1	−1	1	−1	1	z	xyz
E	2	0	−2	0	0	$(x,y)(R_x,R_y)$	(xz,yz)

Additional basis functions for D_{2d}:
- B_1: $z(x^2-y^2)$
- B_2: xyz
- E: $(xz^2, yz^2), [x(x^2-3y^2), y(3x^2-y^2)]$
- E: $(xz^2, yz^2), [x(x^2-3y^2), y(3x^2-y^2)]$

D_{3d}	E	$2C_3$	$3C_2$	i	$2S_6$	$3\sigma_d$		(x axis coincident with C_2)
A_{1g}	1	1	1	1	1	1		x^2+y^2, z^2
A_{2g}	1	1	−1	1	1	−1	R_z	
E_g	2	−1	0	2	−1	0	(R_x, R_y)	$(x^2-y^2, xy)(xz, yz)$
A_{1u}	1	1	1	−1	−1	−1		
A_{2u}	1	1	−1	−1	−1	1	z	
E_u	2	−1	0	−2	1	0	(x,y)	

Additional basis functions for D_{3d}:
- A_{1u}: $x(x^2-3y^2)$
- A_{2u}: $y(3x^2-y^2), z^3$
- E_u: $(xz^2, yz^2), [xyz, z(x^2-y^2)]$

D_{4d}	E	$2S_8$	$2C_4$	$2S_8^3$	C_2	$4C_2'$	$4\sigma_d'$		(x axis coincident with C_2')
A_1	1	1	1	1	1	1	1		x^2+y^2, z^2
A_2	1	1	1	1	1	−1	−1	R_z	
B_1	1	−1	1	−1	1	1	−1		
B_2	1	−1	1	−1	1	−1	1	z	
E_1	2	$(2)^{1/2}$	0	$-(2)^{1/2}$	−2	0	0	(x,y)	
E_2	2	0	−2	0	2	0	0		(x^2-y^2, xy)
E_3	2	$-(2)^{1/2}$	0	$(2)^{1/2}$	−2	0	0	(R_x, R_y)	(xz, yz)

Additional basis functions for D_{4d}:
- B_2: z^3
- E_1: (xz^2, yz^2)
- E_2: $[xyz, z(x^2-y^2)]$
- E_3: $[x(x^2-3y^2), y(3x^2-y^2)]$

D_{5d}	E	$2C_5$	$2C_5^2$	$5C_2$	i	$2S_{10}^3$	$2S_{10}$	$5\sigma_d$			(x axis coincident with C_2)
										x^2+y^2, z^2	
A_{1g}	1	1	1	1	1	1	1	1			
A_{2g}	1	1	1	−1	1	1	1	−1	R_z		
E_{1g}	2	$2\cos 72°$	$2\cos 144°$	0	2	$2\cos 72°$	$2\cos 144°$	0	(R_x, R_y)	(xz, yz)	
E_{2g}	2	$2\cos 144°$	$2\cos 72°$	0	2	$2\cos 144°$	$2\cos 72°$	0		$(x^2−y^2, xy)$	
A_{1u}	1	1	1	1	−1	−1	−1	−1			
A_{2u}	1	1	1	−1	−1	−1	−1	1	z		z^3
E_{1u}	2	$2\cos 72°$	$2\cos 144°$	0	−2	$−2\cos 72°$	$−2\cos 144°$	0	(x, y)		(xz^2, yz^2)
E_{2u}	2	$2\cos 144°$	$2\cos 72°$	0	−2	$−2\cos 144°$	$−2\cos 72°$	0			$[xyz, z(x^2−y^2)],$ $[x(x^2−3y^2), y(3x^2−y^2)]$

D_{6d}	E	$2S_{12}$	$2C_6$	$2S_4$	$2C_3$	$2S_{12}^5$	C_2	$6C_2'$	$6\sigma_d$			(x axis coincident with C_2')
											x^2+y^2, z^2	
A_1	1	1	1	1	1	1	1	1	1			
A_2	1	1	1	1	1	1	1	−1	−1	R_z		
B_1	1	−1	1	−1	1	−1	1	1	−1			
B_2	1	−1	1	−1	1	−1	1	−1	1	z		z^3
E_1	2	$(3)^{1/2}$	1	0	−1	$−(3)^{1/2}$	−2	0	0	(x, y)		(xz^2, yz^2)
E_2	2	1	−1	−2	−1	1	2	0	0		$(x^2−y^2, xy)$	
E_3	2	0	−2	0	2	0	−2	0	0			$[x(x^2−3y^2), y(3x^2−y^2)]$
E_4	2	−1	−1	2	−1	−1	2	0	0			
E_5	2	$−(3)^{1/2}$	1	0	−1	$(3)^{1/2}$	−2	0	0	(R_x, R_y)	(xz, yz)	$[xyz, z(x^2−y^2)]$

8. The S_n groups

S_4	E	S_4	C_2	S_4^3		
A	1	1	1	1	R_z	x^2+y^2, z^2
B	1	−1	1	−1	z	x^2-y^2, xy
E	$\{\begin{matrix}1\\1\end{matrix}$	$\begin{matrix}i\\-i\end{matrix}$	$\begin{matrix}-1\\-1\end{matrix}$	$\begin{matrix}-i\\i\end{matrix}\}$	$(x,y), (R_x, R_y)$	(xz, yz)

							$xyz, z(x^2-y^2)$
							z^3
							$(xz^2, yz^2)[x(x^2-3y^2), y(3x^2-y^2)]$

S_6	E	C_3	C_3^2	i	S_6^5	S_6			$\epsilon = \exp(2\pi i/3)$
A_g	1	1	1	1	1	1	R_z	x^2+y^2, z^2	
E_g	$\{\begin{matrix}1\\1\end{matrix}$	$\begin{matrix}\epsilon\\\epsilon^*\end{matrix}$	$\begin{matrix}\epsilon^*\\\epsilon\end{matrix}$	$\begin{matrix}1\\1\end{matrix}$	$\begin{matrix}\epsilon\\\epsilon^*\end{matrix}$	$\begin{matrix}\epsilon^*\\\epsilon\end{matrix}\}$	(R_x, R_y)	$(x^2-y^2, xy)(xz, yz)$	
A_u	1	1	1	−1	−1	−1	z		$z^3, x(x^2-3y^2), y(3x^2-y^2)$
E_u	$\{\begin{matrix}1\\1\end{matrix}$	$\begin{matrix}\epsilon\\\epsilon^*\end{matrix}$	$\begin{matrix}\epsilon^*\\\epsilon\end{matrix}$	$\begin{matrix}-1\\-1\end{matrix}$	$\begin{matrix}-\epsilon\\-\epsilon^*\end{matrix}$	$\begin{matrix}-\epsilon^*\\-\epsilon\end{matrix}\}$	(x, y)		$(xz^2, yz^2), [xyz, z(x^2-y^2)]$

S_8	E	S_8	C_4	S_8^3	C_2	S_8^5	C_4^3	S_8^7			$\epsilon = \exp(2\pi i/8)$
A	1	1	1	1	1	1	1	1	R_z	x^2+y^2, z^2	
B	1	−1	1	−1	1	−1	1	−1	z		z^3
E_1	$\{\begin{matrix}1\\1\end{matrix}$	$\begin{matrix}\epsilon\\\epsilon^*\end{matrix}$	$\begin{matrix}i\\-i\end{matrix}$	$\begin{matrix}-\epsilon^*\\-\epsilon\end{matrix}$	$\begin{matrix}-1\\-1\end{matrix}$	$\begin{matrix}-\epsilon\\-\epsilon^*\end{matrix}$	$\begin{matrix}-i\\i\end{matrix}$	$\begin{matrix}\epsilon^*\\\epsilon\end{matrix}\}$	(x, y)		(xz^2, yz^2)
E_2	$\{\begin{matrix}1\\1\end{matrix}$	$\begin{matrix}i\\-i\end{matrix}$	$\begin{matrix}-1\\-1\end{matrix}$	$\begin{matrix}-i\\i\end{matrix}$	$\begin{matrix}1\\1\end{matrix}$	$\begin{matrix}i\\-i\end{matrix}$	$\begin{matrix}-1\\-1\end{matrix}$	$\begin{matrix}-i\\i\end{matrix}\}$		(x^2-y^2, xy)	$[xyz, z(x^2-y^2)]$
E_3	$\{\begin{matrix}1\\1\end{matrix}$	$\begin{matrix}-\epsilon^*\\-\epsilon\end{matrix}$	$\begin{matrix}-i\\i\end{matrix}$	$\begin{matrix}\epsilon\\\epsilon^*\end{matrix}$	$\begin{matrix}-1\\-1\end{matrix}$	$\begin{matrix}\epsilon^*\\\epsilon\end{matrix}$	$\begin{matrix}i\\-i\end{matrix}$	$\begin{matrix}-\epsilon\\-\epsilon^*\end{matrix}\}$	(R_x, R_y)	(xz, yz)	$[x(x^2-3y^2), y(3x^2-y^2)]$

9. The cubic groups

T	E	$4C_3$	$4C_3^2$	$3C_2$			$\epsilon = \exp(2\pi i/3)$
A	1	1	1	1		$x^2+y^2+z^2$	
E	$\begin{Bmatrix}1\\1\end{Bmatrix}$	$\begin{matrix}\epsilon\\\epsilon^*\end{matrix}$	$\begin{matrix}\epsilon^*\\\epsilon\end{matrix}$	$\begin{matrix}1\\1\end{matrix}$		$(2z^2-x^2-y^2, x^2-y^2)$	
T	3	0	0	-1	$(R_x, R_y, R_z)(x,y,z)$	(xy, xz, yz)	$(x^3, y^3, z^3), [x(z^2-y^2), y(z^2-x^2), z(x^2-y^2)]$

T_h	E	$4C_3$	$4C_3^2$	$3C_2$	i	$4S_6$	$4S_6^5$	$3\sigma_h$			$\epsilon = \exp(2\pi i/3)$
A_g	1	1	1	1	1	1	1	1		$x^2+y^2+z^2$	
E_g	$\begin{Bmatrix}1\\1\end{Bmatrix}$	$\begin{matrix}\epsilon\\\epsilon^*\end{matrix}$	$\begin{matrix}\epsilon^*\\\epsilon\end{matrix}$	$\begin{matrix}1\\1\end{matrix}$	$\begin{matrix}1\\1\end{matrix}$	$\begin{matrix}\epsilon\\\epsilon^*\end{matrix}$	$\begin{matrix}\epsilon^*\\\epsilon\end{matrix}$	$\begin{matrix}1\\1\end{matrix}$		$(2z^2-x^2-y^2, x^2-y^2)$	
T_g	3	0	0	-1	3	0	0	-1	(R_x, R_y, R_z)	(xy, xz, yz)	
A_u	1	1	1	1	-1	-1	-1	-1			
E_u	$\begin{Bmatrix}1\\1\end{Bmatrix}$	$\begin{matrix}\epsilon\\\epsilon^*\end{matrix}$	$\begin{matrix}\epsilon^*\\\epsilon\end{matrix}$	$\begin{matrix}1\\1\end{matrix}$	$\begin{matrix}-1\\-1\end{matrix}$	$\begin{matrix}-\epsilon\\-\epsilon^*\end{matrix}$	$\begin{matrix}-\epsilon^*\\-\epsilon\end{matrix}$	$\begin{matrix}-1\\-1\end{matrix}$			
T_u	3	0	0	-1	-3	0	0	1	(x,y,z)		$(x^3, y^3, z^3), [x(z^2-y^2), y(z^2-x^2), z(x^2-y^2)]$

T_d	E	$8C_3$	$3C_2$	$6S_4$	$6\sigma_d$			
A_1	1	1	1	1	1		$x^2+y^2+z^2$	
A_2	1	1	1	-1	-1			
E	2	-1	2	0	0		$(2z^2-x^2-y^2, x^2-y^2)$	
T_1	3	0	-1	1	-1	(R_x, R_y, R_z)		xyz
T_2	3	0	-1	-1	1	(x,y,z)	(xy, xz, yz)	$[x(z^2-y^2), y(z^2-x^2), z(x^2-y^2)]$ (x^3, y^3, z^3)

O	E	$6C_4$	$3C_2(=C_4^2)$	$8C_3$	$6C_2$			
A_1	1	1	1	1	1		$x^2+y^2+z^2$	
A_2	1	−1	1	1	−1			xyz
E	2	0	2	−1	0		$(2z^2-x^2-y^2, x^2-y^2)$	
T_1	3	1	−1	0	−1	(R_x, R_y, R_z)		(x^3, y^3, z^3)
T_2	3	−1	−1	0	1		(xy, xz, yz)	$[x(z^2-y^2), y(z^2-y^2), z(x^2-y^2)]$

O_h	E	$8C_3$	$6C_2$	$6C_4$	$3C_2(=C_4^2)$	i	$6S_4$	$8S_6$	$3\sigma_h$	$6\sigma_d$			
A_{1g}	1	1	1	1	1	1	1	1	1	1		$x^2+y^2+z^2$	
A_{2g}	1	1	−1	−1	1	1	−1	1	1	−1			
E_g	2	−1	0	0	2	2	0	−1	2	0		$(2z^2-x^2-y^2, x^2-y^2)$	
T_{1g}	3	0	−1	1	−1	3	1	0	−1	−1			
T_{2g}	3	0	1	−1	−1	3	−1	0	−1	1	(R_x, R_y, R_z)		
A_{1u}	1	1	1	1	1	−1	−1	−1	−1	−1		(xy, xz, yz)	
A_{2u}	1	1	−1	−1	1	−1	1	−1	−1	1			xyz
E_u	2	−1	0	0	2	−2	0	1	−2	0			
T_{1u}	3	0	−1	1	−1	−3	−1	0	1	1	(x, y, z)		(x^3, y^3, z^3)
T_{2u}	3	0	1	−1	−1	−3	1	0	1	−1			$[x(z^2-y^2), y(z^2-x^2), z(x^2-y^2)]$

10. The groups $C_{\infty v}$ and $D_{\infty h}$ for linear molecules

$C_{\infty v}$	E	$2C_\infty^\Phi$	$\cdots$	$\infty\sigma_v$			
$A_1 \equiv \Sigma^+$	1	1	$\cdots$	1	z	x^2+y^2, z^2	
$A_2 \equiv \Sigma^-$	1	1	$\cdots$	-1	R_z		
$E_1 \equiv \Pi$	2	$2\cos\Phi$	$\cdots$	0	$(x,y)(R_x,R_y)$	(xz, yz)	
$E_2 \equiv \Delta$	2	$2\cos 2\Phi$	$\cdots$	0		(x^2-y^2, xy)	
$E_3 \equiv \Phi$	2	$2\cos 3\Phi$	$\cdots$	0			(xz^2, yz^2)
							$[xyz, z(x^2-y^2)]$
							$[x(x^2-3y^2), y(3x^2-y^2)]$
$\cdots$	$\cdots$	$\cdots$	$\cdots$	$\cdots$			z^3

$D_{\infty h}$	E	$2C_\infty^\Phi$	$\cdots$	$\infty\sigma_v$	i	$2S_\infty^\Phi$	$\cdots$	∞C_2			
$A_{1g} \equiv \Sigma_g^+$	1	1	$\cdots$	1	1	1	$\cdots$	1		x^2+y^2, z^2	
$A_{2g} \equiv \Sigma_g^-$	1	1	$\cdots$	-1	1	1	$\cdots$	-1	R_z		
$E_{1g} \equiv \Pi_g$	2	$2\cos\Phi$	$\cdots$	0	2	$-2\cos\Phi$	$\cdots$	0	(R_x, R_y)	(xz, yz)	
$E_{2g} \equiv \Delta_g$	2	$2\cos 2\Phi$	$\cdots$	0	2	$2\cos 2\Phi$	$\cdots$	0		(x^2-y^2, xy)	
$\cdots$	$\cdots$	$\cdots$	$\cdots$	$\cdots$	$\cdots$	$\cdots$	$\cdots$	$\cdots$			
$A_{1u} \equiv \Sigma_u^+$	1	1	$\cdots$	1	-1	-1	$\cdots$	-1			
$A_{2u} \equiv \Sigma_u^-$	1	1	$\cdots$	-1	-1	-1	$\cdots$	1	z		z^3
$E_{1u} \equiv \Pi_u$	2	$2\cos\Phi$	$\cdots$	0	-2	$2\cos\Phi$	$\cdots$	0	(x,y)		(xz^2, yz^2)
$E_{2u} \equiv \Delta_u$	2	$2\cos 2\Phi$	$\cdots$	0	-2	$-2\cos 2\Phi$	$\cdots$	0			$[xyz, z(x^2-y^2)]$
$E_{3u} \equiv \Phi_u$	2	$2\cos 3\Phi$	$\cdots$	0	-2	$2\cos 3\Phi$	$\cdots$	0			$[x(x^2-3y^2), y(3x^2-y^2)]$
$\cdots$	$\cdots$	$\cdots$	$\cdots$	$\cdots$	$\cdots$	$\cdots$	$\cdots$	$\cdots$			

11. The Icosahedral groups

I	E	$12C_5$	$12C_5^2$	$20C_3$	$15C_2$		
							$\eta^{\pm} = 1/2[1 \pm (5)^{1/2}]$
A	1	1	1	1	1		$x^2 + y^2 + z^2$
T_1	3	η^+	η^-	0	-1	$(x, y, z)(R_x, R_y, R_z)$	
T_2	3	η^-	η^+	0	-1		
G	4	-1	-1	1	0		(x^3, y^3, z^3)
H	5	0	0	-1	1		$[x(z^2-y^2), y(z^2-x^2), z(x^2-y^2), xyz]$
							$(2z^2 - x^2 - y^2, x^2 - y^2,$ $xy, xz, yz)$

I_h	E	$12C_5$	$12C_5^2$	$20C_3$	$15C_2$	i	$12S_{10}$	$12S_{10}^3$	$20S_6$	15σ		
												$\eta^{\pm} = 1/2[1 \pm (5)^{1/2}]$
A_g	1	1	1	1	1	1	1	1	1	1		$x^2 + y^2 + z^2$
T_{1g}	3	η^+	η^-	0	-1	3	η^-	η^+	0	-1	(R_x, R_y, R_z)	
T_{2g}	3	η^-	η^+	0	-1	3	η^+	η^-	0	-1		
G_g	4	-1	-1	1	0	4	-1	-1	1	0		
H_g	5	0	0	-1	1	5	0	0	-1	1		$(2z^2 - x^2 - y^2, x^2$ $-y^2, xy, xz, yz)$
A_u	1	1	1	1	1	-1	-1	-1	-1	-1		
T_{1u}	3	η^+	η^-	0	-1	-3	$-\eta^+$	$-\eta^-$	0	1	(x, y, z)	
T_{2u}	3	η^-	η^+	0	-1	-3	$-\eta^-$	$-\eta^+$	0	1		
G_u	4	-1	-1	1	0	-4	1	1	-1	0		(x^3, y^3, z^3) $[x(z^2-y^2), y(z^2-x^2),$ $z(x^2-y^2), xyz]$
H_u	5	0	0	-1	1	-5	0	0	1	-1		

Note: In these groups and others containing C_5, the following relationships may be useful:

$\eta^+ = 1/2[1 + (5)^{1/2}] = 1.61803\ldots = -2\cos 144°$
$\eta^- = 1/2[1 - (5)^{1/2}] = -0.61803\ldots = -2\cos 72°$
$\eta^+ \times \eta^+ = 1 + \eta^+$, $\eta^- \times \eta^- = 1 + \eta^-$, $\eta^+ \times \eta^- = -1$.

Application of Group Theory to Molecular Systems

7

In this chapter, we discuss the various applications of group theory to chemical problems. These include the description of structure and bonding based on hybridization and molecular orbital theories, selection rules in infrared and Raman spectroscopy, and symmetry of molecular vibrations. As will be seen, even though most of the arguments used are qualitative in nature, meaningful results and conclusions can be obtained.

7.1 Molecular orbital theory

As mentioned previously in Chapter 3, when we treat the bonding of a molecule by applying molecular orbital theory, we need to solve the secular determinant

$$\begin{vmatrix} H_{11} - ES_{11} & H_{12} - ES_{12} & \ldots & H_{1n} - ES_{1n} \\ H_{12} - ES_{12} & H_{22} - ES_{22} & \ldots & H_{2n} - ES_{2n} \\ \vdots & \vdots & & \vdots \\ H_{1n} - ES_{1n} & H_{2n} - ES_{2n} & \ldots & H_{nn} - ES_{nn} \end{vmatrix} = 0. \qquad (7.1.1)$$

In eq. (7.1.1), energy E is the only unknown, while energy interaction integrals H_{ij} ($\equiv \int \phi_i \hat{H} \phi_j d\tau$) and overlap integrals S_{ij} ($\equiv \int \phi_i \phi_j d\tau$) are calculated with known atomic orbitals $\phi_1, \phi_2, \ldots \phi_n$. After solving for E (n of them in all), we can substitute each E in the following secular equations to determine the values of coefficients c_i:

$$\begin{cases} (H_{11} - ES_{11})c_1 + (H_{12} - ES_{12})c_2 + \ldots + (H_{1n} - ES_{1n})c_n = 0 \\ (H_{12} - ES_{12})c_1 + (H_{22} - ES_{22})c_2 + \ldots + (H_{2n} - ES_{2n})c_n = 0 \\ \vdots \qquad \vdots \qquad \vdots \qquad \vdots \\ (H_{1n} - ES_{1n})c_1 + (H_{2n} - ES_{2n})c_2 + \ldots + (H_{nn} - ES_{nn})c_n = 0 \end{cases}$$
$$(7.1.2)$$

In molecular orbital theory, the molecular orbitals are expressed as linear combinations of atomic orbitals:

$$\psi = \sum_{i=1}^{n} c_i \phi_i = c_1 \phi_1 + c_2 \phi_2 + \ldots + c_n \phi_n. \qquad (7.1.3)$$

So we have nE's from eq. (7.1.1), and each E value leads to a set of coefficients, or to one molecular orbital. In other words, n atomic orbitals form n molecular orbitals; i.e., the number of orbitals is conserved.

The solution of eq. (7.1.1) is made easier if the secular determinant can be put in block-diagonal form, or block-factored:

$$n \left\{ \begin{array}{c} \overbrace{\begin{vmatrix} k \times k & & & \text{zeroes} \\ & l \times l & & \\ & & \ddots & \\ \text{zeroes} & & & m \times m \end{vmatrix}}^{n} \end{array} \right. = |k \times k| \times |l \times l| \times \ldots \times |m \times m| = 0, \quad \text{with } k + l + \ldots + m = n. \tag{7.1.4}$$

We will then be solving several small determinants ($k \times k$; $l \times l$; ...; $m \times m$) instead of a very large one ($n \times n$). By taking advantage of the symmetry property of the system, group theory can do just that. In this section, we illustrate the reduction of the secular determinant by studying several representative molecular systems.

7.1.1 AH$_n$ ($n = 2$–6) molecules

As mentioned in Chapter 3, to construct the molecular orbitals for an AX$_n$ molecule, we need to combine the atomic orbitals on X and then match the resultant combinations with the atomic orbitals on the central atom A. With group theory, we can derive the linear combinations systematically.

Let us use the simple molecule H$_2$O as the first example. The coordinate system we adopt for this molecule is shown in Fig. 7.1.1. The orientation of the adopted set of axes is similar to that for H$_2$S in Fig. 6.4.2.

If we assume that the 2s and 2p orbitals of oxygen and the 1s orbitals of the hydrogen atoms take part in the bonding, the secular determinant to be solved has the dimensions 6×6. Now we proceed to determine the symmetries of the participating atomic orbitals. From the C_{2v} character table, it can be seen that the $2p_x$, $2p_y$, and $2p_z$ orbitals on oxygen have B_1, B_2, and A_1 symmetries, respectively, while the oxygen 2s orbital, being totally symmetric, has A_1 symmetry. To determine the characters of the representation generated by the hydrogen 1s orbitals, we make use of this simple rule: *the character (of an operation) is equal to the number of objects (vectors or orbitals) unshifted by the operation.* So, for the hydrogen 1s orbitals:

C_{2v}	E	C_2	$\sigma_v(xz)$	$\sigma'_v(yz)$	
Γ_H	2	0	0	2	$\equiv A_1 + B_2$

In other words, the two 1s orbitals will form two linear combinations, one with A_1 symmetry and the other with B_2 symmetry. To deduce these two linear combinations, we need to employ the projection operator, which is defined as

$$P^i = \sum_{j=1}^{h} \chi^i(R_j) R_j, \tag{7.1.5}$$

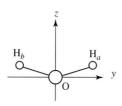

Fig. 7.1.1.
Coordinate system for the H$_2$O molecule. Note that the x axis points toward the reader.

where P^i is the projection operator for representation Γ_i, $\chi^i(R_j)$ is the character of the representation Γ_i for operation R_j, and the summation is over all the symmetry operations (h in total). To derive the linear combinations of the hydrogen 1s orbitals with A_1 symmetry, we apply P^{A_1} on, say, the 1s orbital on H_a (denoted as $1s_a$). So we need to know the result of applying every symmetry operation in the C_{2v} group to $1s_a$:

C_{2v}	E	C_2	$\sigma_v(xz)$	$\sigma'_v(yz)$
$1s_a$	$1s_a$	$1s_b$	$1s_b$	$1s_a$

Now we apply P^{A_1} to $1s_a$:

$$P^{A_1}(1s_a) = [1 \cdot E + 1 \cdot C_2 + 1 \cdot \sigma_v(xz) + 1 \cdot \sigma'_v(yz)]1s_a$$
$$= 1s_a + 1s_b + 1s_b + 1s_a = 2(1s_a) + 2(1s_b)$$
$$\Rightarrow (2)^{-1/2}(1s_a + 1s_b) \quad \text{(after normalization).} \qquad (7.1.6)$$

To obtain the combination with B_2 symmetry:

$$P^{B_2}(1s_a) = [1 \cdot E + (-1)C_2 + (-1)\sigma_v(xz) + 1 \cdot \sigma'_v(yz)]1s_a$$
$$= 1s_a - 1s_b - 1s_b + 1s_a = 2(1s_a) - 2(1s_b)$$
$$\Rightarrow (2)^{-1/2}(1s_a - 1s_b) \quad \text{(after normalization).} \qquad (7.1.7)$$

In Table 7.1.1 we summarize the way in which the molecular orbitals are formed in H_2O. From these results, we can see that the original 6×6 secular determinant is now block-factored into three smaller ones: one 3×3 for functions with A_1 symmetry, one 2×2 with B_2 symmetry, and one 1×1 with B_1 symmetry. A schematic energy diagram for this molecule is shown in Fig. 7.1.2. From this diagram, we can see that there are two bonding orbtals ($1a_1$ and $1b_2$), two other orbitals essentially nonbonding ($2a_1$ and $1b_1$), and two antibonding orbitals ($2b_2$ and $3a_1$). Also, all the bonding and nonbonding orbitals are filled, giving rise to a ground electronic configuration of $(1a_1)^2(1b_2)^2(2a_1)^2(1b_1)^2$ and an electronic state of $^1A_1$. This bonding picture indicates that H_2O has two σ bonds and two filled nonbonding orbitals. Such a result is in qualitative agreement with the familiar valence bond description for this molecule.

In passing, it is of interest to note that, according to Fig. 7.1.2, the first excited electronic configuration is $(1a_1)^2(1b_2)^2(2a_1)^2(1b_1)^1(2b_2)^1$, giving rise to states $^3A_2$ and $^1A_2$. It is easy to show that electronic transition $A_1 \rightarrow A_2$ is not allowed for a molecule with C_{2v} symmetry. In other words, $^1A_1 \rightarrow {}^1A_2$

Table 7.1.1. Formation of the molecular orbitals in H_2O

Symmetry	Orbital on O	Orbitals on H	Molecular orbitals
A_1	2s, $2p_z$	$(2)^{1/2}(1s_a + 1s_b)$	$1a_1, 2a_1, 3a_1$
B_1	$2p_x$	—	$1b_1$
B_2	$2p_y$	$(2)^{1/2}(1s_a - 1s_b)$	$1b_2, 2b_2$

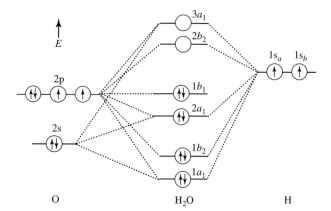

Fig. 7.1.2.
A schematic energy level diagram for H$_2$O.

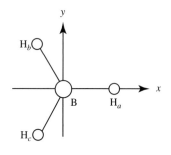

Fig. 7.1.3.
Coordinate system for BH$_3$.

is spin-allowed and symmetry-forbidden, while $^1A_1 \to {}^3A_2$ is both spin- and symmetry-forbidden.

Now let us turn to a slightly more complicated system, that of BH$_3$ with D_{3h} symmetry. Note that BH$_3$ is not a stable species: it dimerizes spontaneously to form diborane, B$_2$H$_6$. A convenient coordinate system for BH$_3$ is shown in Fig. 7.1.3.

From the D_{3h} character table, it is readily seen that boron 2s and 2p$_z$ orbitals have A'_1 and A''_2 symmetry, respectively, while the 2p$_x$ and 2p$_y$ orbitals form an E' set. To determine the symmetry species of the hydrogen 1s orbital combinations, we perform the operations

D_{3h}	E	$2C_3$	$3C_2$	σ_h	$2S_3$	$3\sigma_v$	
Γ_H	3	0	1	3	0	1	$\equiv A'_1 + E'$

So, among the three linear combinations of hydrogen 1s orbitals, one has A'_1 symmetry, while the remaining two form an E' set. To obtain the explicit functions, we require the symmetry operation results

D_{3h}	E	$2C_3$	$3C_2$	σ_h	$2S_3$	$3\sigma_v$
$1s_a$	$1s_a$	$1s_b, 1s_c$	$1s_a, 1s_b, 1s_c$	$1s_a$	$1s_b, 1s_c$	$1s_a, 1s_b, 1s_c$

Now, it is straightforward to obtain the linear combinations

$$P^{A'_1}(1s_a) = 1(1s_a) + 1(1s_b + 1s_c) + 1(1s_a + 1s_b + 1s_c) + 1(1s_a)$$
$$\qquad + 1(1s_b + 1s_c) + 1(1s_a + 1s_b + 1s_c)$$
$$= 4(1s_a + 1s_b + 1s_c)$$
$$\Rightarrow (3)^{-1/2}(1s_a + 1s_b + 1s_c) \quad \text{(after normalization);} \qquad (7.1.8)$$

$$P^{E'}(1s_a) = 2(1s_a) - 1(1s_b + 1s_c) + 2(1s_a) - 1(1s_b + 1s_c)$$
$$= 4(1s_a) - 2(1s_b + 1s_c)$$
$$\Rightarrow (6)^{-1/2}[2(1s_a) - 1s_b - 1s_c] \quad \text{(after normalization).} \qquad (7.1.9)$$

Still one more combination is required to complete the E' set. It is not difficult to see that when we operate $P^{E'}$ on $1s_b$ and $1s_c$, we get

$$P^{E'}(1s_b) = (6)^{-1/2}[2(1s_b) - 1s_a - 1s_c], \quad (7.1.10)$$

$$P^{E'}(1s_c) = (6)^{-1/2}[2(1s_c) - 1s_a - 1s_b]. \quad (7.1.11)$$

Since the two combinations of an E' set must be linearly independent, and summation of eqs. (7.1.10) and (7.1.11) yield eq. (7.1.9) (aside from a normalization factor), we need to take the difference of eqs. (7.1.10) and (7.1.11) to obtain the remaining combination:

$$[2(1s_b) - 1s_a - 1s_c] - [2(1s_c) - 1s_a - 1s_b] = 3(1s_b - 1s_c)$$
$$\Rightarrow (2)^{-1/2}(1s_b - 1s_c) \quad \text{[after normalization]}. \quad (7.1.12)$$

Obviously, there are different ways to choose the combination of an E' set. We choose the ones given by eqs. (7.1.9) and (7.1.12) because these functions overlap with the boron $2p_x$ and $2p_y$ orbitals, respectively, as shown in Fig. 7.1.4.

Table 7.1.2 summarizes how the molecular orbitals in BH_3 are formed. From these results, we can see that the original 7×7 secular determinant (four orbitals from B and three from the H's) is block-factored into three 2×2 and one 1×1 determinants. The 1×1 has A_2'' symmetry, one 2×2 has A_1' symmetry, while the remaining two 2×2 form an E' set. It is important to note that the two 2×2 determinants that form the E' pair have the same pair of roots; i.e., we only need to solve one of these two determinants! A schematic energy level diagram for BH_3 is shown in Fig. 7.1.5. According to this diagram, the ground configuration for BH_3 is $(1a_1')^2(1e')^4$ and the ground state is $^1A_1'$.

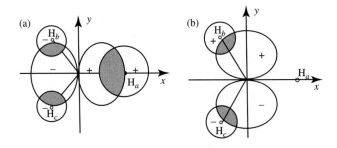

Fig. 7.1.4.
(a) Overlap between the boron $2p_x$ orbital with combination $(6)^{-1/2}[2(1s_a) - 1s_b - 1s_c]$.
(b) Overlap between the boron $2p_y$ orbital with the combination $(2)^{-1/2}[1s_b - 1s_c]$. By symmetry, the total overlaps in (a) and (b) are the same.

Table 7.1.2. Summary of the formation of the molecular orbitals in BH_3

Symmetry	Orbital on B	Orbitals on H	Molecular orbitals
A_1'	2s	$(3)^{-1/2}(1s_a + 1s_b + 1s_c)$	$1a_1', 2a_1'$
E'	$\begin{cases} 2p_x \\ 2p_y \end{cases}$	$\begin{cases} (6)^{-1/2}[2(1s_a) - 1s_b - 1s_c] \\ (2)^{-1/2}(1s_b - 1s_c) \end{cases}$	$1e', 2e'$
A_2''	$2p_z$	—	$1a_2''$

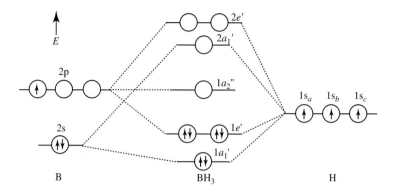

Fig. 7.1.5.
A schematic energy level diagram for BH$_3$.

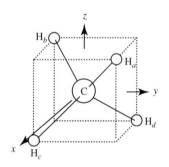

Fig. 7.1.6.
Coordinate system for CH$_4$.

Next we turn to the highly symmetric molecule CH$_4$, which belongs to point group T_d. A coordinate system for this molecule is shown in Fig. 7.1.6. For this molecule, there are eight valence atomic orbitals: 2s and 2p orbitals of carbon and the 1s orbitals of the hydrogens. Regarding the carbon orbitals, 2s has A_1 symmetry, while the $2p_x$, $2p_y$, and $2p_z$ orbitals form a T_2 set. The irreducible representations spanned by the hydrogen 1s orbitals can be readily determined:

T_d	E	$8C_3$	$3C_2$	$6S_4$	$6\sigma_\text{d}$	
Γ_H	4	1	0	0	2	$\equiv A_1 + T_2$

So the four hydrogen 1s orbitals form one linear combination with A_1 symmetry and three other combinations that make up a T_2 set. To obtain these combinations, we make use of the symmetry operation results

T_d	E	$8C_3$	$3C_2$	$6S_4$	$6\sigma_\text{d}$
$1s_a$	$1s_a$	$2(1s_a), 2(1s_b),$ $2(1s_c), 2(1s_d)$	$1s_b, 1s_c,$ $1s_d$	$2(1s_b), 2(1s_c),$ $2(1s_d)$	$3(1s_a), 1s_b, 1s_c,$ $1s_d$

Now the linear combination can be obtained readily:

$$P^{A_1}(1s_a) = 1(1s_a) + 1[2(1s_a) + 2(1s_b) + 2(1s_c) + 2(1s_d)]$$
$$+ 1(1s_b + 1s_c + 1s_d) + 1[2(1s_b) + 2(1s_c) + 2(1s_d)]$$
$$+ 1[3(1s_a) + 1s_b + 1s_c + 1s_d]$$
$$\Rightarrow \frac{1}{2}(1s_a + 1s_b + 1s_c + 1s_d) \quad \text{(after normalization)}.$$
(7.1.13)

$$P^{T_2}(1s_a) = 3(1s_a) - 1(1s_b + 1s_c + 1s_d) - 1[2(1s_b) + 2(1s_c) + 2(1s_d)]$$
$$+ 1[3(1s_a) + 1s_b + 1s_c + 1s_d]$$
$$= 6(1s_a) - 2(1s_b) - 2(1s_c) - 2(1s_d).$$
(7.1.14)

Similarly, when we operate P^{T_2} on $1s_b$, $1s_c$, $1s_d$, we obtain

$$P^{T_2}(1s_b) = 6(1s_b) - 2(1s_a) - 2(1s_c) - 2(1s_d), \quad (7.1.15)$$

$$P^{T_2}(1s_c) = 6(1s_c) - 2(1s_a) - 2(1s_b) - 2(1s_d), \quad (7.1.16)$$

$$P^{T_2}(1s_d) = 6(1s_d) - 2(1s_a) - 2(1s_b) - 2(1s_c). \quad (7.1.17)$$

To obtain the three linear combinations of the T_2 set, we combine eqs. (7.1.14) to (7.1.17) in the following manner:

Sum of eqs. (7.1.14) and (7.1.15):

$$4(1s_a) + 4(1s_b) - 4(1s_c) - 4(1s_d) = \frac{1}{2}(1s_a + 1s_b - 1s_c - 1s_d)$$

$$\text{(after normalization).} \quad (7.1.18)$$

Sum of eqs. (7.1.14) and (7.1.16): $\frac{1}{2}(1s_a - 1s_b + 1s_c - 1s_d). \quad (7.1.19)$

Sum of eqs. (7.1.14) and (7.1.17): $\frac{1}{2}(1s_a - 1s_b - 1s_c + 1s_d). \quad (7.1.20)$

There are many ways of combining eqs. (7.1.14) to (7.1.17) to arrive at the three combinations that form the T_2 set. We choose those ones given by eqs. (7.1.18) to (7.1.20) as these functions overlap effectively with the $2p_z$, $2p_x$, and $2p_y$ orbitals, respectively, as shown in Fig. 7.1.7.

Table 7.1.3 summarizes the formation of the molecular orbitals in CH_4. For this molecule, the original 8×8 secular determinants is reduced to four 2×2 ones, one with A_1 symmetry, while the other three form a T_2 set. In other words, we only need to solve the A_1 2×2 determinant as well as one of the three determinants that form the T_2 set. By symmetry, the three determinants

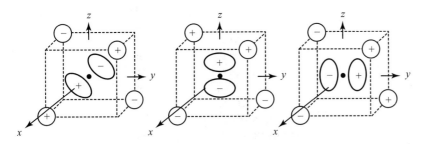

Fig. 7.1.7.
The overlap between $2p_x$, $2p_z$, and $2p_y$ orbitals of carbon with the 1s orbitals of the hydrogens in CH_4

Table 7.1.3. Formation of the molecular orbitals in CH_4

Symmetry	Orbital on C	Orbitals on H	Molecular orbitals
A_1	2s	$\frac{1}{2}(1s_a + 1s_b + 1s_c + 1s_d)$	$1a_1, 2a_1$
T_2	$2p_x$ $2p_y$ $2p_z$	$\frac{1}{2}(1s_a - 1s_b + 1s_c - 1s_d)$ $\frac{1}{2}(1s_a - 1s_b - 1s_c + 1s_d)$ $\frac{1}{2}(1s_a + 1s_b - 1s_c - 1s_d)$	$1t_2, 2t_2$

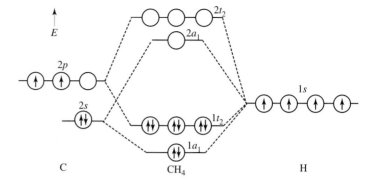

Fig. 7.1.8.
A schematic energy level diagram for CH$_4$.

Table 7.1.4. Formation of the molecular orbitals in AH$_5$

Symmetry	Orbital on A	Orbitals on H	Molecular Orbitals
A_1'	ns	$(2)^{-1/2}(1s_a + 1s_b)$ $(3)^{-1/2}(1s_c + 1s_d + 1s_e)$	$1a_1', 2a_1', 3a_1'$
E'	np_x np_y	$(6)^{-1/2}[2(1s_c) - 1s_d - 1s_e]$ $(2)^{-1/2}(1s_d - 1s_e)$	$1e', 2e'$
A_2''	np_z	$(2)^{-1/2}(1s_a - 1s_b)$	$1a_2'', 2a_2''$

forming the T_2 set have the same roots. A schematic energy level diagram for CH$_4$ is shown in Fig. 7.1.8. According to this diagram, the ground configuration is simply $(1a_1)^2(1t_2)^6$ and the ground state is $^1A_1$.

So far we have illustrated the method for constructing the linear combinations of atomic orbitals, using H$_2$O, BH$_3$, and CH$_4$ as examples. In these simple systems, all the ligand orbitals are equivalent to each other. For molecules with non-equivalent ligand sites, we first linearly combine the orbitals on the equivalent atoms. Then, if the need arises, we can further combine the combinations that have the same symmetry. Take a hypothetical molecule AH$_5$ with trigonal bipyramidal structure (D_{3h} symmetry) as an example. Figure 7.1.9 shows a convenient coordinate system for this molecule. It is clear that these are two sets of hydrogen atoms: equatorial hydrogens H$_c$, H$_d$, and H$_e$ and axial hydrogens H$_a$ and H$_b$.

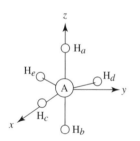

Fig. 7.1.9.
Coordinate system for AH$_5$.

When we linearly combine the orbitals on H$_a$ and H$_b$, two (un-normalized) combinations are obtained: $1s_a + 1s_b$ (A_1' symmetry), $1s_a - 1s_b$ (A_2''). On the other hand, the combinations for the orbitals on the equatorial hydrogens are $1s_c + 1s_d + 1s_e$ (A_1'); $2(1s_c) - 1s_d - 1s_e$ and $1s_d - 1s_e$ (E' symmetry). If we assume central atom A contributes ns and np orbitals to bonding, we can easily arrive at the results summarized in Table 7.1.4. Note that we can further combine the two ligand linear combinations with A_1' symmetry by taking their sum and difference.

Finally, we treat the highly symmetrical octahedral molecule AH$_6$ with O_h symmetry. A coordinate system for this molecule is shown in Fig. 7.1.10. If we assume that central atom A contributes ns, np, and nd orbitals to bonding, the

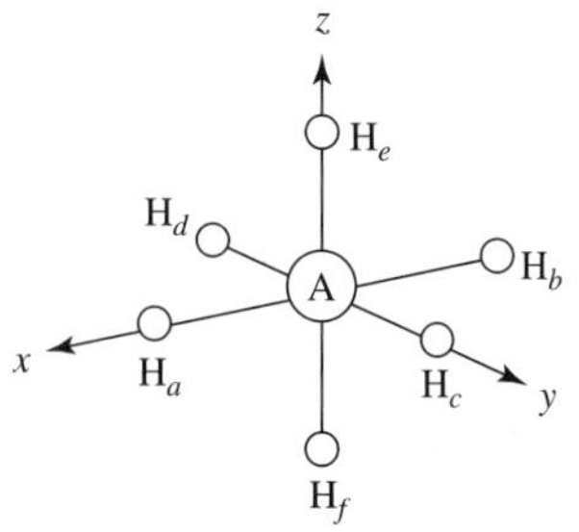

Fig. 7.1.10.
Coordinate system for AH$_6$.

Table 7.1.5. Formation of the molecular orbitals in AH$_6$

Symmetry	Orbital on A	Orbitals on H	Molecular Orbitals
A_{1g}	ns	$(6)^{-1/2}(1s_a + 1s_b + 1s_c + 1s_d + 1s_e + 1s_f)$	$1a_1, 2a_1$
E_g	nd_{z^2} $nd_{x^2-y^2}$	$(12)^{-1/2}[2(1s_e) + 2(1s_f) - 1s_a - 1s_b - 1s_c - 1s_d]$ $\frac{1}{2}(1s_a + 1s_b - 1s_c - 1s_d)$	$1e_g, 2e_g$
T_{2g}	nd_{xy} nd_{yz} nd_{xz}	—	$1t_{2g}$
T_{1u}	np_x np_y np_z	$(2)^{-1/2}(1s_a - 1s_b)$ $(2)^{-1/2}(1s_c - 1s_d)$ $(2)^{-1/2}(1s_e - 1s_f)$	$1t_{1u}, 2t_{1u}$

secular determinant has the dimensions 15×15. To obtain the symmetries of the six 1s orbital linear combinations:

O_h	E	$8C_3$	$6C_2$	$6C_4$	$3C_2 = C_4^2$	i	$6S_4$	$8S_6$	$3\sigma_h$	$6\sigma_d$	
Γ_H	6	0	0	2	2	0	0	0	4	2	$\equiv A_{1g} + E_g + T_{1u}$

To derive the combinations, we need the following tabulation listing the effect of various symmetry operations of O_h on the 1s orbital on hydrogen atom H_a:

O_h	E	$8C_3$	$6C_2$	$6C_4$	$3C_2 = C_4^2$	i	$6S_4$	$8S_6$	$3\sigma_h$	$6\sigma_d$
$1s_a$	$1s_a$	$2(1s_c),$ $2(1s_d),$ $2(1s_e),$ $2(1s_f)$	$2(1s_b),$ $1s_c, 1s_d,$ $1s_e, 1s_f$	$2(1s_a),$ $1s_c, 1s_d,$ $1s_e, 1s_f$	$1s_a,$ $2(1s_b)$	$1s_b$	$2(1s_b),$ $1s_c, 1s_d,$ $1s_e, 1s_f$	$2(1s_c),$ $2(1s_d),$ $2(1s_e),$ $2(1s_f)$	$2(1s_a),$ $1s_b$	$2(1s_a),$ $1s_c, 1s_d,$ $1s_e, 1s_f$

Now it is straightforward to derive the results summarized in Table 7.1.5. It is clear that the 15×15 secular determinant is block-factored into three 1×1 determinants that form a T_{2g} set, one 2×2 with A_{1g} symmetry, two 2×2 that form an E_g set, and three 2×2 that form a T_{1u} set. In other words, we only need to solve one 1×1 and three 2×2 secular determinants for this highly symmetrical molecule. This example shows the great simplification that group theory brings to the solution of a large secular determinant.

7.1.2 Hückel theory for cyclic conjugated polyenes

Previously in Chapter 3 we introduced the Hückel molecular orbital theory and applied it to the π system of a number of conjugated polyene chains. In this section we will apply this approximation to cyclic conjugated polyenes, taking advantage of the symmetry properties of these systems in the process.

Now let us take benzene as an example. The six 2p atomic orbitals taking part in the π bonding are labeled in the manner shown in Fig. 7.1.11. In the

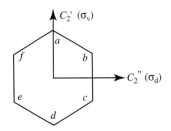

Fig. 7.1.11.
Labeling of the 2p orbitals taking part in the π bonding of benzene. Also shown are the locations of the symmetry elements C_2', C_2'', σ_v and σ_d.

Hückel approximation, the 6×6 secular determinant has the form

$$\begin{vmatrix} \alpha - E & \beta & 0 & 0 & 0 & \beta \\ \beta & \alpha - E & \beta & 0 & 0 & 0 \\ 0 & \beta & \alpha - E & \beta & 0 & 0 \\ 0 & 0 & \beta & \alpha - E & \beta & 0 \\ 0 & 0 & 0 & \beta & \alpha - E & \beta \\ \beta & 0 & 0 & 0 & \beta & \alpha - E \end{vmatrix} = 0. \qquad (7.1.21)$$

If we consider the D_{6h} symmetry of the system, we can determine the symmetries of the six π molecular orbitals:

D_{6h}	E	$2C_6$	$2C_3$	C_2	$3C_2'$	$3C_2''$	i	$2S_3$	$2S_6$	σ_h	$3\sigma_d$	$3\sigma_v$	
Γ_π	6	0	0	0	-2	0	0	0	0	-6	0	2	$\equiv A_{2u} + B_{2g} + E_{1g} + E_{2u}$

Recall that the character of an operation is equal to the number of vectors unshifted by that operation. Previously, for AH$_n$ molecules, in determining Γ_H, the 1s orbitals on the hydrogen atoms are spherically symmetric. In the present case, however, the 2p orbitals have directional properties. Take σ_h as an example. Upon reflection, the direction of the 2p orbitals is reversed. Hence $\chi(\sigma_h) = -6$.

Upon decomposing Γ_π, we can conclude that the secular determinant in eq. (7.1.21) can be factored into six 1×1 blocks: one with A_{2u} symmetry, one with B_{2g} symmetry, two others composing an E_{1g} set (with the same root), and the remaining two forming an E_{2u} set (also with the same root). To derive the six linear combinations, we make use of the following results:

D_{6h}	E	$2C_6$	$2C_3$	C_2	$3C_2'$	$3C_2''$	i	$2S_3$	$2S_6$	σ_h	$3\sigma_d$	$3\sigma_v$
p_a	p_a	p_b, p_f	p_c, p_e	p_d	$-p_a, -p_c,$ $-p_e$	$-p_b, -p_d,$ $-p_f$	$-p_d$	$-p_c, -p_e$	$-p_b, -p_f$	$-p_a$	$p_b, p_d,$ p_f	$p_a, p_c,$ p_e

The derivation of the non-degenerate linear combinations are straightforward:

$$P^{A_{2u}}(p_a) = (6)^{-1/2}(p_a + p_b + p_c + p_d + p_e + p_f) \quad \text{(after normalization).} \qquad (7.1.22)$$

$$P^{B_{2g}}(p_a) = (6)^{-1/2}(p_a - p_b + p_c - p_d + p_e - p_f) \quad \text{(after normalization).} \qquad (7.1.23)$$

The first component of the E_{1g} set can also be derived easily:

$$P^{E_{1g}}(p_a) = (12)^{-1/2}[2(p_a) + p_b - p_c - 2(p_d) - p_e + p_f]$$
$$\text{(after normalization).} \qquad (7.1.24)$$

For the second component, we operate $P^{E_{1g}}$ on p_b an p_c:

$$P^{E_{1g}}(p_b) = p_a + 2(p_b) + p_c - p_d - 2(p_e) - p_f, \qquad (7.1.25)$$

$$P^{E_{1g}}(p_c) = -p_a + p_b + 2(p_c) + p_d - p_e - 2(p_f). \qquad (7.1.26)$$

Since subtracting eq. (7.1.26) from eq. (7.1.25) yields eq. (7.1.24) (aside from a normalization constant), we need to take the sum of these two equations:

Eq. (7.1.25) + Eq. (7.1.26): $\frac{1}{2}(p_b + p_c - p_e - p_f)$ (after normalization). (7.1.27)

Similar manipulation leads to the following two linear combinations forming the E_{2u} set:

$$E_{2u} : \begin{cases} (12)^{-1/2}[2(p_a) - p_b - p_c + 2(p_d) - p_e - p_f], & (7.1.28) \\ \frac{1}{2}(p_b - p_c + p_e - p_f) & (7.1.29) \end{cases}$$

The energy of the six molecular orbitals can now be calculated, using the Hückel approximation as discussed in Chapter 3:

$$E(a_{2u}) = \alpha + 2\beta, \quad (7.1.30)$$
$$E(e_{1g}) = \alpha + \beta, \quad (7.1.31)$$
$$E(e_{2u}) = \alpha - \beta, \quad (7.1.32)$$
$$E(b_{2g}) = \alpha - 2\beta \quad (7.1.33)$$

The π energy levels, along with the molecular orbital wavefunctions, are pictorially displayed in Fig. 7.1.12. Since there are six π electrons in benzene, orbitals a_{2u} and e_{1g} are filled, giving rise to a $^1A_{1g}$ ground state with the total π energy

$$E_\pi = 6\alpha + 8\beta. \quad (7.1.34)$$

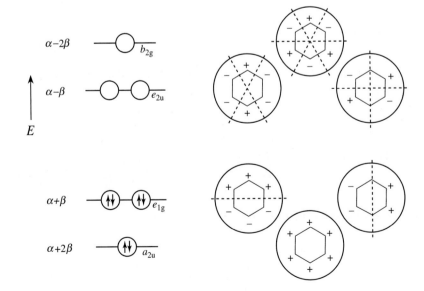

Fig. 7.1.12.
The energy level diagram and the wavefunctions of the six π molecular orbitals in benzene.

If the three π bonds in benzene were independent of each other, i.e., localized, they would have energy $3(2\alpha + 2\beta) = 6\alpha + 6\beta$. Hence the delocalization energy (DE) for benzene is

$$DE = 6\alpha + 8\beta - 6\alpha - 6\beta = 2\beta. \tag{7.1.35}$$

Calculation of β from first principles is a fairly complicated task. On the other hand, it can be approximated from experimental data:

$$|\beta| \sim 18 \text{ kcal mol}^{-1} = 75 \text{ kJ mol}^{-1}. \tag{7.1.36}$$

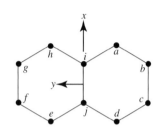

Fig. 7.1.13.
Labelling of the 2p orbitals taking part in the π bonding of naphthalene. The z axis points outward.

Another example of a cyclic conjugated polyene system is naphthalene. A coordinate system for this molecule, along with the labeling of the π atomic orbitals, is shown in Fig. 7.1.13. The 10×10 secular determinant has the form:

$$\begin{vmatrix} \alpha-E & \beta & 0 & 0 & 0 & 0 & 0 & 0 & \beta & 0 \\ \beta & \alpha-E & \beta & 0 & 0 & 0 & 0 & 0 & 0 & 0 \\ 0 & \beta & \alpha-E & \beta & 0 & 0 & 0 & 0 & 0 & 0 \\ 0 & 0 & \beta & \alpha-E & 0 & 0 & 0 & 0 & 0 & \beta \\ 0 & 0 & 0 & 0 & \alpha-E & \beta & 0 & 0 & 0 & \beta \\ 0 & 0 & 0 & 0 & \beta & \alpha-E & \beta & 0 & 0 & 0 \\ 0 & 0 & 0 & 0 & 0 & \beta & \alpha-E & \beta & 0 & 0 \\ 0 & 0 & 0 & 0 & 0 & 0 & \beta & \alpha-E & \beta & 0 \\ \beta & 0 & 0 & 0 & 0 & 0 & 0 & \beta & \alpha-E & \beta \\ 0 & 0 & 0 & \beta & \beta & 0 & 0 & 0 & \beta & \alpha-E \end{vmatrix} = 0. \tag{7.1.37}$$

With the aid of the D_{2h} character table, we can determine the symmetries of the ten molecular orbitals of this system:

D_{2h}	E	$C_2(z)$	$C_2(y)$	$C_2(x)$	i	$\sigma(xy)$	$\sigma(xz)$	$\sigma(yz)$	
Γ_π	10	0	0	-2	0	-10	2	0	$\equiv 2A_u + 3B_{2g} + 2B_{3g} + 3B_{1u}$

In other words, the secular determinant in eq. (7.1.37) can be factored into two 2×2 and two 3×3 blocks. To obtain the explicit forms of these ten combinations, we need the results of each of the eight symmetry operations. Also, since the system now has three types of (structurally non-equivalent) carbon atoms, we need the operation results on the 2p orbitals of these three kinds of atoms:

D_{2h}	E	$C_2(z)$	$C_2(y)$	$C_2(x)$	i	$\sigma(xy)$	$\sigma(xz)$	$\sigma(yz)$
p_a	p_a	p_e	$-p_d$	$-p_h$	$-p_e$	$-p_a$	p_h	p_d
p_b	p_b	p_f	$-p_c$	$-p_g$	$-p_f$	$-p_b$	p_g	p_c
p_i	p_i	p_j	$-p_j$	$-p_i$	$-p_j$	$-p_i$	p_i	p_j

The resultant combinations can be obtained easily:

$$A_u: \quad \phi_1 = \frac{1}{2}(p_a - p_d + p_e - p_h), \tag{7.1.38}$$

$$\phi_2 = \frac{1}{2}(p_b - p_c + p_f - p_g); \tag{7.1.39}$$

$$B_{2g}: \phi_3 = \frac{1}{2}(p_a - p_d - p_e + p_h), \qquad (7.1.40)$$

$$\phi_4 = \frac{1}{2}(p_b - p_c - p_f + p_g), \qquad (7.1.41)$$

$$\phi_5 = (2)^{-1/2}(p_i - p_j); \qquad (7.1.42)$$

$$B_{3g}: \phi_6 = \frac{1}{2}(p_a + p_d - p_e - p_h), \qquad (7.1.43)$$

$$\phi_7 = \frac{1}{2}(p_b + p_c - p_f - p_g); \qquad (7.1.44)$$

$$B_{1u}: \phi_8 = \frac{1}{2}(p_a + p_d + p_e + p_h), \qquad (7.1.45)$$

$$\phi_9 = \frac{1}{2}(p_b + p_c + p_f + p_g), \qquad (7.1.46)$$

$$\phi_{10} = (2)^{-1/2}(p_i + p_j). \qquad (7.1.47)$$

Next we need to set up the four smaller secular determinants. Take the one with A_u symmetry as an example:

$$H_{11} = \int \phi_1 \hat{H} \phi_1 d\tau$$
$$= \frac{1}{4} \int (p_a - p_d + p_e - p_h) \hat{H}(p_a - p_d + p_e - p_h) d\tau = \alpha; \qquad (7.1.48)$$

$$H_{12} = \int \phi_1 \hat{H} \phi_2 d\tau$$
$$= \frac{1}{4} \int (p_a - p_d + p_e - p_h) \hat{H}(p_b - p_c + p_f - p_g) d\tau = \beta; \qquad (7.1.49)$$

$$H_{22} = \int \phi_2 \hat{H} \phi_2 d\tau$$
$$= \frac{1}{4} \int (p_b - p_c + p_f - p_g) \hat{H}(p_b - p_c + p_f - p_g) d\tau = \alpha - \beta. \qquad (7.1.50)$$

Thus the A_u secular determinant has the form

$$A_u: \begin{vmatrix} \alpha - E & \beta \\ \beta & \alpha - \beta - E \end{vmatrix} = 0. \qquad (7.1.51)$$

The other three determinants can be obtained in a similar manner:

$$B_{2g}: \begin{vmatrix} \alpha - E & \beta & (2)^{1/2}\beta \\ \beta & \alpha - \beta - E & 0 \\ (2)^{1/2}\beta & 0 & \alpha - \beta - E \end{vmatrix} = 0. \qquad (7.1.52)$$

$$B_{3g}: \begin{vmatrix} \alpha - E & \beta \\ \beta & \alpha + \beta - E \end{vmatrix} = 0. \qquad (7.1.53)$$

$$B_{1u}: \begin{vmatrix} \alpha - E & \beta & (2)^{1/2}\beta \\ \beta & \alpha + \beta - E & 0 \\ (2)^{1/2}\beta & 0 & \alpha + \beta - E \end{vmatrix} = 0. \qquad (7.1.54)$$

Table 7.1.6. The Hückel energies and wavefunctions of the π molecular orbitals in naphthalene

	Orbital	Energy	Wavefunction
	$3b_{2g}$	$\alpha - 2.303\beta$	$0.3006(p_a - p_d - p_e + p_h) - 0.2307(p_b - p_c - p_f + p_g) - 0.4614(p_i - p_j)$
	$2a_u$	$\alpha - 1.618\beta$	$0.2629(p_a - p_d + p_e - p_h) - 0.4253(p_b - p_c + p_f - p_g)$
	$3b_{1u}$	$\alpha - 1.303\beta$	$0.3996(p_a + p_d + p_e + p_h) - 0.1735(p_b + p_c + p_f + p_g) - 0.3470(p_i + p_j)$
↑	$2b_{2g}$	$\alpha - \beta$	$0.4082(p_b - p_c - p_f + p_g) - 0.4082(p_i - p_j)$
E	$2b_{3g}$	$\alpha - 0.618\beta$	$0.4253(p_a + p_d - p_e - p_h) - 0.2629(p_b + p_c - p_f - p_g)$
	$1a_u$	$\alpha + 0.618\beta$	$0.4253(p_a - p_d + p_e - p_h) + 0.2629(p_b - p_c + p_f - p_g)$
	$2b_{1u}$	$\alpha + \beta$	$0.4082(p_b + p_c + p_f + p_g) - 0.4082(p_i - p_j)$
	$1b_{2g}$	$\alpha + 1.303\beta$	$0.3996(p_a - p_d - p_e + p_h) + 0.1735(p_b - p_c - p_f + p_g) + 0.3470(p_i - p_j)$
	$1b_{3g}$	$\alpha + 1.618\beta$	$0.2629(p_a + p_d - p_e - p_h) + 0.4253(p_b + p_c - p_f - p_g)$
	$1b_{1u}$	$\alpha + 2.303\beta$	$0.3006(p_a + p_d + p_e + p_h) + 0.2307(p_b + p_c + p_f + p_g) + 0.4614(p_i + p_j)$

The solving of these determinants may be facilitated by the substitution $x = (\alpha - E)/\beta$. In any event, the energies of the ten molecular orbitals can be obtained readily. With the energies we can then solve the corresponding secular equations for the coefficients. The energies and the wavefunctions of the ten π molecular orbitals for naphthalene are summarized in Table 7.1.6. From these results, we can arrive at the following ground electronic configuration and state:

$$(1b_{1u})^2 (1b_{3g})^2 (1b_{2g})^2 (2b_{1u})^2 (1a_u)^2, \ {}^1A_g.$$

By adding up the energies for all ten π electrons, we get

$$E_\pi = 10\alpha + 13.684\beta, \tag{7.1.55}$$

and

$$\text{DE} = 3.684\beta. \tag{7.1.56}$$

In addition, we can obtain the following allowed electronic transitions:

$${}^1A_g \rightarrow [\ldots(1a_u)^1(2b_{3g})^1], \ {}^1B_{3u}, \quad x\text{-polarized};$$

$${}^1A_g \rightarrow [\ldots(1a_u)^1(2b_{2g})^1], \ {}^1B_{2u}, \quad y\text{-polarized};$$

$${}^1A_g \rightarrow [\ldots(2b_{1u})^1(1a_u)^2(2b_{3g})^1], \ {}^1B_{2u}, \quad y\text{-polarized}.$$

Note that all three transitions are g ↔ u, in accordance with Laporte's rule.

In the previous section, we discussed the construction of the σ molecular orbitals in AH$_n$ systems. In this section, we confine our treatment to the π molecular orbitals in cyclic conjugated polyenes. In most molecules, there are σ bonds as well as π bonds, and these systems can be treated by the methods introduced in these two sections.

Before concluding this section, it is noted that a fairly user-friendly SHMO (simple Hückel molecular orbital) calculator is now available on the Internet, http://www.chem.ucalgary.ca/shmo/. With this calculator, the Hückel energies and wavefunctions of planar conjugated molecules can be obtained "on the fly."

Application of Group Theory to Molecular Systems

Table 7.1.7. Some useful steps in the derivation of the linear combinations of atomic orbitals for the π system of $(NPX_2)_3$.

D_{3h}	E	$2C_3$	$3C_2$	σ_h	$2S_3$	$3\sigma_v$	
Γ_p	3	0	-1	-3	0	1	$\equiv A_2'' + E''$
Γ_d	3	0	1	-3	0	-1	$\equiv A_1'' + E''$
c	c	$-a, -e$	$a, -c, e$	$-c$	a, e	$-a, c, -e$	
d	d	$-b, -f$	$-b, d, -f$	$-d$	b, f	$b, -d, f$	

7.1.3 Cyclic systems involving d orbitals

Sometimes a cyclic π system also involves d orbitals. An example of such a system is the inorganic phosphonitrilic halide $(NPX_2)_3$, which has D_{3h} symmetry with a pair of out-of-plane halide groups σ-bonded to each phosphorous atom. Now the six atomic orbitals taking part in the π bonding are the three p orbitals on nitrogen and the three d orbitals on phosphorus. These orbitals and their signed lobes above the molecular plane are shown in Fig. 7.1.14. Note that the overlap of the six orbitals is not as efficient as that found in benzene. There is an inevitable "mismatch" of symmetry, here occurring between orbitals a and f, among the six atomic orbitals.

The incorporation of the d orbitals complicates the group theoretic procedure to some extent. Table 7.1.7 gives some useful steps in the derivation of the linear combinations of six atomic orbitals. With the results in Table 7.1.7, the linear combination of the atomic orbitals can be readily derived:

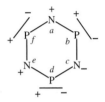

Fig. 7.1.14.
Labeling of the N 2p orbitals and P 3d orbitals taking part in the π bonding of $(NPX_2)_3$. Note that only the signed lobes above the molecular plane are shown. Also, a "mismatch," in this case occurring between orbitals a and f, is inevitable.

$$P^{A_2''}c \Rightarrow (3)^{-1/2}(a - c + e) \quad \text{(after normalization).} \tag{7.1.57}$$

For the degenerate E'' pair, we have

$$P^{E''}a = 2a + c - e$$

$$P^{E''}c = a + 2c + e$$

$$P^{E''}e = -a + c + 2e.$$

Since the first expression is the difference of the last two expressions, we can select the first expression as well as the sum of the last two:

$$E'' : \begin{cases} (6)^{-1/2}(2a + c - e) \\ (2)^{-1/2}(c + e) \end{cases} \tag{7.1.58}$$

Similarly, for the phosphorus 3d orbitals, we have:

$$P^{A_1''}b \Rightarrow (3)^{-1/2}(b - d + f) \quad \text{(after normalization).} \tag{7.1.59}$$

$$E'' : \begin{cases} (6)^{-1/2}(b + 2d + f) \\ (2)^{-1/2}(b - f) \end{cases} \tag{7.1.60}$$

It is not difficult to show that, when we form the 2×2 secular determinants with E'' symmetry, the first component of (7.1.58) interacts with the second component of (7.1.60). Analogously, the second component of (7.1.58) interacts

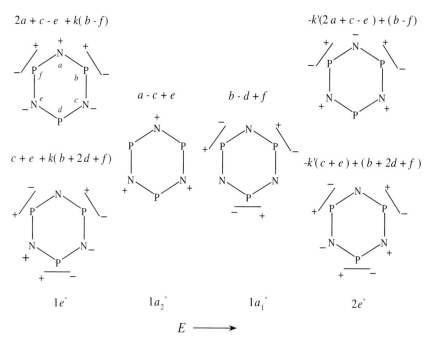

Fig. 7.1.15.
The six *pi* molecular orbitals of $(NPX_2)_3$.

with the first component of (7.1.60). Without further quantitative treatment, it is apparent that the six π golecular orbitals have the energy ordering and nodal characters shown in Fig. 7.1.15.

Molecular orbitals $1e''$ and $2e''$ may be classified as bonding and antibonding, respectively, while $1a_2''$ and $1a_1''$ may be considered as nonbonding orbitals. The ground configuration for this system is $(1e'')^4(1a_2'')^2$. It is clear that the delocalization of the six π electrons of this compound is not as extensive as that in benzene. As a result, the P_3N_3 cycle is not as rigid as the benzene ring. Furthermore, it should be noted that the phosphorus d oribital participation in the bonding of this type of compounds has played an important role in the development of inorganic chemistry. Indeed, the phosphorus d orbitals can participate in the bonding of the $(NPX_2)_3$ molecule in a variety of ways, as discussed in Chapter 15. The presentation here is mainly concerned with the symmetry properties of the π molecular orbitals.

7.1.4 Linear combinations of ligand orbitals for AL_4 molecules with T_d symmetry

In previous discussion, we constructed the symmetry-adapted linear combinations of the hydrogen 1s orbitals for AH_n molecules. In this section, we consider AL_n molecules, where L is a ligand capable of both σ and π bonding. For such systems, the procedure to derive the linear combinations of ligand orbitals can become fairly complicated. As an illustration, we take an AL_4 molecule with T_d symmetry as an example.

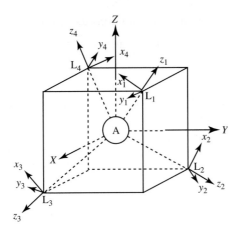

Fig. 7.1.16.
Coordinate system for a tetrahedral AL_4 molecule.

The coordinate systems for all five atoms in AL_4 are shown in Fig. 7.1.16. Note that all of them are right-handed. Also, x_1 and x_4 lie in the AL_1L_4 plane, while x_2 and x_3 lie in the AL_2L_3 plane.

The four linear combinations with A_1 and T_2 symmetries of the four vectors $z_1, \ldots, z_4$ forming σ bonds with orbitals on atom A may be easily obtained by the technique of projection operators. Therefore, only the results are given in Table 7.1.8, where all linear combinations of ligand orbitals will be listed. It is noted that the four combinations of the z_i vectors are identical to the combinations of hydrogen 1s orbitals obtained for methane.

The remaining eight vectors, $x_1, \ldots, x_4$ and $y_1, \ldots, y_4$, will form eight linear combinations having symmetries E, T_1, and T_2. In order to see how these ligand vectors transform under the 24 symmetry operations of the T_d point group, the *direction numbers* of all the ligand vectors are required. They are

$$
\begin{array}{llll}
x_1(-1,-1,2) & x_2(-1,1,2) & x_3(1,-1,2) & x_4(1,1,2) \\
y_1(1,-1,0) & y_2(1,1,0) & y_3(-1,-1,0) & y_4(-1,1,0) \\
z_1(1,1,1) & z_2(-1,1,-1) & z_3(1,-1,-1) & z_4(-1,-1,1).
\end{array}
$$

Among these, the direction numbers of the four z vectors are obvious. From these four vectors, any other vector may be generated by doing the cross product between an appropriate pair of vectors. For instance, $y_1 = z_1 \times z_4$, $x_1 = y_1 \times z_1$, etc.

Next we write down the transformation matrices for all the operations:

$C_3(i)$: A threefold axis passing through atoms A and L_i

$$
C_3(1) = \begin{vmatrix} 0 & 0 & 1 \\ 1 & 0 & 0 \\ 0 & 1 & 0 \end{vmatrix} \quad C_3(2) = \begin{vmatrix} 0 & 0 & 1 \\ -1 & 0 & 0 \\ 0 & -1 & 0 \end{vmatrix} \quad C_3(3) = \begin{vmatrix} 0 & 0 & -1 \\ -1 & 0 & 0 \\ 0 & 1 & 0 \end{vmatrix} \quad C_3(4) = \begin{vmatrix} 0 & 0 & -1 \\ 1 & 0 & 0 \\ 0 & -1 & 0 \end{vmatrix}.
$$

Note that the matrix for $C_3^{-1}(i)$, or $C_3^2(i)$, is simply the transpose of $C_3(i)$.

$C_2(q)$: A twofold axis which coincides with axis q on atom A

$$
C_2(X) = \begin{vmatrix} 1 & 0 & 0 \\ 0 & -1 & 0 \\ 0 & 0 & -1 \end{vmatrix} \quad C_2(Y) = \begin{vmatrix} -1 & 0 & 0 \\ 0 & 1 & 0 \\ 0 & 0 & -1 \end{vmatrix} \quad C_2(Z) = \begin{vmatrix} -1 & 0 & 0 \\ 0 & -1 & 0 \\ 0 & 0 & 1 \end{vmatrix}.
$$

Symmetry in Chemistry

$S_4(q)$: Again, the axis of this operation coincides with axis q on atom A

$$S_4(X) = \begin{vmatrix} -1 & 0 & 0 \\ 0 & 0 & -1 \\ 0 & 1 & 0 \end{vmatrix} \quad S_4(Y) = \begin{vmatrix} 0 & 0 & 1 \\ 0 & -1 & 0 \\ -1 & 0 & 0 \end{vmatrix} \quad S_4(Z) = \begin{vmatrix} 0 & -1 & 0 \\ 1 & 0 & 0 \\ 0 & 0 & 1 \end{vmatrix}.$$

Note that the matrix for $S_4^{-1}(q)$, or $S_4^3(q)$, is also simply the transpose of $S_4(q)$.

$\sigma(ij)$: A symmetry plane passing through atoms A, L_i, and L_j

$$\sigma(12) = \begin{vmatrix} 0 & 0 & 1 \\ 0 & 1 & 0 \\ 1 & 0 & 0 \end{vmatrix} \quad \sigma(13) = \begin{vmatrix} 1 & 0 & 0 \\ 0 & 0 & 1 \\ 0 & 1 & 0 \end{vmatrix} \quad \sigma(14) = \begin{vmatrix} 0 & 1 & 0 \\ 1 & 0 & 0 \\ 0 & 0 & 1 \end{vmatrix}.$$

$$\sigma(23) = \begin{vmatrix} 0 & -1 & 0 \\ -1 & 0 & 0 \\ 0 & 0 & 1 \end{vmatrix} \quad \sigma(24) = \begin{vmatrix} 1 & 0 & 0 \\ 0 & 0 & -1 \\ 0 & -1 & 0 \end{vmatrix} \quad \sigma(34) = \begin{vmatrix} 0 & 0 & -1 \\ 0 & 1 & 0 \\ -1 & 0 & 0 \end{vmatrix}.$$

So now we have the transformation matrices of all the symmetry operations in the T_d point group, except the identity operation E, which is simply a unit matrix with the dimensions 3×3.

When we carry out mathematically a symmetry operation on a ligand vector, we perform a matrix multiplication between the symmetry operation matrix and the vector composed of its directional numbers. For instance, when we operate $C_3(1)$ on x_1, we do the following:

$$C_3(1)x_1 = \begin{vmatrix} 0 & 0 & 1 \\ 1 & 0 & 0 \\ 0 & 1 & 0 \end{vmatrix} \begin{vmatrix} -1 \\ -1 \\ 2 \end{vmatrix} = \begin{vmatrix} 2 \\ -1 \\ -1 \end{vmatrix}.$$

Since operation $C_3(1)$ does not change the position of ligand L_1, the above resultant new vector $(2, -1, -1)$ may be resolved along the x_1 and y_1 directions by taking the dot products

$$(6)^{-1/2}(2, -1, -1) \cdot (6)^{-1/2}(-1, -1, 2) = -\frac{1}{2}x_1$$

$$(6)^{-1/2}(2, -1, -1) \cdot (2)^{-1/2}(1, -1, 0) = \frac{1}{2}(3)^{1/2}y_1.$$

In other words, $C_3(1)x_1 = -\frac{1}{2}x_1 + \frac{1}{2}(3)^{1/2}y_1$. Note that when we resolve the new vector along the x_1 and y_1 directions, all vectors involved need to be normalized first.

Similarly, when we operate $C_3(2)$ on x_1, we get $C_3(2)x_1 = (2, 1, 1)$. Since operation $C_3(2)$ on ligand L_1 yields L_3, the resultant vector $(2, 1, 1)$ is to be resolved along the x_3 and y_3 directions. Now it is easy to show $C_3(2)x_1 = \frac{1}{2}x_3 - \frac{1}{2}(3)^{1/2}y_3$.

When we carry out all 24 symmetry operations in the T_d point group on vector x_1, we get

$$C_3(1): \; -\frac{1}{2}x_1 + \frac{1}{2}(3)^{1/2}y_1, \quad C_3^{-1}(1): \; -\frac{1}{2}x_1 - \frac{1}{2}(3)^{1/2}y_1$$

$$C_3(2): \; \frac{1}{2}x_3 - \frac{1}{2}(3)^{1/2}y_3, \quad C_3^{-1}(2): \; -\frac{1}{2}x_4 - \frac{1}{2}(3)^{1/2}y_4$$

$C_3(3): -\frac{1}{2}x_4 + \frac{1}{2}(3)^{1/2}y_4, \quad C_3^{-1}(3): \frac{1}{2}x_2 + \frac{1}{2}(3)^{1/2}y_2$

$C_3(4): \frac{1}{2}x_2 - \frac{1}{2}(3)^{1/2}y_2, \quad C_3^{-1}(4): \frac{1}{2}x_3 + \frac{1}{2}(3)^{1/2}y_3.$

After summing up, $8C_3x_1 = -x_1 + x_2 + x_3 - x_4$.

$$C_2(X): -x_3, \quad C_2(Y): -x_2, \quad C_2(Z): x_4.$$

After summing up, $3C_2x_1 = -x_2 - x_3 + x_4$.

$S_4(X): -\frac{1}{2}x_4 - \frac{1}{2}(3)^{1/2}y_4, \quad S_4^{-1}(X): \frac{1}{2}x_2 + \frac{1}{2}(3)^{1/2}y_2$

$S_4(Y): \frac{1}{2}x_3 - \frac{1}{2}(3)^{1/2}y_3, \quad S_4^{-1}(Y): -\frac{1}{2}x_4 + \frac{1}{2}(3)^{1/2}y_4$

$S_4(Z): -x_2, \quad S_4^{-1}(Z): -x_3.$

After summing up, $6S_4x_1 = -\frac{1}{2}x_2 - \frac{1}{2}x_3 - x_4 + \frac{1}{2}(3)^{1/2}y_2 - \frac{1}{2}(3)^{1/2}y_3$.

$\sigma(12): -\frac{1}{2}x_1 + \frac{1}{2}(3)^{1/2}y_1, \quad \sigma(13): -\frac{1}{2}x_1 - \frac{1}{2}(3)^{1/2}y_1, \sigma(14): x_1$

$\sigma(23): x_4, \quad \sigma(24): \frac{1}{2}x_3 + \frac{1}{2}(3)^{1/2}y_3, \quad \sigma(34): \frac{1}{2}x_2 - \frac{1}{2}(3)^{1/2}y_2.$

After summing up, $6\sigma_d x_1 = \frac{1}{2}x_2 + \frac{1}{2}x_3 + x_4 - \frac{1}{2}(3)^{1/2}y_2 + \frac{1}{2}(3)^{1/2}y_3$.

Finally, for the identity operation E, we have E, $Ex_1 = x_1$.

If we carry out the same 24 symmetry operations on vector y_1, we get

$Ey_1 = y_1.$

$8C_3y_1 = -y_1 + y_2 + y_3 - y_4.$

$3C_2y_1 = -y_2 - y_3 + y_4.$

$6S_4y_1 = \frac{1}{2}y_2 + \frac{1}{2}y_3 + y_4 + \frac{1}{2}(3)^{1/2}x_2 - \frac{1}{2}(3)^{1/2}x_3.$

$6\sigma_d y_1 = -\frac{1}{2}y_2 - \frac{1}{2}y_3 - y_4 - \frac{1}{2}(3)^{1/2}x_2 + \frac{1}{2}(3)^{1/2}x_3.$

Now we are in a position to apply the projection operator to obtain the linear combinations of ligand orbitals with the desired symmetry. For instance, for the two linear combinations that form the degenerate set with E symmetry, we carry out the operations

$$P^E x_1 = 3x_1 - 3x_2 - 3x_3 + 3x_4, \quad \text{or} \quad \frac{1}{2}(x_1 - x_2 - x_3 + x_4).$$

Similarly, $P^E y_1 = \frac{1}{2}(y_1 - y_2 - y_3 + y_4)$. All linear combinations may be obtained in an analogous manner. Table 7.1.8 lists the 16 linear combinations of ligand orbitals for an AL$_4$ molecule with T_d symmetry.

Table 7.1.8. The symmetry-adapted linear combinations of ligand orbitals for an AL_4 molecule with T_d symmetry

Symmetry	Orbital on A	Orbitals on ligands		Molecular orbitals
A_1	s	$\frac{1}{2}(s_1 + s_2 + s_3 + s_4)$; $\frac{1}{2}(z_1 + z_2 + z_3 + z_4)$		$1a_1, 2a_1, 3a_1$
E	$\begin{cases} d_{z^2} \\ d_{x^2-y^2} \end{cases}$	$\begin{cases} \frac{1}{2}(x_1 - x_2 - x_3 + x_4) \\ \frac{1}{2}(y_1 - y_2 - y_3 + y_4) \end{cases}$		$1e, 2e$
T_1		$\begin{cases} \frac{1}{4}[(3)^{\frac{1}{2}}(x_1 + x_2 - x_3 - x_4) + (y_1 + y_2 - y_3 - y_4)] \\ \frac{1}{4}[(3)^{\frac{1}{2}}(x_1 - x_2 + x_3 - x_4) + (y_1 - y_2 + y_3 - y_4)] \\ \frac{1}{2}(y_1 + y_2 + y_3 + y_4) \end{cases}$		$1t_1$
T_1	$\begin{cases} p_x \\ p_y \\ p_z \\ d_{yz} \\ d_{xz} \\ d_{xy} \end{cases}$	$\begin{cases} \frac{1}{2}(s_1 - s_2 + s_3 - s_4) \\ \frac{1}{2}(s_1 + s_2 - s_3 - s_4) \\ \frac{1}{2}(s_1 - s_2 - s_3 + s_4) \\ \frac{1}{4}[(x_1 + x_2 - x_3 - x_4) + (3)^{\frac{1}{2}}(-y_1 - y_2 + y_3 + y_4)] \\ \frac{1}{4}[(x_1 - x_2 + x_3 - x_4) + (3)^{\frac{1}{2}}(y_1 - y_2 + y_3 - y_4)] \\ \frac{1}{2}(x_1 + x_2 + x_3 + x_4) \end{cases}$	$\begin{cases} \frac{1}{2}(z_1 - z_2 + z_3 - z_4) \\ \frac{1}{2}(z_1 + z_2 - z_3 - z_4) \\ \frac{1}{2}(z_1 - z_2 - z_3 + z_4) \end{cases}$	$1t_2, 2t_2, 3t_2, 4t_2, 5t_2$

For an AL_4 molecule with T_d symmetry, if we assume that nine atomic orbitals on A and four atomic orbitals on each L participate in bonding, there will be 25 molecular orbitals in total. Based on symmetry arguments, as shown in Table 7.1.8, among the molecular orbitals formed, there will be three with a_1 symmetry, two doubly degenerate sets with E symmetry, one triply degenerate set with T_1 symmetry (which is nonbonding, i.e., localized on the ligands), and five triply degenerate set with T_2 symmetry. Clearly, for such a complex system, it is not straightforward to come up with a qualitative energy level diagram.

7.2 Construction of hybrid orbitals

As introduced in Chapter 3, for AX_n systems, the hybrid orbitals are the linear combinations of atomic orbitals on central atom A that point toward the X atoms. In addition, the construction of the sp^n hybrids was demonstrated. In this section, we will consider hybrids that have d–orbital contributions, as well as the relationship between the hybrid orbital coefficient matrix and that of the molecular orbitals, all from the viewpoint of group theory.

7.2.1 Hybridization schemes

As is well known, if we construct four equivalent hybrid orbitals using one s and three p atomic orbitals, the hybrids would point toward the four corners of a tetrahedron. However, is this the only way to construct four such hybrids? If not, what other atomic orbitals can be used to form such hybrid orbitals? To answer these questions, we need to determine the representations spanned by the four hybrid orbitals that point toward the corners of a tetrahedron:

T_d	E	$8C_3$	$3C_2$	$6S_4$	$6\sigma_d$	
Γ_σ	4	1	0	0	2	$\equiv A_1 + T_2$

This result implies that, among the four required atomic orbitals (on the central atom), one must have A_1 symmetry and the other three must form a T_2 set. From Areas III and IV of the T_d character table, we know that the s orbital has A_1 symmetry, while the three p orbitals, or the d_{xy}, d_{yz}, and d_{xz} orbitals, collectively form a T_2 set. In other words, the hybridization scheme can be either the well-known sp^3 or the less familiar sd^3, or a combination of these two schemes.

On symmetry grounds, the sp^3 and sd^3 schemes are entirely equivalent to each other. However, for a particular molecule, we can readily see that one scheme is favored over the other. For instance, in CH$_4$, carbon can use the 2s and three 2p orbitals to form a set of sp^3 hybrids. It is also clear that carbon is unlikely to use its 2s orbital and three 3d orbitals (which lie above the 2p orbitals by about 950 kJ mol^{-1}) to form the hybrids. On the other hand, for tetrahedral transition metal ions such as MnO$_4^-$, it is likely that Mn would use three 3d orbitals, instead of the three higher energy 4p orbitals, for the formation of the hybrids.

We now list the possible hybridization schemes for several important molecular types.

(1) AX$_3$, trigonal planar, D_{3h} symmetry:

D_{3h}	E	$2C_3$	$3C_2$	σ_h	$2S_3$	$3\sigma_v$	
Γ_σ	3	0	1	3	0	1	$\equiv A_1'(s; d_{z^2}) + E'[(p_x, p_y); (d_{xy}, d_{x^2-y^2})]$

Hence the possible schemes include sp^2, sd^2, dp^2, and d^3.

(2) AX$_4$, square planar, D_{4h} symmetry:

D_{4h}	E	$2C_4$	C_2	$2C_2'$	$2C_2''$	i	$2S_4$	σ_h	$2\sigma_v$	$2\sigma_d$	
Γ_σ	4	0	0	2	0	0	0	4	2	0	$\equiv A_{1g}(s; d_{z^2}) + B_{1g}(d_{x^2-y^2}) + E_u(p_x, p_y)$

Hence the possible schemes include dsp^2 and d^{2}p^2.

(3) AX$_5$, trigonal bipyramidal, D_{3h} symmetry:

D_{3h}	E	$2C_3$	$3C_2$	σ_h	$2S_3$	$3\sigma_v$	
Γ_σ	5	2	1	3	0	3	$\equiv 2A_1'(s; d_{z^2}) + A_2''(p_z) + E'[(p_x, p_y); (d_{xy}, d_{x^2-y^2})]$

Hence the possible schemes include dsp^3 and d^{3}sp.

(4) AX$_6$, octahedral, O_h symmetry:

O_h	E	$8C_3$	$6C_2$	$6C_4$	$3C_2$	i	$6S_4$	$8S_6$	$3\sigma_h$	$6\sigma_d$	
Γ_σ	6	0	0	2	2	0	0	0	4	2	$\equiv A_{1g}(s) + E_g(d_{z^2}, d_{x^2-y^2}) + T_{1u}(p_x, p_y, p_z)$

So the only possible scheme is d^{2}sp^3.

Once we have determined the atomic orbitals taking part in the formation of the hybrids, we can employ the method outlined in Chapter 3 to obtain the explicit expressions of the hybrid orbitals.

7.2.2 Relationship between the coefficient matrices for the hybrid and molecular orbital wavefunctions

As has been mentioned more than once already, to construct the hybrid orbitals for an AX$_n$ molecule, we linearly combine the atomic orbitals on A so that the resultant hybrid orbitals point toward the X ligands. On the other hand, to form the molecular orbitals for the AX_n molecule, we linearly combine the orbitals on the ligands such that the combinations match in symmetry with the orbitals on A. Upon studying these two statements, we would not be surprised to find that the coefficient matrices for the hybrid orbitals and for the ligand orbital linear combinations are related to each other. The basis for this relationship is that both matrices are derived by considering the symmetry properties of the molecule. Indeed, this relationship is obvious if we take up a specific example.

From eq. (3.4.31), the sp^2 hybrids for an AX_3 (or AH_3) molecule with D_{3h} symmetry is:

$$\begin{vmatrix} h_1 \\ h_2 \\ h_3 \end{vmatrix} = \begin{vmatrix} (3)^{-1/2} & (2/3)^{1/2} & 0 \\ (3)^{-1/2} & -(6)^{-1/2} & (2)^{-1/2} \\ (3)^{-1/2} & -(6)^{-1/2} & -(2)^{-1/2} \end{vmatrix} \begin{vmatrix} s \\ p_x \\ p_y \end{vmatrix}. \tag{7.2.1}$$

From Table 7.1.2, the linear combinations of ligand orbitals for BH_3 have the form

$$\begin{vmatrix} (3)^{-1/2} & (3)^{-1/2} & (3)^{-1/2} \\ (2/3)^{1/2} & -(6)^{-1/2} & -(6)^{-1/2} \\ 0 & (2)^{-1/2} & -(2)^{-1/2} \end{vmatrix} \begin{vmatrix} 1s_a \\ 1s_b \\ 1s_c \end{vmatrix}. \tag{7.2.2}$$

It is now obvious that the matrix in eq. (7.2.1) is simply the transpose of the matrix in expression (7.2.2), and vice versa. In addition, it can be easily checked that the coefficient matrix for the sp^3 hybrids given in eq. (3.4.35) and the coefficient matrix for the linear combinations of ligand orbitals in CH_4 (Table 7.1.3) have the same relationship.

To conclude, there are two ways to determine the explicit expressions of a set of hybrid orbitals. The first one is that outlined in Chapter 3, taking advantage of the orthonormality relationship among the hybrids as well as the geometry and symmetry of the system. The second method is to apply the appropriate projection operators to the ligand orbitals, which are placed at the ends of the hybrids, to obtain the linear combinations of the ligand orbitals. The coefficient matrix for the hybrids is simply the transpose of the coefficient matrix for the linear combinations. These two methods are naturally closely related to one another. The only difference is that the latter *formally* makes use of group theory techniques, such as projection operator application and decomposition of a reducible representation, whereas the former does not.

7.2.3 Hybrids with d-orbital participation

In Section 7.2.1, we have seen that many hybridization schemes involve d orbitals. In fact, we do not anticipate any technical difficulty in the construction of hybrids that have d orbital participation. Let us take octahedral d^2sp^3 hybrids, directed along Cartesian axes (Fig. 7.1.10), as an example. From Table 7.1.5, we have the coefficient matrix for the linear combinations of ligand orbitals:

$$\begin{vmatrix} (6)^{-1/2} & (6)^{-1/2} & (6)^{-1/2} & (6)^{-1/2} & (6)^{-1/2} & (6)^{-1/2} \\ -(12)^{-1/2} & -(12)^{-1/2} & -(12)^{-1/2} & -(12)^{-1/2} & 2(12)^{-1/2} & 2(12)^{-1/2} \\ 1/2 & 1/2 & -1/2 & -1/2 & 0 & 0 \\ (2)^{-1/2} & -(2)^{-1/2} & 0 & 0 & 0 & 0 \\ 0 & 0 & (2)^{-1/2} & -(2)^{-1/2} & 0 & 0 \\ 0 & 0 & 0 & 0 & (2)^{-1/2} & -(2)^{-1/2} \end{vmatrix} \begin{vmatrix} 1s_a \\ 1s_b \\ 1s_c \\ 1s_d \\ 1s_e \\ 1s_f \end{vmatrix}$$

With these results, we can easily obtain the hybrid wavefunctions

$$\begin{vmatrix} h_a \\ h_b \\ h_c \\ h_d \\ h_e \\ h_f \end{vmatrix} = \begin{vmatrix} (6)^{-1/2} & -(12)^{-1/2} & 1/2 & (2)^{-1/2} & 0 & 0 \\ (6)^{-1/2} & -(12)^{-1/2} & 1/2 & -(2)^{-1/2} & 0 & 0 \\ (6)^{-1/2} & -(12)^{-1/2} & -1/2 & 0 & (2)^{-1/2} & 0 \\ (6)^{-1/2} & -(12)^{-1/2} & -1/2 & 0 & -(2)^{-1/2} & 0 \\ (6)^{-1/2} & 2(12)^{-1/2} & 0 & 0 & 0 & (2)^{-1/2} \\ (6)^{-1/2} & 2(12)^{-1/2} & 0 & 0 & 0 & -(2)^{-1/2} \end{vmatrix} \begin{vmatrix} s \\ d_{z^2} \\ d_{x^2-y^2} \\ p_x \\ p_y \\ p_z \end{vmatrix}.$$

(7.2.3)

In eq. (7.2.3), hybrids $h_a, h_b, \ldots, h_f$ point toward orbitals $1s_a, 1s_b, \ldots, 1s_f$, respectively (Fig. 7.1.10).

Lastly, we consider a system with non-equivalent positions. An example of such a system is the trigonal bipyramidal molecule AX_5 with D_{3h} symmetry. As discussed previously, one possible scheme is the dsp^3 hybridization, where d_{z^2} is the only d orbital participating.

If we use the d_{z^2} and p_z orbitals for the construction of axial hybrids h_a and h_b, and the s, p_x, and p_y orbitals for the equatorial hybrids h_c, h_d, and h_e (Fig. 7.1.8), we can readily write down the wavefunctions of these hybrids:

$$\begin{vmatrix} h_a \\ h_b \\ h_c \\ h_d \\ h_e \end{vmatrix} = \begin{vmatrix} (2)^{-1/2} & (2)^{-1/2} & 0 & 0 & 0 \\ (2)^{-1/2} & -(2)^{-1/2} & 0 & 0 & 0 \\ 0 & 0 & (3)^{-1/2} & 2(6)^{-1/2} & 0 \\ 0 & 0 & (3)^{-1/2} & -(6)^{-1/2} & (2)^{-1/2} \\ 0 & 0 & (3)^{-1/2} & -(6)^{-1/2} & -(2)^{-1/2} \end{vmatrix} \begin{vmatrix} d_{z^2} \\ p_z \\ s \\ p_x \\ p_y \end{vmatrix}$$

(7.2.4)

However, there is no reason at all to assume h_a and h_b are made up of only p_z and d_{z^2} orbitals: The d_{z^2} orbital can also contribute to the equatorial hybrids, and the s orbital can also contribute to the axial hybrids. In fact, if we use only the s and p_z orbitals for the axial hybrids, and the d_{z^2}, p_x, and p_y for the equatorial hybrids, we then have

$$\begin{vmatrix} h_a \\ h_b \\ h_c \\ h_d \\ h_e \end{vmatrix} = \begin{vmatrix} 0 & (2)^{-1/2} & (2)^{-1/2} & 0 & 0 \\ 0 & -(2)^{-1/2} & (2)^{-1/2} & 0 & 0 \\ -(3)^{-1/2} & 0 & 0 & 2(6)^{-1/2} & 0 \\ -(3)^{-1/2} & 0 & 0 & -(6)^{-1/2} & (2)^{-1/2} \\ -(3)^{-1/2} & 0 & 0 & -(6)^{-1/2} & -(2)^{-1/2} \end{vmatrix} \begin{vmatrix} d_{z^2} \\ p_z \\ s \\ p_x \\ p_y \end{vmatrix}.$$

(7.2.5)

Obviously, both of the matrices in eqs. (7.2.4) and (7.2.5) are limiting cases. A general expression encompassing these two cases is

$$\begin{vmatrix} h_a \\ h_b \\ h_c \\ h_d \\ h_e \end{vmatrix} = \begin{vmatrix} (2)^{-1/2}\sin\alpha & (2)^{-1/2} & (2)^{-1/2}\cos\alpha & 0 & 0 \\ (2)^{-1/2}\sin\alpha & -(2)^{-1/2} & (2)^{-1/2}\cos\alpha & 0 & 0 \\ -(3)^{-1/2}\cos\alpha & 0 & (3)^{-1/2}\sin\alpha & 2(6)^{-1/2} & 0 \\ -(3)^{-1/2}\cos\alpha & 0 & (3)^{-1/2}\sin\alpha & -(6)^{-1/2}(2)^{-1/2} & \\ -(3)^{-1/2}\cos\alpha & 0 & (3)^{-1/2}\sin\alpha & -(6)^{-1/2}(2)^{-1/2} & \end{vmatrix} \begin{vmatrix} d_{z^2} \\ p_z \\ s \\ p_x \\ p_y \end{vmatrix}.$$

(7.2.6)

To obtain eqs. (7.2.4) and (7.2.5) from eq. (7.2.6), we only need to set angle α to be 90° and 0°, respectively. It is not difficult to show that the five hybrids form an orthonormal set of wavefunctions. The parameter α in the coefficient matrix in eq. (7.2.6) may be determined in a number of ways, such as by the maximization of overlap between the hybrids and the ligand orbitals, or by the minimization of the energy of the system. In any event, such procedures are clearly beyond the scope of this chapter (or this book) and we will not deal with them any further.

7.3 Molecular vibrations

Molecular vibrations, as detected in infrared and Raman spectroscopy, provide useful information on the geometric and electronic structures of a molecule. As mentioned earlier, each vibrational wavefunction of a molecule must have the symmetry of an irreducible representation of that molecule's point group. Hence the vibrational motion of a molecule is another topic that may be fruitfully treated by group theory.

7.3.1 The symmetries and activities of the normal modes

A molecule composed of N atoms has in general $3N$ degrees of freedom, which include three each for translational and rotational motions, and $(3N-6)$ for the normal vibrations. During a normal vibration, all atoms execute simple harmonic motion at a characteristic frequency about their equilibrium positions. For a linear molecule, there are only two rotational degrees of freedom, and hence $(3N-5)$ vibrations. Note that normal vibrations that have the same symmetry and frequency constitute the equivalent components of a degenerate normal mode; hence the number of normal modes is always equal to or less than the number of normal vibrations. In the following discussion, we shall demonstrate how to determine the symmetries and activities of the normal modes of a molecule, using NH_3 as an example.

Step 1. For a molecule belonging to a certain point group, we first determine the representation $\Gamma(N_0)$ whose characters are the number of atoms that are unshifted by the operations in the group. Taking the NH_3 molecule as an example,

C_{3v}	E	$2C_3$	$3\sigma_v$
$\Gamma(N_0)$	4	1	2

Step 2. Multiply each character of $\Gamma(N_0)$ by the appropriate factor $f(R)$ to obtain Γ_{3N}. Now we show how to determine factor $f(R)$ for operation R. When R is a rotation of angle ϕ, C_ϕ, we use

$$f(C_\phi) = 1 + 2\cos\phi. \qquad (7.3.1)$$

Thus, $f(E) = 3; f(C_2) = -1; f(C_3) = f(C_3^2) = 0; f(C_4) = f(C_4^3) = 1; f(C_6) = f(C_6^5) = 2$; etc. When R is an improper rotation of angle ϕ, S_ϕ, we have

$$f(S_\phi) = -1 + 2\cos\phi. \qquad (7.3.2)$$

Thus, $f(S_1) = f(\sigma) = 1; f(S_2) = f(i) = -3; f(S_3) = -2; f(S_4) = f(S_4^3) = -1; f(S_6) = f(S_6^5) = 0$; etc. So, for NH$_3$,

C_{3v}	E	$2C_3$	$3\sigma_v$
$f(R)$	3	0	1
$\Gamma_{3N} = f(R) \times \Gamma(N_0)$	12	0	2

Note that the values of $f(R)$ can also be obtained from the characters of the symmetry species Γ_{xyz} (representation based on x, y, and z). For point group C_{3v}, $\Gamma_{xyz} = \Gamma_{xy}$ (based on x and y) $+ \Gamma_z$ (based on z) $= E + A_2$.

Upon decomposing Γ_{3N} using e.g. (6.4.3), we get

$$\Gamma_{3N}(\text{NH}_3) = 3A_1 + A_2 + 4E.$$

Step 3. From Γ_{3N} subtract the symmetry species $\Gamma_{\text{trans}}(= \Gamma_{xyz})$ for the translation of the molecule as a whole and Γ_{rot} (based on R_x, R_y, and R_z) for the rotational motion to obtain the representation for molecular vibration, Γ_{vib}:

$$\Gamma_{\text{vib}} = \Gamma_{3N} - \Gamma_{\text{trans}} - \Gamma_{\text{rot}}. \qquad (7.3.3)$$

For NH$_3$, we have

$$\Gamma_{\text{vib}}(\text{NH}_3) = (3A_1 + A_2 + 4E) - (A_1 + E) - (A_2 + E)$$
$$= 2A_1 + 2E.$$

In other words, among the six normal vibrations of NH$_3$, two have A_1 symmetry, two others form a degenerate E set, and the remaining two form another E set.

Step 4. For a vibrational mode to be infrared (IR) active, it must bring about a change in the molecule's dipole moment. Since the symmetry species of the dipole moment's components are the same as Γ_x, Γ_y, and Γ_z, a normal mode having the same symmetry as Γ_x, Γ_y, or Γ_z will be infrared active. The argument employed here is very similar to that used in the derivation of the selection rules for electric dipole transitions (Section 7.1.3). So, of the six vibrations of NH$_3$, all are infrared active, and they comprise four normal modes with distinct fundamental frequencies.

Fig. 7.3.1.
The vibrational modes of NH$_3$ and observed frequencies. Note that only one component is shown for the E modes.

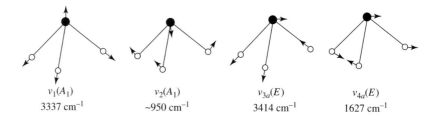

$\nu_1(A_1)$ $\quad\quad$ $\nu_2(A_1)$ $\quad\quad$ $\nu_{3a}(E)$ $\quad\quad$ $\nu_{4a}(E)$
3337 cm^{-1} $\quad$ ~950 cm^{-1} $\quad$ 3414 cm^{-1} $\quad$ 1627 cm^{-1}

On the other hand, for a vibrational mode to be Raman (R) active, it must bring about a change in the polarizability of the molecule. As the components of the polarizability tensor (based on the quadratic products of the coordinates) have the symmetry of Γ_{x^2}, Γ_{y^2}, Γ_{z^2}, Γ_{xy}, Γ_{xz}, and Γ_{yz}, a normal mode having one of these symmetries will be Raman active. So, for NH$_3$, we again will observe four fundamentals in its Raman spectrum. In other words,

$$\Gamma_{\text{vib}}(\text{NH}_3) = 2A_1(\text{R/IR}) + 2E(\text{R/IR}).$$

The four observed frequencies and the pictorial representations of the normal modes for NH$_3$ are displayed in Fig. 7.3.1. From this figure, we can see that ν_1 and ν_3 are stretching modes, while ν_2 and ν_4 are bending modes. Note that the number of stretching vibrations is equal to the number of bonds. Also, the stretching modes have higher frequencies than the bending ones.

Before discussing other examples, we note here that, for a centrosymmetric molecule (one with an inversion center), Γ_x, Γ_y, and Γ_z are "u" (from the German word *ungerade*, meaning odd) species, while binary products of x, y, and z have "g" (*gerade*, meaning even) symmetry. Thus infrared active modes will be Raman forbidden, and Raman active modes will be infrared forbidden. In other words, there are no coincident infrared and Raman bands for a centrosymmetric molecule. This relationship is known as the rule of mutual exclusion.

Another useful relationship: As the totally symmetric irreducible representation Γ_{TS} in every group is always associated with one or more binary products of x, y, and z and it follows that totally symmetric vibrational modes are always Raman active.

Besides being always Raman active, the totally symmetric vibrational modes can also be readily identified in the spectrum. As shown in Fig. 7.3.2, the scattered Raman radiation can be resolved into two intensity components, $I_\perp$ and $I_\parallel$. The ratio of these two intensities is called the depolarization ratio ρ:

$$\rho = I_\perp / I_\parallel. \tag{7.3.4}$$

If the incident radiation is plane-polarized, such as that produced by lasers in Raman spectroscopy, scattering theory predicts that totally symmetric modes will have $0 < \rho < 3/4$ and all other (non-totally symmetric) modes will have $\rho = 3/4$. A vibrational band with $0 < \rho < 3/4$ is said to be polarized, and one with $\rho = 3/4$ is said to be depolarized. In fact, for highly symmetric molecules, the polarized bands often have $\rho \sim 0$, which makes identifying totally symmetric modes relatively simple. For the NH_3 molecule, the two A_1 modes are polarized.

Figure 7.3.2 shows the intensities $I_\perp$ and $I_\parallel$ for the $\nu_1(A_1)$ and $\nu_2(E)$ Raman bands of CCl_4. It is seen that $I_\perp$ of $\nu_1(A_1)$ is essentially zero. Precise measurements yield $\rho = 0.005 \pm 0.002$ for $\nu_1(A_1)$ and $\rho = 0.72 \pm 0.002$ for $\nu_2(E)$. These results are consistent with scattering theory.

7.3.2 Some illustrative examples

In this section, we will attempt to illustrate the various principles and techniques introduced in the previous section with several simple examples. Special emphasis will be on the stretching modes of a molecule.

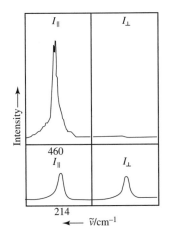

Fig. 7.3.2.
Raman band intensities $I_\perp$ and $I_\parallel$ of CCl_4 for $\nu_1(A_1)$ (top) and $\nu_2(E)$ (bottom). The splitting in the ν_1 band is due to the isotopic effect of Cl.

(1) *trans*-N_2F_2

The molecule N_2F_2 can exist in two geometrical forms, namely, *cis* and *trans*. Here we are only concerned with the *trans* isomer with C_{2h} symmetry. The methodical derivation of the symmetry species of the vibrational modes is best conducted in tabular form, as shown below.

C_{2h}	E	C_2	i	σ_h	Remark
$\Gamma(N_0)$	4	0	0	4	row 1 (atoms not moved by R)
Γ_{3N}	12	0	0	4	row 2 = row 1 × row 3
$\Gamma_{\text{trans}} = f(R)$	3	−1	−3	1	row 3, Γ_{trans} based on (x, y, z)
Γ_{rot}	3	−1	3	−1	row 4, Γ_{rot} based on (R_x, R_y, R_z)
Γ_{vib}	6	2	0	4	row 5 = row 2 − row 3 − row 4

Hence $\Gamma_{3N}(N_2F_2) = 4A_g + 2B_g + 2A_u + 4B_u$, and

$$\Gamma_{\text{vib}}(N_2F_2) = 3A_g(R) + A_u(IR) + 2B_u(IR).$$

So *trans*-N_2F_2 has three bands in both its infrared spectrum and its Raman spectrum. However, none of the bands are coincident, as this is a centrosymmetric molecule. The normal modes, as well as their observed frequencies, of *trans*-N_2F_2 are shown in Fig. 7.3.3. Note that this molecule has three bonds and hence has three stretching modes: $\nu_1(A_g)$ is the symmetric stretch of two N–F bonds, $\nu_2(A_g)$ is the N=N stretching, while $\nu_4(B_u)$ is the asymmetric stretch of the two N–F bonds. As the symmetric N–F stretch and N=N stretch have the same symmetry (A_g), ν_1 and ν_2 are both mixtures of these two types of stretch motion. Among the three modes, $\nu_2(A_g)$ has the highest energy, as N=N is a double bond and N–F is a single bond.

Fig. 7.3.3.
The vibrational modes and the corresponding frequencies of *trans*-N_2F_2. Note that only ν_1, ν_2, and ν_4 are stretching modes.

Fig. 7.3.4.
The vibrational modes and their frequencies of CF_4. Note that only one component is shown for the degenerate modes.

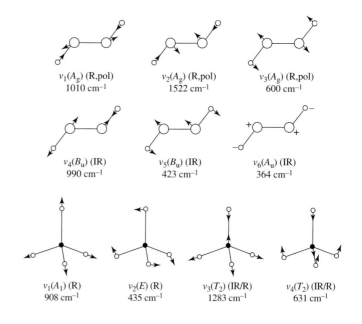

(2) CF_4

This is a tetrahedral molecule with T_d symmetry. The derivation of the vibrational modes is summarized below.

T_d	E	$8C_3$	$3C_2$	$6S_4$	$6\sigma_d$	
$\Gamma(N_0)$	5	2	1	1	3	
$f(R)$	3	0	-1	-1	1	
Γ_{3N}	15	0	-1	-1	3	$\equiv A_1 + E + T_1 + 3T_2$

$$\Gamma_{vib}(CF_4) = A_1(R) + E(R) + 2T_2(IR/R)$$

So CF_4 has two infrared bands and four Raman bands, and there are two coincident absorptions. The normal modes and their respective frequencies are given in Fig. 7.3.4. Note that $\nu_1(A_1)$ and $\nu_3(T_2)$ are the stretching bands. Also, this is an example that illustrates the "rule" that a highly symmetrical molecule has very few infrared active vibrations. The basis of the "rule" is that, in a point group with very high symmetry, x, y, and z often combine to form degenerate representations.

(3) P_4

The atoms in the P_4 molecule occupy the corners of a regular tetrahedron. In point group T_d, following the previous procedure, we have

T_d	E	$8C_3$	$3C_2$	$6S_4$	$6\sigma_d$	
$\Gamma(N_0)$	4	1	0	0	2	
$f(R)$	3	0	-1	-1	1	
Γ_{3N}	12	0	0	0	2	$\equiv A_1 + E + T_1 + 2T_2$

$$\Gamma_{vib}(P_4, T_d) = A_1(R) + E(R) + T_2(IR, R).$$

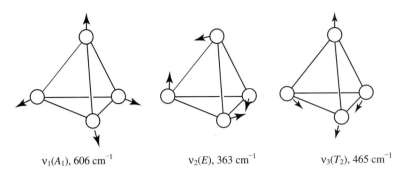

$\nu_1(A_1)$, 606 cm^{-1} $\nu_2(E)$, 363 cm^{-1} $\nu_3(T_2)$, 465 cm^{-1}

Fig. 7.3.5.
The normal modes and Raman frequencies of P$_4$. Only one component is shown for the degenerate modes.

If the molecule were to adopt a hypothetical square-planar structure in point group D_{4h}, the normal modes would be

$$\Gamma_{vib}(P_4, D_{4h}) = A_{1g}(R) + B_{1g}(R) + B_{2g}(R) + B_{2u} + E_u(IR).$$

Both models have the same numbers of infrared, Raman, and polarized Raman bands. However, they can be distinguished by the fact that the T_2 mode in the T_d structure is both IR and Raman active (there is one coincidence), whereas the rule of mutual exclusion holds for the centrosymmetric D_{4h} structure.

The normal modes of P$_4$ and their observed Raman frequencies are shown in Fig. 7.3.5. Here A_1 is called the breathing mode, since all P–P bonds are stretching and contracting in unison. In the $\nu_2(E)$ mode, two bonds are contracting while the other four are lengthening; in the $\nu_3(T_2)$ mode, three bonds are contracting while the other three are lengthening. It is of interest to note that all six vibrations of P$_4$ are stretching motions.

(4) XeF$_4$

This molecule has a square-planar structure with D_{4h} symmetry. With reference to the coordinate system displayed in Fig. 7.3.6, the symmetry of the vibrational modes may be derived in the following manner:

D_{4h}	E	$2C_4$	C_2	$2C_2'$	$2C_2''$	i	$2S_4$	σ_h	$2\sigma_v$	$2\sigma_d$
$\Gamma(N_0)$	5	1	1	3	1	1	1	5	3	1
$f(\mathbf{R})$	3	1	-1	-1	-1	-3	-1	1	1	1
Γ_{3N}	15	1	-1	-3	-1	-3	-1	5	3	1

$\Gamma_{3N}(XeF_4) = A_{1g} + A_{2g} + B_{1g} + B_{2g} + E_g + 2A_{2u} + B_{2u} + 3E_u.$

$$\Gamma_{vib}(XeF_4) = A_{1g}(R) + B_{1g}(R) + B_{2g}(R) + A_{2u}(IR) + B_{2u} + 2E_u(IR). \tag{7.3.5}$$

So XeF$_4$ has three bands in its infrared spectrum as well as in its Raman spectrum. None of the bands are coincident, as this molecule is centrosymmetric. In addition, there is a "silent" mode, with B_{2u} symmetry, which does not show up in either spectrum. The normal modes and their frequencies are given in Fig. 7.3.6.

Fig. 7.3.6.
Normal modes and vibrational frequencies of XeF$_4$. Note that the x and y axes are equivalent, and only one component is shown for the degenerate modes. The coordinate system of locations of secondary twofold axes and mirror planes are shown in the lower part of the figure.

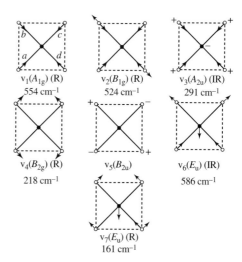

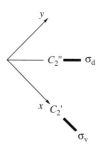

Among the normal modes shown in Fig. 7.3.6, $\nu_1(A_{1g})$, $\nu_2(B_{1g})$, and $\nu_6(E_u)$, are the stretching modes. It is appropriate here to note that it is straightforward to come up with these pictorial representations. If we call the four Xe–F bonds a, b, c, and d, the $\nu_1(A_{1g})$ mode may be denoted as $a + b + c + d$, with "+" indicating contracting motion and "−" indicating stretching motion. The combination $a+b+c+d$ is what we will get if we apply the projection operator $P^{A_{1g}}$ to a. Similarly, if we apply $P^{B_{1g}}$ to a, we will get $a - b + c - d$, which translates to $\nu_2(B_{1g})$ in Fig. 7.3.6. If we apply P^{E_u} to a, we get $a - c$; similarly, applying P^{E_u} to b will yield $b - d$. When we combine these combination further, $(a - c) \pm (b - d)$ are the results. One of these combinations is shown in Fig. 7.3.6, while the other is not. The stretching motions of a molecule can usually be derived in this manner.

If the x and y axes are chosen to bisect the F–Xe–F bond angles, the symmetry of the normal modes is expressed as

$$\Gamma_{\text{vib}}(\text{XeF}_4) = A_{1g}(\text{R}) + B_{1g}(\text{R}) + B_{2g}(\text{R}) + A_{2u}(\text{IR}) + B_{1u} + 2E_u(\text{IR}). \tag{7.3.6}$$

As compared to eq. (7.3.5), the only difference is that the silent mode B_{2u} in that expression is converted to B_{1u} here, and now ν_2 has B_{2g} symmetry and ν_4 has B_{1g} symmetry. All deductions concerning infrared and Raman activities remain unchanged.

Given the illustrations of the normal modes of a molecule, it is possible to identify their symmetry species from the character table. Each non-degenerate normal mode can be regarded as a basis, and the effects of all symmetry operations of the molecular point group on it are to be considered. For instance, the ν_1 mode of XeF$_4$ is invariant to all symmetry operations, i.e. $\boldsymbol{R}(\nu_1) = (1)\nu_1$ for all values of $\boldsymbol{R}$. Since the characters are all equal to 1, ν_1 belongs to symmetry species A_{1g}. For ν_2, the symmetry operation C_4 leads to a character of -1, as shown below:

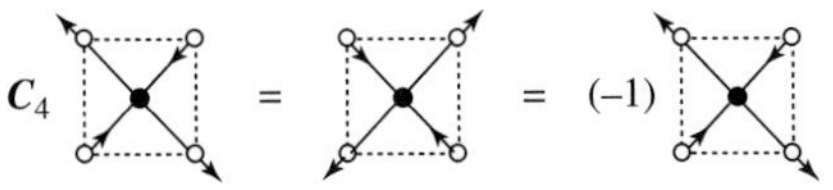

Working through all symmetry operations in the various classes by inspection, we can readily identify the symmetry species of ν_2 as B_{1g}.

For a doubly degenerate normal mode, both components must be used together as the basis of a two-dimensional irreducible representation. For example, the operations C_2 and σ_v on the two normal vibrations that constitute the ν_6 mode lead to the character (sum of the diagonal elements of the corresponding 2×2 matrix) of -2 and 0, respectively, as illustrated below. Working through the remaining symmetry operations, the symmetry species of ν_6 can be identified as E_u.

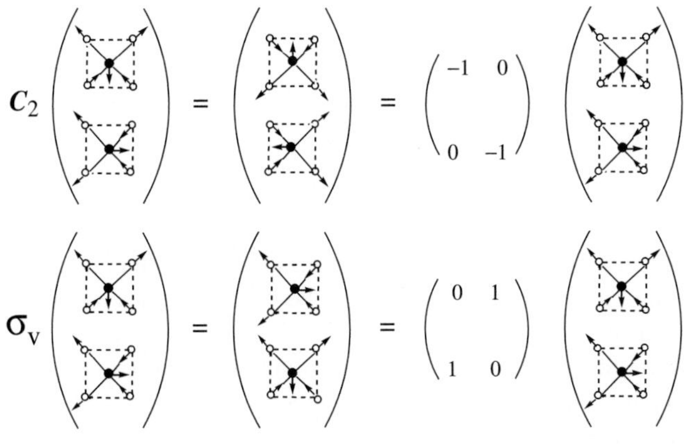

(5) SF$_4$

As shown below, this molecule has C_{2v} symmetry with two types of S–F bonds, equatorial and axial. The vibrational modes can be derived in the following way:

C_{2v}	E	C_2	$\sigma_v(xz)$	$\sigma'_v(yz)$
$\Gamma(N_0)$	5	1	3	3
$f(\boldsymbol{R})$	3	-1	1	1
Γ_{3N}	15	-1	3	3

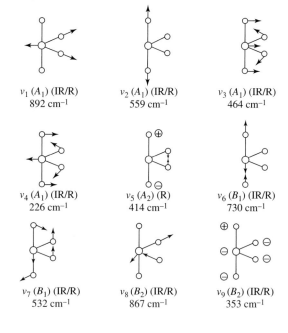

Fig. 7.3.7.
Normal modes of SF$_4$ and observed frequencies. The + and - symbols refer to motions of "coming out" and "going into" the paper, respectively.

$$\Gamma_{3N}(SF_4) = 5A_1 + 2A_2 + 4B_1 + 4B_2.$$

$$\Gamma_{vib}(SF_4) = 4A_1(IR/R) + A_2(R) + 2B_1(IR/R) + 2B_2(IR/R).$$

So there are nine Raman lines and eight infrared lines. All eight infrared bands may be found in the Raman spectrum. The normal modes and their frequencies for SF$_4$ are shown in Fig. 7.3.7. From this figure, it may be seen that stretching modes include ν_1 (symmetric stretch for equatorial bonds), ν_2 (symmetric stretch for axial bonds), ν_6 (asymmetric stretch for axial bonds), and ν_8 (asymmetric stretch for equatorial bonds).

It is worth noting that we have so far discussed the vibrational spectra of three five-atom molecules: CF$_4$, XeF$_4$, and SF$_4$, with T_d, D_{4h}, and C_{2v} symmetry, respectively. In their infrared spectra, two, three, and eight bands are observed, respectively. These results are consistent with the expectation that more symmetrical molecules have less infrared bands.

(6) PF$_5$

This well-known molecule assumes a trigonal bipyramidal structure with D_{3h} symmetry. As in the case of SF$_4$, PF$_5$ also has two types of bonds, equatorial and

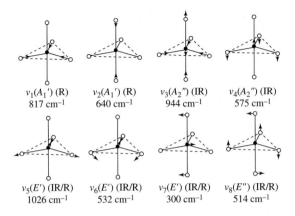

$v_1(A_1')$ (R)
817 cm^{-1}

$v_2(A_1')$ (R)
640 cm^{-1}

$v_3(A_2'')$ (IR)
944 cm^{-1}

$v_4(A_2'')$ (IR)
575 cm^{-1}

$v_5(E')$ (IR/R)
1026 cm^{-1}

$v_6(E')$ (IR/R)
532 cm^{-1}

$v_7(E')$ (IR/R)
300 cm^{-1}

$v_8(E'')$ (IR/R)
514 cm^{-1}

Fig. 7.3.8.
Normal modes of PF$_5$ and their frequencies. Note that only one component is shown for each degenerate mode.

axial P–F bonds. The vibrational modes for this molecule can be determined in the following way:

D_{3h}	E	$2C_3$	$3C_2$	σ_h	$2S_3$	$3\sigma_v$	
$\Gamma(N_0)$	6	3	2	4	1	4	
$f(R)$	3	0	-1	1	-2	1	
Γ_{3N}	18	0	-2	4	-2	4	$\equiv 2A_1' + A_2' + 4E' + 3A_2'' + 2E''$

$\Gamma_{\text{vib}}(\text{PF}_5) = 2A_1'(\text{R}) + 3E'(\text{IR/R}) + 2A_2''(\text{IR}) + E''(\text{R})$.

So there are five infrared and six Raman bands, three of which are coincident. The normal modes and their frequencies are shown in Fig. 7.3.8. From this figure, it is seen that $v_1(A_1')$ and $v_5(E')$ are the stretching modes for the equatorial bonds, while $v_2(A_1')$ and $v_3(A_2'')$ are the symmetric and asymmetric stretches for the axial bonds. Recall that in the derivation of the linear combinations for three equivalent functions a, b, and c, the results are $a+b+c$; $2a-b-c$, and $b-c$, with the last two being degenerate. Mode $v_1(A_1')$ in Fig. 7.3.8 is equivalent to $a+b+c$, while $v_5(E')$ is $2a-b-c$. The remaining combination, $b-c$, is not shown in this figure.

(7) SF$_6$

This highly symmetrical molecule has an octahedral structure with O_h symmetry. Contrary to SF$_4$ and PF$_5$, where there are two types of bonds, all six bonds in SF$_6$ are equivalent to each other. The vibrational modes of this molecule can be determined in the following manner:

O_h	E	$8C_3$	$6C_2$	$6C_4$	$3C_2=C_4^2$	i	$6S_4$	$8S_6$	$3\sigma_h$	$6\sigma_d$
$\Gamma(N_0)$	7	1	1	3	3	1	1	1	5	3
$f(R)$	3	0	-1	1	-1	-3	-1	0	1	1
Γ_{3N}	21	0	-1	3	-3	-3	-1	0	5	3

$\Gamma_{3N}(\text{SF}_6) = A_{1g} + E_g + T_{1g} + T_{2g} + 3T_{1u} + T_{2u}$.

$\Gamma_{\text{vib}}(\text{SF}_6) = A_{1g}(\text{R}) + E_g(\text{R}) + 2T_{1u}(\text{IR}) + T_{2g}(\text{R}) + T_{2u}$.

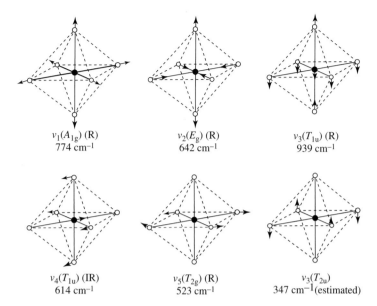

Fig. 7.3.9.
Normal modes of SF₆ and their frequencies. Note that only one component is shown for each degenerate mode.

$\nu_1(A_{1g})$ (R)
774 cm^{-1}

$\nu_2(E_g)$ (R)
642 cm^{-1}

$\nu_3(T_{1u})$ (R)
939 cm^{-1}

$\nu_4(T_{1u})$ (IR)
614 cm^{-1}

$\nu_5(T_{2g})$ (R)
523 cm^{-1}

$\nu_3(T_{2u})$
347 cm^{-1}(estimated)

So there are three Raman and two infrared bands, none of which are coincident. Also, the T_{2u} mode is silent. The normal modes and their frequencies are shown in Fig. 7.3.9. It is clear now that $\nu_1(A_{1g})$, $\nu_2(E_g)$, and $\nu_3(T_{1u})$ are the stretching modes.

If we call the six bonds $a, b, \ldots, f$, with a and b colinear, and so are c and d, as well as e and f, then $\nu_1(A_{1g})$ is equivalent to $a+b+c+d+e+f$; $\nu_2(E_g)$ is equivalent to $2e+2f-a-b-c-d$; $\nu_3(T_{1u})$ is equivalent to $e-f$. The ones not shown in Fig. 7.3.9 are the remaining component of E_g, $a+b-c-d$, and the last two components of T_{1u}, $a-b$ and $c-d$. These six combinations are identical to those listed in Table 7.1.5, the combinations of ligand orbitals in AH₆.

7.3.3 CO stretch in metal carbonyl complexes

In the previous examples for molecules with relatively few atoms, we studied and presented the results of all $3N-6$ vibrations of the molecules. However, for molecules composed of a large number of atoms, often it is convenient to concentrate on a certain type of vibration. The prime example of this kind of investigation is the enumeration of the CO stretching modes in metal carbonyl complexes.

Most metal carbonyl complexes exhibit sharp and intense CO bands in the range 1800–2100 cm^{-1}. Since the CO stretch motions are rarely coupled with other modes and CO absorption bands are not obscured by other vibrations, measurement of the CO stretch bands alone often provides valuable information about the geometric and electronic structures of the carbonyl complexes. As we may recall, free CO absorbs strongly at 2155 cm^{-1}, which corresponds to the stretching motion of a C≡O triple bond. On the other hand, most ketones and aldehyde exhibit bands near 1715 cm^{-1}, which corresponds formally to

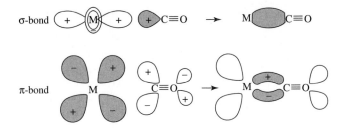

Fig. 7.3.10.
The σ and π bonding in metal carbonyls. Note that electrons flow from the shaded orbital to the unshaded orbital in both instances.

the stretching of a C=O double bond. In other words, the CO bonds in metal carbonyls have a bond order somewhere between 2 and 3. This observation may be rationalized by the simple bonding model illustrated in Fig. 7.3.10.

The M–C σ bond is formed by donating the lone electrons on C to the empty d_{z^2} orbital on M (upper portion of Fig. 7.3.10). The π bond is formed by back donation of the metal dπ electrons to the π* orbital (introduced in Chapter 3) of CO. Populating the π* orbital of CO tends to decrease the CO bond order, thus lowering the CO stretch frequency (lower portion of Fig. 7.3.10). These two components of metal-carbonyl bonding may be expressed by the two resonance structures

$$^-M-C\equiv O^+ \longleftrightarrow M=C=O$$
$$\text{(I)} \hspace{2cm} \text{(II)}$$

Thus structure (I), with higher CO frequencies, tends to build up electron density on M and is thus favored by systems with positive charges accumulated on the metal. On the other hand, structure (II), with lower CO frequencies, is favored by those with negative charge (which enhances back donation) on metal. In other words, net charges on carbonyl compounds have a profound effect on the CO stretch frequencies, as illustrated by the following two isoelectronic series:

	Importance of resonance structure $^-M-C\equiv O^+ \longrightarrow$		
ν (in cm^{-1})	$V(CO)_6^-$	$Cr(CO)_6$	$Mn(CO)_6^+$
$\nu(A_{1g})$	2020	2119	2192
$\nu(E_g)$	1895	2027	2125
$\nu(T_{1u})$	1858	2000	2095
	$Fe(CO)_4^{2-}$	$Co(CO)_4^-$	$Ni(CO)_4$
$\nu(A_1)$	1788	2002	2125
$\nu(T_2)$	1786	1888	2045
	$\longleftarrow$ Importance of resonance structure M=C=O		

The identification of the symmetries of the CO stretching modes and the determination of their activities may be accomplished by the usual group theory techniques. Take the trigonal bipyramidal Fe(CO)$_5$ (with D_{3h} symmetry) as an example. The representation spanned by the five CO stretching motions, Γ_{CO}, is once again derived by counting the number of CO groups unshifted by each operation in the D_{3h} point group.

Fig. 7.3.11.
Carbonyl stretching modes and their frequencies for Fe(CO)$_5$. Note that only one component is shown for $v_{10}(E')$.

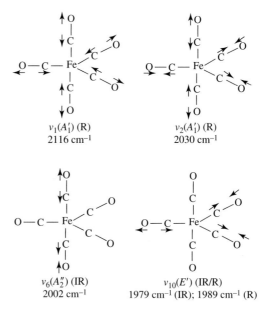

$v_1(A_1')$ (R)
2116 cm^{-1}

$v_2(A_1')$ (R)
2030 cm^{-1}

$v_6(A_2'')$ (IR)
2002 cm^{-1}

$v_{10}(E')$ (IR/R)
1979 cm^{-1} (IR); 1989 cm^{-1} (R)

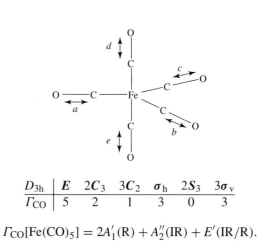

D_{3h}	E	$2C_3$	$3C_2$	σ_h	$2S_3$	$3\sigma_v$
Γ_{CO}	5	2	1	3	0	3

$$\Gamma_{CO}[\text{Fe(CO)}_5] = 2A_1'(\text{R}) + A_2''(\text{IR}) + E'(\text{IR/R}).$$

So there are two and three CO stretch bands in the infrared and Raman spectra, respectively. The CO stretching modes and their frequencies are given in Fig. 7.3.11.

The CO stretching modes shown in Fig. 7.3.11 may be easily obtained using the standard group theory technique. If we call the equatorial CO groups $a, b,$ and c, and the axial groups d and e, the combinations for these five entities are simply $A_1', a+b+c; E', 2a-b-c$ and $b-c; A_1', d+e; A_2'', d-e$; where + denotes stretching and − denotes contracting. Since there are two A_1' combinations, we can combine them further to form $k(a+b+c)+d+e$ and $k'(-a-b-c)+d+e$, where k and k' are the combination coefficients. The stretching modes shown in Fig. 7.3.11 are simply graphical representations of these five linear combinations, except that one component of the E' mode, $b-c$, is not shown. Again, $v_1(A_1')$ is called the breathing mode, since all carbonyl groups are stretching and contracting in unison. For metal-carbonyl complexes, the breathing mode often has the highest energy, as in the case of Fe(CO)$_5$.

Now it is straightforward to show that, for tetrahedral carbonyl complexes M(CO)$_4$ with T_d symmetry,

$$\Gamma_{CO}[M(CO)_4] = A_1(R) + T_2(IR/R).$$

For octahedral carbonyl complexes M(CO)$_6$ with O_h symmetry,

$$\Gamma_{CO}[M(CO)_6] = A_{1g}(R) + E_g(R) + T_{1u}(IR).$$

Since there is a direct relationship between the structure of a metal-carbonyl complex and the number of CO stretching bands, it is often possible to deduce the arrangement of the CO groups in a complex when we compare its spectrum with the number of CO stretch bands predicted for each of the possible structures using group theory techniques. As an illustrative example, consider the *cis* and *trans* isomers of an octahedral M(CO)$_4$L$_2$ complex. For the *trans* isomer, with D_{4h} symmetry, we have

$$\Gamma_{CO}[trans\text{-}M(CO)_4L_2] = A_{1g}(R) + B_{1g}(R) + E_u(IR).$$

So there are two CO stretching bands in the Raman spectrum and only one in infrared. Also, there are no coincident bands. For the *cis* isomer, with C_{2v} symmetry, we have

$$\Gamma_{CO}[cis\text{-}M(CO)_4L_2] = 2A_1(IR/R) + B_1(IR/R) + B_2(IR/R).$$

So there are four infrared/Raman coincident lines for the *cis* isomer. When M is Mo and L is PCl$_3$, the *trans* isomer has indeed only one infrared CO stretch band, while there are four for the *cis* isomer. These results are summarized in Fig. 7.3.12, along with the pictorial illustrations of the CO stretching modes. It appears that the two A_1 modes of the *cis* isomer do not couple strongly. Also, this is another example for the general rule that a more symmetrical molecule will have fewer infrared bands.

In some cases, two isomers may have the same number of CO stretching bands, yet they may still be distinguished by the relative intensities of the bands. Take the *cis*- and *trans*-[(η^5-C$_5$H$_5$)Mo(CO)$_2$PPh$_3$C$_4$H$_6$O]$^+$ and their spectra as an example. The structural formulas of these complex ions and their infrared spectra are shown in Fig. 7.3.13. In both spectra, the two bands are the symmetric and asymmetric CO stretch modes. As shown in the previous example of *cis*-[Mo(CO)$_4$(PCl$_3$)$_2$], the symmetric modes has a slightly higher energy. Also, as shown below, the intensity ratio is related to the angle formed by the two groups.

Referring to Fig. 7.3.14, the dipole vector **R** for a stretching mode is simply the sum or difference of the two individual CO group dipoles. Since the intensities of the symmetric and asymmetric stretches are proportional to $\mathbf{R}^2_{sym}$ and

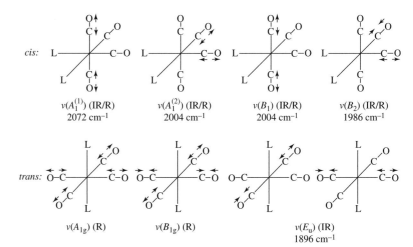

Fig. 7.3.12.
CO stretching modes and their infrared bands of *cis*- and *trans*-[Mo(CO)$_4$(PCl$_3$)$_2$].

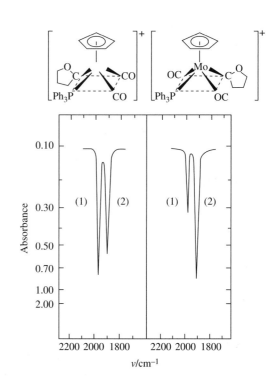

Fig. 7.3.13.
The CO stretching bands and their relative intensities of the *cis* and *trans* isomers of [(η^5–C$_5$H$_5$)Mo(CO)$_2$PPh$_3$C$_4$H$_6$O]$^+$.

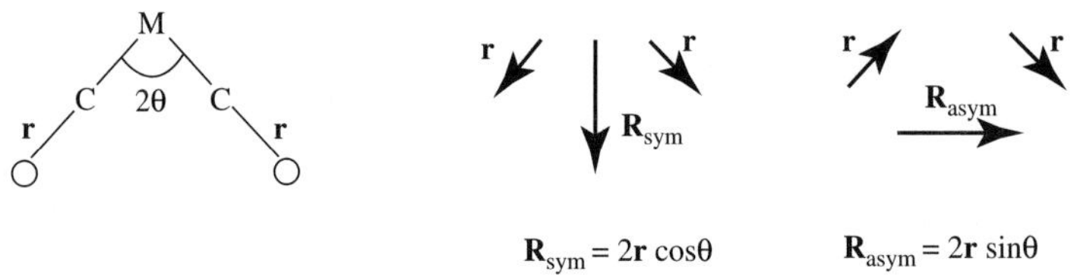

Fig. 7.3.14.
Diagram showing how the two CO dipoles (**r**) are combined to give dipole vector **R** for the M(CO)$_2$ moiety. Different ways of combining lead to **R**$_{sym}$ and **R**$_{asym}$.

$\mathbf{R}^2_{\text{asym}}$, respectively, the ratio of these two intensities is simply

$$I(1)/I(2) = \mathbf{R}^2_{\text{sym}}/\mathbf{R}^2_{\text{asym}}$$
$$= (2\mathbf{r}\cos\theta^2)/(2\mathbf{r}\sin\theta)^2$$
$$= \cot^2\theta,$$

where 2θ is the angle formed by the two CO groups. For the left spectrum in Fig. 7.3.13, $I(1)/I(2) = 1.44$, and 2θ is $79°$, indicating this spectrum is that of the *cis* isomer. For the spectrum on the right, $I(1)/I(2) = 0.32$, and 2θ is $121°$, signifying this spectrum is that of the *trans* isomer.

In polynuclear carbonyls, i.e., those carbonyl complexes with two or more metal atoms, the CO ligands may bond to the metal(s) in different ways. In addition to the terminal carbonyl groups, those we have studied so far, we may also have μ_n bridging carbonyls, which bond to n metal atoms. In the case of μ_2 bridging, it may be assumed that a CO group donates one electron to each metal. Also, both metals back-donate to the CO ligand, thus leading to vibrational frequencies lower than those of terminal COs. Indeed, most μ_2-bridging CO groups absorb in the range of 1700–1860 cm^{-1}. In the following we consider the dinulclear complex Co$_2$(CO)$_8$ which, in the solid state, has the following structure with C_{2v} symmetry:

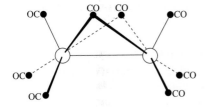

To derive the CO stretching modes, we may consider the bridging and terminal groups separately.

C_{2v}	E	C_2	$\sigma_v(xz)$	$\sigma'_v(yz)$
$(\Gamma_{CO})_{\text{brid}}$	2	0	0	2
$(\Gamma_{CO})_{\text{term}}$	6	0	2	0

$(\Gamma_{CO})_{\text{brid}}[\text{Co(CO)}_8] = A_1(\text{IR/R}) + B_2(\text{IR/R}),$

$(\Gamma_{CO})_{\text{term}}[\text{Co(CO)}_8] = 2A_1(\text{IR/R}) + A_2(\text{R}) + 2B_1(\text{IR/R}) + B_2(\text{IR/R}).$

So there are seven CO stretching bands in the infrared spectrum, as found experimentally. The terminal CO stretching frequencies are at 2075, 2064, 2047, 2035, and 2028 cm^{-1}, and the two bridging CO stretch peaks are at 1867 and 1859 cm^{-1}.

As another example, consider the highly symmetrical Fe$_2$(CO)$_9$, with D_{3h} symmetry, as shown above. The symmetries of the CO stretching modes can be determined readily:

D_{3h}	E	$2C_3$	$3C_2$	σ_h	$2S_3$	$3\sigma_v$
$(\Gamma_{CO})_{\text{brid}}$	3	0	1	3	0	1
$(\Gamma_{CO})_{\text{term}}$	6	0	0	0	0	2

$(\Gamma_{CO})_{\text{brid}}[\text{Fe}_2(\text{CO})_9] = A_1'(\text{R}) + E'(\text{IR/R}),$

$(\Gamma_{CO})_{\text{term}}[\text{Fe}_2(\text{CO})_9] = A_1'(\text{R}) + A_2''(\text{IR}) + E'(\text{IR/R}) + E''(\text{R}).$

So there are three infrared absorption bands: 2066 (A_2''), 2038 (E', terminal), and 1855 (E', bridging) cm^{-1}.

In this section, using purely qualitative symmetry arguments, we have discussed the kind of information regarding structure and bonding we can obtain from vibrational spectroscopy. Obviously, treatments of this kind have their limitations, such as their failure to make reliable assignments for the observed vibrational bands. To carry out this type of tasks with confidence, we need to make use of quantitative methods, which are beyond the scope of this book.

7.3.4 Linear molecules

In this section, we examine the vibrational spectra of a few linear molecules to illustrate the principles we have discussed.

(1) Hydrogen cyanide and carbon dioxide

Both HCN and CO$_2$, with $C_{\infty v}$ and $D_{\infty h}$ symmetry, respectively, have two bonds and, hence, have two stretching modes and one (doubly degenerate) bending mode. In CO$_2$, the two bonds are equivalent and they may couple in a symmetric and an antisymmetric way, giving rise to symmetric and asymmetric stretching modes. However, for HCN, we simply have the C–H and C≡N stretching modes. The observed frequencies and their assignments for these two triatomic molecules are summarized in Tables 7.3.1 and 7.3.2.

Table 7.3.1. The normal modes of HCN and their observed frequencies

Symmetry	cm^{-1}	Normal mode	Activity	Description
Π	712	↑ ↑ H—C≡N ↓	IR/R	bending
Σ^+	2089	← ← → H—C≡N	IR/R	C≡N stretch
Σ^+	3312	← → → H—C≡N	IR/R	C–H stretch

(2) Cyanogen

Cyanogen, N≡C–C≡N, is a symmetric linear molecule. It has three stretching modes and two (doubly degenerate) bending modes. Just as in the case of CO$_2$,

Table 7.3.2. The normal modes of CO_2 and their observed frequencies

Symmetry	cm^{-1}	Normal mode	Activity	Description
Π_u	667	O=C=O (bending: ↑ ↑ ↓)	IR	bending
Σ_g^+	1388	O=C=O (← →)	R	symmetric stretch
Σ_u^+	2349	O=C=O (→ ← →)	IR	asymmetric stretch

there are no coincident bands. Hence, the assignments are fairly straightforward, as summarized in Table 7.3.3.

Table 7.3.3. The normal modes of cyanogen and their observed frequencies

Symmetry	cm^{-1}	Normal mode	Activity	Description
Π_u	226	N≡C—C≡N (↑ ↑ ↓ ↓)	IR	asymmetric C–C–N bending
Π_g	506	N≡C—C≡N (↑ ↑ ↓ ↓)	R	symmetric bending
Σ_g^+	848	N≡C—C≡N (← ← → →)	R	C–C stretch
Σ_u^+	2149	N≡C—C≡N (← → → ←)	IR	asymmetric C≡N stretch
Σ_g^+	2322	N≡C—C≡N (← → ← →)	R	symmetric C≡N stretch

(3) Carbon suboxide

Carbon suboxide, O=C=C=C=O, has four stretching modes and three (doubly degenerate) bending modes. The spectral assignments can be made in a similar manner, as summarized in Table 7.3.4.

While most of the assignments given in Tables 7.3.1 to 7.3.4 may be made by qualitative arguments, it should be stressed that quantitative treatments are indispensable in making certain that the assignments are correct.

Finally, it is useful to mention here a "systematic" way to derive the symmetries of the vibrational modes for a linear molecule. Even though formally carbon suboxide has $D_{\infty h}$ symmetry, its vibrational modes may be derived by using the D_{2h} character table. The procedure is illustrated below for the carbon suboxide molecule:

D_{2h}	E	$C_2(z)$	$C_2(y)$	$C_2(x)$	i	$\sigma(xy)$	$\sigma(xz)$	$\sigma(yz)$
$\Gamma(N_0)$	5	5	1	1	1	1	5	5
$f(R)$	3	−1	−1	−1	−3	1	1	1
Γ_{3N}	15	−5	−1	−1	−3	1	5	5

Table 7.3.4. The normal modes of carbon suboxide and their observed frequencies

Symmetry	cm^{-1}	Normal mode	Activity	Description
Π_u	557	O=C=C=C=O (↑↑↑ / ↓ ↓)	IR	asymmetric C–C–O bending
Π_g	586	O=C=C=C=O (↑ ↑ / ↓ ↓)	R	symmetric bending
Π_u	637	O=C=C=C=O (↑ ↑ / ↓ ↓)	IR	asymmetric C–C–C bending
Σ_g^+	843	O=C=C=C=O (→→ ← ←)	R	symmetric C=C stretch
Σ_u^+	1570	O=C=C=C=O (← ← → ← ←)	IR	asymmetric C=C stretch
Σ_g^+	2200	O=C=C=C=O (→ ← → ←)	R	symmetric C=O stretch
Σ_u^+	2290	O=C=C=C=O (← → → → ←)	IR	asymmetric C=O stretch

$$\Gamma_{3N} = 2A_g + 2B_{2g} + 2B_{3g} + 3B_{1u} + 3B_{2u} + 3B_{3u}.$$

$$\Gamma_{\text{vib}}(C_3O_2, D_{2h}) = 2A_g + B_{2g} + B_{3g} + 2B_{1u} + 2B_{2u} + 2B_{3u}.$$

Note that in deriving these results, we have considered $\Gamma_{\text{vib}}(C_3O_2, D_{2h}) = \Gamma_{3N} - \Gamma_{\text{trans}} - \Gamma(R_x) - \Gamma(R_y)$, which gives rise to a total of 10 vibrations for this *linear* molecule. There is no need to take rotation R_z (with B_{1g} symmetry) into account here as we have assigned the molecular axis to be the z axis.

To convert the symmetry species in point group D_{2h} to those of $D_{\infty h}$, we make use of the correlation diagram shown below, which can be easily deduced from the pair of character tables:

$$
\begin{array}{lll}
D_{2h} & & D_{\infty h} \\
A_g \text{———} z^2 \text{———} & & \Sigma_g^+ \\
B_{2g} \text{———} xz & & \\
 & \searrow & \Pi_g \\
B_{3g} \text{———} yz & \nearrow & \\
B_{1u} \text{———} z \text{———} & & \Sigma_u^+ \\
B_{2u} \text{———} y & & \\
 & \searrow & \Pi_u \\
B_{3u} \text{———} x & \nearrow &
\end{array}
$$

So, in terms of the symmetry species of point group $D_{\infty h}$, we have

$$\Gamma_{\text{vib}}(C_3O_2) = 2\Sigma_g^+ + \Pi_g + 2\Sigma_u^+ + 2\Pi_u.$$

7.3.5 Benzene and related molecules

In the final section of this chapter, we discuss the vibrational spectra of benzene, its isostructural species $C_6O_6^{2-}$ (rhodizonate dianion), and dibenzene metal complexes.

(1) Benzene

Historically, the cyclic structure of benzene with D_{6h} symmetry, as shown in Fig. 7.3.15, was deduced by enumerating the derivatives formed in the mono-, di-, tri-substitution reactions of benzene. The structure can also be established directly using physical methods such as X-ray and neutron diffraction, NMR, and vibrational spectroscopy. We now discuss the infrared and Raman spectral data of benzene.

The symmetries of the 30 normal vibrations can be derived in the usual way. The symmetry elements C_2', C_2'', σ_d, and σ_v are taken as those shown in Fig. 7.3.15.

D_{6h}	E	$2C_6$	$2C_3$	C_2	$3C_2'$	$3C_2''$	i	$2S_3$	$2S_6$	σ_h	$3\sigma_d$	$3\sigma_v$
$\Gamma(N_0)$	12	0	0	0	4	0	0	0	0	12	0	4
$f(R)$	3	2	0	−1	−1	−1	−3	−2	0	1	1	1
Γ_{3N}	36	0	0	0	−4	0	0	0	0	12	0	4

$$\Gamma_{3N} = 2A_{1g} + 2A_{2g} + 2B_{2g} + 2E_{1g} + 4E_{2g} + 2A_{2u} + 2B_{1u} + 2B_{2u}$$
$$+ 4E_{1u} + 2E_{2u}.$$

$$\Gamma_{vib}(C_6H_6) = 2A_{1g}(R) + A_{2g} + 2B_{2g} + E_{1g}(R) + 4E_{2g}(R) + A_{2u}(IR)$$
$$+ 2B_{1u} + 2B_{2u} + 3E_{1u}(IR) + 2E_{2u}.$$

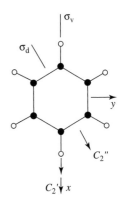

Fig. 7.3.15.
Symmetry elements of the benzene molecule. The x axis is chosen to pass through a pair of carbon atoms, and the z axis points toward the reader.

So four bands are expected in the IR spectrum (one with A_{2u} symmetry and three sets with E_{1u} symmetry), while there should be seven bands in the Raman spectrum (two with A_{2u} symmetry, one set having E_{1g} symmetry, and four sets with E_{2g} symmetry).

The 20 vibrational modes of benzene are pictorially illustrated in Fig. 7.3.16. Also shown are the observed frequencies. In this figure, ν(CH) and ν(CC) represent C–H and C–C stretching modes, respectively, while δ and π denote in-plane and out-of-plane bending modes, respectively. For each E mode, only one component is shown.

The 20 vibrational modes illustrated in Fig. 7.3.16 may be broken down into various types of vibrations. In the following tabulation, the representation generated by the 12 vibrations of the carbon ring skeleton is denoted by Γ_{C6}, and Γ_{C-C} and Γ_{C-H} are the representations for the stretching motions of C–C and C–H bonds, respectively.

D_{6h}	E	$2C_6$	$2C_3$	C_2	$3C_2'$	$3C_2''$	i	$2S_3$	$2S_6$	σ_h	$3\sigma_d$	$3\sigma_v$
$\Gamma(N_0)(C6)$	6	0	0	0	2	0	0	0	0	6	0	2
$f(R)$	3	2	0	−1	−1	−1	−3	−2	0	1	1	1
$\Gamma_{3N}(C6)$	18	0	0	0	−2	0	0	0	0	6	0	2
Γ_{C-C}	6	0	0	0	0	2	0	0	0	6	2	0
Γ_{C-H}	6	0	0	0	2	0	0	0	0	6	0	2

$$\Gamma_{3N}(C6) = A_{1g} + A_{2g} + B_{2g} + E_{1g} + 2E_{2g} + A_{2u} + B_{1u}$$
$$+ B_{2u} + 2E_{1u} + E_{2u}.$$

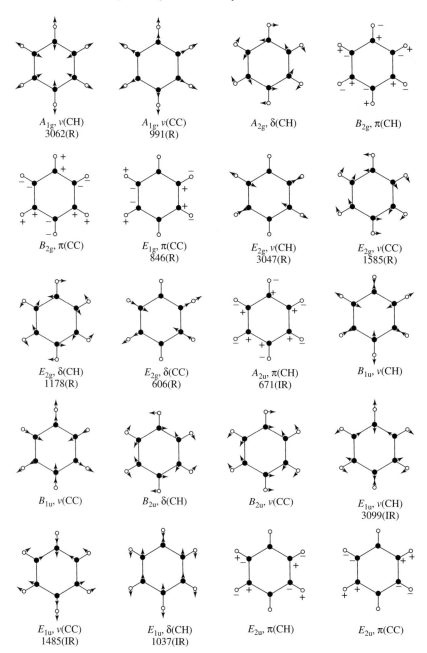

Fig. 7.3.16.
Vibrational modes and their observed frequencies (in cm^{-1}) of benzene.

Now,

$$\Gamma_{C6} \text{ (vibration of the carbon ring)}$$
$$= \Gamma_{3N}(C6) - \Gamma\text{(translation)} - \Gamma\text{(rotation)}$$
$$= A_{1g} + A_{2g} + B_{2g} + E_{1g} + 2E_{2g} + A_{2u} + B_{1u} + B_{2u}$$
$$\quad + 2E_{1u} + E_{2u} - (A_{2u} + E_{1u}) - (A_{2g} + E_{1g})$$
$$= A_{1g} + B_{2g} + 2E_{2g} + B_{1u} + B_{2u} + E_{1u} + E_{2u}.$$

Also,

$$\Gamma_{C-C} = A_{1g} + E_{2g} + B_{2u} + E_{1u}, \quad \text{which are the C–C stretching modes;}$$
$$\Gamma_{C-H} = A_{1g} + E_{2g} + B_{1u} + E_{1u}, \quad \text{which are the C–H stretching modes.}$$

The vibrations that involve the distortion of the carbon ring are therefore $\Gamma_{C6} - \Gamma_{C-C} = B_{2g} + E_{2g} + B_{1u} + E_{2u}$.

Another way to analyse $\Gamma_{\text{vib}}(C_6H_6)$ is to separate the in-plane vibration representation $\Gamma_{\text{vib}}(x,y)$ from the out-of-plane vibration representation $\Gamma_{\text{vib}}(z)$. To do this, we need to first derive the representation based on the x and y coordinates of all 12 atoms, $\Gamma_{2N}(x,y)$:

D_{6h}	E	$2C_6$	$2C_3$	C_2	$3C_2'$	$3C_2''$	i	$2S_3$	$2S_6$	σ_h	$3\sigma_d$	$3\sigma_v$
$\Gamma_{2N}(x,y)$	24	0	0	0	0	0	0	0	0	24	0	0
$\Gamma(x,y) = E_{1u}$	2	1	−1	−2	0	0	−2	−1	1	2	0	0
$R_z = A_{2g}$	1	1	1	1	−1	−1	1	1	1	1	−1	−1

Now, the representation of the in-plane vibrations of benzene is simply $\Gamma_{2N}(x,y) - E_{1u} - A_{2g}$, which is

$$\Gamma_{\text{vib}}(x,y) = 2A_{1g} + A_{2g} + 4E_{2g} + 2B_{1u} + 2B_{2u} + 3E_{1u}.$$

The dimension of $\Gamma_{\text{vib}}(x,y)$ is 24 (total number of degrees of freedom for motion in the xy-plane) − 2 (translation of the molecule in the x and y directions) − 1 (rotation of the molecule about the z axis, R_z) = 21. Furthermore, the representation of the out-of-plane vibrations of benzene can be easily obtained:

$$\Gamma_{\text{vib}}(z) = \Gamma_{\text{vib}}(C_6H_6) - \Gamma_{\text{vib}}(x,y) = 2B_{2g} + E_{1g} + A_{2u} + 2E_{2u}.$$

The infrared and Raman spectra of benzene are shown in Fig. 7.3.17. It is seen that the Raman lines at 991 and 3062 cm^{-1} represent A_{1g} vibrations, with $\rho = I_\perp/I_\parallel \ll 3/4$ while all remaining lines have $\rho = 3/4$. Note that there are also weak bands arising from $2\nu_i$ (first harmonic or overtone), $\nu_i - \nu_j$ (difference band), or $\nu_i + \nu_j$ (combination band) frequencies. These are not to be discussed here.

Fig. 7.3.17.
Infrared (upper) and Raman (lower) spectra of benzene.

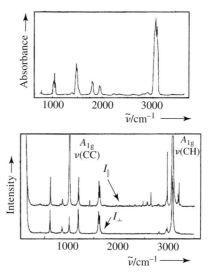

(2) Rhodizonate dianion

Benzene and the rhodizonate dianion ($C_6O_6^{2-}$, structure shown above; also see Section 20.4.4) are isostructural species. Hence all the symmetry arguments and results given in the previous section for benzene are applicable here. The results for $C_6O_6^{2-}$ are summarized in Table 7.3.5.

(3) Dibenzene metal complexes

Dibenzene chromium, $[Cr(C_6H_6)_2]$, has an eclipsed sandwich structure with D_{6h} symmetry. It has $3 \times 25 - 6 = 69$ normal vibrations:

$$\Gamma_{vib}[Cr(C_6H_6)_2] = 4A_{1g}(R) + A_{2g} + 2B_{1g} + 4B_{2g} + 5E_{1g}(R) + 6E_{2g}(R)$$
$$+ 2A_{1u} + 4A_{2u}(IR) + 4B_{1u} + 2B_{2u} + 6E_{1u}(IR) + 6E_{2u}.$$

So it is anticipated that there are 15 Raman and 10 infrared spectral lines for complexes of this type. Listed in Table 7.3.6 are the 10 infrared frequencies for some group VIA dibenzene complexes.

In Table 7.3.6, the term "ring slant" describes a vibration that causes the rings to be inclined (non-parallel) to each other, ν(M-ring) is a vibration that changes the distance between the metal and the rings, while δ(ring-M-ring) is a twisting

Table 7.3.5. The normal modes of the rhodizonate dianion and observed frequencies

Symmetry	Normal Mode	cm^{-1} IR	Raman	Calculated	Activity	Description
A_{1g}	ν(CO)		1669	1594	R(pol)	symmetric CO stretch
A_{1g}	ν(CC)		553	580	R(pol)	breathing mode of the carbon ring
A_{2g}	δ(CO)			854	—	in-plane CO bending
B_{2g}	π(CO)		—	—		ring bending
B_{2g}	π(CC)		—	—		out-of-plane CO bending
E_{1g}	π(CC)		unobs	—	R	out-of-plane ring bending
E_{2g}	ν(CO)		1546	1562	R	CO stretch
E_{2g}	ν(CC)		1252	1222	R	CC stretch
E_{2g}	δ(CO)		436	420	R	in-plane CO bending
E_{2g}	δ(CC)		346	339	R	in-plane ring bending
A_{2u}	π(CO)	235	—		IR	out-of-plane CO bending
B_{1u}	ν(CO)			1551	—	CO stretch
B_{1u}	ν(CC)			489	—	in-plane ring bending
B_{2u}	δ(CO)			451	—	in-plane CO bending
B_{2u}	ν(CC)			1320	—	CC stretch
E_{1u}	ν(CO)	1449		1589	IR	CO stretch
E_{1u}	ν(CC)	1051		1031	IR	CC stretch
E_{1u}	δ(CO)	386		340	IR	in-plane CO bending
E_{2u}	π(CO)		–	–		out-of-plane CO bending
E_{2u}	π(CC)		–	–		out-of-plane ring bending

Table 7.3.6. Observed infrared frequencies (cm^{-1}) for some dibenzene metal complexes

Complex	ν(CH)		ν(CC)	δ(CH)	δ(CC)	π(CH)		Ring slant	ν (M-ring)	δ (ring-M-ring)
[Cr(C$_6$H$_6$)$_2$]	3037	—	1426	999	971	833	794	490	459	140
[Cr(C$_6$H$_6$)$_2$]$^{2+}$	3040	—	1430	1000	972	857	795	466	415	144
[Mo(C$_6$H$_6$)$_2$]	3030	2916	1425	995	966	811	773	424	362	—
[W(C$_6$H$_6$)$_2$]	3012	2898	1412	985	963	882	798	386	331	—

motion that changes the symmetry of the molecule from D_{6h} to D_{6d} and then back to D_{6h}.

References

1. F. A. Cotton, *Chemical Applications of Group Theory*, 3rd edn., Wiley, New York, 1990.
2. S. F. A. Kettle, *Symmetry and Structure: Readable Group Theory for Chemists*, Wiley, Chichester, 1995.
3. S. K. Kim, *Group Theoretical Methods and Applications to Molecules and Crystals*, Cambridge University Press, Cambridge, 1999.
4. D. C. Harris and M. D. Bertolucci, *Symmetry and Spectroscopy: An Introduction to Vibrational and Electronic Spectroscopy*, Oxford University Press, New York, 1978.

5. B. D. Douglas and C. A. Hollingsworth, *Symmetry in Bonding and Spectra: An Introduction*, Academic Press, Orlando, FL, 1985.
6. P. W. M. Jacobs, *Group Theory with Applications in Chemical Physics,* Cambridge University Press, Cambridge, 2005.
7. L. Pauling, *The Nature of the Chemical Bond and the Structure of Molecules and Crystals: An Introduction to Modern Structural Chemistry*, 3rd edn., Cornell University Press, Ithaca, NY, 1960.
8. J. N. Murrell, S. F. A. Kettle and J. M. Tedder, *The Chemical Bond*, Wiley, Chichester, 1978.
9. J. K. Burdett, *Molecular Shapes: Theoretical Models of Inorganic Stereochemistry,* Wiley, New York, 1980.
10. B. M. Gimarc, *Molecular Structure and Bonding: The Quantitative Molecular Orbital Approach*, Academic Press, New York, 1979.
11. T. A. Albright, J. K. Burdett and M.-H. Whangbo, *Orbital Interactions in Chemistry*, Wiley, New York, 1985.
12. T. A. Albright and J. K. Burdett, *Problems in Molecular Orbital Theory,* Oxford University Press, New York, 1992.
13. J. G. Verkade, *A Pictorial Approach to Molecular Bonding and Vibrations,* 2nd edn., Springer, New York, 1997.
14. R. S. Drago, *Physical Methods for Chemists*, 2nd edn., Saunders, Fort Worth, TX, 1992.
15. E. A. B. Ebsworth, D. W. H. Rankin and S. Cradock, *Structural Methods in Inorganic Chemistry*, 2nd edn., CRC Press, Boca Raton, FL, 1991.
16. J. R. Ferraro and K. Nakamoto, *Introductory Raman Spectroscopy*, Academic Press, Boston, MA, 1994.
17. K. Nakamoto, *Infrared and Raman Spectra of Inorganic and Coordination Compounds* (Part A: Theory and Applications in Inorganic Chemistry; Part B: Applications in Coordination, Organometallic, and Bioinorganic Chemistry), 5th edn., Wiley, New York, 1997.
18. R. A. Nyquist, *Interpreting Infrared, Raman, and Nuclear Magnetic Resonance Spectra*, Vols. 1 and 2, Academic Press, San Diego, 2001.

Bonding in Coordination Compounds

8

Coordination compounds are often referred to as *metal complexes*. They consist of one or more *coordination centers* (metal atom or ion), each of which is surrounded by a number of anions (monatomic or polyatomic) or neutral molecules called *ligands*. The set of ligands L constitutes the *coordination sphere* around a metal center M. The term *coordination geometry* is used to describe the spatial arrangement of the ligands, and the number of ligand atoms directly bonded to M is called its *coordination number*. If all ligands are of the same type, the complex is *homoleptic*; otherwise it is *heteroleptic*. A metal complex may be cationic, anionic, or neutral, depending on the sum of the charges on the metal centers and the ligands. It may take the form of a discrete mononuclear, dinuclear, or polynuclear molecule, or exist as a coordination polymer of the chain, layer, or network type. Dinuclear and polynuclear complexes stabilized by bonding interactions between metal centers (i.e., metal–metal bonds) are covered in Chapter 19, and some examples of coordination polymers are described in Chapter 20.

In this chapter, we discuss mostly the bonding in mononuclear homoleptic complexes ML_n using two simple models. The first, called crystal field theory (CFT), assumes that the bonding is ionic; i.e., it treats the interaction between the metal ion (or atom) and ligands to be purely electrostatic. In contrast, the second model, namely the molecular orbital theory, assumes the bonding to be covalent. A comparison between these models will be made.

8.1 Crystal field theory: d-orbital splitting in octahedral and tetrahedral complexes

By considering the electrostatic interaction between the central atom and the surrounding ligands, the CFT shows how the electronic state of a metal atom is affected by the presence of the ligands. Let us first consider the highly symmetrical case of a metal ion M^{m+} surrounded by six octahedrally arranged ligands, as shown in Fig. 8.1.1. If the metal ion has one lone d electron, this electron is equally likely to occupy any of the five degenerate d orbitals, in the absence of the ligands. However, in the presence of the ligands, the d orbitals are no longer degenerate, or equivalent. Specifically, as Fig. 8.1.2 shows, the d_{z^2} and $d_{x^2-y^2}$ orbitals have lobes pointing directly at the ligands, while the lobes of the d_{xy}, d_{yz}, and d_{xz} are pointing between the ligands. As the ligands are negatively

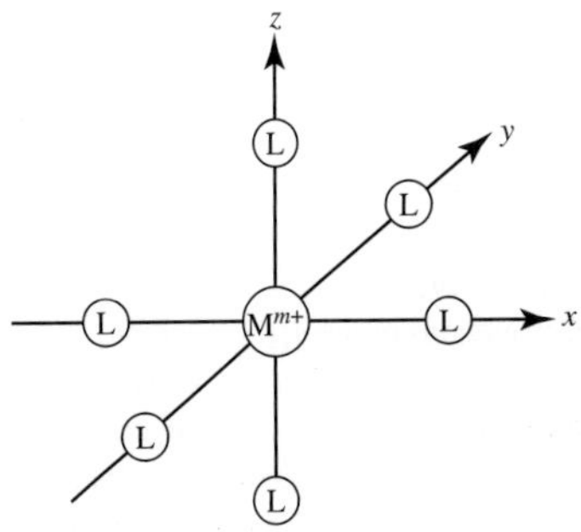

Fig. 8.1.1.
The octahedral arrangement of six ligands surrounding a central metal ion.

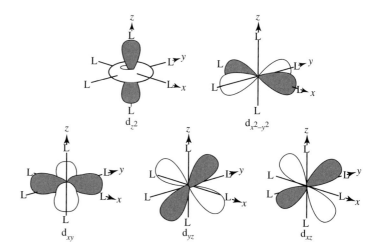

Fig. 8.1.2.
The electron density functions of the five d orbitals relative to the ligand positions in an octahedral complex. The gray and white regions in each orbital bear positive and negative signs, respectively.

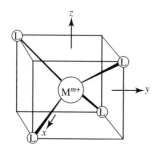

Fig. 8.1.3.
Arrangement of the four ligands in a tetrahedral complex.

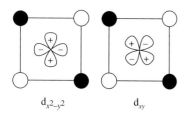

Fig. 8.1.4.
Orientations of the $d_{x^2-y^2}$ and d_{xy} orbitals with respect to the four ligands in a tetrahedral complex, looking down the z axis. Both d orbitals lie in the xy plane. Filled and open circles represent ligands that lie above and below the xy plane, respectively. It is obvious that the $d_{x^2-y^2}$ orbital is more favored for electron occupation in a tetrahedral complex.

charged (or at least have the negative end of the dipole moment pointing at the metal), the d_{xy}, d_{yz}, and d_{xz} orbitals should be more favored for electron occupation than the other two. Moreover, the d_{xy}, d_{yz}, and d_{xz} orbitals are equally likely for electron occupation and hence are degenerate orbitals. On the other hand, the d_{z^2} and $d_{x^2-y^2}$ are less likely to accommodate the single d electron. Furthermore, as pointed out in Chapter 6, it is easy to show that the d_{z^2} is simply the sum of $d_{z^2-x^2}$ and $d_{z^2-y^2}$ functions (aside from a constant), each of which is clearly equivalent to the $d_{x^2-y^2}$ orbital, i.e.,

$$d_{z^2-x^2} + d_{z^2-y^2} \propto z^2 - x^2 + z^2 - y^2 = 3z^2 - r^2 = r^2(3\cos^2\theta - 1) \propto d_{z^2}. \quad (8.1.1)$$

Similarly, when a metal ion is surrounded by four tetrahedrally arranged ligands, as shown in Figs. 8.1.3 and 8.1.4, it is not difficult to see that in this case the electrons tend to occupy the d_{z^2} and $d_{x^2-y^2}$ orbitals in preference to the other three orbitals, d_{xy}, d_{yz}, and d_{xz}. The splitting patterns for a set of d orbitals in octahedral and tetrahedral complexes are summarized in Fig. 8.1.5.

In Fig. 8.1.5(a), it is seen that, in an octahedral complex, the pair of d_{z^2} and $d_{x^2-y^2}$ orbitals are grouped together and called the e_g orbitals, while the d_{xy}, d_{yz}, and d_{xz} orbitals are called the t_{2g} orbitals. The energy difference between these two sets of orbitals is denoted as Δ_o, where the subscript o is the abbreviation for octahedral. In order to retain the "center of gravity" of the d orbitals before and after splitting, i.e., the gain in energy by one set of orbitals (t_{2g} in this case) is offset by the loss in stability by the other set (e_g here), the t_{2g} orbitals lie $(2/5)\Delta_o$ below the set of unsplit d orbitals and the e_g orbitals lie $(3/5)\Delta_o$ above it. Finally, it is noted that t_{2g} and e_g are simply two of the irreducible representations in the O_h group.

In Fig. 8.1.5(b), for tetrahedral complexes, the d_{z^2} and $d_{x^2-y^2}$ pair is now called the e orbitals, and the d_{xy}, d_{yz}, and d_{xz} trio is called the t_2 orbitals. Once again, e and t_2 are two of the irreducible representations of the T_d point group.

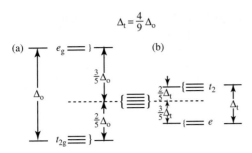

Fig. 8.1.5.
The d orbitals splitting patterns in (a) octahedral and (b) tetrahedral complexes.

The energy difference between these two sets of orbitals is Δ_t (t for tetrahedral). Also, due to the "center of gravity" condition, the t_2 orbitals lie $(2/5)\Delta_t$ above the unperturbed d orbitals and e orbitals lie $(3/5)\Delta_t$ below. When the metal ion, the ligands, and the metal–ligand distance are the same for both the octahedral and tetrahedral systems, it can be shown that

$$\Delta_t = \frac{4}{9}\Delta_o.$$

However, since CFT is a rather crude model, such an exact relationship is seldom of use. Instead, it is instructive and convenient to realize that, with all conditions being equal, the crystal field splitting for a tetrahedral complex is about half of that for an octahedral complex.

Finally, it is of interest to note that, in the purturbative treatment of CFT, Δ_o has a quantum mechanical origin:

$$\Delta_o = 10Dq, \qquad (8.1.2)$$

where

$$D = \frac{35Ze^2}{4a^5}, \qquad (8.1.3)$$

$$q = \frac{2}{105}\int_0^\infty R_{nd}r^4 R_{nd}r^2 dr. \qquad (8.1.4)$$

In the above expressions for D and q, Z is the charge on each ligand, a is the metal–ligand distance, R_{nd} is the radial function for the nd orbital on the metal. The crystal field splittings, Δ_o and Δ_t (or in terms of Dq), are seldom determined by direct calculations. Rather, they are usually deduced from spectroscopic measurements.

8.2 Spectrochemical series, high spin and low spin complexes

We now briefly examine the factors that influence the magnitude of the crystal field splitting Δ_o (or $10Dq$). In general, Δ_o increases for similar transition metals as we go down the Periodic Table, i.e., first row < second row < third row.

Also, Δ_o increases with the charge on the metal ion, i.e., $M^{2+} < M^{3+}$, etc. As far as the ligands are concerned, Δ_o increases according to the spectrochemical series

$$I^- < Br^- < S^{2-} < SCN^- < Cl^- < NO_3^- < F^- < OH^- < C_2O_4^{2-} < H_2O <$$
$$NCS^- < CH_3CN < NH_3 < en(H_2NCH_2CH_2NH_2) < bipy(2,2'\text{-bipyridine}$$
$$\text{or } 2,2'\text{dipyridine}) < phen(o\text{-phenanthroline}) < NO_2^- < CN^- < CO.$$

It is not easy to rationalize this series with the point charge model. For instance, on the basis of ionic bonding, it is difficult to see that a neutral ligand, CO, yields the largest (or at least one of the largest) Δ_o. However, this point may be readily understood once we consider covalent bonding between the metal ion and the ligands.

With the aid of the aufbau principle and Hund's rule, we are now in position to determine the ground electronic configuration of a given octahedral complex. As shown in Fig. 8.2.1 there is only one electronic assignment for d^1–d^3 and d^8–d^{10} complexes, which have the ground configurations

$$d^1 : t_{2g}^1; \ d^2 : t_{2g}^2; \ d^3 : t_{2g}^3; \ d^8 : t_{2g}^6 e_g^2; \ d^9 : t_{2g}^6 e_g^3; \ d^{10} : t_{2g}^6 e_g^4.$$

However, for d^4 to d^7 complexes, there are high-spin and low-spin configurations. In the high-spin complexes, we have Δ_o smaller than the pairing energy; i.e., electrons avoid pairing by occupying both the t_{2g} and e_g orbitals. For low-spin complexes, Δ_o is greater than the pairing energy; i.e., electrons prefer pairing in the t_{2g} orbitals to occupying the less stable e_g orbitals.

For a first transition series octahedral complex, whether it has a high-spin or low-spin configuration depends largely on the nature of the ligand or the charge on the metal ion. For second- or third-row complexes, low-spin configuration is favored, since now Δ_o is larger and pairing energy becomes smaller. In the larger 4d and 5d orbitals, the electronic repulsion, which contributes to pairing energy, is less than that for the smaller 3d orbital.

For tetrahedral complexes, with their crystal field splitting Δ_t being only about half of Δ_o, the high-spin configuration is heavily favored. Indeed, low-spin tetrahedral complexes are rarely observed.

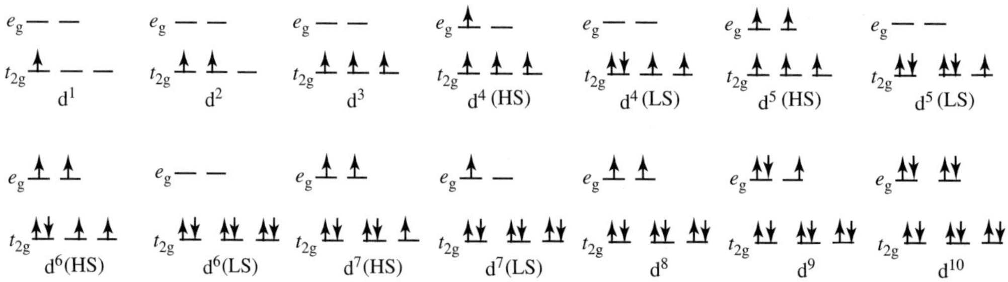

Fig. 8.2.1.
Electronic configurations of octahedral complexes for d^1 – d^{10} metals ions. The high-spin (HS) complexes have Δ_o smaller than pairing energy, while the low-spin (LS) complexes have Δ_o larger than pairing energy.

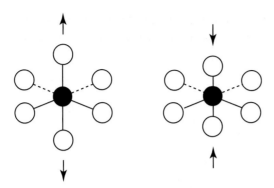

Fig. 8.3.1.
Two examples of tetragonal distortion for an octahedral complex: (a) axial elongation and (b) axial compression. Note that the center of symmetry of the octahedral system is preserved after the distortion.

8.3 Jahn–Teller distortion and other crystal fields

Among the 14 configurations for octahedral complexes shown in Fig. 8.3.1, only five actually exist in the strictest sense. The non-existence of the other nine configurations is due to Jahn–Teller distortion. According to the theory put forth by physicists H. A. Jahn and E. Teller in 1937, any non-linear molecule with a degenerate electronic state, i.e., with orbital degeneracy, will undergo distortion to remove the degeneracy. Furthermore, if the "original" system is centrosymmetric, the center of symmetry will be retained after the distortion. When the 14 configurations shown in Fig. 8.2.1 are examined, it can be readily seen that only those of d^3, d^5 high spin, d^6 low spin, d^8, and d^{10} have non-degenerate ground state and hence are not subjected to Jahn–Teller distortion. On the other hand, the remaining nine systems will distort from its octahedral geometry in order to remove the orbital degeneracy.

In octahedral complexes, the Jahn–Teller distortion is small when t_{2g} orbitals are involved; this distortion becomes larger when there is uneven electronic occupancy in the e_g orbitals. Let us take a d^9 octahedral complex as an example: for configuration $t_{2g}^6 e_g^3$, the orbital degeneracy occurs in the e_g orbitals. In order to preserve the center of symmetry, the complex will undergo a tetragonal distortion: the two ligands lying along the z axis will be either compressed or pulled away from the metal, while the four ligands on the xy plane retain their original positions, as shown in Fig. 8.3.1. The crystal field splitting pattern of a tetragonal distortion on an octahedral complex, deformed by either stretching or compressing two ligands along the z axis, is summarized in Fig. 8.3.2. From this figure, it is clear that either configuration $e_g^4 b_{2g}^2 a_{1g}^2 b_{1g}^1$ (for an axially elongated complex) or $b_{2g}^2 e_g^4 b_{1g}^2 a_{1g}^1$ (for an axially compressed complex) will be energetically more favorable than the $t_{2g}^6 e_g^3$ configuration of an octahedral complex. It is this energy stabilization that leads to the Jahn–Teller distortion. Well-known examples of axial elongation are found in Cu(II) halides. In these systems, each Cu^{2+} ion is surrounded by six halide ions, four of them lying closer to the cation and the remaining two farther away. The structural data for these halides are summarized in Fig. 8.3.3.

Referring to the orbital splitting pattern for an axially elongated tetragonal complex shown in Fig. 8.3.2, if we continue the elongation, the splittings

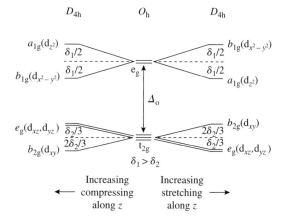

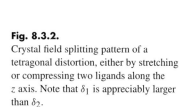

Fig. 8.3.2.
Crystal field splitting pattern of a tetragonal distortion, either by stretching or compressing two ligands along the z axis. Note that δ_1 is appreciably larger than δ_2.

denoted by δ_1 and δ_2 will become larger and larger. Eventually this will lead to the removal of two ligands, forming a square-planar complex in the process. The orbital splitting pattern of such a complex is shown in Fig. 8.3.4.

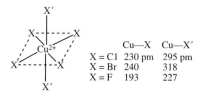

Fig. 8.3.3.
Examples of Jahn–Teller distortion: Cu(II) halides with two long bonds and four short ones.

Most square-planar complexes have the d^8 configuration, and there are also some with the d^9 configuration. Practically all d^8 square-planar complexes are diamagnetic, with a vacant b_{1g} level (cf. Fig. 8.3.4). Also, most d^8 square-planar complexes with first-row transition metals have strong field ligands. Since the energy gap between the b_{1g} and b_{2g} levels is (exactly) Δ_o, strong field ligands would produce a large Δ_o, favoring a low-spin, i.e., diamagnetic, configuration. For example, Ni^{2+} ion forms square-planar complexes with strong field ligands (e.g., $[Ni(CN)_4]^{2-}$), and it forms tetrahedral, paramagnetic complexes with weak field ligands (e.g., $[NiCl_4]^{2-}$). For second- and third-row transition metals, the Δ_o splittings are much larger and square-planar complexes are formed even with relatively weak field ligands (e.g., $[PtCl_4]^{2-}$).

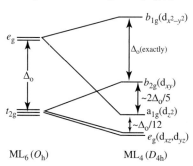

Fig. 8.3.4.
Correlation between the orbital splitting patterns of octahedral ML_6 and square-planar ML_4 complexes.

Table 8.4.1. The character table of the group O

O	E	$6C_4$	$3C_2(=C_4^2)$	$8C_3$	$6C_2$		
A_1	1	1	1	1	1		$x^2+y^2+z^2$
A_2	1	-1	1	1	-1		
E	2	0	2	-1	0		$(2z^2-x^2-y^2, x^2-y^2)$
T_1	3	1	-1	0	-1	$(R_x, R_y, R_z)\,(x,y,z)$	
T_2	3	-1	-1	0	1		(xy, xz, yz)

8.4 Octahedral crystal field splitting of spectroscopic terms

As shown in previous sections, under an octahedral crystal field, the five d orbitals are split into two sets of orbitals, t_{2g} and e_g. In terms of spectroscopic states, we can readily see that the ^2D term arising from configuration d^1 will split into $^2T_{2g}$ and ^2E_g states, arising from configurations t_{2g}^1 and e_g^1, respectively. But what about the spectroscopic terms such as F and G arising from other d^n configurations? In this section, we will see how these terms split in an octahedral crystal field.

Octahedral complex ML$_6$ has O_h symmetry. However, for simplicity, we may work with the O point group, which has only rotations as its symmetry operations and five irreducible representations, $A_1, A_2, \ldots, T_2$. The character table for this group is shown in Table 8.4.1. It is seen that the main difference between the O_h and O groups is that the former has inversion center i, while the latter does not. As a result, the O_h group has ten symmetry species: A_{1g}, $A_{1u}, A_{2g}, A_{2u}, \ldots, T_{2g}, T_{2u}$.

To see how an octahedral crystal field reduces the (2L+1)-fold degeneracy of a spectroscopic term, we first need to determine the character of a rotation operating on an atomic state defined by the orbital angular momentum quantum number L. It can be shown that the character χ for a rotation of angle α about the z axis is simply

$$\chi(\alpha) = \left[\sin\left(L+\frac{1}{2}\right)\alpha\right] / \sin\frac{\alpha}{2}. \qquad (8.4.1)$$

In particular, for the identity operation, i.e., when $\alpha = 0$, we have

$$\chi(\alpha = 0) = 2L+1. \qquad (8.4.2)$$

With these formulas, it is now straightforward to obtain the χ values for all the rotations in the O group and for any L value. Furthermore, we can then reduce the representations with these χ values as characters into the irreducible representations of the O group. All these results are summarized in Table 8.4.2.

To illustrate how the results are obtained, let us take the F term, with $L=3$, as an example. With eq. (8.4.2), we get

$$\chi(E) = 7. \qquad (8.4.3)$$

Table 8.4.2. Representations of atomic terms in the O point group and the states these terms generate

Term	L	E	$6C_4$	$3C_2(=C_4^2)$	$8C_3$	$6C_2$	States generated
S	0	1	1	1	1	1	A_1
P	1	3	1	−1	0	−1	T_1
D	2	5	−1	1	−1	1	$E + T_2$
F	3	7	−1	−1	1	−1	$A_2 + T_1 + T_2$
G	4	9	1	1	0	1	$A_1 + E + T_1 + T_2$
H	5	11	1	−1	−1	−1	$E + 2T_1 + T_2$
I	6	13	−1	1	1	1	$A_1 + A_2 + E + T_1 + 2T_2$

With eq. (8.4.1), we have

$$\chi(C_4) = \left[\sin\left(3 + \frac{1}{2}\right)(90°)\right] \bigg/ \sin 45° = -1. \quad (8.4.4)$$

$$\chi(C_2) = \left[\sin\left(3 + \frac{1}{2}\right)(180°)\right] \bigg/ \sin 90° = -1. \quad (8.4.5)$$

$$\chi(C_3) = \left[\sin\left(3 + \frac{1}{2}\right)(120°)\right] \bigg/ \sin 60° = 1. \quad (8.4.6)$$

Combining the results of eqs. (8.4.3) to (8.4.6), we obtain the following representation for the F term:

O	E	$6C_4$	$3C_2(=C_4^2)$	$8C_3$	$6C_2$
$F, L=3$	7	−1	−1	1	−1

This representation can be easily reduced to the irreducible representations $A_2 + T_1 + T_2$. Note that if this F term arises from a configuration with electrons occupying d orbitals, which are even functions, the electronic states split from this term in an octahedral crystal field are then A_{2g}, T_{1g}, and T_{2g}. (On the other hand, if the F term comes from a configuration with f electron(s) such as f^1, the electronic states in an octahedral crystal field will then be A_{2u}, T_{1u}, and T_{2u}. This is because f orbitals are odd functions.) All the results given in Table 8.4.2 can be obtained in a similar fashion.

8.5 Energy level diagrams for octahedral complexes

After determining what levels are present for an octahedral complex with a given electronic configuration, we are now ready to discuss the energy level diagrams for these spectroscopic terms.

8.5.1 Orgel diagrams

Let us first consider an octahedral complex ML$_6$ with one single d electron. Configuration d^1 has only one term, ^2D, which splits into ^2E_g and $^2T_{2g}$ levels in a crystal field with O_h symmetry. Levels ^2E_g and $^2T_{2g}$ arise from configurations

e_g^1 and t_{2g}^1, respectively. Also, from Fig. 8.1.5(a), we see that the energy of the e_g orbitals are higher than that of the d orbitals by $(3/5)\Delta_o$ (or $6Dq$), and the energy of the t_{2g} orbitals is lower than that of the d orbitals by $(2/5)\Delta_o$ (or $4Dq$). Combining these results we can readily arrive at the energy level diagram shown in Fig. 8.5.1. In this diagram, we plot the energies of the spectroscopic levels against Dq, one tenth of the octahedral crystal field splitting Δ_o. Furthermore, the energy difference between the ^2E_g and $^2T_{2g}$ level is always $10Dq$. Such a diagram is called an Orgel diagram, named after chemist L. E. Orgel, who popularized these diagrams in the 1950's.

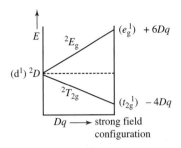

Fig. 8.5.1.
Orgel diagram for a d^1 octahedral complex.

For the d^6 configuration, the ground term is ^5D, which splits into ^5E_g and $^5T_{2g}$ in an O_h field. For a high-spin d^6 complex, the only quintet term arising from configuration $t_{2g}^4 e_g^2$ is $^5T_{2g}$ and that arising from $t_{2g}^3 e_g^3$ is ^5E_g. These states correspond exactly to the d^1 case, except now the states are spin quintets. In other words, for such a complex there is only one spin-allowed transition, $^5T_{2g} \rightarrow {}^5E_g$. Hence the d^1 and high spin d^6 octahedral complexes have the same Orgel diagram. (Note that the d^6 configuration has a total of 16 terms, and ^5D is the only quintet term. The remaining 15 terms are triplets and singlets and they are ignored in the current discussion.)

As mentioned in Chapter 2, both d^9 and d^1 configurations have only one term, ^2D. For a d^9 octahedral complex, the ^2D term once again splits into $^2T_{2g}$ and ^2E_g states, with the energy difference between them being $10Dq$. However, for the d^9 configuration, which is equivalent to the one-hole, or one-positron, case, the energy ordering is just the opposite of that for the d^1 configuration. In other words, for the d^9 octahedral complex, we have the ^2E_g state lower than the $^2T_{2g}$ state, and the energy difference between the states is again $10Dq$.

Similarly, d^4 is the hole-counterpart of d^6, both of which have the ground term ^5D. As have been mentioned many times already, this term splits into $^5T_{2g}$ and ^5E_g states in an O_h crystal field. For the d^4 case, these quintet states arise from the high-spin configurations $t_{2g}^2 e_g^2$ and $t_{2g}^3 e_g^1$, respectively. As may be surmised by the configurations from which they arise, the ^5E_g state is now lower than the $^5T_{2g}$ state for a d^4 complex, which is just the opposite for a high-spin d^6 octahedral complex. The energy difference between the states is once again $10Dq$.

The foregoing results for d^1, high-spin d^4 and d^6, as well as d^9 octahedral complexes can be summarized by the fairly simple Orgel diagram shown in Fig. 8.5.2.

Additionally, we note that octahedral and tetrahedral complexes have opposite energy orderings (Fig. 8.1.5). Hence d^1 octahedral and d^9 tetrahedral have the same Orgel diagram; the same is true for d^6 octahedral and d^4 tetrahedral complexes. In other words, the Orgel diagram in Fig. 8.5.2 is also applicable to d^1, d^4, d^6, and d^9 tetrahedral complexes. Since tetrahedral complexes are not centrosymmetric, their states do not have g or u designations. States in Fig. 8.5.2 do not carry subscripts; they should be put back on for octahedral complexes.

Now we turn our attention to the d^2 octahedral complexes. For this configuration, there are two triplet terms, ^3F (ground term) and ^3P. In an octahedral crystal field, there are three configurations: $t_{2g}^2 < t_{2g}^1 e_g^1 < e_g^2$, in increasing

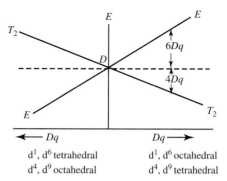

Fig. 8.5.2.
Orgel diagrams for d^1, high-spin d^4 and d^6, and d^9 complex.

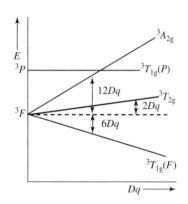

Fig. 8.5.3.
Energy level diagram for a d^2 octahedral complex, ignoring second-order crystal field interaction.

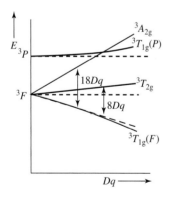

Fig. 8.5.4.
Energy level diagram for a d^2 octahedral complex, including the second-order crystal field interaction.

energy order. From Table 8.4.2, it is clear that ^3P becomes $^3T_{1g}(P)$ in an octahedral field, with no change in energy, while the ^3F term splits into $^3T_{1g}(F)$, $^3T_{2g}$, and $^3A_{2g}$ states. (Since there are two $^3T_{1g}$ states in this system, we call one $^3T_{1g}(P)$ and the other $^3T_{1g}(F)$.) Upon treating the crystal field splitting perturbationally, it is found that the $^3T_{1g}(F)$ state lies $6Dq$ below the energy of ^3F, while the $^3T_{2g}$ and $^3A_{2g}$ states are $2Dq$ and $12Dq$ above ^3F, respectively. (It is seen that the "center of gravity" is also preserved here.) These results are shown in Fig. 8.5.3. Note that the splittings shown in this figure are due to interactions among the components of the same term; these interactions are sometimes called first-order crystal field interactions. In addition to first-order interactions, there are also interactions among components from different terms with the same symmetry, the so-called second-order crystal field interactions. Take the energy levels in Fig. 8.5.3 as an example. Among the four states shown, only $^3T_{1g}(P)$ and $^3T_{1g}(F)$ have the same symmetry and hence they will interact with each other. In quantum mechanics, when two states of the same symmetry interact, the upper level will go still higher and the lower level further below. The results of such a second-order interaction are shown in Fig. 8.5.4. In this figure, it is seen that the (upper) $^3T_{1g}(P)$ level bends upward, while the (lower) $^3T_{1g}(F)$ bends downward.

As we have seen for the d^1 and d^6 systems (Fig. 8.5.2), the d^2 Orgel diagram in Fig. 8.5.4 is also applicable to d^7 octahedral, d^3 and d^8 tetrahedral complexes. Furthermore, the reverse splitting pattern applies to d^3 and d^8 octahedral, as well as to d^2 and d^7 tetrahedral complexes. These results are summarized in Fig. 8.5.5.

The two Orgel diagrams shown in Figs. 8.5.2 and 8.5.5 deal with the spin-allowed transitions for all d^n systems except d^0, d^5, and d^{10}. Let us confine our attention to octahedral complexes. Only one transition is expected for d^1 ($^2T_{2g} \to {}^2E_g$) and d^9 ($^2E_g \to {}^2T_{2g}$) complexes. For the d^1 complex of $[Ti(H_2O)_6]^{3+}$, this band is observed at 20,300 cm^{-1}. This absorption shows a shoulder at ~17,500 cm^{-1}, due to the Jahn-Teller distortion of the excited state ^2E_g (arised from configuration e_g^1). For a d^2 complex, the ground state is $^3T_{1g}(F)$ and three transitions are expected: $^3T_{1g}(F) \to {}^3T_{2g}$, $^3T_{1g}(F) \to {}^3T_{1g}(P)$, and $^3T_{1g}(F) \to {}^3A_{2g}$. For $[V(H_2O)_6]^{3+}$, these Transitions are observed at 17,000, 25,000, and 37,000 cm^{-1} (obscured somewhat by

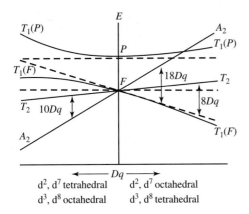

Fig. 8.5.5.
Orgel diagram for d^2, d^3, as well as high-spin d^7 and d^8 complexes. Note that, in the figure on the left, the T_1 states are more bent (than those on the right) because they are closer in energy.

charge transfer transitions), respectively. For a d^3 system, the ground state is $^4A_{2g}$ and again three transitions are expected: $^4A_{2g} \rightarrow {}^4T_{2g}$, $^4A_{2g} \rightarrow {}^4T_{1g}(F)$, $^4A_{2g} \rightarrow {}^4T_{2g}(P)$. For $[Cr(H_2O)_6]^{3+}$, they are observed at $\sim$17,500, 24,700, and 37,000 cm^{-1}, respectively.

Finally, regarding the aforementioned omissions of d^0, d^5, and d^{10} systems, it is noted that d^0 and d^{10} complexes exhibit no d–d band, and there are no spin-allowd transitions for high-spin d^5 complexes. But we shall discuss the electronic spectrum of a d^5 complex, $[Mn(H_2O)_6]^{2+}$, in the next section.

8.5.2 Intensities and band widths of d–d spectral lines

Crystal field, or d–d, transitions are defined as transitions from levels that are exclusively perturbed d orbitals to levels of the same type. In other words, the electron is originally localized at the central metal ion and remains so in the excited state. When the system has O_h symmetry, Laporte's rule says that an electric-dipole allowed transition must be between a "g" state and an "u" state, i.e., u ↔ g. Since all the crystal field electronic states are gerade ("g"), no electric-dipole allowed transitions are possible. In short, all d–d transitions are symmetry forbidden and hence have low intensities. The fact that the d–d transitions are observed at all is due to the interaction between the electronic motion and the molecular vibration. We will discuss this (vibronic) interaction later (Section 8.10).

In addition to the symmetry selection rule, there is also another selection rule on spin: $\Delta S = 0$, i.e., only the transitions between states with the same spin are allowed. The fact that we sometimes do observe spin-forbidden d–d transitions is mainly due to spin–orbit coupling, which will be discussed later in this section.

Spin-forbidden (say, $\Delta S = \pm 1$) transitions, when observed, are usually about 1% as strong as the spin-allowed ones. For the octahedral complexes of the first transition series, the molar extinction coefficients ε (in L mol^{-1}cm^{-1}) of the transitions range from $\sim$0.01 in Mn(II) complexes (see below) to as high as $\sim$25–30 of some non-chelate Co(III) and Ni(II) complexes. The tetrahedral

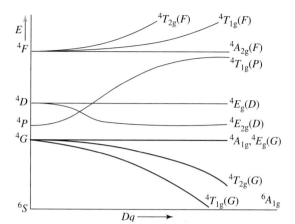

Fig. 8.5.6.
The Orgel diagram for a high-spin d^5 octahedral complex.

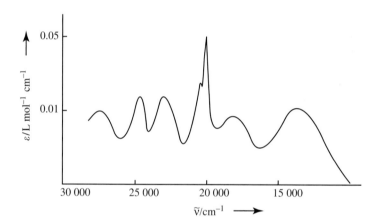

Fig. 8.5.7.
Electronic spectrum of $[Mn(H_2O)_6]^{2+}$. Note that the intensities of all the transitions are very low, since they are both symmetry- and spin-forbidden.

complexes have more intense transitions (by a factor of about 50 to 200) due to their lack of center of symmetry. For examples, the ε value for the $^2T_{2g} \rightarrow {}^2E_g$ transition of $[Ti(H_2O)_6]^{3+}$ is ~5, while those for the d–d transitions of $[V(H_2O)_6]^{2+}$ also have similar intensity.

We now turn to the d–d transitions of the complex $[Mn(H_2O)_6]^{2+}$. For this d^5 configuration, the ground state is ^6S, which becomes $^6A_{1g}$ in an octahedral crystal field. This is the only sextet state for this configuration; the remaining states are either quartets or doublets. An Orgel diagram covering the sextet ground state and the four quartet excited states are shown in Fig 8.5.6. The electronic spectrum of $[Mn(H_2O)_6]^{2+}$ is shown in Fig. 8.5.7.

Note that all transitions for this system are both symmetry- and spin-forbidden and hence have very low ε values, which are responsible for the very pale pink color of this cation. (As a comparison, tetrahedral complexes of Mn(II) are yellow-green, and the color is more intense. The molar absorbance values for these complexes are in the range of 1.0–4.0). The assignment for the spectrum of $[Mn(H_2O)_6]^{2+}$ shown in Fig. 8.5.7 is as follows:

Transition		cm^{-1}
$^6A_{1g}$	$\rightarrow\ ^4T_{1g}(G)$	18,800
	$\rightarrow\ ^4T_{2g}(G)$	23,000
	$\rightarrow\ ^4E_g(G)$	24,900
	$\rightarrow\ ^4A_{1g}(G)$	25,100
	$\rightarrow\ ^4T_{2g}(D)$	28,000
	$\rightarrow\ ^4E_g(D)$	29,700
	$\rightarrow\ ^4T_{1g}(P)$	32,400
	$\rightarrow\ ^4A_{2g}(F)$	35,400
	$\rightarrow\ ^4T_{1g}(F)$	36,900
	$\rightarrow\ ^4T_{2g}(F)$	40,600

Another feature of the $[Mn(H_2O)_6]^{2+}$ spectrum shown in Fig. 8.5.7 is its variation of bandwidths of the spectral lines. Theoretical considerations indicate that the widths of spectral lines is dependent on the relative slopes of the two states involved in the transition. As the ligands vibrate, the value of Dq changes as a result [cf. eqs. (8.1.3) and (8.1.4)]. If the energy of the excited state is a sensitive function of Dq, while that of the ground state is not, the spectral band will be broad. This argument is pictorially illustrated in Fig. 8.5.8. Referring to the Orgel diagram for d^5 complexes shown in Fig. 8.5.6, it is seen that the energies of the ground state $^6A_{1g}$ and the excited states $^4A_{1g}(G)$, $^4E_{1g}(G)$, and $^4E_g(D)$ all happen to be independent of Dq. Hence the spectral bands for the transitions from the ground state to these three excited states should be narrow, as are indeed found to be the case.

Before concluding this section, we briefly discuss how formally spin-forbidden transitions gain their intensity. The mechanism through which this is achieved is the spin–orbit interaction. Here we use the two most stable terms, ^6S and ^4G, of the d^5 configuration for illustration. After spin–orbit (L–S) coupling among the components for each term, ^6S becomes $^6S_{2^1/_2}$, while ^4G splits into $^4G_{5^1/_2}$, $^4G_{4^1/_2}$, $^4G_{3^1/_2}$, and $^4G_{2^1/_2}$. Such "intra-term" coupling may be considered as first-order spin–orbit interaction. Additionally, states with the same J value, namely, $^6S_{2^1/_2}$ and $^4G_{2^1/_2}$ in the present example, will further interact. Such inter-term coupling is called second-order spin–orbit interaction. As a result of the interaction, the ^6S term gains some quartet character; also the ^4G term gains some sextet character. Mathematically, if we call the wavefunctions before and after spin–orbit interaction ψ and ψ', respectively, then we have

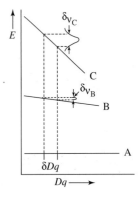

Fig. 8.5.8.
Diagram showing how bandwidth varies with the ratio of the slopes of the ground and excited states. In the present example, the ground state (A) energy is independent of Dq, while excited states B and C have energies which change as Dq changes. Here δDq is the range of variation of Dq due to ligand vibrations, and $\delta\nu_B$ and $\delta\nu_C$ are the bandwidths for transitions A→B and A→C, respectively.

$$\psi'(^6S) = a\psi(^6S) + b\psi(^4G), \quad a \gg b; \quad (8.5.1)$$

$$\psi'(^4G) = c\psi(^4G) + d\psi(^6S), \quad c \gg d. \quad (8.5.2)$$

This minute amount of mixing of the quartet and sextet states leads to a very small intensity for the sextet–quartet transition, as, strictly speaking, the $\Delta S = 0$ selection rule is no longer violated. A more detailed discussion on the spin–orbit interaction in complexes is given in Section 8.7.

8.5.3 Tanabe–Sugano diagrams

The Orgel diagrams shown in Figs. 8.5.2 and 8.5.5 have the distracting feature of decreasing ground state energy with increasing Δ_o or Dq. To remedy this "deficiency," Japanese scientists Y. Tanabe and S. Sugano represented the energy of the ground state of a complex as a horizontal line. The Tanabe–Sugano diagrams for d^2, d^3, and d^8 octahedral complexes are shown in Fig. 8.5.9.

In these figures, states with spin multiplicities different from that of the ground state are also included. Thus, using these diagrams, spin-forbidden transitions may also be considered. The coordinates of the diagrams are E/B and Δ_o/B, both of which are dimensionless (i.e., scalar quantities). Here B is one of the Racah parameters in terms of which the energy of each (free) atomic term is expressed. Hence the diagrams shown can be used for various metal ions with the same d^n configuration and with different ligands.

The Tanabe–Sugano diagrams for d^4–d^7 octahedral complexes are shown in Fig. 8.5.10. In this figure, each diagram has two parts separated by a vertical line. The left part applies to high-spin (or weak field) complexes and the right side to low spin (or strong field) cases. Take the d^6 diagram as an example. For high-spin d^6 complexes, the ground state is $^5T_{2g}$, while that for a low-spin complex is $^1A_{1g}$. This is in agreement with the result shown in Fig. 8.2.1, which indicates that a high-spin d^6 complex has four unpaired electrons (or a quintet spin state) and a low-spin d^6 complex is diamagnetic (or a singlet state).

8.5.4 Electronic spectra of selected metal complexes

In this section, we present some additional spectra in order to illustrate the principles discussed in the previous sections.

(1) $[Co(H_2O)_6]^{2+}$ and $[CoCl_4]^{2-}$

These are d^7 complexes and their spectra are shown in Fig. 8.5.11. From the positions and the intensities of the spectral band, we can readily deduce that $[Co(H_2O)_6]^{2+}$ is pale purple, while $[CoCl_4]^{2-}$ is deep blue. With the help of the Orgel diagram shown in Fig. 8.5.5, and some quantitative results, the only band in each spectrum may be assigned to the following transition: $^4A_2 \rightarrow {}^4T_1(P)$ in $[CoCl_4]^{2-}$ and $^4T_{1g}(F) \rightarrow {}^4T_{1g}(P)$ in $[Co(H_2O)_6]^{2+}$, while the other spin-allowed transitions are not clearly seen.

For the octahedral complex, the $^4T_{1g}(F) \rightarrow {}^4A_{2g}$ should have comparable energy with the observed band. But $^4T_{1g}(F)$ arises from configuration $t_{2g}^5 e_g^2$, while $^4A_{2g}$ is from $t_{2g}^3 e_g^4$. So this is a two-electron process and hence the transition is weak. Meanwhile, the $^4T_{1g}(F) \rightarrow {}^4T_{2g}$ transition occurs in the near-infrared region. For the tetrahedral complex, the $^4A_2 \rightarrow {}^4T_1(F)$ transition also takes place in the near infrared region, while $^4A_2 \rightarrow {}^4T_2$ occurs in a region of even lower energy.

(2) *trans*-$[Cr(en)_2F_2]^+$

This is a d^3 complex with a tetragonal structure with D_{4h} symmetry. (Strictly speaking, this system has D_{2h} symmetry.) Hence the Orgel diagram shown in Fig. 8.5.5 no longer applies. If we assume this cation has an elongated octahedral

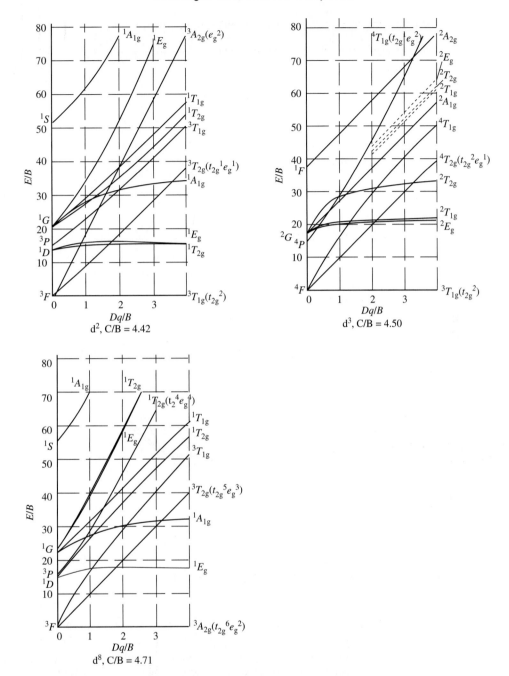

Fig. 8.5.9.
Tanabe–Sugano diagrams for d^2, d^3, and d^8 octahedral complexes. Quantities B and C are Racah parameters, in terms of which the energy of a spectroscopic term is expressed. Hence both E/B and Dq/B are dimensionless.

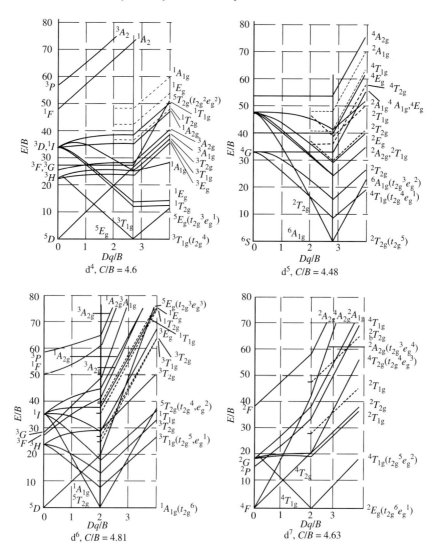

Fig. 8.5.10.
Tanabe–Sugano diagrams for d^4–d^7 octahedral complexes. Both E/B and Dq/B are dimensionless.

geometry, the "original" triply degenerate T states will split into two terms (E with A or B). This splitting pattern is shown in Fig. 8.5.12. With this energy level diagram, the spectral bands of this complex, shown in Fig. 8.5.13, can be assigned in a fairly straightforward manner:

Transition		cm^{-1}
$^4B_{1g}$	$\rightarrow\ ^4E_g$	18,500
	$\rightarrow\ ^4B_{2g}$	21,700
	$\rightarrow\ ^4E_g$	25,300
	$\rightarrow\ ^4A_{2g}$	29,300
	$\rightarrow\ ^4A_{2g}$	41,000 (shoulder)
	$\rightarrow\ ^4E_g$	43,700 (calculated)

Bonding in Coordination Compounds

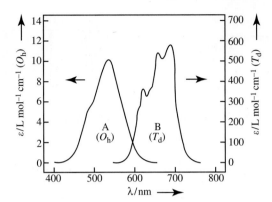

Fig. 8.5.11.
Visible spectra of $[Co(H_2O)_6]^{2+}$ (curve A) and $[CoCl_4]^{2-}$ (curve B). The ε scale on the left side applies to $[Co(H_2O)_6]^{2+}$; the one on the right applies to $[CoCl_4]^{2-}$. As expected, the bands of the tetrahedral complex are much more intense than those of the octahedral complex.

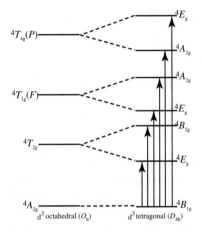

Fig. 8.5.12.
Energy level splitting for a d^3 ion as its crystal field environment changes form octahedral (O_h symmetry) to tetragonal (D_{4h}).

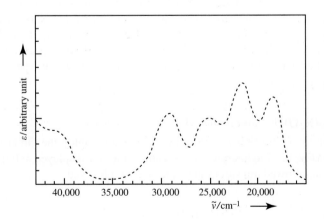

Fig. 8.5.13.
Electronic spectrum of trans-$[Cr(en)_2F_2]^+$.

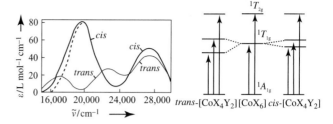

Fig. 8.5.14.
The visible spectrum of *cis*- and *trans*-[Co(en)$_2$F$_2$]$^+$. The broken line signifies that the most intense band actually consists of two overlapping transitions. The asymmetry of this band is caused by the slight splitting of $^1T_{1g}$ state in the *cis* isomer. For the *trans* isomer, the splitting is much more pronounced, and two separate bands are observed. The schematic diagram on the right shows the energy levels involved in these two electronic spectra.

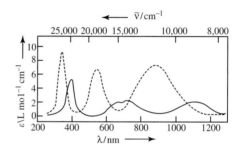

Fig. 8.5.15.
Electronic spectra of [Ni(H$_2$O)$_6$]$^{2+}$ (—) and [Ni(en)$_3$]$^{2+}$ (– – –).

(3) *cis*-[Co(en)$_2$F$_2$]$^+$ and *trans*-[Co(en)$_2$F$_2$]$^+$

These are low-spin d^6 complexes with a $^1A_{1g}$ ground state. According to the Tanabe-Sugano diagram shown in Fig. 8.5.10, the two lowest energy singlet excited states for a d^6 octahedral complex are $^1T_{1g}$ and $^1T_{2g}$. When the symmetry of the complex is lowered by substitution to form *cis*- or *trans*-[CoX$_4$Y$_2$]$^+$, both of the triply degenerate excited states will split. Also, the splitting of the $^1T_{1g}$ state will be more prominent if the ligands X and Y are far apart in the spectrochemical series; understandably this is especially so for the *trans* isomer. These results are summarized schematically in Fig. 8.5.14. Note that, in this figure, the splitting of the $^1T_{2g}$ is so small that it is completely ignored. These expected electronic transitions are observed in the respective spectra. Finally, it is noted that, since the *cis* isomer has no center of symmetry, its spectral bands have higher intensities.

(4) [Ni(H$_2$O)$_6$]$^{2+}$ and [Ni(en)$_3$]$^{2+}$

These are d^8 complexes and we may once again employ the Orgel diagram shown in Fig. 8.5.5 to interpret their spectra, which are displayed in Fig. 8.5.15. Three spin-allowed transitions are expected and observed:

Transition		[Ni(H$_2$O)$_6$]$^{2+}$	[Ni(en)$_3$]$^{2+}$
$^3A_{2g}$	$\to\ ^3T_{2g}$	9,000 cm^{-1}	11,000 cm^{-1}
	$\to\ ^3T_{1g}(F)$	14,000	18,500
	$\to\ ^3T_{1g}(P)$	25,000	30,000

Since ethylenediamine (en) is a stronger ligand than water, all the transitions for $[Ni(en)_3]^{2+}$ have higher energies than the corresponding ones for $[Ni(H_2O)_6]^{2+}$. Also, since $[Ni(H_2O)_6]^{2+}$ is centrosymmetric, its spectral bands have lower intensities. From these spectra, it may be deduced that $[Ni(en)_3]^{2+}$ is purple, while $[Ni(H_2O)_6]^{2+}$ is green.

Finally, it is noted that the splitting of the middle band in the $[Ni(H_2O)_6]^{2+}$ spectrum arises from spin–orbit interaction between the $^3T_{1g}(F)$ and ^1E_g states (cf. Fig. 8.5.9). These two states are close in energy at the Δ_o value generated by six H_2O molecules. But they are far apart at the stronger field of three en molecules. As a result, no splitting is observed in the spectrum of $[Ni(en)_3]^{2+}$.

With the sample spectra discussed in this section, it is clear that the Orgel diagrams and/or Tanabe–Sugano diagrams are useful in making qualitative analysis for the spectra. However, there are also cases where quantitative treatments are indispensable in order to make definitive assignments and interpretations.

8.6 Correlation of weak and strong field approximations

In the Tanabe–Sugano diagrams for d^2 complexes shown in Fig. 8.5.9, it is seen that the $^3T_{1g}$ ground state is identified as derived from the t_{2g}^2 strong field configuration. Actually, such an identification can be made for all states shown. To do that, let us first give a more formal definition for weak and strong field approximations.

For the weak field case, we have the situation where the crystal field interaction is much weaker than the electronic repulsion. In this approximation, the Russell-Saunders terms ^3F, ^3P, ^1G, ^1D, and ^1S for the d^2 configuration are good basis functions. When the crystal field is "turned on," these terms split according to the results given in Table 8.4.2:

$$\begin{aligned}
^3F &\rightarrow {}^3A_{2g} + {}^3T_{1g} + {}^3T_{2g}, \\
^3P &\rightarrow {}^3T_{1g}, \\
^1G &\rightarrow {}^1A_{1g} + {}^1E_g + {}^1T_{1g} + {}^1T_{2g}, \\
^1D &\rightarrow {}^1E_g + {}^1T_{2g}, \\
^1S &\rightarrow {}^1A_{1g}.
\end{aligned}$$

These are the states arising from the weak field approximation.

For the strong field case, we have the reverse situation: the electronic repulsion is much smaller than the crystal field interaction. Now, for a system with two d electrons, we need to consider three crystal field configurations: t_{2g}^2, $t_{2g}^1 e_g^1$, and e_g^2 (in the order of increasing energy). The states derived from these configurations can be obtained with the standard group theoretical methods described in Section 6.4.3. Now it is straightforward to arrive at the following results:

$$\begin{aligned}
t_{2g}^2 &\rightarrow {}^3T_{1g} + {}^1A_{1g} + {}^1E_g + {}^1T_{2g}, \\
t_{2g}^1 e_g^1 &\rightarrow {}^3T_{1g} + {}^3T_{2g} + {}^1T_{1g} + {}^1T_{2g}, \\
e_g^2 &\rightarrow {}^3A_{2g} + {}^1A_{1g} + {}^1E_g.
\end{aligned}$$

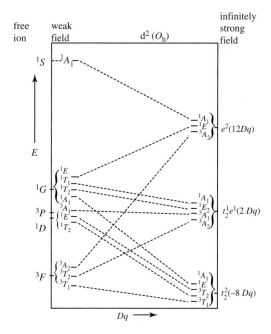

Fig. 8.6.1.
Correlation diagram for a d^2 octahedral complex. The g subscript is omitted in all the state or orbital designations. Note that the lines connecting the triplet states constitute the Orgel diagram shown in Fig. 8.5.4 or Fig. 8.5.5.

Of course, these are the same states obtained in the weak field approximation. The way these two sets of states is correlated is shown in Fig. 8.6.1. In drawing the connection lines, we minimize crossing, even for species with different symmetry. Note that the four lines connecting the triplet states constitute the Orgel diagram for d^2 complexes (Fig. 8.5.4 or Fig. 8.5.8).

8.7 Spin–orbit interaction in complexes: the double group

Previously we stated that, for a state with orbital angular momentum L, the character for a rotation of angle α is given by

$$\chi(\alpha) = [\sin(L + \tfrac{1}{2})\alpha]/\sin \alpha/2 \tag{8.4.1.}$$

When we use this formula to derive crystal field states of a complex, we have made the assumption that spin–orbit interaction is weak and hence is ignored.

When spin–orbit interaction is significant, the quantum number that defines a state is total angular momentum J, instead of L. Quantum number J can be an integer or a half-integer. When J is an integer, we can again make use of eq. (8.4.1) and replace L with J. However, when J is a half-integer, a complication arises: $\chi(\alpha + 2\pi)$ no longer equals to $\chi(\alpha)$, as it should. Mathematically

$$\begin{aligned}\chi(\alpha + 2\pi) &= \sin[(J + \tfrac{1}{2})(\alpha + 2\pi)]/\sin[\tfrac{1}{2}(\alpha + 2\pi)] \\ &= \sin[2\pi + (J + \tfrac{1}{2})\alpha]/\sin\left(\pi + \alpha/2\right) \\ &= \sin[(J + \tfrac{1}{2})\alpha]/\left(-\sin \alpha/2\right) = -\chi(\alpha). \end{aligned} \tag{8.7.1}$$

Bonding in Coordination Compounds

Table 8.7.1. The character table of double group O'

O' (h=48)		E ($\alpha = 4\pi$)	R ($\alpha = 2\pi$)	$4C_3$ $4C_3^2 R$	$4C_3^2$ $4C_3 R$	$3C_2$ $3C_2 R$	$3C_4$ $3C_4^3 R$	$3C_4^3$ $3C_4 R$	$6C_2'$ $6C_2' R$
Γ_1	A_1'	1	1	1	1	1	1	1	1
Γ_2	A_2'	1	1	1	1	1	-1	-1	-1
Γ_3	E'	2	2	-1	-1	2	0	0	0
Γ_4	T_1'	3	3	0	0	-1	1	1	-1
Γ_5	T_2'	3	3	0	0	-1	-1	-1	1
Γ_6	E_2'	2	-2	1	-1	0	$(2)^{1/2}$	$-(2)^{1/2}$	0
Γ_7	E_3'	2	-2	1	-1	0	$-(2)^{1/2}$	$(2)^{1/2}$	0
Γ_8	G'	4	-4	-1	1	0	0	0	0

To get around this difficulty, H. A. Bethe introduced a rotation by 2π, called R, which is different from the identity operation E. To add this operation to any rotation group such as O, we need to expand the group by taking the product of R with all existing rotations; these are now the operations of the new group called the double group. For example, the O group has rotations $E, 8C_3, 3C_2, 6C_4$ and $6C_2'$ (Table 8.4.1) and its corresponding double group O' has operations $E(=R^2), R, (4C_3, 4C_3^2 R), (4C_3^2, 4C_3 R), (3C_2, 3C_2 R), (3C_4, 3C_4^3 R), (3C_4^3, 3C_4 R)$, and $(6C_2', 6C_2' R)$, i.e., a total of 48 symmetry operations forming eight classes. The character table of the O' group is shown in Table 8.7.1. Upon examining this table, it is seen that:

(1) The O' group has eight irreducible representations, with the first five identical to those of the O group (Table 8.4.1). The last three representations of the O' group are new.
(2) There are two systems of notation for the representations. The first one, $\Gamma_1, \Gamma_2, \ldots, \Gamma_8$, is due to Bethe. The other one, $A_1', A_2', \ldots$, etc. is due to Mulliken, who put the primes in to signify that these are the representations of the double group.

We may now use Table 8.7.1 to determine the representations, under O symmetry, spanned by a state with half-integer quantum number J. These results are summarized in Table 8.7.2. Those for states with integral J are included for convenience and completeness. In other words, this table contains all the results listed in Table 8.4.2, in addition to those for half-integer J states.

To illustrate how to obtain the characters for rotation operations listed in Table 8.7.2, we take $\chi(\alpha = 2\pi)$ for $J = 1/2$ as an example:

$$\chi(\alpha = 2\pi) = \lim_{\alpha \to 2\pi} \left[\sin\left(\frac{1}{2} + \frac{1}{2}\right)\alpha \right] / \sin\frac{1}{2}\alpha$$

$$= \lim_{\alpha \to 2\pi} 2\left[\cos\alpha / \cos\frac{1}{2}\alpha \right] = -2. \tag{8.7.2}$$

Now we consider the ^4D term (an excited state) of a d^5 octahedral complex as an example. If spin–orbit interaction is ignored, this term splits into $^4\Gamma_3$ and $^4\Gamma_5$ states (Table 8.7.2) in a crystal field with O_h symmetry. When we "turn on"

Symmetry in Chemistry

Table 8.7.2. Representations, under O or O' symmetry, spanned by a state characterized by an integral or a half-integral J value

J	E $=R^2$	R	$4C_3$ $4C_3^2R$	$4C_3^2$ $4C_3R$	$3C_2$ $3C_2R$	$3C_4$ $3C_4^3R$	$3C_4^3$ $3C_4R$	$6C_2'$ $6C_2'R$	Representations in O or O'
0	1								Γ_1
1/2	2	−2	1	−1	0	$(2)^{1/2}$	$-(2)^{1/2}$	0	Γ_6
1	3								Γ_4
1 1/2	4	−4	−1	1	0	0	0	0	Γ_8
2	5								$\Gamma_3+\Gamma_5$
2 1/2	6	−6	0	0	0	$-(2)^{1/2}$	$(2)^{1/2}$	0	$\Gamma_7+\Gamma_8$
3	7								$\Gamma_2+\Gamma_4+\Gamma_5$
3 1/2	8	−8	1	−1	0	0	0	0	$\Gamma_6+\Gamma_7+\Gamma_8$
4	9								$\Gamma_1+\Gamma_3+\Gamma_4+\Gamma_5$
4 1/2	10	−10	−1	1	0	$(2)^{1/2}$	$-(2)^{1/2}$	0	$\Gamma_6+2\Gamma_8$
5	11								$\Gamma_3+2\Gamma_4+\Gamma_5$
5 1/2	12	−12	0	0	0	0	0	0	$\Gamma_6+\Gamma_7+2\Gamma_8$
6	13								$\Gamma_1+\Gamma_2+\Gamma_3+\Gamma_4+2\Gamma_5$
6 1/2	14	−14	1	−1	0	$-(2)^{1/2}$	$(2)^{1/2}$	0	$\Gamma_6+2\Gamma_7+2\Gamma_8$

the spin–orbit interaction, these states will further split. To do this, it is seen that spin quantum number S is $1\tfrac{1}{2}$ for the present case; this spin state has Γ_8 symmetry (Table 8.7.2). When this spin state interacts with the orbitals parts (Γ_3 and Γ_5), the resultant states are

$$\Gamma_3 \times \Gamma_8 = \Gamma_6 + \Gamma_7 + \Gamma_8;$$
$$\Gamma_5 \times \Gamma_8 = \Gamma_6 + \Gamma_7 + 2\Gamma_8.$$

This means that $^4\Gamma_3$ splits into three states and $^4\Gamma_5$ into four. Note that now spin quantum number S is no longer used to define the resultant states.

On the other hand, if spin–orbit coupling is larger than crystal field interaction, J is the quantum number that defines a state before the crystal field is "turned on." For ^4D, we have J values of $1/2$, $1\tfrac{1}{2}$, $2\tfrac{1}{2}$, and $3\tfrac{1}{2}$. In an octahedral crystal field, with the aid of Table 8.6.2, we can readily see that these states split into

$$J = 1/2 \quad \Gamma_6;$$
$$J = 1\tfrac{1}{2} \quad \Gamma_8;$$
$$J = 2\tfrac{1}{2} \quad \Gamma_7 + \Gamma_8;$$
$$J = 3\tfrac{1}{2} \quad \Gamma_6 + \Gamma_7 + \Gamma_8.$$

The way these two sets of states are correlated is shown in Fig. 8.7.1. It is important to note that, whenever we are dealing with spin–orbit interaction in a system with half-integral J values, we should employ the double group.

8.8 Molecular orbital theory for octahedral complexes

In the crystal field treatment of complexes, the interaction between the central metal ion and the ligands is taken to be purely electrostatic, i.e., ionic bonding. If

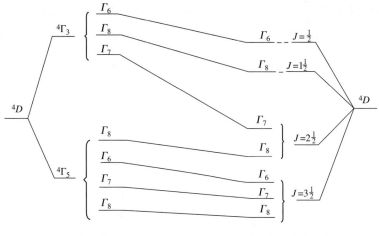

Fig. 8.7.1.
Relative effects of spin-orbit coupling and octahedral crystal field interaction on the electronic state ^4D.

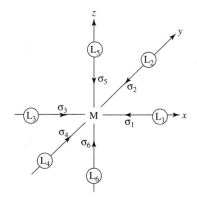

Fig. 8.8.1.
A coordinate system for ML_6, where the ligands have only σ to bond with the metal ion.

we wish to take the overlap between the metal and ligand orbitals, i.e., covalent bonding, into account, we need to turn to molecular orbital theory. The basic concepts of this theory have already been discussed in Chapters 3 and 7.

8.8.1 σ bonding in octahedral complexes

We first treat those complexes that have only σ metal–ligand bonding. Examples of this type of complexes include $[M(H_2O)_6]^{n+}$ and $[M(NH_3)_6]^{n+}$, where M^{n+} is a first transition series metal ion and a typical n value is either 2 or 3. A coordinate system for such a complex is shown in Fig. 8.8.1. The representation generated by the six ligand orbitals is

O_h	E	$8C_3$	$6C_2$	$6C_4$	$3C_2$	i	$6S_4$	$8S_6$	$3\sigma_h$	$6\sigma_d$
Γ_σ	6	0	0	2	2	0	0	0	4	2

Table 8.8.1. Summary of the formation of σ molecular orbitals (Fig. 8.8.2) in ML_6

Symmetry	Metal orbitals	Ligand σ orbitals	Molecular orbitals
A_{1g}	s	$(6)^{-\frac{1}{2}}(\sigma_1 + \sigma_2 + \sigma_3 + \sigma_4 + \sigma_5 + \sigma_6)$	$1a_{1g}, 2a_{1g}$
E_g	$d_{x^2-y^2}$ d_{z^2}	$\frac{1}{2}(\sigma_1 - \sigma_2 + \sigma_3 - \sigma_4)$ $(12)^{-\frac{1}{2}}(2\sigma_5 + 2\sigma_6 - \sigma_1 - \sigma_2 - \sigma_3 - \sigma_4)$	$1e_g, 2e_g$
T_{1u}	p_x p_y p_z	$(2)^{-\frac{1}{2}}(\sigma_1 - \sigma_3)$ $(2)^{-\frac{1}{2}}(\sigma_2 - \sigma_4)$ $(2)^{-\frac{1}{2}}(\sigma_5 - \sigma_6)$	$1t_{1u}, 2t_{1u}$
T_{2g}	d_{xy} d_{yz} d_{zx}	—	$1t_{2g}$

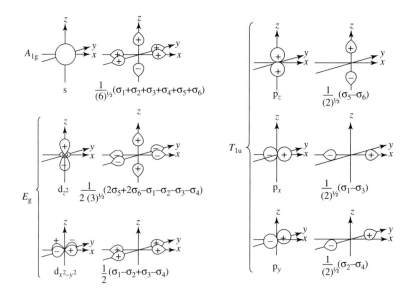

Fig. 8.8.2.
Linear combinations of σ ligand orbitals matching symmetry with the atomic orbitals on the central metal ion in ML_6.

which reduces as $\Gamma_\sigma = A_{1g} + E_g + T_{1u}$. The six linear combinations of ligand orbitals can be readily generated, and they match in symmetry with the suitable metal orbitals. These are summarized in Table 8.8.1 and illustrated in Fig. 8.8.2. The molecular orbital energy level diagram for an octahedral ML_6 complex with only σ bonding is shown in Fig. 8.8.3.

Examining the energy level diagram shown in Fig. 8.8.3, we see that the electron pairs from the six ligands enter into the $1t_{1u}$, $1a_{1g}$, and $1e_g$ orbitals, thus leaving $1t_{2g}$ and $2e_g$ orbitals to accommodate the d electrons from the metal ion. Also note that the $1t_{2g}$ orbitals are more stable than the $2e_g$ orbitals and the energy gap between them is once again called Δ_o. This is in total agreement with the results of crystal field theory (Fig. 8.1.5). But the two theories arrive at their results in different ways. In crystal field theory, the t_{2g} orbitals are more stable than the e_g orbitals because the former have lobes pointing between the ligands and the latter have lobes pointing at the ligands. On the other hand, in molecular orbital theory, the $1t_{2g}$ orbital are more stable than the $2e_g$ orbitals

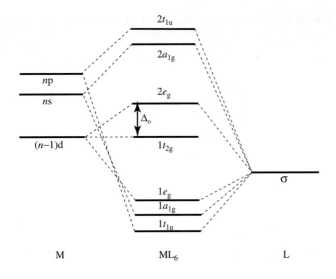

Fig. 8.8.3.
The molecular orbital energy level diagram for an octahedral complex with only σ bonding between the metal and the ligands.

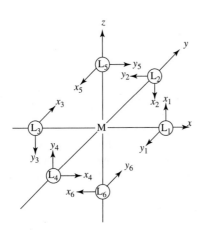

Fig. 8.8.4.
A coordinate system for ML_6, showing the orientations of the twelve π ligand orbitals.

because the former are nonbonding orbitals, while the latter are antibonding σ^* orbitals.

8.8.2 Octahedral complexes with π bonding

When the ligands in complex ML_6 also have π orbitals available for interaction, the bonding picture becomes more complicated. If the ligand π orbitals are oriented in the manner depicted in Fig. 8.8.4, it can be readily shown that the representation generated by the twelve ligand π orbitals is

O_h	E	$8C_3$	$6C_2$	$6C_4$	$3C_2(=C_4^2)$	i	$6S_4$	$8S_6$	$3\sigma_h$	$6\sigma_d$
Γ_π	12	0	0	0	−4	0	0	0	0	0

This reduces to $\Gamma_\pi = T_{1g} + T_{1u} + T_{2g} + T_{2u}$. The twelve linear combinations with the proper symmetry are listed in Table 8.8.2. Since there are no

Table 8.8.2. The symmetry-adapted linear combinations of ligand π orbitals for ML$_6$ complexes (Fig. 8.8.4) with O_h symmetry

Symmetry	Combination	Symmetry	Combination
T_{1g}	$\frac{1}{2}(x_2 + y_4 - x_5 - y_6)$	T_{1u}	$\frac{1}{2}(y_1 - x_3 + x_5 - y_6)$
	$\frac{1}{2}(x_1 + y_3 - y_5 - x_6)$		$\frac{1}{2}(-y_2 + x_4 + y_5 - x_6)$
	$\frac{1}{2}(-y_1 + y_2 - x_3 + x_4)$		$\frac{1}{2}(x_1 - x_2 - y_3 + y_4)$
T_{2g}	$\frac{1}{2}(x_2 + y_4 + x_5 + y_6)$	T_{2u}	$\frac{1}{2}(y_1 - x_3 - x_5 + y_6)$
	$\frac{1}{2}(x_1 + y_3 + y_5 + x_6)$		$\frac{1}{2}(-y_2 + x_4 - y_5 + x_6)$
	$\frac{1}{2}(y_1 + y_2 + x_3 + x_4)$		$\frac{1}{2}(-x_1 - x_2 + y_3 + y_4)$

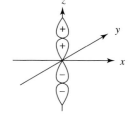

np_z with $\frac{1}{2}(x_1-x_2-x_3+x_4)$ np_z with $(2)^{-\frac{1}{2}}(\sigma_5-\sigma_6)$

Fig. 8.8.5. Comparison of the overlaps between the $T_{1u}\sigma$ and π ligand combinations with the same np_z orbital on the metal. Note that the σ type overlap will lead to more significant interaction between the ligands and the metal.

metal orbitals with either T_{1g} or T_{2u} symmetry, the combinations of ligand orbitals with these symmetries given in Table 8.8.2 are essentially nonbonding orbitals. Also, the linear combinations of ligand π orbitals with T_{1u} symmetry will have less effective overlap with the metal np oritals than the σ combinations (Table 8.8.1) of the same symmetry. The comparison of these two types of overlap is illustrated in Fig. 8.8.5. In other words, the molecular orbitals consisting mostly of the ligand π orbitals with T_{1u} symmetry is essentially nonbonding or at most weakly bonding. So the most important ligand orbitals participating in π bonding are the combinations of T_{2g} symmetry. Figure 8.8.6 shows how these combinations overlap with the metal d orbitals.

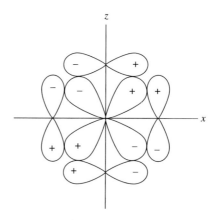

Fig. 8.8.6. The π-type overlap with T_{2g} symmetry between the ligands and the metal.

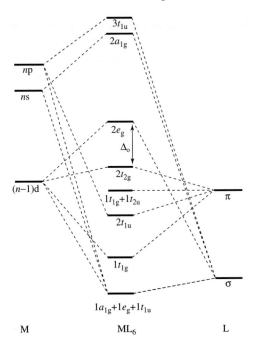

Fig. 8.8.7.
Schematic energy level diagram for an octahedral ML_6 complex in which the ligands have filled π orbitals for bonding. Compared to complexes with only σ bonding, the present system has a smaller Δ_o. See also Fig. 8.8.8(a).

When the ligands have filled π orbitals to bond with the metal, examples of which include F^-, Cl^-, OH^-,..., etc., the schematic energy level diagram for such complexes is shown in Fig. 8.8.7. Note that the orbitals up to and including $1t_{1g}$ and $1t_{2u}$ are filled by electrons from ligands, and we fill in the metal d electrons starting from $2t_{2g}$. Now the $2e_g$ orbitals remain antibonding σ^* in nature, while the $2t_{2g}$ orbitals have become antibonding π^* (recalling that these orbitals, called $1t_{2g}$ in Fig. 8.8.3, are nonbonding in a complex with only σ bonding). As a result, compared to the complexes with only σ bonding, the present system has a smaller Δ_o. As illustrated in Fig. 8.8.8(a), such a decrease in Δ_o implies an electron flow from ligands to metal (L→M).

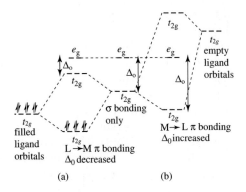

Fig. 8.8.8.
Comparison of the effects of π bonding using (a) filled π ligand orbitals for L→M donation and (b) empty π ligand orbitals for M→L donation. Note that, compared to complexes with only σ bonding, the former leads to a smaller Δ_o and the latter to a larger one.

Conversely, when the ligands have low-lying empty π orbitals for bonding, examples of which include CO, CN^-, PR_3,..., etc., the originally nonbonding t_{2g} orbitals now become π bonding in nature, while the e_g orbitals remain

σ^* antibonding. As a result, compared to complexes with only σ bonding, Δ_o becomes larger. As we feed electrons into these π bonding t_{2g} orbitals, electrons appear to flow from metal to ligands (M→L). This is illustrated in Fig. 8.8.8(b).

To conclude this section, we once again compare and contrast the crystal field theory and molecular orbital theory for metal complexes. Both theories lead to an orbital energy ordering of e_g above t_{2g}. Crystal field theory arrives at these results by arguing that the e_g orbitals point at the ligands and the t_{2g} orbitals point between the ligands. In other words, only electrostatic interaction is considered in this model. In molecular orbital theory, the overlap between the metal and ligand orbitals is taken into account. In complexes with only σ bonding, the e_g orbitals are antibonding σ^* and t_{2g} orbitals are nonbonding. In complexes with ligands having filled π orbitals, the e_g orbitals remain antibonding σ^* and t_{2g} orbitals have become antibonding π^*, leading to a smaller Δ_o. In complexes with ligands having empty π^* orbitals for bonding, the e_g orbitals are still antibonding σ^* and t_{2g} orbitals have become π bonding, thus yielding a larger Δ_o.

8.8.3 The eighteen-electron rule

In studying the compounds of representative elements, we often speak of the octet rule, which ascribes special stability to an electronic configuration of a rare gas atom, i.e., with eight valence electrons. Since transition metals have five d orbitals in their valence shell, in addition to one s and three p orbitals, their stable electronic configuration would have eighteen valence electrons, giving rise to the eighteen-electron rule for transition metal complexes. However, while this rule is often found useful, it is by no means rigorously followed. Indeed, in reference to the eighteen-electron rule, complexes may be broadly classified into three types: (1) those with electronic configurations entirely unrelated to this rule, (2) those with eighteen or less valence electrons, and (3) those with exactly eighteen valence electrons. Employing the energy level diagram shown in Fig. 8.8.3, we will see how these three types of behavior can be accounted for.

For case (1) complexes, examples of which include many first transition series compounds (Table 8.8.3), the $1t_{2g}$ orbitals are essentially nonbonding and Δ_o is small. In other words, the $2e_g$ orbitals are only slightly antibonding and they may be occupied without much energy cost. Hence, there is little or no restriction on the number of d electrons and the eighteen-electron rule has no influence on these complexes.

For case (2) complexes, examples of which include many second- or third-row transition metal complexes (Table 8.8.3), the $1t_{2g}$ orbitals are still nonbonding and Δ_o is large. In other words, the $2e_g$ orbitals are strongly antibonding and there is a strong tendency not to fill them. But there is still no restriction on the number of electrons that occupy the $1t_{2g}$ orbitals. Hence, there are eighteen or less valence electrons for these complexes.

For case (3) complexes, examples of which include many metal carbonyls and their derivatives (Table 8.8.3), the $1t_{2g}$ orbitals are strongly bonding due to back donation and the $2e_g$ orbitals are strongly antibonding. Thus, while it is still imperative not to have electrons occupying the $2e_g$ orbitals, it is equally

Bonding in Coordination Compounds

Table 8.8.3. Three types of complexes in relation to the eighteen-electron rule

Type (1) complex	Number of valence electrons	Type (2) complex	Number of valence electrons	Type (3) complex	Number of valence electrons
$[Cr(NCS)_6]^{3-}$	15	$[WCl_6]^{2-}$	13	$[V(CO)_6]^-$	18
$[Mn(CN)_6]^{3-}$	16	$[WCl_6]^{3-}$	14	$[Mo(CO)_3(PF_3)_3]$	18
$[Fe(C_2O_4)_3]^{3-}$	17	$[TcF_6]^{2-}$	15	$[HMn(CO)_5]$	18
$[Co(NH_3)_6]^{3+}$	18	$[OsCl_6]^{2-}$	16	$[(C_5H_5)Mn(CO)_3]$	18
$[Co(H_2O)_6]^{2+}$	19	$[PtF_6]$	16	$[Cr(CO)_6]$	18
$[Ni(en)_3]^{2+}$	20	$[PtF_6]^-$	17	$[Mo(CO)_6]$	18
$[Cu(NH_3)_6]^{2+}$	21	$[PtF_6]^{2-}$	18	$[W(CO)_6]$	18

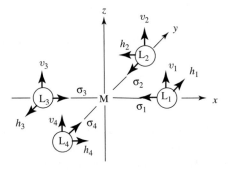

Fig. 8.9.1.
Coordinate system for a square-planar complex ML$_4$, where the ligand σ orbitals are labelled as $\sigma_1, \ldots, \sigma_4$, and the ligand π orbitals as $v_1, \ldots, v_4$ and $h_1, \ldots, h_4$.

important to have the $1t_{2g}$ orbitals fully filled. Removal of electrons from the completely occupied $1t_{2g}$ orbitals would destabilize the complex due to loss of bond energy. Hence these molecules tend to have exactly eighteen electrons.

8.9 Electronic spectra of square planar complexes

In this section, we use the square-planar complex ML$_4$ as an example to discuss the various types of electronic transitions observed in coordination compounds.

8.9.1 Energy level scheme for square-planar complexes

A coordinate system for a square-planar complex ML$_4$ (D_{4h} symmetry) is displayed in Fig. 8.9.1. The linear combinations of ligand orbitals, matched in symmetry with the metal orbitals, and the molecular orbitals they form, are summarized in Table 8.9.1. A schematic energy level diagram for this type of complexes is given in Fig. 8.9.2.

Examining Fig. 8.9.2, we can see that the most stable molecular orbitals are bonding σ orbitals with symmetries B_{1g}, A_{1g}, and E_u; they are mainly located on the ligands. Above them are the bonding and nonbonding π orbitals, also located on the ligands. All these orbitals are fully filled. After these two sets of levels, we come to the five orbitals (in four levels) that are essentially antibonding σ^* and π^* in nature and are located mainly on the metal's d orbitals. The ordering of these four levels may vary from ligand to ligand, but it is well established

Table 8.9.1. Summary of the formation of molecular orbitals in square-planar complexes ML$_4$

Symmetry	Metal orbital	Ligand orbitals	Molecular orbitals
A_{1g}	s d_{z^2}	$\frac{1}{2}(\sigma_1 + \sigma_2 + \sigma_3 + \sigma_4)$	$1a_{1g}, 2a_{1g}, 3a_{1g}$
A_{2g}	—	$\frac{1}{2}(h_1 + h_2 + h_3 + h_4)$	$1a_{2g}$
A_{2u}	p_z	$\frac{1}{2}(v_1 + v_2 + v_3 + v_4)$	$1a_{2u}, 2a_{2u}$
B_{1g}	$d_{x^2-y^2}$	$\frac{1}{2}(\sigma_1 - \sigma_2 + \sigma_3 - \sigma_4)$	$1b_{1g}, 1b_{2g}$
B_{2g}	d_{xy}	$\frac{1}{2}(h_1 - h_2 + h_3 - h_4)$	$1b_{2g}, 2b_{2g}$
B_{2u}	—	$\frac{1}{2}(v_1 - v_2 + v_3 - v_4)$	$1b_{2u}$
E_g	$\begin{cases} d_{xz} \\ d_{yz} \end{cases}$	$\begin{cases} (2)^{-\frac{1}{2}}(v_1 - v_3) \\ (2)^{-\frac{1}{2}}(v_2 - v_4) \end{cases}$	$1e_g, 2e_g$
E_u	$\begin{cases} p_x \\ p_y \end{cases}$	$\begin{cases} (2)^{-\frac{1}{2}}(\sigma_1 - \sigma_3) \\ (2)^{-\frac{1}{2}}(\sigma_2 - \sigma_4) \\ (2)^{-\frac{1}{2}}(h_4 - h_2) \\ (2)^{-\frac{1}{2}}(h_1 - h_3) \end{cases}$	$1e_u, 2e_u, 3e_u$

that the $b_{1g}(d_{x^2-y^2})$ orbital is the most unstable one. In any event, the ordering of these four levels is consistent with the results found for square-planar halides and cyanides. Interestingly, this ordering is also identical to that obtained by crystal field theory (Fig. 8.3.4). So once again we find consistency between the results of crystal field theory and molecular orbital theory, even though these two theories arrive at their results in different manners.

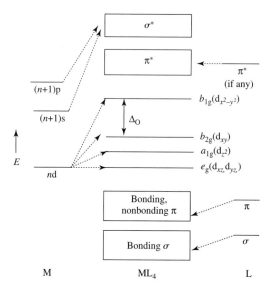

Fig. 8.9.2.
Schematic energy level diagram for a square-planar complex ML$_4$. Note the block of levels labelled π^* is present only for ligands with low-lying π orbitals such as CO, CN$^-$, and PR$_3$.

As mentioned in Section 8.3, many square-planar complexes have the d^8 configuration. This observation is consistent with the energy level scheme shown in Fig. 8.9.2, with eight electrons occupying the relatively more stable e_g, a_{1g}, and b_{2g} orbitals, and leaving the highly unstable b_{1g} level vacant.

Based on the energy level scheme shown in Fig. 8.9.2, we can expect three types of electronic transitions: d–d, ligand to metal (L→M), and metal to ligand (M→L) charge transfers. For d^8 square-planar complexes, there should be three spin-allowed d–d transitions: electron promoted to the b_{1g} orbital from the b_{2g}, a_{1g}, and e_g orbitals. In addition, L→M charge transfer bands originate from symmetry-allowed transitions promoting an electron to the b_{1g} orbital from either the σ or π bonding orbitals. On the other hand, for complexes with π-acceptor ligands, M→L charge transfer bands are due to the promotion of an electron from the three filled "metal" levels to the lowest energy "ligand" orbitals. It is useful to recall here that a charge transfer is an electronic transition from an orbital mostly localized on one atom or group to another orbital mostly localized on another atom or group. The majority of these charge transfer transitions take place in the ultraviolet region; they usually have higher energies, as well as higher intensities, than the d–d transitions. In passing, it is noted that crystal field theory can be employed to make assignments for the d–d transitions. But we need to make use of molecular orbital theory to treat the M→L and L→M charge transfers.

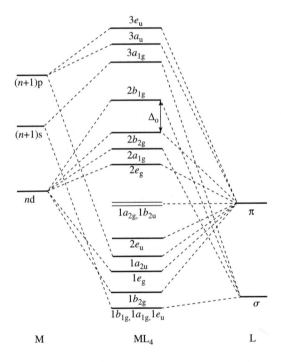

Fig. 8.9.3.
Schematic energy level diagram for square-planar halides.

8.9.2 Electronic spectra of square-planar halides and cyanides

With the energy level scheme shown in Fig. 8.9.2, we can easily derive two types of energy level diagrams: those for halides, as shown in Fig. 8.9.3, and those for cyanides, as shown in Fig. 8.9.4. We will make use of these results to interpret some spectral data for square-planar ML_4 complexes.

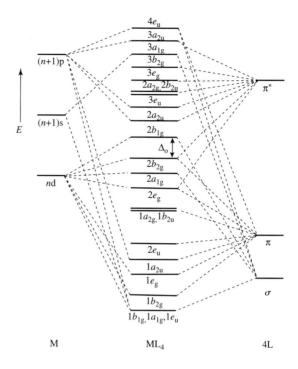

Fig. 8.9.4.
Schematic energy level diagram for square-planar cyanides.

(1) d–d bands

The d–d spectrum of $[PtCl_4]^{2-}$ in solution is shown in Fig. 8.9.5. Note that the bands are relatively weak, signifying that they are due to symmetry-forbidden transitions. As mentioned previously, three d–d bands are expected; indeed, three are observed. Hence the assignment is relatively straightforward:

$$^1A_{1g} \rightarrow {}^1A_{2g}(2b_{2g} \rightarrow 2b_{1g}) \; 21{,}000 \, \text{cm}^{-1}$$
$$\rightarrow {}^1B_{1g}(2a_{1g} \rightarrow 2b_{1g}) \; 25{,}500$$
$$\rightarrow {}^1E_g(2e_g \rightarrow 2b_{1g}) \; 30{,}200.$$

The labeling of the orbitals refers to those given in Fig. 8.9.3.

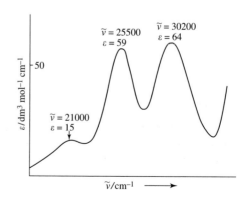

Fig. 8.9.5.
Aqueous solution d–d spectrum of $[PtCl_4]^{2-}$.

(2) L→M charge transfers

Two intense bands are observed in the spectrum of $[PtCl_4]^{2-}$, as shown in Fig 8.9.6. They are charge transfer bands, as characterized by their energies and intensities. Compared to the d–d transitions shown in Fig. 8.9.5, the charge transfer bands are seen at much higher energy and are more intense by a factor of nearly 10^3. Hence these transitions must be symmetry allowed. With the aid of Fig. 8.9.3, as well as some computational treatments, the following assignments have been made:

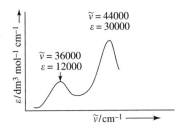

Fig. 8.9.6.
Charge transfer spectrum of $[PtCl_4]^{2-}$ in aqueous solution containing excess Cl^-. Note that these bands are much more intense than the d–d bands observed in $[PtCl_4]^{2-}$ (see Fig. 8.9.5).

$$\left.\begin{array}{r} ^1A_{1g} \rightarrow {}^1E_u(2e_u \rightarrow 2b_{1g}) \\ \rightarrow {}^1A_{2u}(1b_{2u} \rightarrow 2b_{1g}) \end{array}\right\} 36,0000 \text{ cm}^{-1}$$

$$\rightarrow {}^1E_u(1e_u \rightarrow 2b_{1g}) \quad 44,900.$$

(3) M→L charge transfers

Figure 8.9.7 shows that there are three closely spaced bands in the electronic spectrum of $[Ni(CN)_4]^{2-}$. Since CN^- has vacant low-lying π^* orbitals, these three bands have been assigned as the transitions from three closely spaced filled metal d levels ($2e_g$, $2a_{1g}$, and $2b_{2g}$ in Fig. 8.9.4) to the first available ligand level, $2a_{2u}$. Apparently, the $2a_{2u}$ level, mostly localized on the four CN^- groups, has been significantly stabilized by the $4p_z$ orbital on the metal. In any event, these three bands have been assigned as

$$^1A_{1g} \rightarrow {}^1B_{1u}(2b_{2g} \rightarrow 2a_{2u}) \quad 32,300 \text{ cm}^{-1}$$
$$\rightarrow {}^1A_{2u}(2a_{1g} \rightarrow 2a_{2u}) \quad 35,200$$
$$\rightarrow {}^1E_u(2e_g \rightarrow 2a_{2u}) \quad 37,600.$$

It is noted that the $^1A_{1g} \rightarrow {}^1B_{1u}$ transition is formally symmetry forbidden, even though it is Laporte allowed (g ↔ u). Hence the intensity of this transition is somewhere between a d–d band and a symmetry-allowed charge transfer band.

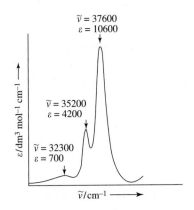

Fig. 8.9.7.
Charge transfer spectrum of $[Ni(CN)_4]^{2-}$ in aqueous solution. Note that these bands are also much more intense than the d–d bands (Fig. 8.9.5).

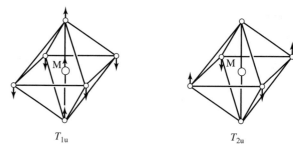

Fig. 8.10.1.
Two vibrational modes in an octahedral complex ML$_6$ that do not preserve the center of symmetry of the molecule.

8.10 Vibronic interaction in transition metal complexes

As mentioned earlier, in a centrosymmetric complex, d–d transitions are Laporte forbidden. The fact that they are observed at all is due to a mechanism called vibronic interaction, which is a mixing of the *vibrational* and *electronic* wavefunctions. Qualitatively, we may imagine that an electronic transition occurs at the very moment some vibrational modes of the complex distort the molecule in such a way that the center of symmetry is destroyed. When such a vibration takes place, the "g" character of the state is lost and the transition becomes (very slightly) allowed. Figure 8.10.1 shows two vibrations, with "u" symmetry, of an octahedral complex which remove the inversion center.

When the vibrational and electronic motions are coupled, the intensity integral for the transition between ψ' (ground state) and ψ'' (excited state) has the form

$$\int (\psi'_e \psi'_v) x (\psi''_e \psi''_v) d\tau, \quad (x \text{ may also be } y \text{ or } z),$$

where ψ_e and ψ_v are the electronic and vibrational wavefunctions for a given state, respectively. To evaluate this integral by symmetry arguments, we first note that ψ'_v is totally symmetric and can thus be ignored. (This is because all vibrational modes are in their ground state, which is totally symmetric.) Now we need to determine whether there exists a vibration (with symmetry Γ_v) such that, even though the product representation of $\psi''_e x \psi'_e$, $\Gamma(\psi''_e x \psi'_e)$, does not contain the totally symmetric representation Γ_{TS}, the product representation $\Gamma(\psi''_e x \psi'_e) \times \Gamma_v$ does. When $\Gamma(\psi''_e x \psi'_e) \times \Gamma_v$ contains Γ_{TS}, the aforementioned intensity integral would not vanish. Also, for $\Gamma(\psi''_e x \psi'_e) \times \Gamma_v$ to contain Γ_{TS}, we need $\Gamma(\psi''_e x \psi'_e)$ to contain Γ_v.

An example will illustrate the above arguments. The ground state of a d^1 octahedral complex ML$_6$ is $^2T_{2g}$ and the only excited state for this complex is ^2E_g. In the O_h point group, $\Gamma_{x,y,z}$ is T_{1u}. Also, this complex has the vibrational modes

$$\Gamma_{vib} = A_{1g} + E_g + 2T_{1u} + T_{2g} + T_{2u}.$$

For the transition $^2T_{2g} \to {}^2E_g$, we have

$$T_{2g} \times T_{1u} \times E_g = A_{1u} + A_{2u} + 2E_u + 2T_{1u} + 2T_{2u}.$$

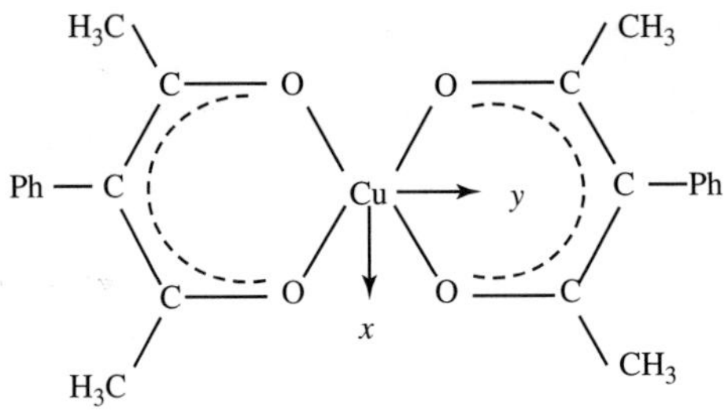

Fig. 8.10.2.
Molecular structure of planar complex [Cu(3-Ph-acac)$_2$] with D_{2h} symmetry.

Therefore, for this electronic transition to have any intensity, it must be accompanied by a vibration of either T_{1u} or T_{2u} symmetry.

We now employ one final example to conclude this discussion. The planar complex [Cu(3-Ph-acac)$_2$] has D_{2h} symmetry and its structure is shown in Fig. 8.10.2. In its crystal spectrum four y-polarized bands are observed and they have been assigned to the transitions $^2B_{1g} \rightarrow {}^2A_g$ (twice), $^2B_{1g} \rightarrow {}^2B_{2g}$, and $^2B_{1g} \rightarrow {}^2B_{3g}$. Given that the "u" vibrational modes of a MA$_2$B$_2$ complex belonging to the D_{2h} point group are of symmetries A_u, B_{1u}, B_{2u}, and B_{3u}, we can readily determine the vibrations responsible for the forbidden transitions to gain intensity.

In the D_{2h} group, $\Gamma_y = B_{2u}$. So for the $^2B_{1g} \rightarrow {}^2A_g$ transition, we have $B_{1g} \times B_{2u} \times A_g = B_{3u}$; i.e., the B_{3u} vibrational mode is needed. For the $^2B_{1g} \rightarrow {}^2B_{2g}$ transition $B_{1g} \times B_{2u} \times B_{2g} = B_{1u}$, or the B_{1u} vibrational mode is required. Finally, for $^2B_{1g} \rightarrow {}^2B_{3g}$, $B_{1g} \times B_{2u} \times B_{3g} = A_u$; i.e., the A_u mode of vibration accompanies this electronic transition.

8.11 The 4f orbitals and their crystal field splitting patterns

To conclude this chapter, we discuss the shapes of the 4f orbitals as well as their splitting patterns in octahedral and tetrahedral crystal fields. The results are of use in studying the complexes of the rare-earth elements.

8.11.1 The shapes of the 4f orbitals

The are seven 4f orbitals that correspond to m_ℓ values of 0, ±1, ±2, and ±3 in their angular parts. Of these the one with $m_\ell = 0$ is real, but the other six are complex, and their linear combinations lead to real functions. However, there is not a unique or conventional way to express the angular functions of the sevenfold degenerate 4f orbitals, all of which have three nodes and ungerade (u) symmetry. If simple linear combinations of the $m_\ell = \pm 1, \pm 2$, and ± 3 functions are taken in pairs, the resulting six Cartesian functions (with abbreviated labels enclosed in square brackets) are $x(4z^2 - x^2 - y^2)$ [xz^2]; $y(4z^2 - x^2 - y^2)$ [yz^2]; xyz; $z(x^2 - y^2)$; $x(x^2 - 3y^2)$; and $y(3x^2 - y^2)$. These six orbitals, together with $z(2z^2 - 3x^2 - 3y^2)$ [z^3] arising from $m_\ell = 0$, constitute the *general set* of 4f orbitals as displayed in Table 2.1.2, and they are also listed in Table 8.11.1.

Table 8.11.1. Angular functions of the 4f orbitals

Orbital*	The general set	The cubic set
z^3	$\frac{1}{4}(7/\pi)^{\frac{1}{2}}(5\cos^3\theta - 3\cos\theta)$	$\frac{1}{4}(7/\pi)^{\frac{1}{2}}(5\cos^3\theta - 3\cos\theta)$
$z(x^2-y^2)$	$\frac{1}{4}(105/\pi)^{\frac{1}{2}}\sin^2\theta\cos\theta\cos 2\phi$	$\frac{1}{4}(105/\pi)^{\frac{1}{2}}\sin^2\theta\cos\theta\cos 2\phi$
xyz	$\frac{1}{4}(105/\pi)^{\frac{1}{2}}\sin^2\theta\cos\theta\sin 2\phi$	$\frac{1}{4}(105/\pi)^{\frac{1}{2}}\sin^2\theta\cos\theta\sin 2\phi$
x^3	—	$\frac{1}{4}(7/\pi)^{\frac{1}{2}}\sin\theta\cos\phi(5\sin^2\theta\cos^2\phi - 3)$
$x(z^2-y^2)$	—	$\frac{1}{4}(105/\pi)^{\frac{1}{2}}\sin\theta\cos\phi(\cos^2\theta - \sin^2\theta\sin^2\phi)$
xz^2	$\frac{1}{8}(42/\pi)^{\frac{1}{2}}\sin\theta - (5\cos^2\theta - 1)\cos\phi$	—
$x(x^2-3y^2)$	$\frac{1}{8}(70/\pi)^{\frac{1}{2}}\sin^3\theta\cos 3\phi$	—
y^3	—	$\frac{1}{4}(7/\pi)^{\frac{1}{2}}\sin\theta\sin\phi(5\sin^2\theta\sin^2\phi - 3)$
$y(z^2-x^2)$	—	$\frac{1}{4}(105/\pi)^{\frac{1}{2}}\sin\theta\sin\phi(\cos^2\theta - \sin^2\theta\cos^2\phi)$
yz^2	$\frac{1}{8}(42/\pi)^{\frac{1}{2}}\sin\theta(5\cos^2\theta - 1)\sin\phi$	—
$y(3x^2-y^2)$	$\frac{1}{8}(70/\pi)^{\frac{1}{2}}\sin^3\theta\sin 3\phi$	—

* z^3 denotes the f_{z^3} orbital, etc.

The shapes of the seven 4f orbitals in the general set are illustrated in Fig. 8.11.1, and their nodal characteristics are shown in Fig. 8.11.2. The number of vertical nodal planes varies from 0 to 3. The z^3, yz^2, and xz^2 orbitals each has two nodes that are the curved surfaces of a pair of cones with a common vertex at the origin.

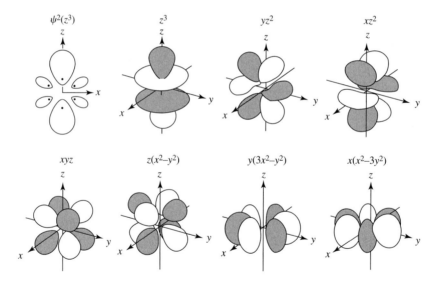

Fig. 8.11.1.
The general set of the seven 4f orbitals. The gray and white regions in each orbital bear positive and negative signs, respectively. Placed to the left of the z^3 orbital is a cross section of $\psi^2(z^3)$, in which dots indicate the "electron-density" maxima, and contour lines are drawn for $\psi^2/\psi^2_{max} = 0.1$.

For systems of cubic symmetry, it is convenient to choose an alternative set of combinations known as the *cubic set*. Their abbreviated Cartesian labels and symmetry species in point group O_h are $xyz(A_{2u})$; x^3, y^3, z^3 (T_{1u}); $z(x^2 - y^2)$, $x(z^2 - y^2)$, $y(z^2 - x^2)$ (T_{2u}). These functions are particularly useful when octahedral and tetrahedral crystal field splitting patterns are considered. The

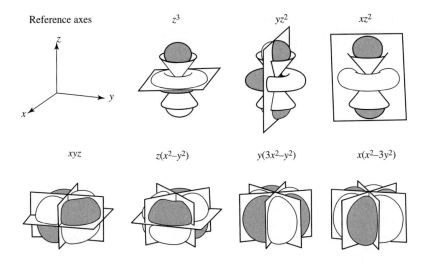

Fig. 8.11.2.
Nodal characteristics of the general set of 4f orbitals. The positive and negative lobes in each orbital are shaded and un-shaded, respectively.

orbitals x^3 and y^3 have the same shape as z^3, but they lie along the x and y axis, respectively. The orbital xyz has eight lobes pointing toward the corners of a cube. The orbitals $x(z^2 - y^2)$, $y(z^2 - x^2)$, and $z(x^2 - y^2)$ are shaped like xyz, but they are rotated by 45° about the x, y, and z axis, respectively.

The two sets of angular functions of the 4f orbitals are compared in Table 8.11.1; it is noted that three of them, namely z^3, $z(x^2 - y^2)$, and xyz, are common to both sets. The remaining four members of the two sets are related by linear transformation. Specifically,

$$f_{x^3} = -\frac{1}{4}[(6)^{\frac{1}{2}} f_{xz^2} - (10)^{\frac{1}{2}} f_{x(x^2-3y^2)}], \qquad (8.11.1)$$

$$f_{y^3} = -\frac{1}{4}[(6)^{\frac{1}{2}} f_{yz^2} + (10)^{\frac{1}{2}} f_{y(3x^2-y^2)}], \qquad (8.11.2)$$

$$f_{x(z^2-y^2)} = \frac{1}{4}[(10)^{\frac{1}{2}} f_{xz^2} + (6)^{\frac{1}{2}} f_{x(x^2-3y^2)}], \qquad (8.11.3)$$

$$f_{y(z^2-x^2)} = \frac{1}{4}[(10)^{\frac{1}{2}} f_{yz^2} - (6)^{\frac{1}{2}} f_{y(3x^2-y^2)}]. \qquad (8.11.4)$$

8.11.2 Crystal field splitting patterns of the 4f orbitals

In an octahedral crystal field, the seven 4f orbitals should split into two groups of triply degenerate sets and one non-degenerate orbital. This information can be readily obtained when we examine the character table of the O_h group, where it is seen that the x^3, y^3, and z^3 orbitals form a T_{2u} set, the $z(x^2 - y^2)$, $y(z^2 - x^2)$, and $x(z^2 - y^2)$ functions constitute a T_{1u} set, while the remaining orbital, xyz, has A_{2u} symmetry. In the following, we discuss the energy ordering of these three sets of orbitals.

If we place six ligands at the centers of the faces of a cube, as shown in Fig. 8.11.3, it can be readily seen that the t_{2u} orbitals (x^3, y^3, z^3) have lobes pointing directly toward the ligands. Hence they have the highest energy. Also,

Fig. 8.11.3.
Splitting of the f orbitals in an octahedral crystal field.

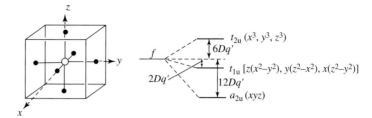

the eight lobes of the a_{2u} orbital (xyz) point directly at the corners of the cube. In other words, the eight lobes of the xyz orbitals are as far away from the ligands as possible. Hence, this orbital should have the lowest energy. Finally, the t_{1u} orbitals have lobes pointing toward the midpoints of the edges of the cube. Hence, they have energy lower than that for the t_{2u} orbitals, but higher than that for the a_{2u} orbital. When this problem is treated quantitatively by perturbation theory, we find that the t_{2u} orbitals are destabilized by $6Dq'$, while the t_{1u} and a_{2u} orbitals are stabilized by $2Dq'$ and $12Dq'$, respectively. These results are summarized in Fig. 8.11.3. It should be noted in the crystal field parameter Dq' given in Fig. 8.11.3 is not the same as that for transition metal complexes [eqs. (8.1.3) and (8.1.4)], as f orbitals, rather than d orbitals, are involved here. Also, examining the splitting pattern shown in Fig. 8.11.3, we can see that the center of gravity is once again preserved, as the energy lost by the t_{2u} orbitals is balanced by the energy gained by the t_{1u} and a_{2u} orbitals.

Fig. 8.11.4.
Splitting pattern of the 4f orbitals in a tetrahedral crystal field.

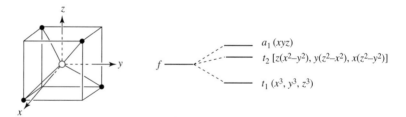

The splitting pattern of the 4f orbitals in a tetrahedral crystal field can be deduced in a similar manner. If we adopt the coordinate system shown in Fig. 8.11.4, we can obtain the splitting pattern shown in the same figure. As expected the 1:3:3 pattern for the tetrahedral crystal field is just the reverse of the 3:3:1 pattern of the octahedral field. Also, the symmetry classification of the orbitals in a tetrahedral complex can be readily obtained from the character table of the T_d group.

References

1. F. A. Cotton, G. Wilkinson, C. A. Murillo and M. Bochmann, *Advanced Inorganic Chemistry*, 6th edn., Wiley, New York, 1999.
2. N. N. Greenwood and A Earnshaw, *Chemistry of the Elements*, 2nd edn., Butterworth-Heinemann, Oxford, 1997.

3. J. E. Huheey, E. A. Keiter and R. L. Keiter, *Inorganic Chemistry: Principles of Structure and Reactivity*, 4th edn., Harper Collins, New York, 1993.
4. W. W. Porterfield, *Inorganic Chemistry: A Unified Approach*, Academic Press, San Diego, CA, 1993.
5. B. E. Douglas, D. H. McDaniel and J. J. Alexander, *Concepts and Models of Inorganic Chemistry*, 3rd edn., Wiley, New York, 1994.
6. C. E. Housecroft and A. G. Sharpe, *Inorganic Chemistry*, 2nd edn., Prentice-Hall, London, 2004.
7. G. L. Miessler and D. A. Tarr, *Inorganic Chemistry*, 3rd edn., Pearson/Prentice Hall, Upper Saddle River, NJ, 2004.
8. J. A. McCleverty and T. J. Meyer (eds.), *Comprehensive Coordination Chemistry II: From Biology to Nanotechnology*, Elsevier Pergamon, Amsterdam, 2004.
9. B. D. Douglas and C. A. Hollingsworth, *Symmetry in Bonding and Spectra: An Introduction*, Academic Press, Orlando, FL, 1985.
10. R. S. Drago, *Physical Methods for Chemists*, 2nd edn., Saunders, Fort Worth, TX, 1992.
11. C. J. Ballhausen, *Introduction to Ligand Field Theory*, McGraw-Hill, New York, 1962.
12. B. N. Figgis and M. A. Hitchman, *Ligand Field Theory and Its Applications*, Wiley–VCH, New York, 2000.
13. C. K. Jørgenson, *Modern Aspects of Ligand Field Theory*, North-Holland, Amsterdam, 1971.
14. T. M. Dunn, D. S. McClure and R. G. Pearson, *Some Aspects of Crystal Field Theory*, Harper and Row, New York, 1965.
15. M. Jacek and Z. Gajek, *The Effective Crystal Field Potential*, Elsevier, New York, 2000.
16. J. S. Griffith, *The Theory of Transition-Metal Ions*, Cambridge University Press, Cambridge, 1961.
17. I. B. Bersuker, *Electronic Structure and Properties of Transition Metal Compounds: Introduction to the Theory*, Wiley, New York, 1996.
18. F. Weinhold and C. Landis, *Valency and Bonding: A Natural Bond Orbital Donor-Acceptor Perspective,* Cambridge University Press, Cambridge, 2005.
19. C. J. Ballhausen and H. B. Gray, *Molecular Orbital Theory*, Benjamin, New York, 1965.
20. T. A. Albright, J. K. Burdett and M.-H. Whangbo, *Orbital Interactions in Chemistry*, Wiley, New York, 1985.
21. T. A. Albright and J. K. Burdett, *Problems in Molecular Orbital Theory,* Oxford University Press, New York, 1992.

9 Symmetry in Crystals

A single crystal is a homogeneous solid, which means that all parts within it have identical properties. However, it is not in general isotropic, so that physical properties such as thermal and electrical conductivity, refractive index, and non-linear optical effect generally vary in different directions.

9.1 The crystal as a geometrical entity

The majority of pure compounds can be obtained in crystalline form, although the individual specimens may often be very small or imperfectly formed. A well-developed crystal takes the shape of a polyhedron with planar faces, linear edges, and sharp vertices. In the simplest terms, a crystal may be defined as a homogeneous, anisotropic solid having the natural shape of a polyhedron.

9.1.1 Interfacial angles

In 1669, Steno discovered that, for a given crystalline material, despite variations in size and shape (i.e., its *habit*), the angles between corresponding faces of individual crystals are equal. This is known as the *Law of Constancy of Interfacial Angles*. The geometrical properties (the *form*) of a crystal may be better visualized by considering an origin inside it from which perpendiculars are drawn to all faces. This radiating set of normals is independent of the size and shape of individual specimens, and is thus an invariant representation of the crystal form. In Fig. 9.1.1, a perfectly developed hexagonal plate-like crystal is shown in part (a), and a set of normals are drawn to its vertical faces from an interior point in part (b). Two other crystal specimens that look very different

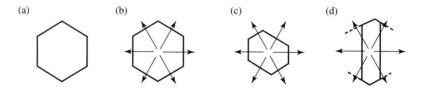

Fig. 9.1.1.
Representation of vertical crystal faces of a hexagonal by a system of radiating normals from a point inside it. The broken lines indicate extensions of the planar faces.

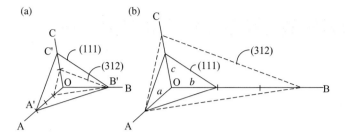

Fig. 9.1.2.
Parametral face, intercepts, and Miller indices.

are shown in parts (c) and (d), and they are seen to possess the same set of radiating normals.

9.1.2 Miller indices

For a mathematical description of crystal faces, take any three non-parallel faces (chosen to be mutually orthogonal, if possible) and take their intersections as reference axes, which are labeled OA, OB, and OC with the origin at O, as shown in Fig. 9.1.2(a). Let another face (the standard face or parametral face A'B'C') meet these axes at A', B', and C', making intercepts $OA' = a$, $OB' = b$ and $OC' = c$, respectively. The ratios $a{:}b{:}c$ are called the *axial ratios*.

If now any face on the crystal makes intercepts of $a/h, b/k$, and c/ℓ on the axes OA, OB, and OC, respectively, it is said to have the *Miller indices* $(hk\ell)$, which have no common divisor. The Miller indices of any face are thus calculated by dividing its intercepts on the axes by a, b, c, respectively, taking the reciprocals, and clearing them of fractions if necessary. If a plane is parallel to an axis, the intercept is at infinity and the corresponding Miller index is zero. The Miller indices of the standard face are (111), and the plane outlined by the dotted lines in Fig. 9.1.2(a) has intercepts $a/3, b, c/2$, which correspond to the Miller indices (312). In Fig. 9.1.2(b) another plane is drawn parallel to the aforesaid plane, making intercepts $a, 3b, 3c/2$; it is obvious that both planes outlined by dotted lines in Fig 9.1.2 have the same orientation as described by the same Miller indices (312).

In 1784, Haüy formulated the *Law of Rational Indices*, which states that all faces of a crystal can be described by Miller indices $(hk\ell)$, and for those faces that commonly occur, h, k, and ℓ are all small integers. The eight faces of an octahedron are (111), ($\bar{1}$11), (1$\bar{1}$1), (11$\bar{1}$), (1$\bar{1}\bar{1}$), ($\bar{1}$1$\bar{1}$), ($\bar{1}\bar{1}$1), and ($\bar{1}\bar{1}\bar{1}$). The *form symbol* that represents this set of eight faces is {111}. The form symbol for the six faces of a cube is {100}. Some examples in the cubic system are shown in Figs. 9.1.3. and 9.1.4.

9.1.3 Thirty-two crystal classes (crystallographic point groups)

The Hermann-Mauguin notation for the description of point group symmetry (in contrast to the Schönflies system used in Chapter 6) is widely adopted in crystallography. An n-fold rotation axis is simply designated as n. An object is said to possess an n-fold inversion axis $\bar{n}$ if it can be brought into an equivalent configuration by a rotation of $360°/n$ in combination with inversion through a

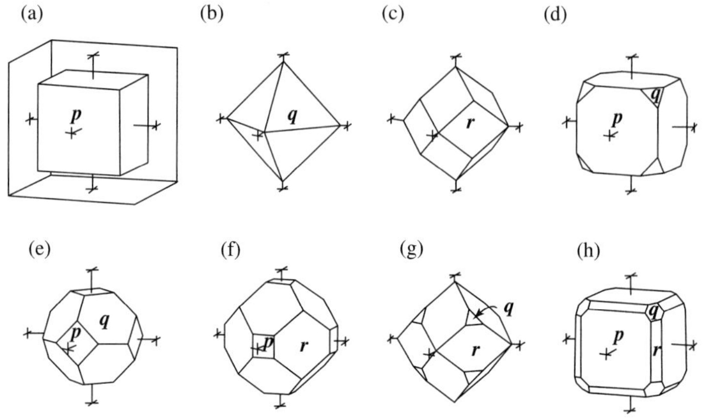

Fig. 9.1.3.
Idealized shapes of cubic crystals with well-developed faces described by form symbols (a) $p\{100\}$, (b) $q\{111\}$, (c) $r\{110\}$, and various combinations (d)–(h).

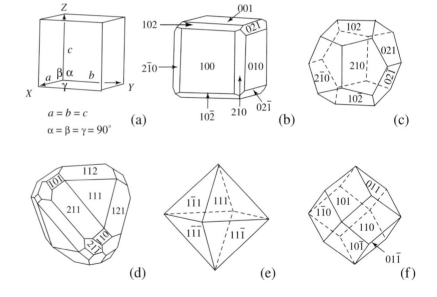

point on the axis. Note that $\bar{1}$ is an inversion center, and $\bar{2}$ is the same as mirror plane m. The symbol $2/m$ means a twofold axis with a mirror plane m lying perpendicular to it.

Fig. 9.1.4.
Some cubic crystals with faces indexed by Miller indices. (a) Cubic unit cell and labels of axes and angles; (b) and (c) two habits of pyrites FeS_2, class $m\bar{3}$; (d) tetrahedrite Cu_3SbS_3, class $\bar{4}3m$; (e) spinel $MgAl_2O_4$, class $m\bar{3}m$; and (f) almandine (garnet) $Fe_3Al_2(SiO_4)_3$, class $m\bar{3}m$.

A single crystal, considered as a finite object, may possess a certain combination of point symmetry elements in different directions, and the symmetry operations derived from them constitute a group in the mathematical sense. The self-consistent set of symmetry elements possessed by a crystal is known as a *crystal class* (or crystallographic point group). Hessel showed in 1830 that there are thirty-two self-consistent combinations of symmetry elements n and $\bar{n}$ ($n = 1, 2, 3, 4$, and 6), namely the thirty-two crystal classes, applicable to the description of the external forms of crystalline compounds. This important

Symmetry in Crystals

result is a natural consequence of the Law of Rational Indices, and the deduction proceeds in the following way.

Consider a crystal having a n-fold rotation axis, often simply referred to as an n-axis. In Fig. 9.1.5, OE, OE', OE'' represent the horizontal projections of three vertical planes generated by successive rotations of $2\pi/n$ about the symmetry axis, which passes through point O. As reference axes, we adopt the n-axis, OE', and OE'', and the latter two are assigned unit length. Next, we consider a plane E'F which is drawn parallel to OE. The intercepts of E'F on the three axes are ∞, 1, and $(\sec 2\pi/n)/2$. The Miller indices of plane E'F are therefore $(0, 1, 2\cos 2\pi/n)$ which must be integers. Since $|\cos 2\pi/n| \le 1$, the only possible values of $2\cos 2\pi/n$ are $\pm 2, \pm 1$, and 0. Hence

$$\cos 2\pi/n = 1, 1/2, 0, -1/2, -1, \text{ leading to } n = 1, 6, 4, 3, 2, \text{ respectively.}$$

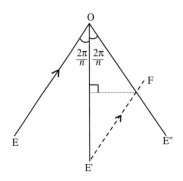

Fig. 9.1.5.
Vertical planes related by a n-fold rotation axis.

Therefore, in compliance with the Law of Rational Indices, only n-axes with $n = 1, 2, 3, 4$ and 6 are allowed in crystals. The occurrence of the inversion center means that the rotation-inversion axes $\bar{1}, \bar{2}(= m), \bar{3}, \bar{4}$ and $\bar{6}$ are also possible.

9.1.4 Stereographic projection

The thirty-two crystal classes are usually represented by stereographic projections of a system of equivalent points. The stereographic projection of a particular crystal class is derived in the manner illustrated in Fig. 9.1.6. Consider an idealized crystal whose planar faces exhibit the full symmetry $\bar{4}$ of its crystallographic point group. Enclose the crystal by a sphere and draw perpendiculars to all faces from a point inside the crystal. The radiating normals intersect the sphere to give a set of points A, B, C and D. From the south pole S of the sphere, draw lines to the points A and B lying in the northern hemisphere, and their intersection in the equatorial plane gives points A' and B' denoted by the small open circles. Similarly, lines are drawn from the north pole N to points C and D in the southern hemisphere to intersect the equatorial plane, yielding points C' and D' denoted by the filled dots. The set of equivalent points A', B', C', and D' in the equatorial circle is a faithful representation of the crystal class $\bar{4}$.

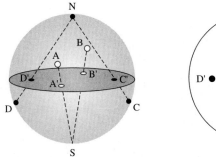
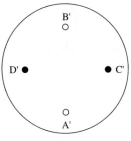

Fig. 9.1.6.
Stereographic projection of crystal class (point group) $\bar{4}$.

The stereographic projection of a crystal class consists of all intersected equivalent points in the equatorial circle and associated symmetry symbols. The location of the symmetry elements and the stereographic projection of point group $\bar{4}3m$ is shown in Fig. 9.1.7. The graphic symbols indicate three $\bar{4}$-axes along the principal axes of the cube, four 3-axes along the body diagonals, and six mirror planes normal to the face diagonals.

All thirty-two crystal classes in stereographic projection are tabulated in Fig. 9.1.8. For each crystal class, two diagrams showing the equipoints and symmetry elements are displayed side by side. Note that the thicker lines indicate mirror planes, and the meaning of the graphic symbols of the symmetry elements are given in Table 9.3.2.

9.1.5 Acentric crystalline materials

Of the thirty-two crystal classes, twenty-two lack an inversion center and are therefore known as non-centrosymmetric, or acentric. Crystalline and polycrystalline bulk materials that belong to acentric crystal classes can exhibit a variety of technologically important physical properties, including optical activity, pyroelectricity, piezoelectricity, and second-harmonic generation (SHG, or frequency doubling). The relationships between acentric crystal classes and physical properties of bulk materials are summarized in Table 9.1.1.

Eleven acentric crystal classes are chiral, i.e., they exist in enantiomorphic forms, whereas ten are polar, i.e., they exhibit a dipole moment. Only five (1, 2, 3, 4, and 6) have both chiral and polar symmetry. All acentric crystal classes except 432 possess the same symmetry requirements for materials to display piezoelectric and SHG properties. Both ferroelectricity and pyroelectricity are related to polarity: a ferroelectric material crystallizes in one of ten polar crystal classes (1, 2, 3, 4, 6, *m*, *mm*2, 3*m*, 4*mm*, and 6*mm*) and possesses a permanent dipole moment that can be reversed by an applied voltage, but the spontaneous polarization (as a function of temperature) of a pyroelectric material is not. Thus all ferroelectric materials are pyroelectric, but the converse is not true.

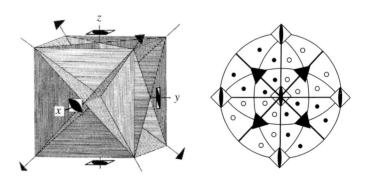

Fig. 9.1.7.
Symmetry elements and stereographic projection of crystal class $\bar{4}3m$.

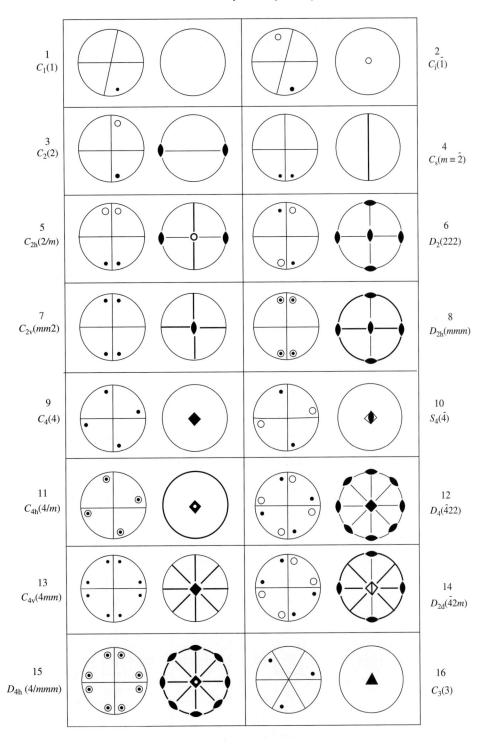

Fig. 9.1.8.

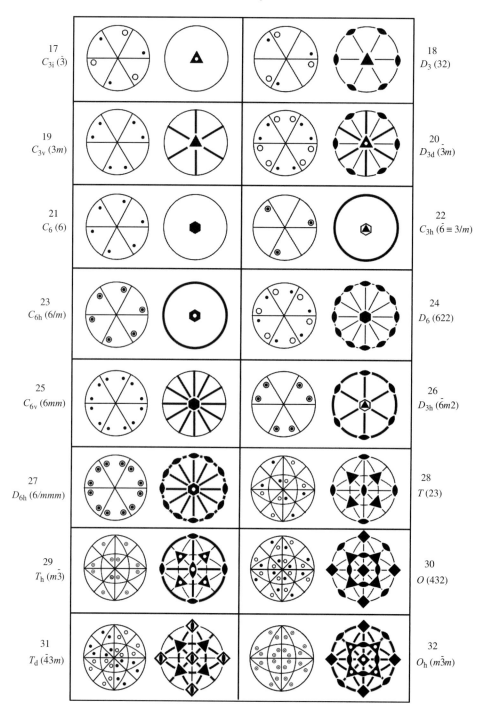

Fig. 9.1.8. (continued)
Stereographic projections showing the general equivalent positions (left figure) and symmetry elements (right figure) of the thirty-two crystal classes (crystallographic point groups). The z axis is normal to the paper in all drawings. Note that the dots and open circles can overlap, and the thick lines represent mirror planes. For each point group, the Hermann–Mauguin symbol is given in parentheses after the Schönflies symbol.

Table 9.1.1. Physical properties of materials found in acentric crystal classes

Crystal system	Crystal class	Chiral (enantiomorphism)	Optical activity (circular dichroism)	Polar (pyroelectric)	Piezoelectric, SHG
Triclinic	1	+	+	+	+
Monoclinic	2	+	+	+	+
	m		+	+	+
Orthorhombic	222	+	+		+
	mm2		+	+	+
Tetragonal	4	+	+	+	+
	$\bar{4}$		+		+
	422	+	+		+
	4mm			+	+
	$\bar{4}2m$		+		+
Trigonal	3	+	+	+	+
	32	+	+		+
	3m			+	+
Hexagonal	6	+	+	+	+
	$\bar{6}$				+
	622	+	+		+
	6mm			+	+
	$\bar{6}m2$				+
Cubic	23	+	+		+
	432	+	+		
	$\bar{4}3m$				+

9.2 The crystal as a lattice

9.2.1 The lattice concept

A study of the external symmetry of crystals naturally leads to the idea that a single crystal is a three-dimensional periodic structure; i.e., it is built of a basic structural unit that is repeated with regular periodicity in three-dimensional space. Such an infinite periodic structure can be conveniently and completely described in terms of a *lattice* (or space lattice), which consists of a set of points (mathematical points that are dimensionless) that have identical environments.

An infinitely extended, linear, and regular system of points is called a *row*, and it is completely described by its repeat distance a. A planar regular array of points is called a *net*, which can be specified by two repeat distances a and b and the angle γ between them. The analogous system of regularly distributed points in three dimensions constitute a lattice (or space lattice), which is described by a set of three non-coplanar vectors (**a**, **b**, **c**) or six parameters: the repeat intervals a, b, c and the angles α, β, γ between the vectors. As far as possible, the angles are chosen to be obtuse, particularly 90° or 120°, and lattices in one, two, and three dimensions are illustrated in Fig. 9.2.1.

9.2.2 Unit cell

Rather than having to visualize an infinitely extended lattice, we can focus our attention on a small portion of it, namely a *unit cell*, which is a parallelopiped (or box) with lattice points located at its eight corners. The entire lattice can then be re-generated by stacking identical unit cells in three dimensions. A unit

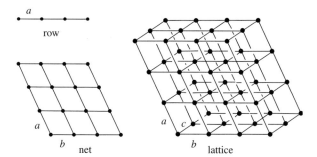

Fig. 9.2.1.
Row, net and lattice.

cell with lattice points only at the corners is described as *primitive*, as it contains only one structural motif (one lattice point). Alternatively, it is also possible to choose a *centered* unit cell that contains more than one lattice point.

Figure 9.2.2(a) shows a net with differently chosen unit cells. The unit cells having sides (a_1, b_1), (a_2, b_1), and (a_2, b_2) are primitive, and the one with sides (a_1, b_2) is C-centered. Note that the unit cell outlined by dotted lines is also a legitimate one, although it is not a good choice.

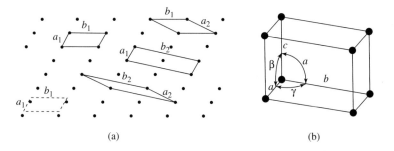

Fig. 9.2.2.
(a) Two-dimensional lattice showing different choices of the unit cell. (b) A primitive unit cell for a three-dimensional lattice.

In general, the size and shape of a unit cell is specified by six parameters: the axial lengths a, b, c and the interaxial angles α, β, γ [Fig.9.2.2(b)]. The volume of a triclinic unit cell is calculated by the formula $V = abc(1 - \cos^2 \alpha - \cos^2 \beta - \cos^2 \gamma + 2 \cos \alpha \cos \beta \cos \gamma)^{1/2}$, which reduces to simpler forms for the other crystal systems.

The choice of a unit cell depends on the best description of the symmetry elements present in the crystal structure. In practice, as far as it is possible, the unit cell is chosen such that some or all of the angles are 90° or 120°. Repetition of the unit cell by translations in three non-coplanar directions regenerates the whole space lattice.

The unit cell is the smallest volume that contains the most essential structural information required to describe the crystal structure. For a particular crystalline compound, each lattice point is associated with a basic structural unit, which may be an atom, a number of atoms, a molecule, or a number of molecules. In other words, if the coordinates of all atoms within the unit cell are known, complete information is obtained on the atomic arrangement, molecular packing, and derived parameters such as interatomic distances and bond angles.

Symmetry in Crystals

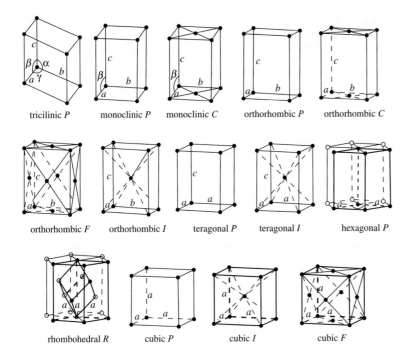

Fig. 9.2.3.
The fourteen Bravais lattices.

9.2.3 Fourteen Bravais lattices

Bravais showed in 1850 that all three-dimensional lattices can be classified into 14 distinct types, namely the fourteen *Bravais lattices*, the unit cells of which are displayed in Fig. 9.2.3. Primitive lattices are given the symbol P. The symbol C denotes a C face centered lattice which has additional lattice points at the centers of a pair of opposite faces defined by the a and b axes; likewise the symbol A or B describes a lattice centered at the corresponding A or B face. When the lattice has all faces centered, the symbol F is used. The symbol I is applicable when an additional lattice point is located at the center of the unit cell. The symbol R is used for a rhombohedral lattice, which is based on a rhombohedral unit cell (with $a = b = c$ and $\alpha = \beta = \gamma \neq 90°$) in the older literature. Nowadays the rhombohedral lattice is generally referred to as a hexagonal unit cell that has additional lattice points at $(2/3, 1/3, 1/3)$ and $(1/3, 2/3, 2/3)$ in the conventional *obverse setting*, or $(1/3, 2/3, 1/3)$ and $(2/3, 1/3, 2/3)$ in the alternative *reverse setting*. In Fig. 9.2.3 both the primitive rhombohedral (R) and obverse triple hexagonal (hR) unit cells are shown for the rhombohedral lattice.

9.2.4 Seven crystal systems

The fourteen Bravais lattices are divided into seven *crystal systems*. The term "system" indicates reference to a suitable set of axes that bear specific relationships, as illustrated in Table 9.2.1. For example, if the axial lengths take arbitrary values and the interaxial angles are all right angles, the crystal system

Table 9.2.1. Characteristics of the seven crystal systems

System	Crystal classes	Unit cell	Standard axes
Triclinic	$1, \bar{1}$	$a \neq b \neq c$ $\alpha \neq \beta \neq \gamma$	**a**, **b**, **c** not coplanar
Monoclinic	$2, m, 2/m$	$a \neq b \neq c$ $\alpha = \gamma = 90°$	principal axis b parallel to 2-axis or perpendicular to m; **a**, **c** smallest lattice vectors perpendicular to **b**
Orthorhombic	$222, mm2, mmm$	$a \neq b \neq c$ $\alpha = \beta = \gamma = 90°$	**a**, **b**, **c** each oriented along a 2-axis, or perpendicular to a mirror plane
Tetragonal	$4, \bar{4}, 4/m, 422,$ $4mm, \bar{4}2m,$ $4/mmm$	$a = b \neq c$ $\alpha = \beta = \gamma = 90°$	**c** oriented parallel to 4- or $\bar{4}$-axis; **a**, **b** smallest lattice vectors perpendicular to **c**
Trigonal	$3, \bar{3}, 32, 3m, \bar{3}m$		
Rhombohedral setting		$a = b = c,$ $\alpha = \beta = \gamma \neq 90°$ and $< 120°$	**a**, **b**, **c** selected to be the smallest non-coplanar lattice vectors related by 3- or $\bar{3}$-axis
Hexagonal setting		$a = b \neq c,$ $\alpha = \beta = 90°$ $\gamma = 120°$	**c** parallel to 3- or $\bar{3}$-axis; **a**, **b** smallest lattice vectors perpendicular to **c**
Hexagonal	$6, \bar{6}, 6/m, 622,$ $6mm, \bar{6}m2,$ $6/mmm$	$a = b \neq c$ $\alpha = \beta = 90°$ $\gamma = 120°$	**c** parallel to 6- or $\bar{6}$-axis; **a**, **b** smallest lattice vectors perpendicular to **c**
Cubic	$23, m\bar{3}, \bar{4}32,$ $\bar{4}3m, m\bar{3}m$	$a = b = c$ $\alpha = \beta = \gamma = 90°$	**a**, **b**, **c** each parallel to a 2-axis (23 and $m\bar{3}$), or to a $\bar{4}$-axis ($\bar{4}3m$), or a 4-axis (432, $m\bar{3}m$)

is orthorhombic; if only one angle is a right angle, the crystal system is monoclinic (meaning one inclined axis). The crystal system is triclinic (meaning three inclined axes) if all six parameters $a, b, c, \alpha, \beta,$ and γ take on unrestricted values. It should be emphasized that the crystal system is determined by the characteristic symmetry elements present in the crystal lattice, and not by the unit-cell parameters.

The thirty-two crystal classes (crystallographic point groups) described in Section 9.1.4 can also be classified into the same seven crystal systems, depending on the most convenient coordinate system used to indicate the location and orientation of their characteristic symmetry elements, as shown in Table 9.2.1.

9.2.5 Unit cell transformation

Consider two different choices of unit cells: a F-centered unit cell with axes (a_1, b_1, c_1) and a primitive one with axes (a_2, b_2, c_2), as shown in Fig. 9.2.4. We can write

$$\mathbf{a}_2 = \tfrac{1}{2}\mathbf{a}_1 + \tfrac{1}{2}\mathbf{c}_1 \qquad \mathbf{a}_1 = \mathbf{a}_2 + \mathbf{b}_2 - \mathbf{c}_2$$
$$\mathbf{b}_2 = \tfrac{1}{2}\mathbf{a}_1 + \tfrac{1}{2}\mathbf{b}_1 \quad \text{and} \quad \mathbf{b}_1 = -\mathbf{a}_2 + \mathbf{b}_2 + \mathbf{c}_2$$
$$\mathbf{c}_2 = \tfrac{1}{2}\mathbf{b}_1 + \tfrac{1}{2}\mathbf{c}_1 \qquad \mathbf{c}_1 = \mathbf{a}_2 - \mathbf{b}_2 + \mathbf{c}_2$$

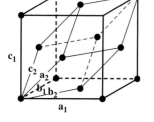

Fig. 9.2.4.
Face-centered and primitive unit cells described by two sets of vectors.

These relationships may be expressed in the form of square matrices:

$$
\begin{array}{c|ccc}
 & a_1 & b_1 & c_1 \\
\hline
h_2 \quad a_2 & 1/2 & 0 & 1/2 \\
k_2 \quad b_2 & 1/2 & 1/2 & 0 \\
l_2 \quad c_2 & 0 & 1/2 & 1/2
\end{array}
\qquad
\begin{array}{c|ccc}
 & a_2 & b_2 & c_2 \\
\hline
h_1 \quad a_1 & 1 & 1 & -1 \\
k_1 \quad b_1 & -1 & 1 & 1 \\
l_1 \quad c_1 & 1 & -1 & 1
\end{array}
$$

The $hk\ell$ indices of a reflection referred to these two unit cells are transformed in the same manner, i.e., $h_2 = 1/2 h_1 + 0 k_1 + 1/2 \ell_1$, etc. The volume ratio is equal to the modulus of the determinant of the transformation matrix; if we take the left matrix, V_2/V_1 is equal to $1/4$.

As a concrete example, consider a rhombohedral lattice and the relationship between the primitive rhombohedral unit cell (in the conventional obverse setting) and the associated triple-sized hexagonal unit cell, as indicated in Fig. 9.2.5.

The triply primitive hexagonal unit cell has lattice points at (0 0 0), (2/3, 1/3, 1/3), and (1/3, 2/3, 2/3). From Fig. 9.2.5, it can be seen that the rhombohedral and hexagonal axes, labeled by subscripts "r" and "h" respectively, are related by vector addition:

$$
\begin{aligned}
\mathbf{a}_r &= 2/3 \mathbf{a}_h + 1/3 \mathbf{b}_h + 1/3 \mathbf{c}_h & \mathbf{a}_h &= \mathbf{a}_r - \mathbf{b} \\
\mathbf{b}_r &= -1/3 \mathbf{a}_h + 1/3 \mathbf{b}_h + 1/3 \mathbf{c}_h \quad \text{and} & \mathbf{b}_h &= \mathbf{b}_r - \mathbf{c}_r \\
\mathbf{c}_r &= -1/3 \mathbf{a}_h - 2/3 \mathbf{b}_h + 1/3 \mathbf{c}_h & \mathbf{c}_h &= \mathbf{a}_r + \mathbf{b}_r + \mathbf{c}_r.
\end{aligned}
$$

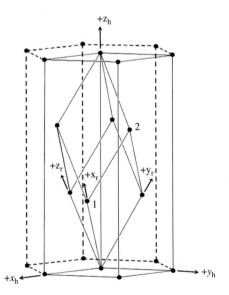
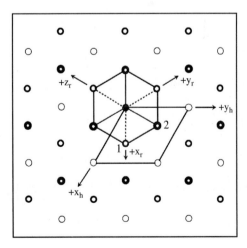

Fig. 9.2.5.
Relationship between rhombohedral (obverse setting) and hexagonal unit cells for a rhombohedral lattice. Note that in the right figure, lattice points at $z = 0, 1/3$, and $2/3$ are differentiated by circles of increasingly darker circumferences, and the lattice point at $z = 1$ is indicated by a filled circle, which obscures the lattice point at the origin.

The pair of matrices can be written down easily, and the volume ratio V_r/V_h worked out to be 1/3.

In the alternative reverse setting (seldom used), the rhombohedral axes are rotated by 60° in an anti-clockwise sense about the c axis of the hexagonal unit cell. The equivalent lattice points are then (0 0 0), (1/3, 2/3, 1/3), and (2/3, 1/3, 2/3). The reader can readily deduce the vectorial and matrix relationships for the two sets of axes and verify that V_r/V_h remains unchanged at 1/3.

9.3 Space groups

An infinitely extended periodic structure can be brought into self-coincidence by point symmetry operations or translations $\ell\mathbf{a}+m\mathbf{b}+n\mathbf{c}$, where $\mathbf{a}, \mathbf{b}, \mathbf{c}$ are the lattice vectors and ℓ, m, n are arbitrary positive and negative integers, including zero. In particular, any symmetry operation that moves a lattice one translation forward (or backward) along an axis leaves it unchanged. This means that, in addition to the point symmetry operations, new types of symmetry operations are also applicable to the periodic arrangement of structural units in a crystal. Such symmetry operations, which are carried out with respect to new symmetry elements called *screw axes* and *glide planes*, are obtained by combining rotations and reflections, respectively, with translations along the lattice directions.

9.3.1   Screw axes and glide planes

The symmetry operation performed by a screw axis n_m is equivalent to a combination of rotation of $2\pi/n$ radians (or $360°/n$) followed by a translation of m/n in the direction of the n-fold axis, where $n = 1, 2, 3, 4$, or 6 is the order of the axis and the subscript m is an integer less than n. There exist totally eleven screw axes: $2_1, 3_1, 3_2, 4_1, 4_2, 4_3, 6_1, 6_2, 6_3, 6_4$, and 6_5 in the crystal lattices. Figure 9.3.1 shows the positions of the equivalent points (also called equipoints for simplicity) around the individual 4-, $\bar{4}$-, 4_1-, 4_2-, and 4_3-axes.

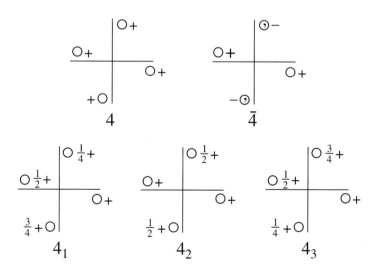

Fig. 9.3.1.
Equipoints related by a 4-, $\bar{4}$-, 4_1-, 4_2-, and 4_3-axis.

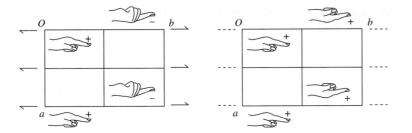

Fig. 9.3.2.
Comparison of the effect of (a) a twofold screw axis and (b) a glide plane on an asymmetric object represented by a left hand. The plus and minus signs indicate that the object lies above and below the plane (001), respectively.

The small circle represents any object (an atom, a group of atoms, a molecule, a group of several molecules, etc.) and the positive sign indicates that it lies above the plane containing the reference axes (x horizontal, y vertical). In the diagram showing a $\bar{4}$-axis, the comma inside the circle emphasizes that the original object has been converted to its mirror image, and the negative sign indicates that it lies below the plane.

The combination of reflection and translation gives a glide plane. If the gliding direction is parallel to the a axis, the symbol for the *axial* glide plane is a and the operation is "reflection in the plane followed by translation parallel to the a axis by $a/2$". Similar axial glide planes b and c have translation components of $b/2$ and $c/2$, respectively.

A screw axis and a glide plane generate equivalent objects in fundamentally different ways. Consider an asymmetric object (such as a chiral molecule, which may be represented by a left hand) located at (x, y, z) close to the origin of the unit cell. The 2_1-axis along b rotates the left hand about the line ($x = 0, z = 0$) to generate the equivalent left hand at $(\bar{x}, 1/2 + y, \bar{z})$, as shown in Fig. 9.3.2(a). In contrast, starting with a left hand at (x, y, z), reflection across the b glide plane at $x = 0$ generates a right hand at $(\bar{x}, 1/2 + y, z)$, as shown in Fig. 9.3.2(b).

If the translation is parallel to a face diagonal by $1/2(a + b)$, $1/2(b + c)$, or $1/2(c + a)$, then the glide plane is called a *diagonal glide* and denoted by n. Less commonly, the n glide operation may involve a translation of $1/2(a + b + c)$ in the tetragonal, rhombohedral, and cubic systems. A third type is the *diamond glide* plane d which has translations of $1/4(a \pm b)$, $1/4(b \pm c)$, or $1/4(c \pm a)$ in the orthogonal and tetragonal systems, and also $1/4(a \pm b \pm c)$ in the cubic system. The recently introduced *double glide* plane has two coexisting $1/2$ translations parallel to it; this type of e glide appears in seven orthorhombic A-, C-, and F-centered space groups, five tetragonal I-centered space groups, and five cubic I- and F-centered space groups.

9.3.2 Graphic symbols for symmetry elements

Different graphic symbols are used to label symmetry axes normal to the plane of projection, symmetry axes in the plane of the figure, symmetry axes inclined to the plane of projection, symmetry planes normal to the plane of projection, and symmetry planes parallel to the plane of projection. These symbols are illustrated in Tables 9.3.1 to 9.3.5, respectively.

Symmetry in Chemistry

Table 9.3.1. Symbols of symmetry axes normal to the plane of projection and inversion center in the plane of the figure

Symmetry element (screw rotation is right-handed)	Graphic symbol	Translation in units of the shortest lattice vector parallel to the axis	Printed symbol
Identity	none	none	1
Twofold rotation axis		none	2
Twofold screw axis: "2 sub 1"		1/2	2_1
Threefold rotation axis		none	3
Threefold screw axis: "3 sub 1"		1/3	3_1
Threefold screw axis: "3 sub 2"		2/3	3_2
Fourfold rotation axis		none	4
Fourfold screw axis: "4 sub 1"		1/4	4_1
Fourfold screw axis: "4 sub 2"		1/2	4_2
Fourfold screw axis: "4 sub 3"		3/3	4_3
Sixfold rotation axis		none	6
Sixfold screw axis: "6 sub 1"		1/6	6_1
Sixfold screw axis: "6 sub 2"		1/3	6_2
Sixfold screw axis: "6 sub 3"		1/2	6_3
Sixfold screw axis: "6 sub 4"		2/3	6_4
Sixfold screw axis: "6 sub 5"		5/6	6_5
Center of symmetry, inversion center: "bar 1"		none	$\bar{1}$
Inversion axis: "bar 3"		none	$\bar{3}$
Inversion axis: "bar 4"		none	$\bar{4}$
Inversion axis: "bar 6"		none	$\bar{6}$
Twofold rotation axis with center of symmetry		none	$2/m$
Twofold screw axis with center of symmetry		1/2	$2_1/m$
Fourfold rotation axis with center of symmetry		none	$4/m$
"4 sub 2" screw axis with center of symmetry		1/2	$4_2/m$
Sixfold rotation axis with center of symmetry		none	$6/m$
"6 sub 3" screw axis with center of symmetry		1/2	$6_3/m$

Symmetry in Crystals

Table 9.3.2. Symmetry axes parallel to the plane of projection

Symmetry axis	Graphic symbol	Symmetry axis	Graphic symbol
Twofold rotation 2		Fourfold screw 4_1	
Twofold screw 2_1		Fourfold screw 4_2	
Fourfold rotation 4		Fourfold screw 4_3	
Inversion axis $\bar{4}$			

Table 9.3.3. Symmetry axes inclined to the plane of projection (cubic space groups only)

Symmetry axis	Graphic symbol	Symmetry axis	Graphic symbol
Twofold rotation 2		Threefold screw 3_1	
Twofold screw 2_1		Threefold screw 3_2	
Threefold rotation 3		Inversion axis $\bar{3}$	

Table 9.3.4. Symmetry planes normal to the plane of projection

Symmetry plane or symmetry line	Graphic symbol	Glide vector in units of lattice translation vectors parallel and normal to the projection plane	Printed symbol
Reflection plane, mirror plane	——	None	m
"Axial" glide plane	- - - -	1/2 lattice vector along line in projection plane	a, b, or c
"Axial" glide plane	········	1/2 lattice vector normal to projection plane	a, b, or c
"Digonal" glide plane	—·—·—	One glide vector with two components: 1/2 along line parallel to projection plane, and 1/2 normal to projection plane	n
"Diamond" glide plane* (pair of planes; in centred cells only)	—·◄·— / —·►·—	1/4 along line parallel to projection plane, combined with 1/4 normal to projection plane (arrow indicates direction parallel to the projection plane for which the normal component is positive)	d
"Double" glide plane (pair of planes; in centered cells only)	—··—··—	Two coexisting glides of 1/2 (related by a centering translation) parallel to and perpendicular to the projection plane	e

* Glide planes d occur only in orthorhombic F space groups, in tetragonal I space groups, and in cubic I and F space groups. They always occur in pairs with alternating glide vectors, for instance $\frac{1}{4}(\mathbf{a}+\mathbf{b})$ and $\frac{1}{4}(\mathbf{a}-\mathbf{b})$. The second power of a glide reflection d is a centering vector.

Table 9.3.5. Symmetry planes parallel to the plane of projection

Symmetry plane	Graphic symbol*	Glide vector in units of lattice translation vectors parallel to the projection plane	Printed symbol
Reflection plane, mirror plane		None	m
"Axial" glide plane		1/2 lattice vector in the direction of the axis	a, b, or c
"Diagonal" glide plane		Latttice vector in the direction of the arrow (glide vector with two 1/2 components)	n
"Diamond" glide plane		1/4 lattice vector in the direction of the arrow (glide vector with two or three components)	d
"Double" glide plane		A pair of 1/2 lattice vectors in the direction of the two arrows	e

* The symbols are given at the upper corner of the space group diagrams. A fraction h attached to a symbol indicates two symmetry planes with "height" h and $h + 1/2$ above the plane of projection; e.g. 1/8 stands for $h = 1/8$ and 5/8. No fraction means $h = 0$ and 1/2.

9.3.3 Hermann–Mauguin space-group symbols

The term *space-group* indicates a self-consistent set of symmetry operations, constituting a group in the mathematical sense, that bring an infinitely extended, three-dimensional periodic structure into self-coincidence. Between the years 1885 and 1894, Fedorow, Schönflies, and Barlow independently showed that there are 230 space groups. The Hermann–Mauguin notation for space groups includes a set of symbols sufficient to completely specify all the symmetry elements and their orientation with respect to the unit-cell axes. It consists of two parts. The first part is a symbol for the type of lattice: P, C (or A or B), I, F, or R (formerly H was also sometimes used for the hexagonal lattice). The second part is a set of symbols for symmetry elements sufficient to fully reflect the space group symmetry without listing other symmetry elements that must necessarily be present. These are written in an order that indicates the orientation of the symmetry elements, in the same manner as for the point groups. In this system a symmetry element associated with a plane has the orientation of the normal to the plane. The symbols 1 and $\bar{1}$ are usually omitted unless they are the only ones or unless they are needed as spacers. When two symmetry elements have the same orientation and both are necessary, their symbols are separated with a slash, for example, $2/m$ indicates a twofold axis with a mirror plane normal to it.

9.3.4 International Tables for Crystallography

The 230 space groups are described in *Volume A: Space-Group Symmetry* of the *International Tables for Crystallography*. Detailed graphic projections along the axial directions showing the symmetry elements present in the unit cell, a list of general and special equivalent positions designated by Wyckoff letters (or positions), multiciplicities, and the symmetry of each local site are given for each space group. This volume is an expanded and updated version of *Volume I: Symmetry Groups* of the *International Tables for X-Ray Crystallography*, which contains a simplified presentation. The 230 space groups are listed in Table 9.3.6.

9.3.5 Coordinates of equipoints

In crystallography the coordinates of the equipoints are each expressed as a fraction of the distance along the corresponding unit-cell edge, and they are therefore actually *fractional coordinates*. For any space group referred to a specific choice of origin (generally at $\bar{1}$, if possible), the coordinates of all equipoints can be derived. Table 9.3.7 shows the derivation of points from those of one prototype with fractional coordinates (x, y, z) by carrying out symmetry operations with respect to various symmetry elements. Note that the coordinates (x, y, z) of a molecule composed of n atoms represent the set of $3n$ atomic coordinates $(x_i, y_i, z_i; i = 1, 2, \ldots, n)$.

Figure 9.3.3 illustrates the derivation of coordinates for equipoints related by a 4-, 3-, and 6-axis along c through the origin. Coordinates related by a $\bar{4}$-, $\bar{3}$-, and $\bar{6}$-axis can be readily written down using this information. If a different choice of origin is made, a different set of coordinates will be obtained.

Given the fractional coordinates of all atoms in the unit cell, bond lengths ℓ_{ij} and bond angles θ_{ijk} can be calculated from their differences ($\Delta x_{ij} = x_i - x_j, \Delta y_{ij} = y_i - y_j, \Delta z_{ij} = z_i - z_j$) and the lattice parameters. In the general triclinic case, the interatomic distance between two atoms labeled 1 and 2 is given by

$$\ell_{12}^2 = (a\Delta x_{12})^2 + (b\Delta y_{12})^2 + (c\Delta z_{12})^2 - 2ab\Delta x \Delta y \cos \gamma \\ - 2ac\Delta x \Delta z \cos \beta - 2bc \Delta y \Delta z \cos \alpha.$$

The θ_{123} angle at atom 2, which forms bonds with 1 and 3, is calculated from the cosine law

$$\cos \theta_{123} = (\ell_{12}^2 + \ell_{13}^2 - \ell_{23}^2)/2\ell_{12}\ell_{13}.$$

For a linear sequence of four bonded atoms A–B–C–D, the torsion angle τ is defined as the angle between the projections of the bond B–A and C–D when viewed along the direction of B–C.

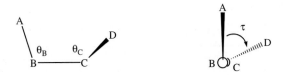

The torsion angle τ an be calculated from the equation

$$\cos \tau = (\mathbf{AB} \times \mathbf{BC}) \cdot (\mathbf{BC} \times \mathbf{CD})/AB(BC)^2 CD \sin \theta_B \sin \theta_C,$$

where **AB** is the vector from A to B, AB is magnitude of **AB**, the symbols · and × indicate scalar and vector products, respectively, and θ_B is the bond

Table 9.3.6. The 230 three-dimensional space groups arranged by crystal systems and classes*

No.	Crystal system and class	Space groups
		Triclinic
1	1 (C_1)	P_1
2	$\bar{1}$ (C_i)	$P\bar{1}$
		Monoclinic
3–5	2 (C_2)	$P2, P2_1, C2$
6–9	$m(C_s)$	Pm, Pc, Cm, Cc
10–15	$2/m(C_{2h})$	$P2/m, P2_1/m, C2/m, P2/c,$ **$P2_1/c$**, $C2/c$
		Orthorhombic
16–24	222 (D_2)	$P222,$ **$P222_1, P2_12_12, P2_12_12_1, C222_1,$** $C222, F222, I222, I2_12_12_1$
25–46	$mm2$ (C_{2v})	$Pmm2, Pmc2_1, Pcc2, Pma2_1, Pca2_1, Pnc2_1, Pmn2_1, Pba2, Pna2_1, Pnn2, Cmm2, Cmc2_1, Ccc2, Amm2, Abm2, Ama2, Aba2, Fmm2,$ **$Fdd2$**, $Imm2, Iba2, Ima2$
47–74	mmm (D_{2h})	$Pmmm,$ **$Pnnn$**, $Pccm,$ **$Pban$**, $Pmma,$ **$Pnna$**, $Pmna, Pcca, Pbam,$ **$Pccn$**, $Pbcm, Pnnm, Pmmn,$ **$Pbcn, Pbca$**, $Pnma, Cmcm, Cmca, Cmmm, Cccm, Cmma,$ **$Ccca$**, $Fmmm,$ **$Fddd$**, $Immm, Ibam,$ **$Ibca$**, $Imma$
		Tetragonal
75–80	4 (C_4)	$P4,$ **$P4_1$**, $P4_2,$ **$P4_3$**, $I4, I4_1$
81–82	$\bar{4}$ (S_4)	$P\bar{4}, I\bar{4}$
83–88	$4/m(C_{4h})$	$P4/m, P4_2/m,$ **$P4/n, P4_2/n$**, $I4/m,$ **$I4_1/a$**
89–98	422 (D_4)	$P422,$ **$P42_12, P4_122, P4_12_12, P4_222, P4_22_12, P4_322, P4_32_12$**, $I422, I4_122$
99–110	$4mm$ (C_{4v})	$P4mm, P4bm, P4_2cm, P4_2nm, P4cc, P4nc, P4_2mc, P4_2bc, I4mm, I4cm, I4_1md,$ **$I4_1cd$**
111–122	$\bar{4}2m(D_{2d})$	$P\bar{4}2m, P\bar{4}2c, P\bar{4}2_1m,$ **$P\bar{4}2_1c$**, $P\bar{4}m2, P\bar{4}c2, P\bar{4}b2, P\bar{4}n2, I\bar{4}m2, I\bar{4}c2, I\bar{4}2m, I\bar{4}2d$
123–142	$4/mmm$ (D_{4h})	$P4/mmm, P4/mcc,$ **$P4/nbm, P4/nnc$**, $P4/mbm, P4/mnc,$ **$P4/nmm, P4/ncc$**, $P4_2/mmc, P4_2/mcm,$ **$P4_2/nbc, P4_2/nnm$**, $P4_2/mbc, P4_2/mnm,$ **$P4_2/nmc, P4_2/ncm$**, $I4/mmm, I4/mcm,$ **$I4_1/amd, I4_1/acd$**
		Trigonal
143–146	3 (C_3)	$P3, P3_1, P3_2, R3$
147–148	$\bar{3}(C_{3i})$	$P\bar{3}, R\bar{3}$
149–155	32 (D_3)	$P312, P321,$ **$P3_112, P3_121, P3_212, P3_221$**, $R32$
156–161	$3m(C_{3v})$	$P3m1, P31m, P3c1, P31c, R3m, R3c$
162–167	$\bar{3}m(D_{3d})$	$P\bar{3}1m, P\bar{3}1c, P\bar{3}m1, P\bar{3}c1, R\bar{3}m, R\bar{3}c$
		Hexagonal
168–173	6 (C_6)	$P6,$ **$P6_1, P6_5, P6_2, P6_4$**, $P6_3$
174	$\bar{6}$ (C_{3h})	$P\bar{6}$
175–176	$6/m(C_{6h})$	$P6/m, P6_3/m$
177–182	622 (D_6)	$P622,$ **$P6_122, P6_522, P6_222, P6_422$**, $P6_322$
183–186	$6mm$ (C_{6v})	$P6mm, P6cc, P6_3cm, P6_3mc$
187–190	$\bar{6}m2$ (D_{3h})	$P\bar{6}m2, P\bar{6}c2, P\bar{6}2m, P\bar{6}2c$
191–194	$6/mmm(D_{6h})$	$P6/mmm, P6/mcc, P6_3/mcm, P6_3/mmc$
		Cubic
195–199	23 (T)	$P23, F23, I23,$ **$P2_13$**, $I2_13$
200–206	$m\bar{3}$ (T_h)	$Pm\bar{3},$ **$Pn\bar{3}$**, $Fm\bar{3},$ **$Fd\bar{3}$**, $Im\bar{3}, Pa\bar{3}, Ia\bar{3}$
207–214	432 (O)	$P432, P4_232, F432,$ **$F4_132$**, $I432,$ **$P4_332, P4_132$**, $I4_132$
215–220	$\bar{4}3m(T_d)$	$P\bar{4}3m, F\bar{4}3m, I\bar{4}3m, P\bar{4}3n, F\bar{4}3c,$ **$I\bar{4}3d$**
221–230	$m\bar{3}m(O_h)$	$Pm\bar{3}m,$ **$Pn\bar{3}n, Pm\bar{3}n, Pn\bar{3}m$**, $Fm\bar{3}m, Fm\bar{3}c,$ **$Fd\bar{3}m, Fd\bar{3}c$**, $Im\bar{3}m, Ia\bar{3}d$

* The 230 space groups include 11 enantiomorphous pairs: $P3_1(P3_2), P3_112 (P3_212), P3_121 (P3_221), P4_1(P4_3), P4_122 (P4_322), P4_12_12 (P4_32_12), P6_1(P6_5), P6_2(P6_4), P6_122 (P6_522), P6_222 (P6_422),$ and $P4_132 (P4_332)$. If the (+)-isomer of an optically active molecule crystallizes in an enantiomorphous space group, the (−)-isomer will crystallize in the related one. Although both members of an enantiomorphous pair are distinguishable by X-ray diffraction (using anomalous dispersion), they can be counted together for classification purposes if the absolute chirality sense (handedness) is unknown or irrelevant, leading to 219 "distinct" space groups. However, once the chirality sense of the asymmetric unit is specified, repetition of that unit in one enantiomorphous space group is quite distinct from repetition in the other. It is therefore virtually impossible for a given chiral compound to crystallize in both enantiomorphous space groups because the resulting packing arrangements (and energies) would be completely different.

Space groups (or enantiomorphous pairs) that are uniquely determined from the symmetry of the diffraction pattern and systematic absences are shown in **boldface** type.

Recent adoption of the double glide plane e as a symmetry element leads to changes in the international symbols for five A- and C-centered orthorhombic space groups: No. 39, $Abm2 \rightarrow Aem2$; No. 41, $Aba2 \rightarrow Aea2$; No. 64, $Cmca \rightarrow Cmce$; No. 67, $Cmma \rightarrow Cmme$; No. 68, **$Ccca \rightarrow Ccce$**.

Table 9.3.7. Derivation of atomic coordinates

Symmetry element	Fractional coordinates
Center at origin (0, 0, 0)	$x, y, z; \bar{x}, \bar{y}, \bar{z};$ or $\pm(x, y, z)$
Center at origin ($1/4$, 0, 0)	$x, y, z; 1/2 - x, \bar{y}, \bar{z};$
Mirror plane at $y = 0$	$x, y, z; x, \bar{y}, z$
Mirror plane at $y = 1/4$	$x, y, z; x, 1/2 - y, z$
c glide plane at $y = 0$	$x, y, z; x, \bar{y}, 1/2 + z$
c glide plane at $y = 1/4$	$x, y, z; x, 1/2 - y, 1/2 + z$
n glide plane at $y = 0$	$x, y, z; 1/2 + x, \bar{y}, 1/2 + z$
n glide plane at $y = 1/4$	$x, y, z; 1/2 + x, 1/2 - y, 1/2 + z$
d glide plane at $x = 1/8$	$x, y, z; 1/4 - x, 1/4 + y, 1/4 + z$
2-axis along line 0, 0, z (i.e. c)	$x, y, z; \bar{x}, \bar{y}, z$
2-axis along line 0, $1/4$, z	$x, y, z; \bar{x}, 1/2 - y, z$
2_1-axis along c	$x, y, z; \bar{x}, \bar{y}, 1/2 + z$
2_1-axis along line $1/4$, 0, z	$x, y, z; 1/2 - x, \bar{y}, 1/2 + z$
4-axis along c	$x, y, z; y, \bar{x}, z; \bar{x}, \bar{y}, z; \bar{y}, x, z$
$\bar{4}$-axis along c with $\bar{1}$ at origin	$x, y, z; y, \bar{x}, \bar{z}; \bar{x}, \bar{y}, z; \bar{y}, x, \bar{z}$
$\bar{4}$-axis along c with $\bar{1}$ at (0, 0, $1/4$)	$x, y, z; y, \bar{x}, 1/2 - z; \bar{x}, \bar{y}, z; \bar{y}, x, 1/2 - z$
4_1-axis along line $1/4$, $1/4$, z with origin on 2	$x, y, z; y, 1/2 - x, 1/4 + z; 1/2 - x, 1/2 - y, 1/2 + z; 1/2 - y, x, 3/4 + z$
3-axis along c	$x, y, z; \bar{y}, x - y, z; y - x, \bar{x}, z$
$\bar{3}$-axis along c with $\bar{1}$ at origin	$\pm (x, y, z; \bar{y}, x - y, z; y - x, \bar{x}, z)$
6-axis along c	$x, y, z; x - y, x, z; \bar{y}, x - y, z; \bar{x},\bar{y}, z; y - x, \bar{x}, z; y, y - x, z$
3-axis along [111] body diagonal in cubic unit cell	$x, y, z; z, x, y; y, z, x$
C-centering (on ab face)	$x, y, z; x, 1/2 + y, 1/2 + z$ or $(0, 0, 0; 1/2, 1/2, 0)$ +
I-centering	$x, y, z; 1/2 + x, 1/2 + y, 1/2 + z;$ or $(0, 0, 0; 1/2, 1/2, 1/2)$+
F-centering	$(0, 0, 0; 1/2, 1/2, 0; 0, 1/2, 1/2; 1/2, 0, 1/2)$+
R-centering,* obverse setting	$x, y, z; 2/3 + x, 1/3 + y, 1/3 + z; 1/3 + x, 2/3 + y, 2/3 + z$ or $(0, 0, 0; 2/3, 1/3, 1/3; 1/3, 2/3, 2/3)$ +
R-centering,* reverse setting	$x, y, z; 1/3 + x, 2/3 + y, 1/3 + z; 2/3 + x, 1/3 + y, 2/3 + z$ or $(0, 0, 0; 1/3, 2/3, 1/3; 2/3, 1/3, 2/3)$ +

* Rhombohedral unit cell based on hexagonal lattice.

angle at atom B. Note that this formula is valid in any coordinate system, but the usual textbook expressions for the scalar and vector products in terms of vector components holds only in a Cartesian coordinate system. The torsion angle τ is not well defined when one or both of θ_B and θ_C is/are 0° or 180°. A positive value of τ indicates that the A–B–C–D link corresponds to the sense

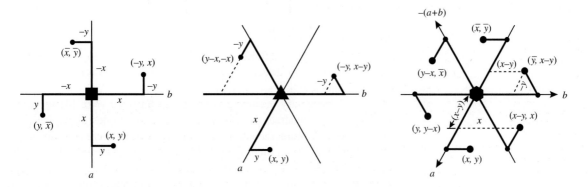

Fig. 9.3.3.
Derivation of coordinates of equipoints related by a 3-, 4-, or 6-axis.

Table 9.3.8. The order of positions in the international symbols for point groups and space groups

System	Position in international symbol		
	1st	2nd	3rd
Triclinic	Only one symbol is used		
Monoclinic	1st setting: 2, 2_1, or $\bar{2}$ along unique **c** axis*		
	2nd setting: 2, 2_1, or $\bar{2}$ along unique **b** axis (conventional)		
Orthorhombic	2, 2_1, or $\bar{2}$ along **a**	2, 2_1, or $\bar{2}$ along **b**	2, 2_1, or $\bar{2}$ along **c**
Tetragonal	4, 4_1, 4_2, 4_3, or $\bar{4}$ along **c**	2, 2_1, or $\bar{2}$ along **a** and **b**	2 or $\bar{2}$ along [110] and [1$\bar{1}$0]
Trigonal	3, 3_1, 3_2, or $\bar{3}$ along **c**	2 or $\bar{2}$ along **a**, **b**, and [110]	2 or $\bar{2}$ perpendicular to **a**, **b**, and [110]
Hexagonal	6, 6_1, 6_2, 6_3, 6_4, 6_5, or $\bar{6}$ along **c**	2 or $\bar{2}$ along **a**, **b**, and [110]	2 or $\bar{2}$ perpendicular to **a**, **b**, and [110]
Cubic	4, 4_1, 4_2, 4_3, $\bar{4}$, 2, 2_1, or $\bar{2}$ along **a**, **b**, and **c**	3 or $\bar{3}$ along <111>	2 or $\bar{2}$ along <110>

* Note that a $\bar{2}$-axis is equivalent to a mirror plane m normal to it; for example, $\bar{2}$ along **a** means m perpendicular to **a**.

of a right-handed screw (or helix). The sign of τ is unaffected by rotation or translation, but is reversed by reflection or inversion.

9.3.6 Space group diagrams

Table 9.3.8 summarizes the order of positions in the point group or space group symbols; note that a $\bar{2}$-axis is equivalent to a mirror plane m perpendicular to it.

For example, in the orthorhombic system, $mm2$ refers to mirror planes *perpendicular to* **a** and **b**, and a twofold axis *along* **c**. In the hexagonal system, $\bar{6}2m$ means a $\bar{6}$-axis along **c**, twofold axes *along* **a**, **b** and **a** + **b** (i.e., [110]), and mirror planes *perpendicular to* **a**, **b** and [110]. In some other books and tables, the equivalent alternative $\bar{6}m2$ may be used in place of $\bar{6}2m$. In the cubic system, $\bar{4}3m$ means three $\bar{4}$-axes *along* the equivalent axes of a cube, four 3-axes *along* the body diagonals <111>, and six mirror planes *perpendicular to* the face diagonals <110>. In $m\bar{3}$ the first position indicates mirror planes *perpendicular to* 0the cubic axes, and the second position means $\bar{3}$-axes *along* the body diagonals. The third position is left blank as there is no symmetry element in the <110> directions. The older symbol for $m\bar{3}$ is $m3$, as the 3- and $\bar{3}$-axis are coincident.

The most frequently occurring space group $P12_1/c1$ (No. 14) serves as a good example to illustrate the use of the space group tables. In the Hermann–Mauguin nomenclature, the lattice symbol is given first; here P means that the lattice is primitive (i.e., there is no lattice centering). The second symbol (1) indicates that there is no symmetry with respect to a, the next symbol ($2_1/c$) shows that both 2_1 and c refer to the b axis (the unique symmetry axis in the conventional second setting), and the last symbol (1) indicates that there is no symmetry with respect to c. The symmetry elements $2_1/c$ mean that the

Laue symmetry is $2/m$ (disregarding the translational components of the 2_1-axis and the c glide), so the space group belongs to the monoclinic system. The simplified symbol $P2_1/c$ normally replaces the full symbol $P12_1/c1$, and the old Schönflies symbol C_{2h}^5 (the fifth space group derived from the point group C_{2h}) is seldom used.

Space group $P2_1/a$ often appears in the older literature in place of $P2_1/c$; in such case the unique axis is still b, but the glide plane is a. If we make a different choice of the axes normal to the 2_1-axis, the glide plane might be changed to a diagonal glide n; and the space group symbol becomes $P2_1/n$. (This choice is sometimes made to make the β angle close to $90°$.) Furthermore, if c were chosen as the unique axis (as in the unconventional first setting), the space group symbol would become $P2_1/a$ or $P2_1/b$. Thus the same space group may be designated by several different symbols depending on the choice of axes and their labeling.

In the following discussion, for simplicity we make use of the pair of symmetry diagrams that constitute a representation of space group $P2_1/c$ (No. 14) in *International Tables for X-Ray Crystallography Volume I*, as displayed in Fig. 9.3.4.

In both diagrams, the origin is located at the upper left corner of the unit cell, with a pointing downward, b horizontally to the right, and c toward the viewer. The right diagram (b) shows the symmetry elements and their location using graphic symbols. Each colinear pair of half arrows represents a 2_1-axis parallel to b, and the label $1/4$ indicates that it lies at a height of $z = 1/4$. The dotted line indicates the c glide with glide component coming out of the plane of the diagram. The small open circle represents a center of symmetry (inversion center) that must be present in $P2_1/c$ but is not explicitly indicated in the space group symbol. Note that all three types of symmetry elements necessarily occur at $1/2$ unit intervals along the lattice translations.

The left diagram (a) shows the locations and coordinates of the general equivalent positions in the space group. To see how this diagram can be derived from the symmetry diagram at the right, we insert a general point labeled **1** in the unit cell and represent it by a large open circle. This position (x, y, z) has an arbitrary fractional height z above the ab plane, which is conventionally designated by the symbol $+$ beside it. For the 2_1 screw axis at $x = 0$, $z = 1/4$, rotation first carries position **1** to a point $(\bar{x}, y, 1/2 - z)$ outside the unit cell at the top, and the

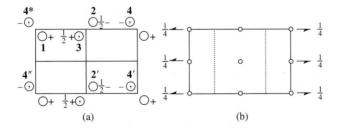

Fig. 9.3.4.
Diagrams showing (a) the general equivalent positions and (b) the symmetry elements in space group $P2_1/c$ as presented in *International Tables for X-Ray Crystallography Volume I*. The bold numerals in the left figure are added for the purpose of discussion.

subsequent $b/2$ translation carries the latter to position **2** marked − (meaning a height of $-z$), and its coordinates are $(\bar{x}, 1/2 + y, 1/2 - z)$. For the c glide at $y = 1/4$, reflection across the plane first brings **1** to the point $(x, 1/2 - y, z)$, and the subsequent $c/2$ translation carries the latter into position **3** marked $1/2+$ (meaning a height of $1/2 + z$), and its coordinates are $(x, 1/2 - y, 1/2 + z)$. The comma inside the circle means that the position, together with its environment, is a mirror image of the original position and its environment. Starting with position **2**, the glide plane at $y = 3/4$ generates the position at $(\bar{x}, 1 - y, \bar{z})$, and we label it **4**. The positions **2′** and **4′** are derived from **2** and **4**, respectively, by unit translation along a. Note that positions **1** and **4′** are related by the inversion center at $(1/2, 1/2, 1/2)$, positions **2** and **3** are related by the inversion center at $(0, 1/2, 1/2)$, and positions **3** and **2′** are related by the inversion center at $(1/2, 1/2, 1/2)$. The point **4*** at $(\bar{x}, \bar{y}, \bar{z})$ is usually chosen instead of **4**, since it is related to **1** by the inversion center at the origin. Equipoints **1**, **4***, **2**, and **3** constitute the set of general equivalent positions in space group $P2_1/c$. Further action of the symmetry operations on them produces no new position, so these four positions reflect the full symmetry of the space group. They are denoted by the Wyckoff notation $4(e)$ in the first row of Table 9.3.9.

The Wyckoff notation for a set of equivalent positions consists of two parts: (i) the multiplicity M, which is the number of equivalent positions per unit cell, and (ii) an italicized small letter a starting at the bottom of the list and moving upward in alphabetical order. For a primitive unit cell, M is equal to the order of the point group from which the space group is derived; for centered cells, M is the product of the order of the point group and the number of lattice points per unit cell.

Note that in Fig. 9.3.4, diagram (b) can be derived from diagram (a), and vice versa, and either one is sufficient to represent the full symmetry of the space group.

Consider the general equivalent positions of space group $P2_1/c$ as shown in Fig. 9.3.4(a). Let position **1** approach the origin of the unit cell; in other words, let the coordinates $x \to 0$, $y \to 0$, and $z \to 0$. As this happens, position **4″** also approaches the origin, while both **2** and **3** simultaneously approach the center of inversion at $(0, 1/2, 1/2)$. When $x = 0$, $y = 0$, and $z = 0$, **1** and **4*** coalesce into one, and **2** and **3** likewise become the same position. There remain only two equivalent positions: $(0, 0, 0)$ and $(0, 1/2, 1/2)$ that occupy sites of symmetry $\bar{1}$, and they constitute the *special* equivalent position $2(a)$, which is designated as Wyckoff position $2(a)$. Other sets of special equivalent positions of site symmetry $\bar{1}$ are obtained by setting $x = 1/2, y = 0, z = 0; x = 0, y = 0,$

Table 9.3.9. General and special positions of space group $P2_1/c$.

No. of positions	Wyckoff letter	Site symmetry	Coordinates of equivalent positions
4	e	1	$x, y, z; \bar{x}, \bar{y}, \bar{z}; \bar{x}, 1/2 + y, 1/2 - z; x, 1/2 - y, 1/2 + z$
2	d	$\bar{1}$	$1/2, 0, 1/2; 1/2, 1/2, 0$
2	c	$\bar{1}$	$0, 0, 1/2; 0, 1/2, 0$
2	b	$\bar{1}$	$1/2, 0, 0; 1/2, 1/2, 1/2$
2	a	$\bar{1}$	$0, 0, 0; 0, 1/2, 1/2$

$z = 1/2$; $x = 1/2$, $y = 0$, $z = 1/2$, giving rise to Wyckoff position $2(b)$, $2(c)$, and $2(d)$, respectively (see Table 9.3.9). The multiplicity of a special position is always a divisor of that of the general position.

To describe the contents of a unit cell, it is sufficient to specify the coordinates of only one atom in each equivalent set of atoms, since the other atomic positions in the set are readily deduced from space group symmetry. The collection of symmetry-independent atoms in the unit cell is called the *asymmetric unit* of the crystal structure. In the *International Tables*, a portion of the unit cell (and hence its contents) is designated as the asymmetric unit. For instance, in space group $P2_1/c$, a quarter of the unit cell within the boundaries $0 \le x \le 1, 0 \le y \le 1/4$, and $0 \le z \le 1$ constitutes the asymmetric unit. Note that the asymmetric unit may be chosen in different ways; in practice, it is preferable to choose independent atoms that are connected to form a complete molecule or a molecular fragment. It is also advisable, whenever possible, to take atoms whose fractional coordinates are positive and lie within or close to the octant $0 \le x \le 1/2, 0 \le y \le 1/2$, and $0 \le z \le 1/2$. Note also that if a molecule constitutes the asymmetric unit, its component atoms may be related by non-crystallographic symmetry. In other words, the symmetry of the site at which the molecule is located may be a subgroup of the idealized molecular point group.

9.3.7 Information on some commonly occurring space groups

Apart from $P2_1/c$, other commonly occurring space groups are $C2/c$ (No. 15), $P2_12_12_1$ (No. 19), $Pbca$ (No. 61), and $Pnma$ (No. 62); the symmetry diagrams and equivalent positions of these space groups are displayed in Table 9.3.10. It should be noted that each space group can be represented in three ways: a diagram showing the location of the general equivalent positions, a diagram showing the location of the symmetry elements, and a list of coordinates of the general positions. The reader should verify that, given any one of these, the other two can be readily deduced.

9.3.8 Using the International Tables

Detailed information on all 230 space groups are given in *Volume A: Space-Group Symmetry* of the *International Tables for Crystallography*. The tabulation for space group $C2/c$ is reproduced in the left column of Table 9.3.11, and explanation of various features are displayed in the right column. This may be compared with the information for $C2/c$ given in Table 9.3.10.

9.4 Determination of space groups

9.4.1 Friedel's law

The intensity of a reflection $(hk\ell)$ is proportional to the square of the structural factor $F(hk\ell)$, which is given by

$$F(hk\ell) = \sum_j f_j \cos 2\pi(hx_j + ky_j + \ell z_j) + i \sum_j f_j \sin 2\pi(hx_j + ky_j + \ell z_j),$$

Table 9.3.10. Symmetry diagrams and equivalent positions of some commonly occurring space groups

Monoclinic 2/m   No. 15 $C2/c$ (C_{2h}^6)

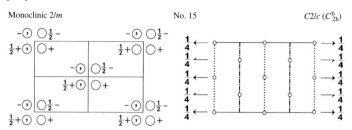

No. of positions	Wyckoff notation	Site symmetry	Coordinates of equivalent positions $(0,0,0; 1/2,1/2,0)+$
8	f	1	$x, y, z; \bar{x}, \bar{y}, \bar{z}; \bar{x}, y, 1/2 - z; x, \bar{y}, 1/2 + z$
4	e	2	$0, y, 1/4; 0, \bar{y}, 3/4$
4	d	$\bar{1}$	$1/4, 1/4, 1/2; 3/4, 1/4, 0$
4	c	$\bar{1}$	$1/4, 1/4, 0; 3/4, 1/4, 1/2$
4	b	$\bar{1}$	$0, 1/2, 0; 0, 1/2, 1/2$
4	a	$\bar{1}$	$0, 0, 0; 0, 0, 1/2$

Orthorhombic 222   No. 19 $P2_12_12_1$ (D_2^4)

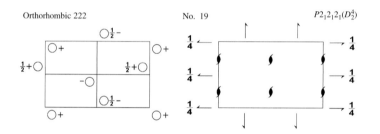

No. of positions	Wyckoff notation	Site symmetry	Coordinates of equivalent positions
4	a	1	$x, y, z; 1/2 - x, \bar{y}, 1/2 + z; 1/2 + x, 1/2 - y, \bar{z}; \bar{x}, 1/2 + y, 1/2 - z$

Orthorhombic mmm No. 61 $Pbca$ (D_{2h}^{15})

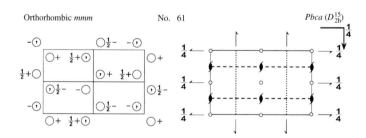

No. of positions	Wyckoff notation	Site symmetry	Coordinates of equivalent positions
8	c	1	$x, y, z; 1/2 + x, 1/2 - y, \bar{z}; \bar{x}, 1/2 + y, 1/2 - z; 1/2 - x,$ $\bar{y}, 1/2 + z; \bar{x}, \bar{y}, \bar{z}; 1/2 - x, 1/2 + y, z; x, 1/2 - y, 1/2 + z;$ $1/2 + x, y, 1/2 - z$
4	b	$\bar{1}$	$0, 0, 1/2; 1/2, 1/2, 1/2; 0, 1/2, 0; 1/2, 0, 0$
4	a	$\bar{1}$	$0, 0, 0; 1/2, 1/2, 0; 0, 1/2, 1/2; 1/2, 0, 1/2$

Table 9.3.10. *Continued*

Orthorhombic *mmm*   No. 62   Pbca (D_{2h}^{16})

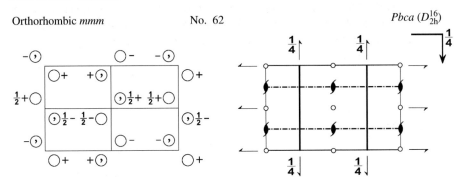

No. of positions	Wyckoff notation	Site symmetry	Coordinates of equivalent positions
8	d	1	x, y, z; $1/2+x, 1/2-y, 1/2-z$; $\bar{x}, 1/2+y, \bar{z}$; $1/2-x, \bar{y}, 1/2+z$; $\bar{x}, \bar{y}, \bar{z}$; $1/2-x, 1/2+y, 1/2+z$; $x, 1/2-y, z$; $1/2+x, y, 1/2-z$
4	c	m	$x, 1/4, z$; $\bar{x}, 3/4, \bar{z}$; $1/2-x, 3/4, 1/2+z$; $1/2+x, 1/4, 1/2-z$
4	b	$\bar{1}$	$0,0,1/2$; $0,1/2,1/2$; $1/2,0,0$; $1/2,1/2,0$
4	a	$\bar{1}$	$0,0,0$; $0,1/2,0$; $1/2,0,1/2$; $1/2,1/2,1/2$

where f_j is the atomic scattering factor of atom j, and the summation is over all atoms in the unit cell. For simplicity, we may write $|F(hk\ell)|^2 = A^2 + B^2$.

$F(hk\ell)$ is in general a complex quantity, which can be expressed as $F(hk\ell) = A + iB$. The structure factor for the reflection with indices $-h, -k, -\ell$ is $F(\bar{h}\bar{k}\bar{\ell})$, and the observed intensity is $|F(\bar{h}\bar{k}\bar{\ell})|^2 = A^2 + B^2$.

The intensities of reflections $(hk\ell)$ and $(\bar{h}\bar{k}\bar{\ell})$ are the same, so that the diffraction pattern has an apparent center of symmetry, even if the crystal structure is not centrosymmetric. This is known as Friedel's law. Note that Friedel's law breaks down under the conditions of anomalous scattering, which happens when the wavelength is such that the X-rays are highly absorbed by the atoms in the crystal.

9.4.2 Laue classes

As a consequence of Friedel's law, the diffraction pattern exhibits the symmetry of a centrosymmetric crystal class. For example, a crystal in class 2, on account of the $\bar{1}$ symmetry imposed on its diffraction pattern, will appear to be in class $2/m$. The same result also holds for crystals in class m. Therefore, it is not possible to distinguish the classes 2, m, and $2/m$ from their diffraction patterns. The same effect occurs in other crystal systems, so that the 32 crystal classes are classified into only 11 distinct Laue groups according to the symmetry of the diffraction pattern, as shown in Table 9.4.1.

Knowledge of the diffraction symmetry of a crystal is useful for its classification. If the Laue group is observed to be $4/mmm$, the crystal system is tetragonal, the crystal class must be chosen from 422, $4mm$, $\bar{4}2m$, and $4/mmm$, and the space group is one of those associated with these four crystallographic point groups.

Table 9.3.11. Detailed information on space group $C2/c$ (No. 15)

$C2/c$	C_{2h}^6	$2/m$	Monoclinic	C_{2h}^6 is Schönflies symbol
No. 15	**C 1 2/c 1**	Patterson symmetry: C 1 2/m 1		Patterson symmetry: C 1 2/m 1

UNIQUE AXIS b, CELL CHOICE 1

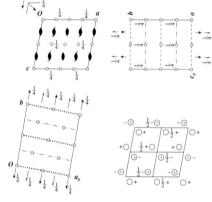

Unique axis is b, which leads to $C2/c$; alternative cell choices are described in other pages. Three projections of the unit cell down b, a, and c (showing the symmetry elements), and a projection down b (showing the general equivalent positions).

c_p is the projection of c

a_p is the projection of a

$\bigcirc$ represents an equipoint

$\odot$ is generated from $\bigcirc$ by an operation of the second kind (inversion, reflection, and glides) $+, -$: coordinate above or below the plane of projection along the unique b axis.

Origin at $\bar{1}$ on glide plane c

Location of origin is specified.

Asymmetric Unit $0 \leq x \leq \frac{1}{2}; 0 \leq y \leq \frac{1}{2}; 0 \leq z \leq \frac{1}{2}$

Symmetry Operations

For $(0,0,0)+$ set

(1) 1   (2) 2 $0, y, \frac{1}{4}$ (3) $\bar{1}$ $0,0,0$ (4) c $x, 0, z$

Numbering (in parentheses) and location of the operations (identity, twofold rotation, inversion, c glide, translation, and n glide).

For $(\frac{1}{2}, \frac{1}{2}, 0)+$ set

(1) $t(\frac{1}{2}, \frac{1}{2}, 0)$ (2) $2(0, \frac{1}{2}, 0)$ $\frac{1}{4}, y, \frac{1}{4}$

(3) $\bar{1}$ $\frac{1}{4}, \frac{1}{4}, 0$ (4) $n(\frac{1}{2}, 0, \frac{1}{2})$ $x\frac{1}{4}, z$

Generators selected   (1); $t(1,0,0)$; $t(0,1,0)$; $t(0,0,1)$; $t(\frac{1}{2}, \frac{1}{2}, 0)$; (2); (3)

Identity, axial and centering translations, twofold rotation, and inversion.

Table 9.3.11. Continued

Positions

Multiplicity Wyckoff letter, Site symmetry	Coordinates $(0,0,0)+$ $(\frac{1}{2},\frac{1}{2},0)+$		Reflection conditions	

Multiplicity, Wyckoff letter, site symmetry, and coordinates given for each set of equivalent positions.

General:

8 f 1   (1) x,y,z (2) $\bar{x},y,\bar{z}+\frac{1}{2}$ $hk\ell: h+k=2n$
         (3) $\bar{x},\bar{y},\bar{z}$ (4) $x,\bar{y},z+\frac{1}{2}$ $h0\ell: h,\ell=2n$
                                                           $0k\ell: k=2n$
                                                           $hk0: h+k=2n$
                                                           $0k0: k=2n$
                                                           $h00: h=2n$
                                                           $00\ell: \ell=2n$

Multiplicity = 8 for general positions
Systematic absences: $hk\ell$ with $(h+k)$ odd indicates C lattice; $h0\ell$ with ℓ odd indicates c glide; the other conditions only provide redundant information.

Special: as above, plus

4 e 2 $0,y,\frac{1}{4},$ $0,\bar{y},\frac{3}{4}$ no extra conditions
4 d $\bar{1}$ $\frac{1}{4},\frac{1}{4},\frac{1}{2},$ $\frac{3}{4},\frac{1}{4},0$ $hk\ell: k+\ell=2n$
4 c $\bar{1}$ $\frac{1}{4},\frac{1}{4},0,$ $\frac{3}{4},\frac{1}{4},\frac{1}{2}$ $hk\ell: k+\ell=2n$
4 b $\bar{1}$ $0,\frac{1}{2},0,$ $0,\frac{1}{2},\frac{1}{2}$ $hk\ell: \ell=2n$
4 a $\bar{1}$ $0,0,0,$ $0,0,\frac{1}{2}$ $hk\ell: \ell=2n$

Special positions are 4(a) to 4(e); 4(a) is always last listed; if an atom occupies 4(c), it only contributes to hkl with $(k+\ell)$ even.

Symmetry of special projections

Along [001] $c2mm$ Along [100] $p2gm$ Along [010] $p2$
$\mathbf{a}' = \mathbf{a}_p$ $\mathbf{b}' = \mathbf{b}$ $\mathbf{a}' = \frac{1}{2}\mathbf{b}$ $\mathbf{b}' = \mathbf{c}_p$ $\mathbf{a}' = \frac{1}{2}\mathbf{c}$ $\mathbf{b}' = \frac{1}{2}\mathbf{a}$
Origin at $0,0,z$ Origin at $x,0,0$ Origin at $0,y,0$

Plane groups and axes of the projections down $\mathbf{c}$, $\mathbf{a}$ and $\mathbf{b}$

Maximal non-isomorphic subgroups

I   [2] $C\,1\,2\,1(C2)$ $(1;2)+$
      [2] $C\,\bar{1}(P\bar{1})$ $(1;3)+$
      [2] $C\,1\,c\,1$ $(1;4)+$
II   [2] $P\,1\,2/c\,1(P2/c)$ $1;2;3;4$
       [2] $P\,1\,2/n\,1(P2/c)$ $1;2;(3;4)+(\frac{1}{2},\frac{1}{2},0)$
       [2] $P\,1\,2_1/n\,1(P2_1/c)$ $1;3;(2;4)+(\frac{1}{2},\frac{1}{2},0)$
       [2] $P\,1\,2_1/c\,1(P2_1/c)$ $1;4;(2;3)+(\frac{1}{2},\frac{1}{2},0)$
IIIb   none

t subgroups type **I** (same lattice translations)
k subgroups type **II** (same crystal class); subdivision **IIa** (same unit cell), **IIb**, **IIc** (larger cell)

Index of the subgroup is enclosed in square brackets.

Maximal isomorphic subgroups of lowest index

IIc   [3] $C\,1\,2\,1(\mathbf{b}' = 3\mathbf{b})(C2/c)$;
       [3] $C\,1\,2\,1(\mathbf{c}' = 3\mathbf{c})(C2/c)$;
       [3] $C\,1\,2\,1(\mathbf{a}' = 3\mathbf{a}$ or $\mathbf{a}' = 3\mathbf{a}, \mathbf{c}' = -\mathbf{a}+\mathbf{c}$ or $\mathbf{a}' = 3\mathbf{a},$
            $\mathbf{c}' = \mathbf{a}+\mathbf{c})(C2/c)$

Index, full international symbol, lattice vectors, and conventional symbol.

Minimal non-isomorphic subgroups

I   [2]$Cmcm$; [2]$Cmca$; [2]$Cccm$; [2]$Ccca$; [2]$Fddd$;
      [2]$Ibam$; [2]$Ibca$; [2]$Imma$; [2]$I4_1/a$; [3]$P\bar{3}12/c$;
      [3]$P\bar{3}2/c1$; [3]$R\bar{3}2/c$
II   [2]$F12/m1(C2/m)$; [2]$C12/m1(2\mathbf{c}' = \mathbf{c})(C2/m)$;
       [2]$P12/c1(2\mathbf{a}' = \mathbf{a}, 2\mathbf{b}' = \mathbf{b})(P2/c)$

t supergroups

k supergroups

328 Symmetry in Chemistry

Table 9.4.1. The 11 Laue groups for diffraction symmetry

Crystal system	Crystal class	Laue group
Triclinic	$1, \bar{1}$	$\bar{1}$
Monoclinic	$2, m, 2/m$	$2/m$
Orthorhombic	$222, mm2, mmm$	mmm
Tetragonal	$4, \bar{4}, 4/m$	$4/m$
	$422, 4mm, \bar{4}2m, 4/mmm$	$4/mmm$
Trigonal	$3, \bar{3}$	$\bar{3}$
	$32, 3m, \bar{3}m$	$\bar{3}m$
Hexagonal	$6, \bar{3}, 6/m$	$6/m$
	$622, 6mm, \bar{6}2m, 6/mmm$	$6/mmm$
Cubic	$23, m\bar{3}$	$m\bar{3}$
	$432, \bar{4}3m, m\bar{3}m$	$m\bar{3}m$

9.4.3 Deduction of lattice centering and translational symmetry elements from systemic absences

Systematic absences (or extinctions) in the X-ray diffraction pattern of a single crystal are caused by the presence of lattice centering and translational symmetry elements, namely screw axes and glide planes. Such extinctions are extremely useful in deducing the space group of an unknown crystal.

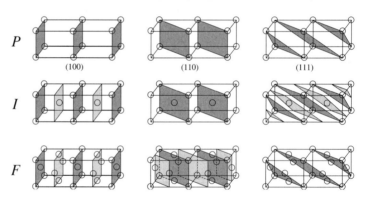

Fig. 9.4.1.
Some lattice planes through primitive (*P*), body-centered (*I*), and face-centered (*F*) lattices.

(1) Effect of lattice centering

In Fig. 9.4.1, the planes (100), (110), and (111) are drawn in heavy shading for pairs of adjoining primitive (*P*), body-centered (*I*), and face-centered (*F*) unit cells. In the *I*-lattice, the centered points lie midway between the (100) and (111) planes, and their contribution to X-ray scattering exactly cancels out that due to points located at the corners of the unit cells. These two reflections are therefore systematically absent. However, the reflections (200), (110), and (222) are observable since each passes through all the lattice points. Similarly, in the *F*-lattice, the (100) and (110) reflections are systematically absent, but the (200), (220), and (111) reflections are observable.

For subsequent discussion of the general case, we make use of the following theorem from geometry:

A line OL from the origin O of a space lattice to a lattice point L, as represented by the vector $\mathbf{OL} = u\mathbf{a} + v\mathbf{b} + w\mathbf{c}$, is divided into $(uh + vk + w\ell)$ parts by the set of planes $(hk\ell)$.

Consider the C-centered unit cell shown in Fig. 9.4.2. Setting $\mathbf{OL} = \mathbf{a} + \mathbf{b}$, the line OL, which runs through the C-centered point, is divided into 5 parts by the planes (320). Thus the points located at the corners of the unit cell lie in the (320) set of planes, but the C-centered point does not. It is easily seen that, in general, the C-centered point lies in the $(hk\ell)$ planes when $(h+k)$ is even, and in between the planes when $(h+k)$ is odd. In the latter case, X-ray scattering by the points at the corners of the unit cell is exactly out of phase with that caused by the C-centered point; in other words, reflections $(hk\ell)$ are systematically absent with $(h+k)$ odd. Conversely, whenever this kind of systematic absences is observed, the presence of a C-centered lattice is indicated.

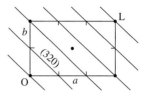

Fig. 9.4.2.
C-centered unit cell showing that the line OL is divided into five parts by the (320) planes, and the C-centered point lies midway between a pair of planes.

For a F-centered lattice, reflections $(hk\ell)$ are observable only when the conditions $(h+k)$ even, $(k+\ell)$ even, and $(\ell+h)$ even are simultaneously satisfied; this means that the indices h, k, ℓ must be either all even or all odd. This result can also be deduced analytically. In a F-centered lattice, the following equivalent positions are present: (x, y, z); $(x, y + 1/2, z + 1/2)$; $(x + 1/2, y, z + 1/2)$; $(x + 1/2, y + 1/2, z)$. Accordingly, the structure factor expression is

$$F_{hk\ell} = \sum_{j=1}^{n/4} f_j \exp[i2\pi(hx_j + ky_j + \ell z_j)]\{1 + \exp[i2\pi(k/2 + \ell/2)]$$
$$+ \exp[i2\pi(k/2 + \ell/2)] + \exp[i2\pi(k/2 + \ell/2)]\}.$$

For reflections of the type $(hk\ell)$, when the indices are all even or all odd (i.e., $h + k = 2n, h + \ell = 2n, k + \ell = 2n$),

$$F_{hk\ell} = 4 \sum_{j=1}^{n/4} f_j \exp[i2\pi(hx_j + ky_j + \ell z_j)].$$

If the h, k, ℓ indices are of mixed parity (i.e., not all odd or all even), then $F_{hk\ell} = 0$. For example, $F_{112} = F_{300} = 0$.

For an I-centered lattice, we consider the vector $\mathbf{OL} = \mathbf{a} + \mathbf{b} + \mathbf{c}$ which passes through the I-centered point. The line OL is divided into $(h + k + \ell)$ parts, which must be even for reflections $(hk\ell)$ to be observable.

Consider a rhombohedral lattice referred to a hexagonal set of axes. In the obverse setting, we consider the vector $\mathbf{OL} = -\mathbf{a}+\mathbf{b}+\mathbf{c}$, which passes through the points at $(-1/3, 1/3, 1/3)$, $(-2/3, 2/3, 2/3)$, and $(-1, 1, 1)$. The line OL is divided into $(-h + k + \ell)$ parts, which must be a multiple of 3 for all lattice points to lie in the planes $(hk\ell)$. In other words, the systematic absences for reflections $(hk\ell)$ are $(-h+k+\ell) \neq 3n$ for a rhombohedral lattice in the obverse setting. In the reverse setting, the systematic absences are $(h - k + \ell) \neq 3n$.

(2) Effect of glide planes and screw axes

Consider the arrangement of asymmetric objects (e.g., left and right hands) generated by a n glide plane normal to the c axis at $z = 0$, as shown in Fig. 9.4.3(a). Note that to the $(hk0)$ reflections, the z coordinate is immaterial, as

Fig. 9.4.3.
Effect of (a) n glide normal to c and (b) b glide normal to c on an asymmetric object represented by a hand. The plus and minus signs indicate that the object lies above and below the plane (001), respectively.

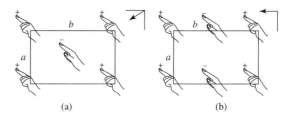

the structure projected onto the ab plane looks exactly like a C-centered lattice. Hence the systematic absences ($hk0$) absent with ($h+k$) odd implies the presence of a n glide plane normal to the c axis. Note that this condition is covered by that for C-centering, and therefore if the lattice is already C-centered, the n glide may or may not be present.

In like manner, for the b glide normal to c shown in Fig. 9.4.3(b), the right and left hands make the same contribution to the ($h00$) reflections. Therefore, the systematic absences of ($hk0$) reflections with k odd indicate the presence of a b glide plane normal to the c axis.

Consider now a 2_1-axis parallel to the b axis. As shown in Fig. 9.4.4, the coordinates x and z are irrelevant for the ($0k0$) planes, and the systematic absences ($0k0$) absent with k odd implies the presence of a 2_1-axis parallel to b. Note that this is a very weak condition as compared to those for lattice centering and glide planes, and is already covered by them. Since the ($0k0$) reflections are few in number, some may be too weak to be observable, and hence the determination of a screw axis from systematic absences is not always reliable.

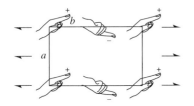

Fig. 9.4.4.
Effect of 2_1-axis along b on an asymmetric object represented by a hand. The plus and minus signs indicate that the object lies above and below the plane (001), respectively.

Analytical deductions can be made in the following way. Suppose the unit cell has a c glide plane perpendicular to the b axis. For an atom at x, y, z, there is an equivalent atom at $x, -y, 1/2+z$. The contribution of these two atoms to the structure factor is

$$F(hk\ell) = f\{\exp[2\pi i(hx + ky + \ell z)] + \exp[2\pi i(hx - ky + \ell/2 + \ell z)]\}.$$

For the special case of $k = 0$,

$$F(h0\ell) = f \exp[2\pi i(hx + \ell z)][1 + \exp(2\pi i\ell)]$$
$$= f \exp[2\pi i(hx + \ell z)][1 + (-1)^\ell].$$

This expression vanishes if ℓ is odd, and is equal to $2f[2\pi i(hx + \ell z)]$ if ℓ is even. Reflections of the type ($h0\ell$) will, therefore, be missing unless ℓ is an even number. The characteristic absence or extinction of ($h0\ell$) reflections with ℓ odd thus indicates a c glide plane perpendicular to the b axis.

For a twofold screw axis running parallel to c, the equivalent positions are x, y, z and $-x, -y, z + 1/2$.

$$F(hk\ell) = \sum_{j=1}^{n} f_j \exp[i2\pi(hx_j + ky_j + \ell z_j)]$$

$$= \sum_{j=1}^{n/2} f_j(\exp\{i2\pi[hx_j + ky_j + \ell z_j]\} + \exp\{i2\pi[-hx_j - ky_j + \ell(z_j + 1/2)]\})$$

When h and k are both zero,

$$F(00\ell) = \sum_{j=1}^{n/2} f_j \exp(i2\pi \ell z_j)[1 + \exp(i2\pi \ell/2)].$$

This expression vanishes if ℓ is odd, and is equal to $2 \sum_{j=1}^{n/2} f_j \exp[i2\pi \ell z_j]$ when ℓ is even. When reflections of the type (00ℓ) are absent with ℓ odd, they suggest the presence of a screw axis in the c direction.

The systematic absences due to the various types of lattice centering, screw axes, and glide planes are given in Table 9.4.2, which is used in the deduction of space groups.

Table 9.4.2. Determination of lattice type, screw axes, and glide planes from systematic absences of X-ray reflections

Lattice type	Reflections affected	Conditions for absence
A-face centered	$hk\ell$	$k + \ell =$ odd
B-face centered	$hk\ell$	$h + \ell =$ odd
C-face centered	$hk\ell$	$h + k =$ odd
F-face centered	$hk\ell$	h, k, ℓ not all odd or even
I-body centered	$hk\ell$	$h + k + \ell =$ odd
R (rhombohedral lattice)		
obverse setting	$hk\ell$	$-h + k + \ell \neq 3n$
reverse setting		$+h - k + \ell \neq 3n$

Screw axis	Reflections affected	Conditions for absence
$2_1, 4_2,$ or 6_3 along a	$h00$	$h =$ odd
b	$0k0$	$k =$ odd
c	00ℓ	$\ell =$ odd
$3_1, 3_2, 6_2,$ or 6_4 along c	00ℓ	$\ell \neq 3n$
4_1 or 4_3 along a	$h00$	$h \neq 4n$
b	$0k0$	$k \neq 4n$
c	00ℓ	$\ell \neq 4n$
6_1 or 6_5 along c	00ℓ	$\ell \neq 6n$

Glide plane	Reflections affected	Conditions for absence
a glide $\perp b$	$h0\ell$	$h =$ odd
$\perp c$	$hk0$	$h =$ odd
b glide $\perp a$	$0k\ell$	$k =$ odd
$\perp c$	$hk0$	$k =$ odd
c glide $\perp a$	$0k\ell$	$\ell =$ odd
$\perp b$	$h0\ell$	$\ell =$ odd
n glide $\perp a$	$0k\ell$	$k + \ell =$ odd
$\perp b$	$h0\ell$	$h + \ell =$ odd
$\perp c$	$hk0$	$h + k =$ odd
d glide $\perp a$	$0k\ell$	$k + \ell \neq 4n$
$\perp b$	$h0\ell$	$h + \ell \neq 4n$
$\perp c$	$hk0$	$h + k \neq 4n$

Table 9.4.3. Examples of deduction of space groups from systematic absences

Crystal system	Symmetry elements deduced by systematic absences with remarks	Possible space group
Monoclinic	(a) no systematic absence	$P2, Pm, P2/m$
	(b) $hk\ell$: $(h+k)$ odd $\Rightarrow$ C-centered	$Cc, C2/c$
	$h0\ell$: h odd $\Rightarrow$ not a new condition	
	$h0\ell$: ℓ odd $\Rightarrow$ c glide $\perp b$ axis	
	$0k0$: k odd $\Rightarrow$ not new	
Orthorhombic	(a) no systematic absence $\Rightarrow P$	$Pccn$
	$0k\ell$: ℓ odd $\Rightarrow$ c glide $\perp a$ axis	
	$hk0$: $(h+k)$ odd $\Rightarrow$ n glide $\perp c$ axis	
	$h0\ell$: ℓ odd $\Rightarrow$ c glide $\perp b$ axis	
	(b) $hk\ell$: $(h+k)$ odd $\Rightarrow C$	$Cmcm, C2cm \equiv$
	$0k\ell$: k odd $\Rightarrow$ not new	$Ama2, Cmc2_1$
	$h0\ell$: h odd $\Rightarrow$ not new	
	$h0\ell$: ℓ odd $\Rightarrow$ c glide $\perp b$ axis	
	$hk0$: $(h+k)$ odd $\Rightarrow$ not new	
Tetragonal	(a) Laue group $4/m$	$P4_2/n$
	$hk\ell$: no systematic absence $\Rightarrow P$	
	$hk0$: $(h+k)$ odd $\Rightarrow$ n glide $\perp c$ axis	
	00ℓ: ℓ odd $\Rightarrow$ 4_2 axis // c axis	
	(b) Laue group $4/m$	$I4, I\bar{4}, I4/m$
	$hk\ell$: $(h+k+\ell)$ odd $\Rightarrow I$	
	$hk0$: $(h+k)$ odd $\Rightarrow$ not new	
	00ℓ: ℓ odd $\Rightarrow$ not new	
	(c) Laue group: $4/mmm$	$I422, I4mm, I\bar{4}m2$
	$hk\ell$: $(h+k+\ell)$ odd $\Rightarrow I$	($m \perp a, b; 2$ //
	$hk0$: $(h+k)$ odd $\Rightarrow$ not new	diagonal of ab
	00ℓ: ℓ odd $\Rightarrow$ not new	plane), $I\bar{4}2m$ (2 //
		$a, b; m \perp$ diagonal of
		ab plane)
Trigonal (based on hexagonal cell)	(a) Laue group: $\bar{3}$	$R3, R\bar{3}$
	$hk\ell$: $(-h+k+\ell) \neq 3n \Rightarrow R$	
	00ℓ: $\ell \neq 3n \Rightarrow$ not new	
	(b) Laue group: $\bar{3}m$	$R32, R3m, R\bar{3}m$
	$hk\ell$: $(-h+k+\ell) \neq 3n \Rightarrow R$	
	00ℓ: $\ell \neq 3n \Rightarrow$ not new	
Hexagonal	(a) Laue group: $6/m$	$P6_3, P6_3/m$
	$hk\ell$: no systematic absence $\Rightarrow P$	
	00ℓ: ℓ odd $\Rightarrow 6_3$ // c axis	
	(b) Laue group: $6mmm$	$P6_322$
	$hk\ell$: no systematic absence $\Rightarrow P$	
	00ℓ: ℓ odd $\Rightarrow 6_3$ // c axis	
Cubic	(a) Laue group: $m\bar{3}$	$Fd\bar{3}$
	$hk\ell$: $(h+k), (k+\ell)$ and $(\ell+h)$ odd $\Rightarrow F$	
	$0k\ell$: $(k+\ell) \neq 4n \Rightarrow$ d glide $\perp a$ axis	
	(b) Laue group: $m\bar{3}m$	$F432, F\bar{4}3m, Fm\bar{3}m$
	$hk\ell$: $(h+k), (k+\ell)$ and $(\ell+h)$ odd $\Rightarrow F$	

(3) Deduction of space groups from systematic absences

Selected examples of the use of systematic absences in the deduction of space groups are illustrated in Table 9.4.3. Note that some conditions are redundant, as they are already covered by more general conditions. Some space groups can be determined uniquely, but others can only be narrowed down to a

9.5 Selected space groups and examples of crystal structures

9.5.1 Molecular symmetry and site symmetry

In a molecular crystal, the idealized symmetry of the molecule is often not fully expressed; in other words, the molecule occupies a site of lower point symmetry. For example, in the crystal structure of naphthalene, the C_8H_{10} molecule (idealized symmetry D_{2h}) is located at a site of symmetry $\bar{1}$. On the other hand, the hexamethylenetetramine molecule, $(CH_2)_6N_4$, retains its T_d symmetry in the crystalline state. Biphenyl, $C_6H_5-C_6H_5$, which exists in a nonplanar conformation with a dihedral angle of $45°$ (symmetry D_2) in the vapor phase, occupies a site of symmetry $\bar{1}$ in the crystalline state and is therefore completely planar.

Crystallization may in rare cases result in chemical transformation of the constituents of a chemical compound. For example, PCl_5 is a trigonal bipyramidal molecule, but its crystal structure is composed of a packing of tetrahedral PCl_4^+ and octahedral PCl_6^- ions. In contrast, PBr_5 in the crystalline state comprises PBr_4^+ and Br^- ions; formation of the PBr_6^- ion is precluded by steric repulsion between the bulky peripheral bromide ions.

9.5.2 Symmetry deductions: assignment of atoms and groups to equivalent positions

The presence of symmetry elements in a crystal imposes restrictions on the numbers of atoms that can be accommodated in a unit cell. If the space group does not have point symmetry elements which can generate special positions, atoms can be placed only in the general position, and their numbers in the cell are restricted to multiples of the multiplicity of that position. For example, space group $P2_12_12_1$ has just one set of equivalent positions of multiplicity equal to 4; accordingly, atoms can occur only in sets of fours in the unit cell of any crystal belonging to this space group. A compound which crystallizes in this space group is codeine hydrobromide dihydrate, $C_{18}H_{21}O_3N \cdot HBr \cdot 2H_2O$. The number of formula units per cell, Z, is equal to 4, so that every atom must be present in multiples of 4.

When the space group has special positions in addition to the general position, there is more latitude to the numbers of each atom type which can be present in the cell. For example, $FeSb_2$ belongs to space group *Pnnm* with $Z = 2$. This space group has one general eightfold, three special fourfold (site symmetry 2 or *m*), and four twofold (site symmetry $2/m$) sets of equipoints. The Fe atoms must occupy one of the twofold sets. The Sb atoms may be assigned to two of the three remaining twofold equipoints or one of the three fourfold equipoints, but not the eightfold equipoint.

In simple inorganic crystals having only a few atoms per unit cell, equipoint considerations alone may lead to a solution of the crystal structure, or to a small number of possible solutions. For example, if a cubic crystal of composition AB has $Z = 4$, its structure is fixed as either the NaCl type or the ZnS type.

Table 9.4.4. Space groups not established uniquely from systematic absences

Laue class	Symbol from systematic absences*	Possible space groups†
Triclinic $\bar{1}$	$P\square$	$P1, P\bar{1}$
Monoclinic $2/m$	$P\square$	$Pm, P2, P2/m$
	$P\square c$	$Pc, P2/c$
	$P2_1\square$	$P2_1, P2_1/m$
	$C\square$	$Cm, C2, C2/m$
	$C\square c$	$Cc, C2/c$
Orthorhombic mmm	$P\square\square\square$	$Pmm2, P222, Pmmm$
	$Pc\square\square$	$Pcm2_1 (= Pmc2_1), Pc2m (= Pma2), Pcmm (= Pmma)$
	$Pn\square\square$	$Pnm2_1 (= Pmn2_1), Pnmm (= Pmmn)$
	$Pcc\square$	$Pcc2, Pccm$
	$Pca\square$	$Pca2_1, Pcam (= Pbcm)$
	$Pba\square$	$Pba2, Pbam$
	$Pnc\square$	$Pnc2, Pncm (= Pmna)$
	$Pna\square$	$Pna2_1, Pnam (= Pnma)$
	$Pnn\square$	$Pnn2, Pnnm$
	$C\square\square\square$	$Cmm2, Cm2m (= Amm2), C222, Cmmm$
	$C\square c\square$	$Cmc2_1, C2cm (= Ama2), Cmcm$
	$C\square\square a$	$Cm2a (= Abm2), Cmma$
	$C\square ca$	$C2ca, Cmca$
	$Ccc\square$	$Ccc2, Cccm$
	$I\square\square\square$	$Imm2, I222, I2_12_12_1, Immm$
	$I\square a\square$	$Ima2, Imam (= Imma)$
	$Iba\square$	$Iba2, Ibam$
	$F\square\square\square$	$Fmm2, F222, Fmmm$
Tetragonal $4/m$	$P\square$	$P\bar{4}, P4, P4/m$
	$P4_2\square$	$P4_2, P4_2/m$
	$I\square$	$I\bar{4}, I4, I4/m$
Tetragonal $P4/mmm$	$P\square\square\square$	$P\bar{4}2m, P\bar{4}m2, P4mm, P422, P4/mmm$
	$P\square 2_1\square$	$P\bar{4}2_1m, P42_12$
	$P\square\square c$	$P\bar{4}2c, P4_2mc, P4_2/mmc$
	$P\square b\square$	$P\bar{4}b2, P4bm, P4/mbm$
	$P\square bc$	$P4_2bc, P4_2/mbc$
	$P\square c\square$	$P\bar{4}c2, P4_2cm, P4_2/mcm$
	$P\square cc$	$P4cc, P4/mcc$
	$P\square n\square$	$P\bar{4}n2, P4_2nm, P4_2/mnm$
	$P\square nc$	$P4nc, P4/mnc$
	$I\square\square\square$	$I\bar{4}m2, I\bar{4}2m, I4mm, I422, I4/mmm$
	$I\square c\square$	$I\bar{4}c2, I4cm, I4/mcm$
	$I\square\square d$	$I\bar{4}2d, I4_1md$
Trigonal $\bar{3}$	$P\square$	$P3, P\bar{3}$
	$R\square$	$R3, R\bar{3}$
Trigonal $\bar{3}m$	$P\square\square\square$	$P3m1, P31m, P312, P321, P\bar{3}1m, P\bar{3}m1$
	$P\square c\square$	$P3c1, P\bar{3}c1$
	$P\square\square c$	$P31c, P\bar{3}1c$
	$R\square\square$	$R3m, R32, R\bar{3}m$
	$R\square c$	$R3c, R\bar{3}c$
Hexagonal $6/m$	$P\square$	$P6, P\bar{6}, P6/m$
	$P6_3$	$P6_3, P6_3/m$
Hexagonal $6/mmm$	$P\square\square\square$	$P\bar{6}m2, P\bar{6}2m, P6mm, P622, P6/mmm$
	$P\square c\square$	$P\bar{6}c2, P6_3cm, P6_3/mcm$
	$P\square\square c$	$P\bar{6}2c, P6_3mc, P6_3/mmc$
	$P\square cc$	$P6cc, P6/mcc$
Cubic $m\bar{3}$	$P\square\square$	$P23, Pm\bar{3}$
	$I\square\square$	$I23, I2_13, Im\bar{3}$
	$F\square\square$	$F23, Fm\bar{3}$
Cubic $m\bar{3}m$	$P\square\square\square$	$P\bar{4}3m, P432, Pm\bar{3}m$
	$P\square\square n$	$P\bar{4}3n, Pm\bar{3}n$
	$I\square\square\square$	$I\bar{4}3m, I432, Im\bar{3}m$
	$F\square\square\square$	$F\bar{4}3m, F432, Fm\bar{3}m$
	$F\square\square c$	$F\bar{4}3c, Fm\bar{3}c$

* The symbol $\square$ indicates a position for a possible point symmetry element.
† For orthorhombic space groups, the standard Hermann–Mauguin symbol for a different choice of axes is enclosed in parentheses.

The first belongs to space group $Fm\bar{3}m$, and the latter to $F\bar{4}3m$, and these two possibilities can be distinguished by comparing the X-ray diffraction intensities expected from the two structure types with those actually observed. This type of deduction is based on consideration of the roles of the individual atoms in a crystal structure, which usually finds application in compounds of inorganic composition such as binary and ternary compounds, alloys, and minerals.

Symmetry considerations are most useful when applied to molecular crystals. In most organic and organometallic crystals, the building units are not the individual atoms but rather sets of atoms already organized into molecules. In such cases the whole chemical molecule must conform to the equipoint restrictions as to both number and symmetry. If the molecule occupies a general position, it need not have any imposed symmetry, because the general position has symmetry 1. But if the molecule is located on a special position, it must have at least the symmetry of that special position. Accordingly, if the space group and the equipoint occupied by the molecule are known, the minimum symmetry which the molecule may have is determined. This information may be useful for a chemist who is trying to decide which, of several configurations that a molecule might conceivably have, is the true one.

It is commonly, but not always, true that all the molecules in a molecular crystal are equivalent. When this is the case, then all the molecules occupy the same set of equipoints; this implies that they are related to one another by the various symmetry operations of the space group. On the other hand, if the molecules occupy two or more Wyckoff positions, whether general or special, they are chemically equivalent but crystallographically distinct. A good example is the 1:1 adduct of hexamethylenetetramine oxide with hydroquinone, $(CH_2)_6N_4 \rightarrow O \cdot p\text{-}C_6H_4(OH)_2$. This compound crystallizes in space group $P\bar{1}$ with $Z = 2$, and the $p\text{-}C_6H_4(OH)_2$ molecules occupy two non-equivalent sites of $\bar{1}$ symmetry, rather than a general position. In other words, the asymmetric unit consists of a $(CH_2)_6N_4 \rightarrow O$ and two "half-molecules" of $p\text{-}C_6H_4(OH)_2$, which are chemically equivalent but crystallographically distinguishable.

The following three compounds all crystallize in space group $P2_1/c$. Useful information on molecular structure can be deduced when Z is less than the space-group multiplicity of 4.

(1) *Naphthalene.* Since $Z = 2$, the molecule occupies a site of symmetry $\bar{1}$; in other words, there is an inversion center in the middle of the C4a–C8a bond. With reference to the atom numbering system shown below, the asymmetric unit can be taken as one of the right (1–4,4a), left (5–8,8a), upper (1, 2, 7, 8, 8a) or lower (3, 4, 4a, 5, 6) half of the molecule. Note that the idealized molecular symmetry D_{2h} of naphthalene is not fully utilized in the solid state, and some of the chemically identical bond distances and bond angles have different measured values.

(2) *Biphenyl*. Since $Z = 2$, an inversion center is present in the middle of the bridging C–C bond between the two C_6H_5 rings. This means that both phenyl rings are coplanar. The planar conformation of biphenyl in the crystalline state, which differs from its non-planar conformation in the solution and gas phases, is attributable to efficient molecular packing stabilized by intermolecular interactions between the phenyl rings.

(3) *Ferrocene*. With $Z = 2$, the molecule must be placed in a special position of symmetry $\bar{1}$. The pair of cyclopentadienyl (Cp) rings are thus expected to be staggered. In fact, the situation is more complicated. A thorough study of the structure of crystalline ferrocene by X-ray and neutron diffraction at temperatures of 173 and 298 K showed that the apparently staggered arrangement of the Cp rings results from the presence of molecules in different orientations randomly distributed at both 173 and 298 K. A disordered model with the two Cp ring rotated about 12° from the eclipsed position is displayed in Fig. 9.5.1. (The rotation angle is 0° for the precisely eclipsed D_{5h} conformation, and 36° for the exactly staggered D_{5d} conformation.)

As a worked example, the assignment of atoms and groups to equivalent positions for basic beryllium acetate is illustrated in detail in the following paragraphs.

Basic beryllium acetate has the properties of a molecular solid rather than an ionic salt. X-ray analysis has shown that it consists of discrete $Be_4O(CH_3COO)_6$ molecules. The crystal structure is cubic, with $a = 1574$ pm and $\rho_x = 1.39$ g cm^{-3}. The molecular weight of $Be_4O(CH_3COO)_6$ is 406, and hence $Z = \rho N_o V/M = (1.39)(6.023 \times 10^{23})(15.74 \times 10^{-8})^3/406 \approx 8$. The diffraction (Laue) symmetry is $m\bar{3}$. The condition for observable $hk\ell$ reflections: h, k, ℓ all even or all odd indicates a F lattice; $0k\ell$ present for $(k + \ell) = 4n$ indicates the presence of a d glide plane. Hence the space group is uniquely established as $Fd\bar{3}$ (No. 203).

Since $Z = 8$, the asymmetric unit consists of one molecule located in special positions $8(a)$ or $8(b)$. The former may be taken for convenience. The molecular symmetry utilized in the space group is therefore 23.

All atomic positional parameters, which define the crystal and molecular structure, can be assigned to the general and special equivalent positions of space group $Fd\bar{3}$ (origin at 23):

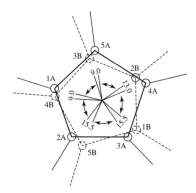

Fig. 9.5.1.
A possible nearly eclipsed configuration for ferrocene in the disordered model. From P. Seiler and J. D. Dunitz, *Acta Crystallogr.*, Sect. B **35**, 1068–74 (1979).

Wyckoff position	Site symmetry	Coordinates $(0, 0, 0; 0, 1/2, 1/2; 1/2, 0, 1/2; 1/2, 1/2, 0) +$
96(g)	1	x, y, z; etc.
48(f)	2	$x, 0, 0$; etc.
32(e)	3	x, x, x; etc.
16(d)	$\bar{3}$	$5/8, 5/8, 5/8$; etc.
16(c)	$\bar{3}$	$1/8, 1/8, 1/8$; etc.
8(b)	23	$1/2, 1/2, 1/2; 3/4, 3/4, 3/4$
8(a)	23	$0, 0, 0; 1/4, 1/4, 1/4$

The atom numbering scheme and molecular structure of $Be_4O(CH_3COO)_6$ is shown in Fig. 9.5.2.

The assignment of atomic positions proceeds as follows. The basic oxygen O(1) is placed in 8(a); Be in 32(e), not in 16(c) + 16(d) because its coordination preference is tetrahedral rather than octahedral, and the molecule possesses symmetry element 3 but not $\bar{3}$; carboxylate carbon C(1) in 48(f) lying on twofold axis; methyl carbon C(2) in 48(f) lying on twofold axis; acetate oxygen O(2) in 96(g). The crystal and molecular structure is therefore determined by six parameters: x of Be; x of C(1); x of C(2); x, y, z of O(2).

The following assignment shows that six positional parameters (neglecting the hydrogen atoms) are required for a description of the crystal structure. The hydrogen atoms are orientationally disordered, and they occupy general position 96(g) with half-site occupancy, as indicated in the last three rows of the table.

Assignment	Parameters	Remarks
$Be_4O(CH_3COO)_6$ in 8(a)	none	Molecular symmetry has 23(T)
O(1) in 8(a)	none	Basic oxygen atom is at the centroid of the tetrahedron formed by Be atoms.
Be(1) in 32(e)	x_{Be}	Beryllium atom lies on C_3 axis.
O(2) in 96(g)	x_{O2}, y_{O2}, z_{O2}	Carboxylate oxygen atom is in general position.
C(1) in 48(f)	x_{C1}	Carboxylate carbon atom lies on C_2 axis.
C(2) in 48(f)	x_{C2}	Methyl carbon lies on the same C_2 axis.
$\frac{1}{2}$H(1) in 96(g)	x_{H1}, y_{H1}, z_{H1}	The methyl group is twofold disordered along
$\frac{1}{2}$H(2) in 96(g)	x_{H2}, y_{H2}, z_{H2}	the C–C bond; the site occupancy factor of
$\frac{1}{2}$H(3) in 96(g)	x_{H3}, y_{H3}, z_{H3}	each methyl hydrogen atom is 1/2.

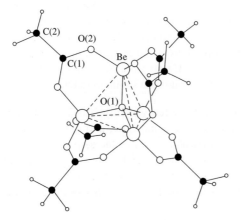

Fig. 9.5.2.
Molecular structure of $Be_4O(CH_3COO)_6$ with atom labels.

In the crystal structure, the symmetry of the Be$_4$O(CH$_3$COO)$_6$ molecule is 23. The central oxygen atom is tetrahedrally surrounded by four beryllium atoms, and each beryllium atom is tetrahedrally surrounded by four oxygen atoms. The six acetate groups are attached symmetrically to the six edges of the tetrahedron. Hydrogen atoms of statistical weight 1/2 are distributed over two sets of positions, corresponding to two equivalent orientations of the methyl group. Although the idealized molecular point group of Be$_4$O(CH$_3$COO)$_6$ is $\bar{4}3m(T_d)$, the molecular symmetry in the crystal is reduced to the subgroup 23 (T), and the symmetry elements $\bar{4}$ and m are not utilized.

9.5.3 Racemic crystal and conglomerate

A racemic crystal contains equal numbers of both enantiomeric forms of a given compound in the unit cell of a centrosymmetric space group. A helicate consists of a linear array of metal ions along a central axis, around which there is a helical arrangement of bridging polytopic ligands. If a helicate crystallizes in a centrosymmetric space group, both right- and left-handed helices must coexist in equal numbers.

A conglomerate contains a 50:50 mixture of single crystals of both enantiomeric forms of a pure compound. A single crystal of either enantiomer may be picked up by chance for X-ray structure analysis. The space group is necessarily non-centrosymmetric.

Note that even for an achiral compound, there is the possibility of its building blocks being organized into chiral structures in the solid state, such as a right-handed (or left-handed) helix or three-leafed propeller. The classical inorganic example is α-quartz; the space group is $P3_121$ (No. 152) for the dextrorotatory form and $P3_221$ (No. 154) for the levorotatory form. An organic example is provided by an isomorphous series of channel inclusion compounds of urea, which crystallize in the hexagonal space group $P6_122$ (No. 178) or $P6_522$ (No. 179). The structure of this type of compounds will be described in Section 9.6.5.

9.5.4 Occurrence of space groups in crystals

Nowadays the number of reported crystal structures is growing at the rate of more than 10,000 per year. There are known compounds in each of the 230 space groups. Several statistical studies have been carried out using the wealth of structural information in three major resources: *Cambridge Structural Database* (for organic compounds, organometallic compounds, and metal complexes containing organic ligands; abbreviated as CSD), the *Inorganic Crystal Structure Database* (ICSD), and the *Metals Data File* (MDF). The principal findings on the distribution of space groups are summarized below:

(1) Inorganic compounds are fairly evenly distributed among the seven crystal systems. Orthorhombic, monoclinic, and cubic crystals (in the order of abundance) comprise about 60% of the total number.
(2) In contrast, organic compounds are heavily concentrated in the lower-symmetry crystal systems. Monoclinic, orthorhombic, and triclinic crystals (in that order) comprise over 95% of the total number.

Table 9.5.1. Distribution of the most common space groups for crystals

Rank	Inorganic		Organic	
	Space group	%	Space group	%
1	***Pnma***	8.25	***$P2_1/c$***	36.57
2	***$P2_1/c$***	8.15	***$P\bar{1}$***	16.92
3	***$Fm\bar{3}m$***	4.42	$P2_12_12_1$	11.00
4	***$P\bar{1}$***	4.35	***C2/c***	6.95
5	***C2/c***	3.82	$P2_1$	6.35
6	***$P6_3/mmc$***	3.61	***Pbca***	4.24
7	***C2/m***	3.40	$Pna2_1$	1.63
8	***I4/mmm***	3.39	***Pnma***	1.57
9	***$Fd\bar{3}m$***	3.03	$P1$	1.23
10	***$R\bar{3}m$***	2.47	***Pbcn***	1.01
11	***Cmcm***	1.95	Cc	0.97
12	***$P\bar{3}m1$***	1.69	C2	0.90
13	***$Pm\bar{3}m$***	1.46	$Pca2_1$	0.75
14	***$R\bar{3}$***	1.44	***$P2_1/m$***	0.64
15	***P6/mmm***	1.44	$P2_12_12$	0.53
16	***Pbca***	1.34	***C2/m***	0.49
17	***$P2_1/m$***	1.33	***P2/c***	0.49
18	***$R\bar{3}c$***	1.10	***$R\bar{3}$***	0.46

* Centrosymmetric space groups are highlighted in bold-face type. From W. H. Baur and D. Kassner, *Acta Crystallogr.*, Sect B **48**, 356–69 (1992).

(3) About 75–80% of known compounds crystallize in centrosymmetric space groups. About 80% of the non-centrosymmetric organic structures occur in space groups without an improper rotation axis $\bar{n}$ (the Sohncke groups).

(4) A 1992 survey of over 86,000 compounds in the three main crystallographic databases showed that about 57% of inorganic compounds are found in 18 most frequent space groups, which are all centrosymmetric. This is compared to about 93% for organic compounds in a different list of 18 space groups, of which eight lack an inversion center, including five possible ones for enantiomerically pure materials. Eight space groups are common to both lists. The principal findings are tabulated in Table 9.5.1.

(5) The three most common space groups for chemical compounds are $P2_1/c, P\bar{1}$, and $C2/c$. In particular, $P2_1/c$ is by far the dominant space group for organic crystals.

(6) Wrong assignment of space groups occurs most frequently for Cc (correct space group is $C2/c$, $Fdd2$, or $R\bar{3}c$). Other incorrect assignments often made are $P1$ (correct space group $P\bar{1}$), $Pna2_1$ (correct space group $Pnma$), and Pc (correct space group $P2_1/c$).

9.6 Application of space group symmetry in crystal structure determination

A precise description of the structure of a crystalline compound necessitates prior knowledge of its space group and atomic coordinates in the asymmetric unit. Illustrative examples of compounds belonging to selected space groups in the seven crystal systems are presented in the following sections.

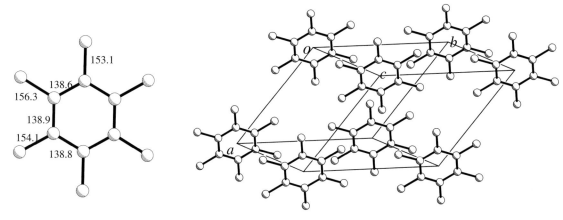

Fig. 9.6.1.
Bond lengths and crystal structure of C_6Me_6.

9.6.1 Triclinic and monoclinic space groups

(1) Space group $P\bar{1}$ (no. 2), multiplicity = 2

Hexamethylbenzene, C_6Me_6, exists as a plastic phase I above 383 K, a room-temperature phase II, and a low-temperature phase III below 117.5 K. Both phase II and phase III crystallize in space group $P\bar{1}$ with $Z = 1$. The molecule is therefore located at an inversion center, and the site symmetry $\bar{1}$ is much lower than the idealized molecular symmetry of D_{6h}. The asymmetric unit consists of one-half of the molecule.

Determination of the crystal structure of phase II by Lonsdale in 1929 unequivocally settled over 70 years of debate concerning the geometry and bonding of aromatic molecular systems. The measured bond lengths and crystal structure of hexamethylbenzene are shown in Fig. 9.6.1. The hexamethylbenzene molecules lie within planes approximately perpendicular to (111). Phase III is structurally very similar to phase II, but differs from it mainly by a shearing process between molecular layers that results in a pseudo-rhombohedral, more densely packed arrangement.

Cisplatin is the commercial name of *cis*-diamminedichloroplatinum(II), which is widely used as an approved drug since 1978 for treating a variety of tumors. In the crystal structure, $Z = 2$, and the *cis*-$PtCl_2(NH_3)_2$ molecule was found to occupy a general position. The measured dimensions and crystal packing are shown in Fig. 9.6.2.

(2) Space group $P2_1$ (no. 4), multiplicity $= 2$

This is one of the most common space groups for non-centrosymmetric molecules, particularly natural products and biological molecules. For example, sucrose is a non-reducing sugar, and X-ray analysis has revealed that its α-D-glucopyranosyl-β-D-fructofuranoside skeleton consists of a linkage between α-glucose and β-fructose moieties. The intramolecular hydrogen bonds and crystal structure are illustrated in Fig. 9.6.3.

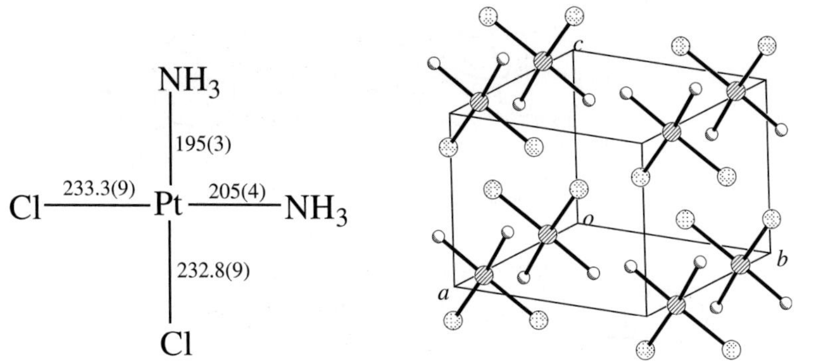

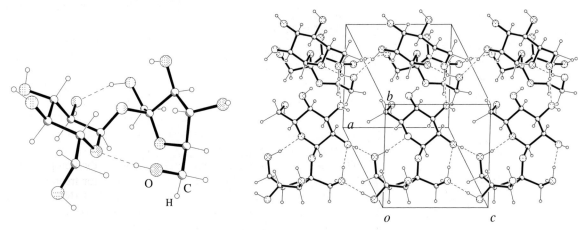

Fig. 9.6.2.
Molecular and crystal structure of cis-PtCl$_2$(NH$_3$)$_2$.

(3) Space group $P2_1/c$ (no. 14), multiplicity = 4

Naphthalene crystallizes in this space group with $Z = 2$; the measured molecular dimensions and molecular packing are displayed in Fig. 9.6.4.

Also crystallizing in space group $P2_1/c$, uranocene has two molecules per unit cell, so that the U(C$_8$H$_8$)$_2$ molecule occupies a special position of site symmetry $\bar{1}$. In other words, the molecule has an eclipsed conformation, and it may be assigned to special position 2(a). Similarly, the two halves of the [Re$_2$Cl$_8$]$^{2-}$ dianion in K$_2$[Re$_2$Cl$_8$]·2H$_2$O ($Z = 2$) are in the eclipsed orientation with respect to each other. The measured molecular dimensions (indicating that the symmetry of the dianion is D_{4h} within experimental error) and crystal structure are shown in Fig. 9.6.5.

Fig. 9.6.3.
Molecular and crystal structure of sucrose.

With $Z = 2$, the iron(II) phthalocyanine molecule occupies a $\bar{1}$ site, and its molecular and crystal structures are shown in Fig. 9.6.6.

N''-Cyano-N,N-diisopropylguanidine crystallizes in space group $P2_1/c$ with $Z = 40$. The asymmetric unit thus contains an exceptionally high number of ten molecules, which are labeled **A** to **J** in Fig. 9.6.7. These crystallographically independent yet structurally similar molecules constitute five

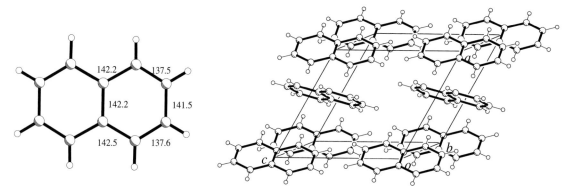

Fig. 9.6.4.
Molecular and crystal structure of naphthalene.

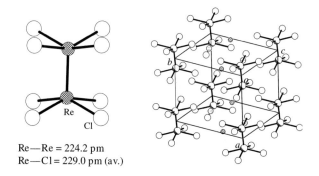

Fig. 9.6.5.
Molecular and crystal structure of [Re$_2$Cl$_8$]$^{2-}$.

Re—Re = 224.2 pm
Re—Cl = 229.0 pm (av.)

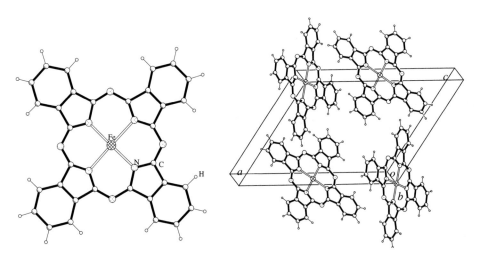

Fig. 9.6.6.
Molecular and crystal structure of iron(II) phthalocyanine.

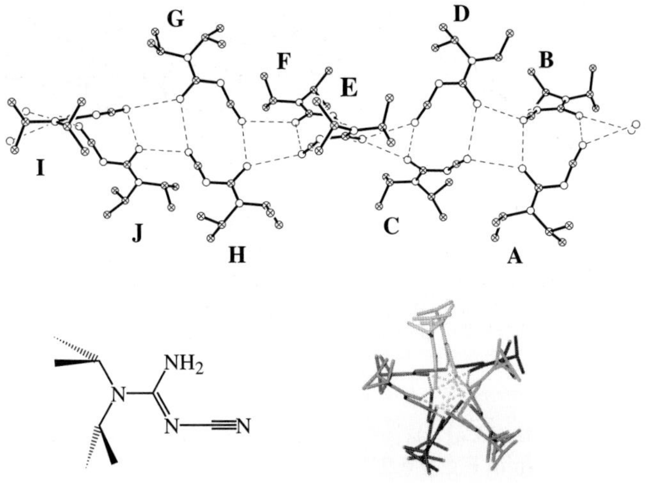

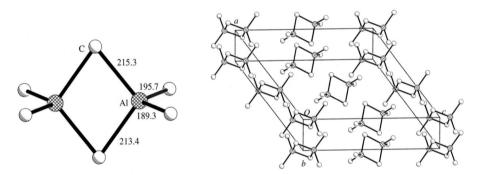

Fig. 9.6.7.
Top: hydrogen-bond ribbon formed from ten independent molecules in the asymmetric unit. Bottom left: structural formula of N''-cyano-N,N-diisopropylguanidine. Bottom right: projection of the asymmetric unit along [101], which appears as a helix.
[Ref. X. Hao, J. Chen, A. Cammers, S. Parkin and C. P. Block, *Acta Crystallogr., Sect B*, **61**, 218–26 (2005).]

hydrogen-bonded dimers (**AB, CD, EF, GH, IJ**), which are connected by additional hydrogen bonds to form a twisted ribbon. Extension of the ribbon along [101] generates an infinite helix with five dimeric units per turn.

Fig. 9.6.8.
Molecular and crystal structure of Al_2Me_6.

(4) Space group $C2/c$ (no. 15), multiplicity = 8

Trimethylaluminum is a dimer with idealized molecular symmetry D_{2h}, so that its molecular formula should be written as Al_2Me_6. There are four dimeric molecules in the unit cell. The molecule may be placed in Wyckoff position $4(a)$ of site symmetry $\bar{1}$ or $4(e)$ of site symmetry 2, and the former has been found. The measured molecular dimensions and crystal structure are shown in Fig. 9.6.8.

9.6.2 Orthorhombic space groups

(1) Space group $P2_12_12_1$ (no. 19), multiplicity = 4

The iron-thiolato complex $(Me_4N)_2[Fe_4S_4(SC_6H_5)_4]$ crystallizes in this space group with $Z = 4$, and the asymmetric unit consists of two Me_4N^+ cations and

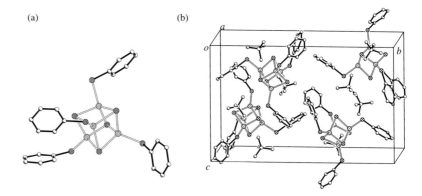

Fig. 9.6.9.
(a) Molecular structure of $[Fe_4S_4(SC_6H_5)_4]^{2-}$. (b) Crystal structure of its Me_4N^+ salt.

a tetranuclear $[Fe_4S_4(SC_6H_5)_4]^{2-}$ cluster dianion. The molecular and crystal structures are shown in Fig. 9.6.9.

(2) Space group *Pnma* (no. 62), multiplicity = 8

Hexachlorocyclophosphazene, $(PNCl_2)_3$, has $Z = 4$. Space group *Pnma* (no. 62) has special positions of site symmetry $\bar{1}$ and m; clearly $\bar{1}$ can be ruled out, and the molecule has a mirror plane passing through a PCl_2 group and an opposite N atom in the ring. The molecular and crystal structure are shown in Fig. 9.6.10.

The Dewar benzene derivative 1′,8′:3,5-naphtho[5.2.2]propella-3,8,10-triene, $C_{18}H_{14}$, also has $Z = 4$, and a crystallographic mirror plane passes through the central naphthalene C–C bond and the midpoints of the bonds C(8)-C(8′), C(9)-C(9′) and C(10)-C(10′). The measured bond lengths in the strained ring system and crystal structure are shown in Fig. 9.6.11.

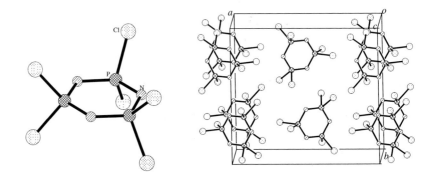

Fig. 9.6.10.
Molecular and crystal structures of $(PNCl_2)_3$.

(3) Space group *Cmca* (no. 64), multiplicity = 16

If the formula of orthorhombic black phosphorus is written simply as P, then $Z = 8$. The special positions of space group *Cmca* are $4(a)$ $2/m$, $4(b)$ $2/m$, $8(c)$ $\bar{1}$, $8(d)$ 2, $8(e)$ 2, $8(f)$ m, and general position $16(g)$ 1. The independent P atom actually is located in $8(f)$, generating a continuous double-layer structure, which is a heavily puckered hexagonal net in which each P atom is covalently bound to three others within the same layer. In Fig. 9.6.12, two puckered layers in

Symmetry in Crystals

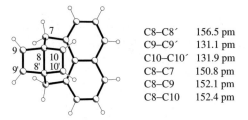
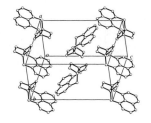

C8–C8′	156.5 pm
C9–C9′	131.1 pm
C10–C10′	131.9 pm
C8–C7	150.8 pm
C8–C9	152.1 pm
C8–C10	152.4 pm

Fig. 9.6.11.
Molecular and crystal structures of 1′,8′:3,5-naphtho[5.2.2]propella-3,8,10-triene.

projection are shown on the left, and a perspective view of the crystal structure is shown on the right.

9.6.3 Tetragonal space groups

(1) Space group $P4_12_12$ (no. 90), multiplicity = 4

The compound 1,1′-binaphthyl exists in its stable conformation as a chiral molecule. In the crystalline state, it exists in higher melting (159°C) and lower melting (145°C) forms. The latter form is a racemic crystal in space group $C2/c$.

In the crystal structure of the chiral higher melting form, a crystallographic twofold axis that bisects the central C(1)–C(1′) bond. The molecule has a *gauche* configuration about the C(1)–C(1′) bond with a dihedral angle of 103.1°. The molecular structure and packing are shown in Fig. 9.6.13.

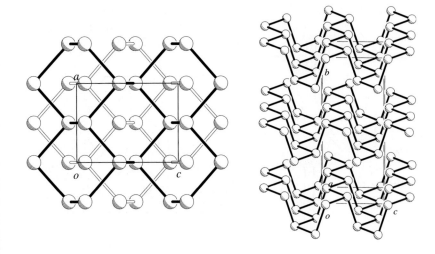

Fig. 9.6.12.
Two views of the crystal structure black phosphorus.

(2) Space group $P\bar{4}2_1m$ (no. 113), multiplicity = 8

Urea was the first organic compound to be synthesized in the laboratory, a feat accomplished by Wöhler in 1828. In the crystal, urea occupies special position 2(c) where its *mm2* molecular symmetry is fully utilized. Every hydrogen atom is involved in a N–H···O hydrogen bond, and this accounts for the fact that urea occurs as a solid at room temperature. A feature of chemical interest in the

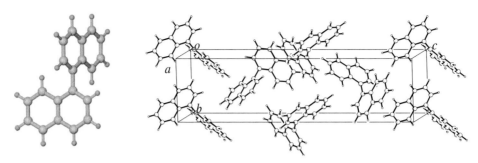

Fig. 9.6.13.
Molecular and crystal structure of 1,1'-binaphthyl.

urea structure is that it provides the first example of a carbonyl O atom which forms four acceptor N–H···O hydrogen bonds. The crystal packing of urea is shown in Fig 9.6.14.

(3) Space group $I4/mmm$ (no. 139), multiplicity = 32

Xenon difluoride XeF_2 and calcium carbide (form I) CaC_2 are isomorphous; with $Z = 2$, the linear triatomic molecule occupies Wyckoff position $2(a)$ of site symmetry $4/mmm$. Hg_2Cl_2 (calomel) has a very similar crystal structure, although it is tetraatomic. The molecular structure and crystal packing of XeF_2 and Hg_2Cl_2 are compared in Fig. 9.6.15.

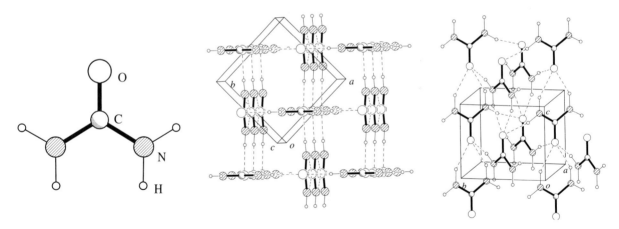

Fig. 9.6.14.
Molecular and crystal structures of urea.

(4) Space group $I4/mcm$ (no. 140), multiplicity = 32

The newly prepared calcium nitridoberyllate $Ca[Be_2N_2]$ crystallizes in this space group with $a = 556.15$ pm, $c = 687.96$ pm, and $Z = 4$. The atoms are located as shown in the following table:

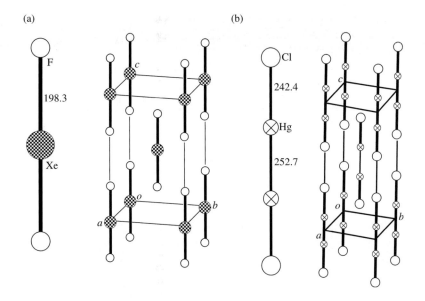

Fig. 9.6.15.
Comparison of the molecular and crystal structure of (a) XeF$_2$ and (b) Hg$_2$Cl$_2$.

Atom	Wyckoff position	Site symmetry	Coordinates $(0,0,0; {}^1\!/_2, {}^1\!/_2, {}^1\!/_2)+$
Be	8(h)	mm	$x, {}^1\!/_2+x, 0; \bar{x}, {}^1\!/_2-x, 0; {}^1\!/_2+x, \bar{x}, 0; {}^1\!/_2-x, x, 0; x = 0.372$
N	8(h)	mm	$x, {}^1\!/_2+x, 0; \bar{x}, {}^1\!/_2-x, 0; {}^1\!/_2+x, \bar{x}, 0; {}^1\!/_2-x, x, 0; x = 0.833$
Ca	4(a)	42	$0, 0, {}^1\!/_4; 0, 0, {}^3\!/_4$

The crystal structure contains planar nitridoberyllate 4.8^2 nets (the notation 4.8^2 indicates that each node in the net is the common vertex of one four-membered and two eight-membered rings), which are stacked in a ...ABAB... sequence but slightly rotated with respect to each other, such that the Be and N atoms aternate along the [001] direction. The slightly irregular octagonal voids between the nitridoberyllate layers are filled by Ca^{2+} ions, as shown in Fig. 9.6.16.

(5) **Space group $I\bar{4}2d$ (no. 122), multiplicity = 16**

meso-(Tetraphenylporphinato)iron(II), [Fe(TPP)], C$_{44}$H$_{28}$N$_4$Fe, has $Z = 4$. The molecule occupies Wyckoff position 4(a) of site symmetry $\bar{4}$; it has a ruffled structure. The dihedral angles between the porphyrin and the pyrrole planes, and between the porphyrin and the benzene planes, are 12.8° and 78.9°, respectively. The molecular and crystal structure of Fe(TPP) are shown in Fig. 9.6.17.

9.6.4 Trigonal and rhombohedral space groups

(1) **Space group $P\bar{3}$ (no. 147), multiplicity = 6**

The crystal structure of the salt [{K(2.2.2-crypt)}$_2$][Pt@Pb$_{12}$] contains a dianion composed of a cage of 12 Pb atoms that encloses a centered Pt atom, denoted

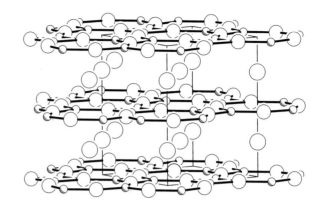

Fig. 9.6.16.
Crystal structure of Ca[Be$_2$N$_2$]; the Be, N and Ca atoms are represented by circles of increasing size. From E. N. Esenturk, J. Fettinger, F. Lam and B. Eichhorn, *Angew. Chem. Int. Ed.* **43**, 2132-4 (2004).

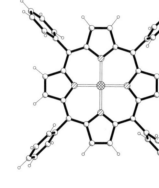 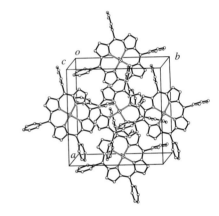

Fig. 9.6.17.
Molecular and crystal structure of [Fe(TPP)].

(a)   (b)

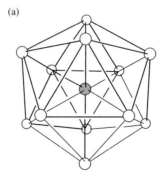 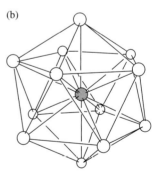

Fig. 9.6.18.
Molecular structure of [Pt@Pb$_{12}$]$^{2-}$: (a) view down the $\bar{3}$-axis; (b) perspective view.

by the symbolic structural formula [Pt@Pb$_{12}$]$^{2-}$, which has near-perfect icosahedral symmetry I_h. With $Z = 1$, the anion occupies a $\bar{3}$ site [Wyckoff position 1(a)], and the [K(2.2.2-crypt)]$^+$ cation lies on a 3-axis [Wyckoff position 2(d)]. The molecular structure of [Pt@Pb$_{12}$]$^{2-}$ is shown in Fig. 9.6.18.

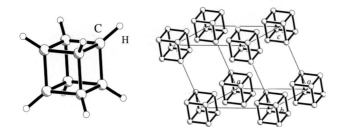

Fig. 9.6.19.
Molecular and crystal structure of cubane.

(2) **Space group $R\bar{3}$ (no. 148), multiplicity = 6 for rhombohedral unit cell**

Cubane was first synthesized by Eaton in 1964. In the crystal, with $Z = 1$, the molecule has symmetry $\bar{3}$, so that there are two independent carbon atoms that occupy general position $6(f)$ and special position $2(c)$ of site symmetry 3. The C–C bond lengths are 155.3(3) and 154.9(3) pm, and the C–C–C bond angles are 89.3°, 89.6°, and 90.5°, all being close to the values expected for a regular cube. Figure 9.6.19 shows the molecular structure and crystal packing of cubane. Historically, the cubane structure was used in the first demonstration of the molecular graphics program ORTEP (Oak Ridge Thermal Ellipsoid Plot) written by C. K. Johnson.

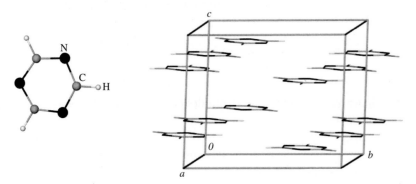

Fig. 9.6.20.
Molecular and crystal structures of s-triazine.

(3) **Space group $R\bar{3}c$ (no. 167), multiplicity = 36 for hexagonal unit cell**

The hexagonal unit cell of s-triazine has $Z = 6$, and the molecule occupies Wyckoff position $2(a)$ of site symmetry 32. The molecule is planar, but it deviates quite markedly from a regular hexagon. From a neutron study, the C–N and C–H bond lengths are 131.5(2) and 93(4) pm, respectively, and the N–C–N angle is 125(1)°. Figure 9.6.20 shows the molecular structure and packing of s-triazine molecules in the unit cell. The molecules are arranged in columns parallel to c with a regular spacing of $c/2$. Besides the pair of neighbors above and below it, each molecule is surrounded by six others at a center-to-center separation of 570 pm.

Crystalline s-triazine undergoes a phase transition from a rhombohedral structure (space group $R\bar{3}c$) at room temperature to a monoclinic structure (space group $C2/c$) below 198 K.

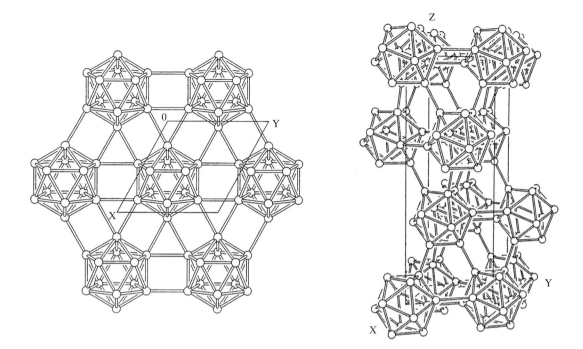

Fig. 9.6.21.
Crystal structure of α-rhombohedral boron: (a) a layer of linked B_{12} icosahedra viewed along the c axis; (b) linkage between layers.

(4) Space group $R\bar{3}m$ (no. 166), multiplicity = 36 for hexagonal unit cell

The hexagonal unit cell of the α-rhombohedral form of elemental boron (α-R12 boron) contains 36 B atoms that form three B_{12} icosahedra. The B_{12} icosahedron occupies Wyckoff position 3(*a*) of site symmetry $\bar{3}m$, so that the asymmetric unit contains two independent B atoms in position 18(*h*) of site symmetry *m*.

In the crystal structure of α-R12 boron, the B_{12} icosahedra are arranged in approximately cubic closest packing and are linked together in a three-dimensional framework. Figure 9.6.21 shows a layer of interlinked icosahedra perpendicular to the $\bar{3}$-axis, and a perspective view of the crystal structure approximately along the *a* axis (note the linkage between layers of icosahedra at $z = 0, 1/3, 2/3, 1$). Further details are given in Section 13.2.

9.6.5 Hexagonal space groups

(1) Space group $P6_122$ (no. 178), multiplicity = 12

Urea forms an isomorphous series of crystalline non-stoichiometric inclusion compounds with *n*-alkanes and their derivatives (including alcohols, esters, ethers, aldehydes, ketones, carboxylic acids, amines, nitriles, thioalcohols, and thioethers), provided that their main chain contains six or more carbon atoms.

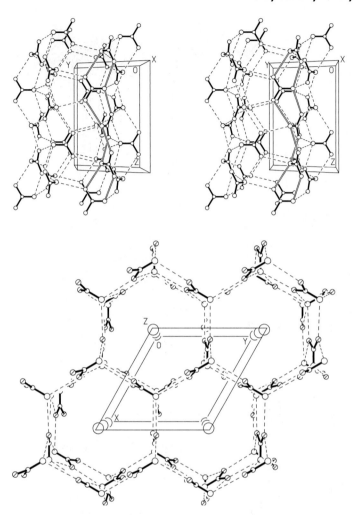

Fig. 9.6.22.
(Top) Stereodrawing of the hydrogen-bonded urea host lattice showing an empty channel extending parallel to the c axis. A shoulder-to-shoulder urea ribbon is highlighted by showing its N–H···O hydrogen bonds as open lines. (Bottom) The channel-type structure of inclusion compounds of urea viewed along the c axis. For clarity the guest molecules are represented by large circles.

The hexagonal unit cell, belonging to space group $P6_122$ or $P6_522$, contains six urea molecules in Wyckoff position $6(a)$ of site symmetry 2. The urea host molecules are arranged in three twined helical spirals, which are interlinked by hydrogen bonds to form the walls of each channel, within which the guest molecules are densely packed (Fig. 9.6.22, top). The host molecules are almost coplanar with the walls of the hexagonal channel, and the channels are packed in a distinctive honeycomb-like manner (Fig. 9.6.22, bottom). The host lattice is stabilized by the maximum possible number of hydrogen bonds that lie virtually within the walls of the hexagonal channels: each NH_2 group forms two donor bonds, and each O atom four acceptor bonds, with their neighbors.

(2) Space group $P6/mcc$ (no. 192), multiplicity = 24

Beryl has the structural formula $Be_3Al_2[Si_6O_{18}]$, with $a = 920.88(5)$, $c = 918.96(7)$ pm and $Z = 2$. The structure of beryl was determined by Bragg

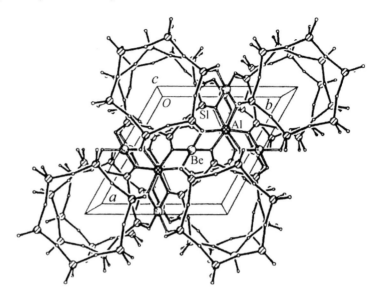

Fig. 9.6.23.
Crystal structure of beryl.

and West in 1926. The atoms are located in the following general and special positions.

Atom	Wyckoff position	Site symmetry	x	y	z
Si	12(ℓ)	m	0.3875	0.1158	0
Be	6(f)	222	1/2	0	1/4
Al	4(c)	32	2/3	1/3	1/4
O(1)	12(ℓ)	m	0.3100	0.2366	0
O(2)	24(m)	1	0.4988	0.1455	0.1453

The crystal structure consists of [SiO$_4$] tetrahedra, which are connected by sharing oxygen atoms of the type O(1) so as to form a sixfold ring. The resulting [Si$_6$O$_{18}$]$^{2-}$ cyclic anions are arranged in parallel layers about the $z = 0$ and $z = 1/4$ planes. They are further linked together by bonding, through oxygen atoms of the type O(2), to Be^{II} and Al^{III} ions, to give a three-dimensional framework, as shown in Fig. 9.6.23. The bond distances are Si–O = 159.2 to 162.0 pm in the [SiO$_4$] tetrahedron, Be–O = 165.3 pm in the [BeO$_4$] tetrahedron, and Al–O = 190.4 pm in the [AlO$_6$] octahedron.

(3) **Space group $P6_3/mmc$ (no. 194), multiplicity = 24**

The crystal structure of cesium iron fluoride, Cs$_3$Fe$_2$F$_9$, comprises a packing of binuclear face-sharing octahedral Fe$_2$F$_9^{3-}$ units and Cs$^+$ ions. With $a = 634.7(1)$, $c = 1480.5(3)$ pm and $Z = 2$, the locations of the atoms and the Fe$_2$F$_9^{3-}$ anion are shown in the following table:

Symmetry in Crystals   353

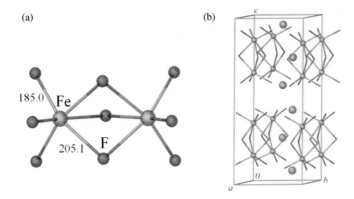

Fig. 9.6.24.
(a) Molecular structure of the $Fe_2F_9^{3-}$ anion and (b) crystal structure of its Cs^+ salt.

Atom/Unit	Wyckoff position	Site symmetry	x	y	z
Cs(1)	2(d)	$\bar{6}m2$	1/3	2/3	3/4
Cs(2)	4(f)	3m	1/3	2/3	0.43271
Fe	4(e)	3m	0	0	0.15153
F(1)	6(h)	mm	0.1312	0.2624	1/4
F(2)	12(k)	1	0.1494	0.2988	0.5940
$Fe_2F_9^{3-}$	2(b)	$\bar{6}m2$	0	0	1/4

The structure of the $Fe_2F_9^{3-}$ anion and the crystal packing in $Cs_3Fe_2F_9$ are shown in Fig. 9.6.24.

9.6.6 Cubic space groups

(1) Space group $Pa\bar{3}$ (no. 205), multiplicity = 24

The alums are a group of hydrated double salts with the general formula $M^IM^{III}(XO_4)_2 \cdot 12H_2O$ in which M^I is a monovalent metal such as $Na^+, K^+, Rb^+, Cs^+, NH_4^+$, or Tl^+; M^{III} is a trivalent metal such as $Al^{3+}, Cr^{3+}, Fe^{3+}, Rh^{3+}, In^{3+}$, or Ga^{3+} and X is S or Se. They crystallize in the space group $Pa\bar{3}$ (no. 205) but occur in three distinct types, α, β, and γ; the order of increasing M^I size is generally γ, α, β, but this is not always true. There are only a few known γ alums with sodium as the monovalent M^I ion. Most cesium sulfate alums adopt the β structure, but exceptions occur when M^{III} is cobalt, rhodium, or iridium, for which the cesium sulfate alums have the α structure. The best known alum is potassium aluminum alum (potash alum), $KAl(SO_4)_2 \cdot 12H_2O$, which belongs to the α type.

The crystal structure of $KAl(SO_4)_2 \cdot 12H_2O$ is cubic, with $a = 1215.8$ pm and $Z = 4$; if the formula is written as $K_2SO_4 \cdot Al_2(SO_4)_3 \cdot 24H_2O$, then $Z = 2$. The Laue symmetry of the diffraction pattern is $m\bar{3}$; the systematic absences are $0k\ell$ reflections with k odd, $h0\ell$ with ℓ odd, and $hk0$ with h odd. From these conditions, the space group can be uniquely established as $Pa\bar{3}$ (no. 205). The space group symbol indicates a primitive lattice, with a glide plane normal to the a axis (the glide direction is 1/2 along b, i.e., a glide component of $b/2$) and four $\bar{3}$-axes along the body diagonals of the unit cell. Since the crystal system is cubic, the symbol a also implies a glide plane normal to b with glide component $c/2$, and a glide plane normal to c with glide component $a/2$.

All atomic positional parameters of the potash alum structure can be assigned to the following general and special equivalent positions of space group $Pa\bar{3}$.

Wyckoff position	Site symmetry	Coordinates
24(d)	1	$x, y, z; \bar{x}, \bar{y}, \bar{z}$; etc.
8(c)	3	$x, x, x; \bar{x}, \bar{x}, \bar{x}$; etc.
4(b)	$\bar{3}$	$1/2, 1/2, 1/2; 1/2, 0, 0; 0, 1/2, 0; 0, 0, 1/2$
4(a)	$\bar{3}$	$0, 0, 0; 0, 1/2, 1/2; 1/2, 0, 1/2; 1/2, 1/2, 0$

There are four K^+ and four Al^{3+} ions in the unit cell, and they can be assigned to the $\bar{3}$ sites 4(a) and 4(b), respectively. The reverse assignment makes no real difference as it merely involves a shift of the origin of the unit cell. Once sites 4(a) and 4(b) have been filled, the eight sulfur atoms must be assigned to 8(c) which means that the SO_4^{2-} ion lies on a 3-axis. The 32 sulfate oxygens can be assigned to $8(c) + 24(d)$; since the sulfate ion is tetrahedral, only one of its oxygen atom can lie on the same 3-axis; there are thus two types of independent sulfate oxygen atoms. The 48 water oxygen atoms can be assigned to $2 \times 24(d)$, and their 96 water hydrogen atoms to $4 \times 24(d)$. In summary, the positional parameters of non-hydrogen atoms can be assigned as follows:

Assignment	Parameters	Remarks
K(1) in 4(a)	none	Arbitrarily placed
Al(1) in 4(b)	none	Must be placed in this position
S(1) in 8(c)	x_S	S atom lying on 3-axis
O(1) in 8(c)	x_{O1}	S-O bond lying on 3-axis
O(2) in 24(d)	x_{O2}, y_{O2}, z_{O2}	The other sulfate O in general position
O(1w) in 24(d)	$x_{O1w}, y_{O1w}, z_{O1w}$	Water molecule coordinated to K^+
O(2w) in 24(d)	$x_{O2w}, y_{O2w}, z_{O2w}$	Water molecule coordinated to Al^{3+}

Therefore, eleven parameters are sufficient to define the crystal structure, and an additional $4 \times 3 = 12$ parameters are required if the hydrogen atoms are also included. When a water molecule is coordinated to a K^+ or Al^{3+} cation, an octahedral coordination environment is generated around the cation by the $\bar{3}$ symmetry operation. Therefore, the crystal structure of $KAl(SO_4)_2 \cdot 12H_2O$ can be regarded as a packing of $[K(H_2O)_6]^+$ and $[Al(H_2O)_6]^{3+}$ octahedra and SO_4^{2-} tetrahedra. If the structural formula of $KAl(SO_4)_2 \cdot 12H_2O$ is rewritten as $K(H_2O)_6 \cdot Al(H_2O)_6 \cdot 2SO_4$, it is easy to see that the three kinds of moieties can be assigned to Wyckoff positions 4(a), 4(b), and 8(c), respectively. Both hydrated cations have site symmetry $\bar{3}$ (a subgroup of $m\bar{3}m$ for a regular octahedron), and the sulfate anion has site symmetry 3 (a subgroup of $\bar{4}3m$ for a regular tetrahedron). The crystal structure of potash alum is shown in Fig. 9.6.25.

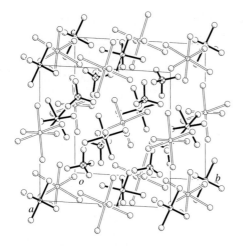

Fig. 9.6.25.
Crystal structure of potash alum.

Representative examples of the β and γ alums are $CsAl(SO_4)_2 \cdot 12H_2O$ and $NaAl(SO_4)_2 \cdot 12H_2O$, respectively. In the α and γ alums, the six water molecules generated from O(1w) constitute a flattened octahedron centered at M^I. In contrast, in a β alum these six water molecules lie almost in a planar hexagon around the large M^I atom. In both α and β alums, the S–O(1) bond points away from the nearest M^I atom, but in the γ form the S–O(1) bond points towards it. Accordingly, the coordination number of M^I in a γ alum is eight (six waters and two sulfate oxygens of type O(1)), which is increased to twelve for the α and β alums (six waters and six sulfate oxygens of type O(2)). In both α and β alums, the $M^{III}(H_2O)_6$ unit is a distorted octahedron whose principal axes are nearly parallel to the unit-cell axes; in contrast, the $M^{III}(H_2O)_6$ octahedron in the γ form is very regular but rotated by approximately 40° about the body diagonal of the unit cell. The crystal structures of the β and γ forms are compared in Fig. 9.6.26.

(2) Space group $I\bar{4}3m$ (no. 217), multiplicity = 48

Hexamethylenetetramine, $(CH_2)_6N_4$, was the first organic molecule to be subjected to crystal structure analysis. The crystals are cubic, with $a = 702$ pm and measured density $\rho = 1.33$ g cm^{-3}. Calculation using the formula $Z = \rho N_o V/M = (1.33)(6.023 \times 10^{23})(7.02 \times 10^{-8})^3/140 \sim 2$ shows that there are two molecules in the unit cell.

From the X-ray photographs, the Laue symmetry is established as cubic $m\bar{3}m$ and the systematic absences are $hk\ell$ with $(h+k+\ell)$ odd. All three point groups 432, $\bar{4}3m$, and $m\bar{3}m$ belong to the same Laue class $m\bar{3}m$, and three space groups $I432$, $I\bar{4}3m$, and $Im\bar{3}m$ are consistent with the observed systematic absences and Laue symmetry. Furthermore, as the $(CH_2)_6N_4$ molecule lacks either a 4-axis or a center of symmetry, the true space group can be unequivocally established as $I\bar{4}3m$, which has the following general and special equivalent positions:

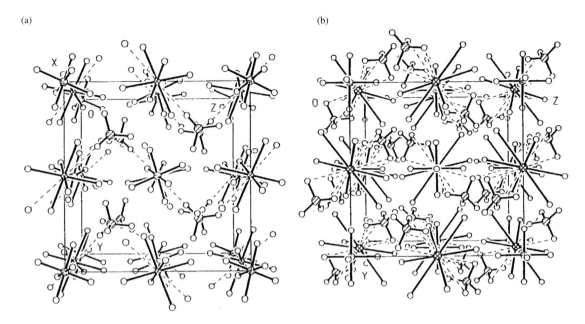

Fig. 9.6.26.
Comparison of the crystal structures of (a) γ alum NaAl(SO$_4$)$_2$·12H$_2$O and (b) β alum CsAl(SO$_4$)$_2$·12H$_2$O. For clarity the bonds between the M^I atoms and the sulfate groups are represented by broken lines.

Wyckoff position	Site symmetry	Coordinates $(0,0,0; 1/2,1/2,1/2)+$
48(h)	1	x, y, z; etc.
24(g)	m	x, x, z; etc.
24(f)	2	$x, 1/2, 0$; etc.
12(e)	mm	$x, 0, 0$; etc.
12(d)	$\bar{4}$	$1/4, 1/2, 0$; etc.
8(c)	3m	x, x, x; etc.
6(b)	$\bar{4}2m$	$0, 1/2, 1/2$; etc.
2(a)	$\bar{4}3m$	$0, 0, 0$

All atomic positions, including hydrogens in the unit cell, can be assigned to the following special positions in space group $I\bar{4}3m$:

Assignment	Parameters	Remarks
(CH$_2$)$_6$N$_4$ in 2(a)	none	molecular symmetry is T_d
N in 8(c)	x_N	N at the intersection of C_3 and three σ_d's
C in 12(e)	x_C	C at the intersection of two σ_d's
H in 24(g)	x_H, z_H	methylene H lies in σ_d

As the structure consists of a body-centered arrangement of (CH$_2$)$_6$N$_4$ molecules, four positional parameters are required for a complete description of the crystal structure. Moreover, the fractional coordinates for carbon and nitrogen can be deduced from relevant chemical data: (a) the van der Waals

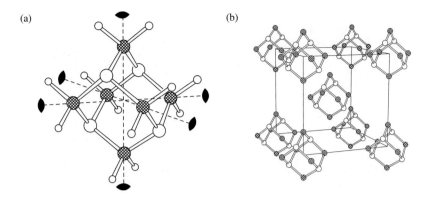

Fig. 9.6.27.
(a) Molecular structure of $(CH_2)_6N_4$, with H atoms represented by small open circles. There are three 2-axes, each passing through a pair of carbon atoms, four 3-axes, each through a nitrogen atom, and six σ_d's, each containing a methylene group. (b) Crystal structure of $(CH_2)_6N_4$; the H atoms are not shown.

separation between carbon atoms in neighboring molecules in an axial direction ∼368 pm and (b) the length of a C–N single bond ∼148 pm.

The carbon atoms having fractional coordinates $(x_C, 0, 0)$ and $(1 - x_C, 0, 0)$ are closest to each other. Therefore, $a[(1 - x_C) - x_C] = 368$, leading to $x_C = 0.238$. Now consider the bond from C at $(x_C, 0, 0)$ to N at (x_N, x_N, x_N), which gives $a[(0.238 - x_N)^2 + x_N^2 + x_N^2)]^{1/2} = 148$. Solution of this equation yields $x_N = 0.127$ or 0.032. Consideration of the molecular dimensions of $(CH_2)_6N_4$ shows that the first answer is the correct one. The molecular and crystal structures of $(CH_2)_6N_4$ are illustrated in Fig. 9.6.27.

Methyllithium, $(CH_3Li)_4$, also crystallizes in space group $I\bar{4}3m$ with $a = 724$ pm and $Z = 2$. The tetrameric unit may be described as a tetrahedral array of Li atoms with a methyl C atom located above the center of each face of the Li_4 tetrahedron or, alternately, as a distorted cubic arrangement of Li atoms and CH_3 groups. The locations of the $(CH_3Li)_4$ unit and its constituent atoms are listed in the following table.

Atom/unit	Wyckoff position	Site symmetry	x	y	z
Li	8(c)	3m	0.131	0.131	0.131
C	8(c)	3m	0.320	0.320	0.320
H	24(g)	m	0.351	0.351	0.192
$(CH_3Li)_4$	2(a)	$\bar{4}3m$	0	0	0

The crystal structure of methyllithium is not built of discrete $(CH_3Li)_4$ molecules. Such tetrameric units are connected together to give a three-dimensional polymeric network, as illustrated in Fig. 9.6.28. The Li–C bond length within a tetrameric unit is 231 pm, which is comparable to the intertetramer Li–C bond distance of 236 pm.

(3) Space group $Fm\bar{3}m$ (no. 225), multiplicity = 192

Adamantane, a structural analog of hexamethylenetetramine, crystallizes at room temperature in the cubic space group $Fm\bar{3}m$ with $a_C = 944.5$ pm and $Z = 4$. The structure is disordered, not ordered in space group $F\bar{4}3m$ (no. 216, multiplicity = 96) as described in a previous report. The cubic close-packed arrangement of $(CH_2)_6(CH)_4$ is favored over the body-centered cubic structure

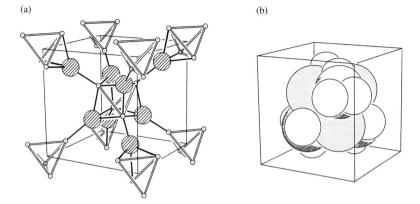

Fig. 9.6.28.
(a) Polymeric crystal structure of methyllithium. (b) Space-filling model of the tetrameric $(CH_3Li)_4$ unit.

of $(CH_2)_6N_4$ because the $(CH_2)_6(CH)_4$ molecule is approximately spherical. The model which gives the best fit with the intensity data consists of hindered reorientations of a rigid molecular skeleton with the CH groups aligned in the [111] directions.

Below $-65°C$, disordered cubic adamantane transforms to an ordered tetragonal form [space group $P\bar{4}2_1c$ (no. 114), multiplicity = 8, $a_T = 660$, $c_T = 881$ pm at $-110°C$, $Z = 2$, molecule occupying site of symmetry $\bar{4}$]. The measured C–C bond lengths have a mean value of 153.6(11) pm. The structure of this low-temperature phase, as illustrated in Fig. 9.6.29, is related to that at room temperature in that a_T lies along the ab face diagonal of the cubic cell $[|a_T| \sim |a|/(2)^{1/2}]$ and c_T corresponds to $c_C = a_C$ in the cubic form.

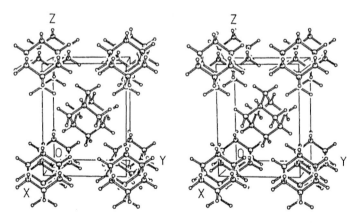

Fig. 9.6.29.
Stereoview of the tetragonal structure of adamantane at $-110°C$.

Many compounds exhibiting the adamantane skeleton can be formally derived by the exchange of "equivalent" atoms or groups. For instance, replacing all four CH groups of $(CH_2)_6(CH)_4$ by nitrogen atoms gives rise to hexamethylenetetramine, $(CH_2)_6N_4$. Some examples of adamantane-like cage compounds containing equivalent constituents include O_6E_4 (E = P, As, Sb), $(AlMe)_6(CMe)_4$, $(BMe)_6(CH)_4$, $(SnR_2)_6P_4$ (R = Me, Bu, Ph), $E_6(SiH)_4$ (E = S, Se), $S_6(EMe)_4$ (E = Si, Sn, Ge), $S_6(PE)_4$ (E = O, S), $(NMe)_6(PE)_4$ (E = O, S, Se) and $[O_6(TaF_3)_4]^{4-}$.

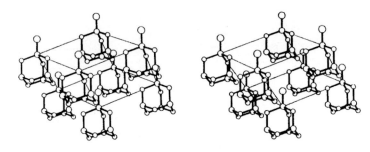

Fig. 9.6.30.
Stereoview of the packing of $(CH_2)_6N_4O$ molecules in a rhombohedral unit cell.

A stereoview of the crystal structure of hexamethylenetetramine N-oxide, $(CH_2)_6N_4O$, which crystallizes in space group $R3m$ (no. 160, multiplicity = 6 with $Z = 1$ in the rhombohedral unit cell), is shown in Fig. 9.6.30. The molecular packing is related to that of $(CH_2)_6N_4$ in a simple way. Starting with a body-centered cubic unit cell of $(CH_2)_6N_4$ (space group $I\bar{4}3m$, $Z = 2$), one can arrive at the $(CH_2)N_4O$ structure in three steps: (i) remove the $(CH_2)_6N_4$ molecule at the body-center, then (ii) introduce the N→O function in the [111] direction to each remaining molecule, and finally (iii) compress the resulting primitive lattice along [111] until the oxygen atom of each $(CH_2)_6N_4O$ molecule nests snugly in a recess formed by three hydrogen atoms of a neighboring molecule.

The borine adduct $(CH_2)_6N_4 \cdot BH_3$ and the hydrohalides $[(CH_2)_6N_4H]^+X^-$ (X = Cl, Br) are isomorphous with $(CH_2)_6N_4O$. Table 9.6.1 shows the relationship of hexamethylenetetramine with its quaternized derivatives and the effect of the different "substituents."

Table 9.6.1. Structural relationship of $(CH_2)_6N_4$ and some quaternized derivatives

Compound	Space group	Z	a_R (pm)	α_R (°)	V_R (pm^3)/Z*
$(CH_2)_6N_4$	$I\bar{4}3m$	2	702.1	90	173.0×10^6
$(CH_2)_6N_4O$	$R3m$	1	591.4	106.3	175.5×10^6
$[(CH_2)_6N_4H]Cl$	$R3m$	1	594.7	97.11	205.0×10^6
$(CH_2)_6N_4 \cdot BH_3$	$R3m$	1	615.0	103.2	210.6×10^6
$[(CH_2)_6N_4H]Br$	$R3m$	1	604.0	96.13	216.2×10^6

* V_R is the volume of the rhombohedral unit cell.

(4) Space group $Fd\bar{3}m$ (No. 227), multiplicity = 192

The structure of spinel, $MgAl_2O_4$, was first analyzed independently by W. H. Bragg and S. Nishikawa in 1915. It has an essentially cubic close-packed array of oxide ions with Mg^{2+} and Al^{3+} ions occupying tetrahedral and octahedral interstices, respectively. With $a = 808.00(4)$ pm and $Z = 8$, the atoms are located in special positions as shown below.

Atom	Wyckoff position	Site symmetry	x	y	z
Mg	8(a)	$\bar{4}3m$	0	0	0
Al	16(d)	$\bar{3}m$	5/8	5/8	5/8
O	32(e)	$3m$	0.387	0.387	0.387

Origin at $\bar{4}3m$, at $(-1/8, -1/8, -1/8)$ from center $(\bar{3}m)$.

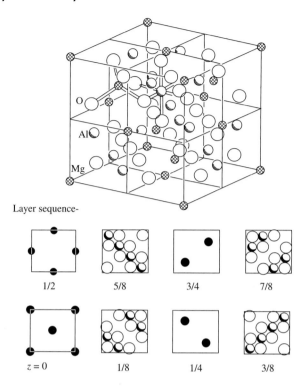

Fig. 9.6.31.
Crystal structure of spinel.

Figure 9.6.31 shows the crystal structure of spinel, the unit cell being divided into octants. The MgO$_4$ tetrahedra and AlO$_6$ octahedra are arranged alternately, so that every oxide ion is the common vertex of one tetrahedron and three octahedra.

The type II clathrate hydrates of the general formula 8G$_L$·16G$_S$·136H$_2$O also crystallize in space group $Fd\bar{3}m$ with $a \sim 1.7$ nm and $Z = 1$. The water molecules are connected by O–H$\cdots$O hydrogen bonds to form a host network composed of a close-packed arrangement of (H$_2$O)$_{20}$ pentagonal dodecahedra (12-hedra bounded by 12 pentagonal faces, idealized symmetry I_h) and (H$_2$O)$_{24}$ hexakaidecahedra (16-hedra composed of 12 pentagons and 4 hexagons, idealized symmetry T_d) in the ratio of 2 to 1. In the crystal structure, both kinds of polyhedra have slightly nonplanar faces with unequal edges and angles. The large guest molecules G$_L$ are accommodated in the 16-hedra, whereas the small guest species G$_S$ occupy the 12-hedra, as illustrated in Fig. 9.6.32(a). Note that here the choice of origin of the space group differs from that used in the structural description of spinel.

Formation of type II clathrate hydrates is facilitated by passing in H$_2$S or H$_2$Se as a "help gas" during the preparation. Crystal structure analyses have been reported for the tetrahydrofuran–H$_2$S hydrate (8C$_4$H$_8$O·16H$_2$S·136H$_2$O) at 250 K, the carbon disulphide–H$_2$S hydrate (8CS$_2$·16H$_2$S·136H$_2$O) at 140 K, and the carbon tetrachloride–Xe hydrate 8CCl$_4$·nXe·136D$_2$O ($n = 3.5$) at 13 K and 100 K. For the last compound, the host and guest molecules are assigned to the special positions in the following table. The CCl$_4$ molecule exhibits orientational disorder, and the Xe atom has a statistical site occupancy of 0.22.

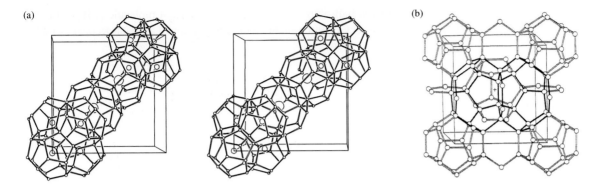

Fig. 9.6.32.
(a) Stereogram illustrating the type II clathrate hydrate host framework, with two 16-hedra centered at (3/8, 3/8, 3/8) and (5/8, 5/8, 5/8), and two cluster units (each consisting of four 12-hedra) centered at (1/8, 1/8, 1/8) and (7/8, 7/8, 7/8). Each line represents a O–H···O hydrogen bond between two host water molecules. The open circles of decreasing sizes represent guest species G_L and G_S, respectively, located inside the corresponding 16- and 12-hedra. (b) The host framework of a type I clathrate hydrate. The 12-hedra occupy the center and corners of the cubic unit cell, and two fused 14-hedra are highlighted by darkened edges.

Atom	Wyckoff position	Site symmetry	x	y	z
$H_2O(1)$	$8(a)$	$\bar{4}3m$	1/8	1/8	1/8
$H_2O(2)$	$32(e)$	$3m$	0.21658	0.21658	0.21658
$H_2O(3)$	$96(g)$	m	0.18215	0.18215	0.36943
$G_L = CCl_4$	$8(b)$	$\bar{4}3m$	3/8	3/8	3/8
$G_S = Xe$	$16(c)$	$3m$	0	0	0

Origin at center $\bar{3}m$, at (1/8, 1/8, 1/8) from $\bar{4}3m$. At 100 K, $a = 1.7240$ nm.

Since small gas species such as O_2, N_2, O_3, and Kr may be trapped inside the 12-hedra, all type II clathrate hydrates usually contain non-stoichiometric amounts of air unless special precautions are taken to avoid its inclusion as guest species.

(5) **Space group $Pm\bar{3}n$ (no. 223), multiplicity = 48**

The type I clathrate hydrates of the general formula $6G_L \cdot 2G_S \cdot 46H_2O$ crystallize in this space group with $a \sim 1.2$ nm and $Z = 1$. The hydrogen-bonded host framework is composed of fused $(H_2O)_{20}$ pentagonal dodecahedra and $(H_2O)_{22}$ tetrakaidecahedra (14-hedra composed of 12 pentagons and 2 hexagons, idealized symmetry D_{6d}) in the ratio of 1 to 3, as illustrated in Fig. 9.6.32(b). The 12-hedra are generally empty or partially filled by air as guest species G_S, and the 14-hedra can accommodate a variety of guest species G_L such as Xe, Cl_2, SO_2, CH_4, CF_4, CH_3Br, and ethylene oxide. The type I clathrates of COS, CH_3SH, and CH_3CHF_2 are reported to change to type II in the presence of H_2S as a help gas.

In a neutron diffraction study of the ethylene oxide clathrate deuterohydrate $6(CH_2)_2O \cdot 46D_2O$, the hydrogen-bonded host framework is defined by three independent deuterated water oxygen atoms occupying special positions, as shown in the table below. The deuterium atoms are located precisely as disordered half-atoms. The ethylene oxide guest molecule $(CH_2)_2O$ is disordered about G_L, and a small amount of entrapped air (site occupancy 0.067) is disordered about G_S.

Atom	Wyckoff position	Site symmetry	x	y	z
O(i)	16(i)	3	0.18375	0.18375	0.18375
O(k)	24(k)	m	0	0.3082	0.1173
O(c)	6(c)	$\bar{4}2m$	0	$1/2$	$1/4$
D(ii)	16(i)	3	0.2323	0.2323	0.2323
D(ck)	24(k)	m	0	0.4340	0.2015
D(kc)	24(k)	m	0	0.3787	0.1588
D(kk)	24(k)	m	0	0.3161	0.2015
D(ki)	48(l)	1	0.0676	0.2648	0.1397
D(ik)	48(l)	1	0.1170	0.2276	0.1588
G_L	6(d)	$\bar{4}2m$	0	$1/4$	$1/2$
G_S	2(a)	$m\bar{3}$	0	0	0

Origin at center $m\bar{3}$. All D atoms have a site occupancy factor of 1/2; D(ck) is a deuterium atom that is covalently bound to O(c) and hydrogen-bonded to O(k). At 80K, $a = 1.187$ nm.

References

The principles and practice of the determination of crystal and molecular structures by X-ray crystallography (single-crystal X-ray analysis) are not covered in this book. The interested reader may consult references [15] to [24], which are arranged in increasing levels of sophistication.

1. W. Borchardt-Ott, *Crystallography*, 2nd edn., Springer-Verlag, Berlin, 1995.
2. D. Schwarzenbach, *Crystallography*, Wiley, Chichester, 1996.
3. J.-J. Rousseau, *Basic Crystallography*, Wiley, Chichester, 1998.
4. C. Hammond, *The Basics of Crystallography and Diffraction*, Oxford University Press, New York, 1997.
5. J. A. K. Tareen and T. R. N. Kutty, *A Basic Course in Crystallography*, Universities Press, Hyderabad, 2001.
6. G. Burns and A. M. Glazer, *Space Groups for Solid State Scientists*, 2nd edn., Academic Press, New York, 1990.
7. T. Hahn (ed.), *Brief Teaching Edition of International Tables for Crystallography, Volume A: Space-group symmetry*, 5th edn. (corrected rpt), Springer, Chichester, 2005.
8. T. Hahn (ed.), *International Tables for Crystallography, Volume A: Space-group symmetry*, 5th edn. (corrected reprint), Kluwer Academic, Dordrecht, 2005.
9. E. Prince (ed.), *International Tables for Crystallography, Volume C: Mathematical, Physical and Chemical Tables*, 3rd edn., Kluwer Academic, Dordrecht, 2004.
10. M. O'Keeffe and B. G. Hyde, *Crystal Structures. I. Patterns and Symmetry*, Mineralogical Society of America, Washington, DC, 1996.
11. J. Bernstein, *Polymorphism in Molecular Crystals*, Oxford University Press, New York, 2002.
12. R. C. Buchanan and T. Park, *Materials Crystal Chemistry*, Marcel Dekker, New York, 1997.
13. B. Douglas and S.-M. Ho, *Structure and Chemistry of Crystalline Solids*, Springer, New York, 2006.
14. J. F. Nye, *Physical Properties of Crystals*, Oxford University Press, Oxford, 1957.

15. J. P. Glusker and K. N. Trueblood, *Crystal Structure Analysis: A Primer*, 2nd edn., Oxford University Press, New York, 1985.
16. W. Clegg, *Crystal Structure Determination*, Oxford University Press, New York, 1998.
17. W. Massa, *Crystal Structure Determination*, 2nd edn., Springer-Verlag, Berlin, 2004.
18. U. Shmueli, *Theories and Techniques of Crystal Structure Determination*, Oxford University Press, Oxford, 2007.
19. P. Müller, R. Herbst-Irmer, A. Spek, T. Schneider and M. Sawaya, *Crystal Structure Refinement: A Crystallographer's Guide to SHELXL*, Oxford University Press, Oxford, 2006.
20. M. Ladd and R. Palmer, *Structure Determination by X-Ray Crystallography*, 4th edn., Kluwer Academic/Plenum Publishers, New York, 2003.
21. J. P. Glusker, M. Lewis and M. Rossi, *Crystal Structure Analysis for Chemists and Biologists*, VCH, New York, 1994.
22. M. M. Woolfson, *An Introduction to X-Ray Crystallography*, 2nd edn., Cambridge University Press, 1997.
23. C. Giacovazzo (ed.), *Fundamentals of Crystallography*, 2nd edn., Oxford University Press, 2002.
24. J. D. Dunitz, *X-Ray Analysis and the Structure of Organic Molecules*, 2nd corrected rpt, VCH, New York, 1995.
25. D. M. Blow, *Outline of Crystallography for Biologists*, Oxford University Press, Oxford, 2002.
26. G. Rhodes, *Crystallography Made Crystal Clear: A Guide for Users of Macromolecular Models*, 3rd edn., Academic Press, San Diego, CA, 2006.
27. D. J. Dyson, *X-Ray and Electron Diffraction Studies in Materials Science*, Maney, London, 2004.
28. W. I. F. David, K. Shankland, L. B. McCusker and Ch. Baerlocher (eds.), *Structure Determination from Powder Diffraction Data*, Oxford University Press, Oxford, 2002.
29. C. C. Wilson, *Single Crystal Neutron Diffraction from Molecular Materials*, World Scientific, Singapore, 2000.
30. G. A. Jeffrey, *An Introduction to Hydrogen Bonding*, Oxford University Press, New York, 1997.
31. I. D. Brown, *The Chemical Bond in Inorganic Chemistry: The Valence Bond Model*, Oxford University Press, New York, 2002.
32. A. F. Wells, *Structural Inorganic Chemistry*, 5th edn., Oxford University Press, Oxford, 1984.
33. T. C. W. Mak and G.-D. Zhou, *Crystallography in Modern Chemistry: A Resource Book of Crystal Structures*, Wiley–Interscience, New York, 1992; Wiley Professional Paperback Edition, 1997.
34. T. C. W. Mak, M. F. C. Ladd and D.C. Povey, Molecular and crystal structure of hexamethylenetetramine oxide. *J. C. S. Perkin II*, 593–5 (1979).
35. T. C. W. Mak and F.-C. Mok, Inorganic cages related to cubane and adamantane. *J. Cryst. Mol. Struct.* **8**, 183–91 (1978).
36. T. C. W. Mak and R. K. McMullan, Polyhedral clathrate hydrates. X. Structure of the double hydrate of tetrahydrofuran and hydrogen sulfide. *J. Chem. Phys.* **42**, 2732–7 (1965).
37. R. K. McMullan and Å. Kvick, Neutron diffraction study of structure II clathrate: $3.5Xe \cdot 8CCl_4 \cdot 136D_2O$. *Acta Crystallogr., Sect. B* **46**, 390–9 (1990).

10 Basic Inorganic Crystal Structures and Materials

Over 25 million compounds are known, but only about one percent of them have their crystal structures elucidated by X-ray and neutron diffraction methods. Many inorganic structures are closely related to each other. Sometimes one basic structural type can incorporate several hundred compounds, and new crystalline compounds may arise from the replacement of atoms, deformation of the unit cell, variation of the stacking order, and the presence of additional atoms in interstitial sites.

This chapter describes some basic inorganic crystal structures and their relationship to the new structural types, which often exhibit different symmetry and characteristic properties that make them useful in materials research and industrial applications.

10.1 Cubic closest packing and related structures

10.1.1 Cubic closest packing (ccp)

Many pure metals and noble gases adopt the cubic closest packing (ccp) structure of identical spheres, which is based on a face-centered cubic (fcc) lattice in space group $O_h^5 - Fm\bar{3}m$. The atomic coordinates and equivalent points of this space group are listed in Table 10.1.1. These data are taken from *International Tables for Crystallography*, Vol. A.

Table 10.1.1. Atomic coordinates* of space group $O_h^5 - Fm\bar{3}m$

Multiplicity	Wyckoff position	Site symmety	Coordinates $\left(0,0,0;\ 0,\tfrac{1}{2},\tfrac{1}{2};\ \tfrac{1}{2},0,\tfrac{1}{2};\ \tfrac{1}{2},\tfrac{1}{2},0\right) +$
32	f	3m	$x,x,x;\ x,\bar{x},\bar{x};\ \bar{x},x,\bar{x};\ \bar{x},\bar{x},x;$ $\bar{x},\bar{x},\bar{x};\ \bar{x},x,x;\ x,\bar{x},x;\ x,x,\bar{x}.$
24	e	4mm	$x,0,0;\ 0,x,0;\ 0,0,x;\ \bar{x},0,0;\ 0,\bar{x},0;\ 0,0,\bar{x}.$
24	d	mmm	$0,\tfrac{1}{4},\tfrac{1}{4};\ \tfrac{1}{4},0,\tfrac{1}{4};\ \tfrac{1}{4},\tfrac{1}{4},0;\ 0,\tfrac{1}{4},\tfrac{3}{4};\ \tfrac{3}{4},0,\tfrac{1}{4};\ \tfrac{1}{4},\tfrac{3}{4},0.$
8	c	$\bar{4}3m$	$\tfrac{1}{4},\tfrac{1}{4},\tfrac{1}{4};\ \tfrac{3}{4},\tfrac{3}{4},\tfrac{3}{4}.$
4	b	$m\bar{3}m$	$\tfrac{1}{2},\tfrac{1}{2},\tfrac{1}{2}$
4	a	$m\bar{3}m$	$0,0,0$

* This table is not complete; equivalent positions with multiplicities > 32 are not listed.

In three-dimensional closest packing, the spherical atoms are located in position 4(a). There are two types of interstices: octahedral and tetrahedral holes which occupy positions 4(b) and 8(c), respectively. The number of tetrahedral holes is twice that of the spheres, while the number of octahedral holes is equal to that of the spheres. The positions of the holes are shown in Fig.10.1.1.

The elements with the ccp structure and their unit-cell parameters are listed in Table 10.1.2.

Some alloys in high-temperature disordered phases adopt the ccp structure. For example, Au and Cu have the same valence electronic configuration. When these two metals are fused and then cooled, the mixture solidifies into a completely disordered solid solution phase, in which the gold atom randomly and statistically replaces the copper atom. Figure 10.1.2(a) shows the structure of $Cu_{1-x}Au_x$ with $x = 0.25$, and the space group is still $O_h^5 - Fm\bar{3}m$. If the $Cu_{0.75}Au_{0.25}$ alloy (or Cu_3Au) is annealed at a temperature lower than 395°C through isothermal ordering, the gold atoms occupy the corners of the

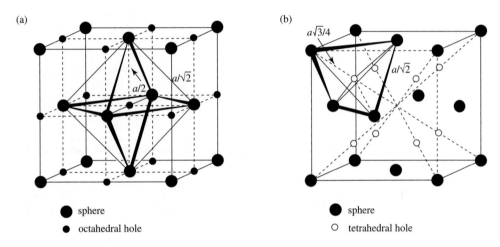

Fig. 10.1.1.
Hole positions in ccp structure.

Table 10.1.2. Elements with the ccp structure and their unit-cell parameters

Element	a (pm)	Element	a (pm)	Element	a (pm)
Ac	531.1	Cu	361.496	Pd	388.98
Ag	408.62	γ-Fe	359.1	Pt	392.31
Al	404.958	In	324.4	Rh	380.31
Am	489.4	Ir	383.94	β-Sc	454.1
Ar	525.6(4.2 K)	Kr	572.1(58K)	α-Sr	608.5
Au	407.825	β-La	530.3	α-Th	508.43
α-Ca	558.2	γ-Mn	385.5	Xe	619.7(58K)
γ-Ce	516.04	Ne	442.9(4.2K)	α-Yb	548.1
β-Co	454.8	Ni	352.387		
β-Cr	368	Pb	495.05		

Fig. 10.1.2.
Structure of (a) disordered $Cu_{1-x}Au_x$ and (b) α-Cu_3Au.

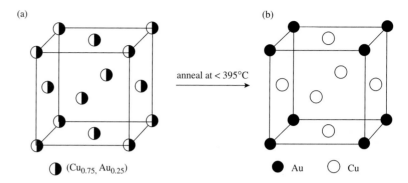

Table 10.1.3. Equivalent positions* of space group $O_h^1 - Pm\bar{3}m$

Multiplicity	Wyckoff position	Site symmetry	Coordinates
8	g	$3m$	$x,x,x;\ x,\bar{x},\bar{x};\ \bar{x},x,\bar{x};\ \bar{x},\bar{x},x;$ $\bar{x},\bar{x},\bar{x};\ \bar{x},x,x;\ x,\bar{x},x;\ x,x,\bar{x}.$
6	f	$4mm$	$x,\frac{1}{2},\frac{1}{2};\ \frac{1}{2},x,\frac{1}{2};\ \frac{1}{2},\frac{1}{2},x;\ \bar{x},\frac{1}{2},\frac{1}{2};\ \frac{1}{2},\bar{x},\frac{1}{2};\ \frac{1}{2},\frac{1}{2},\bar{x}.$
6	e	$4mm$	$x,0,0;\ 0,x,0;\ 0,0,x;\ \bar{x},0,0;\ 0,\bar{x},0;\ 0,0,\bar{x}.$
3	d	$4/mmm$	$\frac{1}{2},0,0;\ 0,\frac{1}{2},0;\ 0,0,\frac{1}{2}.$
3	c	$4/mmm$	$0,\frac{1}{2},\frac{1}{2};\ \frac{1}{2},0,\frac{1}{2};\ \frac{1}{2},\frac{1}{2},0.$
1	b	$m\bar{3}m$	$\frac{1}{2},\frac{1}{2},\frac{1}{2}$
1	a	$m\bar{3}m$	$0,0,0.$

* This table is not complete; equivalent positions with multiplicities greater than 8 are not listed.

cube while the copper atoms occupy the centers of the faces, as shown in Fig. 10.1.2(b). This disordered-to-ordered phase transition gives rise to a simple cubic α-phase in space group $O_h^1 - Pm\bar{3}m$. The equivalent positions of space group $O_h^1 - Pm\bar{3}m$ are listed in Table 10.1.3. In the structure of Cu_3Au in this low-temperature ordered phase, the Au atom is located in position $1(a)$, and the Cu atoms occupy position $3(c)$.

10.1.2 Structure of NaCl and related compounds

(1) Structure type of sodium chloride (rock salt, halite), and its common occurrence

The structure of NaCl can be regarded as being formed from a ccp arrangement of Cl^- anions with the smaller Na^+ cations located in the octahedral interstices. Since spherical Na^+ and Cl^- conform to the $m\bar{3}m$ symmetry of the sites $4(a)$ and $4(b)$, respectively (listed in Table 10.1.1), the space group of NaCl is also $O_h^5 - Fm\bar{3}m$. Either the Na^+ or Cl^- ion can be placed at the origin (site $4a$), and each has coordination number six. There are more than 400 binary compounds that belong to the NaCl structural type, such as alkali metal halides and hydrides, alkaline-earth oxides and chalcogenides, the oxides of divalent early first-row transition metals, the chalcogenides of divalent lanthanides and actinides, the

Table 10.1.4. Compounds with the NaCl structure and parameters of their cubic unit cells

Compound	a (pm)	Compound	a (pm)	Compound	a (pm)
AgBr	577.45	LiH	408.5	ScSb	585.9
AgCl	554.7	LiI	600.0	SnAs	572.48
AgF	492	MgO	421.12	SnSb	613.0
BaO	553.9	MgS	520.33	SnSe	602.3
BaS	638.75	MgSe	545.1	SnTe	631.3
BaSe	660.0	MnO	444.48	SrO	516.02
BaTe	698.6	MnS	522.36	SrS	601.98
CaO	481.05	MnSe	544.8	SrSe	623
CaS	569.03	NaBr	597.324	SrTe	666.0
CaSe	591	NaCl	562.779	TaC	445.40
CaTe	634.5	NaF	462.0	TaO	442.2
CdO	469.53	NaH	488.0	TiC	431.86
CoO	426.67	NaI	647.28	TiN	423.5
CrN	414.0	NbC	446.91	TiO	417.66
CsF	600.8	NiO	416.84	VC	418.2
CsH	637.6	PbS	593.62	VN	412.8
FeO	430.7	PbSe	612.43	YAs	578.6
KBr	660.0	PbTe	645.4	YN	487.7
KCl	629.294	PdH	402	YTe	609.5
KF	534.7	RbBr	685.4	ZrB	465
KH	570.0	RbCl	658.10	ZrC	468.28
KI	706.555	RbF	564	ZrN	456.7
LaN	530	RbH	603.7	ZrO	462
LiBr	550.13	RbI	734.2	ZrP	527
LiCl	512.954	ScAs	548.7	ZrS	525
LiF	401.73	ScN	444		

nitrides, phosphides, arsenides and bismuthides of the lanthanide and actinide elements, silver halides (except the iodide), tin and lead chalcogenides, and many interstitial alloys. Some examples and the parameters of their cubic unit cells are listed in Table 10.1.4.

The data listed in Table 10.1.4 are useful in understanding the structure and properties of many compounds. Two examples are presented below.

(a) Deduction of ionic radii

The effective ionic radii listed in Table 4.2.2 can be calculated from the experimentally determined values of the ionic distances or the parameters of the unit cells. For example, NaH has the structure of NaCl type with $a = 488$ pm, so the $Na^+ \cdots H^-$ distance is 244 pm. Taking the radius of Na^+ as 102 pm, the radius of H^- in the NaH crystal is 142 pm.

(b) The strength of ionic bonding and the melting point of compounds

The alkaline-earth oxides MO are composed of M^{2+} and O^{2-} stabilized by electrostatic forces. The electrostatic Coulombic energies are directly proportional to the product of the two ionic charges and inversely proportional to the interionic distance. The alkaline-earth oxides MO are 2:2 valence compounds and have the same electronic charge for M^{2+}, but the radii of M^{2+} and the distances of $M^{2+}\cdots O^{2-}$ are successively increased with atomic number in the order $Mg^{2+} < Ca^{2+} < Sr^{2+} < Ba^{2+}$. The strength of ionic bonding decreases successively, so the melting points and the hardness also decrease in the same sequence. The experimental melting points and hardness data listed below are consistent with expectation.

	MgO	CaO	SrO	BaO
$M^{2+}\cdots O^{2-}$ distance (pm)	211	241	258	276
mp (K)	3125	2887	2693	2191
Mohs hardness	6.5	4.5	3.5	3.3

(2) Defects in the structure type of NaCl

Various kinds of packing defects exist in the ionic crystals of NaCl type. A pair of cation and anion may be shifted from their stable positions toward the surface of the crystal, thus leaving behind a pair of vacancies. This is called the Schottky defect. The cation may leave its stable position and enter into an interstitial site. The formation of an interstitial cation and a vacancy is called the Frenkel defect. In addition to these two common kinds of defects, the presence of impurity atoms, atoms of varied valence, vacancies, and/or interstitial atoms is also possible. Some other important defects are discussed below.

(a) Color center

A color center or F-center is formed from diffusion of a small quantity of M^+ ion into an ionic crystal MX. Since the crystal must keep its charge neutrality, additional electrons readily move to fill the vacancies normally occupied by anions. Thus the composition of the crystal becomes $(M^+)_{1+\delta}(X^- e_\delta^-)$. The origin of the color is due to electronic motion, and a simple picture of an electron in a vacancy is illustrated by the particle in a three-dimensional box problem, which is discussed in Section 1.5.2.

(b) Koch cluster of $Fe_{1-\delta}O$

The structure of crystalline FeO belongs to the NaCl type. When iron(II) oxide is prepared under normal conditions, the composition of the product (wustite) is always $Fe_{1-\delta}O$. In order to retain overall electric neutrality, part of the Fe^{2+} is oxidized to Fe^{3+}, and the chemical formula becomes $Fe^{3+}_{2\delta}Fe^{2+}_{1-3\delta}O$. Since the radius of Fe^{3+} is small, the Fe^{3+} cations tend to occupy the tetrahedral holes to form a short-range ordered Fe_4O_{10} cluster, which is called the Koch cluster of $Fe_{1-\delta}O$, as shown in Fig. 10.1.3. The Koch clusters are distributed randomly in the crystal structure. To satisfy charge neutrality, the formation of a Koch cluster must be accompanied by the presence of six Fe^{2+} vacancies, one of which is located at the center of the cluster, and the remaining five are distributed randomly at the centers of the edges of the cubic unit cell.

(c) Vacancies in the NbO crystal

The unit cell of niobium monoxide NbO can be derived from the cubic unit cell of NaCl, by deleting the body-central Na^+ (coordinate: $\frac{1}{2},\frac{1}{2},\frac{1}{2}$) and the vertex

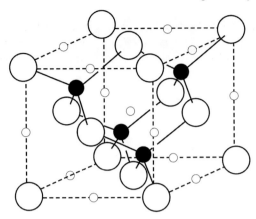

Fig. 10.1.3.
Structure of Koch cluster of $Fe_{1-\delta}O$ (black circle, Fe^{3+}; large white circle, O^{2-}; small white circle represents an octahedral vacancy).

Cl^- (coordinate: 0,0,0), and then replacing Na^+ by Nb^{2+} and Cl^- by O^{2-}. The crystal structure of NbO can be regarded as arising from ordered vacancy defects of the NaCl structural type, as shown in Fig. 10.1.4(a). The NbO crystal has a primitive cubic lattice in space group $O_h^1 - Pm\bar{3}m$, in which the Nb^{2+} and O^{2-} ions occupy sites $3(d)$ and $3(c)$, respectively, as listed in Table 10.1.3. In the NbO crystal, the Nb atoms form octahedral clusters held by Nb–Nb metallic bonds, which result from the overlap of the 5s and 4d orbitals. The Nb_6 octahedral clusters constitute a three-dimensional framework by sharing vertices, and each Nb atom is in square-planar coordination with four O neighbors. Although the geometrical arrangement and coordination environment of the Nb and O atoms are equivalent to each other, an analogously drawn O_6 octahedral cluster has no physical meaning as the O atom lacks d orbitals for bonding.

Some lower oxidation state oxoniobates contain the Nb_6O_{12} cluster shown in Fig. 10.1.4(b) as a basic structural unit. An O atom located above an edge of the Nb_6 octahedron may belong to the cluster or connect with another cluster. Up to six O atoms can be attached to the apices of the Nb_6 octahedron to constitute an outer coordination sphere; such additional atoms may be exclusive to the cluster or shared between clusters.

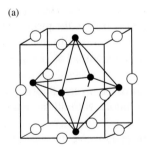
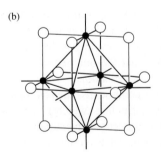

Fig. 10.1.4.
(a) Unit cell of NbO. (b) Nb_6O_{12} cluster with capacity for attachment of six ligands to its metal centers.

(3) Structure of CaC_2 and BaO_2

Calcium carbide CaC_2 exists in at least four crystalline modifications whose occurrence depends on the temperature of formation and the presence of impurities. The tetragonal phase $CaC_2(I)$, which is stable between 298 and 720 K, is

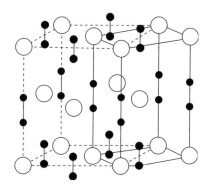

Fig. 10.1.5.
Crystal structure of $CaC_2(I)$.

the common form found in the commercial product. Its structure can be derived from the NaCl structure by replacing the Na^+ cations by Ca^{2+} and the Cl^- anions by C_2^{2-}, thereby reducing the crystal symmetry from cubic $Fm\bar{3}m$ to tetragonal $I4/mmm$, with the C_2^{2-} group lying on the unique 4-axis. The measured C≡C bond length of 120 pm is in agreement with its triple-bond character. Figure 10.1.5 shows the crystal structure of $CaC_2(I)$, in which solid lines represent the body-centered tetragonal unit cell with $a = 389$ pm and $c = 638$ pm, and the broken lines represent a non-primitive face-centered tetragonal unit cell, with $a' = 550$ pm and $c = 638$ pm, which clearly traces its lineage to the NaCl structure.

There are many compounds that are isomorphous with tetragonal $CaC_2(I)$, such as the acetylides of the alkaline- and rare-earth elements (BaC_2, SrC_2, LaC_2, CeC_2, PrC_2, NdC_2, SmC_2), the silicides of molybdenum and tungsten ($MoSi_2$, WSi_2), the superoxides of alkali metals (KO_2, RbO_2, CsO_2), and the peroxides of alkaline-earth metals (CaO_2, BaO_2). In O_2^-, the O–O distance is 128 pm; in O_2^{2-}, the O–O distance is 149 pm; in Si_2^{2-}, the Si–Si distance is 260 pm.

10.1.3 Structure of CaF_2 and related compounds

The structure of fluorite, CaF_2, may be regarded as being composed of cubic closest packed Ca^{2+} cations with all the tetrahedral interstices occupied by F^- anions, as shown in Fig. 10.1.6. The stoichiometric formula is consistent with the fact that the number of tetrahedral holes is twice that of the number of closest packed atoms.

Some compounds, such as Na_2O and K_2S, have the inverse fluorite (or anti-fluorite) structure, which is a ccp structure of anions with all the tetrahedral interstices occupied by cations.

Many inorganic compounds have the fluorite and inverse fluorite structures: (a) all halides of the larger divalent cations except two fluorides in this class; (b) oxides and sulfides of univalent ions; (c) oxides of large quadrivalent cations; and (d) some intermetallic compounds. Table 10.1.5 lists some compounds of the fluorite and inverse fluorite types and their a parameters.

The fluorite structure may also be regarded as being composed of a simple cubic packing of anions, and the number of cubic holes is the same as the

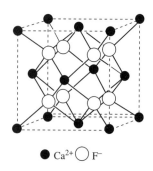

● Ca^{2+} ○ F^-

Fig. 10.1.6.
Structure of fluorite (CaF_2).

Table 10.1.5. Some compounds with fluorite and anti-fluorite structures and a values

Fluorite	a (pm)	Fluorite	a (pm)	Anti-fluorite	a (pm)
$AuAl_2$	600	PrH_2	551.7	Be_2B	467.0
$AuGa_2$	606.3	PrO_2	539.2	Be_2C	433
$AuIn_2$	650.2	$PtAl_2$	591.0	Ir_2P	553.5
$AuSb_2$	665.6	$PtGa_2$	591.1	K_2O	643.6
$BaCl_2$	734	$PtIn_2$	635.3	K_2S	739.1
BaF_2	620.01	$PtSn_2$	642.5	K_2Se	767.6
CaF_2	546.295	RaF_2	636.8	K_2Te	815.2
CdF_2	538.80	ScH_2	478.315	Li_2O	461.9
CeO_2	541.1	$SiMg_2$	639	Li_2S	570.8
$CoSi_2$	535.6	SmH_2	537.6	Li_2Se	600.5
HfO_2	511.5	$SnMg_2$	676.5	Li_2Te	650.4
HgF_2	554	$SrCl_2$	697.67	Mg_2Ge	637.8
$IrSn_2$	633.8	SrF_2	579.96	Na_2O	555
NbH_2	456.3	ThO_2	559.97	Na_2S	652.6
$NiSi_2$	539.5	UN_2	531	Na_2Se	680.9
NpO_2	543.41	YH_2	519.9	Na_2Te	731.4
β-PbF_2	592.732	ZrO_2	507	Rb_2O	674
$PbMg_2$	683.6	$AgAsZn$	591.2*	Rb_2S	765
α-PbO_2	534.9	$NiMgSb$	604.8*	Rb_2P	550.5
α-PoO_2	568.7	$NaYF_4$	545.9+	$LiMgN$	497.0‡

*The last two elements either occupy $1/4, 1/1, 1/4,$; $3/4, 3/4, 3/4$; F.C. separately or are statistically distributed over these positions. †The metal atoms are statistically distributed over the calcium positions of CaF_2. ‡The nitrogen atom takes the place of the calcium atom in CaF_2, and the two kinds of metal atoms are statistically distributed over the fluorine positions.

number of anions. Half of the holes are occupied by cations, and the other half are unoccupied. The center of the unit cell shown in Fig. 10.1.6 is an unoccupied cubic hole.

The structure of zircon, ZrO_2, is of the fluorite type. The Zr^{4+} cations occupy half of the cubic holes, and the O^{2-} anions can easily migrate through the empty cubic holes. Doping of ZrO_2 with Y_2O_3 or CaO stabilizes the fluorite structure and introduces vacant O^{2-} sites. The resulting doped crystalline material shows higher electrical conductivity when the temperature is increased. Hence $ZrO_2(Y_2O_3)$ is a good solid electrolyte of the anionic (O^{2-}) type which can be utilized in devices for measuring the amount of dissolved oxygen in molten steel, and as an electrolyte in fuel cells.

The structure of Li_3Bi is formed by a ccp of Bi atoms with the Li atoms occupying all the tetrahedral and octahedral holes.

10.1.4 Structure of cubic zinc sulfide

The structure of cubic zinc sulfide (zinc blende, sphalerite) may be described as a ccp of S atoms, in which half of the tetrahedral sites are filled with Zn atoms; the arrangement of the filled sites is such that the coordination numbers of S and Zn are both four, as shown in Fig. 10.1.7. The crystal belongs to space group $T_d^2 - F\bar{4}3m$. Note that the roles of the Zn and S atoms can be interchanged by a simple translation of the origin.

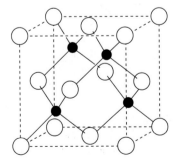

Fig. 10.1.7.
Structure of cubic zinc sulfide.

Many binary compounds adopt the cubic zinc sulfide structure. Some of these compounds and their unit-cell parameter a values are listed in Table 10.1.6.

The ternary iodide Ag_2HgI_4 exists in two forms: the low-temperature form (β) is a yellow tetragonal crystal, whereas the high-temperature form (α) is an orange-red cubic crystal, the transition temperature being 50.7°C. The structure of Ag_2HgI_4 is shown in Fig. 10.1.8. In the β-form, two Ag atoms occupy the sites $(0,0,0)$ and $(\frac{1}{2},\frac{1}{2},0)$, and one Hg atom is disordered over the sites $(0,\frac{1}{2},\frac{1}{2})$ and $(\frac{1}{2},0,\frac{1}{2})$ with occupancy $\frac{1}{2}$. When the β-form undergoes transformation to the α-form, the Ag and Hg atoms are almost randomly distributed over all four sites, so that the average composition at each site is $(Ag_{\frac{1}{2}}Hg_{\frac{1}{4}}\square_{\frac{1}{4}})$. The α-form has the structure type of cubic zinc sulfide and is a good superionic conductor. Therefore, the structure of Ag_2HgI_4 can be considered as a closest packed arrangement of I^- anions, with the Ag^+ and Hg^{2+} cations occupying the tetrahedral holes in either an ordered or disordered fashion.

Table 10.1.6. Some binary compounds with cubic zinc sulfide structure and their a values

Compound	a (pm)	Compound	a (pm)	Compound	a (pm)
γ-AgI	649.5	CdS	583.2	HgTe	646.23
AlAs	566.22	CdSe	605	InAs	605.838
AlP	545.1	CdTe	647.7	InP	586.875
AlSb	613.47	γ-CuBr	569.05	InSb	647.877
BAs	477.7	CuCl	540.57	β-MnS	560.0
BN	361.5	CuF	425.5	β-MnSe	588
BP	453.8	γ-CuI	604.27	SiC	434.8
BePo	583.8	GaAs	565.315	ZnPo	630.9
BeS	486.5	GaP	445.05	ZnS	540.93
BeSe	513.9	GaSb	609.54	ZnSe	566.76
BeTe	562.6	HgS	585.17	ZnTe	610.1
CdPo	666.5	HgSe	608.4		

The ternary iodide Cu_2HgI_4, a structural analog of Ag_2HgI_4 with very similar properties, exists as brown tetragonal crystals at room temperature. Its high-temperature crystalline form (transition temperature is 80°C) belongs to the cubic system and exhibits the red color.

Many ternary, quaternary, and multicomponent metal sulfides also exhibit tetrahedral coordination of atoms and possess structures similar to that of cubic

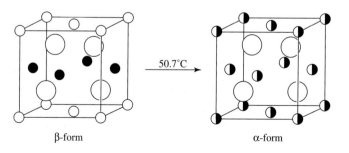

Fig. 10.1.8. Structure of Ag_2HgI_4 [large white circle, I^-; small black circle, $(Hg^{2+})_{\frac{1}{2}}$; small white circle, Ag^+; small black-white circle, $(Ag^+_{\frac{1}{2}}Hg^{2+}_{\frac{1}{4}}\square_{\frac{1}{4}})$].

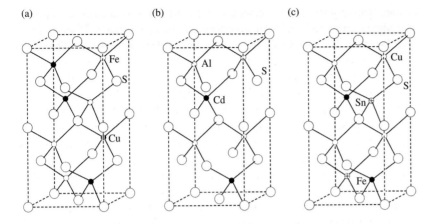

Fig. 10.1.9.
Structure of some multimetal sulfides: (a) $CuFeS_2$, (b) $CdAl_2S_4$, (c) Cu_2FeSnS_4.

zinc sulfide. In these compounds, the Zn atoms are orderly replaced by other atoms or by vacancies. Figure 10.1.9 shows the structures of (a) $CuFeS_2$, (b) $CdAl_2S_4$, and (c) Cu_2FeSnS_4. In these structures, the c axis of the unit cell is twice as long as a and b.

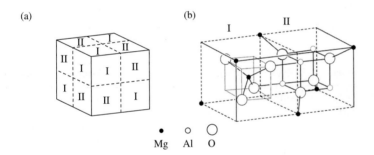

Fig. 10.1.10.
Crystal structure of spinel.

10.1.5 Structure of spinel and related compounds

The mixed-metal oxide spinel, $MgAl_2O_4$, is one of the most important inorganic materials. The structure of spinel can be regarded as a ccp structure of O^{2-} anions with Mg^{2+} ions orderly occupying $1/8$ of the tetrahedral interstices, and Al^{3+} ions orderly occupying half of the octahedral interstices; the remainder $7/8$ tetrahedral interstices and half octahedral interstices are unoccupied. The sites of the three kinds of ions in the face-centered cubic unit cell are displayed in Fig. 9.6.29.

Figure 10.1.10 shows the crystal structure of spinel. The unit cell (represented by solid lines) is divided into eight small cubic octants (represented by broken lines) of two different types, as shown in (a); the structure of type I and type II octants and the connections of atoms are shown in (b). There are four O^{2-} in each octant. The number of O^{2-} in the unit cell is 32 (= 8× 4). The Mg^{2+} ions are located at the centers of type I octants and half of the vertices of type I and type II octants. The number of Mg^{2+} in the unit cell is 8 $[= 4(1+4 \times 1/8) + 4 \times 4 \times 1/8]$. The Mg^{2+} ions have tetrahedral coordination. There are four Al^{3+} ions in the type II octant. The number of Al^{3+} in the unit cell is 16 $(= 4 \times 4)$. The Al^{3+} ions have octahedral coordination.

According to the cation distribution over the $8(a)$ and $16(d)$ sites, the spinels of formula AB_2O_4 can be divided into normal and inverse classes. Examples of normal spinels are $MgAl_2O_4$ and $MgTi_2O_4$, in which the oxidation states of Al and Ti are both +3; their respective structural formulas are $[Mg^{2+}]_t[Al_2^{3+}]_oO_4$ and $[Mg^{2+}]_t[Ti_2^{3+}]_oO_4$, in which the subscripts t and o indicate tetrahedral and octahedral sites, respectively.

In inverse spinels, half of the B ions are in tetrahedral $8(a)$ sites, leaving the remaining B ions and the A ions to fill the $16(d)$ sites. Usually, the occupancy of these $16(d)$ sites is disordered. Examples of inverse spinels are $MgFe_2O_4$ and Fe_3O_4; their structural formulas are $[Fe^{3+}]_t[Mg^{2+}Fe^{3+}]_oO_4$ and $[Fe^{3+}]_t[Fe^{2+}Fe^{3+}]_oO_4$, respectively.

In intermediate spinels any combination of cation arrangement between the extremes of normal and inverse spinels is possible. Table 10.1.7 lists some normal spinels and inverse spinels and the values of their unit-cell parameter a.

Table 10.1.7. Some normal spinels and inverse spinels and their a parameters

Normal Spinel				Inverse Spinel	
Compound	a (pm)	Compound	a (pm)	Compound	a (pm)
$CoAl_2O_4$	810.68	$MgAl_2O_4$	808.0	$CoFe_2O_4$	839.0
$CoCr_2O_4$	833.2	$MgCr_2O_4$	833.3	$CoIn_2S_4$	1055.9
$CoCr_2S_4$	993.4	$MgMn_2O_4$	807	$CrAl_2S_4$	991.4
$CoMn_2O_4$	810	$MgRh_2O_4$	853.0	$CrIn_2S_4$	1059
Co_3O_4	808.3	$MgTi_2O_4$	847.4	$FeCo_2O_4$	825.4
CoV_2O_4	840.7	MgV_2O_4	841.3	$FeGa_2O_4$	836.0
$CdCr_2O_4$	856.7	$MnCr_2O_4$	843.7	Fe_3O_4	839.4
$CdCr_2S_4$	998.3	Mn_3O_4	813	$MgIn_2O_4$	881
$CdCr_2Se_4$	1072.1	$MnTi_2O_4$	860.0	$MgIn_2S_4$	1068.7
$CdFe_2O_4$	869	MnV_2O_4	852.2	$NiCo_2O_4$	812.1
$CdMn_2O_4$	822	$MoAg_2O_4$	926	$NiFe_2O_4$	832.5
$CuCr_2O_4$	853.2	$MoNa_2O_4$	899	$NiIn_2S_4$	1046.4
$CuCr_2S_4$	962.9	$NiCr_2O_4$	824.8	$NiLi_2F_4$	831
$CuCr_2Se_4$	1036.5	$NiRh_2O_4$	836	$NiMn_2O_4$	839.0
$CuCr_2Te_4$	1104.9	WNa_2O_4	899	$SnCo_2O_4$	864.4
$CuMn_2O_4$	833	$ZnAl_2O_4$	808.6	$SnMg_2O_4$	860
$CuTi_2S_4$	988.0	$ZnAl_2S_4$	998.8	$SnMn_2O_4$	886.5
CuV_2S_4	982.4	$ZnCo_2O_4$	804.7	$SnZn_2O_4$	866.5
$FeCr_2O_4$	837.7	$ZnCr_2O_4$	832.7	$TiCo_2O_4$	846.5
$FeCr_2S_4$	999.8	$ZnCr_2S_4$	998.3	$TiFe_2O_4$	850
$GeCo_2O_4$	813.7	$ZnCr_2Se_4$	1044.3	$TiMg_2O_4$	844.5
$GeFe_2O_4$	841.1	$ZnFe_2O_4$	841.6	$TiMn_2O_4$	867
$GeMg_2O_4$	822.5	$ZnMn_2O_4$	808.7	$TiZn_2O_4$	844.5
$GeNi_2O_4$	822.1	ZnV_2O_4	841.4	VCo_2O_4	837.9

10.2 Hexagonal closest packing and related structures

10.2.1 Hexagonal closest packing (hcp)

The crystal structure of many pure metals adopts the hcp of identical spheres, which has a primitive hexagonal lattice in space group $D_{6h}^4 - P6_3/mmc$. There are two atoms in the hexagonal unit cell with coordinates $(0,0,0)$ and $(\frac{2}{3}, \frac{1}{3}, \frac{1}{2})$.

(a)
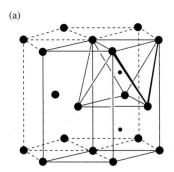

(b)
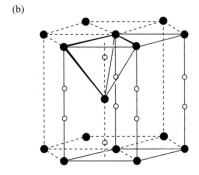

Fig. 10.2.1.
Positions of interstices in the hcp structure: (a) octahedral interstices, (b) tetrahedral interstices.

There are two types of interstices in the hcp structure: octahedral and tetrahedral holes, as in the ccp structure. However, the hcp and ccp structures differ in the linkage of interstices. In ccp, neighboring octahedral interstices share edges, and the tetrahedral interstices behave likewise. In hcp, neighboring octahedral interstices share faces, and a pair of tetrahedral interstices share a common face to form a trigonal bipyramid, so these two tetrahedral interstices cannot be filled by small atoms simultaneously. Figure 10.2.1 shows the positions of the octahedral interstices (a) and tetrahedral interstices (b). Table 10.2.1 lists the pure metals with hcp structure and the parameters of the unit cell.

Table 10.2.1. Metals with hcp structure and the parameters of the hexagonal unit cell

Metal	a (pm)	c (pm)	Metal	a (pm)	c (pm)
Be	228.7	358.3	α-Nd	365.8	1179.9
β-Ca	398	652	Ni	265	433
Cd	297.9	561.8	Os	273.5	431.9
α-Co	250.7	406.9	α-Pr	367.3	1183.5
γ-Cr	272.2	442.7	Re	276.1	445.8
Dy	359.25	565.45	Ru	270.4	428.2
Er	355.90	559.2	α-Sc	330.80	526.53
Gd	363.15	577.7	α-Sm	362.1	2625
He	357	583	β-Sr	432	706
α-Hf	319.7	505.8	Tb	359.90	569.6
Ho	357.61	561.74	α-Ti	250.6	467.88
α-La	377.0	1215.9	α-Tl	345.6	552.5
Li	311.1	509.3	Tm	353.72	556.19
Lu	350.50	554.86	α-Y	364.51	573.05
Mg	320.9	521.0	Zn	266.5	494.7
Na	365.7	590.2	α-Zr	331.2	514.7

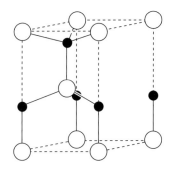

Fig. 10.2.2.
Structure of hexagonal zinc sulfide.

10.2.2 Structure of hexagonal zinc sulfide

In the crystal structure of hexagonal zinc sulfide (wurtzite), the S atoms are arranged in hcp, in which half of the tetrahedral interstices are filled with Zn atoms, and the space group is $C_{6v}^4 - P6_3mc$. The positions of atoms in the hexagonal unit cell are

$$S \quad 0,0,0; \quad 2/3, 1/3, 1/2$$
$$Zn \quad 0,0,3/8; \quad 2/3, 1/3, 7/8$$

The Zn atoms are each tetrahedrally coordinated by four S atoms, and likewise the S atoms are each connected to four Zn atoms, as shown in Fig. 10.2.2. The bonding has both covalent and ionic character.

Note that an equivalent structure is obtained when the positions of the Zn and S atoms are interchanged, but in this case the polar direction of the crystal is reversed. This arises because $P6_3mc$ is a polar space group, and historically the polar sense of a single crystal of ZnS has been used to demonstrate the breakdown of Friedel's law under conditions of anomalous dispersion.

Many binary compounds adopt the structure type of hexagonal zinc sulfide. Table 10.2.2 lists some of these compounds and their unit-cell parameters.

10.2.3 Structure of NiAs and related compounds

In the crystal structure of nickel asenide, NiAs, the As atoms are in hcp with all octahedral interstices occupied by the Ni atoms, as shown in Fig. 10.2.3(a). An important feature of this structure is that the Ni and As atoms are in different coordination environments. Each As atom is surrounded by six equidistant Ni atoms situated at the corners of a regular trigonal prism. Each Ni atom, on the other hand, has eight close neighbors, six of which are As atoms arranged octahedrally about it, while the other two are Ni atoms immediately above and below it at $z = \pm c/2$. The Ni–Ni distance is $c/2 = 503.4/2 = 251.7$ pm, which corresponds to the interatomic distance in metallic nickel. Compound NiAs is semi-metallic, and its metallic property results from the bonding between Ni atoms. In the NiAs structure, the axial ratio $c/a = 503.4/361.9 = 1.39$ is much

Table 10.2.2. Some compounds with hexagonal zinc sulfide structure and their unit-cell parameters

Compound	a (pm)	c (pm)	Compound	a (pm)	c (pm)
AgI	458.0	749.4	MgTe	454	739
AlN	311.1	497.8	MnS	397.6	643.2
BeO	269.8	437.9	MnSe	412	672
CdS	413.48	674.90	MnTe	408.7	670.1
CdSe	430.9	702.1	SiC	307.6	504.8
CuCl	391	642	ZnO	324.95	520.69
CuH	289.3	461.4	ZnS	381.1	623.4
GaN	318.0	516.6	ZnSe	398	653
InN	353.3	569.3	ZnTe	427	699

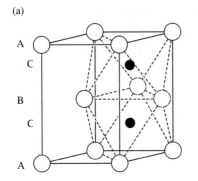
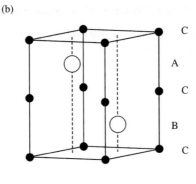

Fig. 10.2.3.
Structure of NiAs: (a) origin at As atom (white circle); (b) origin at Ni atom (black circle).

smaller than the calculated value for hcp of identical spheres (1.633). This fact suggests that, with a decrease in ionic character, metal atoms are less likely to repel each other, and more likely to form metal–metal bonds.

The space group of NiAs is $D_{6h}^4 - P6_3/mmc$. The atomic coordinates are

As   2(a) $\bar{3}m$ 0, 0, 0;   2/3, 1/3, 1/2
Ni   2(c) $\bar{6}m2$ 1/3, 2/3, 3/4,   1/3, 2/3, 1/4

When a translation is applied so that the origin of the unit cell now resides on the Ni atom, as shown in Fig. 10.2.3(b), the atomic coordinates become

As   2(a) $\bar{3}m$ 1/3, 2/3, 1/4;   2/3, 1/3, 3/4
Ni   2(c) $\bar{6}m2$ 0, 0, 0;   0, 0, 1/2

Figure 10.2.3(b) shows clearly that the Ni atoms are arranged in hexagonal layers which exactly eclipse one another, and only half of the large trigonal pyramidal interstices are filled by the As atoms. The large vacant interstices can be occupied easily by other atoms of an additional component. The structure of Ni_2In can be regarded as two Ni atoms forming hexagonal layers that lead to four trigonal pyramidal interstices per unit cell, which are occupied by two In atoms and two Ni atoms.

Some compounds with the NiAs structure and their unit-cell parameters are listed in Table 10.2.3.

10.2.4 Structure of CdI$_2$ and related compounds

The hcp structure consists of a stacking of atomic layers in the sequence ABABAB···. The octahedral holes are located between adjacent layers, as shown in Fig. 10.2.3(a). In the crystal structure of nickel asenide, the As atoms constitute a hcp lattice, and the Ni atoms occupy all the octahedral holes. In contrast, cadmium iodide, CdI_2, may be described as a hcp of I^- anions, in which only half the octahedral holes are occupied by Cd^{2+} ions. The manner of occupancy of the octahedral interstices is such that entire layers of octahedral interstices are filled, and these alternate with layers of empty

Table 10.2.3. Some compounds with the NiAs structure and their unit-cell parameters

Compound	a (pm)	c (pm)	Compound	a (pm)	c (pm)
AuSn	432.3	552.3	NiSb	394.2	515.5
CoS	337.4	518.7	NiSe	366.13	535.62
CoSb	386.6	518.8	NiSn	404.8	512.3
CoSe	362.94	530.06	NiTe	395.7	535.4
CoTe	388.6	536.0	PdSb	407.8	559.3
CrSb	413	551	PdSn	411	544
CrSe	371	603	PdTe	415.2	567.2
CrTe	393	615	PtBi	431.5	549.0
CuSb	387.4	519.3	PtSb	413	548.3
CuSn	419.8	509.6	PtSn	411.1	543.9
FeSb	407.2	514.0	RhBi	407.5	566.9
FeSe	361.7	588	RhSn	434.0	555.3
IrSb	398.7	552.1	RhTe	399	566
IrSn	398.8	556.7	ScTe	412.0	674.8
MnAs	371.0	569.1	TiAs	364	615
MnBi	430	612	TiS	329.9	638.0
MnSb	412.0	578.4	TiSe	357.22	620.5
MnTe	414.29	670.31	VS	333	582
NiAs	361.9	503.4	VSe	366	595
NiBi	407.0	535	VTe	394.2	612.6
NiS	343.92	534.84	ZrTe	395.3	664.7

interstices. The layer stacking sequence along the c axis of the unit cell in CdI_2 is shown schematically in Fig. 10.2.4(a), and the unit cell is shown in Fig. 10.2.4(b).

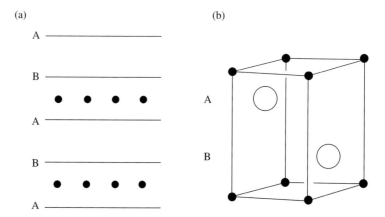

Fig. 10.2.4.
Structure of CdI_2 (black circle, Cd^{2+}): (a) schematic layer structure and (b) unit cell.

A layer structure can be conveniently designated by the following scheme: the capital letters A, B, C indicate the relative positions of anions, lower-case letters a, b, c indicate the positions of cations, and vacant octahedral interstices are represented by □. The structure of CdI_2 can thus be written as

$$A\,c\,B\,\square\,A\,c\,B\ \square\,A\,c\,B\,\square\,A\cdots\cdots$$

Table 10.2.4. Some compounds with the CdI_2 structure and their unit-cell parameters

Compound	a (pm)	c (pm)	Compound	a (pm)	c (pm)
CaI_2	448	696	ZnI_2	425	654
CdI_2	424	684	$IrTe_2$	393.0	539.3
$CoBr_2$	368	612	$NiTe_2$	386.9	530.8
CoI_2	396	665	$PdTe_2$	403.65	512.62
$FeBr_2$	374	617	PtS_2	354.32	503.88
FeI_2	404	675	$PtSe_2$	372.78	508.13
GeI_2	413	679	$PtTe_2$	402.59	522.09
$MgBr_2$	381	626	TiS_2	340.80	570.14
MgI_2	414	688	$TiSe_2$	353.56	600.41
$MnBr_2$	382	619	$TiTe_2$	376.4	652.6
MnI_2	416	682	$Ca(OH)_2$	358.44	489.62
PbI_2	455.5	697.7	$Cd(OH)_2$	348	467
$TiBr_2$	362.9	649.2	$Co(OH)_2$	317.3	464.0
$TiCl_2$	356.1	587.5	$Fe(OH)_2$	325.8	460.5
TiI_2	411.0	682.0	$Mg(OH)_2$	314.7	476.9
VBr_2	376.8	618.0	$Mn(OH)_2$	334	468
VCl_2	360.1	583.5	$Ni(OH)_2$	311.7	459.5
VI_2	400.0	667.0			

Many binary compounds which consist of I^-, Br^-, OH^- and M^{2+} cations, or S^{2-}, Se^{2-}, Te^{2-}, and M^{4+} cations, usually adopt the CdI_2 structure. Table 10.2.4 lists some compounds with the CdI_2 structure and their unit-cell parameters.

The crystal structure of $CdCl_2$ is closely related to that of CdI_2. Both originate from the same type of MXM sandwiches, differing only in the way of stacking. The CdI_2 structure is based on the hcp type, whereas the $CdCl_2$ structure is based on the ccp type. The structure of $CdCl_2$ is represented by the layer sequence:

$$A \, c \, B \, \Box \, C \, b \, A \, \Box \, B \, a \, C \, \Box \, A \, c \, B \, \Box \, C \cdots$$

10.2.5 Structure of α-Al_2O_3

α-Al_2O_3 occurs as the mineral corundum. Because of its great hardness (9 on the Mohs scale), high melting point (2045°C), involatility (0.1 Pa at 1950°C), chemical inertness, and good electrical insulating properties, it is an important inorganic material that finds many applications in abrasives, refractories, and ceramics. Larger crystals of α-Al_2O_3, when colored with metal-ion impurities, are prized as gemstones with names such as ruby (Cr^{3+}, red), sapphire ($Fe^{2+/3+}$ and Ti^{4+}, blue), and oriental amethyst (Cr^{3+}/Ti^{4+}, violet). Many of these crystals are synthesized on an industrial scale and used as laser materials, gems, and jewels.

α-Al_2O_3 belongs to the trigonal system in space group $D_{3d}^6 - R\bar{3}c$. The parameters of the hexagonal unit cell are $a = 476.280$ pm, $c = 1300.320$ pm (31°C). The atomic coordinates (R-centered) are

Al $(12c)(0,0,0;\ ^1/_3, ^2/_3, ^2/_3;\ ^2/_3, ^1/_3, ^1/_3) \pm (0,0,z;\ 0,0, ^1/_2 + z), z = 0.352$

O $(18e)(0,0,0;\ ^1/_3, ^2/_3, ^2/_3;\ ^2/_3, ^1/_3, ^1/_3) \pm (x, 0, ^1/_4;\ 0, x, ^1/_4;\ \bar{x}, \bar{x}, ^1/_4), x = 0.306$

In the crystal structure of α-Al_2O_3, the O atoms are arranged in hcp, while the Al atoms are orderly located in the octahedral interstices. As the number of Al atoms is $2/3$ of the number of O atoms, there are $1/3$ vacant octahedral interstices. The distribution of the interstices can be described in terms of the arrangement of the Al atoms. Between each pair of adjacent closest packed O layers, the Al atoms form a planar hexagonal layer, and the center of each hexagon is an interstice (which is analogous to the arrangement of C atoms in hexagonal graphite). The interstices are evenly distributed in the unit cell at three different sites, as indicated by C', C'', and C''' in Fig. 10.2.5. The Al atoms of two neighboring layers overlap, so that the two AlO_6 octahedral coordination polyhedra are connected by face sharing, as represented by the vertical lines in Fig. 10.2.5(a).

Some trivalent metal oxides adopt the corundum structure. Table 10.2.5 lists these compounds and the parameters of their hexagonal unit cells.

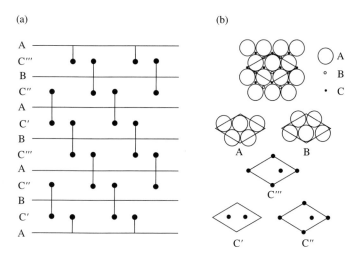

Fig. 10.2.5.
Structure of α-Al_2O_3: (a) schematic distribution of hcp of O atoms and Al atoms in different layers (black circle, Al; vertical lines represent the shared face form of AlO_6), (b) distribution of atoms located in various layers in a unit cell.

10.2.6 Structure of rutile

Rutile, TiO_2, belongs to the tetragonal space group $D_{4h}^{14} - P4_2/mnm$. The parameters of the unit cell are $a = 459.366$ pm, $c = 295.868$ pm (298 K). The atomic coordinates are

Ti $(2a)$ $0, 0, 0$; $1/2, 1/2, 1/2$

O $(4f)$ $\pm (x, x, 0; 1/2 + x, 1/2 - x, 1/2)$, $x = 0.30479$

Table 10.2.5. Compounds with the corundum structure and their lattice parameters

Compound	a (pm)	c (pm)	Compound	a (pm)	c (pm)
α-Al_2O_3	476.3	1300.3	Rh_2O_3	511	1382
Cr_2O_3	495.4	1358.4	Ti_2O_3	514.8	1363
α-Fe_2O_3	503.5	1375	V_2O_3	510.5	1444.9
α-Ga_2O_3	497.9	1342.9			

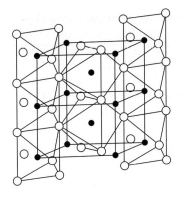
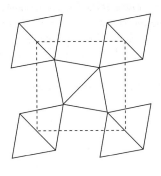

Fig. 10.2.6.
Structure of rutile: (a) connection of TiO$_6$ octahedra and the unit cell, (b) chains share vertices, viewed along the c axis.

In the structure of rutile, the Ti atoms constitute deformed TiO$_6$ octahedra, in which the Ti–O distance are 194.85 pm (four) and 198.00 pm (two). The TiO$_6$ octahedra share edges to form chains, which further share vertices to yield a three-dimensional network, as shown in Fig. 10.2.6.

The structure of rutile may also be described as a nearly hcp of O atoms, with Ti atoms located in half of the octahedral interstices.

Many tetravalent metal oxides and divalent metal fluorides adopt the rutile structure. Table 10.2.6 lists some compounds with the rutile structure and the parameters of their tetragonal unit cells.

Table 10.2.6. Compounds with rutile structure

Compound	a (pm)	c (pm)	Compound	a (pm)	c (pm)
CrO$_2$	441	291	TaO$_2$	470.9	306.5
GeO$_2$	439.5	285.9	TiO$_2$	459.366	295.868
IrO$_2$	449	314	WO$_2$	486	277
β-MnO$_2$	439.5	286	CoF$_2$	469.5	318.0
MoO$_2$	486	279	FeF$_2$	469.7	330.9
NbO$_2$	477	296	MgF$_2$	462.3	305.2
OsO$_2$	451	319	MnF$_2$	487.3	331.0
PbO$_2$	494.6	337.9	NiF$_2$	465.1	308.4
RuO$_2$	451	311	PdF$_2$	493.1	336.7
SnO$_2$	473.7	318.5	ZnF$_2$	470.3	313.4

10.3 Body-centered cubic packing and related structures

10.3.1 Body-centered cubic packing (bcp)

Body-centered cubic packing (bcp) is a common structure type among metals. The space group is O_h^9-$Im\bar{3}m$, and the atomic coordinates of equivalent positions are listed in Table 10.3.1.

In the bcp structure, the cubic unit cell has two atoms in site $2(a)$: $(0, 0, 0; 1/2, 1/2, 1/2)$. Figure 10.3.1(a) shows that at the center of every face and the edge of the unit cell there are octahedral interstices, each surrounded by

Table 10.3.1. Atomic coordinates of equivalent positions in cubic space group O_h^9 – $Im\bar{3}m$*

Multiplicity	Wyckoff position	Site symmetry	Coordinates $\left(0,0,0;\ \tfrac{1}{2},\tfrac{1}{2};\ \tfrac{1}{2}\right)+$
24	h	mm2	$0,y,y;\ 0,\bar{y},y;\ 0,y,\bar{y};\ 0,\bar{y},\bar{y};\ y,0,y;\ y,0,\bar{y};$ $\bar{y},0,y;\ \bar{y},0,\bar{y};\ y,y,0;\ \bar{y},y,0;\ y,\bar{y},0;\ \bar{y},\bar{y},0.$
24	g	mm2	$x,0,\tfrac{1}{2};\ \bar{x},0,\tfrac{1}{2};\ \tfrac{1}{2},x,0;\ \tfrac{1}{2},\bar{x},0;\ 0,\tfrac{1}{2},x;\ 0,\tfrac{1}{2},\bar{x};$ $0,x,\tfrac{1}{2};\ 0,\bar{x},\tfrac{1}{2};\ x,\tfrac{1}{2},0;\ \bar{x},\tfrac{1}{2},0;\ \tfrac{1}{2},0,\bar{x};\ \tfrac{1}{2},0,x.$
16	f	3m	$x,x,x;\ \bar{x},\bar{x},x;\ \bar{x},x,\bar{x};\ x,\bar{x},\bar{x}.$ $x,x,\bar{x};\ \bar{x},\bar{x},\bar{x};\ x,\bar{x},x;\ \bar{x},x,x.$
12	e	4mm	$x,0,0;\ \bar{x},0,0;\ 0,x,0;\ 0,\bar{x},0;\ 0,0,x;\ 0,0,\bar{x}.$
12	d	$\bar{4}m2$	$\tfrac{1}{4},0,\tfrac{1}{2};\ \tfrac{3}{4},0,\tfrac{1}{2};\ \tfrac{1}{2},\tfrac{1}{4},0;\ \tfrac{1}{2},\tfrac{3}{4},0;\ 0,\tfrac{1}{2},\tfrac{1}{4};\ 0,\tfrac{1}{2},\tfrac{3}{4}.$
8	c	$\bar{3}m$	$\tfrac{1}{4},\tfrac{1}{4},\tfrac{1}{4};\ \tfrac{1}{4},\tfrac{3}{4},\tfrac{3}{4};\ \tfrac{3}{4},\tfrac{1}{4},\tfrac{3}{4};\ \tfrac{3}{4},\tfrac{3}{4},\tfrac{1}{4}.$
6	b	4/mmm	$0,\tfrac{1}{2},\tfrac{1}{2};\ \tfrac{1}{2},0,\tfrac{1}{2};\ \tfrac{1}{2},\tfrac{1}{2},0.$
2	a	$m\bar{3}m$	$0,0,0$

* This table is not complete and only lists eight sets.

six spheres. Each sphere of radius R corresponds to three such octahedral interstices. These octahedra are not regular but are compressed along one axis. The shortest dimension of an interstice dictates the radius r of a small sphere that can be accommodated in it, and the ratio r/R is 0.154.

Distorted tetrahedral interstices also occur on the faces of the unit cell. Each face has four such distorted tetrahedral interstices with a radius ratio $r/R = 0.291$, as shown in Fig. 10.3.1(b). Each sphere corresponds to six of these tetrahedral interstices.

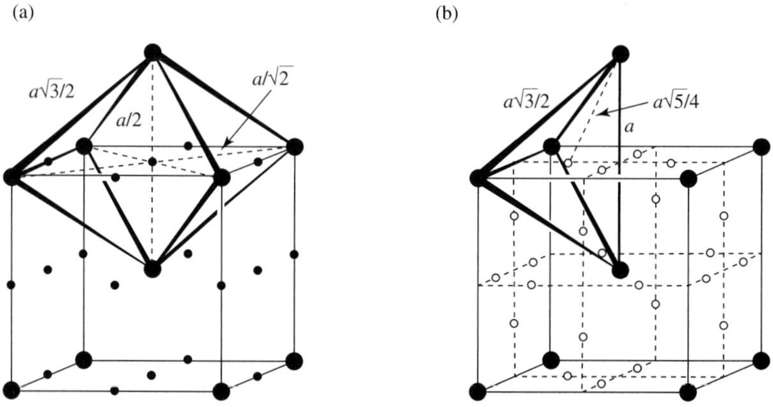

Fig. 10.3.1.
The positions of interstices in the bcp structure: (a) octahedral interstices (small black spheres), (b) tetrahedral interstices (small open spheres).

These two types of interstitial polyhedra are not mutually exclusive; i.e., a particular part of volume in the unit cell does not belong to one particular polyhedron. In addition, distorted trigonal-bipyramidal interstitial polyhedra are each formed from two face-sharing tetrahedra. The number of such interstices that corresponds to one sphere is 12. Therefore, each sphere in the bcp structure is associated with three octahedral, six tetrahedral, and twelve trigonal-bipyramidal interstices, i.e., a total of 21 interstices. The size and distribution

Table 10.3.2. Some metals with the bcp structure and their a values

Metal	a (pm)	Metal	a (pm)	Metal	a (pm)
Ba	502.5	Mo	314.73	β-Th	411
γ-Ca	438	Na	429.06	β-Ti	330.7
δ-Ce	412	Nb	330.04	β-Tl	388.2
α-Cr	288.39	β-Nd	413	γ-U	347.4
Cs	606.7	γ-Np	352	V	302.40
Eu	457.8	β-Pr	413	W	316.496
α-Fe	286.65	ε-Pu	363.8	β-Y	411
K	524.7	Rb	560.5	β-Yb	444
γ-La	426	β-Sn	407	β-Zr	362
Li	350.93	γ-Sr	485		
δ-Mn	307.5	Ta	330.58		

of these interstices directly affect the properties of the structure. Table 10.3.2 lists some metals with the bcp structure and their a values.

10.3.2 Structure and properties of α-AgI

Silver iodide, AgI, exists in several polymorphic forms. In the α-AgI crystal, the I^- ions adopt the bcp structure, and the Ag^+ cations are distributed statistically among the $6(b)$, $12(d)$, and $24(h)$ sites of space group $O_h^9 - Im\bar{3}m$, as listed in Table 10.3.1, and also partially populate the passageways between these positions. The cubic unit cell, with $a = 504$ pm, provides 42 possible positions for two Ag^+ cations, and the $Ag^+ \cdots I^-$ distances are listed below:

$6(b)$ positions having 2 I^- neighbors at 252 pm,

$12(d)$ positions having 4 I^- neighbors at 282 pm, and

$24(h)$ positions having 3 I^- neighbors at 267 pm.

Figure 10.3.2 shows the crystal structure of α-AgI and the possible positions of the Ag^+ ions, the mobility of which accounts for the prominent ionic transport properties of α-AgI as a solid electrolyte.

The structure of AgI varies at different temperatures and pressures. The stable form of AgI below 409 K, γ-AgI, has the zinc blende (cubic ZnS) structure. On the other hand, β-AgI, with the wurtzite (hexagonal ZnS) structure, is the stable form between 409 and 419 K. Above 419 K, β-AgI undergoes a phase change to cubic α-AgI. Under high pressure, AgI adopts the NaCl structure. Below room temperature, γ-AgI obtained from precipitation from an aqueous solution exhibits prominent covalent bond character, with a low electrical conductivity of about 3.4×10^{-4} ohm^{-1}cm^{-1}. When the temperature is raised, it undergoes a phase change to α-AgI, and the electrical conductivity increases ten-thousand-fold to 1.3 ohm^{-1}cm^{-1}. Compound α-AgI is the prototype of an important class of ionic conductors with Ag^+ functioning as the carrier.

The transformation of β-AgI to α-AgI is accompanied by a dramatic increase in the ionic electrical conductivity of the solid, which leaps by a factor of nearly 4000 from 3.4×10^{-4} to 1.3 ohm^{-1}cm^{-1}. This arises because in β-AgI the Ag

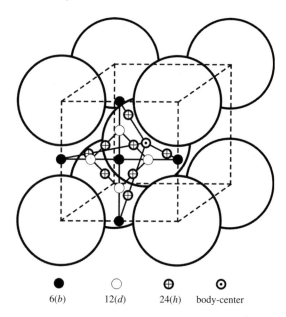

Fig. 10.3.2.
Structure of α-AgI.

● 6(b)   ○ 12(d)   ⊕ 24(h)   ⊙ body-center

atoms have tetrahedral coordination, in which covalent bonding is the dominant factor, but ionic bonding becomes important in α-AgI.

The reaction of AgI with RbI forms RbAg$_4$I$_5$, which has been found to have the highest specific electrolytic conductivity at room temperature (0.27 ohm^{-1}cm^{-1}) among known solids, and this property is retained at low temperature.

The AgI-type superionic solids are important inorganic materials. Several generalizations on the relationship between the conductivity of solid electrolytes and their structures can be made:

(a) Almost all solid electrolytes of any kind have a network of passageways formed from the face-sharing of anionic polyhedra.
(b) In structures in which the sites available to the current carriers are not crystallographically equivalent, the distribution of carriers over different sites is markedly non-uniform.
(c) The conductivity is associated with the nature of the passageways: more open channels lead to higher conductivity. In general, three-dimensional networks exhibit higher average conductivities than two-dimensional networks. A larger number of available sites and/or a larger crystal volume available to the conduction passageways tends to give higher conductivities.
(d) The stability of the Ag$^+$ ions in both four- and three-coordination and their univalent character account for the fact that they constitute the mobile phase in most good solid electrolytes.

10.3.3 Structure of CsCl and related compounds

The cesium chloride structure may be regarded as derived from the two equivalent atoms of the cubic bcp unit cell by changing one to Cs$^+$ and the other to

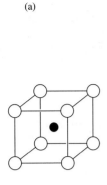 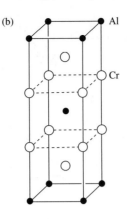

Fig. 10.3.3.
Unit cell of (a) CsCl and (b) Cr$_2$Al.

Cl$^-$. The lattice type is reduced from face-centered cubic to primitive cubic, and the space group of CsCl is $O_h^1 - Pm\bar{3}m$. Figure 10.3.3(a) shows a unit cell in the crystal structure of CsCl.

In the CsCl structure, the coordination number of either kind of ion is eight, and the interionic distance is $[\frac{1}{2}(3)^{1/2}]a$. Two principal kinds of compounds crystallize with the CsCl structure. In one group are the halides of the largest univalent ions, and in the other are intermetallic compounds. Some examples are listed in Table 10.3.3.

Table 10.3.3. Compounds of CsCl structure type and their unit-cell parameters

Compound	a (pm)	Compound	a (pm)	Compound	a (pm)
AgCd	333	CsBr	428.6	MgTl	362.8
AgCe	373.1	CsCl	412.3	NH$_4$Cl	386
AgLa	376.0	CsI	456.67	NH$_4$Br	405.94
AgMg	328	CuPd	298.8	NH$_4$I	437
AgZn	315.6	CuZn	294.5	NiAl	288.1
AuMg	325.9	FeTi	297.6	NiIn	309.9
AuZn	319	LiAg	316.8	NiTi	301
BeCo	260.6	LiHg	328.7	SrTl	402.4
BeCu	269.8	LiTl	342.4	TlBi	398
BePd	281.3	MgCe	389.8	TlBr	397
CaTl	384.7	MgHg	344	TlCl	383.40
CoAl	286.2	MgLa	396.5	TlI	419.8
CoTi	298.6	MgSr	390.0	TlSb	384

The structures of some intermetallic compounds are formed from the stacking of blocks, each resembling a unit cell of the CsCl type. For example, the unit cell of Cr$_2$Al may be regarded as a stack of three blocks, as shown in Fig. 10.3.3(b).

10.4 Perovskite and related compounds

10.4.1 Structure of perovskite

The mineral CaTiO$_3$ is named perovskite, the structure of which is related to a vast number of inorganic crystals, and it plays a central role in the development of the inorganic functional materials.

Fig. 10.4.1.
(a) A-type and (b) B-type unit cells of perovskite (large black circles, Ca^{2+}; small black circles, Ti^{4+}; white circles, O^{2-}).

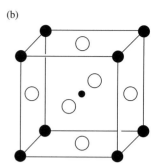

Perovskite crystallizes in the cubic space group $O_h^1 - Pm\bar{3}m$. The Ti^{4+} ions are located at the corners of the unit cell, a Ca^{2+} ion at the body center, and O^{2-} ions at the mid-points of the edges; this so-called A-type cell is shown in Fig. 10.4.1(a). When the origin of the cubic unit cell is taken at the Ca^{2+} ion, the Ti^{4+} ion occupies the body center and the O^{2-} ions are located at the face centers; this B-type unit cell is shown in Fig. 10.4.1(b).

Each Ca^{2+} is thus twelve-coordinated and each Ti^{4+} six-coordinated by oxygen neighbors, while each O^{2-} is linked to four Ca^{2+} and two Ti^{4+} ions. As expected, it is the larger metal ion that occupies the site of higher coordination. Geometrically the structure can be regarded as a ccp of (O^{2-} and Ca^{2+}) ions, with the Ti^{4+} ions orderly occupying ¼ of the octahedral interstices.

The basic perovskite structure ABX_3 forms the prototype for a wide range of other structures related to it by combinations of topological distortions, substitution of the A, B and X ions, and intergrowth with other structure types. These compounds exhibit a range of magnetic, electrical, optical, and catalytic properties of potential application in solid state physics, chemistry, and materials science.

Many ABX_3 compounds have true cubic symmetry, whereas some owing to strain or to small departures from perfect cubic symmetry have appreciably distorted atomic arrangements. In the idealized structure, a simple relationship exists between the radii of the component ions,

$$r_A + r_X = (2)^{1/2}(r_B + r_X),$$

where A is the larger cation. It is found that, in practice, the real structure conforms to the condition

$$r_A + r_X = t(2)^{1/2}(r_B + r_X),$$

where t is a "factor of tolerance" which lies within the approximate limiting range of 0.7–1.0. If t lies outside this range, other structures are obtained. Table 10.4.1 lists some compounds with the perovskite structure.

Only a small fraction of the enormous number of perovskite-type compounds is listed in Table 10.4.1. For instance, substitution of the lanthanide element

Table 10.4.1. Some compounds with the perovskite structure

Compound	a (pm)	Compound	a (pm)	Compound	a (pm)
$AgZnF_3$	398	KIO_3	441.0	$RbMnF_3$	425.0
$BaCeO_3$	439.7	$KMgF_3$	397.3	$SmAlO_3$	373.4
$BaFeO_3$	401.2	$KMnF_3$	419.0	$SmCoO_3$	375
$BaMoO_3$	404.04	$KNbO_3$	400.7	$SmCrO_3$	381.2
$BaPbO_3$	427.3	$KTaO_3$	398.85	$SmFeO_3$	384.5
$BaSnO_3$	411.68	$LaCoO_3$	382.4	$SmVO_3$	389
$BaTiO_3$	401.18	$LaCrO_3$	387.4	$SrFeO_3$	386.9
$BaZrO_3$	419.29	$LaFeO_3$	392.0	$SrHfO_3$	406.9
$CaSnO_3$	392	$LaGaO_3$	387.5	$SrMoO_3$	397.51
$CaTiO_3$	384	$LaVO_3$	399	$SrSnO_3$	403.34
$CaVO_3$	376	$LiBaF_3$	399.6	$SrTiO_3$	390.51
$CaZrO_3$	402.0	$LiBaH_3$	402.3	$SrZrO_3$	410.1
$CsCaF_3$	452.2	$LiWO_3$	372	$TaSnO_3$	388.0
$CsCdF_3$	520	$MgCNi_3$	381.2	$TlCoF_3$	413.8
$CsPbBr_3$	587.4	$NaAlO_3$	373	$TlIO_3$	451.0
$EuTiO_3$	390.5	$NaNbO_3$	391.5	$YCrO_3$	376.8
$KCdF_3$	429.3	$NaWO_3$	386.22	$YFeO_3$	378.5
$KCoF_3$	406.9	$RbCaF_3$	445.2		
$KFeF_3$	412.2	$RbCoF_3$	406.2		

La for Sm will generate another isostructural series. In view of the common occurrence of perovskite-type compounds, it is worthwhile to elaborate their structural characteristics and important trends.

(1) **Relative sizes of ion A and ion B in ABX_3**

In all the perovskite compounds, the A ions are large and comparable in size to the oxygen or fluoride ion. The B ions must have a radius for six-coordination by oxygen (or fluorine). Thus, the radii of A and B ions must lie within the ranges 100 to 140 pm and 45 to 75 pm, respectively.

(2) **Valences of A and B cations**

Oxides with the perovskite structure must have a pair of cations of valences +1 and +5 (e.g. $KNbO_3$), +2 and +4 (e.g. $CaTiO_3$), or +3 and +3 (e.g. $LaCrO_3$). Sometimes, the A and/or B sites are not all occupied by atoms of the same kind, as in $(K_{0.5}La_{0.5})TiO_3$, $Sr(Ga_{0.5}Nb_{0.5})O_3$, $(Ba_{0.5}K_{0.5})(Ti_{0.5}Nb_{0.5})O_3$. The components of these compounds need to satisfy overall charge neutrality.

(3) **Perovskite as a complex oxide**

Despite the resemblance between the empirical formulas $CaCO_3$ and $CaTiO_3$, these two compounds are structurally entirely distinct. While $CaCO_3$ is a salt that contains the anion CO_3^{2-}, $CaTiO_3$ has no discrete TiO_3^{2-} species in its crystal structure, in which each Ti^{4+} ion is coordinated symmetrically by six O^{2-}. Thus $CaTiO_3$ should more properly be regarded as a complex oxide.

(4) **Perovskite compounds are not limited to oxides and fluorides**

The compound $MgCNi_3$ has the perovskite structure, as determined by neutron diffraction, and is a superconducting material with $T_C = 8$ K.

Table 10.4.2. Structure and properties of BaTiO$_3$ in different temperature ranges

Temperature	> 393 K	278–393 K	193–278 K	< 193 K
System	cubic	tetragonal	orthorhombic	hexagonal
Space group	$O_h^1 - Pm\bar{3}m$	$C_{4v}^1 - P4mm$	$C_{2v}^{14} - Amm2$	$D_{6h}^4 - P6_3/mmc$
Parameters of unit cell (pm)	$a = 401.18$	$a = 399.47$ $c = 403.36$	$a = 399.0$ $b = 566.9$ $c = 568.2$	$a = 573.5$ $c = 1405$
Atomic coordinates	Ba: 1(a) 0,0,0 Ti: 1(b) $\frac{1}{2},\frac{1}{2},\frac{1}{2}$ O: 3(d) $\frac{1}{2},\frac{1}{2},0$	Ba: 1(a) 0,0,0 Ti: 1(b) $\frac{1}{2},\frac{1}{2},0.512$ O(1): 1(b) $\frac{1}{2},\frac{1}{2},0.023$ O(2): 2(c) $\frac{1}{2},0,0.486$	Ba: 2(a) 0,0,z ($z=0$) Ti: 2(b) $\frac{1}{2},0,z$ ($z=0.510$) O(1): 2(a) 0,0,z ($z=0.490$) O(2): 4(e) $\frac{1}{2},y,z$ ($y=0.253$, $z=0.237$)	Ba(1): 2(b) 0,0,$\frac{1}{4}$ Ba(2): 4(f) $\frac{1}{3},\frac{2}{3},z$ ($z=0.097$) Ti(1): 2(a) 0,0,0 Ti(2): 4(f) $\frac{1}{3},\frac{2}{3},z$ ($z=0.845$) O(1):(6h) $x,2x,\frac{1}{4}$ ($x=0.522$) O(2): 12(k) $x,2x,z$ ($x=0.836$, $z=0.076$)
Ferroelectric property	no	yes	yes	no
Structure in Fig.10.4.2	(a)	(b)	(c)	(d)

10.4.2 Crystal structure of BaTiO$_3$

Mixed solid solutions of the general formula (Ba,Sr,Pb,Ca)(Ti,Zr,Sn)O$_3$ are the most commonly used ferroelectric materials.

In different temperature ranges, barium titanate BaTiO$_3$ exists in several stable phases. Table 10.4.2 lists the crystal data and properties of the polymorphic forms of BaTiO$_3$, and Fig. 10.4.2 shows their structures.

Above 393 K, BaTiO$_3$ has the perovskite structure, as shown in Fig. 10.4.2(a). This cubic crystal has a center of symmetry and does not exhibit ferroelectric properties.

In the temperature range 278–393 K, the BaTiO$_3$ lattice is transformed to the tetragonal system with local C_{4v} symmetry. The tetragonal unit cell has axial ratio $c/a = 1.01$. The Ti^{4+} ion departs from the center and moves toward the upper face; the O^{2-} ions are divided into two sets, one moving downward and other moving upward, as shown in Fig. 10.4.2(b). The arrows in the figure indicate the direction of the movement. The TiO$_6$ octahedra are no longer regular, which is the result of spontaneous polarization, the strength of which depends on temperature (thermoelectric effect) and pressure (piezoelectric effect).

In the temperature range 193–278 K, the atoms in the BaTiO$_3$ crystal undergo further shifts, and the lattice symmetry is reduced to the orthorhombic system and C_{2v} point group, as shown in Fig. 10.4.2(c). This crystalline modification has spontaneous polarization with ferroelectric properties.

(a)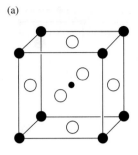

(b)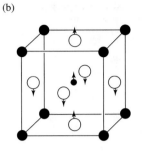

(c)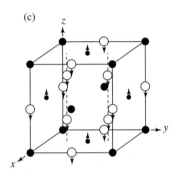

(d)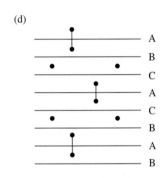

Fig. 10.4.2.
Structure of BaTiO$_3$: (a) cubic, (b) tetragonal, (c) orthorhombic, (d) hexagonal (white circles, O^{2-}; large black circles, Ba^{2+}; small black circles, Ti^{4+}).

When the temperature is below 193 K, the orthorhomic form further transforms to the hexagonal system in space group $D_{6h}^4 - P6_3/mmc$. The O^{2-} and Ba^{2+} ions together form close packed layers and stack in the sequence ABCACB and the Ti^{4+} ions occupy the octahedral interstices composed only of O^{2-} anions. In this layer stacked structure, the A layers have horizontal mirror symmetry as shown in Fig. 10.4.2(d). Two-thirds of the TiO$_6$ octahedra share faces to form the Ti$_2$O$_9$ group, in which the Ti–Ti distance is 267 pm, as represented by a line. This centrosymmetrical crystal is non-ferroelectric.

10.4.3 Superconductors of perovskite structure type

The crystal structure of many oxide superconductors can be regarded as based on the perovskite structure, which can be modified by atomic replacements, displacements, vacancies, and changes in the order of stacking layers.

(a) Structure of La$_2$CuO$_4$

The first-discovered oxide superconductor is La$_2$CuO$_4$ and it has the K$_2$NiF$_4$ structure type, as shown in Fig. 10.4.3(a). This structure can be derived from one B-type unit cell of the perovskite structure in combination with two A-type unit cells, each with one layer removed. In this manner, a tetragonal unit cell with $c \sim 3a$ is generated. The resulting structure belongs to space group $D_{4h}^{17} - I4/mmm$, in which the Cu atom has octahedral coordination and the La atom has nine-coordination. The CuO$_6$ octahedra share vertices and form infinite layers.

Fig. 10.4.3.
Structures of three oxide superconductors with perovskite structure type: (a) La$_2$CuO$_4$, (b) YBa$_2$Cu$_3$O$_6$, (c) YBa$_2$Cu$_3$O$_7$ (small white circles, O; small black circles, Cu; large black and open circles, M).

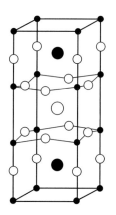
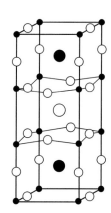

(b) **Structure of YBa$_2$Cu$_3$O$_6$ and YBa$_2$Cu$_3$O$_7$**

These two compounds are both high-temperature superconductors with similar structures, as shown in Figs. 10.4.3(b) and (c). YBa$_2$Cu$_3$O$_6$ belongs to the tetragonal system in space group $D_{4h}^1 - P4/mmm$, with $a = 385.70$ pm and $c = 1181.94$ pm. The crystals of YBa$_2$Cu$_3$O$_7$ are orthorhombic, space group $D_{2h}^1 - Pmmm$, with $a = 381.98$ pm, $b = 388.49$ pm, and $c = 1167.62$ pm. These two structures are conveniently described as oxygen-deficient perovskites with triple A-type unit cells owing to Ba–Y ordering along the c axis. In YBa$_2$Cu$_3$O$_7$, one set of Cu atoms from linear chains of corner-linked squares orientated along the b axis, and the other set of Cu atoms form layers of corner-shared square pyramids. The structure of YBa$_2$Cu$_3$O$_6$ is derived from that of YBa$_2$Cu$_3$O$_7$ by removal of the oxygen atoms at the base. The coordination geometry of the Cu atom at the origin then becomes linear two-coordinate, but the square-pyramidal environment of the other Cu atoms remains unchanged. In these two crystals, the Cu and O atoms that constitute the basal components of a square-pyramidal Cu–O layer are not coplanar.

10.4.4 ReO$_3$ and related compounds

(1) Structure of ReO$_3$

The rhenium trioxide ReO$_3$ structure consists of regular octahedra sharing vertices to form a large unoccupied interstice surrounded by 12 O atoms, as shown in Fig. 10.4.4(a). The structure of ReO$_3$ can also be regarded as derived from the perovskite structure by deleting all Ca^{2+} ions and keeping the TiO$_3$ framework. It belongs to space group $O_h^1 - Pm\bar{3}m$, the same one for perovskite.

Some trivalent metal fluorides, hexavalent metal oxides, and copper(I) nitride adopt the ReO$_3$ structure:

MoF$_3$, $a = 389.85$ pm;   TaF$_3$, $a = 390.12$ pm;   NbF$_3$, $a = 390.3$ pm;

UO$_3$, $a = 415.6$ pm;   ReO$_3$, $a = 373.4$ pm;   Cu$_3$N, $a = 380.7$ pm.

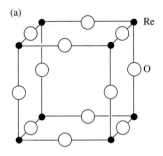 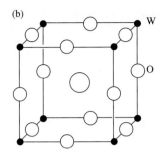

Fig. 10.4.4.
Crystal structure of (a) ReO₃, and (b) Na$_x$WO$_3$ (central large circle, Na$^+$ of fractional site occupancy).

(2) Structure of Na$_x$WO$_3$

The structure of WO$_3$ is of the ReO$_3$ type, as shown in Fig. 10.4.4(a). The tungsten bronzes of the formula M$_x$WO$_3$ ($0 < x < 1$) are derived from the three-dimensional network of WO$_3$, in which the W atoms adopt variable valence states, and the large unoccupied interstices conveniently accommodate other cationic species. When a portion of the W^{6+} cations is converted to W^{5+}, the requisite cations M (normally Na or K; but also Ca, Sr, Ba, Al, In, Tl, Sn, Pb, Cu, Ag, Cd, lanthanides, H$^+$, and NH$_4^+$) are incorporated into the structure to maintain electrical neutrality. The additional cations M are partially located at the centers of the unit cells to give the perovskite structure type, as shown in Fig. 10.4.4(b).

The name "tungsten bronzes" originates from their characteristic properties: intense color, metallic luster, metallic conductivity or semiconductivity, a range of variable composition, and resistance to attack by non-oxidizing acids. The bronzes Na$_x$WO$_3$ exhibit colors that change with the occupancy factor x as follows:

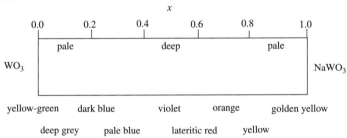

10.5 Hard magnetic materials

10.5.1 Survey of magnetic materials

There are four different classes of hard magnetic materials:

(a) Steel and iron-based alloys, such as carbon steel and the alnicos;
(b) Transition metal oxides, such as Fe$_3$O$_4$, γ-Fe$_2$O$_3$, and M-type ferrites [MO·6Fe$_2$O$_3$ (M = Ba, Sr, Pb)];
(c) Rare-earth intermetallics, such as SmCo$_5$, Sm$_2$Co$_{17}$, and Nd$_2$Fe$_{14}$B; and
(d) Metal complexes containing organic ligands, such as V(TCNE)$_x$ · y(CH$_2$Cl$_2$).

Table 10.5.1. Trends in $[BH]_{max}$ at 300 K of hard magnetic materials

Years	Materials	$[BH]_{max}$ (kJ m^{-3})
1960s	Magnetic materials of transition metals oxides:	
	Ba-ferrite (anis.)	70
	Sr-ferrite (anis.)	80
	Iron-based alloys: alnico 8, 9	100
1970s	Rare-earth intermetallics:	
	First generation: SmCo$_5$	150
	Second generation: Sm(Co, Cu, Fe, Zr)$_{7.4}$	250
1980s	Third generation: Nd$_2$Fe$_{14}$B	300
1990s	Improved third generations: Nd$_2$Fe$_{14}$B	400
	Organic ferromagnets: V(TCNE)$_x \cdot$y(CH$_2$Cl$_2$)	~ 5

For many centuries, carbon steels were the only synthetic permanent magnetic materials superior to loadstone. Starting in about 1890, improved understanding of the metallurgy of alloy steels, and later, control of solid state precipitation in iron-based alloys via heat treatment led to substantial advances. Today the alnicos (Fe–Co–Ni–Al–Cu alloys) corner a large fraction of the market on magnets. However, increasing availability of rare-earth metals has led to a new development path based on compounds containing 3d (transition) and 4f (rare-earth) metals. The trend in the maximum energy product $[BH]_{max}$ at 300 K for different classes of hard magnetic materials that have been studied over the past four decades is displayed in Table 10.5.1.

Transition metal oxides having the spinel structure are typical magnetic materials. In a magnetic solid, the spin interaction between neighboring cations through oxygen as a bridging element plays a major role in determining the magnitude of the magnetic moment. This spin interaction is most effective when the adjacent cations and bridging oxygens are in a straight line. In spinel AB$_2$O$_4$, the A(tetrahedral cation)–O–B(octahedral cation) angle is 125°, while the angles for A–O–A and B–O–B are 79° and 90°, respectively. Since the A–B interaction is the most effective, the magnetic properties of spinel structures can be fine-tuned by controlling the normal and inverse structures through composition changes and heat treatment.

The alnicos are a very important group of permanent magnetic alloys, which are used in a wide range of applications. They are essentially heat-treated alloys of variable Fe–Co–Ni–Al–Cu composition:

Alnico-5 (wt %): Ni 12–15, Al 7.8–8.5, Co 23–25, Cu 2–4, Ti 4–8, Nb 0–1, Fe as remainder
Alnico-9 (wt %): Ni 14–16, Al 7–8, Co 32–36, Cu 3–4, Nb 0–1, S 0.3, Fe as remainder

The alnicos are polycrystalline solids, in which very complex structural disorder generally occurs, due to the wide variety of deviations from periodicity that may be present. Physical disorder corresponds to the displacement of atoms, and/or the presence of amorphous regions and non-uniform crystallite sizes, while chemical disorder corresponds to site occupation by impurity atoms, and/or the presence of vacancies accompanied by variable valence states of

10.5.2 Structure of SmCo$_5$ and Sm$_2$Co$_{17}$

Most of the permanent magnets based on cobalt-rare-earth alloys exhibit several times higher magnetic energy than conventional magnets such as alnicos and hard ferrites. They also exhibit very high stability of magnetic flux density even at elevated temperatures, a property unattainable with any other existing permanent-magnet materials.

Among the rare-earths, samarium is the most commonly used because it provides the best permanent-magnet property. Other rare-earth elements are sometimes employed in combination with samarium to meet special requirements. The compounds SmCo$_5$ and Sm$_2$Co$_{17}$ are the most important magnetic materials among the cobalt-lanthanide alloys.

Compound SmCo$_5$ belongs to the hexagonal system, space group *P6/mmm*, with $a = 498.9$ pm and $c = 398.1$ pm. Its crystal structure, as shown in Fig. 10.5.1, consists of two different layers of atoms. One layer is composed of two kinds of atoms in the ratio of 1:2 for samarium to cobalt, with the Sm atoms arranged in a closed packed layer and the Co atoms at the center of each triangular interstice. This layer alternates with another layer consisting of only cobalt atoms.

The crystal structure of Sm$_2$Co$_{17}$ is closely related to the SmCo$_5$ structure. One Sm atom is substituted by a pair of Co atoms at one-third of the Sm sites in the SmCo$_5$ structure. This pair of Co atoms lie along the c axis. The positions of the substitution in the neighboring (but similar) layer are different. The order of the substitution can be either (a) ABABAB… or (b) ABCABC… along the c-axis, as shown in Fig. 10.5.2.

In the substitution order ABABAB…, Sm$_2$Co$_{17}$ forms a hexagonal structure, with $a = 836.0$ pm and $c = 851.5$ pm [Fig. 10.5.2(a)]. The substitution order ABCABC… yields a rhombohedral form of Sm$_2$Co$_{17}$ with $a = 837.9$ pm and $c = 1221.2$ pm [Fig. 10.5.2(b)].

10.5.3 Structure of Nd$_2$Fe$_{14}$B

Permanent magnets based on the Nd$_2$Fe$_{14}$B phase were introduced in 1983 and are currently the strongest magnets available.

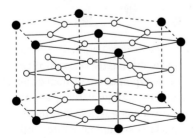

Fig. 10.5.1.
Crystal structure of SmCo$_5$.

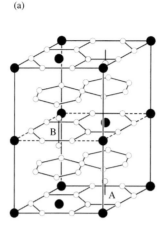
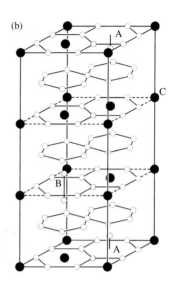

Fig. 10.5.2.
Crystal structure of Sm_2Co_{17}: (a) hexagonal form and (b) rhombohedral form.

The $Nd_2Fe_{14}B$ crystal is tetragonal, with space group $P4_2/mnm$, unit cell $a = 880$ pm, $c = 1220$ pm, and $Z = 4$. There is a six-layer stacking sequence along the c axis. The first and fourth layers are mirror planes that contain Fe, Nd, and B atoms, and the other layers are puckered nets containing only Fe atoms. Each B atom is at the center of a trigonal prism formed by six Fe atoms, three above and three below the Fe–Nd–B planes. Figure 10.5.3 shows the structure of $Nd_2Fe_{14}B$.

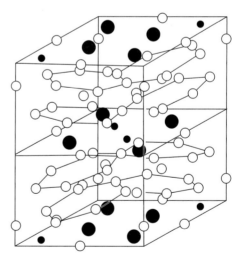

Fig. 10.5.3.
Crystal structure of $Nd_2Fe_{14}B$ (large black circles, Nd; small black circles, B; open circles, Fe).

An effective strategy of modifying the magnetic properties of iron-rich rare-earth intermetallics is to incorporate small interstitial atoms into the crystal lattice. Besides hydrogen, only boron, carbon, and nitrogen atoms are small enough to enter the structure in this way, and they preferentially occupy interstitial sites surrounded by the largest number of rare-earth atoms. For example, in

Sm_2Fe_{17} (iso-structural to Sm_2Co_{17}) there are three octahedral interstitial sites which may be filled by nitrogen atoms to form $Sm_2Fe_{17}N_3$. The effect of interstitial atoms on the intrinsic magnetic properties of iron-based intermetallics can be further exploited in materials design.

References

1. A. F. Wells, *Structural Inorganic Chemistry*, 5th edn., Oxford University Press, Oxford, 1984.
2. T. C. W. Mak and G.-D. Zhou, *Crystallography in Modern Chemistry: A Resource Book of Crystal Structures*, Wiley, New York, 1992.
3. B. K. Vainshtein, V. M. Fridkin, and V. L. Indenbom, *Structures of Crystals*, 2nd edn., Springer Verlag, Berlin, 1995.
4. R. J. D. Tilley, *Crystals and Crystal Structures*, Wiley, Chichester, 2006.
5. U. Müller, *Inorganic Structural Chemistry*, 2nd edn., Wiley, Chichester, 2006.
6. F. S. Galasso, *Structure and Properties of Inorganic Solids*, Pergamon, Oxford, 1970.
7. B. G. Hyde and S. Andersson, *Inorganic Crystal Structures*, Wiley, New York, 1989.
8. L. E. Smart and E. A. Moore, *Solid State Chemistry: An Introduction*, 3rd edn., CRC Press, Boca Raton, Fl, 2005.
9. A. R. West, *Basic Solid State Chemistry*, 2nd edn., Wiley, New York, 1999.
10. A. K. Cheetham and P. Day (eds.), *Solid State Chemistry: Techniques*, Oxford University Press, New York, 1987.
11. A. K. Cheetham and P. Day (eds.), *Solid State Chemistry: Compounds*, Oxford University Press, New York, 1991.
12. G. S. Rohrer, *Structure and Bonding in Crystalline Materials*, Cambridge University Press, Cambridge, 2001.
13. J. I. Gersten, and F. W. Smith, *The Physics and Chemistry of Materials*, Wiley, New York, 2001.
14. R. C. Buchanan and T. Park, *Materials Crystal Chemistry*, Marcel Dekker, New York, 1997.
15. B. Douglas and S.-M. Ho, *Structure and Chemistry of Crystalline Solids*, Springer, New York, 2006.
16. C. N. R. Rao and Raveau, *Transition Metal Oxides: Structure, Properties, and Synthesis of Ceramic Oxides*, 2nd edn., Wiley–VCH, New York, 1998.
17. W. M. Meier and D. H. Olson, *Atlas of Zeolite Structure Types,* 2nd edn., Butterworths, London, 1987.
18. H. Yanagida, K. Koumoto and M. Miyayama, *The Chemistry of Ceramics*, 2nd edn., Wiley, Chichester, 1996.
19. J. Evetts (ed.), *Concise Encyclopedia of Magnetic and Superconducting Materials*, Pergamon Press, Oxford, 1992.
20. R. Skomski and J. M. D. Coey, *Permanent Magnetism,* Institute of Physics Publishing, Bristol, 1999.

Structural Chemistry of Selected Elements

Part III comprises a succinct account of the structural chemistry of the elements in the Periodic Table.

In general, each chemical element or group has its own characteristic structural chemistry. In Part III, seven chapters are devoted to discussing the structural chemistry of hydrogen, alkali, and alkaline-earth metals, and the elements in Group 13 to Group 18. These chapters are followed by one chapter each on the rare-earth elements, transition metal clusters, and supramolecular structural chemistry. The compounds described include organometallics, metal–metal bonded systems, coordination polymers, host–guest compounds, and supramolecular assemblies. Basic organic crystal structures are discussed in Chapter 14 under the chemistry of carbon. The structural chemistry of bioinorganic compounds is not covered here as, in the writers' opinion, the subject would be better taught in a separate course on "Bioinorganic Chemistry," for which several excellent texts are available.

The presentation in Part III attempts to convey the message that inorganic synthesis is inherently less organized than organic synthesis, and serendipitous discoveries are being made from time to time. Selected examples illustrating interesting aspects of structure and bonding, generalizations of structural patterns, and highlights from the current literature are discussed in the light of the theoretical principles presented in Parts I and II. The most up-to-date resources and references are used in the compilation of tables of structural data. It is hoped that the immense impact of chemical crystallography in the development of modern chemistry comes through naturally to the reader.

Structural Chemistry of Hydrogen

11.1 The bonding types of hydrogen

Hydrogen is the third most abundant element (after oxygen and silicon in terms of the number of atoms) in the earth's crust and oceans. Hydrogen atom exists as three isotopes: ^{1}H (ordinary hydrogen, protium, H), ^{2}H (deuterium, D) and ^{3}H (tritium, T). Some of the important physical properties of the isotopes of hydrogen are listed in Table 11.1.1. Hydrogen forms more compounds than any other element, including carbon. This fact is a consequence of the electronic structure of the hydrogen atom.

Hydrogen is the smallest atom with only one electron and one valence orbital, and its ground configuration is $1s^1$. Removal of the lone electron from the neutral atom produces the H$^+$ ion, which bears some similarity with an alkali metal cation, while adding one electron produces the H$^-$ ion, which is analogous to a halide anion. Hydrogen may thus be placed logically at the top of either Group 1 or Group 17 in the Periodic Table. The hydrogen atom forms many bonding types in a wide variety of compounds, as described below.

(1) Covalent bond

The hydrogen atom can use its 1s orbital to overlap with a valence orbital of another atom to form a covalent bond, as in the molecules listed in Table 11.1.2. The covalent radius of hydrogen is 37 pm.

Table 11.1.1. Properties of hydrogen, deuterium, and tritium

Property	H	D	T
Abundance (%)	99.985	0.015	$\sim 10^{-16}$
Relative atomic mass	1.007825	2.014102	3.016049
Nuclear spin	1/2	1	1/2
Nuclear magnetic moment (μ_N)	2.79270	0.85738	2.9788
NMR frequency (at 2.35 T) (MHz)	100.56	15.360	104.68
Properties of diatomic molecules	H$_2$	D$_2$	T$_2$
mp (K)	13.957	18.73	20.62
bp (K)	20.30	23.67	25.04
T_c (K)	33.19	38.35	40.6 (calc.)
Enthalpy of dissociation (kJ mol^{-1})	435.88	443.35	446.9
Zero point energy (kJ mol^{-1})	25.9	18.5	15.1

Table 11.1.2. Molecules containing covalently bonded hydrogen

Molecules	H_2	HCl	H_2O	NH_3	CH_4
Bond	H–H	H–Cl	H–O	H–N	H–C
Bond lengths (pm)	74.14	147.44	95.72	101.7	109.1

(2) Ionic bond

(a) Hydrogen present as H^-

The hydrogen atom can gain an electron to form the hydride ion H^- with the helium electronic configuration

$$H + e^- \rightarrow H^- \quad \Delta H = 72.8 \,\text{kJ mol}^{-1}.$$

With the exception of beryllium, all the elements of Groups 1 and 2 react spontaneously when heated in hydrogen gas to give white solid hydrides $M^I H$ or $M^{II} H_2$. All MH hydrides have the sodium chloride structure, while MgH_2 has the rutile structure and CaH_2, SrH_2, and BaH_2 have the structure of distorted $PbCl_2$. The chemical and physical properties of these solid hydrides indicate that they are ionic compounds.

The hydride anion H^- is expected to be very polarizable, and its size changes with the partner in the MH or MH_2 compounds. The experimental values of the radius of H^- cover a rather wide range:

Compounds	LiH	NaH	KH	RbH	CsH	MgH_2
$r(H^-)$ (pm)	137	142	152	154	152	130

From NaH, which has the NaCl structure with cubic unit-cell parameter $a = 488$ pm, the radius of H^- has been calculated as 142 pm.

(b) Hydrogen present as H^+

The loss of an electron from a H atom to form H^+ is an endothermic process:

$$H(g) \rightarrow H^+(g) + e^- \quad \Delta H = 1312.0 \,\text{kJ mol}^{-1} = 13.59 \,\text{eV}.$$

In reality H^+, a bare proton, cannot exist alone except in isolation inside a high vacuum. The radius of H^+ is about 1.5×10^{-15} m, which is 10^5 times smaller than other atoms. When H^+ approaches another atom or molecule, it can distort the electron cloud of the latter. Therefore, other than in a gaseous ionic beam, H^+ must be attached to another atom or molecule that possesses a lone pair of electrons. The proton as an acceptor can be stabilized as in pyramidal hydronium (or hydroxonium) ion H_3O^+, tetrahedral NH_4^+, and linear H_2F^+. These cations generally combine with various anions through ionic bonding to form salts.

The total hydration enthalpy of H^+ is highly exothermic and much larger than those of other simple charged cations:

$$H^+(g) + nH_2O(\ell) \rightarrow H_3O^+(aq) \quad \Delta H = -1090 \,\text{kJ mol}^{-1}$$

(3) Metallic bond

At ultrahigh pressure and low temperature, such as 250 GPa and 77 K, solid molecular hydrogen transforms to a metallic phase, in which the atoms are held together by the metallic bond, which arises from a band-overlap mechanism. Under such extreme conditions, the H_2 molecules are converted into a linear chain of hydrogen atoms (or a three-dimensional network). This polymeric H_n structure with a partially filled band (conduction band) is expected to exhibit metallic behavior. Schematically, the band-overlap mechanism may be represented in the following manner:

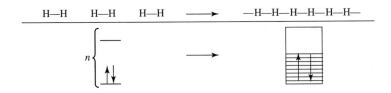

The physical properties of H_n are most interesting. This material has a nearly opaque appearance and exhibits metallic conduction, and has been suggested to be present in several planets.

(4) The hydrogen bond

The hydrogen bond is usually represented as X–H···Y, where X and Y are highly electronegative atoms such as F, O, N, Cl, and C. Note that the description of "donor" and "acceptor" in a hydrogen bond refers to the proton; in the hydrogen bond X–H···Y, the X–H group acts as a hydrogen donor to acceptor atom Y. In contrast, in a coordination bond the donor atom is the one that bears the electron pair.

Hydrogen bonding can be either intermolecular or intramolecular. For an intramolecular X–H···Y hydrogen bond to occur, X and Y must be in favorable spatial configuration in the same molecule. This type of hydrogen bond will be further elaborated in Section 11.2.

(5) Multicenter hydrogen bridged bonds

(a) B–H–B bridged bond

Boranes, carboranes, and metallocarboranes are electron-deficient compounds in which B–H–B three-center two-electron (3c-2e) bridged bonds are found. The B–H–B bond results from the overlap of two boron sp^3 hybrid orbitals and the 1s orbital of the hydrogen atom, which will be discussed in Chapter 13.

(b) M–H–M and M–H–B bridged bonds

The 3c-2e hydrogen bridged bonds of type M–H–M and M–H–B can be formed with main-group metals (such as Be, Mg and Al) and transition metals (such as Cr, W, Fe, Ta, and Zr). Some examples are shown in Fig. 11.1.1.

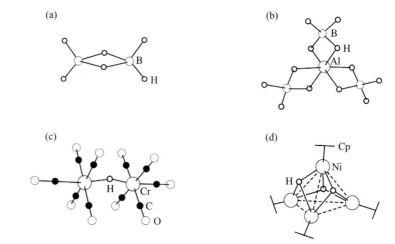

Fig. 11.1.1.
Structures containing multiple hydrogen bridged bonds: (a) B_2H_6, (b) $Al(BH_4)_3$, (c) $[(CO)_5Cr-H-Cr(CO)_5]^-$, (d) $H_3Ni_4Cp_4$.

(c) Triply bridged (μ_3-H)M$_3$ bond

In this bond type, a H atom is covalently bonded simultaneously to three metal atoms, as shown in Fig. 11.1.1(d).

(6) Hydride coordinate bond

The hydride ion H^- as a ligand can donate a pair of electrons to a metal to form a metal hydride. The crystal structures of many metal hydride complexes, such as Mg_2NiH_4, Mg_2FeH_6, and K_2ReH_9, have been determined. In these compounds, the M–H bonds are covalent σ coordinate bonds.

(7) Molecular hydrogen coordinate bond

Molecular hydrogen can coordinate to a transition metal as an intact ligand. The bonding of the H_2 molecule to the transition metal atom appears to involve the transfer of σ bonding electrons of H_2 to a vacant metal d orbital, coupled with synergistic back donation of metal d electrons to the vacant σ^* antibonding orbital of the H_2 molecule. This type of coordinate bond formation weakens the H–H covalent bond of the H_2 ligand and, in the limit, leads to its cleavage to two H atoms.

(8) Agostic bond

The existence of the agostic bond C–H→M has been firmly established by X-ray and neutron diffraction methods. The symbolic representation C–H→M indicates formal donor interaction of a C–H bond with an electron-deficient metal atom M. As in all 3c-2e bridging systems involving only three valence orbitals, the bonded C–H→M fragment is bent. The agostic bond will be further discussed in Section 11.5.3.

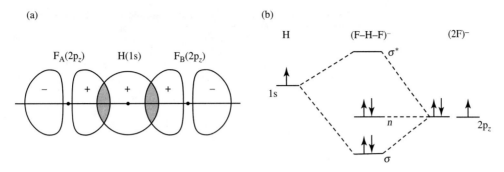

Fig. 11.2.3.
Bonding in HF_2^-: (a) orbital overlap; (b) qualitative MO energy level diagram.

B), as shown in Fig. 11.2.3(a), to form three molecular orbitals:

$$\psi_1(\sigma) = N_1[2p_z(A) + 2p_z(B) + c1s]$$
$$\psi_2(n) = N_2[2p_z(A) - 2p_z(B)]$$
$$\psi_3(\sigma^*) = N_3[2p_z(A) + 2p_z(B) - c1s],$$

where c is a weighting coefficient and N_1, N_2, N_3 are normalization constants. The ordering of the molecular orbitals is shown qualitatively in Fig. 11.2.3(b). Since there are four valence electrons, the bonding (ψ_1) and nonbonding (ψ_2) molecular orbitals are both occupied to yield a $3c$-$4e$ bond. The bond order and force constant of each F–H link in HF_2^- can be compared with those in the HF molecule:

Molecule	Bond order	d (pm)	k (N m^{-1})
HF	1	93	890
HF_2^-	0.5	113	230

In the $\{[(NH_2)_2CO]_2H\}(SiF_6)$ crystal, there are symmetrical hydrogen bonds of the type O–H–O, with bond lengths 242.4 and 244.3 pm in two independent $\{[(NH_2)_2CO]_2H\}^+$ cations. Figure 11.2.4 shows the structure of the cation.

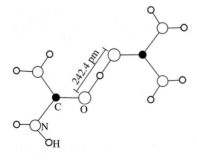

Fig. 11.2.4.
Structure of $\{[(NH_2)_2CO]_2H\}^+$.

Fig. 11.2.5.
The hydrogen bond W–C≡O···H–O in crystalline W(CO)$_3$(P^{i}Pr$_3$)$_2$(H$_2$O)(THF).

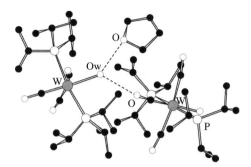

11.2.4 Hydrogen bonds in organometallic compounds

In transition metal carbonyls, the M–C=O group acts as a proton acceptor, which interacts with appropriate donor groups to form one or more C=O···H–X (X is O or N) hydrogen bonds. The CO ligand can function in different μ_1, μ_2, and μ_3 modes, corresponding to a formal C–O bond order of 3, 2, and 1, respectively. Examples of hydrogen bonding formed by metal carbonyl complexes are shown below:

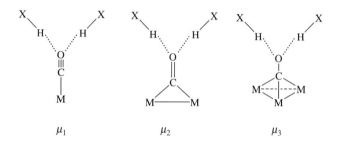

When the M atom has a strong back-donation to the CO π^* orbital, the basic property of CO increases, and the O···H distance shortens. These results are consistent with the sequence of shortened distances from terminal μ^1 to μ^3 coordinated forms. In such hydrogen bonds, the bond angles of C–O···X are all about 140°. Figure 11.2.5 shows the hydrogen bond W–C≡O···H–O in the crystal of W(CO)$_3$(P^{i}Pr$_3$)$_2$(H$_2$O)(THF), in which the O···O bond length is 279.2 pm. Note that the aqua ligand also forms a donor hydrogen with a THF molecule.

In organometallic compounds, the μ_3-CH and μ_2-CH$_2$ ligands can act as proton donors to form hydrogen bonds, as shown below:

Since the acidity of μ_3-CH is stronger than that of μ_2-CH$_2$, the length of a μ_3-CH hydrogen bond is shorter than that of μ_2-CH$_2$.

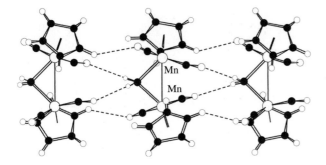

Fig. 11.2.6.
C–H···O hydrogen bonds between μ_2-CH$_2$ and Mn–CO groups in [CpMn(CO)$_2$]$_2$(μ_2-CH$_2$).

Neutron diffraction studies have shown that the –CH ligand in the cluster [Co(CO)$_3$]$_3$(μ_3-CH) forms three hydrogen bonds, with H···O distances of 250, 253, and 262 pm. In the compound [CpMn(CO)$_2$]$_2$(μ_2-CH$_2$), the μ_2-CH$_2$ group forms hydrogen bonds with the O atoms of Mn–CO groups, as shown in Fig. 11.2.6.

11.2.5 The universality and importance of hydrogen bonds

Hydrogen bonds exist in numerous compounds. The reasons for its universal appearance are as follows:

(a) *The abundance of H, O, N, C and halogen elements:* Many compounds are composed of H, O, N, C, and halogens, such as water, HX, oxyacids, and organic compounds. These compounds generally contain functional group such as –OH, –NH$_2$, and >C=O, which readily form hydrogen bonds.
(b) *Geometrical requirement of hydrogen bonding:* A hydrogen bond does not require rigorous conditions for its formation as in the case for covalent bonding. The bond lengths and group orientations allow for more flexibility and adaptability.
(c) *Small bond energy:* The hydrogen bond is intermediate in strength between the covalent bond and van der Waals interaction. The small bond energy of the hydrogen bond requires low activation energy in its formation and cleavage. Its relative weakness permits reversibility in reactions involving its formation and a greater subtlety of interaction than is possible with normal covalent bonds.
(d) *Intermolecular and intramolecular bonding modes:* Hydrogen bonds can form between molecules, within the same molecule, or in a combination of both varieties. In liquids, hydrogen bonds are continuously being broken and reformed at random. Figures 11.2.7 and 11.2.8, respectively, display some structures featuring intermolecular and intramolecular hydrogen bonds.

Hydrogen bonds are important because of the effects they produce:

(a) Hydrogen bonds, especially the intramolecular variety, dictate many chemical properties, influence the conformation of molecules, and often

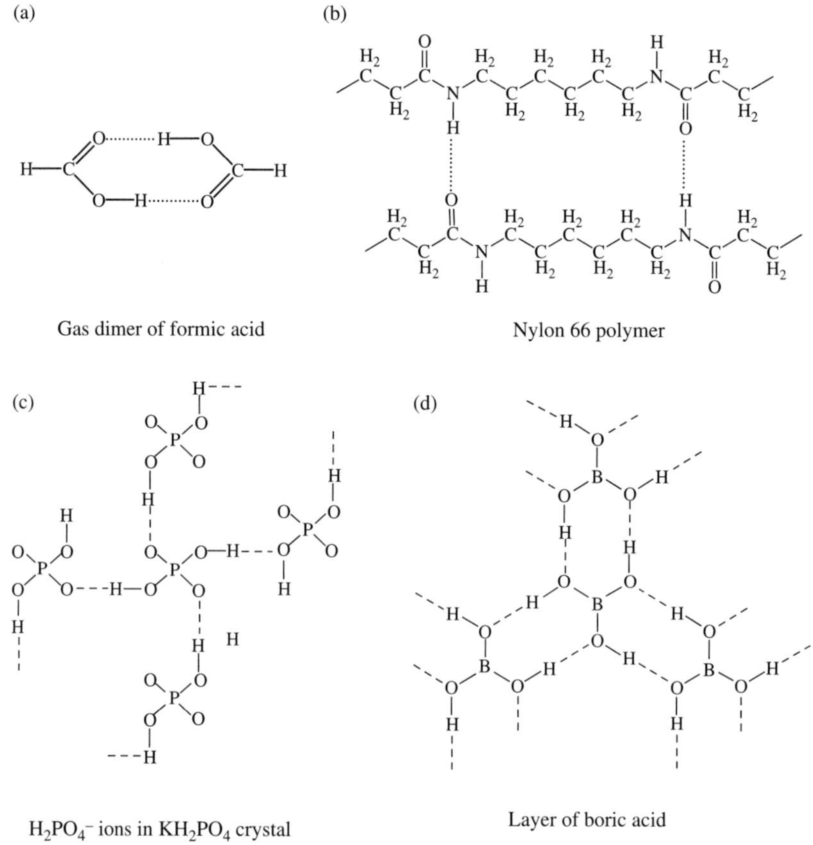

Fig. 11.2.7.
Examples of intermolecular hydrogen bonds.

(a) Gas dimer of formic acid

(b) Nylon 66 polymer

(c) $H_2PO_4^-$ ions in KH_2PO_4 crystal

(d) Layer of boric acid

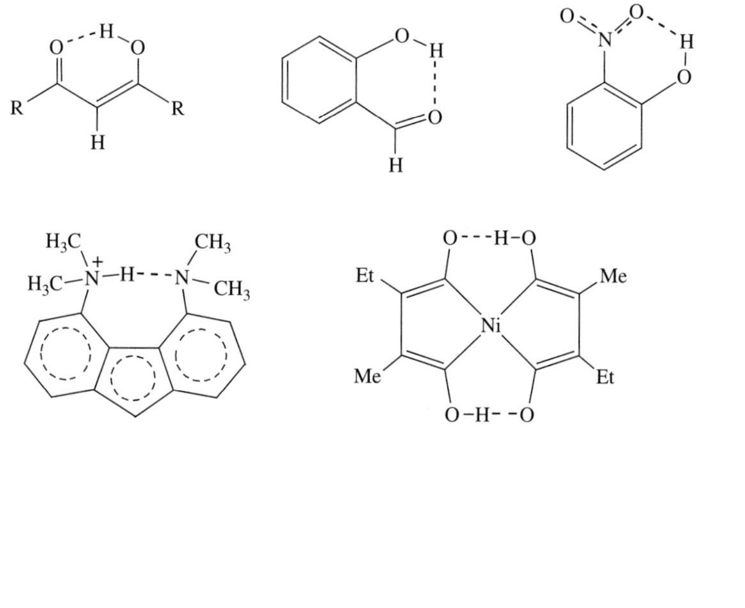

Fig. 11.2.8.
Examples of intramolecular hydrogen bonds.

play a critical role in determining reaction rates. Hydrogen bonds are responsible for stabilization of the three-dimensional architecture of proteins and nucleic acids.
(b) Hydrogen bonds affect IR and Raman frequencies. The ν(X–H) stretching frequency shifts to a lower energy (caused by weakening the X–H bond) but increases in width and intensity. For N–H$\cdots$F, the change of frequency is less than 1000 cm^{-1}. For O–H$\cdots$O and F–H$\cdots$F, the change of frequency is in the range 1500–2000 cm^{-1}. The ν(X–H) bending frequency shifts to higher wavenumbers.
(c) Intermolecular hydrogen bonding in a compound raises the boiling point and frequently the melting point.
(d) If hydrogen bonding is possible between solute and solvent, solubility is greatly increased and often results in infinite solubility. The complete miscibility of two liquids, for example water and ethanol, can be attributed to intermolecular hydrogen bonding.

11.3 Non-conventional hydrogen bonds

The conventional hydrogen bond X–H$\cdots$Y is formed by the proton donor X–H with a proton acceptor Y, which is an atom with a lone pair (X and Y are all highly electronegative atoms such as F, O, N, and Cl). Some non-conventional hydrogen bonds that do not conform to this condition are discussed below:

11.3.1 X–H$\cdots$$\pi$ hydrogen bond

In a X–H$\cdots$$\pi$ hydrogen bond, π bonding electrons interact with the proton to form a weakly bonded system. The phenyl ring and delocalized π system as proton acceptors interact with X–H to form X–H$\cdots$$\pi$ hydrogen bonds. The phenyl group is by far the most important among π-acceptors, and the X–H$\cdots$Ph hydrogen bond is termed an "aromatic hydrogen bond." The N–H and phenyl groups together form aromatic hydrogen bonds which stabilize the conformation of polypeptide chains. Calculations show that the bond energy of a N–H$\cdots$Ph bond is about 12 kJ mol^{-1}. Two major types of N–H$\cdots$Ph hydrogen bonds generally occur in the polypeptide chains of biomolecules:

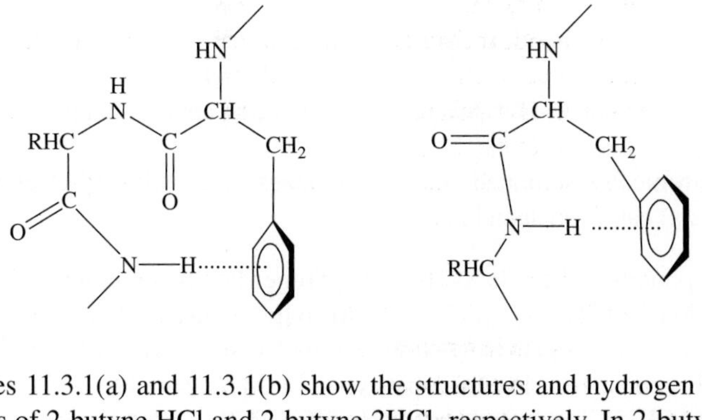

Figures 11.3.1(a) and 11.3.1(b) show the structures and hydrogen bonding distances of 2-butyne·HCl and 2-butyne·2HCl, respectively. In 2-butyne·HCl,

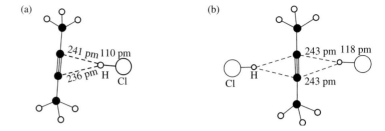

Fig. 11.3.1.
The Cl–H···π hydrogen bonds in (a) 2-butyne·HCl and (b) 2-butyne·2HCl.

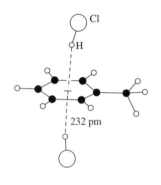

Fig. 11.3.2.
Cl–H···π hydrogen bonds in the crystal of toluene·2HCl.

the distance from Cl to the center of the C≡C bond is 340 pm. In 2-butyne·2HCl, the distance from Cl to the center of the C≡C bond is 347 pm.

The Cl–H···π hydrogen bond in the crystal structure of toluene·2HCl has been characterized. In this case the π bonding electrons of the aromatic ring of toluene serve as the proton acceptor. The distance of the H atom to the ring center is 232 pm, as shown in Fig. 11.3.2.

In addition to N–H···π and Cl–H···π hydrogen bonds, there are also O–H···π and C–H···π hydrogen bonds in many compounds. Two examples are shown in Figs. 11.3.3(a) and (b). A neutron diffraction study of the crystal structure of 2-acetylenyl-2-hydroxyl adamatane [Fig. 11.3.3(c)] has shown that intermolecular O–H···O and C–H···O hydrogen bonds co-exist [Fig. 11.3.3(d)] with the O–H···π hydrogen bonds [Fig. 11.3.3(e)].

11.3.2 Transition metal hydrogen bond X–H···M

The X–H···M hydrogen bond is analogous to a conventional hydrogen bond and involves an electron-rich transition metal M as the proton acceptor in a 3c-4e interaction. Several criteria that serve to characterize a 3c-4e X–H···M hydrogen bond are as follows:

(a) The bridging hydrogen is covalently bonded to a highly electronegative atom X and is protonic in nature, enhancing the electrostatic component of the interaction.
(b) The metal atom involved is electron-rich, i.e., typically a late transition metal, with filled d-orbitals that can facilitate the 3c-4e interaction involving the H atom.
(c) The ^{1}H NMR chemical shift of the bridging H atom is downfield of TMS and shifted downfield relative to the free ligand.
(d) Intermolecular X–H···M interactions have an approximately linear geometry.
(e) Electronically saturated metal complexes (e.g., with 18-electron metal centers) can form such interactions.

Two compounds with 3c-4e X–H···M hydrogen bonds are shown in Fig. 11.3.4. The dianion $\{(PtCl_4) \cdot cis\text{-}[PtCl_2(NH_2Me)_2]\}^{2-}$ consists of two square-planar d^8-Pt centers held together by short intermolecular N–H···Pt and N–H···Cl hydrogen bonds: H···Pt 226.2 pm, H···Cl 231.8 pm. The N–H···Pt bond angle is 167.1°. The presence of the filled Pt d_{z^2} orbital oriented towards

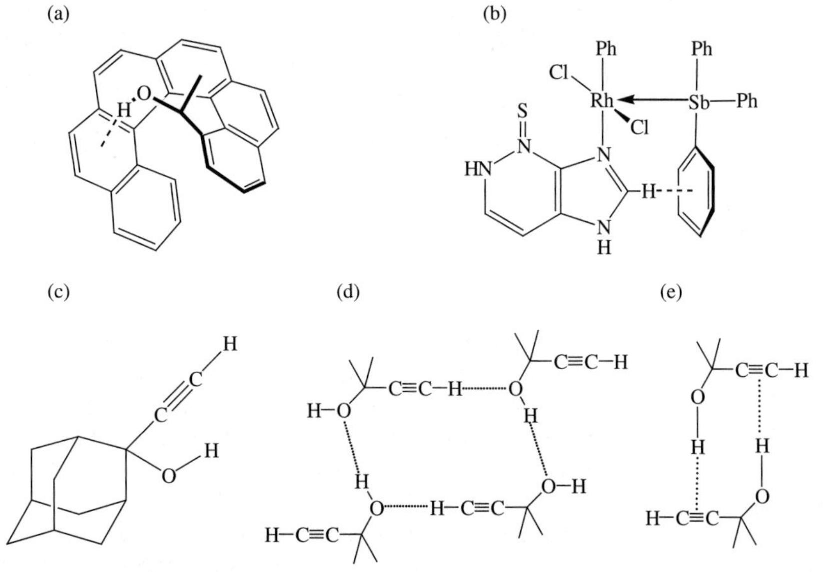

Fig. 11.3.3.
Some compounds containing O–H···π and C–H···π hydrogen bonds.

the amine N–H group favors a 3c-4e interaction. Figure 11.3.4(a) shows the structure of this dianion. Figure 11.3.4(b) shows the molecular structure of PtBr(1-C$_{10}$H$_6$NMe$_2$)(1-C$_{10}$H$_5$NHMe$_2$) with Pt···N = 328 pm and Pt···H–N = 168°.

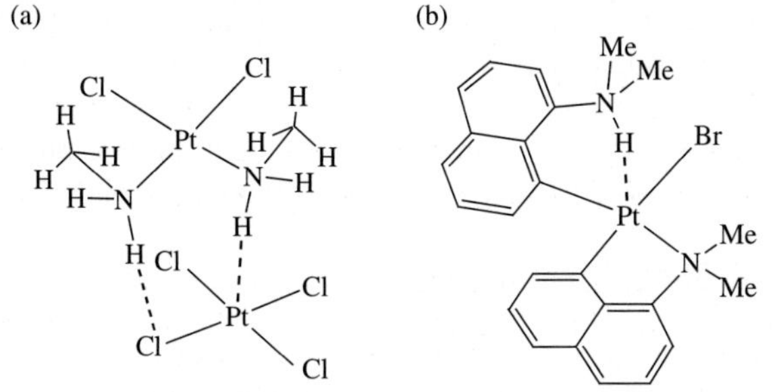

Fig. 11.3.4.
Molecular structures of compounds with X–H···M hydrogen bond:
(a) {(PtCl$_4$)·*cis* − [PtCl$_2$(NH$_2$Me)$_2$]}$^{2-}$,
(b) PtBr(1-C$_{10}$H$_6$NMe$_2$)(1-C$_{10}$H$_5$NHMe$_2$).

11.3.3 Dihydrogen bond X–H···H–E

Conventional hydrogen bonds are formed between a proton donor, such as an O–H or N–H group and a proton acceptor, such as oxygen or nitrogen lone pair. In all such cases a nonbonding electron pair acts as the weak base component.

A wide variety of E–H σ bonds (E = boron or transition metal) act unexpectedly as efficient hydrogen bond acceptors toward conventional proton donors, such as O–H and N–H groups. The resulting X–H···H–E systems have close H···H contacts (175-190 pm) and are termed "dihydrogen bonds."

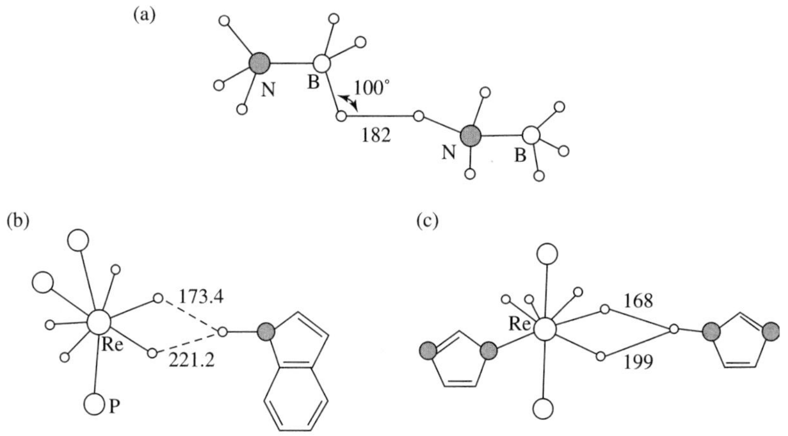

Fig. 11.3.5.
Structure of compounds containing
X–H···H–E bonds: (a)
H₃N–BH₃ ··· H₃N–BH₃,
(b) ReH₅(PPh₃)₃·indole, and
(c) ReH₅(PPh₃)₂(imidazole).

Their enthalpies of formation are substantial, 13–30 kJ mol^{-1}, and lie in the range for conventional H bonds. Some examples are listed below.

(1) N–H···H–B bond

A comparison of the melting points of the isoelectronic species H₃C–CH₃ (−181°C), H₃C–F (−141°C) and H₃N–BH₃ (104 °C) suggests the possibility that unusually strong intermolecular interactions are present in H₃N–BH₃. As compared to the H₃C–F molecule, the less polar H₃N–BH₃ molecule has a smaller dipole-dipole interaction, and both slack a lone pair to form a conventional hydrogen bond. The abnormally high melting point of H₃N–BH₃ originates from the presence of N–H···H–B bonding in the crystalline state.

Many close intermolecular N–H···H–B contacts in the range 170-220 pm have been found in various crystal structures. In these dihydrogen bonds, the (NH)···H–B angle is strongly bent, falling in the range 95–120°. Figure 11.3.5(a) shows the (NH)···H–B bond between H₃N–BH₃ molecules.

(2) X–H···H–M bonds

X-Ray and neutron diffraction studies have established the presence of transition metal N–H···H–M and O–H···H–M dihydrogen bonds, in which the hydride ligand acts as a proton acceptor. Figure 11.3.5(b) shows the dihydrogen bond N–H···H–Re in the crystal structure of ReH₅(PPh₃)₃·indole·C₆H₆. The H···H distances are 173.4 and 221.2 pm. Figure 11.3.5(c) shows a similar dihydrogen bond with measured H···H distances of 168 and 199 pm in the crystal structure of ReH₅(PPh₃)₂(imidazole).

In the crystal structure of [K(1,10-diaza-18-crown-6)][IrH₄(P^{i}Pr₃)₂], the two kinds of ions form an infinite chain held together by N–H···H–Ir bonds, as shown schematically in Fig. 11.3.6. The observed distance of H···H is 207 pm, and the observed N–H bond length of 77 pm is likely to be less than the true value. The corrected N–H and H···H distances are 100 and 185 pm, respectively.

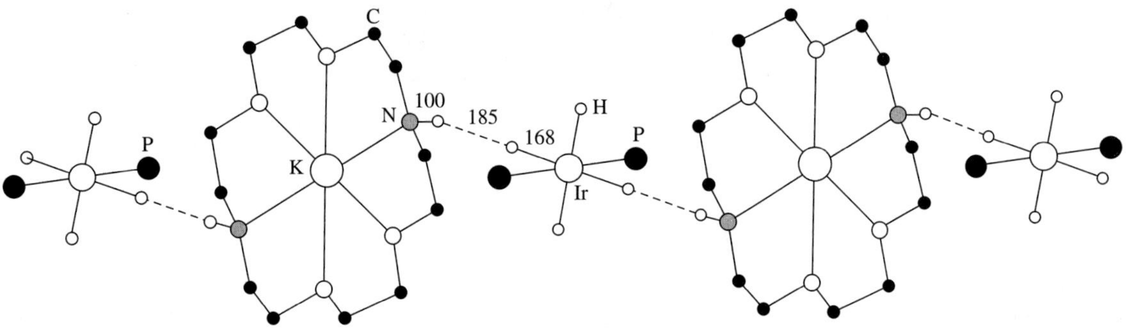

Fig. 11.3.6.
Chain structure of [K(1,10-diaza-18-crown-6)][IrH$_4$(P^{i}Pr$_3$)$_2$] and relevant dihydrogen bond lengths in pm. P^{i}Pr$_3$C and H atoms and non-essential crown H atoms have been omitted for clarity.

The X–H···H–M systems generally tend to lose H$_2$ readily, and indeed such H···H bonded intermediates may be involved whenever a hydride undergoes protonation.

11.3.4 Inverse hydrogen bond

In the normal hydrogen bond X–H···Y, the H atom plays the role of electron acceptor, while the Y atom is the electron donor. The interaction is of the kind

$$X-H\cdots Y.$$

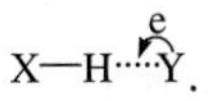

In the inverse hydrogen bond, the H atom plays the role of electron donor, while the Y atom becomes the electron acceptor, and the interaction is of the kind

$$X-H\cdots Y.$$

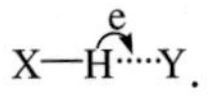

Some examples of inverse hydrogen bonds are presented below:

(1) The so-called "lithium hydrogen bond," Li–H···Li–H, occurs in the hypothetical linear (LiH)$_2$ dimer. The inner Li atom is electron-deficient, and the inner H atom is sufficiently electron-rich to act as a donor in the formation of an inverse hydrogen bond. The calculated bond lengths (in pm) and electron donor–acceptor relationship are illustrated below:

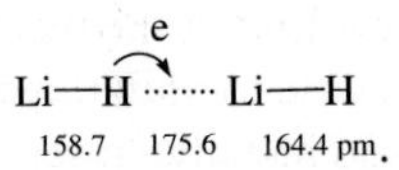

$$\text{Li—H} \cdots \text{Li—H}$$
$$158.7 \quad 175.6 \quad 164.4 \text{ pm}.$$

The distance H···Li is shorter than the sum of the atomic van der Waals radii of H and Li, and the linkage of Li–H···Li is almost linear.

Fig. 11.3.7.
Structure of the adduct Nb$_2$(hpp)$_4$·2NaEt$_3$BH.

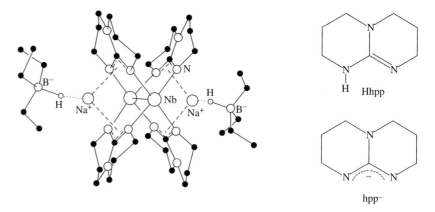

(2) The dihydrogen bond X–H···H–M may be formally regarded as a normal hydrogen bond

$$X-H\overset{e}{\cdots}Y.$$

where Y stands for the hydridic H atom as the electron donor, or an inverse hydrogen bond

$$M-H\overset{e}{\cdots}Y.$$

where Y is the H atom bonded to X.

(3) The inverse hydrogen bond B$^-$–H···Na$^+$ is found in the adduct of Na(Et$_3$BH) and Nb$_2$(hpp)$_4$, where hpp is the anion of 1,3,4,6,7,8-hexahydro-2H-pyrimido-[1,2-a]-pyrimidine (Hhpp). Figure 11.3.7 shows the B$^-$–H···Na$^+$ interaction in the compound Nb$_2$(hpp)$_4$·2NaEt$_3$BH.

The concept of inverse hydrogen bond is a relatively recent development, and additional varieties of this novel type of interaction may be uncovered in the future.

11.4 Hydride complexes

Hydrogen combines with many metals to form binary hydrides MH$_x$. The hydride ion H$^-$ has two electrons with the noble gas configuration of He. Binary metal hydrides have the following characteristics:

(1) Most of these hydrides are non-stoichiometric, and their composition and properties depend on the purity of the metals used in the preparation.
(2) Many of the hydride phases exhibit metallic properties such as high electrical conductivity and metallic luster.
(3) They are usually produced by the reaction of metal with hydrogen. Besides the formation of true hydride phases, hydrogen also dissolves in the metal to give a solid solution.

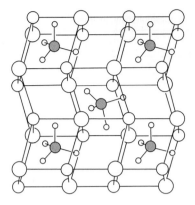

Fig. 11.4.1.
Crystal structure of CaMgNiH$_4$ (large circles represent Ca, small circles represent Mg, and tetrahedral groups represent NiH$_4$).

Metal hydrides may be divided into two types: covalent and interstitial. They are discussed in the following two sections.

11.4.1 Covalent metal hydride complexes

In recent years, many crystal structures of transition metal hydride complexes have been determined. In compounds such as CaMgNiH$_4$, Mg$_2$NiH$_4$, Mg$_2$FeH$_6$, and K$_2$ReH$_9$, H$^-$ behaves as an electron-pair donor covalently bonded to the transition metals. The transition metal in the NiH$_4^{2-}$, FeH$_6^{4-}$, and ReH$_9^{2-}$ anions generally has the favored noble gas electronic configuration with 18 valence electrons. Figure 11.4.1 shows the crystal structure of CaMgNiH$_4$.

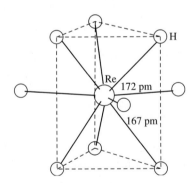

Fig. 11.4.2.
Structure of ReH$_9^{2-}$.

The anion ReH$_9^{2-}$ is a rare example of a central metal atom forming nine 2c-2e bonds, which are directed toward the vertices of a tricapped trigonal prism. Figure 11.4.2 shows the structure of ReH$_9^{2-}$, and Table 11.4.1 lists the structures of a number of transition metal hydride complexes. In these complex anions, the distances of transition metal to hydrogen are in the range 150–160 pm for 3d metals and 170–180 pm for 4d and 5d metals, except for Pd (160–170 pm) and Pt (158–167 pm).

Low-temperature neutron diffraction analysis of H$_4$Co$_4$(C$_5$Me$_4$Et)$_4$ shows that the molecule consists of four face-bridging hydrides attached to a

Structural Chemistry of Selected Elements

Table 11.4.1. Structures of transition metal hydrides

Geometry	Example	Bond length* (pm)
Tricapped trigonal-prismatic	K_2ReH_9	ReH_9^{2-}: Re(1)–H (3×)172,(6×) 167 Re(2)–H (3×)161,(6×) 170
Octahedral	Na_3RhH_6 Mg_2FeH_6 Mg_3RuH_6	RhH_6^{3-}: Rh–H 163 – 168 FeH_6^{4-}: Fe–H 156 RuH_6^{6-}: Ru–H (4×) 167, (2×) 173
Square-pyramidal	Mg_2CoH_5 Eu_2IrH_5	CoH_5^{4-}: Co–H (4×) 152, (1×) 159 IrH_5^{4-}: Ir–H (6×, disorder) 167
Square-planar	Na_2PtH_4 Li_2RhH_4	PtH_4^{2-}: Pt–H (4×) 164 RhH_4^{3-}: Rh–H (2×) 179, (2×) 175
Tetrahedral	Mg_2NiH_4	NiH_4^{4-}: Ni–H 154–157
Saddle	Mg_2RuH_4	RuH_4^{4-}: Ru–H (2×) 167, (2×) 168
T-shaped	Mg_3RuH_3	RuH_3^{6-}: Ru–H 171
Linear	Na_2PdH_2 $MgRhH_{1-x}$	PdH_2^{2-}: Pd–H (2×) 168 $Rh_4H_4^{8-}$: Rh–H (2×) 171

*The bond lengths are determined by neutron diffraction for M–D.

tetrahedral cobalt metal core, as shown in Fig. 11.4.3. The average distances (in pm) and angles in the core of the molecule are as follows:

Co–Co 257.1   Co–H 174.9   Co–C 215.8
H···H 236.6   Co–H–Co 94.6°   H–Co–H 85.1°

The hydride ligands are located off the Co–Co–Co planes by an average distance of 92.3 pm.

Polynuclear platinum and palladium carbonyl clusters containing the bulky tri-*tert*-butylphosphine ligand are inherently electron-deficient at the metal centers. The trigonal bipyramidal cluster [$Pt_3Re_2(CO)_6(P^tBu_3)_3$], as shown in Fig. 11.4.4(a), is electronically unsaturated with a deficit of 10 valence electrons, as it needs 72 valence electrons to satisfy an 18-electron configuration at

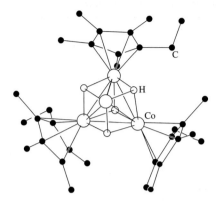

Fig. 11.4.3.
Molecular structure of H$_4$Co$_4$(C$_5$Me$_4$Et)$_4$ by neutron diffraction analysis at 20 K.

each metal vertex. The cluster reacts with 3 equivalents of molecular hydrogen at room temperature to give the addition product [Pt$_3$Re$_2$(CO)$_6$(P^{t}Bu$_3$)$_3$(μ-H)$_6$] in 90% yield. Single-crystal X-ray analysis showed that the hexahydrido complex has a similar trigonal bipyramidal structure with one hydrido ligand bridging each of the six Pt–Re edges of the cluster core, as illustrated in Fig. 11.4.4(b). The Pt–Pt bond lengths are almost the same in both complexes (average values 272.25 pm versus 271.27 pm), but the Pt–Re bonds are much lengthened in the hexahydrido complex (average values 264.83 pm versus 290.92 pm). The Pt–H bonds are significantly shorter than the Re–H bonds (average values 160 pm versus 189 pm).

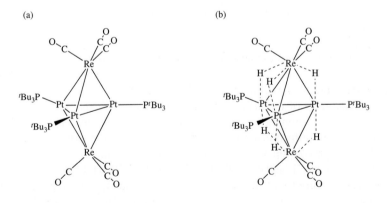

Fig. 11.4.4.
Molecular structure of
(a) [Pt$_3$Re$_2$(CO)$_6$(P^{t}Bu$_3$)$_3$] and its hydrogen adduct (b)
[Pt$_3$Re$_2$(CO)$_6$(P^{t}Bu$_3$)$_3$(μ-H)$_6$]. From R. D. Adams and B. Captain, *Angew. Chem. Int. Ed.* **44**, 2531–3 (2005).

11.4.2 Interstitial and high-coordinate hydride complexes

Most interstitial metal hydrides have variable composition, for example, PdH$_x$ with $x < 1$. The hydrogen atoms are assumed to have lost their electrons to the d

Table 11.4.2. Some hydrogen storage systems

Storage medium	Hydrogen percent by weight	Hydrogen density (kg dm^{-1})
MgH_2	7.6	0.101
Mg_2NiH_4	3.16	0.081
VH_2	2.07	0.095
$FeTiH_{1.95}$	1.75	0.096
$LaNi_5H_6$	1.37	0.089
Liquid H_2	100	0.070
Gaseous H_2 (10 MPa)	100	0.008

orbitals of the metal atoms and behave as mobile protons. This model accounts for the mobility of hydrogen in PdH_x, the fact that the magnetic susceptibility of palladium falls as hydrogen is added, and that if an electric potential is applied across a filament of PdH_x, hydrogen migrates toward the negative electrode.

A striking property of many interstitial metal hydrides is the high rate of hydrogen diffusion through the solid at slightly elevated temperatures. This mobility is utilized in the ultra-purification of H_2 by diffusion through a palladium-silver alloy tube.

Hydrogen has the potential to be an important fuel because it has an extremely high energy density per unit weight. It is also a non-polluting fuel, the main combustion product being water. The metallic hydrides, which decompose reversibly to give hydrogen gas and the metals, can be used for hydrogen storage. Table 11.4.2 lists the capacities of some hydrogen storage systems. It is possible to store more hydrogen in the form of these hydrides than in the same volume of liquid hydrogen.

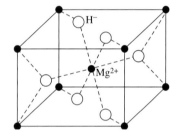

Fig. 11.4.5.
Crystal structure of MgH_2.

The crystal structure of MgH_2 has been determined by neutron diffraction. It has the rutile structure, space group $P4_2/mnm$, with $a = 450.25$ pm, $c = 301.23$ pm, as shown in Fig. 11.4.5. The Mg^{2+} ion is surrounded octahedrally by six H^- anions at 194.8 pm. Taking the radius of Mg^{2+} as 72 pm (six-coordinate, Table 4.2.2), the radius of H^- (three-coordinate) is calculated to be 123 pm.

Figure 11.4.6 shows the pressure (P) versus composition (x) isotherms for the hydrogen–iron–titanium system. This system is an example of the formation of a ternary hydride from an intermetallic compound.

In the majority of its metal complexes, the hydride ligand normally functions in the μ^1 (terminal), μ^2 (edge-bridging), and μ^3 (triangular face-capping) modes. Research efforts in recent years have led to the syntheses of an increasing number of high-coordinate (μ^4, μ^5, and μ^6) hydride complexes.

The compounds $[Li_8(H)\{N(2\text{-Py})Ph\}_6]^+[Li(Me_2Al^tBu_2)_2]^-$ and $Li_7(H)[N(2\text{-Py})Ph]_6$ are examples of molecular species that contain a μ^6-hydride ligand surrounded by an assembly of main-group metal ions. The cation $[Li_8(H)\{N(2\text{-Py})Ph\}_6]^+$ encapsulates a H^- ion within an octahedron composed of Li atoms, the average Li–H distance being 201.5 pm, as shown in Fig. 11.4.7(a). In molecular $Li_7(H)[N(2\text{-Py})Ph]_6$, the H^- ion is enclosed in a distorted octahedral coordination shell, as shown in Fig. 11.4.7(b). The average Li–H distance is 206 pm. The seventh Li atom is located at a much longer Li··· H distance of 249 pm.

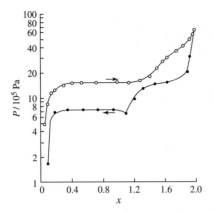

Fig. 11.4.6.
P–x isotherms for the FeTiH$_x$ alloy. The upper curve corresponds to the equilibrium pressure as hydrogen is added stepwise to the alloy; the lower curve corresponds to the equilibrium pressure as hydrogen is removed stepwise from the hydride.

Neutron diffraction analysis of the hydrido cluster complex [H$_2$Rh$_{13}$(CO)$_{24}$]$^{3-}$ at low temperature has revealed a hexagonal close-packed metal skeleton in which each surface Rh atom is coordinated by one terminal and two bridging carbonyls. The two hydride ligands occupy two of the six square-pyramidal sites on the surface, each being slightly displaced from the plane of four basal Rh atoms toward the central Rh atom of the Rh$_{13}$ cluster core (Fig. 11.4.8(a)). For either μ^5-hydride, the axial Rh(central)–H distance is shorter (average 184 pm) than the four Rh(surface)–H distances (average 197 pm).

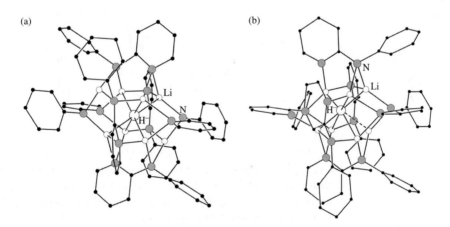

Fig. 11.4.7.
Structure of main-group metal hydride complexes: (a) [Li$_8$(H){N(2-Py)Ph}$_6$]$^+$, (b) Li$_7$(H)[N(2-Py)Ph]$_6$; the broken line indicates a weak interaction.

DFT calculations, as well as X-ray and neutron diffraction studies, have established the first existence of a four-coordinate interstitial hydride ligand. In the isomorphous tetranuclear lanthanide polyhydride complexes [C$_5$Me$_4$(SiMe$_3$)]$_4$Ln$_4$H$_8$ (Ln = Lu, Y), the tetrahedral Ln$_4$H$_8$ cluster core adopts a pseudo C_{3v} configuration with a body-centered μ^4, one face-capping μ^3, and six edge-bridging μ^2 hydride ligands. The molecular structure of the yttrium(III) complex is shown in Fig. 11.4.8(b).

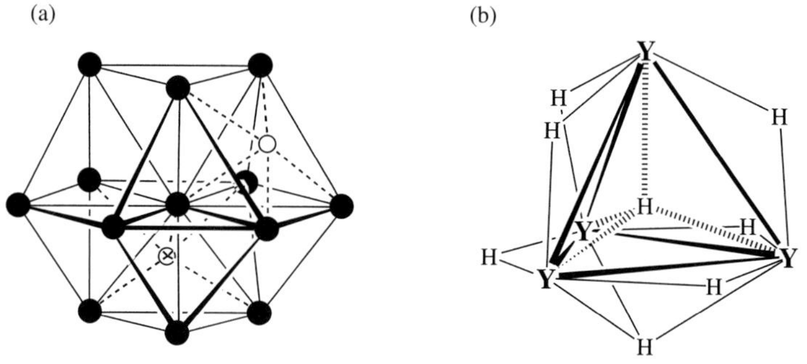

Fig. 11.4.8.
(a) Molecular structure of $[H_2Rh_{13}(CO)_{24}]^{3-}$; the carbonyl groups are omitted for clarity. (b) Molecular structure of $[C_5Me_4(SiMe_3)]_4Y_4H_8$; the $C_5Me_4(SiMe_3)^-$ ligands, each capping a metal center, are omitted for clarity.

11.5 Molecular hydrogen (H_2) coordination compounds and σ-bond complexes

11.5.1 Structure and bonding of H_2 coordination compounds

The activation of hydrogen by a metal center is one of the most important chemical reactions. The H–H bond is strong (436 kJ mol^{-1}), so that H_2 addition to unsaturated organic and other compounds must be mediated by metal centers whose roles constitute the basis of catalytic hydrogenation. In catalytic mechanisms, hydride complexes formed by the cleavage of H_2 are regarded as key intermediates.

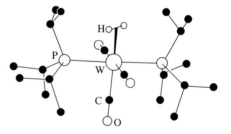

Fig. 11.5.1.
Molecular structure of $W(CO)_3[P(CHMe_2)_3]_2(H_2)$.

The first isolable transition metal complex containing a coordinated H_2 molecule is $W(CO)_3[P(CHMe_2)_3]_2(H_2)$. X-ray and neutron diffraction studies and a variety of spectroscopic methods have confirmed that it possesses a η^2-H_2 ligand, as shown in Fig. 11.5.1.

The resulting geometry about the W atom is that of a regular octahedron. The H_2 molecule is symmetrically coordinated in an η^2 mode with an average W–H distance of 185 pm (X-ray) and 175 pm (neutron) at $-100\,°C$. The H–H distance is 75 pm (X-ray) and 82 pm (neutron), slightly longer than that of the free H_2 molecule (74 pm).

The side-on bonding of H_2 to the metal involves the transfer of σ bonding electrons of H_2 to a vacant metal d orbital (or hybrid orbital), together with the transfer of electrons from a filled metal d orbital into the empty σ^* orbital of H_2, as shown in Fig. 11.5.2. This synergistic (mutually assisting) bonding mode is similar to that of CO and ethylene with metal atoms. The π back donation from

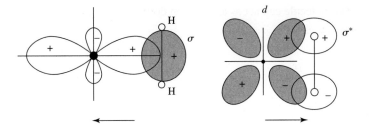

Fig. 11.5.2.
Bonding in M-η^2-H_2. Arrow indicates direction of donation.

metal to σ^* orbital on H_2 is consistent with the fact that H–H bond cleavage is facilitated by electron-rich metals.

Table 11.5.1. H$\cdots$H bond distances (d_{HH}) determined by neutron diffraction for H_2 coordination compounds

H_2 Coordination compound	d^*_{HH} (pm)
Mo(Co)(dppe)$_2$(H_2)	73.6 (84)
[FeH(H_2)(dppe)$_2$]$^+$	81.6
W(CO)$_3$(P^{i}Pr$_3$)$_2$(H_2)	82
[RuH(H_2)(dppe)$_2$]$^+$	82 (94)
FeH$_2$(H_2)(PEtPh$_2$)$_3$	82.1
[OsH(H_2)(dppe)$_2$]$^+$	97
[Cp*OsH$_2$(H_2)(PPh$_3$)]$^+$	101
[Cp*OsH$_2$(H_2)(AsPh$_3$)]$^+$	108
[Cp*Ru(dppm)(H_2)]$^+$	108 (110)
cis-IrCl$_2$H(H_2)(P^{i}Pr$_3$)$_2$]	111
trans-[OsCl(H_2)(dppe)$_2$]$^+$	122
[Cp*OsH$_2$(H_2)(PCy$_3$)]$^+$	131
[Os(en)$_2$(H_2)(acetate)]$^+$	134
ReH$_5$(H_2)(P(p-tolyl)$_3$)$_2$	135.7
[OsH$_3$(H_2)(PPhMe$_2$)$_3$]$^+$	149

* Uncorrected for effects of vibrational motion; the corrected values are given in parentheses.

The structure of trans-[Fe(η^2-H_2)(H)(PPh$_2$CH$_2$CH$_2$PPh$_2$)]BPh$_4$ has been determined by neutron diffraction at 20 K. The coordination environment about the Fe atom is shown in Fig. 11.5.3. This is the first conclusive demonstration of the expected difference between a hydride and a H_2 molecule in coordination to the same metal center. The H–H bond distance is 81.6 pm and the H–Fe bond length is 161.6 pm, which is longer than the terminal H–Fe distance of 153.5 pm.

A large number of H_2 coordination compounds have been identified and characterized. Table 11.5.1 lists the H$\cdots$H distances (d_{HH}) determined by neutron diffraction for some of these complexes. The H$\cdots$H distances ranges from 82 to 160 pm, beyond which a complex is generally regarded to be a classical hydride. A "true" dihydrogen complex can be considered to have $d_{HH} < 100$ pm, and the complexes with $d_{HH} > 100$ pm are more hydride-like in their properties and have highly delocalized bonding.

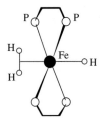

Fig. 11.5.3.
Coordination of the Fe atom in trans-[Fe(η^2-H_2)(H)(PPh$_2$CH$_2$CH$_2$PPh$_2$)]BPh$_4$.

The H_2 molecule can act as an electron donor in combination with a proton to yield a H_3^+ species:

$$\begin{array}{c} H \\ | \\ H \end{array} + H^+ \implies \begin{array}{c} H \\ | \\ H \end{array} \rightarrow H^+ \implies \left[\begin{array}{c} H \\ \diagup \diagdown \\ H \cdots H \end{array} \right]^+$$

The existence of H_3^+ has been firmly established by its mass spectrum: H_3^+ has a relative molecular mass of 3.0235, which differs from those of HD (3.0219) and T (3.0160). The H_3^+ species has the shape of an equilateral triangle and is stabilized by a 3c-2e bond.

The higher homologs H_5^+, H_7^+, and H_9^+ have also been identified by mass spectroscopy. These species are 10^2 to 10^3 times more abundant than the even-member species H_{2n}^+ ($n = 2, 3, 4, \ldots$) in mass spectral measurements. The structures of H_5^+, H_7^+, and H_9^+ are shown below:

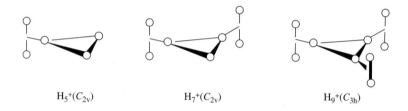

$H_5^+(C_{2v})$ $H_7^+(C_{2v})$ $H_9^+(C_{3h})$

11.5.2 X–H σ-bond coordination metal complexes

By analogy to the well characterized η^2-H_2 complexes, methane and silane can coordinate to the metal in an μ^2-fashion via a C–H σ bond and a Si–H σ bond:

$$L_nM \leftarrow \begin{array}{c} H \\ | \\ H \end{array} \qquad L_nM \leftarrow \begin{array}{c} H \\ | \\ CH_3 \end{array} \qquad L_nM \leftarrow \begin{array}{c} H \\ | \\ SiH_3 \end{array}$$

Transition metal complexes with a variety of σ-coordinated silane ligands have been extensively investigated. The molecular structure of $Mo(\eta^2$-H-$SiH_2Ph)(CO)(Et_2PCH_2CH_2PEt_2)_2$, as shown in Fig. 11.5.4(a), exhibits η^2-coordination of the Si–H bond at a coordination site cis to the CO ligand. The distances (in pm) and angles of the Si–H coordinate bond are Si–H 177, Mo–H 170, Mo–Si 250.1, Mo–H–Si 92°, and Mo–Si–H 42.6°. The SiH_2Ph_2 analog, $Mo(\eta^2$-H–$SiHPh_2)(CO)(Et_2PCH_2CH_2PEt_2)_2$, as shown in Fig. 11.5.4(b), exhibits a similar structure with Si–H 166, Mo–H 204, and Mo–Si 256.4.

Two molecules can be combined to form an ion-pair through a σ coordination bond, in which one molecule provides its X–H (X = B, C, N, O, Si) σ bonding electrons to a transition metal atom (such as Zr) of another molecule. A good example is $[(C_5Me_5)_2Zr^+Me][B^-Me(C_6F_5)_3]$, whose structure is shown in Fig. 11.5.5. This bonding type is called an intermolecular pseudo-agostic (IPA) interaction.

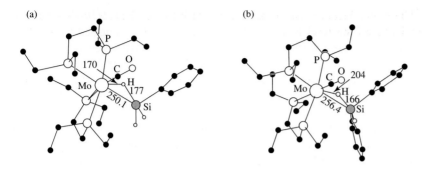

Fig. 11.5.4.
Structure of (a) Mo(η^2-H–SiH$_2$Ph)(CO)(Et$_2$PCH$_2$CH$_2$PEt$_2$)$_2$ and (b) Mo(η^2-H–SiHPh$_2$)(CO)(Et$_2$PCH$_2$CH$_2$PEt$_2$)$_2$ (bond lengths in pm).

11.5.3 Agostic bond

The agostic bond is an intramolecular 3c-2e C–H→M bond, with the vacant metal orbital accepting an electron pair from the C–H σ bonding orbital. The term "agostic" denotes the intramolecular coordination of C–H bonds to transition metals and is derived from a Greek word meaning "to clasp, to draw toward, to hold to oneself." The term "agostic" should not be used to describe external ligand binding solely through a σ bond, which is best referred to as σ-bonded coordinate binding. Agostic bonds may be broadly understood in the following way. Many transition-metal compounds have less than an 18-electron count at the metal center, and thus are formally unsaturated. One way in which this deficiency may be alleviated is to increase the coordination number at the transition metal by clasping a H atom of a coordinated organic ligand. Therefore, the agostic bond generally occurs between carbon–hydrogen groups and transition-metal centers in organometallic compounds, in which a H atom is covalently bonded simultaneously to both a C atom and to a transition-metal atom. An agostic bond is generally written as C–H→M, where the "half arrow"

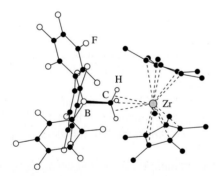

Fig. 11.5.5.
Structure of [(C$_5$Me$_5$)$_2$Zr$^+$Me][B$^-$Me(C$_6$F$_5$)$_3$].

indicates formal donation of two electrons from the C–H bond to the M vacant orbital. As in all 3c-2e bridging systems involving only three valence orbitals, the C–H→M fragment is bent and the agostic bond can be more accurately represented by

The agostic C–H distance is in the range 113–19 pm, about 5-10% longer than a non-bridging C–H bond; the M–H distance is also elongated by 10–20% relative to a normal terminal M–H bond. Usually NMR spectroscopy can be used for the diagnosis of static agostic systems which exhibit low $J(C-H_\alpha)$ values owing to the reduced C–H_α bond order. Typical values of $J(C-H_\alpha)$ are in the range of 60–90 Hz, which are significantly lower than those (120–30 Hz) expected for $C(sp^3)$–H bonds.

The C–H→M agostic bond is analogous to a X–H···Y hydrogen bond, in the sense that the strength of the M–H linkage, as measured by its internuclear distance, is variable, and correlates with the changes in C–H distance on the ligand. There are, however, two important differences. Firstly, in a hydrogen bond the H atom is attracted by an electronegative acceptor atom, but the strength of an agostic bond is stronger for more electropositive metals. Secondly, the hydrogen bond is a $3c$-$4e$ system, but the agostic bond is a $3c$-$2e$ system.

The agostic interaction is considered to be important in such reactions as α elimination, β-hydrogen elimination, and orthometalation. For example, in a β-hydrogen elimination, a H atom on the β-C atom of an alkyl group is transferred to the metal atom and an alkene is eliminated. This reaction proceeds through an agostic-bond intermediate, as shown below:

The discovery of agostic bonding has led to renewed interest in the ligand behavior of simple organic functional groups, such as the methyl group. The C–H group used to be regarded as an inert spectator, but now C–H containing hydrocarbon ligand systems are recognized as being capable of playing important roles in complex stereochemical reactions.

Agostic bonding has been extended to include general X–H→M systems, in which X may be B, N, Si, as well as C. The principal types of compounds with agostic bonds are shown in Fig. 11.5.6.

Figure 11.5.7 shows the molecular structures of some transition metal complexes in which agostic bonding has been characterized by either X-ray or neutron diffraction. Structural data for these compounds are listed in Table 11.5.2.

Many complexes containing P-, N-, or O-containing ligands can also form agostic bonds:

Agostic interactions involving phosphines are quite significant because they possess residual binding sites for other small molecules such as H_2, and phosphine groups play crucial roles in many homogeneous catalysts. Figure 11.5.8(a)

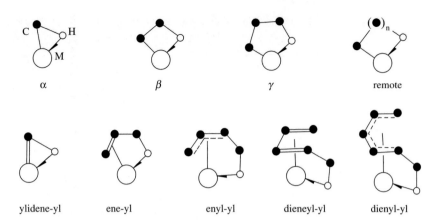

Fig. 11.5.6.
Structural types of agostic alkyl and unsaturated hydrocarbonyl complexes.

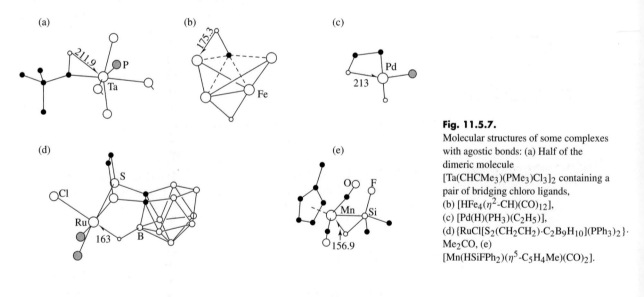

Fig. 11.5.7.
Molecular structures of some complexes with agostic bonds: (a) Half of the dimeric molecule [Ta(CHCMe$_3$)(PMe$_3$)Cl$_3$]$_2$ containing a pair of bridging chloro ligands, (b) [HFe$_4$(η^2-CH)(CO)$_{12}$], (c) [Pd(H)(PH$_3$)(C$_2$H$_5$)], (d) {RuCl[S$_2$(CH$_2$CH$_2$)·C$_2$B$_9$H$_{10}$](PPh$_3$)$_2$}·Me$_2$CO, (e) [Mn(HSiFPh$_2$)(η^5-C$_5$H$_4$Me)(CO)$_2$].

Table 11.5.2. Structural data of some agostic-bonded X–H→M compounds

Compound	M–X (pm)	X–H (pm)	M–H (pm)	X–H–M(°)	Fig. 11.5.7
[Ta(CHCMe$_3$)(PMe$_3$)Cl$_3$]$_2$	Ta–C 189.8	C–H 113.1	Ta–H 211.9	C–H–Ta 84.8	(a)
[HFe$_4$(η^2-CH)(CO)$_{12}$]	Fe–C 182.7–194.9	C–H 119.1	Fe–H 175.3	C–H–Fe 79.4	(b)
[Pd(H)(PH$_3$)(C$_2$H$_5$)]	Pd–C 208.5	C–H 113	Pd–H 213	C–H–Pd 88	(c)
{RuCl[S$_2$(CH$_2$CH$_2$)C$_2$B$_9$H$_{10}$]-(PPh$_3$)$_2$}·Me$_2$CO		B–H 121	Ru–H 163		(d)
[Mn(HSiFPh$_2$)(η^5-C$_5$H$_4$Me)(CO)$_2$]	Mn–Si 235.2	Si–H 180.2	Mn–H 156.9		(e)

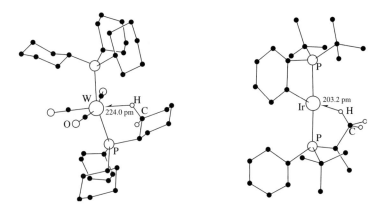

Fig. 11.5.8.
Structure of (a) W(CO)$_3$(PCy$_3$)$_2$ and
(b) [IrH(η^2-C$_6$H$_4$P^{t}Bu$_2$)(P^{t}Bu$_2$Ph)]$^+$.

shows the structure of W(CO)$_3$(PCy$_3$)$_2$, in which the W←H distance is 224.0 pm and the W···C distance is 288.4 pm. Figure 11.5.8(b) shows the structure of [IrH(η^2-C$_6$H$_4$P^{t}Bu$_2$)(P^{t}Bu$_2$Ph)]$^+$ in its [BAr$_4$]$^-$ salt, in which the Ir←H distance is 203.2 pm and the Ir···C distance is 274.5 pm.

11.5.4 Structure and bonding of σ complexes

Recent studies have established the following generalizations regarding σ (sigma-bond) complexes:

(1) The X–H (X = B, C, Si) bonds in some ligand molecules, like the H–H bond in H$_2$, provide their σ bonding electron pairs to bind the ligands to metal centers. They form intermolecular coordination compounds (i.e., sigma-bond complexes) or intramolecular coordination compounds (i.e., agostic-bond complexes). All such compounds involve non-classical 3c-2e bonding and are collectively termed σ complexes.

(2) An understanding of the bonding nature in the σ complexes has extended the coordination concept to complement the classical Werner-typer donation of a lone pair and the π-electron donation of unsaturated ligands.

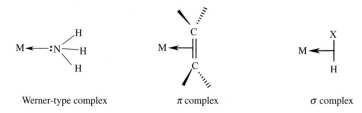

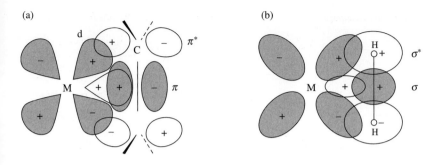

Fig. 11.5.9.
Electron donation in (a) π complexes and (b) σ complexes. The filled orbitals are shaded.

(3) In σ complexes, the σ ligand is side-on (η^2 mode) bonded to the metal (M) to form a 3c-2e bond. The electron donation in σ complexes is analogous to that in π complexes. The transition metals can uniquely stabilize the σ ligands and π ligands due to back donation from their d orbitals, as shown in Fig. 11.5.9. The main-group metals, lacking electrons in their outer d orbitals, do not form stable σ complexes.

(4) H–H and X–H are the strong covalent bonds. These bonds can be weakened and even broken by σ coordination to the transition metal and back donation from the metal to the σ* orbitals. The lengthening and eventual cleavage of a σ bond by binding it to M are dependent on the electronic character of M, which is influenced by the other ligands on M. In the σ complexes, the H···H and X···H distances can vary greatly. The gradual conversion of a metal dihydrogen σ complex to a dihydride is indicated by the following scheme:

$$\begin{array}{ccccc}
\text{H} & \text{H} & \text{H} & \text{H} & \text{H} \\
| \quad \text{M} & \text{M}—\| & \text{M}\cdots\| & \text{M}\diagup & \text{M}\diagup \\
\text{H} & \text{H} & \text{H} & \text{H} & \text{H} \\
\\
\text{H}_2 \text{ molecule} & \text{"true" H}_2 \text{ } \sigma \text{ molecule} & \text{elongated H}_2 \text{ complex} & & \text{hydride} \\
74 \text{ pm} & 80\text{–}90 \text{ pm} & 100\text{–}120 \text{ pm} \quad 130\text{–}150 \text{ pm} & & > 160 \text{ pm}
\end{array}$$

(5) There are two completely different pathways for the cleavage of H–H bonds: oxidative addition and heterolytic cleavage. Both pathways have been identified in catalytic hydrogenation and may also be applicable to other types of X–H σ bond activation such as C–H cleavage.

Back donation in the σ complexes is the crucial component in both aiding the binding of H_2 to M and activating the H–H bond toward cleavage. If the back donation becomes too strong, the σ bond breaks to form a dihydride due to over-population of the H_2 antibonding orbital. It is notable that back donation controls σ-bond activation toward cleavage and a σ bond cannot be broken solely by sharing its electron pair with a vacant metal d orbital. Although σ interaction is generally the predominant bonding component in a σ complex, it is unlikely to be stable at room temperature without at least a small amount of back donation.

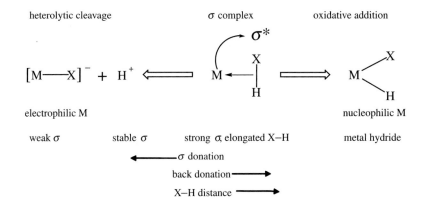

References

1. G. A. Jeffery, *An Introduction to Hydrogen Bonding*, Oxford University Press, Oxford, 1997.
2. R. B. King (ed.), *Encyclopedia of Inorganic Chemistry*, Wiley, New York, 1994: (a) K. Yvon, Hydrides: solid state transition metal complexes, pp. 1401–20; (b) M. Kakiuchi, Hydrogen: inorganic chemistry, pp. 1444–71.
3. F. D. Manchester (ed.), *Metal Hydrogen Systems: Fundamentals and Applications*, vols. I and II, Elsevier Sequoia, Lausanne, 1991.
4. G. R. Desiraju and T. Steiner, *The Weak Hydrogen Bond in Structural Chemistry and Biology*, Oxford University Press, Oxford, 1999.
5. G. A. Jeffrey and W. Saenger, *Hydrogen Bonding in Biological Structures*, Springer-Verlag, Berlin, 1991.
6. R. H. Crabtree, *The Organometallic Chemistry of the Transition Metals*, 4th edn., Wiley, New York, 2005.
7. C. C. Wilson, *Single Crystal Neutron Diffraction from Molecular Materials*, World Scientific, Singapore, 2000.
8. M. Peruzzini and R. Poli (eds.), *Recent Advances in Hydride Chemistry*, Elsevier, Amsterdam, 2001.
9. G. J. Kubas, *Metal Dihydrogen and σ-Bond Complexes,* Kluwer/Plenum, New York, 2001.
10. C.-K. Lam and T. C. W. Mak, Carbonate and oxalate dianions as prolific hydrogen-bond acceptors in supramolecular assembly. *Chem. Commun.*, 2660–61 (2003).
11. G. R. Desiraju, Hydrogen bridges in crystal engineering: interactions without borders. *Acc. Chem. Res.* **35**, 565–73 (2002).
12. R. H. Crabtree, P. E. M. Siegbahn, O. Eisensten, A. L. Rheingold and T. F. Koetzle, A new intermolecular interaction: unconventional hydrogen bonds with element-hydride bonds as proton acceptor. *Acc. Chem. Res.* **29**, 348–54 (1996); C. J. Cramer and W. L. Gladfelter, Ab initio characterization of $[H_3N \cdot BH_3]_2$, $[H_3N \cdot AlH_3]_2$ and $[H_3N \cdot GaH_3]_2$. *Inorg. Chem.* **36,** 5358–62 (1997).
13. I. Alkorta, I. Rozas and J. Elguero, Non-conventional hydrogen bonds. *Chem. Soc. Rev.* **27**, 163–70 (1998).
14. I. Rozas, I. Alkorta and J. Elguero, Inverse hydrogen-bonded complexes. *J. Phys. Chem.* **101**, 4236–44 (1997).
15. S. E. Landau, K. E. Groh, A. J. Lough and R. H. Morris, Large effects of ion pairing and protonic-hydridic bonding on the stereochemistry and basicity of crown-, azacrown-, and cryptand-222-potassium salts of anionic tetrahydride complexes of iridium(III). *Inorg. Chem.* **41**, 2995–07 (2002).

16. J. J. Schneider, Si–H and C–H activition by transition metal complexes: a step towards isolable alkane complexes? *Angew. Chem. Int. Ed.* **35,** 1069–76 (1996).
17. D. R. Armstrong, W. Clegg, R. P. Davies, S. T. Liddle, D. J. Linton, P. R. Raithby, R. Snaith and A. E. H. Wheatley, The first molecular main group metal species containing interstitial hydride. *Angew. Chem. Int. Ed.* **38**, 3367–90 (1999).
18. R. Bau, N. N. Ho, J. J. Schneider, S. A. Mason and G. J. McIntyre, Neutron diffraction study of $H_4Co_4(C_5Me_4Et)_4$. *Inorg. Chem.* **43**, 555–8 (2004).
19. F. A. Cotton, J. H. Matonic and C. A. Murillo, Triply bonded Nb_2^{4+} tetragonal lantern compounds: some accompanied by novel $B-H \cdots Na^+$ interactions. *J. Am. Chem. Soc.* **120**, 6047–52 (1998).

12 Structural Chemistry of Alkali and Alkaline-Earth Metals

12.1 Survey of the alkali metals

The alkali metals — lithium, sodium, potassium, rubidium, cesium, and francium — are members of Group 1 of the Periodic Table, and each has a single ns^1 valence electron outside a rare gas core in its ground state. Some important properties of alkali metals are given in Table 12.1.1.

With increasing atomic number, the alkali metal atoms become larger and the strength of metallic bonding, enthalpy of atomization (ΔH_{at}), melting point (mp), standard enthalpy of fusion (ΔH_{fus}) and boiling point (bp) all progressively decrease. The elements show a regular gradation in physical properties down the group.

The radii of the metals increase with increasing atomic number and their atomic sizes are the largest in their respective periods. Such features lead to relatively small first ionization energy (I_1) for the atoms. Thus the alkali metals are highly reactive and form M^+ ions in the vast majority of their compounds. The very high second ionization energy (I_2) prohibits formation of the M^{2+} ions. Even though the electron affinities (Y) indicate only mild exothermicity, M^- ions can be produced for all the alkali metals (except Li) under carefully controlled conditions.

Table 12.1.1. Some properties of alkali metals*

Property	Li	Na	K	Rb	Cs
Atomic number, Z	3	11	19	37	55
Electronic configuration	[He]2s^1	[Ne]3s^1	[Ar]4s^1	[Kr]5s^1	[Xe]6s^1
ΔH^0_{at} (kJ mol^{-1})	159	107	89	81	76
mp (K)	454	371	337	312	302
ΔH^0_{fus} (kJ mol^{-1})	3.0	2.6	2.3	2.2	2.1
bp (K)	1620	1156	1047	961	952
r_M (CN = 8) (pm)	152	186	227	248	265
r_M (CN = 6) (pm)	76	102	138	152	167
I_1 (kJ mol^{-1})	520.3	495.8	418.9	403.0	375.7
I_2 (kJ mol^{-1})	7298	4562	3051	2633	2230
Electron affinity (Y) (kJ mol^{-1})	59.8	52.7	48.3	46.9	45.5
Electronegativity (χ_S)	0.91	0.87	0.73	0.71	0.66

* Data are not available for francium, of which only artificial isotopes are known.

Alkali and Alkaline-Earth Metals

The chemistry of the alkali metals has in the past attracted little attention as the metals have a fairly restricted coordination chemistry. However, interesting and systematic study has blossomed over the past 25 years, largely prompted by two major developments: the growing importance of lithium in organic synthesis and materials science, and the exploitation of macrocyclic ligands in the formation of complexed cations. Section 12.4 deals with the use of complexed cations in the generation of alkalides and electrides.

12.2 Structure and bonding in inorganic alkali metal compounds

12.2.1 Alkali metal oxides

The following types of binary compounds of alkali metals (M) and oxygen are known:

(a) Oxides: M_2O;
(b) Peroxides: M_2O_2;
(c) Superoxides: MO_2 (LiO_2 is stable only in matrix at 15 K);
(d) Ozonides: MO_3 (except Li);
(e) Suboxides (low-valent oxides): Rb_6O, Rb_9O_2, Cs_3O, Cs_4O, Cs_7O, and $Cs_{11}O_3$; and
(f) Sesquioxides: M_2O_3: These are probably mixed peroxide–superoxides in the form of $M_2O_2 \cdot 2MO_2$.

The structures and properties of the alkali metal oxides are summarized in Table 12.2.1.

Suboxides exist for the larger alkali metals. The crystal structures of all binary suboxides (except Cs_3O) as well as the structures of $Cs_{11}O_3Rb$, $Cs_{11}O_3Rb_2$, and $Cs_{11}O_3Rb_7$ have been determined by single-crystal X-ray analysis. The structural data are listed in Table 12.2.2. All structures conform to the following rules:

(a) Each O atom occupies the center of an octahedron composed of Rb or Cs atoms.
(b) Face sharing of two such octahedra results in the cluster Rb_9O_2, and three equivalent octahedra form the cluster $Cs_{11}O_3$, as shown in Fig. 12.2.1.
(c) The O–M distances are near the values expected for M^+ and O^{2-} ions. The ionic character of the M atoms is reflected by the short intracluster M–M distances.
(d) The intercluster M–M distances are comparable to the distances in metallic Rb and Cs.
(e) The Rb_9O_2 or $Cs_{11}O_3$ clusters and additional alkali metals atoms form compounds of new stoichiometries. Some crystal structures have the intercluster space filled by metal atoms of the same kind, such as $(Cs_{11}O_3)Cs_{10}$ shown in Fig. 12.2.2(a), or by metal atoms of a different kind, such as $(Cs_{11}O_3)Rb$ and $(Cs_{11}O_3)Rb_7$ illustrated in Figs. 12.2.2(b) and (c), respectively.

The bonding within the alkali metal suboxides is illustrated by comparing the interatomic distances in the compounds Rb_9O_2 and $Cs_{11}O_3$ with those in

Table 12.2.1. Structures and properties of alkali metal oxides

	Li	Na	K	Rb	Cs
Oxides M_2O	Colorless mp 1843 K anti-CaF$_2$ structure	Colorless mp 1193 K anti-CaF$_2$ structure	Pale-yellow mp > 763 K anti-CaF$_2$ structure	Yellow mp > 840 K anti-CaF$_2$ structure	Orange mp 763 K anti-CdCl$_2$ structure
Peroxides M_2O_2	Colorless dec. > 473 K	Pale yellow dec. ≈ 948 K	Yellow dec. ≈ 763 K	Yellow dec. ≈ 843 K	Yellow dec. ≈ 863 K
Superoxides MO_2	—	Orange dec. ≈573 K NaCl-structure	Orange mp 653 K dec. ≈ 673 K CaC$_2$ structure	Orange mp 685 K CaC$_2$ structure	Orange mp 705 K CaC$_2$ structure
Ozonides MO_3	—	Red dec.<room temp.	Dark red dec. at room temp. CsCl structure	Dark red dec. ≈ room temp. CsCl structure	Dark red dec. > 323 K CsCl structure
Suboxides	—	—	—	Rb$_6$O Bronze color dec. 266 K Rb$_9$O$_2$ Copper color mp 313 K	Cs$_3$O Blue green dec. 439 K Cs$_4$O Red-violet dec. 284 K Cs$_7$O Bronze mp 277 K Cs$_{11}$O$_3$ Violet mp 326 K

the "normal" oxides and in metallic Rb and Cs. The M–O distances nearly match the sum of ionic radii. The large intercluster M–M distances correspond to the distances in elemental M (Rb and Cs). Therefore, the formulations $(Rb^+)_9(O^{2-})_2(e^-)_5$ and $(Cs^+)_{11}(O^{2-})_3(e^-)_5$, where e^- denotes an electron, represent a rather realistic description of the bonding in these clusters. Their

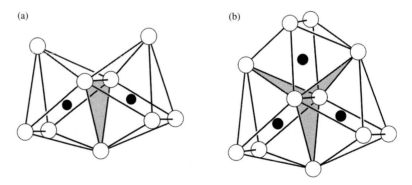

Fig. 12.2.1. Clusters in alkali suboxides: (a) Rb$_9$O$_2$ and (b) Cs$_{11}$O$_3$. The shared faces are shaded.

stability is attributable to consolidation by strong O–M and weak M–M bonds. All the alkali metal suboxides exhibit metallic luster and are good conductors.

Alkali and Alkaline-Earth Metals

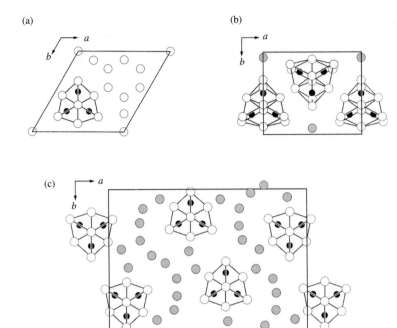

Fig. 12.2.2.
Unit-cell projections showing the atomic arrangements in some low-valent alkali metal oxides: (a) $(Cs_{11}O_3)Cs_{10}$, (b) $(Cs_{11}O_3)Rb$, (c) $(Cs_{11}O_3)Rb_7$. The open, blackened, and shaded circles represent Cs, O, and Rb atoms, respectively.

Thus the electrons in M–M bonding states do not stay localized within the clusters but are delocalized throughout the crystal due to the close contacts between the clusters.

12.2.2 Lithium nitride

Among the alkali metals, only lithium reacts with N_2 at room temperature and normal atmospheric pressure to give red-brown, moisture-sensitive lithium nitride Li_3N (α-form), which has a high ionic conductivity.

Table 12.2.2. Structural data of alkali metal suboxides

Compounds	Space group	Structural feature
Rb_9O_2	$P2_1/m$	Rb_9O_2 cluster, Fig. 12.1.1(a)
Rb_6O	$P6_3/m$	$(Rb_9O_2)Rb_3$; three Rb atoms per Rb_9O_2 cluster
$Cs_{11}O_3$	$P2_1/c$	$Cs_{11}O_3$ cluster, Fig. 12.1.1(b)
Cs_4O	$Pna2_1$	$(Cs_{11}O_3)Cs$; one Cs atom per $Cs_{11}O_3$ cluster
Cs_7O	$P6m2$	$(Cs_{11}O_3)Cs_{10}$; ten Cs atoms per $Cs_{11}O_3$ cluster, Fig. 12.2.2(a)
$Cs_{11}O_3Rb$	$Pmn2_1$	One Rb atom per $Cs_{11}O_3$ cluster, Fig. 12.2.2(b)
$Cs_{11}O_3Rb_2$	$P2_1/c$	Two Rb atoms per $Cs_{11}O_3$ cluster
$Cs_{11}O_3Rb_7$	$P2_12_12_1$	Seven Rb atoms per $Cs_{11}O_3$ cluster, Fig. 12.2.2(c)

The α form of Li_3N is made up of planar Li_2N layers, in which the Li atoms form a simple hexagonal arrangement, as in the case of the carbon layer in graphite, with a nitrogen atom at the center of each ring, as shown in Fig. 12.2.3(a). These layers are stacked and are linked by additional Li atoms midway between the N atoms of adjacent overlapping layers. Figure 12.2.3(b)

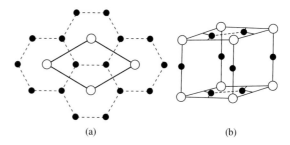

Fig. 12.2.3.
Crystal structure of Li_3N: (a) Li_2N layer, (b) the unit cell. The small black circle and large open circle represent Li and N atoms, respectively.

shows a hexagonal unit cell, with $a = 364.8$ pm, $c = 387.5$ pm. The Li–N distance is 213 pm within a layer, and 194 pm between layers.

α-Li_3N is an ionic compound, in which the N^{3-} ion has a much greater size than the Li^+ ion [$r_{N^{3-}}$ 146 pm, r_{Li^+} 59 pm (CN = 4)]. The crystal structure is very loosely packed, and conductivity arises from Li^+ vacancies within the Li_2N layers. The interaction between the carrier ions (Li^+) and the fixed ions (N^{3-}) is relatively weak, so that ion migration is quite effective.

12.2.3 Inorganic alkali metal complexes

(1) Structural features of alkali metal complexes

A growing number of alkali metal complexes $(MX \cdot xL)_n$ (M = Li, Na, K; X = halogens, OR, $NR_2, \cdots$; L = neutral molecules) are known. It is now generally recognized that the M–X bonds in these complexes are essentially ionic or have a high degree of ionic character. This bonding character accounts for the following features of these complexes:

(a) Most alkali metal complexes have an inner ionic core, and at the same time frequently form molecular species. The ionic interaction leads to large energies of association that stabilize the formation of complexes. The outer molecular fragments prevent the continuous growth of MX to form an ionic lattice. This characteristic has been rationalized in terms of the relatively large size and low charge of the cations M^+. According to this view, the stability of alkali metal complexes should diminish in the sequence Li > Na > K > Rb > Cs, and this is frequently observed.

(b) The electrostatic association favors the formation of the M^+X^- ion pairs, and it is inevitable that such ion pairs will associate, so as to delocalize the electronic charges. The most efficient way of accomplishing this is by ring formation. If the X^- groups are relatively small and for the most part coplanar with the ring, then stacking of rings can occur. For example, two dimers are connected to give a cubane-like tetramer, and two trimers lead to a hexagonal prismatic hexamer.

(c) In the ring structures, the bond angles at X^- are acute. Hence the formation of larger rings would reduce repulsion between X^- ions and between M^+ ions.

(d) Since the alkali metal complexes are mostly discrete molecules with an ionic core surrounded by organic peripheral groups, they have relatively low melting points and good solubility in weakly polar organic solvents. These properties have led to some practical applications; for example, lithium

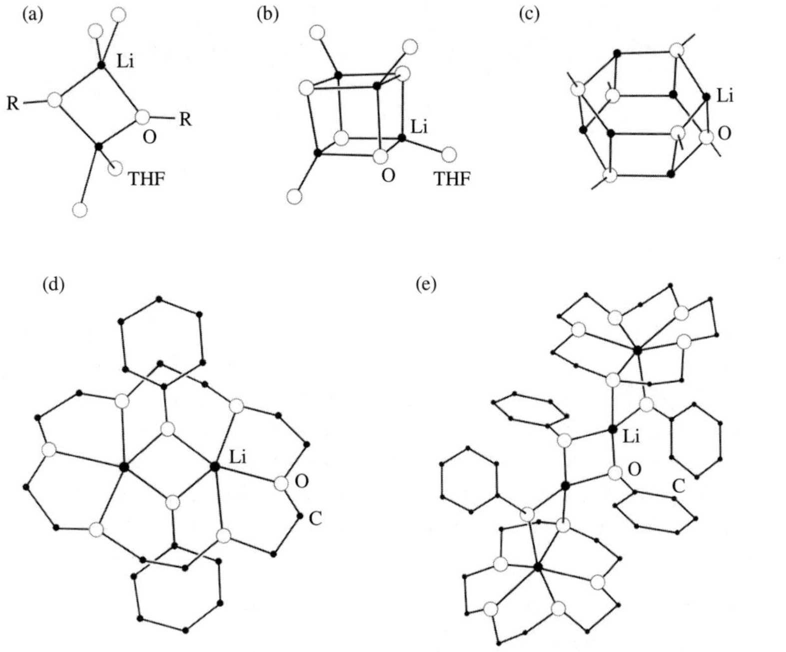

Fig. 12.2.4.
Core structures of some lithium alkoxides and aryloxides:
(a) [Li(THF)$_2$OC(NMe$_2$)(c-2,4,6-C$_7$H$_6$)]$_2$,
(b) [Li(THF)OC(CH^{t}Bu)OMe]$_4$,
(c) [LiOC(CH$_2$)^{t}Bu]$_6$,
(d) (LiOPh)$_2$(18-crown-6), and
(e) [(LiOPh)$_2$(15-crown-5)]$_2$.

complexes are used as low-energy electrolytic sources of metals, in the construction of portable batteries, and as halogenating agents and specific reagents in organic syntheses.

(2) Lithium alkoxides and aryloxides

Anionic oxygen donors display a strong attraction for Li$^+$ ions and usually adopt typical structures of their aggregation state, as shown in Figs 12.2.4(a)-(c). Figure 12.2.4(a) shows the dimeric structure of [Li(THF)$_2$OC(NMe$_2$)(c-2,4,6-C$_7$H$_6$)]$_2$, in which Li$^+$ is four-coordinated by the oxygen atoms from two THF ligands and two alkoxides OC(NMe$_2$)(c-2,4,6-C$_7$H$_6$). The Li–O distances are 188 and 192 pm for the bridging oxygen of the enolate, and 199 pm for THF. Figure 12.2.4(b) shows the tetrameric cubane-type structure of [Li(THF)OC(CH^{t}Bu)OMe]$_4$. The Li–O distances are 196 pm (enolate) and 193 pm (THF). Figure 12.2.4(c) shows the hexagonal-prismatic structure of [LiOC(CH$_2$)^{t}Bu]$_6$. The Li–O distances are 190 pm (av.) in the six-membered ring of one stack, and 195 pm (av.) between the six-membered rings.

Figures 12.2.4(d) and (e) show the structures of (LiOPh)$_2$(18-crown-6) and [(LiOPh)$_2$(15-crown-5)]$_2$. In the former, there is a central dimeric (Li–O)$_2$ unit. Owing to its larger size, 18-crown-6 can coordinate both Li$^+$ ions, each through three oxygen atoms (Li–O 215 pm, av.). These Li$^+$ ions are bridged by phenoxide groups (Li–O 188 pm). The structure of the latter has a very similar (LiOPh)$_2$ core (Li–O 187 and 190 pm). Each Li$^+$ ion is coordinated to another –OPh group as well as one ether oxygen from a 15-crown-5 donor. This ether oxygen atom also connected to a further Li$^+$ ion whose coordination is completed by the remaining four oxygen donors of the 15-crown-5 ligands.

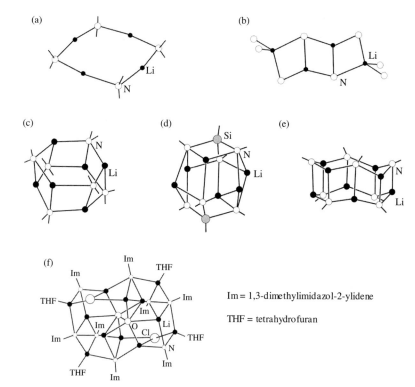

Fig. 12.2.5.
Core structures of some lithium amides and imides: (a) (LiTMP)$_4$, (b) Li$_4$(TMEDA)$_2$[c-N(CH$_2$)$_4$], (c) [Li{c-N(CH$_2$)$_6$}]$_6$, (d) [{LiN(SiMe$_3$)}$_3$SiR]$_2$, (e) [LiNH^{t}Bu]$_8$, and (f) [Li$_{12}$O$_2$Cl$_2$(ImN)$_8$(THF)$_4$]·8(THF).

Im = 1,3-dimethylimidazol-2-ylidene

THF = tetrahydrofuran

(3) Lithium amides

The structures of lithium amides are characterized by a strong tendency for association. Monomeric structures are observed only in the presence of very bulky groups at nitrogen and/or donor molecules that coordinate strongly to lithium. In the dimeric lithium amides, the Li$^+$ ion is usually simultaneously coordinated by N atoms and other atoms. The trimer [LiN(SiMe$_3$)$_2$]$_3$ has a planar cyclic arrangement of its Li$_3$N$_3$ core. The Li–N bonds are 200 pm long with internal angles of 148° at Li and 92° at N (average values). Figure 12.2.5 shows the structures of some higher aggregates of lithium amides.

(a) The tetramer (LiTMP)$_4$ (TMP = 2,2,6,6-tetramethylpiperidinide) has a planar Li$_4$N$_4$ core with a nearly linear angle (168.5°) at Li and an internal angle of 101.5° at N. The average Li–N distance is 200 pm, as shown in Fig. 12.2.5(a).

(b) The tetrameric molecule Li$_4$(TMEDA)$_2$[c-N(CH$_2$)$_4$] (TMEDA = N,N'-tetramethylenediamine) has a ladder structure in which adjacent Li$_2$N$_2$ rings share edges with further aggregation blocked by TMEDA coordination, as shown in Fig. 12.2.5(b).

(c) The hexameric [Li{c-N(CH$_2$)$_6$}]$_6$ has a stacked structure, as shown in Fig. 12.2.5(c). This involves the association of two [Li-c-N(CH$_2$)$_6$]$_3$ units to form a hexagonal-prismatic Li$_6$N$_6$ unit, in which all N atoms have the relatively rare coordination number of five.

(d) The basic framework in [{LiN(SiMe$_3$)}$_3$SiR]$_2$ (R = Me, ^{t}Bu and Ph) is derived from the dimerization of a trisamidosilane. It exhibits molecular

D_{3d} symmetry, as shown in Fig. 12.2.5(d). Each of the three Li$^+$ ions in a monomeric unit is coordinated by two N atoms in a chelating fashion, and linkage of units occurs through single Li–N contacts. This causes all N atoms to be five-coordinate, which leads to weakening of the Li–N and Si–N bonds.

(e) Figure 12.2.5(e) shows the core structure of [LiNH^{t}Bu]$_8$, in which planar Li$_2$N$_2$ ring units are connected to form a discrete prismatic ladder molecule.

(f) Figure 12.2.5(f) shows the centrosymmetric core structure of [Li$_{12}$O$_2$Cl$_2$(ImN)$_8$(THF)$_4$]·8(THF), which consists of a folded Li$_4$N$_2$O$_2$ ladder in which the central O$_2^{2-}$ ion is connected to two Li centers. The two adjacent Li$_4$ClN$_3$ ladders are connected with the central Li$_4$N$_2$O$_2$ unit via Li–O, Li–Cl and Li–N interactions. The bond distances are

> Li–N   195.3–218.3 pm,   Li–O   191.7–259.3 pm,
>
> Li–Cl 238.5–239.6 pm,   O–O   154.4 pm.

In this structure, Li atoms are four-coordinated, and N atoms are four- or five-coordinated.

(4) Lithium halide complexes

In view of the very high lattice energy of LiF, there is as yet no known complex that contains LiF and a Lewis base donor ligand. However, a range of crystalline fluorosilyl-amide and -phosphide complexes that feature significant Li···F contacts have been synthesized and characterized. For example, {[^{t}Bu$_2$Si(F)]$_2$N}Li·2THF has a heteroatomic ladder core, as shown in Fig. 12.2.6(a), and dimeric [^{t}Bu$_2$SiP(Ph)(F)Li·2THF]$_2$ contains an eight-membered heteroatomic ring.

A large number of complexes containing the halide salts LiX (X = Cl, Br, and I) have been characterized in the solid state. The structures of some of these complexes are shown in Figs. 12.2.6(b)-(f).

In the center of the cubane complex [LiCl·HMPA]$_4$ (HMPA is (Me$_2$N)$_3$P=O), each Li$^+$ is bonded to three Cl$^-$ and one oxygen of HMPA, as shown in Fig. 12.2.6(b). The complex (LiCl)$_6$(TMEDA)$_2$ has a complicated polymeric structure based on a (LiCl)$_6$ core, as shown in Fig. 12.2.6(c). Tetrameric [LiBr]$_4$·6[2,6-Me$_2$Py] has a staggered ladder structure, as shown

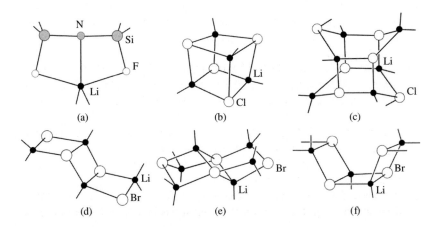

Fig. 12.2.6.
Structures of some lithium halide complexes: (a) {[^{t}Bu$_2$Si(F)]$_2$N}Li·2THF, (b) [LiCl·HMPA]$_4$, (c) (LiCl)$_6$(TMEDA)$_2$, (d) [LiBr]$_4$·6[2,6-Me$_2$Py], (e) [Li$_6$Br$_4$(Et$_2$O)$_{10}$]$^{2+}$, and (f) [LiBr·THF]$_n$.

in Fig. 12.2.6(d). In the complex $[Li_6Br_4(Et_2O)_{10}]^{2+} \cdot 2[Ag_3Li_2Ph_6]^-$, the hexametallic dication contains a Li_6Br_4 unit which can be regarded as two three-rung ladders sharing their terminal Br anions, as shown in Fig. 12.2.6(e). The structure of oligomeric $[LiBr \cdot THF]_n$ is that of a unique corrugated ladder, in which $[LiBr]_2$ units are linked together by μ_3-Br bridges, as shown in Fig. 12.2.6(f). Polymeric ladder structures have also been observed for a variety of other alkali metal organometallic derivatives.

(5) Lithium salts of heavier heteroatom compounds

Lithium can be coordinated by heavier main-group elements, such as S, Se, Te, Si, and P atoms, to form complexes. In the complex $[Li_2(THF)_2Cp^*TaS_3]_2$, the monomeric unit consists of a pair of four-membered Ta–S–Li–S rings sharing a common Ta–S edge. Dimerization occurs through further Li–S linkages, affording a hexagonal-prismatic skeleton, as shown in Figure 12.2.7(a). The bond lengths are Li–S 248 pm(av.) and Ta–S 228 pm(av.). Figure 12.2.7(b) shows the core structure of $Li(THF)_3SeMes^*$ ($Mes^* = 2,4,6-C_6H_2R_3$, where R is a bulky group such as Bu^t) which has Li–Se bond length 257 pm. Figure 12.2.7(c) shows the core structure of the dimeric complex $[Li(THF)_2TeSi(SiMe_3)_3]_2$, in which the distances of bridging Li–Te bonds are 282 and 288 pm.

Some lithium silicide complexes have been characterized. The complex $Li(THF)_3SiPh_3$ displays a terminal Li–Si bond that has a length of 267 pm, which is the same as that in $Li(THF)_3Si(SiMe_3)$. Figure 12.2.7(d) shows the core structure of $Li(THF)_3SiPh_3$. There is no interaction between Li and the phenyl rings. The C–Si–C bond angles are smaller (101.3° av.) than the ideal tetrahedral value, while the Li–Si–C angles are correspondingly larger (116.8° av.). This indicates the existence of a lone pair on silicon.

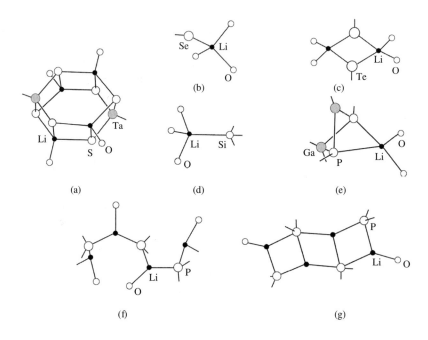

Fig. 12.2.7.
Core structures of some lithium complexes bearing heavier heteroatom ligands: (a) $[Li_2(THF)_2Cp^*TaS_3]_2$, (b) $Li(THF)_3SeMes^*$, (c) $[Li(THF)_2TeSi(SiMe_3)_3]_2$, (d) $Li(THF)_3SiPh_3$, (e) $Li(Et_2O)_2(P^tBu)_2(Ga_2^tBu_3)$, (f) $[Li(Et_2O)PPh_2]_n$, (g) $[LiP(SiMe_3)_2]_4(THF)_2$.

Figures 12.2.7(e)–(g) show the core structures of three lithium phosphide complexes. Li(Et$_2$O)$_2$(P^{t}Bu)$_2$(Ga$_2^t$Bu$_3$) has a puckered Ga$_2$P$_2$ four-membered ring, in which one Ga atom is four- and the other three-coordinated, as shown in Fig. 12.2.7(e). The Li–P bond length is 266 pm(av.).

The polymeric complex [Li(Et$_2$O)PPh$_2$]$_n$ has a backbone of alternating Li and P chain, as shown in Fig. 12.2.7(f). The bond length of Li–P is 248 pm (av.) and Li–O is 194 pm (av.). Figure 12.2.7(g) shows the core structure of [LiP(SiMe$_3$)$_2$]$_4$(THF)$_2$, which has a ladder structure with Li–P bond length 253 pm(av.).

(6) Sodium coordination complexes

Recent development in the coordination chemistry of Group 1 elements has extended our knowledge of their chemical behavior and provided many interesting new structural types. Three examples among known sodium coordination complexes are presented below:

(a) A rare example of a bimetallic imido complex is the triple-stacked Li$_4$Na$_2$[N=C(Ph)(^{t}Bu)]$_6$. This molecule has six metal atoms in a triple-layered stack of four-membered M$_2$N$_2$ rings, with the outer rings containing lithium and the central ring containing sodium. In the structure, lithium is three-coordinated and sodium is four-coordinated, as shown in Fig. 12.2.8(a).

(b) The large aggregate [Na$_8$(OCH$_2$CH$_2$OCH$_2$CH$_2$OMe)$_6$(SiH$_3$)$_2$] has been prepared and characterized. The eight sodium atoms form a cube, the faces of which are capped by the alkoxo oxygen atoms of the six (OCH$_2$CH$_2$OCH$_2$CH$_2$OMe) ligands, which are each bound to four sodium atoms with Na–O distances in the range 230–242 pm. The sodium and oxygen atoms constitute the vertices of an approximate rhombododecahedron. Six of eight sodium atoms are five-coordinated by oxygen atoms, and each of the other two Na atoms is bonded by a SiH$_3^-$ group, which has inverted C_{3v} symmetry with Na–Si–H bond angles of 58°–62°, as shown in Fig. 12.2.8(b).

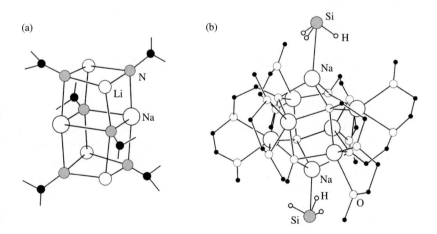

Fig. 12.2.8. Structures of (a) Li$_4$Na$_2$[N=C(Ph)(^{t}Bu)]$_6$ (the Ph and ^{t}Bu groups are not shown), and (b) [Na$_8$(OCH$_2$CH$_2$OCH$_2$CH$_2$OMe)$_6$(SiH$_3$)$_2$].

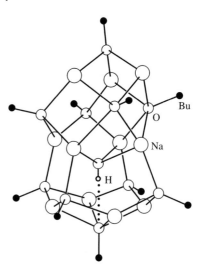

Fig. 12.2.9.
Structure of [Na$_{11}$(O^{t}Bu)$_{10}$(OH)].

Despite the strange appearance of the SiH$_3^-$ configuration in this complex, *ab initio* calculations conducted on the simplified model compound (NaOH)$_3$NaSiH$_3$ indicate that the form with inverted hydrogens is 6 kJ mol^{-1} lower in energy than the un-inverted form. The results suggest that electrostatic interaction, rather than agostic interaction between the SiH$_3$ group and its three adjacent Na neighbors, stabilizes the inverted form.

(c) The complex [Na$_{11}$(O^{t}Bu)$_{10}$(OH)] is obtained from the reaction of sodium *tert*-butanolate with sodium hydroxide. Its structure features a 21-vertex cage constructed from eleven sodium cations and ten *tert*-butanolate anions, with an encapsulated hydroxide ion in its interior, as shown in Fig. 12.2.9. In the lower part of the cage, eight sodium atoms constitute a square antiprism, and four of the lower triangular faces are each capped by a μ_3-O^{t}Bu group. The Na$_4$ square face at the bottom is capped by a μ_4-O^{t}Bu group, and the upper Na$_4$ square face is capped in an inverted fashion by the μ_4-OH$^-$ group. Thus the OH$^-$ group resides within a Na$_8$-core, and it also forms an O–H$\cdots$O hydrogen bond of length 297.5 pm. The Na–O distances in the 21-vertex cage are in the range of 219–43 pm.

12.3 Structure and bonding in organic alkali metal compounds

12.3.1 Methyllithium and related compounds

Lithium readily interacts with hydrocarbon π systems such as olefins, arenes, and acetylenes at various sites simultaneously. Lithium reacts with organic halides to give alkyllithium or aryllithium derivatives in high yield. Therefore, a wide variety of organolithium reagents can be made. These organolithium compounds are typically covalent species, which can be sublimed, distilled in a vacuum, and dissolved in many organic solvents. The Li–C bond is a strong polarized covalent bond, so that organolithium compounds serve as sources

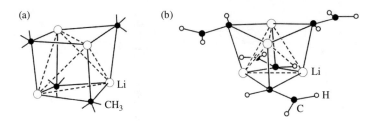

Fig. 12.3.1.
Molecular structure of (a) methyllithium tetramer (Li$_4$Me$_4$) (H atoms have been omitted); and (b) vinyllithium·THF tetramer (LiHC=CH$_2$·THF)$_4$ (THF ligands have been omitted).

of anionic carbon, which can be replaced by Si, Ge, Sn, Pb, Sb, or Bi. When Sn(C$_6$H$_5$)$_3$ or Sb(C$_6$H$_{11}$)$_2$ groups are used to replace CH$_3$, CMe$_3$ or C(SiMe$_3$)$_3$ in organolithium compounds, new species containing Li–Sn or Li–Sb bonds are formed.

Methyllithium, the simplest organolithium compound, is tetrameric. The Li$_4$(CH$_3$)$_4$ molecule with T_d symmetry may be described as a tetrahedral array of four Li atoms with a methyl C atom located above each face of the tetrahedron, as shown in Fig. 12.3.1(a). The bond lengths are Li–Li 268 pm and Li–C 231 pm, and the bond angle Li–C–Li is 68.3°.

Vinyllithium, LiHC=CH$_2$, has a similar tetrameric structure in the crystalline state with a THF ligand attached to each lithium atom [Fig. 12.3.1(b)]. In THF solution vinyllithium is tetrameric as well. The internuclear Li–H distances are in agreement with those calculated from the two-dimensional NMR ^{6}Li, ^{1}H HOSEY spectra.

When steric hindrance is minimal, organolithium compounds tend to form tetrameric aggregates with a tetrahedral Li$_4$ core. Figures 12.3.2(a)–(d) show the structures of some examples: (a) (LiBr)$_2$·(CH$_2$CH$_2$CHLi)$_2$·4Et$_2$O, (b) (PhLi·Et$_2$O)$_3$·LiBr, (c) [C$_6$H$_4$CH$_2$N(CH$_3$)$_2$Li]$_4$, and (d) (^{t}BuC≡CLi)$_4$(THF)$_4$. In these complexes, the Li–C bond lengths range from 219 to 253 pm, with an average value of 229 pm, while the Li–Li distances lie in the range 242–263 pm (average 256 pm). The Li–C bond lengths are slightly longer than those for the terminally bonded organolithium compounds. This difference may be attributed to multicenter bonding in the tetrameric structures.

12.3.2 π-Complexes of lithium

The interactions between Li atom and diverse π systems yield many types of lithium π complexes. Figures 12.3.3(a)–(c) show the schematic representation of the core structures of some π complexes of lithium, and Fig. 12.3.3(d) shows the molecular structure of [C$_2$P$_2$(SiMe$_3$)$_2$]$^{2-}$·2[Li$^+$(DME)] (DME = dimethoxyethane).

(a) Li[Me$_2$N(CH$_2$)$_2$NMe$_2$]·[C$_5$H$_2$(SiMe$_3$)$_3$]: In the cyclopentadienyllithium complex, the Li atom is coordinated by the planar Cp ring and by one chelating TMEDA ligand. The Li–C distances are 226 to 229 pm (average value 227 pm).

(b) Li$_2$[Me$_2$N(CH$_2$)$_2$NMe$_2$]·C$_{10}$H$_8$: In the naphthalene–lithium complex, each Li atom is coordinated by one η^6 six-membered ring and a chelating TEMDA ligand. The two Li atoms are not situated directly opposite to

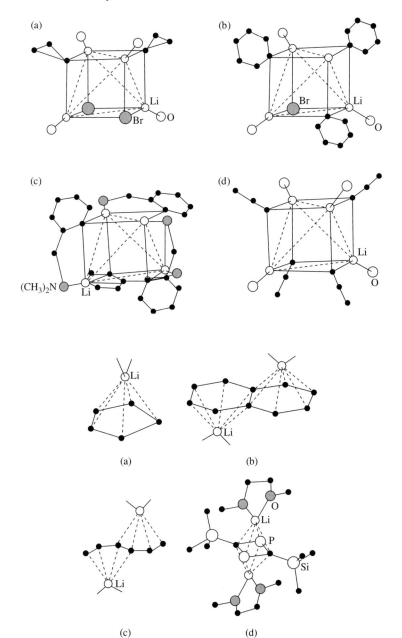

Fig. 12.3.2.
Structures of some organolithium complexes:
(a) (LiBr)$_2$·(CH$_2$CH$_2$CHLi)$_2$·4Et$_2$O,
(b) (PhLi·Et$_2$O)$_3$·LiBr,
(c) [C$_6$H$_4$CH$_2$N(CH$_3$)$_2$Li]$_4$, and
(d) (^{t}BuC≡CLi)$_4$(THF)$_4$.

Fig. 12.3.3.
Structures of some lithium π complexes:
(a) Li[Me$_2$N(CH$_2$)$_2$NMe$_2$]·[C$_5$H$_2$(SiMe$_3$)$_3$],
(b) Li$_2$[Me$_2$N(CH$_2$)$_2$NMe$_2$]·C$_{10}$H$_8$,
(c) Li$_2$[Me$_2$N(CH$_2$)$_2$NMe$_2$][H$_2$C=CH–CH=CH–CH=CH$_2$], and
(d) [C$_2$P$_2$(SiMe$_3$)$_2$]$^{2-}$·2[Li$^+$(DME)].

each other. The Li–C distances are in the range 226–66 pm; the average value is 242 pm.

(c) Li$_2$[Me$_2$N(CH$_2$)$_2$NMe$_2$][H$_2$C=CH–CH=CH–CH=CH$_2$]: In this complex, each Li atom is coordinated by the bridging bistetrahaptotriene and by one chelating TMEDA ligand. The Li–C distances are 221 to 240 pm (average value 228 pm).

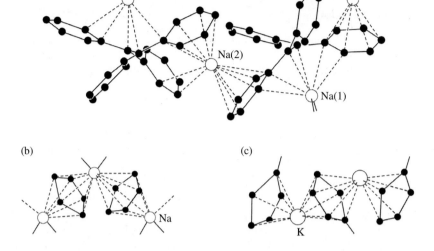

Fig. 12.3.4.
Structures of some π complexes of sodium and potassium:
(a) $Na_2[Ph_2C=CPh_2]\cdot 2Et_2O$,
(b) $Na(C_5H_5)(en)$, and (c) $K[C_5H_4(SiMe_3)]$.

(d) $[C_2P_2(SiMe_3)_2]^{2-}\cdot 2[Li^+(DME)]$: In this complex, the C_2P_2 unit forms a planar four-membered ring, which exhibits aromatic properties with six π electrons. The two Li atoms are each coordinated to the central η^4-C_2P_2 ring. The distances of Li–C and Li–P are 239.1 and 245.8 pm, respectively. Each Li atom is also coordinated by one chelating DME ligand.

12.3.3 π-Complexes of sodium and potassium

Many π complexes of sodium and potassium have been characterized. The Na and K atoms of these complexes are polyhapto-bonded to the ring π systems, and frequently attached to two or more rings to give infinite chain structures. Figure 12.3.4 shows the structures of some π complexes of sodium and potassium.

In the structure of $Na_2[Ph_2C=CPh_2]\cdot 2Et_2O$, the two halves of the $[Ph_2C=CPh_2]^{2-}$ dianion twist through 56° relative to each other, and the central C=C bond is lengthened to 149 pm as compared to 136 pm in $Ph_2C=CPh_2$. The coordination of Na(1) involves the >C=C< π bond and two adjacent π bonds. The two Et_2O ligands are also coordinated to it. The Na(2) is sandwiched between two phenyl rings in a bent fashion, and further interacts with a π bond in the third ring, as shown in Fig. 12.3.4(a). The Na(1)–C and Na(2)–C distances lie in the ranges 270–282 pm and 276–309 pm, respectively.

Figure 12.3.4(b)–(c) show the structures of $Na(C_5H_5)(en)$ and $K[C_5H_4(SiMe_3)]$, respectively. In these complexes, each metal (Na or K) atom is centrally positioned between two bridging cyclopentadienyl rings, leading to a bent sandwich polymeric zigzag chain structure. The sodium atom is further coordinated by an ethylenediamine ligand, and the potassium atom interacts with an additional Cp ring of a neighboring chain.

12.4 Alkalides and electrides

12.4.1 Alkalides

The alkali metals can be dissolved in liquid ammonia, and also in other solvents such as ethers and organic amines. Solutions of the alkali metals (except Li) contain solvated M^- anions as well as solvated M^+ cations:

$$2M(s) \rightleftharpoons M^+(solv) + M^-(solv).$$

Successful isolation of stable solids containing these alkalide anions depends on driving the equilibrium to the right and then on protecting the anion from the polarizing effects of the cation. Both goals have been realized by using macrocyclic ethers (crown ethers and cryptands). The oxa-based cryptands such as 2.2.2-crypt (also written as C222) are particularly effective in encapsulating M^+ cations.

Alkalides are crystalline compounds that contain the alkali metal anions, M^- (Na^-, K^-, Rb^-, or Cs^-). The first alkalide compound $Na^+(C222) \cdot Na^-$ was synthesized and characterized by Dye in 1974.

Elemental sodium dissolves only very slightly ($\sim 10^{-6}$ mol L^{-1}) in ethylamine, but when C222 is added, the solubility increases dramatically to 0.2 mol L^{-1}, according to the equation

$$2Na(s) + C(222) \longrightarrow Na^+(C222) + Na^-.$$

When cooled to –15 °C or below, the solution deposits shiny, gold-colored thin hexagonal crystals of $[Na^+(C222)] \cdot Na^-$. Single-crystal X-ray analysis shows that two sodium atoms are in very different environments in the crystal. One is located inside the cryptand at distances from the nitrogen and oxygen atoms characteristic of a trapped Na^+ cation. The other is a sodium anion (natride), Na^-, which is located far away from all other atoms. Figure 12.4.1 shows the cryptated sodium cation with its six nearest natrides in the crystals. The close analogy of the natride ion with an iodide ion is brought out clearly by comparing $[Na^+(C222)] \cdot Na^-$ with $[Na^+(C222)] \cdot I^-$.

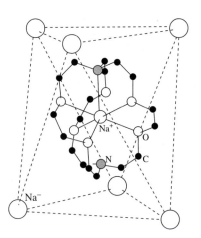

Fig. 12.4.1.
The cryptated sodium cation surrounded by six natride anions in crystalline $[Na^+(C222)] \cdot Na^-$.

Table 12.4.1. Radii of alkali metal anions from structure of alkalides and alkali metals*

Compound	$r_{M^-(min)}$ (pm)	$r_{M^-(av)}$ (pm)	d_{atom} (pm)	r_{M^+} (pm)	r_{M^-} (pm)
Na metal	—	—	372	95	277
$K^+(C222) \cdot Na^-$	255	273 (14)			
$Cs^+(18C6) \cdot Na^-$	234	264 (16)			
$Rb^+(15C5)_2 \cdot Na^-$	260	289 (16)			
$K^+(HMHCY) \cdot Na^-$	248	277 (10)			
$Cs^+(HMHCY) \cdot Na^-$	235	279 (8)			
K metal	—	—	463	133	330
$K^+(C222) \cdot K^-$	294	312 (10)			
$Cs^+(15C5)_2 \cdot K^-$	277	314 (16)			
Rb metal	—	—	485	148	337
$Rb^+(C222) \cdot Rb^-$	300	321 (14)			
$Rb^+(18C6) \cdot Rb^-$	299	323 (9)			
$Rb^+(15C5)_2 \cdot Rb^-$	264	306 (16)			
Cs metal	—	—	527	167	360
$Cs^+(C222) \cdot Cs^-$	317	350 (15)			
$Cs^+(18C6)_2 \cdot Cs^-$	309	346 (15)			

* HMHCY stands for hexamethyl hexacyclen, which is the common name of 1,4,7,10,13,16-hexaaza-1,4,7,10,13,16-hexamethyl cyclooctadecane. Here $r_{M^-(min)}$ is the distance between an anion and its nearest hydrogen atoms minus the van der Waals radius of hydrogen (120 pm), while $r_{M^-(av)}$ is the average radius over the nearest hydrogen atoms; the numbers in the brackets are the numbers of hydrogen atoms for averaging. Similarly, d_{atom} is the interatomic distance in the metal; r_{M^-} is equal to d_{atom} minus r_{M^+}.

More than 40 alkalide compounds that contain the anions Na^-, K^-, Rb^-, or Cs^- have been synthesized, and their crystal structures have been determined. Table 12.4.1 lists the calculated radii of alkali metal anions from structures of alkalides and alkali metals. The values of $r_{M^-(av)}$ derived from alkalides and r_{M^-} from the alkali metals are in good agreement.

12.4.2 Electrides

The counterparts to alkalides are electrides, which are crystalline compounds with the same type of complexed M^+ cations, but the M^- anions are replaced by entrapped electrons that usually occupy the same sites. The crystal structures of several electrides are known: $Li^+(C211)e^-$, $K^+(C222)e^-$, $Rb^+(C222)e^-$, $Cs^+(18C6)_2e^-$, $Cs^+(15C5)_2e^-$, and $[Cs^+(18C6)(15C5)e^-]_6(18C6)$. Comparison of the structures of the complexed cation in the natride $Li^+(C211)Na^-$ and the electride $Li^+(C211)e^-$ shows that the geometric parameters are virtually identical, as shown in Fig. 12.4.2, supporting the assumption that the "excess" electron density in the electride does not penetrate substantially into the cryptand cage. In $Cs^+(18C6)_2Na^-$ and $Cs^+(18C6)_2e^-$, not only are the complexed cation geometries the same, but the crystal structures are also very similar, the major difference being a slightly larger anionic site for Na^- than for e^-.

Electrides made with crown ethers and oxa-based cryptands are generally unstable above −40 °C, since the ether linkages are vulnerable to electron capture and reductive cleavage. Using a specifically designed pentacyclic tripiperazine cryptand TripPip222 [molecular structure displayed

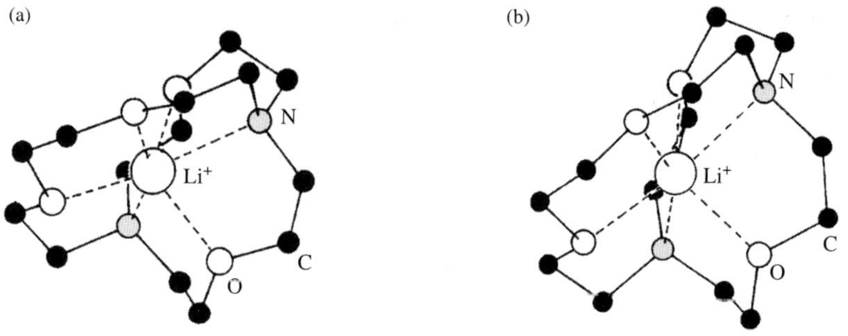

Fig. 12.4.2.
Comparison of the structures of the complexed cation in (a) Li$^+$(C211)Na$^-$ and (b) Li$^+$(C211)e$^-$.

in Fig. 12.4.3(a)], the pair of isomorphous compounds Na$^+$(TripPip222)Na$^-$ and Na$^+$(TripPip222)e$^-$ have been synthesized and fully characterized. This air-sensitive electride is stable up to about 40 °C before it begins to decompose into a mixture of the sodide and the free complexant.

While the trapped electron could be viewed as the simplest possible anion, there is a significant difference between alkalides and electrides. Whereas the large alkali metal anions are confined to the cavities, only the probability density of a trapped electron can be defined. The electronic wavefunction can extend into all regions of space, and electron density tends to seek out the void spaces provided by the cavities and by intercavity channels.

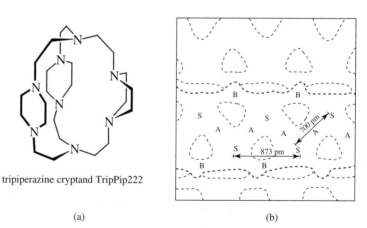

Fig. 12.4.3.
(a) Structural formula of the complexant 1,4,7,10,13,16,21,24-octaazapentacyclo [8.8.8.2^{4,7}.2^{13,16}.2^{21,24}]dotriacontane (TriPip222). (b) The "ladder-like" zigzag chain of cavities S and channels A and B, along the a direction (horizontal), in the crystal structure of electride Rb$^+$(C222)e$^-$. Connection along b (near perpendicular to the plane of the figure) between chains involves channel C (not labeled). The c axis is in the plane at 72° to the horizontal.

tripiperazine cryptand TripPip222

Figure 12.4.3(b) displays the cavity-channel geometry in the crystal structure of Rb$^+$(C222)e$^-$, in which the dominant void space consists of parallel zigzag chains of cavities S and large channels A (diameter 260 pm) along the a-direction, which are connected at the "corners" by narrower channel B of diameter 115 pm to form "ladder-like" one-dimensional chains along a. Adjacent chains are connected along b by channels C of diameter 104 pm and along c by channels with diameter of 72 pm. Thus the void space in the crystal comprises a three-dimensional network of cavities and channels. The inter-electron coupling in adjacent cavities depends on the extent of overlap of the electronic wavefunction. The dimensionalities, diameters, and lengths of the channels that connect the cavities play major roles. These structural features are consistent with the nature of wave–particle duality of the electron.

Removal of enclathrated oxygen ions from the cavities in a single crystal of $12CaO \cdot 7Al_2O_3$ leads to the formation of the thermally and chemically stable electride $[Ca_{24}Al_{28}O_{64}]^{4+}(e^-)_4$.

12.5 Survey of the alkaline-earth metals

Beryllium, magnesium, calcium, strontium, barium, and radium constitute Group 2 in the Periodic Table. These elements (or simply the Ca, Sr, and Ba triad) are often called alkaline-earth metals. Some important properties of group 2 elements are summarized in Table 12.5.1.

All group 2 elements are metals, but an abrupt change in properties between Be and Mg occurs as Be shows anomalous behavior in forming mainly covalent compounds. Beryllium most frequently displays a coordination number of four, usually tetrahedral, in which the radius of Be^{2+} is 27 pm. The chemical behavior of magnesium is intermediate between that of Be and the heavier elements, and it also has some tendency for covalent bond formation.

With increasing atomic number, the group 2 metals follow a general trend in decreasing values of I_1 and I_2 except for Ra, whose relatively high I_1 and I_2 are attributable to the 6s inert pair effect. Also, high I_3 values for all members of the group preclude the formation of the +3 oxidation state.

As the atomic number increases, the electronegativity decreases. In the organic compounds of group 2 elements, the polarity of the M–C bond increases in the order

$$BeR_2 < MgR_2 < CaR_2 < SrR_2 < BaR_2 < RaR_2$$

This is due to the increasing difference between the χ_s of carbon (2.54) and those of the metals. The tendency of these compounds to aggregate also increases in the same order. The BeR_2 and MgR_2 compounds are linked in the solid phase

Table 12.5.1. Properties of group 2 elements

Property	Be	Mg	Ca	Sr	Ba	Ra
Atomic number, Z	4	12	20	38	56	88
Electronic configuration	$[He]2s^2$	$[Ne]3s^2$	$[Ar]4s^2$	$[Kr]5s^2$	$[Xe]6s^2$	$[Rn]7s^2$
ΔH^0_{at} (kJ mol^{-1})	309	129	150	139	151	130
mp (K)	1551	922	1112	1042	1002	973
ΔH_{fuse}(mp) (kJ mol^{-1})	7.9	8.5	8.5	7.4	7.1	—
bp (K)	3243	1363	1757	1657	1910	1413
r_M(CN = 12) (pm)	112	160	197	215	224	—
$r_{M^{2+}}$ (CN = 6) (pm)	45	72	100	118	135	148
I_1 (kJ mol^{-1})	899.5	737.3	589.8	549.5	502.8	509.3
I_2 (kJ mol^{-1})	1757	1451	1145	1064	965.2	979.0
I_3 (kJ mol^{-1})	14850	7733	4912	4138	3619	3300
χ_s	1.58	1.29	1.03	0.96	0.88	—
mp of MCl_2 (K)	703	981	1045	1146	1236	—

via M–C–M $3c$-$2e$ covalent bonds to form polymeric chains, while CaR_2 to RaR_2 form three-dimensional network structures in which the M–C bonds are largely ionic.

All the M^{2+} ions are smaller and considerably less polarizable than the isoelectronic M^+ ions, as the higher effective nuclear charge binds the remaining electrons tightly. Thus the effect of polarization of cations on the properties of their salts are less important. Ca, Sr, Ba, and Ra form a closely allied series in which the properties of the elements and their compounds vary systematically with increasing size in much the same manner as in the alkali metals.

The melting points of group 2 metal chlorides MCl_2 increase steadily, and this trend is in sharp contrast to the alkali metal chlorides: LiCl (883 K), NaCl (1074 K), KCl (1045 K), RbCl (990 K), and CsCl (918 K). This is due to several subtle factors: (a) the nature of bonding varies from covalent (Be) to ionic (Ba); (b) from Be to Ra the coordination number increases, so the Madelung constants, the lattice energies, and the melting points also increase; (c) the radius of Cl^- is large (181 pm), whereas the radii of M^{2+} are small, and in the case of metal coordination, an increase in the radius of M^{2+} reduces the $Cl^- \cdots Cl^-$ repulsion.

12.6 Structure of compounds of alkaline-earth metals

12.6.1 Group 2 metal complexes

The coordination compounds of the alkaline-earth metals are becoming increasingly important to many branches of chemistry and biology. A considerable degree of structural diversity exists in these compounds, and monomers up to nonametallic clusters and polymeric species are known.

Beryllium, in view of its small size and simple set of valence orbitals, almost invariably exhibits tetrahedral four-coordination in its compounds. Figure 12.6.1(a) shows the structure of $Be_4O(NO_3)_6$. The central oxygen atom is tetrahedrally surrounded by four Be atoms, and each Be atom is in turn tetrahedrally surrounded by four O atoms. The six nitrate groups are attached symmetrically to the six edges of the tetrahedron. This type of structure also appears in $Be_4O(CH_3COO)_6$.

Magnesium shows a great tendency to form complexes with ligands which have oxygen and nitrogen donor atoms, often displaying six-coordination, and its compounds are more polar than those of beryllium. The structure of hexameric $[MgPSi^tBu_3]_6$ is based on a Mg_6P_6 hexagonal drum, with Mg–P distances varying between 247 and 251 pm in the six-membered Mg_3P_3 ring, and 250 and 260 pm between the two rings, as shown in Fig. 12.6.1(b).

Calcium, strontium, and barium form compounds of increasing ionic character with higher coordination numbers, of which six to eight are particularly common. The complex $Ca_9(OCH_2CH_2OMe)_{18}(HOCH_2CH_2OMe)_2$ has an interesting structure: the central $Ca_9(\mu_3\text{-}O)_8(\mu_2\text{-}O)_8O_{20}$ skeleton is composed of three six-coordinate Ca atoms and six seven-coordinate Ca atoms that can be viewed as filling the octahedral holes in two close-packed oxygen layers, as shown in Fig. 12.6.1(c). The distances of Ca–O are Ca–(μ_3-O) 239 pm, Ca–(μ_2-O) 229 pm and Ca–O_{ether} 260 pm.

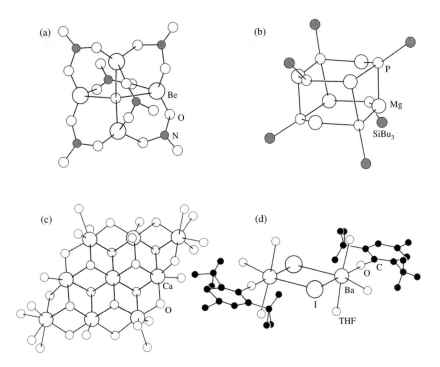

Fig. 12.6.1.
Structures of some coordination compounds of alkaline-earth metals: (a) Be$_4$O(NO$_3$)$_6$, (b) [MgPSi^{t}Bu$_3$]$_6$, (c) Ca$_9$(OCH$_2$CH$_2$OMe)$_{18}$(HOCH$_2$CH$_2$OMe)$_2$ and (d) [BaI(BHT)(THF)$_3$]$_2$.

In the dimeric complex [BaI(BHT)(THF)$_3$]$_2$ [BHT = 2,6-di-t-butyl-4-methylphenol, C$_6$H$_2$Me^{t}Bu$_2$(OH)], the coordination geometry around the Ba atom is distorted octahedral, with the two bridging iodides, the BHT, and a THF ligand in one plane, and two additional THF molecules lying above and below the plane, as shown in Fig. 12.6.1(d). The bond distances are Ba–I 344 pm, Ba–OAr 241 pm (av).

Many interesting coordination compounds of Group 2 elements have been synthesized in recent years. For example, the cation [Ba(NH$_3$)$_n$]$^{2+}$ is generated in the course of reducing the fullerenes C$_{60}$ and C$_{70}$ with barium in liquid ammonia. In the [Ba(NH$_3$)$_7$]C$_{60}$·NH$_3$ crystal, the Ba^{2+} cation is surrounded by seven NH$_3$ ligands at the vertices of a monocapped trigonal antiprism, and the C$_{60}$ dianion is well ordered. In the [Ba(NH$_3$)$_9$]C$_{70}$·7NH$_3$ crystal, the coordination geometry around Ba^{2+} is a distorted tricapped trigonal prism, with Ba–N distances in the range 289–297 pm; the fullerene C$_{70}$ units are linked into slightly zigzag linear chains by single C–C bonds with length 153 pm.

12.6.2 Group 2 metal nitrides

The crystals of M[Be$_2$N$_2$] (M = Mg, Ca, Sr) are composed of complex anion layers with covalent bonds between Be and N atoms. For example, Mg[Be$_2$N$_2$] contains puckered six-membered rings in the chair conformation. These rings are condensed into single nets and are further connected to form double layers,

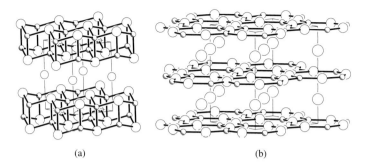

Fig. 12.6.2.
Crystal structures of (a) Mg[Be$_2$N$_2$] and (b) Ca[Be$_2$N$_2$]. Small circles represent Be, large circles N, while circles between layers are either Mg or Ca.

in which the Be and N atoms are alternately linked. In this way, each beryllium atom is tetrahedrally coordinated by nitrogens, while each nitrogen occupies the apex of a trigonal pyramid whose base is made of three Be atoms, and a fourth Be atom is placed below the base. Thus the nitrogen atoms are each coordinated by an inverse tetrahedron of Be atoms and occupy positions at the outer boundaries of a double layer, as shown in Fig. 12.6.2(a). The Be–N bond lengths are 178.4 pm (3×) and 176.0 pm. The magnesium atoms are located between the anionic double layers, and each metal center is octahedrally surrounded by nitrogens with Mg–N 220.9 pm.

The ternary nitrides Ca[Be$_2$N$_2$] and Sr[Be$_2$N$_2$] are isostructural. The crystal structure contains planar layers which consist of four- and eight-membered rings in the ratio of 1:2. The layers are stacked with successive slight rotation of each about the common normal, forming octagonal prismatic voids in the interlayer region. The eight-coordinated Ca^{2+} ions are accommodated in the void space, as shown in Fig. 12.6.2(b). The Be–N bond lengths within the layers are 163.2 pm (2×) and 165.7 pm. The Ca–N distance is 268.9 pm.

12.6.3 Group 2 low-valent oxides and nitrides

(1) (Ba$_2$O)Na

Similar to the alkali metals, the alkaline-earth metals can also form low-valent oxides. The first crystalline compound of this type is (Ba$_2$O)Na, which belongs to space group *Cmma* with $a = 659.1, b = 1532.7, c = 693.9$ pm, and $Z = 4$. In the crystal structure, the O atom is located inside a Ba$_4$ tetrahedron. Such [Ba$_4$O] units share *trans* edges to form an infinite chain running parallel to the a axis. The [Ba$_{4/2}$O]$_\infty$ chains are in parallel alignment and separated by the Na atoms, as illustrated in Fig. 12.6.3. Within each chain, the Ba–O bond length is 252 pm, being shorter than the corresponding distance of 277 pm in crystalline barium oxide. The interatomic distances between Ba and Na atoms lie in the range 421–433 pm, which are comparable with the values in the binary alloys BaNa and BaNa$_2$ (427 and 432 pm, respectively).

(2) [Ba$_6$N] cluster and related compounds

When metallic barium is dissolved in liquid sodium or K/Na alloy in an inert atmosphere of nitrogen, an extensive class of mixed alkali metal–barium

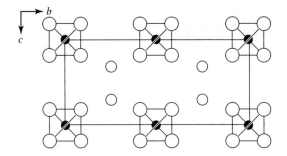

Fig. 12.6.3.
Crystal structure of $(Ba_2O)Na$. The $[Ba_{4/2}O]_\infty$ chains are seen end-on.

subnitrides can be prepared. They contain $[Ba_6N]$ octahedra that exist either as discrete entities or are condensed into finite clusters and infinite arrays.

Discrete $[Ba_6N]$ clusters are found in the crystal structure of $(Ba_6N)Na_{16}$, as shown in Fig. 12.6.4. $(Ba_6N)Na_{16}$ crystallizes in space group $Im\bar{3}m$ with $a = 1252.7$ pm and $Z = 2$. The Na atoms are located between the cluster units, in a manner analogous to that in the low-valent alkali metal suboxides.

The series of barium subnitrides Ba_3N, $(Ba_3N)Na$, and $(Ba_3N)Na_5$ are characterized by parallel infinite $(Ba_{6/2}N)_\infty$ chains each composed of *trans* face-sharing octahedral $[Ba_6N]$ clusters. Figure 12.6.5(a) shows the crystal structure of Ba_3N projected along a sixfold axis. In the crystal structure of $(Ba_3N)Na$, the sodium atoms are located in between the $(Ba_{6/2}N)_\infty$ chains, as shown in Fig. 12.6.5(b). In contrast, since $(Ba_3N)Na_5$ has a much higher sodium content, the $(Ba_{6/2}N)_\infty$ chains are widely separated from one another by the additional Na atoms, as illustrated in Fig. 12.6.5(c).

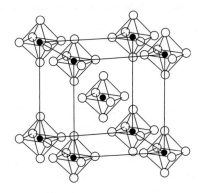

Fig. 12.6.4.
Arrangement of the (Ba_6N) clusters in the crystal structure of $(Ba_6N)Na_{16}$.

(3) $[Ba_{14}CaN_6]$ cluster and related compounds

The $[Ba_{14}CaN_6]$ cluster can be considered as a Ca-centered Ba_8 cube with each face capped by a Ba atom, and the six N atoms are each located inside a Ba_5 tetragonal pyramid, as shown in Fig. 12.6.6. The structure can be viewed alternatively as a cluster composed of the fusion of six N-centered Ba_5Ca octahedra sharing a common Ca vertex. In the synthesis of this type of cluster compounds, variation of the atomic ratio of the Na/K binary alloy system leads to a series of compounds of variable stoichiometry: $[Ba_{14}CaN_6]Na_x$ with $x = 7, 8, 14, 17, 21$

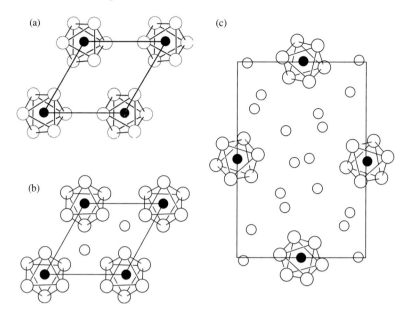

Fig. 12.6.5.
Crystal structures of (a) Ba_3N, (b) $(Ba_3N)Na$ and (c) $(Ba_3N)Na_5$. The $(Ba_{6/2}N)_\infty$ chains are seen end-on.

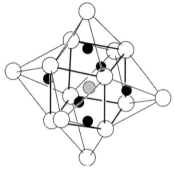

Fig. 12.6.6.
Structure of the $[Ba_{14}CaN_6]$ cluster.

and 22. This class of compounds can be considered as composed of an ionically bonded $[Ba_8^{2+}Ca^{2+}N_6^{3-}]$ nucleus surrounded by Ba_6Na_x units, which interact with the nucleus mainly via metallic bonding.

12.7 Organometallic compounds of group 2 elements

There are numerous organometallic compounds of Group 2 elements, and in this section only the following three typical species are described.

12.7.1 Polymeric chains

Dimethylberyllium, $Be(CH_3)_2$, is a polymeric white solid containing infinite chains, as shown in Fig. 12.7.1(a). Each Be center is tetrahedrally coordinated and can be considered to be sp^3 hybridized. As the CH_3 group only contributes one orbital and one electron to the bonding, there are insufficient electrons to form normal $2c$-$2e$ bonds between the Be and C atoms. In this electron-deficient system, the BeCBe bridges are $3c$-$2e$ bonds involving sp^3 hybrid atomic orbitals from two Be atoms and one C atom, as shown in Fig. 12.7.1(b).

The polymeric structures of BeH_2 and $BeCl_2$ are similar to that of $Be(CH_3)_2$. The $3c$-$2e$ bonding in BeH_2 is analogous to that in $Be(CH_3)_2$. But in $BeCl_2$ there are sufficient valence electrons to form normal $2c$-$2e$ Be–Cl bonds.

12.7.2 Grignard reagents

Grignard reagents are widely used in organic chemistry. They are prepared by the interaction of magnesium with an organic halide in ethers. Grignard reagents

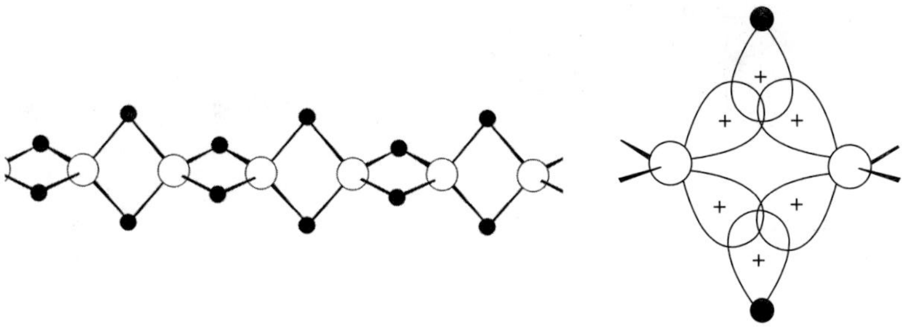

Fig. 12.7.1.
(a) Linear structure of Be(CH$_3$)$_2$; (b) its BeCBe 3c-2e bonding in the polymer.

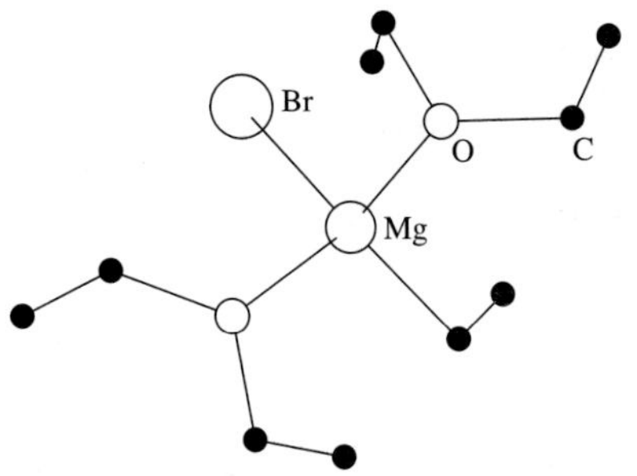

Fig. 12.7.2.
Molecular structure of Grignard reagent (C$_2$H$_5$)MgBr·2(C$_2$H$_5$)$_2$O.

are normally assigned the simple formula RMgX, but this is an oversimplification as solvation is important. The isolation and structural determination of several crystalline RMgX compounds have demonstrated that the essential structure is RMgX·(solvent)$_n$. Figure 12.7.2 shows the molecular structure of (C$_2$H$_5$)MgBr·2(C$_2$H$_5$)$_2$O, in which the central Mg atom is surrounded by one ethyl group, one bromine atom, and two diethyl ether molecules in a distorted tetrahedral configuration. The bond distances are Mg–C 215 pm, Mg–Br 248 pm, and Mg–O 204 pm. The latter is among the shortest Mg–O distances known.

In a Grignard compound the carbon atom bonded to the electropositive Mg atom carries a partial negative charge. Consequently a Grignard reagent is an extremely strong base, with its carbanion-like alkyl or aryl portion acting as a nucleophile.

12.7.3 Alkaline-earth metallocenes

The alkaline-earth metallocenes exhibit various structures, depending on the sizes of the metal atoms. In Cp$_2$Be, the Be^{2+} ion is η^5/η^1 coordinated between two Cp rings, as shown in Fig. 12.7.3(a). The mean Be–C distances are 193 and 183 pm for η^5 and η^1 Cp rings, respectively. In contrast, Cp$_2$Mg adopts a typical sandwich structure, as shown in Fig. 12.7.3(b). The mean Mg–C distance is 230 pm. In the crystalline state, the two parallel rings have a staggered conformation.

The compounds Cp$_2$Ca, Cp*Sr and Cp*Ba exhibit a different coordination mode, as the two Cp rings are not aligned in a parallel fashion like Cp$_2$Mg, but

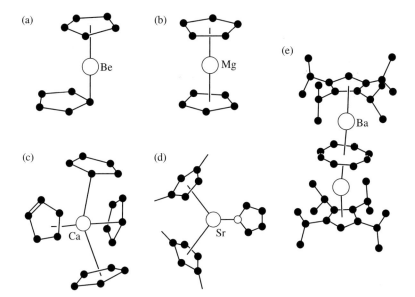

Fig. 12.7.3.
Structures of some alkaline-earth metallocenes: (a) Cp$_2$Be, (b) Cp$_2$Mg, (c) Cp$_2$Ca, (d) Sr[C$_5$H$_3$(SiMe$_3$)$_2$]$_2$·THF, (e) Ba$_2$(COT)[C$_5$H(CHMe$_2$)$_4$]$_2$.

are bent with respect to each other making a Cp$_{(centroid)}$–M–Cp$_{(centroid)}$ angle of 147°–154°. The positive charge of the M^{2+} ion is not only shared by two η^5-bound Cp rings, but also by other ligands. In Cp$_2$Ca, apart from the two η^5-Cp ligands, there are significant Ca–C contacts to η^3- and η^1-Cp rings, as shown in Figure 12.7.3(c). In Sr[C$_5$H$_3$(SiMe$_3$)$_2$]$_2$·THF, apart from the two η^5-Cp rings, a THF ligand is coordinated to the Sr^{2+} ion, as shown in Fig. 12.7.3(d). The Ba^{2+} ion has the largest size among the alkaline-earth metals, and accordingly it can be coordinated by a planar COT (C$_8$H$_8$) ring bearing two negative charges. Figure 12.7.3(e) shows the structure of a triple-decker sandwich complex of barium, Ba$_2$(COT)[C$_5$H(CHMe$_2$)$_4$]$_2$, in which the mean Ba–C distances are 296 and 300 pm for the η^5-Cp ring and η^8-COT ring, respectively.

12.8 Alkali and alkaline-earth metal complexes with inverse crown structures

The term "inverse crown ether" refers to a metal complex in which the roles of the central metal core and the surrounding ether oxygen ligand sites in a conventional crown ether complex are reversed. An example of an inverse crown ether is the organosodium-zinc complex Na$_2$Zn$_2$\{N(SiMe$_3$)$_2$\}$_4$(O) shown in Fig. 12.8.1(a). The general term "inverse crown" is used when the central core of this type of complex comprises non-oxygen atoms, or more than just a single O atom. Two related inverse crowns with cationic (Na–N–Mg–N–)$_2$ rings that encapsulate peroxide and alkoxide groups, respectively, are shown in Figs. 12.8.1(b) and (c). The octagonal macrocycle in Na$_2$Mg$_2$\{N(SiMe$_3$)$_2$\}$_4$(OOct^n)$_2$ takes the chair form with the Na–OOct^n groups displaced on either side of the plane defined by the N and Mg atoms. Each alkoxide O atom interacts with both Mg atoms to produce a perfectly planar

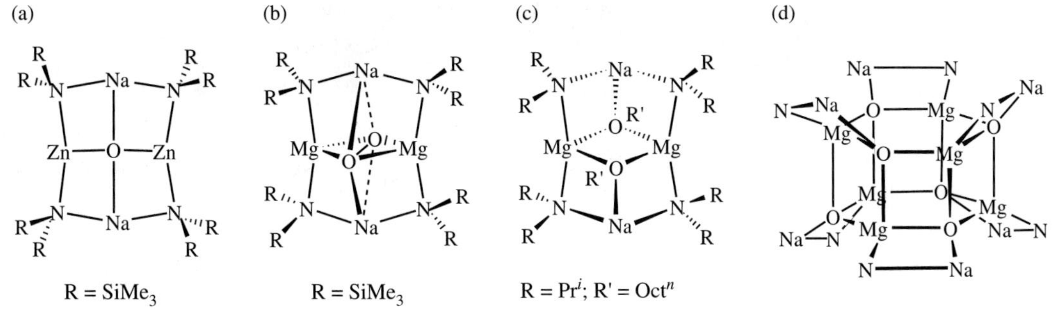

Fig. 12.8.1.
Examples of some inverse crown compounds:
(a) Na$_2$Zn$_2${N(SiMe$_3$)$_2$}$_4$(O),
(b) Na$_2$Mg$_2${N(SiMe$_3$)$_2$}$_4$(O$_2$),
(c) Na$_2$Mg$_2${N(SiMe$_3$)$_2$}$_4$(OOct^n)$_2$,
(d) cluster core of {(THF)NaMg(Pr$_2^i$N)(O)}$_6$; the THF ligand attached to each Na atom and the isopropyl groups are omitted for clarity.

(MgO)$_2$ ring. The Mg–O and Na–O$_{syn}$ bond lengths are 202.7(2)-203.0(2) and 256.6(1)-247.2(2) pm, respectively; the non-bonded Na$\cdots$O$_{anti}$ distances lie in the range 310.4(1)-320.4(2) pm.

Figure 12.8.1(d) shows the structure of a hexameric "super" inverse crown ether {(THF)NaMg(Pr$_2^i$N)(O)}$_6$, which consists of a S_6-symmetric hexagonal prismatic Mg$_6$O$_6$ cluster with six external four-membered rings constructed with Na–THF and Pr$_2^i$N appendages.

Table 12.8.1. Compositions of some inverse crowns containing organic guest species

Group 1 metal	Group 2 metal	Amide ion	Core moiety	Host ring size
4Na	2Mg	6TMP	C$_6$H$_3$Me^{2-}	12
4Na	2Mg	6TMP	C$_6$H$_4^{2-}$	12 [Fig. 2.7.5(a)]
6K	6Mg	12TMP	6C$_6$H$_5^-$	24
6K	6Mg	12TMP	6C$_6$H$_4$Me$^-$	24
4Na	4Mg	8Pr$_2^i$N	Fe(C$_5$H$_3$)$_2^{4-}$	12 [Fig. 2.7.5(b)]

A remarkable series of inverse crown compounds featuring organic guest moieties encapsulated within host-like macrocyclic rings composed of sodium/potassium and magnesium ions together with anionic amide groups, such as TMP$^-$ (TMPH = 2,2,6,6-tetramethylpiperidine) and Pr$_2^i$N$^-$ (Pr$_2^i$NH = diisopropylamine), have been synthesized and characterized. Table 12.8.1 lists some examples of this class of inverse crown molecules.

Figure 12.8.2(a) shows the molecular structure of Na$_4$Mg$_2$(TMP)$_6$(C$_6$H$_4$), in which the N atom of each tetramethylpiperidinide is bonded to two metal atoms to form a cationic 12-membered (Na–N–Na–N–Mg–N–)$_2$ ring. The mixed metal macrocyclic amide acts as a host that completely encloses a 1,4-deprotonated benzenediide guest species whose naked carbon atoms are each stabilized by a covalent Mg–C bond and a pair of Na$\cdots$C π interactions.

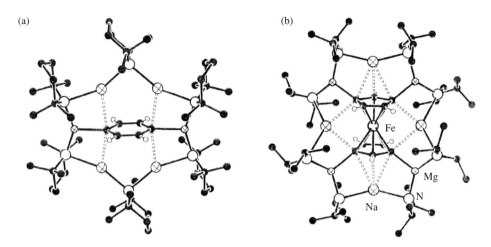

Fig. 12.8.2.
Molecular structures of two centrosymmetric inverse crown molecules containing organic cores:
(a) $Na_4Mg_2(TMP)_6(C_6H_4)$,
(b) $Na_4Mg_4(^iPr_2N)_8[Fe(C_5H_3)_2]$.
The Na···C π interactions are indicated by broken lines.

Figure 12.8.2(b) shows the molecular structure of $Na_4Mg_4(Pr^i_2N)_8$ $[Fe(C_5H_3)_2]$. Each amido N atom is bound to one Na and one Mg atom to form a 16-membered macrocyclic ring, which accommodates the ferrocene-1,1',3,3'-tetrayl residue at its center. The deprotonated 1,3-positions of each cyclopentadienyl ring of the ferrocene moiety are each bound to a pair of Mg atoms by Mg–C covalent bonds, and the 1,2,3-positions further interact with three Na atoms via three different kinds of Na···C π bonds.

References

1. N. Wiberg, *Inorganic Chemistry*, 1st English edn. (based on *Holleman-Wiberg: Lehrbuch der Anorganischen Chemie*, 34th edn., Walter de Gruyter, Berlin, 1995), Academic Press, San Diego, CA, 2001.
2. F. A. Cotton, G. Wilkinson, C. A. Murillo and M. Bochmann, *Advanced Inorganic Chemistry*, 6th edn., Wiley-Interscience, New York, 1999.
3. A. G. Massey, *Main Group Chemistry*, 2nd edn., Wiley, Chichester, 2000.
4. C. E. Housecroft and A. G. Sharpe, *Inorganic Chemistry*, 2nd edn., Prentice-Hall, Harlow, 2004.
5. D. F. Shriver, P. W. Atkins, T. L. Overton, J. P. Rourke, M. T. Weller and F. A. Armstrong, *Inorganic Chemistry*, 4th edn., Oxford University Press, Oxford, 2006.
6. A.-M. Sapse and P. von R. Schleyer, *Lithium Chemistry: A Theoretical and Experimental Overview*, Wiley, New York, 1995.
7. T. C. W. Mak and G.-D. Zhou, *Crystallography in Modern Chemistry: A Resource Book of Crystal Structures*, Wiley, New York, 1992.
8. M. Gielen, R. Willen, and B. Wrackmeyer (eds.), *Unusual Structures and Physical Properties in Organometallic Chemistry*, Wiley, West Sussex, 2002.
9. J. A. McCleverty and T. J. Meyer (editors-in-chief), *Comprehensive Coordination Chemistry: From Biology to Nanotechnology*, vol. 3, G. F. R. Parkin (volume ed.),

Coordination Chemistry of the s, p, and f Metals, Elsevier-Pergamon, Amsterdam, 2004.
10. A. Simon, Alkali and alkaline earth metal suboxides and subnitrides. In M. Driess and H. Nöth (eds.), *Molecular Clusters of the Main Group Elements*, Wiley–VCH, Weinheim, 2004, pp. 246–66.
11. M. Driess, R. E. Mulvey and M. Westerhausen, Cluster growing through ionic aggregation: synthesis and structural principles of main group metal-nitrogen, phosphorus and arsenic rich clusters. In M. Driess and H. Nöth (eds.), *Molecular Clusters of the Main Group Element*s, Wiley–VCH, Weinheim, 2004, pp. 246–66.
12. J. L. Dye, Electrides: From LD Heisenberg chains to 2D pseudo-metals. *Inorg. Chem.* **36**, 3816–26 (1997).
13. Q. Xie, R. H. Huang, A. S. Ichimura, R. C. Phillips, W. P. Pratt Jr. and J. L. Dye, Structure and properties of a new electride, Rb^+(cryptand[2.2.2])e^-. *J. Am. Chem. Soc.* **122**, 6971–8 (2000).
14. M. Y. Redko, J. E. Jackson, R. H. Huang and J. L. Dye, Design and synthesis of a thermally stable organic electride. *J. Am. Chem. Soc.* **127**, 12416–22 (2005).
15. H. Sitzmann, M. D. Walter and G. Wolmershäuser, A triple-decker sandwich complex of barium. *Angew. Chem. Int. Ed.* **41**, 2315–16 (2002).
16. S. Matsuishi, Y. Toda, M. Miyakawa, K. Hayashi, T. Kamiya, M. Hirano, I. Tanaka and H. Hosono, High-density electron anions in a nanoporous single crystal: $[Ca_{24}Al_{28}O_{64}]^{4+}(4e^-)$. *Science* **301**, 626–9 (2003).
17. M. Sebastian, M. Nieger, D. Szieberth, L. Nyulaszi and E. Niecke, Synthesis and structure of a 1,3-diphospha-cyclobutadienediide. *Angew. Chem. Int. Ed.* **43**, 637–41 (2004).
18. D. Stalke, The lithocene anion and "open" calcocene — new impulses in the chemistry of alkali and alkaline earth metallocenes. *Angew. Chem. Int. Ed.* **33**, 2168–71(1994).
19. J. Geier and H. Grützmacher, Synthesis and structure of $[Na_{11}(O^tBu)_{10}(OH)]$. *Chem. Commun.* 2942–3 (2003).
20. M. Somer, A. Yarasik, L. Akselrud, S. Leoni, H. Rosner, W. Schnelle and R. Kniep, $Ae[Be_2N_2]$: Nitridoberyllates of the heavier alkaline-earth metals. *Angew. Chem. Int. Ed.* **43**, 1088–92 (2004).
21. R. E. Mulvey, s-Block metal inverse crowns: synthetic and structural synergism in mixed alkali metal-magnesium (or zinc) amide chemistry. *Chem. Commun.* 1049–56 (2001).

13

Structural Chemistry of Group 13 Elements

13.1 Survey of the group 13 elements

Boron, aluminum, gallium, indium, and thallium are members of group 13 of the Periodic Table. Some important properties of these elements are given in Table 13.1.1.

From Table 13.1.1, it is seen that the inner electronic configurations of the group 13 elements are not identical. The ns^2np^1 electrons of B and Al lie outside a rare gas configuration, while the ns^2np^1 electrons of Ga and In are outside the d^{10} subshell, and those of Tl lie outside the $4f^{14}5d^{10}$ core. The increasing effective nuclear charge and size contraction, which occur during successive filling of d and f orbitals, combine to make the outer s and p electrons of Ga, In, and Tl more strongly held than expected by simple extrapolation from B to Al. Thus there is a small increase in ionization energy between Al and Ga, and between In and Tl. The drop in ionization energy between Ga and In mainly

Table 13.1.1. Table 13.1.1 Some properties of group 13 elements

Property	B	Al	Ga	In	Tl
Atomic number, Z	5	13	31	49	81
Electronic configuration	[He]$2s^22p^1$	[Ne]$3s^23p^1$	[Ar]$3d^{10}4s^24p^1$	[Kr]$4d^{10}5s^25p^1$	[Xe]$4f^{14}5d^{10}6s^26p^1$
ΔH^θ_{at} (kJ mol^{-1})	582	330	277	243	182
mp (K)	2453*	933	303	430	577
ΔH^θ_{fuse} (kJ mol^{-1})	50.2	10.7	5.6	3.3	4.1
bp (K)	4273	2792	2477	2355	1730
I_1 (kJ mol^{-1})	800.6	577.5	578.8	558.3	589.4
I_2 (kJ mol^{-1})	2427	1817	1979	1821	1971
I_3 (kJ mol^{-1})	3660	2745	2963	2704	2878
$I_1 + I_2 + I_3$ (kJ mol^{-1})	6888	5140	5521	5083	5438
χ_s	2.05	1.61	1.76	1.66	1.79
r_M (pm)	—	143	153	167	171
r_{cov} (pm)	88	130	122	150	155
$r_{ion, M^{3+}}$ (pm)	—	54	62	80	89
r_{ion, M^+} (pm)	—	—	113	132	140
$D_{0, M-F}$ (in MF$_3$) (kJ mol^{-1})	613	583	469	444	439 (in TlF)
$D_{0, M-Cl}$ (in MCl$_3$) (kJ mol^{-1})	456	421	354	328	364 (in TlCl)

* For β-rhombohedral boron

reflects the fact that both elements have a completely filled inner shell and the outer electrons of In are further from the nucleus.

The +3 oxidation state is characteristic of group 13 elements. However, as in the later elements of this groups, the trend in I_2 and I_3 shows increases at Ga and Tl, leading to a marked increase in stability of their +1 oxidation state. In the case of Tl, this is termed the 6s inert-pair effect. Similar effects are observed for Pb (group 14) and Bi (group 15), for which the most stable oxidation states are +2 and +3, respectively, rather than +4 and +5.

Because of the 6s inert-pair effect, many Tl^+ compounds are found to be more stable than the corresponding Tl^{3+} compounds. The progressive weakening of M^{3+}–X bonds from B to Tl partly accounts for this. Furthermore, the relativistic effect (as described in Section 2.4.3) is another factor which contributes to the inert-pair effect.

13.2 Elemental Boron

The uniqueness of structure and properties of boron is a consequence of its electronic configuration. The small number of valence electrons (three) available for covalent bond formation leads to "electron deficiency," which has a dominant effect on boron chemistry.

Boron is notable for the complexity of its elemental crystalline forms. There are many reported allotropes, but most are actually boron-rich borides. Only two have been completely elucidated by X-ray diffraction, namely α-R12 and β-R105; here R indicates the rhombohedral system, and the numeral gives the number of atoms in the primitive unit cell. In addition, α-T50 boron (T denotes the tetragonal system) has been reformulated as a carbide or nitride, $B_{50}C_2$ or $B_{50}N_2$, and β-T192 boron is still not completely elucidated. These structures all contain icosahedral B_{12} units, which in most cases are accompanied by other boron atoms lying outside the icosahedral cages, and the linkage generates a three-dimensional framework.

The α-R12 allotrope consists of an approximately cubic closest packing arrangement of icosahedral B_{12} units bound to each other by covalent bonds. The parameters of the rhombohedral unit cell are $a = 505.7$ pm, $\alpha = 58.06°$ (60° for regular ccp). In space group $R\bar{3}m$, the unit cell contains one B_{12} icosahedron. A layer of interlinked icosahedra perpendicular to the threefold symmetry axis is illustrated in Fig. 13.2.1. There are 12 neighboring icosahedra for each icosahedron: six in the same layer, and three others in each of the upper and lower layers, as shown in Fig. 9.6.21(b).

The B_{12} icosahedron is a regular polyhedron with 12 vertices, 30 edges, and 20 equilateral triangular faces, with B atoms located at the vertices, as shown in Fig. 13.2.2. The B_{12} icosahedron is a basic structural unit in all isomorphic forms of boron and in some polyhedral boranes such as $B_{12}H_{12}^{2-}$. It has 36 valence electrons; note that each line joining two B atoms (mean B–B distance 177 pm) in Fig. 13.2.2 does not represent a normal two-center two-electron ($2c$-$2e$) covalent bond.

When the total number of valence electrons in a molecular skeleton is less than the number of valence orbitals, the formation of normal $2c$-$2e$ covalent

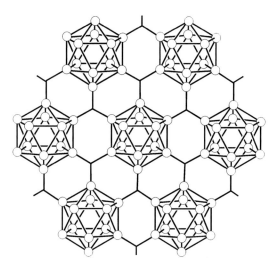

Fig. 13.2.1.
A close-packed layer of interlinked icosahedra in the crystal structure of α-rhombohedral boron. The lines meeting at a node (not a B atom) represents a BBB 3c-2e bond linking three B_{12} icosahedra.

bonds is not feasible. In this type of electron-deficient compound, three-center two-electron (3c-2e) bonds generally occur, in which three atoms share an electron pair. Thus one 3c-2e bond serves to compensate for the shortfall of four electrons and corresponds to a bond valence value of 2, as illustrated for a BBB bond in Fig. 13.2.3.

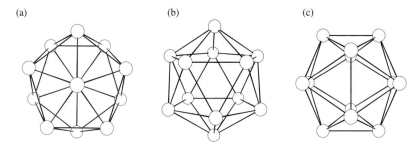

Fig. 13.2.2.
From left to right, perspective view of the B_{12} icosahedron along one of its (a) 6 fivefold, (b) 10 threefold, and (c) 15 twofold axes.

In a closo-B_n skeleton, there are only three 2c-2e covalent bonds and $(n-2)$ BBB 3c-2e bonds. (This will be derived in Section 13.4.) For an icosahedral B_{12} unit, there are three B–B 2c-2e bonds and ten BBB 3c-2e bonds, as shown in Fig. 13.2.4.

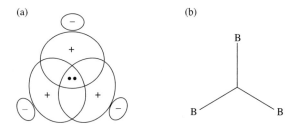

Fig. 13.2.3.
The BBB 3c-2e bond: (a) three atoms share an electron pair, and (b) simplified representation of 3c-2e bond.

The 36 electrons in the B_{12} unit may be partitioned as follow: 26 electrons are used to form bonds within an icosahedron, and the remaining ten

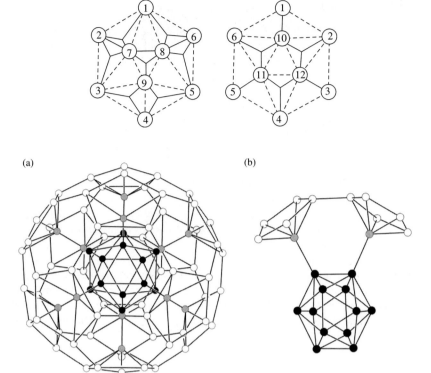

Fig. 13.2.4.
Chemical bonds in the B_{12} icosahedron (one of many possible canonical forms): three $2c$-$2e$ bonds between 1–10, 3–12, and 5–11; ten BBB $3c$-$2e$ bonds between 1–2–7, 2–3–7, 3–4–9, ….

Fig. 13.2.5.
Structure of β-R105 boron: (a) perspective view of the B_{84} unit [B_{12a}@B_{12}@B_{60}]. Black circles represent the central B_{12a} icosahedron, shaded circles represent the B atoms connecting the surface 60 B atoms to the central B_{12a} unit; (b) each vertex of inner B_{12a} is connected to a pentagonal pyramidal B_6 unit (only two adjoining B_6 units are shown).

electrons to form the intericosahedral bonds. In the structure of α-R12 boron, each icosahedron is surrounded by six icosahedra in the same layer and forms six $3c$-$2e$ bonds [see Fig. 9.6.21(b)], in which every B atom contributes 2/3 electron, so each icosahedron uses $6 \times 2/3 = 4$ electrons. Each icosahedron is bonded by normal $2c$-$2e$ B–B bonds to six icosahedra in the upper and lower layers, which need six electrons. The number of bonding electrons in an icosahedron of α-R12 boron is therefore $26 + 4 + 6 = 36$.

The β-R105 boron allotrope has a much more complex structure with 105 B atoms in the unit cell (space group $R\bar{3}m$, $a = 1014.5$ pm, $\alpha = 65.28°$). A basic building unit in the crystal structure is the B_{84} cluster illustrated in Fig. 13.2.5(a); it can be considered as a central B_{12a} icosahedron linked radially to 12 B_6 half-icosahedra (or pentagonal pyramids), each attached like an inverted umbrella to an icosahedral vertex, as shown in Fig. 13.2.5(b).

The basal B atoms of adjacent pentagonal pyramids are interconnected to form a B_{60} icosahedron that resembles fullerene-C_{60}. The large B_{60} icosahedron encloses the B_{12a} icosahedron, and the resulting B_{84} cluster can be formulated as B_{12a}@B_{12}@B_{60}. Additionally, a six-coordinate B atom lies at the center of symmetry between two adjacent B_{10} condensed units, and the crystal structure is an intricate coordination network resulting from the linkage of B_{10}–B–B_{10} units and B_{84} clusters, such that all 105 atoms in the unit cell except one are the vertices of fused icosahedra. Further details and the electronic structure of β-R105 boron are described in Section 13.4.6.

13.3 Borides

Solid borides have high melting points, exceptionally high hardness, excellent wear resistance, and good immunity to chemical attack, which make them industrially important with uses as refractory materials and in rocket cones and turbine blades. Some metal borides have been found to exhibit superconductivity.

13.3.1 Metal borides

A large number of metal borides have been prepared and characterized. Several hundred binary metal borides M_xB_y are known. With increasing boron content, the number of B–B bonds increases. In this manner, isolated B atoms, B–B pairs, fragments of boron chains, single chains, double chains, branched chains, and hexagonal networks are formed, as illustrated in Fig. 13.3.1. Table 13.3.1 summarizes the stoichiometric formulas and structures of metal borides.

In boron-rich compounds, the structure can often be described in terms of a three-dimensional network of boron clusters (octahedra, icosahedra, and cubo-octahedra), which are linked to one another directly or via non-cluster atoms.

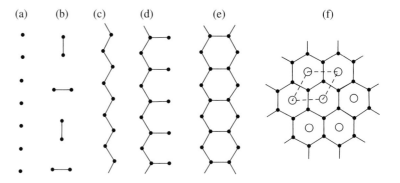

Fig. 13.3.1.
Idealized patterns of boron catenation in metal-rich borides: (a) isolated boron atoms, (b) B–B pairs, (c) single chain, (d) branched chain, (e) double chain, (f) hexagonal net in MB_2.

Table 13.3.1. Stoichiometric formulas and structure of metal borides

Formula	Example	Catenation of boron and figure showing structure
M_4B	Mn_4B, Cr_4B	
M_3B	Ni_3B, Co_3B	
M_5B_2	Pd_5B_2	isolated B atom, Fig. 13.3.1(a)
M_7B_3	Ru_7B_3, Re_7B_3	
M_2B	Be_2B, Ta_2B	
M_3B_2	V_3B_2, Nb_3B_2	B_2 pairs, Fig. 13.3.1(b)
MB	FeB, CoB	single chains, Fig. 13.3.1(c)
$M_{11}B_8$	$Ru_{11}B_8$	branched chains, Fig. 13.3.1(d)
M_3B_4	Ta_3B_4, Cr_3B_4	double chains, Fig. 13.3.1(e)
MB_2	MgB_2, AlB_2	hexagonal net, Fig. 13.3.1(f) and Fig. 13.3.2
MB_4	LaB_4, ThB_4	B_6 octahedra and B_2 Fig. 13.3.3(a)
MB_6	CaB_6	B_6 octahedra, Fig. 13.3.3(b)
MB_{12}	YB_{12}, ZrB_{12}	B_{12} cubooctahedra, Fig. 13.3.4(a)
MB_{15}	NaB_{15}	B_{12} icosahedra, Fig. 13.3.4(b) and B_3 unit
MB_{66}	YB_{66}	$B_{12} \subset (B_{12})_{12}$ giant cluster

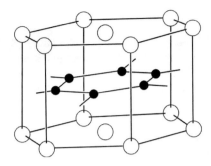

Fig. 13.3.2.
Crystal structure of MgB$_2$.

The metal atoms are accommodated in cages between the octahedra or icosahedra, thereby providing external bonding electrons to the electron-deficient boron network.

The structures of some metal borides in Table 13.3.1 are discussed below.

(1) MgB$_2$ and AlB$_2$

Magnesium boride MgB$_2$ was discovered in 2001 to behave as a superconductor at $T_c = 39$ K, and its physical properties are similar to those of Nb$_3$Sn used in the construction of high-field superconducting magnets in NMR spectrometers. MgB$_2$ crystallizes in the hexagonal space group *P6/mmm* with $a = 308.6$, $c = 352.4$ pm, and $Z = 1$. The Mg atoms are arranged in a close-packed layer, and the B atoms form a graphite-like layer with a B–B bond length of $a/(3)^{1/2} = 178.2$ pm. These two kinds of layers are interleaved along the c axis, as shown in Fig. 13.3.1(f). In the resulting crystal structure, each B atom is located inside a Mg$_6$ trigonal prism, and each Mg atom is sandwiched between two planar hexagons of B atoms that constitute a B$_{12}$ hexagonal prism, as illustrated in Fig. 13.3.2. AlB$_2$ has the same crystal structure with a B–B bond length of 175 pm.

(2) LaB$_4$

Figure 13.3.3(a) shows a projection of the tetragonal structure of LaB$_4$ along the c axis. The chains of B$_6$ octahedra are directly linked along the c axis and joined laterally by pairs of B$_2$ atoms in the ab plane to form a three-dimensional skeleton. In addition, tunnels accommodating the La atoms run along the c axis.

(3) CaB$_6$

The cubic hexaboride CaB$_6$ consists of B$_6$ octahedra, which are linked directly in all six orthogonal directions to give a rigid but open framework, and the Ca atoms occupy cages each surrounded by 24 B atoms, as shown in Fig. 13.3.3(b).

The number of valence electrons for a B$_6$ unit can be counted as follows: six electrons are used to form six B–B bonds between the B$_6$ units, and the *closo*-B$_6$ skeleton is held by three B–B 2c-2e bonds and four BBB 3c-2e bonds, which require 14 electrons. The total number is $6+14 = 20$ electrons per B$_6$ unit. Thus in the CaB$_6$ structure each B$_6$ unit requires the transfer of two electrons from metal atoms. However, the complete transfer of 2e per B$_6$ unit is not mandatory for a three-dimensional crystal structure, and theoretical calculations for MB$_6$ (M = Ca, Sr, Ba) indicate a net transfer of only 0.9e to 1.0e. This also explains

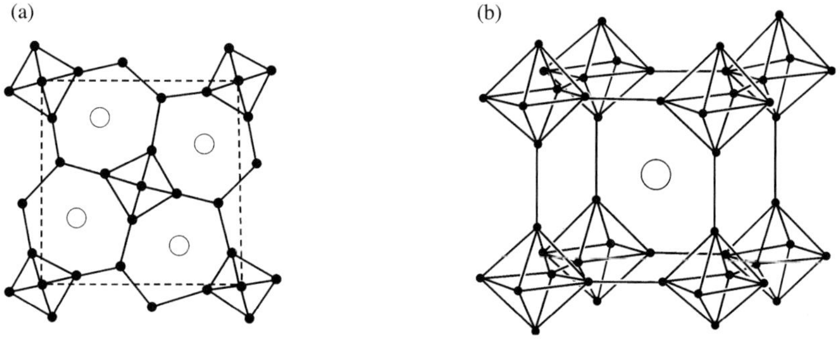

Fig. 13.3.3.
Structures of (a) LaB$_4$ and (b) CaB$_6$.

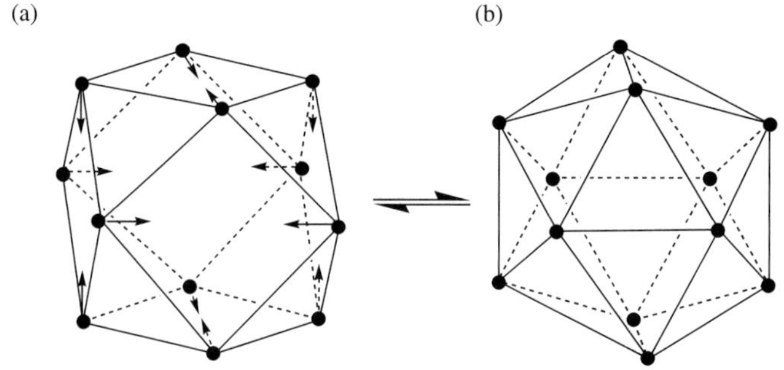

Fig. 13.3.4.
Structures of (a) cubo-octahedral B$_{12}$ unit and (b) icosahedral B$_{12}$ unit. These two units can be interconverted by the displacements of atoms. The arrows indicate shift directions of three pairs of atoms to form additional linkages (and triangular faces) during conversion from (a) to (b), and concurrent shifts of the remaining atoms are omitted for clarity.

why metal-deficient phases M$_{1-x}$B$_6$ remain stable and why the alkali metal K can also form a hexaboride.

(4) High boron-rich metal borides

In high boron-rich metal borides, there are two types of B$_{12}$ units in their crystal structures: one is cubo-octahedral B$_{12}$ found in YB$_{12}$, as shown in Fig. 13.3.4(a); the other is icosahedral B$_{12}$ in NaB$_{12}$, as shown in Fig. 13.3.4(b). These two structural types can be interconverted by small displacements of the B atoms. The arrows shown in Fig. 13.3.4(a) represent the directions of the displacements.

(5) Rare-earth metal borides

The ionic radii of the rare-earth metals have a dominant effect on the structure types of their crystalline borides, as shown in Table 13.3.2.

The "radius" of the 24-coordinate metal site in MB$_6$ is too large (M–B distances are in the range of 215-25 pm) to be comfortably occupied by the later (smaller) lanthanide elements Ho, Er, Tm, and Lu, and these elements form MB$_4$ compounds instead, where the M–B distances vary within the range of 185-200 pm.

YB$_{66}$ crystallizes in space group $Fm\bar{3}c$ with $a = 2344$ pm and $Z = 24$. The basic structural unit is a thirteen-icosahedra B$_{156}$ giant cluster in which a central B$_{12}$ icosahedron is surrounded by twelve B$_{12}$ icosahedra; the B–B bond distances within the icosahedra are 171.9–185.5 pm, and the intericosahedra

Table 13.3.2. Variation of structural types with the ionic radii of rare-earth metals

Cation	radii (pm)	Structural type				
		YB$_{12}$	AlB$_2$	YB$_{66}$	ThB$_4$	CaB$_6$
Eu^{2+}	130					+
La^{3+}	116.0				+	+
Ce^{3+}	114.3				+	+
Pr^{3+}	112.6				+	+
Nd^{3+}	110.9			+	+	+
Sm^{3+}	107.9			+	+	+
Gd^{3+}	105.3		+	+	+	+
Tb^{3+}	104.0	+	+	+	+	
Dy^{3+}	102.7	+	+	+	+	
Y^{3+}	101.9	+	+	+	+	
Ho^{3+}	101.5	+	+	+	+	
Er^{3+}	100.4	+	+	+	+	
Tm^{3+}	99.4	+	+	+	+	
Yb^{3+}	98.5	+	+	+	+	
Lu^{3+}	97.7	+	+	+	+	

distances vary between 162.4 and 182.3 pm. Packing of the eight B$_{12}$ ⊂ (B$_{12}$)$_{12}$ giant clusters in the unit cell generates channels and non-icosahedral bulges that accommodate the remaining 336 boron atoms in a statistical distribution. The yttrium atom is coordinated by twelve boron atoms belonging to four icosahedral faces and up to eight boron atoms located within a bulge.

13.3.2 Non-metal borides

Boron forms a large number of non-metal borides with oxygen and other non-metallic elements. The principal oxide of boron is boric oxide, B$_2$O$_3$. Fused B$_2$O$_3$ readily dissolves many metal oxides to give borate glasses. Its major application is in the glass industry, where borosilicate glasses find extensive use because of their small coefficient of thermal expansion and easy workability. Borosilicate glasses include Pyrex glass, which is used to manufacture most laboratory glassware. The chemistry of borates and related oxo complexes will be discussed in Section 13.5.

(1) Boron carbides and related compounds

The potential usefulness of boron carbides has prompted intensive studies of the system B$_{1-x}$C$_x$ with $0.1 \leq x \leq 0.2$. The carbides of composition B$_{13}$C$_2$ and B$_4$C, as well as the related compounds B$_{12}$P$_2$ and B$_{12}$As$_2$, all crystallize in the rhombohedral space group $R\bar{3}m$. The structure of B$_{13}$C$_2$ is derived from α-R12 boron with the addition of a linear C–B–C linking group along the [111] direction, as shown in Fig. 13.3.5. The structure of B$_4$C ($a = 520$ pm, $\alpha = 66°$) has been shown to be B$_{11}$C(CBC) rather than B$_{12}$(CCC), with statistical distribution of a C atom over the vertices of the B$_{11}$C icosahedron. For boron phosphide and boron arsenide, a two-atom P–P or As–As link replaces the three-atom chain in the carbides.

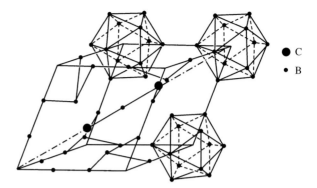

Fig. 13.3.5.
Structure of $B_{13}C_2$. The small dark circles represent B atoms and large circles represent C atoms.

(2) Boron nitrides

The common form of BN has an ordered layer structure containing hexagonal rings, as shown in Fig. 13.3.6(a). The layers are arranged so that a B atom in one layer lies directly over a N atom in the next, and vice versa. The B–N distance within a layer is 145 pm, which is shorter than the distance of 157 pm for a B–N single bond, implying the presence of π-bonding. The interlayer distance of 330 pm is consistent with van der Waals interactions. Boron nitride is a good lubricant that resembles graphite. However, unlike graphite, BN is white and an electrical insulator. This difference can be interpreted in terms of band theory, as the band gap in boron nitride is considerably greater than that in graphite because of the polarity of the B–N bond.

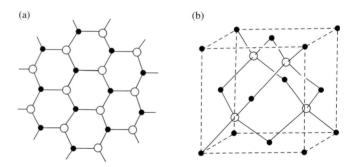

Fig. 13.3.6.
Structures of BN: (a) graphite-like; (b) diamond-like (borazon).

Heating the layered form of BN at ~2000 K and >50 kbar pressure in the presence of a catalytic amount of Li_3N or Mg_3N_2 converts it into a more dense polymorph with the zinc blende structure, as shown in Fig. 13.3.6(b). This cubic form of BN is called borazon, which has a hardness almost equal to that of diamond and is used as an abrasive.

As the BN unit is isoelectronic with a C_2 fragment, replacement of the latter by the former in various organic compounds leads to azaborane structural analogs. Some examples are shown below. Planar borazine (or borazole) $B_3N_3H_6$ is stabilized by π-delocalization, but it is much more reactive than benzene in view of the partial positive and negative charges on the N and B atoms, respectively.

(3) Boron halides

Boron forms numerous binary halides, of which the monomeric trihalides BX_3 are volatile and highly reactive. These planar molecules differ from aluminum halides, which form Al_2X_6 dimers with a complete valence octet on Al. The difference between BX_3 and Al_2X_6 is due to the smallness of the B atom. The B atom uses its sp^2 hybrid orbitals to form σ bonds with the X atoms, leaving the remaining empty $2p_z$ orbital to overlap with three filled p_z orbitals of the X atoms. This generates delocalized π orbitals, three of which are filled with electrons. Thus there is some double-bond character in the B–X bonding that stabilizes the planar monomer. Table 13.3.3 lists some physical properties of BX_3 molecules. The energy of the B–F bond in BF_3 is 645 kJ mol^{-1}, which is consistent with its multiple-bond character.

All BX_3 compounds behave as Lewis acids, the strength of which is determined by the strength of the aforementioned delocalized π bonding; the stronger it is, the molecule more easily retains its planar configuration and the acid strength is weaker. Because of the steadily weakening of π bonding in going from BF_3 to BI_3, BF_3 is a weaker acid and BI_3 is stronger. This order is the reverse of that predicated when the electronegativity and size of the halogens are considered: the high electronegativity of fluorine could be expected to make BF_3 more receptive to receiving a lone pair from a Lewis base, and the small size of fluorine would least inhibit the donor's approach to the boron atom.

Salts of tetrafluoroborate, BF_4^-, are readily formed by adding a suitable metal fluoride to BF_3. There is a significant lengthening of the B–F bond from 130 pm in planar BF_3 to 145 pm in tetrahedral BF_4^-.

Among the boron halides, B_4Cl_4 is of interest as both B_4H_4 and $B_4H_4^{2-}$ do not exist. Figure 13.3.7 shows the molecular structure of B_4Cl_4, which has a tetrahedral B_4 core consolidated by four terminal B–Cl 2c-2e bonds and four BBB 3c-2e bonds on four faces. This structure is further stabilized by σ–π interactions between the lonepairs of Cl atoms and BBB 3c-2e bonds. The

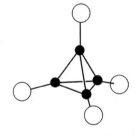

Fig. 13.3.7.
Molecular structure of B_4Cl_4

Table 13.3.3. Some physical properties of BX_3

Molecule	B–X bond length (pm)	$(r_B + r_X)^*$ (pm)	Bond energy (kJ mol^{-1})	mp (K)	bp (K)
BF_3	130	152	645	146	173
BCl_3	175	187	444	166	286
BBr_3	187	202	368	227	364
BI_3	210	221	267	323	483

*Quantities r_B and r_X represent the covalent radii of B and X atoms, respectively.

bond length of B–Cl is 170 pm, corresponding to a single bond, and the bond length of B–B is also 170 pm.

13.4 Boranes and carboranes

13.4.1 Molecular structure and bonding

Boranes and carboranes are electron-deficient compounds with interesting molecular geometries and characteristic bonding features.

(1) Molecular structure of boranes and carboranes

Figures 13.4.1 and 13.4.2 show the structures of some boranes and carboranes, respectively, that have been established by X-ray and electron diffraction methods.

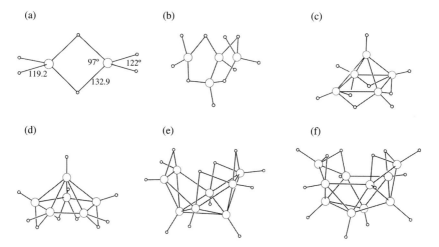

Fig. 13.4.1.
Structures of boranes: (a) B_2H_6, (b) B_4H_{10}, (c) B_5H_9, (d) B_6H_{10}, (e) B_8H_{12}, (f) $B_{10}H_{14}$.

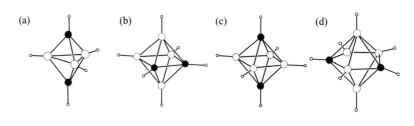

Fig. 13.4.2.
Structures of carboranes:
(a) $1,5\text{-}C_2B_3H_5$, (b) $1,2\text{-}C_2B_4H_6$,
(c) $1,6\text{-}C_2B_4H_6$, (d) $2,4\text{-}C_2B_5H_7$,
(e)–(g) three isomers of $C_2B_{10}H_{12}$
(H atoms are omitted for clarity).

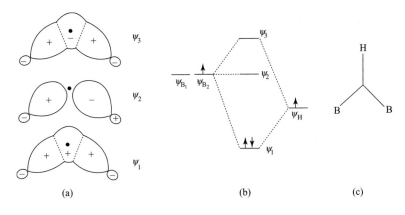

Fig. 13.4.3.
The 3c-2e BHB bond: (a) orbitals overlap, (b) relative energies, and (c) simple representation.

(2) Bonding in boranes

In simple covalent bonding theory, there are four bonding types in boranes: (a) normal 2c-2e B–B bond, (b) normal 2c-2e B–H bond, (c) 3c-2e BBB bond (described in Fig. 13.2.4), and (d) 3c-2e BHB bond.

The 3c-2e BHB bond is constructed from two orbitals of B_1 and B_2 atoms (ψ_{B_1} and ψ_{B_2}, which are sp^x hybrids) and ψ_H (1s orbital) of the H atom. The combinations of these three atomic orbitals result in three molecular orbitals:

$$\psi_1 = \frac{1}{2}\psi_{B_1} + \frac{1}{2}\psi_{B_2} + \frac{1}{(2)^{1/2}}\psi_H$$

$$\psi_2 = \frac{1}{(2)^{1/2}}(\psi_{B_1} - \psi_{B_2})$$

$$\psi_3 = \frac{1}{2}\psi_{B_1} + \frac{1}{2}\psi_{B_2} - \frac{1}{(2)^{1/2}}\psi_H.$$

The orbital overlaps and relative energies are illustrated in Fig. 13.4.3. Note that ψ_1, ψ_2, and ψ_3 are bonding, nonbonding, and antibonding molecular orbitals, respectively. For the 3c-2e BHB bond, only ψ_1 is filled with electrons.

(3) Topological description of boranes

The overall bonding in borane molecules or anions is sometimes represented by a 4-digit code introduced by Lipscomb, the so-called *styx* number, where

s is the number of 3c-2e BHB bonds,

t is the number of 3c-2e BBB bonds,

y is the number of normal B–B bonds, and

x is the number of BH_2 groups.

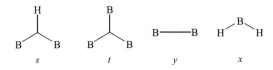

The molecular structural formulas of some boranes and their *styx* codes are shown in Fig. 13.4.4.

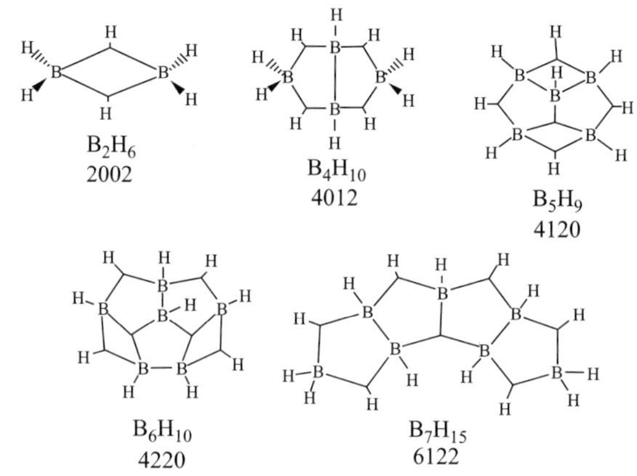

Fig. 13.4.4.
Molecular structural formulas and *styx* codes of some boranes.

The following simple rules must hold for the borane structures:

(a) Each B atom has at least a terminal H atom attached to it.
(b) Every pair of boron atoms which are geometric neighbors must be connected by a B–B, BHB, or BBB bond.
(c) Every boron atom uses four valence orbitals in bonding to achieve an octet configuration.
(d) No two boron atoms may be bonded together by both 2c-2e B–B and 3c-2e BBB bonds, or by both 2c-2e B–B and 3c-2e BHB bonds.

13.4.2 Bond valence in molecular skeletons

The geometry of a molecule is related to the number of valence electrons in it. According to the valence bond theory, organic and borane molecules composed of main-group elements owe their stability to the filling of all four valence orbitals (ns and np) of each atom, with eight valence electrons provided by the atom and those bonded to it. Likewise, transition-metal compounds achieve stability by filling the nine valence orbitals [$(n-1)$d, ns, and np] of each transition-metal atom with 18 electrons provided by the metal atom and the surrounding ligands. In molecular orbital language, the octet and 18-electron rules are rationalized in terms of the filling of all bonding and nonbonding molecular orbitals and the existence of a large HOMO-LUMO energy gap.

Consider a molecular skeleton composed of n main-group atoms, M_n, which takes the form of a chain, ring, cage, or framework. Let g be the total number of valence electrons of the molecular skeleton. When a covalent bond is formed between two M atoms, each of them effectively gains one electron in its valence shell. In order to satisfy the octet rule for the whole skeleton, $\frac{1}{2}(8n-g)$ electron pairs must be involved in bonding between the M atoms. The number of these bonding electron pairs is defined as the "bond valence" b of the molecular skeleton:

$$b = \tfrac{1}{2}(8n - g). \tag{13.4.1}$$

When the total number of valence electrons in a molecular skeleton is less than the number of valence orbitals, the formation of normal 2c-2e covalent bonds is insufficient to compensate for the lack of electrons. In this type of electron-deficient compound there are usually found 3c-2e bonds, in which three atoms share an electron pair. Thus one 3c-2e bond serves to compensate for the lack of four electrons and corresponds to a bond valence value of 2, as discussed in 13.2.

There is an enormous number of compounds that possess metal–metal bonds. A metal cluster may be defined as a polynuclear compound in which there are substantial and directed bonds between the metal atoms. The metal atoms of a cluster are also referred to as skeletal atoms, and the remaining non-metal atoms and groups are considered as ligands.

According to the 18-electron rule, the bond valence of a transition metal cluster is

$$b = \tfrac{1}{2}(18n - g). \tag{13.4.2}$$

If the bond valence b calculated from (13.4.1) and (13.4.2) for a cluster M_n matches the number of connecting lines drawn between pairs of adjacent atoms in a conventional valence bond structural formula, the cluster is termed "electron-precise."

For a molecular skeleton consists of n_1 transition-metal atoms and n_2 main-group atoms, such as transition-metal carboranes, its bond valence is

$$b = \tfrac{1}{2}(18n_1 + 8n_2 - g). \tag{13.4.3}$$

The bond valence b of a molecular skeleton can be calculated from expressions (13.4.1) to (13.4.3), in which g represents the total number of valence electrons in the system. The g value is the sum of:

(1) the number of valence electrons of the n atoms that constitute the molecular skeleton M_n,
(2) the number of electrons donated by the ligands to M_n, and
(3) the number of net positive or negative charges carried by M_n, if any.

The simplest way to count the g value is to start from uncharged skeletal atoms and uncharged ligands. Ligands such as NH_3, PR_3, and CO each supply two electrons. Non-bridging halogen atoms, H atom, and CR_3 and SiR_3 groups are one-electron donors. A μ_2-bridging halogen atom contributes three electrons, and a $\mu_3$bridging halogen atom donates five electrons. Table 13.4.1 lists the number of electrons contributed by various ligands, depending on their coordination modes.

13.4.3 Wade's rules

For boranes and carboranes, the terms *closo*, *nido*, *arachno*, and *hypho* are used to describe their molecular skeletons with reference to a series of increasingly open deltahedra. Boranes of the *nido*, *arachno*, and *hypho* types are compounds based on formulas B_nH_{n+4}, B_nH_{n+6}, and B_nH_{n+8}, respectively. The structures

Table 13.4.1. Number of electrons supplied by ligands to a molecular skeleton

Ligand	Coordinate mode*	No. of electrons	Ligand	Coordinate mode*	No. of electrons
H	μ_1, μ_2, μ_3	1	NR_3, PR_3	μ_1	2
B	int	3	NCR	μ_1	2
CO	μ_1, μ_2, μ_3	2	NO	μ_1	3
CR	μ_3, μ_4	3	OR, SR	μ_1	1
CR_2	μ_1	2	OR, SR	μ_2	3
CR_3, SiR_3	μ_1, μ_2	1	O, S, Se, Te	μ_2	2
η^2–C_2R_2	μ_1	2	O, S, Se, Te	μ_3	4
η^2–C_2R_4	μ_1	2	O, S	int	6
η^5–C_5R_5	μ_1	5	F, Cl, Br, I	μ_1	1
η^6–C_6R_6	μ_1	6	F, Cl, Br, I	μ_2	3
C, Si	int	4	Cl, Br, I	μ_3	5
N, P, As, Sb	int	5	PR	μ_3, μ_4	4

The skeletal atoms are considered to be uncharged.
*μ_1 = terminal ligand, μ_2 = ligand bridging two atoms, μ_3 = ligand bridging three atoms, int = interstitial atom.

of *nido*-boranes are analogous to the corresponding *closo* compounds with the exception that one vertex is missing. For example, in B_5H_9, five B atoms are located at the vertices of an octahedron whereas the sixth vertex is empty. Likewise, the arrangement of *n* skeletal atoms of *arachno* compounds is referred to a deltahedron having $(n + 2)$ vertices; for instance, the B atoms in B_4H_{10} lie at four vertices of an octahedron in which two adjacent vertices are not occupied. For hypothetical B_nH_{n+8} and its isoelectronic equivalent $B_nH_{n+6}^{2-}$, the cluster skeleton is a *n*-vertex deltahedron with three missing vertices.

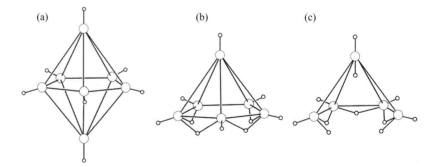

Fig. 13.4.5.
Structural relationship between (a) *closo*-$B_7H_7^{2-}$, (b) *nido*-B_6H_{10} and (c) *arachno*-B_5H_{11}.

Figure 13.4.5 shows the relations between (a) *closo*-$B_7H_7^{2-}$, (b) *nido*-B_6H_{10}, and (c) *arachno*-B_5H_{11}. The structure of *closo*-$B_7H_7^{2-}$ is a pentagonal bipyramid. In *nido*-B_6H_{10}, the B atoms are located at six vertices of the pentagonal bipyramid, and one vertex on the C_5 axis is not occupied. In *arachno*-B_5H_{11}, two vertices of the pentagonal bipyramid, one axial and one equatorial, are not occupied.

Historically, a scheme of skeletal electron-counting was developed to rationalize the structures of boranes and their derivatives, to which the following Wade's rules are applicable.

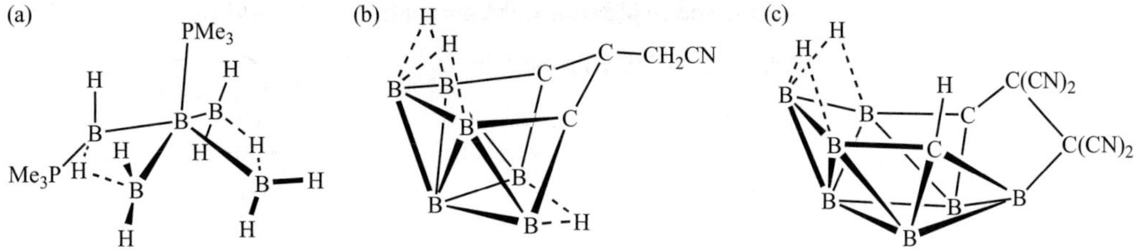

Fig. 13.4.6.
Molecular structures of some hypho cluster systems: (a) $B_5H_9(PMe_3)_2$; (b) $(NCCH_2)C_3B_6H_{12}^-$; (c) $[\mu^2-\{C(CN)_2\}_2]C_2B_7H_{12}^-$. In carboranes (b) and (c), for clarity only bridging and extra H atoms are included, and all *exo* H atoms attached to the vertices are omitted.

(1) A *closo*-deltahedral cluster with n vertices is held together by $(n+1)$ bonding electron pairs.
(2) A *nido*-deltahedral cluster with n vertices is held together by $(n+2)$ bonding electron pairs.
(3) An *arachno*-deltahedral cluster with n vertices is held together by $(n+3)$ bonding electron pairs.
(4) A *hypho*-deltahedral cluster with n vertices is held together by $(n+4)$ bonding electron pairs.

Boranes and carboranes of the *hypho* type are quite rare. The prototype *hypho*-borane $B_5H_{11}^{2-}$ has been isolated, and its proposed structure is analogous to that of the isoelectronic analog $B_5H_9(PMe_3)_2$ (Fig. 13.4.6(a)), which was established by X-ray crystallography. The 9-vertex *hypho* tricarbaborane cluster anion $(NCCH_2)$-1,2,5-$C_3B_6H_{12}^-$ shown in Fig. 13.4.6(b) has 13 skeletal pairs, and accordingly it may be considered as derived from an icosahedron with three missing vertices. Similarly, in *endo*-6-*endo*-7-$[\mu^2-\{C(CN)_2\}_2]$-*arachno*-6,8-$C_2B_7H_{12}^-$ [Fig. 13.4.6(c)], the C_2B_7-fragment of the cluster anion can be viewed as a 13-electron pair, 9-vertex *hypo* system with an *exo*-polyhedral TCNE substituent bridging a C atom and a B atom at the open side.

13.4.4 Chemical bonding in *closo*-boranes

Boranes of the general formula $B_nH_n^{2-}$ and isoelectronic carboranes such as $CB_{n-1}H_n^-$ and $C_2B_{n-2}H_n$ have *closo* structures, in which n skeletal B and C atoms are located at the vertices of a polyhedron bounded by trianglular faces (deltahedron). For the $B_nH_n^{2-}$ *closo*-boranes

$$b = \tfrac{1}{2}[8n - (4n+2)] = 2n - 1. \qquad (13.4.4)$$

From the geometrical structures of *closo*-boranes, s and x of the *styx* code are both equal to 0. Each 3c–2e BBB bond has a bond valence value of 2. Thus,

$$b = 2t + y. \qquad (13.4.5)$$

Combining expressions (13.4.4) and (13.4.5) leads to

$$2t + y = 2n - 1. \tag{13.4.6}$$

When the number of electron pairs of the molecular skeleton is counted, one gets

$$t + y = n + 1. \tag{13.4.7}$$

Hence from (13.4.6) and (13.4.7),

$$t = n - 2, \tag{13.4.8}$$

$$y = 3, \tag{13.4.9}$$

so that each *closo*-borane $B_nH_n^{2-}$ has exactly three B–B bonds in its valence bond structural formula.

Table 13.4.2. Bond valence and other parameters for $B_nH_n^{2-}$

n in $B_nH_n^{2-}$	styx code	Bond valence b	No. of edges = $3t$	No. of faces = $2t$	$t + y$ (No. of skeletal electron pairs)
5	0330	9	9	6	6
6	0430	11	12	8	7
7	0530	13	15	10	8
8	0630	15	18	12	9
9	0730	17	21	14	10
10	0830	19	24	16	11
11	0930	21	27	18	12
12	0,10,3,0	23	30	20	13

Table 13.4.2 lists the bond valence of $B_nH_n^{2-}$ and other parameters; n starts from 5, not from 4, as $B_4H_4^{2-}$ would need $t = 2$ and $y = 3$, which cannot satisfy rule (d) for a tetrahedron. Figure 13.4.7 shows the localized bonding

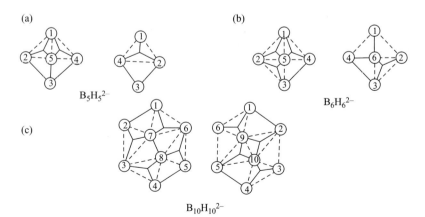

Fig. 13.4.7.
The B–B and BBB bonds in (a) $B_5H_5^{2-}$, (b) $B_6H_6^{2-}$, and (c) $B_{10}H_{10}^{2-}$; for each molecule both the front and back sides of a canonical structure are displayed. The molecular skeleton of $B_{12}H_{12}^{2-}$ is shown in Fig.13.2.2, and the bonding is shown in Fig.13.2.4.

description (in one canonical form) for some *closo*-boranes. The charge of -2 in $B_nH_n^{2-}$ is required for consolidating the n BH units into a deltahedron. Except for $B_5H_5^{2-}$, the bond valence b is always smaller than the number of edges of the polyhedron.

In the B_n skeleton of *closo*-boranes, there are $(n-2)$ 3c-2e BBB bonds and three normal B–B bonds, which amounts to $(n-2)+3 = n+1$ electron pairs.

Some isoelectronic species of $C_2B_{10}H_{12}$, such as $CB_{11}H_{12}^-$, $NB_{11}H_{12}$, and their derivatives, are known. Since they have the same number of skeletal atoms and the same bond valence, these compounds are isostructural.

13.4.5 Chemical bonding in *nido*- and *arachno*-boranes

Using the *styx* code and bond valence to describe the structures of *nido*-boranes or carboranes, the bond valence b is equal to

$$b = \tfrac{1}{2}[8n - (4n+4)] = 2n - 2, \quad (13.4.10)$$

$$b = s + 2t + y. \quad (13.4.11)$$

Combining expressions (13.4.10) and (13.4.11) leads to

$$s + 2t + y = 2n - 2. \quad (13.4.12)$$

The number of valence electron pairs for *nido* B_nH_{n+4} is equal to $\tfrac{1}{2}(4n+4) = 2n + 2$, in which $n+4$ pairs are used to form B–H bonds and the remaining electron pairs used to form 2c–2e B–B bonds (y) and 3c–2e BBB bonds (t). Thus,

$$t + y = (2n+2) - (n+4) = n - 2. \quad (13.4.13)$$

The number of each bond type in *nido*-B_nH_{n+4} are obtained from expressions (13.4.12) and (13.4.13):

$$t = n - s, \quad (13.4.14)$$

$$y = s - 2, \quad (13.4.15)$$

$$x + s = 4. \quad (13.4.16)$$

The stability of a molecular species with open skeleton may be strengthened by forming 3c-2e bond(s). In a stable *nido*-borane, neighboring H–B–H and B–H groups always tend to be converted into the H–BHB–H system:

Table 13.4.3. The *styx* codes of some *nido*-boranes

	Borane						
	B_4H_8	B_5H_9	B_6H_{10}	B_7H_{11}	B_8H_{12}	B_9H_{13}	$B_{10}H_{14}$
s	4	4	4	4	4	4	4
t	0	1	2	3	4	5	6
y	2	2	2	2	2	2	2
x	0	0	0	0	0	0	0
$b = s + 2t + y$	6	8	10	12	14	16	18

Thus, in *nido*-borane, $x = 0$, $s = 4$, and

$$t = n - 4 \qquad (13.4.17)$$
$$y = 2. \qquad (13.4.18)$$

Table 13.4.3 lists the *styx* code and bond valence of some *nido*-boranes, and Fig. 13.4.8 shows the chemical bonding in their skeletons.

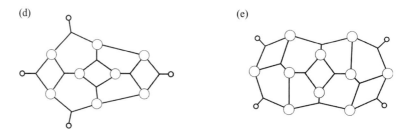

Fig. 13.4.8.
Chemical bonding in the skeletons of some *nido*-boranes (large circles represent BH group, small circles represent H atoms in BHB bonds):
(a) B_4H_8 (4020), (b) B_5H_9 (4120),
(c) B_6H_{10} (4220), (d) B_8H_{12} (4420),
(e) $B_{10}H_{14}$ (4620).

For the *arachno*-boranes, B_nH_{n+6}, the relations between *styx* code and n are as follows:

$$t = n - s \qquad (13.4.19)$$
$$y = s - 3 \qquad (13.4.20)$$
$$x + s = 6. \qquad (13.4.21)$$

According to the octet and 18-electron rules, when one C atom replaces one BH group in a borane, the g value and b value are unchanged, and the structure

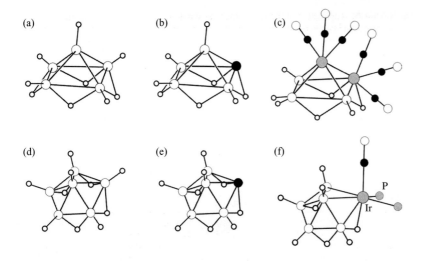

Fig. 13.4.9.
Structures of corresponding boranes, carboranes, and metalloboranes having the same bond valence: (a) B_5H_9, (b) CB_4H_8, and (c) $B_3H_7[Fe(CO)_3]_2$, $b = 8$; (d) B_6H_{10}, (e) CB_5H_9, and (f) $B_5H_8Ir(CO)(PPh_3)_2$, $b = 10$.

of the carborane is the same as that of the borane. Likewise, when one transition-metal atom replaces one BH group, the bond valence counting of (13.4.3) also remains unchanged. Figure 13.4.9 shows two series of compounds, which have the same b value in each series.

13.4.6 Electron-counting scheme for macropolyhedral boranes: *mno* rule

A generalized electron-counting scheme, known as the *mno* rule, is applicable to a wide range of polycondensed polyhedral boranes and heteroboranes, metallaboranes, metallocenes, and any of their combinations. According to this *mno* rule, the number of electron pairs N necessary for a macropolyhedral system to be stable is

$$N = m + n + o + p - q,$$

where

m = the number of polyhedra,
n = the number of vertices,
o = the number of single-vertex-sharing connections,
p = the number of missing vertices, and
q = the number of capping vertices.

For a *closo* macropolyhedral borane cluster, the rule giving the required number of electron pairs is $N = m + n$.

Some examples illustrating the application of the *mno* rule are given in Table 13.4.4, which makes reference to the structures of compounds shown in Fig. 13.4.10.

$B_{20}H_{16}$ is composed of two polyhedra sharing four atoms. The $(m + n)$ electron pair count is $2 + 20 = 22$, which is appropriate for a *closo* structure stabilized by 22 skeletal electron pairs (16 from 16 BH groups and 6 from four B

Table 13.4.4. Application of the *mno* rule for electron counting in some condensed polyhedral boranes and related compounds*

Formula	Structure in Fig. 3.4.10	m	n	o	p	q	N	BH	B	CH	H_b	α	β	x
$B_{20}H_{16}$	(a)	2	20	0	0	0	22	16	6	0	0	0	0	0
$(C_2B_{10}H_{11})_2$	(b)	2	24	0	0	0	26	18	2	6	0	0	0	0
$[(C_2B_9H_{11})_2Al]^-$	(c)	2	23	1	0	0	26	18	0	6	0	1.5	0	0.5
$[B_{21}H_{18}]^-$	(d)	2	21	0	0	0	23	18	4.5	0	0	0	0	0.5
Cp_2Fe	(e)	2	11	1	2	0	16	0	0	15	0	1	0	0
$Cp^*IrB_{18}H_{20}$	(f)	3	24	1	3	0	31	15	4.5	7.5	2.5	1.5	0	0
$[Cp^*IrB_{18}H_{19}S]^-$	(g)	3	25	1	3	0	32	16	3	7.5	1.5	1.5	2	0.5
$Cp_2^*Rh_2S_2B_{15}H_{14}(OH)$	(h)	4	29	2	5	0	40	13	3	15	2	3	4	0
$(CpCo)_3B_4H_4$	(i)	4	22	3	3	1	31	4	0	22.5	0	4.5	0	0

*BH = number of electron pairs from BH groups, B = number of electron pairs from B atoms, CH = number of electron pairs from CH groups, α = number of electron pairs from metal center(s), β = number of electron pairs from main-group hetero atom, H_b = number of electron pairs from bridging hydrogen atoms, x = number of electron pairs from the charge. Also, BH (or CH) can be replaced by BR (or CR).

atoms). In $(C_2B_{10}H_{11})_2$, which consists of two icosahedral units connected by a B–B single bond, the electron count is simply twice that of a single polyhedron. The 26 skeletal electron pairs are provided by 18 BH groups (18 pairs), 2 B atoms (2 pairs, B–B bond not involved in cluster bonding), and 4 CH groups (6 pairs, 3 electrons from each CH group). The structure of $[(C_2B_9H_{11})_2Al]^-$ is constructed from condensation of two icosahedra through a common vertex. The required skeletal electron pairs are contributed by 18 BH groups (18 pairs), 4 CH groups (6 pairs), the Al atom (1.5 pairs), and the anionic charge (0.5 pair). The polyhedral borane anion $B_{21}H_{18}^-$ has a shared triangular face. Its 18 BH groups, 3 B atoms, and the negative charge provide $18 + 4.5 + 0.5 = 23$ skeletal pairs, which fit the *mno* rule with $m+n = 2+21$. For ferrocene, which has 16 electron pairs (15 from 10 CH groups and 1 from Fe), the *mno* rule suggests a molecular skeleton with two open (*nido*) faces ($m + n + o + p = 2 + 11 + 1 + 2 = 16$). The metalloborane $[Cp^*IrB_{18}H_{20}]$ is condensed from three *nido* units. The *mno* rule gives 31 electron pairs, and the molecular skeleton is stabilized by electron pairs from 15 BH groups (15), 5 CH groups (7.5), 3 shared B atoms (4.5), 5 bridging H atoms (2.5), and the Ir atom $[\frac{1}{2}(9-6)] = 1.5$; 6 electrons occupying 3 nonbonding metal orbitals]. Addition of a sulfur atom as a vertex to the above cluster leads to $[Cp^*IrB_{18}H_{19}S]^-$, which requires 32 electron pairs. The electrons are supplied by the S atom (4 electrons, as the other 2 valence electrons occupy an *exo*-cluster orbital), 16 BH groups, 2 shared B atoms, 3 bridging H atoms, the Ir atom, and the negative charge. The cluster compound $[Cp_2^*Rh_2S_2B_{15}H_{14}(OH)]$ contains a *nido* {$RhSB_8H_7$} unit and an *arachno* {$RhSB_9H_8(OH)$} unit conjoined by a common B–B edge. The *mno* rule leads to 40 electron pairs, which are supplied by 13 BH groups (13), 2 B atoms (3), 10 CH groups (15), 4 bridging H atoms (2), 2 Rh atoms (3), and 2 S atoms (4). In the structure of $(CpCo)_3B_4H_4$, 1 B atom caps the Co_3 face of a Co_3B_3 octahedron, and the 31 electron pairs are supplied by 4 BH groups (4), 15 CH groups (22.5), and 3 Co atoms (4.5).

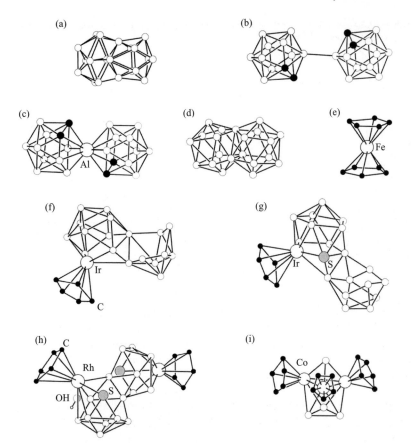

Fig. 13.4.10.
Structure of condensed polyhedral boranes, metallaboranes, and ferrocene listed in Table 13.4.4: (a) $B_{20}H_{16}$, (b) $(C_2B_{10}H_{11})_2$, (c) $[(C_2B_9H_{11})_2Al]^-$, (d) $[B_{21}H_{18}]^-$, (e) Cp^*_2Fe, (f) $Cp^*IrB_{18}H_{20}$, (g) $[Cp^*IrB_{18}H_{19}S]^-$, (h) $Cp^*_2Rh_2S_2B_{15}H_{14}(OH)$, and (i) $(CpCo)_3B_4H_4$. (The Me group of Cp^* and H atoms is not shown.)

13.4.7 Electronic structure of β-rhombohedral boron

In Section 13.2 and Fig. 13.2.5, the structure of β-R105 boron is described in terms of large B_{84} ($B_{12a}@B_{12}@B_{60}$) clusters and B_{10}–B–B_{10} units. Each B_{10} unit of C_{3v} symmetry is fused with three B_6 half-icosahedra from three adjacent B_{84} clusters, forming a B_{28} cluster composed of four fused icosahedra, as shown in Fig. 13.4.11(a). A pair of such B_{28} clusters are connected by a six-coordinate B atom to form a B_{57} (B_{28}–B–B_{28}) unit.

To understand the subtlety and electronic structure of the complex covalent network of β-R105 boron, it is useful to view the contents of the unit cell [Fig. 13.4.11(b)] as assembled from B_{12a} (at the center of a B_{84} cluster) and B_{57} (lying on a C_3 axis) fragments connected by 2c-2e bonds.

The electronic requirement of the B_{57} unit can be assessed by saturating the dangling valencies with hydrogen atoms, yielding the molecule $B_{57}H_{36}$. According to the *mno* rule, $8 + 57 + 1 = 66$ electron pairs are required for stability, but the number available is 67.5 [36 (from 36 BH) + (21 × 3)/2 (from 21B)]. Thus the $B_{57}H_{36}$ polyhedral skeleton needs to get rid of three electrons to achieve stability, whereas the $B_{12a}H_{12}$ skeleton requires two additional electrons for sufficiency. As the unit cell contains four B_{12a} and one B_{57} fragments, the idealized structure of β-rhombohedral boron has a net deficiency of five electrons.

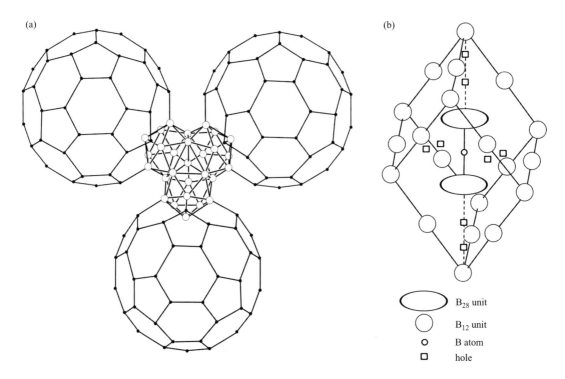

Fig. 13.4.11.
Structure of β-R105 boron. (a) Three half-icosahedra of adjacent B_{84} clusters fuse with a B_{10} unit to form a B_{28} unit. (b) Unit cell depicting the B_{12} units at the vertices and edge centers, leaving the B_{28}–B–B_{28} unit along the main body diagonal. Holes of three different types are also indicated.

X-ray studies have established that β-R105 boron has a very porous (only 36% of space is filled in the idealized model) and defective structure with the presence of interstitial atoms and partial occupancies. The B_{57} fragment can dispose of excess electrons by removal of some vertices to form *nido* or *arachno* structures, and individual B_{12a} units can gain electrons by incorporating capping vertices that are accommodated in interstitial holes (see Fig. 13.4.11(b)).

13.4.8 Persubstituted derivatives of icosahedral borane $B_{12}H_{12}^{2-}$

The compound $Cs_8[B_{12}(BSe_3)_6]$ is prepared from the solid-state reaction between Cs_2Se, Se, and B at high temperature. In the anion $[B_{12}(BSe_3)_6]^{8-}$, each trigonal planar BSe_3 unit is bonded to a pair of neighboring B atom in the B_{12} skeleton, as shown in Fig. 13.4.12. It is the first example of bonding between a chalcogen and an icosahedral B_{12} core. The point group of $[B_{12}(BSe_3)_6]^{8-}$ is D_{3d}.

Icosahedral $B_{12}H_{12}^{2-}$ can be converted to $[B_{12}(OH)_{12}]^{2-}$, in which the 12 hydroxy groups can be substituted to form carboxylic acid esters $[B_{12}(O_2CMe)_{12}]^{2-}$, $[B_{12}(O_2CPh)_{12}]^{2-}$, and $[B_{12}(OCH_2Ph)_{12}]^{2-}$. The sequential two-electron oxidation of $[B_{12}(OCH_2Ph)_{12}]^{2-}$ with Fe^{3+} in ethanol affords *hyper closo*-$B_{12}(OCH_2Ph)_{12}$ as a dark-orange crystal. An X-ray diffraction study of *hypercloso*-$B_{12}(OCH_2Ph)_{12}$ revealed that the molecule has distorted icosahedral

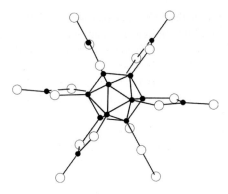

Fig. 13.4.12.
Structure of $[B_{12}(BSe_3)_6]^{8-}$.

geometry (D_{3d}). The term *hypercloso* refers to an *n*-vertex polyhedral structure having fewer skeletal electron pairs than $(n + 1)$ as required by Wade's rule.

13.4.9 Boranes and carboranes as ligands

Many boranes and carboranes can act as very effective polyhapto ligands to form metallaboranes and metallacarboranes. Metallaboranes are borane cages containing one or more metal atoms in the skeletal framework. Metallacarboranes have both metal and carbon atoms in the cage skeleton. In contrast to the metallaboranes, syntheses of metallacarboranes via low- or room-temperature metal insertion into carborane anions in solution are more controllable, usually occurring at a well-defined C_2B_n open face to yield a single isomer.

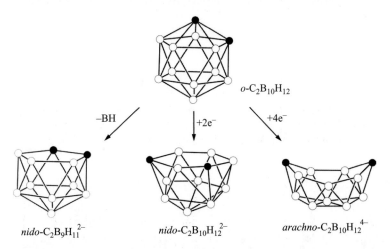

Fig. 13.4.13.
Conversion of icosahedral carborane o-$C_2B_{10}H_{12}$ to nido-$C_2B_9H_{11}^{2-}$, nido-$C_2B_{10}H_{12}^{2-}$, and arachno-$C_2B_{10}H_{12}^{4-}$.

The icosahedral carborane cage $C_2B_{10}H_{12}$ can be converted to the nido-$C_2B_9H_{11}^{2-}$, nido-$C_2B_{10}H_{12}^{2-}$, or arachno-$C_2B_{10}H_{12}^{4-}$ anions, as shown in Fig. 13.4.13. Nido-$C_2B_9H_{11}^{2-}$ has a planar five-membered C_2B_3 ring with delocalized π orbitals, which is similar to the cyclopentadienyl ring. Likewise, nido-$C_2B_{10}H_{12}^{2-}$ has a nearly planar C_2B_4 six-membered ring and delocalized π orbitals, which is analogous to benzene. Arachno-$C_2B_{10}H_{12}^{4-}$ has a boat-like C_2B_5 bonding face in which the five B atoms are coplanar and the two C atoms lie approximately 60 pm above this plane.

The similarity between the C_2B_3 open face in $nido$-$C_2B_9H_{11}{}^{2-}$ and the $C_5H_5{}^-$ anion implies that the dicarbollide ion can act as an η^5-coordinated ligand to form sandwich complexes with metal atoms, which are analogous to the metallocenes. Likewise, the C_2B_4 open face in $nido$-$C_2B_{10}H_{12}{}^{2-}$ is analogous to benzene and it may be expected to function as an η^6-ligand.

A large number of metallacarboranes and polyhedral metallaboranes of s-, p-, d-, and f-block elements are known. In these compounds, the carboranes and boranes act as polyhapto ligands. The domain of metallaboranes and metallacarboranes has grown enormously and engenders a rich structural chemistry. Some new advances in the chemistry of metallacarboranes of f-block elements are described below.

The full-sandwich lanthanacarborane $[Na(THF)_2][(\eta^5\text{-}C_2B_9H_{11})_2La(THF)_2]$ has been prepared by direct salt metathesis between $Na_2[C_2B_9H_{11}]$ and $LaCl_3$ in THF. The structure of the anion is shown in Fig. 13.4.14(a). The average La-cage atom distance is 280.4 pm, and the ring centroid–La–ring centroid angle is 132.7°.

The structure of samaracarborane $[\eta^5\!:\!\eta^6\text{-}Me_2Si(C_5H_4)(C_2B_{10}H_{11})]Sm(THF)_2$ is shown in Fig. 13.4.14(b), in which the Sm^{3+} ion is surrounded by an η^5-cyclopentadienyl ring, η^6-hexagonal C_2B_4 face of the $C_2B_{10}H_{11}$ cage, and two THF molecules in a distorted tetrahedral manner. The ring centroid–Sm–ring centroid angle is 125.1° and the average Sm–cage atom and Sm–C(C_5 ring) distances are 280.3 and 270.6 pm, respectively.

Figure 13.4.14(c) shows the coordination environment of the Er atom in the $\{\eta^7\text{-}[(C_6H_5CH_2)_2(C_2B_{10}H_{10})]Er(THF)\}_2\cdot\{Na(THF)_3\}_2\cdot 2THF$. The Er atom is coordinated by η^7-$[arachno\text{-}(C_6H_5CH_2)_2C_2B_{10}H_{10}]^{4-}$ and σ-bonded to two

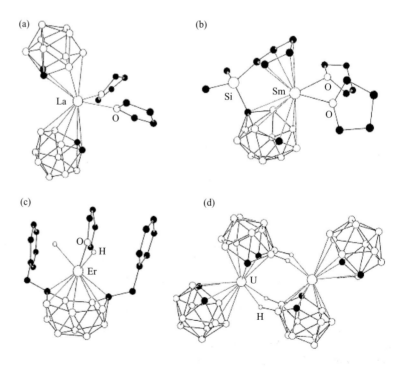

Fig. 13.4.14.
Structure of some metallacarboranes of f-block elements:
(a) $[\eta^5\text{-}C_2B_9H_{11}]_2La(THF)_2]^-$,
(b) $[\eta^5\!:\!\eta^6\text{-}Me_2Si(C_5H_4)(C_2B_{10}H_{11})]Sm(THF)_2$,
(c) $\{[\eta^7\text{-}C_2B_{10}H_{10}(CH_2C_6H_5)_2]Er(THF)^-\}$,
(d) $[(\eta^7\text{-}C_2B_{10}H_{12})(\eta^6\text{-}C_2B_{10}H_{12})U]_2{}^{4-}$.

B–H groups from a neighboring [$arachno$-(C$_6$H$_5$CH$_2$)$_2$C$_2$B$_{10}$H$_{10}$]$^{4-}$ ligand and one THF molecule. The average Er–B(cage) distance is 266.5 pm, and that for the Er–C(cage) is 236.6 pm.

The compound [{(η^7-C$_2$B$_{10}$H$_{12}$)(η^6-C$_2$B$_{10}$H$_{12}$)U}$_2${K$_2$(THF)$_5$}]$_2$ has been prepared from the reaction of o-C$_2$B$_{10}$H$_{12}$ with excess K metal in THF, followed by treatment with a THF suspension of UCl$_4$. In this structure each U^{4+} ion is bonded to η^6-$nido$-C$_2$B$_{10}$H$_{12}^{2-}$ and η^7-$arachno$-C$_2$B$_{10}$H$_{12}^{4-}$, and coordinated by two B–H groups from the C$_2$B$_5$ bonding face of a neighboring $arachno$-C$_2$B$_{10}$H$_{12}^{4-}$ ligand, as shown in Fig. 13.4.14(d). This is the first example of an actinacarborane bearing a η^6- C$_2$B$_{10}$H$_{12}^{2-}$ ligand.

13.4.10 Carborane skeletons beyond the icosahedron

Recent synthetic studies have shown that carborane skeletons larger than the icosahedron can be constructed by the insertion of additional BH units. The reaction scheme and structure motifs of these carborane species are shown in Fig. 13.4.15. Treatment of 1,2-(CH$_2$)$_3$-1,2-C$_2$B$_{10}$H$_{10}$ (**1**) with excess lithium metal in THF at room temperature gave {[(CH$_2$)$_3$-1,2-C$_2$B$_{10}$H$_{10}$][Li$_4$(THF)$_5$]}$_2$ (**2**) in 85% yield. X-ray analysis showed that, in the crystalline state, both the hexagonal and pentagonal faces of the "carbon-atoms-adjacent" $arachno$-carborane tetra-anion in **2** are capped by lithium ions, which may in principle be substituted by BH groups.

The reaction of **2** with 5.0 equivalents of HBBr$_2$·SMe$_2$ in toluene at −78 °C to 25 °C yielded a mixture of a 13-vertex $closo$-carborane (CH$_2$)$_3$C$_2$B$_{11}$H$_{11}$ (**3**, 32%), a 14-vertex $closo$-carborane (CH$_2$)$_3$C$_2$B$_{12}$H$_{12}$ (**4**, 7%); **1** (2%). $Closo$-carboranes **3** and **4** can be converted through reduction reaction into their $nido$-carborane salts {[(CH$_2$)$_3$C$_2$B$_{11}$H$_{11}$][Na$_2$(THF)$_4$]}$_n$ (**5**) and {[(CH$_2$)$_3$C$_2$B$_{12}$H$_{12}$][Na$_2$(THF)$_4$]}$_n$ (**6**), respectively. Compound **4** can also be prepared from the reaction of **5** with HBBr$_2$·SMe$_2$, whose cage skeleton is a bicapped hexagonal antiprism.

Quite different from the planar open faces in $nido$-C$_2$B$_9$H$_{11}^{2-}$ and $nido$-C$_2$B$_{10}$H$_{12}^{2-}$, the open faces of the 13- and 14-vertex $nido$ cages in **5** and **6** are

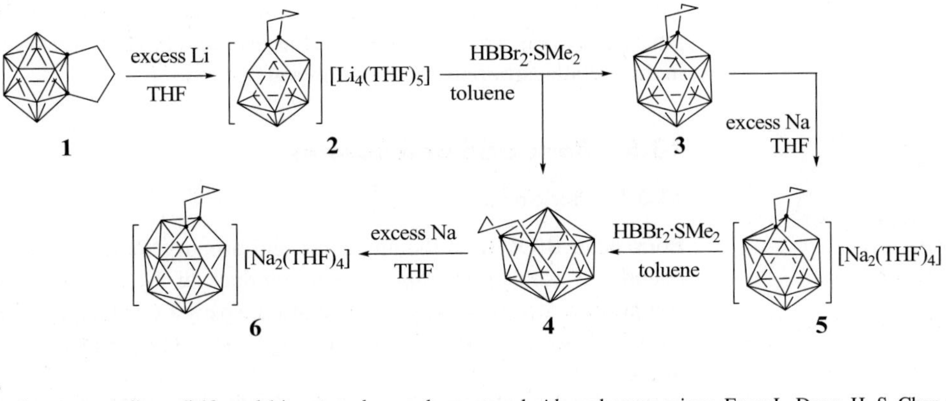

Fig. 13.4.15.
Synthesis of "carbon-atoms-adjacent" 13- and 14-vertex $closo$-carboranes and $nido$-carborane anions. From L. Deng, H.-S. Chan and Z. Xie, *Angew. Chem. Int. Ed.* **44**, 2128–31 (2005).

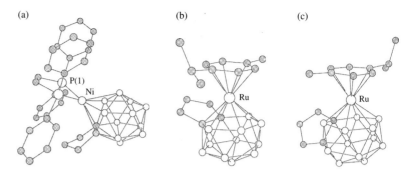

Fig. 13.4.16.
Structure of some metallacarboranes with 14 and 15 vertices:
(a) [η^5-(CH$_2$)$_3$C$_2$B$_{11}$H$_{11}$]Ni(dppe); (b) [η^6-(CH$_2$)$_3$C$_2$B$_{11}$H$_{11}$]Ru(p-cymene);
(c) [η^6-(CH$_2$)$_3$C$_2$B$_{12}$H$_{12}$]Ru(p-cymene). From L. Deng, H.-S. Chan and Z. Xie, *J. Am. Chem. Soc.* **128**, 5219–30 (2006); and L. Deng, J. Zhang, H.-S. Chan and Z. Xie, *Angew. Chem. Int. Ed.* **45**, 4309–13 (2006).

bent five-membered rings. Despite this, they can be capped by metal atoms to give metallacarboranes with 14 and 15 vertices, respectively. Figure 13.4.16 shows some examples of these novel metallacarboranes.

The complexes [η^5-(CH$_2$)$_3$C$_2$B$_{11}$H$_{11}$]Ni(dppe) and [η^6-(CH$_2$)$_3$C$_2$B$_{11}$H$_{11}$]Ru(p-cymene) were prepared by the reactions of **5** with (dppe)NiCl$_2$ and [(p-cymene)RuCl$_2$]$_2$, respectively. Although both of them are 14-vertex metallacarboranes having a similar bicapped hexagonal antiprismatic cage geometry, the coordination modes of their carborane anions are quite different: the carborane anion [(CH$_2$)$_3$C$_2$B$_{11}$H$_{11}$]$^{2-}$ is η^5-bound to the Ni atom in the nickelacarborane [Fig. 13.4.16(a)], whereas an η^6-bonding fashion is observed in the ruthenacarborane with an average Ru-cage atom distance of 226.6 pm [Fig. 13.4.16(b)].

Figure 13.4.16(c) shows the structure of a 15-vertex ruthenacarborane, [(CH$_2$)$_3$C$_2$B$_{12}$H$_{12}$]Ru(p-cymene), which was synthesized by the reaction of **6** with [(p-cymene)RuCl$_2$]$_2$. Similar to that in the above-mentioned 14-vertex ruthenacarborane, the carborane moiety [(CH$_2$)$_3$C$_2$B$_{12}$H$_{12}$]$^{2-}$ in this 15-vertex ruthenacarborane is also η^6-bound to the Ru atom, forming a hexacosahedral structure. The distances of Ru–arene(cent) and Ru–CB$_5$(cent) are 178 and 141 pm, respectively. This 15-vertex ruthenacarborane has the largest vertex number in the metallacarborane family known at present.

13.5 Boric acid and borates

13.5.1 Boric acid

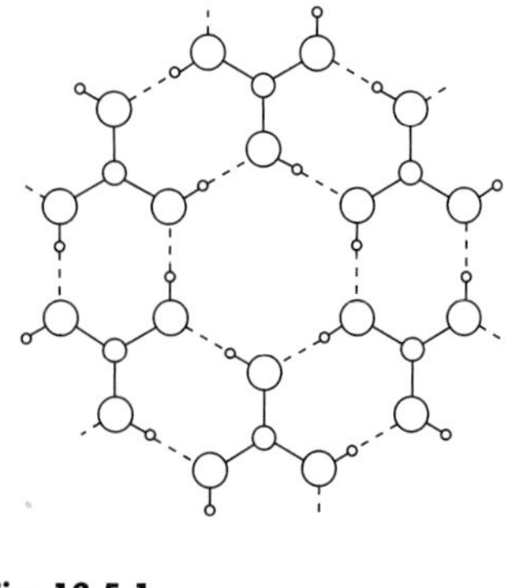

Boric acid, B(OH)$_3$, is the archetype and primary source of oxo-boron compounds. It is also the normal end product of hydrolysis of most boron compounds. It forms flaky, white and transparent crystals, in which the BO$_3$ units are joined to form planar layers by O–H$\cdots$O hydrogen bonds, as shown in Fig. 13.5.1.

Fig. 13.5.1.
Structure of layer of B(OH)$_3$ (B–O 136.1 pm, O–H$\cdots$O 272 pm).

Boric acid is a very weak monobasic acid (pK_a = 9.25). It generally behaves not as a Brønsted acid with the formation of the conjugate-base [BO(OH)$_2$]$^-$, but rather as a Lewis acid by accepting an electron pair from an OH$^-$ anion to form the tetrahedral species $[B(OH)_4]^-$:

$$(HO)_3B + :OH^- \longrightarrow B(OH)_4^-.$$
$$[sp^2+p(unoccupied)] \qquad sp^3$$

Thus this reaction is analogous to the acceptor–donor interaction between BF_3 and NH_3:

$$F_3B + :NH_3 \longrightarrow F_3B \leftarrow NH_3.$$
$$[sp^2+p(unoccupied)] \qquad sp^3$$

In the above two reactions, the hybridization of the B atom changes from sp^2(reactant) to sp^3(product).

Partial dehydration of $B(OH)_3$ above 373 K yields metaboric acid, HBO_2, which consists of trimeric units $B_3O_3(OH)_3$. Orthorhombic metaboric acid is built of discrete molecules $B_3O_3(OH)_3$, which are linked into layers by O–H$\cdots$O bonds, as shown in Fig. 13.5.2.

In dilute aqueous solution, there is an equilibrium between $B(OH)_3$ and $B(OH)_4^-$:

$$B(OH)_3 + 2H_2O \rightleftharpoons H_3O^+ + B(OH)_4^-.$$

At concentrations above 0.1 M, secondary equilibria involving condensation reactions of the two dominant monomeric species give rise to oligomers such as the triborate monoanion $[B_3O_3(OH)_4]^-$, the triborate dianion $[B_3O_3(OH)_5]^{2-}$, the tetraborate $[B_4O_5(OH)_4]^{2-}$, and the pentaborate $[B_5O_6(OH)_4]^-$.

In $(Et_4N)_2[BO(OH)_2]_2\cdot B(OH)_3\cdot 5H_2O$, the conjugate acid–base pair $B(OH)_3$ and dihydrogen borate $[BO(OH)_2]^-$ coexist in the crystalline state. In the crystal structure of the inclusion compound $Me_4N^+[BO(OH)_2]^- \cdot 2(NH_2)_2CO\cdot H_2O$, the $BO(OH)_2^-$, $(NH_2)_2CO$, and H_2O molecules are linked by N–H$\cdots$O and O–H$\cdots$O bonds to form a host lattice featuring a system of parallel channels, which accommodate the guest Me_4N^+ cations (see Section 20.4.4 for further details). The measured dimensions of the $[BO(OH)_2]^-$ anion, which has crystallographically imposed symmetry m, show that the valence tautomeric form **I** with a formal B=O double bond makes a more important contribution than **II** to its electronic structure (Fig. 13.5.3).

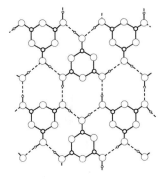

Fig. 13.5.2.
Structure of layer of orthorhombic metaboric acid.

13.5.2 Structure of borates

(1) Structural units in borates

The many structural components that occur in borates consist of planar triangular BO_3 units and/or tetrahedral BO_4 units with vertex-sharing linkage, and may be divided into three kinds. Some examples are given below:

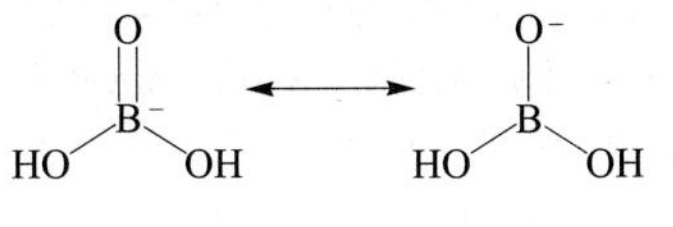

	I	II
B–O	132.3(5) pm	O–B–OH 124.3(2)°
B–OH	139.5(3) pm	HO–B–OH 111.3(3)°

Fig. 13.5.3.
Valence-bond structural formulas for the dihydrogen borate ion, $[BO(OH)_2]^-$, and its measured dimensions in $[(CH_3)_4N]^+[BO(OH)_2]^- \cdot 2(NH_2)_2CO\cdot H_2O$.

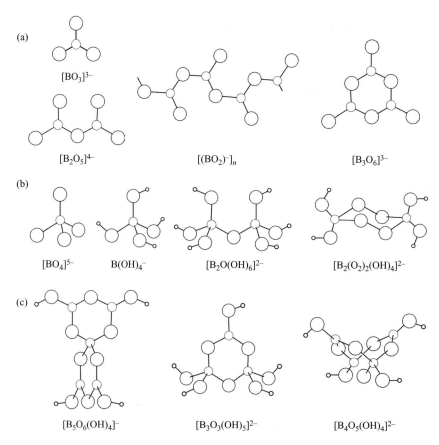

Fig. 13.5.3.
Structural units in borates.

(a) Units containing B in planar BO₃ coordination only [structures show in Fig. 13.5.3(a)]:
BO_3^{3-}: $Mg_3(BO_3)_2$, $LaBO_3$
$[B_2O_5]^{4-}$: $Mg_2B_2O_5$, $Fe_2B_2O_5$
$[B_3O_6]^{3-}$: $K_3B_3O_6$, $Ba_3(B_3O_6)_2 \equiv BaB_2O_4$ (BBO)
$[(BO_2)^-]_n$: $Ca(BO_2)_2$

(b) Units containing B in tetrahedral BO₄ coordination only [structure shown in Fig. 13.5.3(b)]:
BO_4^{5-}: $TaBO_4$
$B(OH)_4^-$: $Na_2[B(OH)_4Cl]$
$[B_2O(OH)_6]^{2-}$: $Mg[B_2O(OH)_6]$
$[B_2(O_2)_2(OH)_4]^{2-}$: $Na_2[B_2(O_2)_2(OH)_4] \cdot 6H_2O$

(c) Units containing B in both BO₃ and BO₄ coordination [Structure shown in Fig. 13.5.3(c)]:
$[B_5O_6(OH)_4]^-$: $K[B_5O_6(OH)_4] \cdot 2H_2O$
$[B_3O_3(OH)_5]^{2-}$: $Ca[B_3O_3(OH)_5] \cdot H_2O$
$[B_4O_5(OH)_4]^{2-}$: $Na_2[B_4O_5(OH)_4] \cdot 8H_2O$

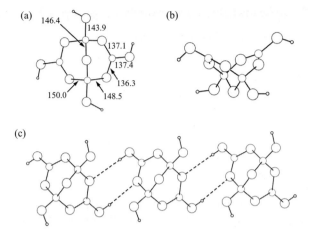

Fig. 13.5.4.
Structure of borax: (a) bond lengths in $[B_4O_5(OH)_4]^{2-}$ (in pm), (b) perspective view of the anion, and (c) $[B_4O_5(OH)_4^{2-}]_n$ chain.

(2) Structure of borax

The main source of boron comes from borax, which is used in the manufacture of borosilicate glass, borate fertilizers, borate-based detergents, and flame-retardants. The crystal structure of borax was determined by X-ray diffraction in 1956, and by neutron diffraction in 1978. It consists of $B_4O_5(OH)_4^{2-}$, Na^+ ions, and water molecules, so that its chemical formula is $Na_2[B_4O_5(OH)_4]\cdot 8H_2O$, not $Na_2B_4O_7\cdot 10H_2O$ as given in many older books. The structure of $[B_4O_5(OH)_4]^{2-}$ is shown in Figs. 13.5.4(a) and (b). In the crystal, the anions are linked by O–H$\cdots$O hydrogen bonds to form an infinite chain, as shown in Fig. 13.5.4(c). The sodium ions are octahedrally coordinated by water molecules, and the octahedra share edges to form another chain. These two kinds of chains are linked by weaker hydrogen bonds between the OH groups of $[B_4O_5(OH)_4]^{2-}$ anions and the aqua ligands, and this accounts for the softness of borax.

(3) Structural principles for borates

Several general principles have been formulated in connection with the structural chemistry of borates:

(a) In borates, boron combines with oxygen either in trigonal planar or tetrahedral coordination. In addition to the mononuclear anions BO_3^{3-} and BO_4^{5-} and their partially or fully protonated forms, there exists an extensive series of polynuclear anions formed by corner-sharing of BO_3 and BO_4 units. The polyanions may form separate boron–oxygen complexes, rings, chains, layers, and frameworks.
(b) In boron–oxygen polynuclear anions, most of the oxygen atoms are linked to one or two boron atoms, and a few can bridge three boron atoms.
(c) In borates, the hydrogen atoms do not attach to boron atoms directly, but always to oxygens not shared by two borons, thereby generating OH groups.
(d) Most boron–oxygen rings are six-membered, comprising three B atoms and three oxygen atoms.
(e) In a borate crystal there may be two or more different boron–oxygen units.

13.6 Organometallic compounds of group 13 elements

13.6.1 Compounds with bridged structure

The compounds $Al_2(CH_3)_6$, $(CH_3)_2Al(\mu\text{-}C_6H_5)_2Al(CH_3)_2$, and $(C_6H_5)_2Al(\mu\text{-}C_6H_5)_2Al(C_6H_5)_2$ all have bridged structures, as shown in Figs. 13.6.1 (a), (b) and (c) respectively. The bond distances and angles are listed in Table 13.6.1.

The bonding in the Al–C–Al bridges involves a 3c-2e molecular orbital formed from essentially sp^3 orbitals of the C and Al atoms. This type of 3c-2e interaction in the bridge necessarily results in an acute Al–C–Al angle (75° to 78°), which is consistent with the experimental data. This orientation is sterically favored and places each *ipso*-carbon atom in an approximately tetrahedral configuration.

In the structure of $Me_2Ga(\mu\text{-}C{\equiv}C\text{-}Ph)_2GaMe_2$ [Fig. 13.6.1(d)], each alkynyl bridge leans over toward one of the Ga centers. The organic ligand

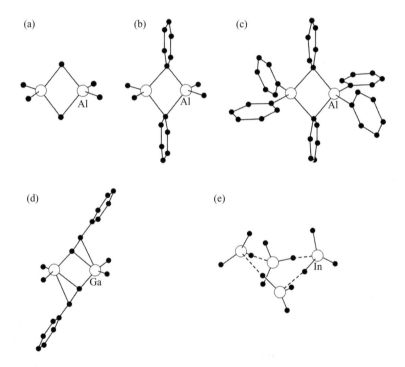

Fig. 13.6.1.
Structure of some group 13 organometallic compounds:
(a) $Al_2(Me)_6$, (b) $Al_2(Me)_4(Ph)_2$,
(c) $Al_2(Ph)_6$, (d) $Ga_2Me_4(C_2Ph)_2$, and
(e) $In(Me)_3$ tetramer.

Table 13.6.1. Structural parameters found in some carbon-bridged organoaluminum compounds

Compound	Distance (pm)			Angle (°)		
	Al–C$_b$	Al–C$_t$	Al···Al	Al–C–Al	C–Al–C (internal)	C–Al–C (external)
Al_2Me_6	214	197	260	74.7	105.3	123.1
$Al_2Me_4Ph_2$	214	197	269	77.5	101.3	115.4
Al_2Ph_6	218	196	270	76.5	103.5	121.5

forms an Ga–C σ bond and interacts with the second Ga center by using its C≡C π bond. Thus each alkynyl group is able to provide three electrons for bridge bonding, in contrast to one electron as normally supplied by an alkyl or aryl group.

Compounds In(Me)$_3$ and Tl(Me)$_3$ are monomeric in solution and the gas phase. In the solid state, they also exist as monomers essentially, but close intermolecular contacts become important. In crystalline In(Me)$_3$, significant In$\cdots$C intermolecular interactions are observed, suggesting that the structure can be described in terms of cyclic tetramers, as shown in Fig. 13.6.1(e), with interatomic distances of In–C 218 pm and In$\cdots$C 308 pm. The corresponding bond distances in isostructural Tl(Me)$_3$ are Tl–C = 230 pm and Tl$\cdots$C = 316 pm.

13.6.2 Compounds with π bonding

Many organometallic compounds of group 13 are formed by π bonds. The π ligands commonly used are olefins, cyclopentadiene, or their derivatives.

The structure of the dimer [ClAl(Me–C=C–Me)AlCl]$_2$ is of interest. There are 3c-2e π bonds between the Al atoms and the C=C bonds, as shown in Fig. 13.6.2(a). The Al–C distances in the Al-olefin units are relatively long, approximately 235 pm, but the sum of the four interactions lead to remarkable stability of the dimer. The energy gained through each Al–olefin interaction is calculated to be about 25 to 40 kJ mol^{-1}.

The [Al(η^5-Cp*)$_2$]$^+$ (Cp* = C$_5$Me$_5$) ion is at present the smallest sandwich metallocene of any main-group element. The Al–C bond length is 215.5 pm. Figure 13.6.2(b) shows the structure of this ion.

The molecular structure of Al$_4$Cp$_4^*$ in the crystal is shown in Fig. 13.6.2(c). The central Al$_4$ tetrahedron has Al–Al bond length 276.9 pm, which is shorter

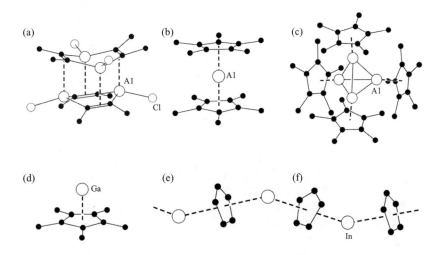

Fig. 13.6.2.
Structure of some organometallic compounds formed by π bonds: (a) [ClAl(Me–C=C–Me)AlCl]$_2$, (b) [Al(Cp*)$_2$]$^+$, (c) Al$_4$Cp$_4^*$, (d) GaCp*, and (e) InCp(or TlCp).

than those in Al metal (286 pm). Each Cp* ring is η^5-coordinated to an Al atom, whereby the planes of the Cp* rings are parallel to the opposite base of the tetrahedron. The average Al–C distances is 233.4 pm.

The half-sandwich structure of GaCp* in the gas phase, as determined by electron diffraction, is shown in Fig. 13.6.2(d). This compound has a pentagonal pyramidal structure with an η^5-C$_5$ ring. The Ga–C distance is 240.5 pm.

Both CpIn and CpTl are monomeric in the gas phase, but in the solid they possess a polymeric zigzag chain structure, in which the In (or Tl) atoms and Cp rings alternate, as shown in Fig. 13.6.2(e).

13.6.3 Compounds containing M–M bonds

The compounds $R_2Al–AlR_2$, $R_2Ga–GaR_2$, and $R_2In–InR_2$ [R=CH(SiMe$_3$)$_2$] have been prepared and characterized. The bond lengths in these compounds are Al–Al 266.0 pm, Ga–Ga 254.1 pm, and In–In 282.8 pm; the M_2C_4 frameworks are planar as shown in Fig. 13.6.3(a). In Al$_2$(C$_6$H$_2{}^i$Pr$_3$)$_4$, Ga$_2$(C$_6$H$_2{}^i$Pr$_3$)$_4$, and In$_2$[C$_6$H$_2$(CF$_3$)$_3$]$_4$, the bond lengths are Al–Al 265 pm, Ga–Ga 251.5 pm, and In–In 274.4 pm, and here the M_2C_4 framework are nonplanar, as shown in Fig. 13.6.3(b). In these structures the Ga–Ga bond lengths are shorter than the Al–Al bond lengths. The anomalous behavior of gallium among Group 13 elements is due to the insertion of electrons into the d orbitals of the preceding 3d elements in the Periodic Table and the associated contraction of the atomic radius.

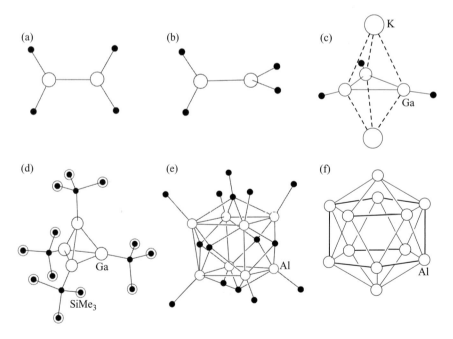

Fig. 13.6.3.
Structures of some compounds containing M–M bonds: (a) planar M_2R_4 (M = Al, Ga, In), (b) nonplanar M_2R_4 (M = Al, Ga, In), (c) K_2Ga_3 core of $K_2[Ga_3(C_6H_3Mes_2)_3]$, (d) [GaC(SiMe$_3$)$_3$]$_4$, (e) (AlMe)$_8$(CCH$_2$Ph)$_5$(C≡C–Ph), and (f) icosahedral Al$_{12}$ core in $K_2[Al_{12}{}^i Bu_{12}]$.

The structures of compounds $Na_2[Ga_3(C_6H_3Mes_2)_3]$ and $K_2[Ga_3(C_6H_3Mes_2)_3]$ are worthy of note. In these compounds, the Na_2Ga_3 and K_2Ga_3 cores have trigonal bipyramidal geometry, as shown in Fig. 13.6.3(c). The bond lengths are Ga–Ga 244.1 pm (Na salt), 242.6 pm (K salt), Ga–Na 322.9 pm, and Ga–K 355.4 pm. The planarity of the Ga_3 ring and the very short Ga–Ga bonds implies electron delocalization in the three-membered ring, the requisite two π electrons being provided by the two K (or Na) atoms (one electron each) to the empty p_z orbitals of the three sp^2-hybridized Ga atoms.

Figures 13.6.3(d)–(f) show the structures of three compounds which contain metallic polyhedral cores. In $[GaC(SiMe_3)_3]_4$, the Ga–Ga bond length is 268.8 pm, in $(AlMe)_8(CCH_2Ph)_5(C\equiv C-Ph)$ the Al–Al bond lengths are observed to be close to two average values: 260.9 and 282.9 pm, and in $K_2[Al_{12}{}^iBu_{12}]$ the Al–Al bond length is 268.5 pm.

$Al_{50}Cp^*_{12}$ (or $Al_{50}C_{120}H_{180}$) is a giant molecule, whose crystal structure shows a distorted square-antiprismatic Al_8 moiety at its center, as shown in Fig. 13.6.4. This Al_8 core is surrounded by 30 Al atoms that form an icosidodecahedron with 12 pentagonal faces and 20 trigonal faces. Each pentagonal faces is capped by an $AlCp^*$ unit, and the set of 12 Al atoms forms a very regular icosahedron, in which the Al···Al average distance is 570.2 pm. Each of the peripheral 12 Al atoms is coordinated by 10 atoms (5 Al and 5 C) in the form of a staggered "mixed sandwich." In the molecule, the average Al–Al bond length is 277.0 pm (257.8–287.7 pm). The 60 CH_3 groups of the 12 Cp^* ligands at the surface exhibits a topology resembling that of the carbon atoms in fullerene-C_{60}. The average distance between a pair of nearest methyl groups of neighboring Cp^* ligands (386 pm) is approximately twice the van der Waals radius of a methyl group (195 pm). The entire $Al_{50}Cp^*_{12}$ molecule has a volume which is about five times large than that of a C_{60} molecule.

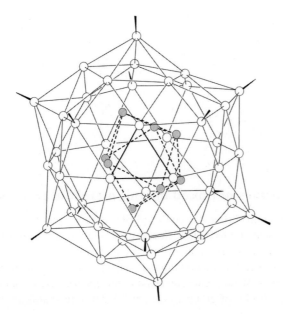

Fig. 13.6.4.
Molecular structure of $Al_{50}Cp^*_{12}$ (Cp^* = pentamethylcyclopentadienide). For clarity, the Cp^* units are omitted, and only the bonds from the outer 12 Al atoms toward the ring centers are shown. The shaded atoms represent the central distorted square-antiprismatic Al_8 moiety.

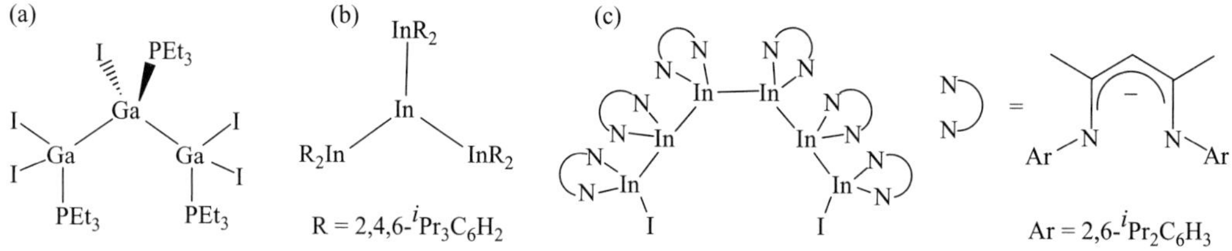

494 Structural Chemistry of Selected Elements

Fig. 13.6.5.
Structural formulas of some open-chain homocatenated compounds of heavier group 13 elements.

13.6.4 Linear catenation in heavier group 13 elements

For the heavier congeners of boron, the occurrence of well-characterized compounds possessing a linear extended skeleton containing two or more unsupported two-electron E–E bonds is quite rare. Some discrete molecules exhibiting this feature include the open-chain trigallium subhalide complex $I_2(PEt_3)Ga–GaI(PEt_3)–GaI_2(PEt_3)$ (Fig. 13.6.5(a)) and the trigonal tetranuclear indium complex $\{In[In(2,4,6-^iPr_3C_6H_2)_2]_3\}$ (Fig. 13.6.5(b)).

Using a chelating ligand of the β-diketiminate class, a novel linear homocatenated hexanuclear indium compound has been synthesized (Fig. 13.6.5(c)). The four internal indium atoms are in the +1 oxidation state, and the terminal indium atoms, each carrying an iodo ligand, are both divalent. The coordination geometry at each indium center is distorted tetrahedral. As shown in Fig. 13.6.6, the mixed-valent molecule has a pseudo C_2 axis with a β-diketiminate ligand chelated to each metal center, and its zigzag backbone is held together by five unsupported In–In single bonds constructed from sp^3 hybrid orbitals.

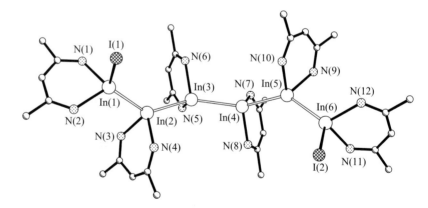

Fig. 13.6.6.
Molecular structure of a linear homocatenated compound containing six indium centers. The 3,5-dimethylphenyl groups of the β-diketiminate ligands have been omitted for clarity. Bond lengths (pm): In(1)–In(2) 281.2, In(2)–In(3) 283.5, In(3)–In(4) 285.4, In(4)–In(5) 284.1, In(5)–In(6) 282.2; In(1)–I(1) 279.8, In(6)–I(2) 278.0; standard deviation 0.1 pm. Average In–In–In bond angle at In(2) to In(5) is 139.4°.

13.7 Structure of naked anionic metalloid clusters

The term metalloid cluster is used to describe a multinuclear molecular species in which the metal atoms exhibit closest packing (and hence delocalized intermetallic interactions) like that in bulk metal, and the metal–metal contacts outnumber the peripheral metal–ligand contacts. Most examples are found in the field of precious-metal cluster chemistry. In recent years, an increasing number of cluster species of group 13 elements have been synthesized with cores

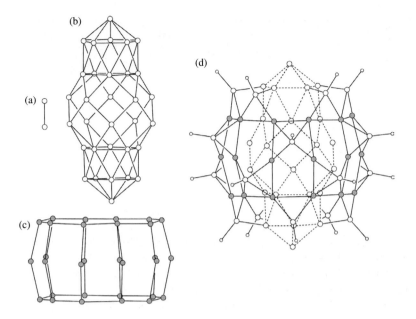

Fig. 13.7.1.
Structure of $Ga_{84}[N(SiMe_3)_2]_{20}{}^{4-}$ [the larger circles (shaded and unshaded) represent Ga atom, and the small circles represent N atoms]: (a) central Ga_2 unit, (b) Ga_{32} shell, (c) a "belt" of 30 Ga atoms, and (d) the front of $Ga_{84}[N(SiMe_3)_2]_{20}{}^{4-}$. The Ga_2 and N atoms at the back are not shown, the Ga_{32} shell is emphasized by the broken lines, the "belt" of 30 Ga atoms are shaded, and for the ligands $N(SiMe_3)_2$ only the N atoms directly bonded to the Ga atoms are shown.

consisting of Al_n (n = 7, 12, 14, 50, 69) and Ga_m (m = 9, 10, 19, 22, 24, 26, 84) atoms, for example $[Al_{77}\{N(SiMe_3)_2\}_{20}]^{2-}$ and $[Ga_{84}\{N(SiMe_3)_2\}_{20}]^{4-}$.

13.7.1 Structure of $Ga_{84}[N(SiMe_3)_2]_{20}{}^{4-}$

The largest metalloid cluster characterized to date is $Ga_{84}[N(SiMe_3)_2]_{20}^{4-}$, the structure of which is illustrated in Fig. 13.7.1. It comprises four parts: a Ga_2 unit, a Ga_{32} shell, a Ga_{30} "belt," and 20 $Ga[N(SiMe_3)_2]$ groups. The Ga_2 unit [as shown in (a)] is located at the center of the 64 naked Ga atoms; the Ga–Ga bond distance is 235 pm, which is almost as short as the "normal" triple Ga–Ga bond (232 pm). The Ga_2 unit is encapsulated by a Ga_{32} shell in the form of a football with icosahedral caps, as shown in (b). The $Ga_2@Ga_{32}$ aggregate is surrounded by a belt of 30 Ga atoms that are also naked, as shown in (c). Finally the entire Ga_{64} framework is protected by 20 $Ga[N(SiMe_3)_2]$ groups to form $Ga_{84}[N(SiMe_3)_2]_{20}{}^{4-}$, as shown in (d). This large anionic cluster has a diameter of nearly 2 nm.

13.7.2 Structure of NaTl

The structure of sodium thallide NaTl can be understood as a diamond-like framework of Tl atoms, whose vacant sites are completely filled with Na atoms. Figure 13.7.2(a) shows the structure of NaTl, in which the Tl–Tl covalent bonds are represented by solid lines. The Tl atom has three valence electrons, which are insufficient for the construction of a stable diamond framework. The deficit can be partially compensated by the introduction of Na atoms. The effective radius of the Na atom is considerably smaller than that in pure metallic sodium.

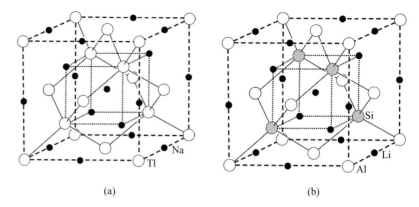

Fig. 13.7.2.
Structure of (a) NaTl and (b) Li$_2$AlSi.

Therefore, the chemical bonding in NaTl is expected to be a mixture of covalent, ionic, and metallic interactions.

The NaTl-type structure is the prototype for Zintl phases, which are intermetallic compounds which crystallize in typical "non-metal" crystal structures. Binary AB compounds LiAl, LiGa, LiIn and NaIn are both isoelectronic (isovalent) and isostructural with NaTl. In the Li$_2$AlSi ternary compound, Al and Si form a diamond-like framework, in which the octahedral vacant sites of the Al sublattice are filled by Li atoms, as shown in Fig. 13.7.2(b).

From the crystal structure, physical measurements, and theoretical calculations, the nature of the chemical bond in the NaTl-type compound AB can be understood in the following terms:

(a) Strong covalent bonds exist between the B atoms (Al, Ga, In, Tl, Si).
(b) The alkali atoms (A) are not in bonding contact.
(c) The chemical bond between the A and B framework is metallic with a small ionic component.
(d) For the upper valence/conduction electron states a partial metal-like charge distribution can be identified.

13.7.3 Naked Tl$_n^{m-}$ anion clusters

The heavier elements of Group 13, in particular thallium, are able to form discrete naked clusters with alkali metals. Table 13.7.1 lists some examples of Tl$_n^{m-}$ naked anion clusters, and Fig. 13.7.3 shows their structures.

The bonding in these Tl$_n^{m-}$ anion clusters is similar to that in boranes. For example, the bond valences (b) of Tl$_5^{7-}$ and centered Tl$_{13}^{11-}$ clusters are equal to those of B$_5$H$_5^{2-}$ and B$_{12}$H$_{12}^{2-}$, respectively. Note that, in Tl$_{13}^{11-}$, the central Tl atom contributes all three valence electrons to cluster bonding, so that the total number of bonding electrons is $(3 + 12 \times 1 + 11) = 26$. This cluster is thus consolidated by ten Tl–Tl–Tl 3c-2e and three Tl–Tl 2c-2e bonds. The Tl$_9^{9-}$ cluster has 36 valence electrons, and its bond valence b is 18. The cluster is stabilized by three Tl–Tl–Tl 3c-2e bonds, as labeled by the three shaded faces in Fig. 13.7.3(e), and the remaining twelve edges represent 12 Tl–Tl 2c-2e bonds.

Group 13 Elements

Table 13.7.1. Some examples of Tl_n^{m-} anion clusters

Composition	Anion	Structure in Fig. 13.7.3	Cluster symmetry	Bond valence (b)	Bonding
Na$_2$Tl	Tl_4^{8-}	(a)	T_d	6	6 Tl–Tl 2c-2e bonds
Na$_2$K$_{21}$Tl$_{19}$	$2Tl_5^{7-}$	(b)	D_{3h}	9	3Tl–Tl–Tl 3c-2e bonds 3Tl–Tl 2c-2e bonds
	Tl_9^{9-}	(e)	defective I_h	18	3Tl–Tl–Tl 3c-2e bonds 12Tl–Tl 2c-2e bonds
KTl	Tl_6^{6-}	(c)	D_{4h}	12	12 Tl–Tl 2c-2e bonds
K$_{10}$Tl$_7$	Tl_7^{7-}, 3e$^-$	(d)	~ D_{5h}	14	5Tl–Tl–Tl 3c-2e bonds 4Tl–Tl 2c-2e bonds
K$_8$Tl$_{11}$	Tl_{11}^{7-}, e$^-$	(f)	~ D_{3h}	24	3Tl–Tl–Tl 3c-2e bonds 18Tl–Tl 2c-2e bonds
Na$_3$K$_8$Tl$_{13}$	Tl_{13}^{11-}	(g)	Centered ~ I_h	23	10Tl–Tl–Tl 3c-2e bonds 3Tl–Tl 2c-2e bonds

The compound K$_{10}$Tl$_7$ is composed of ten K$^+$, a Tl_7^{7-} and three delocalized electrons per formula unit and exhibits metallic properties. The Tl_7^{7-} cluster is an axially compressed pentagonal bipyramid conforming closely to D_{5h} symmetry [Fig. 13.7.3(d)]. The apex–apex bond distance of 346.2 pm is slightly longer than the bonds in the pentagonal waist (318.3–324.7 pm). Comparison between the structures of Tl_7^{7-} and $B_7H_7^{2-}$ [Fig. 13.4.6(a)] shows that both are pentagonal bipyramidal but Tl_7^{7-} is compressed along the C_5 axis for the formation of a coaxial 2c-2e Tl–Tl bond, so the bond valences of Tl_7^{7-} and $B_7H_7^{2-}$ are 14 and 13, respectively.

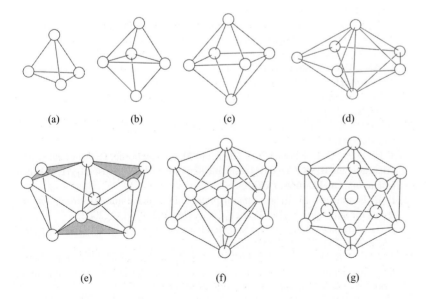

Fig. 13.7.3. Structures of some Tl_n^{m-} anion clusters: (a) Tl_4^{8-}, (b) Tl_5^{7-}, (c) Tl_6^{6-}, (d) Tl_7^{7-}, (e) Tl_9^{9-} (shaded faces represent three Tl–Tl–Tl 3c-2e bonds), (f) Tl_{11}^{7-}, (g) Tl_{13}^{11-}.

References

1. N. N. Greenwood and A. Earnshaw, *Chemistry of the Elements*, 2nd edn., Butterworth Heinemann, Oxford, 1997.
2. C. E. Housecroft and A. G. Sharpe, *Inorganic Chemistry*, 2nd edn., Prentice-Hall, London, 2004.
3. D. F. Shriver, P. W. Atkins, T. L. Overton, J. P. Rourke, M. T. Weller and F. A. Armstrong, *Inorganic Chemistry*, 4th edn., Oxford University Press, Oxford, 2006.
4. D. M. P. Mingos, *Essential Trends in Inorganic Chemistry*, Oxford University Press, Oxford, 1998.
5. G. E. Rodgers, *Introduction to Coordination, Solid State, and Descriptive Inorganic Chemistry*, McGraw-Hill, New York, 1994.
6. G. Meyer, D. Naumann and L. Wesemann (eds.), *Inorganic Chemistry Highlights*, Wiley–VCH, Weinheim, 2002.
7. C. E. Housecroft, *Boranes and Metallaboranes: Structure, Bonding and Reactivity*, 2nd edn., Ellis Horwood, New York, 1994.
8. C. E. Housecroft, *Cluster Molecules of the* p-*Block Elements*, Oxford University Press, Oxford, 1994.
9. D. M. P. Mingos and D. J. Wales, *Introduction to Cluster Chemistry*, Prentice-Hall Englewood Cliffs, NJ, 1990.
10. W. Siebert (ed.), *Advances in Boron Chemistry*, Royal Society of Chemistry, Cambridge, 1997.
11. G. A. Olah, K. Wade and R. E. Williams (eds.), *Electron Deficient Boron and Carbon Clusters*, Wiley, New York, 1991.
12. M. Driess and H. Nöth (eds.), *Molecular Clusters of the Main Group Elements*, Wiley–VCH, Weinheim, 2004.
13. R. A. Beaudet, "The molecular structures of boranes and carboranes," in J.F. Liebman, A. Greenberg and R. E. Williams (eds.), *Advances in Boron and the Boranes*, VCH, New York, 1988.
14. W. Siebert (ed.), *Advances in Boron Chemistry*, Royal Society of Chemistry, Cambridge, 1997.
15. I. D. Brown, *The Chemical Bond in Inorganic Chemistry: The Bond Valence Model*, Oxford University Press, New York, 2002.
16. U. Müller, *Inorganic Structural Chemistry*, 2nd edn., Wiley, Chichester, 2006.
17. T. C. W. Mak and G.-D. Zhou, *Crystallography in Modern Chemistry: A Resource Book of Crystal Structures*, Wiley–Interscience, New York, 1992.
18. E. Abel, F. G. A. Stone and G. Wilkinson (eds.), *Comprehensive Organometallic Chemistry II*, Vol. 1, Pergamon Press, Oxford, 1995.
19. S. M. Kauzlarich (ed.), *Chemistry, Structure, and Bonding of Zintl Phases and Ions*, VCH, New York, 1996.
20. G. D. Zhou, Bond valence and molecular geometry. *University Chemistry* (in Chinese), **11**, 9–18 (1996).
21. T. Peymann, C. B. Knobler, S. I. Khan and M. F. Hawthorne, Dodeca(benzyloxy)-dodecaborane $B_{12}(OCH_2Ph)_{12}$: A stable derivative of hypercloso-$B_{12}H_{12}$. *Angew. Chem. Int. Ed.* **40**, 1664–7 (2001).
22. E. D. Jemmis, M. M. Balakrishnarajan and P. D. Pancharatna, Electronic requirements for macropolyhedral boranes. *Chem. Rev.* **102**, 93–144 (2002).
23. Z. Xie, Advances in the chemistry of metallacarboranes of f-block elements. *Coord. Chem. Rev.* **231**, 23–46 (2002).
24. J. Vollet, J. R. Hartig and H. Schnöckel, $Al_{50}C_{120}H_{180}$: A pseudofullerene shell of 60 carbon atoms and 60 methyl groups protecting a cluster core of 50 aluminium atoms. *Angew. Chem. Int. Ed.* **43**, 3186-9 (2004).

25. H. W. Roesky and S. S. Kumar, Chemistry of aluminium(I). *Chem. Commun.* 4027–38 (2005).
26. C. Dohmeier, D. Loos and H. Schnöckel, Aluminum(I) and gallium(I) compounds: syntheses, structures, and reactions. *Angew. Chem. Int. Ed.* **35**, 129–49 (1996).
27. A. Schnepf and H. Schnöckel, Metalloid aluminum and gallium clusters: element modifications on the molecular scale?. *Angew. Chem. Int. Ed.* **41**, 3532–52 (2002).
28. W. Uhl, Organoelement compounds possessing Al–Al, Ga–Ga, In–In, and Tl–Tl single bonds. *Adv. Organomet. Chem.* **51**, 53–108 (2004).
29. S. Kaskel and J. D. Corbett, Synthesis and structure of $K_{10}Tl_7$: the first binary trielide containing naked pentagonal bipyramidal Tl_7 clusters. *Inorg. Chem.* **39**, 778–82 (2000).
30. M. S. Hill, P. B. Hitchcock and R. Pongtavornpinyo, A linear homocatenated compound containing six indium centers. *Science* **311**, 1904–7 (2006).

14 Structural Chemistry of Group 14 Elements

14.1 Allotropic modifications of carbon

With the exception of the gaseous low-carbon molecules: $C_1, C_2, C_3, C_4, C_5, \ldots$, the carbon element exists in the diamond, graphite, fullerene, and amorphous allotropic forms.

14.1.1 Diamond

Diamond forms beautiful, transparent, and highly refractive crystals, and has been used as a noble gem since antiquity. It consists of a three-dimensional network of carbon atoms, each of which is bonded tetrahedrally by covalent C–C single bonds to four others, so that the whole diamond crystal is essentially a "giant molecule." Nearly all naturally occurring diamonds exist in the cubic form, space group $Fd\bar{3}m$ (no. 227), with $a = 356.688$ pm and $Z = 8$. The C–C bond length is 154.45 pm, and the C–C–C bond angle is 109.47°. In the crystal structure, the carbon atoms form six-membered rings that take the all-chair conformation [Fig. 14.1.1(a)]. The mid-point of every C–C bond is located at an inversion center, so that the six nearest carbon atoms about it are in a staggered arrangement, which is the most stable conformation.

In addition to cubic diamond, there is a metastable hexagonal form, which has been found in aerorite and can be prepared from graphite at 13 GPa above 1300 K. Hexagonal diamond crystallizes in space group $P6_3/mmc$ (no. 194) with $a = 251$ pm, $c = 412$ pm and $Z = 4$, as shown in Fig. 14.1.1(b). The bond type and the C–C bond length are the same as in cubic diamond. The difference between the two forms is the orientation of two sets of tetrahedral bonds about neighboring carbon atoms: cubic diamond has the staggered arrangement

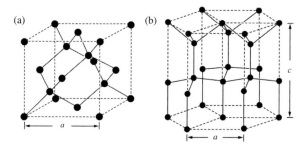

Fig. 14.1.1.
Crystal structure of diamond: (a) cubic form and (b) hexagonal form.

about each C–C bond at an inversion center, but in hexagonal diamond some bonds take the eclipsed arrangement related by mirror symmetry. As repulsion between nonbonded atoms in the eclipsed arrangement is greater than that in the staggered arrangement, hexagonal diamond is much rarer than cubic diamond.

Diamond is the hardest natural solid known and has the highest melting point, 4400 ± 100 K (12.4 GPa). It is an insulator in its pure form. Since the density of diamond (3.51 g cm^{-3}) far exceeds that of graphite (2.27 g cm^{-3}), high pressures can be used to convert graphite into diamond even though graphite is thermodynamically more stable by 2.9 kJ mol^{-1}. To attain commercially viable rates for the pressure-induced conversion of graphite to diamond, a transition-metal catalyst such as iron, nickel, or chromium is generally used. Recently, the method has been employed to deposit thin films of diamond onto a metallic or other material surface. The extreme hardness and high thermal conductivity of diamond find applications in numerous areas, notably the hardening of surfaces of electronic devices and as cutting and/or grinding materials.

Elemental silicon, germanium, and tin have the cubic diamond structure with unit-cell edge $a = 543.072$, 565.754, and 649.12 pm (α-Sn), respectively.

14.1.2 Graphite

Graphite is the common modification of carbon that is stable under normal conditions. Its crystal structure consists of planar layers of hexagonal carbon rings. Within the layer each carbon atom is bonded covalently to three neighboring carbon atoms at 141.8 pm. The σ bonds between neighbors within a layer are formed from the overlap of sp^2 hybrids, and overlap involving the remaining electron and perpendicular p$_z$ orbital on every atom generates a network of π bonds that are delocalized over the entire layer. The relatively free movement of π electrons in the layers leads to abnormally high electrical conductivity for a non-metallic substance.

The layers are stacked in a staggered manner with half of the atoms of one layer situated exactly above atoms of the layer below, and the other half over the rings centers. There are two crystalline forms that differ in the sequence of layer stacking.

(1) *Hexagonal graphite* or α-*graphite*: The layers are arranged in the sequence …ABAB…, as shown in Fig. 14.1.2(a). The space group is $P6_3/mmc$

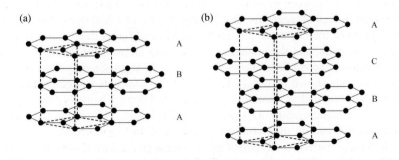

Fig. 14.1.2.
Structure of graphite: (a) α-graphite and (b) β-graphite.

(no. 194), and the unit cell has dimensions of $a = 245.6$ pm and $c = 669.4$ pm, so that the interlayer distance is $c/2 = 334.7$ pm.

(2) *Rhombohedral graphite* or *β-graphite*: The layers are arranged in the sequence ...ABCABC..., as shown in Fig. 14.1.2(b). The space group is $R\bar{3}m$ (no. 166), and the unit cell has dimensions of $a = 246.1$ pm and $c = 1006.4$ pm. The interlayer separation $c/3 = 335.5$ pm is similar to that of α-graphite.

The enthalpy difference between hexagonal and rhombohedral graphite is only 0.59 ± 0.17 kJ mol^{-1}. The two forms are interconvertible by grinding (hexagonal → rhombohedral) or heating above 1025 °C (rhombohedral → hexagonal). Partial conversion leads to an increase in the average spacing between layers; this reaches a maximum of 344 pm for turbostratic graphite in which the stacking sequence of the parallel layers is completely random.

In graphite the layers are held together by van der Waals forces. The relative weak binding between layers is consistent with its softness and lubricity, as adjacent layers are able to easily slide by each other. Graphite mixed with clay constitutes pencil "lead," which should not be confused with metallic lead or dark-gray lead sulfide.

14.1.3 Fullerenes

The third allotropic modification of carbon, the fullerenes, consists of a series of discrete molecules of closed cage structure bounded by planar faces, whose vertices are made up of carbon atoms. If the molecular cage consists of pentagons and hexagons only, the number of pentagons must always be equal to 12, while the number of hexagons may vary. Fullerenes are discrete globular molecules that are soluble in organic solvents, and their structure and properties are different from those of diamond and graphite.

Fullerenes can be obtained by passing an electric arc between two graphite electrodes in a controlled atmosphere of helium, or by controlling the helium/oxygen ratio in the incomplete combustion of benzene, followed by evaporation of the carbon vapor and re-crystallization from benzene. The main product of the preparation is fullerene-C_{60}, and the next abundant product is fullerene-C_{70}.

The structure of C_{60} has been determined by single-crystal neutron diffraction and electron diffraction in the gas phase. It has icosahedral (I_h) symmetry with 60 vertices, 90 edges, 12 pentagons, and 20 hexagons. The C–C bond lengths are 139 pm for 6/6 bonds (fusion of two six-membered rings) and 144 pm for 6/5 bonds (fusion between five- and six-membered rings). The 60 carbon atoms all lie within a shell of mean diameter 700 pm. Figure 14.1.3(a) shows the soccer-ball shape of the C_{60} molecule. Each carbon atom forms three σ bonds with three neighbors, and the remaining orbitals and electrons of the 60 C atoms form delocalized π bonds. This structure may be formulated by the valence-bond formula shown in Fig. 14.1.3(b). The C_{60} molecule can be chemically functionalized, and an unambiguous atom-numbering scheme is required for systematic nomenclature of its derivatives. Figure 14.1.3(c) shows the planar formula with carbon atom-numbering scheme of the C_{60} molecule.

In the C_{60} molecule, the sum of the σ bond angles at each C atom is 348° (= 120° + 120° + 108°), and the mean C–C–C angle is 116°. The π atomic orbital

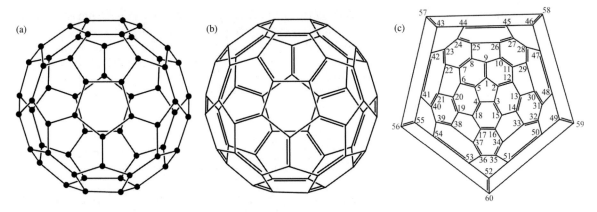

Fig. 14.1.3.
Structure of fullerene-C_{60}: (a) molecular shape, (b) valence-bond formula, and (c) planar formula with carbon atom-numbering scheme.

lies normal to the convex surface, the angle between the σ and π orbitals being 101.64°. It may be approximately calculated that each σ orbital has s component 30.5% and p 69.5%, and each π orbital has s 8.5% and p 91.5%.

Table 14.1.1. Some physical properties of fullerene-C_{60}

Density (g cm^{-3})	1.65
Bulk modulus (GPa)	18
Refractive index (630 nm)	2.2
Heat of combustion (crystalline C_{60}) (kJ mol^{-1})	2280
Electron affinity (eV)	2.6–2.8
First ionization energy (eV)	7.6
Band gap (eV)	1.9
Solubility (303 K) (g dm^{-3})	
$\quad CS_2$	5.16
$\quad$ Toluene	2.15
$\quad$ Benzene	1.44
$\quad CCl_4$	0.45
$\quad$ Hexane	0.04

Fullerene-C_{60} is a brown-black crystal, in which the nearly spherical molecules rotate continuously at room temperature. The structure of the crystal can be considered as a stacking of spheres of diameter 1000 pm in cubic closest packing ($a = 1420$ pm) or hexagonal closest packing ($a = 1002$ pm, $c = 1639$ pm). Figure 14.1.4 shows the crystal structure of fullerene-C_{60}.

Below 249 K, the molecules are orientated in an ordered fashion, and the symmetry of the crystal is reduced from a face-centered cubic lattice to a primitive cubic lattice. At 5 K, the crystal structure determined by neutron diffraction yielded the following data: space group $Pa\bar{3}$ (no. 205), $a = 1404.08(1)$ pm; C–C bond lengths: (6/6) 139.1 pm, (6/5) 144.4 pm, and 146.6 pm (mean 145.5 pm).

In the C_{60} molecule, the mean distance from the center to every C vertex is 350 pm, so the molecule has a spherical skeleton with diameter 700 pm. Allowing for the van der Waals radius of the C atom (170 pm), the C_{60} molecule has a central cavity of diameter 360 pm, which can accommodate a foreign atom. Some physical properties of C_{60} are summarized in Table 14.1.1.

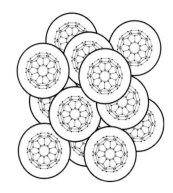

Fig. 14.1.4.
The cubic face-centered structure of fullerene-C_{60}.

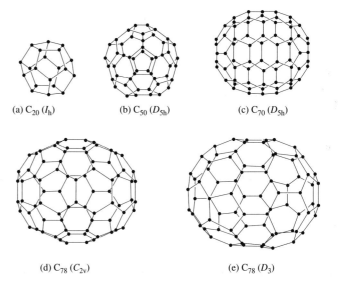

Fig. 14.1.5.
Structures of some fullerenes. Note that (b) shows the carbon skeleton of $C_{50}Cl_{10}$, whose ten equatorial chloro substituents have been omitted.

(a) C_{20} (I_h) (b) C_{50} (D_{5h}) (c) C_{70} (D_{5h})

(d) C_{78} (C_{2v}) (e) C_{78} (D_3)

In addition to C_{60}, many other higher homologs have been prepared and characterized. Several synthetic routes to fullerenes have yielded gram quantities of pure C_{60} and C_{70}, whereas C_{76}, C_{78}, C_{80}, C_{82}, C_{84} and other fullerenes have been isolated as minor products. Figure 14.1.5 shows the structures of the following fullerenes: C_{20}, C_{50}, C_{70}, and two C_{78}-isomers.

In general, the polyhedral closed cages of fullerenes are made up entirely of n three-coordinate carbon atoms that constitute 12 pentagonal and $(n/2 - 10)$ hexagonal faces. The larger fullerenes synthesized so far do faithfully satisfy the isolated pentagon rule (IPR), which governs the stability of fullerenes comprising hexagons and exactly 12 pentagons. On the other hand, smaller fullerenes do not obey the IPR, and are so labile that their properties and reactivity have only been studied in the gas phase. The smallest fullerene that can exist theoretically is C_{20}. It has been synthesized from dodecahedrane $C_{20}H_{20}$ by replacing the hydrogen atoms with relatively weakly bound bromine atoms to form a triene precursor of average composition $[C_{20}HBr_{13}]$, which was then subjected to debromination in the gas phase. The bowl isomer of C_{20} is likewise generated by gas-phase debromination of a $[C_{20}HBr_9]$ precursor prepared from bromination of corannulene $C_{20}H_{10}$. Identification of the two C_{20} isomers was achieved by mass-selective anion-photoelectron spectroscopy. Figure 14.1.5(a) shows the molecular structure of fullerene C_{20}.

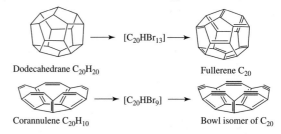

Dodecahedrane $C_{20}H_{20}$ → $[C_{20}HBr_{13}]$ → Fullerene C_{20}

Corannulene $C_{20}H_{10}$ → $[C_{20}HBr_9]$ → Bowl isomer of C_{20}

[From H. Prinzbach, A. Weiler, P. Landenberger, F. Wahl, J. Worth, L. T. Scott, M. Gelmont, D. Olevano and B. v. Issendorff, *Nature* **407**, 60–3 (2000).]

Fullerene-C_{50} has been trapped as its perchloro adduct $C_{50}Cl_{10}$, which was obtained in milligram quantity from the addition of CCl_4 to the usual graphite arc-discharge process for the synthesis of C_{60} and larger fullerenes. The D_{5h} structure of $C_{50}Cl_{10}$, with all chlorine atoms lying in the equatorial plane and attached to sp^3 carbon atoms, was established by mass spectrometry, ^{13}C NMR, and other spectroscopic methods. The idealized structure of C_{50} is displayed in Fig. 14.1.5(b).

The fullerene-C_{70} molecule has D_{5h} symmetry and an approximately ellipsoidal shape, as shown in Fig. 14.1.5(c). As in C_{60}, the 12 five-membered rings in C_{70} are not adjacent to one another. In contrast to C_{60}, C_{70} has an equatorial phenylene belt comprising 15 fused hexagons and two polar caps, each assembled from a pentagon that shares its edges with 5 hexagons. The curvature at the polar region is very similar to that of C_{60}. There are five sets of geometrically distinct carbon atoms, and the observed ^{13}C NMR signals have the intensity ratios of 1:1:2:2:1. The measured carbon–carbon bond lengths (Fig. 14.1.6) in the crystal structure of the complex $(\eta^2\text{-}C_{70})Ir(CO)Cl(PPh_3)_2$ suggest that two equivalent lowest-energy Kekulé structures per equatorial hexagon are required to describe the structure and reactivity properties of C_{70}.

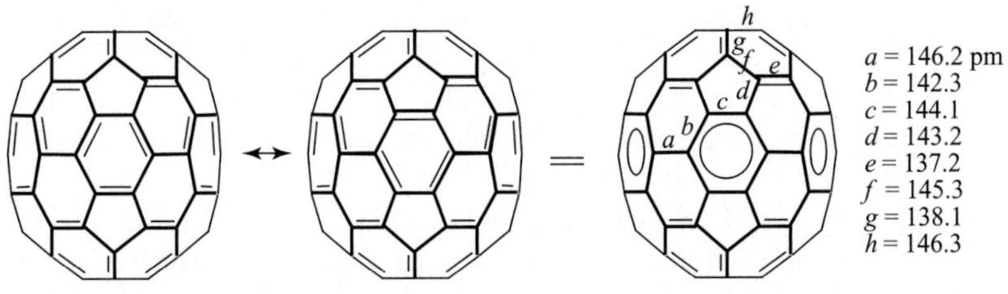

$a = 146.2$ pm
$b = 142.3$
$c = 144.1$
$d = 143.2$
$e = 137.2$
$f = 145.3$
$g = 138.1$
$h = 146.3$

Fig. 14.1.6.
Two equivalent valence bond structures of C_{70} and measured bond lengths of the polyhedral cage in $(\eta^2\text{-}C_{70})Ir(CO)Cl(PPh_3)_2$.

The structures of two geometric isomers of fullerene-C_{78} have been elucidated by ^{13}C NMR spectroscopy: one has C_{2v} symmetry, as shown in Fig. 14.1.5(d); the other has D_3 symmetry, as shown in Fig. 14.1.5(e). Structural assignments with qualities ranging from reliable to absolutely certain have also been made for $C_{74}(D_{3h})$, $C_{76}(D_2)$, $C_{78'}$ [a new isomer of C_{2v} symmetry, which has a more spherical shape compared to the $C_{78}(C_{2v})$ isomer shown in Fig. 14.1.5(d)], $C_{80}(D_2)$, $C_{82}(C_2)$, $C_{84}(D_2)$, and $C_{84}(D_{2d})$.

In addition to the single globular species, fullerenes can be formed by joining two or more carbon cages, as found for the dimer C_{120} shown in Fig. 14.1.7. X-ray diffraction showed that the dimer is connected by a pair of C–C bonds linking the edges of hexagonal faces (6/6 bonds) in two C_{60} units to form a central four-membered ring, in which the bond lengths are 157.5 pm (connecting the two cages) and 158.1 pm (6/6 bonds).

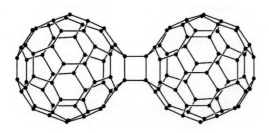

Fig. 14.1.7.
Molecular structure of the dimer $[C_{60}]_2$.

Fullerene-C_{60} was selected as 'Molecule of the Year 1991' by the journal *Science (Washington)*. Discovery of this modification of carbon in the mid-1980s created a great excitement in the scientific community and popular press. Some of this interest undoubtedly stemmed from the fact that carbon is a common element that had been studied since ancient times, but still an entirely new field of fullerene chemistry suddenly emerged with great potential for exciting research and practical applications.

14.1.4 Amorphous carbon

Amorphous carbon is a general term that covers non-crystalline forms of carbon such as coal, coke, charcoal, carbon black (soot), activated carbon, vitreous carbon, glassy carbon, carbon fiber, carbon nanotubes, and carbon onions, which are important materials and widely used in industry. The arrangements of the carbon atoms in amorphous carbon are different from those in diamond, graphite, and fullerenes, but the bond types of carbon atoms are the same as in these three crystalline allotropes. Most forms of amorphous carbon consist of graphite scraps in irregularly packing.

Coal is by far the world's most abundant fossil fuel, with a total recoverable resource of about 1000 billion (10^{12}) tons. It is a complex mixture of many compounds that contain a high percentage of carbon and hydrogen, but many other elements are also present as impurities. The composition of coal varies considerably depending on its age and location. A typical bituminous coal has the approximate composition 80% C, 5% H, 8% O, 3% S, and 2% N. The manifold structures of coal are very complex and not clearly defined.

The high-temperature carbonization of coal yields coke, which is a soft, poorly graphitized form of carbon, most of which is used in steel manufacture.

Activated carbon is a finely divided form of amorphous carbon manufactured from the carbonization of an organic precursor, which possesses a microporous structure with a large internal surface area. The ability of the hydrophobic surface to adsorb small molecules accounts for the widespread applications of activated carbon as gas filters, decoloring agents in the sugar industry, water purification agents, and heterogeneous catalysts.

Carbon black (soot) is made by the incomplete combustion of liquid hydrocarbons or natural gas. The particle size of carbon black is exceedingly small, only 0.02 to 0.3 μm, and its principal application is in the rubber industry where it is used to strengthen and reinforce natural rubber.

Carbon fibers are filaments consisting of non-graphitic carbon obtained by carbonization of natural or synthetic organic fibers, or fibers drawn from organic precursors such as resins or pitches, and subsequently heat-treated up to temperatures of about 300 °C. Carbon fibers are light and exceeding strong, so that they are now important industrial materials that have gained increasing applications ranging from sport equipment to aerospace strategic uses.

From observations in transmission electron micrography (TEM), carbon soot particles are found to have an idealized onion-like shelled structure, as shown in Fig. 14.1.8. Carbon onions varying from 3 to 1000 nm in diameter have been observed experimentally. In an idealized model of a carbon onion, the first shell is a C_{60} core of I_h symmetry, the second shell is C_{240} (comprising $2^2 \times 60$

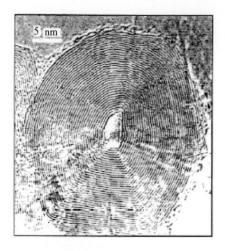

Fig. 14.1.8.
Section of a carbon onion.

atoms), and in general the number of carbon atom in the nth shell is $n^2 \times 60$. The molecular formula of a carbon onion is $C_{60}@C_{240}@C_{540}@C_{960}@\ldots$, and the intershell distance is always ~ 350 pm.

Recent work based on HRTEM (high-resolution transmission electron microscopy) and simulations has established that carbon onions are spherical rather than polyhedral, and the intershell spacing increases gradually from a value well below that of graphite at the center to the expected value at the outermost pair. The individual shells are not aligned in any regular fashion and do not rotate relative to each other. The innermost core can be a smaller fullerene (e.g. C_{28}) or a diamond fragment varying in size from 2 to 4.5 nm, and the internal cavity can be as large as 2 nm in diameter.

14.1.5 Carbon nanotubes

Carbon nanotubes were discovered in 1991 and consist of elongated cages bounded by cylindrical walls constructed from rolled graphene (graphite-like) sheets. In contrast to the fullerenes, nanotubes possess a network of fused six-membered rings, and each terminal of the long tube is closed by a half-fullerene cap. Single-walled carbon nanotubes (SWNTs) with diameters in the range 0.4 to 3.0 nm have been observed experimentally; most of them lie within the range 0.6 to 2.0 nm, and those with diameter 0.7, 0.5 and 0.4 nm correspond to the fullerenes C_{60}, C_{36} and C_{20}, respectively. Electron micrographs of the smallest 0.4 nm nanotubes prepared by the arc-discharge method showed that each is capped by half of a C_{20} dodecahedron and has an antichiral structure.

A carbon SWNT can be visualized as a hollow cylinder formed by rolling a planar sheet of hexagonal graphite (unit-cell parameters $a = 0.246$, $c = 0.669$ nm). It can be uniquely described by a vector $\mathbf{C} = n\mathbf{a}_1 + m\mathbf{a}_2$, where $\mathbf{a}_1$ and $\mathbf{a}_2$ are reference unit vectors as defined in Fig. 14.1.9. The SWNT is generated by rolling up the sheet such that the two end-points of the vector $\mathbf{C}$ are superimposed. The tube is denoted as (n, m) with $n \geq m$, and its diameter D

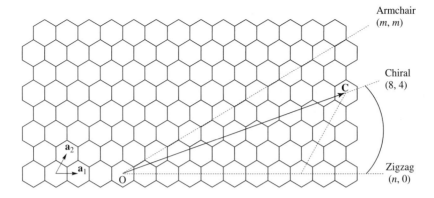

Fig. 14.1.9.
Generation of the chiral (8,4) carbon nanotube by rolling a graphite sheet along the vector $\mathbf{C} = n\mathbf{a}_1 + m\mathbf{a}_2$, and definition of the chiral angle θ. The reference unit vectors $\mathbf{a}_1$ and $\mathbf{a}_2$ are shown, and the broken lines indicate the directions for generating achiral zigzag and armchair nanotubes.

is given by

$$D = |\mathbf{C}|/\pi = a(n^2 + nm + m^2)^{\frac{1}{2}}/\pi$$

The tubes with $m = n$ are called "armchair" and those with $m = 0$ are referred to as "zigzag." All others are chiral with the chiral angle θ defined as that between the vectors $\mathbf{C}$ and $\mathbf{a}_1$; θ can be calculated from the equation

$$\theta = \tan^{-1}[3^{\frac{1}{2}}m/(m+2n)].$$

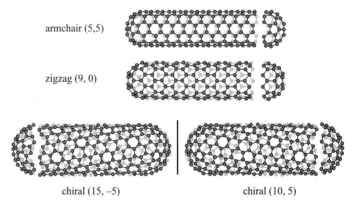

Fig. 14.1.10.
Lateral view of three kinds of carbon nanotubes with end caps: (a) armchair (5,5) capped by one-half of C_{60}, (b) zigzag (9,0) capped by one-half of C_{60}, and (c) an enantiomorphic pair of chiral SWNTs each capped by a hemisphere of icosahedral fullerene C_{140}.

The values of θ lies between 0° (for a zigzag tube) and 30° (for an armchair tube). Note that the mirror image of a chiral (n, m) nanotube is specified by $(n + m, -m)$. The three types of SWNTs are illustrated in Fig. 14.1.10.

There are multiwalled carbon nanotubes (MWNTs), each consisting of ten inner tubes or more. In a carbon MWNT, the spacing between two adjacent coaxial zigzag tubes $(n_1, 0)$ and $(n_2, 0)$ is $\Delta d/2 = (0.123/\pi)(n_2 - n_1)$. However, this cannot be made to be close to $c/2 = 0.335$ nm (the interlayer separation

Fig. 14.1.11.
A helical multi-walled carbon nanotube.

in graphite) by any reasonable combination of n_2 and n_1, and hence no zigzag nanotube can exist as a component of a MWNT. On the other hand, a MWNT can be constructed for all armchair tubes $(5m, 5m)$ with $m = 1, 2, 3$, etc., for which the intertube spacing is $(0.123/\pi)3^{1/2}(5) = 0.339$ nm, which satisfies the requirement.

In practice, defect-free coaxial nanotubes rarely occur in experimental preparations. The observed structures include the capped, bent, and toroidal SWNTs, as well as the capped and bent, branched, and helical MWNTs. Figure 14.1.11 shows the HRTEM micrograph of a helical multiwalled carbon nanotube which incorporates a small number of five- and seven-membered rings into the graphene sheets of the nanotube surfaces.

14.2 Compounds of carbon

More than twenty million compounds containing carbon atoms are now known, the majority of which are organic compounds that contain carbon–carbon bonds.

From the perspective of structural chemistry, the modes of bonding, coordination, and the bond parameters of a particular element in its allotropic modifications may be further extended to its compounds. Thus organic compounds can be conveniently divided into three families that originate from their prototypes: aliphatic compounds from diamond, aromatic compounds from graphite, and fullerenic compounds from fullerenes.

14.2.1 Aliphatic compounds

Aliphatic compounds comprise hydrocarbons and their derivatives in which the molecular skeletons consist of tetrahedral carbon atoms connected by C–C single bonds. These tetrahedral carbon atoms can be arranged as chains, rings, or finite frameworks, and often with an array of functional groups as substituents on various sites. The alkanes C_nH_{2n+2} and their derivatives are typical examples of aliphatic compounds.

Some frameworks of alicyclic compounds are derived from fragments of diamond, as shown in Fig. 14.2.1. In these molecules, all six-membered carbon rings have the chair conformation. Diamantane $C_{14}H_{20}$ is also named

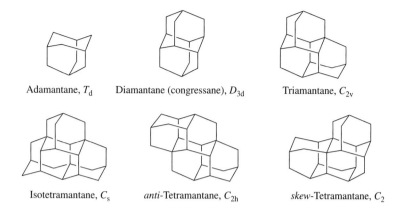

Fig. 14.2.1.
Some frameworks of alicyclic compounds as fragments of diamond.

congressane as it was chosen as the logo of the XIXth Conference of IUPAC in London in 1963 as a challenging target for the participants; the successful synthesis was accomplished two years later. There are three structural isomers of tetramantane $C_{22}H_{28}$. X-ray analysis of *anti*-tetramantane has revealed an interesting bond-length progression: $CH–CH_2 = 152.4$ pm, $C–CH_2 = 152.8$ pm, $CH–CH = 153.7$ pm, and $C–CH = 154.2$ pm, approaching the limit of 154.45 pm in diamond as the number of bonded H atoms decreases.

14.2.2 Aromatic compounds

Graphite typifies the basic structural unit present in aromatic compounds, in which the planar carbon skeletons of these molecules and their derivatives can be considered as fragments of graphite, each consisting of carbon atoms which use their sp^2 hybrids to form σ bonds to one another, and overlap between the remaining parallel p_z orbitals gives rise to delocalized π bonding. The aromatic compounds may be divided into the following four classes.

(1) Benzene and benzene derivatives

Up to six hydrogen atoms in benzene can be mono- or poly-substituted by other atoms or groups to give a wide variety of derivatives. Up to six sterically bulky groups such as $SiMe_3$ and ferrocenyl ($C_5H_5FeC_5H_4$, Fc) groups can be substituted into a benzene ring. Hexaferrocenylbenzene, C_6Fc_6, is of structural interest as a supercrowded arene, a metalated hexakis(cyclopentadienylidene)radialene, and the core for the construction of "Ferris wheel" supermolecules. In the crystalline solvate $C_6Fc_6 \cdot C_6H_6$, the C_6Fc_6 molecule adopts a propeller-like configuration with alternating up and down Fc groups around the central benzene ring. Perferrocenylation causes the benzene ring to take a chair conformation with alternating C–C–C–C dihedral angles of $\pm14°$, and the elongated C–C bonds exhibit noticeable bond alternation averaging 142.7/141.1 pm. The $C_{ar}–C_{Fc}$ bonds average 146.9(5) pm.

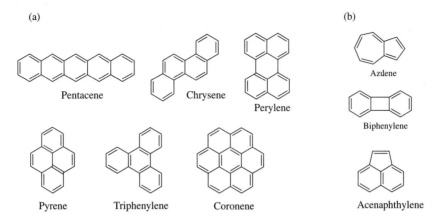

Fig. 14.2.2.
Carbon skeletons of some polycyclic aromatic compounds of the (a) benzenoid type and (b) non-benzenoid type.

(2) Polycyclic benzenoid aromatic compounds

These compounds consist of two or more benzene rings fused together, and the number of delocalized π electrons conforms to the Hückel $(4n + 2)$ rule for aromaticity. Figure 14.2.2(a) shows the carbon skeletons of some typical examples.

(3) Non-benzenoid aromatic compounds

Many aromatic compounds have considerable resonance stabilization but do not possess a benzene nucleus, or in the case of a fused polycyclic system, the molecular skeleton contains at least one ring that is not a benzene ring. The cyclopentadienyl anion $C_5H_5^-$, the cycloheptatrienyl cation $C_7H_7^+$, the aromatic annulenes (except for [6]annulene, which is benzene), azulene, biphenylene and acenaphthylene (see Fig. 14.2.2(b)) are common examples of non-benzenoid aromatic hydrocarbons. The cyclic oxocarbon dianions $C_nO_n^{2-}$ ($n = 3, 4, 5, 6$) constitute a class of non-benzenoid aromatic compounds stabilized by two delocalized π electrons. Further details are given in Section 20.4.4.

(4) Heterocyclic aromatic compounds

In many cyclic aromatic compounds an element other than carbon (commonly N, O, or S) is also present in the ring. These compounds are called heterocycles. Figure 14.2.3 shows some nitrogen heterocycles commonly used as ligands, as well as planar, sunflower-like octathio[8]circulene, $C_{16}S_8$, which can be regarded as a novel form of carbon sulfide.

14.2.3 Fullerenic compounds

The derivatives of fullerenes are called fullerenic compounds, which are now mainly prepared from C_{60} and, to a lesser extent, from C_{70} and C_{84}.

Since efficient methods for the synthesis and purification of gram quantities of C_{60} and C_{70} became available in the early 1990s, fullerene chemistry has developed at a phenomenal pace. There are many reactions which can generate

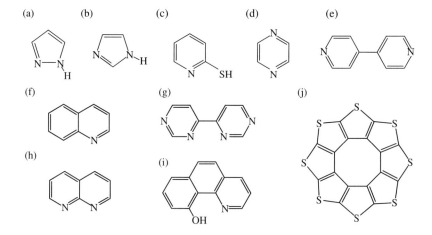

Fig. 14.2.3.
Some commonly used heterocyclic nitrogen ligands: (a) pyrazole, (b) imidazole, (c) pyridine-2-thiol, (d) pyrazine, (e) 4,4'-bipyridine, (f) quinoline, (g) 4,4'-bipyrimidine, (h) 1,8-naphthyridine, (i) 1,10-phenanthroline. Compounds (a), (b), and (c) occur in metal complexes in the anionic (deprotonated) form with delocalization of the negative charge. (j) Octathio[8]circulene has a highly symmetric planar structure.

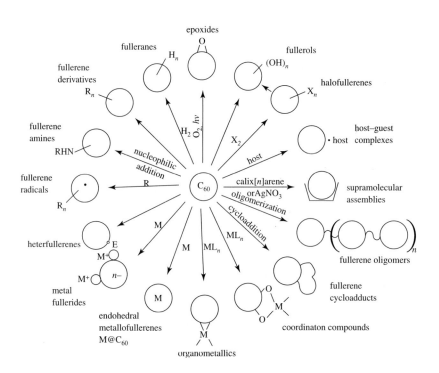

Fig. 14.2.4.
The principal reactions of C_{60}.

fullerenic compounds, as shown in Fig. 14.2.4. Unlike the aromatics, fullerenes have no hydrogen atoms or other groups attached, and so are unable to undergo substitution reactions. However, the globular fullerene carbon skeleton gives rise to an unprecedented diversity of derivatives. This unique feature leads to a vast number of products that may arise from addition of just one reagent. Substitution reactions can take place on derivatives, once these have been formed by addition. Some fullerenic compounds are briefly described below.

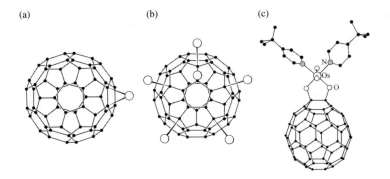

Fig. 14.2.5.
Fullerenes bonded to non-metallic elements: (a) $C_{60}O$, (b) $C_{60}Br_6$, and (c) $C_{60}[OsO_4(pyBu)_2]$.

(1) Fullerenes bonded to non-metallic elements

This class of compounds consists of fullerene adducts with covalent bonds formed between the fullerene carbon atoms and non-metallic elements. Since all carbon atoms lie on the globular fullerene surface, the number of sites, as well as their positions, where additions take place vary from case to case. Examples of these compounds include $C_{60}O$, $C_{60}(CH_2)$, $C_{60}(CMe_3)$, $C_{60}Br_6$, $C_{60}Br_8$, $C_{60}Br_{24}$, $C_{50}Cl_{10}$, and $C_{60}[OsO_4(pyBu)_2]$. In the first two examples, the oxygen atom and the methylene carbon atom are each bonded to two carbon atoms on a (6/6) edge in the fullerene skeleton. The structure of $C_{60}O$ is shown in Fig. 14.2.5(a). In the next five examples, the C atom of CMe_3, as well as Br and Cl, is each bonded to only one carbon atom of fullerene. The structure of $C_{60}Br_6$ is displayed in Fig. 14.2.5(b). In the final example, the six-coordinate osmium(VI) moiety $OsO_4(pyBu)_2$ is linked to C_{60} through the formation of a pair of O–C single bonds with two neighboring carbon atoms in the fullerene skeleton, as shown in Fig. 14.2.5(c).

(2) Coordination compounds of fullerene

This class of coordination compounds features direct covalent bonds between complexed metal groups and the carbon atoms of fullerene systems. Monoadducts such as $C_{60}Pt(PPh_3)_2$ each have only one group bonded to a fullerene, as shown in Fig. 14.2.6(a). Multiple adducts are formed when several groups are attached to the same fullerene nucleus. Typical examples are $C_{60}[Pt(PPh_3)_2]_6$ and $C_{70}[Pt(PPh_3)_2]_4$, whose structures are displayed in Figs. 14.2.6(b) and 14.2.6(c), respectively.

(3) Fullerenes as π-ligands

In this class of metal complexes, there is delocalized π bonding between fullerene and the metal atom. The structures of $(\eta^5\text{-}C_5H_5)Fe(\eta^5\text{-}C_{60}Me_5)$ and $(\eta^5\text{-}C_5H_5)Fe(\eta^5\text{-}C_{70}Me_3)$, each containing a fused ferrocene moiety, are shown in Fig. 14.2.7. The shared pentagonal carbon ring of the C_{60} (or C_{70}) skeleton acts as a 6π-electron donor ligand to the Fe(II) atom of the $Fe(C_5H_5)$ fragment. In $(\eta^5\text{-}C_5H_5)Fe(\eta^5\text{-}C_{60}Me_5)$, the five methyl groups attached to five sp^3 carbon atoms protrude outward at an angle of 42° relative to the symmetry axis of the molecule. The C_5H_5 group and cyclopentadienide in $Fe(C_{60}Me_5)$ are arranged in a staggered manner; the C–C bond lengths are 141.1 pm (averaged for C_5H_5) and 142.5 pm (averaged for $C_{60}Me_5$). The Fe–C distances are 203.3 pm for C_5H_5 and 208.9 pm for $C_{60}Me_5$, which are comparable to those in known ferrocene derivatives. The structural features of $(\eta^5\text{-}C_5H_5)Fe(\eta^5\text{-}C_{70}Me_3)$ are similar to those of $(\eta^5\text{-}C_5H_5)Fe(\eta^5\text{-}C_{60}Me_5)$. The C–C bond lengths in the shared pentagon are 141–3 pm. The Fe–C bond distances are 205.4 pm (averaged for C_5H_5) and 208.3 pm (averaged for $C_{70}Me_3$).

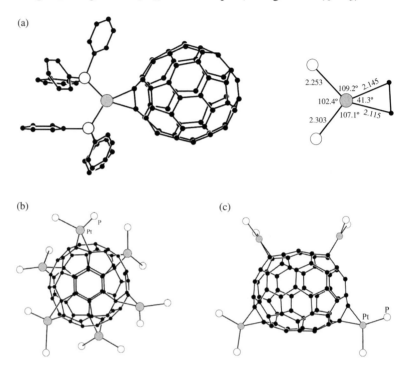

Fig. 14.2.6.
Molecular structure of (a) $C_{60}Pt(PPh_3)_2$, (b) $C_{60}[Pt(PPh_3)_2]_6$, and (c) $C_{70}[Pt(PPh_3)_2]_4$.

(4) Metal fullerides

Fullerenes exhibit an electron-accepting nature and react with strong reducing agents, such as the alkali metals, to yield metal fulleride salts. The compounds $Li_{12}C_{60}$, $Na_{11}C_{60}$, M_6C_{60} (M = K, Rb, Cs), K_4C_{60} and M_3C_{60} (M_3 = K_3, Rb_3, $RbCs_2$) have been prepared.

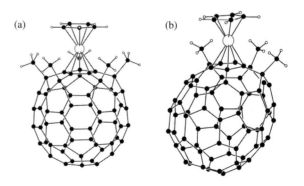

Fig. 14.2.7.
Structure of
(a) $(\eta^5\text{-}C_5H_5)Fe(\eta^5\text{-}C_{60}Me_5)$ and
(b) $(\eta^5\text{-}C_5H_5)Fe(\eta^5\text{-}C_{70}Me_3)$.

The M_3C_{60} compounds are particularly interesting as they become superconducting materials at low temperature, with transition temperatures (T_c) of 19 K for K_3C_{60}, 28 K for Rb_3C_{60}, and 33 K for $RbCs_2C_{60}$.

Fulleride K_3C_{60} is a face-centered cubic crystal, space group $Fm\bar{3}m$, with $a = 1424(1)$ pm and $Z = 4$. The C_{60}^{3-} ions form a ccp structure with K^+ ions filling all the tetrahedral interstices (radius 112 pm) and octahedral interstices (radius 206 pm), as shown in Fig. 14.2.8.

Both K_4C_{60} and Rb_4C_{60} are tetragonal, and the structure of M_6C_{60} at room temperature is body-centered cubic. These metal fullerides are all insulators.

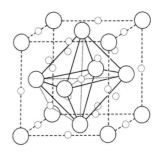

Fig. 14.2.8.
Crystal structure of K_3C_{60} (large circles represent C_{60} and small circles represent K).

(5) Fullerenic supramolecular adducts

Fullerenes C_{60} and C_{70} form supramolecular adducts with a variety of molecules, such as crown ethers, ferrocene, calixarene, and hydroquinone. In the solid state, the intermolecular interactions may involve ionic interaction, hydrogen bonding, and van der Waals forces. Figure 14.2.9 shows a part of the structure of $[K(18C6)]_3 \cdot C_{60} \cdot (C_6H_5CH_3)_3$, in which C_{60}^{3-} is surrounded by a pair of $[K^+(18C6)]$ complexed cations.

(6) Fullerene oligomers and polymers

This class of compounds contain two or more fullerenes and they may be further divided into the following three categories.

(a) Dimers and polymers containing two or more fullerenes are linked together through C–C covalent bonds formed by the carbon atoms on the globular surface. Figure 14.2.10(a) shows the structure of dimeric $[C_{60}(^tBu)]_2$. In addition, chain-like polymeric fullerenes with the following structure have been proposed, although none has yet been prepared:

$$-\!\!\!\!-\!\!\!\!\{C_{60}H_2\!\!-\!\!C_{60}H_2\!\!-\!\!C_{60}H_2\}_n\!\!-\!\!\!\!-$$

(b) Two or more fullerenes are linked together through the formation of C–C covalent bonds with an organic functional group. Figure 14.2.10(b) shows the structure of $[C_{60}]_3(C_{14}H_{14})$, in which three fullerenes form an adduct through C-C linkage with the central $C_{14}H_{14}$ unit.

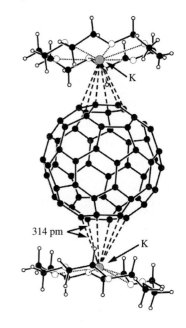

Fig. 14.2.9.
A part of the structure of $[K(18C6)]_3 \cdot C_{60} \cdot (C_6H_5CH_3)_3$.

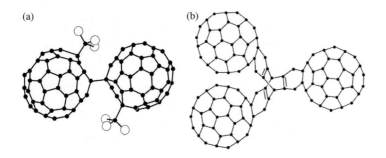

Fig. 14.2.10.
Structures of (a) $[C_{60}(^tBu)]_2$ and (b) $[C_{60}]_3(C_{14}H_{14})$.

(c) Fullerenes are linked as pendants at regular intervals to a skeleton of a polymeric chain, such as:

$$\sim\sim\sim\text{CH--CR=CH--CH}_2\sim\sim\sim\text{CH--CR=CH--CH}_2\sim\sim\sim$$
$$\qquad\qquad |\qquad\qquad\qquad\qquad\qquad |$$
$$\qquad\quad C_{60}H\qquad\qquad\qquad\qquad\quad C_{60}H$$

(7) Heterofullerenes

Heterofullerenes are fullerenes in which one or more carbon atoms in the cage are replaced by other main-group atoms. Compounds containing boron, such as $C_{59}B$, $C_{58}B_2$, $C_{69}B$, and $C_{68}B_2$, have been detected in the mass spectra of the products obtained using boron/graphite rods in arc discharge synthesis. Since nitrogen has one more electron than carbon, the azafullerenes are radicals ($C_{59}N\cdot$, $C_{69}N\cdot$), which can either dimerise to give $(C_{59}N)_2$ and $(C_{69}N)_2$, or take up hydrogen to give $C_{59}NH$.

(8) Endohedral fullerenes (incar-fullerenes)

Fullerenes can encapsulate various atoms within the cages, and these compounds have been referred to as endohedral fullerenes. For example, the symbolic representations La@C_{60} and La$_2$@C_{80} indicate that the fullerene cage encapsulates one and two lanthanum atom(s), respectively. The IUPAC description refers to these fullerenes species as incar-fullerenes, and the formulas are written as iLaC$_{60}$ and iLa$_2$C$_{80}$, (i is derived from $incarcerane$). Some metal endohedral fullerenes are listed in Table 14.2.1. The endohedral fullerenes are expected to have interesting and potentially very useful bulk properties as well as a fascinating chemistry. Some non-metallic elements, such as N, P, and noble gases, can be incarcerated into fullerenes to form N@C_{60}, P@C_{60}, N@C_{70}, Sc$_3$N@C_{80}, Ar@C_{60}, etc.

Table 14.2.1. Endohedral fullerenes

Fullerene	Metallic atom	Fullerene	Metallic atom
C_{36}	U	C_{56}	U_2
C_{44}	K, La, U	C_{60}	Y_2, La$_2$, U_2
C_{48}	Cs	C_{66}	Sc$_2$
C_{50}	U, La	C_{74}	Sc$_2$
C_{60}	Li, Na, K, Rb, Cs, Ca, Ba, Co, Y, La, Ce, Pr, Nd, Sm, Eu, Gd, Tb, Dy, Ho, Lu, U	C_{76}	La$_2$
		C_{79}N	La$_2$
		C_{80}	La$_2$, Ce$_2$, Pr$_2$
C_{70}	Li, Ca, Y, Ba, La, Ce, Gd, Lu, U	C_{82}	Er$_2$, Sc$_2$, Y$_2$, La$_2$, Lu$_2$, Sc$_2$C$_2$
C_{72}	U		
C_{74}	Ca, Sc, La, Gd, Lu	C_{84}	Sc$_2$, La$_2$, Sc$_2$C$_2$
C_{76}	La	C_{68}	Sc$_3$N
C_{80}	Ca, Sr, Ba	C_{78}	Sc$_3$N
C_{81}N	La	C_{80}	Sc$_3$N, ErSc$_2$N, Sc$_2$LaN, ScLa$_2$N, La$_3$N
C_{82}	Ca, Sr, Ba, Sc, Y, La, Ce, Pr, Nd, Sm, Eu, Gd, Tb, Dy, Ho, Er, Tm, Yb, Lu	C_{82}	Sc$_3$
		C_{84}	Sc$_3$
C_{84}	Ca, Sr, Ba, Sc, La	C_{82}	Sc$_4$

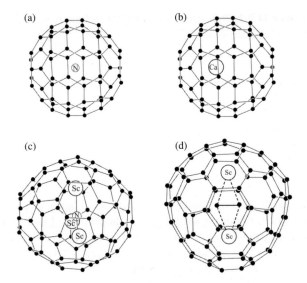

Fig. 14.2.11.
Structures of some endofullerenes: (a) N@C_{60}, (b) Ca@C_{60}, (c) Sc_3N@C_{78} and (d) Sc_2C_2@C_{84}.

Theoretical and experimental studies of the structures and electronic properties of endohedral fullerenes have yielded many interesting results. The enclosed N and P atoms of N@C_{60} and P@C_{60} retain their atomic ground state configuration and are localized at the center of the fullerenes, as shown in Fig. 14.2.11(a). The atoms are almost freely suspended inside the respective molecular cages and exhibit properties resembling those of ions in electromagnetic traps. In Ca@C_{60}, the Ca atom lies at an off-center position by 70 pm, as shown in Fig. 14.2.11(b). The symmetry of Ca@C_{60} is predicted to be C_{5v}, implying that the Ca atom lies on a fivefold rotation axis of the C_{60} cage. The predicted distances between the Ca atom and the first and second set of nearest C atoms are 279 and 293 pm, respectively.

A single-crystal X-ray diffraction study of [Sc_3N@C_{78}]·[Co(OEP)] ·1.5(C_6H_6)·0.3($CHCl_3$) (OEP is the dianion of octaethylporphyrin) showed that the fullerene is embraced by the eight ethyl groups of the porphyrin macrocycle. The structure of [Sc_3N@C_{78}] is shown in Fig. 14.2.11(c). The N–Sc distances range from 198 to 212 pm, and the shortest C–Sc distances fall within a narrow range of 202–11 pm. The flat Sc_3N unit is oriented so that it lies near the equatorial mirror plane of the C_{78} cage.

A synchrotron X-ray powder diffraction study of (Sc_2C_2)@C_{84} showed that the lozenge-shaped Sc_2C_2 unit is encapsulated by the D_{2d}-C_{84} fullerene, as shown in Fig. 14.2.11(d). The Sc–Sc, Sc–C and C–C distances in the Sc_2C_2 unit are 429, 226 and 142 pm, respectively.

14.3 Bonding in carbon compounds

14.3.1 Types of bonds formed by the carbon atom

The capacity of the carbon atom to form various types of covalent bonds is attributable to its unique characteristics as an element in the Periodic Table.

518   Structural Chemistry of Selected Elements

Table 14.3.1. Hybridization schemes of carbon atom

	sp	sp^2	sp^3
Number of orbitals	2	3	4
Interorbital angle	180°	120°	109.47°
Geometry	Linear	Trigonal	Tetrahedral
% s character	50	33	25
% p character	50	67	75
Electronegativity of carbon	3.29	2.75	2.48
Remaining p orbitals	2	1	0

The electronegativity of the carbon atom is 2.5, which means that the carbon atom cannot easily gain or lose electrons to form an anion or cation. As the number of valence orbitals is exactly equal to the number of valence electrons, the carbon atom cannot easily form a lone pair or electron-deficient bonds. Carbon has a small atomic radius, so its orbitals can overlap effectively with the orbitals of neighbor atoms in a molecule.

For simplicity, we use the conventional concept of hybridization to describe the bond types, but bonding is frequently more subtle and more extended than implied by this localized description. The parameters of typical hybridization schemes of the carbon atom are listed in Table 14.3.1.

The hybrid orbitals of carbon always overlap with orbitals of other atoms in a molecule to form σ bonds. The remaining p orbitals can then be used to form π bonds, which can be classified into two categories: localized and delocalized. The localized π bonds of carbon form double and triple bonds as illustrated below:

The carbon atom can also form σ and π bonds to metal atoms in various fashions. For example:

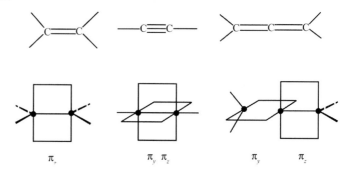

(a) >C—M   single bond

(b) polycenter metal-carbon bonds

(c) >C=M   double bond

(d) —C≡M   triple bond

A particularly interesting example is the tungsten complex

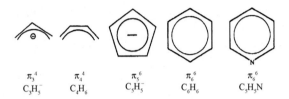

which contains C≡W, C=W, and C–W bonds with lengths 179, 194, and 226 pm, respectively.

The delocalized π bonds involve three or more carbon atoms or heteroatoms. For instance:

π_3^4 $C_3H_5^-$ π_4^4 C_4H_6 π_5^6 $C_5H_5^-$ π_6^6 C_6H_6 π_6^6 C_5H_5N

An enormous variety of π-bonded systems, whether they be neutral or ionic, cyclic or linear, odd or even in the number of carbon atoms, serve as ligands that coordinate to transition metals. Figure 14.3.1 shows some representatives of the innumerable organometallic coordination compounds stabilized by metal-π bonding.

Larger aromatic rings and polycyclic aromatic hydrocarbons can also function as π ligands in forming sandwich-type metal complexes. For example, the planar cyclooctatetraenyl dianion $C_8H_8^{2-}$ functions as a η^8-ligand to form the sandwich compound uranocene, $U(C_8H_8)_2$, which takes the eclipsed configuration.

Recently the "sandwich" motif has been extended to the case of two cyclic aromatic ligands flanking a small planar aggregate of metal atoms. In $[Pd_3(\eta^7\text{-}C_7H_7)_2Cl_3](PPh_4)$, a triangular unit of palladium(0) atoms each coordinated by

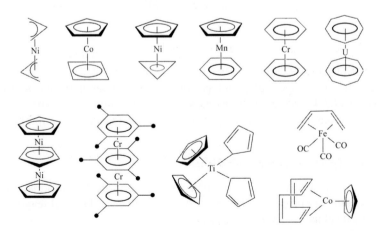

Fig. 14.3.1.
Some examples of organometallic compounds formed with π bonding ligands.

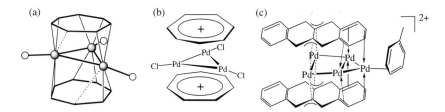

Fig. 14.3.2.
(a) Molecular geometry and (b) structural formula of the $[Pd_3(\eta^7\text{-}C_7H_7)_2Cl_3]^-$ ion; the Pd–C and Pd···C bond lengths are in the ranges 215–47 and 253–61 pm, respectively. (c) Structure of the $[Pd_5(C_{18}H_{12})_2(C_7H_8)]^{2+}$ ion; coordination of an electronically delocalized C_3 fragment to a Pd center is represented by a broken line.

a terminal chloride ligand is sandwiched between two planar cycloheptatrienyl $C_7H_7^+$ rings, as shown in Figs. 14.3.2(a, b). The measured Pd–Pd (274.5 to 278.9 pm) and Pd–Cl (244.2 to 247.1 pm) bonds are within the normal ranges.

In $[Pd_5(\text{naphthacene})_2(\text{toluene})][B(Ar_f)_4]_2 \cdot 3\text{toluene}$, where $Ar_f = 3,5\text{-}(CF_3)_2C_6H_3$, the sheet-like pentapalladium(0) core is composed of a triangle sharing an edge with a trapezoid; the innermost Pd–Pd distance (291.6 pm) is relatively long. This metal monolayer is sandwiched between two naphthacene radical cations, each of which coordinates to the Pd_5 sheet through 12 carbons via the $\mu_5\text{-}\eta^2{:}\eta^2{:}\eta^2{:}\eta^3{:}\eta^3$ mode, as illustrated in Fig. 14.3.2(c). One of the four independent toluene molecules in the unit cell is located near the apex Pd atom at a closest contact of 252 pm, but its coordination mode (either η^2 or η^1) cannot be definitively assigned owing to disorder.

14.3.2 Coordination numbers of carbon

Carbon is known with all coordination numbers from 0 to 8. Some typical examples are given in Table 14.3.2, and their structures are shown in Fig. 14.3.3. In these examples, the compounds with the high coordination numbers, such as ≥ 5, do not belong to the class of hypervalent compounds, but rather to electron-deficient systems. Hypervalent molecules usually have a central atom which requires the presence of more than an octet of electrons to form more than four $2c\text{-}2e$ bonds, such as the S atom in SF_6.

14.3.3 Bond lengths of C–C and C–X bonds

The bond lengths of carbon–carbon bonds are listed in Table 14.3.3. The bond lengths of some important bond types of carbon–heteroatom bonds are given in Table 14.3.4. The values given here are average values from experimental determinations and do not necessarily exactly apply to a particular compound.

14.3.4 Factors influencing bond lengths

The measured bond lengths in a molecule provide valuable information on its structure and properties. Some factors that influence the bond lengths are discussed below.

Group 14 Elements

Table 14.3.2. Coordination numbers of carbon

Coordination number	Examples	Structure in Fig. 14.3.2
0	C atoms, gas phase	
1	CO, stable gas	(a)
2	CO_2, stable gas	(b) Linear
2	HCN, stable gas	(c) Linear
2	:CX_2 (carbene), X = H, F, OH	(d) Bent
3	C=OXY (oxohalides, ketones)	(e) Planar
3	CH_3^-, CPh_3^-	(f) Pyramidal
3	Ta(=CHCMe$_3$)$_2$(Me$_3$C$_6$H$_2$)(PMe$_3$)$_2$	(g) T-shaped*
4	CX_4 (X = H, F, Cl)	(h) Tetrahedral
4	$Fe_4C(CO)_{13}$	(i) C capping Fe_4
5	Al_2Me_6	(j) Bridged dimer
5	$(Ph_3PAu)_5C^+ \cdot BF_4^-$	(k) Trigonal bipyramidal
6	$[Ph_3PAu]_6C^{2+}$	(l) Octahedral
6	$C_2B_{10}H_{12}$	(m) Pentagonal pyramidal
7	[LiMe]$_4$ crystal	(n) †
8	$[(Co_8C(CO)_{18}]^{2-}$	(o) Cubic

* The unique H is equatorial and angle Ta=C–CMe$_3$ is 169°.
† The distance of intramolecular C–Li is 231 pm, (a C atom caps 3 Li atoms in each face of the Li$_4$ tetrahedron), C–H is 96 pm, and intermolecular C–Li is 236 pm.

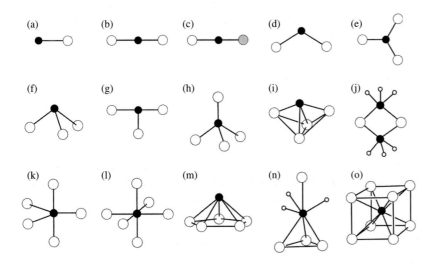

Fig. 14.3.3. Coordination numbers of carbon in its compounds.

(1) Bonds between atoms of different electronegativities

It is generally found that the greater the difference in electronegativity between the bonding atoms, the greater the deviation from the predicted bond distance based on covalent radii or the mean values given in Tables 14.3.3 and 14.3.4. Empirical methods for the adjustment of covalent bond lengths by a factor dependent on the electronegativity differences between the atoms have been proposed.

Table 14.3.3. Bond lengths (in pm) of carbon–carbon bonds

Bond	Bond length	Example
C–C		
sp^3–sp^3	153	Ethane (H_3C–CH_3)
sp^3–sp^2	151	Propene (H_3C–CH=CH_2)
sp^3–sp	147	Propyne (H_3C–C≡CH)
sp^2–sp^2	148	Butadiene (H_2C=CH–CH=CH_2)
sp^2–sp	143	Vinylacetylene (H_2C=CH–C≡CH)
sp–sp	138	Butadiyne (HC≡C–C≡CH)
C=C		
sp^2–sp^2	132	Ethylene (H_2C=CH_2)
sp^2–sp	131	Allene (H_2C=C=CH_2)
sp–sp	128	Butatriene (H_2C=C=C=CH_2)
C≡C		
sp–sp	118	Acetylene (HC≡CH)

Table 14.3.4. Bond lengths (in pm) of C–X bonds

C–H			C–N		C–S	
			sp^3–N	147	sp^3–S	182
	sp^3–H	109	sp^2–N	138	sp^2–S	175
	sp^2–H	108			sp–S	168
	sp–H	108	C=N			
C–O			sp^2–N	128	C=S	
					sp^2–S	167
	sp^3–O	143	C≡N			
	sp^2–O	134	sp–N	114	C–Si	
C=O			C–P		sp^3–Si	189
	sp^3–O	121	sp^3–P	185	C=Si	
	sp^2–O	116	C=P		sp^2–Si	170
			sp^2–P	166		
			C≡P			
			sp–P	154		

C–X	X =	F	Cl	Br	I
	sp^3–X	140	179	197	216
	sp^2–X	134	173	188	210
	sp–X	127	163	179	199

(2) Steric strain

Steric strain exists in a molecule when bonds are forced to make abnormal angles. There are in general two kinds of structural features that result in sterically-caused abnormal bond angles. One of these is common to small ring compounds, where the bond angles must be less than those resulting from normal orbital overlap. Such strain is called small-angle strain. The other arises when nonbonded atoms are forced into close proximity by the geometry of the molecule. This effect is known as steric overcrowding.

Steric strain generally leads to elongated bond lengths as compared to the expected values given in Tables 14.3.3 and 14.3.4.

(3) Conjugation

A hybrid atomic orbital of higher s content has a smaller size and lies closer to the nucleus. Accordingly, carbon–carbon bonds are shortened by increasing s character of the overlapping hybrid orbitals. The C(sp^3)–C(sp^3) single bond is generally longer than other single bonds involving sp^2 and sp carbon atoms. This general rule arises mainly from conjugation between the bonded atoms. When mean values are used to estimate bond lengths, the conjugation factor must be taken into account. For example, the C–C bond length in benzene is 139.8 pm, which is virtually equal to the mean value of the C(sp^2)–C(sp^2) and C=C bond lengths: $\frac{1}{2}(148 + 132) = 140$ pm. Hence the conjugation of alternating single and double bonds in benzene can be expressed as resonance between two limiting valence bond (canonical) structures:

(4) Hyperconjugation

The overlap of a C–H σ orbital with the π (or p) orbital on a directly bonded carbon atom is termed hyperconjugation. This interaction has a shortening effect on the C–C bond length, a good example being the structure of the *tert*-butyl cation [C(CH$_3$)$_3$]$^+$.

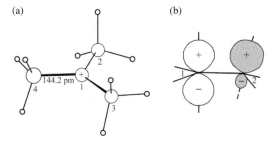

Fig. 14.3.4.
Hyperconjugation in [C(CH$_3$)$_3$]$^+$: (a) structure of [C(CH$_3$)$_3$]$^+$ and (b) hyperconjugation between central carbon atom C(1) and one of the C–H σ bonds.

The structure of [C(CH$_3$)$_3$]$^+$ in the crystalline salt [C(CH$_3$)$_3$]Sb$_2$F$_{11}$ is shown in Fig. 14.3.4(a). The carbon skeleton is planar with an average C–C bond length of 144.2 pm and approximate D_{3h} molecular symmetry. The experimental C–C bond length is shorter by 6.8 pm than the normal C(sp^3)-C(sp^2) bond length of 151 pm. This is due to hyperconjugative interaction between three filled C–H σ bond orbitals and the empty p orbital on the central carbon atom that leads to partial C–C π bonding, as shown in Fig. 14.3.4 (b).

(5) Surroundings of bonding atoms

The geometry and connected groups of bonding atoms usually influence the bond lengths. For example, an analysis of C–OR bond distances in more than 2000 ethers and carboxylic esters (all with sp^3 carbon) showed that this distance increases with increasing electron-withdrawing power of the R group, and also when the C atom changes from primary through secondary to tertiary. For such compounds mean C–O bond lengths range from 141.8 to 147.5 pm.

As an illustrative example taken from the current literature, consider the variation of C–C and C–O bond lengths in the deltate species $C_3O_3^{2-}$ held within a dinuclear organometallic uranium(IV) complex. In a remarkable synthesis, this cyclic aromatic oxocarbon dianion is generated by the metal-mediated reductive cyclotrimerization of carbon monoxide, as indicated in the reaction scheme

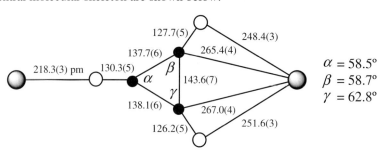

A strongly reducing organouranium(III) complex stabilized by η^5-cyclopentadienide and η^8-cyclooctatetraenediide ligands, together with tetrahydrofuran, is used to crack the robust C≡O triple bond at room temperature and ambient pressure. Low-temperature X-ray analysis revealed the presence of a reductively homologated CO trimer, $C_3O_3^{2-}$, as a $\eta^1{:}\eta^2$ bridging ligand between two organouranium(IV) centers. The measured dimensions of the central molecular skeleton are shown below:

The noticeable distortion of the C_3 ring and measured C–C, C–O and U–O bond lengths reflect the conjugation effect, the chemical environment of individual atoms, steric congestion around the respective uranium centers, and variation caused by experimental errors.

14.3.5 Abnormal carbon–carbon single bonds

(1) Unusually long C–C bonds

Figure 14.3.5 shows some organic molecules which have abnormally long C–C bonds. Rationalization of these unusual structures is discussed below.

(a) Oxalic acid, HOOC–COOH

In the oxalic acid molecule, there is a O=C–C=O system which satisfies the condition for conjugation, and the existence of delocalized π bonding has been

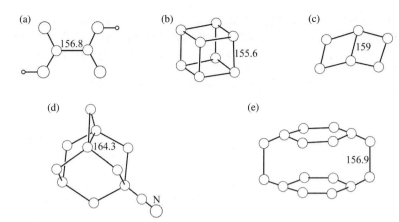

Fig. 14.3.5.
Some molecules with abnormally long C–C bonds (lengths in pm): (a) oxalic acid, (b) cubane, (c) skeleton of Dewar benzene derivative, (d) 1-cyano-tetracyclodecane, and (e) paracyclophane.

substantiated by deformation density studies. The central C(sp^2)–C(sp^2) bond has a length of 156.8 pm. A MNDO computation showed that the four fully occupied π MOs are alternately bonding and antibonding orbitals between the two C atoms. Thus the π MOs contribute little to bonding. The σ bond order is less than unity mainly because each carbon atom is bonded to two highly electronegative oxygen atoms, leading to a reduction of charge density on the carbon atom. The bond order in the C–C bond is made up of a σ component of 0.815 and a π component of only 0.015. The strongly reducing properties of oxalic acid are associated with its relatively weak C–C bond which easily cleaves during a reaction, and the resulting fragments are then oxidized to carbon dioxide.

(b) Cubane, C$_8$H$_8$

Several cubane structures have been determined. The mean C–C bond length is 155.6 pm from 42 independent measurements of the edges. This long bond length can be understood by considering the lesser overlap of endocyclic orbitals that are richer in p character than the typical sp^3 hybrid.

(c) Dewar benzene derivatives

In derivatives of Dewar benzene, the bond length of the bridging C–C bond is found to lie in the range 156–159 pm. The elongation of bond length is mainly due to the strain caused by the fusion of two four-membered rings.

(d) 1-Cyano-tetracyclodecane, C$_{10}$H$_{13}$CN

In this molecule, the pair of bridgehead atoms each has all four bonds directed to the same side as its partner atom, and the central bond length is 164.3 pm. Thus the bridgehead atoms have an "inverted" bond configuration and serious C–C bond strain.

(e) [2.2]Paracyclophane, C$_{16}$H$_{16}$

In the cyclophanes, the bridging bond lengths are generally the longest and in some instances have been found to be near or greater than 160 pm. The

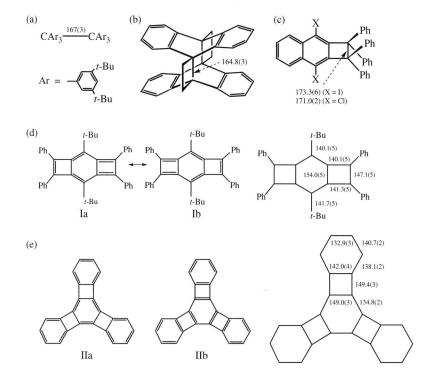

Fig. 14.3.6.
Molecules containing very long C–C single bonds. The bond lengths are shown in pm.

elongation of the C–C single bond in the bridging –CH$_2$–CH$_2$– group of [2.2]paracyclophane is attributable to the inherent steric strain.

Further examples of organic molecules containing very long carbon–carbon single bonds have been reported in recent years. The extended C(sp^3)–C(sp^3) bond in the hexaarylethane shown in Fig. 14.3.6(a) is caused by severe steric repulsion between the bulky substituted phenyl groups. In the bi(anthracene-9,10-dimethylene) photodimer shown in Fig. 14.3.6(b), the bridge bond of the cyclobutane ring has a length longer than the rest. Extremely long C(sp^3)–C(sp^3) bonds have been shown to exist in the naphthocyclobutenes shown in Fig. 14.3.6(c).

The bond length variation in the substituted benzodicyclobutadiene shown in Fig. 14.3.6(d), as determined from X-ray analysis, can be fairly well accounted for by resonance between the canonical formulas Ia and Ib. The central six-membered ring contains a pair of extremely long C(sp^2)–C(sp^2) bonds at 154.0(5) pm, which significantly exceed the reference bond distance of 149.0(3) pm observed for tris(benzocyclobutadieno)benzene [Fig. 14.3.6(e)]. The bond length pattern in the latter compound indicates that its structure is better described by the radiallene formula IIa in preference to formula IIb, which contains anti-aromatic cyclobutadiene moieties.

(2) Unusually short bonds between tetracoordinate carbon atoms

Figure 14.3.7 shows some organic molecules containing abnormally short single bonds between two four-coordinate carbon atoms.

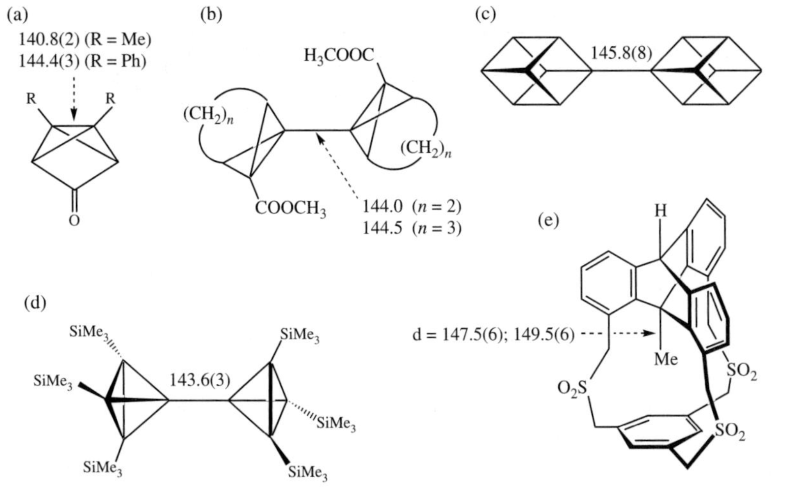

Fig. 14.3.7.
Molecules containing very short C–C single bonds. The bond lengths are shown in pm. Values are given for two independent molecules of compound (e).

The bond between two bridgehead carbon atoms in bicyclo[1.1.0]butane exhibits the properties of a carbon–carbon multiple bond, although it is formally a single bond. The dihedral angle δ between the three-membered rings in 1,5-dimethyltricyclo[2.1.0.0]pentan-3-one is made small by the short span of the carbonyl linkage, as shown in Fig. 14.3.7(a). The bridgehead bond has a pronounced π character with a length of 140.8(2) pm. In the 1,5-diphenyl analog, the two aromatic rings are oriented almost perpendicular (at 93.6° and 93.6°) to the plane bisecting the angle δ. There is optimal conjugation between the phenyl groups via the π-population of the bridge bond, and the conjugation effect leads to its lengthening of the latter to 144.4(3) pm. The central exocyclic $C(sp^3)$–$C(sp^3)$ bond connecting two bicyclobutane moieties is quite short [Fig. 14.3.7(b)], as are the related linkages in bicubyl [Fig. 14.3.7(c)] and hexakis(trimethylsilyl)bitetrahedryl [Fig. 14.3.7(d)]. The calculated s character of the linking C–C bond in the bitetrahedryl molecule is $sp^{1.53}$, which is consistent with its significant shortening.

In the crystal structure of the *in*-isomer of the methylcyclophane shown in Fig. 14.3.7(e), there are two independent molecules with measured C–Me bond distances of 147.5(6) and 149.5(6) pm. The inward-pointing methyl group is forced into close contact with the basal aromatic ring, and the steric congestion accounts for compression of the C–Me bond.

14.3.6 Complexes containing a naked carbon atom

In the realm of all-carbon ligands in the formation of transition-metal complexes, the naked carbon atom holds a special position. Based on the geometry of metal–carbon interaction, these compounds can be divided into four classes: terminal carbide (I), 1,3-dimetallaallene (II), C-metalated carbyne (III), and carbido cluster (IV):

$$:C\equiv M \qquad M=C=M \qquad M\equiv C-M \qquad C@M_n$$
$$(I) \qquad\quad (II) \qquad\quad (III) \qquad\quad (IV)$$

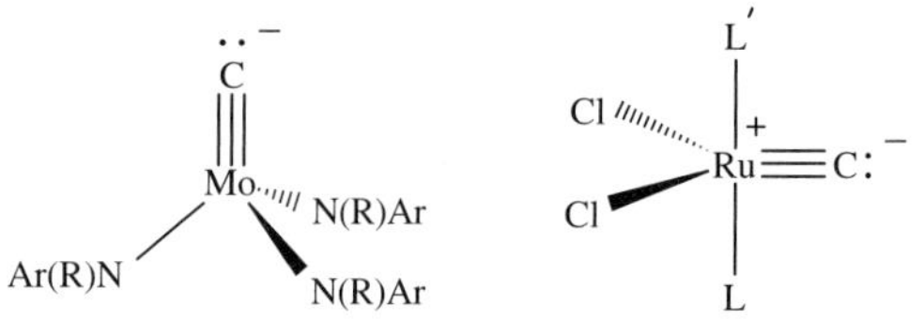

Fig. 14.3.8.
Complexes containing a metal–terminal carbon triple bond.

There are two well-characterized examples of a naked carbon atom bound by a triple bond to a metal center (Fig 14.3.8). The molybdenum carbide anion $[CMo\{N(R)Ar\}_3]^-$ (R = $C(CD_3)_2(CH_3)$, Ar = $C_6H_3Me_2$-3,5), an isoelectronic analog of $NMo\{N(R)Ar\}_3$, can be prepared in a multistep procedure via deprotonation of the d^0 methylidyne complex $HCMo\{N(R)Ar\}_3$. The Mo≡C distance of 171.3(9) pm is at the low end of the known range for molybdenum–carbon multiple bonds. In the diamagnetic, air-stable terminal ruthenium carbide complex $Ru(\equiv C:)Cl_2(LL')(L = L' = PCy_3$, or L = PCy_3 and L' = 1,3-dimesityl-4,5-dihydroimidazol-2-ylidene), the measured Ru–C distance of 165.0(2) pm is consistent with the existence of a very short Ru≡C triple bond.

Many complexes containing a M–C–M' bridge have been reported. The earliest know example of a 1,3-dimetallallene complex, $\{Fe(tpp)\}_2C$ (tpp = tetraphenylporphyrin), was synthesized by the reaction of Fe(tpp) with CI_4, CCl_3SiMe_3, CH_2Cl_2 and BuLi. The single carbon atom bridges the two Fe(tpp) moieties with Fe–C 167.5(1) pm in the linear Fe–C–Fe unit. Thermal decomposition of the olefin metathesis catalyst $(IMesH_2)(PCy_3)(Cl)_2Ru=CH_2$ ($IMesH_2$ = 1,3-dimesityl-4,5-dihydroimidazol-2-ylidene) results in the formation of a C-bridged dinuclear ruthenium complex, as shown in the following scheme. The Ru≡C–Ru bond angle is 160.3(2)°. The measured Ru≡C bond distance of 169.8(4) pm is slightly longer than those in reported μ-carbide ruthenium complexes such as $(PCy_3)_2(Cl)_2Ru\equiv C-Pd(Cl)_2(SMe_2)$ (166.2(2) pm), and the Ru–C distance of 187.5(4) pm is much shorter than the usual R–C single bonds in ruthenium complexes such as $(Me_3CO)_3W\equiv C-Ru(CO)_2(Cp)$ (209(2) pm).

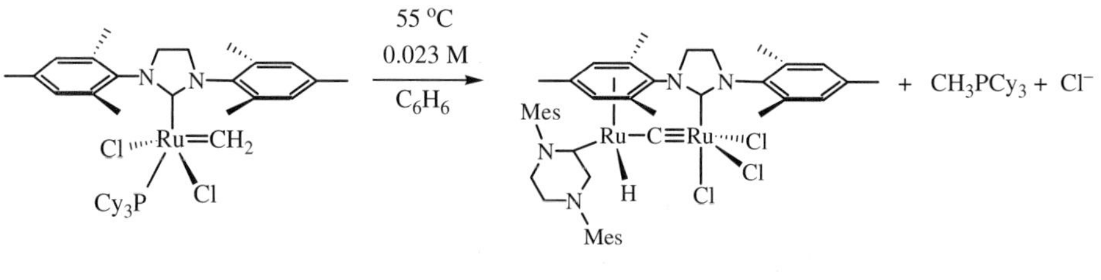

The carbide-centered polynuclear transition-metal carbonyl clusters exhibit a rich variety of structures. A common feature to this class of carbide complexes is that the naked carbon is wholly or partially enclosed in a metal cage composed of homo/hetero metal atoms, and there is also a subclass that can be considered as tetra-metal-substituted methanes. The earliest known compound of this kind is $Fe_5C(CO)_{15}$, in which the carbon atom is located at the center

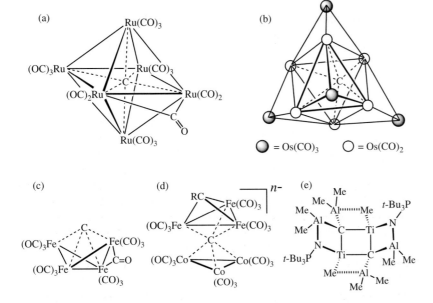

Fig. 14.3.9. Molecular structures of some transition-metal clusters containing a naked carbon atom.

of the base of a square pyramid with Fe(CO)$_3$ groups occupying its five vertices. Carbido carbonyl clusters of Ru and Os are well documented. Octahedral Ru$_6$C(CO)$_{17}$ is composed of four Ru(CO)$_3$ and two *cis*-Ru(CO)$_2$ fragments, the latter being bridged by a carbonyl group [Fig. 14.3.9(a)]. Os$_{10}$C(CO)$_{24}$ is built of an octahedral arrangement of Os(CO)$_2$ groups with four of its eight faces each capped by an Os(CO)$_3$ group, as shown in Fig. 14.3.9(b). The exposed carbon atom in Fe$_4$C(CO)$_{13}$ shows the greatest chemical activity, and the Fe$_4$C system serves as a plausible model for a surface carbon atom in heterogeneous catalytic processes [Fig. 14.3.9(c)]. Addition of Co$_3$(μ_3-CCl)(CO)$_9$ to a solution of (PPh$_4$)$_2$[Fe$_3$(μ-CCO)(CO)$_9$] (CCO is the ketenylidene group C=C=O) in CH$_2$Cl$_2$, in the presence of thallium salt, generates the species [{Co$_3$(CO)$_9$}C{Fe$_3$(CO)$_9$(μ-CCO)}]$^-$ containing a single carbon atom linking two different trimetallic clusters; subsequent addition of ethanol yielded the complex [{Co$_3$(CO)$_9$}C{Fe$_3$(CO)$_9$(μ-C–CO$_2$Et)}]$^{2-}$. The structures of these two hexanuclear hetereometallic anions are shown in Fig. 14.3.9(d).

The reaction of AlMe$_3$ with (*t*-Bu$_3$PN)$_2$TiMe$_2$ leads to the formation of two Ti complexes, of which [(μ^2-*t*-Bu$_3$PN)Ti(μ-Me)(μ^4-C)(AlMe$_2$)$_2$]$_2$ is the major product. Single-crystal X-ray analysis revealed that it has a saddle-like structure, with two phosphinimide ligands lying on one side and four AlMe$_2$ groups on the other [Fig. 14.3.9(e)]. The Ti and carbide C atoms in the central Ti$_2$C$_2$ ring both exhibit distorted tetrahedral coordination geometry.

14.3.7 Complexes containing naked dicarbon ligands

There is an interesting series of heterobinuclear complexes in which the Ru and Zr centers are connected by three different types of C$_2$ bridges: C–C, C=C, and C≡C.

$Cp(Me_3P)_2RuC{\equiv}CH \xrightarrow{[Cp_2Zr(Cl)(NMe_2)]} Cp(Me_3P)_2Ru\text{—}C{\equiv}C\text{—}ZrCp_2Cl$

$\downarrow [Cp_2Zr(Cl)(H)]$

[Structure: Cp(Me₃P)₂Ru and ZrCp₂Cl connected via a C=C unit with H substituents]
$\xrightarrow{[Cp_2Zr(Cl)(H)]}$ [Structure: four-membered metallacycle with Cp(Me₃P)₂Ru, ZrCp₂Cl bridged by CH₂–CH]

In polynuclear metal complexes bearing a naked C_2 species, multiple metal–carbon interactions generally occur, and the measured carbon–carbon bond distances indicate that the C_2 ligand may be considered to originate from fully deprotonated ethane, ethylene, or acetylene, which is stabilized in a "permetallated" coordination environment. Singly and doubly bonded dicarbon moieties are found in some polynuclear transition-metal carbonyl complexes (carbon–carbon bond length in pm): $Rh_{12}(C_2)(CO)_{25}$, 148(2); $[Co_6Ni_2(C_2)_2(CO)_{16}]^{2-}$, 149(1); $Fe_2Ru_6(\mu_6\text{-}C_2)_2(\mu\text{-}CO)_3(CO)_{14}Cp_2$, 133.4(8) and 135.4(7); $Ru_6(\mu_6\text{-}C_2)(\mu\text{-}SMe_2)_2(\mu\text{-}PPh_2)_2(CO)_{14}$, 138.1(8). The following discussion is concerned only with anionic species derived from acetylene.

Acetylene is a Brønsted acid ($pK_a \sim 25$). Its chemistry is associated with its triple-bond character and the labile hydrogen atoms. It can easily lose one proton to form the acetylide monoanion $HC{\equiv}C^-$ (IUPAC name acetylenide) or release two to give the acetylide dianion $^-C{\equiv}C^-$ (C_2^{2-}, IUPAC name acetylenediide). The acetylenide $H\text{–}C{\equiv}C^-$ and substituted derivatives $R\text{–}C{\equiv}C^-$ form organometallic compounds with the alkali metals. In these compounds, the interactions of the π orbitals of the $C{\equiv}C$ fragment with metal orbitals may lead to many structural types, e.g.,

$$\begin{array}{c} R\text{–}C{\equiv}C\text{–}Li \\ |\quad\quad| \\ Li\text{–}C{\equiv}C\text{–}R \end{array}$$

The acetylenediide C_2^{2-} can combine with alkali and alkaline-earth metals to form ionic salts, which are readily decomposed by water. Four modifications of CaC_2 (commonly known as calcium carbide) are known: room-temperature tetragonal CaC_2 **I**, high-temperature cubic CaC_2 **IV** (in which the C_2^{2-} dianion exhibits orientational disorder), low-temperature CaC_2 **II**, and a fourth modification CaC_2 **III** (Fig. 14.3.10). MgC_2, SrC_2, and BaC_2 adopt the tetragonal CaC_2 **I** structure, which consists of a packing of Ca^{2+} and discrete C_2^{2-} ions in a distorted NaCl lattice, with the C_2^{2-} dumbbell (bond length 119.1 pm from neutron powder diffraction) aligned parallel to the c axis.

Ternary metal acetylenediides of composition AM^IC_2 (A = Li to Cs and M^I = Ag(I), Au(I); or A = Na to Cs and M^I = Cu(I)) have been prepared; $NaAgC_2$, $KAgC_2$, and $RbAgC_2$ are isomorphous [Fig. 14.3.11(a)], but $LiAgC_2$ [Fig. 14.3.11(b)] and $CsAgC_2$ [Fig. 14.3.11(c)] belong to different structural types. The ternary acetylenediides A_2MC_2 (A = Na to Cs, M = Pd, Pt) crystallize in the same structure type, which is characterized by $[M(C_2)_{2/2}^{2-}]_\infty$ chains

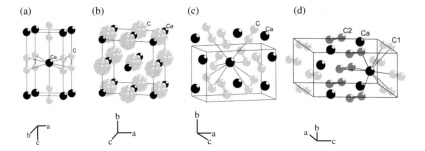

Fig. 14.3.10.
Crystal structure of (a) tetragonal CaC$_2$ **I** (*I4/mmm*, $Z = 2$); (b) cubic CaC$_2$ **IV** (*Fm$\bar{3}$m*, $Z = 4$), the C$_2^{2-}$ dumbbell exhibits orientational disorder; (c) low-temperature CaC$_2$ **II** (*C2/c*, $Z = 4$); (d) meta-stable CaC$_2$ **III** (*C2/m*, $Z = 4$).

separated by alkali metal ions [Fig. 14.3.11(d)]. The ternary alkaline-earth acetylenediide Ba$_3$Ge$_4$C$_2$ can be synthesized from the elements or by the reaction of BaC$_2$ with BaGe$_2$ at 1530 K; it consists of slightly compressed tetrahedral [Ge$_4$]$^{4-}$ anions inserted into a twisted octahedral Ba$_{6/2}$ three-dimensional framework. The Ba$_6$ octahedra are centered by C$_2^{2-}$ dumbbells (C–C bond length 120(6) pm), which are statistically oriented in two directions.

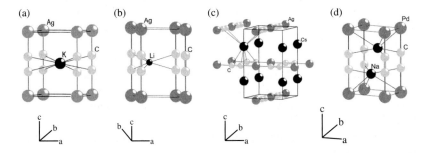

Fig. 14.3.11.
Crystal structure of (a) KAgC$_2$ (*P4/mmm*, $Z = 1$); (b) LiAgC$_2$ (*P$\bar{6}$m2*, $Z = 1$); (c) CsAgC$_2$ (*P4$_2$/mmc*, $Z = 2$); (d) Na$_2$PdC$_2$ (*P$\bar{3}$m1*, $Z = 1$).

The group 11 (Cu$_2$C$_2$, Ag$_2$C$_2$, and Au$_2$C$_2$) and group 12 (ZnC$_2$, CdC$_2$ and Hg$_2$C$_2$·H$_2$O, and Hg$_2$C$_2$) acetylenediides exhibit properties characteristic of covalent polymeric solids, but their tendency to detonate upon mechanical shock and insolubility in common solvents present serious difficulties in structural characterization. The earliest known and most studied non-ionic acetylenediide is Ag$_2$C$_2$ (commonly known as silver acetylide or silver carbide), which forms a series of double salts of the general formula Ag$_2$C$_2$·mAgX, where X$^-$ = Cl$^-$, I$^-$, NO$_3^-$, H$_2$AsO$_4^-$, or $\frac{1}{2}$EO$_4^{2-}$ (E = S, Se, Cr, or W), and m is the molar ratio. From 1998 onward, systematic studies have yielded a wide range of double, triple, and quadruple salts of silver(I) containing silver acetylenediide as a component, e.g., Ag$_2$C$_2$·mAgNO$_3$ (m = 1, 5, 5.5, and 6), Ag$_2$C$_2$·8AgF, Ag$_2$C$_2$·2AgClO$_4$·2H$_2$O, Ag$_2$C$_2$·AgF·4AgCF$_3$SO$_3$·RCN (R = CH$_3$, C$_2$H$_5$) and 2Ag$_2$C$_2$·3AgCN·15AgCF$_3$CO$_2$·2AgBF$_4$·9H$_2$O. A structural feature common to these compounds is that the C$_2^{2-}$ species is fully encapsulated inside a polyhedron with Ag(I) at each vertex, which may be represented as C$_2$@Ag$_n$. Note that each C$_2$@Ag$_n$ cage carries a charge of $(n-2)+$, and such cages are linked by anionic (and co-existing neutral) ligands to form a two- or three-dimensional coordination network. Some polyhedral Ag$_n$ cages with encapsulated C$_2^{2-}$ species found in various silver(I) double salts are shown in Fig. 14.3.12. In Ag$_2$C$_2$·6AgNO$_3$, the dumbbell-like C$_2^{2-}$ moiety is located inside a rhombohedral silver cage whose edges lie

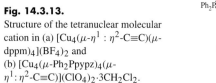

Fig. 14.3.12.
Polyhedral Ag_n (n = 6-9) cages with encapsulated C_2^{2-} species found in various silver(I) double salts:
(a) $Ag_2C_2 \cdot 2AgClO_4 \cdot 2H_2O$;
(b) $Ag_2C_2 \cdot AgNO_3$;
(c) $Ag_2C_2 \cdot 5.5AgNO_3 \cdot 0.5H_2O$;
(d) $Ag_2C_2 \cdot 5AgNO_3$;
(e) $Ag_2C_2 \cdot 6AgNO_3$ (the C_2^{2-} group is shown in one of its three possible orientations); (f) $Ag_2C_2 \cdot 8AgF$. Other ligands bonded to the silver vertices are not shown.
The dotted lines represent polyhedral edges that exceed 340 pm (twice the van der Waals radius of the Ag atom).

in the range of 295–305 pm, but it exhibits orientational disorder about a crystallographic threefold axis [Fig. 14.3.12(e)]. The utilization of $C_2@Ag_n$ polyhedra as building blocks for the supramolecular assembly of new coordination frameworks has resulted in a series of discrete, 1-D, 2-D, and 3-D complexes bearing interesting structural motifs; further details are presented in Section 20.4.5.

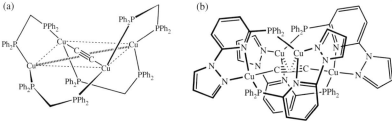

Fig. 14.3.13.
Structure of the tetranuclear molecular cation in (a) $[Cu_4(\mu-\eta^1:\eta^2-C\equiv C)(\mu\text{-dppm})_4](BF_4)_2$ and (b) $[Cu_4(\mu\text{-Ph}_2\text{Ppypz})_4(\mu-\eta^1:\eta^2-C\equiv C)](ClO_4)_2 \cdot 3CH_2Cl_2$.

To date, there are only two well-characterized copper(I) acetylenediide complexes. In $[Cu_4(\mu-\eta^1:\eta^2-C\equiv C)(\mu\text{-dppm})_4](BF_4)_2$ (dppm = $Ph_2PCH_2PPh_2$), the cation contains a saddle-like $Cu_4(\mu\text{-dppm})_4$ system, with the C_2 unit surrounded by a distorted rectangular Cu_4 array and interacting with the Cu atoms in η^1 and η^2 modes [Fig. 14.3.13(a)]. In comparison, the tetranuclear, C_2-symmetric cation of $[Cu_4(\mu\text{-Ph}_2\text{Ppypz})_4(\mu-\eta^1:\eta^2-C\equiv C)](ClO_4)_2 \cdot 3CH_2Cl_2$ [Ph_2Ppypz = 2-(diphenylphosphino-6-pyrazolyl)pyridine] consists of a butterfly-shaped Cu_4C_2 core in which the acetylenediide anion bridges a pair of Cu_2 subunits in both η^1 and η^2 bonding modes; the $C\equiv C$ bond length is 126(1) pm [Fig. 14.3.13(b)].

14.4 Structural chemistry of silicon

After oxygen (approx 45.5 wt%), silicon is the next most abundant element in the earth's crust (approx 27 wt%). Elemental Si does not occur naturally, but it combines with oxygen to form a large number of silicate minerals.

14.4.1 Comparison of silicon and carbon

Silicon and carbon command dominant positions in inorganic chemistry (silicates) and organic chemistry (hydrocarbons and their derivatives), respectively. Although they have similar valence electronic configurations, $[He]2s^22p^2$ for C and $[Ne]3s^23p^2$ for Si, their properties are not similar. The reasons for the difference between the chemistry of the two elements are elaborated below.

(1) Electronegativity

The electronegativity of C is 2.54, as compared with 1.92 for Si. Carbon is strictly nonmetallic whereas Si is essentially a non-metallic element with some metalloid properties.

(2) Configuration of valence shell and multiplicity of bonding

Unlike carbon, the valence shell of the silicon atom has available d orbitals. In many silicon compounds, the d orbitals of Si contribute to the hybrid orbitals and Si forms more than four 2c-2e covalent bonds. For example, SiF_5^- uses sp^3d hybrid orbitals to form five Si–F bonds, and SiF_6^{2-} uses sp^3d^2 hybrid orbitals to form six Si–F bonds.

Furthermore, silicon can use its d orbitals to form $d\pi$–$p\pi$ multiple bonds, whereas carbon only uses its p orbitals to form $p\pi$–$p\pi$ multiple bonds. Trisilylamines such as $N(SiH_3)_3$ is a planar molecule, as shown in Fig. 14.4.1(a), which differs from the pyramidal $N(CH_3)_3$ molecule as shown in Fig. 14.4.1(b). In the planar configuration of $N(SiH_3)_3$, the central N atom uses trigonal planar sp_xp_y hybrid orbitals to form N–Si bonds, with the nonbonding electron pair of N residing in the $2p_z$ orbital. Silicon has empty, relatively low-lying 3d orbitals able to interact appreciably with the $2p_z$ orbital of the N atom. This additional delocalized $p\pi$–$d\pi$ interaction shown in Fig 14.4.1(c) enhances the strength of the bonding that causes the NSi_3 skeleton to adopt a planar configuration.

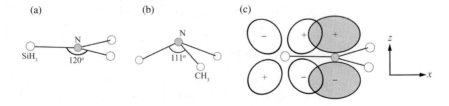

Fig. 14.4.1.
(a) Structure of planar $N(SiH_3)_3$, (b) pyramidal $N(CH_3)_3$, and (c) the $d\pi$–$p\pi$ bonding in $N(SiH_3)_3$. The occupied $2p_z$ orbital of the N atom is shaded, and only one Si $3d_{xz}$ orbital is shown for clarity.

Table 14.4.1. Comparison of chemical bonds formed by C and Si

Bond	Bond length (pm)	Bond energy (kJ mol^{-1})
C–C	154	356
C–H	109	413
C–O	143	336
Si–Si	235	226
Si–H	148	318
Si–O	166	452

(3) Catenation and silanes

The term catenation is used to describe the tendency for covalent bond formation between atoms of a given element to form chains, cycles, layers, or 3D frameworks. Catenation is common in carbon compounds, but it only occurs to a limited extent in silicon chemistry. The reason can be deduced from the data listed in Table 14.4.1.

Inspection of Table 14.4.1 shows that E(C–C) > E(Si–Si), E(C–H) > E(Si–H) and E(C–C) > E(C–O), but E(Si–Si) << E(Si–O). Thus alkanes are much more stable than silanes, and silanes react readily with oxygen to convert the Si–Si bonds to stronger Si–O bonds. Although silanes do not exist in nature, some compounds with Si–Si and Si=Si bonds have been synthesized in the absence of air and in non-aqueous solvents. The silanes Si$_n$H$_{2n+2}$ ($n = 1 - 8$), cyclic silanes Si$_n$H$_{2n}$ ($n = 5, 6$), and some polyhedral silanes are known. The structures of tetrahedral Si$_4$(Si^{t}Bu$_3$)$_4$, trigonal-prismatic Si$_6$(2,6-^{i}PrC$_6$H$_3$)$_6$, and cubane-like Si$_8$(2,6-Et$_2$C$_6$H$_3$)$_8$ are shown below:

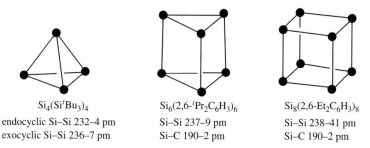

Si$_4$(Si^{t}Bu$_3$)$_4$
endocyclic Si–Si 232–4 pm
exocyclic Si–Si 236–7 pm

Si$_6$(2,6-^{i}Pr$_2$C$_6$H$_3$)$_6$
Si–Si 237–9 pm
Si–C 190–2 pm

Si$_8$(2,6-Et$_2$C$_6$H$_3$)$_8$
Si–Si 238–41 pm
Si–C 190–2 pm

The structures of other examples of oligocyclosilanes are shown in Fig. 14.4.2. The adamantane-like cluster (SiMe$_2$)$_6$(Si–SiMe$_3$)$_4$ with pendant methyl and trimethylsilyl groups was prepared in 2005.

14.4.2 Metal silicides

There exist a number of metal silicides, in which silicon anions formally occur. The structural types of the silicides in the solid state are diverse:

(1) Isolated units

(a) Si^{4-}: Examples are found in Mg$_2$Si, Ca$_2$Si, Sr$_2$Si, Ba$_2$Si.

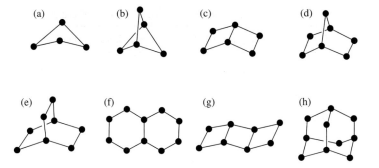

Fig. 14.4.2.
The structures of the molecular skeletons of some oligocyclosilanes:
(a) $Si_4{}^tBu_2(2,6\text{-}Et_2C_6H_3)_4$,
(b) $Si_5H_2{}^tBu_4(2,4,6\text{-}{}^tBu^iPr_2C_6H_2)_2$,
(c) $Si_6{}^iPr_{10}$, (d) Si_7Me_{12}, (e) Si_8Me_{14},
(f) $Si_{10}Me_{18}$, (g) $Si_8{}^iPr_{12}$,
(h) $Si_{10}Me_{16}$.

(b) Si_2 units: The $Si_2{}^{6-}$ ion occurs in U_3Si_2 (Si–Si distance is 230 pm) and in the series Ca_5Si_3, Sr_5Si_3 and Ba_5Si_3 [as $(M^{2+})_5(Si_2)^{6-}(Si)^{4-}$].

(c) Si_4 units: The $Si_4{}^{6-}$ ion has a butterfly structure; examples are found in Ba_3Si_4, as shown in Fig. 14.4.3(a). On the other hand, $Si_4{}^{4-}$ is tetrahedral and examples are found in NaSi, KSi, CsSi, $BaSi_2$, as shown in Fig. 14.4.3(b). Finally, in K_7LiSi_8, pairs of Si_4 units are connected by Li^+, as shown in Fig. 14.4.3(c), with additional interactions involving K^+ ions.

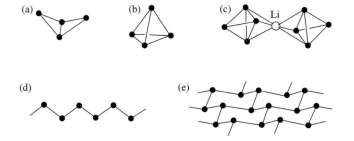

Fig. 14.4.3.
Structure of silicon anions in metal silicides: (a) $Si_4{}^{6-}$, (b) $Si_4{}^{4-}$, (c) $[LiSi_8]^{7-}$, (d) $[Si^{2-}]_n$, (e) $[Si^-]_n$.

(2) Si_n chains: Polymeric unit $[Si^{2-}]_n$ has a planar zigzag shape. Examples are found in USi and CaSi, as shown in Fig. 14.4.3(d). The Si–Si distances are 236 (USi) and 247 pm (CaSi).

(3) Si_n layers: Polymeric unit $[Si^-]_n$ in $CaSi_2$ consists of corrugated layers of six-membered rings, as shown in Fig. 14.4.3(e). In $\beta\text{-}ThSi_2$, the $[Si^-]_n$ units form planar hexagonal nets, which are similar to the B_n layers in AlB_2 [Fig. 13.3.1(f)].

(4) Si_n three-dimensional networks: Examples are $SrSi_2$, $\alpha\text{-}USi_2$, in which $[Si^-]_n$ constitutes a network with metal cations occupying the interstitial sites.

14.4.3 Stereochemistry of silicon

(1) Silicon molecular compounds with coordination numbers from three to six

The compounds formed by one Si atom and one to two other non-metallic atoms, such as SiX or SiX_2 (X = H, F, O, and Cl), are not stable species.

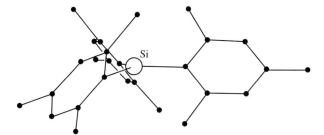

Fig. 14.4.4.
Structure of silylium ion (mes)$_3$Si$^+$.

A number of silenes (R$_2$Si=CR$_2$) and disilenes (R$_2$Si=SiR$_2$) containing double bonds, where each Si atom is three-coordinated, are known.

The structure of the free silylium ion (mes)$_3$Si$^+$ has been characterized by X-ray analysis of [(mes)$_3$Si][HCB$_{11}$Me$_5$Br$_6$]·C$_6$H$_6$. The (mes)$_3$Si$^+$ cation is well separated from the carborane anion and benzene solvate molecule. As shown in Fig. 14.4.4, the Si atom has trigonal planar coordination geometry expected of a sp^2 silylium center, with Si–C bond length 181.7 pm (av.), and the mesityl groups have a propeller-like arrangement around the Si center with an averaged twist angle $\tau = 49.2°$.

In the vast majority of its compounds, Si is tetrahedrally coordinated, which include SiH$_4$, SiX$_4$, SiR$_4$, SiO$_2$, and silicates. Five-coordinate Si can be either trigonal bipyramidal or square pyramidal, the former being more stable than the latter. In [Et$_4$N][SiF$_5$] the anion SiF$_5^-$ is trigonal bipyramidal with Si–F$_{(ax)}$ 165 pm and Si–F$_{(eq)}$ 159 pm (av). Among the stable compounds of penta-coordinate silicon, the most studied are the silatranes, an example of which is shown below:

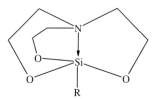

In this molecule, the axial nitrogen atom is linked through three (CH$_2$)$_2$ tethers to oxygen atoms at the triangular base. The tetradentate tripodal ligand occupies four coordination positions around the central Si atom, forming an intramolecular transannular dative bond of the type N→Si. The observed N→Si bond distances in about 50 silatranes containing three condensed rings vary within the range 200–40 pm. The longer the N→Si bond, the more planar the configuration of the NC$_3$ moiety, and the structure of the RSiO$_3$ fragment approaches tetrahedral.

Numerous compounds of octahedral hexa-coordinate Si are known. Some examples are shown below.

In the salt [C(NH$_2$)$_3$]$_2$[SiF$_6$], the Si–F bond length in octahedral SiF$_6^{2-}$ is 168 pm, which is longer than the Si–F bonds in SiF$_5^-$.

(2) Structure of Si(IV) compounds with SiO$_5$ and SiO$_6$ skeletons

The chemistry of silicon oxygen compounds with SiO$_5$ and SiO$_6$ skeletons in aqueous solution is of special interest. It has been speculated that such Si(IV) complexes with ligands derived from organic hydroxy compounds (such as pyrocatechol derivatives, hydroxycarboxylic acids, and carbohydrates) may play a significant role in silicon biochemistry by controlling the transport of silicon.

Several zwitterionic neutral compounds and anionic species with the SiO$_5$ skeleton have been synthesized and characterized by X-ray diffraction. The Si coordination polyhedra of these compounds are typically distorted trigonal bipyramids, with the carboxylate oxygen atoms of the two bidentate ligands in the axial positions. Figure 14.4.5 shows the structures of (a) Si[C$_2$O$_3$(Me)$_2$]$_2$[O(CH$_2$)$_2$NHMe$_2$] with Si-O$_{(ax)}$ 177.3 and 179.8 pm, and Si-O$_{(eq)}$ 164.3 to 165.9 pm, and (b) the anion in crystalline [Si(C$_2$O$_3$Ph$_2$)$_2$(OH)]$^-$[H$_3$NPh]$^+$, where Si–O$_{(ax)}$ 179.8 pm and Si–O$_{(eq)}$ 165.0 to 166.0 pm.

Some neutral compounds, cationic species, and anionic species with distorted octahedral SiO$_6$ skeletons have been established. Figure 14.4.6 shows the structures of (a) Si[C$_2$O$_3$Ph$_2$][C$_3$HO$_2$Ph$_2$]$_2$ in which the Si–O bond lengths are 169.3 to 182.1 pm, and (b) the dianion in [Si(C$_2$O$_4$)$_3$]$^{2-}$ [HO(CH$_2$)$_2$NH(CH$_2$)$_4^+$]$_2$, in which the Si–O bond lengths vary from 176.7 to 178.8 pm.

(3) Structural features of Si=Si double bond in disilenes

The first reported disilene is (mes)$_2$Si=Si(mes)$_2$, which was isolated and characterized in 1981. The characteristic structural features of disilenes are the

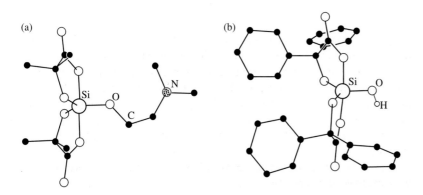

Fig. 14.4.5. Structures of (a) Si[C$_2$O$_3$Me$_2$]$_2$[O(CH$_2$)$_2$NHMe$_2$] and (b) [Si(C$_2$O$_3$Ph$_2$)$_2$(OH)]$^-$.

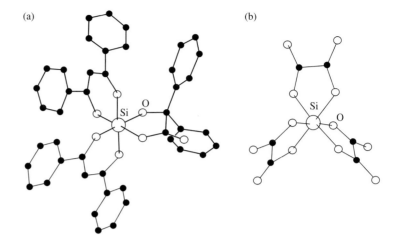

Fig. 14.4.6.
Structures of
(a) $Si[C_2O_3Ph_2][C_3HO_2Ph_2]_2$ and (b) $[Si(C_2O_4)_3]^{2-}$.

length of the Si=Si double bond d, the twist angle τ, and the *trans*-bent angle θ, as shown below.

In contrast to the C=C double bond of sterically crowded alkenes, in which variations of the bond lengths are small, the Si=Si bond lengths of disilenes vary between 214 and 229 pm. The twist angle τ of the two SiR_2 planes varies between 0° and 25°. A further peculiarity of the disilenes, not observed in alkenes, is the possibility of *trans*-bending of the substituents, which is described by the *trans*-bent angle θ between the SiR_2 planes and the Si=Si vector. The θ values can reach up to 34°. These differences can be

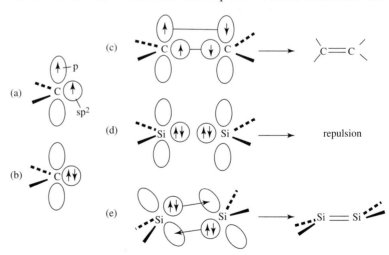

Fig. 14.4.7.
Double-bond formation from two triplet carbenes and singlet silylenes.

rationalized as follows: carbenes have either a triplet ground state (T), as shown in Fig. 14.4.7(a), or a singlet ground state (S), as shown in Fig. 14.4.7(b), with relatively low S→T transition energies. The familiar picture of a C=C double-bond results from the approach of two triplet carbenes [Fig. 14.4.7(c)]. In contrast, silylenes have singlet ground states with a relative large S→T excitation energy. Approach of two singlet silylenes should usually result in repulsion rather than bond formation [Fig. 14.4.7(d)]. However, if the two silylene molecules are tilted with respect to each other, interaction between the doubly occupied sp^2-type orbitals and the vacant p orbitals can form a double donor–acceptor bond. This is accompanied by a *trans*-bending of the substituents about the Si=Si vector, as illustrated in Fig. 14.4.7(e).

(4) Stable silyl radicals

Silyl radicals stabilized with bulky trialkylsilyl groups can be isolated in crystalline form. X-ray analysis showed that the $(^tBu_2MeSi)_3Si^{\cdot}$ radical has trigonal planar geometry about the central sp^2 Si atom, with an averaged Si–Si bond length of 242(1) pm. Interestingly, all the methyl substituents at the α-Si atoms lie in the plane of the polysilane skeleton so that steric hindrance is minimized. Reaction of this radical with lithium metal in hexane at room temperature afforded $[(^tBu_2MeSi)_3Si]Li$, the crystal structure of which showed a planar configuration (sum of Si–Si–Si angles = 119.7° for the central anionic Si atom) and an averaged Si–Si bond length of 236(1) pm.

Crystal structure analysis of the disilene $(^tBu_2MeSi)_2Si=Si(SiMe^tBu_2)_2$ has revealed a highly twisted configuration about the central Si=Si double bond, as shown in Fig. 14.4.8(a). The Si(1) and Si(2) atoms forming the double bond are both sp^2 hybridized, and the twist angle τ has an exceptionally large value of 54.5° (see Fig. 14.4.8(b)). The disilene reacts with ^tBuLi to produce the $[Li(THF)_4]^+$ salt of the silyl anion radical $(^tBu_2MeSi)_2Si^--Si^{\cdot}(SiMe^tBu_2)_2$, in which the anionic $Si(1)^-$ atom has flattened pyramidal geometry (sum of bond angles = 352.7°) whereas the radical $Si(2)^{\cdot}$ atom retains its sp^2 configuration. The mean planes of the Si(3)–Si(1)–Si(4) and Si(5)–Si(2)–Si(6) fragments are nearly orthogonal, with a τ angle of 88° about the central Si–Si single bond, as illustrated in Fig. 14.4.8(c).

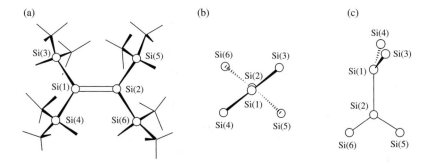

Fig. 14.4.8.
(a) Molecular structure of disilene $(^tBu_2MeSi)_2Si=Si(SiMe^tBu_2)_2$. Si(m)–Si(n) bond lengths (in pm): 1–2, 226.0(1); 1–3, 241.6(2); 1–4, 241.9(2); 2–5, 241.3(2); 2–6, 242.3(2). (b) Twisted configuration of the $Si_2Si=SiSi_2$ skeleton. (c) Nearly orthogonal configuration of the silyl anion radical $(^tBu_2MeSi)_2Si^--Si^{\cdot}(SiMe^tBu_2)_2$. Bond lengths Si(m)–Si(n) (in pm): 1–2, 234.1(5); 1–3, 239.0(5); 1–4, 239.2(5); 2–5, 241.2(1); 2–6, 240.1(1).

14.4.4 Silicates

(1) Classification of silicates

Silicates constitute the largest part of the earth's crust and mantle. They play a dominant role as raw materials as well as products in technological processes, such as building materials, cements, glasses, ceramics, and refractory materials.

In the silicates, the tetrahedral SiO_4 units are either isolated or share corners with other tetrahedra, giving rise to an enormous variety of structures. In many silicates, silicon may be replaced to a certain extent by other elements, such as aluminium, so the structures of silicates are further extended to cover cases where such partial substitution occurs.

Table 14.4.2. Classification of natural silicates

Number of shared O atoms in SiO_4 unit	Structure	Name
0 (no O atom shared)	Discrete SiO_4 unit	Neso-silicates
1	Discrete Si_2O_7 unit	Soro-silicates
2	Cyclic $(SiO_3)_n$ structures	Cyclo-silicates
2	Infinite chains or ribbons	Ino-silicates
3	Infinite layers	Phyllo-silicates
4	Infinite 3D frameworks	Tecto-silicates

The kind and the degree of linkage of SiO_4 tetrahedra constitute a basis for the classification of natural silicates, as shown in Table 14.4.2.

(a) Silicates containing discrete SiO_4^{4-} or $Si_2O_7^{6-}$

The discrete SiO_4^{4-} anion [as shown in Fig. 14.4.9(a)] occurs in the orthosilicates, such as phenacite Be_2SiO_4, olivine $(Mg,Fe,Mn)_2SiO_4$, and zircon $ZrSiO_4$. Another important group of orthosilicates is the garnets,

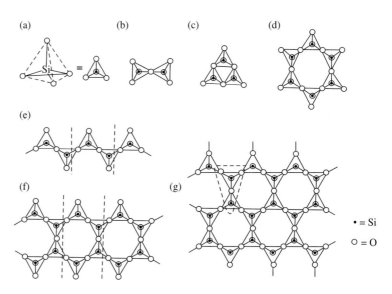

Fig. 14.4.9.
Structures of some silicates: (a) SiO_4^{4-}, (b) $Si_2O_7^{6-}$, (c) cyclic $[Si_3O_9]^{6-}$, (d) cyclic $[Si_6O_{18}]^{12-}$, (e) single chain $[SiO_3]_n^{2n-}$, (f) double chain $[Si_4O_{11}]_n^{6n-}$, (g) layer $[Si_2O_5]_n^{2n-}$.

• = Si
○ = O

$M^{II}_3M^{III}_2(SiO_4)_3$, in which M^{II} are eight-coordinate (e.g. Ca, Mg, Fe) and M^{III} are six-coordinate (e.g., Al, Cr, Fe). Orthosilicates are also vital components of Portland cements: β-Ca_2SiO_4 has a discrete [SiO_4] group with Ca in six- or eight-coordination.

An example of a disilicate containing the discrete $Si_2O_7^{6-}$ ion [as shown in Fig. 14.4.9(b)] is the mineral thortveitite $Sc_2Si_2O_7$, in which Sc^{III} is octahedral and Si–O–Si is linear with staggered conformation. There is also a series of lanthanoid disilicates $Ln_2Si_2O_7$, in which the Si–O–Si angle decreases progressively from 180° to 133°, and the coordination number of Ln increases from 6 through 7 to 8 as the size of Ln increases from six-coordinated Lu^{III} to eight-coordinated Nd^{III}. In the Zn mineral hemimorphite the angle is 150°, but the conformation of the two tetrahedra is eclipsed rather than staggered.

Discrete chains of three or four linked SiO_4 tetrahedra are extremely rare, but they exist in aminoffite $Ca_3(BeOH)_2[Si_3O_{10}]$, kinoite $Cu_2Ca_2[Si_3O_{10}]\cdot 2H_2O$, and vermilion $Ag_{10}[Si_4O_{13}]$.

(b) Silicates containing cyclic $[SiO_3]_n^{2n-}$ ion

Cyclic $[SiO_3]_n^{2n-}$ having 3, 4, 6 or 8 linked tetrahedra are known, though 3 and 6 [as shown in Figs 14.4.9(c)–(d)] are the most common. These anions are exemplified by the minerals benitoite $BaTi[Si_3O_9]$, catapleite $Na_2Zr[Si_3O_9]\cdot 2H_2O$, beryl $Be_3Al_2[Si_6O_{18}]$, tourmaline $(Na,Ca)(Li,Al)_3Al_6(OH)_4(BO_3)_3[Si_6O_{18}]$, and murite $Ba_{10}(Ca,Mn,Ti)_4[Si_8O_{24}](Cl,OH,O)_{12}\cdot 4H_2O$.

(c) Silicates with infinite chain or ribbon structure

Single chains formed by tetrahedral SiO_4 groups sharing two vertices adopt various configurations in the crystal. There are different numbers of tetrahedra in the repeat unit for different minerals; for example, both diopside $CaMg[SiO_3]_2$ and enstatite $MgSiO_3$ have two repeat units, as shown in Fig. 14.4.9(e).

In the double chains or ribbons, there are different kinds of tetrahedra sharing two and three vertices. The most numerous amphiboles and asbestos minerals, such as tremolite, $Ca_2Mg_5(Si_4O_{11})_2(OH)_2$, adopt the $[Si_4O_{11}]_n^{6n-}$ double chain structure, as shown in Fig. 14.4.9(f).

(d) Silicates with layer structures

Silicates with layer structures include clay minerals, micas, asbestos, and talc. The simplest single layer has composition $[Si_2O_5]_n^{2n-}$, in which each SiO_4 tetrahedron shares three vertices with three others, as shown in Fig. 14.4.9(g). Muscovite $KAl_2[AlSi_3O_{10}](OH)_2$ is a layer silicate of the mica group. There are many and complex variations of structure based on one or more of the layer types. The vertices of the tetrahedra can form part of a hydroxide-like layer. Their bases can either be directly opposed, or can be fitted to hydroxyl or water layers. When layers of different types succeed each other with more or less regularity, the changes in stacking these layers together occur in an endless variety of ways. This irregularity of structure is no doubt related to their capacity to take up or lose water and mediate ion exchange. In many cases the structures swell or contract in different stages of hydration, and can take up replaceable ions.

(e) Silicates with framework structures

In silicates, the most important and widespread substitution is that of Al for Si in tetrahedral coordination to form (Si, Al)O_4 tetrahedra, so that most of the "silicates" that occur in nature are in fact aluminosilicates. This substitution must be accompanied by the incorporation of cations to balance the charge on the Si–O framework.

Sharing of all vertices of each (Si,Al)O_4 tetrahedron leads to infinite 3D framework aluminosilicates, such as felspars, ultramarines, and zeolites. The felspars are the most abundant of all minerals and comprise about 60% of the earth's crust, and are subdivided into two groups according to the symmetry of their structures. Typical members of the groups are: (a) orthoclase KAlSi$_3$O$_8$ and celsian BaAl$_2$Si$_2$O$_8$, and (b) the plagioclase felspar: albite NaAlSi$_3$O$_8$ and anorthite CaAl$_2$Si$_2$O$_8$. Examples of ultramarine are sodalite Na$_8$Cl$_2$[Al$_6$Si$_6$O$_{24}$] and ultramarine Na$_8$(S$_2$)[Al$_6$Si$_6$O$_{24}$].

(2) Zeolites

Zeolites are crystalline, hydrated aluminosilicates that possess framework structures containing regular channels and/or polyhedral cavities. Zeolites generally contain loosely held water, which can be removed by heating the crystals and subsequently regained on exposure to a moist atmosphere. When occluded water is removed, it can be replaced by other small molecules. There is a close relationship between the size of molecules that can diffuse through the framework and the dimensions of the aperture or "bottle-neck" connecting one cavity to the next. This capacity of zeolites to admit molecules below a certain limiting size, while blocking passage to larger molecules, has led to their being termed "molecular sieves."

There are more than 40 natural zeolites and over 100 synthesized zeolites. Table 14.4.3 lists the ideal composition and the member of rings of some zeolites. Figure 14.4.10(a) shows the construction of a truncated cubo-octahedron (β-cage) formed from 24 linked (Si,Al)O$_4$ tetrahedra, and Fig. 14.4.10(b) shows a simplified representation of this cavity formed by joining the Si(Al) atom positions. Several other types of polyhedra have also been observed.

Zeolite A is a synthetic zeolite that has not been found in nature. In the framework structure of dehydrated zeolite 4A, Na$_{12}$[Al$_{12}$Si$_{12}$O$_{48}$], cubo-octahedra lie at the corners of the unit cell, generating small cubes at the centers of the cell edges and eight-membered rings at the face centers. Access through these eight-membered rings to a very large cavity (truncated cubo-octahedron, α-cage) at

Table 14.4.3. Ideal composition and member of rings in some zeolites

Zeolite	Ideal composition	Member of rings
Zeolite A	Na$_{12}$[Al$_{12}$Si$_{12}$O$_{48}$]·27H$_2$O	4,6,8
Faujasite (Zeolite X and Y)	Na$_{58}$[Al$_{58}$Si$_{134}$O$_{384}$]·240H$_2$O	4,6,12
Zeolite ZSM-5	Na$_3$[Al$_3$Si$_{93}$O$_{192}$]·16H$_2$O	4,5,6,7,8,10
Chabazite	Na$_4$Ca$_8$[Al$_{20}$Si$_{52}$O$_{144}$]·56H$_2$O	4,5,6,8,10
Mordenite	Na$_8$[Al$_8$Si$_{40}$O$_{96}$]·24H$_2$O	4,5,6,8,12

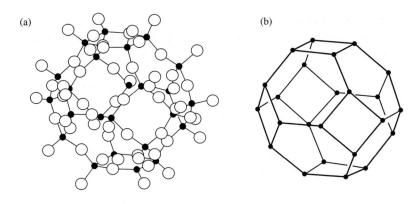

Fig. 14.4.10.
Structure of cubo-octahedral cage (β-cage): (a) arrangement of $(Si,Al)O_4$ (black circles represent Si and Al), (b) simplified representation of the polyhedral cavity.

the center of the unit cell is facilitated. Figure 14.4.11(a) shows the structure of zeolite A framework.

Faujasite or zeolite X and Y exists in nature. Its framework can be formed as follows. The cubo-octahedra are placed at the positions of the carbon atoms in the diamond structure, and they are joined by hexagonal prisms through four of the eight hexagonal faces of each cubo-octahedron, as shown in Fig. 14.4.11(b).

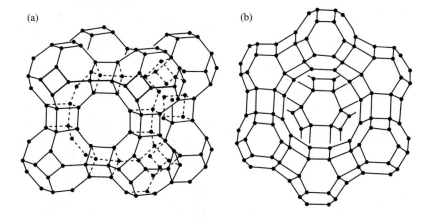

Fig. 14.4.11.
Structure of the framework of (a) zeolite A and (b) faujasite.

(3) General rules governing the structures of natural silicates

The structures of natural silicates obey the following general rules:

(a) With some exceptions, such as stishovite, the Si atoms form SiO_4 tetrahedra with little deviation of bond lengths and bond angles from the mean values: Si–O = 162 pm, O–Si–O = 109.5°.
(b) The most important and widespread substitution is that of Al for Si in tetrahedral coordination, which is accompanied by the incorporation of cations to balance the charge. The mean Al–O bond length in AlO_4 is 176 pm. In almost all silicates that occur as minerals, the anions are tetrahedral $(Si,Al)O_4$ groups. The Al atoms also occupy positions of octahedral coordination in aluminosilicates.
(c) The $(Si,Al)O_4$ tetrahedra are linked to one another by sharing common corners rather than edges or faces. Since two Si–O–Al groups have a lower

energy content than one Al–O–Al plus one Si–O–Si group, the AlO$_4$ tetrahedra do not share corners in tetrahedral framework structures if this can be avoided.
(d) One oxygen atom can belong to no more than two SiO$_4$ tetrahedra.
(e) If s is the number of oxygen atoms of a SiO$_4$ tetrahedron shared with other SiO$_4$ tetrahedra, then for a given silicate anion the difference between the s values of all SiO$_4$ tetrahedra tends to be small.

14.5 Structures of halides and oxides of heavier group 14 elements

14.5.1 Subvalent halides

Subvalent halides of heavier group 14 elements MX$_2$ (M = Ge, Sn, and Pb; X = F, Cl, Br, and I) and their complexes exhibit the following structural features and properties.

(1) Stereochemically active lone pair

In the dihalides of Ge, Sn, and Pb and their complexes, the metal atom always has one lone pair. The discrete, bent MX$_2$ molecules are only present in the gas phase with bond angles less than 120°. Figure 14.5.1(a) shows the structure of SnCl$_2$, which has bond angle 95° and bond distance 242 pm.

In the crystalline state, the coordination numbers of metal atoms are usually increased to three or four. Because of the existence of the lone pair, the MX$_3^-$ or MX$_4^{2-}$ units all adopt the trigonal pyramidal or square-pyramidal configuration, as shown in Figs. 14.5.1(b) to 14.5.1(d). Figure 14.5.1(b) shows the structure of Sn$_2$F$_5^-$ in NaSn$_2$F$_5$, in which each Sn is trigonal pyramidal with two close F$_t$ (Sn–F$_t$ 207 and 208 pm) and one F$_b$ (Sn–F$_b$ 222 pm). Figure 14.5.1(c) shows the structure of (SnCl$_2$F)$^-$ in [Co(en)$_3$](SnCl$_2$F)[Sn$_2$F$_5$]Cl. Figure 14.5.1(d) shows the structure of (SnF$_3^-$)$_\infty$ in KSnF$_3 \cdot \frac{1}{2}$H$_2$O. In this compound, the Sn atoms are square pyramidal and each bridging F atom is connected to two Sn atoms to give an infinite chain with Sn–F$_b$ 227 pm and Sn–F$_t$ 201 and 204 pm.

(2) Oligomers and polymers

The structural chemistry of MX$_2$ is complex, partly because of the stereochemical activity (or nonactivity) of the lone pair of electrons and partly because of

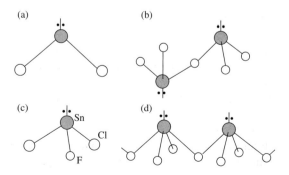

Fig. 14.5.1.
Structure of subhalides of tin: (a) SnCl$_2$, (b) (Sn$_2$F$_5$)$^-$ in NaSn$_2$F$_5$, (c) (SnCl$_2$F)$^-$ in [Co(en)$_3$][SnCl$_2$F][Sn$_2$F$_5$]Cl, and (d) (SnF$_3^-$)$_\infty$ in KSnF$_3 \cdot \frac{1}{2}$H$_2$O.

the propensity of M^{II} to increase its coordination numbers by polymerization into larger structural units, such as rings, chains or layers. The M^{II} center rarely adopts structures typical of spherically symmetrical ions because the lone pair of electrons, which is ns^2 in the free gaseous ion, is readily distorted in the condensed phase. This can be described in terms of ligand-field distortions or the adoption of some p-orbital character. The lone pair can act as a donor to vacant orbitals, and the vacant np orbitals and nd orbitals can act as acceptors in forming extra covalent bonds.

In the solid state, GeF_2 has a unique structure in which trigonal pyramidal GeF_3 units share two F atoms to form an infinite spiral chain. Reaction between GeF_2 and F^- gives GeF_3^-, which is a trigonal pyramidal ion.

Crystalline SnF_2 is composed of Sn_4F_8 tetramers, which are puckered eight-membered rings of alternating Sn and F atoms, with Sn–F_b 218 pm and Sn–F_t 205 pm. The tetramers are interlinked by weaker Sn–F interactions. Bridge formation is observed in $Na(Sn_2F_5)$, $Na_4(Sn_3F_{10})$, and other salts such as $[Co(en)_3][SnCl_2F][Sn_2F_5]Cl$.

The structure of $SnCl_2$ is shown in Fig. 14.5.1(a). In the crystalline state it has a layer structure with chains of corner-shared, trigonal-pyramidal $SnCl_3$ units. The commercially available solid hydrate $SnCl_2·2H_2O$ also has a puckered-layer structure.

The dihalides PbX_2 is much more stable thermally and chemically than PbX_4. Structurally, α-PbF_2, $PbCl_2$ and $PbBr_2$ all form colorless orthorhombic crystals in which Pb^{II} is surrounded by nine X ligands (seven closer and two farther away) at the corners of a tricapped trigonal prism.

(3) Stability

The stability of the dihalides of Ge, Sn, and Pb steady increases in the sequence $GeX_2 < SnX_2 < PbX_2$. Thus PbX_2 is more stable than PbX_4, whereas GeX_4 is more stable than GeX_2. At 978 K, SnF_4 sublimes to give a vapor containing SnF_4 molecules, which is thermally stable, but PbF_4 (prepared by the action of F_2 on Pb compounds) decomposes into PbF_2 and F_2 when heated.

The preference for the +2 over +4 oxidation state increases down the group, the change being due to relativistic effects that make an important contribution to the inert pair. The "inert pair" concept holds only for the lead ion Pb^{2+}(aq), which could have a $6s^2$ configuration. In more covalent Pb^{II} compounds and most Sn^{II} compounds there are stereochemically active lone pairs. In some MX_2 (M = Ge or Sn) compounds Ge and Sn can act as donor ligands.

(4) Mixed halides and mixed-valence complexes

Many mixed halides are known, such as all ten trihalogenostannate(II) anions $[SnCl_xBr_yI_z]^-$ (xyz = 300, 210, 201, 120, 102, 111, 021, 012, 030, 003), have been observed and characterized by ^{119}Sn NMR spectroscopy. An example $(SnCl_2F)^-$ is shown in Fig. 14.5.1(c).

Many mixed halides of Pb^{II} have also been characterized, including PbFCl, PbFBr, PbFI, and $PbX_2·4PbF_2$. Of these, PbFCl has an important tetragonal layer structure, which is frequently adopted by large cations in the presence of two anions of differing size; its sparing solubility in water forms the basis of a gravimetric method for the determination of F.

Some mixed-valence halide complexes are known, such as $Ge_5F_{12} = (Ge^{II}F_2)_4(Ge^{IV}F_4)$, $\alpha\text{-}Sn_2F_6 = Sn^{II}Sn^{IV}F_6$, and $Sn_3F_8 = Sn_2^{II}Sn^{IV}F_8$. In these compounds, the M^{II}–X bond distances are longer than the corresponding M^{IV}–X bond distances.

The coordination geometries of M^{IV} in compounds or mixed-valence complexes, in contrast to M^{II}, tend to exhibit high symmetries; the MX_4 molecules are tetrahedral and $[MX_6]^{2-}$ units adopt the octahedral (or slightly distorted) configuration.

14.5.2 Oxides of Ge, Sn, and Pb

(1) Subvalent oxides

Both SnO and PbO exist in several modifications. The most common blue-black modification of SnO and red PbO (litharge) have a tetragonal layer structure, in which the M^{II} (Sn^{II} and Pb^{II}) atom is bonded to four oxygen atoms arranged in a square to one side of it, with the lone pair of electrons presumably occupying the apex of the tetragonal pyramid (Sn–O 221 pm and Pb–O 230 pm). Each oxygen atom is surrounded tetrahedrally by four M^{II} atoms. Figure 14.5.2 shows the crystal structure of SnO (and PbO).

Dark-brown crystalline GeO is obtained when germanium powder and GeO_2 are heated or $Ge(OH)_2$ is dehydrated, but this compound is not well characterized.

(2) Dioxides

Germanium dioxide, GeO_2, closely resembles SiO_2 and exists in both α-quartz and tetragonal rutile (Fig. 10.2.6) forms. In the latter, Ge^{IV} is octahederally coordinated with Ge–O 188 pm (mean). SnO_2 occurs naturally as cassiterite, adopting the rutile structure with Sn–O 209 pm (mean). There are two modifications of PbO_2: the tetragonal maroon form has the rutile structure with Pb–O 218 pm (mean), whereas $\alpha\text{-}PbO_2$ is an orthorhombic black form whose structure is derived from hcp layers with half of the octahedral sites filled.

(3) Mixed-valence oxides

Pb_3O_4 ($Pb_2^{II}Pb^{IV}O_4$) and Pb_2O_3 ($Pb^{II}Pb^{IV}O_3$) are the two well-known mixed-valence oxides of lead. Red lead Pb_3O_4 is important commercially as a pigment and primer. Its tetragonal crystal structure consists of chains of $Pb^{IV}O_6$ octahedra sharing opposite edges (mean Pb^{IV}–O 214 pm). These chains are aligned

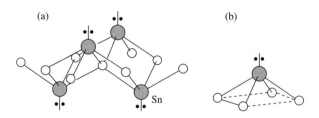

Fig. 14.5.2.
Crystal structure of SnO (and PbO):
(a) viewed from the side of a part of one layer, (b) the arrangement of bonds and lone pair at a metal atom.

parallel to the c axis and linked by the Pb^{II} atoms, each being pyramidally coordinated by three oxygen atoms ($Pb^{II}O_3$ unit, Pb^{II}–O two at 218 pm and one at 213 pm). Figure 14.5.3 shows a portion of the tetragonal unit cell of Pb_3O_4.

The sesquioxide Pb_2O_3 is a black monoclinic crystal, which consists of $Pb^{IV}O_6$ octahedra (Pb^{IV}–O 218 pm, mean) and very irregular six-coordinate $Pb^{II}O_6$ units (Pb^{II}–O 231, 243, 244, 264, 291 and 300 pm). The Pb^{II} atoms are situated between layers of distorted $Pb^{IV}O_6$ octahedra.

Among the various mixed-valence oxides of tin that have been reported, the best characterized is Sn_3O_4, i.e., $Sn_2^{II}Sn^{IV}O_4$, the structure of which is similar to that of Pb_3O_4. Mixed oxides of Pb^{IV} (or Pb^{II}) with other metals find numerous applications in technology and industry. Specifically, $M^{II}Pb^{IV}O_3$ and $M^{II}Pb^{IV}O_4$ (M^{II} = Ca, Sr, Ba) have important uses; for example, $CaPbO_3$ is a priming pigment to protect steel corrosion by salt water. Mixed oxides of Pb^{II} are also important. $PbTiO_3$, $PbZrO_3$, $PbHfO_3$, $PbNb_2O_6$ and $PbTi_2O_6$ are ferroelectric materials. The high Curie temperature of many Pb^{II} ferroelectrics makes them particularly useful for high-temperature applications.

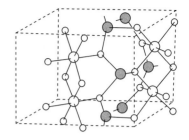

Fig. 14.5.3.
Crystal structure of Pb_3O_4 (large circles represent Pb atom, white Pb^{IV} and shaded Pb^{II}; small circles represent O atom).

14.6 Polyatomic anions of Ge, Sn, and Pb

Germanium, tin, and lead dissolve in liquid ammonia in the presence of alkali metals to give highly colored anions, which are identified as polyatomic species (e.g., Sn_5^{2-}, Pb_5^{2-}, Sn_9^{4-}, Pb_9^{4-}, and Ge_{10}^{2-}) in salts containing cryptated cations, e.g., $[Na(C222)]_2Pb_5$, $[Na(C222)]_4Sn_9$, and $[K(C222)]_3Ge_9$.

In an unusual synthesis, the compound $[K(C222)]_2Pb_{10}$ was obtained by oxidation of a solution of Pb_9^{4-} ions in ethylenediamine with Au^I in the form of $[P(C_6H_5)_3AuCl]$ as the oxidizing agent:

$$2Pb_9^{4-} + 6Au^+ \rightarrow Pb_{10}^{2-} + 8Pb^0 \downarrow + 6Au^0.$$

A single-crystal X-ray structure analysis revealed that the intensely brown colored Pb_{10}^{2-} anion has a bicapped square-antiprismatic structure of nearly perfect D_{4d} symmetry.

Since the homopolyatomic anions and cations (Zintl ions) are devoid of ligand attachment, they are sometimes referred to as "naked" clusters. The structures of some naked anionic clusters are shown in Fig. 14.6.1: (a) tetrahedral Ge_4^{4-}, Sn_4^{4-}, Pb_4^{4-}, $Pb_2Sb_2^{2-}$; (b) trigonal bipyramidal Sn_5^{2-}, Pb_5^{2-}; (c) octahedral Sn_6^{2-}; (d) tricapped-trigonal prismatic Ge_9^{2-} and paramagnetic Ge_9^{3-}, Sn_9^{3-}; (e) monocapped square-antiprismatic Ge_9^{4-}, Sn_9^{4-}, Pb_9^{4-}; (f) bicapped square-antiprismatic Ge_{10}^{2-}, Pb_{10}^{2-}, Sn_9Tl^{3-}.

The bonding in these polyatomic anions is delocalized, and for the diamagnetic species the bond valence method (see Section 13.4) can be used to rationalize the observed structures. This method is to be applied according to the following rules:

(a) Each pair of neighbor M atoms is connected by a M–M 2c-2e bond or MMM 3c-2e bond to form the cluster.
(b) Each M atom use its four valence orbitals to form chemical bonds to attain the octet configuration.

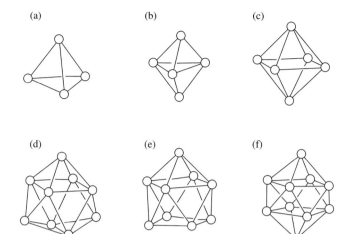

Fig. 14.6.1.
Structure of polyatomic anions of Ge, Sn, and Pb: (a) Ge_4^{4-}, Sn_4^{4-}, Pb_4^{4-}, and $Pb_2Sb_2^{2-}$; (b) Sn_5^{2-}, Pb_5^{2-}; (c) Sn_6^{2-}; (d) Ge_9^{2-}; (e) Ge_9^{4-}, Sn_9^{4-}, Pb_9^{4-}; (f) Ge_{10}^{2-}, Pb_{10}^{2-}, Sn_9Tl^{3-}.

(c) Two M atoms cannot participate in forming both a M–M 2c-2e bond and a MMM 3c-2e bond.

For example, the bond valence (b) of Ge_9^{4-} is $b = \frac{1}{2}(8n - g) = \frac{1}{2}[9 \times 8 - (9 \times 4 + 4)] = 16$, and the monocapped square-antiprismatic anion is stabilized by six MMM 3c-2e bonds and four M–M 2c-2e bonds. Bonding within the polyhedron involves a resonance hybrid of several canonical forms of the type shown in Fig. 14.6.2. The structures of the polyatomic anions are summarized in Table 14.6.1.

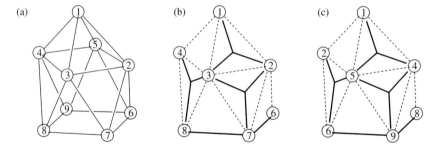

Fig. 14.6.2.
(a) Structure and atom numbering of Ge_9^{4-}. Bonding in Ge_9^{4-}: (b) front view of a canonical structure, in which a M–M 2c-2e bond is represented by a solid line joining two atoms, and a MMM 3c-2e bond by three solid lines connecting atoms to the bond center; (c) rear view of the canonical structure.

The polystannide cluster Sn_9^{4-} (or the related Pb_9^{4-}) reacts with $Cr(CO)_3(mes)$ to form a bicapped square-antiprismatic cluster $[Sn_9Cr(CO)_3]^{4-}$;

$$Cr(CO)_3(mes) + K_4Sn_9 + 4\,C222 \xrightarrow{en/toluene} [K(C222)]_4[Sn_9Cr(CO)_3].$$

The nonastannide cluster is slightly distorted to accommodate the chromium carbonyl fragment at the open face. Figure 14.6.3 shows the structure of $[M_9Cr(CO)_3]^{4-}$ (M = Sn, Pb).

Group 14 Elements

Table 14.6.1. Structure of polyatomic anions

Anion	Polyhedron	b	Bonding
Ge_4^{4-}, Sn_4^{4-}, Pb_4^{4-}, $Pb_2Sb_2^{2-}$	Tetrahedral	6	6[M–M 2c-2e]
Ge_5^{2-}, Pb_5^{2-}, Sn_5^{2-}	Trigonal bipyramidal	9	9[M–M 2c-2e]
Sn_6^{2-}	Octahedral	11	4[MMM 3c-2e]
			3[M–M 2c-2e]
Ge_9^{2-} (Sn_9^{3-}, Ge_9^{3-})	Tricapped-trigonal prismatic	17	7[MMM 3c-2e]
			3[M–M 2c-2e]
Ge_9^{4-}, Pb_9^{4-}, Sn_9^{4-} *	Monocapped Square-antiprismatic	16	6[MMM 3c-2e]
			4[M–M 2c-2e]
Ge_{10}^{2-}, Pb_{10}^{2-}, Sn_9Tl^{3-}	Bicapped Square-antiprismatic	19	8[MMM 3c-2e]
			3[M–M 2c-2e]

*The cation Bi_9^{5+} is isoelectronic and isostructural with these anions, and has the same kind of bonding.

The bond valence (b) in the ten-atom cluster $[Sn_9Cr(CO)_3]^{4-}$ is

$$b = \frac{1}{2}[8n_1 + 18n_2 - g]$$

$$= \frac{1}{2}[8 \times 9 + 18 - (9 \times 4 + 6 + 6 + 4)]$$

$$= 19.$$

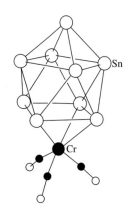

Fig. 14.6.3.
Structure of $[M_9Cr(CO)_3]^{4-}$ (M = Sn, Pb).

This anion has eight MMM 3c–2e bonds and three M–M 2c–2e bonds, just as in $B_{10}H_{10}^{2-}$, and the scheme is illustrated in Fig. 13.4.5(c).

14.7 Organometallic compounds of heavier group 14 elements

Elements Ge, Sn, and Pb form many interesting varieties of organometallic compounds, some of which are important commercial products. For example, Et_4Pb is an anti-knock agent, and organotin compounds are employed as polyvinylchloride (PVC) stabilizers against degradation by light and heat. Some examples and advances are discussed in the present section.

14.7.1 Cyclopentadienyl complexes

Cyclopentadienyl complexes of Ge, Sn, and Pb exist in a wide variety of composition and structure, such as half-sandwiches CpM and CpMX, bent and parallel sandwiches Cp_2M, and polymeric $(Cp_2M)_x$.

(1) $(\eta^5\text{-}C_5Me_5)Ge^+$ and $(\eta^5\text{-}C_5Me_5)GeCl$

In crystalline $[\eta^5\text{-}(C_5Me_5)Ge]^+ [BF_4]^-$ the cation has a half-sandwich structure, as shown in Fig. 14.7.1(a). In the chloride $(\eta^5\text{-}C_5Me_5)GeCl$, the Ge atom is bonded to the Cl atom in a bent configuration, as shown in Fig. 14.7.1(b).

(2) $(\eta^5\text{-}C_5R_5)_2M$ (R=H, Me, Ph)

The compounds $(\eta^5\text{-}C_5H_5)_2M$ (M = Ge, Sn and Pb) are angular molecules in the gas phase, as shown in Fig. 14.7.1(c). The ring centroid-M-ring centroid

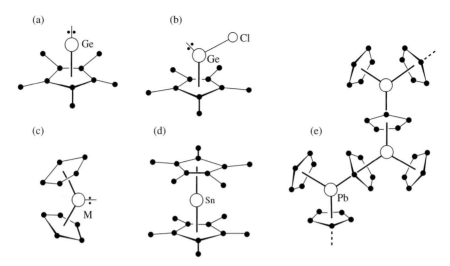

Fig. 14.7.1.
Structures of some cyclopentadienyl complexes: (a) $(\eta^5\text{-}C_5Me_5)Ge^+$, (b) $(\eta^5\text{-}C_5Me_5)GeCl$, (c) $(\eta^5\text{-}C_5H_5)_2M$, (M=Ge, Sn and Pb), (d) $(\eta^5\text{-}C_5Ph_5)_2Sn$ (only one C atom of each phenyl group is shown), and (e) $[(\eta^5\text{-}C_5H_5)_2Pb]_x$ chain.

angles of $(\eta^5\text{-}C_5H_5)_2M$ are –Ge– 130°, –Sn– 134°, and –Pb– 135°. However, as the H atoms in Cp is substituted by other groups, the angles will be changed. For example, the angle in $(\eta^5\text{-}C_5Me_5)_2$ Pb is 151°. For $(\eta^5\text{-}C_5Ph_5)_2Sn$, the two planar C_5 rings are exactly parallel and staggered, as shown in Fig. 14.7.1(d), and the opposite canting of the phenyl rings with respect to the C_5 rings results in overall S_{10} molecular symmetry.

(3) **Polymeric $[(\eta^5\text{-}C_5H_5)_2Pb]_x$**

There are three known crystalline modifications of $(\eta^5\text{-}C_5H_5)_2Pb$. When crystals are grown by sublimation, the structure features a zigzag chain with alternating bridging and nonbridging C_5H_5 groups; however, crystallization from toluene results in modification of the conformation of the polymeric chain [as shown in Fig. 14.7.1(e)], and also cyclization into a hexamer $[(\eta^5\text{-}C_5H_5)_2Pb]_6$.

14.7.2 Sila- and germa-aromatic compounds

Sila- and germa-aromatic compounds are Si- and Ge-containing $(4n + 2)$ π electron ring systems that represent the heavier congeners of carbocyclic aromatic compounds. Several kinds of sila- and germa-aromatic compounds bearing a bulky hydrophobic 2,4,6-tris[bis(trimethylsilyl)methyl]phenyl group [Tbt = 2,4,6,-(CH(SiMe$_3$)$_2$)$_3$C$_6$H$_2$] on the sp^2 Si or Ge atom have been prepared and isolated as stable crystalline solids, as shown in Figs. 14.7.2(a)–(e).

X-ray diffraction analyses showed that all these compounds have an almost planar aromatic ring, and the geometry around the central Si or Ge atom is completely trigonal planar. The measured bond lengths of Si–C or Ge–C are

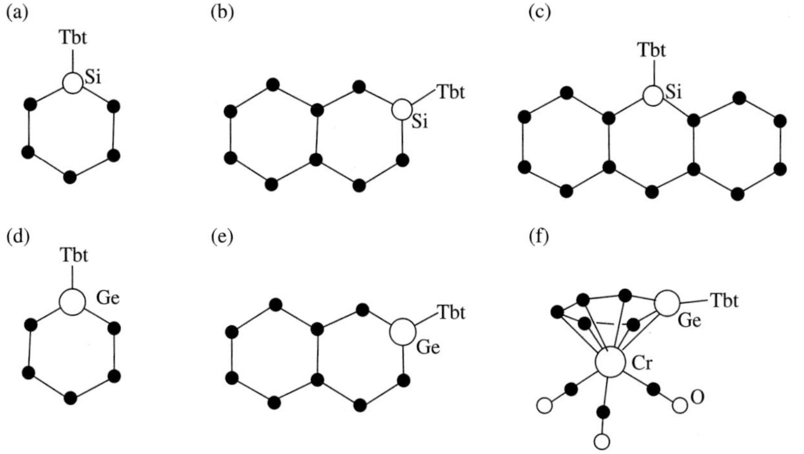

Fig. 14.7.2.
Structure of some sila- and germa-aromatic compounds: (a) silabenzene, (b) 2-silanaphthalene, (c) 9-silaanthracene, (d) germabenzene, (e) 2-germanaphthalene, and (f) a part of (η^6-C$_5$H$_5$GeTbt)Cr(CO)$_3$.

almost equal, the average values being Si–C 175.4 pm and Ge–C 182.9 pm, which lie between those of typical double and single bonds.

The germabenzene shown in Fig. 14.7.2(d) serves as an η^6-arene ligand in reacting with [M(CH$_3$CN)$_3$(CO)$_3$] (M = Cr, Mo, W) to form sandwich complexes [(η^6-C$_5$H$_5$GeTbt)M(CO)$_3$]. Figure 14.7.2(f) shows a part of the structure of [(η^6-C$_5$H$_5$GeTbt)Cr(CO)$_3$]. This complexation of germa-aromatic compounds indicates that sila- and germa-benzenes have a π electron delocalized structure with considerable aromaticity.

14.7.3 Cluster complexes of Ge, Sn, and Pb

Extensive research over the past decade has greatly increased the number of cluster complexes of Ge, Sn, and Pb. The properties of M–M bonds and structures of these compounds are similar to the C–C bonds and carbon skeletons in organic compounds. Table 14.7.1 lists some of these cluster compounds that have been synthesized, isolated, and characterized by X-ray crystallography. In

Table 14.7.1. Cluster complexes of Ge and Sn

Compound	Bond length / pm	Bond valence	Fig. 14.7.3
[Ge(Si^{t}Bu$_3$)]$_3^+$[BPh$_4$]$^-$	Ge ≡ Ge, 232.6	1^2/$_3$	(a)
[Ge(SiMe$_3$)$_3$]$_2$SnCl$_2$	Ge—Sn, 263.1	1	(b)
[Ge(Cl)Si(SiMe$_3$)$_3$]$_4$	Ge—Ge, 253.4	1	(c)
Ge$_4$(Si^{t}Bu$_3$)$_4$	Ge—Ge, 244	1	(d)
Ge$_6$[CH(SiMe$_3$)$_2$]$_6$	Ge—Ge, 256	1	(e)
[Sn(Ph$_2$)]$_6$	Sn—Sn, 277.5	1	(f)
Sn$_5$(2,6-Et$_2$C$_6$H$_3$)$_6$	Sn—Sn, 285.8	1	(g)
	Sn---Sn, 336.7		
Sn$_7$(2,6-Et$_2$C$_6$H$_3$)$_8$	Sn—Sn, 284.5	1	(h)
	Sn---Sn, 334.8		
Ge$_8$(CMeEt$_2$)$_8$	Ge—Ge, 249.0	1	(i)
Ge$_8^t$Bu$_8$Br$_2$	Ge—Ge, 248.4	1	(j)
Sn$_{10}$(2,6-Et$_2$C$_6$H$_3$)$_{10}$	Sn—Sn, 285.6	1	(k)

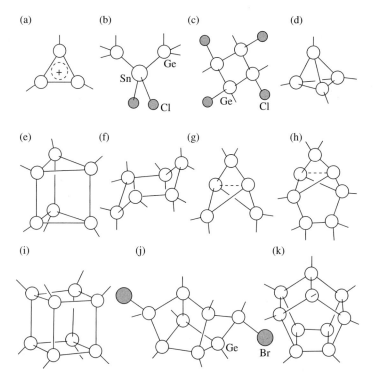

Fig. 14.7.3.
Structures of metallic cluster complexes of Ge and Sn (for clarity only the metal and halogen atoms are shown, and the organic groups have been omitted):
(a) [Ge(Si^{t}Bu$_3$)]$_3^+$,
(b) [Ge(SiMe$_3$)$_3$]$_2$SnCl$_2$,
(c) [Ge(Cl)Si(SiMe$_3$)$_3$]$_4$,
(d) Ge$_4$(Si^{t}Bu$_3$)$_4$,
(e) Ge$_6$[CH(SiMe$_3$)$_2$]$_6$, (f) [SnPh$_2$]$_6$,
(g) Sn$_5$(2,6-Et$_2$C$_6$H$_3$)$_6$,
(h) Sn$_7$(2,6-Et$_2$C$_6$H$_3$)$_8$,
(i) Ge$_8$(CMeEt$_2$)$_8$, (j) Ge$_8^t$Bu$_8$Br$_2$,
(k) Sn$_{10}$(2,6-Et$_2$C$_6$H$_3$)$_{10}$.

the table, the bond lengths are average values. The b value of M$_n$ is calculated according to formula (13.4.1), and the bond valence per M–M bond is listed for each compound.

In [GeSi^{t}Bu$_3$]$_3^+$[BPh$_4$]$^-$, the cyclotrigermanium cation has a π delocalized bond π_3^2 similar to that of the carbon analog, the cyclopropenium ion. The Ge–Ge bond in this cation has bond valence 5/3. Figure 14.7.3(a) shows the structure of the cluster.

The mixed-metal compound [Ge(SiMe$_3$)$_3$]$_2$SnCl$_2$ is a bent molecule [as shown in Fig. 14.7.3(b)], which forms two Ge–Sn bonds with bond lengths 263.6 and 262.6 pm.

The tetramer [Ge(Cl)Si(SiMe$_3$)$_3$]$_4$ forms a four-membered ring with Ge–Ge bond lengths 255.8 and 250.9 pm [as shown in Fig. 14.7.3(c)], while Ge$_4$(Si^{t}Bu$_3$)$_4$ is the first molecular germanium compound with a Ge$_4$ tetrahedron, as shown in Fig. 14.7.3(d).

The hexamer Ge$_6$[CH(SiMe$_3$)$_2$]$_6$ has a prismane structure as shown in Fig. 14.7.3(e). The Ge–Ge distances within the two triangular faces are 258 pm, being longer than those in the quadrilateral edges, which are 252 pm. On the other hand, the hexamer (SnPh$_2$)$_6$ exists in the chair conformation, (as shown in Fig. 14.7.3(f)) with Sn–Sn distances 277 to 278 pm.

Both Sn$_5$(2,6-Et$_2$C$_6$H$_3$)$_6$ and Sn$_7$(2,6-Et$_2$C$_6$H$_3$)$_8$ have a pentastanna[1.1.1] propellane structure, which consists of three fused three-membered rings, as shown in Figs. 14.7.3(g) and (h). In these two compounds, the bridgehead Sn atoms each have all four bonds directed to one side of the relevant atom, thus forming "inverted configuration" Sn⋅⋅⋅Sn bonds, the distances of which are 336.7 and 334.8 pm, respectively. These distances far exceed the longest known Sn–Sn single-bond length of 280 pm. These structures are also similar to that of 1-cyanotetracyclo-decane [Fig. 14.3.5(d)], in which the two bridgehead C atoms form a very long (164.3 pm) C–C bond.

The octagermacubanes $Ge_8(CMeEt_2)_8$ and $Ge_8(2,6-Et_2C_6H_3)_8$ have a cubic structure, as shown in Fig. 14.7.3(i), with Ge–Ge bond lengths 247.8 to 250.3 pm, with a mean value of 249.0 pm. The polycyclic octagermane $Ge_8{}^tBu_8Br_2$ forms a chiral C_2 symmetric skeleton, as shown in Fig. 14.7.3(j). The average Ge–Ge bond length is 248.4 pm.

The pentaprismane skeleton of $Sn_{10}(2,6-Et_2C_6H_3)_{10}$ is composed of five four-membered rings and two five-membered rings, as shown in Fig. 14.7.3(k). The average Sn–Sn bond length is 285.6 pm.

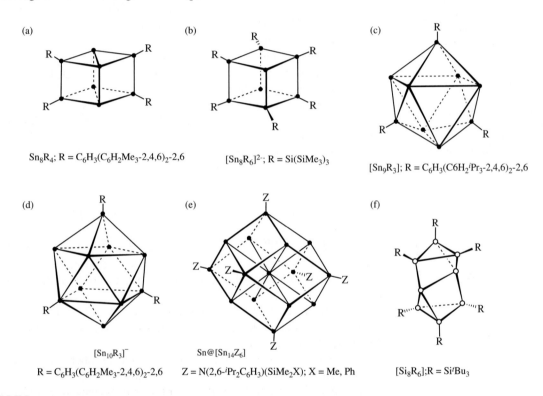

Fig. 14.7.4.
Structures of some metalloid clusters of tin and the isoleptic compound $Si_8(Si^tBu_3)_6$.

14.7.4 Metalloid clusters of Sn

Literature information is available for high-nuclearity metalloid clusters $[Sn_xR_y]$ ($x > y$), including $[Sn_8R_4]$, $[Sn_8R_6]^{2-}$, $[Sn_9R_3]$, and $[Sn_{10}R_3]^-$, in which R represents a bulky aryl or silyl ligand, and their structures are displayed in Figs. 14.7.4(a)–(d), respectively.

The largest tin clusters known to date are the pair of isoleptic amido complexes $[Sn_{15}Z_6]$ (Z = N(2,6-^{i}Pr$_2$C$_6$H$_3$)(SiMe$_2$X); X = Me, Ph). Both possess a common polyhedral core consisting of fourteen tin atoms that fully encapsulates a central tin atom. The eight ligand-free peripheral tin atoms constitute the corners of a distorted cube, with each of its six faces capped by a Sn{N(2,6-^{i}Pr$_2$C$_6$H$_3$)(SiMe$_2$X)} moiety, as shown in Fig. 14.7.4(e). The edges of the Sn$_{14}$ polyhedron have an average length of 302 pm. The Sn(central)–Sn bonds in this high-nuclearity Sn@Sn$_{14}$ metalloid cluster have an average length of 315 pm, which is close to the corresponding value of 310 pm in gray tin (α-Sn, diamond lattice) but considerably longer than that in white tin (β-Sn).

Interestingly, the isoleptic compound Si$_8$R$_6$ (R = Si^{t}Bu$_3$; "supersilyl") does not exhibit a cubanoid cluster skeleton, but instead contains an unsubstituted Si$_2$ dumbbell with a short Si–Si single bond of 229(1) pm, which is sandwiched between two almost parallel Si$_3$R$_3$ rings, as illustrated in Fig. 14.7.4(f). Each terminal of the Si$_2$ dumbbell acts as a bridgehead with an "inverted" tetrahedral bond configuration analogous to that found in 1-cyano-tetracyclodecane (see Section 14.3.5), forming two normal Si–Si bonds of 233 pm to one Si$_3$R$_3$ ring and a longer one of 257 pm to the other. In the crystal structure, the atoms of the Si$_2$ dumbbell are disordered over six sites.

An intermetalloid cluster is a metalloid cluster composed of more than one metal, for example [Ni$_2$@Sn$_{17}$]$^{4-}$. Known intermetalloid clusters involving other Group 14 elements and various transition metals include [Pd$_2$@Ge$_{18}$]$^{4-}$, [Ni(Ni@Ge$_9$)$_2$]$^{4-}$, [Ni@Pb$_{10}$]$^{2-}$, and [Pt@Pb$_{12}$]$^{2-}$; in the latter complex, the Pt atom is encapsulated by a Pb$_{12}$ icosahedron (see Fig. 9.6.18).

14.7.5 Donor–acceptor complexes of Ge, Sn and Pb

The system R$_2$Sn→SnCl$_2$, with R = CH(SiMe$_3$)C$_9$H$_6$N), provides the first example of a stable donor–acceptor complex between two tin centers, as shown in Fig. 14.7.5. The alkyl ligand R is bonded in a C,N-chelating fashion to a Sn atom which adopts a pentacoordinate square-pyramidal geometry. This Sn atom is bonded directly to the other Sn atom of the SnCl$_2$ fragment with a Sn–Sn distance of 296.1 pm, which is significantly longer than the similar distance of 276.8 pm in R$_2$Sn=SnR$_2$ [R = CH(SiMe$_3$)$_2$], and much shorter than the similar distance of 363.9 pm in Ar$_2$Sn–SnAr$_2$ [Ar = 2,4,6-(CF$_3$)$_3$C$_6$H$_2$]. The fold angle defined as the angle between the Sn–Sn vector and the SnCl$_2$ plane is 83.3°.

Some donor–acceptor complexes containing Ge–Ge and Sn–Sn bonds are listed in Table 14.7.2.

The compounds listed in Table 14.7.2 have the formal formula: which indicate the presence of a M=M double bond, but the measured M–M bond lengths vary over a wide range. The M$_2$C$_4$ (or M$_2$Si$_4$) skeleton is not planar, in contrast with that of an olefin. Thus the properties of the formal M=M bonds are diverse and interesting. Two examples listed in this table are discussed below.

As reference data, the "normal" values M–M single-bond lengths are calculated from the covalent radii (Table 3.4.3) and the "normal" bond lengths of M=M double bonds are estimated as 0.9 × (single-bond lengths). Thus the

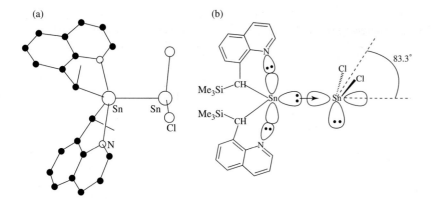

Fig. 14.7.5.
Structure of the R$_2$Sn→SnCl$_2$ (R = CH(SiMe$_3$)C$_9$H$_6$N). (a) Molecular structure (the SiMe$_3$ groups have been omitted for clarity), (b) the lone pair forming a donor–acceptor bond.

"normal" bond lengths are

$$\text{Ge–Ge } 244 \text{ pm, } \text{Ge}=\text{Ge } 220 \text{ pm,}$$
$$\text{Sn–Sn } 280 \text{ pm, } \text{Sn}=\text{Sn } 252 \text{ pm.}$$

Compound A in Table 14.7.2, (Me$_3$SiN=PPh$_2$)$_2$C=Ge→Ge=C(Ph$_2$P=NSiMe$_3$)$_2$, comprises two germavinylidene units Ge=C(Ph$_2$P=NSiMe$_3$)$_2$ bonded together in a head-to-head manner. The molecule is asymmetrical with

Table 14.7.2. Some donor–acceptor complexes with Ge–Ge and Sn–Sn bonds

Compound*	M–M bond length (pm)
[Ge(2,6-Et$_2$C$_6$H$_3$)$_2$]$_2$	Ge–Ge, 221.3
[Ge(2,6-i-C$_3$H$_7$)$_2$C$_6$H$_3$]{2,4,6-Me$_3$C$_6$H$_2$}$_2$	Ge–Ge, 230.1
[Ge{CH(SiMe$_3$)$_2$}$_2$]$_2$	Ge–Ge, 234.7
A	Ge–Ge, 248.3
[Sn{CH(SiMe$_3$)$_2$}$_2$]$_2$	Sn–Sn, 276.8
[Sn{Si(SiMe$_3$)$_3$}$_2$]$_2$	Sn–Sn, 282.5
[Sn(4,5,6-Me$_3$-2-^{t}BuC$_6$H)$_2$]$_2$	Sn–Sn, 291.0
B	Sn–Sn, 300.9
C	Sn–Sn, 308.7
[Sn{2,4,6-(CF$_3$)$_3$C$_6$H$_2$}$_2$]$_2$	Sn–Sn, 363.9

*A, B, C structures shown below.

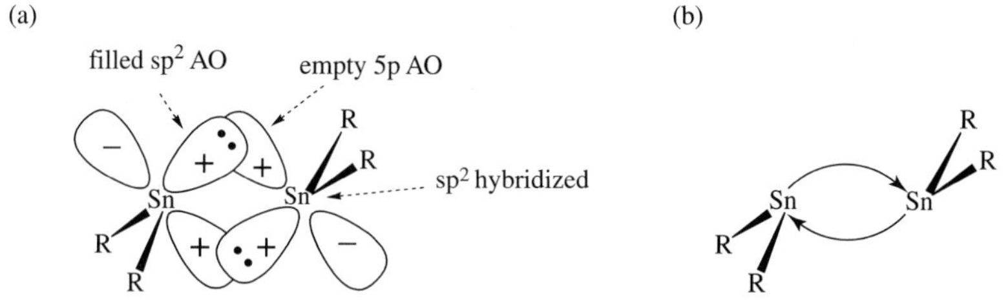

Fig. 14.7.6.
Sn–Sn bonding model in [Sn{CH(SiMe$_3$)$_2$}$_2$]$_2$: (a) overlap of atomic orbitals, (b) representation of the donor–acceptor bonding.

two different Ge environments: four-coordinate Ge and two-coordinate Ge, and shows *trans* linkage of the C=Ge–Ge=C skeleton, the torsion angle about the Ge–Ge axis being 43.9°. The Ge–Ge bond distance of 248.3 pm is consistent with a single bond, and both observed Ge–C bond lengths 190.5 and 190.8 pm are between those of standard Ge–C and Ge=C bonds. Therefore, the Ge–Ge bond in compound A is more approximately described as a donor–acceptor interaction similar to that of R$_2$Sn→SnCl$_2$ (as shown in Fig. 14.7.5), and the four-coordinate Ge behaves as the donor and the two-coordinate Ge as a Lewis acid center.

The compound [Sn{CH(SiMe$_3$)$_2$}$_2$]$_2$ does not possess a planar Sn$_2$C$_4$ framework, and the Sn–Sn bond length (276.8 pm) is too long to be consistent with a "normal" double bond. A bonding model involving overlap of filled sp^2 hybrids and vacant 5p atomic orbitals has been suggested, as shown in Fig. 14.7.6. The structure is similar to that in Fig. 14.4.7(e), but the *trans*-bent angle is larger.

The digermanium alkyne analog 2,6-Dipp$_2$H$_3$C$_6$Ge≡GeC$_6$H$_3$-2,6-Dipp$_2$ (Dipp = C$_6$H$_3$-2,6-^{i}Pr$_2$) was synthesized by the reaction of Ge(Cl)C$_6$H$_3$-2,6-Dipp$_2$ with potassium in THF or benzene. It was isolated as orange-red crystals and fully characterized by spectroscopic methods and X-ray crystallography. The digermyne molecule is centrosymmetric with a planar *trans*-bent C(*ipso*)–Ge–Ge–C(*ipso*) skeleton, and the central aryl ring of the terphenyl ligand is virtually coplanar with the molecular skeleton, each flanking aryl ring being oriented at ∼82° with respect to it. The structural parameters are C–Ge, 199.6 pm; Ge–Ge, 228.5 pm (considerably shorter than the Ge–Ge single-bond distance of approx. 244 pm); C–Ge–Ge, 128.67°. The measured Ge–Ge distance lies on the short side of the known range (221–46 pm) for digermenes, the digermanium analogs of alkenes.

Fig. 14.7.7.
Ge–Ge bonding model in Ar'Ge≡GeAr' (Ar' = 2,6-(C$_6$H$_3$-2,6-^{i}Pr$_2$)$_2$C$_6$H$_3$): (a) overlap of atomic orbitals, showing that donor–acceptor σ bonding occurs in the molecular plane, and (b) representation of the multiple bonding.

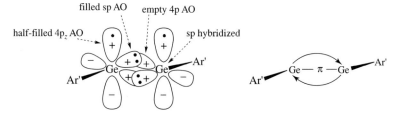

By analogy to the model used to describe a Sn=Sn double bond, the observed geometry of the digermyne molecule and its multiple bonding character can be rationalized as shown in Fig. 14.7.7. Each germanium atom is considered to be sp hybridized in the C–Ge–Ge–C plane; a singly filled sp hybrid orbital is used to form a covalent bond with the terphenyl ligand, and the other sp hybrid is completely filled and forms a donor bond with an empty in-plane 4p orbital on the other germanium atom. Two half-filled $4p_z$ orbitals, one on each germanium atom and lying perpendicular to the molecular skeleton, then overlap to form a much weaker π bond. This simplified bonding description is consistent with the *trans*-bent molecular geometry and the fact that the formal Ge≡Ge bond length is not much shorter than that of the Ge=Ge bond.

The distannyne Ar'SnSnAr' and diplumbyne Ar*PbPbAr* (Ar* = 2,4,6-$(C_6H_3-2,6-{}^iPr_2)_3C_6H_2$) have also been synthesized as crystalline solids and fully characterized. They are isostructural with digermyne, and their structural parameters are Sn–Sn, 266.75 pm; C–Sn–Sn, 125.24°; Pb–Pb, 318.81 pm; C–Pb–Pb, 94.26°. Recent studies of the chemical reactivities of digermynes showed that it has considerable diradical character, which can be represented by a canonical form with an unpaired electron on each germanium atom, i.e., negligible overlap between the half-filled $4p_z$ orbitals. The reactivity of the alkyne analogs decreases in the order Ge > Sn > Pb. In the diplumbyne, the Pb–Pb bond length is significantly longer than the value of approximately 290 pm normally found in organometallic lead–lead bonded species, e.g., $Me_3PbPbMe_3$. This suggests that a lone pair resides in the 6s orbital of each lead atom in the dilead compound, and a single σ bond results from head-to-head overlap of 6p orbitals.

In summary, it is noted that multiple bonding between the heavier Group 14 elements E (Ge, Sn, Pb) differs in nature in comparison with the conventional σ and π covalent bonds in alkenes and alkynes. In an E=E bond, both components are of the donor–acceptor type, and a formal E≡E bond involves two donor–acceptor components plus a p–p π bond. There is also the complication that the bond order may be lowered when each E atom bears an unpaired electron or a lone pair. The simple bonding models provide a reasonable rationale for the marked difference in molecular geometries, as well as the gradation of bond properties in formally single, double and triple bonds, in compounds of carbon versus those of its heavier congeners.

References

1. J. March, *Advanced Organic Chemistry*, 4th edn., Wiley, New York, 1992.
2. N. N. Greenwood and A. Earnshaw, *Chemistry of the Elements*, 2nd edn., Butterworth-Heinemann, Oxford, 1997.
3. C. E. Housecroft and A. G. Sharpe, *Inorganic Chemistry*, 2nd edn., Prentice-Hall, Harlow, 2004.
4. D. F. Shriver, P. W. Atkins, T. L. Overton, J. P. Rourke, M. T. Weller and F. A. Armstrong, *Inorganic Chemistry*, 4th edn., Oxford University Press, Oxford, 2006.
5. F. H. Allen, O. Kennard, D. G. Watson, L. Brammer, A. G. Orpen and R. Taylor, in A. J. C. Wilson (ed.), *International Tables for Crystallography*, vol. C, pp. 685–06, Kluwer Academic Publishers, Dordrecht, 1992.
6. J. Baggott, *Perfect Symmetry, The Accidental Discovery of Buckminster Fullerene*, Oxford University Press, Oxford, 1994.
7. H. W. Kroto, J. E. Fischer and D. E. Cox (eds.), *Fullerenes*, Pergamon Press, Oxford, 1993.
8. R. Taylor, *Lecture Notes on Fullerene Chemistry: A Handbook for Chemists*, Imperial College Press, London, 1999.

9. P. W. Fowler and D. E. Manolopoulos, *An Atlas of Fullerenes*, Oxford University Press, Oxford, 1995.
10. A. Hirsch and M. Brettreich, *Fullerenes: Chemistry and Reactions*, Wiley–VCH, Weinheim, 2005.
11. F. Langa and J.-F. Nierengarten (eds.), *Fullerenes: Principles and Applications*, Royal Society of Chemistry, Cambridge, 2007.
12. T. Akasaka and S. Nagase (eds.), *Endofullerenes: A New Family of Carbon Clusters*, Kluwer, Dordrecht, 2002.
13. M. S. Dresselhaus, G. Dresselhaus and P. C. Eklund, *Science of Fullerenes and Carbon Nanotubes*, Academic Press, San Diego, CA, 1996.
14. P. J. F. Harris, *Carbon Nanotubes and Related Structures*, Cambridge University Press, Cambridge, 1999.
15. M. Meyyappan (ed.), *Carbon Nanotubes: Science and Applications*, CRC Press, Boca Raton, FL, 2005.
16. L. R. Radovic (ed.), *Chemistry and Physics of Carbon*, vol. 28, Marcel Dekker, New York, 2003.
17. J. Y. Corey, E. R. Corey and P. P. Gaspar (eds.), *Silicon Chemistry*, Ellis Horword, Chichester, 1988.
18. M. A. Brook, *Silicon in Organic, Organometallic, and Polymer Chemistry*, Wiley, New York, 2000.
19. P. Jutzi and U. Schubert (eds.), *Silicon Chemistry: From the Atom to Extended Systems,* Wiley-VCH, Weinheim, 2003.
20. F. Liebau, *Structural Chemistry of Silicates*, Springer, New York, 1985.
21. D. T. Griffen, *Silicate Crystal Chemistry*, Oxford University Press, New York, 1992.
22. R. Szostak, *Handbook of Molecular Sieves*, Van Nostrand Reinhold, New York, 1992.
23. W. M. Meier and D. H. Olson, *Atlas of Zeolite Structure Types*, Butterworths, London, 1987.
24. R. West and F. G. A. Stone (eds.), *Multiply Bonded Main Group Metals and Metalloids*, Academic Press, San Diego, CA, 1996.
25. S. M. Kauzlarich (ed.), *Chemistry, Structure and Bonding of Zintl Phases and Ions*, VCH, New York, 1996.
26. J. Sloan, A. I. Kirkland, J. L. Hutchison and M. L. H. Green, Integral atomic layer architectures of 1D crystals inserted into single walled carbon nanotubes. *Chem. Commun.*, 1319–32 (2002).
27. J. Sloan, A. I. Kirkland, J. L. Hutchison and M. L. H. Green, Aspects of crystal growth within carbon nanotubes. *C. R. Physique* **4**, 1063–74 (2003).
28. Y. Yu, A. D. Bond, P. W. Leonard, U. J. Lorenz, T. V. Timofeeva, K. P. C. Vollhardt, G. D. Whitener and A. D. Yakovenko, Hexaferrocenylbenzene. *Chem. Commun.*, 2572–74 (2006).
29. K. Yu. Chernichenko, V. V. Sumerin, R. V. Shpanchenko, E. S. Balenkova and V. G. Nenajdenko, "Sunflower": a new form of carbon sulfide. *Angew. Chem. Int. Ed.* **45**, 7367–70 (2006).
30. M. Sawamura, Y. Kuninobu, M. Toganoh, Y. Matsuo, M. Yamanaka and E. Nakamura, Hybrid of ferrocene and fullerene. *J. Am. Chem. Soc.* **124**, 9354–55 (2002).
31. S. Y. Xie, F. Gao, X. Lu, R.-B. Huang, C.-R. Wang, X. Zhang, M.-L. Liu, S.-L. Deng and L.-S. Zheng, Capturing the labile fullerene[50] as $C_{50}Cl_{10}$. *Science* **304**, 699 (2004).

32. T. A. Murphy, T. Pawlik, A. Weidinger, M. Höhne, R. Alcala and J. M. Spaeth, Observation of atom-like nitrogen in nitrogen-implanted solid C_{60}. *Phys. Rev. Lett.* **77**, 1075–78 (1996).
33. S. Stevenson, G. Rice, T. Glass, K. Harich, F. Cromer, M. R. Jordan, J. Craft, A. Hadju, R. Bible, M. M. Olmsted, K. Maitra, A. J. Fisher, A. L. Balch and H. C. Dorn, Small-bandgap endohedral metallofullerenes in high yield and purity. *Nature* **401**, 55–7 (1999).
34. T. Murahashi, M. Fujimoto, M.-a. Oka, Y. Hashimoto, T. Uemura, Y. Tatsumi, Y. Nakao, A. Ikeda, S. Sakaki and H. Kurosawa, Discrete sandwich compounds of monolayer palladium sheets. *Science* **313**, 1104–07 (2006).
35. F. Toda, Naphthocyclobutenes and benzodicyclobutadienes: synthesis in the solid state and anomalies in the bond lengths. *Eur. J. Org. Chem.* 1377–86 (2000).
36. O. T. Summerscales, F. G. N. Cloke, P. B. Hitchcock, J. C. Green and N. Hazari, Reductive cyclotrimerization of carbon monoxide to the deltate dianion by an organometallic uranium complex. *Science* **311**, 829–31 (2006).
37. D. R. Huntley, G. Markopoulos, P. M. Donovan, L. T. Scott and R. Hoffmann, Squeezing C–C bonds. *Angew. Chem. Int. Ed.* **44**, 7549–53 (2005).
38. F. R. Lemke, D. J. Szalda and R. M. Bullock, Ruthenium/zirconium complexes containing C_2 bridges with bond orders of 3, 2, and 1. Synthesis and structures of $Cp(PMe_3)_2RuCH_nCH_nZrClCp_2$ ($n = 0, 1, 2$). *J. Am. Chem. Soc.* **113**, 8466–77 (1991).
39. U. Ruschewitz, Binary and ternary carbides of alkali and alkaline-earth metals. *Coord. Chem. Rev.* **244**, 115–36 (2003).
40. G.-C. Guo, G.-D. Zhou and T. C. W. Mak, Structural variation in novel double salts of silver acetylide with silver nitrate: fully encapsulated acetylide dianion in different polyhedral silver cages. *J. Am. Chem. Soc.* **121**, 3136–41 (1999).
41. H.-B. Song, Q.-M. Wang, Z.-Z. Zhang and T. C. W. Mak, A novel luminescent copper(I) complex containing an acetylenediide-bridged, butterfly-shaped tetranuclear core. *Chem. Commun.*, 1658–59 (2001).
42. K. C. Kim, C. A. Reed, D. W. Elliott, L. J. Mueller, F. Tham and J. B. Lambert, Crystallographic evidence for a free silylium Ion. *Science* **297**, 825–7 (2002).
43. A. Sekiguchi and H. Sakurai, Cage and cluster compounds of silicon, germanium and tin. *Adv. Organometal. Chem.* **45**, 1–38 (1995).
44. A. Sekiguchi, S. Inoue, M. Ichinoche and Y. Arai, Isolable anion radical of blue disilene $(^tBu_2MeSi)_2Si=Si(SiMe^tBu_2)_2$ formed upon one-electron reduction: synthesis and characterization. *J. Am. Chem. Soc.* **126**, 9626–29 (2004).
45. S. Nagase, Polyhedral compounds of the heavier group 14 elements: silicon, germanium, tin and lead. *Acc. Chem. Res.* **28**, 469–76 (1995).
46. M. Brynda, R. Herber, P. B. Hitchcock, M. F. Lappert, I. Nowik, P. P. Power, A. V. Protchenko, A. Ruzicka and J. Steiner, Higher-nuclearity group 14 metalloid clusters: $[Sn_9\{Sn(NRR')\}_6]$. *Angew. Chem. Int. Ed.* **45**, 4333–37 (2006).
47. G. Fischer, V. Huch, P. Maeyer, S. K. Vasisht, M. Veith and N. Wiberg, $Si_8(Si^tBu_3)_6$: a hitherto unknown cluster structure in silicon chemistry. *Angew. Chem. Int. Ed.* **44**, 7884–87 (2005).
48. W. P. Leung, Z. X. Wang, H. W. Li and T. C. W. Mak, Bis(germavinylidene)-$[(Me_3SiN=PPh_2)_2C=Ge\rightarrow Ge=C(Ph_2P=NSiMe_3)]$ and 1,3-dimetallacyclobutanes $[M\{\mu^2\text{-}C(Ph_2PSiMe_3)_2\}]_2$ (M = Sn, Pb). *Angew. Chem. Int. Ed.* **40**, 2501–3 (2001).
49. W. P. Leung, W. H. Kwok, F. Xue and T. C. W. Mak, Synthesis and crystal structure of an unprecedented tin(II)-tin(II) donor-acceptor complex, $R_2Sn\rightarrow SnCl_2$, $[R=CH(SiMe_3)C_9H_6N-8]$. *J. Am. Chem. Soc.* **119**, 1145–46 (1997).

50. W. P. Leung, H. Cheng, R.-B. Huang, Q.-C. Yang and T.C.W. Mak, Synthesis and structural characterization of a novel asymmetric distannene [{1-{N(Bu^t)C(SiMe$_3$)C(H)}-2-{N(Bu^t)(SiMe$_3$)CC(H)}-C$_6$H$_4$}Sn → Sn{1,2-{N(Bu^t)(SiMe$_3$)CC(H)}$_2$C$_6$H$_4$}]. *Chem. Commun.*, 451–2 (2000).
51. N. Tokitoh, New progress in the chemistry of stable metalla-aromatic compounds of heavier group 14 elements. *Acc. Chem. Res.* **37**, 86–94 (2004).
52. M. Stender, A. D. Phillips, R. J. Wright and P. P. Power, Synthesis and characterization of a digermanium analogue of an alkyne. *Angew. Chem Int. Ed.* **41**, 1785–87 (2002).
53. P. P. Power, Synthesis and some reactivity studies of germanium, tin and lead analogues of alkyne. *Appl. Organometal. Chem.* **19**, 488–93 (2005).

Structural Chemistry of Group 15 Elements

15.1 The N_2 molecule, all-nitrogen ions and dinitrogen complexes

15.1.1 The N_2 molecule

Nitrogen is the most abundant uncombined element in the earth's surface. It is one of the four essential elements (C, H, O, N) that support all forms of life. It constitutes, on the average, about 15% by weight in proteins. The industrial fixation of nitrogen in the production of agricultural fertilizers and other chemical products is now carried out on a vast scale.

The formation of a triple bond, $N\equiv N$, comprising one σ and two π components with bond length 109.7 pm, accounts for the extraordinary stability of the dinitrogen molecule N_2; the energy level diagram for N_2 is shown in Fig. 3.3.3(a). Gaseous N_2 is rather inert at room temperature mainly because of the great strength of the $N\equiv N$ bond and the large energy gap between the HOMO and LUMO ($\sim$8.6 eV), as well as the absence of bond polarity. The high bond dissociation energy of the N_2 molecule, 945 kJ mol^{-1}, accounts for the following phenomena:

(a) Molecular nitrogen constitutes 78.1% by volume (about 75.5% by weight) of the earth's atmosphere.
(b) It is difficult to "fix" nitrogen, i.e., to convert molecular nitrogen into other nitrogen compounds by means of chemical reactions.
(c) Chemical reactions that release N_2 as a product are highly exothermic and often explosive.

15.1.2 Nitrogen ions and catenation of nitrogen

There are three molecular ions which consist of nitrogen atoms only.

(1) Nitride ion N^{3-}

The N^{3-} ion exists in salt-like nitrides of the types M_3N and M_3N_2. In M_3N compounds, M is an element of group 1 (Li) or Group 11 (Cu, Ag). In M_3N_2 compounds, M is an element of group 2 (Be, Mg, Ca, Sr, Ba) or Group 12 (Zn, Cd, Hg). In these compounds, the bonding interactions between N and M atoms are essentially ionic, and the radius of N^{3-} is 146 pm. Nitride Li_3N exhibits

(2) Azide N_3^-

The azide ion N_3^- is a symmetrical linear group that can be formed by neutralization of hydrogen azide HN_3 with alkalis. The bent configuration of HN_3 is shown below:

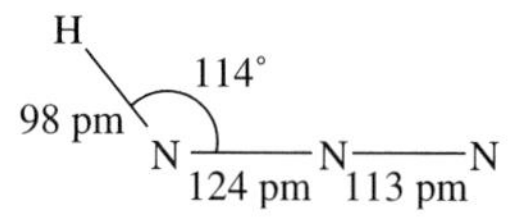

$$\underset{98\,\text{pm}}{H}\diagdown\underset{124\,\text{pm}}{N}\!\!-\!\!\underset{113\,\text{pm}}{N}\!\!-\!\!N \quad 114°$$

The Group 1 and 2 azides NaN_3, KN_3, $Sr(N_3)_2$ and $Ba(N_3)_2$ are well-characterized colorless crystalline salts which can be melted with little decomposition. The corresponding Group 11 and 12 metal azides such as AgN_3, $Cu(N_3)_2$ and $Pb(N_3)_2$, are shock-sensitive and detonate readily, and they are far less ionic with more complex structures. The salt $(PPh_4)^+(N_3HN_3)^-$ has been synthesized from the reaction of Me_3SiN_3 with $(PPh_4)(N_3)$ in ethanol; the $(N_3HN_3)^-$ anion has a nonplanar bent structure consisting of distinct N_3^- and HN_3 units connected by a hydrogen bond, as shown in Fig. 15.1.1(a). In $Cu_2(N_3)_2(PPh_3)_4$ and $[Pd_2(N_3)_6]^{2-}$, N_3^- acts either as a terminal or a bridging ligand, as shown in Figs. 15.1.1(b) and (c), respectively. The azide ion N_3^- can also be combined with a metalloid ion such as As^{5+} to form $[As(N_3)_6]^-$, as shown in Fig. 15.1.1(d). Table 15.1.1 lists the coordination modes of N_3^- in its metal complexes; the highest-ligation μ-1,1,1,3,3,3 mode (h) was found to exist in the double salt $AgN_3 \cdot 2AgNO_3$.

The uranium(IV) heptaazide anion $U(N_3)_7^{3-}$ has been synthesized as the *n*-tetrabutylammonium salt. This is the first homoleptic azide of an actinide

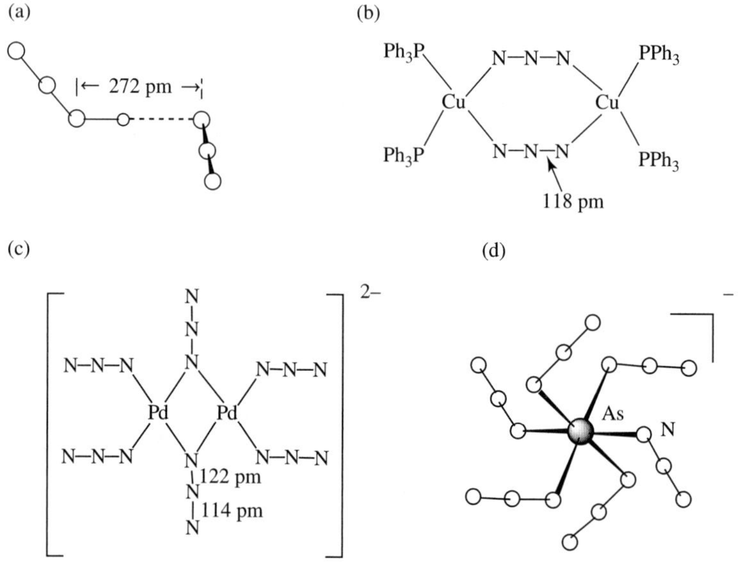

Fig. 15.1.1.
Structures of some azide compounds: (a) $(N_3HN_3)^-$, (b) $Cu(N_3)_2(PPh_3)_4$, (c) $[Pd_2(N_3)_6]^{2-}$, and (d) $[As(N_3)_6]^-$.

Table 15.1.1. Coordination modes of N_3^- in metal complexes

M—N=N=N	M—N=N=N—M	M\N=N=N/M
(a)	(b)	(c)
M\N—N—N≡M / M	M\N=N=N—M / M	M\N=N=N/M M/ \M
(d)	(e)	(f)
M\ M—N—N—N≡M / M	M\ M—N=N=N—M / M / M	
(g)	(h)	

as well as the first structurally characterized heptaazide. Two polymorphs of $(Bu_4N)_3[U(N_3)_7]$ were obtained from crystallization in $CH_3CN/CFCl_3$ (form A) or CH_3CH_2CN (form B). Form A belongs to space group $Pa\bar{3}$ with $Z = 8$, and hence the U atom and one of the three independent azide groups are located on a crystallographic 3-axis. This results in a 1:3:3 monocapped octahedral arrangement of the azide ligands around the central uranium atom, as illustrated in Fig. 15.1.2(a). Form B crystallizes in space group $P2_1/c$ with $Z = 4$, and the azide ligands exhibit a distorted 1:5:1 pentagonal-bipyramidal coordination mode, as shown in Fig. 15.1.2(b). The $U-N_\alpha$ bond lengths range from 232 to 243 pm.

Two isomeric polymorphs of $[(C_5Me_5)_2U(\mu\text{-}N)U(\mu\text{-}N_3)(C_5Me_5)_2]_4$ have been synthesized and structurally characterized by X-ray crystallography. In the tetrameric macrocycle, eight $(\eta^5\text{-}C_5Me_5)_2U^{2+}$ units are charge-balanced by four N_3^- (azide) ligands and four formally N^{3-} (nitride) ligands. Isomer A has a $(UNUN_3)_4$ ring in a pseudo-crown (or chair) conformation, whereas that of isomer B exhibits a pseudo-saddle (or boat) geometry, as shown in Fig. 15.1.3. The observed U–N(azide) bond distances in the range 246.7-252.5 pm are longer than typical U(IV)–N single bonds. The U–N(nitride) bond lengths in the range

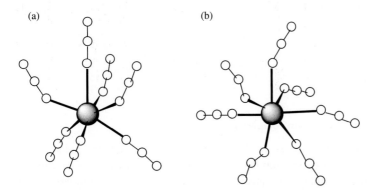

Fig. 15.1.2.
Molecular structure of $U(N_3)_7^{3-}$ in (a) form A and (b) form B of the Bu_4N^+ salt. [Ref. M.-J. Crawford, A. Ellern and P. Meyer, *Angew. Chem. Int. Ed.* **44**, 7874–78 (2005).]

Fig. 15.1.3.
Molecular skeleton of the $(UNUN_3)_4$ ring in two crystalline polymorphs of $[(C_5Me_5)_2U(\mu\text{-}N)U(\mu\text{-}N_3)(C_5Me_5)_2]_4$: pseudo-crown conformation observed for isomeric form A and pseudo-saddle observed for form B. The pair of η^5-C_5Me_5 groups bonded to each uranium(IV) atom is not shown. From W. J. Evans, S. A. Kozimor and J. W. Ziller, *Science* **309**, 1835–8 (2005).

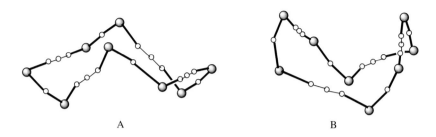

A   B

204.7-209.0 pm are consistent with a bond order of 2 in a symmetrical bonding scheme for each UNU segment, rather than an alternating single-triple-bond pattern, as shown below:

$$U=\!\!=\!\!N=\!\!=U \leftrightarrow U=\!\!=N=\!\!=U \qquad U-\!\!\!-N\equiv U$$

symmetrical nitride bridging   asymmetrical nitride bridging

(3) Pentanitrogen cation N_5^+

The pentanitrogen cation N_5^+ was first synthesized in 1999 as the white solid $N_5^+AsF_6^-$, which is not very soluble in anhydrous HF but stable at $-78°C$. The N_5^+ cation has a closed-shell singlet ground state, and its C_{2v} symmetry was characterized by Raman spectroscopy and NMR of ^{14}N and ^{15}N nuclei. A series of 1:1 salts of N_5^+ have also been prepared with the monoanions HF_2^- (as the adduct $N_5^+HF_2^- \cdot nHF$), BF_4^-, PF_6^-, SO_3F^-, $(Sb_2F_{11})^-$, $B(N_3)_4^-$ and $P(N_3)_6^-$. A determination of the crystal structure of $N_5^+(Sb_2F_{11})^-$, which is stable at room temperature, yielded the structural parameters shown in Fig. 15.1.4. Note that each terminal N–N–N segment in N_5^+ deviates from exact linearity. A description of the bonding in this cation is given in Chapter 5, Section 5.8.3.

The catenation of nitrogen refers to the tendency of N atoms to be connected to each other, and is far lower than those of C and P. This is because the repulsion of lone pairs on adjacent N atoms weakens the N–N single bond, and the lone pairs can easily react with electrophilic species. The structures and examples of known compounds that contain chains and rings of N atoms are listed in Table 15.1.2. Note that none of these has a linear configuration of nitrogen atoms.

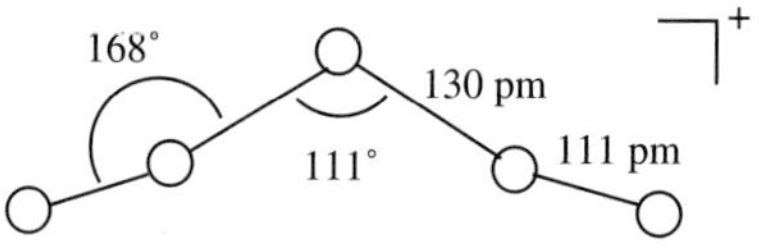

Fig. 15.1.4.
Molecular dimensions of the pentanitrogen cation N_5^+ from crystal structure analysis.

15.1.3 Dinitrogen complexes

Molecular nitrogen can react directly with some transition-metal compounds to form dinitrogen complexes, the structure and properties of which are of considerable interest because they may serve as models for biological nitrogen

Table 15.1.2. Species containing catenated N atoms

Number of N atoms	Chain or ring	Example
3	$[N-N-N]^+$	$[H_2NNMe_2NH_2]Cl$
3	$N-N=N$	$MeHN-N=NH$
4	$N-N=N-N$	$H_2N-N=N-NH_2$, $Me_2N-N=N-NMe_2$
4	$N-N-N-N$	$(CF_3)_2N-N(CF_3)-N(CF_3)-N(CF_3)_2$
5	$N=N-N-N=N$	$PhN=N-N(Me)-N=NPh$
6	$N=N-N-N-N=N$	$PhN=N-N(Ph)-N(Ph)-N=NPh$
8	$N=N-N-N=N-N-N=N$	$PhN=N-N(Ph)-N=N-N(Ph)-N=NPh$
5	![ring structure] N=N, N, N=N, N—	N=N, N, N=N, N–Ph

fixation and as intermediates in synthetic applications. The known coordination modes of dinitrogen are discussed below and summarized in Table 15.1.3. The skeletal views of dinitrogen coordination modes in some metal complexes are shown in Fig. 15.1.5.

(1) η^1-N_2

The first complex containing molecular nitrogen as a ligand, $[Ru(NH_3)_5(N_2)]Cl_2$, was synthesized and identified in 1965 [Fig. 15.1.5(a)]. The complex $(Et_2PCH_2CH_2PEt_2)_2Fe(N_2)$ exhibits trigonal bipyramidal coordination geometry, with N_2 lying in the equatorial plane, as shown in Fig. 15.1.5(a'). Most examples of stable dinitrogen complexes have been found to belong to the η^1–N_2 category, in which the dinitrogen ligand binds in a linear, end-on mode with only a slightly elongated N–N bond length (112–24 pm) as compared to that in gaseous dinitrogen (109.7 pm). The stable monomeric titanocene complexes $\{(PhMe_2Si)C_5H_4\}_2TiX$ (X = N_2, CO) are isomorphous, with a crystallographic C_2 axis passing through the Ti atom and the η^1–X ligand. The measured bond distances (in pm) are Ti–N = 201.6(1), N–N = 111.9(2) for the dinitrogen complex, and Ti–C = 197.9(2), C–O = 115.1(2) for the carbonyl complex.

(2) μ-(bis-η^1)-N_2

As a bridging ligand in dinuclear systems, dinitrogen may formally be classified into three types.

(a) M–N≡N–M: This dinitrogen ligand corresponds to a neutral N_2 molecule, which uses its lone pairs to coordinate to two M atoms. Complexes of this type show a relatively short N–N distance of 112–20 pm. In $[(\eta^5\text{-}C_5Me_5)_2Ti]_2(N_2)$, the binuclear molecular skeleton consists of two $(\eta^5\text{-}C_5Me_5)_2Ti$ moieties bridged by the N_2 ligand in an essentially linear Ti–N≡N–Ti arrangement, as shown in Fig. 15.1.5(b). The N–N distance is 116 pm (av.).

(b) M=N=N=M: This dinitrogen ligand corresponds to diazenido(-2). The $(N_2)^{2-}$ anion coordinates to two M atoms. In $(Mes)_3Mo(N_2)Mo(Mes)_3$

(where Mes = 2,4,6-Me$_3$C$_6$H$_2$), Mo=N=N=Mo forms a linear chain. The length of the N–N distance is 124.3 pm.

(c) M≡N–N≡M: This dinitrogen ligand corresponds to hydrazido(-4). The (N$_2$)$^{4-}$ anion coordinates to two M atoms. In [PhP(CH$_2$SiMe$_2$NPh$_2$)$_2$NbCl]$_2$ (N$_2$), the Nb⇐N–N⇒Nb moiety is linear, and the distance of N–N is 123.7 pm.

(3) μ_3-η^1: η^1: η^2-N$_2$

In the [(C$_{10}$H$_8$)(C$_5$H$_5$)$_2$Ti$_2$](μ_3–N$_2$)[(C$_5$H$_4$)(C$_5$H$_5$)$_3$Ti$_2$] complex, the dinitrogen ligand is coordinated simultaneously to three Ti atoms, as shown in Fig. 15.1.5(c). The N–N distance is 130.1 pm.

(4) μ_3-η^1: η^1: η^1-N$_2$

In the mixed-metal complex [WCl(py)(PMePh)$_3$(μ_3-N$_2$)]$_2$(AlCl$_2$)$_2$, both WNN linkages are essentially linear, and the four metal atoms and two μ_3–N$_2$ ligands almost lie in the same plane, as shown in Fig. 15.1.5(d).

(5) μ-(η^1: η^2)-N$_2$

In [PhP(CH$_2$SiMe$_2$NPh)$_2$]$_2$Ta$_2$(μ-H)$_2$(N$_2$), the dinitrogen moiety is end-on bound to one Ta atom and side-on bound to the other, as shown in Fig. 15.1.5(e). The N–N distance of 131.9 pm is consistent with a formal assignment of the bridging dinitrogen moiety as (N$_2$)$^{4-}$. The shortest distance of the end-on Ta–N bond is 188.7 pm, which is consistent with its considerable double-bond character.

(6) μ-(bis-η^2)-N$_2$ (planar)

Figure 15.1.5(f) shows a planar, side-on bonded N$_2$ ligand between two Zr atoms in the compound [Cp″$_2$Zr]$_2$(N$_2$) [Cp″ = 1,3-(SiMe$_2$)$_2$C$_5$H$_3$]. The N–N bond length is 147 pm. Side-on coordination of the dinitrogen ligand appears to be important for its reduction.

(7) μ-(bis-η^2)-N$_2$ (nonplanar)

In the compound Li{[(SiMe$_3$)$_2$N]$_2$Ti(N$_2$)}$_2$, each Ti atom is side-on bound to two N$_2$ molecules, as shown in Fig. 15.1.5(g). The N–N distance is 137.9 pm.

(8) μ_4-η^1: η^1: η^2: η^2-N$_2$

In [Ph$_2$C(C$_4$H$_3$N)$_2$Sm]$_4$(N$_2$), the N$_2$ moiety is end-on bound to two Sm atoms and side-on bound to the other two, as shown in Fig. 15.1.5(h). The bond lengths are: N–N 141.2 pm, Sm(terminal)–N 217.7 pm and Sm(bridging)–N 232.7 pm.

(9) μ_5-η^1: η^1: η^2: η^2: η^2-N$_2$

In {[(–CH$_2$–)$_5$]$_4$calixtetrapyrrole}$_2$Sm$_3$Li$_2$ (N$_2$)[Li(THF)$_2$]·(THF), the N$_2$ moiety is end-on bound to two Li atoms and side-on bound to three Sm atoms, as shown in Fig. 15.1.5(i). The bond lengths are N–N 150.2 pm, Sm–N 233.3 pm(av.), and Li–N 191.0 pm(av.).

Group 15 Elements

Table 15.1.3. Coordination modes of dinitrogen

Coordination mode	Example	d_{N-N} (pm)	Structure in Fig.15.1.5
(1) η^1-N_2 M—N≡N	$[Ru(NH_3)_5(N_2)]^{2+}$	112	(a)
	$(depe)_2Fe(N_2)$	113.9	(a')
(2) μ-(bis-η^1)-N_2 (a) M—N≡N—M (b) M=N=N=M (c) M≡N—N≡M	$[(C_5Me_5)_2Ti]_2(N_2)$ $(Mes)_3Mo(N_2)Mo(Mes)_3$ $[PhP(CH_2SiMe_2NPh)_2NbCl]_2(N_2)$	116 (av.) 124.3 123.7	(b)
(3) μ_3-η^1:η^1:η^2-N_2 M—N═N—M M	$[(C_{10}H_8)(C_5H_5)_2Ti_2]$- $[(C_5H_4)(C_5H_5)_3Ti_2](N_2)$	130.1	(c)

(4) μ_3-η^1:η^1:η^1-N_2
       M
 M—N—N—N—M
       M
	$[WCl(py)(PMePh)_3(N_2)]_2(AlCl_2)_2$	125	(d)

(5) μ-(η^1:η^2)-N_2
    M M
    N—N
	$[PhP(CH_2SiMe_2NPh)_2]_2Ta_2(\mu\text{-}H)_2(N_2)$	131.9	(e)

(6) μ-(bis-η^2)-N_2
      N
 M   M
      N
 (planar)
	$[Cp''_2Zr]_2(N_2)$	147	(f)

(7) μ-(bis-η^2)-N_2 N M—N—M (nonplanar)	$Li\{[(SiMe_3)_2N]_2Ti(N_2)\}_2$	137.9	(g)

(8) μ_4-η^1:η^1:η^2:η^2-N_2
       M
 M—N—N—M
       M
	$[Ph_2C(C_4H_3N)_2Sm]_4(N_2)$	141.2	(h)

(9) μ_5-η^1:η^1:η^2:η^2:η^2-N_2
         M
 M'—N═N—M'
      M   M
	$\{[(-CH_2-)_5]_4\text{calixtetrapyrrole}\}_2Sm_3Li_2(N_2)$	150.2	(i)

(10) μ_6-η^1:η^1:η^2:η^2:η^2:η^2-N_2
       M M'
 M'—N═N—M'
    M'   M
	$[(THF)_2Li(OEPG)Sm]_2Li_4(N_2)$	152.5	(j)

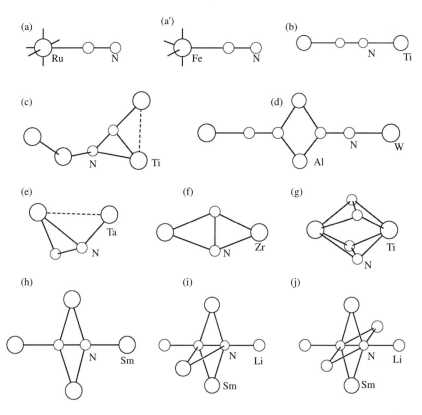

Fig. 15.1.5.
Dinitrogen coordination modes in metal complexes.

(10) μ_6-η^1: η^1: η^2: η^2: η^2: η^2-**N$_2$**

In [(THF)$_2$Li(OEPG)Sm]$_2$Li$_4$(N$_2$) (OEPG = octaethylporphyrinogen), the N$_2$ moiety is end-on bound to two Li atoms and side-on bound to two Sm atoms and two Li atoms, as shown in Fig. 15.1.5(j). The bond lengths are N–N 152.5 pm, Sm–N 235.0 pm (av.), and Li–N 195.5 pm(av.).

The bonding of dinitrogen to transition metals may be divided into the "end-on" and "side-on" categories. In the end-on arrangement, the coordination of the N$_2$ ligand is accomplished by a σ bond between the 2σ_g orbital of the nitrogen molecule and a hybrid orbital of the metal, and by π back-bonding from a doubly degenerate metal dπ orbitals (d$_{xz}$, d$_{yz}$) to the vacant 1π_g^* orbitals (π_{xz}^*, π_{yz}^*) of the nitrogen molecule, as shown in Fig. 15.1.6.

In the side-on arrangement, the bonding is considered to arise from two interdependent components. In the first part, σ overlap between the filled π orbital of N$_2$ and a suitably directed vacant hybrid metal orbital forms a donor bond. In the second part, the M atom and N$_2$ molecule are involved in two back-bonding interactions, one having π symmetry as shown in Fig. 15.1.7(a), and the other with δ symmetry as shown in Fig. 15.1.7(b). These π and δ-back bonds synergically reinforce the σ bond.

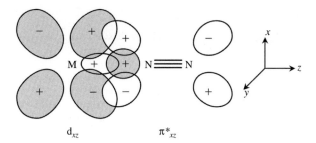

Fig. 15.1.6.
Orbital interactions of N_2 and transition-metal M in end-on coordinated dinitrogen complexes. The $d\pi$-$1\pi^*_g$ overlap shown here also occurs in the yz plane.

Since the overlap of a δ bond should be less effective than that of a π bond, the end-on mode is generally preferred over the side-on form.

The transition-metal dinitrogen complexes have been investigated theoretically, and the results lead to the following generalizations:

(a) Both σ donation and π or δ back donation are related to the formation of the metal-nitrogen bond, the former interaction being more important.
(b) The N–N bond of the side-on complex is appreciably weakened by electron donation from the bonding π and σ orbitals of the N_2 ligand to the vacant orbitals of the metal.
(c) The end-on coordination mode takes precedence over the side-on one. The weak N–N bond in the side-on complex indicates that the N_2 ligand in this type of compound is fairly reactive. The reduction of the coordinated nitrogen molecule may proceed through this activated form.

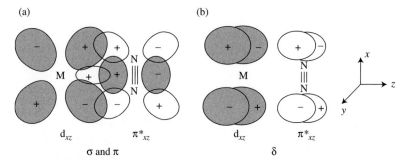

Fig. 15.1.7.
Orbital interactions of N_2 and transition-metal M in side-on coordinate dinitrogen complexes.

15.2 Compounds of nitrogen

15.2.1 Molecular nitrogen oxides

Nitrogen displays nine oxidation states ranging from -3 to $+5$. Since it is less electronegative than oxygen, nitrogen forms oxides and oxidized compounds with an oxidation number between $+1$ and $+5$. Eight oxides of nitrogen, N_2O, NO, N_2O_2, N_2O_3, NO_2, N_2O_4, N_2O_5 and N_4O, are known, and the ninth, NO_3, exists as an unstable intermediate in various reactions involving nitrogen oxides. Their structures and some properties are presented in Table 15.2.1 and Fig. 15.2.1.

All nitrogen oxides have planar structures. Nitrogen displays all its positive oxidation states in these compounds, and in N_2O, N_2O_3 and N_4O the N atoms in

Structural Chemistry of Selected Elements

Table 15.2.1. Structure and properties of the nitrogen oxides

Formula	Name	Oxidation number	Structure (in Fig.15.2.1)	ΔH_f^0 (kJ mol^{-1})	Properties
N_2O	Dinitrogen monoxide (dinitrogen oxide, nitrous oxide, laughing gas)	+1 or (0,+2)	(a) linear $C_{\infty v}$	82.0	mp 182.4 K, bp 184.7 K, colorless gas, fairly unreactive with pleasuring odor and sweet taste
NO	Nitrogen oxide (nitric oxide, nitrogen monoxide)	+2	(b) linear $C_{\infty v}$	90.2	mp 109 K, bp 121.4 K, colorless, paramagnetic gas
$(NO)_2$	Dimer of nitrogen oxide	+2	(c), (d)	—	—
N_2O_3	Dinitrogen trioxide	+3 or (+2,+4)	(e) planar C_s	80.2	mp 172.6 K, dec. 276.7 K, dark blue liquid, pale blue solid, reversibly dissociates to NO and NO_2
NO_2	Nitrogen dioxide	+4	(f) C_{2v}	33.2	orange brown, paramagnetic gas, reactive
N_2O_4	Dinitrogen tetroxide	+4	(g) planar D_{2h}	9.16	mp 262.0 K, bp 294.3 K, colorless liquid, reversibly dissociates to NO_2
N_2O_5	Dinitrogen pentoxide	+5	(h) planar C_{2v}	11.3 (gas) −43.1 (cryst.)	sublimes at 305.4 K, colorless, volatile solid consisting of NO_2^+ and NO_3^-; exists as N_2O_5 in gaseous state
N_4O	Nitrosyl azide	*	(i) planar C_s	−297.3	pale yellow solid (188 K)

* The assignment of an oxidation number is not appropriate as only one of the four nitrogen atoms is bonded to the oxygen atom in the N_4O molecule.

molecule have different oxidation states. In the gaseous state, six stable nitrogen oxides exist, each with a positive heat of formation primarily because the N≡N bond is so strong. The structure and properties of nitrogen oxides are presented in Table 15.2.1.

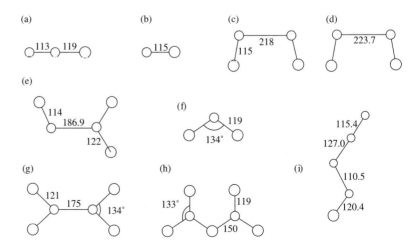

Fig. 15.2.1.
Structure of nitrogen oxides (bond length in pm): (a) N_2O, (b) NO, (c) N_2O_2 (in crystal), (d) N_2O_2 (in gas), (e) N_2O_3, (f) NO_2, (g) N_2O_4, (h) N_2O_5, (i) N_4O.

(1) N_2O

Dinitrogen oxide (nitrous oxide), N_2O, is a linear unsymmetrical molecule with a structure similar to its isoelectronic analog CO_2:

$$\overset{-}{N} \leftarrow \overset{+}{N} = O \longleftrightarrow N \equiv \overset{+}{N} - \overset{-}{O}.$$

Dinitrogen oxide is unstable and undergoes dissociation when heated to about 870 K:

$$N_2O \longrightarrow N_2 + \tfrac{1}{2}O_2.$$

The activation energy for this process is high (~ 520 kJ mol^{-1}), so N_2O is relatively unreactive at room temperature. Dinitrogen oxide has a pleasant odor and sweet taste, and its past use as an anaesthetic with undesirable side effect accounts for its common name as "laughing gas."

(2) NO and $(NO)_2$

The simplest thermally stable odd-electron molecule known is nitrogen monoxide, NO, which is discussed in the next section. High-purity nitrogen monoxide partially dimerizes when it liquefies to give a colorless liquid. The heat of dissociation of the dimer is 15.5 kJ mol^{-1}. The structure of the $(NO)_2$ dimer in the crystalline and vapor states are shown in Figs. 15.2.1(c) and 15.2.1(d), respectively. The $(NO)_2$ dimer adopts the *cis* arrangement with C_{2v} symmetry. In the crystalline state, the N–N distance is 218 pm (223.7 pm in the gas phase), and the O···O distance is 262 pm. The very long N–N distance has not been accounted for satisfactorily in bonding models.

(3) N_2O_3

Dinitrogen trioxide is formed by the reaction of stoichiometric quantities of NO and O_2:

$$2NO + \tfrac{1}{2}O_2 \rightarrow N_2O_3.$$

At temperatures below 172.6 K, N_2O_3 crystallizes as a pale blue solid. On melting it forms an intensely blue liquid which, as the temperature is raised, is increasingly dissociated into NO and an equilibrium mixture of NO_2 and N_2O_4. This dissociation occurs significantly above 243 K and the liquid assumes a greenish hue resulting from the brown color of NO_2 mixed with the blue. The N–N distance of the N_2O_3 molecule, 186.9 pm, is considerably longer than the typical N–N single bond (145 pm) in hydrazine, H_2N–NH_2.

(4) NO_2 and N_2O_4

Nitrogen dioxide and dinitrogen tetroxide are in rapid equilibrium that is highly dependent on temperature. Below the melting point (262.0 K) the oxide consists entirely of colorless, diamagnetic N_2O_4 molecules. As the temperature is raised to the boiling point (294.3 K), the liquid changes to an intense red-brown colored, highly paramagnetic phase containing 0.1% NO_2. At 373 K, the proportion of NO_2 increases to 90%. The configuration, bond distances, and bond angles of NO_2 and N_2O_4 are given in Figs. 15.2.1(f) and 15.2.1(g). In view of the large bond angle of NO_2 and its tendency for dimerization, bonding in the molecule can be described in terms of resonance between the following canonical structures:

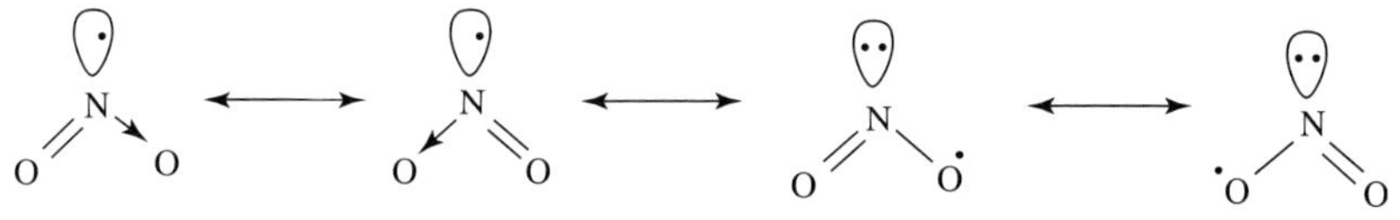

The N–N distance in planar N_2O_4 is 175 pm, with a rotation barrier of about 9.6 kJ mol^{-1}. Although the N–N bond is of the σ type, it is lengthened because the bonding electron pair is delocalized over the entire N_2O_4 molecule with a large repulsion between the doubly occupied MOs on the two N atoms.

(5) N_2O_5 and NO_3

Nitrogen pentoxide, N_2O_5, is a colorless, light- and heat-sensitive crystalline compound that consists of linear NO_2^+ cations (N–O 115.4 pm) and planar NO_3^- anions (N–O 124 pm). In the gas phase N_2O_5 is molecular but its configuration and dimensions have not been reliably measured. Also, N_2O_5 is the anhydride of nitric acid and can be obtained by carefully dehydrating the concentrated acid with P_4O_{10} at low temperatures:

$$4HNO_3 + P_4O_{10} \xrightarrow{-10°C} 2N_2O_5 + 4HPO_3.$$

The existence of the fugitive, paramagnetic trioxide NO_3 is also implicated in the N_2O_5-catalyzed decomposition of ozone, and its concentration is sufficiently high for its absorption spectrum to be recorded. It has not been isolated as a pure compound, but probably has a symmetrical planar structure like that of NO_3^-.

(6) N_4O

Nitrosyl azide, N_4O, is a pale yellow solid formed by the reaction of activated, anhydrous NaN_3 and $NOCl$, followed by low-temperature vacuum sublimation. The Raman spectrum and *ab initio* calculation characterized the structure of N_4O as that shown in Fig. 15.2.1(i). The bonding in the N_4O molecule can be represented by the following resonance structures:

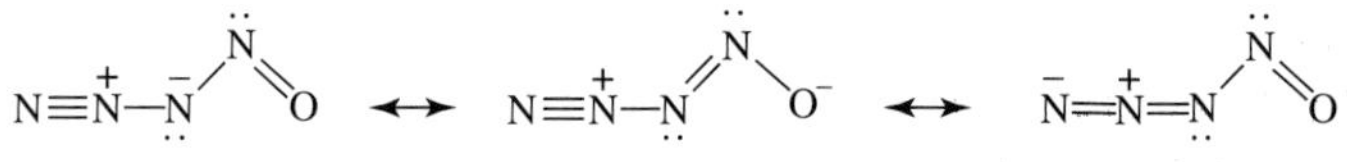

(7) Molecule of the year for 1992: nitric oxide

Nitric oxide, NO, was named molecule of the year for 1992 by the journal *Science*. It is one of the most extensively investigated molecules in inorganic and bioinorganic chemistry.

Nitric oxide is biosynthesized in animal species, and its polarity and small molecular dimensions allow it to readily diffuse through cell walls, acting as a messenger molecule in biological systems. It plays an important role in the normal maintenance of many important physiological functions, including neurotransmission, blood clotting, regulation of blood pressure, muscle relaxation, and annihilation of cancer cells.

The valence shell electron configuration of NO in its ground state is $(1\sigma)^2(1\sigma^*)^2(1\pi)^4(2\sigma)^2(1\pi^*)^1$, which accounts for the following properties:

(a) The NO molecule has a net bond order of 2.5, bond energy 627.5 kJ mol^{-1}, bond length 115 pm, and an infrared stretching frequency of 1840 cm^{-1}.
(b) The molecule is paramagnetic in view of its unpaired π^* electron.
(c) NO has a much lower ionization energy (891 kJ mol^{-1} or 9.23 eV) than N_2 (15.6 eV) or O_2 (12.1 eV).
(d) NO has a dipole moment of 0.554×10^{-30} C m, or 0.166 Debye.
(e) Employing its lone-pair electrons, NO serves as a terminal or bridging ligand in forming numerous coordination compounds.
(f) NO is thermodynamically unstable ($\Delta G^o = 86.57$ kJ mol^{-1}, $\Delta S^o = 217.32$ kJ mol^{-1} K^{-1}) and decomposes to N_2 and O_2 at high temperature. It is capable of undergoing a variety of redox reactions. The electrochemical oxidation of NO around 1.0 V has been used to devise NO-selective amperometric microprobe electrodes to detect its release in biological tissues.

As a result of its unique structure and properties, the NO molecule exhibits a wide variety of reactions with various chemical species. In particular, it readily releases an electron in the antibonding π^* orbital to form the stable nitrosyl cation NO$^+$, increasing the bond order from 2.5 to 3.0, so that the bond distance decreases by 9 pm. The NO species in an aqueous solution of nitrous acid, HONO, is NO$^+$:

$$HONO + H^+ \longrightarrow NO^+ + H_2O.$$

There are many nitrosyl salts, including (NO)HSO$_4$, (NO)ClO$_4$, (NO)BF$_4$, (NO)FeCl$_4$, (NO)AsF$_6$, (NO)PtF$_6$, (NO)PtCl$_6$, and (NO)N$_3$.

Nitrosyl halides are formed when NO reacts with F$_2$, Cl$_2$, and Br$_2$, and they all have a bent structure:

X–N=O
X = F, X–N 152 pm, N=O 114 pm, X–N=O 110°
X = Cl, X–N 197 pm, N=O 114 pm, X–N=O 113°
X = Br, X–N 213 pm, N=O 114 pm, X–N=O 117°

Because of its close resemblance to O_2 and its paramagnetism, NO has been used extensively as an O_2 surrogate to probe the metal environment of various metalloproteins. In particular, NO has been shown to bind to Fe^{2+} in heme-containing oxygen-transporting proteins, in a fashion nearly identical to O_2, with the important consequence of rendering the complex paramagnetic and thus detectable by EPR spectroscopy. Since the unpaired electron assumes appreciable iron d-orbital character, analysis of the EPR characteristics of hemo–protein nitrosyl complexes yields valuable information on both the ligand-binding environment around the heme and on the conformational state of the protein.

Nitric oxide can bind to metals in both terminal and bridging modes to give metal nitrosyl complexes. Depending upon the stereochemistry of the complexes, NO may exhibit within one given complex either NO^+ or NO^- character, as illustrated in Table 15.2.2.

Nitric oxide contains one more electron than CO and generally behaves as a three-electron donor in metal nitrosyl complexes. Formally this may be regarded as the transfer of one electron to the metal atom, thereby reducing its oxidation state by one, followed by coordination of the resulting NO^+ to the metal atom as a two-electron donor. This is in accordance with the general rule that three terminal CO groups in a metal carbonyl compound may be replaced by two NO groups. In this type of bonding the M–N–O bond angle would formally be 180°. However, in many instances this bond angle is somewhat less than 180°, and slight bent M–N–O groups with angles in the range 165° to 180° are frequently found.

In a second type of bonding in nitrosyl coordination compounds, the M–N–O bond angle lies in the range of 120° to 140° and the NO molecule acts as a one-electron donor. Here the situation is analogous to XNO compounds and the M–N bond order is one.

It should be emphasized that the NO^+ and NO^- "character" of NO and the linear or bent angle in a nitrosyl complex do not necessarily imply that NO would be released from the complex in the free form NO or as NO^+ or NO^-.

Table 15.2.2. Two types of MNO coordination geometry.

	M–N–O	M—N⟍O
Bond angle	165° – 180°	120° – 140°
M–N distance	~ 160 pm	> 180 pm
NO frequency	1650 – 1985 cm^{-1}	1525 – 1590 cm^{-1}
Chemical properties	Electrophilic; NO^+ similar to CO: $\overset{-}{M}=\overset{+}{N}=O$	Nucleophilic; NO^- similar to O_2: M—N⟍O
Ligand behavior	three-electron donor	one-electron donor

Group 15 Elements

Table 15.2.3. Oxo-acids of nitrogen

Formula	Name	Properties	Structure
HON=NOH	Hyponitrous acid	Weak acid; salts are known	trans form, Fig. 15.2.3 (a)
H_2NNO_2	Nitramide	Isomeric with hyponitrous acid	Fig. 15.2.2 (a)
HNO	Nitroxyl	Reactive intermediate; salt known	Fig. 15.2.2 (b)
$H_2N_2O_3$	Hyponitric acid	Known only in solution and as salts	
HNO_2	Nitrous acid	Unstable, weak acid	Fig. 15.2.2 (c)
HNO_3	Nitric acid	Stable, strong acid	Fig. 15.2.2 (d)
H_3NO_4	Orthonitric acid	Acid unknown; Na_3NO_4 and K_3NO_4 have been prepared	Tetrahedral NO_4^{3-}, Fig. 15.2.3(f)

15.2.2 Oxo-acids and oxo-ions of nitrogen

The oxo-acids of nitrogen known either as the free acids or in the form of their salts are listed in Tables 15.2.3 and 15.2.4. The structures of these species are shown in Figs. 15.2.2 and 15.2.3.

(1) Hyponitrous acid, $H_2N_2O_2$

Spectroscopic data indicate that hyponitrous acid, HON=NOH, has the planar *trans* configuration. Its structural isomer nitramide, H_2N-NO_2, is a weak acid. The structure of nitramide is shown in Fig. 15.2.2(a); in this molecule, the angle between the NNO_2 plane and the H_2N plane is 52°. Both *trans* and *cis* forms of the hyponitrite ion, $(ONNO)^{2-}$, are known. The *trans* isomer, as illustrated in Fig. 15.2.3(a), is the stable form with considerable π bonding over the molecular skeleton; in the *cis* isomer, the N=N and N–O bond lengths are 120 and 140 pm, respectively.

(2) Nitroxyl, HNO

This is a transient species whose structure is shown in Fig. 15.2.2(b). The existence of the NO^- anion has been established by single-crystal X-ray analysis of $(Et_4N)_5[(NO)(V_{12}O_{32})]$, in which the NO^- group lies inside the cage of

Table 15.2.4. The nitrogen oxo-ions

Oxiation number	Formula	Name	Structure (Fig. 15.2.3)	Properties
+1	$N_2O_2^{2-}$	Hyponitrite	trans, C_{2h}, (a) cis, C_{2v} (not shown)	Reducing agent
+2	$N_2O_3^{2-}$	Hyponitrate	trans, C_s, (b) cis, C_s, (c)	Reducing agent
+3	NO_2^-	Nitrite	Bent, C_{2v}, (d), bond angle 115°	Oxidizing or reducing agent
+5	NO_3^-	Nitrate	Planar, D_{3h}, (e)	Oxidizing agent
+5	NO_4^{3-}	Orthonitrate	Tetrahedral, T_d, (f)	Oxidizing agent
+3	NO^+	Nitrosonium	$C_{\infty v}$, (g)	Oxidizing agent
+5	NO_2^+	Nitronium	$D_{\infty v}$, (h)	Oxidizing agent

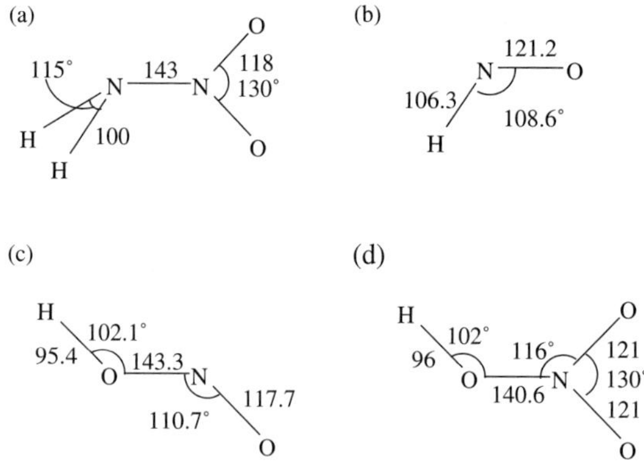

Fig. 15.2.2.
Structures of two amides and two oxo-acids of nitrogen: (a) $H_2N_2O_2$, (b) HNO, (c) HNO_2, and (d) HNO_3.

$(V_{12}O_{32})^{4-}$. The bond length of NO^- is 119.8 pm, which is longer than that of the neutral molecule NO, 115.0 pm.

(3) Hyponitric acid, $H_2N_2O_3$

Hyponitric acid has not been isolated in pure form, but its salts are known. In the hyponitrate anion $N_2O_3^{2-}$, both N atoms each bears a lone pair $\left(O = \ddot{N} - \ddot{N} \begin{smallmatrix} O^- \\ O^- \end{smallmatrix} \right)$. The structure of the anion adopts a non-planar configuration, which can exist in both *cis* and *trans* forms, as shown in Figs. 15.2.3(b) and 15.2.3(c).

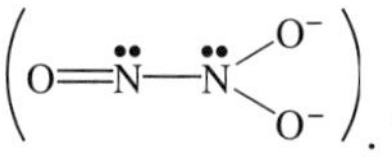

Fig. 15.2.3.
Structure of nitrogen oxo-ions (bond lengths in pm): (a) *trans*-$(ON=NO)^{2-}$, (b) *cis*-$N_2O_3^{2-}$, (c) *trans*-$N_2O_3^{2-}$, (d) NO_2^-, (e) NO_3^-, (f) NO_4^{3-}, (g) NO^+, (h) NO_2^+.

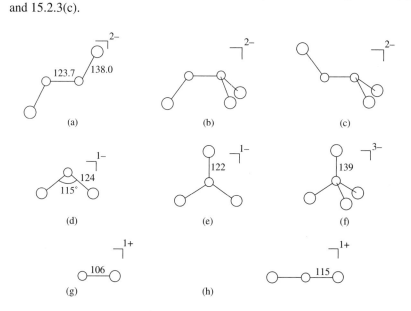

(4) Nitrous acid, HNO_2

Although nitrous acid has never been isolated as a pure compound, its aqueous solution is a widely used reagent. Nitrous acid is a moderately weak acid with pKa = 3.35 at 291 K. In the gaseous state, it adopts the *trans*-planar structure, as shown in Fig. 15.2.2(c).

The nitrite ion, NO_2^- is bent with C_{2v} symmetry, and its structure can be represented by two simple resonance formulas:

Many stable metal nitrites (Li^+, Na^+, K^+, Rb^+, Cs^+, Ag^+, Tl^+, Ba^{2+}, NH_4^+) contain the bent $(O–N–O)^-$ anion with N–O bond length in the range of 113-123 pm, and the angle 116°–132°, as shown in Fig. 15.2.3(d).

(5) Nitric acid, HNO_3

Nitric acid is one of the three major inorganic acids in the chemical industry. Its structural parameters in the gaseous state are shown in Fig. 15.2.2(d). The crystals of nitric acid monohydrate consist of H_3O^+ and NO_3^-, which are connected by strong hydrogen bonds. In the acid salts, HNO_3 molecules are bound to nitrate ions by strong hydrogen bonds. For example, the structures of $[H(NO_3)_2]^-$ in $K[H(NO_3)_2]$ and $[H_2(NO_3)_3]^-$ in $(NH_4)H_2(NO_3)_3$ are as follows:

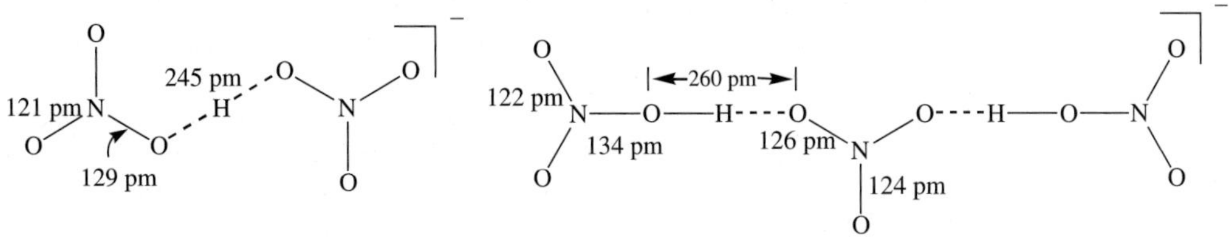

The nitrate ion NO_3^- has D_{3h} symmetry, as shown in Fig. 15.2.3(e), and its bonding can be represented by the following resonance structures:

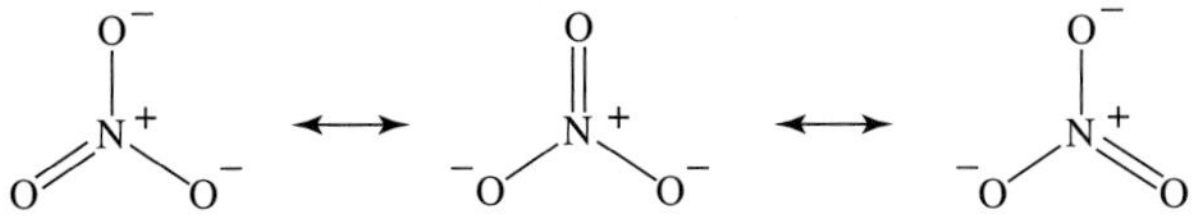

(6) Orthonitric acid, H_3NO_4

Orthonitric acid is still unknown, but its salts Na_3NO_4 and K_3NO_4 have been characterized by X-ray crystallography. The NO_4^{3-} ion has regular T_d symmetry, and its bonding can be represented by resonance structures:

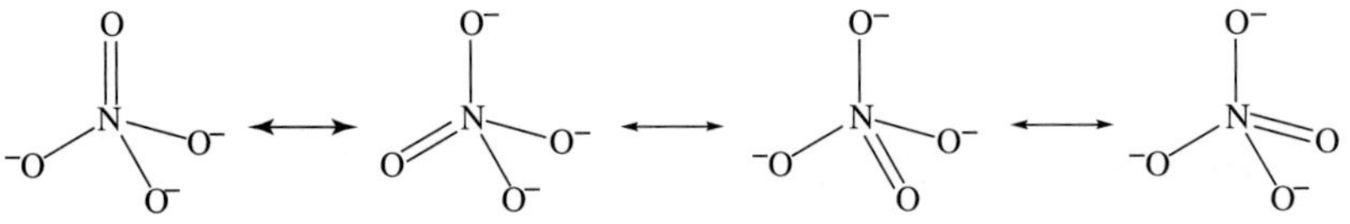

(7) Oxo-cations, NO^+ and NO_2^+

The nitrosonium NO^+ and nitronium NO_2^+ ions, which have been mentioned in previous sections, have $C_{\infty v}$ and $D_{\infty h}$ symmetry, respectively. Their structures shown in Figs. 15.2.3(g) and 15.2.3(h) can be represented by the familiar Lewis structures: $N{\equiv}O^+$ and $O{=}N^+{=}O$.

15.2.3 Nitrogen hydrides

(1) Ammonia, NH_3

The most important nitrogen hydride is ammonia, which is a colorless, alkaline gas with a unique odor. Its melting point is 195 K. Its boiling point 240 K is far higher than that of PH_3 (185.4 K). This is due to the strong hydrogen bonds between molecules in liquid ammonia. Liquid ammonia is an excellent solvent and a valuable medium for chemical reactions, as its high heat of vaporization (23.35 kJ mol^{-1}) makes it relatively easy to handle. As its dielectric constant ($\varepsilon = 22$ at 239 K) and self-ionization are both lower than those of water, liquid ammonia is a poorer ionizing solvent but a better one for organic compounds. The self-ionization of ammonia is represented as

$$2NH_3 \rightleftharpoons NH_4^+ + NH_2^-, K = 10^{-33} \text{ (223 K)},$$

and ammonium compounds behave as Lewis acids while amides are bases.

Ammonia is an important industrial chemical used principally (over 80%) as fertilizers in various forms, and is employed in the production of many other compounds such as urea, nitric acid, and explosives.

(2) Hydrazine, H_2NNH_2

Hydrazine is an oily, colorless liquid in which nitrogen has an oxidation number of -2. The length of the N–N bond is 145 pm, and there is a lone pair on each N atom. Its most stable conformer is the *gauche* form, rather than the *trans* or *cis* form. The rotational barrier through the *trans* or staggered position is 15.5 kJ mol^{-1}, and through the *cis* or eclipsed position is 49.7 kJ mol^{-1}. These numbers reflect the modest repulsion of the nonbonding electrons for a neighboring bond pair and the significantly greater repulsion of the nonbonding pairs for each other in the *cis* arrangement. The melting point of hydrazine is 275 K, and the boiling point is 387 K. The very high exothermicity of its combustion makes it a valuable rocket fuel.

(3) Diazene, HN=NH

Diazene (or diimide) is a yellow crystalline compound that is unstable above 93 K. In the molecule, each N atom uses two sp^2 hybrids for σ bonding with the neighboring N and H atoms, and the lone pair occupies the remaining sp^2 orbital. The molecule adopts the *trans* configuration:

$$\text{H}-\ddot{\text{N}}=\ddot{\text{N}}-\text{H}$$

(4) Hydroxylamine, NH_2OH

Anhydrous NH_2OH is a colorless, thermally unstable hygroscopic compound which is usually handled as an aqueous solution or in the form of its salts. Pure hydroxylamine melts at 305 K and has a very high dielectric constant (77.6–77.9). Aqueous solutions are less basic than either ammonia or hydrazine:

$$NH_2OH(aq) + H_2O \rightleftharpoons NH_3OH^+ + OH^-, K = 6.6 \times 10^{-9} \text{ (298 K)}.$$

Hydroxylamine can exist as two configurational isomers (*cis* and *trans*) and in numerous intermediate *gauche* conformations. In the crystalline form, hydrogen bonding tends to favor packing in the *trans* conformation. The N–O bond length is 147 pm, consistent with its formulation as a single bond. Above room temperature the compound decomposes by internal oxidation–reduction reactions into a mixture of N_2, NH_3, N_2O, and H_2O. Aqueous solutions are much more stable, particularly acid solutions in which the protonated species $[NH_3(OH)]^+$ is generally used as a reducing agent.

15.3 Structure and bonding of elemental phosphorus and P_n groups

Homonuclear aggregates of phosphorus atoms exist in many forms: discrete molecules, covalent networks in crystals, polyphosphide anions, and phosphorus fragments in molecular compounds.

15.3.1 Elemental phosphorus

(1) P_4 and P_2 molecules

Elemental phosphorus is known in several allotropic forms. All forms melt to give the same liquid which consists of tetrahedral P_4 molecules, as shown in Fig. 15.3.1(a). The same molecular entity exists in the gas phase, the P–P bond length being 221 pm. At high temperature (> 800°C) and low pressure P_4 is in equilibrium with P_2 molecules, in which the P≡P bond length is 189.5 pm.

The bonding between phosphorous atoms in the P_4 molecule can be described by a simple bent bond model, which is formed by the overlap of sp^3 hybrids of the P atoms. Maximum overlap of each pair of sp^3 orbitals does not occur along an edge of the tetrahedron. Instead, the P–P bonds are bent, as shown in Fig. 15.3.1(b). In a more elaborate model, the P_4 molecule is further stabilized by the d orbitals of P atoms which also participate in the bonding.

(2) White phosphorus

White phosphorus (or yellow phosphorus when impure) is formed by condensation of phosphorus vapor. Composed of P_4 molecules, it is a soft, waxy, translucent solid, and is soluble in many organic solvents. It oxidizes spontaneously in air, often bursting into flame. It is a strong poison and as little as 50 mg can be fatal to humans.

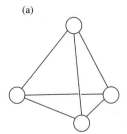
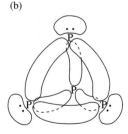

Fig. 15.3.1.
(a) Structure of P_4 molecule; (b) bent bonds in P_4 molecule.

At normal temperature, white phosphorus exists in the cubic α-form, which is stable from –77°C to its melting point (44.1°C). The crystal data of α-white phosphorus are $a = 1.851$ nm, $Z = 56$ (P$_4$), and $D = 1.83$ g cm^{-3}, but its crystal structure is still unknown. At –77°C the cubic α-form transforms to a hexagonal β-form with a density of 1.88 g cm^{-3}.

(3) Black phosphorus

Black phosphorus is thermodynamically the most stable form of the element and exists in three known crystalline modifications: orthorhombic, rhombohedral, and cubic, as well as in an amorphous form. Unlike white phosphorus, the black forms are all highly polymeric, insoluble, and practically non-flammable, and have comparatively low vapor pressures. The black phosphorus varieties represent the densest and chemically the least reactive of all known forms of the element.

Under high pressure, orthorhombic black phosphorus undergoes reversible transitions to produce denser rhombohedral and cubic forms. In the rhombohedral form the simple hexagonal layers are not as folded as in the orthorhombic form, and in the cubic form each atom has an octahedral environment, as shown in Figs. 15.3.2(a)–(c).

(4) Violet phosphorus

Violet phosphorus (Hittorf's phosphorus) is a complex three-dimensional polymer in which each P atom has a pyramidal arrangement of three bonds linking it to neighboring P atoms to form a series of interconnected tubes, as shown in Fig. 15.3.3. These tubes lie parallel to each other, forming double layers, and in the crystal structure one layer has its tubes packed at right angles to those in adjacent layers.

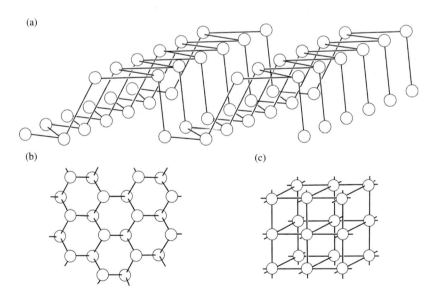

Fig. 15.3.2.
Structure of black phosphorus: (a) orthorhombic, (b) rhombohedral, and (c) cubic black phosphorus.

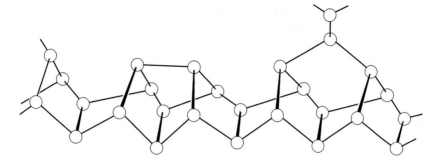

Fig. 15.3.3.
Structure of the tube in violet phosphorus

(5) Red phosphorus

Red phosphorus is a term used to describe a variety of different forms, some of which are crystalline and all are more or less red in color. They show a range of densities from 2.0 to 2.4 g cm^{-3} with melting points in the range of 585–610°C. Red phosphorus is a very insoluble species. It behaves as a high polymer that is inflammable and almost non-toxic.

15.3.2 Polyphosphide anions

Almost all metals form phosphides, and over 200 different binary compounds are now known. In addition, there are many ternary mixed-metal phosphides. These phosphides consist of metal cations and phosphide anions. In addition to some simple anions (P^{3-}, P_2^{4-}, P_3^{5-}), there are many polyphosphide anions that exist in the form of rings, cages, and chains, as shown in Fig. 15.3.4.

In some metal phosphides, the polyphosphide anions constitute infinite chains and sheets, as shown in Fig. 15.3.5.

15.3.3 Structure of P_n groups in transition-metal complexes

Transition-metal complexes with phosphorus bonded to metal atoms have been investigated extensively. They include single P atoms encapsulated in cages of metal atoms, and various P_n groups where $n = 2$ to at least 12. These P_n groups can be chains, rings, or fragments which are structurally related to their valence electron numbers. Figure 15.3.6 shows some skeletal structures of the transition-metal complexes with P_n groups.

Fig. 15.3.6(a) shows the coordination of a pair of P_2 groups to metal centers in the dinuclear complex (Cp″Co)$_2$(P$_2$)$_2$. In Figs. 15.3.6(b), (f), (j), and (n), the P_n groups P_3, P_4, P_5, and P_6 form planar three-, four-, five-, and six-membered rings, respectively. They can be considered as isoelectronic species of planar (CH)$_3$, (CH)$_4$, (CH)$_5$, and (CH)$_6$ molecules. The [(P$_5$)$_2$Ti]$^{2-}$ anion [Fig. 15.3.6(j)] is the first entirely inorganic metallocene, the structure of which has a pair of parallel and planar P_5 rings symmetrically positioned about the central Ti atom. The average P–P bond distance is 215.4 pm, being intermediate

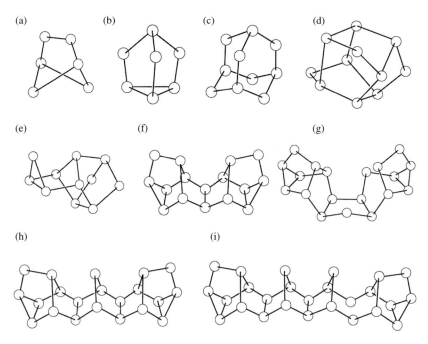

Fig. 15.3.4.
The structures of some polyphosphide anions: (a) P_6^{4-} in [Cp"Th(P$_6$)ThCp"] (Cp"=1,3-Bu$_2^+$C$_5$H$_3$), (b) P_7^{3-} in Li$_3$P$_7$, (c) P_{10}^{6-} in Cu$_4$SnP$_{10}$, (d) P_{11}^{3-} in Na$_3$P$_{11}$, (e) P_{11}^{3-} in [Cp$_3$(CO)$_4$Fe$_3$]P$_{11}$, (f) P_{16}^{2-} in (Ph$_4$P)$_2$P$_{16}$, (g) P_{19}^{3-} in Li$_3$P$_{19}$, (h) P_{21}^{3-} in K$_4$P$_{21}$I, and (i) P_{26}^{4-}.

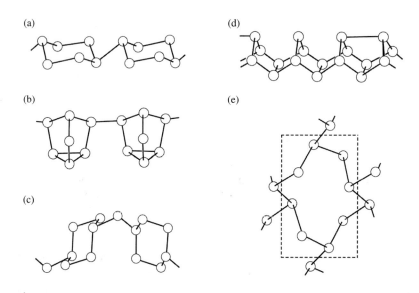

Fig. 15.3.5.
Infinite chain and sheet structures of polyphosphide anions: (a) $[P_6^{4-}]_n$ in BaP$_3$, (b) $[P_7^-]_n$ in RbP$_7$, (c) $[P_7^{5-}]_n$ in Ag$_3$SnP$_7$, (d) $[P_{15}^-]_n$ in KP$_{15}$, and (e) $[P_8^{4-}]_n$ in CuP$_2$.

between those of P–P single (221 pm) and P=P double (202 pm) bonds. The average Ti–P distance is 256 pm. The Ti–P$_5$ (center) distance is 179.7 pm.

The P_n groups in the complexes shown in Figs. 15.3.6(e), (k), and (o) are four-, five- and six-membered rings, respectively. Open chains P$_4$ and P$_5$ as

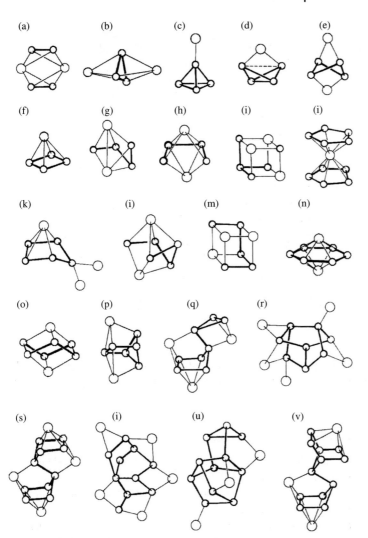

Fig. 15.3.6.
Structure of P_n groups bonded to metal atoms in transition-metal complexes: (a) $(Cp''Co)_2(P_2)_2$, (b) $(Cp_2Th)_2P_3$, (c) $W(CO)_3(PCy_3)_2P_4$, (d) $RhCl(PPh_3)_2P_4$, (e) $[Cp*Co(CO)]_2P_4$, (f) $Cp*Nb(CO)_2P_4$, (g) $[Ni(CO)_2Cp]_2P_4$, (h) $(CpFe)_2P_4$, (i) $[(Cp*Ni)_3P]P_4$, (j) $[Ti(P_5)_2][K(18\text{-}C\text{-}6)]$, (k) $[Cp*Fe]P_5[Cp*Ir(CO)_2]$, (l) $[Cp*Fe]P_5[TaCp'']$, (m) $[Cp*Fe]P_5[TaCp'']_2$, (n) $Cp*Mo_2P_6$, (o) $(Cp*Ti)_2P_6$, (p) $(Cp'_2Th)_2P_6$, (q) $[Cp'''Co(CO)_2]_3P_8$, (r) $[Cp*Ir(CO)]_2P_8[Cr(CO)_5]_3$, (s) $(Cp'Rh)_4P_{10}$, (t) $[CpCr(CO)_2]_5P_{10}$, (u) $Cp^{Pr}Fe(CO)_2]P_{11}[Cp^{Pr}Fe(CO)]_2$, and (v) $[CpCo(CO)_2]_3P_{12}$.

multidentate ligands are exemplified by the structures shown in Figs. 15.3.6(g), (h), (l), and (m). Metal complexes containing bi- and tricyclic P_n rings are shown in Figs. 15.3.6(d) and (p), respectively. The skeletons of metal complexes of polyphosphorus P_n ligands with $n > 6$ are shown in Figs. 15.3.6(q) to (v).

The reaction of Cp*FeP$_5$ with CuCl in CH$_2$Cl$_2$/CH$_3$CN solvent leads to the formation of [Cp*FeP$_5$]$_{12}$(CuCl)$_{10}$(Cu$_2$Cl$_3$)$_5${Cu(CH$_3$CN)$_2$}$_5$. In this large molecule, the *cyclo*-P$_5$ rings of Cp*FeP$_5$ are surrounded by six-membered P$_4$Cu$_2$ rings that result from the coordination of each of the P atomic lone pairs to CuCl metal centers, which are further coordinated by P atoms of other *cyclo*-P$_5$ rings. Thus five- and six-membered rings are fused in a manner reminiscent of the formation of the fullerene-C$_{60}$ molecule. Figure 15.3.7 shows the structure of a hemisphere of this globular molecule. The two hemispheres are joined by [Cu$_2$Cl$_3$]$^-$ as well as by [Cu(CH$_3$CN)$_2$]$^+$ units,

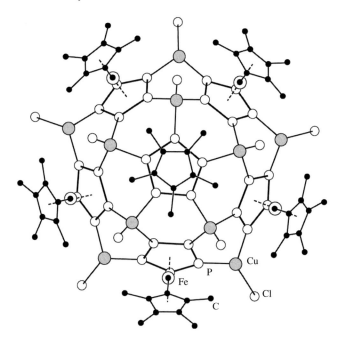

Fig. 15.3.7.
Structure of a hemisphere of
[Cp*FeP$_5$]$_{12}$(CuCl)$_{10}$(Cu$_2$Cl$_3$)$_5$[Cu(CH$_3$CN)$_2$]$_5$.
The central Fe atom is omitted for clarity.

and this inorganic fullerene-like molecule has an inner diameter of 1.25 nm and an outer diameter of 2.13 nm, making it about three times as large as C$_{60}$.

15.3.4 Bond valence in P$_n$ species

The bonding in P$_n$ species can be expressed by their bond valence b, which corresponds to the number of P–P bonds. Let g be the total number of valence electrons in P$_n$. When a covalent bond is formed between two P atoms, each of them gains one electron in its valence shell. In order to satisfy the octet rule for P$_n$, $\frac{1}{2}(8n - g)$ electron pairs must be involved in bonding between the P atoms. The number of these bonding electron pairs is defined as the bond valence b of the P$_n$ species:

$$b = \tfrac{1}{2}(8n - g).$$

Applying this simple formula:

$$P_4 : \quad b = \tfrac{1}{2}(4 \times 8 - 4 \times 5) = 6, \quad \text{6 P–P bonds;}$$
$$P_7^{3-} : \quad b = \tfrac{1}{2}[7 \times 8 - (7 \times 5 + 3)] = 9, \quad \text{9 P–P bonds;}$$
$$P_{11}^{3-} : \quad b = \tfrac{1}{2}[11 \times 8 - (11 \times 5 + 3)] = 15, \quad \text{15 P–P bonds;}$$
$$P_{16}^{2-} : \quad b = \tfrac{1}{2}[16 \times 8 - (16 \times 5 + 2)] = 23, \quad \text{23 P–P bonds.}$$

In these P$_n$ species, the b value is exactly equal to the bond number in the structural formula, as shown in Figs. 15.3.1, 15.3.4, and 15.3.5. But for the planar ring P_6 [Fig. 15.3.6(n)],

$$b = \tfrac{1}{2}[6 \times 8 - (6 \times 5)] = 9,$$

which implies the existence of three P–P bonds and three P=P bonds, and hence aromatic behavior as in the benzene molecule.

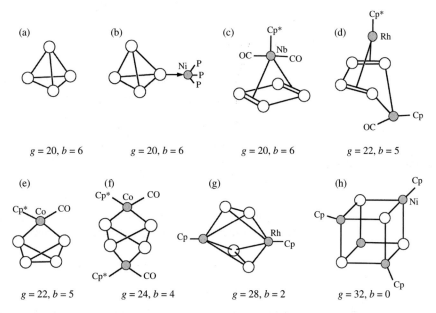

Fig. 15.3.8.
Structure of P_4 group in some transition-metal complexes (large circle represents P atom and arrow → represents dative bond): (a) P_4 molecule, (b) $(P_4)Ni(PPh_2CH_2)_3CCH_3$, (c) $(P_4)Nb(CO)_2Cp^*$, (d) $(P_4)Rh_2(CO)(Cp)(Cp^*)$, (e) $(P_4)Co(CO)Cp^*$, (f) $(P_4)[Co(CO)Cp^*]_2$, (g) $(P_2)_2Rh_2(Cp)_2$, and (h) $(P)_4(NiCp)_4$.

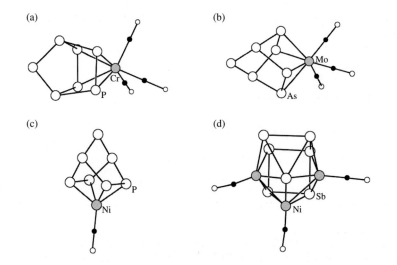

Fig. 15.3.9.
Structure of transition-metal complexes of P_7^{3-}, As_7^{3-}, and Sb_7^{3-}: (a) $[P_7Cr(CO)_3]^{3-}$, (b) $[As_7Mo(CO)_3]^{3-}$, (c) $[P_7Ni(CO)]^{3-}$, and (d) $[Sb_7Ni_3(CO)_3]^{3-}$.

The bond valence of a P_n group changes with its number of valence electrons. The P_4 species is a good example, as shown in Fig. 15.3.8. In these structures, each transition metal atom also conforms to the 18-electron rule.

The bonding structure of transition-metal complexes with P_n group can be classified into four types:

(1) Covalent P–M σ bond: each atom donates one electron to bonding and the g value of P_n increases by one, as shown in Fig. 15.3.8(e).
(2) P $\to$ M dative bond: the g value of P_n species does not change, as shown in Fig. 15.3.8(b).
(3) $\begin{matrix}P\\||\\P\end{matrix} \to M$ π dative bond: the g value of P_n species also does not change, as shown in Fig. 15.3.8 (c).
(4) $(\eta^n-P_n) \to$ M dative bond: the P_n ring (n = 3–6) donates its delocalized π electrons to the M atom, and the g value of P_n group does not change.

Some molecular transition-metal complexes of P_7^{3-}, As_7^{3-}, and Sb_7^{3-} have been isolated as salts of cryptated alkali metal ions. The structures of the complex anions are shown in Fig. 15.3.9. The bond valence b and bond number of these complex anions are as follows:

(a) $[P_7Cr(CO)_3]^{3-}$ in $[Rb\cdot crypt]_3[P_7Cr(CO)_3]$: b = 12, eight P–P and four P–Cr bonds.
(b) $[As_7Mo(CO)_3]^{3-}$ in $[Rb\cdot crypt]_3[As_7Mo(CO)_3]$: b = 12, eight As–As and four As–Mo bonds.
(c) $[P_7Ni(CO)]^{3-}$ in $[Rb\cdot crypt]_3[P_7Ni(CO)]$: b = 12, eight P–P and four P–Ni bonds.
(d) $[Sb_7Ni_3(CO)_3]^{3-}$ in $[K\cdot crypt]_3[Sb_7Ni_3(CO)_3]$: b = 18, four Sb–Sb, five SbSbNi 3c-2e, and two SbNiNi 3c-2e bonds.

15.4 Bonding type and coordination geometry of phosphorus

15.4.1 Potential bonding types of phosphorus

To illustrate the potential diversity of structure and bonding of phosphorus, the classic Lewis representations for each possible coordination number one to six are shown in Fig. 15.4.1. Many of these bonding types have been observed in stable compounds, which are discussed in the following sections.

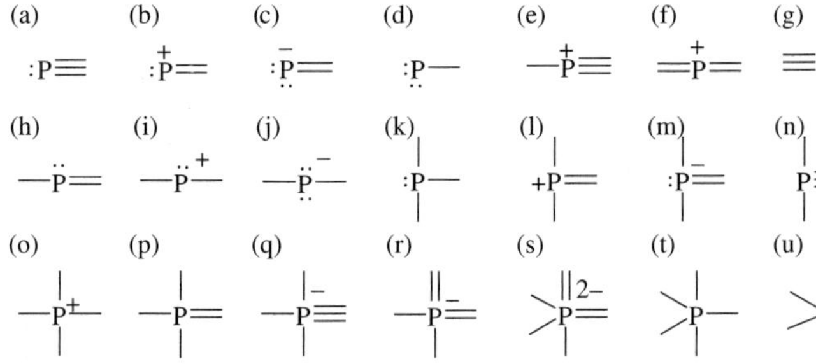

Fig. 15.4.1. Potential bond types of phosphorus.

Phosphines are classical Lewis bases or ligands in transition-metal complexes, but the cationic species shown in Fig. 15.4.1(i) are likely to exhibit Lewis acidity by virtue of the positive charge. Despite their electron-rich nature, an extensive coordination chemistry has been developed for Lewis acidic phosphorus. For example, the compound shown below has a coordinatively unsaturated Ga(I) ligand bonded to a phosphenium cation; it can be considered as a counter-example of the traditional coordinate bond since the metal center (Ga) behaves as a Lewis donor (ligand) and the non-metal center (P) behaves as a Lewis acceptor.

$$\begin{bmatrix} OTf = OTeF_5^- \\ Dipp = (2,6\text{-}^iC_3H_7)C_6H_3 \end{bmatrix}$$

15.4.2 Coordination geometries of phosphorus

Phosphorus forms various compounds with all elements except Sb, Bi, and the inert gases for the binary compounds. The stereochemistry and bonding of phosphorus are very varied. Some typical coordination geometries are summarized in Table 15.4.1 and illustrated in Fig. 15.4.2. Many of these compounds will be discussed below.

Table 15.4.1. Coordination geometries of phosphorus atoms in compounds

CN	Coordination geometry	Example	Structure (in Fig. 15.4.1)
1	Linear	P≡N, F–C≡P	(a)
2	Bent	$[P(CN)_2]^-$	(b)
3	Pyramidal	PX_3 (X = H, F, Cl, Br, I)	(c)
	Planar	$PhP\{Mn(C_5H_5)(CO)_2\}_2$	(d)
4	Tetrahedral	P_4O_{10}	(e)
	Square	$[P\{Zr(H)Cp_2\}_4]^+$	(f)
5	Square pyramidal	$Os_5(CO)_{15}(\mu_4\text{-POMe})$	(g)
	Trigonal bipyramidal	PF_5	(h)
6	Octahedral	PCl_6^-	(i)
	Trigonal prismatic	$(\mu_6\text{-}P)[Os(CO)_3]_6^-$	(j)
7	Monocapped trigonal prismatic	Ta_2P	(k)
8	Cubic	Ir_2P	(l)
	Bicapped trigonal prismatic	Hf_2P	(m)
9	Monocapped square antiprismatic	$[Rh_9(CO)_{21}P]^{2-}$	(n)
	Tricapped trigonal prismatic	Cr_3P	(o)
10	Bicapped square antiprismatic	$[Rh_{10}(CO)_{22}P]^{3-}$	(p)

(1) Coordination number 1

Coordination number 1 is represented by the compounds P≡N, P≡C–H, P≡C–X (X = F, Cl), and P≡C–Ar (Ar = $^tBu_3C_6H_2$). In these compounds, the P atom

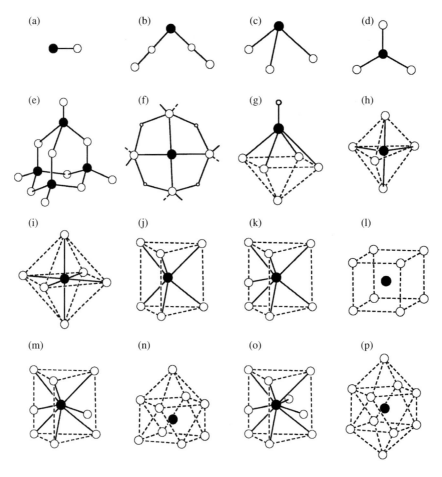

Fig. 15.4.2.
Coordination geometries of the phosphorus atom in some compounds (black circles represent P, open circles represent other atoms): (a) P≡N, (b) [P(CN)$_2$]$^-$, (c) PX$_3$, (d) PhP{Mn(C$_5$H$_5$)(CO)$_2$}$_2$, (e) P$_4$O$_{10}$, (f) [P{Zr(H)Cp$_2$}$_4$]$^+$, (g) Os$_5$(CO)$_{15}$(μ_4–POMe), (h) PF$_5$, (i) [PCl$_6$]$^-$, (j) [Os(CO)$_3$]$_6$P, (k) Ta$_2$P, (l) Ir$_2$P, (m) Hf$_2$P, (n) [Rh$_9$(CO)$_{21}$P]$^{2-}$, (o) Cr$_3$P, and (p) [Rh$_{10}$(CO)$_{22}$P]$^{3-}$.

forms a triple bond with the N or C atom. The bond length are P≡N 149 pm, P≡C 154 pm.

(2) Coordination number 2

Compounds of this coordination number have three bond types:

P(CN)$_2^-$ R$_2$N–P=NR

In P(CN)$_2^-$, the bond length P–C is 173 pm, C≡N is 116 pm, and bond angle C–P–C is 95°.

(3) Coordination number 3

In compounds of this coordination number, the pyramidal configuration is the most common type, and the planar one is much less favored. In pyramidal PX_3, the three X groups are either the same or different, X = F, Cl, Br, I, H, OR, OPh, Ph, ^tBu, etc. The observed data of some PX_3 compounds are listed below:

PX_3	PH_3	PF_3	PCl_3	PBr_3	PI_3
P–X bond length (pm)	144	157	204	222	252
X–P–X bond angle	94°	96°	100°	101°	102°

Since PX_3 has a lone pair at the P atom and the X groups can be varied, the molecules PX_3 are important ligands. The strength of the coordinate bond $X_3P{\rightarrow}M$ is influenced by different X groups in three aspects:

(a) σ P$\rightarrow$M bonding

The stability of the σ P$\rightarrow$M interaction, which uses the lone pair of electrons on P atom and a vacant orbital on M atom, is influenced by different X groups in the sequence:

$$P^tBu_3 > P(OR)_3 > PR_3 \approx PPh_3 > PH_3 > PF_3 > P(OPh)_3.$$

(b) π back donation

The possibility of synergic π back donation from a nonbonding d_π pair of electrons on M into a vacant $3d_\pi$ orbital on P varies in the sequence:

$$PF_3 > P(OPh)_3 > PH_3 > P(OR)_3 > PPh_3 \approx PR_3 > P^tBu_3.$$

(c) Steric interference

The stability of the P–M bonds are influenced by steric interference of the X groups, in accordance with the sequence:

$$P^tBu_3 > PPh_3 > P(OPh)_3 > PMe_3 > P(OR)_3 > PF_3 > PH_3.$$

In the $PhP\{Mn(C_5H_5)(CO)_2\}_2$ molecule, the P atom is bonded to two Mn atoms and one phenyl C atom by single bonds in a planar configuration, as shown in Fig. 15.4.2(d). The bond angle Mn–P–Mn is 138°.

(4) Coordination number 4

This very common coordination number usually leads to a tetrahedral configuration for phosphoric acid, phosphates, and many phosphorus(V) compounds. In these compounds, the P(V) atom forms one double bond and three single bonds with other atoms. Figure 15.4.2(e) shows the structure of P_4O_{10}, whose symmetry is T_d. The length of the terminal P=O bond is 143 pm, and the bridging P–O bond is 160 pm. The bond angle O–P–O is 102°, and P–O–P is 123°.

The compound P_4O_{10}, known as "phosphorus pentoxide," is the most common and most important oxide of phosphorus. This compound exists in three modifications. When phosphorus burns in air and condenses rapidly from the vapor, the common hexagonal (H) form of P_4O_{10} is obtained. The H form is metastable and can be transformed into a metastable orthorhombic (O) form by heating for 2 h at 400°C, and into a stable orthorhombic (O′) form by heating for 24 h at 450°C. All three modifications undergo hydrolysis in cold water to give phosphoric acid, H_3PO_4.

Square-planar geometry of the P atom occurs in the compound $[P\{Zr(H)Cp_2\}_4][BPh_4]$. Figure 15.4.2(f) shows the structure of $P\{Zr(H)Cp_2\}_4^+$, in which the Zr–P–Zr angles are very close to 90°, and the bridging hydrogens and the Zr atoms form a nearly planar eight-membered ring that encircles the central P atom.

(5) Coordination number 5

Figure 15.4.2(g) shows the structure of $Os_5(CO)_{15}(\mu_4\text{-POMe})$, in which five Os atoms constitute a square-pyramidal cluster. Each Os atom is coordinated by three CO ligands, and only the four basal Os atoms are bonded to the apical P atom.

Various structures have been found for phosphorus pentahalides:

(a) The molecular structure of PF_5 is trigonal bipyramidal, as shown in Fig. 15.4.2(h). The axial P–F bond length is 158 pm, which is longer than the equatorial P–F bond length of 153 pm.
(b) In the gaseous phase, PCl_5 is trigonal bipyramidal with axial P–Cl_{ax} bond length 214 pm and equatorial P–Cl_{eq} bond length 202 pm. In the crystalline phase, PCl_5 consists of a packing of tetrahedral $[PCl_4]^+$ and octahedral $[PCl_6]^-$ ions; in the latter anion shown in Fig. 15.4.2(i), the P–Cl bond length is 208 pm.
(c) In the crystalline phase, PBr_5 consist of a packing of $[PBr_4]^+$ and Br^-. Owing to steric overcrowding, $[PBr_6]^-$ cannot be formed by grouping six bulky Br atoms surrounding a relatively small P atom.
(d) There is as yet no evidence for the existence of PI_5.

(6) Coordination number ≥ 6

In each of the structures shown in Figs. 15.4.2(j)–(p), the transition-metal atoms form a polyhedron around the P atom, which donates five valence electrons to stabilize the metal cluster.

15.5 Structure and bonding in phosphorus–nitrogen and phosphorus–carbon compounds

15.5.1 Types of P–N bonds

When phosphorus and nitrogen atoms are directly bonded, they form one of the most intriguing and chemically diverse linkages in inorganic chemistry. A convenient classification of phosphorus–nitrogen compounds can be made on the basis of formal bonding. Azaphosphorus compounds containing the P–N group are known as phosphazanes, those containing the P=N group are

Table 15.5.1. The common types of phosphorus–nitrogen bonds*

	Phosphazanes	Phosphazenes	Phosphazynes
P^{III}	>P—N< (sp²) σ², λ²	>P=N< (sp²) σ², λ³	:P≡N σ¹, λ³
	>P—N< (sp³) σ³, λ³		
P^V	>P—N< (sp³) σ⁴, λ⁴		
	>P—N< (sp³) σ⁴, λ⁵	>P=N< (sp³) σ⁴, λ⁵	>P≡N (sp²) σ³, λ⁵
	>P—N< (sp³d) σ⁵, λ⁵	>P—N< (sp³) σ⁴, λ⁵	
	>P—N< (sp³d) σ⁵, λ⁵	>P=N< (sp³) σ⁴, λ⁵	
	>P—N< (sp³d²) σ⁶, λ⁶	>P=N< (sp²) σ³, λ⁵	

* The hybridization of the P atom is enclosed in parentheses. The symbols σ and λ represent the coordination number and bonding number of P atoms.

phosphazenes, and those with the P≡N group are phosphazynes. The common types of phosphorus–nitrogen bonds of these three classes of compounds are displayed in Table 15.5.1.

Phosphazynes are rare. The bond length of the diatomic molecule P≡N is 149 pm. The first stable compound containing the P≡N group is [P≡N–R]$^+$[AlCl$_4$]$^-$, [R = C$_6$H$_2$(2,4,6-^{t}Bu$_3$)], in which the length of the P≡N bond is 147.5 pm.

15.5.2 Phosphazanes

Some examples of phosphazanes are described below according to the types of P–N bonds.

(1) >P—N<

In this bonding type, the P atom uses its sp² hybrid orbitals, one of which accommodates a lone pair. A bent molecular structure of this type is found for $(^iPr_2N)_2P^+$:

$$R_2\ddot{N}-\overset{+}{\overset{..}{P}}-\ddot{N}R_2 \longleftrightarrow R_2\overset{+}{N}=\overset{..}{\overset{..}{P}}-\ddot{N}R_2 \longleftrightarrow R_2\ddot{N}-\overset{..}{\overset{..}{P}}=\overset{+}{N}R_2$$

The bond length of P–N is 161.2 pm, shorter than the standard P–N single-bond distance of 177 pm observed in H_3N–PO_3, which does not have a lone pair on the N atom. The bond angle N–P–N is 115°, smaller than idealized 120°, due to the repulsion of the lone pairs.

(2) $\overset{..}{P}-N\overset{\diagup}{\diagdown}$

The configuration of most R_2P–NR_2 compounds is represented by F_2P–NMe_2, which features a short P–N distance (162.8 pm) and trigonal planar arrangement (sp^2 hybridization) at the N atom. The plane defined by the C_2N unit bisects the F–P–F angle. This configuration minimizes steric repulsion between the substituents on P and N and also orients the lone pairs on the P and N atoms at a dihedral angle of about 90°, as shown in Fig. 15.5.1(a).

The geometry of the molecule and the short P–N distance indicate that there is π bonding between the P and N atoms. In the molecule, the z axis is perpendicular to the PNC_2 plane and the x axis lies parallel to the P–N bond. The $p_z(N)$ orbital with a lone pair and the empty $d_{xz}(P)$ orbital overlap to form a π bond, as shown in Fig. 15.5.1(b).

The structure of F_2PNH_2 resembles that of F_2PNMe_2 with a P–NH_2 bond distance of 166 pm.

(3) $\overset{\diagup}{\diagdown}\overset{+}{P}-N\overset{\diagup}{\diagdown}$

Many compounds contain formally a single-bonded P–N group, such as the cation $[PCl_2(NMe_2)_2]^+$ in crystalline $[PCl_2(NMe_2)_2](SbCl_6)$, the cation $[P(NH_2)_4]^+$, and the anion $[P(NR)_4]^{3-}$. The bond length of P–N in $[P(NH_2)_4]^+$ is 160 pm, and that in $[P(NR)_4]^{3-}$ is 164.5 pm.

(4) $\overset{\diagup}{\diagdown}\overset{\Vert}{P}-N\overset{\diagup}{\diagdown}$

An important group of compounds such as $(Me_2N)_2POCl$, $(Me_2N)POCl_2$, and $(R_2N)_3PO$ possess this bonding type. For example, $(Me_2N)_3PO$ is a colorless

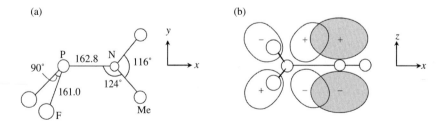

Fig. 15.5.1.
(a) Molecular structure of F_2P–NMe_2, (b) overlap between the d_{xz} orbital of the P atom and the p_z orbital of the N atom.

mobile liquid which is miscible with water in all proportions. It forms an adduct with $HCCl_3$, dissolves ionic compounds, and can dissolve alkali metals to give blue paramagnetic solutions which are strong reducing agents.

(5) $>\!\!\overset{|}{\underset{|}{P}}\!\!-\!\!N\!\!<$

The compound F_4P-NEt_2 has this bonding type, in which the P atom uses sp^3d hybrid orbitals.

(6) $>\!\!\overset{|}{\underset{}{P}}\!\!^+\!\!\leftarrow\!\!N\!\!<$

The phosphatranes that have this bonding type are analogs of silatranes. The cage molecule shown on the right is a trigonal bipyramidal five-coordinated phosphazane with a rather long P–N bond.

(7) $>\!\!\overset{|}{\underset{|}{P}}\!\!\leftarrow\!\!N\!\!<$

The adduct $F_5P\leftarrow NH_3$ is an octahedral six-coordinated phosphanane, in which the P–N bond length is 184.2 pm. In another example, $Cl_5P \leftarrow N\!\!\!\diagup\!\!\!\diagdown\!\!\!N$, the P–N bond length is 202.1 pm, which is longer than the former. This difference is due to the high electronegativity of F, which renders the F_5P group a better acceptor as compared to Cl_5P.

15.5.3 Phosphazenes

Phosphazenes, formerly known as phosphonitrilic compounds, are characterized by the presence of the group P=N. Known compounds, particularly those containing the $\searrow\!\!P\!\!=\!\!N\!\!-\!\!\nearrow$ group, are very numerous and they have important potential applications.

(1) Bonding types of phosphazenes

(a) Containing –P=N– bonding type

In this bonding type, the P atom uses sp^2 hybrid orbitals, one of which contains the lone-pair electrons. This type of phosphazene compounds exists in a bent configuration. For example, the structure of $(SiMe_3)_2NPN(SiMe_3)$ is shown below:

$$(Me_3Si)_2N\underset{108°}{\overset{167.4\text{ pm}}{-\!\!\!-\!\!\!-}}P\overset{154.5\text{ pm}}{\diagdown}N-SiMe_3$$

(b) Containing $\diagdown$P=N— bonding type

The compound $(Me_3Si)_2N-P(NSiMe_3)_2$ belongs to this type, as shown below:

$$Me_3SiN\diagdown \atop Me_3SiN\diagup \overset{150.3\ pm}{P}\overset{164.6\ pm}{\underset{113°}{\longrightarrow}}N(SiMe_3)_2$$

(c) Containing $\diagdown\!\!\diagup$P=N— bonding type

The simplest compound of this type is iminophosphorane, $H_3P=NH$, whose derivatives are very numerous, including $R_3P=NR'$, $Cl_3P=NR$, $(RO)_3P=NR'$, and $Ph_3P=NR$. In these compounds the P atom uses its sp^3 hybrid orbitals to form four σ bonds, and is also strengthened by $d\pi-p\pi$ overlap with N and other atoms.

(2) Structure and bonding of cyclic phosphazenes

The main products of refluxing a mixture of PCl_5 and NH_4Cl using tetrachloroethane as solvent are the cyclic trimer $(PNCl_2)_3$ and tetramer $(PNCl_2)_4$, which are stable white crystalline compounds that can be isolated and purified by recrystallization from nonpolar solvents.

The trimer $(PNCl_2)_3$ has a planar six-membered ring structure of D_{3h} symmetry with the six Cl atoms disposed symmetrically above and below the plane of the ring. The P–N bonds within the ring are of the same length, 158 pm, and the interior angles are all close to 120°. The Cl–P–Cl planes are perpendicular to the plane of the central ring, and the Cl–P–Cl angle is 120°. Figure 15.5.2 shows the structures of $(PNCl_2)_3$ and chair-like $(PNCl_2)_4$.

The shortness and equality of the P–N bond lengths in $(PNCl_2)_3$ arise from electron delocalization involving the d orbitals of the P atoms and the p orbitals of the N atoms. In such a system, the σ bonds formed from phosphorus sp^3 orbitals overlapping with the nitrogen sp^2 orbitals are enhanced by π bonding between the nitrogen p_z orbitals and the phosphorus d orbitals.

In $(PNCl_2)_3$, the π bonding occurs over the entire ring. Firstly, $d\pi-p\pi$ overlap occurs between the nitrogen p_z orbital (z axis perpendicular to the ring plane) and the d_{xz} orbital of phosphorus. Figure 15.5.3(a) shows the orientation of one d_{xz} orbital of the P atom and two p_z orbitals of neighbor N atoms. Figure 15.5.3(b) shows a projection of the overlap of d_{xz} with the p_z orbitals. Secondly, "in-plane" electron delocalization probably arises from overlap of the

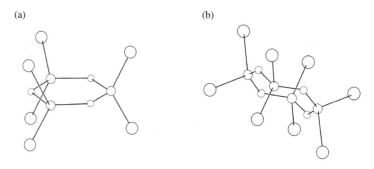

Fig. 15.5.2.
Structure of (a) $(PNCl_2)_3$ and (b) $(PNCl_2)_4$ (circles of decreasing sizes represent Cl, P, and N atoms, respectively).

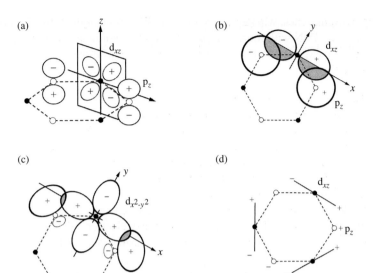

Fig. 15.5.3.
dπ–pπ Bonding in the ring of $(PNCl_2)_3$: (a) orientation of d_{xz} of P atom and p_z of N atom; (b) overlap of d_{xz} and p_z; (c) overlap of $d_{x^2-y^2}$ of P atom and lone pair (sp^2) of N atom; (d) possible mismatch of the d_{xz} orbital of P atom and p_z orbital of the N atom.

lone-pair orbitals on nitrogen with the $d_{x^2-y^2}$ orbitals on phosphorus, forming additional π' bonds in the plane of the ring. Fig. 15.5.3(c) shows the $d_{x^2-y^2}$ of P overlapping with the lone-pair (sp^2) orbitals of two adjacent N atoms. Thirdly, the d_{z^2} orbitals of P atoms overlap with the p orbitals of the exocyclic Cl atoms. Fig. 15.5.3(d) shows that electron delocalization around the ring is hindered by a mismatch of orbital symmetry. A simplified description of the bonding in this molecule based on the group-theoretic method is given in Chapter 7.

The geometric disposition of the d orbitals in dπ–pπ systems allows puckering and accounts for the variety of ring conformations found among larger cyclic phosphazene compounds, such as $(PNCl_2)_4$.

Of the five binary phosphorus–nitrogen molecules described in the literature, namely P_4N_4, $P(N_3)_3$, $P(N_3)_5$, the anion in the ionic compound $(N_5)^+[P(N_3)_6]^-$, and the phosphazene derivative $[PN(N_3)_2]_3$, only the last has been fully structurally characterized. These compounds are difficult to isolate and handle owing to their highly endothermic character and extremely low energy barriers, which often lead to uncontrollable explosive decomposition. Single-crystal X-ray analysis of $[PN(N_3)_2]_3$ conducted in 2006 showed that it is a structural analog of $[PNCl_2]_3$, with three azide groups oriented nearly parallel to the phosphazene ring and the other three nearly perpendicular to the ring.

A hybrid borazine-phosphazine ring system has been found in [ClBNMePCl$_2$NPCl$_2$NMe](GaCl$_4$); an X-ray study revealed that the 6π aromatic cation (structural formula shown on the right) is virtually planar with B–N bond lengths of 143.6(9) and 142.2(10) pm, which are close to those found in borazines (143 pm).

All phosphazenes, whether cyclic or chain-like, contain the formally unsaturated group P=N with four-coordinate P and two-coordinate N atoms. Based on available experimental data, the following generalizations may be made in regard to their structure and properties:

(a) The rings and chains are very stable.
(b) The skeletal interatomic distances are equal within the ring or along the chain, unless there is different substitution at various P atoms.
(c) The P–N distances are shorter than expected for a covalent single bond (∼177 pm) and are usually in the range 158±2 pm.
(d) The N–P–N angles are usually in the range 120±2°; but the P–N–P angles in various compounds span the range 120°–148°.
(e) Skeletal N atoms are weakly basic and can coordinate to metals or be protonated, especially when the P atoms carry electron-releasing groups.
(f) Unlike many aromatic systems, the phosphazene skeleton is difficult to reduce electrochemically.
(g) Spectral effects associated with organic π systems are not exhibited.

15.5.4 Bonding types in phosphorus–carbon compounds

Phosphorus and carbon are diagonal relatives in the Periodic Table. The diagonal analogy stresses the electronegativity of the element (C 2.5 vs P 2.2) which governs its ability to release or accept electrons. This property controls the reactivity of any species containing the element. This section covers the types of phosphorus–carbon bonds and the structures of representative species.

(1) Phosphinidenes and phosphinidene complexes, >C—P̈

Phosphinidenes (recommended IUPAC name: phosphanylidenes) are unstable species and analogous to the carbenes. The parent compound H–P is a six-electron species that is still unknown, but its organic derivatives can give rise to seven different types of complexes, as listed in Table 15.5.2.

(2) Phosphaalkenes, $R^1R^2C=PR^3$

Phosphaalkenes are tervalent phosphorus derivatives with a double bond between carbon and phosphorus. The observed P=C bond lengths range from 161 to 171 pm (average 167 pm), appreciably shorter than the single P–C bond length of 185 pm.

Phosphaalkenes may coordinate to transition-metal fragments in various ways:

(a) η^1 mode via the lone pair on P atom. An example is

P=C 167.9 pm

Group 15 Elements

Table 15.5.2. Types of phosphinidene complexes

Types	Structure and properties
Two-electron complexes	R–P⃛→M η^1-bent, electrophilic
	R–P=M η^1-bent, nucleophilic*
	R–P(⃛)(M)(M) bridging μ_2-pyramidal
Four-electron complexes	R–P≡M η^1-linear†
	R–P(=M)(=M) μ_2-planar
	R–P(M)(M)(M) μ_3-tetrahedral
	R–P(M)(M)(M)(M) μ_4-bipyramidal

* In the complex Mes–P=Mo(Cp)$_2$, P=Mo 237.0 pm, Mo –P –C 115.8°.
† In the complex Mes–P≡WCl$_2$(CO)(PMePh$_2$), P≡W 216.9 pm, C–P–W 168.2°.

(b) η^2 mode via the π bond electrons. An example is

Tms–C(Cp*)=P(Cp*)→Ni(PEt$_3$)$_2$

In η^2 complexes the P–C bond length is longer than that in η^1 complexes or the free ligands.

(c) η^1, η^2 mode via both the lone pair and π bond electrons. An example is

H$_2$C=P(Mes*)→Fe(CO)$_4$, Fe(CO)$_4$ P=C 173.7 pm

(3) Phosphaalkynes, RC≡P

Phosphaalkynes are compounds of tervalent phosphorus which contain a P≡C triple bond. Figure 15.5.4 shows the structure of ^{t}Bu–C≡P, whose P≡C bond

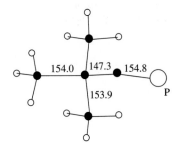

Fig. 15.5.4. Structure of ^{t}Bu–C≡P (bond lengths in pm).

Structural Chemistry of Selected Elements

Table 15.5.3. Coordination modes of phosphaalkynes

Coordination mode		Example and structure
η^1	R—C≡P→M	[^{t}Bu–C≡P–Fe(H)(dppe)$_2$][BPh$_4$] C≡P 151.2 pm
η^2	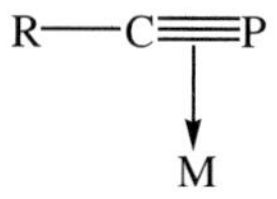 R—C≡P ↓ M	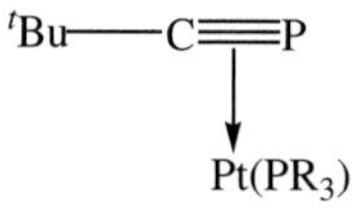 ^{t}Bu—C≡P ↓ Pt(PR$_3$)
η^1, η^2	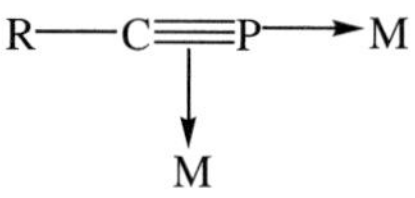 R—C≡P→M ↓ M	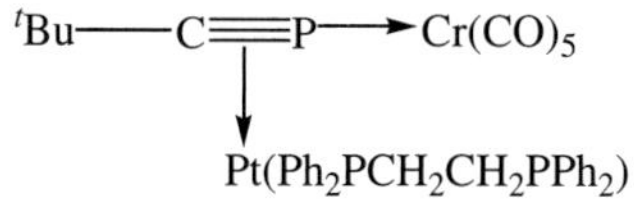 ^{t}Bu—C≡P→Cr(CO)$_5$ ↓ Pt(Ph$_2$PCH$_2$CH$_2$PPh$_2$)
4e	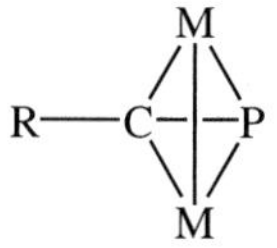 R—C⫫P with 2 M	^{t}BuCP[Fe$_2$(CO)$_5$(PPh$_2$CH$_2$PPh$_2$)] [Fig. 15.5.5(a)]
6e	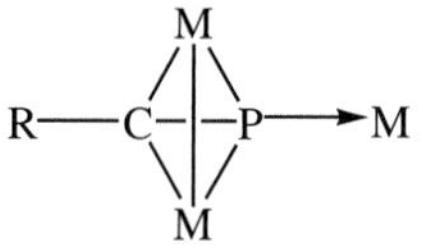 R—C⫫P→M with 2 M	^{t}BuCP[W(CO)$_5$][Co$_2$(CO)$_6$] [Fig. 15.5.5(b)]

is very short (154.8 pm), and the electronic ionization energies [$I_1(\pi\text{MO})$ = 9.61 eV, I_2(P lone pair) = 11.44 eV] are low, suggesting that it may have a chemistry closely related to that of the alkynes.

Phosphaalkynes have a rich coordination chemistry, in which both the triple bond and the lone pair of the P atom can participate. Table 15.5.3 lists the coordination modes of phosphaalkynes, and two structures are shown in Fig. 15.5.5.

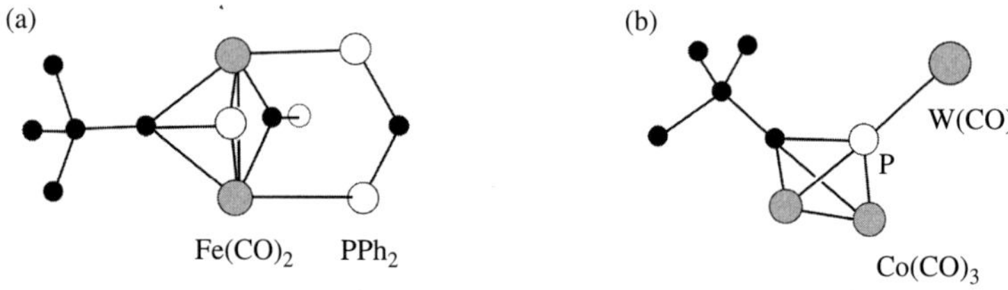

Fig. 15.5.5.
Structure of
(a) ^{t}Bu-CP[Fe$_2$(CO)$_5$(PPh$_2$CH$_2$PPh$_2$)]
and (b) ^{t}Bu-CP[W(CO)$_5$][Co$_2$(CO)$_6$].

Some oligomers of the phosphaalkynes ^{t}BuCP have been characterized. The phosphaalkyne cyclotetramer exists in several isomeric forms, whose structures are shown in Fig. 15.5.6(a) to Fig. 15.5.6(e). In cubane-like P$_4$C$_4$ ^{t}Bu$_4$, the P–C bond lengths are all identical (188 pm), and are typical for single bonds. The angle at P is reduced from the idealized 90° to 85.6°, while that at C is widened to 94.4°.

The phosphaalkyne pentamer P$_5$C$_5$ ^{t}Bu$_5$ has the cage structure shown in Fig. 15.5.6(f), which can be derived from the tetramer by replacing one corner C atom of the "cube" by a C$_2$P triangular fragment.

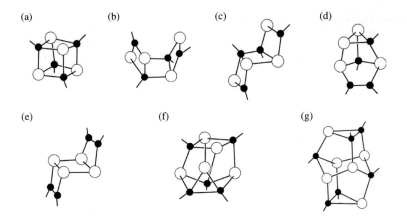

Fig. 15.5.6.
Structure of oligomers of phosphaalkyne: (a)–(e) cyclotetramer, (f) pentamer, (g) hexamer.

The phosphaalkyne hexamer $P_6C_6{}^tBu_6$ consists of a lantern-like cage constructed from the linkage of a chair-like P_4C_2 ring with a pair of C_2P rings above and below it, as shown in Fig. 15.5.6(g).

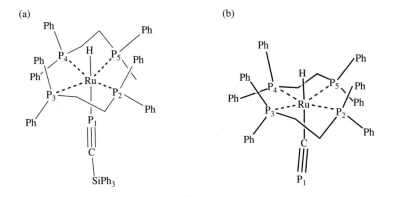

Fig. 15.5.7.
Molecular structures and bond lengths (pm) of (a) $[RuH(dppe)_2(Ph_3SiC\equiv P)]^+$, $C\equiv P_1$ 153.0(3), Ru–P_1 224.85(8), Ru–P_2 238.11(7), Ru–P_3 237.21(7), Ru–P_4 235.59(7), Ru–P_5 236.94(7); (b) $[RuH(dppe)_2(C\equiv P)]$, $C\equiv P_1$ 157.3(2), Ru–C 205.7(2), Ru–P_2 233.42(5), Ru–P_3 233.15(5), Ru–P_4 232.22(5), Ru–P_5 233.96(4).

(4) Cyaphide P≡C⁻

The quest for cyaphide, the phosphorus homolog of cyanide, as a ligand in a stable metal complex came to a satisfactory conclusion in 2006. The pair of related complexes $[RuH(dppe)_2(Ph_3SiC\equiv P)]OTf$ and $[RuH(dppe)_2(C\equiv P)]$ were obtained via a new synthetic route and structurally characterized by X-ray crystallography. Their molecular geometries are displayed in Fig. 15.5.7. As expected, the Si–C≡P and P≡C⁻ ligands are *P*- and *C*-coordinated to the Ru(II) center, respectively. The long C≡P bond in the cyaphide complex likely arises from back donation from Ru to the π^* orbitals of the ligand.

Table 15.5.4. Coordination modes of diphosphenes

Coordination mode		Example and structure	
η^1	R₂P=P-M(R)	Ar₂P=P-Mo(CO)₅(Ar)	(Ar = 2,4,6-^{t}Bu₃C₆H₂)
η^2	R₂P=P(R)→M	(η^2-^{t}BuP=P^{t}Bu)Zr(Cp*)₂	
$\mu(\eta^1: \eta^1)$	M-P(R)=P(R)-M	(CO)₅Cr-P(Ph)=P(Ph)-Cr(CO)₅	
$\mu(\eta^1: \eta^2)$	M-P(R)=P(R)-M (η²)	—	
$\mu_3(\eta^1: \eta^1: \eta^2)$	M, M, P(R)=P(R)-M	(CO)₅W, W(CO)₅, P(Ph)=P(Ph)-W(CO)₅	
$\mu(\eta^2: \eta^2)$	R-P-P-R with M, M (butterfly)	Ph-P-P-Ph with Mo[(CO)₂Cp], Mo[(CO)₂Cp]	Mo₂P₂ in a butterfly configuration
$\mu(\eta^2: \eta^2)$	R, R, P, P, M-M (tetrahedral)	^{t}Bu, ^{t}Bu, PH-PH, (OC)₃Fe-Fe(CO)₃	Fe₂P₂ in a tetrahedral configuration

15.5.5 π-Coordination complexes of phosphorus–carbon compounds

(1) Diphosphenes (R–P=P–R)

Diphosphenes contain the –P=P– group, the majority of which adopt a *trans*-configuration. For example, the stable compound

$$\text{Ar-P=P-Ar} \quad (\text{Ar} = 2,4,6\text{-}^t\text{Bu}_3\text{C}_6\text{H}_2)$$

exhibits the *trans* form with a P=P bond length of 203.4 pm.

Various coordination modes of diphosphenes toward transition metals involve σ and π interaction with the metal center, as listed in Table 15.5.4.

(2) η^3-Phosphaallyl and η^3-phosphirenes

The phosphaallyl anions (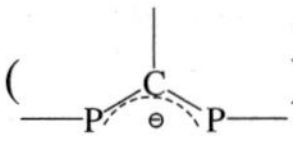) and the phosphirenes which contain the group (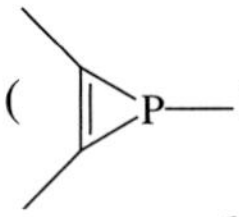) are η^3-ligands that can form complexes with transition metals. Some examples are shown below:

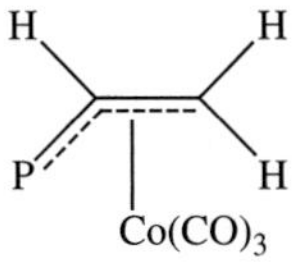

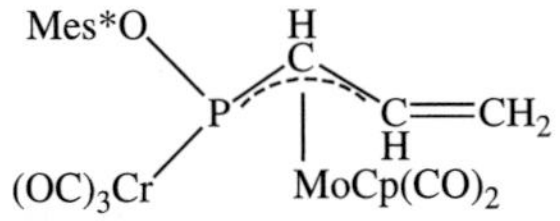

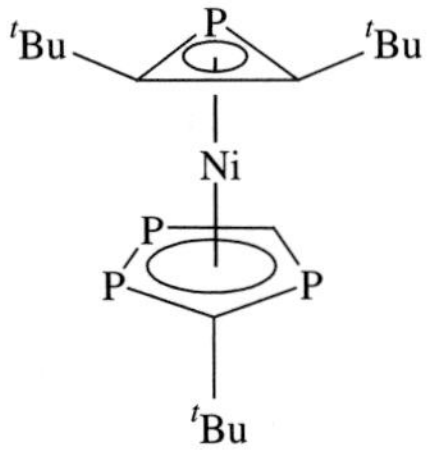

(3) η^4-Phosphadienes and diphosphacyclobutadienes

Phosphadienes, , are η^4-ligands that can form complexes with transition metals, for example:

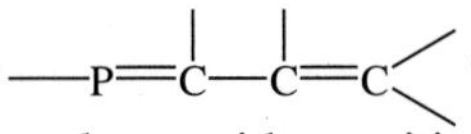

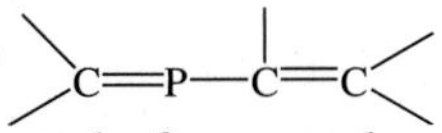

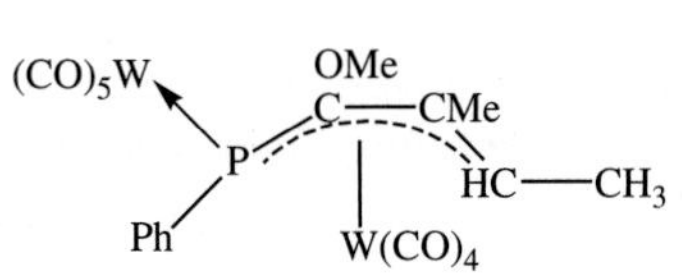

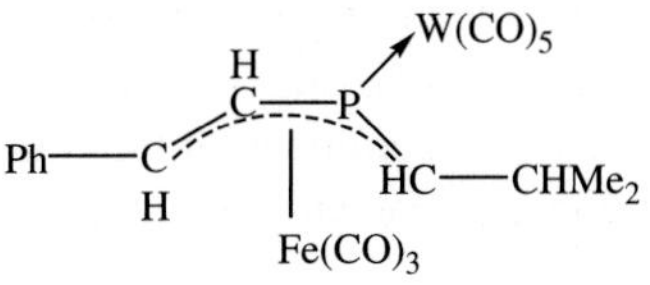

Many metal complexes containing the 1,2 or 1,3-diphosphacyclobutadiene ring have been characterized, and Fig. 15.5.8 shows the structures of some examples.

Interestingly, it has proved impossible to displace the η^4-ligated (P_2C_2 $^tBu_2$) rings from any of the above complexes, in contrast with the behavior of the analogous η^4-ligated cyclobutadiene ring complexes. This may be attributed to the significantly stronger π interaction between the metal and the phosphorus-containing ring system.

(4) η^5-Phospholyl complexes

The phospholyl unit, which contains one to five P atoms, is an analog of the cyclopentadienyl ligand. Figures 15.5.9(a) to 15.5.9(f) show some phosphametallocenes, including sandwich, half-sandwich, and tilted structures. Figures 15.5.9(g) to 15.5.9(j) show some complex phosphametallocenes, in which only essential parts of the structure are displayed. Figure 15.5.9(k) shows the supramolecular structure of [Sm(η^5-PC$_4$Me$_4$)$_2$(η^1-PC$_4$Me$_4$)K(η^6-C$_6$H$_5$Me)]Cl.

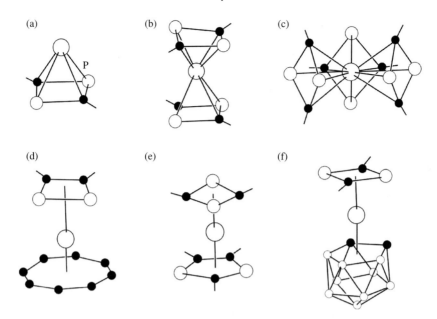

Fig. 15.5.8.
Structure of diphosphacyclobutadiene complexes: (a) Fe(CO)$_3$[η^4-P$_2$C$_2{}^t$Bu$_2$], (b) Ni[η^4-P$_2$C$_2{}^t$Bu$_2$]$_2$, (c) Mo[η^4-P$_2$C$_2{}^t$Bu$_2$]$_3$, (d) Ti(η^4-P$_2$C$_2{}^t$Bu$_2$)(η^8-COT), (e) Co(η^4-P$_2$C$_2{}^t$Bu$_2$)(η^5-P$_2$C$_3{}^t$Bu$_3$), and (f) Rh(η^4-P$_2$C$_2{}^t$Bu$_2$)(η^5-C$_2$B$_9$H$_{11}$).

(5) η^6-Phosphinine complexes

The phosphinine rings (PC$_5$R$_5$, P$_2$C$_4$R$_4$,... P$_6$) are analogs of benzene and form sandwich structures with the transition metals. In η^6-phosphinine complexes, the phosphinine rings are all planar.

15.6 Structural chemistry of As, Sb, and Bi

15.6.1 Stereochemistry of As, Sb, and Bi

The series As, Sb, and Bi show a gradation of properties from non-metallic to metallic, but the discrete molecules and ions of these elements exhibit similar stereochemistry, as listed in Table 15.6.1 and shown in Fig. 15.6.1. The presence of a lone pair (denoted by E in the table) in these atoms implies M^{III}; otherwise it is M^V.

Many compounds of the MX$_3$E type have been prepared, and all 12 trihalides of As, Sb, and Bi are well known and available commercially. In either the gaseous or solid state, the lone pair causes the bond angles to be less than the ideal tetrahedral angle in every case. For example, SbCl$_3$ in the gas phase has bond length 233 pm and bond angle 97.1°, and in the crystal it has three short Sb–Cl 236 pm and three long Sb$\cdots$Cl $\geq$ 350 pm, and the bond angle Cl–Sb–Cl is 95°. The cations of MX$_4^+$ are all tetrahedral. The anion SbF$_4^-$ is known in the monomeric form and has the MX$_4$E type disphenoidal geometry. In the dimer Sb$_2$F$_7^-$ both Sb atoms have a disphenoidal geometry with the bridging fluorine in one axial position.

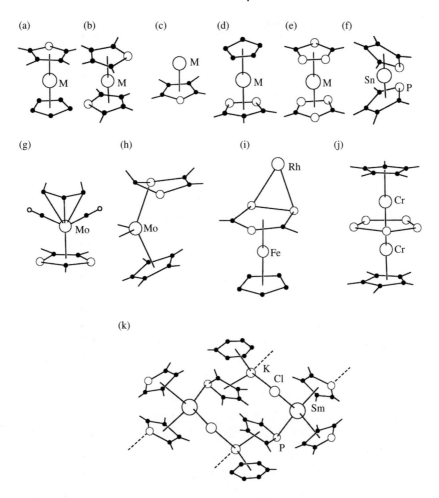

Fig. 15.5.9.
Structure of phospholyl π complexes: (a)–(e) sandwich-type phosphametallocenes, (f) Sn[η^5-PC$_4$(TMS)$_2$Cp$_2$]$_2$, (g) (η^3-C$_9$H$_7$)Mo(CO)$_2$(η^5-P$_2$C$_3{}^t$Bu$_3$), (h) (η^3-P$_2$C$_3{}^t$Bu$_3$)Mo(CO)$_2$(η^5-Cp*), (i) [(η^5-Cp*)(CO)]Rh[η^5-P$_3$C$_2{}^t$Bu$_2$]Fe(η^5-Cp), (j) (η^5-Cp*)Cr(η^5-P$_5$)Cr(η^5-Cp*), (k) [Sm(η^5-PC$_4$Me$_4$)$_2$(η^1-PC$_4$Me$_4$)K(η^6-C$_6$H$_5$Me)]Cl.

Table 15.6.1. Stereochemistry of As, Sb, and Bi

Total number of electron pairs	General formula*	Geometry	Example (refer to Fig. 15.6.1)
4	MX$_3$E	Trigonal pyramidal	AsCl$_3$, SbCl$_3$, BiCl$_3$ (a)
4	MX$_4$	Tetrahedral	AsCl$_4^+$, SbCl$_4^+$ (b)
5	MX$_4$E	Disphenoidal	SbF$_4^-$ (c)
5	MX$_5$	Trigonal bipyramidal	AsF$_5$, SbCl$_5$, BiF$_5$ (d)
5	MX$_5$	Square pyramidal	Sb(C$_6$H$_5$)$_5$, Bi(C$_6$H$_5$)$_5$ (e)
6	MX$_5$E	Square pyramidal	SbCl$_5^{2-}$ (f)
6	MX$_6$	Octahedral	SbBr$_6^-$ (g)
7	MX$_6$E	Octahedral	SbBr$_6^{3-}$, BiBr$_6^{3-}$ (g)

*M = As, Sb, or Bi, X = ligand atom or group, E = lone pair.

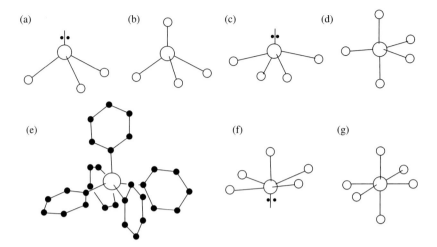

Fig. 15.6.1.
Stereochemistry of As, Sb, and Bi:
(a) $AsCl_3$, (b) $AsCl_4^+$, (c) SbF_4^-, (d) $SbCl_5$, (e) $Bi(C_6H_5)_5$, (f) $SbCl_5^{2-}$, and (g) $SbBr_6^{3-}$ and $BiBr_6^{3-}$.

Compounds of the MX_5 type exhibit two geometries, trigonal bipyramidal (more common) and square pyramidal. In the trigonal bipyramidal molecules, the axial bonds are longer than the equatorial bonds. If there are different ligands, the more electronegative ones usually occupy the axial positions. The compounds $Bi(C_6H_5)_5$ and $Sb(C_6H_5)_5$ have a square-pyramidal shape, as shown in Fig. 15.6.1(e), for which the bond lengths and bond angles are as follows:

$Bi(C_6H_5)_5$: Bi–C_{ax} 222.1 pm   Bi–C_{ba} 233.6 pm C_{ax}–Bi–C_{ba} 101.6°
$Sb(C_6H_5)_5$: Sb–C_{ax} 211.5 pm   Sb–C_{ba} 221.6 pm C_{ax}–Sb–C_{ba} 105.4°

The molecules of MX_5E type, such as SbF_5^{2-}, $BiCl_5^{2-}$, and some oligolymeric anions $(SbF_4)_4^{4-}$ and $(BiCl_4)_2^{2-}$, have square-pyramidal geometry at each M atom. In these cases, the four ligands in the base of the square pyramid lie in a plane slightly above the central M atom, and the bond angles are all less than 90° caused by the greater lone-pair repulsions.

The anions of the type MX_6, such as SbF_6^-, $SbBr_6^-$, and $Sb(OH)_6^-$, have the expected octahedral geometry. The anions of the type MX_6E, such as $SbBr_6^{3-}$ and $BiBr_6^{3-}$, frequently also have a regular octahedral structure. The undistorted nature of the $SbBr_6^{3-}$ octahedral suggests that the lonepair is predominantly $5s^2$, but in a sense it is still stereochemically active since the Sb–Br distance in $Sb^{III}Br_6^{3-}$ is 279.5 pm, which is longer than the distance 256.4 pm in $Sb^VBr_6^-$.

The bismuthonium ylide 4,4-dimethyl-2,6-dioxo-1-triphenylbismuthoniocyclohexane exhibits a distorted tetrahedral geometry with a Bi–C_{ylide} bond length of 215.6 pm, Bi–C_{Ph} bond lengths in the range 221–2 pm, and a weak Bi···O interaction of 301.9 pm with one of the carbonyl oxygen atoms (the other Bi···O separation is 335.2 pm). The X-ray data are consistent with the expectation that the negative charge resides mainly on a deprotonated enolic oxygen atom rather than on the ylidic carbon atom, whose 2p orbital does not overlap effectively with the 6d orbital of bismuth. Accordingly, formula I is a faithful representation of the structure in preference to II, III, and IV, as displayed below:

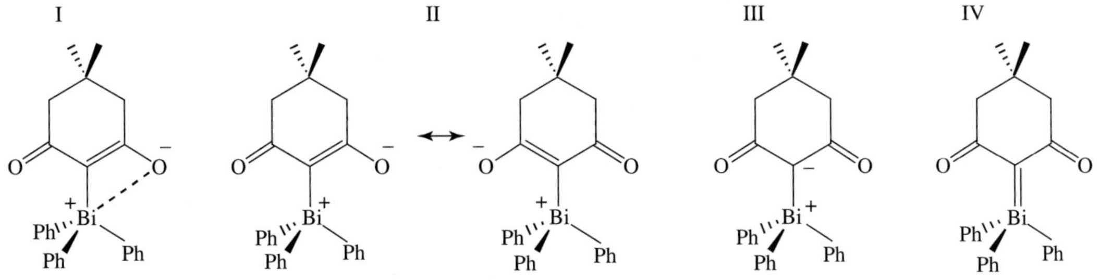

I   II   III   IV

From 1995 onward, studies have led to the synthesis and structural characterization of stable tungsten complexes with a heavier Group 15 element functioning as a triple-bonded terminal ligand. In the series of complexes [(CH$_2$CH$_2$NSiMe$_3$)$_3$N]W≡E (E = P, As, Sb), the tungsten atom exhibits a distorted trigonal bipyramidal coordination geometry with three equatorial N atoms and one N atom and the E atom occupying the axial positions. The molecular structure and W≡E bond distances are shown below:

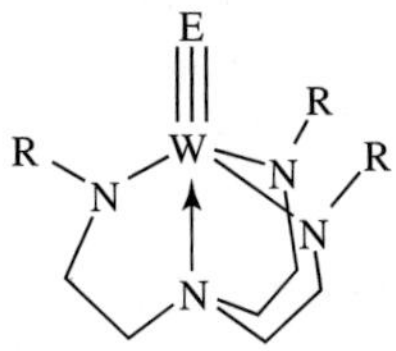

R = SiMe$_3$

W≡P   216.2(4) pm
W≡As 229.0(1) pm
W≡Sb 252.6(2) pm

15.6.2 Metal–metal bonds and clusters

Many compounds containing M–M bonds or stable rings and clusters of Group 15 elements are known. Figures 15.6.2(a) to 15.6.2(c) show the structures of some organometallic compounds of As, Sb, and Bi, which contain M–M bonds, and Figs. 15.6.2(d) to 15.6.2(e) show the structures of naked cluster cations Bi$_n^{m+}$, which are the components of some complex salts of bismuth. The structure of As$_6$(C$_6$H$_5$)$_6$ illustrates the typical trigonal pyramidal environment of the As atom; the As–As bond length is 246 pm, in which the As$_6$ ring adopts a chair conformation. In Sb$_4$(η^1-C$_5$Me$_4$)$_4$, Sb$_4$ forms a twisted ring, with Sb–Sb 284 pm, and all Sb–Sb–Sb bond angles are acute. Tetrameric bis(trimethylsilyl)methylbismuthine [(Me$_3$Si)$_2$CHBi]$_4$ contains a folded four-membered metallacycle with fold angles of 112.6° and 112.9°, Bi–Bi bond lengths in the range 297.0–304.4 pm, and Bi–Bi–Bi angles in the range 79.0–79.9°. The discrete molecules As$_2$Ph$_4$, Sb$_2$Ph$_4$, and Bi$_2$Ph$_4$ all adopt the staggered conformation, and the bond lengths are As–As 246 pm, Sb–Sb 286 pm, and Bi–Bi 299 pm.

Dibismuthenes of the general formula LBi=BiL can be synthesized when L is a bulky aryl ligand. The LBi=BiL molecule is centrosymmetric and therefore exists in the *trans* configuration. For L = 2,4,6-tris[bis(trimethylsilyl)methyl]phenyl, X-ray analysis of the dibismuthene

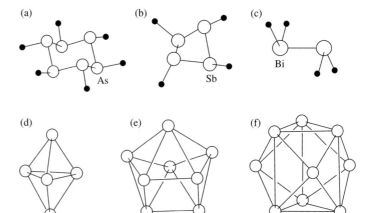

Fig. 15.6.2.
Structures of (a) $As_6(C_6H_5)_6$, (b) $Sb_4(\eta^1\text{-}C_5Me_4)_4$, (c) $Bi_2(C_6H_5)_4$, (d) Bi_5^{3+}, (e) Bi_8^{2+}, and (f) Bi_9^{5+}. In (a)–(c) the phenyl ligands are represented by C atoms bonded to the metal skeleton.

yielded a Bi=Bi double-bond length of 282.1 pm and a Bi=Bi–C angle of 100.5°.

The structures of cationic bismuth clusters Bi_5^{3+}, Bi_8^{2+}, and Bi_9^{5+} are listed in Table 15.6.2.

Table 15.6.2. Cationic bismuth clusters

Cation	Crystal	Structure	Symmetry
Bi_5^{3+}	$Bi_5(AlCl_4)_3$	Trigonal bipyramidal	D_{3h}
Bi_8^{2+}	$Bi_8(AlCl_4)_2$	Square antiprism	D_{4h}
Bi_9^{5+}	$Bi_{24}Cl_{28}$, or $(Bi_9^{5+})_2(BiCl_5^{2-})_4(Bi_2Cl_8^{2-})$	Tricapped trigonal prism	$C_{3h}(\sim D_{3h})$

Metalloid and intermetalloid clusters of Group 13 and 14 elements have been described in the two preceding chapters. For Group 15 elements, the ligand-free intermetalloid clusters $[As@Ni_{12}@As_{20}]^{3-}$ and $[Zn@Zn_8Bi_4@Bi_7]^{5-}$ are known. In $[As@Ni_{12}@As_{20}]^{3-}$, a central As atom is located inside a Ni_{12} icosahedron, which is in turn enclosed by a As_{20} pentagonal dodecahedron, as shown in Fig. 15.6.3(a).

The first example of an intermetalloid cluster that almagamates 11 Bi and 9 Zn atoms has been found in the complex $[K(2.2.2\text{-crypt})]_5[Zn_9Bi_{11}]\cdot$ 2en·toluene (2.2.2-crypt = 4,7,13,16,21,24-hexaoxa-1,10-diazabicyclo[8.8.8] hexacosane). The naked $[Zn@Zn_8Bi_4@Bi_7]^{5-}$ cluster consists of a central Zn atom trapped inside a distorted icosahedron whose vertices are 8 Zn atoms and 4 Bi atoms, with 7 of the 20 triangular faces each capped by a Bi atom, as shown in Fig. 15.6.3(b).

A simplified model may be used to rationalize the bonding in the heteroatomic species $[Zn@Zn_8Bi_4@Bi_7]^{5-}$. According to the electron counting theory proposed by Wade, the formation of a *closo* deltahedra of 12 vertices is stabilized by 13 skeletal electron pairs. The total of 26 electrons required for skeletal bonding may be considered to be provided as follows: 2 from the interstitial Zn atom, $(8 \times 0 + 4 \times 3 = 12)$ from the Zn_8Bi_4 icosahedral unit (each vertex atom carries an *exo* lone pair or bond pair), 7 from the capping Bi atoms, and 5 from

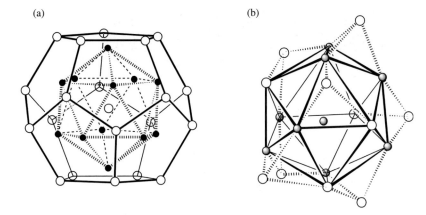

Fig. 15.6.3.
(a) Structure of the $[As@Ni_{12}@As_{20}]^{3-}$ cluster; the solid and open circles represent Ni and As atoms, respectively. (b) Structure of the $[Zn@Zn_8Bi_4@Bi_7]^{5-}$ cluster; the shaded and open circles represent Zn and Bi atoms, respectively. Interatomic distances: Zn_{cent}–Zn_{ico} 283.2(2)–359.4(3), Zn_{cent}–Bi_{ico} 282.2(2)–292.8(2), Zn_{ico}–Bi_{ico} 287.6(2)–375.5(2), Bi_{ico}–Bi_{ico} 320.5(1), Zn_{ico}–Zn_{ico} 289.4(2)–376.2(2), Bi_{cap}–Zn_{ico} 259.8(1)–276.8(1), Bi_{cap}–Bi_{ico} 310.7(1)–331.2(1) pm; the subscripts cent, ico, and cap denote atoms occupying the central, icosahedral, and capping sites, respectively.

the negative charges. Note that each vertex Bi atom is considered to retain one lone pair, whereas each capping Bi atom withholds two lone pairs.

15.6.3 Intermolecular interactions in organoantimony and organobismuth compounds

Intermolecular interactions have long been known to exist in inorganic compounds of antimony and bismuth. For example, $SbCl_3$ has three Sb–Cl bonds of length 234 pm and five Sb$\cdots$Cl distances in the range 346–74 pm, thus enlarging the coordination sphere from trigonal pyramidal to (3 + 5) bicapped trigonal prismatic geometry.

The organoantimony and organobismuth compounds in oxidation states I–III also exhibit strong intermolecular interactions, which lead to secondary bonds between molecules in forming chain-like, layer-type, and three-dimensional supramolecular structures. Figure 15.6.4 shows the structures of three organometallic compounds of Sb and Bi.

(1) MeSbCl$_2$

Crystalline $MeSbCl_2$ contains alternating layers, each consisting of double chains of $MeSbCl_2$ molecules. Along a $MeSbCl_2$ chain, the bond lengths and angles are Sb–Cl, 236.8 and 243.0 pm; Sb$\cdots$Cl, 333.7 and 386.5 pm; Sb–Cl$\cdots$Sb (within chain), 102.9°.

(2) Me$_2$SbI

In the Me_2SbI chain, the bond lengths are Sb–I 279.9 pm and Sb$\cdots$I 366.6 pm, so that the short and long Sb–I bonds are clearly different. The chain atoms lie

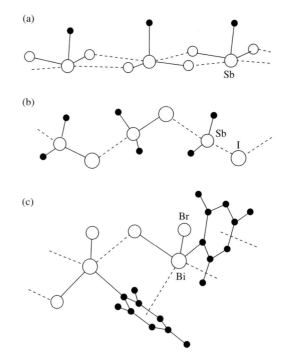

Fig. 15.6.4.
Structures of organometallic compounds of Sb and Bi: (a) MeSbCl$_2$, (b) Me$_2$SbI, (c) MesBiBr$_2$ (Mes = mesityl).

almost in a plane, while the methyl groups are directed to one side of this plane. The reverse side is exposed to neighboring chains with very weak interchain Sb···I contacts (402.4–416.7 pm) close to the van der Waals separation. This type of chain packing leads to the formation of double layers.

(3) MesBiBr$_2$

In the chain structure of MesBiBr$_2$, the Bi and bridging Br atoms constitute a zigzag chain, with the nonbridging Br atoms lying on one side and the mesityl groups on the other side of the plane as defined by the positions of the chain atoms. The structure is stabilized by π interaction between the Bi atom and the noncoordinated mesityl group. The bond lengths are Bi–Br, 261.9 and 281.8 pm; Bi···Bi, 301.7 and 302.2 pm; Bi– mesityl ring centroid, 319.5 and 330.1 pm.

References

1. N. N. Greenwood and A. Earnshaw, *Chemistry of the Elements*, 2nd edn., Butterworth-Heinemann, Oxford, 1997.
2. R. B. King (ed.), *Encyclopedia of Inorganic Chemistry*, Wiley, New York, 1994: (a) H. H. Sisler, Nitrogen: Inorganic chemistry, pp. 2516–57; (b) J. R. Lancaster, Nitrogen oxides in biology, pp. 2482–98; (c) J. Novosad, Phosphorus: Inorganic chemistry, pp. 3144–80; (d) R. H. Neilson, Phosphorus–nitrogen compounds, pp. 3180–99.
3. Y. A. Henry, A. Guissani and B. Ducastel, *Nitric Oxide Research from Chemistry to Biology: EPR Spectroscopy of Nitrosylated Compounds*, Springer, New York, 1997.
4. D. E. C. Corbridge, *Phosphorus: An Outline of its Chemistry, Biochemistry and Technology*, 5th edn., Elsevier, Amsterdam, 1995.

5. M. Regitz and O. J. Scherer (eds.), *Multiple Bonds and Low Coordination in Phosphorus Chemistry*, Verlag, Stuttgart, 1990.
6. A. Durif, *Crystal Chemistry of Condensed Phosphate*, Plenum Press, New York, 1995.
7. K. B. Dillon, F. Mathey and J. F. Nixon, *Phosphorus: The Carbon Copy*, Wiley, Chichester, 1998.
8. D. E. C. Corbridge, *The Structural Chemistry of Phosphorus*, Elsevier, Amsterdam, 1974.
9. J. Donohue, *The Structure of Elements*, Wiley, New York, 1974.
10. M.-T. Averbuch-Pouchot and A. Durif, *Topics in Phosphate Chemistry*, World Scientific, Singapore, 1996.
11. I. Haiduc and D. B. Sowerby (eds.), *The Chemistry of Inorganic Homo- and Heterocycles*, vol. 1–2, Academic Press, London, 1987.
12. J.-P. Majoral (ed.), *New Aspects in Phosphorus Chemistry I*, Springer, Berlin, 2002.
13. G. Meyer, D. Neumann and L. Wesemann (eds.), *Inorganic Chemistry Highlights*, Wiley–VCH, Weimheim, 2002.
14. S. M. Kauzlarich (ed.), *Chemistry, Structure and Bonding of Zintl Phases and Ions*, VCH, New York, 1996.
15. M. Gielen, R. Willem and B. Wrackmeyer (eds.), *Unusual Structures and Physical Properties in Organometallic Chemistry*, Wiley, West Sussex, 2002.
16. H. Suzuki and Y. Matano (eds.), *Organobismuth Chemistry*, Elsevier, Amsterdam, 2001.
17. K. O. Christe, W. W. Wilson, J. A. Sheehy and J. A. Boatz, N_5^+: a novel homoleptic polynitogen ion. *Angew. Chem. Int. Ed.* **38**, 2004–9 (1999).
18. B. A. Mackay and M. D. Fryzuk, Dinitrogen coordination chemistry: on the biomimetic borderlands. *Chem. Rev.* **104**, 385–401 (2004).
19. N. Burford and P. J. Ragogna, New synthetic opportunities using Lewis acidic phosphines. *Dalton Trans.*, 4307–15 (2002).
20. E. Urnezius, W. W. Brennessel, C. J. Cramer, J. E. Ellis and P. von R. Schleyer, A carbon-free sandwich complex $[(P_5)_2Ti]^{2-}$. *Science* **295**, 832–4 (2002).
21. M. Peruzzini, L. Gonsalvia and A. Romerosa, Coordination chemistry and functionalization of white phosphorus via transition metal complexes. *Chem. Soc. Rev.* **34**, 1038–47 (2005).
22. J. G. Cordaro, D. Stein, H. Rüegger and H. Grützmacher, Making the true "CP" ligand. *Angew. Chem. Int. Ed.* **45**, 6159–62 (2006).
23. J. Bai, A. V. Virovets and M. Scheer, Synthesis of inorganic fullerene-like molecules. *Science* **300**, 781–3 (2003).
24. M. J. Moses, J. C. Fettinger and B. W. Eichhorn, Interpenetrating As_{20} fullerene and Ni_{12} icosahedra in the onion-skin $[As@Ni_{12}@As_{20}]^{3-}$ ion. *Science* **300**, 778–80 (2003).
25. J. M. Goicoechea and S. C. Sevov, $[Zn_9Bi_{11}]^{5-}$: a ligand-free intermetalloid cluster. *Angew. Chem. Int. Ed.* **45**, 5147–50 (2006).

16

Structural Chemistry of Group 16 Elements

16.1 Dioxygen and ozone

Oxygen is the most abundant element on the earth's surface. It occurs both in the free state and as a component in innumerable compounds. The common allotrope of oxygen is dioxygen (O_2) or oxygen gas; the other allotrope is ozone (O_3).

16.1.1 Structure and properties of dioxygen

Molecular oxygen (or dioxygen) O_2 and related species are involved in many chemical reactions. The valence molecular orbitals and electronic configurations of the homonuclear diatomic species O_2 and O_2^- are shown in Fig. 16.1.1.

In the ground state of O_2, the outermost two electrons occupy a doubly degenerate set of antibonding π^* orbitals with parallel spins. Dioxygen is thus a paramagnetic molecule with a triplet ground state ($^3\Sigma$), and its formal double bond has a length of 120.752 pm.

The first electronic excited state ($^1\Delta$) is 94 kJ mol^{-1} above the ground state and is a singlet with no unpaired electrons. This very reactive species, commonly called singlet oxygen, has a half-life of 2 μs in water. Its bond length is 121.563 pm. Another singlet ($^1\Sigma$) species, with paired electrons in the π_g orbitals, is higher in energy than the ground state by 158 kJ mol^{-1}, and its bond length is 122.688 pm. Of the two singlet states only the lower $^1\Delta$ survives long enough to participate in chemical reactions. The higher $^1\Sigma$ state is deactivated to the O_2 ($^1\Delta$) state so rapidly that it has no significant chemistry.

Simple reduction of dioxygen to the superoxide ion, O_2^-, is thermodynamically unfavorable as it involves adding an electron to a half-filled antibonding orbital. Reduction to the peroxide ion, O_2^{2-}, however, is favorable from either superoxide or dioxygen. Owing to the increased population of antibonding orbitals in the reduced O_2 species (O_2^- and O_2^{2-}), both the bond order and O–O IR stretching frequency decrease, while the O–O distance increases. Table 16.1.1 lists relevant data on the bond properties of dioxygen species.

Owing to its nonpolar nature, O_2 is more soluble in organic solvents than in H_2O. Since the $^3\Sigma$ ground state of dioxygen is a spin triplet, concerted oxygenation reaction is subject to spin restriction. As a result, many types of dioxygen reactions proceed slowly, even in cases where such reactions are

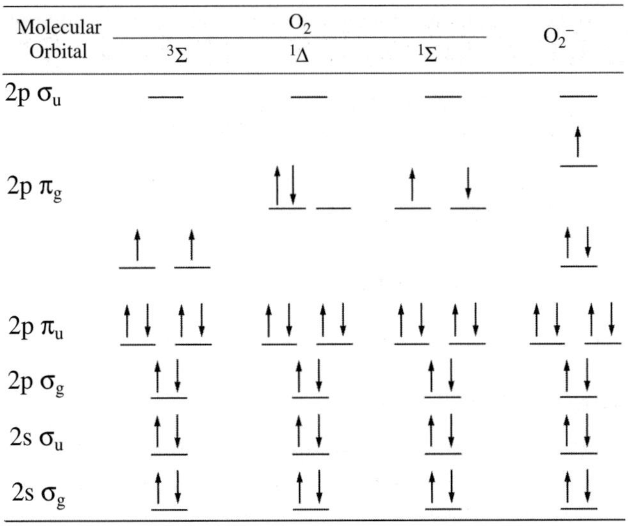

Fig. 16.1.1.
MO diagrams for O_2 and O_2^-.

strongly favored thermodynamically, such as a mixture of O_2 and H_2. The reactivity of dioxygen with singlet molecules can be increased by exciting it to its $^1\Delta$ singlet state, thereby removing the spin restriction. This energy corresponds to an infrared wavelength of 1270 nm. As singlet oxygen in solution deactivates by transferring its electronic energy to vibration of solvent molecules, its lifetime depends strongly on the medium. Solvents with high vibrational frequencies provide the most efficient relaxation. For this reason, the lifetime (about 2–4 μs) is shortest in water, which has a strong OH vibration near 3600 cm^{-1}. Solvents with CH groups ($\approx$ 3000 cm^{-1}) are the next most efficient, with lifetime in the range of 30–100 μs.

Table 16.1.1. Bond properties of dioxygen species

Species	Compound	Bond order	O–O (pm)	Bond energy (kJ mol^{-1})	ν(O–O) (cm^{-1})
Oxygenyl O_2^+	$O_2[AsF_6]$	2.5	112.3	625.1	1858
Triplet $O_2(^3\Sigma)$	$O_2(g)$	2	120.752	490.4	1554.7
Singlet $O_2(^1\Delta)$	$O_2(g)$	2	121.563	396.2	1483.5
Singlet $O_2(^1\Sigma)$	$O_2(g)$	2	122.688	—	—
Superoxide O_2^-	KO_2	1.5	128	—	1145
Peroxide O_2^{2-}	Na_2O_2	1	149	204.2	842
—O–O—	H_2O_2(cryst.)	1	145.8	213	882

There are two major sources of singlet oxygen O_2 ($^1\Delta$):

(1) *Photochemical sources.* One of the most common sources of singlet oxygen is energy transfer from an excited "sensitizer" (Sens), such as a dye or natural pigment, to the ground state of the oxygen molecule. The role of the sensitizer is to absorb irradiation and be converted to an electronically excited state (Sens*). This state, normally a triplet, then transfers its

energy to the oxygen molecule, producing singlet oxygen and regenerating the sensitizer, as indicated below:

$$\text{Sens} \xrightarrow{h\nu} \text{Sens}^* \xrightarrow{O_2} \text{Sens} + O_2(^1\Delta).$$

(2) *Chemical sources.* An early chemical source is the reaction of HOCl with H_2O_2, which produces singlet $O_2(^1\Delta)$ in nearly quantitative yield:

$$H_2O_2 + HOCl \rightarrow O_2(^1\Delta) + H_2O + HCl.$$

Another source is derived from phosphite ozonides as in the reaction

In addition to the above sources, a number of other chemical reactions have been suggested for the production of singlet oxygen.

Singlet oxygen $O_2(^1\Delta)$ is a useful synthetic reagent, which allows stereospecific and regiospecific introduction of O_2 into organic substrates. Since singlet oxygen $O_2(^1\Delta)$ has a vacant π_g orbital, similar to the frontier MO of ethylene, it behaves as an electrophilic reagent, with a reactivity pattern reminiscent of that exhibited by ethylene with electronegative substituents. For example, it undergoes [4 + 2] cycloadditions with 1,3-dienes, and [2 + 2] cycloadditions and ene reactions with isolated double bonds. Some examples are given below:

16.1.2 Crystalline phases of solid oxygen

X-ray studies of solid oxygen began in the 1920s, and at present six distinct crystallographic forms have been unequivocally identified. The α, β, and γ phases exist at ambient pressure and low temperature. At $T = 295$ K, oxygen solidifies to form the β phase at 5.4 GPa, which transforms into the orange δ

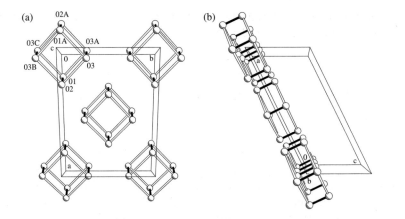

Fig. 16.1.2.
Molecular packing in the crystal structure of ε-oxygen; the solid and open lines indicate covalent and weak bonds, respectively. (a) C-centered layer of $(O_2)_4$ clusters viewed along the c axis; (b) the same layer viewed along the b axis. Symmetry codes: A $\bar{x}, y, \bar{z}$; B $x, \bar{y}, z$; C $\bar{x}, \bar{y}, \bar{z}$.

phase at 9.6 GPa, which in turn is converted to the red ε phase at 10 GPa. Above 96 GPa, the ε phase undergoes transformation to the metallic ς phase.

The monoclinic α, rhombohedral β, and orthorhombic γ phases all have layered structures composed of O_2 molecules. Recently, single-crystal X-ray analysis has established that in the crystal structure of the ε phase at 13.2–17.6 GPa (space group $C2/m$ with $Z = 8$), four O_2 molecules associate into a rhombohedral $(O_2)_4$ cluster by serving as the short, parallel edges of a rhombic prism located at a site of symmetry $2/m$, which is held together by weak chemical bonds. The layer-type crystal structure is shown in Fig. 16.1.2. The O$\cdots$O$\cdots$O angles at the rhombic face O2–O3–O1A–O3B are 84.5° and 95.8°. The two independently measured O–O bond lengths are 120 and 121 pm, which are in agreement with that in the gas phase. The shortest intra-cluster, intercluster, and interlayer O$\cdots$O contacts are 218, 259, and 250 pm, respectively.

16.1.3 Dioxygen-related species and hydrogen peroxide

The oxygenyl cation, O_2^+, is well characterized in both the gas phase and in salts with non-oxidizable anions. Removal of one of the antibonding π_g electrons from O_2 gives O_2^+ with a bond order of 2.5 and, consequently, a shorter bond length of 112.3 pm. The chemistry of O_2^+ is quite limited in scope because of the high ionization potential of O_2 (1163 kJ mol^{-1}).

The superoxide ion O_2^- is formed by adding one electron into the antibonding orbital π_g of O_2, thereby reducing the bond order to 1.5 and increasing the bond length to 128 pm, as shown in Table 16.1.1.

The peroxide ion, O_2^{2-}, has two more electrons than neutral dioxygen, and these additional electrons completely fill the π_g MO, resulting in a diamagnetic molecule with an O–O bond order of 1. The single-bond character is evident from the longer bond length (about 149 pm) and from the lower rotational barrier of the peroxide bond. The peroxide anion is found in ionic salts with the alkali metals and heavier alkaline-earth metals such as calcium, strontium, and barium. These ionic peroxides are powerful oxidants, and hydrogen peroxide, H_2O_2, is formed when they are dissolved in water. Figure 16.1.3 shows the molecular structure of H_2O_2.

Fig. 16.1.3.
Molecular structure of H₂O₂: (a) in the crystal and (b) in the gas phase (bond lengths in pm).

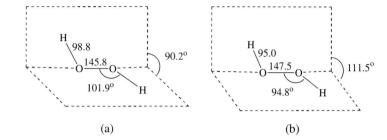

Table 16.1.2 lists the values of the dihedral angle of H₂O₂ in some crystalline phases (a value of 180° corresponds to a planar *trans* configuration). This large range of values indicates that the rotational barrier is low and the molecular configuration of H₂O₂ is very sensitive to its surroundings.

Table 16.1.2. Dihedral angle of H₂O₂ in some crystalline phases

Compound	Dihedral angle	Compound	Dihedral angle
H_2O_2 (s)	90.2°	$Li_2C_2O_4 \cdot H_2O_2$	180°
$K_2C_2O_4 \cdot H_2O_2$	101.6°	$Na_2C_2O_4 \cdot H_2O_2$	180°
$Rb_2C_2O_4 \cdot H_2O_2$	103.4°	$NH_4F \cdot H_2O_2$	180°
$H_2O_2 \cdot 2H_2O$	129°	theoretical	90° – 120°

Hydrogen peroxide has a rich and varied chemistry which arises from its versatility:

(a) It can act either as an oxidizing or a reducing agent in both acid and alkaline media.
(b) It undergoes proton acid/base reactions to form peroxonium salts $(H_2OOH)^+$, hydroperoxides $(OOH)^-$, and peroxides (O_2^{2-}).
(c) It gives rise to peroxometal complexes and peroxoacid anions.
(d) It can form crystalline addition compounds with other molecules through hydrogen bonding, such as $Na_2C_2O_4 \cdot H_2O_2$, $NH_4F \cdot H_2O_2$, and $H_2O_2 \cdot 2H_2O$.

The O–O bond length of 134 pm for the superoxide ion, O_2^-, was determined accurately for the first time in its $[1,3\text{-}(Me_3N)_2C_6H_4][O_2]_2 \cdot 3NH_3$ salt. Orientational disorder of the O_2^- ion that often occurs in crystals is avoided by using the bulky, nonspherical organic counter cation, which imposes order on the anionic sites.

16.1.4 Ozone

Ozone is a diamagnetic gas with a distinctive strong odor. The O_3 molecule [Fig. 16.1.4(a)] is angular, with an O–O bond length of 127.8 pm, which is longer than that of O_2 (120.8 pm); the bond energy of 297 kJ mol⁻¹ is lesser than that of O_2 (490 kJ mol⁻¹). The central O atom uses its sp^2 hybrid orbitals to form σ bonds with the terminal O atoms and to accommodate a lone pair, leaving the remaining p orbital for π bonding, as shown in Fig. 16.1.4(b). Each O–O bond has a bond order of 1.5. Ozone has a measured dipole moment of

1.73×10^{-30} C·m (0.52 Debye) due to the uneven electron distribution over the central and terminal O atoms.

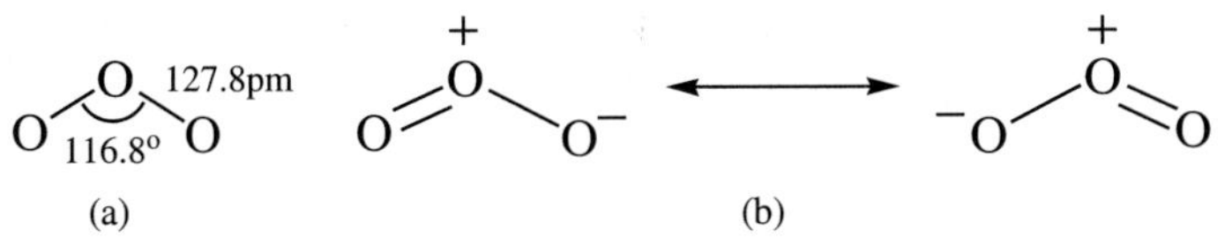

Fig. 16.1.4.
The O_3 molecule: (a) geometry and (b) bonding.

The deep red ozonide ion O_3^- has a bent structure, and sodium ozonide is isostructural with sodium nitrite ($NaNO_2$). The X-ray structural data of the alkali metal ozonides show that an increase in cationic size corresponds to a decrease in O–O bond length and an increase in O–O–O angle, as summarized in the following table:

MO_3	O–O (pm)	O–O–O(°)
NaO_3	135.3(3)	113.0(2)
KO_3	134.6(2)	113.5(1)
RbO_3	134.3(7)	113.7(5)
CsO_3	133.3(9)	114.6(6)

Since the additional electron enters an antibonding π^* orbital, the bond order of O–O in O_3^- is 1.25 and the bond length increases. Ionic ozonides such as KO_3 are thermodynamically metastable and extremely sensitive to moisture and carbon dioxide.

Ozone is a major atmospheric pollutant in urban areas. In addition to its damaging effect on lung tissue and even on exposed skin surfaces, ozone attacks the rubber of tires, causing them to become brittle and crack. But in the stratosphere, where ozone absorbs much of the short-wavelength UV radiation from the sun, it provides a vital protective shield for life forms on earth.

Ozone has proven to be useful as a reagent to oxygenate organic compounds, a good example being the conversion of olefins to cleaved carbonyl products (Fig. 16.1.5). When O_3 reacts with olefins, the first product formed

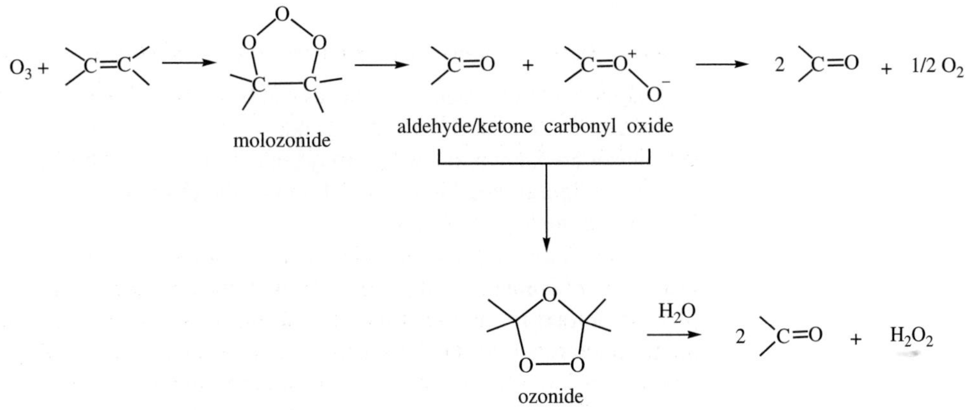

Fig. 16.1.5.
Mechanism of reaction between O_3 and an olefin.

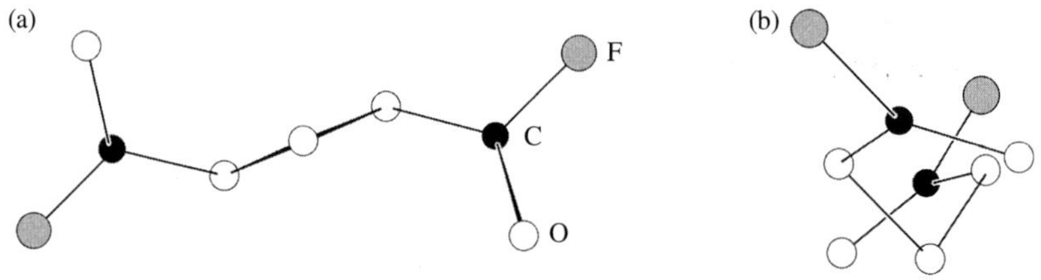

Fig. 16.1.6.
Structure of the molecule FC(O)OOOC(O)F showing (a) its C_2 symmetry axis and (b) left-handed chiral conformation.

is a molozonide which in turn undergoes an O–O bond cleavage to give an aldehyde or a ketone plus a carbonyl oxide. The carbonyl oxide then decomposes to yield the products. But under appropriate conditions, the aldehyde or ketone will recombine with the carbonyl oxide to generate an ozonide, which then decomposes in the presence of H_2O to give two carbonyl compounds and hydrogen peroxide.

There exist only a few compounds which contain a linear chain of three oxygen atoms. The unstable compound bis(fluoroformyl)trioxide, F–C(O)–O–O–O–C(O)–F, exists as the *trans-syn-syn* rotamer with C_2 symmetry in the crystalline state (with respect to the central O_3 fragment, the C–O bonds are *trans* and both C=O bonds are *cis*). The C–O–O–O torsion angle and O–O–O bond angle are 99.0(1)° and 104.0(1)°, respectively, as shown in Fig. 16.1.6. In the stable molecule bis(trifluoromethyl)trioxide F_3C–O–O–O–CF_3, the corresponding values are 95.9(8) and 106.4(1)°, respectively.

16.2 Oxygen and dioxygen metal complexes

16.2.1 Coordination modes of oxygen in metal–oxo complexes

The structures of inorganic crystals are usually described by the packing of anions and metal coordination geometries. In this section, we place particular emphasis on the coordination environment of the oxygen atom. The oxygen atom (oxo ligand) exhibits various ligation modes in binding to metal centers, as shown in Table 16.2.1.

16.2.2 Ligation modes of dioxygen in metal complexes

In general, dioxygen metal complexes can be classified in terms of their mode of binding. With nonbridging O_2 ligands, both η^1 (superoxo) and η^2 (peroxo) complexes are known. Bridging dioxygen groups can adopt $\eta^1:\eta^1$, $\eta^1:\eta^2$, and $\eta^2:\eta^2$ configurations. Figure 16.2.1 shows the observed coordination modes in dioxygen metal complexes.

Another commonly adopted classification of dioxygen metal complexes is based on O–O bond lengths (referring to Table 16.2.2): superoxo complexes of types I*a* and I*b*, in which the O–O distance is roughly constant and close to the value reported for the superoxide anion (~130 pm); and types II*a* and II*b*, in which the O–O distance is close to the values reported for H_2O_2 and O_2^{2-} (~148 pm). The *a* or *b* classification distinguishes complexes in which the dioxygen is bound to one metal atom (type *a*) or bridging two metal atoms (type *b*).

Group 16 Elements

Table 16.2.1. Coordination geometry of oxo ligand

Coordination geometry	Examples and notes
M=O	$[VO]^{2+}$, V=O bond length (155 to 168 pm) has partial triple-bond character.
M—O—M	Linear, ReO_3: Re–O–Re forms the edges of cubic unit cell.
M–O–M (bent)	Bent, $Cr_2O_7^{2-}$: Cr–O bond length 177 pm, Cr–O–Cr bond angle 123°.
O=M—O—M=O	Linear (in the presence of strong π bonding) and bent: these types exist in the compounds of Re and Tc; bond length M=O 165 to 170 pm, and M–O 190 to 192 pm.
M₃-O pyramidal/trigonal planar	Pyramidal, as in H_3O^+ and $[O(HgCl)_3]^+$. Trigonal planar, as in $[Fe_3(\mu_3\text{-}O)(O_2CCMe_3)_6(MeOH)_3]^+$.
M₄-O tetrahedral	Tetrahedral, as in Cu_2O and $Be_4(\mu_4\text{-}O)(O_2CMe)_6$.
M₄-O square planar	Square planar, as in $[Fe_8(\mu_4\text{-}O)(\mu_3\text{-}O)_4(OAc)_8(tren)_4]^{6+}$.
M₅-O trigonal bipyramidal	Trigonal bipyramidal, as in $[Fe_5(\mu_5\text{-}O)(O_2CMe)_{12}]^+$.
M₆-O octahedral	Octahedral, as in $[Fe_6(\mu_6\text{-}O)(\mu_2\text{-}OMe)_{12}(OMe)_6]^{2-}$.

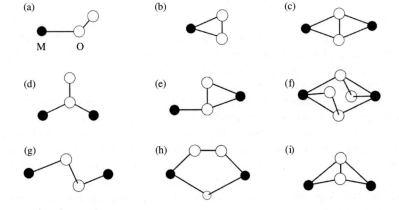

Fig. 16.2.1. Coordination modes of dioxygen metal complexes: (a) η^1-superoxo, (b) η^2-superoxo, (c) $\mu\text{-}\eta^2:\eta^2$ peroxo, symmetric, (d) $\mu\text{-}\eta^1$-superoxo, symmetrical, (e) $\mu\text{-}\eta^1:\eta^2$ peroxo, (f) $\mu\text{-}\eta^1:\eta^2$-peroxo, centrosymmetric, (g) *trans* $\mu\text{-}\eta^1:\eta^1$, superoxo or peroxo, (h) *cis* $\mu\text{-}\eta^1:\eta^1$, superoxo or peroxo, (i) $\mu\text{-}\eta^2:\eta^2$ peroxo, unsymmetric.

Table 16.2.2. Structural data of dioxygen metal complexes

Complex	Type	O_2:M ratio	Structure	O–O (pm) (normal range)	v(O–O) (cm^{-1}) (normal range)
Superoxo	Ia	1:1	M—O⋯O	125 – 135	1130 – 1195
Superoxo	Ib	1:2	M–O⋯O–M	126 – 136	1075 – 1122
Peroxo	IIa	1:1	M(O–O) triangle	130 – 155	800 – 932
Peroxo	IIb	1:2	M–O–O–M	144 – 149	790 – 884

The stretching frequencies attributed to the O–O vibration are closely related to the structural type. Type I complexes show O–O stretching vibrations around 1125 cm^{-1} and type II around 860 cm^{-1}. This sharp difference enables the O–O stretching frequency as measured by infrared or Raman spectroscopy to be used for structure type classification. Table 16.2.2 lists the structural data of dioxygen metal complexes.

16.2.3 Biological dioxygen carriers

The proteins hemoglobin (Hb) and myoglobin (Mb) serve to transport and to store oxygen, respectively, in all vertebrates. The active site in both proteins is a planar heme group (Fig. 16.2.2). Myoglobin is composed of one globin (globular protein molecule) and one heme group. Hemoglobin consists of four myoglobin-like subunits, two α and two β. In each subunit, the heme group is partially embedded in the globin, and an oxygen molecule can bond to the iron atom on the side of the porphyrin opposite the proximal histidine, thus forming a hexa-coordinate iron complex, as shown in Fig. 16.2.3.

Myoglobin contains Fe^{2+} (d^6 configuration) in a high-spin state, which has a radius of about 78 pm in a pseudo-octahedral environment. The Fe^{2+} ion is too large to fit into the central cavity of the porphyrin ring, and lies some 42 pm above the plane of the N donor atoms. When a dioxygen molecule binds to Fe^{2+}, the Fe^{2+} becomes low-spin d^6 with a shrunken radius of only 61 pm, which enables it to slip into the porphyrin cavity.

In hemoglobin, O_2 binding to Fe^{2+} does not oxidize it to Fe^{3+}, as it is protected by protein units around the heme group. A nonaqueous environment is required for reversible O_2 binding.

The $O_2(^3\Sigma)$ and the high-spin Fe^{2+} atom of the heme combine to form a spin-paired diamagnetic system. The resulting Fe–O–O is bent, with a bond angle varying from 115° to 153°. The stronger σ interaction is between the d_{z^2}(Fe) and $\pi_g(O_2)$ orbitals, and the weak π interaction is between d_{xz}(Fe) and $\pi_g^*(O_2)$. The increased ligand effect results in pairing of electrons and a weakened O–O bond.

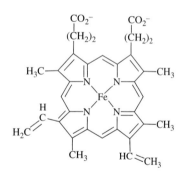

Fig. 16.2.2.
The heme group.

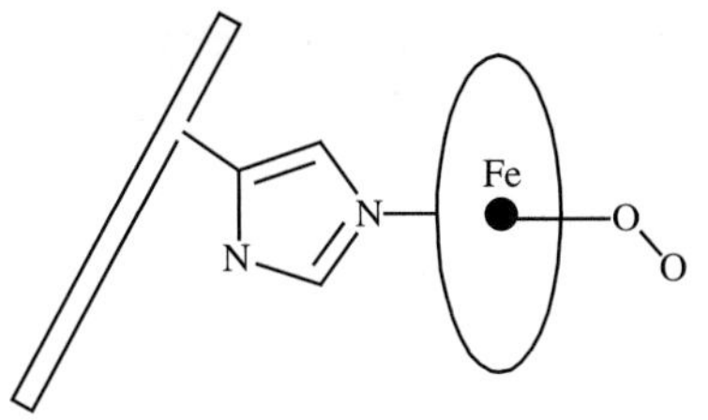

Fig. 16.2.3.
The binding of dioxygen by heme.

Oxygen enters the blood in the lungs, where the partial pressure of O_2 is relatively high (2.1×10^4 Pa) under ideal conditions; in the lungs with mixing of inhaled and nonexhaled gases, the O_2 partial pressure is $\sim 1.3 \times 10^4$ Pa. It is then carried by red blood cells to the tissues where the O_2 partial pressure is lowered to $\sim 4 \times 10^3$ Pa. The reactions that occur are represented as follows:

Lungs: $Hb + 4O_2 \rightarrow Hb(O_2)_4$
Tissues: $Hb(O_2)_4 + 4Mb \rightarrow 4Mb(O_2) + Hb$.

The four subunits ($2\alpha + 2\beta$) of hemoglobin function cooperatively. When one O_2 is bound to one heme group, the conformation of Hb subtly changes to make binding of additional oxygen molecules easier. Thus the four irons can each carry one O_2 with steadily increasing equilibrium constants. As a result, as soon as some O_2 has been bound to the molecule, all four irons readily become oxygenated (oxyHb). In a similar fashion, initial removal of oxygen triggers the release of the remainder, and the entire load of O_2 is delivered at the required site. This effect is also facilitated by pH changes caused by increased CO_2 concentration in the capillaries. As the CO_2 concentration increases, the pH decreases due to formation of HCO_3^-, and the increased acidity favors release of O_2 from oxyHb. This is known as the Bohr effect.

Hemoglobin carries oxygen via the circulatory system from the lungs to body tissues. There the oxygen can be transferred to myoglobin, where it is stored for oxidation of foodstuffs. The properties of Hb and Mb can be represented by the oxygen-binding curves of Fig. 16.2.4. At high oxygen concentrations, Hb and Mb bind O_2 with approximately equal affinity. However, at low oxygen concentrations, such as in muscle tissue during or immediately after muscular activity, Hb is a poor oxygen binder and therefore can deliver its oxygen to Mb. The difference in O_2-binding ability between Mb and Hb is accentuated at low pH values; hence the transfer from Hb to Mb has a great driving force where it is most needed, viz., in tissues where oxygen has been converted to CO_2.

16.3 Structure of water and ices

Water is the most common substance on the earth's surface, and it constitutes about 70% of the human body and the food it consumes. It is also the smallest molecule with the greatest potential for hydrogen bonding. Water is the most

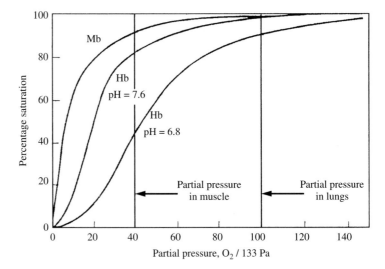

Fig. 16.2.4.
The oxygen binding curves for Mb and Hb, showing the pH dependence for the latter.

important species in chemistry because a majority of chemical reactions can be carried out in its presence.

16.3.1 Water in the gas phase

The H_2O molecule has a bent structure, with O–H bond length 95.72 pm and bond angle H–O–H 104.52°; the O–H bonds and the lone pairs form a tetrahedral configuration, as shown in Fig. 16.3.1(a). They are capable of forming donor O–H$\cdots$X and acceptor O$\cdots$H–X hydrogen bonds (X is a highly electronegative atom).

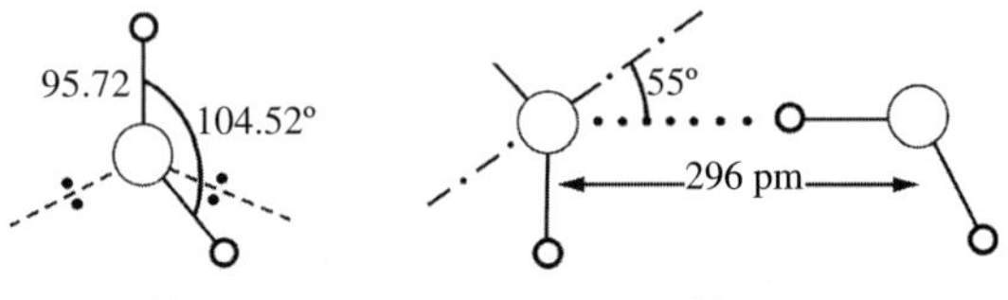

Fig. 16.3.1.
Structures of (a) gaseous H_2O molecule and (b) water dimer (bond length in pm).

In the gas phase, the adducts of H_2O with HF, HCl, HBr, HCN, HC≡CH, and NH_3 have been studied by microwave spectroscopy to determine the relationship of hydrogen-bond donor and acceptor. The structures of the adducts are $H_2O{:}\cdots H{-}F$, $H_2O{:}\cdots H{-}Cl$, and $H_2O{:}\cdots H{-}CN$. Only with NH_3 is water a H donor: $H_3N{:}\cdots H{-}OH$.

Water forms a hydrogen-bonded dimer, the structure of which has been determined by microwave spectroscopy, as shown in Fig. 16.3.1(b). The experimental value of the binding energy of this dimeric system is 22.6 kJ mol^{-1}.

16.3.2 Water in the solid phase: ices

Solid H_2O has 11 known crystalline forms, as listed in Table 16.3.1.

Ordinary hexagonal ice I_h and cubic ice I_c are formed at normal pressures, the latter being stable below $-120°$ C. Their hydrogen bonding patterns are

Table 16.3.1. Crystal structures of ice polymorphs

Ice	Space group	Parameters of unit cell (pm)	Z	Hydrogen positions	Density (g cm^{-3})	Coordination number*	Hydrogen bond distance (pm)	Nearest nonbonding distance (pm)
I_h (273 K)	$P6_3/mmc$ (no. 194)	$a = 451.35(14)$ $c = 735.21(12)$	4	Disordered	0.92	4	275.2–276.5	450
I (5 K)	$Cmc2_1$ (no. 36)	$a = 450.19(5)$ $b = 779.78(8)$ $c = 732.80(2)$	12	Ordered	0.93	4	273.7	450
I_c (143 K)	$Fd\bar{3}m$ (no. 227)	$a = 635.0(8)$	8	Disordered	0.93	4	275.0	450
II	$R\bar{3}$ (no. 148)	$a = 779(1)$ $\alpha = 113.1(2)°$	12	Ordered	1.18	4	275–84	324
III	$P4_12_12$ (no. 92)	$a = 673(1)$ $c = 683(1)$	12	Disordered	1.16	4	276–80	343
IV	$R\bar{3}c$ (no. 167)	$a = 760(1)$ $\alpha = 70.1(2)°$	16	Disordered	1.27	4	279–92	314
V	$A2/a$ (no. 15)	$a = 922(2)$ $b = 754(1)$ $c = 1035(2)$ $\beta = 109.2(2)°$	28	Disordered	1.23	4	276–87	328
VI	$P4_2/nmc$ (no. 137)	$a = 627$ $c = 579$	10	Disordered	1.31	4	280–2	351
VII	$Pn\bar{3}m$ (no. 224)	$a = 343$	2	Disordered	1.49	8	295	295
VIII	$I4_1/amd$ (no. 141)	$a = 467.79(5)$ $c = 680.29(10)$	8	Ordered	1.49	6	280–96	280
IX	$P4_12_12$ (no. 92)	$a = 673(1)$ $c = 683(1)$	12	Ordered	1.16	4	276–80	351

*No. of nearest neighbors closer than 300 pm.

very similar, and only the arrangements of the oxygen atoms differ. In I_c, the oxygens are arranged like the carbon atoms in cubic diamond, whereas in I_h the atomic arrangement corresponds to that in hexagonal diamond. Figure 16.3.2 shows the structure of I_h.

In both I_h and I_c, the hydrogen-bonded systems contain cyclohexane-like buckled hexagonal ring motifs: chair form in I_c and boat form in I_h. An important distinction is that in I_c all four hydrogen bonds are equivalent by symmetry. In I_h, the hydrogen bond in the direction of the hexagonal axis can be differentiated from the other three.

A neutron diffraction analysis of H_2O and D_2O ice I_h at different temperatures has shown that, in addition to the disordered hydrogens, the oxygen atoms are also disordered about sites 6 pm off the hexagonal axis.

In the high-pressure ices the hydrogen-bonding patterns become progressively more complex with increasing pressure until the ultra-high-pressure ice polymorphs VII and VIII are formed. Ice II contains nearly planar hexagonal rings of hydrogen-bond O atoms arranged in columns. The O···O···O

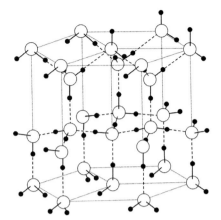

Fig. 16.3.2.
Structure of ice I_h. (Large circle represents O atom, small black circle represents H atom, and broken line represents a disordered hydrogen bond, for which only one of the two arrangements, O–H···O and O···H–O, is shown).

angles are greatly distorted from tetrahedral, with some close nonbonded O···O separation of 324 pm.

Ice III is composed mainly from hydrogen-bonded pentagons with some heptagons, but with no hexagonal rings. Ice IX has the same structure as ice III, except that the hydrogen atoms are ordered in IX and disordered in III. The transformation III → IX occurs between 208 and 165 K.

Ice IV, which is metastable, is even more distorted with O···O···O angles between 88° and 128° and a large number of non-hydrogen-bonded O···O contacts between 314 and 329 pm. Some oxygen atoms lie at the centers of hexagonal rings. Ice V contains quadrilaterals, pentagons, and hexagons, in which respect it resembles some of the clathrate hydrate host frameworks.

It is of interest to note that ice VI is the first of the ice structures composed of two independent interpenetrating hydrogen-bonded frameworks (e.g., it is a self-clathrate).

Ice VII and VIII are constructed from much more regular interpenetrating frameworks, each with the I_c structure with little distortion from tetrahedral bonding configuration. In ice VII, each oxygen has eight nearest neighbors, four hydrogen-bonded at O···O = 288 pm, and four nonbonded at a shorter distance of 274 pm, demonstrating that it is easier to compress a van der Waals O···O distance than a O–H···O distance. Until the formation of interpenetrating frameworks, the principal effect of increasing pressure is to produce more complex hydrogen-bond patterns with large departures from tetrahedral coordination around the oxygen atoms; with the interpenetrating hydrogen bond frameworks, the patterns become much simpler.

A very-high-pressure form, ice X, which was predicated to have a H atom located midway between every pair of hydrogen-bonded O atoms, was identified by infrared spectroscopy with a transition at 44 GPa.

In addition to the various forms of ice, there are a large number of ice-like structures with four-coordinated water molecules that constitute a host framework, which are only stable when voids in the host framework are occupied by other guest molecules. Such compounds are known as clathrate hydrates.

16.3.3 Structural model of liquid water

Water is a unique substance in view of its unusual physical and chemical properties:

(a) The density of liquid water reaches a maximum at 4 °C.
(b) Water has a high dielectric constant associated with the distortion or breaking of hydrogen bonds.
(c) Water has relatively high electrical conductivity associated with the transfer of H_3O^+ and OH^- ions through the hydrogen-bonded structure.
(d) Water can be super-cooled and its fluidity increases under pressure.
(e) Water is an essential chemical for all life processes.

Many models for the structure of liquid water have been proposed. One of the most useful is the polyhedral model.

In liquid water, the thermal motions of molecules are perpetual, and the relative positions of the molecules are changing all the time. Although the structure of liquid water has no definite pattern, the hydrogen bonds between molecules still exist in large numbers. Thus liquid water is a dynamic system in which the H_2O molecules self-assemble in perfect, imperfect, isolated, linked and fused polyhedra (Fig. 16.3.3), among which the pentagonal dodecahedron takes precedence.

The ideal process of the melting of ice may be visualized as a partial (∼15%) breakage of hydrogen bonds to form "ice fragments," which are then converted into a polyhedral system, as indicated in Fig. 16.3.4.

The polyhedral model is useful for understanding the properties of liquid water.

(1) Hydrogen bonds in liquid water

The geometry of a pentagonal dodecahedron is well suited to the formation of hydrogen bonds between molecules. The interior angle of a five-membered ring

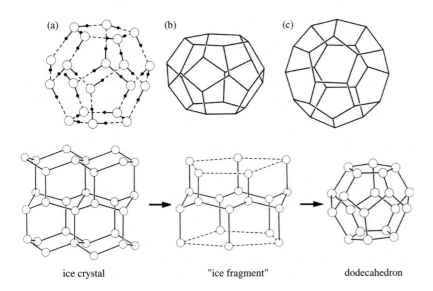

Fig. 16.3.3.
Structures of some polyhedra for liquid water: (a) pentagonal dodecahedron (5^{12}), (b) 14-hedron ($5^{12}6^2$), and (c) 15-hedron ($5^{12}6^3$).

Fig. 16.3.4.
Schematic diagram for melting of ice to form water.

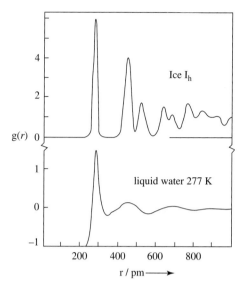

Fig. 16.3.5.
The "$g(r)$ vs r" diagrams of ice I_h (top) and liquid water at 277 K (bottom). [Here $g(r)$ represents the oxygen atom-pair correlation function and r represents the distance of an oxygen atom-pair; taken from A. H. Narten, C. A. Venkatesh and S. A. Rice, *J. Chem Phys.* **64**, 1106 (1976).]

(108°) is close to the H–O–H bond angle (104.5°) of the water molecule and its tetrahedral charge distribution.

(2) Density of water

When "ice fragments" convert to the dodecahedral form, the packing of molecules becomes compact, so liquid water is denser than solid ice. In other circumstances, the density of water decreases when the temperature is raised. Water has its highest density at 4 °C.

(3) X-ray diffraction of liquid water

The "$g(r)$ vs r" diagrams from X-ray diffraction of ice and liquid water are shown in Fig. 16.3.5, in which $g(r)$ represents the oxygen atom-pair correlation function and r represents the distance between an oxygen atom-pair. The diagram of liquid water shows that the strongest peak occurs at $r = 280$ pm, which corresponds to the distance of the hydrogen bond in the dodecahedral system. The next peak is at 450 pm, and the others lie in the range of 640 to 780 pm. There is no peak in the ranges of 280–450 pm and 450–640 pm. In the pentagonal dodecahedron the distances between a vertex and its neighbors are 1 (length of the edge, which is taken as the unit length), τ, $(2)^{1/2}\tau$, τ^2, $(3)^{1/2}\tau$, with $\tau = 1.618$. The calculated distances of oxygen atom-pairs are 280, 450, 640, 730, and 780 pm, which fit the X-ray diffraction data of liquid water.

The diagram of ice I_h has a distinct peak at $r = 520$ pm, which does not occur in liquid water. At 77 K, the "$g(r)$ vs r" diagrams of amorphous water solid and liquid water are very similar to each other.

(4) Thermal properties

The relations between the thermal properties of ice, liquid water, and steam are indicated below:

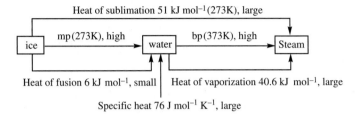

These properties show that in liquid water there still exist a large number of intermolecular hydrogen bonds, thus lending support to the polyhedral model.

(5) Formation of clathrate hydrates

Many gases, such as Ar, Kr, Xe, N_2, O_2, Cl_2, CH_4 and CO, can be crystallized with water to form ice-like clathrate hydrates. The basic structural components of these hydrates are the $(H_2O)_{20}$ pentagonal dodecahedron and other larger polyhedra bounded by five- and six-membered hydrogen-bonded rings, which can accommodate the small neutral molecules. The inclusion properties of water imply that such polyhedra are likely to be present in liquid water as its structural components.

Recently there have been extensive studies on the structure and property of water. In December, 2004, upon assessing the major achievements of the past twelve months, *Science* declared that the study of water as one of the Breakthroughs of the Year. (Specifically, it was listed as No. 8.)

In the structural study of gaseous protonated water clusters, $H^+(H_2O)_n$, with $n = 6 - 27$, it was found that the mass peak with $n = 21$ is particularly intense, signifying a "magic number" in the spectrum. Furthermore, infrared spectroscopic studies revealed that this cluster resembles the pentagonal dodecahedral cage that encapsulates a methane molecule in a Type I clathrate hydrate. In the $H^+(H_2O)_{21}$ cluster, the hydronium ion H_3O^+ can either occupy the (methane) position inside the clathrate cage or take up a site on the cage's surface, thereby pushing a neutral water molecule to the center of the cage.

Another ongoing debate is concerned with the structure of liquid water. In the conventional century-old picture, liquid water is simply an extended network where each water molecule is linked (or hydrogen-bonded) to four others in a tetrahedral pattern. Contrary to this, there are recent synchrotron X-ray results which suggest that many water molecules are linked to only two neighbors. Yet there are also more-recent X-ray data which support the "original" structure, and so this controversy is by no means resolved. Water clearly plays indispensable roles in essentially all fields of science, most notably in chemistry and the life sciences. These roles will be better understood once the structure of liquid water is unequivocally determined.

16.3.4 Protonated water species, H_3O^+ and $H_5O_2^+$

The oxonium (or hydronium) ion H_3O^+ is a stable entity with a pyramidal structure, in which the O–H bond has the same length as that in ice, and the apical angle lies between 110° and 115°. Figure 16.3.6(a) shows the structure of H_3O^+.

The $H_5O_2^+$ ion is a well-documented chemical entity in many crystal structures. Some relevant data are listed in Table 16.3.2. The common characteristic of these compounds is a very short O–H···O hydrogen bond (240–5 pm). Figure 16.3.6(b) shows the structure of $H_5O_2^+$ in the crystalline dihydrate of hydrochloric acid $H_5O_2^+ \cdot Cl^-$.

Besides H_3O^+ and $H_5O_2^+$, several other hydronium ion complexes have been proposed, for example $H_7O_3^+$, $H_9O_4^+$, $H_{13}O_6^+$ and $H_{14}O_6^{2+}$. If 245 pm is taken as the upper limit for the O–H···O distance within a hydronium ion, longer O–H···O distances represent hydrogen bonds between a hydronium ion and neighboring water molecules. According to this criterion, reasonable and unambiguous structural formulas can be assigned to many acid hydrates, as listed in Table 16.3.3.

(a) (b)

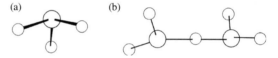

Fig. 16.3.6.
Structure of (a) H_3O^+ and (b) $H_5O_2^+$.

Table 16.3.2. Some compounds containing the $H_5O_2^+$ ion

Compound	Diffraction method	O···O (pm)	Geometry of hydrogen bond
$(H_5O_2)Cl \cdot H_2O$	X-ray	243.4	
$(H_5O_2)ClO_4$	X-ray	242.4	
$(H_5O_2)Br$	Neutron	240	117 122 O—H—O 174.7 °
$(H_5O_2)_2SO_4 \cdot 2H_2O$	X-ray	243	
(H_5O_2)[*trans*-$Co(en)_2Cl_2]_2$	Neutron	243.1	
$(H_5O_2)[C_6H_2(NO_2)_3SO_3] \cdot 2H_2O$	Neutron	243.6	Centered
$(H_5O_2)[C_6H_4(COOH)SO_3] \cdot H_2O$	X-ray, neutron	241.4	112.8 131.0 O—H—O 175° 120.1 121.9 O—H—O
$(H_5O_2)[PW_{12}O_{40}]$	X-ray, neutron	241.4	Centered
$(H_5O_2)[Mn(H_2O)_2(SO_4)_2]$	X-ray	242.6	Centered

Table 16.3.3. Structural formulas for some acid hydrates

Acid hydrate	Empirical formula	Structural formula
$HNO_3 \cdot 3H_2O$	$(H_7O_3)NO_3$	$(H_3O)NO_3 \cdot 2H_2O$
$HClO_4 \cdot 3H_2O$	$(H_7O_3)ClO_4$	$(H_3O)ClO_4 \cdot 2H_2O$
$HCl \cdot 3H_2O$	$(H_7O_3)Cl$	$(H_5O_2)Cl \cdot H_2O$
$HSbCl_6 \cdot 3H_2O$	$(H_7O_3)SbCl_6$	$(H_5O_2)SbCl_6 \cdot H_2O$
$HCl \cdot 6H_2O$	$(H_9O_4)Cl \cdot 2H_2O$	$(H_3O)Cl \cdot 5H_2O$
$CF_3SO_3H \cdot 4H_2O$	$(H_9O_4)CF_3SO_3$	$(H_3O)(CF_3SO_3) \cdot 3H_2O$
$2[HBr \cdot 4H_2O]$	$(H_9O_4)(H_7O_3)Br_2 \cdot H_2O$	$2[(H_3O)Br \cdot 3H_2O]$
$[(C_9H_{18})_3(NH)_2Cl] \cdot Cl \cdot HCl \cdot 6H_2O$	$[(C_9H_{18})_3(NH)_2Cl] \cdot H_{13}O_6 \cdot Cl_2$	$[(C_9H_{18})_3(NH)_2Cl \cdot (H_5O_2) \cdot Cl_2 \cdot 4H_2O]$
$HSbCl_6 \cdot 3H_2O$	$(H_{14}O_6)_{0.5}(SbCl_6)$	$(H_5O_2)(SbCl_6) \cdot H_2O$

16.4 Allotropes of sulfur and polyatomic sulfur species

16.4.1 Allotropes of sulfur

The allotropy of sulfur is far more extensive and complex than that of any other element. This arises from the following factors.

(1) A great variety of molecules can be generated by –S–S– catenation. This becomes evident by comparing the experimental bond enthalpies of O_2 and S_2, which are 498 and 431 kJ mol^{-1}, respectively. On the other hand, the S–S single-bond enthalpy, 263 kJ mol^{-1}, is much greater than the value of 142 kJ mol^{-1} for the O–O single bond. In view of the difference in atomic size, the repulsion energies of lone pairs on two O atoms are larger than that on two S atoms. Thus elemental sulfur readily forms rings or chains with S–S single bonds, whereas oxygen exists primarily as a diatomic molecule.

(2) The S–S bonds are very variable and flexible, generating large S_n molecules or extended structures, and hence the relevant compounds are solids at room temperature. The interatomic distances cover the range of 198 to 218 pm, depending mainly on the extent of multiple bonding. Bond angles S–S–S are in the range 101° to 111° and the dihedral angles S–S–S–S vary from 74° to 100°.

(3) The S_n molecules can be packed in numerous ways in the crystal lattice.

All allotropic forms of sulfur that occur at room temperature consist of S_n rings, with $n = 6$ to 20 (Table 16.4.1). Many homocyclic sulfur allotropes have been characterized by X-ray crystallography: S_6, S_7 (two modifications), S_8 (three modifications), S_9, S_{10}, $S_6 \cdot S_{10}$, S_{11}, S_{12}, S_{13}, S_{14}, S_{18} (two forms), S_{20}, and polymer chain S_∞. Their structural data are listed in Table 16.4.2, and the molecular structures are shown in Fig. 16.4.1.

The most common allotrope of sulfur is orthorhombic α-S_8. At about 95.3°C, α-S_8 transforms to monoclinic β-S_8, such that the packing of S_8 molecules is altered and their orientation becomes partly disordered. This leads to a lower density of 1.94 to 2.01 g cm^{-3}, but the dimensions of S_8 rings in the two allotropes are very similar. Monoclinic γ-S_8 also comprises cyclo-S_8 molecules, but the packing is more efficient and leads to a higher density of 2.19 g cm^{-3}. It reverts slowly to α-S_8 at room temperature, but rapid heating gives a melting point of 106.8°C.

Structural Chemistry of Selected Elements

Table 16.4.1. Some properties of sulfur allotropes

Allotrope	Color	Density (g cm)$^{-3}$	mp or dp (°C)
S_2(g)	Blue-violet	—	Very stable at high temperature
S_3(g)	Cherry-red	—	Stable at high temperature
S_6	Orange-red	2.209	dp > 50
S_7	Yellow	2.182(−110° C)	dp > 39
S_8 (α)	Yellow	2.069	112.8
S_8 (β)	Yellow	1.94–2.01	119.6
S_8 (γ)	Light-yellow	2.19	106.8
S_9	Intense yellow	—	Stable below room temp.
S_{10}	Pale yellow green	2.103 (−110° C)	dp > 0
S_{11}	—	—	—
S_{12}	Pale yellow	2.036	148
S_{14}	Yellow	—	113
S_{18}	Lemon yellow	2.090	mp 128 (dec)
S_{20}	Pale yellow	2.016	mp 124 (dec)
S_x	Yellow	2.01	104 (dec)

mp = melting point; dp = decomposition temperature; dec = with decomposition

Allotrope S_7 is known in four crystalline modifications, one of which (δ form) is obtained by crystallization from CS_2 at −78°C. It features one long bond (bond S^6–S^7 in Fig. 16.4.1) of 218.1 pm, which probably arises from the virtually coplanar set of atoms S^4–S^6–S^7–S^5, which leads to maximum repulsion between nonbonding lone pairs on adjacent S atoms. As a result of this weakening of the S^6–S^7 bond, the adjacent S^4–S^6 and S^5–S^7 bonds are strengthened (199.5 pm), and there are further alternations of bond lengths (210.2 and 205.2 pm) throughout the molecule.

Table 16.4.2. Structural data of S_n molecules in the sulfur allotropes

Molecule	Bond length (pm)	Bond angle (°)	Torsion angle (°)
S_2 (matrix at 20 K)	188.9	—	—
S_6	206.8	102.6	73.8
γ–S_7	199.8 – 217.5	101.9 – 107.4	0.4 – 108.8
δ–S_7	199.5 – 218.2	101.5 – 107.5	0.3 – 108.0
α–S_8	204.6 – 205.2	107.3 – 109.0	98.5
β–S_8	204.7 – 205.7	105.8 – 108.3	96.4 – 101.3
γ–S_8	202.3 – 206.0	106.8 – 108.5	97.9 – 100.1
α–S_9	203.2 – 206.9	103.7 – 109.7	59.7 – 115.6
S_{10}	203.3 – 207.8	103.3 – 110.2	75.4 – 123.7
S_{11}	203.2 – 211.0	103.3 – 108.6	69.3 – 140.5
S_{12}	204.8 – 205.7	105.4 – 107.4	86.0 – 89.4
S_{13}	197.8 – 211.3	102.8 – 111.1	29.5 – 116.3
S_{14}	204.7 – 206.1	104.0 – 109.3	72.5 – 101.7
α–S_{18}	204.4 – 206.7	103.8 – 108.3	79.5 – 89.0
β–S_{18}	205.3 – 210.3	104.2 – 109.3	66.5 – 87.8
S_{20}	202.3 – 210.4	104.6 – 107.7	66.3 – 89.9
catena-S_x	206.6	106.0	85.3

cyclo-S_9 crystallizes in two allotropic forms: α and β. α-S_9 belongs to space group $P2_1/n$ with two independent molecules of similar structure in the unit

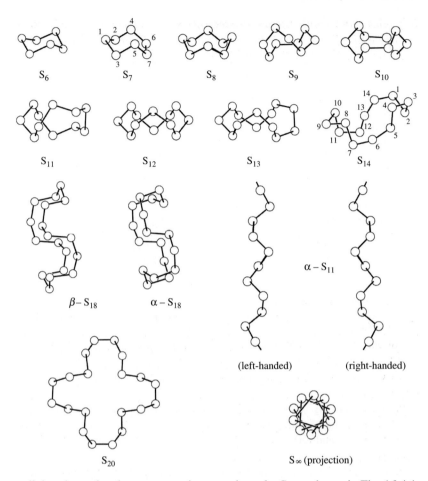

Fig. 16.4.1.
Structures of S_n molecules.

cell, but the molecular symmetry is approximately C_2, as shown in Fig. 16.4.1. The bond lengths lie between 203.2 and 206.9 pm. The bond angles vary in the range of 103.7° to 109.7°, and torsion angles in the range of 59.7 to 115.6°. The conformation of a sulfur homocycle is best described by its "motif", i.e., the order of the signs of the torsion angles around the ring. The motif of α-S_9 is $++--++-+-$.

A new allotrope of sulfur, *cyclo*-S_{14}, has been isolated as yellow rod-like crystals (mp 113°C) from the reaction of [(tmeda)ZnS$_6$] (tmeda = N,N,N′,N′-tetra-methylethylenediamine) with S$_8$Cl$_2$. The bond distances vary from 204.7 to 206.1 pm, the bond angles from 104.0° to 109.3°, and the torsion angles in the range 72.5° to 101.7°. The motif of *cyclo*-S_{14} is $++--++--++--+-$, in which the first twelve signs are the same as those of *cyclo*-S_{12}, as shown in Fig. 16.4.1. Hence the structure of S_{14} can be obtained by opening a bond in S_{12} and inserting an S_2 fragment (i.e., atoms S^9 and S^{10} in the figure).

Solid polycatena sulfur comes in many forms as described by their names: rubbery S, plastic S, lamina S, fibrous S, polymeric S, and insoluble S which is also termed S_μ or S_ω. Fibrous S consists of parallel helical chains of sulfur atoms whose axes are arranged on a hexagonal close-packed net 463 pm apart. The structure contains both left- and right-handed helices of radius 95 pm

(Fig. 16.4.1, S_∞) with a repeat distance of 1380 pm comprising 10 S atoms in three turns. Within each helix the metric parameters are bond distance S–S 206.6 pm, bond angle S–S–S 106.0°, and dihedral angle S–S–S–S 85.3°

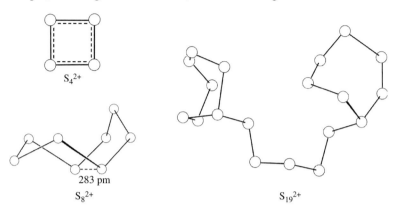

Fig. 16.4.2.
Structures of polyatomic sulfur cations.

16.4.2 Polyatomic sulfur ions

(1) Cations

Sulfur may be oxidized by different oxidizing agents to yield a variety of polyatomic cations. The currently known species that have been characterized by X-ray diffraction are S_4^{2+}, S_8^{2+}, and S_{19}^{2+}, the structures of which are shown in Fig. 16.4.2.

In crystalline $(S_4^{2+})(S_7I^+)_4(AsF_6^-)_6$, the cation S_4^{2+} takes the form of a square-planar ring of edge 198 pm. As this is significantly shorter than a typical S–S single bond (206 pm), the S–S bonds in S_4^{2+} have some double-bond character.

In the salt $(S_8^{2+})(AsF_6^-)_2$, the eight-membered ring of S_8^{2+} has an *exo-endo* conformation and a rather long transannular bond of 283 pm. This structure may be regarded as being halfway between those of the cage molecule S_4N_4 and the crown-shaped ring of S_8. Since S_4N_4 has two electrons less than S_8^{2+} and is isoelectronic with the unknown S_8^{4+}, it is conceivable that, as two electrons are removed from S_8, one end folds up to generate a transannular bond. Then, with the removal of two more electrons, the other end also folds up and another bond is formed to give the S_4N_4 structure, as shown in Fig. 16.4.3.

In crystalline $(S_{19}^{2+})(AsF_6^-)_2$, the S_{19}^{2+} cation consists of two seven-membered rings joined by a five-atom chain. One of the rings has a boat conformation while the other is disordered, existing as a 4:1 mixture of chair and boat conformations. The S–S distances vary greatly from 187 to 239 pm, and S–S–S angles from 91.9° to 127.6°.

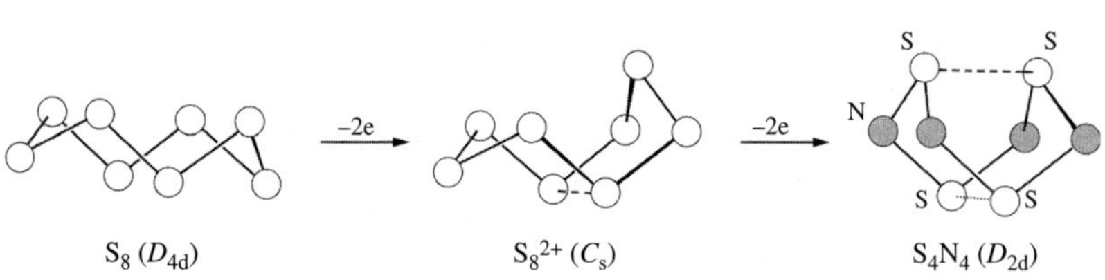

Fig. 16.4.3.
Relationship between the structure of S_8, S_8^{2+} and S_4N_4.

$S_8\ (D_{4d})$ →(−2e)→ $S_8^{2+}\ (C_s)$ →(−2e)→ $S_4N_4\ (D_{2d})$

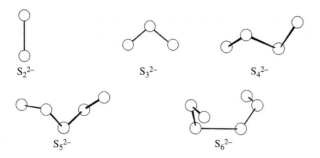

Fig. 16.4.4. Structures of S_n^{2-}.

(2) Anions

The majority of polysulfide anions have acyclic structures. The configurations of S_n^{2-} species are in accord with their bond valence (b):

$$b = \frac{1}{2}(8n - g)$$
$$= \frac{1}{2}[8n - (6n + 2)]$$
$$= n - 1.$$

Figure 16.4.4 shows the puckered chain motifs of some S_n^{2-} species, and their structural data are listed in Table 16.4.3.

Table 16.4.3. Structural data of S_n^{2-}

S_n^{2-}	Compound	Bond length (pm)	Bond angle (central)	Symmetry
S_2^{2-}	pyrites	208 – 215	—	$D_{\infty h}$
S_3^{2-}	BaS$_3$	207.6	144.9°	C_{2v}
S_4^{2-}	Na$_2$S$_4$	206.1 – 207.4	109.8°	C_2
S_5^{2-}	K$_2$S$_5$	203.7 – 207.4	106.4°	C_2
S_6^{2-}	Cs$_2$S$_6$	201 – 211	108.8°	C_2

16.5 Sulfide anions as ligands in metal complexes

Sulfur has an extensive coordination chemistry involving the S^{2-} and S_n^{2-} anions, which exhibit an extremely versatile variety of coordination modes.

16.5.1 Monosulfide S^{2-}

Ligands in which S acts as a donor atom are usually classified as soft Lewis bases, in contrast to oxygen donor-atom ligands which tend to be hard Lewis bases. The large size of the S atom and its easily deformed electron cloud account for this difference, and the participation of metal $d\pi$ orbitals in bonding to sulfur is well established. The monosulfide dianion S^{2-} can act either as a terminal or a bridging ligand. As a terminal ligand, S^{2-} donates a pair of electron to its

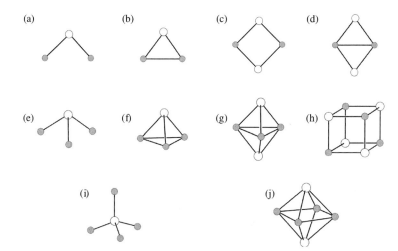

Fig. 16.5.1.
Coordination modes of S^{2-} (open circle): (a) $(\mu_2\text{-S})[\text{Au(PEt}_3)]_2$, (b) $(\mu_2\text{-S})[\text{Pt(PPh}_3)_2]_2$, (c) $(\mu_2\text{-S})_2\text{Mo(S)}_2\text{Fe(SPh)}_2]^{2-}$, (d) $(\mu_2\text{-S})_2\text{Fe}_2(\text{NO})_4]^{2-}$, (e) $(\mu_3\text{-S})[\text{Au(PPh}_3)]_3$, (f) $(\mu_3\text{-S})[\text{Co(CO)}_3]_3$, (g) $(\mu_3\text{-S})_2(\text{CoCp})_3$, (h) $(\mu_3\text{-S})_4[\text{Fe(NO)}]_4$, (i) $(\mu_4\text{-S})[\text{Zn}_4(\text{S}_2\text{AsMe}_2)_6]$, and (j) $(\mu_4\text{-S})_2\text{Co}_4(\text{CO})_{10}$.

bonded atom. In the μ_2 bridging mode S is usually regarded as a two-electron donor.

Figures 16.5.1(a) to Fig. 16.5.1(d) shows four μ_2 bridging modes. In the μ_3 triply bridging mode S can be regarded as a four-electron donor, using both its two unpaired electrons and one lone pair. The pseudo-cubane structure adopted by some of the μ_3-S compounds is a crucial structural unit in many biologically important systems, e.g., the [(RS)MS]$_4$ (M = Mo, Fe) units which cross-link the polypeptide chains in nitrogenase and ferredoxins. Figures 16.5.1(e) to Fig. 16.5.1(h) shows four μ_3 bridging modes. In the μ_4 bridging mode S can be regarded as a four- or six-electron donor, depending on the geometry of the mode. Figures 16.5.1(i) to Fig. 16.5.1(j) show two μ_4 bridging modes. As yet there is no molecular compound in which S bridges six or eight metal atoms, but interstitial sulfur is well known.

16.5.2 Disulfide S_2^{2-}

The coordination modes of disulfide S_2^{2-} in some representative compounds are listed in Table 16.5.1. Their molecular structures are shown in Fig. 16.5.2.

16.5.3 Polysulfides S_n^{2-}

The S_n^{2-} ($n = 3 - 9$) anions generally have a chain structure. The average S–S bond length is smaller in S_n^{2-} ions ($n > 2$) than in S_2^{2-}, and the length of the S–S terminal bond decreases from S_3^{2-} (215 pm) to S_7^{2-} (199.2 pm). These data indicate that the negative charge (filling a π^* antibonding MO) is delocalized over the entire chain, but the delocalization along the chain is less in higher polysulfides. These considerations are important in comparing the S–S bond lengths of the free ions with those of their metal complexes. Fig. 16.5.3 shows the structures of some polysulfide coordination compounds.

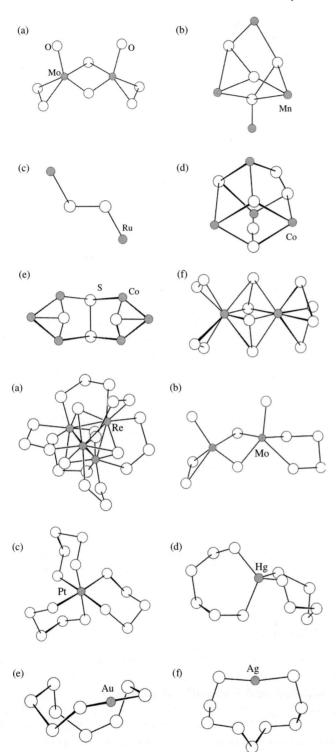

Fig. 16.5.2. Structure of S_2^{2-} coordination compounds: (a) $[Mo_2O_2S_2(S_2)_2]^{2-}$, (b) $Mn_4(CO)_{15}(S_2)_2$, (c) $[Ru_2(NH_3)_{10}S_2]^{4+}$, (d) $Co_4Cp_4(\mu_3\text{-}S)_2(\mu_3\text{-}S_2)_2$, (e) $[S(Co_3(CO)_7]_2S_2$, and (f) $[Mo_2(S_2)_6]^{2-}$.

Fig. 16.5.3. Structures of some S_n^{2-} coordination compounds: (a) $[Re_4(\mu_3\text{-}S)_4(S_3)_6]^{4-}$, (b) $[Mo_2(S)_2(\mu\text{-}S)_2(\eta^2\text{-}S_2)(\eta^2\text{-}S_4)]^{2-}$, (c) $[Pt(\eta^2\text{-}S_5)_3]^{2-}$, (d) $[Hg(\eta^2\text{-}S_6)_2]^{2-}$, (e) $[Au(\eta^2\text{-}S_9)]^-$, and (f) $[Ag(\eta^2\text{-}S_9)]^-$.

Table 16.5.1. Types of disulfide coordination compounds

Type		Example	d(S–S) (pm)	Structure in Fig. 16.5.2
Ia	M(η²-S₂) (side-on)	[Mo₂O₂S₂(S₂)₂]²⁻	208	(a)
Ib	μ-η²:η¹ bridging (two M on one S, side-on)	[Mo₄(NO)₄(S₂)₅(S)₃]⁴⁻	204.8	—
Ic	μ-η²:η¹:η¹ (M on each S plus side-on)	Mn₄(CO)₁₅(S₂)₂	207	(b)
Id	μ₄ type with three M on S₂ unit	Mo₄(CO)₁₅(S₂)₂	209	(b)
IIa	μ-η¹:η¹ bridging (end-on, one M each S)	[Ru₂(NH₃)₁₀S₂]⁴⁺	201.4	(c)
IIb	μ₃ type (two M on one S, one M on other)	Co₄Cp₄(μ₃-S)₂(μ₃-S₂)₂	201.3	(d)
IIc	μ₄ type (two M on each S)	[SCo₃(CO)₇]₂S₂	204.2	(e)
III	S₂ bridging two M forming rhombus	[Mo₂(S₂)₆]²⁻	204.3	(f)

16.6 Oxides and oxoacids of sulfur

16.6.1 Oxides of sulfur

More than ten oxides of sulfur are known, and the most important industrially are SO_2 and SO_3. The six homocyclic polysulfur monoxides S_nO ($5 \leq n \leq 10$) are prepared by oxidizing the appropriate cyclo-S_n with trifluoroperoxoacetic

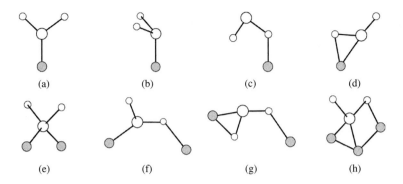

Fig. 16.6.1.
Coordination modes of SO$_2$ to metal atoms (shaded circles).

acid, CF$_3$C(O)OOH, at −30°C. The dioxides S$_7$O$_2$ and S$_6$O$_2$ are also known. In addition, there are the thermally unstable acylclic oxides S$_2$O, S$_2$O$_2$, SO and the elusive species SOO and SO$_4$. The S$_2$O molecule has a bent structure with C_s symmetry, and its structural parameters are S=O bond length 145.6 pm, S=S 188.4 pm, and bond angle S–S–O is 117.9°.

(1) Sulfur dioxide, SO$_2$

When sulfide ores (such as pyrite), sulfur-rich organic compounds, and fossil fuels (such as coal) are burned in air, the sulfur therein is mostly converted to SO$_2$.

Sulfur dioxide is a colorless, toxic gas with a choking odor. Gaseous SO$_2$ neither burns nor supports combustion. It is readily soluble in water (3927 cm^3 SO$_2$ in 100 g H$_2$O at 20°C) and forms "sulfurous acid."

Sulfur dioxide is a component of air pollutants, and is capable of causing severe damage to human and other animal lungs, particularly in the sulfate form. It is also an important precursor to acid rain. The SO$_2$ molecule survives for a few days in the atmosphere before it is oxidized to SO$_3$. The direct reaction

$$2SO_2(g) + O_2(g) \rightarrow 2SO_3(g)$$

is very slow, but it speeds up markedly in the presence of metal cations such as Fe^{3+}.

Sulfur dioxide is a bent molecule with C_{2v} symmetry; the bond angle is 119°, and the bond length is 143.1 pm, which is shorter than the expected value of a single S–O (176 pm) or S=O double bond (154 pm) calculated from the sum of covalent radii. The S–O bond energy in SO$_2$ is 548 kJ mol^{-1}, which is larger than that of the unstable molecule SO (524 kJ mol^{-1}). These data indicate that the S–O bond order in SO$_2$ exceeds 2, a consequence of the participation of the 3d orbitals of S in bonding. Sulfur dioxide is homologous with O$_3$. In the bent molecule O$_3$ the O–O bond length is 127.8 pm, which is longer than that in O$_2$ (120.7 pm).

There are lone pairs in the S and O atoms of SO$_2$ molecule, which as a ligand forms many coordination modes with metal atoms, as shown in Fig. 16.6.1.

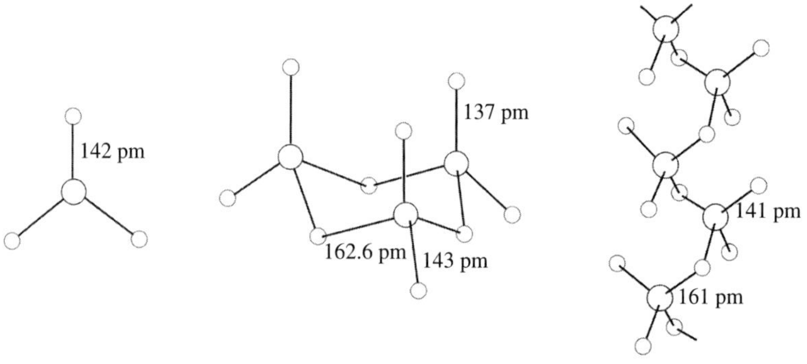

Fig. 16.6.2.
Structures of SO_3: (a) SO_3(g), (b) γ–SO_3, and (c) β–SO_3.

(a) 142 pm

(b) 137 pm, 162.6 pm, 143 pm

(c) 141 pm, 161 pm

(2) Sulfur trioxide, SO_3

Sulfur trioxide is made on a huge scale by the catalytic oxidation of SO_2. It is usually not isolated but is immediately converted to H_2SO_4.

In the gas phase, monomeric SO_3 has a D_{3h} planar structure with bond length S–O 142 pm, as shown in Fig. 16.6.2(a). The cyclic trimer $(SO_3)_3$ occurs in colorless orthorhombic γ-SO_3 (mp 16.9°C), and its structure is shown in Fig. 16.6.2(b). The helical chain structure of β-SO_3 is shown in Fig. 16.6.2(c). A third and still more stable form, α-SO_3 (mp 62°C), involves cross-linkage between the chains to give a complex layer structure. The standard enthalpies of formation of the four forms of SO_3 at 298 K are listed below:

	SO_3(g)	α-SO_3	β-SO_3	γ-SO_3
ΔH_f^0 / kJ mol^{-1}	−395.2	−462.4	−449.6	−447.4

Sulfur trioxide reacts vigorously and extremely exothermically with water to give H_2SO_4. Anhydrous H_2SO_4 has an unusually high dielectric constant, and also a very high electrical conductivity which results from the ionic self-dissociation of the compound coupled with a proton-switch mechanism for the rapid conduction of current through the viscous hydrogen-bonded liquid:

$$2H_2SO_4 \rightleftharpoons H_3SO_4^- + HSO_4^-.$$

Photolysis of SO_3 and O_3 mixtures yields monomeric SO_4, which can be isolated by inert-gas matrix techniques at low temperatures (15 to 78 K). Vibration spectroscopy indicates a sulfuryl group together with either an open SOO branch (C_s structure) or a closed three-membered peroxo ring (C_{2v} structure), the latter being preferred on the basis of the infrared absorption bands. The structures of these two species are illustrated pictorially below.

C_s (planar), or C_1 (non-planar) C_{2v}

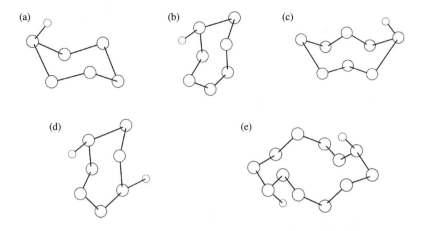

Fig. 16.6.3.
Molecular structure of oxides of cyclic poly-sulfur: (a) S_6O, (b) S_7O, (c) S_8O, (d) S_7O_2, (e) $S_{12}O_2$.

(3) Oxides of cyclic poly-sulfur, S_nO and S_nO_2

When cyclo-S_8, −S_9, and −S_{10} are dissolved in CS_2 and oxidized by freshly prepared $CF_3C(O)O_2H$ at temperatures below −10°C, modest yields (10–20%) of the corresponding crystalline monoxides S_nO are obtained. Of these compounds, S_5O has not yet been isolated, but it has been prepared in solution by the same method. On the other hand, S_6O has two crystalline modifications: orange α-S_6O with mp 39°C (dec) and dark orange β-S_6O with mp 34°C (dec), and its molecular structure is shown in Fig. 16.6.3(a). Also, S_7O is a orange crystal with mp 55°C (dec), and its molecular structure is shown in Fig. 16.6.3(b). Finally, S_8O is a orange-yellow crystal with mp 78°C (dec), and its structure is shown in Fig. 16.6.3(c).

In the S_nO_2 compounds, S_7O_2 is a dark orange crystal which decomposes above room temperature. Its molecular structure is shown in Fig. 16.6.3(d). The structure of $S_{12}O_2$ as determined from the orange adduct $S_{12}O_2 \cdot 2SbCl_5 \cdot 3CS_2$ is shown in Fig. 16.6.3(e).

16.6.2 Oxoacids of sulfur

Sulfur forms many oxoacids, though few of them can be isolated in pure form. Most are prepared in aqueous solution or as crystalline salts of the corresponding oxoacid anions. Table 16.6.1 lists the common oxoacids of sulfur.

(1) Sulfuric acid and disulfuric acid

Sulfuric acid is the most important chemical of all sulfur compounds. Anhydrous sulfuric acid is a dense, viscous liquid which is readily miscible with water in all proportions. Sulfuric acid forms hydrogen sulfate (also known as bisulfate, HSO_4^-) and sulfate (SO_4^{2-}) salts with many metals, which are frequently very stable and are important mineral compounds. Figures 16.6.4(a)–(c) shows the molecular structures of H_2SO_4, HSO_4^-, and SO_4^{2-}.

The crystal structure of sulfuric acid consists of layers of $O_2S(OH)_2$ tetrahedra connected via hydrogen bonds involving the donor OH groups and acceptor O atoms. Figure 16.6.5 shows a layer of hydrogen-bonded H_2SO_4 molecules.

Table 16.6.1. The common oxoacids of sulfur

Formula	Name	Oxidation state of S	Schematic structure	Salt
H_2SO_4	Sulfuric	6		Sulfate SO_4^{2-}
$H_2S_2O_7$	Disulfuric	6		Disulfate $O_3SOSO_3^{2-}$
$H_2S_2O_3$	Thiosulfuric	6, −2		Thiosulfate $S_2O_3^{2-}$
H_2SO_5	Peroxo-monosulfuric	6		Peroxo-monosulfate $OOSO_3^{2-}$
$H_2S_2O_8$	Peroxo-disulfuric	6		Peroxo-disulfate $O_3SOOSO_3^{2-}$
$H_2S_2O_6$	Dithionic*	5		Dithionate $O_3SSO_3^{2-}$
$H_2S_{n+2}O_6$	Polythionic	5, 0		Polythionate $O_3S(S)_nSO_3^{2-}$
H_2SO_3	Sulfurous*	4		Sulfite SO_3^{2-}
$H_2S_2O_5$	Disulfurous*	5, 3		Disulfite $O_3SSO_2^{2-}$
$H_2S_2O_4$	Dithionous*	3		Dithionite O_2SSO_2

* Acids only exist as salts.

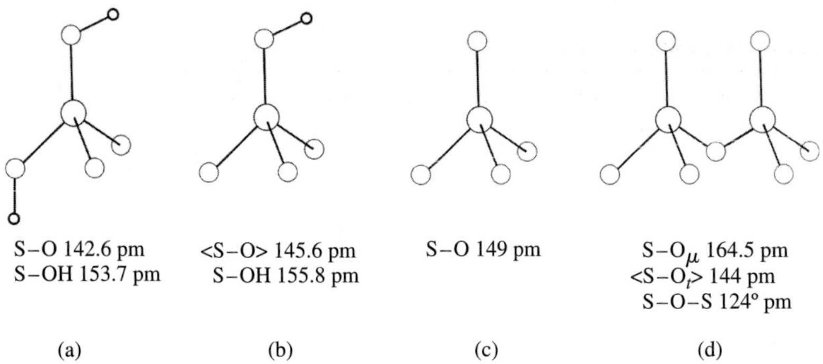

S–O 142.6 pm <S–O> 145.6 pm   S–O 149 pm   S–O$_\mu$ 164.5 pm
S–OH 153.7 pm S–OH 155.8 pm   <S–O$_t$> 144 pm
                                             S–O–S 124° pm

(a)   (b)   (c)   (d)

Fig. 16.6.4.
Molecular structure of (a) H_2SO_4, (b) HSO_4^-, (c) SO_4^{2-} and (d) $S_2O_7^{2-}$.

The O–H$\cdots$O hydrogen bond length is 264.8 pm and the bond angle O–H$\cdots$O is 170°.

The hydrogen sulfate (or bisulfate) anion HSO_4^- exists in crystalline salts such as $(H_3O)(HSO_4)$, $K(HSO_4)$ and $Na(HSO_4)$. The bond lengths of HSO_4^- in $(H_3O)(HSO_4)$ are S–O = 145.6 pm and S–OH = 155.8 pm.

Disulfuric acid (also known as pyrosulfuric acid), $H_2S_2O_7$, which is the major constituent of "fuming sulfuric acid", is formed from sulfur trioxide and sulfuric acid:

$$SO_3 + H_2SO_4 \rightarrow H_2S_2O_7.$$

Figure 16.6.4(d) shows the structure of the $S_2O_7^{2-}$ anion.

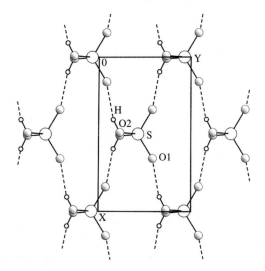

Fig. 16.6.5.
A layer of hydrogen-bonded H_2SO_4 molecules.

(2) Sulfurous acid and disulfurous acid

Sulfurous acid, H_2SO_3, and disulfurous acid, $H_2S_2O_5$, are examples of sulfur oxoacids that do not exist in the free state, although numerous salts derived from them containing the HSO_3^-, SO_3^{2-}, $HS_2O_5^-$, and $S_2O_5^{2-}$ anions are stable

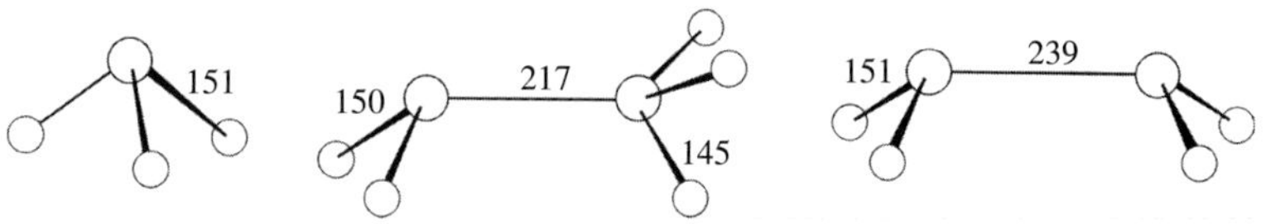

Fig. 16.6.6.
Structure of (a) SO_3^{2-}, (b) $S_2O_5^{2-}$, and (c) $S_2O_4^{2-}$ (bond lengths in pm).

solids. An aqueous solution of SO_2, though acidic, contain negligible quantities of the free acid H_2SO_3. The apparent hexahydrate $H_2SO_3 \cdot 6H_2O$ is actually the gas hydrate $6SO_2 \cdot 46H_2O$, in which the SO_2 molecules are enclosed in cages within a host framework constructed from hydrogen-bonded water molecules. Figure 16.6.6 shows the structures of the anions SO_3^{2-}, $S_2O_5^{2-}$, and $S_2O_4^{2-}$. The hydrogen sulfite (bisulfite) ion HSO_3^- has been found to exist in two isomeric forms: $HO-SO_2^-$ and $H-SO_3^-$.

(3) Thiosulfate, SSO_3^{2-}

In the thiosulfate ion, a terminal S atom replaces an O atom of the sulfate ion. The S–S bond length is 201.3 pm, which indicates essentially single-bond character, while the mean S–O bond length is 146.8 pm, which indicates considerable π bonding between the S and O atoms.

Single-crystal X-ray analysis has shown that the structure of $SeSO_3^{2-}$ is isostructural with the $S_2O_3^{2-}$ ion with a Se–S bond length of 217.5(1) pm.

The thiosulfate ion, in which the terminal S and O atoms can function as ligand sites, is a polyfunctional species in various coordination modes with metal atoms. Figure 16.6.7 shows the coordination modes of $S_2O_3^{2-}$.

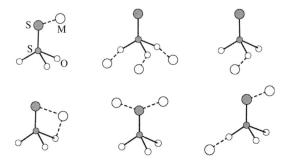

Fig. 16.6.7.
The coordination modes of $S_2O_3^{2-}$.

(4) Peroxoacids of sulfur

The peroxoacids of sulfur and their salts all contain the –O–O– group. The salts of $S_2O_8^{2-}$, such as $K_2S_2O_8$, are very convenient and powerful oxidizing agents. Peroxomonosulfuric acid (Caro's acid), H_2SO_5, is a colorless, explosive solid (mp 45°C), and salts of HSO_5^- are known. In HSO_5^- and $S_2O_8^{2-}$, the S–O (peroxo) and S–O (terminal) bond distances are different. The S–O (peroxo) bond length is about 160 pm, which corresponds to a single bond, and the S–O (terminal) bond length is about 145 pm, which corresponds to a double bond.

The structural formulas of HSO_5^- and $S_2O_8^{2-}$ are shown below:

[Structural formulas of HSO_5^- and $S_2O_8^{2-}$]

16.7 Sulfur–nitrogen compounds

Sulfur and nitrogen are diagonally related elements in the Periodic Table and might therefore be expected to have similar electronic charge densities for similar coordination numbers, and to form cyclic, acyclic, and polycyclic molecules through extensive covalent bonding. Some sulfur–nitrogen compounds exhibit interesting chemical bonding and have unusual properties, as discussed below.

16.7.1 Tetrasulfur tetranitride, S_4N_4

Nitride S_4N_4 is an air-stable compound that can be prepared by passing NH_3 gas into a warm solution of S_2Cl_2 in CCl_4 or benzene. It is a thermochromic crystal: colorless at 83 K, pale yellow at 243 K, orange at room temperature, and deep red above 373 K.

The D_{2d} molecular structure of S_4N_4 is shown in Fig. 16.7.1(a). The atoms of S_4N_4 are arranged so that the electropositive S atoms occupy the vertices of a tetrahedron, while the electronegative N atoms constitute a square that intersects the tetrahedron. All the S–N distances are equal; for gaseous S_4N_4 they are 162.3 pm, which is intermediate between the distances of the S–N single bond (174 pm) and S=N double bond (154 pm). The N–S–N bond angle is 105.3°, S–N–S is 114.2°, and S–S–N is 88.4°. A peculiarity of the S_4N_4 structure is the short distance between the two S atoms connected by a broken line in Fig. 16.7.1(a); at 258 pm, it lies between the S–S single-bond length (208 pm) and van der Waals contacting distance $S \cdots S$ 360 pm. This may imply that the S atom uses two electrons for two σ S–N bonds, one electron for delocalized π bonding, and one electron for $S \cdots S$ weak bonding.

Tetraselenium tetranitride, Se_4N_4, forms red, hygroscopic crystals and is highly explosive. The structure of Se_4N_4 resembles that of S_4N_4 with Se–N bond length of 180 pm and cross-cage $Se \cdots Se$ distance of 276 pm (note that $2r_{cov}(Se) = 2 \times 117 pm = 234 pm$).

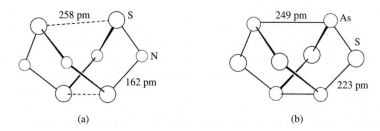

Fig. 16.7.1.
Molecular structure of (a) S_4N_4 and (b) As_4S_4.

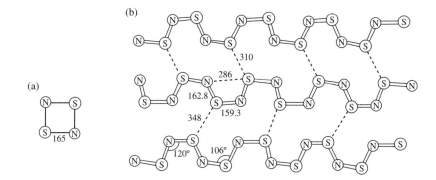

Fig. 16.7.2.
Structure of (a) S$_2$N$_2$ and (b) polymeric chains in one layer of (SN)$_x$ and the important structural parameters (bond lengths in pm).

The homolog realgar, As$_4$S$_4$, has an analogous but different structure with the electronegative S atoms at the vertices of a square and the electropositive As atoms at the vertices of a tetrahedron. The As atoms are linked by normal single bonds, as shown by the solid lines between them in Fig. 16.7.1(b). The As–As distance is 249 pm, which is nearly equal to the calculated value of 244 pm (Table 3.4.3).

16.7.2 S$_2$N$_2$ and (SN)$_x$

When the heated vapor of S$_4$N$_4$ is passed over silver wool at 520 to 570 K, the unstable cyclic dimer S$_2$N$_2$ is obtained. It forms large colorless crystals which are insoluble in water but soluble in many organic solvents.

The molecular structure of S$_2$N$_2$, as shown in Fig. 16.7.2(a), is a D_{2h} square-planar ring with S–N edge 165 pm, somewhat analogous to the isoelectronic cation S$_4^{2+}$ (Fig. 16.4.2). The valence-bond representations of the S$_2$N$_2$ molecule are as follows:

When colorless S$_2$N$_2$ crystals are allowed to stand at room temperature, golden (SN)$_x$ crystals are gradually formed. The (SN)$_x$ chain can conceivably be generated from adjacent square-planar S$_2$N$_2$ molecules, and a free radical mechanism has been proposed. Since polymerization can take place with only minor movements of the atoms, the starting material and product are pseudomorphs without alteration of the crystallinity. Figure 16.7.2 shows the configuration of the (SN)$_x$ chains and the packing of the chains in the crystal.

Polymeric (SN)$_x$ has some unusual properties. For example, it has a bronze color and metallic luster, and its electrical conductivity is about that of mercury metal. Values of the conductivity of (SN)$_x$ depend on the purity and crystallinity of the polymer and on the direction of measurement, being much greater along the fibers than across them. A conjugated single-bond/double-bond system can be formulated, in which every S–N unit has one antibonding π^* electron. The half-filled overlapping π^* orbitals combine to form a half-filled conduction band, in much the same way as the half-filled ns orbitals of alkali metal atoms

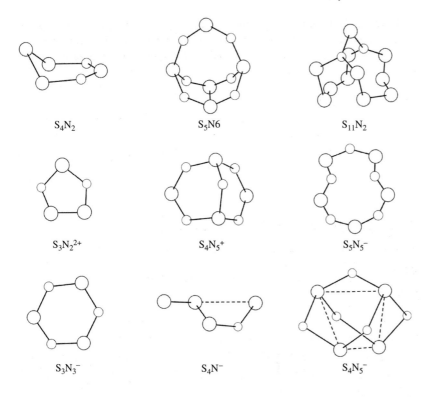

Fig. 16.7.3.
Structure of some cyclic sulfur–nitrogen compounds (large circle represents S atom and small circle represents N atom).

form a conduction band. However, in (SN)$_x$, the conduction band lies only in the direction of the (SN)$_x$ fibers, so the polymer behaves as an "one-dimensional metal".

16.7.3 Cyclic sulfur–nitrogen compounds

Many cyclic sulfur–nitrogen compounds are known, some of which are shown in Fig. 16.7.3. The general structural features of sulfur–nitrogen compounds are formulated as follows:

(a) The S atom has the ability to form various types of S–S and S–N bonds, some of which contain catenated –S–S– chains that can insert into the cyclic chains as a fragment in molecules. For example, the S$_{11}$N$_2$ molecule has two S$_5$ chain fragments, and S$_4$N$_2$ and S$_4$N$^-$ each has one S$_3$ chain fragment. The S$\cdots$S interactions can vary in strength: 314 pm in S$_4$N$^-$ as compared to 271–5 pm in S$_4$N$_5^-$.

(b) The S–N bond distances are in the range of 155 to 165 pm, which are shorter than the calculated single-bond length, so the S–N bonds have some double-bond character. The bond angles vary over a large range: for example, in $S_5N_5^+$ the bond angles are between 138° and 151°. The wide variations of bond distances and angles indicate that the bond types are quite complex.

(c) Normally the S and N atoms are each bonded to two adjacent atoms, but in some cases they are also three-connected to form polycyclic molecules. Some examples are as S_5N_6, $S_{11}N_2$, $S_4N_5^+$ and $S_4N_5^-$, the structures of which are shown in Fig. 16.7.3.

(d) The conformations of the cyclic S–N molecules exhibit diversity. For example, $S_3N_2^{2+}$ and $S_3N_3^-$ adopt planar conformations which are stabilized by delocalized π bonding, while the majority of these cyclic molecules are nonplanar, in which the d orbitals of S atoms also participate in bonding.

16.8 Structural chemistry of selenium and tellurium

16.8.1 Allotropes of selenium and tellurium

Selenium forms several allotropes but tellurium forms only one. The thermodynamically stable form of selenium (α-selenium or gray selenium) and the crystalline form of tellurium are isostructural. In both Te and gray Se, the atoms form infinite, helical chains having three atoms in every turn, the axes of which lie parallel to each other in the crystal, as shown in Fig. 16.8.1.

The distance of two adjacent atoms within the chain are Se–Se 237 pm and Te–Te 283 pm. Each atom has four adjacent atoms from three different chains at an average distance of Se$\cdots$Se 344 pm, Te$\cdots$Te 350 pm. The interchain distance is significantly shorter than expected from the van der Waals separation (380 pm for Se and 412 pm for Te).

Red monoclinic selenium exists in three forms, each containing Se_8 rings with the crown conformation of S_8 (Fig. 16.4.1). Vitreous black selenium, the ordinary commercial form of the element, comprises an extremely complex and irregular structure of large polymeric rings.

16.8.2 Polyatomic cations and anions of selenium and tellurium

Like their sulfur congener, selenium and tellurium can form polyatomic cations and anions in many compounds.

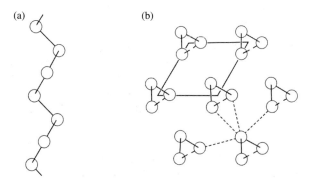

Fig. 16.8.1.
Structure of α-selenium (or tellurium): (a) side view of Se_x (or Te_x) helical chain; (b) viewed along the helices; the hexagonal unit cell and the coordination environment about one atom is indicated.

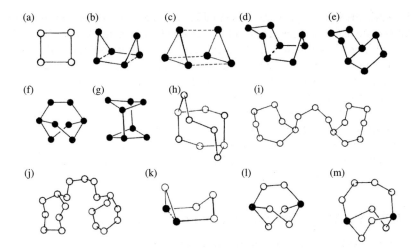

Fig. 16.8.2.
Structures of some polyatomic cations of Se and Te. (a) Se_4^{2+} in $Se_4(H_2S_2O_7)_2$; Te_4^{2+} in $Te_4(AsF_6)_2$; (b) Te_6^{2+} in $Te_6(MOCl_4)_2$ (M = Nb, W); (c) Te_6^{4+} in $Te_6(AsF_6)_4 \cdot 2AsF_3$; (d) Te_8^{2+} in $Te_8(ReCl_6)_2$; Se_8^{2+} in $Se_8(AlCl_4)_2$; (e) Te_8^{2+} in $Te_8(WCl_6)_2$; (f) Te_8^{2+} in $(Te_6)(Te_8)(WCl_6)_4$; (g) Te_8^{4+} in $(Te_8)(VOCl_4)_2$; (h) Se_{10}^{2+} in $Se_{10}(SbF_6)_2$; (i) Se_{17}^{2+} in $Se_{17}(NbCl_6)_2$; (j) Se_{19}^{2+} in $Se_{19}(SbF_6)_2$; (k) $(Te_2Se_4)^{2+}$ in $(Te_2Se_4)(SbF_6)_2$; (l) $(Te_2Se_6)^{2+}$ in $(Te_2Se_6)(Te_2Se_8)(AsF_6)_4$; (m) $(Te_2Se_8)^{2+}$ in $(Te_2Se_8)(AsF_6)_2$.

(1) Polyatomic cations

Figure 16.8.2 show the structures of some polyatomic cations of Se and Te that exist in crystalline salts. Figure 16.8.2(a) shows the square-planar geometry of the Se_4^{2+} and Te_4^{2+} cations. In $Se_4(HS_2O_7)_2$ the Se–Se distance is 228 pm, and in $Te_4(AsF_6)_2$ the Te–Te distance is 266 pm. These two distances are shorter than those in the respective elemental forms, 237 and 284 pm, respectively, being consistent with the effect of some multiple bonding.

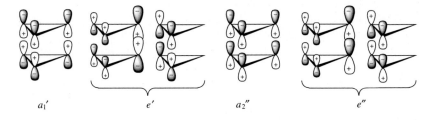

Fig. 16.8.3.
Molecular orbitals in Te_6^{4+}.

The structure of Te_6^{4+} is shown in Fig. 16.8.2(c). In this trigonal prismatic cation, the average Te–Te bond length within a triangular face is 268 pm, and the average Te$\cdots$Te distance between the parallel triangular faces is 313 pm. The Te_6^{4+} cation can be considered as a dimer of two Te_3^{2+} units consolidated by a $\pi^* - \pi^*$ 6c-4e bonding interaction. As shown in Fig. 16.8.3, the tellurium $5p_z$ orbitals give rise to six molecular orbitals of a_1', e', a_2'', and e'' symmetry, the first three being used to accommodate eight valence electrons. The e' orbitals are nonbonding within the individual Te_3^{2+} units but form bonding interaction between them, and the bonding a_1' and antibonding a_2'' orbitals cancel each other. The formal bond order along each prism edge is therefore 2/3.

Formal addition of two electrons to Te_6^{4+} gives Te_6^{2+}, which takes the shape of a boat-shaped six-membered ring, as shown in Fig. 16.8.2(b). The average length of the pair of weak transannular interactions is 329 pm, which indicates that the two positive charges are delocalized over all four Te atoms in the rectangular base.

Figure 16.8.2(d) shows the conformation of and weak central transannular interaction in the Se_8^{2+} and Te_8^{2+} cations, which are isostructural with S_8^{2+}. Their common bicyclic structure can be regarded as being derived from a crown-shaped eight-membered ring by flipping one atom from an *exo* to an *endo* position, with the formally positively charged atoms interacting in transannular linkage. In $Te_8(ReCl_6)_2$, the Te–Te bond length indicates a normal single bond, and the Te$\cdots$Te distance is 315 pm. In $Se_8(AlCl_4)_2$, the Se–Se bond lengths lie in the range 229-36 pm, and Se$\cdots$Se is 284 pm.

Figures 16.8.2(e) and (f) show two other isomeric forms of Te_8^{2+}. In $Te_8(WCl_6)_2$, the Te_8^{2+} cation is composed of two five-membered rings, each taking an envelope conformation, with an average Te–Te bond length of 275 pm and a relatively short transannular Te$\cdots$Te bond of 295 pm. In $(Te_6)(Te_8)(WCl_6)_4$, Te_8^{2+} exhibits a bicyclo[2.2.2]octane geometry with two bridge-head Te atoms.

Figure 16.8.2(g) shows the structure of Te_8^{4+}; this 44 valence electron cluster takes a cube shape with two cleaved edges, the positive charges being located on four three-coordinate Te atoms. The Te_8^{4+} cation can be viewed as two planar Te_4^{2+} ions that have dimerized via the formation of a pair of Te–Te bonds with simultaneous loss of electronic delocalization and distortion from planarity.

Figures 16.8.2(h), (i), and (j) show the structures of Se_{10}^{2+}, Se_{17}^{2+}, and Se_{19}^{2+}, respectively. They all consist of seven- or eight-membered rings connected by short chains. Each homopolyatomic cation has two three-coordinate atoms that formally carry the positive charges. Se_{10}^{2+} has a bicyclo[2.2.4]decane geometry. The Se–Se bond distances vary between 225 and 240 pm, and the Se–Se–Se angles range from 97° to 106°. Se_{17}^{2+} and Se_{19}^{2+} comprise a pair of seven-membered rings connected by a three- and four-atom chain, respectively.

Figures 16.8.2(k), (l), and (m) show the structures of $(Te_2Se_4)^{2+}$, $(Te_2Se_6)^{2+}$ and $(Te_2Se_8)^{2+}$, respectively. In these heteropolyatomic cations, the heavier Te atoms generally have a higher coordination number of three and serve as positive charge bearers, which is consistent with the lower electronegativity of Te compared to Se. As expected and confirmed by experiment, a weak transannular Te$\cdots$Te bond exists in the boat-shaped $(Te_2Se_4)^{2+}$ cation. The $(Te_2Se_6)^{2+}$ and $(Te_2Se_8)^{2+}$ cations have bicyclo[2.2.2]octane and bicyclo[2.2.4]decane geometries, respectively, with the Te atoms located at the bridge-head positions.

Figure 16.8.4 shows the structures of some polymeric cations of Se and Te. In most of these systems, the Te–Te bonds link the Te atoms to form an infinite polymeric chain. The coordination numbers of the Te atoms are normally two or three, but some may attain the value of four in forming hypervalent structures. Various polymeric cations contain four-, five-, or six-membered rings. The four-membered ring is planar, but the larger rings are nonplanar. The rings are directly connected or linked by short fragments of one, two, three atoms. In the heteroatom polymeric cations, the Te atoms invariably occupy the three-coordinate sites.

Figures 16.8.4(a) and (j) show the structures of the coexisting $(Te_4^{2+})_\infty$ and $(Te_{10}^{2+})_\infty$ in $(Te_4)(Te_{10})(Bi_4Cl_{16})$. $(Te_4^{2+})_\infty$ is composed of planar squares of Te atoms connected by Te–Te bonds to form an infinite zigzag chain. There are

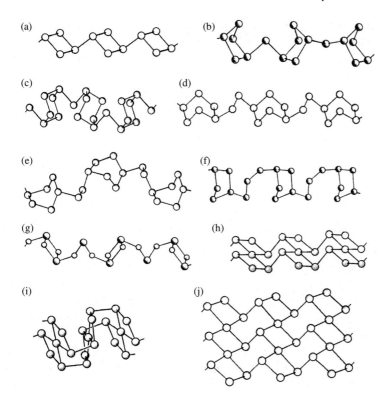

Fig. 16.8.4.
Structures of some polymeric cations of Se and Te. (a) $(Te_4^{2+})_\infty$ in $(Te_4)(Te_{10})(Bi_4Cl_{16})$, (b) $(Te_6^{2+})_\infty$ in $(Te_6)(HfCl_6)$, (c) $(Te_7^{2+})_\infty$ in $(Te_7)(AsF_6)_2$, (d) $(Te_8^{2+})_\infty$ in $(Te_8)(U_2Br_{10})$, (e) $(Te_8^{2+})_\infty$ in $(Te_8)(Bi_4Cl_{14})$, (f) $(Te_{3.15}Se_{4.85}^{2+})_\infty$ in $(Te_{3.15}Se_{4.85})(WOCl_4)_2$, (g) $(Te_3Se_4^{2+})_\infty$ in $(Te_3Se_4)(WOCl_4)_2$, (h) $(Te_7^{2+})_\infty$ in $(Te_7)(Bi_2Cl_6)$, (i) $(Te_7^{2+})_\infty$ in $(Te_7)(NbOCl_4)_2$, and (j) $(Te_{10}^{2+})_\infty$ in $(Te_4)(Te_{10})(Bi_4Cl_{16})$.

equal numbers of two- and three-coordinate Te atoms, so that the latter carry the positive charges. The bond lengths within each square ring are 275 and 281 pm, and the interring bond distance is 297 pm.

Figures 16.8.4(b) to (e) show the structures of $(Te_6^{2+})_\infty$ in $(Te_6)(HfCl_6)$, $(Te_7^{2+})_\infty$ in $(Te_7)(AsF_6)_2$, and $(Te_8^{2+})_\infty$ in $(Te_8)(U_2Br_{10})$ and $(Te_8)(Bi_4Cl_{14})$, respectively. These polymeric zigzag chains are composed of five- or six-membered rings linked by one or two atoms.

Figures 16.8.4(f) and (g) show the structures of two related heteroatom polymeric cationic chains composed of Te and Se. In $(Te_{3.15}Se_{4.85})(WOCl_4)_2$, the non-stoichiometric, disordered $(Te_{3.15}Se_{4.85}^{2+})_\infty$ cationic chains are constructed from the linkage of five-membered rings by nonlinear three-atom fragments. In $(Te_3Se_4)(WOCl_4)_2$, the $(Te_3Se_4^{2+})_\infty$ chain is composed of planar Te_2Se_2 rings connected by nonlinear Se–Te–Se fragments.

Figures 16.8.4(h) to (j) show the structures of polymeric cations that contain hypervalent Te atoms. Two kinds of $(Te_7^{2+})_\infty$ chains are found separately in $(Te_7)(Bi_2Cl_6)$ and $(Te_7)(NbOCl_4)_2$, and $(Te_{10}^{2+})_\infty$ exists in $(Te_4)(Te_{10})(Bi_4Cl_{16})$. In these polymeric cations, the hypervalent Te atoms each exhibits square-planar coordination to form a $TeTe_4$ unit with Te–Te bond lengths in the range 292–7 pm. In $(Te_7^{2+})_\infty$, the $TeTe_4$ unit and a pair of terminal Te atoms constitute an enlarged Te_7 unit composed of two planar squares sharing a common vertex. In $(Te_{10}^{2+})_\infty$, the basic Te_{10} structural unit consists of a linear arrangement of three corner-sharing planar squares, and such Te_{10}

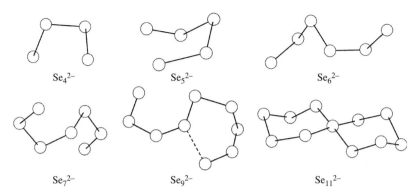

Fig. 16.8.5.
Structures of some dianions Se_x^{2-}. For Se_9^{2-}, the longest bond is represented by a broken line.

units are laterally connected by Te–Te bonds to generate a corrugated polymeric ribbon that also contain chair-like six-membered rings.

(2) Polyatomic anions

The chemistry of polyselenides, polytellurides, and their metal complexes is very well established. Typical structures of polyselenide dianions are shown in Fig. 16.8.5. In these species, the Se–Se bond distances vary from 227 to 236 pm, and the bond angles from 103° to 110°. The tethered monocyclic structure of Se_9^{2-} in the complex $Sr(15\text{-}C5)_2(Se_9)$ has a three-connected Se atom forming two long and one normal Se–Se bonds at 295, 247, and 231 pm (anticlockwise in Fig. 16.8.5, with the longest bond represented by a broken line). The other Se–Se bonds are in the range 227–39 pm.

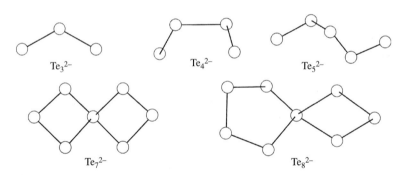

Fig. 16.8.6.
Structures of some dianions Te_x^{2-}.

The dianion Se_{11}^{2-} has a centrosymmetric spiro-bicyclic structure involving a central square-planar Se atom common to the two chair-shaped rings. The shared atom forms four long Se–Se bonds of length 266–8 pm, and the structure may be described as a central Se^{2+} chelated by two $\eta^2\text{-}Se_5^{2-}$ ligands.

Some typical structures of polytelluride dianions, Te_x^{2-}, are shown in Fig. 16.8.6. In these species, the Te–Te bond distances vary from 265 to 284 pm. In the bicyclic polytellurides Te_7^{2-} and Te_8^{2-}, the central Te atom each has four long bonds with bond lengths from 292 to 311 pm.

The Se_x^{2-} and Te_x^{2-} are effective chelating ligands for both main group and transition metals, giving rise to complexes such as $Sn(\eta^2\text{-}Se_4)_3^{2-}$, $[M(\eta^2\text{-}Se_4)_2]^{2-}$ (M = Zn, Cd, Hg, Ni, Pb), $Ti(\eta^5\text{-}C_5H_5)_2(\eta^2\text{-}Se_5)$, $[Hg(\eta^3\text{-}Te_7)]^{2-}$, and $M_2(\mu_2\text{-}Te_4)(\eta^2\text{-}Te_4)_2$ (M = Cu, Ag).

16.8.3 Stereochemistry of selenium and tellurium

Selenium and tellurium exhibit a great variety of molecular geometries as a consequence of the number of stable oxidation states. Various observed structures are summarized in Table 16.8.1, in which A is a central Se or Te atom, X is an atom bonded to A, and E represents a lone pair.

Table 16.8.1. Molecular geometries of Se and Te

Type	Molecular geometry	Example (structure in Fig. 16.8.7)
AX_2E	Bent	SeO_2
AX_2E_2	Bent	$SeCl_2$, $Se(CH_3)_2$, $TeCl_2$, $Te(CH_3)_2$
AX_3	Trigonal planar	SeO_3, TeO_3
AX_3E	Trigonal pyramidal	$(SeO_2)_x$ (a), $OSeF_2$ (b)
AX_3E_2	T-shaped	$[SeC(NH_2)_2I_3]^{2+}$ (c), $C_6H_5TeBr(SC_3N_2H_6)$ (d)
AX_4	Tetrahedral	O_2SeF_2, $(SeO_3)_4$
AX_4E	Disphenoidal	$Se(C_6H_5)_2Cl_2$, $Se(C_6H_5)_2Br_2$, $Te(CH_3)_2Cl_2$, $Te(C_6H_5)_2Br_2$
AX_5E	Square pyramidal	TeF_5^- (e), $(TeF_4)_x$ (f)
AX_6	Octahedral	SeF_6, TeF_6, $(TeO_3)_x$, $F_5TeOTeF_5$ (g)
AX_6E	Octahedral	$SeCl_6^{2-}$, $TeCl_6^{2-}$
AX_7	Pentagonal bipyramidal	TeF_7^- (h)

With reference to Table 16.8.1, the stereochemistries of Se and Te compounds are briefly described below.

(1) AX_2E type

The SeO_2 molecule in the vapor phase has a bent configuration with Se–O 160.7 pm and O–Se–O 114°. Crystalline SeO_2 is built of infinite chains, in which each Se atom is bonded to three oxygen atoms (AX_3E type) in a trigonal pyramidal configuration. The bond lengths are Se–O_b 178 pm, Se–O_t 173 pm, as shown in Fig. 16.8.7(a).

(2) AX_2E_2 type

Many AX_2E_2 type molecules, such as $Se(CH_3)_2$, $SeCl_2$, $Te(CH_3)_2$, and $TeBr_2$, all exhibit a bent configuration, in which repulsion of lone pairs makes the interbond angles smaller than the ideal tetrahedral angle:

	$Se(CH_3)_2$	$SeCl_2$	$Te(CH_3)_2$	$TeCl_2$
A–X (pm)	194.5	215.7	214.2	232.9
X–A–X	96.3°	99.6°	94°	97°

(3) AX_3 type

Monomeric selenium trioxide (SeO_3) and tellurium trioxide (TeO_3) have a trigonal planar structure in the gas phase. In the solid state, SeO_3 forms cyclic tetramers $(SeO_3)_4$, in which each Se atom connects two bridging O atoms and two terminal O atoms, with Se–O_b 177 pm and Se–O_t 155 pm (AX_4 type).

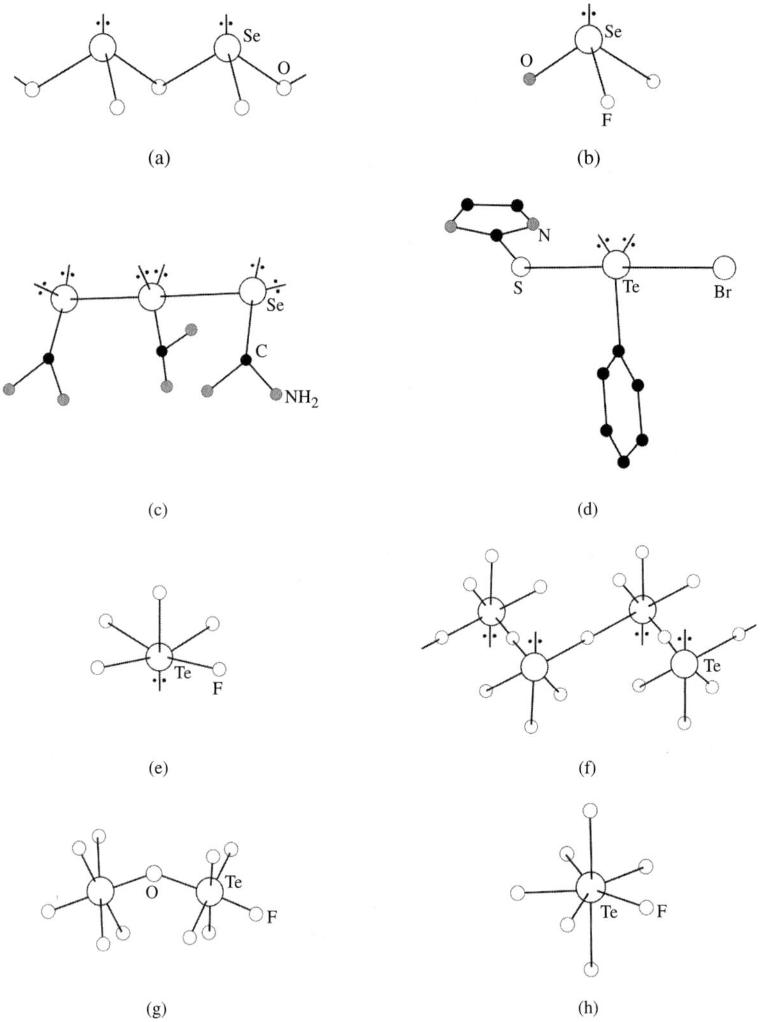

Fig. 16.8.7.
Stereochemistry of Se and Te compounds.

The solid-state structure of TeO₃ is a three-dimensional framework, in which Te(VI) forms TeO₆ octahedra (AX₆ type) sharing all vertices.

(4) AX₃E type

Pyramidal molecule SeOF₂ has bond lengths Se=O 158 pm and Se–F 173 pm, bond angles F–Se–F 92° and F–Se–O 105°, as shown in Fig. 16.8.7(b). Its dipole moment (2.62 D in benzene) and dielectric constant (46.2 at 20°C) are both high, and accordingly it is a useful solvent.

(5) AX₃E₂ type

The cation $[SeC(NH_2)_2]_3^{2+}$ adopts T-shaped geometry, as shown in Fig. 16.8.7(c). In this structure, the central Se atom must bear a formal negative charge to have two lone pairs at the equatorial positions of a trigonal bipyramid.

The valence-bond structural formula of this cation is given below:

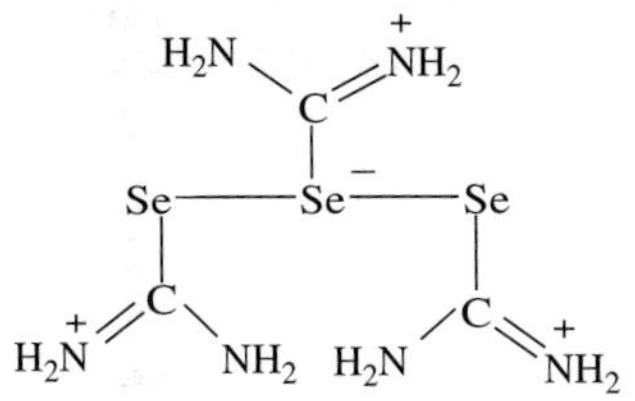

In the molecule C$_6$H$_5$TeBr(SC$_3$N$_2$H$_6$), the Te atom has a similar T-shaped configuration, as shown in Fig. 16.8.7(d). Bonding can be described in terms of resonance between a pair of valence-bond structures:

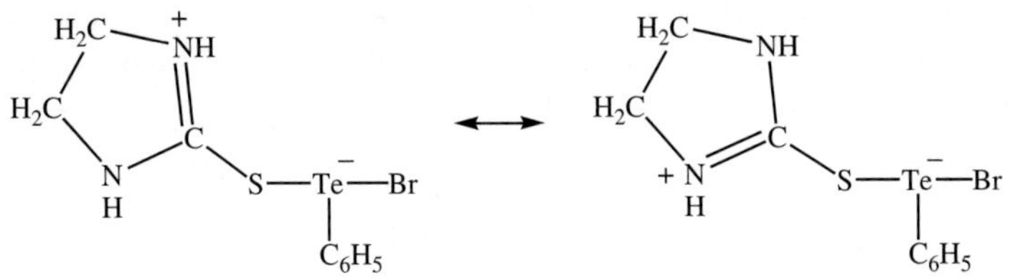

(6) AX$_4$ type

The molecule SeO$_2$F$_2$ and analogous compounds have tetrahedral geometry with three different bond angles: O–Se–O 126.2°, O–Se–F 108.0°, and F–Se–F 94.1°.

(7) AX$_4$E type

The molecules Se(C$_6$H$_5$)$_2$Cl$_2$, Se(C$_6$H$_5$)$_2$Br$_2$, Te(CH$_3$)$_2$Cl$_2$, and Te(C$_6$H$_5$)$_2$Br$_2$ constitute this structure type. In all these molecules, the halogen atoms occupy the axial positions, as shown below:

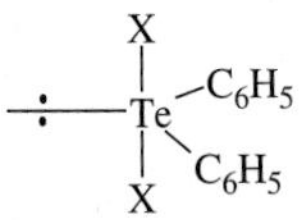

(8) AX$_5$E type

The anion TeF$_5^-$ and polymeric (TeF$_4$)$_x$ belong to the AX$_5$E type with square-pyramidal configuration. In TeF$_5^-$, the bonds in the square base (196 pm) are longer than the axial bond (185 pm), and the bond angles (79°) are smaller than 90°, as shown in Fig. 16.8.7(e). Crystalline (TeF$_4$)$_x$ has a chain structure, in which TeF$_5$ groups are linked by bridging F atoms in such a way that alternate pyramids are oriented in opposite directions, as shown in Fig. 16.8.7(f).

(9) AX$_6$ type

The hexafluorides SeF$_6$ and TeF$_6$ have the expected regular octahedral configuration. In F$_5$SeOSeF$_5$ and F$_5$TeOTeF$_5$, the Se and Te atoms take the octahedral configuration, and the four equatorial bonds in each case are bent away from the bridging O atom, as shown in Fig. 16.8.7(g).

(10) AX$_6$E type

The anions SeX$_6^{2-}$ and TeX$_6^{2-}$ (X = Cl, Br) adopt regular octahedral geometry, apparently indicating that the lone pair in the valence shell is stereochemically inactive. The observed result can be explained as follows.

(a) With increasing size of the central atom the tendency for the lone pair to spread around the core is enhanced. It is drawn inside the valence shell, behaving like an s-type orbital and effectively becoming the outer shell of the core.
(b) This tendency is also enhanced by the presence of six bonding pairs in the valence shell, which leaves rather little space for the lone pair.
(c) With the addition of the nonbonding electron pair, the core size increases and the core charge decreases from $+6$ to $+4$, with the result that the bond pairs move farther from the central nucleus, thus increasing the bond lengths. In accordance with this, the observed bond lengths of SeX$_6^{2-}$, TeX$_6^{2-}$, and SbX$_6^{2-}$ (X = Cl, Br) ions are considerably longer than those expected from the sum of the covalent radii by about 20-5 pm.

(11) AX$_7$ type

The TeF$_7^-$ ion, an isoelectronic and isostructural analog of IF$_7$, has a pentagonal bipyramidal structure with Te–F$_{ax}$ 179 pm and Te–F$_{eq}$ 183-90 pm, as shown in Fig. 16.8.5(h). The equatorial F atoms deviate slightly from the mean equatorial plane.

References

1. N. N. Greenwood and A. Earnshaw, *Chemistry of the Elements*, 2nd edn., Butterworth-Heinemann, Oxford, 1997.
2. D. F. Shriver, P. W. Atkins, T. L. Overton, J. P. Rourke, M. T. Weller and F. A. Armstrong, *Inorganic Chemistry*, 4th edn., Oxford University Press, Oxford, 2006.
3. R. B. King, *Inorganic Chemistry of Main Group Elements*, VCH, New York, 1995.
4. J. Gillespie and I. Hargittai, *The VSEPR Model of Molecular Geometry*, Allyn and Bacon, Boston, 1991.
5. J. Donohue, *The Structure of the Elements*, Wiley, New York, 1974.
6. U. Müller, *Inorganic Structural Chemistry*, 2nd edn., Wiley, Chichester, 2006.
7. K. Kuchitsu (ed.), *Structure of Free Polyatomic Molecules: Basic Data*, Springer, Berlin, 1998.
8. R. B. Kings (eds.), *Encyclopedia of Inorganic Chemistry*, Wiley, New York, 1994: (a) R. R. Conry and K. D. Karlin, Dioxygen and related ligands, pp. 1036–40; (b) D. T. Sawyer, Oxygen: inorganic chemistry, pp. 2947–88; (c) J. D. Woolins, Sulfur: inorganic chemistry, pp. 3954–88; (d) T. Chivers, Sulfur-nitrogen compounds, pp. 3988–4009.
9. D. T. Sawyer, *Oxygen Chemistry*, Oxford University Press, Oxford, 1991.
10. C. S. Foote, J. S. Valentine, A. Greenberg and J. F. Liebman (eds.), *Active Oxygen in Chemistry*, Blackie, London, 1995.
11. A. E. Martell and D. T. Sawyer (eds.), *Oxygen Complexes and Oxygen Activation by Transition Metal Complexes*, Plenum Press, New York, 1988.

12. I. Haiduc and D. B. Sowerby (eds.), *The Chemistry of Inorganic Homo- and Heterocycles*, Vol. 1-2, Academic Press, London, 1987.
13. R. Steudel (ed.), *The Chemistry of Inorganic Ring Systems*, Elsevier, Amsterdam, 1992.
14. T. Chivers, *A Guide to Chalcogen-Nitrogen Chemistry*, World Scientific, Singapore, 2005.
15. E. I. Stiefel and K. Matsumoto (eds.), *Transitional Metal Sulfur Chemistry: Biological and Industrial Significance*, American Chemical Society, Washington, DC, 1996.
16. J.-X. Lu (ed.), *Some New Aspects of Transitional-Metal Cluster Chemistry*, Science Press, Beijing/New York, 2000.
17. L. F. Lundegaard, G. Weck, M. I. McMahon, S. Desgreniers and P. Loubeyre, Observation of an O_8 molecular lattice in the ε phase of solid oxygen. *Nature* **443**, 201–4 (2006).
18. H. Pernice, M. Berkei, G. Henkel, H. Willner, G. A. Argüello, H. L. McKee and T. R. Webb, Bis(fluoroformyl)trioxide, FC(O)OOOC(O)F. *Angew. Chem. Int. Ed.* **43**, 2843–6 (2004).
19. T. S. Zwier, The structure of protonated water clusters. *Science* **304**, 1119–20 (2004).
20. M. Miyazaki, A. Fujii, T. Ebata and N. Mikami, Infrared spectroscopic evidence for protonated water clusters forming nanoscale cages. *Science* **304**, 1134–7 (2004).
21. J.-W. Shin, N. I. Hammer, E. G. Diken, M. A. Johnson, R. S. Walters, T. D. Jaeger, M. A. Duncan, R. A. Christie and K. D. Jordan, Infrared signature of structures associated with the $H^+(H_2O)_n$ ($n = 6$ to 27) clusters. *Science* **304**, 1137–40 (2004).
22. R. Steudel, K. Bergemann, J. Buschmann and P. Luger, Application of dicyanohexasulfane for the synthesis of *cyclo*-nonasulfur. Crystal and molecular structures of $S_6(CN)_2$ and of α-S_9. *Inorg. Chem.* **35**, 2184–8 (1996).
23. R. Steudel, O. Schumann, J. Buschmann and P. Luger, A new allotrope of elemental sulfur: convenient preparation of *cyclo*-S_{14} from S_8. *Angew. Chem. Int. Ed.* **37**, 2377–8 (1998).
24. J. Beck, Polycationic clusters of the heavier group 15 and 16 elements, in G. Meyer, D. Naumann and L. Wesemann (eds.), *Inorganic Chemistry in Focus II*, Wiley–VCH, Weinheim, 2005, pp. 35–52.
25. W. S. Sheldrick, Cages and clusters of the chalcogens, in M. Driess and H. Nöth (eds.), *Molecular Clusters of the Main Group Elements*, Wiley-VCH, Weinheim, 2004, pp. 230–45.

17 Structural Chemistry of Group 17 and Group 18 Elements

17.1 Elemental halogens

17.1.1 Crystal structures of the elemental halogens

The halogens are diatomic molecules, whose color increases steadily with atomic number. Fluorine (F_2) is a pale yellow gas, bp 85.0 K. Chlorine (Cl_2) is a greenish-yellow gas, bp 239.1 K. Bromine (Br_2) is a dark-red liquid, bp 331.9 K. Iodine (I_2) is a lustrous black crystalline solid, mp 386.7 K, which sublimes and boils readily at 458.3 K. Actually solid iodine has a vapor pressure of 41 Pa at 298 K and 1.2×10^4 Pa at the melting point. In the solid state, the halogen molecules are aligned to give a layer structure. Fluorine exists in two crystalline modifications: a low-temperature α-form and a higher temperature β-form, neither of which resembles the orthorhombic layer structure of the isostructural chlorine, bromine, and iodine crystals. Figure 17.1.1 shows the crystal structure of iodine. Table 17.1.1 gives the interatomic distances in gaseous and crystalline halogens.

The molecules F_2, Cl_2, and Br_2 in the crystalline state have intramolecular distances (X–X) which are nearly the same as those in the gaseous state. In crystalline iodine, the intramolecular I–I bond distance is longer than that in a gaseous molecule, and the lowering of the bond order is offset by the intermolecular bonding within each layer. The closest interatomic distance between neighboring I_2 molecules is 350 pm, which is considerably shorter than twice the van der Waals radius (430 pm). It therefore seems that appreciable secondary bonding interactions occur between the iodine molecules, giving rise to the semiconducting properties and metallic luster; under very high pressure iodine becomes a metallic conductor. The distance between layers in the iodine crystal (427 and 434 pm) corresponds to the van der Waals distance.

17.1.2 Homopolyatomic halogen anions

The homopolyhalogen anions are formed mainly by iodine, which exhibits the highest tendency to form stable catenated anionic species. Numerous examples of small polyiodides, such as I_3^-, I_4^{2-} and I_5^-, and extended discrete oligomeric anionic polyiodides, such as I_7^-, I_8^{2-}, I_9^-, I_{12}^{2-}, I_{16}^{2-}, I_{16}^{4-}, I_{22}^{4-} and I_{29}^{3-}, and polymeric $(I_7^-)_n$ networks have been reported. These polyiodides are all formed by the relatively loose association of several I_2 molecules with several I^- and/or I_3^-

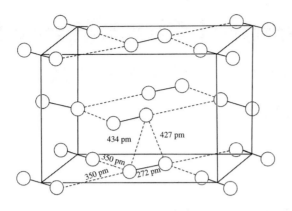

Fig. 17.1.1.
Crystal structure of iodine.

anions. In order to assess the association of such species, the following equation between the bond length (d) and bond order (n) has been proposed:

$$d = d_o - c \log n = 267 \text{ pm} - (85 \text{ pm}) \log n.$$

In this equation, the reference I–I single-bond length d_o is taken to be 267 pm; when the distance d between two iodine atoms is ≤ 293 pm, corresponding to bond order $n \geq 0.50$, there is a relatively strong bond between them, which is represented by a solid line. The distance d between 293 and 352 pm corresponds to bond order n between 0.50 and 0.10, indicating a relatively weak bond, which is represented by a broken line. When the distance is longer than 352 pm, there is only van der Waals interaction between the two molecular/ionic species, and no discrete polyiodide is formed.

Table 17.1.1. Interatomic distances in gaseous and crystalline halogens

X	X–X (pm)		X··· X (pm)		Ratio	X··X (Shortest)
	Gas	Solid	Within layer	Between layers		X-X (Solid)
F	143.5	149	324	284		1.91
Cl	198.8	198	332	374		1.68
Br	228.4	227	331	399		1.46
I	266.6	272	350	427		1.29

Figure 17.1.2 shows some polyiodides, which have been characterized structurally. All polyiodine anions consist of units of I^-, I_2, and I_3^-. The bond length of the structural components of the polyiodides are often characteristic: 267 to 285 pm in I_2 molecular fragments, whereas those of symmetrical triiodide I_3^- are about 292 pm.

The formation and stability of an extended polyiodide species are dependent on the size, shape, and charge of its accompanying cation. In the solid state, the polyiodide species are assembled around a central cation to form a discrete or one-, two-, or three-dimensional structure.

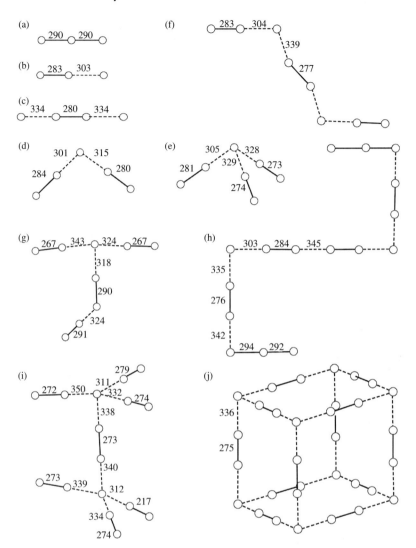

Fig. 17.1.2.
Structures of some polyiodides: (a) I_3^- (symmetric) in Ph_4AsI_3, (b) I_3^- (asymmetric) in CsI_3, (c) I_4^{2-} in $Cu(NH_3)_4I_4$, (d) I_5^- in $Fe(S_2CNEt_2)_3I_5$, (e) I_7^- in Ph_4PI_7, (f) I_8^{2-} in $[(CH_2)_6N_4Me]_2I_8$, (g) I_9^- in Me_4NI_9, (h) I_{16}^{4-} in $[(C_7H_8N_4O_2)H]_4I_{16}$, (i) I_{16}^{2-} in $(Cp^*{}_2Cr_2I_3)_2I_{16}$, and (j) $(I_7^-)_n$ network in a unit cell of the $\{Ag[18]aneS_6\}I_7$ complex. Bond lengths are in pm.

17.1.3 Homopolyatomic halogen cations

The structures of the following homopolyatomic halogen cations have been determined by X-ray analysis:

X_2^+:	Br_2^+ and I_2^+ (in $Br_2^+[Sb_3F_{16}]^-$ and $I_2^+[Sb_2F_{11}]^-$)
X_3^+:	Cl_3^+, Br_3^+ and I_3^+ (in X_3AsF_6)
X_4^{2+}:	I_4^{2+} (in $I_4^{2+}[Sb_3F_{16}]^-[SbF_6]^-$)
X_5^+:	Br_5^+ and I_5^+ (in X_5AsF_6)
X_{15}^+:	I_{15}^+ (in $I_{15}AsF_6$)

X_2^+: The bond lengths of Br_2^+ and I_2^+, with a formal bond order of 1½, are 215 and 258 pm, respectively, which are shorter than the bond lengths of molecular Br_2(228 pm) and I_2(267 pm). This is consistent with the loss of an electron from an antibonding orbital.

X_3^+: These cations have a bent structure (Fig. 17.1.3(a)). The X–X bond lengths are similar to those in gaseous X_2, being consistent with their single-bond character. The bond angles are between the 101° and 104°.

X_4^{2+}: Compound $I_4[Sb_3F_{16}][SbF_6]$ contains an I_4^{2+} cation, which has the shape of a planar rectangle with I–I bond lengths of 258 and 326 pm, as shown in Fig. 17.1.3(b).

X_5^+: Cations Br_5^+ and I_5^+ are iso-structural, as shown in Fig. 17.1.3(c).

X_{15}^+: Compound $I_{15}AsF_6$ contains an I_{15}^+ cation, which has the shape of a centrosymmetric zigzag chain. This cation may be considered to be a finite zigzag chain composed of three connected I_5^- units, as shown in Fig. 17.1.3(d).

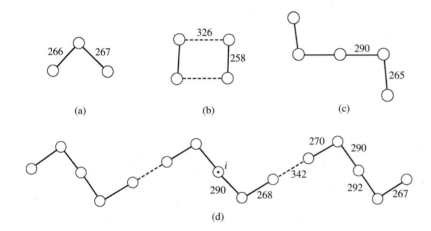

Fig. 17.1.3.
Structures of some polyiodine cations: (a) I_3^+, (b) I_4^{2+}, (c) I_5^+ and (d) I_{15}^+. Bond lengths are in pm. Both I_5^+ and I_{15}^+ are centrosymmetric.

17.2 Interhalogen compounds and ions

The halogens form many compounds and ions that are binary or ternary combinations of halogen atoms. There are three basic types: (a) neutral interhalogen compounds, (b) interhalogen cations, and (c) interhalogen anions.

17.2.1 Neutral interhalogen compounds

The halogens react with each other to form binary interhalogen compounds XY, XY_3, XY_5 and XY_7, where X is the heavier halogen. A few ternary compounds are also known, e.g., $IFCl_2$ and IF_2Cl. All interhalogen compounds contain an even number of halogen atoms. Table 17.2.1 lists the physical properties of some XY_n compounds.

XY: All six possible diatomic interhalogen compounds between F, Cl, Br and I are known, but IF is unstable, and BrCl cannot be isolated free from Br_2 and Cl_2. In general, the diatomic interhalogens exhibit properties intermediate

Table 17.2.1. Physical properties of some interhalogen compounds

Compound	Appearance at 298 K	mp (K)	bp(K)	Bond length*(pm)	
ClF	Colorless gas	117	173	163	
BrF	Pale brown gas	240	293	176	
BrCl	Red brown gas	—	—	214	
ICl (α)	Ruby red crystal	300	~373	237, 244	
ICl (β)	Brownish red crystal	287	—	235, 244	
IBr	Black crystal	314	~389	249	
ClF$_3$	Colorless gas	197	285	160 (eq)	170 (ax)
BrF$_3$	Yellow liquid	282	399	172 (eq)	181 (ax)
IF$_3$	Yellow solid	245 (dec)	—	—	
(ICl$_3$)$_2$	Orange solid	337 (sub)	—	238 (t)	268 (b)
ClF$_5$	Colorless gas	170	260	172 (ba)	162 (ap)
BrF$_5$	Colorless liquid	212.5	314	172 (ba)	168 (ap)
IF$_5$	Colorless liquid	282.5	378	189 (ba)	186 (ap)
IF$_7$	Colorless gas	278 (sub)	—	186 (eq)	179 (ax)

* The XY$_3$ molecule has a T-shaped structure: axial (ax), equatorial (eq); (ICl$_3$)$_2$ is a dimer: bridging (b), terminal (t); XY$_5$ forms a square-based pyramid: apical (ap), basal (ba); XY$_7$ has the shape of a pentagonal bipyramid: equatorial (eq), axial (ax).

between their parent halogens. However, the electronegativities of X and Y differ significantly, so the X–Y bond is stronger than the mean of the X–X and Y–Y bond strengths, and the X–Y bond lengths are shorter than the mean of d(X–X) and d(Y–Y). The dipole moments for polar XY molecules in the gas phase are ClF 0.88 D, BrF 1.29 D, BrCl 0.57 D, ICl 0.65 D, and IBr 1.21 D.

Iodine monochloride ICl is unusual in forming two modifications: the stable α-form and the unstable β-form, both of which have infinite chain structures and significant I···Cl intermolecular interactions of 294 to 308 pm. Figure 17.2.1(a) shows the chain structure of β-ICl.

XY$_3$: Both ClF$_3$ and BrF$_3$ have a T-shaped structure, being consistent with the presence of 10 electrons in the valence shell of the central atom, as shown in Fig. 17.2.1(b). The relative bond lengths of d(X–Y$_{ax}$) > d(X–Y$_{eq}$) and the bond angle of Y$_{ax}$–X–Y$_{eq}$ < 90° (ClF$_3$ 87.5° and BrF$_3$ 86°) reflect the greater electronic repulsion of the nonbonding pair of electrons in the equatorial plane of the molecule.

Iodine trichloride is a fluffy orange powder that is unstable above room temperature. Its dimer (ICl$_3$)$_2$ has a planar structure, as shown in Fig. 17.2.1(c), that contains two I–Cl–I bridges (I–Cl distances in the range of 268–272 pm) and four terminal I–Cl bonds (238–9 pm).

XY$_5$: The three fluorides ClF$_5$, BrF$_5$, and IF$_5$ are the only known interhalogens of the XY$_5$ type, and they are extremely vigorous fluorinating reagents. All three compounds occur as a colorless gas or liquid at room temperature. Their structure has been shown to be square pyramidal with the central atom slightly below the plane of the four basal F atoms [Fig. 17.2.1(d)]. The bond angles F$_{(ap)}$–X–F$_{(ba)}$ are ~90° (ClF$_5$), 85° (BrF$_5$) and 81° (IF$_5$).

XY$_7$: IF$_7$ is the sole representative of this structural type. Its structure, as shown in Fig. 17.2.1(e), exhibits a slight deformation from pentagonal

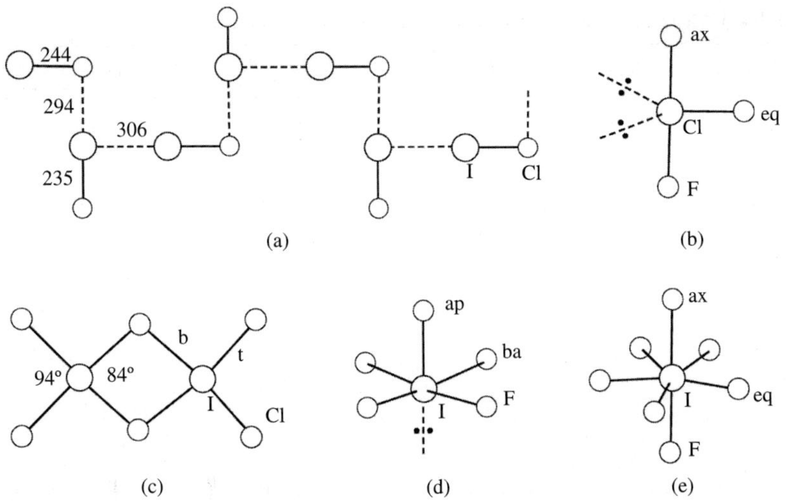

Fig. 17.2.1.
Structures of some interhalogen molecules: (a) β-ICl, (b) ClF$_3$, (c) I$_2$Cl$_6$, (d) IF$_5$, and (e) IF$_7$.

bipyramidal D_{5h} symmetry due to a 7.5° puckering and a 4.5° axial bending displacement. Bond length I–F$_{(eq)}$ is 185.5 pm and I–F$_{(ax)}$ is 178.6 pm.

17.2.2 Interhalogen ions

These ions have the general formulas XY_n^+ and XY_n^-, where n can be 2, 4, 5, 6, and 8, and the central halogen X is usually heavier than Y. Table 17.2.2 lists many of the known interhalogen ions.

Table 17.2.2. Some interhalogen ions

	XY$_2$		XY$_4$		XY$_5$	XY$_6$	XY$_8$
Cations	ClF$_2^+$	I$_2$Cl$^+$	ClF$_4^+$		—	ClF$_6^+$	—
	Cl$_2$F$^+$	IBr$_2^+$	BrF$_4^+$			BrF$_6^+$	
	BrF$_2^+$	I$_2$Br$^+$	IF$_4^+$			IF$_6^+$	
	IF$_2^+$	IBrCl$^+$	I$_3$Cl$_2^+$				
	ICl$_2^+$						
Anions	BrCl$_2^-$	ClICl$^-$	ClF$_4^-$	I$_2$Cl$_3^-$	IF$_5^{2-}$	ClF$_6^-$	IF$_8^-$
	Br$_2$Cl$^-$	ClIBr$^-$	BrF$_4^-$	I$_2$BrCl$_2^-$		BrF$_6^-$	
	I$_2$Cl$^-$	BrIBr$^-$	IF$_4^-$	I$_2$Br$_2$Cl$^-$		IF$_6^-$	
	FClF$^-$		ICl$_3$F$^-$	I$_2$Br$_3^-$			
	FIBr$^-$		ICl$_4^-$	I$_4$Cl$^-$			
			IBrCl$_3^-$				

The structures of these ions normally conform to those predicted by the VSEPR theory, as shown in Fig. 17.2.2. Since the anion XY$_n^-$ has two more electrons than the cation XY$_n^+$, they have very different shapes. The anion IF$_5^{2-}$ is planar with lone pairs occupying the axial positions of a pentagonal bipyramid. In [Me$_4$N](IF$_6$), IF$_6^-$ is a distorted octahedron (C_{3v} symmetry) with a sterically active lone pair, whereas both BrF$_6^-$ and ClF$_6^-$ are octahedral. The anion IF$_8^-$ has the expected square antiprismatic structure.

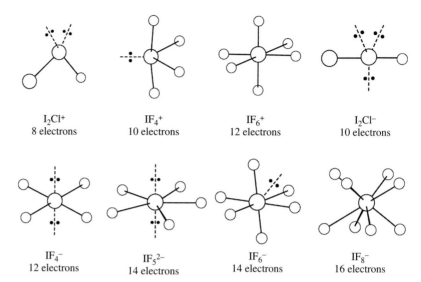

Fig. 17.2.2.
Structures of some interhalogen ions. The number of electrons in the valence shell of the central atom is given for each ion.

17.3 Charge-transfer complexes of halogens

A charge-transfer (or donor–acceptor) complex is one in which a donor and an acceptor species interact weakly with some net transfer of electronic charge, usually facilitated by the acceptor. The diatomic halogen molecule X_2 has HOMO π^* and LUMO σ^* molecular orbitals, and the σ^* orbital is antibonding and acts as an acceptor. If the X_2 molecule is dissolved in a solvent such as ROH, H_2O, pyridine, or CH_3CN that contains N, O, S, Se, or π electron pairs, the solvent molecule can function as a donor through the interaction of one of its σ or π electron pairs with the σ^* orbital of X_2. This donor–acceptor interaction leads to the formation of a charge-transfer complex between the solvent (donor) and X_2 (acceptor) and alters the optical transition energy of X_2, as shown in Fig. 17.3.1.

Let us take I_2 as an example. The normal violet color of gaseous iodine is attributable to the allowed $\pi^* \rightarrow \sigma^*$ transition. When iodine is dissolved in a solvent, the interaction of I_2 with a donor solvent molecule causes an increase in the energy separation of the π^* to σ^* orbitals from E_1 to E_2, as shown in Fig. 17.3.1. The color of this solution is then changed to brown. (The absorption maximum for the violet solution occurs at 520 to 540 nm, and that of a typical brown solution at 460 to 480 nm.) The electron transition in these $I_2 \cdot$ solvent complexes is called a charge-transfer transition. However, the most direct evidence for the formation of a charge-transfer complex in solution comes from the appearance of an intense new charge-transfer band occurring in the near ultraviolet spectrum in the range 230–330 nm.

The structures of many charge-transfer complexes have been determined. All the examples shown in Fig. 17.3.2 share the following common structural characteristics:

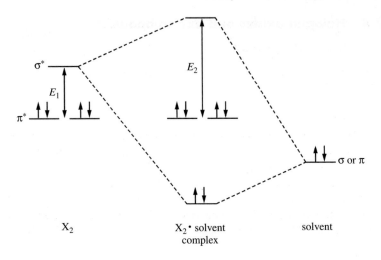

Fig. 17.3.1.
Interaction between the σ^* orbital of X_2 and a donor orbital of solvent.

(a) The donor atom (D) or π orbital (π) and X_2 molecule are essentially linear: $D \cdots X-X$ or $\pi \cdots X-X$.

(b) The bond lengths of the X–X groups in the complexes are all longer than those in the corresponding free X_2 molecules; the $D \cdots X$ distance is invariably shorter than the sum of their van der Waals radii.

(c) Each X_2 molecule in an infinite chain structure is engaged by donors (D) at both ends, and the $D \cdots X-X \cdots D$ unit is essentially linear, as expected for a σ-type acceptor orbital.

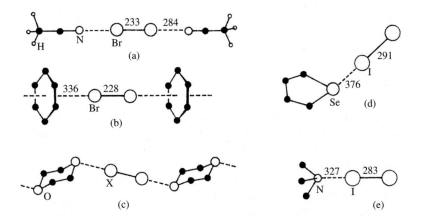

Fig. 17.3.2.
Structures of some charge-transfer complexes (bond lengths in pm). (a) $(H_3CCN)_2 \cdot Br_2$, (b) $C_6H_6 \cdot Br_2$, (c) $C_4H_8O_2 \cdot X_2$ (X = Cl, Br, X–X distance: 202 pm for X = Cl and 231 pm for X = Br), (d) $C_4H_8Se \cdot I_2$, and (e) $Me_3N \cdot I_2$.

In the extreme case, complete transfer of charge may occur, as in the formation of $[I(py)_2]^+$:

$$2I_2 + 2\left(\underset{}{\bigcirc}N\right) \longrightarrow \underset{}{\bigcirc}N \rightarrow I \leftarrow N\underset{}{\bigcirc} + I_3^-$$

17.4 Halogen oxides and oxo compounds

17.4.1 Binary halogen oxides

The structures of some binary halogen oxides are listed in Table 17.4.1.

(1) X_2O molecules

(a) Since fluorine is more electronegative than oxygen, the binary compounds of F_2 and O_2 are named oxygen fluorides, rather than fluorine oxides. F_2O is a colorless, highly toxic, and explosive gas. The molecule has C_{2v} symmetry, as expected for a molecule with 20 valence electrons and two normal single bonds.

(b) Dichlorine monoxide Cl_2O is a yellow-brown gas that is stable at room temperature. There are two linkage isomers, Cl–Cl–O and Cl–O–Cl, but

Table 17.4.1. Structure of halogen oxides

Oxide	Molecular structure	Structural formula	Formal oxidation state of X	Bond length / pm and bond angle
X_2O		Cl–O–Cl	−1 (F) +1	X = F, a = 141, θ = 103° X = Cl, a = 169, θ = 111° X = Br, a = 185, θ = 112°
X_2O_2		F–O–O–F	−1	X = F, a = 122, b = 158, θ = 109°
XO_2		O–Cl–O	+3	X = Cl, a = 147, θ = 118°
X_2O_3		O=Br–O–Br(=O)(=O)	+1, +5	X = Br, a = 161, b = 185, θ = 112°
X_2O_4		O=Br–O–Br(=O)(=O)	+1, +7	X = Br, a = 161, b = 186, θ = 110°
X_2O_5		O=I–O–I=O	+5	X = Br, a = 188, b = 161, θ = 121° X = I, a = 193(av), b = 179(av), θ = 139°
X_2O_6		O=Cl(–O)–O–Cl(=O)(=O)	+5, +7	X = Cl, a = 144 (av), b = 141, θ = 119°
X_2O_7		O=Cl(=O)–O–Cl(=O)=O	+7	X = Cl, a = 172, b = 142, θ = 119°

Group 17 and Group 18 Elements

only the latter is stable. The stable form Cl–O–Cl is bent (111°) with Cl–O 169 pm.

(c) Dibromine monoxide Br_2O is a dark-brown crystalline solid stable at 213 K (mp 255.6 K with decomposition). The molecule has C_{2v} symmetry in both the solid and vapor phases with Br–O 185 pm and angle Br–O–Br 112°.

(2) X_2O_2 molecules

Only F_2O_2 is known. This is a yellow-orange solid (mp 119 K) that decomposes above 223 K. Its molecular shape resembles that of H_2O_2, although the internal dihedral angel is smaller (87°). The long O–F bond (157.5 pm) and short O–O bond (121.7 pm) in F_2O_2 can be rationalized by resonance involving the following valence bond representations:

(3) XO_2 molecule

Only ClO_2 is known. Chlorine dioxide is an odd-electron molecule. Theoretical calculations suggest that the odd electron is delocalized throughout the molecule, and this probably accounts for the fact that there is no evidence of dimerization in solution, or even in the liquid or solid phase. Its important Lewis structures are shown below:

(4) X_2O_3 molecules

(a) Cl_2O_3 is a dark-brown solid which explodes even below 273 K. Its structure has not been determined.

(b) Br_2O_3 is an orange crystalline solid and has been shown by X-ray analysis to be syn-$BrOBrO_2$ with Br^I–O 184.5 pm, Br^V–O 161.3 pm, and angle Br–O–Br 111.6°. It is thus, formally, the anhydride of hypobromous and bromic acid.

(5) X_2O_4 molecules

(a) Chlorine perchlorate, Cl_2O_4, is most likely $ClOClO_3$. Little is known of the structure and properties of this pale-yellow liquid; it is even less stable than ClO_2 and decomposes at room temperature to Cl_2, O_2, and Cl_2O_6.

(b) Br_2O_4 is a pale yellow crystalline solid, whose structure has been shown by EXAFS to be bromine perbromate, $BrOBrO_3$, with Br^I–O 186.2 pm, Br^{VII}–O 160.5 pm and angle Br–O–Br 110°.

(6) X_2O_5 molecules

(a) Ozonization of Br_2 generates Br_2O_3 and eventually Br_2O_5:

$$Br_2 \xrightarrow{O_3, 195\,K} Br_2O_3 \xrightarrow{O_3, 195\,K} Br_2O_5$$

$$\text{brown} \qquad\qquad \text{orange} \qquad\qquad \text{colorless}$$

The end product can be crystallized from propionitrile as $Br_2O_5 \cdot EtCN$, and crystal structure analysis has shown that Br_2O_5 is $O_2BrOBrO_2$ with each Br atom pyramidally surrounded by three O atoms, and the terminal O atoms are eclipsed with respect to each other.

(b) I_2O_5 is the most stable oxide of the halogens. Crystal structure analysis has shown that the molecule consists of two pyramidal IO_3 groups sharing a common oxygen. The terminal O atoms have a staggered conformation, as shown in Table 17.4.1.

(7) X_2O_6 molecule

Cl_2O_6 is actually a mixed-valence ionic compound $ClO_2^+ ClO_4^-$, in which the angular ClO_2^+ and tetrahedral ClO_4^- ions are arranged in a distorted CsCl-type crystal structure. Cation ClO_2^+ has Cl–O 141 pm, angle O–Cl–O 119°; ClO_4^- has Cl–O(av) 144 pm.

(8) X_2O_7 molecule

Cl_2O_7 is a colorless liquid at room temperature. The molecule has C_2 symmetry in both gaseous and crystalline states, the ClO_3 groups being twisted from the staggered (C_{2v}) configuration, with Cl–O(bridge) 172.3 pm and Cl–O(terminal) 141.6 pm.

In addition to the compounds mentioned above, other unstable binary halogen oxides are known. The structures of the short-lived gaseous XO radicals have been determined. For ClO, the interatomic distance $d = 156.9$ pm, dipole moment $\mu = 1.24$ D, and bond dissociation energy $D_o = 264.9$ kJ mol^{-1}. For BrO, $d = 172.1$ pm, $\mu = 1.55$ D, and $D_o = 125.8$ kJ mol^{-1}. For IO, $d = 186.7$ pm and $D_o = 175$ kJ mol^{-1}.

The structures of the less stable oxides I_4O_9 and I_2O_4 are still unknown, but I_4O_9 has been formulated as $I^{3+}(I^{5+}O_3)_3$, and I_2O_4 as $(IO)^+(IO_3)^-$.

17.4.2 Ternary halogen oxides

The ternary halogen oxides are mainly compounds in which a heavier X atom (Cl, Br, I) is bonded to both O and F. These compounds are called halogen oxide fluorides. The structures of the ternary halogen oxides are summarized in Fig. 17.4.1.

The geometries of these molecules are consistent with the VSEPR model. (a) FClO is bent with C_s symmetry (two lone pairs). (b) FXO_2 is pyramidal with C_s symmetry (one lone pair). (c) FXO_3 has C_{3v} symmetry. (d) F_3XO is an incomplete trigonal bipyramid with F, O, and a lone pair in the equatorial plane, having C_s symmetry. (e) F_3ClO_2 is a trigonal bipyramid with one fluorine and both oxygens in the equatorial plane; owing to the strong repulsion from two

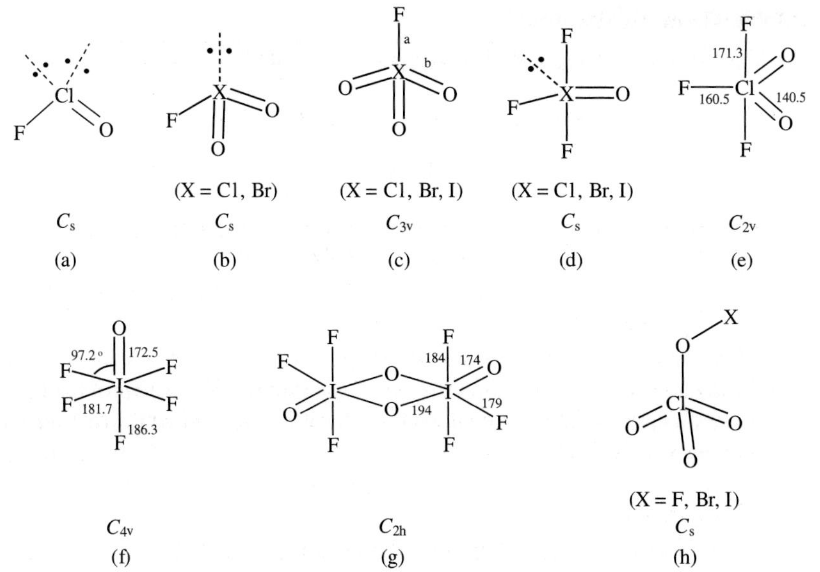

Fig. 17.4.1.
Structures of ternary halogen oxides. The bond lengths are in pm. For (c), X = Cl, $a = 162$ pm, $b = 140$ pm; X = Br, $a = 171$ pm, $b = 158$ pm.

Cl=O double bonds in the equatorial plane, the axial Cl–F bonds are longer than the equatorial Cl–F bonds, as confirmed by experimental data. (f) F_5IO has C_{4v} symmetry, in which the bond lengths are I–F$_{ax}$ 186.3 pm, I–F$_{eq}$ 181.7 pm, and I=O 172.5 pm. Also, the I atom lies above the plane of four equatorial F atoms, and bond angle O–I–F$_{eq}$ is 97.2°. (g) F_3IO_2 forms oligomeric species, and the dimer $(F_3IO_2)_2$ has C_{2h} symmetry. (h) O_3ClOX are "halogen perchlorate salts".

A variety of cations and anions are derived from some of the above neutral molecules by gaining or losing one F^-, as shown in Fig. 17.4.2. Their structures are again in accord with predictions by the VSEPR model.

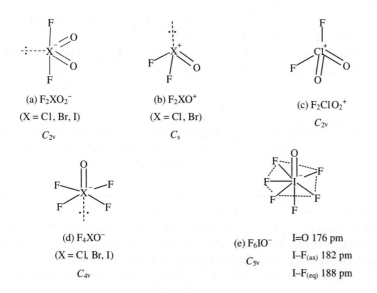

Fig. 17.4.2.
Structures of cations and anions of ternary halogen oxides.

Structural Chemistry of Selected Elements

Table 17.4.2. Halogen oxoacids

Generic name	Fluorine	Chlorine	Bromine	Iodine
Hypohalous acids	HOF*	HOCl	HOBr	HOI
Halous acids	—	HOClO	—	—
Halic acids	—	HOClO$_2$	HOBrO$_2$	HOIO$_2$*
Perhalic acids	—	HOClO$_3$*	HOBrO$_3$	HOIO$_3$*, (HO)$_5$IO*

* Isolated as pure compounds; others are stable only in aqueous solutions.

17.4.3 Halogen oxoacids and anions

Numerous halogen oxoacids are known, though most of them cannot be isolated as pure species and are stable only in aqueous solution or in the form of their salts. Anhydrous hypofluorous acid (HOF), perchloric acid (HClO$_4$), iodic acid (HIO$_3$), orthoperiodic acid (H$_5$IO$_6$), and metaperiodic acid (HIO$_4$) have been isolated as pure compounds. Table 17.4.2 lists the halogen oxoacids.

(1) Hypofluorous and other hypohalous acids

At room temperature hypofluorous acid HOF is a gas. Its colorless solid (mp = 156 K) melts to a pale yellow liquid. In the crystal structure, O–F = 144.2 pm, angle H–O–F = 101°, and the HOF molecules are linked by O–H···O hydrogen bonds (bond length 289.5 pm and bond angle O–H···O 163°) to form a planar zigzag chain, as shown below:

In general, fluorine has a formal oxidation state of −1, but in HOF and other hypohalous acids HOX, the formal oxidation state of F and X is +1.

Hypochlorous acid HOCl is more stable than HOBr and HOI, with Cl–O 169.3 pm and angle H–O–Cl 103° in the gas phase.

(2) Chlorous acid HOClO and chlorite ion ClO$_2^-$

Chlorous acid, HOClO, is the least stable of the oxoacids of chlorine. It cannot be isolated in pure form, but exists in dilute aqueous solution. Likewise, HOBrO and HOIO are even less stable, showing only a transient existence in aqueous solution.

The chlorite ion (ClO$_2^-$) in NaClO$_2$ and other salts has a bent structure (C_{2v} symmetry) with Cl–O 156 pm, O–Cl–O 111°.

(3) Halic acids HOXO$_2$ and halate ions XO$_3^-$

Iodic acid, HIO$_3$, is a stable white solid at room temperature. In the crystal structure, trigonal HIO$_3$ molecules are connected by extensive hydrogen bonding with I–O 181 pm, I–OH 189 pm, angle O–I–O 101°, O–I–(OH) 97°.

Halate ions are trigonal pyramidal, with C_{3v} symmetry, as shown below:

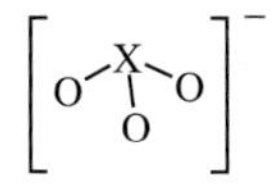

$\begin{bmatrix} O-\overset{X}{\underset{O}{|}}-O \end{bmatrix}^-$ X = Cl, Cl–O 149 pm
X = Br, Br–O 165 pm
X = I, I–O 184 pm
Angles O–X–O 106° to 107°

Note that in the solid state, some metal halates do not consist of discrete ions. For example, in the iodates there are three short I–O distances 177–90 pm and three longer distances 251–300 pm, leading to distorted pseudo-sixfold coordination and piezoelectric properties.

(4) Perchloric acid HClO$_4$ and perhalates XO$_4^-$

Perchloric acid is the only oxoacid of chlorine that can be isolated. The crystal structure of HClO$_4$ at 113 K exhibits three Cl–O distances of 142 pm and a Cl–OH distance of 161 pm. The structure of HClO$_4$, as determined by electron diffraction in the gas phase, shows Cl–O 141 pm, Cl–OH 163.5 pm, and angles O–Cl–(OH) 106°, O–Cl–O 113°.

The hydrates of perchloric acid exist in at least six crystalline forms. The monohydrate is composed of H$_3$O$^+$ and ClO$_4^-$ connected by hydrogen bonds.

All perhalate ions XO$_4^-$ are tetrahedral, with T_d symmetry, as shown below:

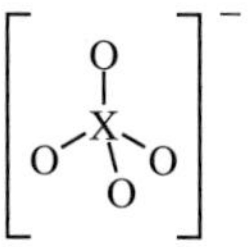

X = Cl, Cl–O 144 pm
X = Br, Br–O 161 pm
X = I, I–O 179 pm

(5) Periodic acids and periodates

Several different periodic acids and periodates are known.

(a) **Metaperiodic acid HIO$_4$**

Acid HIO$_4$ consists of one-dimensional infinite chains built up of distorted *cis*-edge-sharing IO$_6$ octahedra, as shown in Fig. 17.4.3. Until now no discrete HIO$_4$ molecule has been found.

(b) **Orthoperiodic (or paraperiodic) acid (HO)$_5$IO**

The crystal structure of orthoperiodic acid, commonly written as H$_5$IO$_6$, consists of axially distorted octahedral (HO)$_5$IO molecules linked into a three-dimensional array by O–H $\cdots$ O hydrogen bonds (10 for each molecule,

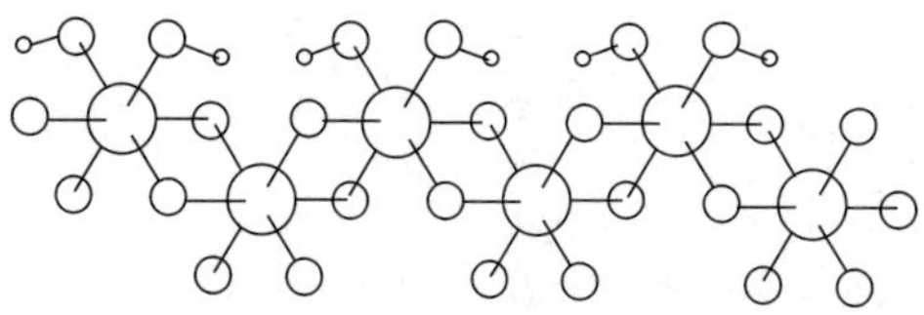

I–(OH) 184 pm
I–O(bridge) 201 pm
I–O(terminal) 191 pm

Fig. 17.4.3.
Structure of metaperiodic aicd.

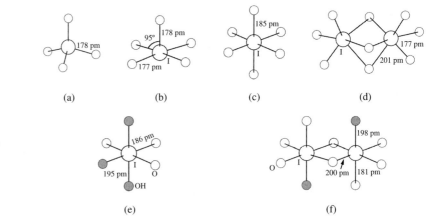

Fig. 17.4.4.
Structures of periodates (a) to (d) and hydrogen periodates (e) to (f); small shaded circles represent OH group.
(a) IO_4^- in $NaIO_4$, (b) IO_5^{3-} in K_3IO_5, (c) IO_6^{5-} in K_5IO_6, (d) $I_2O_9^{4-}$ in $K_4I_2O_9$, (e) $[IO_3(OH)_3]^{2-}$ in $(NH_4)_2IO_3(OH)_3$, and (f) $[I_2O_8(OH)_2]^{4-}$ in $K_4[I_2O_8(OH)_2]\cdot 8H_2O$.

260–78 pm). The $(HO)_5IO$ molecule has the maximum number of –OH groups surrounding the $I(+7)$ atom and hence it is called orthoperiodic acid:

$$\begin{array}{c} O \\ \parallel \\ HO \diagup I \diagdown OH \\ HO \diagup \mid \diagdown OH \\ OH \end{array} \qquad \begin{array}{l} I=O\ 178\ pm \\ \\ I-OH\ 189\ pm \end{array}$$

The structure of $H_7I_3O_{14}$ does not show any new type of catenation, because this compound exists in the solid state as a stoichiometric phase containing orthoperiodic and metaperiodic acids according to the formula $(HO)_5IO\cdot 2HIO_4$.

The structures of several periodates and hydrogen periodates are shown in Fig. 17.4.4.

17.4.4 Structural features of polycoordinate iodine compounds

Iodine differs in many aspects from the other halogens. Because of the large atomic size and the relatively low ionization energy, it can easily form stable polycoordinate, multivalent compounds. Interest in polyvalent organic iodine compounds arises from several factors: (a) the similarity of the chemical properties and reactivity of I(III) species to those of Hg(+2), Tl(+3), and Pb(+4), but without the toxic and environmental problems of these heavy metal congeners; (b) the recognition of similarities between organic transition-metal complexes and polyvalent main-group compounds such as organoiodine species; and (c) the commercial availability of key precursors, such as $PhI(OAc)_2$.

Six structural types of polyvalent iodine species are commonly encountered, as shown below:

The first two types, (a) and (b), called iodanes, are conventionally considered as derivatives of trivalent iodine I(+3). The next two, (c) and (d), periodanes, represent the most typical structural types of pentavalent iodine I(+5). The structural types (e) and (f) are typical of heptavalent iodine I(+7).

The most important structural features of polyvalent iodine compounds may be summarized as follows:

(1) The iodonium ion (type a) generally has a distance of 260–80 pm between iodine and the nearest anion, and may be considered as having pseudotetrahedral geometry about the central iodine atom.
(2) Species of type (b) has an approximately T-shaped structure with a collinear arrangement of the most electronegative ligands. Including the nonbonding electron pairs, the geometry about iodine is a distorted trigonal bipyramid with the most electronegative groups occupying the axial positions and the least electronegative group and both electron pairs residing in equatorial positions.
(3) The I–C bond lengths in both iodonium salts (type a) and iodoso derivatives (type b) are approximately equal to the sum of the covalent radii of I and C atoms, ranging generally from 200 to 210 pm.
(4) For type (b) species with two heteroligands of the same electronegativity, both I–L bonds are longer than the sum of the appropriate covalent radii, but shorter than purely ionic bonds. For example, the I–Cl bond lengths in PhICl$_2$ are 245 pm, whereas the sum of the covalent radii of I and Cl is 232 pm. Also, the I–O bond lengths in PhI(OAc)$_2$ are 215–16 pm, whereas the sum of the covalent radii of I and O is 199 pm.

(5) The geometry of the structural types (c) and (d) can be square pyramidal, pesudo-trigonal bipyramidal, and pesudo-octahedral. The bonding in I(+5) compounds IL$_5$ with a square-pyramidal structure may be described in terms of a normal covalent bond between iodine and the ligand in the apical position, and two orthogonal, hypervalent 3c-4e bonds accommodating four ligands. The carbon ligand and unshared electron pair in this case should occupy the apical positions, with the most electronegative ligands residing at equatorial positions.
(6) The typical structures of I(+7) involve a distorted octahedral configuration (type e) about iodine in most periodates and oxyfluoride, IOF$_5$, and the heptacoordinated, pentagonal bipyramidal species (type f) for the IF$_7$ and IOF$_6^-$ anions. The pentagonal bipyramidal structure can be described as two covalent collinear axial bonds between iodine and ligands in the apical positions and a coplanar, hypervalent 6c-10e bond system for the five equatorial bonds.

17.5 Structural chemistry of noble gas compounds

17.5.1 General survey

All the Group 18 elements (He, Ne, Ar, Kr, Xe, and Rn; rare gases or noble gases) have the very stable electronic configurations ($1s^2$ or ns^2np^6) and are monoatomic gases. The nonpolar, spherical nature of the atoms leads to physical properties that vary regularly with atomic number. The only interatomic interactions are weak van der Waals forces, which increase in magnitude as the polarizabilities of the atoms increase and the ionization energies decrease. In other words, interatomic interactions increase with atomic size.

In Section 5.8, we have encountered the noble gas compound HArF and complexes NgMX. The discussion there is mainly concerned with the electronic structures of these linear triatomic species. In recent years, the use of matrix-isolation techniques has led to the generation of a wide range of compounds of the general formula HNgY, where Ng = Kr or Xe, and Y is an electronegative atom or group such as H, halides, pseudohalides, OH, SH, C≡CH, and C≡C–C≡CH. These molecules can be easily detected by the extremely strong intensity of the H–Ng stretching vibration. The observed ν(H–Ng) values for some HXeY molecules are Y = H, 1166, 1181; Y = Cl, 1648; Y = Br, 1504; Y = I, 1193; Y = CN, 1623.8, Y = NC, 1851.0 cm^{-1}. The electronic structure of HNgY is best described in terms of the ion pair HNg$^+$Y$^-$, in which the HNg fragment is held by a covalent bond, and the interaction between Ng and Y is mostly ionic.

In this section, we turn our attention to the structural chemistry of those noble gas compounds that can be isolated in bulk quantities.

Xenon compounds with direct bonds to the electronegative main-group elements F, O, N, C, and Cl are well established. The first noble gas compound, a yellow-orange solid formulated as XePtF$_6$, was prepared by Neil Bartlett in 1962. Noting that the first ionization energy of Xe (1170 kJ mol^{-1}) is very similar to that of O$_2$ (1175 kJ mol^{-1}), and that PtF$_6$ and O$_2$ can combine to form O$_2$PtF$_6$, Bartlett replaced O$_2$ by a molar quantity of Xe in the reaction with PtF$_6$ and produced XePtF$_6$. It is now established that the initial product he obtained was a mixture containing diamagnetic Xe^{II}Pt^{IV}F$_6$ as the major product, which is most likely a XeF$^+$ salt of (PtF$_5^-$)$_n$ with a polymeric chain structure, as illustrated below, by analogy with the known crystal structure of XeCrF$_6$.

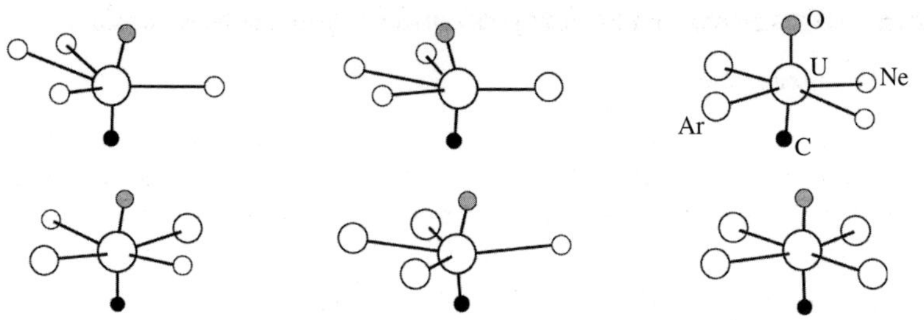

Fig. 17.5.1.
Structures of CUO(Ne)$_{4-n}$(Ar)$_n$ ($n = 0, 1, 2, 3, 4$).

If xenon is mixed with a large excess of PtF$_6$ vapor, further reactions proceed as follows:

$$XePtF_6 + PtF_6 \rightarrow XeF^+PtF_6^- + PtF_5 \text{(non-crystalline)}$$

$$XeF^+PtF_6^- + PtF_5 \text{(warmed} \geq 60°C) \rightarrow XeF^+Pt_2F_{11}^- \text{(orange} - \text{red solid)}.$$

Evidently the Xe(I) oxidation state is not a viable one, and Xe(II) is clearly favored.

Until now, no compound of helium has been discovered, and as radon has intense α-radioactivity, information about its chemistry is very limited. Compounds of the other rare gas elements, Ne, Ar, Kr, and Xe, have been reported. For example, experimental investigation of CUO(Ng)$_n$ (Ng = Ar, Kr, Xe; $n = 1, 2, 3, 4$) complexes in solid neon have provided evidence of their formation. The computed structures of CUO(Ne)$_{4-n}$(Ar)$_n$ ($n = 0, 1, 2, 3, 4$) complexes are illustrated in Fig. 17.5.1.

17.5.2 Stereochemistry of xenon

Xenon reacts directly with fluorine to form fluorides. Other compounds of Xe can be prepared by reactions using xenon fluorides as starting materials, which fall into four main types:

(1) In combination with F$^-$ acceptors, yielding fluorocations of xenon, as in the formation of (XeF$_5$)(AsF$_6$) and (XeF$_5$)(PtF$_6$);
(2) In combination with F$^-$ donors, yielding fluoroanions of xenon, as in the formation of Cs(XeF$_7$) and (NO$_2$)(XeF$_8$);
(3) F/H metathesis between XeF$_2$ and an anhydrous acid, such as

$$XeF_2 + HOClO_3 \rightarrow F\text{–}Xe\text{–}OClO_3 + HF;$$

(4) Hydrolysis, yielding oxofluorides, oxides, and xenates, such as

$$XeF_6 + H_2O \rightarrow XeOF_4 + 2HF.$$

Xenon exhibits a rich variety of stereochemistry. Some of the more important compounds of xenon are listed in Table 17.5.1, and their structures are shown in Fig. 17.5.2. The structural description of these compounds depends on whether

Table 17.5.1. Structures of some compounds of xenon with fluorine and oxygen

Compound	Geometry/ symmetry	Xe–F (pm)	Xe–O (pm)	Arrangement in the bonded and nonbonded electron pairs in Fig. 17.5.2	
XeF_2	Linear, $D_{\infty h}$	200		Trigonal bipyramidal	(a)
XeO_3	Pyramidal, C_{3v}		176	Tetrahedral	(b)
XeF_3^+ in $(XeF_3)(SbF_5)$	T-shaped, C_{2v}	184–91		Trigonal bipyramidal	(c)
$XeOF_2$	T-shaped, C_{2v}			Trigonal bipyramidal	(d)
XeF_4	Square planar, D_{4h}	193		Octahedral	(e)
XeO_4	Tetrahedral, T_d		174	Tetrahedral	(f)
XeO_2F_2	See-saw, C_{2v}	190	171	Trigonal bipyramidal	(g)
$XeOF_4$	Square pyramidal, C_{4v}	190	170	Octahedral	(h)
XeO_3F_2	Trigonal bipyramidal, D_{3h}			Trigonal bipyramidal	(i)
XeF_5^+ in $(XeF_5)(PtF_6)$	Square pyramidal, C_{4v}	179–85		Octahedral	(j)
XeF_5^- in $(NMe_4)(XeF_5)$	Pentagonal planar, D_{5h}	189–203		Pentagonal bipyramidal	(k)
XeF_6	Distorted octahedral, C_{3v}	189(av)		Capped octahedral	(l)
XeO_6^{4-} in $K_4XeO_6 \cdot 9H_2O$	Octahedral, O_h		186	Octahedral	(m)
XeF_7^- in $CsXeF_7$	Capped octahedral, C_s	193–210		Capped octahedral	(n)
XeF_8^{2-} in $(NO)_2XeF_8$	Square antiprismatic, D_{4d}	196–208		Square antiprismatic	(o)

only nearest neighbor atoms are considered or whether the electron lone pairs are also taken into consideration.

Figure 17.5.2 shows that the known formal oxidation state of xenon ranges from +2 (XeF_2) to +8 (XeO_4, XeO_3F_2 and XeO_6^{4-}), and the structures of the xenon compounds are all consistent with the VSEPR model.

17.5.3 Chemical bonding in xenon fluorides

(1) Xenon difluoride and xenon tetrafluoride

The XeF_2 molecule is linear. A simple bonding description takes the $5p_z$ AO of the xenon atom and the $2p_z$ AO of each fluorine atom to construct the MOs of XeF_2: bonding (σ), nonbonding (σ^n), and antibonding (σ^*), as shown in Fig. 17.5.3. The four valence electrons fill σ and σ^n, forming a 3c-4e σ bond extending over the entire F–Xe–F system. Hence the formal bond order of the Xe–F bond can be taken as 0.5. The remaining 5s, $5p_x$ and $5p_y$ AOs of the Xe

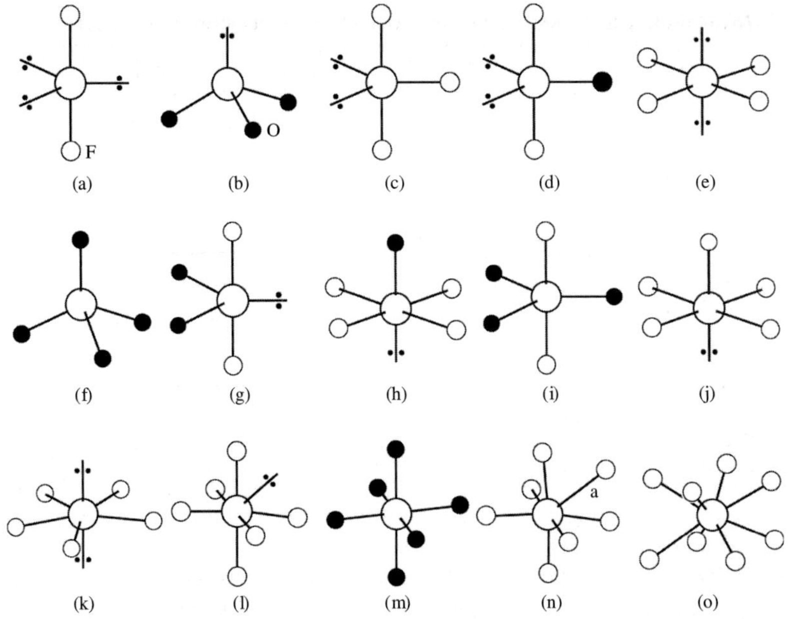

Fig. 17.5.2.
Structures of xenon compounds (small shaded circles represent the O atoms): (a) XeF_2, (b) XeO_3, (c) XeF_3^+, (d) $XeOF_2$, (e) XeF_4, (f) XeO_4, (g) XeO_2F_2, (h) $XeOF_4$, (i) XeO_3F_2, (j) XeF_5^+, (k) XeF_5^-, (l) XeF_6, (m) XeO_6^{4-}, (n) XeF_7^-, and (o) XeF_8^{2-}. The electron lone pairs in XeF_7^- and XeF_8^{2-} are not shown in the figure. The XeF_6 molecule (l) has no static structure but is continually interchanging between eight possible C_{3v} structures in which the lone pair caps a triangular face of the octahedron, but in the figure the lone pair is shown in only one possible position. Also, connecting these C_{3v} structures are the transition states with C_{2v} symmetry in which the lone pair pokes out from an edge of the octahedron.

atom are hybridized to form sp² hybrid orbitals to accommodate the three lone pairs, as shown in Fig. 17.5.2(a).

A similar treatment, involving two 3c-4e bonds located in X and Y directions, accounts satisfactorily for the planar structure of XeF_4. The remaining 5s and $5p_z$ AOs of Xe are hybridized to form sp hybrid orbitals that accommodate the two lone-pair electrons, as shown in Fig. 17.5.2(e). The crystal structure of XeF_2 is analogous to that of α-KrF_2, which is illustrated in Fig. 17.5.8(a) (see below).

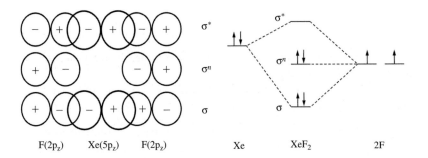

Fig. 17.5.3.
Molecular orbitals of the 3c-4e F–Xe–F σ bond of XeF_2: (a) the possible combinations of the AOs of Xe and F atoms, and (b) the schematic energy levels of XeF_2 molecule.

(2) Xenon hexafluoride

There are two possible theoretical models for the isolated XeF_6 molecule: (i) regular octahedral (O_h symmetry), with three 3c-4e bonds and a sterically inactive lone pair of electrons occupying a spherically symmetric s orbital, or (ii) distorted octahedral (C_{3v} symmetry), where the lone pair is sterically active and lies above the center of one face, but the molecule is readily converted into other configurations. All known experimental data are consistent with the following nonrigid model. Starting from the C_{3v} structure, the electron pair can move across an edge between two F atoms (C_{2v} symmetry for the intermediate configuration) to an equivalent position surrounded by three F atoms. This continuous molecular rearrangement, designated as a $C_{3v} \to C_{2v} \to C_{3v}$ transformation, involves only modest changes in bond angles and virtually no change in bond lengths.

(3) Perfluoroxenates XeF_8^{2-} and XeF_7^-

There are nine electron pairs in XeF_8^{2-}, which are to be accommodated around the Xe atom. From experimental data, the XeF_8^{2-} ion has a square antiprismatic structure [Fig. 17.5.2(o)] showing no distortion that could reveal a possible position for the nonbonding pair of electrons. The lone pair presumably resides in the spherical 5s orbital.

There are eight electron pairs to be placed around the Xe atom in XeF_7^-. Six F atoms are arranged octahedrally around the central Xe atom. The approach of the seventh F atom [labelled "a" for this F atom in Fig. 17.5.2(n)] towards the midpoint of one of the faces causes severe distortion of the basic octahedral shape. Such a distortion could easily mask similar effects arising from a nonspherical, sterically active lone pair. The bond between Xe atom and the capping fluorine atom $F_{(a)}$ has the length Xe–$F_{(a)}$ 210 pm, or about 17 pm longer than the other Xe–F bonds, which might suggest some influence of a lone pair in the $F_{(a)}$ direction.

17.5.4 Structures of some inorganic xenon compounds

The structure and bonding of the $[AuXe_4]^{2+}$ cation in $(AuXe_4)(Sb_2F_{11})_2$ have been discussed in Section 2.4.3. Some other interesting inorganic xenon compounds are described below.

(1) $Xe_2Sb_4F_{21}$

Although the Xe_2^+ cation is known to exist from Raman spectroscopy, with $\nu(Xe–Xe) = 123$ cm^{-1}, its bond length was only determined in 1997 from the crystal structure of $Xe_2Sb_4F_{21}$. The Xe–Xe bond length is 309 pm, the longest recorded homonuclear bond between main-group elements.

(2) Complexes of xenon fluorides

The pentafluoride molecules, such as AsF_5 and RuF_5, act essentially as fluoride ion acceptors, so that their complexes with xenon fluorides can be formulated as salts containing cationic xenon species, e.g., $[XeF][AsF_6]$, $[XeF][RuF_6]$, $[Xe_2F_3][AsF_6]$, $[XeF_3][Sb_2F_{11}]$, and $[XeF_5][AgF_4]$.

In a number of adducts of XeF_2 with metal complexes, the configuration at the Xe atom remains virtually linear, with one F atom lying in the coordination sphere of the metal atom. The ion Xe–F$^+$ does not ordinarily occur as a discrete ion, but rather is attached covalently to a fluorine atom on the anion. Figure 17.5.4(a) shows the structure of $[XeF][RuF_6]$. The terminal Xe–$F_{(t)}$ bond is appreciably shorter than that in XeF_2 (200 pm), while the bridging Xe–$F_{(b)}$ bond is longer. Table 17.5.2 lists the Xe–F bond lengths in some adducts of this type.

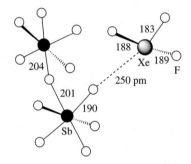

Fig. 17.5.4. Structure of (a) [XeF][RuF$_6$] and (b) [Xe$_2$F$_{11}$]$_2$[NiF$_6$].

In the crystal structure of [XeF$_3$][Sb$_2$F$_{11}$], the T-shaped XeF$_3^+$ cation is nearly coplanar with a F atom of a Sb$_2$F$_{11}^-$ group in a close Xe···F contact, as shown below. The terminal Sb–F bond lengths of the anion lie in the range 183–9 pm.

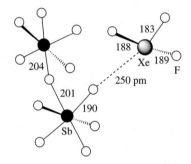

The crystal structure of [XeF$_5$][AgF$_4$] consists of alternate stacks of double layers of square-pyramidal XeF$_5^+$ cations and approximately square-planar AgF$_4^-$ anions (site symmetry D_{2h}, Ag–F = 190.2 pm). The XeF$_5^+$ ion has C_{4v} symmetry with Xe–F(ax) = 185.2, Xe–F(eq) = 182.6 pm, and F(ax)–Xe–F(eq) = 77.7°. Each cation lying on a 4-axis interacts with one bridging F ligand of each of four anions at 263.7 pm, as illustrated below:

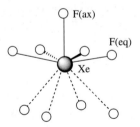

Table 17.5.2. Xe–F bond lengths in some complexes containing XeF$^+$

Compound	Xe–F$_{(t)}$ (pm)	Xe–F$_{(b)}$ (pm)
F–Xe–FAsF$_5$	187	214
F–Xe–FRuF$_5$	187	218
F–Xe–FWOF$_4$	189	204
F–Xe–Sb$_2$F$_{10}$	184	235

The [Xe$_2$F$_3$]$^+$ cation in [Xe$_2$F$_3$][AsF$_6$] is V-shaped, as shown below:

The [Xe$_2$F$_{11}$]$^+$ cation can be considered as [F$_5$Xe $\cdots$ F $\cdots$ XeF$_5$]$^+$ by analogy to [Xe$_2$F$_3$]$^+$. The compounds [Xe$_2$F$_{11}^+$]$_2$ [NiF$_6^{2-}$] and [Xe$_2$F$_{11}^+$] [AuF$_6^-$] contain Ni(IV) and Au(V), respectively. Figure 17.5.4(b) shows the structure of [Xe$_2$F$_{11}$]$_2$[NiF$_6$].

(3) FXeOSO$_2$F

When XeF$_2$ reacts with an anhydrous acid, such as HOSO$_2$F, elimination of HF occurs to yield FXeOSO$_2$F, which contains a linear F–Xe–O group, as shown in Fig. 17.5.5(a).

(4) FXeN(SO$_2$F)$_2$

This compound is produced by the replacement of a F atom in XeF$_2$ by a N(SO$_2$F)$_2$ group from HN(SO$_2$F)$_2$. Its molecular structure has a linear F–Xe–N fragment and a planar configuration at the N atom, as shown in Fig. 17.5.5(b).

(5) Cs$_2$(XeO$_3$Cl$_2$)

This salt is obtained from the reaction of XeO$_3$ with CsCl in aqueous HCl solution, and contains an anionic infinite chain. Figure 17.5.5(c) shows the structure of the chain [Xe$_2$O$_6$Cl$_4$]$_n^{4n-}$.

(6) M(CO)$_5$E (M = Cr, Mo, W; E = Ar (W only), Kr, Xe)

These complexes are transient species generated from photolysis of M(CO)$_6$ in supercritical noble gas solution at room temperature. Their octahedral structure (symmetry C_{4v}) has been characterized by IR spectroscopy and solution NMR. The noble gas atom E can be formally considered as a neutral two-electron donor ligand, and the bonding involves interactions between the p orbitals of E and orbitals on the equatorial CO groups. The stabilities of the complexes decrease in the order W > Mo ∼ Cr and Xe > Kr > Ar. Organometallic noble gas complexes

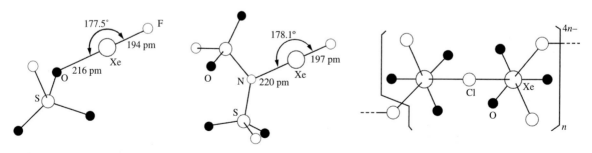

Fig. 17.5.5.
Structures of some inorganic xenon compounds: (a) FXeOSO$_2$F, (b) FXeN(SO$_2$F)$_2$, and (c) [Xe$_2$O$_6$Cl$_4$]$_n^{4n-}$.

that can be generated using matrix isolation techniques include Fe(CO)$_4$Xe, Rh(η^5-C$_5$R$_5$)(CO)E (R = H, Me; E = Kr, Xe), M(η^5-C$_5$R$_5$)(CO)$_2$E (M = Mn, Re; R = H, Me, Et (Mn only); E = Ar (only with Re and C$_5$H$_5$), Kr, Xe), and M(η^5-C$_5$H$_5$)(CO)$_3$Xe (M = Nb, Ta).

17.5.5 Structures of some organoxenon compounds

More than ten organoxenon compounds which contain Xe–C bonds have been prepared and characterized. The first structural characterization of a Xe–C bond was performed on [MeCN–Xe–C$_6$H$_5$]$^+$[(C$_6$H$_5$)$_2$BF$_2$]$^-$ · MeCN. The structure of the cation [MeCN–Xe–C$_6$H$_5$]$^+$ is shown in Fig. 2.4.4. The Xe–C bond length is 209.2 pm, and the Xe–N bond length is 268.1 pm, with the Xe atom in a linear environment. This compound has no significant fluorine bridge between cation and anion, as the shortest intermolecular Xe $\cdots$ F distance is 313.5 pm.

Other examples of compounds containing Xe–C bonds are shown in Fig. 17.5.6 and described below.

(1) Xe(C$_6$F$_5$)$_2$

This is the first reported homoleptic organoxenon(II) compound with two Xe–C bonds of length 239 and 235 pm, longer than the corresponding bond in other compounds by about 30 pm. The C–Xe–C unit is almost linear (angle 178°) and the two C$_6$F$_5$ rings are twisted by 72.5° with respect to each other.

(2) [(C$_6$F$_5$Xe)$_2$Cl][AsF$_6$]

This is the first isolated and unambiguously characterized xenon(II) chlorine compound. The cation [(C$_6$F$_5$Xe)$_2$Cl]$^+$ consists of two C$_6$F$_5$Xe fragments bridged through a chloride ion. Each linear C–Xe–Cl linkage can be considered to involve an asymmetric hypervalent 3c-4e bond. Thus a shorter Xe–C distance (mean value 211.3 pm) occurs and is accompanied by a longer Xe–Cl distance (mean value 281.6 pm). The Xe–Cl–Xe angle is 117°.

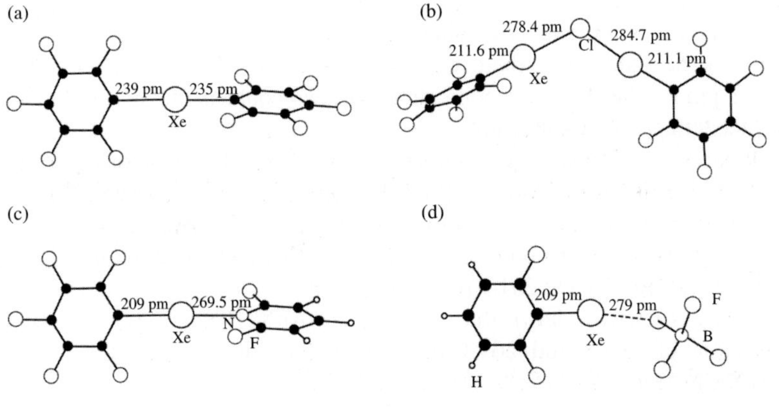

Fig. 17.5.6. Structures of some organoxenon compounds: (a) Xe(C$_6$F$_5$)$_2$, (b) the cation [(C$_6$F$_5$Xe)$_2$Cl]$^+$, (c) the cation (2,6-F$_2$H$_3$C$_5$N–Xe–C$_6$F$_5$)$^+$, and (d) 2,6-F$_2$H$_3$C$_6$–Xe–FBF$_3$.

(3) [2,6-F$_2$H$_3$C$_5$N–Xe–C$_6$F$_5$][AsF$_6$]

In the structure of the cation (2,6-F$_2$H$_3$C$_5$N–Xe–C$_6$F$_5$)$^+$, the Xe atom is (exactly) linearly bonded to C and N atoms; the bond lengths are Xe–C 208.7 pm and Xe–N 269.5 pm.

(4) 2,6-F$_2$H$_3$C$_6$–Xe–FBF$_3$

In this compound the Xe atom is linearly bonded to C and F atoms; bond lengths are Xe–C 209.0 pm and Xe–F 279.36 pm and the bond angle C–Xe–F is 167.8°.

The structural data of three other organoxenon compounds are listed below:

Compound	Xe–C (pm)	Xe–E (pm)	C–Xe–E
(C$_6$F$_5$–Xe)(AsF$_6$)	(1) 207.9	271.4	170.5°
	(2) 208.2	267.2	174.2°
(C$_6$F$_5$–Xe)(OCC$_6$F$_5$) [O double bond]	212.2	236.7	178.1°
(F$_2$H$_3$C$_6$–Xe)(OSO$_2$CF$_3$)	(1) 207.4	268.7	173.0°
	(2) 209.2	282.9	165.1°

In the known organoxenon compounds, the C–Xe–E angle (E = F, O, N, Cl) deviates only slightly from 180°, indicating a hypervalent 3c-4e bond in all cases. The Xe–C bond lengths, except in the compound Xe(C$_6$F$_5$)$_2$, vary within the range 208–12 pm. All Xe···E contacts are significantly shorter than the sum of the van der Waals radii of Xe and E, indicating at least a weak secondary Xe···E interaction.

17.5.6 Gold–xenon complexes

Xenon can act as a complex ligand to form M–Xe bonds, especially with gold, which exhibits significant relativistic effects in view of its electronic structure, as discussed in Section 2.4.3. Some gold–xenon complexes have been prepared and characterized, and their structures are shown in Fig. 17.5.7.

The compound [AuXe$_4^{2+}$][Sb$_2$F$_{11}^-$]$_2$ is mentioned in Section 2.4.3 in the context of relativistic effects. In fact it exists in two crystallographically distinct modifications: triclinic and tetragonal. The cation [AuXe$_4$]$^{2+}$ is square planar with Au–Xe bond lengths ranging from 267.0 to 277.8 pm (Fig. 17.5.7(a)). Around the gold atom, there are three weak Au···F contacts of 267.1 to 315.3 pm for the triclinic modification, and two contacts of Au···F 292.8 pm for the tetragonal modification.

Figures 17.5.7(b) and (c) compare the structure of trans-[AuXe$_2$][SbF$_6$]$_2$ with that of the trans-[AuXe$_2$F][SbF$_6$] moiety in crystalline [AuXe$_2$F][SbF$_6$][Sb$_2$F$_{11}$]. Each Au atom resides in a square-planar environment, with two Xe atoms and two F atoms bonded to it. The Au–Xe bond length is 270.9 pm for the former and 259.3 to 261.9 pm for the latter.

Figure 17.5.7(d) shows the structure of [AuXe$_2$][Sb$_2$F$_{11}$]$_2$. The Au–Xe distances of 265.8 and 267.1 pm are slightly shorter than those in the [AuXe$_4$]$^{2+}$ ion.

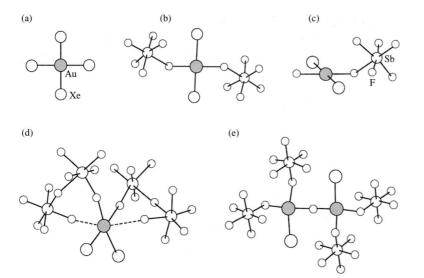

Fig. 17.5.7.
Structures of some gold–xenon complexes: (a) [AuXe$_4$]$^{2+}$, (b) AuXe$_2$(SbF$_6$)$_2$, (c) (AuXe$_2$F)(SbF$_6$), (d) (AuXe$_2$)(Sb$_2$F$_{11}$)$_2$, and (e) (Au$_2$Xe$_2$F)$^{3+}$ and its immediate anion environment in [Au$_2$Xe$_2$F][SbF$_6$]$_3$.

Figure 17.5.7(e) shows the structure of the Z-shaped binuclear [Xe–Au–F–Au–Xe]$^{3+}$ ion and its immediate anion environment in [Au$_2$Xe$_2$F][SbF$_6$]$_3$. The Au atom is coordinated by one Xe atom and three F atoms in a square-planar configuration, with Au–Xe bond length 264.7 pm.

17.5.7 Krypton compounds

The known compounds of krypton are limited to the +2 oxidation state, and the list includes the following:

(a) KrF$_2$
(b) Salts of KrF$^+$ and Kr$_2$F$_3^+$:
 [KrF][MF$_6$] (M = P, As, Sb, Bi, Au, Pt, Ta, Ru)
 [KrF][M$_2$F$_{11}$] (M = Sb, Ta, Nb)
 [Kr$_2$F$_3$][MF$_6$] (M = As, Sb, Ta)
 [KrF][AsF$_6$]·[Kr$_2$F$_3$][AsF$_6$]
(c) Molecular adducts:
 KrF$_2$·MOF$_4$ (M = Cr, Mo, W)
 KrF$_2$ · nMoOF (n = 2, 3)
 KrF$_2$·VF$_5$
 KrF$_2$·MnF$_4$
 KrF$_2$·[Kr$_2$F$_3$][SbF$_5$]$_2$
 KrF$_2$·[Kr$_2$F$_3$][SbF$_6$]
(d) Other types
 salts of [RCN–KrF]$^+$ (R = H, CF$_3$, C$_2$F$_5$, nC$_3$F$_7$)
 Kr(OTeF$_5$)$_2$

(1) Structure of KrF$_2$

Krypton difluoride KrF$_2$ exists in two forms in the solid state: α-KrF$_2$ and β-KrF$_2$, whose crystal structures are shown in Fig. 17.5.8. Specifically, α-KrF$_2$

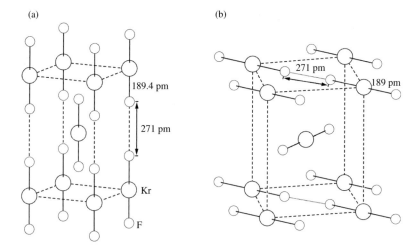

Fig. 17.5.8.
Crystal structure of (a) α-KrF$_2$ and (b) β-KrF$_2$.

crystallizes in a body-centered tetragonal lattice, with space group *I*4/*mmm*, and all the KrF$_2$ molecules are aligned parallel to the *c* axis. In contrast, β-KrF$_2$ belongs to tetragonal space group *P*4$_2$/*mnm*, in which the KrF$_2$ molecules located at the corners of the unit cell all lie in the *ab* plane and are rotated by 45° with respect to the *a* axis. The central KrF$_2$ molecule also lies in the *ab* plane, but its molecular axis is orientated perpendicular to those of the corner KrF$_2$ molecules.

The Kr–F bond length in α-KrF$_2$ is 189.4 pm and is in excellent agreement with those determined for β-KrF$_2$, 189 pm, by X-ray diffraction and for gaseous KrF$_2$ by electron diffraction, 188.9 pm. The interatomic F···F distance between collinearly orientated KrF$_2$ molecules is 271 pm in both structures.

(2) Structures of [KrF][MF$_6$] (M = As, Sb, Bi)

These three compounds form an isomorphous series, in which the [KrF]$^+$ cation strongly interacts with the anion by forming a fluorine bridge with the pseudo-octahedral anion bent about F$_b$, as shown in Fig. 17.5.9. The terminal Kr–F$_t$ bond lengths in these salts (176.5 pm for [KrF][AsF$_6$] and [KrF][SbF$_6$], 177.4 pm for [KrF][BiF$_6$]) are shorter, and the Kr–F$_b$ bridge bond lengths (213.1 pm for [KrF][AsF$_6$], 214.0 pm for [KrF][SbF$_6$], and 209.0 pm for [KrF][BiF$_6$]) are longer, than the Kr–F bonds of α-KrF$_2$ (189.4 pm).

The Kr–F$_b$–M bridge bond angles (133.7° for [KrF][AsF$_6$], 139.2° for [KrF][SbF$_6$], and 138.3° for [KrF][BiF$_6$]) are consistent with the bent geometry predicted by the VSEPR arrangements at their respective F$_b$ atoms, but are more open than the ideal tetrahedral angle.

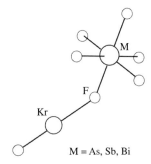

Fig. 17.5.9.
Structure of [KrF][MF$_6$] (M = AS, Sb, Bi).

References

1. N. N. Greenwood and A. Earnshaw, *Chemistry of the Elements*, 2nd edn., Butterworth-Heinemann, Oxford, 1997.
2. A. G. Massey, *Main Group Chemistry*, 2nd edn., Wiley, Chichester, 2000.

3. C. E. Housecroft and A. G. Sharpe, *Inorganic Chemistry*, 2nd edn., Prentice-Hall, Harlow, 2004.
4. D. F. Shriver, P. W. Atkins, T. L. Overton, J. P. Rourke, M. T. Weller and F. A. Armstrong, *Inorganic Chemistry*, 4th edn., Oxford University Press, Oxford, 2006.
5. G. Meyer, D. Naumann and L. Wesemann (eds.), *Inorganic Chemistry Highlights*, Wiley–VCH, Weimheim, 2002.
6. M. Pettersson, L. Khriachtchev, J. Lundell and M. Räsänen, Noble gas hydride compounds. In G. Meyer, D. Naumann and L. Wesemann (eds.), *Inorganic Chemistry in Focus II*, Wiley–VCH, Weinheim, 2005, pp. 15–34.
7. N. Bartlett (ed.), *The Oxidation of Oxygen and Related Chemistry: Selected Papers of Neil Bartlett*, World Scientific, Singapore, 2001.
8. K. Akiba (ed.), *Chemistry of Hypervalent Compounds*, Wiley–VCH, New York, 1999.
9. A. J. Blake, F. A. Devillanova, R. O. Gould, W.-S. Li, V. Lippolis, S. Parsons, C. Radek and M. Schröder, Template self-assembly of polyiodide networks. *Chem. Soc. Rev.* **27**, 195–205 (1998).
10. J. Li, S. Irle, and W. H. E. Schwarz, Electronic structure and properties of trihalogen X_3^+ and XY_2^+. *Inorg. Chem.* **35**, 100–9 (1996).
11. D. B. Morse, T. B. Rauchfuss and S. R. Wilson, Main-group-organotransition metal chemistry: the cyclopentadienylchromium polyiodides including [$(C_5Me_5)_2Cr_2I_3^+$]$_2$[I_{16}^{2-}]. *J. Am. Chem. Soc.* **112**, 1860–4 (1990).
12. T. Kraft and M. Jansen, Crystal structure determination of metaperiodic acid, HIO_4, with combined X-ray and neutron diffraction. *Angew. Chem. Int. Ed.* **36**, 1753–4 (1997).
13. M. Jansen and T. Kraft, The structural chemistry of binary halogen oxides in the solid state. *Chem. Ber,* **130**, 307–15 (1997).
14. J. H. Holloway and E. G. Hope, Recent advances in noble-gas chemistry. *Adv. Inorg. Chem.* **46**, 51–100 (1999).
15. T. Drews and K. Seppelt, The Xe_2^+ ion – preparation and structure. *Angew. Chem. Int. Ed.* **36**, 273–4 (1997).
16. O. S. Jina, X. Z. Sun and M. W. George, Do early and late transition metal noble gas complexes react by different mechanisms? A room temperature time-resolved infrared study of (η^5-C_5H_5)Rh(CO)$_2$ (R = H or Me) in supercritical noble gas solution at room temperature. *Dalton Trans.*, 1773–8 (2003).
17. H. Bock, D. Hinz-Hübner, U. Ruschewitz and D. Naumann, Structure of bis(pentafluorophenyl)xenon, $Xe(C_6F_5)_2$. *Angew. Chem. Int. Ed.* **41**, 448–50 (2002).
18. J. F. Lehmann, D. A. Dixon and G. J. Schrobilgen, X-ray crystal structures of α-KrF_2, [KrF][MF_6] (M = As, Sb, Bi), [Kr_2F_3][SbF_6]·KrF_2, [Kr_2F_3]$_2$[SbF_6]$_2$·KrF_2, and [Kr_2F_3][AsF_6]·[KrF][AsF_6]; Synthesis and characterization of [Kr_2F_3][PF_6]·$nKrF_2$; and theoretical studies of KrF_2, KrF^+, $Kr_2F_3^+$, and the [KrF][MF_6] (M = P, As, Sb, Bi) ion pairs. *Inorg. Chem.* **40**, 3002–17 (2001).
19. T. Drews, S. Seidal and K. Seppelt, Gold–xenon complexes. *Angew. Chem. Int. Ed.* **41**, 454–6 (2002).
20. B. Liang, L, Andrews, J. Li and B. E. Bursten, On the noble-gas-induced intersystem crossing for the CUO molecule. *Inorg. Chem.* **43**, 882–94 (2004).

18 Structural Chemistry of Rare-Earth Elements

18.1 Chemistry of rare-earth metals

The lanthanides (Ln) include lanthanum (La) and the following fourteen elements—Ce, Pr, Nd, Pm, Sm, Eu, Gd, Tb, Dy, Ho, Er, Tm, Yb and Lu—in which the 4f orbitals are progressively filled. These fifteen elements together with scandium (Sc) and yttrium (Y) are termed the rare-earth metals. The designation of *rare earths* arises from the fact that these elements were first found in rare minerals and were isolated as oxides (called earths in the early literature). In fact, their occurrence in nature is quite abundant, especially in China, as reserves have been estimated to exceed 84×10^6 tons. In a broader sense, even the actinides (the 5f elements) are sometimes included in the rare-earth family.

The rare-earth metals are of rapidly growing importance, and their availability at quite inexpensive prices facilitates their use in chemistry and other applications. Much recent progress has been achieved in the coordination chemistry of rare-earth metals, in the use of lanthanide-based reagents or catalysts, and in the preparation and study of new materials. Some of the important properties of rare-earth metals are summarized in Table 18.1.1. In this table, r_M is the atomic radius in the metallic state and $r_{M^{3+}}$ is the radius of the lanthanide(III) ion in an eight-coordinate environment.

18.1.1 Trends in metallic and ionic radii: lanthanide contraction

The term *lanthanide contraction* refers to the phenomenon of a steady decrease in the radii of the Ln^{3+} ions with increasing atomic number, from La^{3+} to Lu^{3+}, amounting overall to 18 pm (Table 18.1.1). A similar contraction occurs for the metallic radii and is reflected in many smooth and systematic changes, but there are marked breaks at Eu and Yb. Lanthanide contraction arises because the 4f orbitals lying inside the 4d, 5s, and 5p orbitals provide only incomplete shielding of the outer electrons from the steadily increasing nuclear charge. Therefore the outer electron cloud as a whole steadily shrinks as the 4f subshell is filled. In recent years, theoretical work suggests that relativistic effects also plays a significant role. Further details are given in Section 2.4.3.

The spectacular irregularity in the metallic radii of Eu and Yb occurs because they have only two valence electrons in the conduction band, whereas the other lanthanide metals have three valence electrons in the 5d/6s conduction band. Therefore, lanthanide contraction is not manifested by Eu and Yb in the metallic

Table 18.1.1. Properties of the rare-earth metals

	Symbol name	Electronic configuration		r_M (pm) (CN = 12)	$r_{M^{3+}}$ (pm) (CN = 8)	Crystal structure	Lattice parameters	
		Metal	M^{3+}				a (pm)	c (pm)
Sc	Scandium	[Ar]$3d^1 4s^2$	[Ar]	164.1	87.1	hcp	330.9	526.8
Y	Yttrium	[Kr]$4d^1 5s^2$	[Kr]	180.1	101.9	hcp	364.8	573.2
La	Lanthanum	[Xe]$5d^1 6s^2$	[Xe]	187.9	116.0	dhcp	377.4	1217.1
Ce	Cerium	[Xe]$4f^1 5d^1 6s^2$	[Xe]$4f^1$	182.5	114.3	ccp	516.1	—
Pr	Praseodymium	[Xe]$4f^3 6s^2$	[Xe]$4f^2$	182.8	112.6	dhcp	367.2	1183.3
Nd	Neodymium	[Xe]$4f^4 6s^2$	[Xe]$4f^3$	182.1	110.9	dhcp	365.8	1179.7
Pm	Promethium	[Xe]$4f^5 6s^2$	[Xe]$4f^4$	(181.0)	(109.5)	dhcp	365.0	1165
Sm	Samarium	[Xe]$4f^6 6s^2$	[Xe]$4f^5$	180.4	107.9	rhom	362.9	2620.7
Eu	Europium	[Xe]$4f^7 6s^2$	[Xe]$4f^6$	204.2	106.6	bcp	458.3	—
Gd	Gadolinium	[Xe]$4f^7 5d^1 6s^2$	[Xe]$4f^7$	180.1	105.3	hcp	363.4	578.1
Tb	Terbium	[Xe]$4f^9 6s^2$	[Xe]$4f^8$	178.3	104.0	hcp	360.6	569.7
Dy	Dysprosium	[Xe]$4f^{10} 6s^2$	[Xe]$4f^9$	177.4	102.7	hcp	359.2	565.0
Ho	Holmium	[Xe]$4f^{11} 6s^2$	[Xe]$4f^{10}$	176.6	101.5	hcp	357.8	561.8
Er	Erbium	[Xe]$4f^{12} 6s^2$	[Xe]$4f^{11}$	175.7	100.4	hcp	355.9	558.5
Tm	Thulium	[Xe]$4f^{13} 6s^2$	[Xe]$4f^{12}$	174.6	99.4	hcp	353.8	555.4
Yb	Ytterbium	[Xe]$4f^{14} 6s^2$	[Xe]$4f^{13}$	193.9	98.5	ccp	548.5	—
Lu	Lutetium	[Xe]$4f^{14} 5d^1 6s^2$	[Xe]$4f^{14}$	173.5	97.7	hcp	350.5	554.9

state or in some compounds such as the hexaborides, where the lower valence of these two elements is evident from the large radii compared with those of neighboring elements.

Lanthanide contraction has important consequences for the chemistry of the third-row transition metals. The reduction in radius caused by the poor shielding ability of the 4f electrons means that the third-row transition metals are approximately the same size as their second-row congeners, and consequently exhibit similar chemical behavior. For instance, it has been shown that the covalent radius of gold (125 pm) is less than that of silver (133 pm).

18.1.2 Crystal structures of the rare-earth metals

The 17 rare-earth metals are known to adopt five crystalline forms. At room temperature, nine exist in the hexagonal closest packed structure, four in the double c-axis hcp (dhcp) structure, two in the cubic closest packed structure and one in each of the body-centered cubic packed and rhombic (Sm-type) structures, as listed in Table 18.1.1. This distribution changes with temperature and pressure as many of the elements go through a number of structural phase transitions. All of the crystal structures, with the exception of bcp, are closest packed, which can be defined by the stacking sequence of the layers of close-packed atoms, and are labeled in Fig. 18.1.1.

hcp: AB···   dhcp: ABAC···
ccp: ABC···   Sm-type: ABABCBCAC···

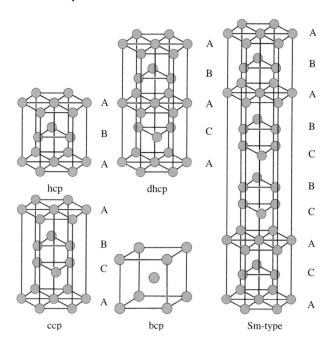

Fig. 18.1.1.
Structure types of rare-earth metals.

If Ce, Eu, and Yb are excluded, then the remaining 14 rare-earth metals can be divided into two major subgroups:

(a) The heavy rare-earth metals Gd to Lu, with the exception of Yb and the addition of Sc and Y—these metals adopt the hcp structure.
(b) The light rare-earth metals La to Sm, with the exception of Ce and Eu—these metals adopt the dhcp structure. (The Sm-type structure can be viewed as a mixture of one part of ccp and two parts of hcp.)

Within each group, the chemical properties of the elements are very similar, so that they invariably occur together in mineral deposits.

18.1.3 Oxidation states

Because of stability of the half-filled and filled 4f subshell, the electronic configuration of the Ln atom is either $[Xe]4f^n5d^06s^2$ or $[Xe]4f^n5d^16s^2$.

The most stable and common oxidation state of Ln is +3. The principal reason is that the fourth ionization energy I_4 of a rare-earth atom is greater than the sum of the first three ionization energies ($I_1 + I_2 + I_3$), as listed in Table 18.1.2. The energy required to remove the fourth electron is so great that in most cases it cannot be compensated by chemical bond formation, and thus the +4 oxidation state rarely occurs.

Although the +3 oxidation state dominates lanthanide chemistry, other oxidation states are accessible, especially if a $4f^0$, $4f^7$ or $4f^{14}$ configuration is generated. The most common 2+ ions are Eu^{2+} ($4f^7$) and Yb^{2+} ($4f^{14}$), and the most common 4+ ions are Ce^{4+} ($4f^0$) and Tb^{4+} ($4f^7$). Of the five lanthanides that exhibit tetravalent chemistry, Nd^{4+} and Dy^{4+} are confined to solid-state

Table 18.1.2. Ionization energies of rare-earth elements (kJ mol^{-1})

Element	I_1	I_2	I_3	$(I_1 + I_2 + I_3)$	I_4
Sc	633	1235	2389	4257	7091
Y	616	1181	1980	3777	5963
La	538	1067	1850	3455	4819
Ce	527	1047	1949	3523	3547
Pr	523	1018	2086	3627	3761
Nd	530	1035	2130	3695	3899
Pm	536	1052	2150	3738	3970
Sm	543	1068	2260	3871	3990
Eu	547	1085	2404	4036	4110
Gd	593	1167	1990	3750	4250
Tb	565	1112	2114	3791	3839
Dy	572	1126	2200	3898	4001
Ho	581	1139	2204	3924	4100
Er	589	1151	2194	3934	4115
Tm	597	1163	2285	4045	4119
Yb	603	1176	2415	4194	4220
Lu	524	1340	2022	3886	4360

fluoride complexes, while Pr^{4+} (4f^1) and Tb^{4+} also form the tetrafluoride and dioxide. The most extensive lanthanide(IV) chemistry is that of Ce^{4+}, for which a variety of tetravalent compounds and salts are known (e.g., CeO_2, $CeF_4 \cdot H_2O$). The common occurrence of Ce^{4+} is attributable to the high energy of the 4f orbitals at the start of the lanthanide series, such that Ce^{3+} is not sufficiently stable to prevent the loss of an electron.

The crystallographic ionic radii of the rare-earth elements in oxidation states +2 (CN = 6), +3 (CN = 6), and +4 (CN = 6) are presented in Table 18.1.3. The data provide a set of conventional size parameters for the calculation of hydration energies. It should be noted that in most lanthanide(III) complexes the Ln^{3+} center is surrounded by eight or more ligands, and that in aqueous solution the primary coordination sphere has eight and nine aqua ligands for light and heavy Ln^{3+} ions, respectively. The crystal radii of Ln^{3+} ions with CN = 8 are listed in Table 18.1.1.

18.1.4 Term symbols and electronic spectroscopy

Atomic and ionic energy levels are characterized by a term symbol of the general form $^{2S+1}L_J$. The values of S, L, and J of lanthanide ions Ln^{3+} in the ground state can be deduced from the arrangement of the electrons in the 4f subshell, which are determined by Hund's rules and listed in Table 18.1.4.

Three types of electronic transition can occur for lanthanide compounds. These are f→f transition, $nf \rightarrow (n+1)d$ transition and ligand → metal f charge-transfer transition.

In Ln^{3+} ions, the 4f orbitals are radially much more contracted than the d orbitals of transition metals, to the extent that the filled 5s and 5p orbitals largely shield the 4f electrons from the ligands. The result is that vibronic coupling is much weaker in Ln^{3+} compounds than in transition-metal compounds, and hence the intensities of electronic transitions are much lower. As many of

Structural Chemistry of Selected Elements

Table 18.1.3. Crystallographic ionic radii (pm) and hydration entropies (kJ mol^{-1}) of the rare-earth elements in oxidation states +2, +3 and +4

Element	$r_{M^{2+}}$ (CN = 6)	$-\Delta_{hyd}H°$ (M^{2+})	$r_{M^{3+}}$ (CN = 6)	$-\Delta_{hyd}H°$ (M^{3+})	$r_{M^{4+}}$ (CN = 8)	$-\Delta_{hyd}H°$ (M^{4+})
Sc	—	—	74.5	—	—	—
Y	—	—	90.0	3640	—	—
La	130.4	—	103.2	3372	—	—
Ce	127.8	—	101.0	3420	96.7	6390
Pr	125.3	1438	99.0	3453	94.9	6469
Nd	122.5	1459	98.3	3484	93.6	6528
Pm	120.6	1474	97.0	3520	92.5	6579
Sm	118.3	1493	95.8	3544	91.2	6639
Eu	116.6	1507	94.7	3575	90.3	6682
Gd	114.0	—	93.8	3597	89.4	6726
Tb	111.9	1546	92.1	3631	88.6	6765
Dy	109.6	1566	91.2	3661	87.4	6824
Ho	107.5	1585	90.1	3692	86.4	6875
Er	105.6	1602	89.0	3718	85.4	6926
Tm	103.8	1619	88.0	3742	84.4	6978
Yb	102.6	1631	86.8	3764	83.5	7026
Lu	—	—	86.1	3777	82.7	7069

these electronic transitions lie in the visible region of the electromagnetic spectrum, the colors of Ln^{3+} compounds are typically less intense than those of the transition metals. The colors of the Ln^{3+} ions in hydrated salts are given in Table 18.1.4. The lack of 4f orbital and ligand interaction means that the f→f

Table 18.1.4. Electronic configuration, ground state term symbol, and magnetic properties of Ln^{3+} ions

Ln^{3+}	4f electronic configuration	Ground state term symbol	Color of Ln^{3+}	Magnetic moment, μ(298 K)/μ_B Calculated	Observed
La^{3+}	4f^0	$^1S_0$	Colorless	0	0
Ce^{3+}	4f^1	$^2F_{5/2}$	Colorless	2.54	2.3–2.5
Pr^{3+}	4f^2	$^3H_4$	Green	3.58	3.4–3.6
Nd^{3+}	4f^3	$^4I_{9/2}$	Lilac	3.62	3.5–3.6
Pm^{3+}	4f^4	$^5I_4$	Pink	2.68	2.7
Sm^{3+}	4f^5	$^6H_{5/2}$	Pale yellow	0.85	1.5–1.6
Eu^{3+}	4f^6	$^7F_0$	Colorless	0	3.4–3.6
Gd^{3+}	4f^7	$^8S_{7/2}$	Colorless	7.94	7.8–8.0
Tb^{3+}	4f^8	$^7F_6$	Very pale pink	9.72	9.4–9.6
Dy^{3+}	4f^9	$^6H_{15/2}$	Pale yellow	10.65	10.4–10.5
Ho^{3+}	4f^{10}	$^5I_8$	Yellow	10.60	10.3–10.5
Er^{3+}	4f^{11}	$^4I_{15/2}$	Pink	9.58	9.4–9.6
Tm^{3+}	4f^{12}	$^3H_6$	Pale green	7.56	7.1–7.4
Yb^{3+}	4f^{13}	$^2F_{7/2}$	Colorless	4.54	4.4–4.9
Lu^{3+}	4f^{14}	$^1S_0$	Colorless	0	0

transition energies for a given Ln^{3+} change little between compounds, and hence the colors of Ln^{3+} are often characteristic. In view of the small interaction of the Ln^{3+} 4f orbitals with the surrounding ligands, the f→f transition energies in Ln^{3+} compounds are well defined, and thus the bands in their electronic absorption spectra are much sharper.

Since f→d transitions are Laporte allowed, they have much higher intensity than f→f transitions. Ligand-to-metal charge-transfer transitions are also Laporte allowed and also have high intensity. These two types of transitions generally fall in the ultraviolet region, so they do not affect the colors of Ln^{3+} compounds. For easily reduced Ln^{3+} (Eu and Yb), they are at lower energy than the f→d transitions, and for easily oxidized ligands they may tail into the visible region of the spectrum, giving rise to much more intensely colored complexes.

The Ln^{2+} ions are often highly colored. This arises because the 4f orbitals in Ln^{2+} are destabilized with respect to those in Ln^{3+}, and hence lie closer in energy to the 5d orbitals. This change in orbital energy separation causes the f→d transitions to shift from the ultraviolet into the visible region of the spectrum.

The fluorescence which arises from f→f transitions within the Ln^{3+} ion is employed in color television sets, the screens of which contain three phosphor emitters. The red emitter is Eu^{3+} in Y_2O_2S or Eu^{3+}:Y_2O_3. The main emissions for Eu^{3+} are between the $^5D_o \to {}^7F_n$ ($n = 4$ to 0) levels. The green emitter is Tb^{3+} in Tb^{3+}:La_2O_2S. The main emissions for Tb^{3+} are between the $^5D_4$ and ^7F_n ($n = 6$ to 0) levels. The best blue emitter is Ag, Al:ZnS, which has no Ln^{3+} component.

18.1.5 Magnetic properties

The paramagnetism of Ln^{3+} ions arises from their unpaired 4f electrons which interact little with the surrounding ligands in Ln^{3+} compounds. The magnetic properties of these compounds are similar to those of the free Ln^{3+} ions. For most Ln^{3+} the magnitude of the spin–orbital interaction in f orbital is sufficiently large, so that the excited levels are thermally inaccessible, and hence the magnetic behavior is determined entirely by the ground level. The effective magnetic moment μ_{eff} of this level is given by the equation

$$\mu_{eff} = g_J \sqrt{J(J+1)},$$

where

$$g_J = \frac{3}{2} + \frac{S(S+1) - L(L+1)}{2J(J+1)}.$$

The calculated μ_{eff} and observed values from experiments are listed in Table 18.1.4 and shown in Fig. 18.1.2. There is good agreement in all cases except for Sm^{3+} and Eu^{3+}, both of which have low-lying excited states ($^6H_{31/2}$ for Sm^{3+}, and $^7F_1$ and $^7F_2$ for Eu^{3+}) which are appreciably populated at room temperature.

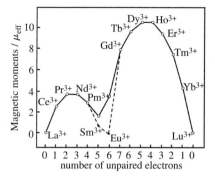

Fig. 18.1.2.
Measured and calculated effective magnetic moments (μ_{eff}) of Ln^{3+} ions at 300 K (broken lines represents the calculated values).

18.2 Structure of oxides and halides of rare-earth elements

The bonding in the oxides and halides of rare-earth elements is essentially ionic. Their structures are determined almost entirely by steric factors and gradual variation across the series, and can be correlated with changes in ionic radii.

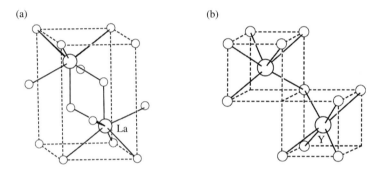

Fig. 18.2.1.
(a) Structure of La_2O_3 and (b) the two coordination environments of Y^{3+} in the Y_2O_3 structure.

18.2.1 Oxides

The sesquioxides M_2O_3 are the stable oxides for all rare-earth elements except Ce, Pr, and Tb, and are the final product of calcination of many salts such as oxalates, carbonates, and nitrates. The sesquioxides M_2O_3 adopt three structural types:

The type-A (hexagonal) structure consists of MO_7 units which approximate to capped octahedral geometry, and is favored by the lightest lanthanides (La, Ce, Pr, and Nd). Figure 18.2.1(a) shows the structure of La_2O_3.

The type-B (monoclinic) structure is related to type-A, but is more complex as it contains three kinds of non-equivalent M atoms, some with octahedral and the remainder with monocapped trigonal prismatic coordination. In the latter type of coordination geometry, the capping O atom is appreciably more distant than those at the vertices of the prism. For example, in Sm_2O_3, the seventh atom is at 273 pm (mean) and the others are at 239 pm (mean). This type is favored by the middle lanthanides (Sm, Eu, Gd, Tb, and Dy).

The type-C (cubic) structure is related to the fluorite structure (Fig. 10.1.6), but with one-quarter of the anions removed in such a way as to reduce the metal coordination number from 8 to 6, resulting in two different coordination geometries, as shown in Fig. 18.2.1(b). This type is favored by Sc, Y, and heavy lanthanides from Nd to Lu.

The dioxides CeO_2 and PrO_2 adopt the fluorite structure with cubic unit-cell parameters a = 541.1 and 539.2 pm, respectively, and Tb_4O_7 is closely related to fluorite.

18.2.2 Halides

(1) Fluorides

Trifluorides are known for all the rare-earth elements. The structure of ScF_3 is close to the cubic ReO_3 structure [Fig. 10.4.4(a)], in which Sc^{3+} has octahedral coordination. The YF_3 structure has a nine-coordinate Y^{3+} in a distorted tricapped trigonal prism, eight F^- at approximately 230 pm, and the ninth at 260 pm. The YF_3 structure type is adopted by the 4f trifluorides from SmF_3 to LuF_3.

The early light trifluorides from LaF_3 to HoF_3 adopt the LaF_3 structure, in which the La^{3+} is 11-coordinate with a fully capped, distorted trigonal prismatic coordination geometry. The neighbors of La^{3+} are seven F^- at 242-8 pm, two F^- at 264 pm, and a further two F^- at 300 pm.

Tetrafluorides LnF_4 are known for Ce, Pr, and Tb. In the structure for LnF_4, the Ln^{4+} ion is surrounded by eight F^- forming a slightly distorted square antiprism, which shares its vertices with eight others.

(2) Chlorides

Chlorides $ScCl_3$, YCl_3, and later $LnCl_3$ (from $DyCl_3$ to $LuCl_3$) adopt the YCl_3 layer structure, in which the small M^{3+} ion is surrounded by an octahedron of Cl^- neighbors. The early trichlorides (from $LaCl_3$ to $GdCl_3$) adopt the $LaCl_3$ structure, in which the large La^{3+} ion is surrounded by nine approximately equidistant Cl^- neighbors in a tricapped trigonal prismatic arrangement. Figure 18.2.2 shows the structure of $LaCl_3$.

Chlorides are also known in oxidation state +2 for Nd, Sm, Eu, Dy, and Tm. The structure of the sesquichloride Gd_2Cl_3 is best formulated as

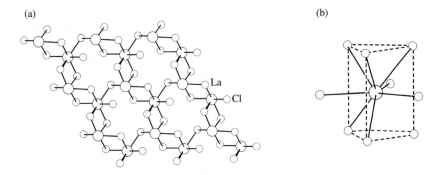

Fig. 18.2.2.
Crystal structures of $LaCl_3$: (a) viewed along the c axis, and (b) coordination geometry of La^{3+}.

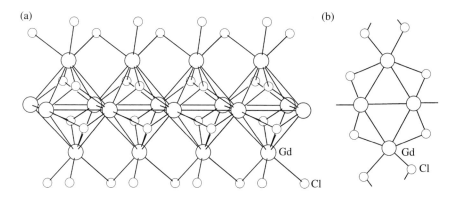

Fig. 18.2.3.
Structure of Gd_2Cl_3: (a) perspective view of a chain aligned along the b axis, and (b) view perpendicular to the b axis.

$[Gd_4]^{6+}[Cl^-]_6$ and is made up of infinitive chains of Gd_6 octahedra sharing opposite edges, with chlorine atoms capping the triangular faces, as shown in Fig. 18.2.3.

(3) Bromides and iodides

The trihalides MBr_3 and MI_3 are known for all the lanthanide elements. The early lanthanide tribromide (La to Pr) adopt the $LaCl_3$ structure, while the later tribromides (from Nd to Lu) and the early triiodides (from La to Nd) form a layer structure with eight-coordinate lanthanide ions.

Ionic dibromides and diiodides are known for Nd, Sm, Eu, Dy, Tm, and Yb. SmI_2, EuI_2, and YbI_2 are useful starting materials for organometallic compounds of these elements in their +2 oxidation states. Iodide SmI_2 is a popular one-electron reducing agent for organic synthesis. The diiodides of La, Ce, Pr, and Gd exhibit metallic properties and are best formulated as $Ln^{3+}(I^-)_2 e^-$ with delocalized electrons. The reduction chemistry of lanthanide(II) compounds will be discussed in Section 18.5.

(4) Oxohalides

Many oxohalides of rare-earth elements have been characterized. The crystal of γ-LaOF has a tetragonal unit cell with $a = 409.1$ pm and $c = 583.6$ pm, space group $P4/nmm$. In this structure, each La^{3+} is coordinated by four O^{2-} and four F^- anions, forming a distorted cube, as shown in Fig. 18.2.4. The distance of La–O is 261.3 pm and La–F 242.3 pm.

18.3 Coordination geometry of rare-earth cations

In lanthanide complexes, the Ln ions are hard Lewis acids, which prefer to coordinate hard bases, such as F, O, N ligands. The f-orbitals are not involved to a significant extent in M–L bonds, so their interaction with ligands is almost electrostatic in nature. Table 18.3.1 lists some examples of the various coordination

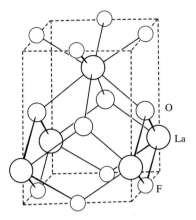

Fig. 18.2.4.
Structure of γ-LaOF.

Table 18.3.1. Coordination number and geometry of rare-earth cations

Oxidation state	CN	Coordination geometry*	Example (structure)
+2	6	Octahedral	Yb(PPh$_2$)$_2$(THF)$_4$, SmO, EuTe
	7	Pentagonal bipyramidal	SmI$_2$(THF)$_5$
	8	Cubic	SmF$_2$
+3	3	Pyramidal	La[N(SiMe$_3$)$_2$]$_3$ [Fig. 18.3.1(a)]
	4	Tetrahedral	[Lu^{t}Bu$_4$]$^-$
	5	Trigonal bipyramidal	Nd[N(SiHMe$_2$)$_2$]$_3$(THF)$_2$
	6	Octahedral	GdCl$_4$(THF)$_2^-$, ScCl$_3$, YCl$_3$, LnCl$_3$ (Dy–Lu)
	6	Trigonal prismatic	Pr[S$_2$P(C$_6$H$_{11}$)$_2$]$_3$
	7	Monocapped trigonal prismatic	Gd$_2$S$_3$, Y(acac)$_3$·H$_2$O
	7	Monocapped octahedral	La$_2$O$_3$ [Fig. 18.2.1(a)]
	8	Square antiprismatic	Nd(CH$_3$CN)(CF$_3$SO$_3$)$_3$L† [Fig. 18.3.1(b)]
	8	Dodecahedral	Lu(S$_2$CNEt$_2$)$_4^-$
	8	Cubic	[La(bipyO$_2$)$_4$]$^{3+}$
	8	Bicapped trigonal prismatic	Gd$_2$S$_3$
	9	Tricapped trigonal prismatic	Nd(H$_2$O)$_9^{3+}$, LaCl$_3$ [Fig. 18.2.2]
	9	Capped square antiprismatic	LaCl$_3$(18C6) [Fig. 18.3.1(c)]
	10	Bicapped dodecahedral	Lu(NO$_3$)$_5^{2-}$ [Fig. 18.3.1(d)]
	10	Irregular	Eu(NO$_3$)$_3$(12C4)
	11	Fully capped trigonal prismatic	LaF$_3$
	11	Irregular	La(NO$_3$)$_3$(15C5) [Fig. 18.3.1(e)]
	12	Icosahedral	La(NO$_3$)$_6^{3-}$ [Fig. 18.3.1(f)]
+4	6	Octahedral	Cs$_2$CeCl$_6$
	8	Square antiprismatic	Ce(acac)$_4$
	8	Cubic	CeO$_2$
	10	Irregular	Ce(NO$_3$)$_4$(OPPh$_3$)$_2$
	12	Icosahedral	[Ce(NO$_3$)$_6$]$^{2-}$

* The polyhedron may exhibit regular or distorted geometry.
† L = 1-methyl-1,4,7,10-tetraazacyclododecane

numbers and geometries of rare-earth cations, which are determined by three factors:

(1) *Steric bulk*. The ligands are packed around the metal ion in such a way as to minimize interligand repulsion, and the coordination number is determined

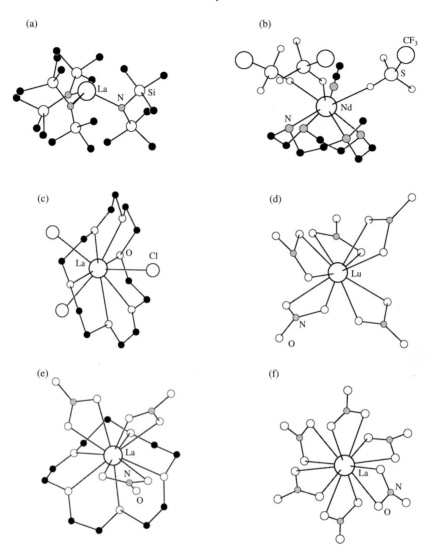

Fig. 18.3.1.
Structures of some lanthanide complexes: (a) La[N(SiMe$_3$)$_2$]$_3$, CN = 3;
(b) Nd(CH$_3$CN)(CF$_3$SO$_3$·)$_3$[CH$_3$—$\overline{\text{N—(CH}_2\text{—CH}_2\text{—NH})_3\text{—CH}_2\text{—CH}_2}$], CN = 8;
(c) LaCl$_3$(18C6), CN = 9; (d) [Lu(NO$_3$)$_5$]$^{2-}$, CN = 10; (e) La(15C5)(NO$_3$)$_3$, CN = 11;
(f) [La(NO$_3$)$_6$]$^{3-}$, CN = 12.

by the steric bulk of the ligands. The coordination number varies over a wide range from 3 to 12, and lower coordination numbers can be achieved with very bulky ligand such as hexamethyldisilylamide. In La[N(SiMe$_3$)$_2$]$_3$, the coordination number is only three, as shown in Fig. 18.3.1(a).

(2) *Ionic size*. The large sizes of lanthanide ions lead to high coordination numbers; eight or nine is very common, and several complexes are known with coordination number 12. For example the NO$_3^-$ ligand, which has a small bite angle, forms a 12-coordinate La^{3+} complex. Figures 18.3.1(b) to 18.3.1(f) show

the structures of Ln^{3+} complexes with coordination numbers ranging from 8 to 12, respectively. The coordination geometries of high coordination complexes are often irregular.

(3) *Chelate effect*. Higher coordination numbers are usually achieved with chelating ligands, such as crown ethers, EDTA and NO_3^- or CO_3^{2-}. Cations Ln^{3+} form a range of complexes with crown ethers. In the crystal structure of $La(18C6)(NO_3)_3$, the La^{3+} ion is coplanar with the six O-donors of the crown ether. The flexibility of 18C6 allows it to pucker and bind effectively to the smaller later lanthanides. Dibenzo-18C6 is much less flexible than 18C6, and can only form complexes with large early Ln^{3+} ions (La to Nd) in the presence of the strongly coordinating bidentate NO_3^- counterion. Complexes of Ln^{3+} with 15C5 or 12C4 are known for La–Lu with NO_3^- counterions. The larger Ln^{3+} ion cannot fit within the cavity of these ligands, so that it sits above the plane of the macrocycle.

In the coordination chemistry of rare-earth metals, compounds containing M–M bonds are very rare, but the complexes often exist as dimers or oligomers linked by bridging ligands. Figure 18.3.2 shows the structures of some dimeric and oligomeric coordination compounds: (a) in $Rb_5Nd_2(NO_3)_{11} \cdot H_2O$, one NO_3^- ligand bridges two Nd atoms to form the dimeric anion $Nd_2(NO_3)_{11}^{5-}$; (b) in the complex $[Y(\eta^5: \eta^1\text{-}C_5Me_4SiMe_2N^tBu)(\mu\text{-}C_4H_3S)]_2$, a pair of $\mu\text{-}C_4H_3S$ ligands bridge two Y atoms; (c) in $Yb_2(OC_6H_3Ph_2\text{-}2,6)_4(PhMe)_{1.5}$, two $OC_6H_3Ph_2\text{-}2,6$ ligands bridge two Yb atoms; and (d) in $K^+(THF)_6\{[MePhC(C_4H_3N)_2]Sm\}_5(\mu_5\text{-}I^-)$, the pentameric anion is formed

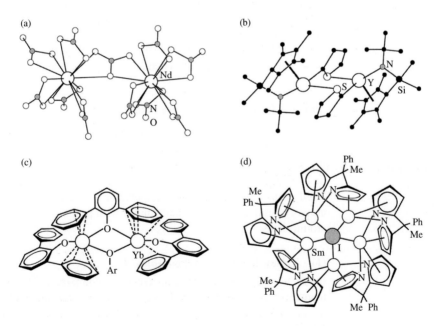

Fig. 18.3.2.
Structures of some dimeric and oligomeric coordination compound of rare-earth metals: (a) $[Nd_2(NO_3)_{11}]^{5-}$, (b) $[Y(\eta^5: \eta^1\text{-}C_5Me_4SiMe_2N^tBu)(\mu\text{-}C_4H_3S)]_2$, (c) $Yb_2(2,6\text{-}Ph_2C_6H_3O)_4$, $OAr = (2,6\text{-}Ph_2C_6H_3O)$; (d) $\{[MePhC(C_4H_3N)_2]Sm\}_5(\mu_5\text{-}I^-)$.

by five bridging [MePhC(C$_4$H$_3$N)$_2$] ligands and consolidated by a central μ_5-I$^-$ anion.

The lanthanide complexes exhibit a number of characteristic features in their structures and properties:

(a) There is a wide range of coordination numbers, generally 6 to 12, but 3 to 5 are known, as listed in Table 18.3.1.
(b) Coordination geometries are determined by ligand steric factors rather than crystal field effects. For example, the donor oxygen atoms in different ligands coordinate to the La^{3+} ions in different geometries: monocapped octahedral (CN = 7), irregular (CN = 11), and icosahedral (CN = 12).
(c) The lanthanides prefer anionic ligands with hard donor atoms of rather high electronegativity (e.g., oxygen and fluorine) and generally form labile "ionic" complexes that undergo facile ligand exchange.
(d) The lanthanides do not form Ln=O or Ln≡N multiple bonds of the type known for many transition metals and certain actinides.
(e) The 4f orbitals in the Ln^{3+} ions do not participate directly in bonding. Their spectroscopic and magnetic properties are thus largely unaffected by the ligands.

18.4 Organometallic compounds of rare-earth elements

In contrast to the extensive carbonyl chemistry of the d-transition metals, lanthanide metals do not form complexes with CO under normal conditions. All organolanthanide compounds are highly sensitive to oxygen and moisture, and in some cases they are also thermally unstable.

Organolanthanide chemistry has mainly been developed using the cyclopentadienyl C$_5$H$_5$(Cp) group and its substituted derivatives, such as C$_5$Me$_5$ (Cp*), largely because their size allows some steric protection of the large metal center.

18.4.1 Cyclopentadienyl rare-earth complexes

The properties of cyclopentadienyl lanthanide compounds are influenced markedly by the relationship between the size of the lanthanide atoms and the steric demand of the Cp group. The former varies from La to Lu according to lanthanide contraction, while the latter varies from the least bulky Cp to highly substituted Cp*, which is appreciably larger. The Sc and Y complexes are very similar to those of lanthanides with proper allowance for the relative atomic sizes.

(1) Triscyclopentadienyl complexes

In LaCp$_3$, the coordination requirements of the large La^{3+} ion are satisfied by formation of a polymer, where each La atom is coordinated to three η^5-C$_5$H$_5$ ligands with an additional η^2-C$_5$H$_5$ interaction, as shown in Fig. 18.4.1(a). The intermediate-size Sm atom forms a simple Sm(η^5-C$_5$H$_5$)$_3$ monomeric species, as shown in Fig. 18.4.1(b). The smallest lanthanide, Lu, is unable to accommodate three η^5-C$_5$H$_5$ ligands, so that LuCp$_3$ adopts a polymeric structure with

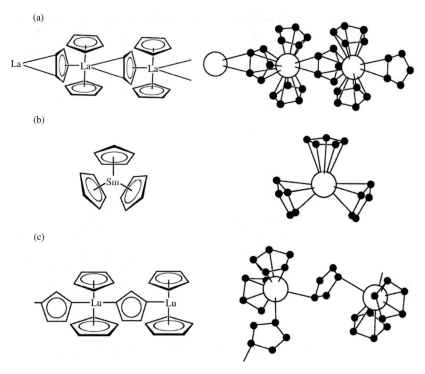

Fig. 18.4.1.
Structural formulas and geometries of LnCp$_3$ complexes: (a) LaCp$_3$, (b) SmCp$_3$, and (c) LuCp$_3$.

each Lu coordinated to two η^5-C$_5$H$_5$ ligands and two μ_2-Cp ligands, as shown in Fig. 18.4.1(c).

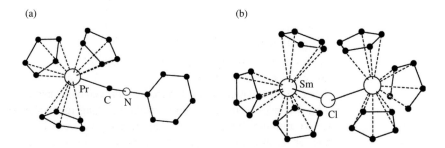

Fig. 18.4.2.
Structure of (a) Cp$_3$Pr(CNC$_6$H$_{11}$) and (b) [Cp$_3$SmClSmCp$_3$]$^-$.

Many adducts of LnCp$_3$ with neutral Lewis bases (such as THF, esters, phosphines, pyridines, and isocyanides) have been prepared and characterized. These complexes usually exhibit pseudo-tetrahedral geometry. Figure 18.4.2(a) shows the structure of Cp$_3$Pr(CNC$_6$H$_{11}$) and Fig. 18.4.2(b) shows the structure of the anion of [Li(DME)$_3$]$^+$[Cp$_3$SmClSmCp$_3$]$^-$, in which two Cp$_3$Ln fragments are bridged by one anionic ligand.

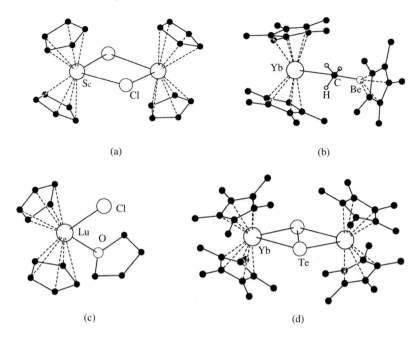

Fig. 18.4.3.
Structures of some biscyclopentadienyl rare-earth complexes: (a) $(Cp_2ScCl)_2$, (b) $Cp_2^*Yb(MeBeCp^*)$, (c) $Cp_2LuCl(THF)$, and (d) $(Cp_2^*Yb)_2Te_2$.

18.4.2 Biscyclopentadienyl complexes

Many crystal structures of the biscyclopentadienyl rare-earth compounds have been determined; four examples are shown in Fig. 18.4.3.

(a) $(Cp_2LnCl)_2$

The configuration of dimers of this type is dependent on the relative importance of the steric interaction of the Cp rings with the halide bridge versus the interaction between the Cp rings on both metal centers. In $(Cp_2ScCl)_2$, the Cp ring is relatively small in comparison with the halide ligand; accordingly all Cp centroids lie in one plane, and the Sc_2Cl_2 plane is perpendicular to it, as shown in Fig. 18.4.3(a).

(b) $Cp_2^*Yb(MeBeCp^*)$

The Cp_2^*Yb unit can be coordinated by saturated hydrocarbons, as in MeBeCp*. The positions of the hydrogens on the bridging methyl group show that this is not a methyl bridge having a $3c$-$2e$ bond between Yb–C–Be, as shown in Fig. 18.4.3(b).

(c) $Cp_2LuCl(THF)$

This adduct is monomeric with one coordinated THF, forming a pseudo-tetrahedral arrangement (each Cp^- ligand is considered to be tridentate), as shown in Fig. 18.4.3(c).

(d) $(Cp_2^*Yb)_2Te_2$

In the chalcogenide complex $Cp_2^*Yb(Te_2)YbCp_2^*$, the Te_2 unit serves as a bridging ligand, as shown in Fig. 18.4.3(d).

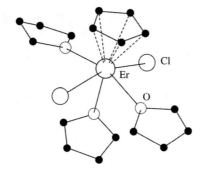

Fig. 18.4.4.
Structure of CpErCl$_2$(THF)$_3$.

(3) Monocyclopentadienyl complexes

Since monocyclopentadienyl rare-earth complexes require four to six σ-donor ligands to reach the stable coordination numbers 7 to 9, monomeric complexes must bind several neutral donor molecules. Generally, some of these donor molecules are easily lost, and dinuclear and polynuclear complexes are formed. This accounts for the fact that monocyclopentadienyl rare-earth complexes exhibit rich structural complexity. Figure 18.4.4 shows the structure of CpErCl$_2$(THF)$_3$.

18.4.3 Benzene and cyclooctatetraenyl rare-earth complexes

All the lanthanides and Y are able to bind two substituted benzene rings, such as Gd(η^6-C$_6$H$_3{}^t$Bu$_3$)$_2$ and Y(η^6-C$_6$H$_6$)$_2$. Figure 18.4.5(a) shows the sandwich structure of Gd(η^6-C$_6$H$_3{}^t$Bu$_3$)$_2$. In this complex, the Gd center is zero-valent.

The large lanthanide ions are able to bind a planar cyclooctatetraene dianion ligand, as shown in Fig. 18.4.5(b) to Fig. 18.4.5(d). In [(C$_8$H$_8$)$_2$Ce]K[CH$_3$O(CH$_2$CH$_2$O)$_2$CH$_3$], Ce(III) is sandwiched by two (η^8-C$_8$H$_8$) dianions, as shown in Fig. 18.4.5(b). The structure of (C$_8$H$_8$)$_3$Nd$_2$(THF)$_2$ shows a [(C$_8$H$_8$)$_2$Nd]$^-$ anion coordinating to the [(C$_8$H$_8$)Nd(THF)$_2$]$^+$ cation via two carbon atoms of the C$_8$H$_8^{2-}$ ligand, as shown in Fig. 18.4.5(c).

In the sandwich complex (C$_8$H$_8$)Lu(Cp*), as shown in Fig. 18.4.5(d), the molecular structure is slightly bent with a ring centroid–Lu–ring centroid angle of 173°. The methyl groups of Cp* are bent away from the Lu center by 0.7°-2.3° from the idealized planar configuration. The averaged Lu–C bond distances for the C$_8$H$_8^{2-}$ and Cp*$^-$ ligands are 243.3 and 253.6 pm, respectively.

18.4.4 Rare-earth complexes with other organic ligands

A variety of "open" π complexes, such as allyl complexes, are known. Generally, the allyl group is η^3-bound to the rare-earth center. Figure 18.4.6(a) shows the structure of the anion in [Li$_2$(μ-C$_3$H$_5$)(THF)$_3$]$^+$[Ce(η^3-C$_3$H$_5$)$_4$]$^-$.

Using bulky ligands or chelating ligands to saturate the complexes coordinatively or sterically, the rare-earth complexes without π ligands have been obtained and characterized. The La and Sm complexes with bulky alkyl ligands

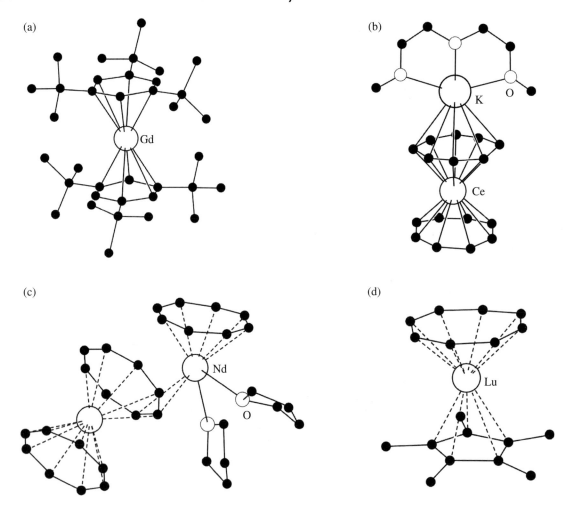

Fig. 18.4.5.
Structures of (a) Gd(η^6-C$_6$H$_3$^{t}Bu$_3$)$_2$, (b) [(C$_8$H$_8$)$_2$Ce]K[CH$_3$O(CH$_2$CH$_2$O)$_2$CH$_3$], (c) (C$_8$H$_8$)$_3$Nd$_2$(THF)$_2$, and (d) (C$_8$H$_8$)Lu(Cp*).

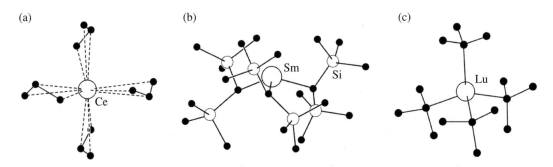

Fig. 18.4.6.
Structures of some rare-earth complexes: (a) [Ce(η^3-C$_3$H$_5$)$_4$]$^-$, (b) Sm[CH(SiMe$_3$)$_2$]$_3$, and (c) [Lu(^{t}Bu)$_4$]$^-$.

[CH(SiMe$_3$)$_2$] are pyramidal, being stabilized by agostic interactions of methyl groups with the highly Lewis acidic metal center, as shown in Fig. 18.4.6(b).

The second half of the lanthanide series react with the more bulky ^{t}BuLi reagent to form tetrahedrally coordinated complexes, such as [Li(TMEDA)$_2$]$^+$[Lu(^{t}Bu)$_4$]$^-$. Figure 18.4.6(c) shows the structure of the anion [Lu(^{t}Bu)$_4$]$^-$.

18.5 Reduction chemistry in oxidation state +2

Only the elements samarium, europium, and ytterbium have significant "normal" chemistry based on true Ln^{2+} ions, although SmCl$_2$ and EuCl$_2$ are well-characterized and have been known for over a century. The compounds of Pr, Nd, Dy, Ho, and Tm in oxidation state +2 are unstable in aqueous solution. Lanthanide(II) compounds are of current interest as reductants and coupling agents in organic and main-group chemistry.

18.5.1 Samarium(II) iodide

SmI$_2$, which can be prepared conveniently from samarium powder and 1,2-diiodoethane in THF, finds application as a versatile one-electron reducing agent in organic synthesis. Two typical synthetic procedures mediated by SmI$_2$ are the pinacol coupling of aldehydes and the Barbier reaction, as shown in the following schemes:

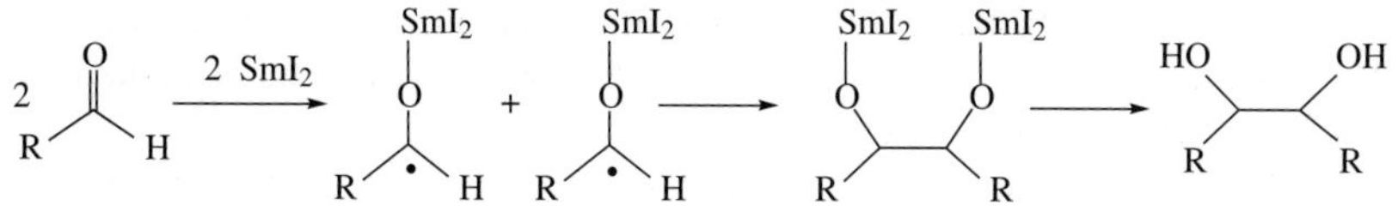

pinacol coupling

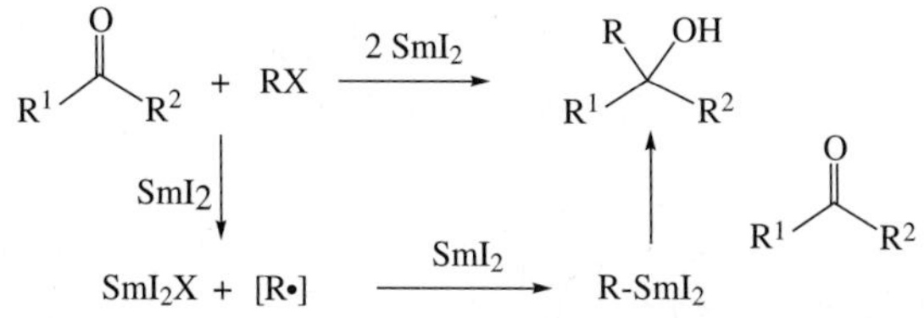

Barbier reaction

SmI$_2$ reacts with a variety of donor solvents to form crystalline solvates. SmI$_2$(Me$_3$CCN)$_2$ is an iodo-bridged polymeric solid containing six-coordinate Sm^{2+} centers. In contrast, SmI$_2$(HMPA)$_4$, where HMPA is (Me$_3$N)$_3$P=O, is a discrete six-coordinate molecule with a linear I–Sm–I unit. The diglyme solvate SmI$_2$[O(CH$_2$CH$_2$OMe)$_2$]$_2$ exhibits eight-coordination.

18.5.2 Decamethylsamarocene

The organosamarium(II) complex $Sm(\eta^5\text{-}C_5Me_5)_2$ has the bent-metallocene structure, which can be attributed to polarization effects; its THF solvate $Sm(\eta^5\text{-}C_5Me_5)_2(THF)_2$ exhibits the expected pseudo-tetrahedral coordination geometry. $[Sm(\eta^5\text{-}C_5Me_5)_2]$ is a very powerful reductant, and its notable reactions include the reduction of aromatic hydrocarbons such as anthracene and of dinitrogen, as illustrated in the following scheme:

The N–N distance in the dinuclear N_2 complex is 129.9 pm, as compared with the bond length of 109.7 pm in dinitrogen, and is indicative of a considerable decrease in bond order. An analogous planar system comprising a side-on bonded $\mu\text{-}(bis\text{-}\eta^2)\text{-}N_2$ ligand between two metal centers, with a much longer N–N bond length of 147 pm, occurs in $[Cp''_2Zr]_2(N_2)$, as described in Section 15.1.3.

(1) Diiodides of thulium, dysprosium, and neodymium

The molecular structures of $LnI_2(DME)_3$ (Ln = Tm, Dy; DME = dimethoxyethane) are shown in Fig. 18.5.1(a) and (b). The larger ionic size of Dy^{2+} compared with Tm^{2+} accounts for the fact that the thulium(II) complex is seven-coordinate, whereas the dysprosium(II) complex is eight-coordinate. The complex $NdI_2(THF)_5$ has pentagonal bipyramidal geometry with the iodo ligands occupying the axial positions.

Thulium(II) complexes are stabilized by phospholyl or arsolyl ligands that can be regarded as derived from the cyclopentadienyl group by replacing one CH group by a P or As atom. Their decreased π-donor capacity relative to the parent cyclopentadienyl system enhances the stability of the Tm(II) center, and stable complexes of the bent-sandwiched type have been isolated.

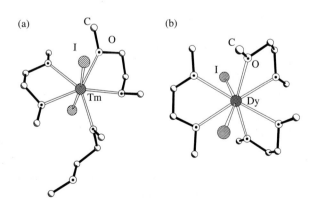

Fig. 18.5.1.
Comparison of the molecular structures of (a) hepta-coordinate $TmI_2(DME)_3$ and (b) octa-coordinate $DyI_2(DME)_3$.

The LnI$_2$ (Ln = Tm, Dy, Nd) species have been used in a wide range of reactions in organic and organometallic syntheses, including the reduction of aromatic hydrocarbons and the remarkable reductive coupling of MeCN to form a tripodal N$_3$-ligand, as shown in the following schemes:

References

1. H. C. Aspinall, *Chemistry of the f-Block Elements*, Gordon and Breach, Amsterdam, 2001.
2. S. Cotton, *Lanthanide and Actinide Chemistry*, Wiley, Chichester, 2006.
3. N. Kaltsoyannis and P. Scott, *The f Elements*, Oxford University Press, Oxford, 1999.
4. S. D. Barrett and S. S. Dhesi, *The Structure of the Rare-Earth Metal Surfaces*, Imperial College Press, London, 2001.
5. A. F. Wells, *Structural Inorganic Chemistry*, 5th edn., Oxford University Press, Oxford, 1984.
6. C. E. Housecroft and A. G. Sharpe, *Inorganic Chemistry*, 2nd edn., Prentice-Hall, Harlow, 2004.
7. D. F. Shriver, P. W. Atkins, T. L. Overton, J. P. Rourke, M. T. Weller and F. A. Armstrong, *Inorganic Chemistry*, 4th edn., University Press, Oxford, 2006.
8. C. H. Huang, *Coordination Chemistry of Rare-Earth Elements* (in Chinese), Science Press, Beijing, 1997.
9. R. B. King (editor-in-chief), *Encyclopedia of Inorganic Chemistry*, Wiley, Chichester, 1994.
10. K. A. Gschneidner, Jr., L. Eyring, G. R. Choppin and G. H. Lander (eds.), *Handbook on the Physics and Chemistry of Rare Earths*, Vol. 18: *Lanthanides/Actinides: Chemistry*, North-Holland, Amsterdam, 1994.

11. J. A. McCleverty and T. J. Meyer (editors-in-chief), *Comprehensive Coordination Chemistry II*, vol. 3 (G. F. R. Parkin, volume editor), Elsevier-Pergamon, Amsterdam, 2004.
12. E. W. Abel, F. G. A. Stone and G. Wilkinson (editors-in-chief), *Comprehensive Organometallic Chemistry II*, vol. 4 (M. F. Lappert, volume editor), Pergamon, Oxford, 1995.
13. Thematic issue on "Frontiers in Lathanide Chemistry," *Chem. Rev.* **102**, no. 6, 2002.

Metal–Metal Bonds and Transition-Metal Clusters

19

19.1 Bond valence and bond number of transition-metal clusters

A dinuclear transition-metal complex contains two transition-metal atoms each surrounded by a number of ligands. A transition-metal cluster has a core of three or more metal atoms directly bonded with each other to form a discrete molecule containing metal–metal (M–M) bonds. The first organometallic complex reported to possess a M–M bond was $Fe_2(CO)_9$. Since this Fe–Fe bond is supported by three bridging CO ligands, as shown in Fig. 19.1.1(a), it cannot be taken as definitive proof of direct metal–metal interaction. In contrast, the complexes $Re_2(CO)_{10}$, $Mn_2(CO)_{10}$, and $[MoCp(CO)_3]_2$, with no bridging ligands, provide unequivocal examples of unsupported M–M bonding. In these species the metal atoms are considered to be bonded through a single bond of the $2c$-$2e$ type. Figure 19.1.1(b) shows the molecular structure of $Mn_2(CO)_{10}$.

Since polynuclear complexes and cluster compounds are in general rather complicated species, the application of quantitative methods for describing bonding is not only difficult but also impractical. Qualitative approaches and empirical rules often play an important role in treating such cases. We have used the octet rule and bond valence to describe the structure and bonding of boranes and their derivatives (Sections 13.3 and 13.4). Now we use the 18-electron rule and bond valence to discuss the bonding and structure of polynuclear transition-metal complexes and clusters.

The metal atoms in most transition-metal complexes and clusters obey the 18-electron rule. The bond valence, b, of the skeleton of complex $[M_nL_p]^{q-}$ can be calculated from the formula

$$b = 1/2(18n - g),$$

where g is the total number of valence electrons in the skeleton of the complex, which is the sum of the following three parts:

(a) n times the number of valence electrons of metal atom M;
(b) p times the number of valence electrons donated to the metal atoms by ligand L; and
(c) q electrons from the net charge of the complex.

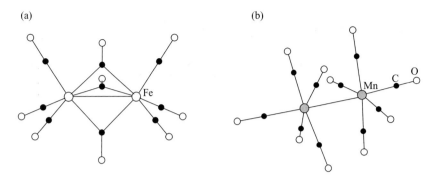

Fig. 19.1.1.
Structure of dinuclear transition metal complexes: (a) $Fe_2(CO)_9$ and (b) $Mn_2(CO)_{10}$.

The bond valence b of the skeleton of a complex or cluster corresponds to the sum of the bond numbers of the metal–metal bonds. For a M–M single bond, the bond number is equal to 1; similarly, the bond number is 2 for a M=M double bond, 3 for a M≡M triple bond, 4 for a M≣M quadruple bond, and 2 for a 3c-2e MMM bond.

The majority of cluster compounds have carbonyl ligands coordinated to the metal atoms. Some clusters bear NO, CNR, PR_3, and H ligands, while others contain interstitial C, N, and H atoms. The compounds can be either neutral or anionic, and the common structural metal building blocks are triangles, tetrahedra, octahedra, and condensed clusters derived from them. The carbonyl ligand has two special features. First, the carbonyl ligand functions as a two-electron donor in the terminal-, edge-, or face-bridging mode, and neither changes the valence of the metal skeleton nor requires the involvement of additional ligand electrons. In contrast, the Cl ligand may change from a one- (terminal) to a three- (edge-bridging) or five- (face-bridging) electron donor. Second, the synergic bonding effects which operate in the alternative bridging modes of the carbonyl ligand lead to comparable stabilization energies and therefore readily favor the formation of transition-metal cluster compounds.

As mentioned in Sections 13.3 and 13.4, the structures of most known boranes and carboranes are based on the regular deltahedron and divided into three types: *closo*, *nido*, and *arachno*. In a *closo*-structure the skeletal B and C atoms occupy all the vertices of the polyhedron. In the cases of *nido*- and *arachno*-structures, one and two of the vertices of the appropriate polyhedron remain unoccupied, respectively. On each skeletal B and C atom, there is always a H atom (or some other simple terminal ligand such as halide) that points away from the center of the polyhedron and is also linked to the skeleton by a radial 2c-2e single bond. In a polyhedron of n vertices, there are $4n$ atomic orbitals. Among them, n orbitals are used up for n terminal B–H bonds, the remaining $3n$ atomic orbitals being available for skeletal bonding. In the skeleton of *closo*-borane $B_nH_n^{2-}$, there are $4n+2$ valence electrons (n B atoms contribute $3n$ electrons, n H atoms donate n electrons, and the charge accounts for 2 electrons), and hence $2n+2$ electrons or $n+1$ electron pairs are available for skeletal bonding. These numbers are definite. For the transition-metal cluster compounds, however, the electron counts vary with different bonding types. For example, octahedral M_6 clusters of different compounds form various structures and bond types. Some examples are discussed below and shown in Fig 19.1.2.

Figure 19.1.2(a) shows the structure of $[Mo_6(\mu_3\text{-Cl})_8Cl_6]^{2-}$. In this structure, eight μ_3-Cl and six terminal Cl atoms are coordinated to the Mo_6 cluster. Each μ_3-Cl donates five electrons and each terminal Cl donates one electron. Thus the g value is

$$g = 6 \times 6 + (8 \times 5 + 6 \times 1) + 2 = 84.$$

The bond valence of the Mo_6 cluster is

$$b = {}^1/_2(6 \times 18 - 84) = 12.$$

The bond valence b precisely matches 12 2c-2e Mo–Mo bonds, as shown in the front and back views of the Mo_6 cluster on the right side of Fig. 19.1.2(a).

Figure 19.1.2(b) shows the structure of $[Nb_6(\mu_2\text{-Cl})_{12}Cl_6]^{4-}$. In this structure, there are 12 μ_2-Cl atoms each donating 3 electrons, plus six terminal μ_1-Cl atoms, each donating one electron, to the Nb cluster. The g value is

$$g = 6 \times 5 + (12 \times 3 + 6 \times 1) + 4 = 76.$$

The bond valence of the Nb_6 cluster is

$$b = {}^1/_2(6 \times 18 - 76) = 16.$$

In this Nb_6 cluster, each edge is involved in bonding with a μ_2-Cl ligand, so the edges do not correspond to 2c-2e Nb–Nb bonds. Each face of the Nb_6 cluster forms a 3c-2e NbNbNb bond, and has bond number 2. The sum of bond number, 16, is just equal to the bond valence of 16, as shown in the front and back views of the Nb cluster in Fig. 19.1.2(b).

Figure 19.1.2(c) shows the structure of $Rh_6(\mu_3\text{-CO})_4(CO)_{12}$. The CO group as a ligand always donates two electrons to the M_n cluster. The g value and bond valence of the Rh_6 cluster are

$$b = {}^1/_2(6 \times 18 - 86) = 11$$

The b value of Rh_6 is 11, which just matches the b value of $B_6H_6^{2-}$. The Rh_6 cluster is stabilized by four 3c-2e RhRhRh bonds and three 2c-2e Rh-Rh bonds. Various bond formulas can be written for the Rh_6 cluster, one of which is shown on the right side of Fig. 19.1.2(c). These formulas are equivalent by resonance and the octahedron has overall O_h symmetry.

19.2 Dinuclear complexes containing metal–metal bonds

Studies on dinuclear transition-metal compounds containing metal–metal bonds have deeply enriched our understanding of chemical bonding. The nature and

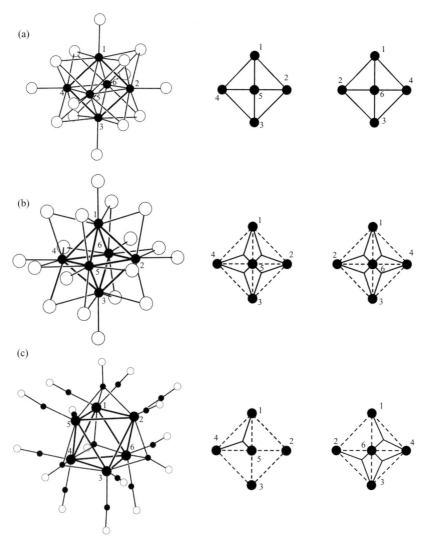

Fig. 19.1.2.
Structure and bonding of three octahedral clusters: (a) $[Mo_6(\mu_3\text{-}Cl)_8Cl_6]^{2-}$; (b) $[Nb_6(\mu_2\text{-}Cl)_{12}Cl_6]^{4-}$; and (c) $Rh_6(\mu_3\text{-}CO)_4(CO)_{12}$.

implication of M–M bonds are richer and more varied than the covalent bonds of the second-period representative elements. The following points may be noted.

(a) The d atomic orbitals of transition metals are available for bonding, in addition to the s and p atomic orbitals.
(b) The number of valence orbitals increases from 4 for the second-period elements to 9 for the transition-metal atoms. Thus the octet rule suits the former and eighteen-electron rule applies to the latter.
(c) The use of d-orbitals for bonding leads to the possible formation of a quadruple bond (and even a quintuple bond) between two metal atoms, in addition to the familiar single, double, and triple bonds.

(d) The transition-metal atoms interact with a large number of ligands in coordination compounds. The various geometries and electronic factors have significant effects on the properties of metal–metal bonds. The 18-electron rule is not strictly valid, thus leading to varied M–M bond types.

19.2.1 Dinuclear transition-metal complexes conforming to the 18-electron rule

When the structure of a dinuclear complex has been determined, the structural data may be used to enumerate the number of valence electrons available in the complex, to determine its bond valence, and to understand the properties of the metal–metal bond. For a dinuclear complex, the bond valence $b = 1/2(18 \times 2 - g)$. Table 19.2.1 lists the data for some dinuclear complexes.

In the complex $Ni_2(Cp)_2(\mu_2\text{-}PPh_2)_2$, with two bridging ligands ($\mu_2\text{-}PPh_2$) linking the Ni atoms, the bond valence equals zero, indicating that there is no bonding interaction between them.

$Mn_2(CO)_{10}$ is one of the simplest dimeric compounds containing a metal–metal single bond unsupported by bridging ligands, and its diamagnetic behavior is accounted for in compliance with the 18-electron rule. The structures of $Tc_2(CO)_{10}$, $Re_2(CO)_{10}$, and $MnRe(CO)_{10}$, which are isomorphous with $Mn_2(CO)_{10}$, have also been determined. The measured M–M bond lengths (pm) are Tc–Tc 303.6, Re–Re 304.1, and Mn–Re 290.9. The structure of $Fe_2(CO)_9$, shown in Fig. 19.1.1(a), has an Fe–Fe single bond of 252.3 pm, which is further stabilized by the bridging carbonyl ligands.

Many dinuclear complexes have a M=M double bond. Some examples are

$$Co_2(CO)_2Cp_2^*, Co=Co\ 233.8\,pm;$$

$$Fe_2(NO)_2Cp_2, Fe=Fe\ 232.6\,pm;$$

$$Re_2(\mu\text{-}Cl)_2Cl_4(dppm)_2, Re=Re\ 261.6\,pm;$$

$$Mo_2(OR)_8\ (R = {}^iPr, {}^tBu), Mo=Mo\ 252.3\,pm.$$

Most compounds with triple and quadruple bonds are formed by Re, Cr, Mo, and W. The ligands in such compounds are in general relatively hard Lewis bases such as halides, carboxylic acids, and amines. Nevertheless, in some cases π-acceptor ligands such as carbonyl, phosphines and nitrile are also present.

Table 19.2.1. Dinuclear complexes

Complex	g	b	M–M(pm)	Bond properties
$Ni_2(Cp)_2(\mu_2\text{-}PPh_2)_2$	36	0	336	Ni$\cdots$Ni, no bonding
$(CO)_5Mn_2(CO)_5$	34	1	289.5	Mn–Mn, single bond
$Co_2(\mu_2\text{-}CH_2)(\mu_2\text{-}CO)(Cp^*)_2$	32	2	232.0	Co=Co, double bond
$Cr_2(CO)_4(Cp)_2$	30	3	222	Cr≡Cr, triple bond
$[Mo_2(\mu_2\text{-}O_2CMe)_2(MeCN)_6]^{2+}$	28	4	213.6	Mo≣Mo, quadruple bond

19.2.2 Quadruple bonds

The recognition and understanding of the quadruple bond is one of the most important highlights in modern inorganic chemistry. The overlap of d atomic orbitals can generate three types of molecular orbitals: σ, π, and δ. These molecular orbitals can be used to form a quadruple bond between two transition-metal atoms under appropriate conditions. In a given compound, however, not all of these orbitals are always available for multiple metal–metal bonding. Thus the 18-electron rule does not always hold for the dinuclear complexes and needs to be modified according to their structures.

The most interesting aspect of the crystal structure of $K_2[Re_2Cl_8]\cdot 2H_2O$ is the presence of the dianion $Re_2Cl_8^{2-}$ (Fig. 19.2.1), which possesses an extremely short Re–Re bond distance of 224.1 pm, as compared with an average Re–Re distance of 275 pm in rhenium metal. Another unusual feature is the eclipsed configuration of the Cl atoms with a Cl$\cdots$Cl separation of 332 pm. As the sum of van der Waals radii of two Cl atoms is 360 pm, the staggered configuration would normally be expected for $Re_2Cl_8^{2-}$. These two features are both attributable to the formation of a Re≣Re quadruple bond.

The bonding in the skeleton of the $[Re_2Cl_8]^{2-}$ ion can be formulated as follows: each Re atom uses its square-planar set of dsp^2 ($d_{x^2-y^2}$, s, p_x, p_y) hybrid orbitals to overlap with ligand Cl p orbitals to form Re–Cl bonds. The p_z atomic orbital of Re is not available for bonding. The remaining d_{z^2}, d_{xz}, d_{yz}, and d_{xy} atomic orbitals on each Re atom overlap with the corresponding orbitals on the other Re atom to generate the following MOs:

$$d_{z^2} \pm d_{z^2} \rightarrow \sigma \text{ and } \sigma^* \text{ MO}$$
$$d_{xz} \pm d_{xz} \rightarrow \pi \text{ and } \pi^* \text{ MO}$$
$$d_{yz} \pm d_{yz} \rightarrow \pi \text{ and } \pi^* \text{ MO}$$
$$d_{xy} \pm d_{xy} \rightarrow \delta \text{ and } \delta^* \text{ MO}$$

Figure 19.2.2 shows the pairing up of d AO's of two Re atoms to form MO's and the ordering of the energy levels. In the $[Re_2Cl_8]^{2-}$ ion, the two Re atoms have 16 valence electrons including two from the negative charge. Eight valence electrons are utilized to form eight Re–Cl bonds, and the remaining eight occupy four metal–metal bonding orbitals to form a quadruple bond: one σ bond, two π bonds, and one δ bond, leading to the $\sigma^2\pi^4\delta^2$ configuration.

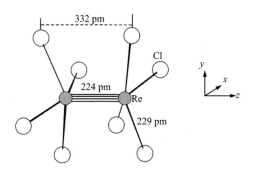

Fig. 19.2.1.
Structure of $[Re_2Cl_8]^{2-}$.

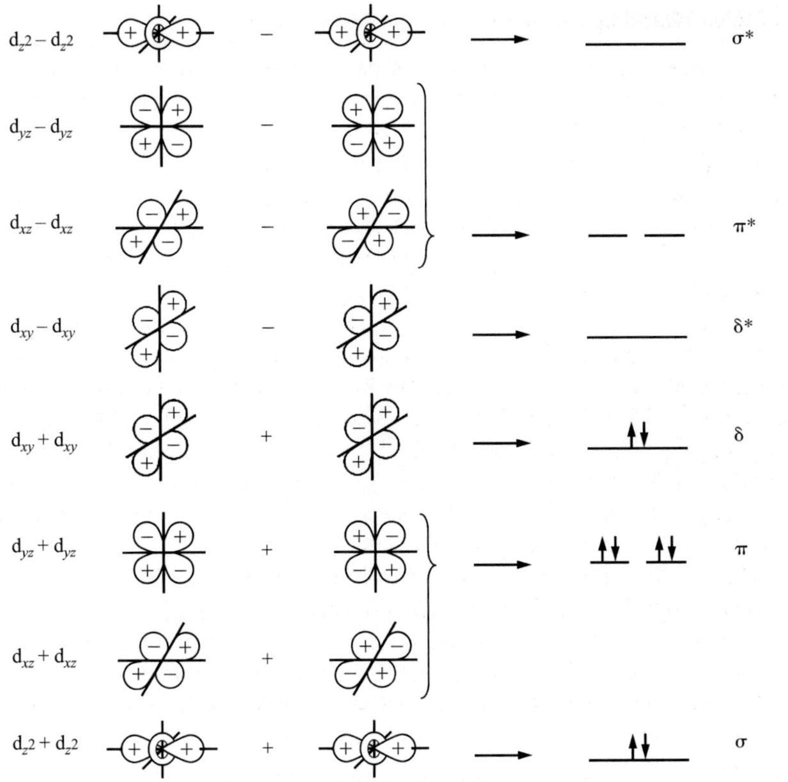

Fig. 19.2.2.
Overlap of d orbitals leading to the formation of a quadruple bond between two metal atoms. Note that the z axis of each metal atom is taken to point toward the other, such that if a right-handed coordinate system is used for the atom on the left, a left-handed coordinate system must be used for the atom on the right.

The structures of two transition-metal complexes that contain a Mo≣Mo quadruple bond are shown in Fig. 19.2.3. In the compound $[Mo_2(O_2CMe)_2(NCMe)_6](BF_4)_2$, each Mo atom is six-coordinate, and all nine valence AO's are used for bonding according to the 18-electron rule. In the cation $[Mo_2(O_2CMe)_2(NCMe)_6]^{2+}$,

$$g = 2 \times 6 + 2 \times 3 + 6 \times 2 - 2 = 28,$$

and

$$b = {}^1\!/_2(2 \times 18 - 28) = 4.$$

The Mo–Mo bond distance of 213.6 pm is in accord with the quadruple Mo≣Mo bond. In the molecule $Mo_2(O_2CMe)_4$, each Mo atom is five-coordinate and has one empty $AO(p_z)$ which is not used for bonding. It obeys the 16-electron rule with $g = 2 \times 6 + 4 \times 3 = 24$ and $b = {}^1\!/_2(2 \times 16 - 24) = 4$. The observed Mo–Mo distance of 209.3 pm is consistent with that expected of a quadruple bond.

A large number of dinuclear complexes containing a M≣M quadruple bond formed by the Group 16 and 17 metals Cr, Mo, W, Re, and Tc have been reported in the literature. Selected examples are listed in Table 19.2.2.

Table 19.2.2. Compounds containing a M≡M quadruple bond

Compound	M≡M	Distance(pm)	Rule	g
Cr_2(2-MeO-5-Me-C_6H_3)$_4$	Cr≡Cr	182.8	16e	24
Cr_2[MeNC(Ph)NMe]$_4$	Cr≡Cr	184.3	16e	24
$Cr_2(O_2CMe)_4$	Cr≡Cr	228.8	16e	24
$Cr_2(O_2CMe)_4(H_2O)_2$	Cr≡Cr	236.2	18e	28
Mo_2(hpp)$_4$*	Mo≡Mo	206.7	16e	24
$K_4[Mo_2(SO_4)_4]\cdot 2H_2O$	Mo≡Mo	211.0	18e	28
$[Mo_2(O_2CCH_2NH_3)_4]Cl_4\cdot 3H_2O$	Mo≡Mo	211.2	16e	24
$Mo_2[O_2P(OPh)_2]_4$	Mo≡Mo	214.1	16e	24
W_2(hpp)$_4\cdot$ 2NaHBEt$_3$	W≡W	216.1	16e	24
$W_2(O_2CPh)_4(THF)_2$	W≡W	219.6	18e	28
$W_2(O_2CCF_3)_4$	W≡W	222.1	16e	24
$W_2Cl_4(PBu_3^n)_4\cdot C_7H_8$	W≡W	226.7	16e	24
$(Bu_4N)_2Tc_2Cl_8$	Tc≡Tc	214.7	16e	24
$K_2[Tc_2(SO_4)_4]\cdot 2H_2O$	Tc≡Tc	215.5	16e	24
$Tc_2(O_2CCMe_3)_4Cl_2$	Tc≡Tc	219.2	18e	28
$(Bu_4N)_2Re_2F_8\cdot 2Et_2O$	Re≡Re	218.8	16e	24
$Na_2[Re_2(SO_4)_4(H_2O)_2]\cdot 6H_2O$	Re≡Re	221.4	18e	28
$[Re_2(O_2CMe)_2Cl_4(\mu\text{-pyz})]_n$	Re≡Re	223.6	18e	28

* hpp is the anion of 1,3,4,6,7,8-hexahydro-2H-pyrimido-[1,2-a]-pyrimidine (Hhpp); pyz is pyrazine.

The quadruple bond can undergo a variety of interesting reactions, as outlined in Fig. 19.2.4.

(1) There is a rich chemistry in which ligands are exchangable, and virtually every type of ligand can be used except the strong π acceptors.
(2) Addition of a mononuclear species to an M≡M bond can yield a trinuclear cluster.
(3) Two quadruple bonds can combine to form a metallacyclobutadiyne ring.
(4) Oxidative addition of acids to generate a M≡M bond (particularly W≡W) is a key part of molybdenum and tungsten chemistry.
(5) Phosphines can act as reducing agents as well as ligands to give products with triple bonds of the $\sigma^2\pi^4\delta^2\delta^{*2}$ type, as in $Re_2Cl_4(PEt_3)_4$.
(6) Photo excitation by the $\delta \to \delta^*$ transition can lead to reactive species which are potentially useful in photosensitizing various reactions, including the splitting of water.

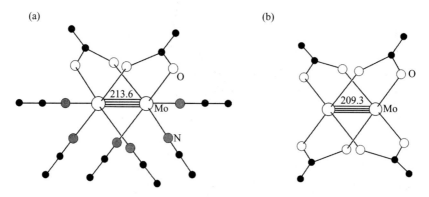

Fig. 19.2.3.
Structure of (a) $[Mo_2(O_2CMe)_2(NCMe)_6]^{2+}$ and (b) $Mo_2(O_2CMe)_4$.

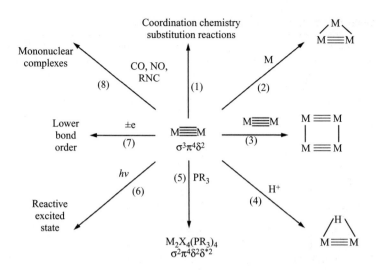

Fig. 19.2.4.
Some reaction types of dimetal compounds containing a M≡M quadruple bond.

(7) Electrochemical oxidation or reduction reduces the bond order and generates reactive intermediates.
(8) With π-acceptor ligands, the M≡M bonds are usually cleaved to give mononuclear products, which are sometimes inaccessible by any other synthetic route.

19.2.3 Bond valence of metal–metal bond

In dinuclear complexes, the bond valence of the metal–metal (M–M) bond can be calculated from its number of bonding electrons, g_M. A simple procedure for counting the metal–metal bond valence is as follows:

(a) Calculate the g value in the usual manner.
(b) Calculate the number of valence electron used for M–L bonds, g_L.
(c) Calculate the number of valence electron used for M–M bonds, $g_M = g - g_L$.
(d) Assign the g_M electrons to the following orbitals according to the energy sequence: σ, (π_x, π_y), δ, δ^*, (π_x^*, π_y^*), and σ^*, as shown in Fig. 19.2.2.

Some examples are presented below:

(1) $Mo_2(O_2CMe)_4$

$g = 2 \times 6 + 4 \times 3 = 24, \quad g_L = 8 \times 2 = 16, \quad g_M = g - g_L = 24 - 16 = 8.$

The electronic configuration is $\sigma^2\pi^4\delta^2$, and the bond order is 4; i.e., the bond valence is 4. For this Mo≡Mo bond, the bond length is 209.3 pm.

(2) $Mo_2(O_2CMe)_2(MeCN)_6]^{2+}$

$g = 2 \times 6 + 2 \times 3 + 6 \times 2 - 2 = 28, \quad g_L = 2 \times 5 \times 2 = 20,$

$g_M = 28 - 20 = 8.$

The electronic configuration is $\sigma^2\pi^4\delta^2$, and the bond valence is 4. So again this is a Mo≡Mo bond, and the bond length is 213.6 pm.

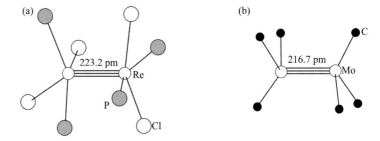

Fig. 19.2.5.
Structure of (a) $Re_2Cl_4(PEt_3)_4$ and (b) $Mo_2(CH_2SiMe_3)_6$

(3) $Re_2Cl_4(PEt_3)_4$

$$g = 2 \times 7 + 4 \times 1 + 4 \times 2 = 26, \quad g_L = 8 \times 2 = 16, \quad g_M = 26 - 16 = 10.$$

The structure of the molecule is shown in Fig. 19.2.5(a). The electronic configuration is $\sigma^2\pi^4\delta^2\delta^{*2}$. The bond valence is 3, indicating a Re≡Re triple bond, and the bond length is 223.2 pm.

(4) $Mo_2(CH_2SiMe_3)_6$

The molecule has D_{3d} symmetry, as shown in Fig. 19.2.5(b).

$$g = 2 \times 6 + 6 \times 1 = 18, \quad g_L = 6 \times 2 = 12, \quad g_M = 18 - 12 = 6.$$

The electronic configuration is $\sigma^2\pi^4$. The length of the Mo≡Mo triple bond is 216.7 pm.

(5) $Mo_2(OR)_8$, ($R = {}^iPr, {}^tBu$)

$$g = 2 \times 6 + 8 \times 1 = 20, \quad g_L = 8 \times 2 = 16, \quad g_M = 20 - 16 = 4.$$

The electronic configuration is $\sigma^2\pi_x^1\pi_y^1$, indicating a Mo=Mo double bond. The experimental bond distance is 252.3 pm, and the molecule is a paramagnetic species.

19.2.4 Quintuple bonding in a dimetal complex

Recently, structural evidence for the first quintuple bond between two metal atoms was found in the dichromium(I) complex Ar′CrCrAr′ (where Ar′ is the sterically encumbering monovalent 2,6-bis[(2,6-diisopropyl)phenyl]phenyl ligand). This complex exists as air- and moisture-sensitive dark-red crystals that remain stable up to 200°C. X-ray diffraction revealed a centrosymmetric molecule with a planar *trans*-bent C–Cr–Cr–C backbone with measured structural parameters Cr–Cr 183.51(4) pm, Cr–C 213.1(1) pm, and C–Cr–Cr 102.78(3)°. Characterization of the compound is further substantiated by magnetic and spectroscopic data, as well as theoretical computations. Figure 19.2.6 shows the molecular geometry and structural formula of Ar′CrCrAr′.

In principle, a homodinuclear transition-metal species can form up to six bonds using the ns and five $(n-1)d$ valence orbitals. In a simplified bonding description of Ar′CrCrAr′, the planar C–Cr–Cr–C skeleton has idealized molecular symmetry C_{2h} with reference to a conventional z axis lying perpendicular to it. For each chromium atom, a local z axis is chosen to be directed toward

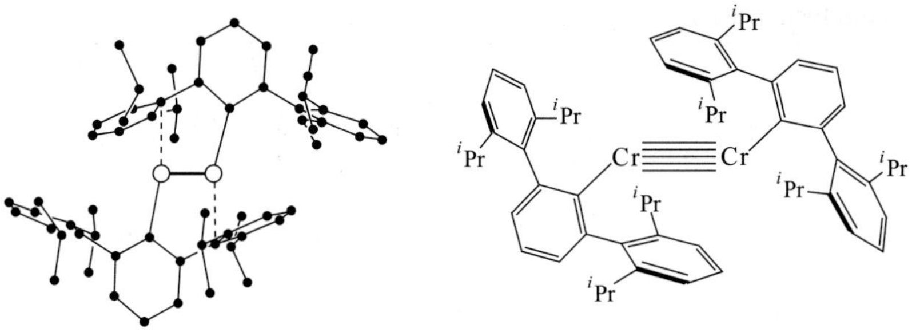

Fig. 19.2.6.
Molecular geometry and structural formula of the dinuclear complex Ar'CrCrAr' (Ar' = C_6H_3-2,6(C_6H_3-2,6-^{i}Pr$_2$)$_2$).

the other chromium atom, and the local x axis to lie in the skeletal plane. Each chromium atom (electronic configuration $3d^54s^1$) uses it 4s orbital to overlap with a sp^2 hybrid orbital on the *ipso* carbon atom of the terphenyl ligand to form a Cr–C σ bond. That leaves five d orbitals at each Cr(I) center for the formation of a fivefold (i.e., quintuple) metal–metal bond, which has one σ ($d_{z^2} + d_{z^2}$; symmetry species A_g), two π ($d_{yz} + d_{yz}$, $d_{xz} + d_{xz}$; A_u, B_u) and two δ ($d_{x^2-y^2} + d_{x^2-y^2}$, $d_{xy} + d_{xy}$; A_g, B_g) components. The situation is actually more complex as mixing of the chromium 4s, $3d_{z^2}$ and $3d_{x^2-y^2}$ orbitals (all belonging to A_g) can occur. In an alternative bonding scheme, each chromium atom may be considered to be d_{z^2}s hybridized. The outward-extended (s − d_{z^2}) hybrid orbital is used to form a Cr–C σ bond. Note that this type of overlap results in a C–Cr–Cr angle of 90° and allows free rotation of the C–Cr bond about the Cr–Cr axis, and it is the steric repulsion between the pair of bulky Ar' ligands that accounts for the obtuse C–Cr–Cr bond angle and the *trans*-bent geometry of the central C–Cr–Cr–C core. The pair of chromium (s + d_{z^2}) hybrid orbitals aligned along the common local z axis overlap to form the metal–metal σ bond, while the π and δ bonds are formed in the manner described above. In either bonding description, there is a formal bond order of 5 between the chromium(I) centers.

19.3 Clusters with three or four transition-metal atoms

19.3.1 Trinuclear clusters

Table 19.3.1 lists the structural data and bond valences of some trinuclear clusters. In $Os_3(CO)_9(\mu_3$-$S)_2$, the Os_3 unit is in a bent configuration with two Os–Os bonds of average length 281.3 pm, and the other Os⋯Os distance (366.2 pm) is significantly longer. The cluster $(CO)_5Mn$–$Fe(CO)_4$–$Mn(CO)_5$ adopts a linear configuration. The other clusters are all triangular. The Fe_3 skeleton of $Fe_3(CO)_{12}$ has 48 valence electrons, and the Fe_3 unit contains three Fe–Fe single bonds. The Os_3 skeleton of $Os_3H_2(CO)_{10}$, a 46-electron triangular cluster, has one Os=Os double bond and two Os–Os single bonds. The length of the Os=Os double bond is 268.0 pm, and the Os–Os single bonds are 281.8 and 281.2 pm.

The remaining three clusters $[Mo_3(\mu_3$-$S)_2(\mu_2$-$Cl)_3Cl_6]^{3-}$, $[Mo_3(\mu_3$-$O)(\mu_2$-$O)_3F_9]^{5-}$, and $Re_3(\mu_2$-$Cl)_3(CH_2SiMe_3)_6$ all have nearly equilateral M_3 skeletons. According to the calculated bond valences of 5, 6, and 9 for the

Structural Chemistry of Selected Elements

Table 19.3.1. Some trinuclear clusters

Cluster	g	b	M–M(pm)	Figure
$Os_3(CO)_9(\mu_3\text{-}S)_2$	50	2	Os–Os, 281.3	(a)
$Mn_2Fe(CO)_{14}$	50	2	Mn–Fe, 281.5	(b)
$Fe_3(CO)_{12}$	48	3	Fe–Fe, 281.5	(c)
$Os_3H_2(CO)_{10}$	46	4	two Os–Os, 281.5 Os=Os, 268.0	(d)
$[Mo_3(\mu_3\text{-}S)_2(\mu_2\text{-}Cl)_3Cl_6]^{3-}$	44	5	Mo---Mo, 261.7*	(e)
$[Mo_3(\mu_3\text{-}O)(\mu_2\text{-}O)_3F_9]^{5-}$	42	6	Mo=Mo, 250.2	(f)
$Re_3(\mu_2\text{-}Cl)_3(CH_2SiMe_3)_6$	36	9	Re≡Re, 238.7	(g)

* Bond order of $1\frac{2}{3}$.

Table 19.3.2. Some tetranuclear clusters

Cluster	g	b	M–M(pm)	Figure
$Re_4(\mu_3\text{-}H)_4(CO)_{12}$	56	8	6 Re---Re, 291*	(a)
$Ir_4(CO)_{12}$	60	6	6 Ir–Ir, 268	(b)
$Re_4(CO)_{16}^{2-}$	62	5	5 Re–Re, 299	(c)
$Fe_4(CO)_{13}C$	62	5	5 Fe–Fe, 263	(d)
$Co_4(CO)_{10}(\mu_4\text{-}S)_2$	64	4	4 Co–Co, 254	(e)
$Re_4H_4(CO)_{15}^{2-}$	64	4	4 Re–Re, 302	(f)
$Co_4(\mu_4\text{-}Te)_2(CO)_{11}$	66	3	3 Co–Co, 262	(g)
$Co_4(CO)_4(\mu\text{-}SEt)_8$	68	2	2 Co–Co, 250	(h)

* Bond order of $1\frac{1}{3}$.

metal–metal bonds in these compounds, the bond types are Mo---Mo (bond order $1\frac{2}{3}$), Mo=Mo, and Re≡Re, respectively.

19.3.2 Tetranuclear clusters

Table 19.3.2 lists the bond valence and structural data of some tetranuclear clusters.

The bond valence of $Re_4(\mu_3\text{-}H)_4(CO)_{12}$ is 8, and the tetrahedral Re_4 skeleton can be described in two ways: (I) resonance between valence-bond structures, leading to a formal bond order of $1\frac{1}{3}$, and (II) four 3c-2e ReReRe bonds. Since there are already four μ_3-H capping the faces, description (II) is not as good as (I).

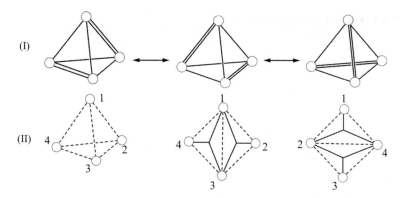

In other examples listed in Table 19.3.2, the calculated bond valence b is just equal to the number of edges of the corresponding metal skeleton, signifying $2c$-$2e$ M–M single bonds.

19.4 Clusters with more than four transition-metal atoms

19.4.1 Pentanuclear clusters

Selected examples of pentanuclear metal clusters are listed in Table 19.4.1. In these examples, the calculated bond valence b is exactly the same as the number of edges of the metal skeleton, indicating $2c$-$2e$ M–M bonds. In general, electron-rich species have lower bond valence and more open structures than the electron-deficient ones.

The structures and skeletal bond valences of $Os_5(CO)_{16}$ and $B_5H_5^{2-}$ are similar as a pair, as are also $Fe_5C(CO)_{15}$ and B_5H_9. But the bonding types in the boranes and the metal clusters are not the same. Since every B atom in a polyhedral borane has three AO's for bonding of the B_n skeleton, any vertex more than three-connected must involve multicenter bonds. In the transition-metal skeleton, the M_n atoms form either $2c$-$2e$ single bonds or $3c$-$2e$ multicenter bonds.

Some clusters have 76 valence electrons based on a trigonal bipyramidal skeleton, such as $[Ni_5(CO)_{12}]^{2-}$, $[Ni_3Mo_2(CO)_{16}]^{2-}$, $Co_5(CO)_{11}(PMe_2)_3$, and $[FeRh_4(CO)_{15}]^{2-}$, which are not shown in Table 19.4.1. The additional four valence electrons compared to $Os_5(CO)_{16}$ have a significant effect on its geometry, and the bond lengths to the apical metal atoms are increased, and the bond valence b is decreased.

The cluster $[Os_5C(CO)_{14}(O_2CMe)I]$ has 78 valence electrons, which is not shown in Table 19.4.1, and its bond valence is equal to 6: [$b = \frac{1}{2}(5 \times 18 - 78) = 6$]. This cluster has a deformed trigonal bipyramidal geometry with no bonding in the equatorial plane of the bipyramid.

19.4.2 Hexanuclear clusters

Table 19.4.2 lists the bond valence and structural data of some selected examples of hexanuclear clusters. The first three clusters have different b values, yet they

Table 19.4.1. Some pentanuclear clusters

Cluster	g	b	No. of edges	Figure
$Os_5(CO)_{16}$	72	9	9	(a)
$Fe_5C(CO)_{15}$	74	8	8	(b)
$Os_5H_2(CO)_{16}$	74	8	8	(c)
$Ru_5C(CO)_{15}H_2$	76	7	7	(d)
$Os_5(CO)_{18}$	76	7	7	(e)
$Os_5(CO)_{19}$	78	6	6	(f)
$Re_2Os_3H_2(CO)_{20}$	80	5	5	(g)

(a) (b) (c) (d) (e) (f) (g)

are all octahedral with 12 edges. There are three stable types of bonding schemes for an octahedron, as shown in Fig. 19.1.2:

(a) $Mo_6Cl_{14}{}^{2-}$, $g = 84$, $b = 12$; in this cluster there are 12 2c-2e bonds at the 12 edges.
(b) $Nb_6Cl_{18}{}^{4-}$, $g = 76$, $b = 16$; in this cluster there are eight 3c-2e bonds on the eight faces of an octahedron.
(c) $Rh_6(CO)_{16}$, $g = 86$, $b = 11$; in this cluster there are four 3c-2e bonds on four faces and three 2c-2e bonds at three edges, as in the case of $B_6H_6^{2-}$.

Table 19.4.2. Some hexanuclear clusters

Cluster	g	b	No. of edges	Figure
$Mo_6(\mu_3\text{-Cl})_8Cl_6^{2-}$	84	12	12	(a)
$Nb_6(\mu_2\text{-Cl})_{12}Cl_6^{4-}$	76	16	12	(a)
$Rh_6(CO)_{16}$	86	11	12	(a)
$Os_6(CO)_{18}$	84	12	12	(b)
$Os_6(CO)_{18}H_2$	86	11	11	(c)
$Os_6C(CO)_{16}(MeC\equiv CMe)$	88	10	10	(d)
$Ru_6C(CO)_{15}^{2-}$	90	9	9	(e)
$Os_6(CO)_{20}[P(OMe)_3]$	90	9	9	(f)
$Co_6(\mu_2\text{-}C_2)(\mu_4\text{-}S)(CO)_{14}$	92	8	8	(g)

(a) (b) (c) (d) (e) (f) (g)

Other clusters listed in Table 19.4.2 have the property that their b value equals the number of edges of their M_n skeletons.

19.4.3 Clusters with seven or more transition-metal atoms

Table 19.4.3 listed the bond valence and structural data of some selected examples of high-nuclearity clusters, each consisting of seven or more transition-metal atoms. The skeletal structures of these clusters are shown in Fig. 19.4.1.

Table 19.4.3. Clusters with more than six transition metal atoms

Cluster	g	b	Structure (Fig. 19.4.1)	Remark
$Os_7(CO)_{21}$	98	14	(a)	Capped octahedron
$[Os_8(CO)_{22}]^{2-}$	110	17	(b)	Para-bicapped octahedron
$[Rh_9P(CO)_{21}]^{2-}$	130	16	(c)	Capped square antiprism; iso-bond valence with B_9H_{13}
$[Rh_{10}P(CO)_{22}]^-$	142	19	(d)	Bicapped square antiprism; iso-bond valence with $B_{10}H_{10}^{2-}$
$[Rh_{11}(CO)_{23}]^{3-}$	148	25	(e)	Three face-sharing octahedra
$[Rh_{12}Sb(CO)_{27}]^{3-}$	170	23	(f)	Icosahedron; iso-bond valence with $B_{12}H_{12}^{2-}$

When the number of metal atoms in a cluster increases, the geometries of the clusters become more complex, and some are often structurally better described in terms of capped or decapped polyhedra and condensed polyhedra. For example, the first and second clusters listed in Table 19.4.3 are a capped octahedron and a bicapped octahedron, respectively. Consequently, capping or decapping with a transition-metal fragment to a deltapolyhedral cluster leads to an increase or decrease in the cluster valence electron count of 12. When a transition-metal atom caps a triangular face of the cluster, it forms three M–M bonds with the vertex atoms, so according to the 18-electron rule, the cluster needs an additional $18 - 6 = 12$ electrons. The parent octahedron of $[Os_6(CO)_{18}]^{2-}$ has $g = 86$, the monocapped octahedron $Os_7(CO)_{21}$ has $g = 98$, and the bicapped octahedron $[Os_8(CO)_{22}]^{2-}$ has $g = 110$.

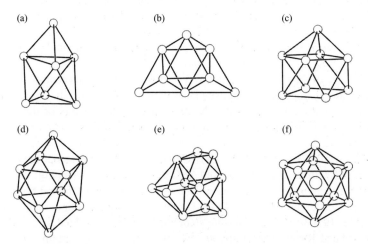

Fig. 19.4.1. Structures of some transition-metal clusters: (a) $Os_7(CO)_{21}$, (b) $[Os_8(CO)_{22}]^{2-}$, (c) $[Rh_9P(CO)_{21}]^{2-}$, (d) $[Rh_{10}P(CO)_{22}]^-$, (e) $[Rh_{11}(CO)_{23}]^{3-}$, and (f) $[Rh_{12}Sb(CO)_{27}]^{3-}$.

The metal cluster of $[Rh_{10}P(CO)_{22}]^-$ forms a deltapolyhedron, which has $g = 142$, as shown in Fig. 19.4.1(d). The skeleton of $[Rh_9P(CO)_{21}]^{2-}$ is obtained by removal of a vertex transition-metal fragment. The skeletal valence electron count of $[Rh_9P(CO)_{21}]^{2-}$ gives $g = 142 - 12 = 130$.

The metal cluster of $[Rh_{11}(CO)_{23}]^{3-}$ is composed of three face-sharing octahedra, as shown in Fig. 19.4.1(e). The metal cluster of $[Rh_{12}Sb(CO)_{27}]^{3-}$ consists of an icosahedron with an encapsulated Sb atom at its center.

Generally, capped or decapped deltapolyhedral clusters are characterized by the number of skeletal valence electrons g

$$g = (14n + 2) \pm 12m,$$

where n is the number of M atoms in the parent deltapolyhedron and m is the number of capped (+) or decapped (−) metal fragments.

A classical example of correlation of structure with valence electron count of transition-metal clusters is shown in Fig. 19.4.2. There the structures of a series of osmium clusters are systematized by applying the capping and decapping procedures.

19.4.4 Anionic carbonyl clusters with interstitial main-group atoms

There is much interest in transition-metal carbonyl clusters containing interstitial (or semi-interstitial) atoms in view of the fact that insertion of the encapsulated atom inside the metallic cage increases the number of valence electrons but leaves the molecular geometry essentially unperturbed. The clusters are generally anionic, and the most common interstitial heteroatoms are carbon, nitrogen, and phosphorus. Some representative examples are displayed in Fig. 19.4.3.

The core of the anionic carbonyl cluster $[Co_6Ni_2(C)_2(CO)_{16}]^{2-}$ consists of two trigonal prisms sharing a rectangular face [Fig. 19.4.3(a)]. All four vertical edges and two horizontal edges, one on the top face and the other on the bottom face, are each bridged by a carbonyl group. Each of the two Co* atoms has two terminal carbonyl groups, and each of the Ni and Co atoms has one.

The $[Os_{18}Hg_3(C)_2(CO)_{42}]^{2-}$ cluster is composed of two tricapped octahedral $Os_9(C)(CO)_{21}$ units sandwiching a Hg_3 triangle (Fig. 19.4.3(b)). Each corner Os atom in the top and bottom faces has three terminal carbonyl groups, and the remaining Os atoms each have two.

The core of the $[Fe_6Ni_6(N)_2(CO)_{24}]^{2-}$ cluster comprises a central Ni_6 octahedron that shares a pair of opposite faces with two Ni_3Fe_3 octahedra, as shown in Fig. 19.4.3(c). The interstitial N atoms occupy the centers of the Ni_3Fe_3 octahedra. Each Fe atom has two terminal carbonyl groups, and each Ni atom has one.

The metallic core of the $[Rh_{28}(N)_4(CO)_{41}H_x]^{4-}$ cluster is composed of three layers of Rh atoms in a close-packed ABC sequence, as shown in Fig. 19.4.3(d). Four N and an unknown number of H atoms occupy the octahedral holes.

The metal atoms in $[Ru_8(P)(CO)_{22}]^-$ constitute a square antiprismatic assembly [Fig. 19.4.3(e)]. Two opposite slant edges are each bridged by a carbonyl group. Each Ru bridged atoms has two terminal carbonyl groups, and each

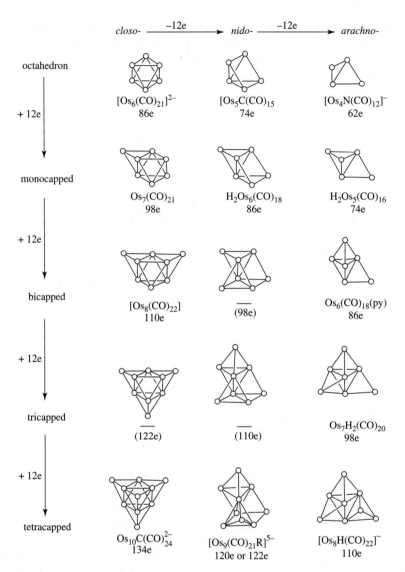

Fig. 19.4.2.
The structures of osmium carbonyl compounds vary with an increase or decrease of the valence electrons.

of the remaining four has three. Comparison of this cluster core with those of $[Rh_9(P)(CO)_{21}]^{2-}$ [Fig. 19.4.3(f)] and $[Rh_{10}(S)(CO)_{22}]^{2-}$ [Fig. 19.4.3(g)] shows that the latter two are derived from successive capping of the rectangular faces of the square antiprism.

19.5 Iso-bond valence and iso-structural series

For a cluster consisting of n_1 transition-metal atoms and n_2 main-group atoms, the bond valence b is evaluated as

$$b = \tfrac{1}{2}(18n_1 + 8n_2 - g),$$

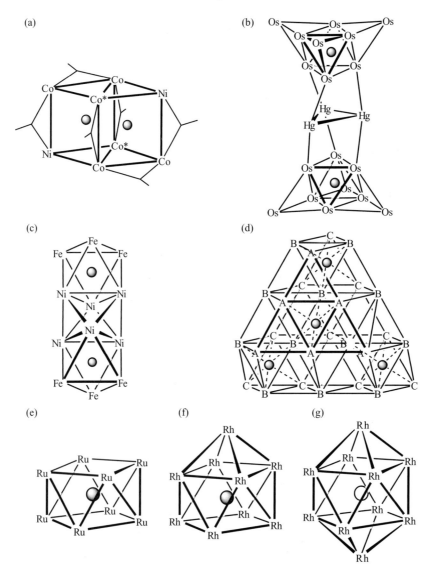

Fig. 19.4.3.
Molecular structure of some carbonyl cluster anions containing encapsulated heteroatoms; all terminal CO groups are omitted for clarity;
(a) $[Co_6Ni_2(C)_2(CO)_{16}]^{2-}$; only the bridging CO groups are shown; (b) $[Os_{18}Hg_3(C)_2(CO)_{42}]^{2-}$; (c) $[Fe_6Ni_6(N)_2(CO)_{24}]^{2-}$;
(d) $[Rh_{28}(N)_4(CO)_{41}H_x]^{4-}$; to avoid clutter, not all Rh–Rh bonds are included; (e) $[Ru_8(P)(CO)_{22}]^-$; (f) $[Rh_9(P)(CO)_{21}]^{2-}$;
(g) $[Rh_{10}(S)(CO)_{22}]^{2-}$.

where g is the number of valence electrons of the skeleton formed by the n_1 transition-metals and n_2 main-group atoms.

When a BH group of the octahedral cluster $(BH)_6^{2-}$ is replaced by a CH^+ group, both g and b retain their values and the structure of the cluster anion $(BH)_5CH^-$ remains octahedral. On the other hand, when a BH group of $(BH)_6^{2-}$ is replaced by a $Ru(CO)_3$ group, the b value still remains the same. But g

Metal–Metal Bonds

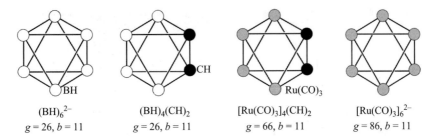

Fig. 19.5.1.
Iso-bond valence and iso-structural series of $B_6H_6^{2-}$.

increases its value by 10, as a BH group contributes 4 electrons to the skeleton, while a Ru(CO)$_3$ group contributes 14 (8 from Ru and 2 from each CO). Therefore, replacement of one or more BH groups in (BH)$_6^{2-}$ by either CH$^+$ or Ru(CO)$_3$ groups results in a series of iso-bond valence and iso-structural clusters. The structures of some members of this series, (BH)$_6^{2-}$, (BH)$_4$(CH)$_2$, [Ru(CO)$_3$]$_4$(CH)$_2$, and [Ru(CO)$_3$]$_6^{2-}$, are shown in Fig. 19.5.1. Similar substitutions by either transition-metal or representative-element groups give rise to a variety of cluster compounds, and five iso-structural series of clusters containing both transition-metal and main-group element components are displayed in Fig. 19.5.2.

Clearly the structure of a given cluster depends on electronic, geometric, and other factors. Hence the structure of a compound cannot be predicted until it has been determined experimentally. Still, based on the bond valence and structural principle illustrated above, an educated guess on the structure of a cluster becomes feasible. In addition, the bond valence concept provides a useful link between apparently dissimilar clusters such as (BH)$_6^{2-}$ and [Ru(CO)$_3$]$_4$(CH)$_2$.

19.6 Selected topics in metal–metal interactions

Since the 1980s, studies on metal clusters and metal string complexes have revealed unusual interactions between metal atoms, some of which are discussed in this section.

19.6.1 Aurophilicity

The term aurophilicity (or aurophilic attraction) refers to the formally non-bonding but attractive interaction between gold(I) atoms in gold cluster compounds. The Au(I) atom has a closed-shell electronic configuration: [Xe] $4f^{14}5d^{10}6s^0$. Normally, repulsion exists between the nonbonding homoatoms. However, there is extensive crystallographic evidence of attractive interaction between gold(I) cations. Figure 19.6.1 shows the structures of three Au(I) compounds.

(1) O[AuP(o-tol)$_3$]$_3^+$ (o-tol = C$_6$H$_4$Me-2): In O[AuP(o-tol)$_3$]$_3$(BF$_4$), the O atom forms covalent bonds with three Au atoms in a OAu$_3$ pyramidal configuration, as shown in Fig. 19.6.1(a). The Au atoms are linearly

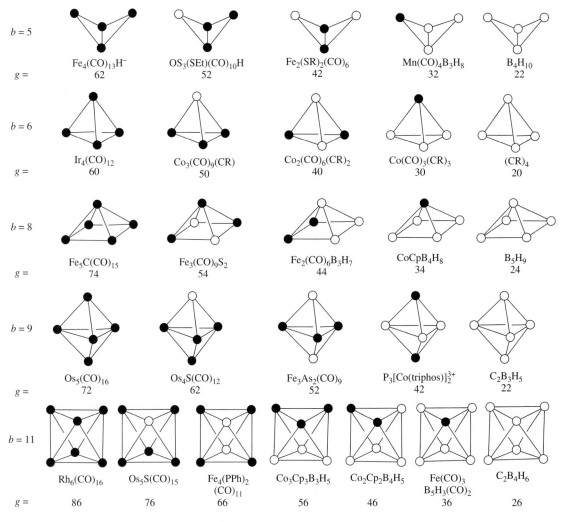

Fig. 19.5.2.
Iso-structural series of transition-metal and main-group clusters.

coordinated by O and P atoms. In this structure, the mean Au···Au distance is 308.6 pm, which is shorter than the sum of van der Waals radii, $2 \times 166 = 332$ pm.

(2) $S[AuP(o\text{-tol})_3]_4^{2+}$: In $S[AuP(o\text{-tol})_3]_4(ClO_4)_2$, the SAu_4 unit takes a square-pyramidal configuration, as shown in Fig. 19.6.1(b). The Au atoms are near linearly coordinated by S and P atoms. The Au···Au distances lie in the range of 288.3 to 293.8 pm, and the mean distance is 293.0 pm. The distance of the S atom to the center of the Au_4 basal plane is 130 pm.

(3) $Au_{11}I_3[P(p\text{-}C_6FH_4)_3]_7$: The structure of the molecule is shown in Fig. 19.6.1(c). In the Au_{11} cluster, the central Au atom is surrounded by ten Au atoms, which form an incomplete icosahedron (lacking two vertices) with each vertex carrying one terminal iodo ligand or $P(p\text{-}C_6FH_4)_3$ group.

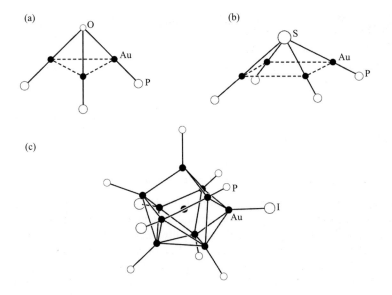

Fig. 19.6.1.
Molecular structure of gold cluster compounds: (a) O[AuP(o-tol)$_3$]$_3^+$, (b) S[AuP(o-tol)$_3$]$_4^{2+}$, (c) Au$_{11}$I$_3$[P(p-C$_6$FH$_4$)$_3$]$_7$.

The mean Au···Au distance from the central atom to the surrounding atoms is 268 pm, and the mean distance between the ten Au vertices is 298 pm.

In other gold(I) cluster compounds, such as tetrahedral [(AuL)$_4$(μ_4-N)] and [(AuL)$_4$(μ_4-O)]$^{2+}$, trigonal-bipyramidal [(AuL)$_5$(μ_5-C)]$^+$, [(AuL)$_5$(μ_5-N)]$^{2+}$ and [(AuL)$_5$(μ_5-P)]$^{2+}$, and octahedral [(AuL)$_6$(μ_6-C)]$^{2+}$ and [(AuL)$_6$(μ_6-N)]$^{3+}$ (L = PPh$_3$ or PR$_3$), the Au···Au distances lie in the range 270 to 330 pm. These data substantiate that aurophilicity is a common phenomenon among gold cluster complexes.

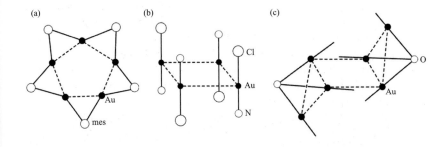

Fig. 19.6.2.
Structure of some gold oligomeric aggregate molecules: (a) Au$_5$(mes)$_5$ (mes = C$_6$H$_3$Me$_3$-2,4,6); (b) [LAuCl]$_4$, L = $\overline{\text{HN(CH}_2\text{)}_4\text{CH}_2}$; and (c) [O(AuPPh$_3$)$_3$]$_2$.

Aurophilicity presumably arises from relativistic modification of the gold valence AOs energies, which brings the 5d and 6s orbitals into close proximity in the energy level diagram. In more recent theoretical studies, the effect is primarily attributed to electron correlation, which takes precedence over 6s/5d hybridization. To date, the origin of the aurophilicity has not yet been unambiguously established.

Making use of the concept of aurophilicity, simple gold compounds can be combined to yield complicated oligomeric aggregates in designed synthesis. Figure 19.6.2 shows the structures of three oligomeric molecules.

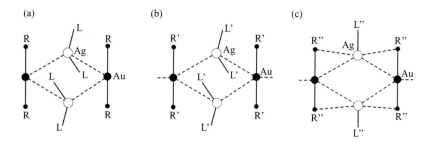

Fig. 19.6.3.
Some mixed Au and Ag metal complexes (filled circle, Au; open circle, Ag): (a) [Au(CH$_2$PPh$_2$)$_2$]$_2$[Ag(OClO$_3$)$_2$]$_2$, (b) [Au(C$_6$F$_5$)]$_2$[Ag(C$_6$H$_6$)]$_2$, and (c) [Au(C$_6$F$_5$)]$_2$[Ag(COMe$_2$)]$_2$.

19.6.2 Argentophilicity and mixed metal complexes

By analogy to aurophilicity, argentophilicity has been demonstrated to exist in silver cluster complexes. In the crystal structures of a variety of silver(I) double and multiple salts containing a fully encapsulated acetylide dianion C_2^{2-} (IUPAC name acetylenediide) in different polyhedral silver cages (see Fig. 14.3.11), there exist many Ag···Ag contacts shorter than twice the van der Waals radius of silver ($2 \times 170 = 340$ pm). Further details are given in Chapter 20.

Taking advantage of both aurophilicity and argentophilicity, tetranuclear mixed-metal complexes which contain pairs of Au and Ag atoms have been prepared, as shown in Fig. 19.6.3. In the preparation of mixed gold/silver polynuclear complexes, aurophilicity and argentophilicity have been utilized to promote cluster formation. Figure 19.6.4 shows the cores of several mixed gold-silver clusters.

The [Au$_{13}$Ag$_{12}$] molecular skeletons (a) and (b) of [(Ph$_3$P)$_{10}$Au$_{13}$Ag$_{12}$Br$_8$]SbF$_6$ and [(p-tol$_3$P)$_{10}$Au$_{13}$Ag$_{12}$Br$_8$]Br, respectively, can each be considered as two centered icosahedra sharing a common vertex. In an alternative description, the 25 metal atoms constitute three fused icosahedra, with a common pentagon shared between an adjacent pair of fused icosahedra. Structure (a) adopts the **ses** (staggered-eclipsed-staggered) configuration, and the sequence of relative positioning of atoms is ABBA. Structure (b) adopts the **sss** (staggered-staggered-staggered) configuration, and the sequence is ABAB. The structure of (c) consists of the three centered icosahedra, each of which uses one edge to form a central triangle, with an additional Ag atom lying above and below it. The structure of (d) consists of six Au atoms that form a planar six-membered ring, with one Ag atom located at the center; each edge of the Au$_6$ hexagon is bridged by a bridging C atom, with three C atoms lying above the plane and three below it.

19.6.3 Metal string molecules

A metal string molecule contains a linear metal-atom chain in its structure. In a molecule of this type, all or a part of the neighboring metal atoms are involved in metal–metal bonding interactions. A general strategy of synthesizing metal string molecules is to design bridging ligands possessing multiple donor sites arranged in a linear sequence such that they can coordinate simultaneously to metal centers.

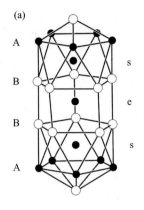
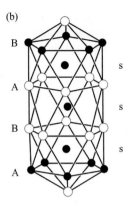
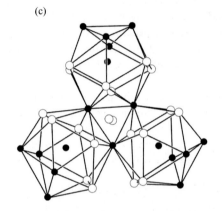
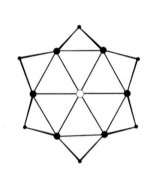

Fig. 19.6.4.
Skeletal structure of some mixed Au/Ag clusters (filled circle, Au; open circle, Ag): (a) [$Au_{13}Ag_{12}$] in [$(Ph_3P)_{10}Au_{13}Ag_{12}Br_8$]$SbF_6$, (b) [$Au_{13}Ag_{12}$] in [$(p\text{-tol}_3P)_{10}Au_{13}Ag_{12}Br_8$]Br, (c) [$Au_{18}Ag_{20}$] in [$(p\text{-tol}_3P)_{12}Au_{18}Ag_{20}Cl_{14}$], and (d) [$AgAu_6C_6$] in [$Ag(AuC_6H_2(CHMe_2)_3)_6CF_3SO_3$].

(1) Metal string molecules constructed with oligo(α-pyridyl)amido ligand

The common formula of a series of polypyridylamines is shown below:

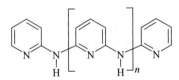

$n = 0$, Hdpa
$n = 1$, H_2tpda
$n = 2$, H_3teptra
$n = 3$, H_4peptea

Four fully deprotonated polypyridylamines, or oligo(α-pyridyl)amido ligands, can coordinate simultaneously to metal atoms from the upper, lower, front, and back directions to form a metal string molecule:

n	M (II)
0	Cr, Ru, Co, Rh, Ni, Cu
1	Cr, Co, Ni
2	Cr, Ni
3	Cr, Ni

As an oligo(α-pyridyl)amido ligand has an odd number of donor sites, the corresponding metal string molecule has the same number of metal atoms.

From the reactions of Ni(II) salts with polypyridylamines, a series of metal string molecules, in which the number of nickel atom varies from 3 (i.e., $n =$

0) to 9 (i.e., $n = 3$), have been prepared. The structures of these molecules are very similar. Figure 19.6.5 shows the structure of [Ni$_9$(μ_9-peptea)$_4$Cl$_2$], (H$_4$peptea = pentapyridyltetramine). The nine Ni atoms are in a straight line, and the distances between atoms are approximately equal. Every N atom in the four (α-pyridyl)amido ligands coordinates to one Ni atom. Each Ni atom is coordinated by four N atoms to form a square oriented perpendicular to the metal string. Because of steric repulsion between H atoms of the neighboring pyridine rings, the (α-pyridyl)amido ligands are helically distributed around the string axis, as shown in Fig. 19.6.5.

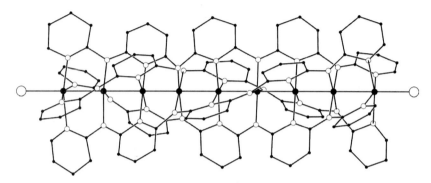

Fig. 19.6.5.
Structure of [Ni$_9$(μ_9-peptea)$_4$Cl$_2$] (filled circle, Ni; large open circle, Cl; small open circle, N; and small filled circle, C).

The Ni–Ni and Ni–N distances in a family of related metal string molecules, in which the numbers of Ni atoms are 3, 5, 7, and 9, are shown in Fig. 19.6.6. The Ni–Ni distances increase from the center toward each terminal, and the Ni–N distances are nearly equal except for the outermost ones. These effects

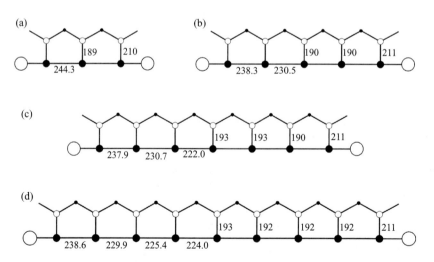

Fig. 19.6.6.
Bond lengths in metal string molecules (in pm): (a) Ni$_3$(μ_3-dpa)$_4$Cl$_2$, (b) Ni$_5$(μ_5-tpda)$_4$Cl$_2$, (c) Ni$_7$(μ_7-teptra)$_4$Cl$_2$, and (d) Ni$_9$(μ_9-peptea)$_4$Cl$_2$.

are attributable to the fact that, unlike the inner metal atoms, each terminal metal atom has square-pyramidal coordination.

The metal string molecule [Cr$_5$(tpda)$_4$Cl$_2$]· 2Et$_2$O·4CHCl$_3$ contains alternately long and short metal–metal distances, as shown in Fig. 19.6.7. The short Cr–Cr distances are 187.2 and 196.3 pm, which correspond to the quadruple bond Cr≣Cr. The long Cr···Cr distances are 259.8 and 260.9 pm, which are indicative of nonbonding interaction.

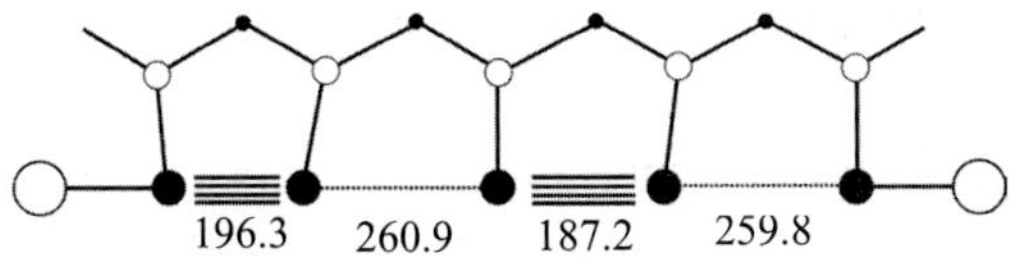

Fig. 19.6.7.
Structure of Cr$_5$(tpda)$_4$Cl$_2$ (bond length in pm) (large filled circle, Cr; small filled circle, C; large open circle, Cl; small open circle, N).

(2) Hexanuclear metal string cationic complexes

Modification of the H$_2$tpda ligand by substitution of the central pyridyl group with a naphthyridyl group gives rise to the new ligand 2,7-bis (α-pyridylamino)-1,8-naphthyridine (H$_2$bpyany), which has been used to generate a series of hexanuclear metal string complexes of the general formula [M$_6$(μ_6-bpyany)$_4$X$_2$]Y$_n$ (M=Co, Ni; X$^-$ = terminal monoanionic ligand; Y$^-$ = counter monoanion; n = 1, 2). All compounds contain a linear hexanuclear cation helically supported by four bpyany^{2-} ligands and conforms approximately to idealized D_4 molecular symmetry if the axial terminal ligands are ignored. The structure of a representative example is shown in Fig. 19.6.8.

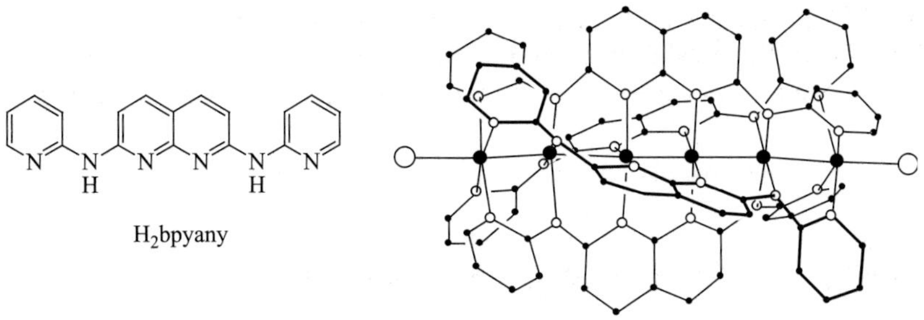

Fig. 19.6.8.
Structural formula of the H$_2$bpyany ligand and molecular geometry of the hexanuclear monocation in crystalline [Co$_6$(μ_6-bpyany)$_4$Cl$_2$]PF$_6$.

The averaged metal–metal bond lengths in the series of linear M$_6^{12+}$ (M = Co, Ni) complexes and their M$_6^{11+}$ one-electron reduction products are

Table 19.6.1. Hexanuclear metal string complexes and average M–M bond distances

M	X	Y	n	Bond distance (pm) averaged for D_4 symmetry		
				Outermost	Mid-way	Innermost
Co	NCS	PF$_6$	2	231.3(1)	225.5(1)	224.5(1)
Co	NCS	PF$_6$	1	231.3(1)	227.3(1)	225.6(1)
Co	CF$_3$SO$_3$	CF$_3$SO$_3$	2	228.3(1)	224.3(1)	226.7(1)
Co	CF$_3$SO$_3$	CF$_3$SO$_3$	1	228.4(1)	224.9(1)	225.1(1)
Ni	NCS	BPh$_4$	2	240.3(1)	231.4(1)	229.6(1)
Ni	Cl	PF$_6$	1	241.1(3)	228.5(3)	220.2(3)

tabulated in Table 19.6.1. In all complexes the outermost M–M distance is in general slightly longer than the inner bond distances, but neither the nature of the axial ligands nor the addition of one electron to the Co_6^{12+} system results in significant structural changes. In contrast, the innermost Ni–Ni bond shows a substantial decrease of 9.4(3) pm upon one-electron reduction of the Ni_6^{12+} system. The crystallographic data are consistent with the proposed model of a delocalized electronic structure for the Co_6^{n+} (n = 11, 12) complexes, whereas the extra electron in the Ni_6^{12+} system partakes in a δ bond constructed from $d_{x^2-y^2}$ orbitals of the naphthyridyl-coordinated nickel atoms.

(3) Mixed-valence metal string complexes of gold

The structures of two gold metal string molecules are shown in Fig. 19.6.9. The formula of molecule (a) in this figure is:

$$\left[\begin{array}{c} \text{Ph}_2 \quad\quad \text{Ph}_2 \\ \text{P} \quad R \quad \text{P} \\ \text{R-Au-Au-Au-Au-Au-R} \\ \text{P} \quad R \quad \text{P} \\ \text{Ph}_2 \quad\quad \text{Ph}_2 \end{array} \right]^+ (\text{AuR}_4)^-, R = C_6F_5$$

From the Au–Au bond lengths shown and related theoretical calculations, the valence states of the Au atoms are in the sequence of Au(III)–Au(I)–Au(I)–Au(I)–Au(III). The molecular cation is composed of the central unit $[Au(C_6F_5)_2]^-$ and two outer dinuclear gold cations $[Au_2\{(CH_2)_2(PPh_2)\}_2C_6F_5]^+$, with the central unit donating electrons to the outer units.

The formula of molecule (b) in Fig. 19.6.9 is

$$\left[\begin{array}{c} \text{Ph}_2 \quad \text{Ph}_2 \quad \text{Ph}_2 \\ \text{P} \quad \text{P} \quad \text{P} \\ \text{R-Au-Au-Au-Au-Au-Au-R} \\ \text{P} \quad \text{P} \quad \text{P} \\ \text{Ph}_2 \quad \text{Ph}_2 \quad \text{Ph}_2 \end{array} \right]^{2+} (\text{ClO}_4^-)_2, R = C_6F_3H_2$$

(a)
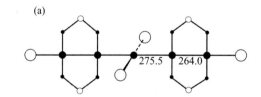

(b)
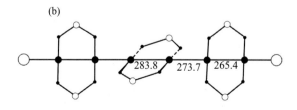

Fig. 19.6.9.
Structures of two mixed-valence metal string molecules (large filled circle, Au; large open circle, R group; small filled circle, C; small open circle, P; bond lengths in pm).

From the Au–Au bond lengths shown and theoretical calculations, the valence states of the Au atoms are identified as Au(III)–Au(I)–Au(I)–Au(I)–Au(III).

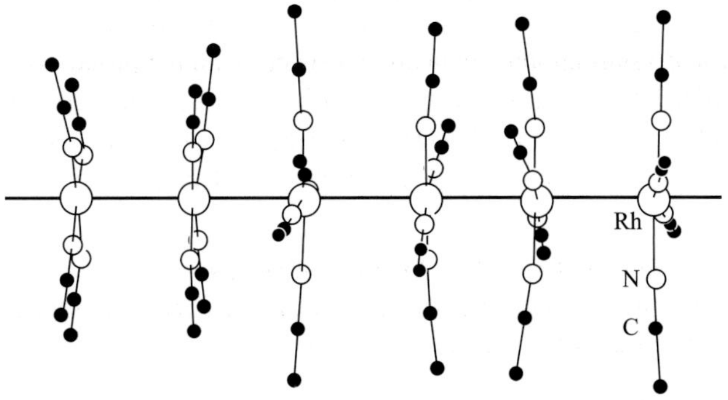

Fig. 19.6.10.
Structure of a section of the infinite rhodium chain, $[Rh(CH_3CN)_4^{1.5+}]_\infty$.

19.6.4 Metal-based infinite chains and networks

Infinite metal-based chains are expected to be much more promising as conducting inorganic "molecular wires" than short-chain oligomers. The infinite rhodium chain, $[Rh(CH_3CN)_4^{1.5+}]_\infty$, consists of alternating Rh–Rh distances of 284.42 and 292.77 pm, and is a semiconductor. Figure 19.6.10 shows a section of the infinite cationic chain in the polymer $[\{Rh(CH_3CN)_4\}(BF_4)_{1.5}]_\infty$.

A series of polymeric complexes featuring the metallophilic interaction between gold(I) and thallium(I) has been synthesized employing acid–base strategy. For example, the treatment of $Bu_4N[Au(C_6Cl_5)_2]$ with $TlPF_6$ in THF gave $[AuTl(C_6Cl_5)_2]_n$, which consists of an infinite linear $(Tl\cdots Au\cdots)_\infty$ chain consolidated by unsupported $Au^I\cdots Tl^I$ interactions, as shown in Fig. 19.6.11(a).

In the presence of triphenylphosphine oxide, and also depending on the nature of the pentahalophenyl group employed, similar synthetic reactions afforded $[AuTl(C_6F_5)_2(Ph_3P=O)_2]_n$ and $[AuTl(C_6Cl_5)_2(Ph_3P=O)_2(THF)]_n$ with the

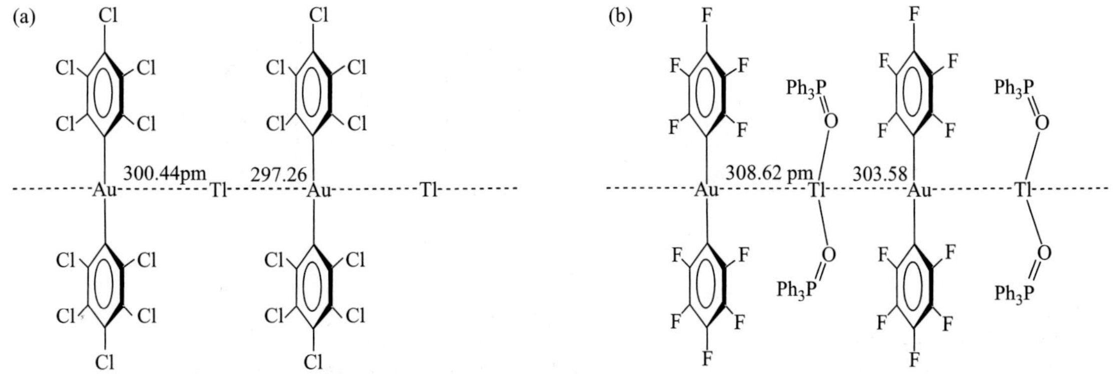

Fig. 19.6.11.
Structure of (a) $[AuTl(C_6Cl_5)_2]_n$ and (b) $[AuTl(C_6F_5)_2(Ph_3P=O)_2]_n$.

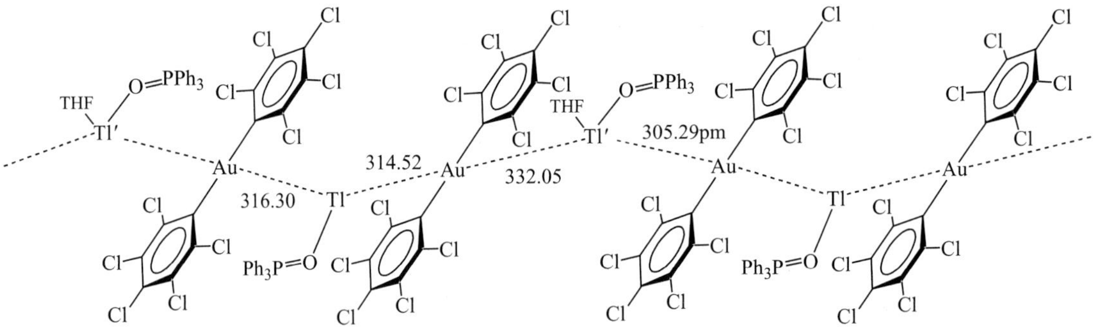

Fig. 19.6.12.
Structure of [AuTl(C$_6$Cl$_5$)$_2$(Ph$_3$P=O)$_2$(THF)]$_n$.

phosphine oxide incorporated into their polymeric structures. In the pentafluorophenyl complex, thallium(I) adopts a distorted trigonal-bipyramidal geometry with one equatorial coordination site filled by a stereochemically active lone pair, forming a linear (Tl···Au···)$_\infty$ chain, as shown in Fig. 19.6.11(b). In contrast, the pentachlorophenyl complex comprises an infinite zigzag (Tl···Au···Tl'···Au···)$_\infty$ chain constructed from two kinds of thallium(I) centers with distorted trigonal-pyramidal and pseudo-tetrahedral geometries for Tl and Tl' (each possessing a stereochemically active lone pair), respectively, as shown in Fig. 19.6.12.

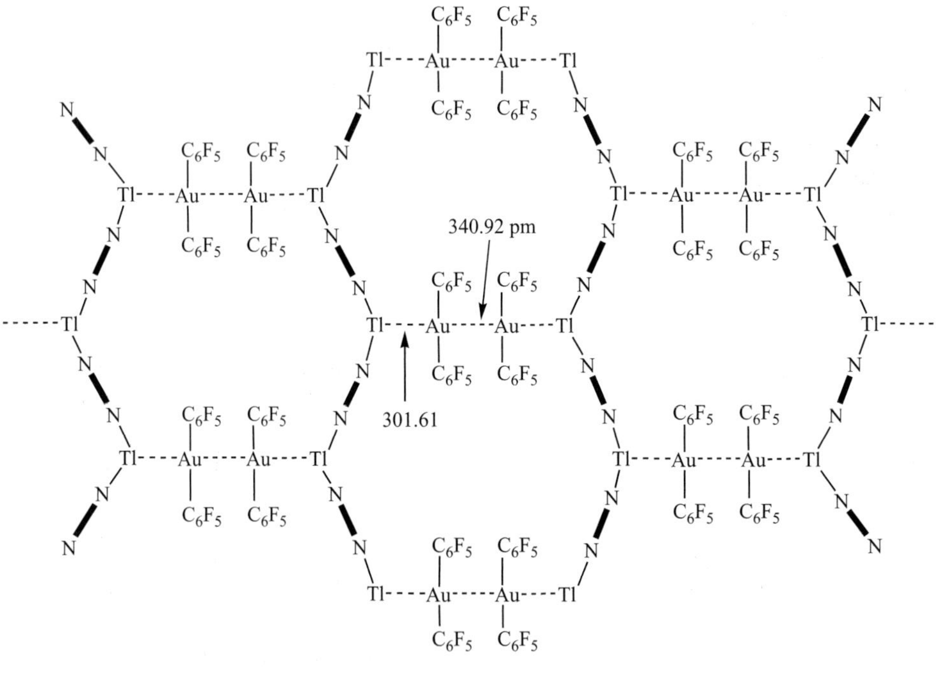

Fig. 19.6.13.
Layer structure of [AuTl(C$_6$F$_5$)$_2$(bipy)]$_n$. The bridging 4,4'-bypyridine ligand is represented by a thick rod joining its two terminal N atoms.

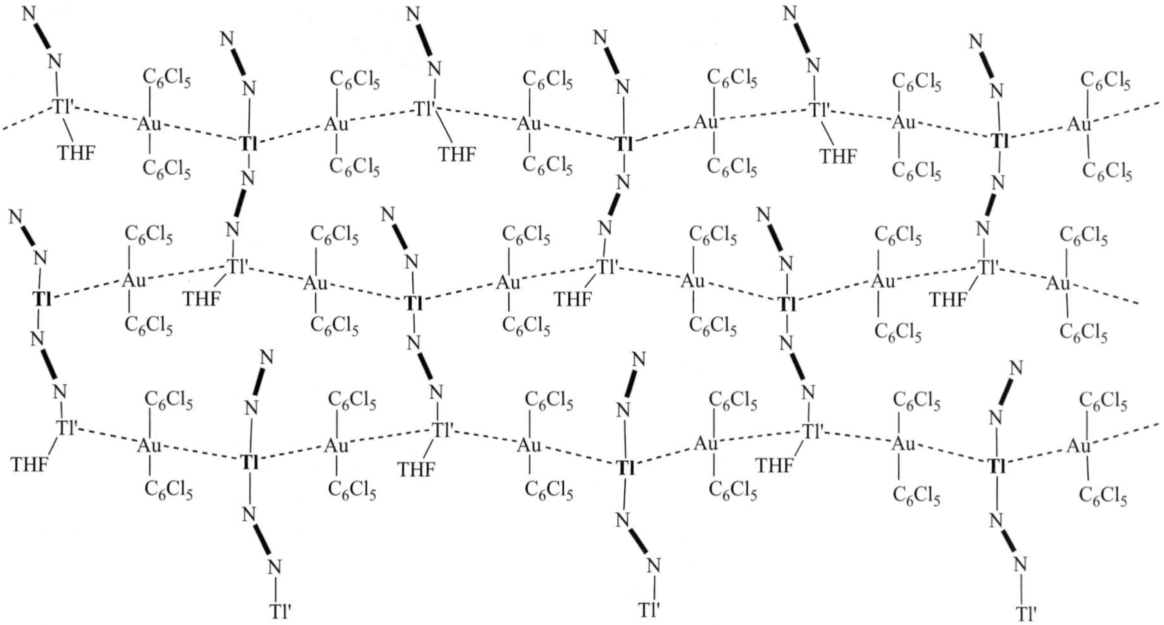

Fig. 19.6.14.
Layer structure of $[Au_2Tl_2(C_6Cl_5)_4(bipy)_{1.5}(THF)]_n$. Atom **Tl** is shown in boldface type to distinguish it from atom Tl$'$. Each bridging 4,4$'$-bypyridine ligand is represented by a thick rod joining the terminal N atoms.

Similar acid–base reactions with the introduction of the *exo*-bidentate bridging ligand 4,4$'$-bipyridine led to the formation of $[AuTl(C_6F_5)_2(bipy)]_n$ and $[Au_2Tl_2(C_6Cl_5)_4(bipy)_{1.5}(THF)]_n$, which exhibit higher-dimensional polymeric structures. In the pentafluorophenyl complex, linear tetranuclear Tl···Au···Au···Tl units are linked through bipyridine bridges to form a honeycomb-like network, as shown in Fig. 19.6.13.

The pentachlorophenyl complex contains two kinds of thallium(I) centers: atom Tl is coordinated by two bipyridyl N atoms, whereas atom Tl$'$ is bound to a THF ligand and a bipyridyl N atom, as shown in Fig. 19.6.14. The asymmetric unit contains two non-equivalent bridging 4,4$'$-bipyridine ligands, one of which occupies a $\bar{1}$ site in the unit cell and necessarily exists in the planar configuration. Hetero-metallophilic interaction generates infinite $(Tl···Au···Tl'···Au···)_\infty$ zigzag chains linked by the first kind of bridging bipyridine ligand (located in a general position and nonplanar) across Tl and Tl$'$ to form a brick-like layer. The centrosymmetric bipyridine, which is bound to Tl and represented by a dangling rod in the figure, connects the Tl atoms in adjacent layers to form a double layer.

References

1. F. A. Cotton, G. Wilkinson, C. A. Murillo and M. Bochmann, *Advanced Inorganic Chemistry*, 6th edn., Wiley, New York, 1999.

2. F. A. Cotton, C. A. Murillo and R. A. Walton (eds.), *Multiple Bonds between Metal Atoms*, 3rd edn., Springer, New York, 2005.
3. F. P. Purchnik, *Organometallic Chemistry of the Transition Elements*, Plenum Press, New York, 1990.
4. D. M. P. Mingos and D. J. Wales, *Introduction to Cluster Chemistry*, Prentice Hall, Englewood Clifts, NJ, 1990.
5. D. F. Shriver, H. D. Karsz and R. D. Adams (eds.), *The Chemistry of Metal Cluster Complexes*, VCH, New York, 1990.
6. P. J. Dyson and J. S. McIndoe, *Transition Metal Carbonyl Cluster Chemistry*, Gordon and Breach, Amsterdam, 2000.
7. M. H. Chisholm (ed.), *Early Transition Metal Clusters with π -Donor Ligands*, VCH, New York, 1995.
8. M. Gielen, R. Willem and B. Wrackmeyer (eds.), *Unusual Structures and Physical Properties in Organometallic Chemistry*, Wiley, West Sussex, 2002.
9. T. P. Fehlner (ed.), *Inorganometallic Chemistry*, Plenum Press, New York, 1992.
10. J.-X. Lu (ed.), *Some New Aspects of Transition-Metal Cluster Chemistry*, Science Press, Beijing/New York, 2000.
11. P. Braunstein, L. A. Oro and P. R. Raithby (eds.), *Metal Clusters in Chemistry*: vol. 1 *Molecular Metal Cluster*; vol. 2 *Catalysis and Dynamics and Physical Properties of Metal Clusters*; vol. 3 *Nanomaterials and Solid-state Cluster Chemistry*, Wiley–VCH, Weinheim, 1999.
12. A. J. Welch and S. K. Chapman (eds.), *The Chemistry of the Copper and Zinc Triads*, Royal Society of Chemistry, Cambridge, 1993.
13. J. P. Collman, R. Boulatov and G. B. Jameson, The first quadruple bond between elements of different groups. *Angew. Chem. Int. Ed.* **40**, 1271–4 (2001).
14. T. Nguyen, A. D. Sutton, M. Brynda, J. C. Fettinger, G. J. Long and P. P. Power, Synthesis of a stable compound with fivefold bonding between two chromium(I) centers. *Science* **310**, 844–7 (2005).
15. G.-D. Zhou, Bond valence and molecular geometry. *University Chemistry* (in Chinese) **11**, 9–18 (1996).
16. P. Pyykkö, Strong closed-shell interaction in inorganic chemistry. *Chem. Rev.* **97**, 579–636 (1997).
17. N. Kaltsoyannis, Relativistic effects in inorganic and organometallic chemistry. *J. Chem. Soc. Dalton Trans.*, 1–11 (1997).
18. S.-M. Peng, C.-C. Wang, Y.-L. Jang, Y.-H. Chen, F.-Y. Li, C.-Y. Mou and M.-K. Leung, One-dimensional metal string complexes. *J. Magn. Magn. Mater.* **209**, 80–3 (2000).
19. C.-H. Chien, J.-C. Chang, C.-Y. Yeh, G.-H. Lee, J.-M. Fang, Y. Song and S.-M. Peng, *Dalton Trans.*, 3249–56 (2006).
20. J. K. Bera and K. R. Dunbar, Chain compounds based on transition metal backbones: new life for an old topic. *Angew. Chem. Int. Ed.* **41**, 4453–7 (2002).
21. E. J. Fernández, A. Laguna, J. M. López-de-Luzuriaga, F. Mendizábal, M. Monge, M. E. Olmos and J. Pérez, Theoretical and photoluminescence studies on the $d^{10} - s^2$ Au^I-Tl^I interaction in extended unsupported chains. *Chem. Eur. J.* **9**, 456–65 (2003).

Supramolecular Structural Chemistry

20.1 Introduction

Supramolecular chemistry is a highly interdisciplinary field of science covering the chemical, physical, and biological features of molecular assemblies that are organized and held together by intermolecular interactions. The basic concepts and terminology were introduced by J.-M. Lehn, who together with D. J. Cram and C. J. Pedersen were awarded the 1987 Nobel Prize in Chemistry. In the words of Lehn, supramolecular chemistry may be defined as chemistry beyond the molecule, i.e., the study of organized entities of higher complexity (supermolecule) resulting from the association of two or more chemical species consolidated by intermolecular forces. The relationship of supermolecules to molecules and intermolecular binding is analogous to that of molecules to atoms and covalent bonds (Fig. 20.1.1).

A clarification about vocabulary in the chemical literature: the prefix in the word supermolecule (a noun) is derived from the Latin *super*, meaning "more than" or "above"; it should not be used interchangeably with the prefix *supra* in the word supramolecular (an adjective), which means "beyond" or "at a higher level than."

20.1.1 Intermolecular interactions

Intermolecular interactions constitute the core of supramolecular chemistry. The design of supermolecules requires a clear understanding of the nature, strength, and spatial attributes of intermolecular bonding, which is a generic term that includes ion pairing (Coulombic), hydrophobic and hydrophilic interactions, hydrogen bonding, host–guest complementarity, π–π stacking, and van der Waals interactions. For inorganic systems, coordination bonding is included in this list if the metal acts as an attachment template. Intermolecular interactions in organic compounds can be classified as (a) isotropic, medium-range forces that define molecular shape, size, and close packing and (b) anisotropic, long-range forces, which are electrostatic and involve heteroatom interactions. In general, isotropic forces (van der Waals interactions) usually mean dispersive and repulsive forces, including $C \cdots C$, $C \cdots H$, and $H \cdots H$ interactions, while most interactions involving heteroatoms (N, O, Cl, Br, I, P, S, Se, etc.) with one another or with carbon and hydrogen are anisotropic in character, including ionic forces, strongly directional hydrogen bonds ($O–H \cdots O$, $N–H \cdots O$),

Structural Chemistry of Selected Elements

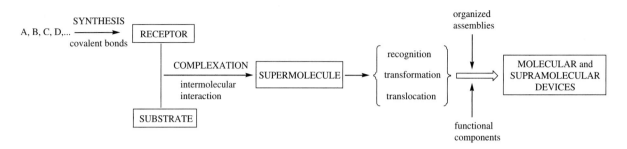

Fig. 20.1.1.
Conceptual development from molecular to supramolecular chemistry: molecules, supermolecules, molecular devices, and supramolecular devices.

weakly directional hydrogen bonds (C–H···O, C–H···N, C–H···X, where X is a halogen, and O–H···π), and other weak forces such as halogen···halogen, nitrogen···nitrogen, and sulfur···halogen interactions. In a crystal, various strong and weak intermolecular interactions coexist (sometimes in delicate balance, as is demonstrated by the phenomenon of polymorphism) and consolidate the three-dimensional scaffolding of the molecules.

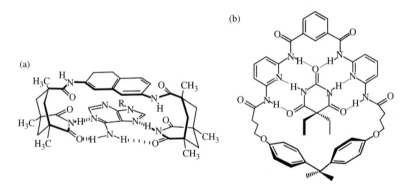

Fig. 20.1.2.
Molecular recognition through hydrogen bonding of (a) adenine in a cleft, and (b) barbituric acid in a macrocyclic receptor (right).

20.1.2 Molecular recognition

The concept of molecular recognition has its origin in effective and selective biological functions, such as substrate binding to a receptor protein, enzyme reactions, assembly of protein–DNA complexes, immunological antigen–antibody association, reading of the genetic code, signal induction by neurotransmitters, and cellular recognition. Many of these functions can be performed by artificial receptors, whose design requires an optimal match of the steric and electronic features of the non-covalent intermolecular forces between substrate and receptor. Some of the recognition processes that have been well studied by chemists include spherical recognition of metal cations by cryptates, tetrahedral recognition by macrotricyclic cryptands, recognition of specific anions, and the binding and recognition of neutral molecules through Coulombic, donor–acceptor, and in particular hydrogen-bonding interactions (see Fig. 20.1.2).

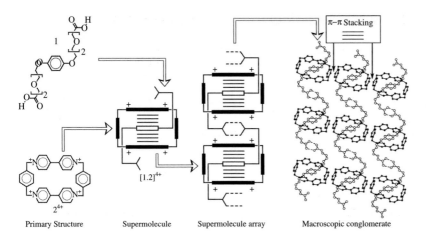

Fig. 20.1.3.
Aspects of supramolecular hierarchy in increasing superstructural complexity.

Molecular recognition studies are typically carried out in solution, and the effects of intermolecular interactions are often probed by spectroscopic methods.

20.1.3 Self-assembly

The term "self-assembly" is used to designate the evolution toward spatial confinement through spontaneous connection of molecular components, resulting in the formation of discrete or extended entities at either the molecular or the supramolecular level. Molecular self-assembly yields covalent structures, while in supramolecular self-assembly several molecules spontaneously associate into a single, highly structured supramolecular aggregate. In practice, self-assembly can be achieved if the molecular components are loaded with recognition features that are mutually complementary; i.e., they contain two or more interaction sites for establishing multiple connections. Thus well-defined molecular and supramolecular architectures can be spontaneously generated from specifically "engineered" building blocks. For example, self-assembly occurs with interlocking of molecular components using $\pi-\pi$ interactions (Fig. 20.1.3) and the formation of capsules with some curved molecules bearing complementary hydrogen bonding sites (Fig. 20.1.4).

The cyclic octapeptide *cyclo*-[–(D-Ala–L-Glu–D-Ala–L-Gln)$_2$–] has been designed by Ghadiri and co-workers to generate a hydrogen-bonded organic nanotube having an internal diameter of approximately 0.7–0.8 nm (Fig. 20.1.5).

20.1.4 Crystal engineering

Structural chemists and crystallographers rightfully regard an organic crystal as the "supermolecule *par excellence*," being composed of Avogadro's number of molecules self-assembled by mutual recognition at an amazing level of precision. In contrast to a molecule, which is constructed by connecting atoms with covalent bonds, a molecular crystal (solid-state supermolecule) is built by connecting molecules with intermolecular interactions. The process of

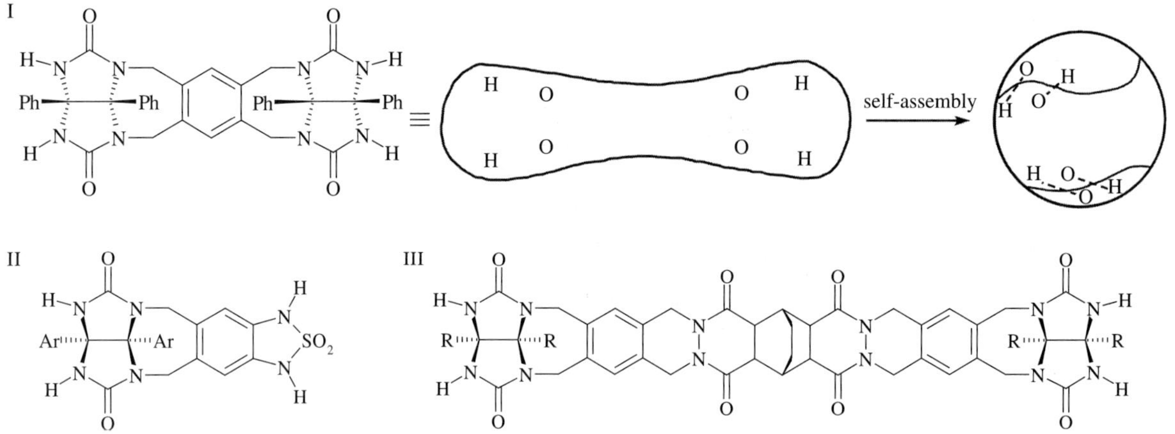

Fig. 20.1.4.
A tennis-ball-shaped molecular aggregate can be constructed by the self-assembly of curved molecule I. Tetrameric assembly of II generates a pseudo-spherical capsule. Dimeric assembly of III can be induced by the encapsulation of smaller molecules of appropriate size and shape at the center of a spherical complex.

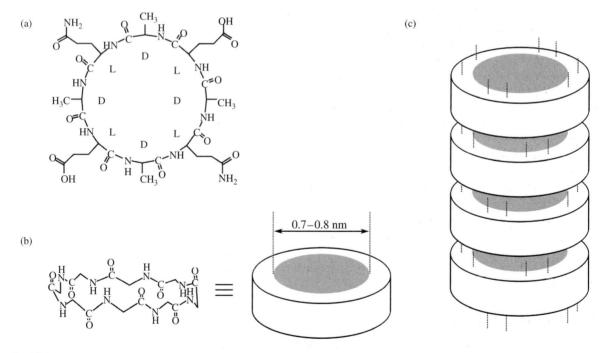

Fig. 20.1.5.
(a) Structural formula of *cyclo*-[–(D-Ala–L-Glu–D-Ala–L-Gln)$_2$–]; Ala = alanine, Glu = glutamic acid, Gln = glutamine, D or L indicates chirality at the carbon atom. (b) Perspective view of the backbone of the flat, ring-shaped octapeptide. (c) Tubular architecture generated from a stack of octapeptide molecules held by intermolecular hydrogen bonding.

Table 20.1.1. Comparison of crystal engineering and molecular recognition

	Crystal engineering	Molecular recognition
(1)	Concerned with the solid state	Concerned mainly with solution phase
(2)	Considers both convergent and divergent binding of molecules	Most cases only focus on convergent binding of molecules
(3)	Intermolecular interaction are examined directly in terms of their geometrical features obtained from X-ray crystallography	Intermolecular interactions are studied indirectly in terms of association constants obtained from various spectroscopic (NMR, UV, etc.) methods
(4)	Design strategies involve the control of the three-dimensional arrangement of molecules in the crystal; such an arrangement ideally results in desired chemical and physical properties	Design strategies are confined to the mutual recognition of generally two species: the substrate and the receptor; such recognition is expected to mimic some biological functionality
(5)	Both strong and weak interactions are considered independently or jointly in the design strategy	Only strong interactions such as hydrogen bonding are generally used for the recognition event
(6)	The design may involve either single-component species or multicomponent species; a single-component molecular crystal is a prime example of self-recognition	The design usually involves two distinct species: the substrate and the receptor; ideas concerning self-recognition are poorly developed
(7)	In host–guest complexes, the host cavity is composed of several molecules whose synthesis may be fairly simple; the geometry and functionality of the guest molecules are often of significance in the complexation	In host–guest complexes, the host cavity is often a single macrocyclic molecule whose synthesis is generally tedious; the host framework rather than the guest molecule plays a critical role in the complexation
(8)	Systematic retrosynthetic pathways may be deduced with the Cambridge Structural Database (CSD) to design new recognition patterns using both strong and weak interactions	There is no systematic set of protocols for the identification of new recognition patterns; much depends on individual style and preferences

crystallization is one of the most precise and spectacular examples of molecular recognition.

The determination of crystal structures by X-ray crystallography provides precise and unambiguous data on intermolecular interactions. Crystal engineering has been defined by Desiraju as "the understanding of intermolecular interactions in the context of crystal packing and in the utilization of such knowledge in the design of new solids with desired physical and chemical properties."

Crystal engineering and molecular recognition are twin tenets of supramolecular chemistry that depend on multiple matching of functionalities among molecular components. Historically, crystal engineering has been developed by structural and physical chemists with a view to design new materials and solid-state reactions, whereas molecular recognition has been developed by physical organic chemists interested in mimicking biological processes. The methodologies and goals of these two related fields are summarized in Table 20.1.1.

20.1.5 Supramolecular synthon

In the context of organic synthesis, the term "synthon" was introduced by Corey in 1967 to refer to "structural units within molecules which can be formed and/or assembled by known or conceivable synthetic operations." This general definition was modified by Desiraju for supramolecular chemistry: "*Supramolecular synthons* are structural units within supermolcules which can be formed and/or assembled by known or conceivable synthetic operations involving intermolecular interactions." The goal of crystal engineering is to recognize and design synthons sufficiently robust to be carried over from one network structure to another, which ensures generality and predictability. Some common examples of supramolecular synthons are shown in Fig. 20.1.6.

It should be emphasized that supramolecular synthons are derived from designed combinations of interactions and are not identical to the interactions. A supramolecular synthon incorporates both chemical and geometrical recognition features of two or more molecular fragments, i.e., both explicit and implicit involvement of intermolecular interactions. However, in the simplest cases, a single interaction may be regarded as a synthon, for instance $Cl \cdots Cl$, $I \cdots I$ or $N \cdots Br$ [Figs. 20.1.6(**23–25**)]. Besides the strong hydrogen bonds ($N-H \cdots O$ and $O-H \cdots O$), which are expected to be frequently involved in supramolecular synthons [Figs. 20.1.6(**1–5**)], weak hydrogen bonds of the $C-H \cdots X$ variety and $\pi-\pi$ interactions may also be significant. Although such weaker interactions have low energies in the range of 2 to 20 kJ mol^{-1}, their cumulative effects on molecular association and crystal structure and packing are just about as predictable as the effects of conventional hydrogen bonding.

The nature of the $X \cdots X$ interaction in trimer synthon **44** is illustrated in Fig. 20.1.7. The C–X bond in a halo-substituted phenyl ring is polarized, so that there are regions of positive and negative electrostatic potentials around the X atom. The cyclic interaction of three C–X groups optimizes electrostatic potential overlap in the halogen trimer system.

20.2 Hydrogen-bond directed assembly

Hydrogen bonding is an indispensable tool for designing molecular aggregates within the fields of supramolecular chemistry, molecular recognition, and crystal engineering. It is well recognized that in organic crystals certain building blocks or supramolecular synthons have a clear pattern preference, and molecules that contain these building blocks tend to crystallize in specific arrangements with efficient close packing. As already mentioned in Section 11.2.1, three important rules are generally applicable to the formation of hydrogen bonds between functional groups in neutral organic molecules:

(a) all strong donor and acceptor sites are fully utilized;
(b) intramolecular hydrogen bonds giving rise to a six-membered ring will form in preference to intermolecular hydrogen bonds; and
(c) the remaining proton donors and acceptors not used in (b) will form intermolecular hydrogen bonds to one another.

For additional rules and a full discussion the reader is referred to the papers of Etter [*Acc. Chem. Res.* **23**, 120–6 (1990); *J. Phys. Chem.* **95**, 4601–10 (1991)]. Figure 20.2.1 shows a hydrogen-bonded polar sheet where 3,5-dinitrobenzoic acid and 4-aminobenzoic acid are co-crystallized.

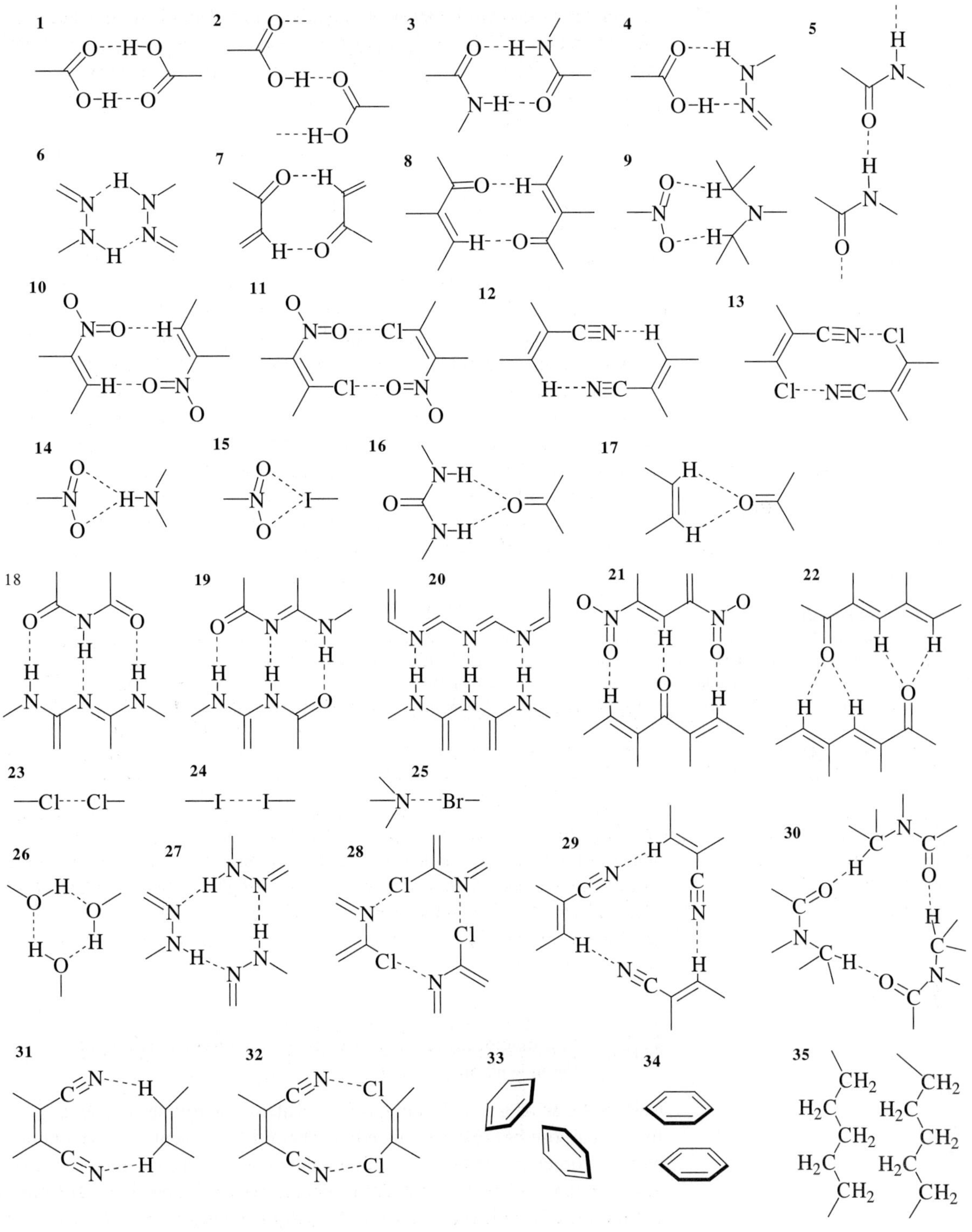

Fig. 20.1.6.
(continued on next page)

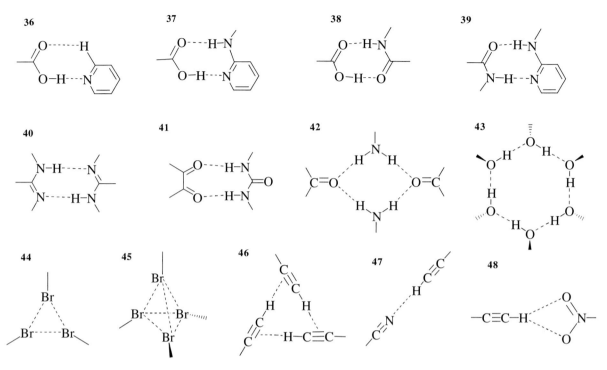

Fig. 20.1.6. (continued)
Representative supramolecular synthons. Synthon Nos. 1–35 are taken from G. R. Desiraju, *Angew. Chem. Int. Ed.* **34**, 2311 (1995). Nos. **33**: known as EF (edge-to-face). and **34**, OFF (offset face-to-face) phenyl-phenyl interactions from, M. L. Scudder and I. G. Dance, *Chem. Eur. J.* **8**, 5456 (2002); I. Dance, Supramolecular inorganic chemistry, in G. R. Desiraju (ed.), *The Crystal as a Supramolecular Entity, Perspectives in Supramolecular Chemistry*, vol. 2, Wiley, New York, 1996, pp. 137–233; **36** from T. Steiner, *Angew. Chem. Int. Ed.* **41**, 48 (2002); **37** from A. Nangia, *CrystEngComm* **17**, 1 (2002); **38** from P. Vishweshwar, A. Nangia and V. M. Lynch, *CrystEngComm* **5**, 164 (2003); **39** from F. H. Allen, W. D. S. Motherwell, P. R. Raithby, G. P. Shields and R. Taylor, *New J. Chem.*, 25 (1999); **40** from R. K. Castellano, V. Gramlich and F. Diederich, *Chem. Eur. J.* **8**, 118 (2002); **41** from C.-K. Lam and T. C. W. Mak, *Angew. Chem. Int. Ed.* **40**, 3453 (2001); **42** from M. D. Hollingsworth, M. L. Peterson, K. L. Pate, B. D. Dinkelmeyer and M. E. Brown, *J. Am. Chem. Soc.* **124**, 2094 (2002); **43**, observed in classical hydroquinone clathrates and phenolic compounds, from T. C. W. Mak and B. R. F. Bracke, Hydroquinone clathrates and diamondoid host lattices, in D. D. MacNicol, F. Toda and R. Bishop (eds.), *Comprehensive Supramolecular Chemistry*, vol. 6, Pergamon Press, New York, 1996, pp. 23–60; **44** from C. K. Broder, J. A. K. Howard, D. A. Keen, C. C. Wilson, F. H. Allen, R. K. R. Jetti, A. Nangia and G. R. Desiraju, *Acta Crystallogr.* **B56**, 1080 (2000); **45** from D. S. Reddy, D. C. Craig and G. R. Desiraju, *J. Am. Chem. Soc.* **118**, 4090 (1996); **46** from B. Goldfuss, P. v. R. Schleyer and F. Hampel, *J. Am. Chem. Soc.* **119**, 1072 (1997); **47** from P. J. Langley, J. Hulliger, R. Thaimattam and G. R. Desiraju, *New J. Chem.*, 307 (1998); **48** from B. Moulton and M. J. Zaworotko, *Chem. Rev.* **101**, 1629 (2001).

20.2.1 Supramolecular architectures based on the carboxylic acid dimer synthon

Carboxylic acids are commonly used as pattern-controlling functional groups for the purpose of crystal engineering. The most prevalent hydrogen bonding patterns formed by carboxylic acids are the dimer and the catemer. Carboxylic acids containing small substituent groups (formic acid, acetic acid) form the catemer motif, while most others (especially aromatic carboxylic acids) form

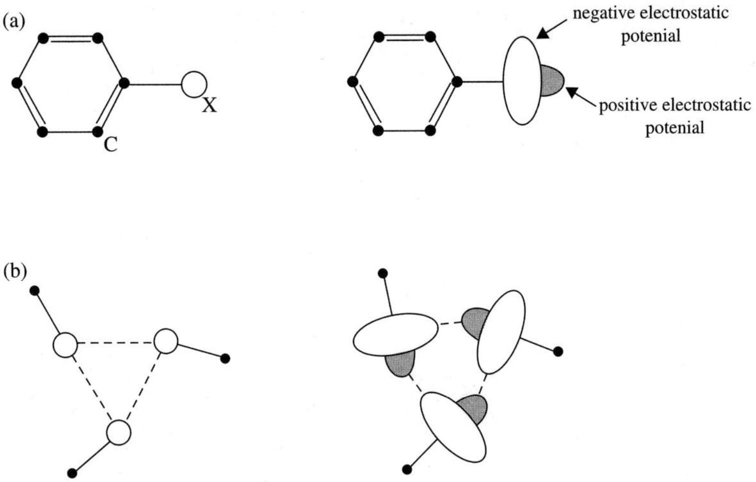

Fig. 20.1.7.
(a) Areas of positive and negative electrostatic potentials at the halogen substituent of a phenyl ring and (b) stabilization of the X···X trimer supramolecular synthon **44**.

dimers, although not exclusively. In the case of di- and polycarboxylic acids, terephthalic acid, and isophthalic acid form linear and zigzag ribbons (or tapes), respectively, trimesic acid (1,3,5-benzenetricarboxylic acid) with its threefold molecular symmetry forms a hydrogen-bonded sheet, and adamantane-1,3,5,7-tetracarboxylic acid forms a diamondoid network (Fig. 20.2.2).

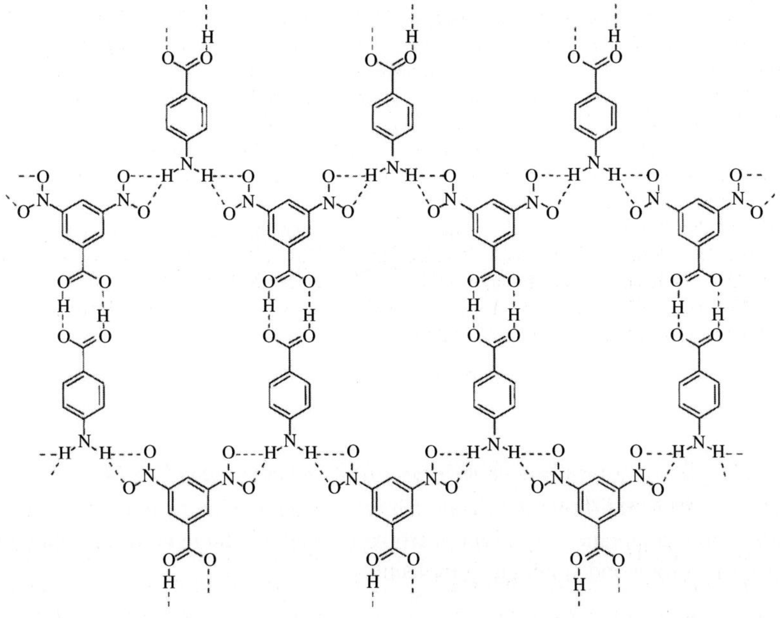

Fig. 20.2.1.
Polar sheet formed by 3,5-dinitrobenzoic acid and 4-aminobenzoic acid using the nitro-amine and carboxylic acid dimer motif.

If a bulky hydrophobic group is introduced at the 5-position of isophthalic acid, a cyclic hexamer (rosette) is generated. In the crystal structure of trimesic acid, the voids are filled through interpenetration of two honeycomb

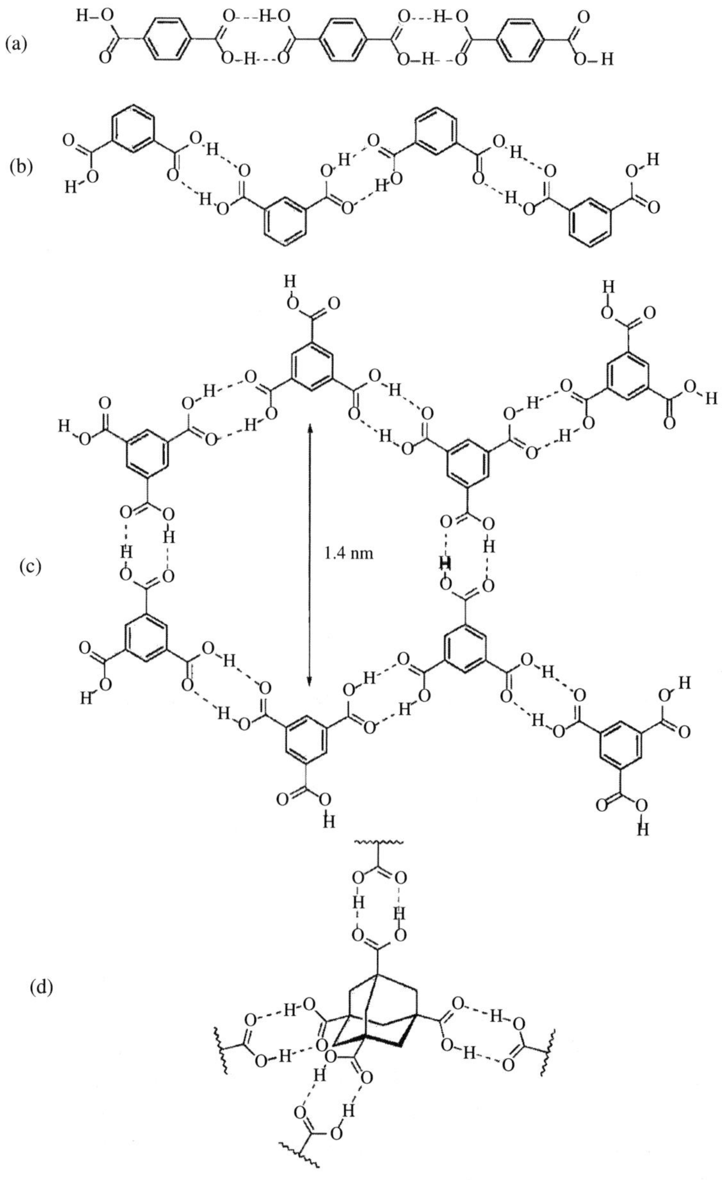

Fig. 20.2.2.
(a) One-dimensional linear tape and
(b) crinkled (or zigzag) tape,
(c) two-dimensional sheet (or layer), and
(d) three-dimensional network (or framework) held together by the carboxylic acid dimer synthon.

networks with additional stabilization by $\pi-\pi$ stacking interactions. In the host lattice of adamantane-1,3,5,7-tetracarboxylic acid, the large voids are filled by interpenetration and small guest molecules.

20.2.2 Graph-set encoding of hydrogen-bonding pattern

The robust intermolecular motifs found in organic systems can be used to direct the synthesis of supramolecular complexes in crystal engineering. In the interest

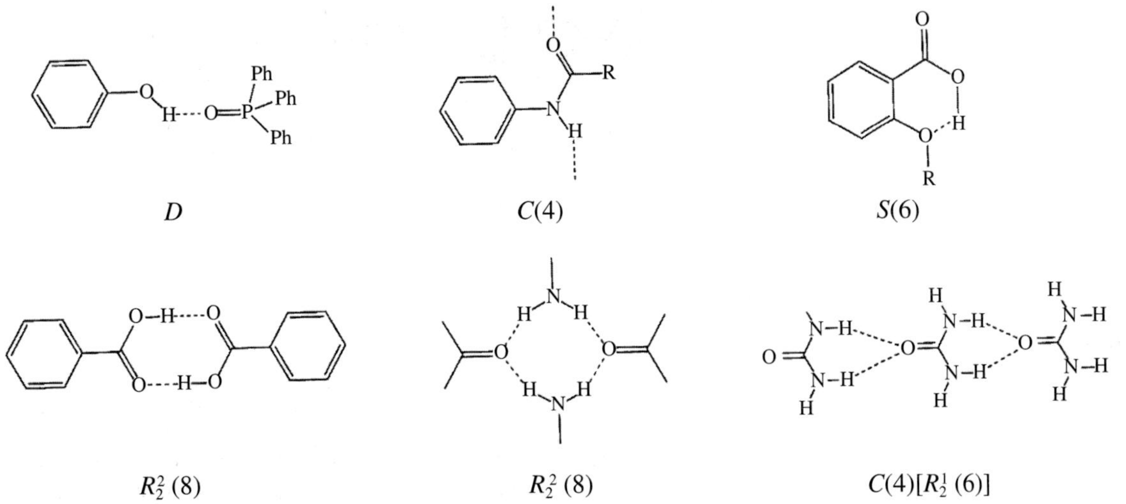

Fig. 20.2.3.
Some examples of graph-set descriptors of hydrogen-bonded structural motifs.

of adopting a systematic notation for the topology of hydrogen-bonded motifs and networks, a graph-set approach has been suggested by Etter and Ward. This provides a description of hydrogen-bonding schemes in terms of four pattern designators (G), i.e., infinite chain (C), ring (R), discrete complex (D), and intramolecular (self-associating) ring (S), which together with the degree of the pattern (n, the number of atoms comprising the pattern), the number of donors (d), and the number of acceptors (a) are combined to form the quantitative graph-set descriptor $G_d^a(n)$. Examples of the use of these quantitative descriptors are given in Fig. 20.2.3.

Preferable hydrogen-bonding patterns of a series of related compounds containing a particular type of functional group can be obtained by graph-set analysis. For example, most primary amides prefer forming cyclic dimers and chains with a common hydrogen-bonding pattern $C(4)[R_2^1(6)]$. Furthermore, important insights may be gained by graph-set analysis of two seemingly unrelated organic crystals, which may lead to similar hydrogen-bonding patterns involving different functional groups. Some of the most common supramolecular synthons found in the Cambridge Structure Database are illustrated below:

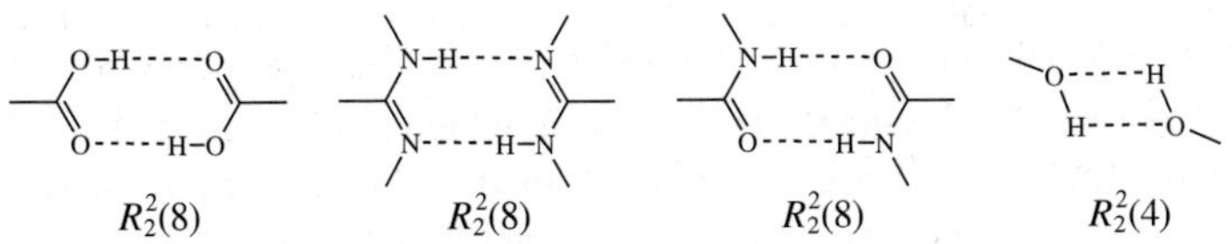

The same set of hydrogen bond donors and acceptors may be connected in alternate ways to generate distinguishable motifs, giving rise to different

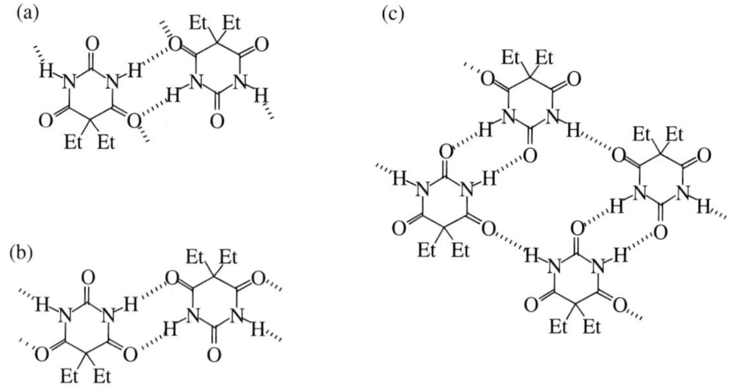

Fig. 20.2.4.
Patterns of hydrogen bonding found in three polymorphic forms of 5,5-diethylbarbituric acid. The graph-set descriptors are (a) $C(6)[R_2^2(8)R_4^2(12)]$, (b) $C(10)[R_2^2(8)]$, and (c) $C(6)[R_2^2(8)R_4^4(16)]$.

polymorphic forms. A good example is 5,5-diethylbarbituric acid, for which three crystalline polymorphs that exhibit polymeric ribbon structures are shown in Fig. 20.2.4.

20.2.3 Supramolecular construction based on complementary hydrogen bonding between heterocycles

The simple heterocyclic compounds melamine and cyanuric acid possess perfectly matched sets of donor/acceptor sites, 6/3 and 3/6 respectively, for complementary hydrogen bonding to form a planar hexagonal network (Fig. 20.2.5). Three distinct structural motifs can be recognized in this extended array: (a) linear tape,, (b) crinkled tape and (c) cyclic hexameric aggregate (rosette). Using barbituric acid derivatives and 2,4,6-triaminopyrimidine derivatives, the group of Whitesides has successfully synthesized all three preconceived systems. The conceptual design involves disruption of N–H···O and N–H···N hydrogen bonding in specific directions by introducing suitable hydrophobic bulky groups (Fig. 20.2.6).

The group of Reinhoudt has reported the construction of a D_3 hydrogen-bonded assembly of three calix[4]arene bismelamine and six barbituric acid derivatives (Fig. 20.2.7).

20.2.4 Hydrogen-bonded networks exhibiting the supramolecular rosette pattern

Ward has shown that the self-assembly of cations and anions in a guanidinium sulfonate salt, through precise matching of donor and acceptor sites, gives rise to a layer structure displaying the rosette motif, which is not planar but corrugated since the configuration at the S atom is tetrahedral (Fig. 20.2.8). If the sulfonate R group is small, a bilayer structure with interdigitated substituents is formed [structural motif (a)]. When R is large, a single layer structure with substituents alternating on opposite sides of the hydrogen-bonded layer is obtained [motif (b)]. Alternatively, the use of disulfonates provides covalent linkage between

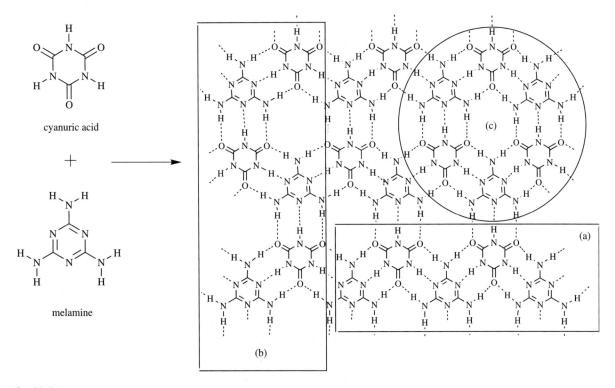

Fig. 20.2.5.
Hexagonal layer structure of the 1:1 complex of melamine and cyanuric acid. Three kinds of assembly in lower dimensions are possible: (a) linear tape, (b) crinkled tape, and (c) rosette.

adjacent layers to generate a pillared three-dimensional network, which may enclose a variety of guest species G [motif (c)].

The design and construction of hydrogen-bonded "supramolecular rosettes" from guanidinium/organic sulfonate, trimesic acid, or cyanuric acid/melamine depend on utilization of their topological equivalence, i.e., equal numbers of donor and acceptor hydrogen bonding sites and C_3 symmetry of the component moieties. As a modification of this strategy, a new kind of "fused-rosette ribbon" can be constructed with the guanidinium cation (GM^+) and hydrogen carbonate dimer $(HC^-)_2$ in the ratio of 1:1 (Fig. 20.2.9).

Each supramolecular rosette comprises a quasi-hexagonal assembly of two GM^+ and four HC^- units connected by strong $N_{GM}-H \cdots O_{HC}$ and $O_{HC}-H \cdots O_{HC}$ hydrogen bonds. The $(HC^-)_2$ dimer is shared as a common edge of adjacent rosettes and makes full use of its remaining acceptor sites in linking with GM^+. On the other hand, each GM^+ in the resulting linear ribbon (or tape) still possesses a pair of free donor sites, and it is anticipated that some "molecular linker" with suitable acceptor sites may be used to bridge an array of parallel ribbons to form a sheet-like network. This design objective has been realized in the synthesis and characterization of the inclusion compound $5[C(NH_2)_3^+] \cdot 4(HCO_3^-) \cdot 3[(n\text{-}Bu)_4N^+] \cdot 2[1,4\text{-}C_6H_4\text{-}(COO^-)_2] \cdot 2H_2O$ with the terephthalate (TPA^{2-}) anion functioning as a linker.

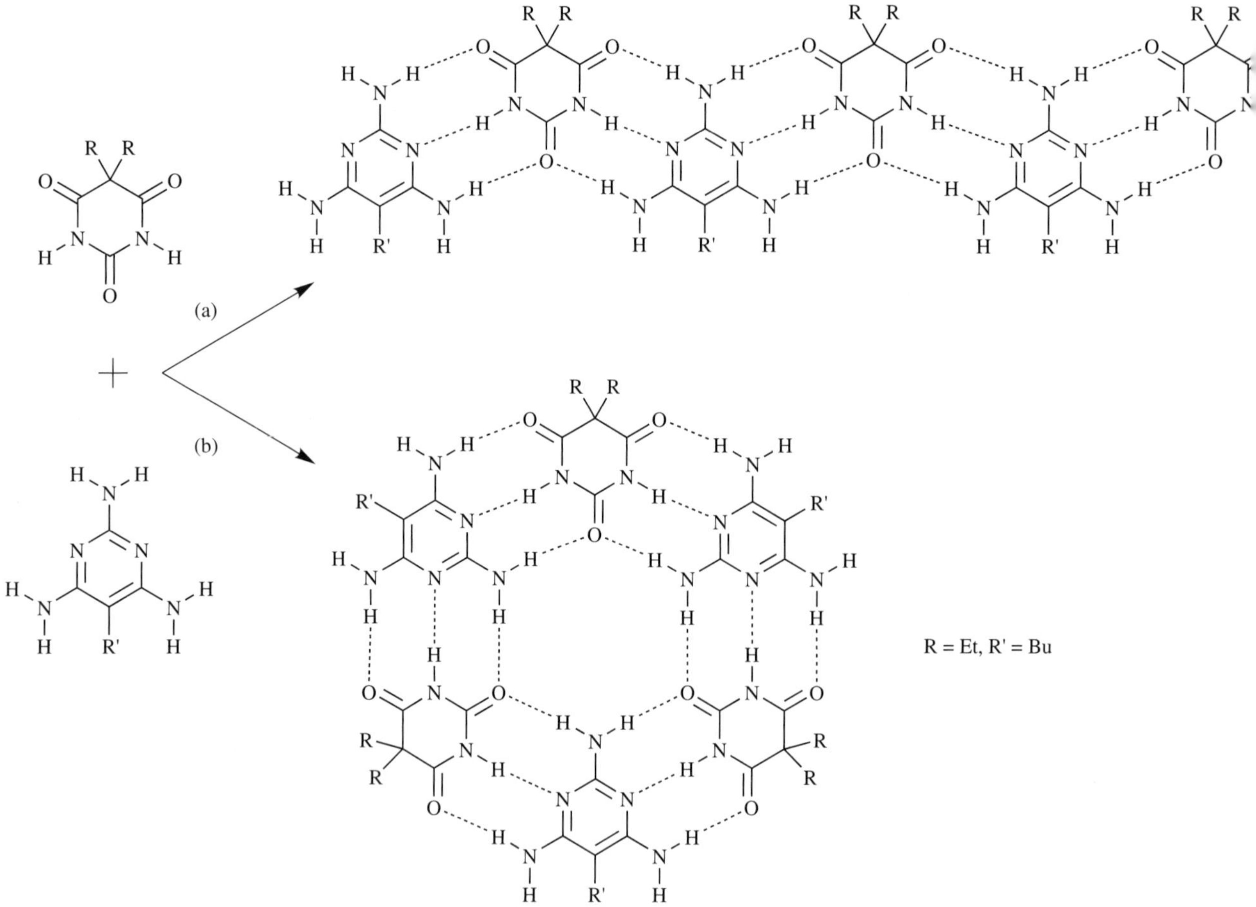

Fig. 20.2.6.
(a) Linear tape and (b) rosette 1:1 complexes constructed with barbituric acid derivatives and 2,4,6-triaminopyrimidine derivatives.

As shown in Fig. 20.2.10, each HC$^-$ provides one donor and one acceptor site to form a planar dimer motif [**A**, $R_2^2(8)$]. The remaining eight acceptor sites of each (HC$^-$)$_2$ dimer are topologically complemented by four GM$^+$ units, such that each GM$^+$ connects two (HC$^-$)$_2$ dimers through two pairs of N–H$_{syn}\cdots$O hydrogen bonds [**B**, $R_2^2(8)$]. Thus two GM$^+$ units and two (HC$^-$)$_2$ dimers constitute a planar, pseudo-centrosymmetric, quasi-hexagonal supramolecular rosette [**C**, $R_6^4(12)$] with inner and outer diameters of approximately 0.55 and 0.95 nm, respectively. In the resulting fused-rosette ribbon, the remaining two *exo*-orientated donor sites of each GM$^+$ unit form a pair of N–H$_{anti}\cdots$O hydrogen bonds [**D**, $R_2^2(8)$] with a TPA^{2-} carboxylate group. Thus two types of ladders are developed: type (I) [TPA^{2-} composed of C(10) to C(17) and O(13) to O(16)] is consolidated by two independent water molecules that alternately bridge carboxylate oxygen atoms of neighboring steps by pairs of donor O$_w$–H$\cdots$O hydrogen bonds, generating a centrosymmetric ring motif [**E**, $R_4^4(22)$] and pentagon pattern [**F**, $R_6^3(12)$]; in the type (II) ladder [TPA^{2-} composed of C(18) to C(25) and O(17) to O(20)], carboxylate oxygen atoms O(18)′

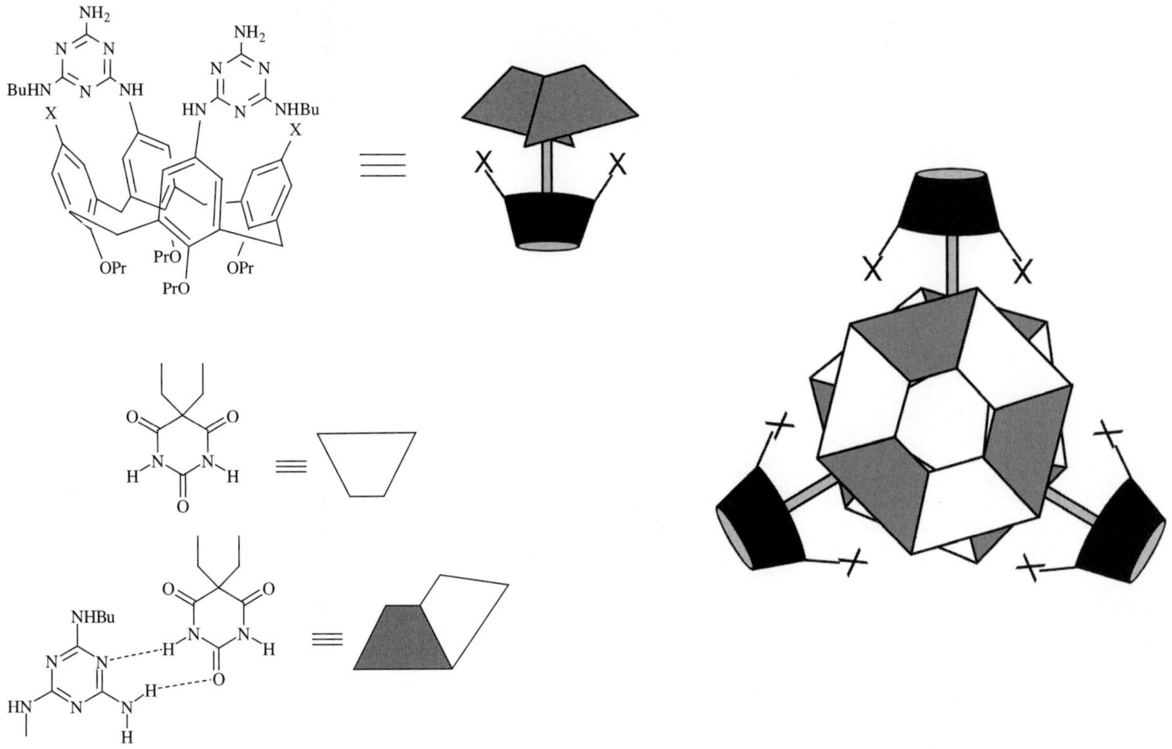

Supramolecular Structural Chemistry

Fig. 20.2.7.
D_3-symmetric hydrogen-bonded assembly of three calix[4]arene bismelamine and six barbituric acid derivatives.

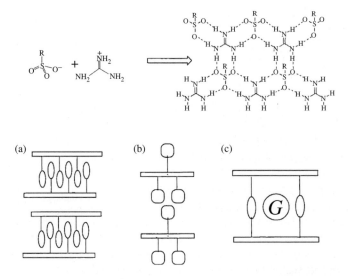

Fig. 20.2.8.
Top: assembly of guanidinium and sulfonate groups by N–H···O hydrogen bonds to give a corrugated rosette layer. Bottom: structural motifs (a), (b), and (c), which are formed depending on the size of R and the nature of the sulfonate group used.

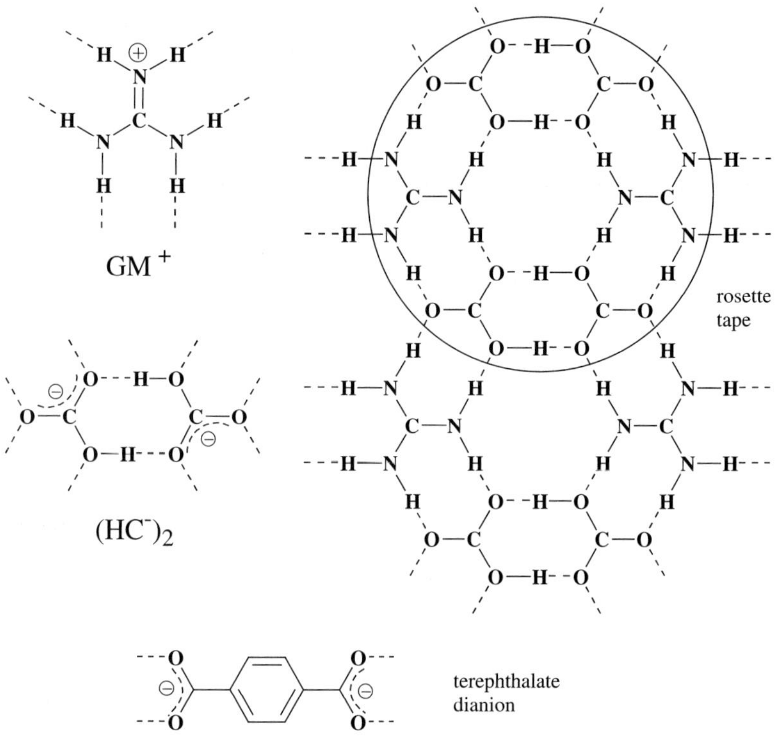

Fig. 20.2.9.
Design of supramolecular rosette tape and linker. From T. C. W. Mak and F. Xue, *J. Am. Chem. Soc.* **122**, 9860–1 (2000).

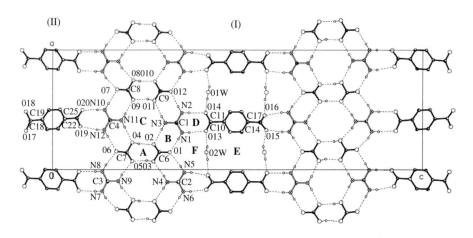

Fig. 20.2.10.
Projection on (010) showing the two-dimensional network. Hydrogen bonds are represented by broken lines, and ring motifs **A** to **F** are shown in boldface type. The TPA^{2-} ladders linking the rosette tapes are labeled (I) and (II).

and O(19) belonging to adjacent steps are connected by the remaining GM$^+$ ions, which are not involved in rosette formation, via two pairs of donor hydrogen bonds [**G**, $R_2^1(6)$] (Fig. 20.2.11(a)). The remaining two donor sites of each free GM$^+$ ion are linked to a carboxylate oxygen atom and a water molecule of TPA^{2-} column (I) of an adjacent layer to form a pentagon motif [**H**, $R_3^2(8)$], thus yielding a three-dimensional pillared layer structure (Fig. 20.2.11(b)). The large voids in the pillar region generate nanoscale channels extending along the [100] direction. The dimensions of the cross section of each channel are approximately 0.8 × 2.2 nm, within which three independent [(n-Bu)$_4$N]$^+$ cations are aligned in separate columns in a well-ordered manner.

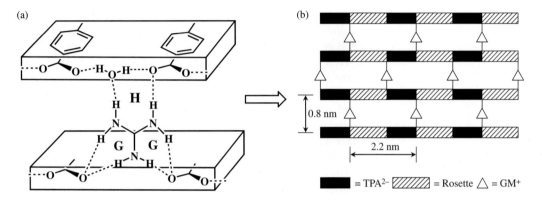

Fig. 20.2.11.
(a) Hydrogen-bonding motifs **G** and **H** involving linkage of the free guanidinium ion to neighboring rosette ribbon-terephthalate layers. (b) Schematic presentation of the pillared layer structure.

Drawing upon the above successful design of a linear "fused-rosette ribbon" assembled from the (HC$^-$)$_2$ dimer and GM$^+$ in 1:1 molar ratio, it would be challenging to attempt the hydrogen-bond mediated construction of two premeditated anionic rosette-layer architectures using guanidinium and ubiquitous C_3-symmetric oxo-anions that carry *unequal* charges, namely guanidinium-carbonate **I** and guanidinium-trimesate **II**, as illustrated in Fig. 20.2.12.

In principle, the negatively charged, presumably planar network **I** can be combined with one molar equivalent of tetraalkylammonium ion R$_4$N$^+$ of the right size as interlayer template to yield a crystalline inclusion compound of stoichiometric formula (R$_4$N$^+$)[C(NH$_2$)$_3^+$]CO$_3^{2-}$ that is reminiscent of the graphite intercalates. Anionic network **II**, on the other hand, needs twice as many monovalent cations for charge balance, and furthermore possesses honeycomb-like host cavities of diameter ~700 pm that must be filled by

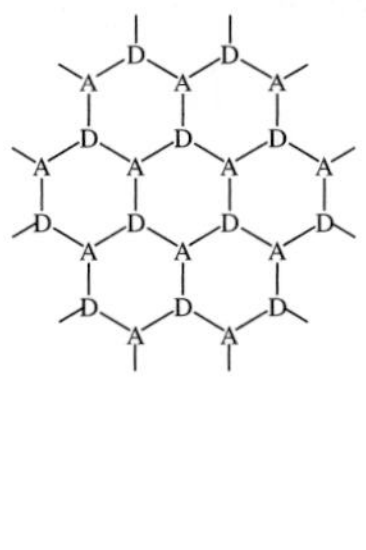

	Synthon D—A	Rosette Network	D	A
	N—H⋯O —C⟨ ⟩ N—H⋯O	I	[C(NH$_2$)$_3^+$]	CO$_3^{2-}$
		II	[C(NH$_2$)$_3^+$]	1,3,5-C$_6$H$_3$(COO$^-$)$_3$

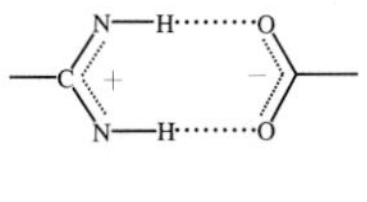

Fig. 20.2.12.
Design of supramolecular rosette layers. From C.-K. Lam and T. C. W. Mak, *J. Am. Chem. Soc.* **127**, 11536–7 (2005).

suitable guest species. The expected formula of the corresponding inclusion compound is $(R_4N^+)_2[C(NH_2)_3^+] [1,3,5-C_6H_3(COO^-)_3] \cdot G$, where G is an entrapped guest moiety with multiple hydrogen-bond donor sites to match the nearly planar set of six carboxylate oxygens that line the inner rim of each cavity.

Crystallization of $(R_4N^+)[C(NH_2)_3^+]CO_3^{2-}$ by variation of R, based on conceptual network **I**, was not successful. Taking into account the fact that the guanidinium ion can function as a pillar between layers and the carbonate ion is capable of forming up to twelve acceptor hydrogen bonds, as observed in crystalline bis(guanidinium) carbonate and $[(C_2H_5)_4N^+]_2 \cdot CO_3^{2-} \cdot 7(NH_2)_2CS$ [see Fig. 11.2.2(b)], the synthetic strategy was modified by incorporating a second guanidinium salt $[C(NH_2)_3]X$ as an extra component. After much experimentation with various combinations of R and X, the targeted construction of network **I**, albeit in undulating form, was realized through the isolation of crystalline $4[(C_2H_5)_4N^+] \cdot 8[C(NH_2)_3^+] \cdot 3(CO_3)^{2-} \cdot 3(C_2O_4)^{2-} \cdot 2H_2O$ (**1**).

In the asymmetric unit of (**1**), there are two independent carbonate anions and five independent guanidinium cations, which are henceforth conveniently referred to by their carbon atom labels in bold type. Carbonate **C(1)** and guanidinium **C(3)** each has one bond lying in a crystallographic mirror plane; together with carbonate **C(2)** and guanidinium **C(4)**, they form a nonplanar zigzag ribbon running parallel to the b axis, neighboring units being connected by a pair of strong N^+–H···O^- hydrogen bonds (Fig. 20.2.13). Adjacent anti-parallel $\{[C(NH_2)_3]^+ \cdot CO_3^{2-}\}_\infty$ ribbons are further cross-linked by strong N^+–H···O^- hydrogen bonds to generate a highly corrugated rosette layer, which is folded into a plane-wave pattern by guanidinium **C(5)** (Fig. 20.2.14). Guanidinium **C(6)** and **C(7)** protruding away from carbonate **C(1)** and **C(2)** are hydrogen-bonded to the twofold disordered oxalate ion containing **C(8)** and **C(9)** (Fig. 20.2.14), forming a pouch that cradles the disordered $(C_2H_5)_4N^+$ ion. The carbonate ions **C(1)** and **C(2)** each form eleven acceptor hydrogen bonds, only one fewer than the maximum number. The resulting composite hydrogen-bonded layers at $x = 1/4$ and $3/4$ are interconnected by $[(C_2O_4^{2-} \cdot (H_2O)_2]_\infty$

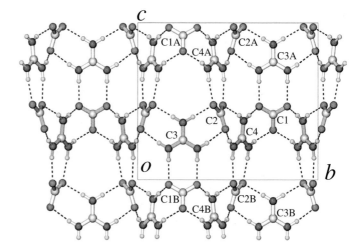

Fig. 20.2.13.
Projection diagram showing a portion of the nonplanar anionic rosette network **I** concentrated at $a = 1/4$ in the crystal structure of (**1**). The atom types are differentiated by size and shading, and hydrogen bonds are indicated by dotted lines. Adjacent antiparallel $\{[C(NH_2)_3]^+ \cdot CO_3^{2-}\}_\infty$ ribbons run parallel to the b axis. Symmetry transformations: A ($1/2 - x, 1 - y, 1/2 + z$); B ($1/2 - x, y - 1/2, z - 1/2$).

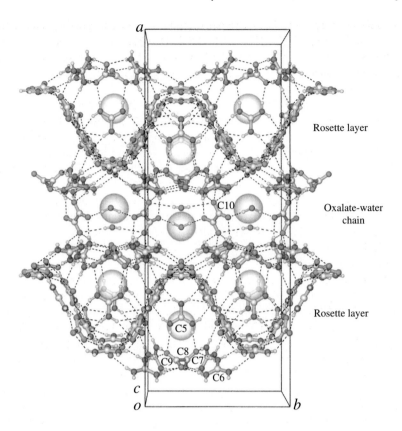

Fig. 20.2.14.
Perspective view of the crystal structure of (**1**) along [001]. The undulating guanidinium-carbonate rosette network **I** appears as a sinusoidal cross section. Adjacent composite hydrogen-bonded layers are interconnected by a $[C_2O_4^{2-} \cdot (H_2O)_2]_\infty$ chain. The disordered oxalate is shown in one possible orientation, and the two different types of disordered Et_4N^+ ions (represented by large semi-transparent spheres) are included in the pouches and the zigzag channels running parallel to the [010] direction, respectively.

chains derived from centrosymmetric oxalate **C(10)** and water molecules O1w and O2w via strong $^+N-H\cdots O^-$ hydrogen bonds to generate a complex three-dimensonal host framework, within which the second kind of disordered $(C_2H_5)_4N^+$ ions are accommodated in a zigzag fashion within channels extending along the [010] direction.

The predicted assembly of guanidinium-trimesate network **II** was achieved through the crystallization of $[(C_2H_5)_4N^+]_2 \cdot [C(NH_2)_3^+] \cdot [1,3,5-C_3H_3(COO^-)_3] \cdot 6H_2O$ (**2**). The guanidinium and trimesate ions are connected together by pairs of strong charge-assisted $^+N-H\cdots O^-$ hydrogen bonds to generate an essentially planar rosette layer with large honeycomb cavities [Fig. 20.2.15 (left)]. Three independent water molecules constitute a cyclic $(H_2O)_6$ cluster of symmetry 2, which is tightly fitted into each host cavity by adopting a flattened-chair configuration in an out-of-plane orientaion, with $O\cdots O$ distances comparable to 275.9 pm in deuterated ice I_h. Each water molecule has its ordered hydrogen atom pointing outward to form a strong $O-H\cdots O^-$ hydrogen bond with a carboxylate oxygen on the inner rim of the cavity. The well-ordered $(C_2H_5)_4N^+$ guests, represented by large spheres, are sandwiched between anionic rosette host layers with an interlayer spacing of ~ 7.5 Å [Fig. 20.2.15 (right)].

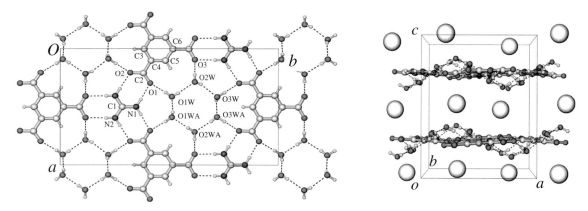

Fig. 20.2.15.
(Left) Projection diagram showing the hydrogen-bonding scheme in the infinite rosette layer **II** of (**2**). Only one of the two cyclic arrangements of disordered H atoms lying on the edges of each $(H_2O)_6$ ring is displayed. Symmetry transformation: A $(1 - x, y, 1/2 - z)$. (Right) Sandwich-like crystal structure of (**2**) viewed along the b axis.

The malleability of guanidinium-carbonate network **I**, rendered possible by the prolific hydrogen-bond accepting capacity of its carbonate building block, opens up opportunities for further exploration of supramolecular assembly. The flattened-chair $(H_2O)_6$ guest species, filling the cavity within robust guanidinium-trimesate layer **II** and being comparable to that in the host lattice of bimesityl-3,3′-dicarboxylic acid, may conceivably be replaced by appropriate hydrogen-bond donor molecules. The present anionic rosette networks are unlike previously reported neutral honeycomb lattices of the same (6,3) topology, thus expanding the scope of *de novo* engineering of charge-assisted hydrogen-bonded networks using ionic modular components, from which discrete molecular aggregates bearing the rosette motif may be derived.

20.3 Supramolecular chemistry of the coordination bond

The predictable coordination geometry of transition metals and the directional characteristics of interacting sites in a designed ligand provide the blueprint (or programmed instructions) for the rational synthesis of a wide variety of supramolecular inorganic and organometallic systems. Current research is concentrated in two major areas: (a) the construction of novel supermolecules from the intermolecular association of a few components and (b) the spontaneous organization (or self-assembly) of molecular units into one-, two-, and three-dimensional arrays. In the solid state, the supermolecules and supramolecular arrays can further associate with one another to yield gigantic macroscopic conglomerates, i.e., supramolecular structures of higher order.

20.3.1 Principal types of supermolecules

The supermolecules that have been synthesized include large metallocyclic rings, helices, host–guest complexes, and interlocked structures such as catenanes, rotaxanes, and knots.

Supramolecular Structural Chemistry

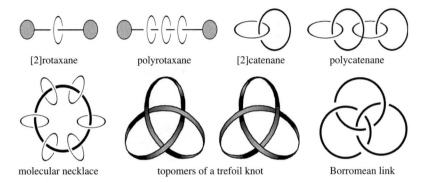

Fig. 20.3.1.
Catenanes, rotaxanes, molecular necklace, knots, and Borromean link.

The simplest catenane is [2]catenane which contains two interlocked rings. A polycatenane has three or more rings interlocked in a one-to-one linear fashion. A rotaxane has a ring component (or a bead) threaded by a linear component (or a string) with a stopper at each end. A polyrotaxane has several rings threaded onto the same string. A molecular necklace is a cyclic oligorotaxane with several rings threaded by a closed loop. There are two topological isomers for a trefoil knot. The Borromean link is composed of three interlocked rings such that the scission of any one ring unlocks the other two. These topologies are shown in Fig. 20.3.1.

20.3.2 Some examples of inorganic supermolecules

(1) Ferric wheel

The best known example of a large metallocyclic ring is the "ferric wheel" $[Fe(OMe)_2(O_2CCH_2Cl)]_{10}$ prepared from the reaction of oxo-centered trinuclear $[Fe_3O(O_2CCH_2Cl)_6(H_2O)_3](NO_3)$ and $Fe(NO_3)_3 \cdot 9H_2O$ in methanol. An X-ray analysis showed that it is a decameric wheel having a diameter of about 1.2 nm with a small and unoccupied hole in the middle (Fig. 20.3.2). Each pair of iron(III) centers are bridged by two methoxides and one O,O'-chloroacetate group.

(2) Hemicarceplex

A carcerand is a closed-surface, globular host molecule with a hollow interior that can enclose guest species such as small organic molecules and inorganic ions to form a carceplex. A hemicarcerand is a carcerand that contains portals large enough for the imprisoned guest molecule to escape at high temperatures, but otherwise remains stable under normal laboratory conditions. The hemicarcerand host molecule designed by Cram shown in Fig. 20.3.3 has idealized D_{4h} symmetry with its fourfold axis roughly coincident with the long axis of the ferrocene guest, which lies at an inversion center and hence adopts a fully staggered D_{5d} conformation. The 1,3-diiminobenzene groups connecting the northern and southern hemispheres of the hemicarceplex are arranged like paddles in a paddle wheel around the circumference of the central cavity.

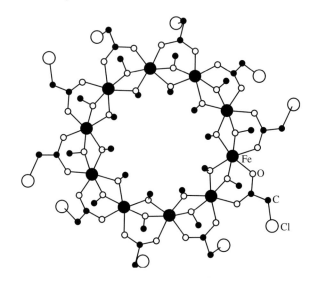

Fig. 20.3.2.
Molecular structure of the ferric wheel.

(3) [2]Catenane Pt(II) complex

The coordination bond between a pyridyl N atom and Pt(II) is normally quite stable, but it becomes labile in a highly polar medium at high concentration. Figure 20.3.4 shows the overall one-way transformation of a binuclear cyclic Pt(II) complex into a dimeric [2]catenane framework, which can be isolated as its nitrate salt upon cooling.

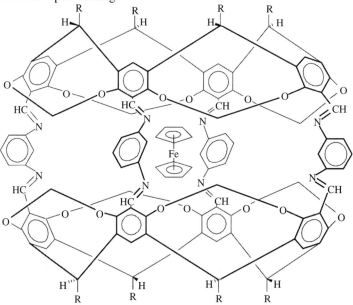

Fig. 20.3.3.
A hemicarceplex consisting of a hemicarcerand host molecule enclosing a ferrocene guest molecule.

(4) Helicate

In a helicate the polytopic ligand winds around a linear array of metal ions lying on the helical axis, so that its ligation sites match the coordination requirements

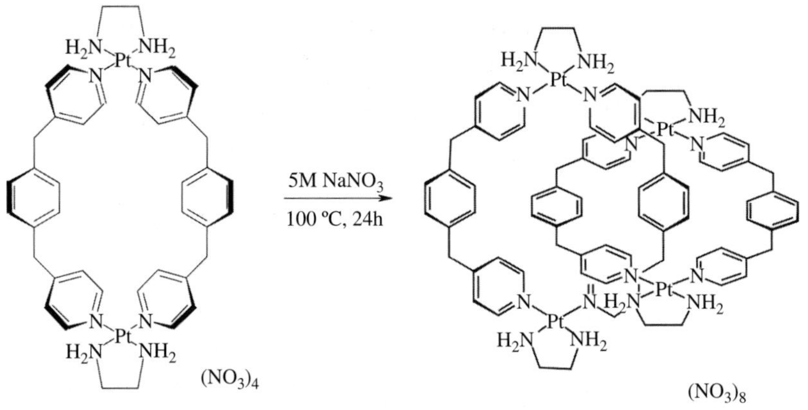

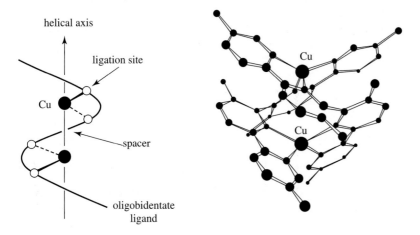

Fig. 20.3.4.
Self-assembly of a [2]catenane Pt(II) complex.

of the metal centers. The two enantiomers of a helicate are designated P- for a right-handed helix and M- for a left-handed helix [Fig. 20.3.5(a)]. An example of a M-type Cu(I) double helicate is shown in Fig. 20.3.5(b). Double helicates containing up to five Cu(I) centers and triple helicates involving the wrapping of oligobidentate ligands around octahedral Co(II) and Ni(II) centers have been synthesized.

Fig. 20.3.5.
(a) Structural features of a M-type helicate (only one strand is shown). (b) A Cu(I) double helicate containing an oligopyridine ligand.

(5) Molecular trefoil knot

The first molecular trefoil knot was successfully synthesized by Dietrich-Buchecker and Sauvage according to the scheme shown in Fig. 20.3.6. A specially designed ligand consisting of two diphenolic 1,10-phenanthroline units tied together by a tetramethylene tether was reacted with [Cu(MeCN)$_4$]BF$_4$ to give a dinuclear double helix. This precursor was then treated under high dilution conditions with two equivalents of the diiodo derivative of hexaethylene glycol in the presence of Cs$_2$CO$_3$ to form the cyclized complex in low yield. Finally, demetallation of this helical dicopper complex yielded the desired free trefoil knot with retention of topological chirality. It should be noted that in the crucial cyclization step, the double helical precursor is in equilibrium with a

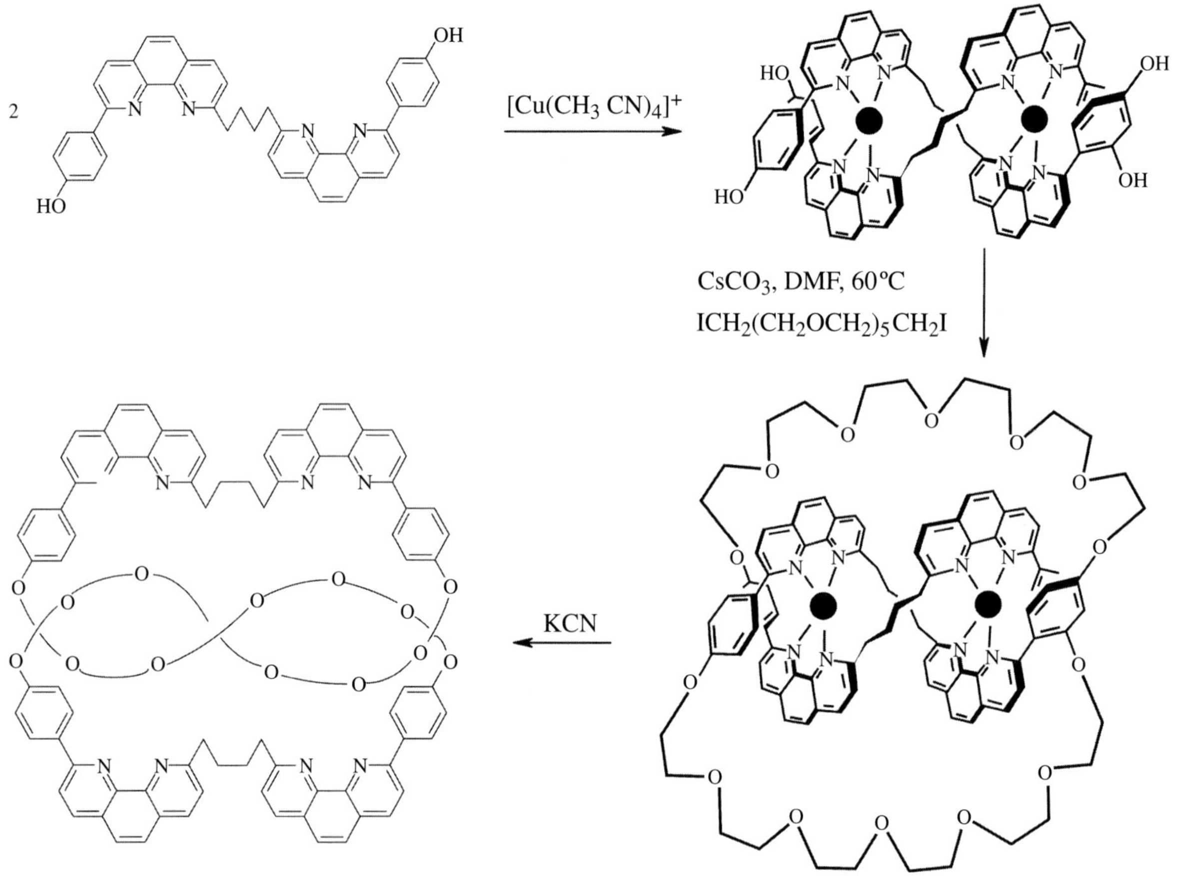

Fig. 20.3.6.
Synthetic scheme for dicopper and free trefoil knots. From C. O. Dietrich-Buchecker, J. Guilhem, C. Pascard and J.-P. Sauvage, *Angew. Chem. Int. Ed.* **29**, 1154–6 (1990).

non-helical species, which leads to an unknotted dicopper complex consisting of two 43-membered rings arranged around two Cu(I) centers in a face-to-face manner (Fig. 20.3.7).

(6) Molecular Borromean link

The connection between chemistry and topology is exemplified by the elegant construction of a molecular Borromean link that consists of three identical interlocked rings. In the designed synthesis, each component ring comprises two short and two long segments. The programmed, one-step self-assembly process is indicated by the scheme illustrated in Fig. 20.3.8.

Self-assembly of the molecular Borromean link (a dodecacation **BR**$^{12+}$) is achieved by a template-directed cooperative process that results in over 90% yield. Each of the three component rings (L) in **BR**$^{12+}$ is constructed from [2+2] macrocyclization involving two DFP (2,6-diformylpyridine) and two DAB (diamine containing a 2,2'-bipyridyl group) molecules. Kinetically labile

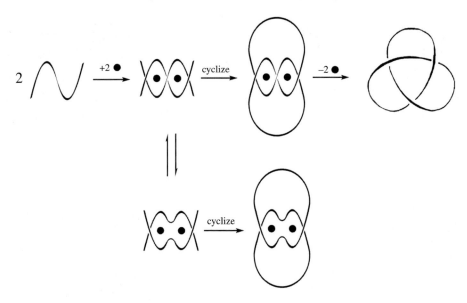

Fig. 20.3.7.
Template synthesis of a free trefoil knot and an unknotted product.

zinc(II) ions serve efficiently as templates in multiple molecular recognition that results in precise interlocking of three macrocyclic ligands:

$$6\text{DFB} + 6[\text{DABH}_4](\text{CF}_3\text{CO}_2)_4 + 9\text{Zn}(\text{CF}_3\text{CO}_2)_2 \xrightarrow{\text{CH}_3\text{OH}} [\text{L}_3(\text{ZnCF}_3\text{CO}_2)_6] \cdot 3[\text{Zn}(\text{CF}_3\text{CO}_2)_4] + 12\text{H}_2\text{O}$$

$$|||$$

$$[\mathbf{BR}(\text{CF}_3\text{CO}_2)_6] \cdot 3[\text{Zn}(\text{CF}_3\text{CO}_2)_4]$$

An X-ray analysis of $[\mathbf{BR}(\text{CF}_3\text{CO}_2)_6]\cdot 3[\text{Zn}(\text{CF}_3\text{CO}_2)_4]$ revealed that the hexacation $[\mathbf{BR}(\text{CF}_3\text{CO}_2)_6]^{6+}$ has S_6 symmetry with each macrocyclic ligand L adopting a chair-like conformation (Fig. 20.3.8). The interlocked rings are consolidated by six Zn(II) ions, each being coordinated in a slightly distorted octahedral geometry by the *endo*-tridentate diiminopyridyl group of one ring, the *exo*-bidentate bipyridyl group of another ring, and an oxygen atom of a CF_3CO_2^- ligand. Each bipyridyl group is sandwiched unsymmetrically between a pair of phenolic rings at π–π stacking distance of 361 and 366 pm in different directions.

In the crystal packing, the $[\mathbf{BR}(\text{CF}_3\text{CO}_2)_6]^{6+}$ ions are arranged in hexagonal arrays with intermolecular π–π stacking interactions of 331 pm and C–H$\cdots$O=C hydrogen bonds (H$\cdots$O 252 pm), generating columns along c that accommodate the $[\text{Zn}(\text{CF}_3\text{CO}_2)_4]^{2-}$ counterions.

20.3.3 Synthetic strategies for inorganic supermolecules and coordination polymers

Two basic approaches for the synthesis of inorganic supermolecules and one-, two-, and three-dimensional coordination polymers have been developed:

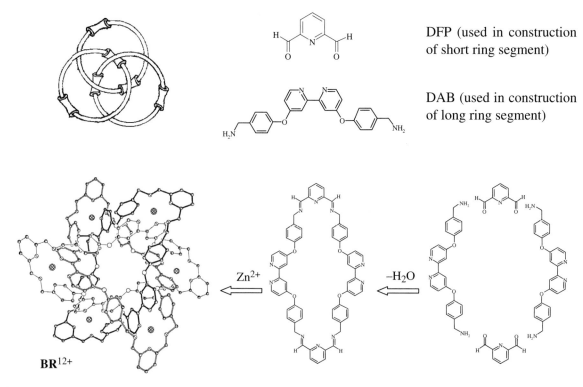

Fig. 20.3.8.
Synthetic scheme for one-step supramolecular assembly of the molecular Borromean link $[L_3Zn_6]^{12+} \equiv \mathbf{BR}^{12+}$; its three component interlocked rings are differentiated by different degrees of shading. Each octahedral zinc(II) ion is coordinated by an *endo*-N_3 ligand set and a chelating bipyridyl group belonging to a different ring, the monodentate acetate ligand being omitted for clarity. From K. S. Chichak, S. J. Cantrill, A. R. Pease, S.-H. Chiu, G. W. V. Cave, J. L. Atwood and J. F. Stoddart, *Science (Washington)* **304**, 1308–12 (2004).

(1) Transition-metal ions are employed as nodes and bifunctional ligands as spacers. Commonly used spacer ligands are pseudohalides such as cyanide, thiocyanate, and azide, and N-donor ligands such as pyrazine, 4,4′-bipyridine, and 2,2′-bipyrimidine. Besides discrete supermolecules, some one-, two-, and three-dimensional architectural motifs generated from this strategy are shown in Fig. 20.3.9.

If all nodes at the boundary of a portion of a motif are bound by terminal ligands, a discrete molecule will be formed. An example is the square grid shown in Fig. 20.3.10. Reaction of the tritopic ligand 6,6′-bis[2-(6-methylpyridyl)]-3,3′-bipyridazine (Me$_2$bpbpz) with silver triflate in 2:3 molar ratio in nitromethane results in self-assembly of a complex of the formula $[Ag_9(Me_2bpbpz)_6](CF_3SO_3)_9$. X-ray analysis showed that the $[Ag_9(Me_2bpbpz)_6]^{9+}$ cation is in the form of a 3 × 3 square grid, with two sets of Me$_2$bpbpz ligands positioned above and below the mean plane of the silver centers, as shown in Fig. 20.3.10(a), so that each Ag(I) atom is in a distorted tetrahedral environment. The grid is actually distorted into a diamond-like shape due to the curved nature of the ligand, and the

Supramolecular Structural Chemistry

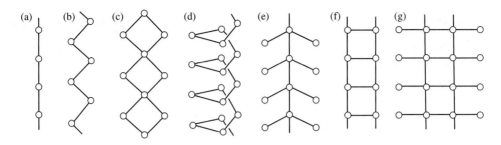

One-dimensional

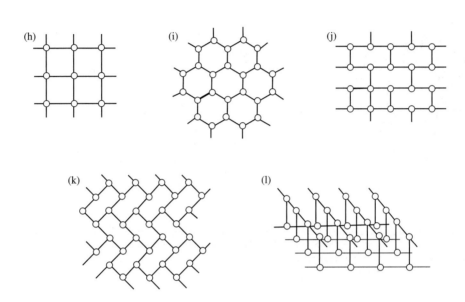

Two-dimensional

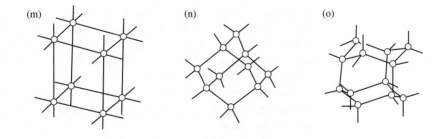

Three-dimensional

Fig. 20.3.9.
Schematic representation of the motifs generated from the connection of transition metals by bifunctional spacer ligands. One-dimensional: (a) linear chain, (b) zigzag chain, (c) double chain, (d) helical chain, (e) fishbone, (f) ladder, and (g) railroad. Two-dimensional: (h) square grid, (i) honeycomb, (j) brick wall, (k) herringbone, and (l) bilayer. Three-dimensional: (m) six-connected network, (n) four-connected diamondoid network, and (o) four-connected ice-I_h network.

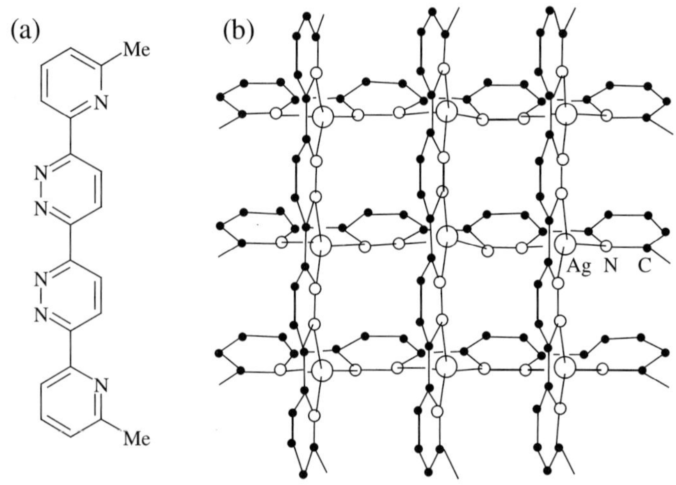

Fig. 20.3.10.
The 3 × 3 square molecular grid [Ag$_9$(Me$_2$bpbpz)$_6$]$^{9+}$: (a) structural formula and (b) molecular structure. From [P. N. W. Baxter, J-M. Lehn, J. Fischer and M-T. Youinou. *Angew. Chem. Int. Ed.* **33**, 2284–7 (1994).

angle between the mean planes of the two sets of ligands is about 72° (Fig. 20.3.10(b)).

(2) Exodentate multitopic ligands are used to link transition metal ions into building blocks. Some examples of such ligands are 2,4,6-tris(4-pyridyl)-1,3,5-triazine, oligopyridines, and 3- and 4-pyridyl-substituted porphyrins.

In the following sections, selected examples from the recent literature are used to illustrate the creative research activities in transition-metal supramolecular chemistry.

20.3.4 Molecular polygons and tubes

(1) Nickel wheel

The reaction of hydrated nickel acetate with excess 6-chloro-2-pyridone (Hchp) produces in 60% yield a dodecanuclear nickel complex, which can be recrystallized from tetrahydrofuran as a solvate of stoichiometry [Ni$_{12}$(O$_2$CMe)$_{12}$(chp)$_{12}$(H$_2$O)$_6$(THF)$_6$]. X-ray structure analysis revealed that this wheel-like molecule (Fig. 20.3.11) lies on a crystallographic threefold axis. There are two kinds of nickel atoms in distorted octahedral coordination: Ni(1) is bound to three O atoms from acetate groups, two O atoms from chp ligands, and an aqua ligand, whereas Ni(2) is surrounded by two acetate O atoms, two chp O atoms, an aqua ligand, and the terminal THF ligand. All ligands, other than THF, are involved in bridging pairs of adjacent nickel atoms. The structure of this nickel metallocycle resembles that of the decanuclear "ferric wheel" [Fe(OMe)$_2$(O$_2$CCH$_2$Cl)]$_{10}$ (see Fig. 20.3.2). Both complexes feature a closed chain of intersecting M$_2$O$_2$ rings, with each ring additionally bridged by an acetate ligand. They differ in that in the ferric wheel the carboxylate ligands are all exterior to the ring, whereas in the nickel wheel half of the acetate ligands lie within the central cavity.

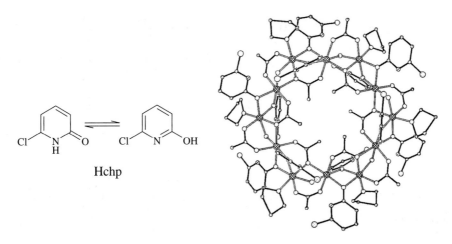

Fig. 20.3.11.
Molecular structure of the nickel wheel. The bridging pyridone and terminal THF ligands point alternately above and below the mean plane of the twelve nickel atoms. From A. J. Blake, C. M. Grant, S. Parsons, J. M. Rawson and R. E. P. Winpenny. *Chem. Commun.*, 2363–4 (1994).

(2) Nano-sized tubular section

The ligand 2,4,6-tris[(4-pyridyl)methylsulfanyl]-1,3,5-triazine (tpst) possesses nine possible binding sites to transition metals for the assembly of supramolecular systems. Its reaction with $AgNO_3$ in a 1:2 molar ratio in DMF/MeOH followed by addition of $AgClO_4$ produces $Ag_7(tpst)_4(ClO_4)_2(NO_3)_5(DMF)_2$. In the crystal structure, two tpst ligands coordinate to three silver(I) ions to form a bicyclic ring. Two such rings are fitted together by Ag–N and Ag–S bonds involving a pair of bridging silver ions to generate a nano-sized tubular section $[Ag_7(tpst)_4]$ with dimensions of $1.34 \times 0.96 \times 0.89$ nm, which encloses two perchlorate ions and two DMF molecules (Fig. 20.3.12). The tubular sections are further linked by additional Ag–N and Ag–S bonds to form an infinite chain. The nitrate ions are located near the silver ions and imbedded in between the linear polymers.

The two independent tpst ligands are each bound to four silver atoms but in different coordination modes: three pyridyl N plus one thioether S, or three pyridyl N plus one triazine N. The silver atoms exhibit two kinds of coordination modes: normal linear AgN_2 and a very unusual AgN_2S_2 mode of distorted square-planar geometry.

(3) Infinite square tube

The reaction of 2-aminopyrimidine (apym), $Na[N(CN)_2]$, and $M(NO_3)_2 \cdot 6H_2O$ (M = Co, Ni) gives $M[N(CN)_2]_2(apym)$, which consists of a packing of infinite molecular tubes of square cross section. In each tube, the metal atoms constitute the edges and three connecting $N(CN)_2^-$ (dicyanamide) ligands form the sides (Fig. 20.3.13). The octahedral coordination of the metal atoms is completed by chains of two-connecting ligands (with the amide nitrogen uncoordinated) which occupy the outside of each edge, as well as monodentate apym ligands. The length of a side of the molecular tube is 486.4 pm for M = Co and 482.1

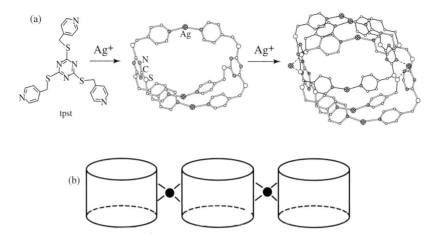

Fig. 20.3.12.
(a) Formation and structure of the nano-sized tubular section [Ag$_7$(tpst)$_4$]. Bonds formed by bridging silver ions are indicated by broken lines. (b) Linkage of tubular sections to form a linear polymer. From M. Hong, Y. Zhao, W. Su, R. Cao, M. Fujita, Z. Zhou and A. S. C. Chan. *Angew. Chem. Int. Ed.* **39**, 2468–70 (2000).

pm for M = Ni. In the crystal structure extensive hydrogen bonding between the tubes occurs via the terminal apym ligands.

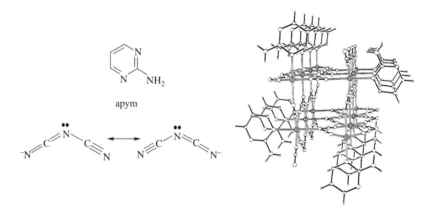

Fig. 20.3.13.
Structure of a square molecular tube in M[N(CN)$_2$]$_2$(apym) (M=Co, Ni). From P. Jensen, S. R. Batten, B. Moubaraki and K. S. Murray, *Chem. Commun.*, 793–4 (2000).

20.3.5 Molecular polyhedra

(1) Tetrahedral iron(II) host–guest complex

The structural unit M(tripod)$^{n+}$ (tripod = CH$_3$C(CH$_2$PPh$_2$)$_3$) has been used as a template for the construction of many supramolecular complexes. For example, the host–guest complex [BF$_4$ ⊂ {(tripod)$_3$Fe}$_4$(*trans*-NCCH=CHCN)$_6$(BF$_4$)$_4$](BF$_4$)$_3$ has been synthesized from the reaction of tripod, Fe(BF$_4$)$_3$·6H$_2$O, and fumaronitrile in a 4:4:6 molar ratio in CH$_2$Cl$_2$/EtOH at 20°C. The tetranuclear Fe(II) complex has idealized symmetry T, with a crystallographic twofold axis passing through the midpoints of a pair of edges of the Fe$_4$ tetrahedron (Fig. 20.3.14). The B–F bonds of the encapsulated BF$_4^-$ ion point toward the corner iron atoms. Each face of the Fe$_4$ tetrahedron is

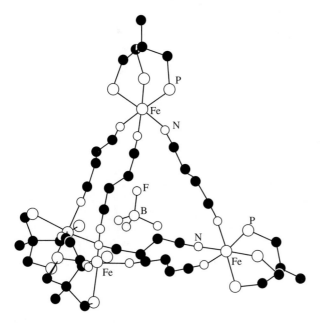

Fig. 20.3.14.
Molecular structure of the tetrahedral Fe(II) host–guest complex cation. The capping BF_4^- groups and the phenyl rings of the tripod ligands have been omitted for clarity. From S. Mann, G. Huttner, L. Zsolnai and K. Heinze, *Angew. Chem. Int. Ed.* **35**, 2808–9 (1996).

capped by a BF_4^- group, and the remaining three are located in voids between the complex cations.

(2) Cubic molecular box

The cyanometalate box $\{Cs \subset [Cp^*Rh(CN)_3]_4[Mo(CO)_3]_4\}^{3-}$ is formed in low yield from the reaction of $[Cp^*Rh(CN)_3]^-$ ($Cp^* = C_5Me_5$) and $(\eta^6\text{-}C_6H_3Me_3)Mo(CO)_3$ in the presence of cesium ions, and it can be crystallized as a Et_4N^+ salt. The Cs^+ ion serves as a template in the self-assembly of the anionic molecular box, which has a cubic $Rh_4Mo_4(\mu\text{-}CN)_{12}$ core with three exterior carbonyl ligands attached to each Mo and a Cp^* group to each Rh. The encapsulated Cs^+ ion has a formal coordination number of 12 if interaction with the centers of cyano groups is considered (Fig. 20.3.15).

(3) Lanthanum square antiprism

A tris-bidentate pyrazolone ligand, 4-(1,3,5-benzenetricarbonyl)-tris(3-methyl-1-phenyl-2-pyrazoline-5-one) (H_3L), has been designed and synthesized. When this rigid C_3-symmetric ligand was reacted with $La(acac)_3$ in dimethylsufoxide (DMSO), the complex $La_8L_8(DMSO)_{24}$ was obtained in 81% yield. An X-ray analysis revealed a square-antiprismatic structure with idealized D_{4d} symmetry, with each L^{3-} ligand occupying one of the eight triangular faces (Fig. 20.3.16). In the coordination sphere of the La^{3+} ion, six sites are filled by O atoms from three L^{3-} ligands and the remaining three by DMSO molecules which point into the central cavity. It was found that the DMSO ligands could be replaced partially by methanol in recrystallization.

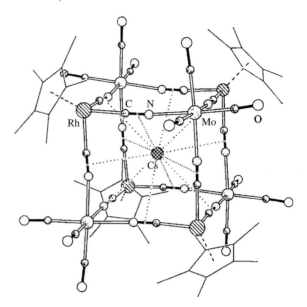

Fig. 20.3.15.
Structure of the molecular cube that encloses a Cs$^+$ ion. From K. K. Klausmeyer, S. R. Wilson and T. B. Rauchfuss. *J. Am. Chem. Soc.* **121**, 2705–11 (1999).

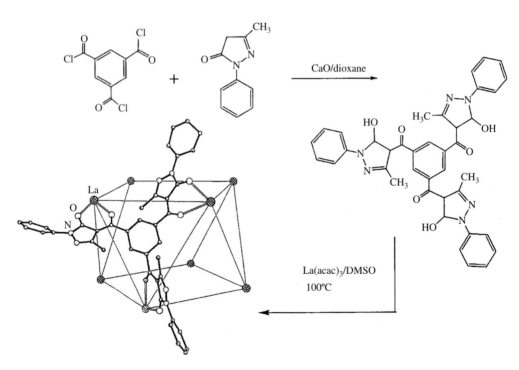

Fig. 20.3.16.
Synthesis and structure of the La$_8$L$_8$ cluster. Only one L^{3-} ligand is shown, and the coordinating DMSO moleclues have been omitted for clarity. From K. N. Raymond and J. Xu. *Angew. Chem. Int. Ed.* **39**, 2745–7 (2000).

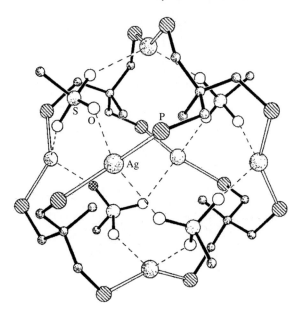

Fig. 20.3.17.
The super-adamantoid core of the [Ag$_6$(triphos)$_4$(CF$_3$SO$_3^-$)$_4$]$^{2+}$ cage. The broken lines indicate one of the two adamantane cores formed by silver ions and triflate ligands, whose F atoms have been omitted for clarity. From S. L. James, D. M. P. Mingos, A. J. P. White and D. Williams, *Chem Commun.*, 2323–4 (2000).

(4) Super-adamantoid cage

A 2:3 molar mixture of MeC(CH$_2$PPh$_2$)$_3$ (triphos) and silver triflate gives [Ag$_6$(triphos)$_4$(CF$_3$SO$_3$)$_4$](CF$_3$SO$_3$)$_2$ in high yield. The inorganic "super-adamantoid" cage [Ag$_6$(triphos)$_4$(CF$_3$SO$_3$)$_4$]$^{2+}$ exhibiting approximate T molecular symmetry is formed with the CF$_3$SO$_3^-$ ion as a template. In the cage structure (Fig. 20.3.17), an octahedron of silver(I) ions is bound by two sets of tritopic triphos and triflate ligands, each occupying four alternating faces of the octahedron. Thus six silver ions and four triphos ligands constitute one adamantane core, and likewise the silver ions and triflate ligands form a second adamantane core. A novel feature in this structure is the "endo-methyl" conformation of the triphos ligand, leaving only a small cavity at the center.

(5) Chemical reaction in a coordination cage

Flat panel-like ligands with multiple interacting sites have been used for metal-directed self-assembly of many fascinating supramolecular 3D structures. A simple triangular "molecular panel" is 2,4,6-tris(4-pyridyl)-1,3,5-triazine (L), which has been employed by Fujita to assemble a discrete [{Pt(bipy)}$_6$L$_4$]$^{12+}$ (bipy = 2,2′-bipyridine) coordination cage in quantitative yield by treating Pt(bipy)(NO$_3$)$_2$ with L in a 3:2 molar ratio. In this complex cation, the Pt(II) atoms constitute an octahedron, and the triangular panels (L ligands) are located at four of the eight faces (Fig. 20.3.18). The Pt(bipy)$^{2+}$ fragment thus serves as a *cis*-protected coordination block, each linking a pair of molecular panels. The nano-sized central cavity with a diameter of ∼1 nm is large enough to accommodate several guest molecules. Different types of guest species such as adamantane, adamantane carboxylate, *o*-carborane, and anisole have been used. In particular, C-shaped molecules such as *cis*-azobenzene and *cis*-stilbene derivatives can be encapsulated in the cavity as a dimer stabilized by the "phenyl-embrace" interaction.

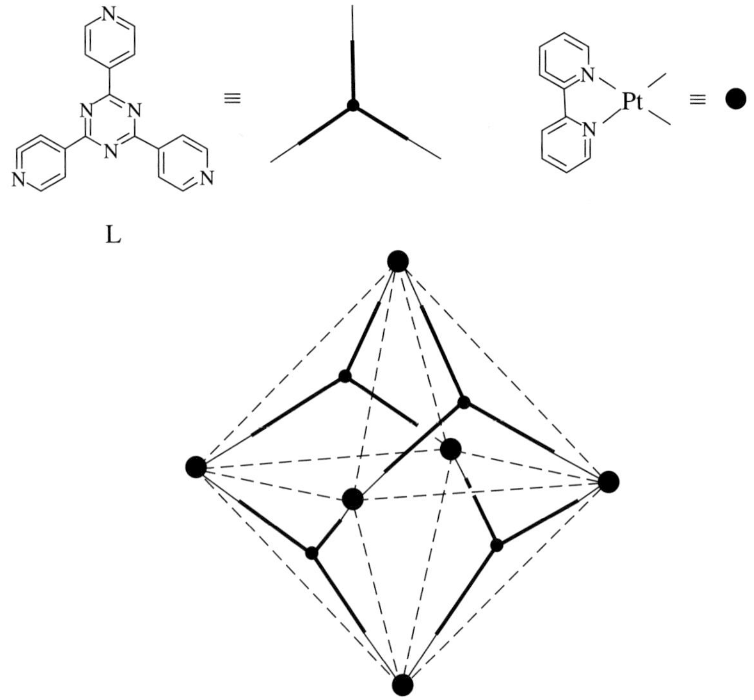

Fig. 20.3.18.
Self-assembly of [{Pt(bipy)}$_6$L$_4$]$^{12+}$ cage.

Useful chemical reactions have been carried out in the nano-sized cavity, as illustrated by the *in situ* isolation of a labile cyclic siloxane trimer (Fig. 20.3.19). In the first step, three to four molecules of phenyltrimethoxysilane enter the cage and are hydrolyzed to siloxane molecules. Next, condensation takes place in the confined environment to generate the cyclic trimer {SiPh(OH)O–}$_3$, which is trapped and stabilized in a pure form. The overall reaction yields an inclusion complex [{SiPh(OH)O–}$_3$ ⊂ {Pt(bipy)}$_6$L$_4$](NO$_3$)$_{12}$·7H$_2$O, which can be crystallized from aqueous solution in 92% yield. The all-*cis* configuration of the cyclic siloxane trimer and the structure of the inclusion complex have been determined by NMR and ESI-MS.

(6) Nanoscale dodecahedron

The dodecahedron is a Platonic solid that contains 12 fused pentagons formed from 20 vertices and 30 edges. An organic molecule of this exceptionally high icosahedral (I_h) symmetry is the hydrocarbon dodecahedrane C$_{20}$H$_{20}$, which was first synthesized by Paquette in 1982. Recently an inorganic analog has been obtained from edge-directed self-assembly of a metallocyclic structure (Fig. 20.3.20) in a remarkably high 99% yield. The tridentate ligand at each vertex is tris(4′-pyridyl)methanol, and the linear bidentate subunit at each edge is bis[4,4′-(*trans*-Pt(PEt$_3$)$_2$(CF$_3$SO$_3$))]benzene. The dodecahedral molecule carries 60 positive charges and encloses 60 CF$_3$SO$_3^-$ anions, and its estimated diameter d along the threefold axis is about 5.5 nm. Using bis[4,4′-(*trans*-Pt(PPh$_3$)$_2$(CF$_3$SO$_3$))]biphenyl as a longer linear linker, d for the resulting enlarged dodecahedron increases to about 7.5 nm.

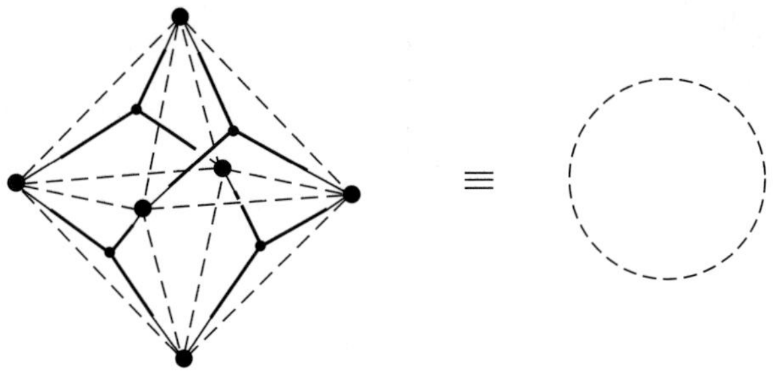

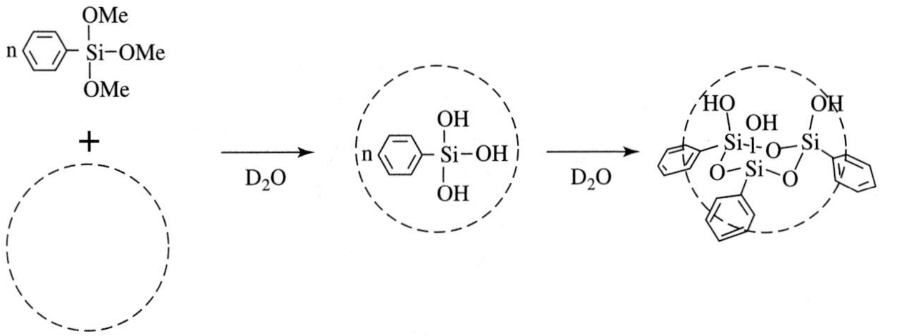

Fig. 20.3.19.
Generation and stabilization of cyclic siloxane trimer in a self-assembled coordination cage. From M. Yoshizawa, T. Kusukawa, M. Fujita, and K. Yamaguchi. *J. Am. Chem. Soc.* **122**, 6311–12, (2000).

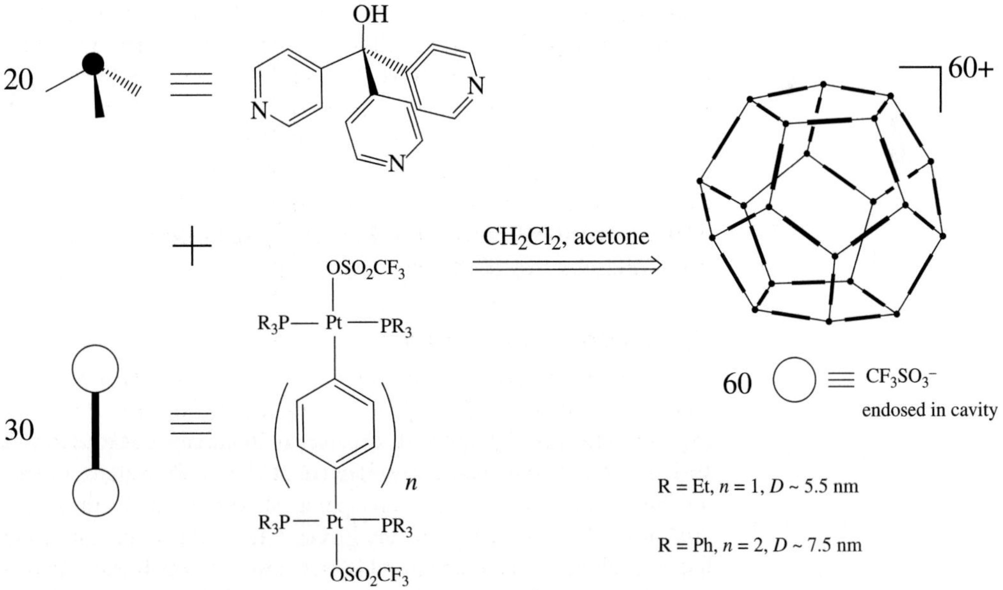

R = Et, n = 1, D ~ 5.5 nm
R = Ph, n = 2, D ~ 7.5 nm

Fig. 20.3.20.
Self-assembly of a nanoscale dodecahedron. The OH group attached to each quaternary C atom of the tris(4′-pyridyl)methanol molecule has been omitted for clarity. From B. Olenyuk, M. D. Levin, J. A. Whiteford, J. E. Shield and P. J. Stang, *J. Am. Chem. Soc.* **121**, 10434–5 (2000).

Fig. 20.3.21.
The [Mn$_9$(μ-CN)$_{30}$Mo$_6$] core of the high-spin cluster. From J. Larionova, M. Gross, M. Pilkington, H. Andres, H. Stoeckli-Evans,
H. U. Güdel and S. Decurtins, *Angew. Chem. Int. Ed.* **39**, 1605–9 (2000).

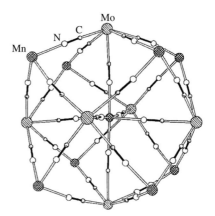

(7) High-spin rhombic dodecahedron

The cluster [Mn^{II} {Mn^{II}(MeOH)$_3$}$_8$(μ-CN)$_{30}$ {Mo(CN)$_3$}$_6$]·5MeOH·2H$_2$O has a pentadecanuclear core of idealized O_h symmetry, as shown in Fig. 20.3.21. The nine Mn(II) ions constitute a body-centered cube, the six Mo(V) ions define an octahedron, and the two polyhedra interpenetrate each other so that the peripheral atoms exhibit the geometry of a rhombic dodecahedron. Each pair of adjacent metal centers is linked by a μ-cyano ligand with Mo bonded to C and Mn bonded to N. Each outer Mn(II) atom is surrounded by three methanol ligands, leading to octahedral coordination. Similarly, three terminal cyano ligands are bound to each Mo(V) to establish an eight-coordination environment.

Actually the cluster has a lower symmetry with a crystallographic C_2 axis passing through the central Mn atom and the midpoints of two opposite Mo···Mo edges. The resulting neutral cluster has a high-spin ground state with $S = 25^1/_2$. [Note that $S = 2^1/_2$ for Mn(II) and $^1/_2$ for Mo(V); thus the total spin of the system is $9 \times 2^1/_2 + 6 \times {^1/_2} = 25^1/_2$.]

20.4 Selected examples in crystal engineering

Examples from the recent literature that illustrate various approaches in the rational design of novel crystalline materials are given in this section.

20.4.1 Diamondoid networks

The three-dimensional network structure of diamond can be considered as constructed from the linkage of nodes (C atoms) with rods (C–C bonds) in a tetrahedral pattern. From the viewpoint of crystal engineering, in a diamondoid network the node can be any group with tetrahedral connectivity, and the linking rods (or linker) can be all kinds of bonding interactions (ionic, covalent, coordination, hydrogen bond, and weak interactions) or molecular fragment.

The molecular skeletons of adamantane, (CH$_2$)$_6$(CH)$_4$, and hexamethylenetetramine, (CH$_2$)$_6$N$_4$ (Fig. 20.4.1) constitute the characteristic structural units of diamondoid networks containing one and two kinds of four-connected nodes, respectively. If the rod is long, the resulting diamondoid network

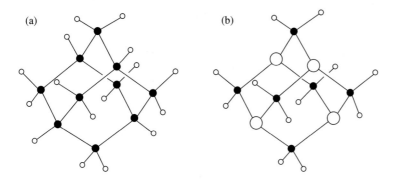

Fig. 20.4.1.
Molecular structure of (a) adamantane and (b) hexamethylenetetramine.

becomes quite porous, and stability can only be achieved by interpenetration. If two diamondoid networks interpenetrate to form the crystal structure, the degree of interpenetration ρ is equal to 2. Examples in which ρ ranges from 2 to 9 are known.

Some crystalline compounds that exhibit diamondoid structures are listed in Table 20.4.1. The rod linking a pair of nodes can be either linear or nonlinear.

In the crystal structure of Cu_2O, each O atom is surrounded tetrahedrally by four Cu atoms, and each Cu atom is connected to two O atoms in a linear fashion. Hence the node is the O atom, and the rod is O–Cu–O. Figure 20.4.2 shows a single Cu_2O diamondoid network, and the crystal structure is composed of two interpenetrating networks.

In the crystal structure of ice-VII, which is formed under high pressure, the node is the O atom, and the rod is a hydrogen bond. Since the H atoms are disordered, the hydrogen bond is written as O···H···O in Table 20.4.1, indicating equal population of O–H···O and O···H–O. The degree of interpenetration is two, as shown in Fig. 20.4.3.

Figures 20.4.4(a) and 20.4.4(b) illustrate the crystal structure of the 1:1 complex of tetraphenylmethane and carbon tetrabromide. The nodes comprise $C(C_6H_5)_4$ and CBr_4 molecules, and the each linking rod is the weak interaction between a Br atom and a phenyl group. The hexamethylenetetramine-like structural unit is outlined by broken lines. Figures 20.4.4(c) and 20.4.4(d) show the crystal structure of tetrakis(4-bromophenyl)methane, which has a distorted diamondoid network based on the hexamethylenetetramine building unit. If the synthon composed of the aggregation of four Br atoms is considered as a node, then two kinds of nodes (Br_4 synthon and quatenary C atom) are connected by rods consisting of p-phenylene moeities.

Figure 20.4.5 shows the crystal structure of $[Cu\{1,4-C_6H_4(CN)_2\}_2]BF_4$. The nodes are the four-connected Cu atoms, each being coordinated tetrahedrally by N≡C–C_6H_4–C≡N ligands as rods. The adamantane-like structure unit is shown in Fig. 20.4.5(a), and repetition of such units along a twofold axis leads to fivefold interpenetration. The remaining space is filled by the BF_4^- ions.

Figure 20.4.6 shows the crystal structure of $C(C_6H_4C_2C_5NH_4O)\cdot 8C_2H_5COOH$. The whole $C(C_6H_4C_2C_5NH_4O)_4$ molecule serves as a node, and

Table 20.4.1. Diamondoid networks

Compound	Node	Linking rod	Bond type between nodes	Degree of interpenetration ρ	Remarks
Diamond	C	C–C	Covalent	None	
M_2O (M = Cu, Ag, Pb)	O	O–M–O	Ionic/covalent	Two	Fig. 20.4.2
Ice-VII	O	O···H···O	H bond	Two	H atom disordered Fig. 20.4.3
KH_2PO_4	$H_2PO_4^-$	O–H···O	H bond	Two	K^+ in channel
Methane-tetraacetic acid	$C(CH_2COOH)_4$	(cyclic dimeric H-bond motif)	Cyclic dimeric H bond	Three	
$(CH_2)_6N_4 \cdot CBr_4$	$(CH_2)_6N_4$, CBr_4	N···Br	N···Br interaction	Two	Fig. 20.4.4(a) Fig. 20.4.4(b)
$C(C_6H_5)_4 \cdot CBr_4$	$C(C_6H_5)_4$, CBr_4	–Br···phenyl–	Br···phenyl interaction	None	
$C(4\text{-}C_6H_4Br)_4$	$C(C_6H_4)_4$ unit, Br_4 synthon	C–Br	Covalent	Three	Tetrahedral Br_4 synthon consolidated by weak Br···Br interaction Fig. 20.4.4(c) Fig. 20.4.4(d)
$M(CN)_2$ (M = Zn, Cd)	M	M←C≡N→M	Coordination	Two	
$[Cu(L)_2]BF_4$ (L = p-$C_6H_4(CN)_2$)	Cu	Ag←NCC$_6$H$_4$CN→Ag	Coordination	Five	BF_4^- in channel Fig. 20.4.5
$C(C_6H_4C_2C_5NH_4O)_4$ ·8CH_3CH_2COOH	$C(C_6H_4C_2C_5NH_4O)_4$	Double N–H···O	Cyclic dimeric H bond	Seven	Fig. 20.4.6
$[Ag(L)_2]XF_6$ (L = 4,4'-NCC$_6$H$_4$–C$_6$H$_4$CN, X = P, As, Sb)	Ag	Ag←NCC$_6$H$_4$–C$_6$H$_4$CN→Ag	Coordination	Nine	XF_6^- in channel
$[Ag(L)]BF_4 \cdot xPhNO_2$ (L = C(4-C_6H_4mCN)$_4$)	Ag, C(4-C_6H_4CN)$_4$	CN→Ag	Coordination	None	BF_4^- and $PhNO_2$ guest species in cavity
$[Mn(CO)_3(\mu\text{-}OH)]_4 \cdot$ (H$_2$NCH$_2$CH$_2$NH$_2$)	$[Mn(CO)_3\text{-}(\mu\text{-}OH)]_4$	O–H···NH$_2$CH$_2$–CH$_2$NH$_2$···H–O	H bond	Three	

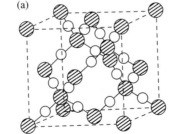 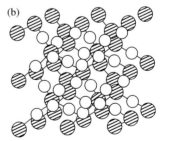

Fig. 20.4.2.
Crystal structure of Cu_2O: (a) single diamondoid network and (b) two interpenetrating networks.

Supramolecular Structural Chemistry

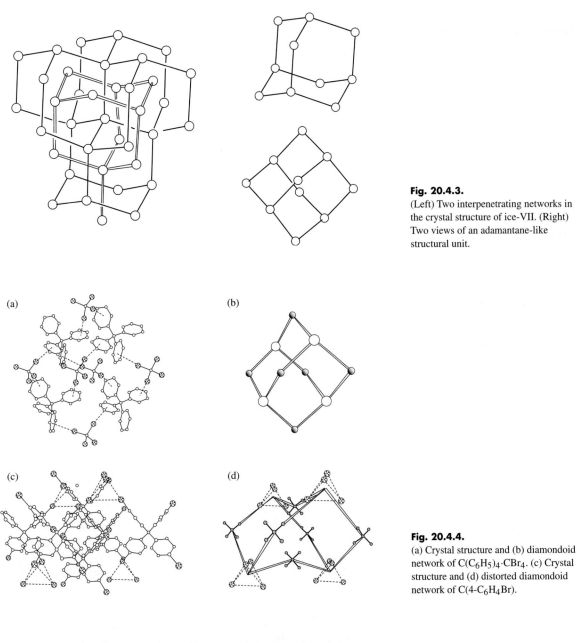

Fig. 20.4.3.
(Left) Two interpenetrating networks in the crystal structure of ice-VII. (Right) Two views of an adamantane-like structural unit.

Fig. 20.4.4.
(a) Crystal structure and (b) diamondoid network of $C(C_6H_5)_4 \cdot CBr_4$. (c) Crystal structure and (d) distorted diamondoid network of $C(4\text{-}C_6H_4Br)$.

Fig. 20.4.5.
(a) Single diamond network of $[Cu\{1,4\text{-}C_6H_4(CN)_2\}_2]$ and (b) fivefold interpenetration in the crystal structure.

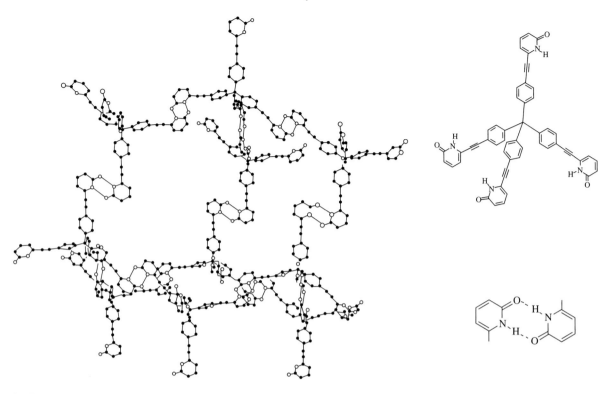

Fig. 20.4.6.
Crystal structure of C(C$_6$H$_4$C$_2$C$_5$NH$_4$O)$_4$·8CH$_3$CH$_2$COOH. The structure of the host molecule is shown in the upper right and the pairwise linkage of pyridone groups is shown at the lower right.

such nodes are connected by rods each comprising a pair of N–H···O hydrogen bonds between pyridine groups. As the resulting diamondoid network is quite open, sevenfold interpenetration occurs, and the remaining space is used to accommodate the ethanol guest molecules.

In the crystal structure of [Ag{C(4-C$_6$H$_4$CN)$_4$}]BF$_4$·xPhNO$_2$, the structural unit is of the hexamethylenetetramine type (Fig. 20.4.7). The nodes are Ag atoms and C(4-C$_6$H$_4$CN)$_4$ molecules, and the rods are CN→Ag coordination bonds. The void space is filled by the nitrobenzene guest molecules and BF$_4^-$ ions.

20.4.2 Interlocked structures constructed from cucurbituril

Cucurbituril is a hexameric macrocyclic compound with the formula (C$_6$H$_6$N$_4$O$_2$)$_6$ shaped like a pumpkin which belongs to the botanical family *Cucurbitaceae*. This macrocyclic cavitand has idealized symmetry D_{6h} with a hydrophobic internal cavity of about 0.55 nm. The two portals, which are each laced by six hydrophilic carbonyl groups, have a diameter of 0.4 nm [Fig. 20.4.8(a)].

Like a molecular bead, cucurbituril can be threaded with a linear diammonium ion to form an inclusion complex [Fig. 20.4.8(b)]. This is stabilized by the fact that each protonated amino N atom forms hydrogen bonds to three of the

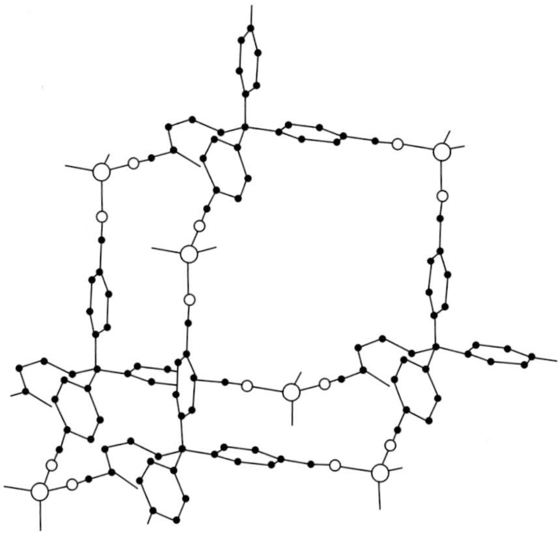

Fig. 20.4.7.
Hexamethylenetetramine-like structure unit in the diamondoid network of [Ag{C(4-C$_6$H$_4$CN)$_4$}]BF$_4$.

six carbonyl groups at its adjacent portal. Since each rotaxane unit has two terminal pyridyl groups, it can serve as an *exo*-bidentate ligand to link up transition metals to form a polyrotaxane which may take the form of a linear coordination polymer, a zigzag polymer, a molecular necklace, or a puckered layer network [Fig. 20.4.8(c)]. The structures of the linear and zigzag polyrotaxane polymers are shown in Fig. 20.4.9.

The two-dimensional network is constructed from the fusion of chair-like hexagons with Ag(I) ions at the corners and rotaxane units forming the edges. The nitrate ions lie above and below the puckered layer such that each Ag(I) ion is coordinated by three rotaxanes and a nitrate ion in a distorted tetrahedral geometry. In the crystal structure, two sets of parallel two-dimensional networks stacked in different directions make a dihedral angle of 69°, and they interpenetrate in such a way that a hexagon belonging to one set interlocks with four hexagons of the other set, and *vice versa*.

The rotaxane building unit can be modified by replacing the 4-pyridyl group by another functional group such as 3-cyanobenzyl. When a rotaxane unit built in this way is treated with Tb(NO$_3$)$_3$ under hydrothermal conditions, the cyano group is converted to the carboxylate group to generate a three-dimensional coordination polymeric network. The basic building block of the framework consists of a binuclear Tb^{3+} center and two types of rotaxane units: type I having bridging 3-phenylcarboxylate terminals and type II having chelating carboxylate terminals (Fig. 20.4.10).

The binuclear terbium centers and type I rotaxanes form a two-dimensional layer. Stacked layers are further interconnected via type II rotaxanes to form a three-dimensional polyrotaxane network, which has an inclined α-polonium topology with the binuclear terbium centers behaving as six-connected nodes (Fig. 20.4.10). The void space in the crystal packing is filled by a free rotaxane unit, NO$_3^-$ and OH$^-$ counter ions, and water molecules.

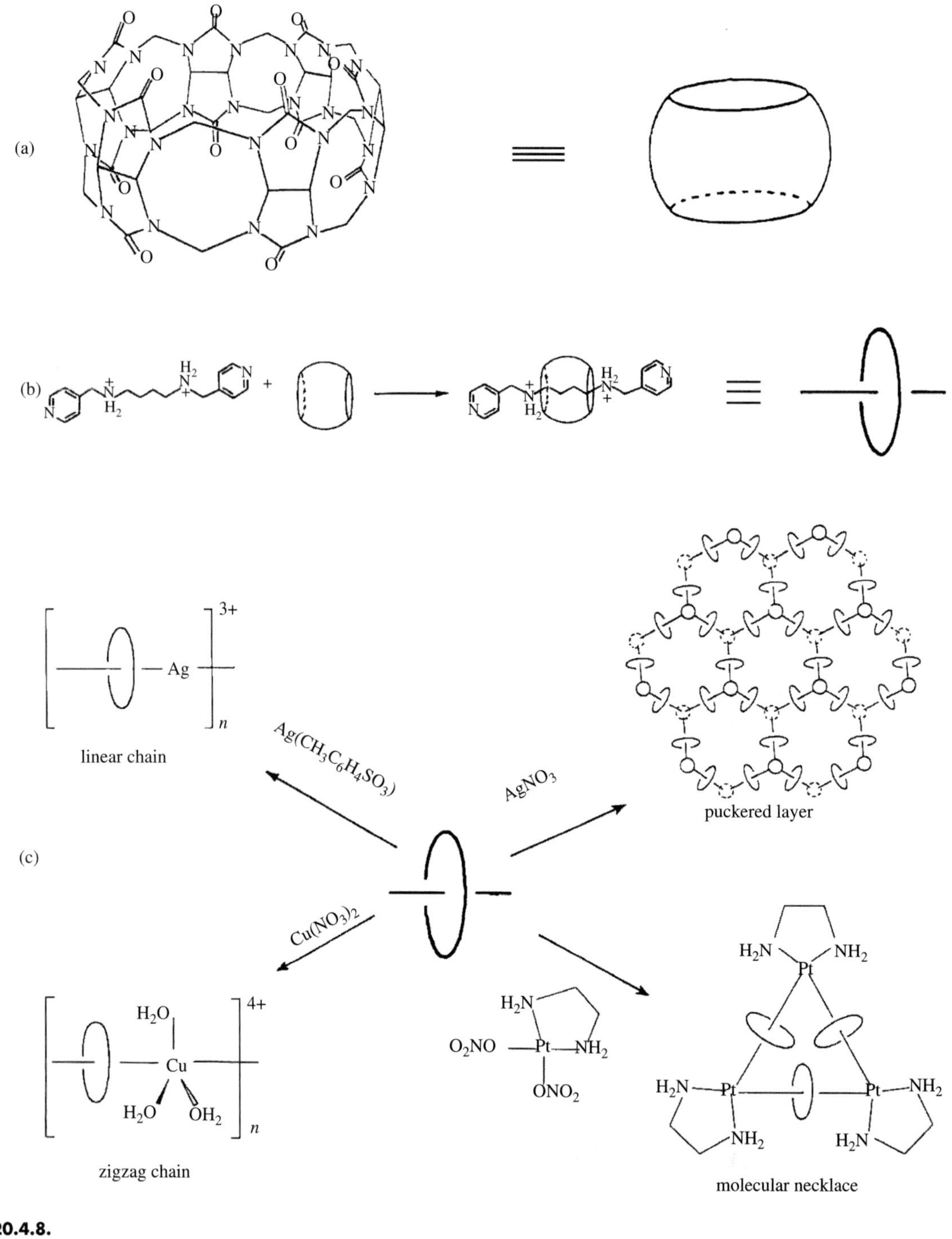

Fig. 20.4.8.
(a) Molecular structure of cucurbituril and its representation as a molecular bead or barrel, (b) cucurbituril threaded with a linear diammonium ion to form an inclusion complex, and (c) rotaxane building unit obtained by threading cucurbituril with diprotonated N,N'-bis(4-pyridylmethyl)-1,4-diaminobutane, and its subsequent reactions to yield linear polymers, a puckered layer, and a molecular triangle.

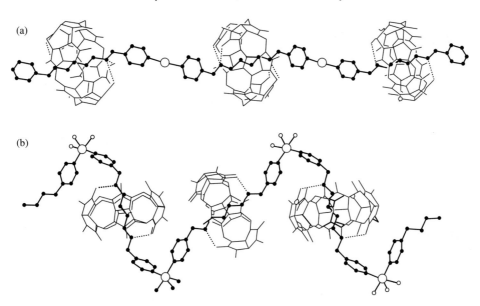

Fig. 20.4.9.
Structure of (a) a linear polyrotaxane and (b) a zigzag polyrotaxane.

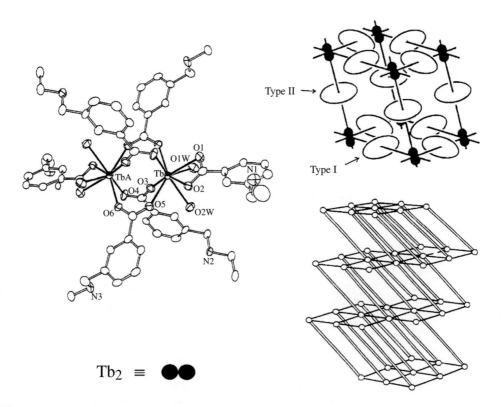

Fig. 20.4.10.
(a) Coordination geometry around the centrosymmetric binuclear terbium center. (b) Unit cell (top) and schematic representation (bottom) of the α-polonium-type network. Two contacting black circles stand for a binuclear terbium center, and the solid and open rods represent type I and type II rotaxane, respectively. From E. Lee, J. Heo and K. Kim, *Angew. Chem. Int. Ed.* **39**, 2699–701 (2000).

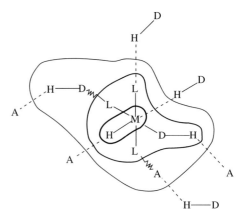

Fig. 20.4.11.
Domain model for hydrogen bonding involving metal complexes. Here D and A represent donor and acceptor atom, respectively. Note that both M–H and D–H units remain intact.

20.4.3 Inorganic crystal engineering using hydrogen bonds

The utilization of hydrogen bonding in inorganic crystal design has gained prominence in recent years. A conceptual framework for understanding supramolecular chemistry involving metals and metal complexes is provided by the domain model according to Dance and Brammer (Fig. 20.4.11).

The central *metal domain* of a metal complex consists of the metal atom M, a metal hydride group, or a number of metal atoms if a metal cluster complex is considered. The *ligand domain* is composed of ligand atoms L directly bonded with the metal center(s). The *periphery domain* is the outmost part of the complex, consisting of those parts of the ligand not strongly influenced by electronic interaction with the metal center.

Hydrogen bonding arising from donor groups (M)O–H and (M)N–H is commonly observed. The σ-type coordinated ligands are hydroxy (OH), aqua (OH$_2$), alcohol (ROH), and amines (NH$_3$, NRH$_2$, NR$_2$H). Acceptors include halide-type (M–X with X = F, Cl, Br, I), hydride-type (D–H$\cdots$H–M and D–H$\cdots$H–E with E = B, Al, Ga) and carbonyl-type (D–H$\cdots$OC–M).

Some examples of the coordination polymers consolidated by hydrogen bonding are discussed below.

(1) Chains

In the crystal structure of Fe[η^5-CpCOOH]$_2$, a hydrogen-bonded chain with carboxyl groups interacting via the $R_2^2(8)$ dimer synthon is formed [Fig. 20.4.12(a)]. The complex [Ag(nicotinamide)$_2$]CF$_3$SO$_3$ has a ladder structure propagated via the N–H$\cdots$O catemer with rungs comprising $R_2^2(8)$ amide dimer interactions [Fig. 20.4.12(b)]. Two CF$_3$SO$_3^-$ ions (not shown) lie inside each centrosymmetric macrocyclic ring, and their O atoms form N–H$\cdots$O–S–O$\cdots$H–N hydrogen-bonded and weak Ag$\cdots$O$\cdots$Ag bridges. The crystal structure of [Ru(η^5-Cp)(η^5-1-p-tolyl-2-hydroxyindenyl)] features a zigzag chain linked by O–H$\cdots \pi$(Cp) hydrogen bonds [Fig. 20.4.12(c)].

The crystal structures of [Pt(NCN–OH)Cl] and [Pt(SO$_2$)(NCN–OH)Cl], where NCN is the tridentate pincer ligand {C$_6$H$_2$-4-(OH)-2,6-(CH$_2$NMe$_2$)$_2$}$^-$, are compared in Fig. 20.4.13. The colorless complex [Pt(NCN–OH)Cl]

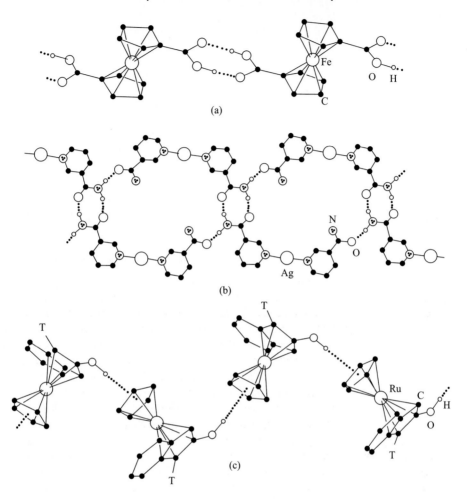

Fig. 20.4.12.
Hydrogen-bonded chain in (a) Fe[η^5-CpCOOH]$_2$, (b) [Ag(nicotinamide)$_2$]$^+$, and (c) [Ru(η^5-Cp)(η^5-1-p-tolyl-2-hydroxyindenyl)] (T stands for p-tolyl group).

consists of a sheet-like array of parallel zigzag chains connected via O–H$\cdots$Cl(Pt) hydrogen bonds. Reversible uptake of SO$_2$ is accompanied by a color change, resulting in an orange complex [Pt(SO$_2$)(NCN–OH)Cl] in which the coordinated SO$_2$ ligand is involved in a donor–acceptor S$\cdots$Cl interaction.

(2) Two-dimensional networks

The O–H$\cdots$O$^-$ hydrogen-bonded square grid found in [Pt(L$_2$)(HL)$_2$]·2H$_2$O (HL = isonicotinic acid) is shown in Fig. 20.4.14(a). Water molecules (not shown) occupy channels in the threefold interpenetrated network. The crystal structure of [Zn(SC(NH$_2$)NHNH$_2$)$_2$(OH)$_2$][1,4-O$_2$CC$_6$H$_4$CO$_2$]·2H$_2$O has a brick-wall sheet structure [Fig. 20.4.14(b)]. Each N,S-chelating thiosemicarbazide ligand forms two donor hydrogen bonds with one carboxylate group

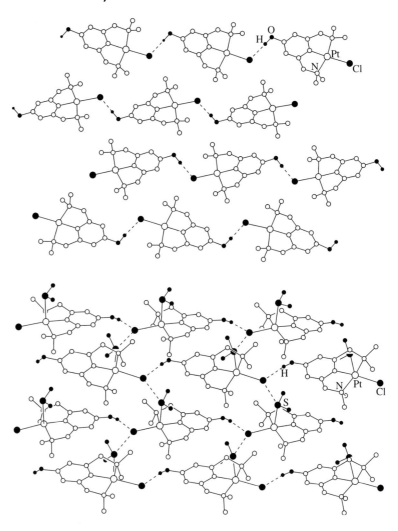

Fig. 20.4.13.
(a) Zigzag chain of [Pt(NCN–OH)Cl] held by O–H···Cl hydrogen bonds. (b) Two-dimensional network of [Pt(SO$_2$)(NCN–OH)Cl] consolidated by additional S···Cl donor–acceptor interactions.

of a terephthalate ion and one donor hydrogen bond with another terephthalate ion. The layers are linked via N–H···O and O–H···O hydrogen bonds to the water molecules (not shown). The complex [Ag(nicotinamide)$_2$]PF$_6$ has a cationic herringbone layer constructed from amide N–H···O hydrogen bonds [Fig. 20.4.14(c)]. Such layers are further cross-linked by N–H···F and C–H···F hydrogen bonds involving the PF$_6^-$ ions (not shown).

(3) Three-dimensional networks

In the complex [Cu(L)$_4$]PF$_6$ (L = 3-cyano-6-methylpyrid-2(1H)-one), the hydrogen-bonded linkage involves the amido $R_2^2(8)$ supramolecular synthon [Fig. 20.4.15(a)]. Each Cu(I) center serves as a tetrahedral node in a fourfold interpenetrated cationic diamondoid network. The tetrahedral metal cluster [Mn(μ_3-OH)(CO)$_3$]$_4$ can be used to construct a diamondoid network with the linear 4,4′-bipyridine spacers [Fig. 20.4.15(b)].

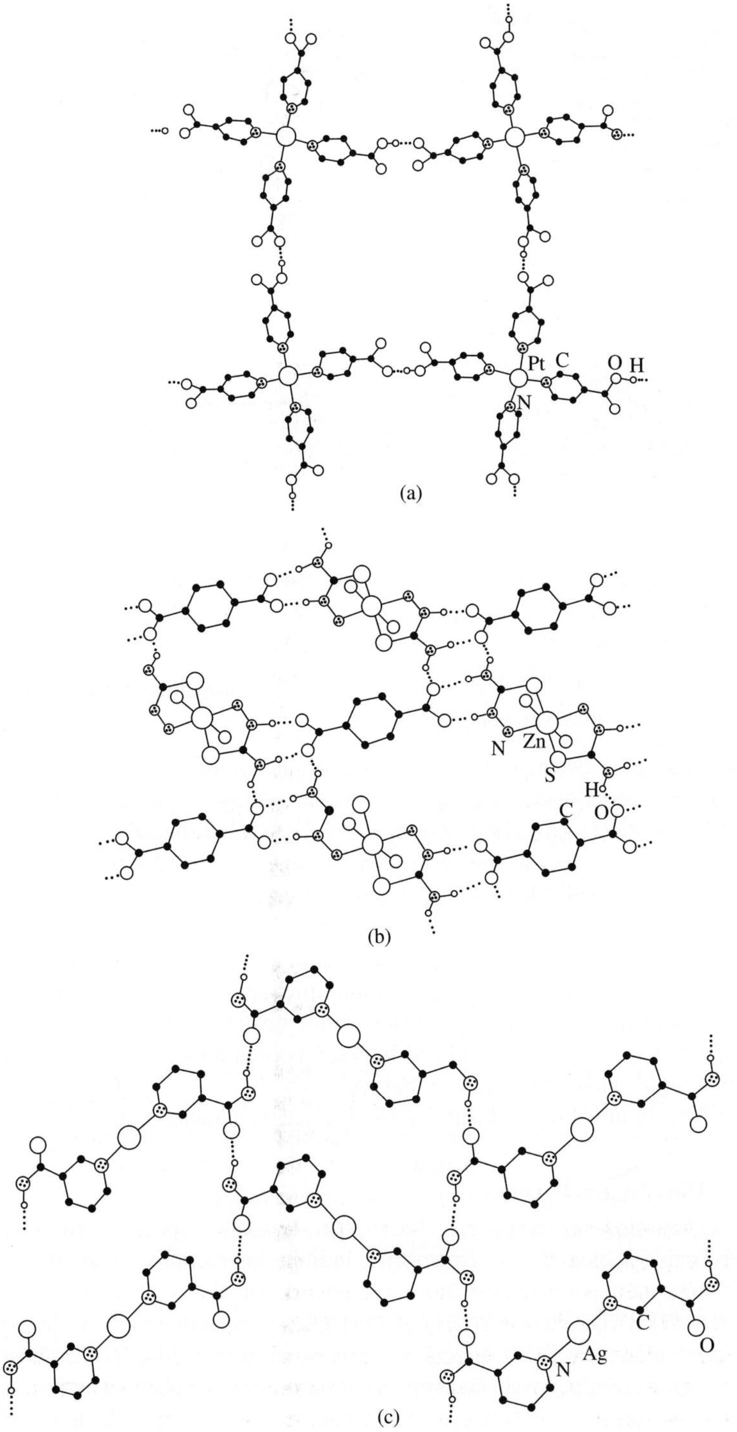

Fig. 20.4.14.
(a) Square grid in [Pt(L$_2$)(HL)$_2$]·2H$_2$O (HL = isonicotinic acid). (b) Brick-wall sheet in [Zn(thiosemicarbazide)(OH)$_2$](terephthate)·2H$_2$O (c) Cationic layer structure of [Ag(nicotinamide)$_2$]PF$_6$.

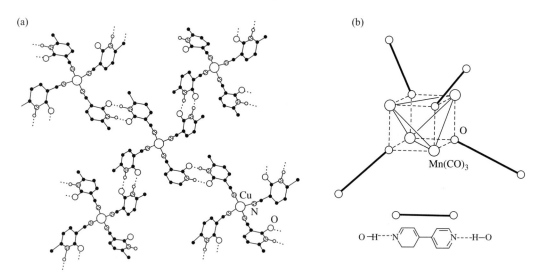

Fig. 20.4.15.
(a) Part of the cationic diamondoid network in [Cu(L)$_4$]PF$_6$ (L = pyridone ligand). (b) Molecular components and linkage mode of the diamondoid network in [Mn(μ_3-OH)(CO)$_3$]$_4$·2(bipy)·2CH$_3$CN.

20.4.4 Generation and stabilization of unstable inorganic/organic anions in urea/thiourea complexes

Urea, thiourea, or their derivatives are often employed as useful building blocks for supramolecular architectures because they contain amido functional group which can form moderately strong N–H···X hydrogen bonds with rather well-defined and predictable hydrogen-bonding patterns. Furthermore, in the presence of anions, the hydrogen bond is strengthened by two to three times (40 to 190 kJ mol^{-1}) compared with the bond strength involving uncharged molecular species (10 to 65 kJ mol^{-1}). Hence, making use of this kind of charge-assisted N–H···X$^-$ hydrogen bonding interactions and bulky tetraalkylammonium cations as guest templates, some unstable organic anions A$^-$ can be generated *in situ* and stabilized in urea/thiourea-anion host frameworks. A series of inclusion compounds of the type R$_4$N$^+$A$^-$·m(NH$_2$)$_2$CX (where X = O or S) with novel topological features has been characterized.

(1) Dihydrogen borate

As mentioned in Section 13.5.1, the transient species [BO(OH)$_2$]$^-$ has been stabilized by hydrogen-bonding interactions with the nearest urea molecules in the host framework of the inclusion compound [(CH$_3$)$_4$N]$^+$[BO(OH)$_2$]$^-$·2(NH$_2$)$_2$CO·H$_2$O. A perspective view of the crystal structure along the [010] direction is presented in Fig. 20.4.16. The host lattice consists of a parallel arrangement of unidirectional channels whose cross section has the shape of a peanut. The diameter of each spheroidal half is about 704 pm, and the separation between two opposite walls at the waist of the channel is about 585 pm. The well-ordered tetramethylammonium cations are accommodated in double columns within each channel.

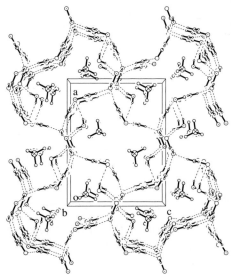

Fig. 20.4.16.
Crystal structure of
$[(CH_3)_4N]^+[BO(OH)_2]^- \cdot 2(NH_2)_2CO \cdot H_2O$
showing the channels extending parallel to the b axis and the enclosed cations. Broken lines represent hydrogen bonds, and the atoms are shown as points for clarity. From Q. Li, F. Xue and T. C. W. Mak, *Inorg. Chem.* **38**, 4142–5 (1999).

(2) **Allophanate and 3-thioallophanate**

Allophanate esters $H_2NCONHCOOR$ are among the oldest organic compounds recorded in the literature. The parent allophanic acid, $H_2NCONHCOOH$, is not known in the free state, whereas inorganic allophanate salts are unstable and readily hydrolyzed by water to carbon dioxide, urea, and carbonate. However, the elusive allophanate anion can be generated *in situ* and stabilized in the following three inclusion compounds:

$[(CH_3)_4N]^+[NH_2CONHCO_2]^- \cdot 5(NH_2)_2CO$
$[(n\text{-}C_3H_7)_4N]^+[NH_2CONHCO_2]^- \cdot 3(NH_2)_2CO$
$[(CH_3)_3N^+CH_2CH_2OH][NH_2CONHCO_2]^- \cdot (NH_2)_2CO$

A part of the host framework in $[(CH_3)_4N]^+[NH_2CONHCO_2]^- \cdot 5(NH_2)_2CO$ is shown in Fig. 20.4.17. Two neighboring allophanate anions are arranged in a head-to-tail fashion and bridged by a urea molecule with N–H$\cdots$O and charge-assisted N–H$\cdots$O$^-$ hydrogen bonds to generate a zigzag ribbon. This ribbon is further joined to another ribbon related to it by an inversion center via pairs of N–H$\cdots$O$^-$ hydrogen bonds to form a double ribbon.

The hitherto unknown 3-thioallophanate anion has been trapped in the host lattice of $[(n\text{-}C_4H_9)_4N]^+[H_2NCSNHCO_2]^- \cdot (NH_2)_2CS$. The cyclic structure and molecular dimensions of the allophanate and 3-thioallophanate ions are compared in Fig. 20.4.18. As expected, the C–O bond involved in intramolecular hydrogen bonding in the 3-thioallophanate anion is longer than that in the allophanate anion.

(3) **Valence tautomers of the rhodizonate dianion**

The rhodizonate dianion $C_6O_6^{2-}$ (Fig. 20.4.19) is a member of a series of planar monocyclic oxocarbon dianions $C_nO_n^{2-}$ ($n = 3$, deltate; $n = 4$, squarate; $n = 5$, croconate; $n = 6$; rhodizonate) which have been recognized as nonbenzenoid aromatic compounds. However, this six-membered ring species

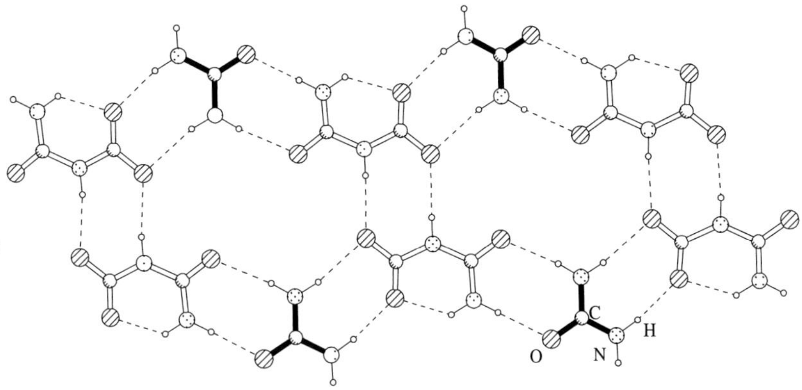

Fig. 20.4.17.
Part of the host framework in $[(CH_3)_4N]^+[H_2NCONHCO_2]^-\cdot 5(NH_2)_2CO$ showing the double ribbon constructed by allophanate anions and one of the independent urea molecules. From T. C. W. Mak, W.-H. Yip and Q. Li, *J. Am. Chem. Soc.* **117**, 11995–6 (1995).

is not stable in aqueous solution as it readily undergoes oxidative ring contraction reaction to the croconate dianion, and the decomposition is catalyzed by alkalis. Recently, this relatively unstable species has been generated *in situ* and stabilized by hydrogen bonding in two novel inclusion compounds $[(n\text{-}C_4H_9)_4N^+]_2C_6O_6^{2-}\cdot 2(m\text{-}OHC_6H_4NHCONH_2)\cdot 2H_2O$ (Fig. 20.4.20) and $[(n\text{-}C_4H_9)_4N^+]_2C_6O_6^{2-}\cdot 2(NH_2CONHCH_2CH_2NHCONH_2)\cdot 3H_2O$ (Fig. 20.4.21), respectively.

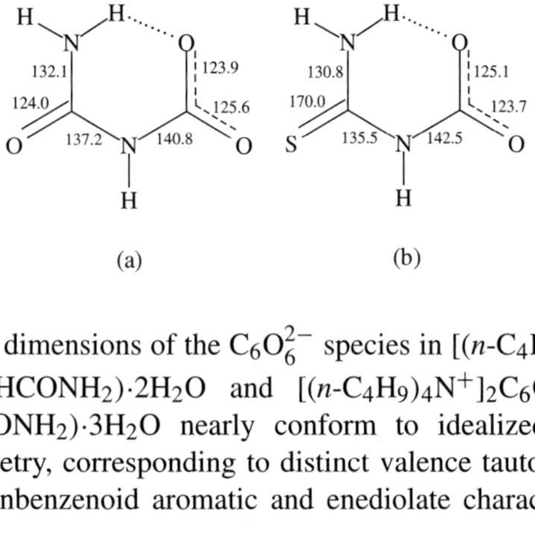

Fig. 20.4.18.
Bond lengths (pm) of the (a) allophanate and (b) 3-thioallophanate anion. From C.-K. Lam, T.-L. Chan and T. C. W. Mak, *CrystEngComm*. **6**, 290–2 (2004).

The measured dimensions of the $C_6O_6^{2-}$ species in $[(n\text{-}C_4H_9)_4N^+]_2C_6O_6^{2-}\cdot 2(m\text{-}OHC_6H_4NHCONH_2)\cdot 2H_2O$ and $[(n\text{-}C_4H_9)_4N^+]_2C_6O_6^{2-}\cdot 2(NH_2CONHCH_2CH_2NHCONH_2)\cdot 3H_2O$ nearly conform to idealized D_{6h} and C_{2v} molecular symmetry, corresponding to distinct valence tautomeric structures that manifest nonbenzenoid aromatic and enediolate character, respectively (Fig. 20.4.22).

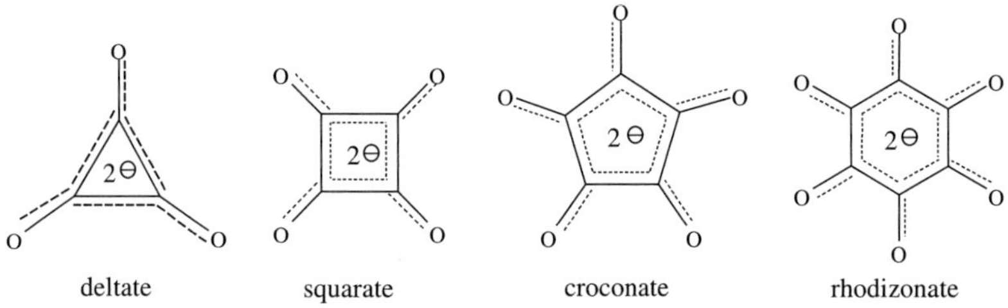

Fig. 20.4.19.
Structural formulas of cyclic oxocarbon dianions.

deltate   squarate   croconate   rhodizonate

Occurrence of the charge-localized structure of $C_6O_6^{2-}$ in $[(n\text{-}C_4H_9)_4N^+]_2C_6O_6^{2-}\cdot 2(NH_2CONHCH_2CH_2NHCONH_2)\cdot 3H_2O$, as well as its

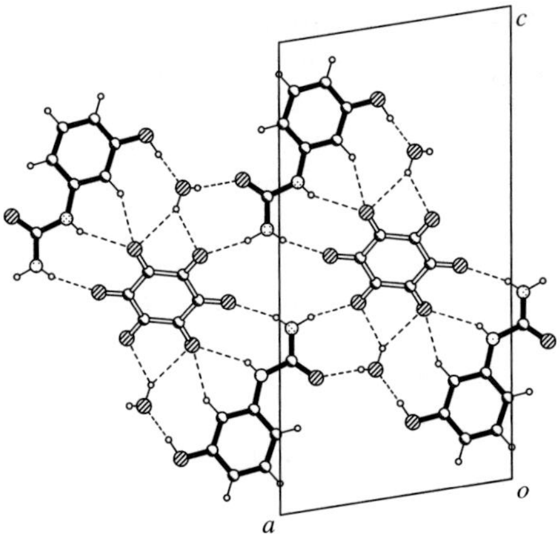

Fig. 20.4.20.
Projection along the *b* axis showing extensive hydrogen-bonding interactions around the centrosymmetric rhodizonate dianion in the host lattice of $[(n\text{-}C_4H_9)_4N^+]_2 C_6O_6^{2-} \cdot 2(m\text{-}OHC_6H_4NHCONH_2) \cdot 2H_2O$. From C.-K. Lam and T. C. W. Mak, *Angew. Chem. Int. Ed.* **40**, 3453–5 (2001).

noticeable deviation from idealized C_{2v} molecular symmetry, can be attributed to unequal hydrogen-bonding interaction with its two neighboring bisurea donors and a pair of water molecules (Fig. 20.4.21).

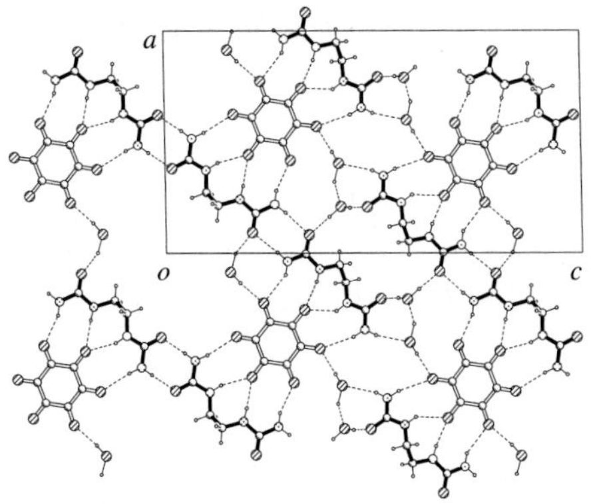

Fig. 20.4.21.
Projection along the *b* axis showing the hydrogen-bonding interactions within the puckered rhodizonate-bisurea-water layer of $[(n\text{-}C_4H_9)_4N^+]_2 C_6O_6^{2-} \cdot 2(NH_2CONHCH_2CH_2NHCONH_2) \cdot 3H_2O$. From C.-K. Lam and T. C. W. Mak, *Angew. Chem. Int. Ed.* **40**, 3453–5 (2001).

(4) Valence tautomers of the croconate dianion

In the host lattice of $[(n\text{-}C_3H_7)_4N^+]_2 C_5O_5^{2-} \cdot 3(NH_2)_2CO \cdot 8H_2O$, the croconate anion resides in a rather symmetrical hydrogen-bonding environment (Fig. 20.4.23) and thus its measured dimensions are consistent with its expected charge-delocalized D_{5h} structure in the ground state, as shown in

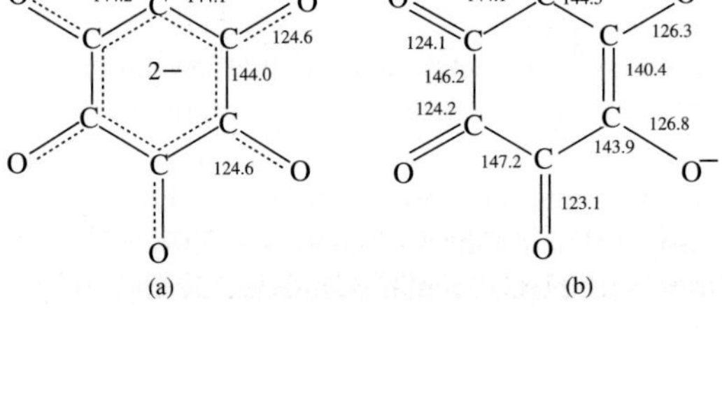

Fig. 20.4.22.
Bond lengths (pm) of (a) D_{6h} (note that in this case the dianion is located at a site of symmetry) and (b) C_{2v} valence tautomers of the rhodizonate dianion.

Fig. 20.4.24(a). In the host lattice of $[(C_2H_5)_4N^+]_2C_5O_5^{2-}\cdot 3(CH_3NH)_2CO$, the croconate dianion resides on a twofold axis, being directly linked to three 1,3-dimethylurea molecules through pairs of N–H···O hydrogen bonds to form a semi-circular structural unit (Fig. 20.4.25). In particular, each type O1 oxygen atom forms two strong acceptor hydrogen bonds with the N–H donors from a pair of 1,3-dimethylurea molecules, while each type O2 oxygen atom forms only one N–H···O hydrogen bond. In contrast, the solitary type O3 oxygen atom is stabilized by two weak C–H···O hydrogen bonds with neighboring tetra-*n*-butylammonium cations. Such a highly unsymmetrical environment engenders a sharp gradient of hydrogen-bonding donor strength around the croconate ion, which is conducive to stabilization of its C_{2v} valence tautomer with significantly different C–C and C–O bond lengths around the cyclic system, as shown in Fig. 20.4.24(b).

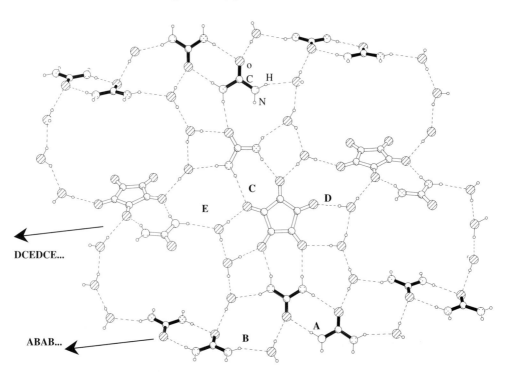

Fig. 20.4.23.
Projection diagram on the $(1\bar{1}0)$ plane showing the hydrogen-bonding scheme for a portion of the host lattice in $[(n\text{-}C_3H_7)_4N^+]_2C_5O_5^{2-}\cdot 3(NH_2)_2CO\cdot 8H_2O$. The slanted vertical (urea dimer–croconate–urea)$_\infty$ chain constitutes a side wall of the [110] channel system. The parallel [urea dimer–$(H_2O)_2]_\infty$ **ABAB...** and [croconate–urea–$(H_2O)_4]_\infty$ **DCEDCE...** ribbons define the channel system in the *c* direction. From C.-K. Lam, M.-F. Cheng, C.-L. Li J.-P. Zhang, X.-M. Chen, W.-K. Li and T. C. W. Mak, *Chem. Commun.*, 448–9 (2004).

The examples given in this section show that crystal engineering provides a viable route to breaking the degeneracy of canonical forms of a molecular species, and an elusive anion can be generated *in situ* and stabilized in a crystalline inclusion compound through hydrogen-bonding interactions with neighboring hydrogen-bond donors, such as urea/thiourea or their derivatives, which can function as supramolecular stabilizing agents.

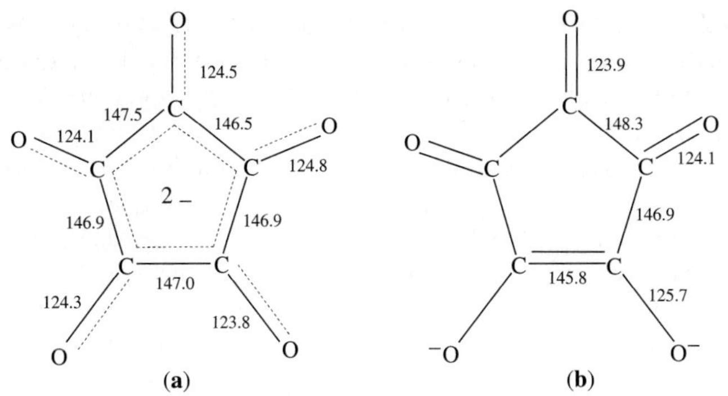

Fig. 20.4.24.
Bond lengths (pm) and angles (°) of D_{5h} (a) and C_{2v} (b) valence tautomers of the croconate dianion.

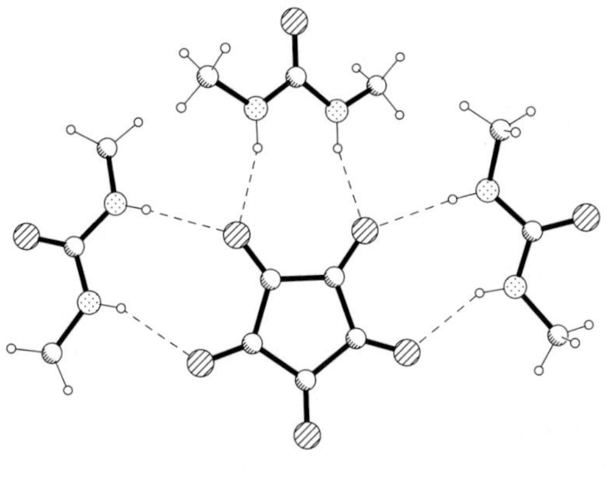

Fig. 20.4.25.
Hydrogen-bonding environment of the croconate dianion in the crystal structure of $[(C_2H_5)_4N^+]_2C_5O_5^{2-} \cdot 3(CH_3NH)_2CO$.

20.4.5 Supramolecular assembly of silver(I) polyhedra with embedded acetylenediide dianion

Since 1998, a wide range of double, triple, and quadruple salts containing Ag_2C_2 (IUPAC name silver acetylenediide, commonly known as silver acetylide or silver carbide in the older literature) as a component have been synthesized and characterized (see Section 14.3.7 and the comprehensive review by Bruce and Low cited at the end of this chapter). Such Ag_2C_2-containing double, triple, and quadruple salts can be formulated as $Ag_2C_2 \cdot nAgX$, $Ag_2C_2 \cdot mAgX \cdot nAgY$, and $Ag_2C_2 \cdot lAgX \cdot mAgY \cdot nAgZ$, respectively. The accumulated experimental data indicate that the acetylenediide dianion C_2^{2-} preferentially resides inside a silver(I) polyhedron of the type $C_2@Ag_n$ (n = 6-10), which is jointly stabilized by ionic, covalent (σ, π, and mixed), and argentophilic interactions (see Section 19.6.2). However, the $C_2@Ag_n$ cage is quite labile and its formation can be influenced by various factors such as solvent, reaction temperature, and the coexistence of anions, crown ethers, tetraaza macrocycles, organic cations, neutral ancillary ligands, and *exo*-bidentate bridging ligands, so that cage size and geometry are in general unpredictable.

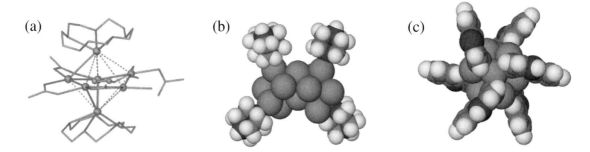

Fig. 20.4.26.
(a) Crown-sandwiched structure of the discrete molecule in [$Ag_2C_2 \cdot 5CF_3CO_2Ag \cdot 2(15C5) \cdot H_2O$]·$3H_2O$. The F and H atoms have been omitted for clarity. (b) Space-filling drawing of the discrete supermolecule in [$Ag_{14}(C_2)_2(CF_3CO_2)_{14}(dabcoH)_4(H_2O)_{1.5}$]·$H_2O$. (dabco = 1,4-diazabicyclo[2.2.2]octane). The trifluoroacetate and aqua ligands have been omitted for clarity. (c) Space-filling drawing of the discrete supermolecule [$Ag_8(C_2)(CF_3CO_2)_6(L^1)_6$] viewed along its $\bar{3}$ symmetry axis (L^1 = 4-hydroxyquinoline). The trifluoroacetate ligands have been omitted for clarity. From Q.-M. Wang and T. C. W. Mak, *Angew. Chem. Int. Ed.* **40**, 1130–3 (2001); Q.-M. Wang and T. C. W. Mak, *Inorg. Chem.* **42**, 1637–43 (2003); X.-L. Zhao, Q.-M. Wang and T. C. W. Mak, *Inorg. Chem.* **42**, 7872–6 (2003).

(1) Discrete molecules

To obtain discrete molecules, one effective strategy is to install protective cordons around the $C_2@Ag_n$ moiety with neutral, multidentate ligands that can function as blocking groups or terminal stoppers.

Small crown ethers have been introduced as structure-directing agents into the Ag_2C_2-containing system to prevent catenation and interlinkage of silver polyhedra. Judging from the rather poor host–guest complementarity of Ag(I) (soft cation) with a crown ether (hard O ligand sites), the latter is expected not to affect the formation of $C_2@Ag_n$, but to act as a capping ligand to an apex of the polyhedral silver cage.

[$Ag_2C_2 \cdot 5CF_3CO_2Ag \cdot 2(15C5) \cdot H_2O$]·$3H_2O$ (15C5 = [15]crown-5) can be obtained from an aqueous solution containing silver acetylenediide, silver trifluoroacetate, and 15C5. The discrete $C_2@Ag_7$ moiety in [$Ag_2C_2 \cdot 5CF_3CO_2Ag \cdot 2(15C5) \cdot H_2O$]·$3H_2O$ is a pentagonal bipyramid, with four equatorial edges bridged by $CF_3CO_2^-$ groups while the two apical Ag atoms are each attached to a 15C5, as shown in Fig. 20.4.26(a). In this discrete molecule, one equatorial Ag atom is coordinated by a monodentate $CF_3CO_2^-$ and the other by an aqua ligand.

In [$Ag_{14}(C_2)_2(CF_3CO_2)_{14}(dabcoH)_4(H_2O)_{1.5}$]·$H_2O$, the core is a Ag_{14} double cage constructed from edge-sharing of two triangulated dodecahedra. Apart from trifluoroacetate and aqua ligands, there are four monoprotonated dabco ligands surrounding the core unit, each being terminally coordinated to a silver(I) vertex, as shown in Fig. 20.4.26(b).

For the discrete molecule [$Ag_8(C_2)(CF_3CO_2)_6(L^1)_6$] ($L^1$ = 4-hydroxyquinoline) of $\bar{3}$ symmetry displayed in Fig. 20.4.26(c), the encapsulated acetylenediide dianion in the rhombohedral Ag_8 core is disordered about a crystallographic threefold axis that bisects the C≡C bond and passes through two opposite corners of the rhombohedron. Apart from the μ_2-O,O' trifluoroacetate liagnds, there are six keto-L^1 ligands surrounding the polynuclear core, each bridging an edge in the μ-O mode.

In the structures shown in Fig. 20.4.26, the hydrophobic tails of perfluorocarboxylates, together with the bulky ancillary ligands, successfully prevent linkage between adjacent silver polyhedra, thus leading to discrete supermolecules.

(2) Chains and columns

Free betaines are zwitterions bearing a carboxylate group and a quaternary ammonium group, and the prototype of this series is the trimethylammonio derivative commonly called betaine ($Me_3N^+CH_2COO^-$, IUPAC name trimethylammonioacetate, hereafter abbreviated as Me_3bet). Owing to their permanent bipolarity and overall charge neutrality, betaine and its derivatives (considered as carboxylate-type ligands) have distinct advantages over most carboxylates as ligands in the formation of coordination polymers:

(1) synthetic access to water-soluble metal carboxylates;
(2) generation of new structural varieties, such as complexes with metal centers bearing additional anionic ligands, and those with variable metal to carboxylate molar ratios; and
(3) easy synthetic modification of ligand property by varying the substituents on the quaternary nitrogen atom or the backbone between the two polar terminals.

The introduction of such ligands into the Ag_2C_2-containing system has led to isolation of supramolecular complexes showing chain-like or columnar structures.

$[(Ag_2C_2)_2(AgCF_3CO_2)_9(L^2)_3]$ has a columnar structure composed of fused silver(I) double cages: a triangulated dodecahedron and a bicapped trigonal prism, each encapsulating an acetylenediide dianion. Such a neutral column is coated by a hydrophobic sheath composed of trifluoroacetate and L^2 ligands, as shown in Fig. 20.4.27(a).

The core in $[(Ag_2C_2)_2(AgCF_3CO_2)_{10}(L^3)_3] \cdot H_2O$ is a double cage generated from edge-sharing of a square-antiprism and a distorted bicapped trigonal-prism, with each single cage encapsulating an acetylenediide dianion. Double cages of this type are fused together to form a helical column, which is surrounded by a hydrophobic sheath composed of trifluoroacetate and L^3 ligands, as shown in Fig. 20.4.27(b).

The building unit in $[(Ag_2C_2)(AgC_2F_5CO_2)_6(L^4)_2]$ is a centrosymmetric double cage, in which each half encapsulates an acetylenediide dianion. Each single cage is an irregular monocapped trigonal prism with one appended atom. The L^4 ligand acting in the μ_2-O,O' coordination mode links a pair of double cages to form an infinite chain, as shown in Fig. 20.4.27(c).

In $[(Ag_2C_2)(AgCF_3CO_2)_7(L^5)_2(H_2O)]$, the basic building unit is a distorted monocapped cube. The trifluoroacetate and L^5 ligands act as μ_3-bridges across adjacent single cage blocks to form a bead-like chain [Fig, 20.4.27(d)].

(3) Two-dimensional structures

The incorporation of ancillary N,N'- and N,O-donor ligands into the Ag_2C_2-containing system has led to a series of two-dimensional structures.

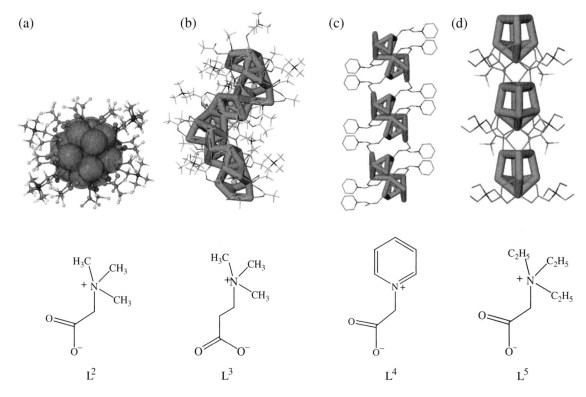

Fig. 20.4.27.
(a) Projection along an infinite silver(I) column with enclosed C_2^{2-} species and hydrophobic sheath in $[(Ag_2C_2)_2(AgCF_3CO_2)_9(L^2)_3]$. (b) Perspective view of the infinite silver(I) helical column with C_2^{2-} species embedded in its inner core and an exterior coat comprising anionic and zwitterionic carboxylates in $[(Ag_2C_2)_2(AgCF_3CO_2)_{10}(L^3)_3]\cdot H_2O$. (c) Infinite chain generated from the linkage of $(C_2)_2@Ag_{16}$ double cages by μ_2-O,O' L^4 ligands in $[(Ag_2C_2)(AgC_2F_5CO_2)_6(L^4)_2]$. (d) Infinite chain constructed from $C_2@Ag_9$ polyhedra connected by L^5 and trifluoroacetate bridges in $[(Ag_2C_2)(AgCF_3CO_2)_7(L^5)_2(H_2O)]$. From X.-L. Zhao, Q-M, Wang and T. C. W. Mak, *Chem. Eur. J.* **11**, 2094–102 (2005).

In the synthesis of $[(Ag_2C_2)(AgCF_3CO_2)_4(L^6)(H_2O)]\cdot H_2O$ under hydrothermal reaction condition, the starting ligand 4-cyanopyridine undergoes hydrolysis to form 4-pyridine-carboxamide (L^6). The basic building unit is a $C_2@Ag_8$ single cage in the shape of a distorted triangulated dodecahedron. Such dodecahedra share edges to form a zigzag composite chain, which are further linked via L^6 to generate a two-dimensional network Fig. 20.4.28(a).

The L^1 ligand, H_3O^+ species, and the anionic polymeric system $\{[Ag_{11}(C_2)_2(C_2F_5CO_2)_9(H_2O)_2]^{2-}\}_\infty$ comprise $(L^1\cdot H_3O)_2[Ag_{11}(C_2)_2(C_2F_5CO_2)_9(H_2O)_2]\cdot H_2O$. The basic building unit in the latter is a Ag_{12} double cage composed of two irregular monocapped trigonal antiprisms sharing an edge. The double cages are fused together to generate an infinite, sinuous anionic column. The oxygen atoms of L^1 and the water molecule are bridged by a proton to give the cationic aggregate $L^1\cdot H_3O^+$, which links the columns into a layer structure via hydrogen bonds with the pentafluoropropionate groups (Fig. 20.4.28(b)).

$[Ag_8(C_2)(CF_3CO_2)_8(H_2O)_2]\cdot(H_2O)_4\cdot(L^7H_2)$ represents a rare example of a hydrogen-bonded layer-type host structure containing $C_2@Ag_n$ that features the inclusion of organic guest species. The core is a centrosymmetric $C_2@Ag_8$ single cage in the shape of a slightly distorted cube. The trifluoroacetate ligands functioning in the μ_3-coordination mode further interlink the single cages into an infinite zigzag silver(I) chain. Of the three independent water molecules, one forms an acceptor hydrogen bond with a terminal of the L^7H_2 dication, the other serves as an aqua ligand bonded to a silver atom, and the third lying on a two fold axis functions as a bridge between aqua ligands belonging to adjacent silver(I) chains. The centrosymmetric L^7H_2 ions each hydrogen-bonded to a pair of terminal water molecules are accommodated between adjacent layers, forming an inclusion complex, as shown in Fig. 20.4.28(c).

(4) Three-dimensional structures

The strategy of using $C_2@Ag_n$ polyhedra as building blocks for the assembly of new coordination frameworks via introduction of potentially *exo*-bidentate nitrogen/oxygen-donor bridging ligands between agglomerated components has led to the isolation of three-dimensional supramolecular complexes exhibiting interesting crystal structures.

In $(Ag_2C_2)(AgCF_3CO_2)_8(L^8)_2(H_2O)_4$, square-antiprismatic $C_2@Ag_8$ cores are linked by trifluoroacetate groups to generate a columnar structure. Hydrogen bonds with the amino group of L^8 and aqua ligands serving as donors and the oxygen atoms of the trifluoroacetate group as acceptors further connect the columns into a three-dimensional scaffold [Fig. 20.4.29(a)].

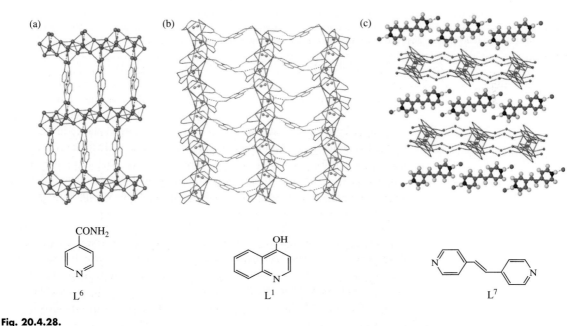

Fig. 20.4.28.
(a) Ball-and-stick drawing of the two-dimensional structure in $[(Ag_2C_2)(AgCF_3CO_2)_4(L^6)(H_2O)]\cdot H_2O$. (b) Schematic showing of the layer structure in $(L^1\cdot H_3O)_2[Ag_{11}(C_2)_2(C_2F_5CO_2)_9(H_2O)_2]\cdot H_2O$. (c) Two-dimensional host layer structure of $[Ag_8(C_2)(CF_3CO_2)_8(H_2O)_2]\cdot(H_2O)_4\cdot(L^7H_2)$ constructed from hydrogen bonds linking the silver(I) chains, with hydrogen-bonded $(bpeH_2\cdot 2H_2O)^{2+}$ moieties being accommodated between the host layers. From X.-L. Zhao and T. C. W. Mak, *Dalton Trans.*, 3212–7 (2004); X.-L. Zhao, Q.-M. Wang and T. C. W. Mak, *Inorg. Chem.*, **42**, 7872–6 (2003); X.-L. Zhao, and T. C. W. Mak, *Polyhedron* **24**, 940–8 (2005).

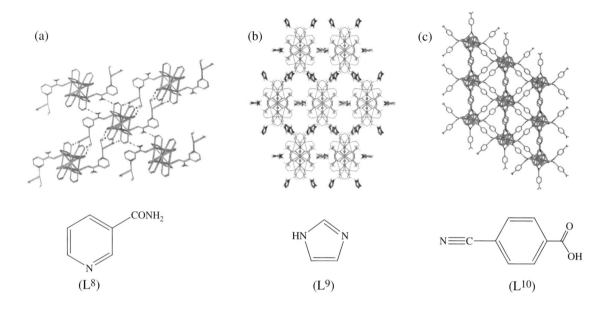

Fig. 20.4.29.
(a) Three-dimensional architecture of $(Ag_2C_2)(AgCF_3CO_2)_8(L^8)_2(H_2O)_4$ resulting from the linkage of silver columns via hydrogen bonds. (b) Three-dimensional architecture in $(L^9H)_3 \cdot [Ag_8(C_2)(CF_3CO_2)_9] \cdot H_2O$ generated from covalent silver chains linked by hydrogen bonds. (c) The (3,6) covalent network in $[Ag_7(C_2)(CF_3CO_2)_2(L^{10})_3]$ constructed by the linkage of L^{10} with silver(I) columns. From X.-L.Zhao and T. C. W. Mak, *Dalton Trans.* 3212-7 (2004); X.-L. Zhao and T. C. W. Mak, *Polyhedron* **25**, 975–82 (2006).

The building block in $(L^9H)_3 \cdot [Ag_8(C_2)(CF_3CO_2)_9] \cdot H_2O$ is a $C_2@Ag_8$ single cage in the shape of triangulated dodecahedron located on a twofold axis. Silver cages of such type are connected by μ_3-O,O,O' trifluoroacetate ligands to form a zigzag anionic silver(I) column along the *a* direction. All three independent L^9 molecules are protonated to satisfy the overall charge balance in the crystal structure. Notably, the resulting L^9H cations play a key role in the construction of the three-dimensional architecture. As shown in Fig. 20.4.29(b), the silver(I) columns are interconnected by hydrogen bonding with the protonated L^9 serving as donors and O, F atoms of trifluoroacetate ligands as acceptors to form the three-dimensional network.

In $[Ag_7(C_2)(CF_3CO_2)_2(L^{10})_3]$, the basic building block is a centrosymmetric $(C_2)_2@Ag_{14}$ double cage, with each half taking the shape of a distorted bicapped trigonal prism. Such double cages are fused together to form an infinite column. Each silver(I) column is linked to six other radiative silver(I) columns via L^{10}, and every three neighboring silver columns encircle a triangular hole, thus resulting in a (3,6) (or 3^6) topology, as displayed in Fig. 20.4.29(c).

(5) Mixed-valent silver(I,II) compounds containing Ag_2C_2

To investigate the effect of coexisting metal ions on the assembly of polyhedral silver(I) cages, macrocyclic *N*-donor ligand 1,4,8,11-tetramethyl-1,4,8,11-tetraazacyclotetradecane (tmc) has been used for *in situ* generation of [Ag^{II}(tmc)]. Mixed-valent silver complexes [Ag^{II}(tmc)(BF$_4$)][Ag$^{I}_{6}$(C$_2$)(CF$_3$CO$_2$)$_5$(H$_2$O)]· H$_2$O and [Ag^{II}(tmc)][Ag^{II}(tmc)(H$_2$O)]$_2$[Ag$^{I}_{11}$(C$_2$)(CF$_3$CO$_2$)$_{12}$(H$_2$O)$_4$]$_2$ have been isolated and structurally characterized.

In [Ag^{II}(tmc)(BF$_4$)][Ag$^{I}_{6}$(C$_2$)(CF$_3$CO$_2$)$_5$(H$_2$O)]· H$_2$O, the addition of tmc leads to disproportionation of silver(I) to give elemental silver and complexed silver(II), the latter being stabilized by tmc to form [Ag^{II}(tmc)]$^{2+}$. Weak axial interactions of the d^9 silver(II) center with adjacent BF$_4^-$ serve to link the complexed Ag(II) cations into a [Ag^{II}(tmc)(BF$_4$)]$^{+1}_{\infty}$ column, which further induces the assembly of a novel anionic zigzag chain constructed from edge-sharing of silver(I) triangulated dodecahedra, each enclosing a C$_2^{2-}$ species (Fig. 20.4.30(a)).

In [Ag^{II}(tmc)][Ag^{II}(tmc)(H$_2$O)]$_2$[Ag$^{I}_{11}$(C$_2$)(CF$_3$CO$_2$)$_{12}$(H$_2$O)$_4$]$_2$, which lacks the participation of BF$_4^-$ ions, the cations do not line up in a one-dimensional array and instead a dimeric supramolecular cluster anion is generated [Fig. 20.4.30(b)].

(6) Ligand-induced disruption of polyhedral C$_2$@Ag$_n$ cage assembly

Attempts to interfere with the assembly process to open the C$_2$@Ag$_n$ cage or construct a large single cage for holding two or more C$_2^{2-}$ species were carried out via the incorporation of the multidentate ligand pyzCONH$_2$ (pyrazine-2-carboxamide) into the reaction system. Pyrazine-2-carboxamide was selected as a structure-directing component by virtue of its very short spacer length and chelating capacity, and the introduction of the amide functionality could conceivably disrupt the assembly of C$_2$@Ag$_n$ via the formation of hydrogen bonds. The ensuing study yielded two silver(I) complexes Ag$_{12}$(C$_2$)$_2$(CF$_3$CO$_2$)$_8$(2-pyzCONH$_2$)$_3$ and Ag$_{20}$(C$_2$)$_4$(C$_2$F$_5$CO$_2$)$_8$(2-pyzCOO)$_4$(2-pyzCONH$_2$)(H$_2$O)$_2$ exhibiting novel C$_2$@Ag$_n$ motifs.

The basic structural unit of Ag$_{12}$(C$_2$)$_2$(CF$_3$CO$_2$)$_8$(2-pyzCONH$_2$)$_3$ comprises the fusion of a distorted triangulated dodecahedral Ag$_8$ cage containing an embedded C$_2^{2-}$ dianion and an open fish-like Ag$_6$(μ_6-C$_2$) motif

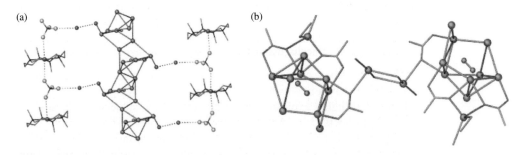

Fig. 20.4.30.
(a) Crystal structure of [Ag^{II}(tmc)(BF$_4$)][Ag$^{I}_{6}$(C$_2$)(CF$_3$CO$_2$)$_5$(H$_2$O)]· H$_2$O. (b) Perspective view of the structure of the dimeric supramolecular anion in
[Ag^{II}(tmc)][Ag^{II}(tmc)(H$_2$O)]$_2$[Ag$^{I}_{11}$(C$_2$)(CF$_3$CO$_2$)$_{12}$(H$_2$O)$_4$]$_2$. The F atoms of the
CF$_3$CO$_2^-$ ligands and some CF$_3$CO$_2$ are omitted for clarity. From Q.-M. Wang and T. C. W. Mak, *Chem. Commun.*, 807-8 (2001); Q.-M. Wang, H. K. Lee and T. C. W. Mak, *New J. Chem.* **26**, 513-5 (2002).

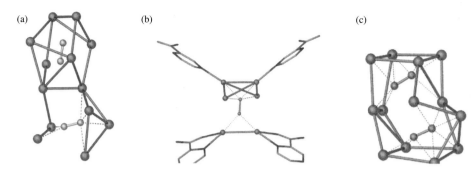

Fig. 20.4.31.
(a) Basic building unit in $Ag_{12}(C_2)_2(CF_3CO_2)_8$(2-pyzCONH$_2$)$_3$. (b) Open fish-like $Ag_6(\mu_6\text{-}C_2)$ motif coordinated by four pyrazine-2-carboxamide ligands in $Ag_{12}(C_2)_2(CF_3CO_2)_8$(2-pyzCONH$_2$)$_3$. (c) The $(C_2)_2@Ag_{13}$ in $Ag_{20}(C_2)_4(C_2F_5CO_2)_8$(2-pyzCOO)$_4$(2-pyzCONH$_2$)(H$_2$O)$_2$. From X.-L. Zhao and T. C. W. Mak, *Organometallics* **24**, 4497–9 (2005).

[Fig. 20.4.31(a)]. In the $Ag_6(\mu_6\text{-}C_2)$ motif, one carbon atom is embraced by four silver atoms in a butterfly arrangement and the other bonds to two silver atoms. Its existence can be rationalized by the fact that it is stabilized by four surrounding pyrazine-2-carboxamide ligands so that steric overcrowding obstructs the aggregation of silver(I) into a closed cage [Fig. 20.4.31(b)].

The basic building block in $Ag_{20}(C_2)_4(C_2F_5CO_2)_8$(2-pyzCOO)$_4$(2-pyzCONH$_2$)(H$_2$O)$_2$ is an aggregate composed of three polyhedral units: an unprecedented partially opened cage $(C_2)_2@Ag_{13}$ [Fig.20.4.31(c)] and two similar distorted $C_2@Ag_6$ trigonal prisms. A pair of C_2^{2-} dianions are completely encapsulated in the Ag_{13} cage. For simplicity, this single cage can be visualized as composed of two distorted cubes sharing a common face, with cleavage of four of the edges and capping of a lateral face. The two embedded C_2^{2-} dianions retain their triple-bond character with similar C–C bond lengths of 118(2) pm.

20.4.6 Supramolecular assembly with the silver(I)-ethynide synthon

In 2004, the silver carbide (Ag_2C_4) was synthesized as a light gray powder, which behaves like its lower homologue Ag_2C_2, being insoluble in most solvents and highly explosive in the dry state when subjected to heating or mechanical shock. Using Ag_2C_4 and the crude polymeric silver ethynide complexes $[R–(C{\equiv}CAg)_m]_\infty$ (R = aryl; $m = 1$ or 2) as starting materials, a variety of double and triple silver(I) salts containing 1,3-butadiynediide and related carbon-rich ethynide ligands have been synthesized. Investigation of the coordination modes of the ethynide moiety in these compounds led to the recognition of a new class of supramolecular synthons R–C≡C⊃Ag$_n$ (n = 4, 5), which can be utilized to assemble a series of one-, two-, and three-dimensional networks together with argentophilic interactions, π–π stacking, silver-aromatic interactions and hydrogen bonding.

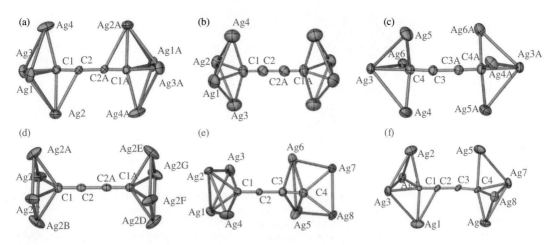

Fig. 20.4.32.
Observed μ_8-coordination modes of C_4^{2-} dianion. (a) Symmetrical mode with two butterfly-shaped baskets in $Ag_2C_4 \cdot 6AgNO_3 \cdot nH_2O$ ($n = 2$, 3). From L. Zhao and T. C. W. Mak, *J. Am. Chem. Soc.* **126**, 6852-3 (2004). (b) Symmetrical mode only with Ag—C σ bonds in $Ag_2C_4 \cdot 16AgC_2F_5CO_2 \cdot 6CH_3CN \cdot 8H_2O$. (c) Barb-like μ_8-coordination mode with two linearly coordinated C≡C—Ag bonds in triple salt $Ag_2C_4 \cdot AgF \cdot 3AgNO_3 \cdot 0.5H_2O$. (d) Symmetrical mode with two parallel planar Ag_4 aggregates in $Ag_2C_4 \cdot 16AgC_2F_5CO_2 \cdot 24H_2O$. (e) Unsymmetrical μ_8-coordination mode with one butterfly-shaped Ag_4 basket and one planar Ag_4 aggregate in $Ag_2C_4 \cdot 6AgCF_3CO_2 \cdot 7H_2O$. (f) Unsymmetrical μ_8-coordination mode with two butterfly-shaped Ag_4 baskets in quadruple salt $Ag_2C_4 \cdot 4AgNO_3 \cdot Ag_3PO_4 \cdot AgPF_2O_2$.

(1) Silver(I) complexes containing the C_4^{2-} dianion

In all of its silver(I) complexes, the linear $^-C{\equiv}C{-}C{\equiv}C^-$ dianion exhibits an unprecedented μ_8-coordination mode, each terminal being capped by four silver(I) atoms (Fig. 20.4.32). However, the σ- and π-type silver–ethynide interactions play different roles in symmetrical and unsymmetrical μ_8-coordination. Furthermore, coexisting ancillary anionic ligands, nitrile groups, and aqua molecules also influence the coordination environment around each terminal ethynide, which takes the form of a butterfly-shaped, barb-like, or planar Ag_4 basket. The carbon–carbon triple- and single-bond lengths in C_4^{2-} are in good agreement with those observed in transition-metal 1,3-butadiyne-1,4-diyl complexes. The Ag$\cdots$Ag distances within the Ag_4 baskets are all shorter than 340 pm, suggesting the existence of significant Ag$\cdots$Ag interactions.

The [$Ag_4C_4Ag_4$] aggregates *vide supra* can be further linked by other anionic ligands such as nitrate and perfluorocarboxylate groups, and/or water molecules, to produce various two- or three-dimensional coordination networks. In the structure of $Ag_2C_4 \cdot 6AgNO_3 \cdot 2H_2O$, the [$Ag_4C_4Ag_4$] aggregates arranged in a pseudo-hexagonal array are connected by one nitrate group acting in the μ_3-O,O',O'' plus O,O'-chelating mode to form a thick layer normal to [100] [Fig. 20.4.33(a)]. Linkage of adjacent layers by the remaining two independent nitrate groups, abetted by O–H$\cdots$O(nitrate) hydrogen bonding involving the aqua ligand, then generates a three-dimensional network.

In the crystal structure of $Ag_2C_4 \cdot 16AgC_2F_5CO_2 \cdot 24H_2O$, each [$Ag_4C_4Ag_4$] unit connects with eight such units by eight [$Ag_2(\mu\text{-}O_2CC_2F_5)_4$] bridging ligands to form a (4,4) coordination network [Fig. 20.4.33(b)]. Through the

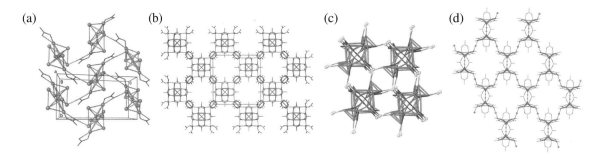

Fig. 20.4.33.
Some examples of two- and three-dimensional coordination networks of Ag_2C_4: (a) pseudohexagonal array of [$Ag_4C_4Ag_4$] aggregates linked by an independent nitrate group in $Ag_2C_4 \cdot 6AgNO_3 \cdot 2H_2O$: (b) (4,4) network in ab plane of $Ag_2C_4 \cdot 16AgC_2F_5CO_2 \cdot 24H_2O$, which is composed of [$Ag_4C_4Ag_4$] aggregates connected by bridging [$Ag_2(\mu\text{-}O_2CC_2F_5)_4$] groups: (c) three-dimensional coordination network in $Ag_2C_4 \cdot AgF \cdot 3AgNO_3 \cdot 0.5H_2O$ through the coordination of μ_3-fluoride ligands, (d) Rosette layer in $Ag_2C_4 \cdot 10AgCF_3CO_2 \cdot 2[(Et_4N)CF_3CO_2]$ $\cdot 4(CH_3)_3CCN$ composed of [$Ag_4C_4Ag_4$] aggregates linked by one external silver atom and two trifluoroacetate groups.

linkage of the C_4 carbon chains perpendicular to this network, an infinite channel is aligned along the [001] direction, and each accommodates a large number of pentafluoroethyl groups.

In the crystal structure of $Ag_2C_4 \cdot AgF \cdot 3AgNO_3 \cdot 0.5H_2O$, [$Ag_4C_4Ag_4$] aggregates are mutually connected through the linkage of nitrate groups and sharing of some silver atoms to form a silver column. The fluoride ions bridge these silver columns in the μ_3-mode to produce a structurally robust three-dimensional coordination network [(Fig. 20.4.33(c)]. With an external silver atom and two carboxylato oxygen atoms as bridging groups, the [$Ag_4C_4Ag_4$] aggregates in $Ag_2C_4 \cdot 10AgCF_3CO_2 \cdot 2[(Et_4N)CF_3CO_2] \cdot 4(CH_3)_3CCN$ are linked to form a two-dimensional rosette layer, in which the C_4^{2-} dianion acts as the shared border of two metallacycles [Fig. 20.4.33(d)].

(2) Silver(I) complexes of isomeric phenylenediethynides with the supramolecular synthons $Ag_n \subset C_2$–x-C_6H_4–$C_2 \supset Ag_n$ ($x = p, m, o$; $n = 4, 5$)

The above study of silver(I) 1,3-butadiynediide complexes suggests that the $Ag_4 \subset C_2$–R–$C_2 \supset Ag_4$ moiety may be conceived as a synthon for the assembly of coordination networks, by analogy to the plethora of well-known supramolecular synthons that involve hydrogen bonding and other weak intermolecular interactions (see Section 20.1.5). With reference to 1,3-butadiynediide as a standard, the p-phenylene ring was introduced as the bridging R group in the supramolecular synthon $Ag_n \subset C_2$–R–$C_2 \supset Ag_n$ with a lengthened linear π-conjugated backbone, and the aromatic ring of the resulting p-phenylenediethynide dianion could presumably partake in π–π stacking and silver–aromatic interaction. The isomeric m- and o-phenylenediethynides were also investigated in order to probe the influence of varying the relative orientation of the pair of terminal ethynide groups.

In $2[Ag_2(p\text{-}C{\equiv}CC_6H_4C{\equiv}C)] \cdot 11AgCF_3CO_2 \cdot 4CH_3CN \cdot 2CH_3CH_2CN$, the p-phenylenediethynide ligand exhibits the highest ligation number reported to date for the ethynide moiety by adopting an unprecedented μ_5-η^1 mode.

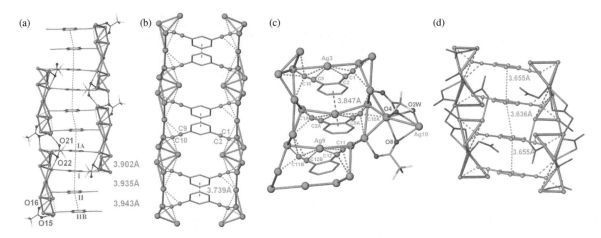

Fig. 20.4.34.
(a) Broken silver(I) double chain in $2[Ag_2(pC\equiv CC_6H_4C\equiv C)]\cdot 11AgCF_3CO_2\cdot 4CH_3CN\cdot 2CH_3CH_2CN$ stabilized by continuous $\pi-\pi$ stacking between parallel p-phenylene rings. (b) Silver double chain in $Ag_2(m\text{-}C\equiv CC_6H_4C\equiv C)\cdot 6AgCF_3CO_2\cdot 3CH_3CN\cdot 2.5H_2O$ assembled by argentophilic interaction and $\pi-\pi$ interaction between adjacent pairs of m-phenylene rings. (c) Widened silver chain in $3[Ag_2(o\text{-}C\equiv CC_6H_4C\equiv C)]\cdot 14AgCF_3CO_2\cdot 2CH_3CN\cdot 9H_2O$ constructed through the cross-linkage of two narrow silver chains by bridging silver atoms with pairwise $\pi-\pi$ interaction between the o-phenylene rings. (d) Broken silver(I) double chain in $Ag_2(m\text{-}C\equiv CC_6H_4C\equiv C)]\cdot 5AgNO_3\cdot 3H_2O$ stabilized by continuous $\pi-\pi$ stacking between parallel m-phenylene rings. From L. Zhao and T. C. W. Mak, *J. Am. Chem. Soc.* **127**, 14966–7 (2005).

The Ag_{14} aggregate, being constructed essentially from two independent $Ag_n \subset C_2-(p\text{-}C_6H_4)-C_2 \supset Ag_n$ ($n = 4, 5$) synthons through argentophilic interaction and continuous $\pi-\pi$ stacking, is connected to its symmetry equivalents to form a broken silver(I) double chain along the a-axis. Adjacent Ag_{14} segments within a single chain are bridged by the oxygen atoms of two independent trifluoroacetate groups, and the pair of single chains are arranged in interdigitated fashion [Fig. 20.4.34(a)].

In contrast to the popular μ_1,μ_1-coordination mode of m-phenylenediethynide in most transition-metal complexes, this ligand exhibits two different terminal ethynide bonding modes, namely, $\mu_4\text{-}\eta^1,\eta^1,\eta^1,\eta^1$ and $\mu_4\text{-}\eta^1,\eta^1,\eta^1,\eta^2$, in the crystal structure of $Ag_2(m\text{-}C\equiv CC_6H_4C\equiv C)\cdot 6AgCF_3CO_2\cdot 3CH_3CN\cdot 2.5H_2O$. Through inversion centers located between successive pairs of m-phenylene rings, the $Ag_n \subset C_2-(m\text{-}C_6H_4)-C_2 \supset Ag_n$ ($n = 4$) synthon is extended along the a direction by Ag$\cdots$Ag interactions to form a silver double chain [Fig. 20.4.34(b)], which are further consolidated by $\pi - \pi$ interaction between adjacent m-phenylene rings protruding alternatively on either side.

In the crystal structure of $3[Ag_2(o\text{-}C\equiv CC_6H_4C\equiv C)]\cdot 14AgCF_3CO_2 \cdot 2CH_3CN\cdot 9H_2O$, the ethynide moieties bond to silver atoms via three different coordination modes: $\mu_4\text{-}\eta^1,\eta^1,\eta^1,\eta^2$, $\mu_4\text{-}\eta^1,\eta^1,\eta^2,\eta^2$, and $\mu_5\text{-}\eta^1,\eta^1,\eta^1,\eta^1,\eta^2$ [Fig. 20.4.34(c)]. The two independent $Ag_n \subset C_2-(o\text{-}C_6H_4)-C_2 \supset Ag_n$ ($n = 4, 5$) synthons mutually associate to generate an undulating silver chain consolidated by argentophilic interaction, pairwise $\pi-\pi$ interaction between the o-phenylene rings, and linkage by silver atoms Ag3 and Ag9 across two silver single chains. Bridged by silver atom Ag10 through Ag$\cdots$Ag interaction and three oxygen atoms (O2W, O4, O8), the silver columns are linked to form a wave-like layer with o-phenylene groups protruding on both sides.

Fig. 20.4.35.
Schematic diagram showing the structural relationship between the supramolecular synthons Ag$_4$⊂C$_2$—C$_2$⊃Ag$_4$, Ag$_n$⊂ C$_2$—R—C$_2$⊃Ag$_n$ (R = p-, m-, o-C$_6$H$_4$; n = 4, 5), and C$_2$@Ag$_n$ (n = 6–10). The circular arc represents a Ag$_n$ (n = 4, 5) basket.

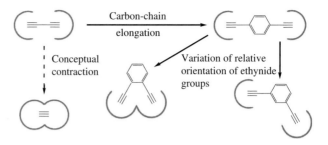

The decreasing separation of the pair of ethynide groups in the above three complexes is accompanied by strengthened argentophilic interaction at the expense of weakened π–π stacking, yielding a broken double chain, a double chain and a silver layer, respectively. When the pair of ethynide vectors make an angle of 60°, sharing of a common silver atom for the Ag$_n$ caps occurs in the o-phenylenediethynide complex.

The crystal structure of Ag$_2$(m-C≡CC$_6$H$_4$C≡C)]·5AgNO$_3$·3H$_2$O features a broken silver(I) double chain analogous to that of the p-phenylenediethynide complex [see Fig. 20.4.34(a). However, the single chains containing the Ag$_{14}$ aggregates are arranged in parallel fashion, each matching Ag$_{14}$ pair being connected by two m-phenylenediethynide ligands [Fig. 20.4.34(d)]. Linkage of a series of Ag$_n$⊂C$_2$—(m-C$_6$H$_4$)—C$_2$⊃Ag$_n$ (n = 4) fragments by intrachain bridging nitrate groups with continuous π–π interaction between adjacent m-phenylene rings engenders the broken silver(I) double chain.

The structural correlation between various silver–ethynide supramolecular synthons affords a rationale for the preponderant existence of C$_2$@Ag$_n$ (n = 6–10) polyhedra in Ag$_2$C$_2$ complexes (see Section 20.4.5). If the linear $^-$C≡C–C≡C$^-$ chain were contracted to a C$_2^{2-}$ dumbbell, further overlap between atoms of the terminal Ag$_n$ caps would conceivably yield a closed cage with 6–10 vertices (Fig. 20.4.35).

(3) **Silver(I) arylethynide complexes containing R-C$_2$ ⊃ Ag$_n$ (R = C$_6$H$_5$, C$_6$H$_4$Me-4, C$_6$H$_4$Me-3, C$_6$H$_4$Me-2, C$_6$H$_4^t$Bu-4; n = 4, 5)**

π–π Stacking or π–π interactions are important noncovalent intermolecular interactions, which contribute much to self-assembly when extended structures are formed from building blocks with aromatic moieties. In relation to the rich variety of π–π stacking in the crystal structure of silver(I) complexes of phenylenediethynide, related silver complexes of phenylethynide and its homologues with different substituents (–CH$_3$, –C(CH$_3$)$_3$) or the –CH$_3$ group in different positions (o-, m-, p-) are investigated.

In the crystal structure of 2AgC≡CC$_6$H$_5$·6AgC$_2$F$_5$CO$_2$·5CH$_3$CN, the ethynide group composed of C1 and C2 is capped by a square-pyramidal Ag$_5$ basket in an unprecedented μ_5-$\eta^1,\eta^1,\eta^1,\eta^1,\eta^2$ coordination mode and the other one comprising C9 and C10 by a butterfly-shaped Ag$_4$ basket in a μ_4-$\eta^1,\eta^1,\eta^1,\eta^2$ coordination mode, as shown in Fig. 20.4.36(a). With an inversion center located at the center of the Ag1···Ag1A bond, two Ag$_5$ baskets share an edge to engender a Ag$_8$ aggregate, whereas another Ag$_8$ aggregate results from fusion of a pair of inversion-related Ag$_4$ baskets. Two adjacent Ag$_8$ agregates

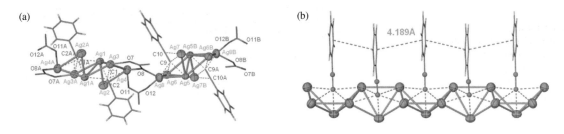

Fig. 20.4.36.
(a) Coordination modes of the independent phenylethynide ligands in $2AgC{\equiv}CC_6H_5{\cdot}6AgC_2F_5CO_2{\cdot}5CH_3CN$. (b) Coordination mode of the $C_6H_5C{\equiv}C^-$ ligand in $AgC{\equiv}CC_6H_5{\cdot}3AgCF_3CO_2{\cdot}CH_3CN$. The silver column is connected by edge-sharing between adjacent square-pyramidal Ag_5 aggregates, and continuous π–π stacking of phenyl rings occurs on one side of the column. From L. Zhao, W.-Y. Wong and T. C. W. Mak, *Chem. Eur. J.* **12**, 4865–72 (2006).

are linked by two pentafluoropropionate groups via μ_3-O, O′, O′ and μ_2-O, O′ coordination modes, respectively, to generate an infinite column along the [111] direction. No π–π interaction is observed in this complex.

An infinite array of parallel phenyl rings stabilized by π–π stacking (center-to-center distance 418.9 pm) occurs in the complex $AgC{\equiv}CC_6H_5{\cdot}3AgCF_3CO_2{\cdot}CH_3CN$ [Fig. 20.4.36(b)]. The capping square-planar Ag_5 baskets are fused through argentophilic interactions via edge-sharing to form an infinite coordination column along the [100] direction, with continuous π–π stacking of phenyl rings lying on the same side of the column.

When substituents are introduced into the phenyl group, the π–π stacking between consecutive aromatic rings is affected by the size of the substituted groups and their positions. In the crystal structure of $2AgC{\equiv}CC_6H_4Me\text{-}4{\cdot}6AgCF_3CO_2{\cdot}1.5CH_3CN$ [Fig. 20.4.37(a)], the methyl group has a little influence on the formation of a silver column stabilized by π–π stacking, which is almost totally identical with the structure of $AgC{\equiv}CC_6H_5{\cdot}3AgCF_3CO_2{\cdot}CH_3CN$ [see Fig. 20.4.36(b)]. However, when a more bulky *tert*-butyl group is employed, the π–π stacking system is interrupted despite the formation of a similar silver chain [Fig. 20.4.37(b)]. The entire $C_6H_4^t Bu\text{-}4$ moiety rotates around the $C(^tBu)$–C(phenyl) single bond to generate a highly disordered structure. On the other hand, putting a *meta*-methyl group on the phenyl ring can form C–H···π interaction, but the constitution of the silver chain is changed from edge-sharing to vertex-sharing [Fig. 20.4.37(c)]. Finally, use of the 2-methyl-substituted phenyl ligand entirely destroys the π–π stacking and even breaks the Ag···Ag interactions between Ag_n caps to form a silver chain connected by trifluoroacetate groups [Fig. 20.4.37(d)].

20.4.7 Self-assembly of nanocapsules with pyrogallol[4]arene macrocycles

Recent studies have shown that the bowl-shaped *C*-alkyl substituted pyrogallol[4]arene macrocycles readily self-assemble to form a gobular hexameric cage, which is structurally robust and remains stable even in aqueous media (Fig. 20.4.38). Slow evaporation of a solution of *C*-heptylpyrogallol[4]arene in ethyl acetate gives crystalline [(*C*-heptylpyrogallol[4]arene)$_6$(EtOAc)$_6$

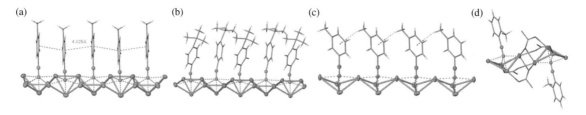

Fig. 20.4.37.
(a) Silver column in 2AgC≡CC$_6$H$_4$Me-4 · 6AgCF$_3$CO$_2$ · 1.5CH$_3$CN connected by the fusion of square-pyramidal Ag$_5$ baskets and stabilized by continuous π–π stacking of phenyl rings. (b) Similar silver column in AgC≡CC$_6$H$_4^t$Bu-4·3AgCF$_3$CO$_2$·CH$_3$CN connected only by argentophilic interaction. (c) Silver chain in AgC≡CC$_6$H$_4$Me-3·2AgCF$_3$SO$_3$ through atom sharing. (d) Silver chain in AgC≡CC$_6$H$_4$Me-2·4AgCF$_3$CO$_2$·H$_2$O through the connection of trifluoroacetate groups.

(H$_2$O)]·6EtOAc, and X-ray analysis revealed that the large spheroidal supermolecule is stabilized by a total of 72 O–H···O hydrogen bonds (four intramolecular and eight intermolecular per macrocycle building block). The nano-sized molecular capsule, having an internal cavity volume of about 1.2 nm^3, contains six ethyl acetate molecules and one water molecule; the methyl terminal of each encapsulated ethyl acetate guest molecule is orientated toward a bulge on the surface, and the single guest water molecule resides at the center of the capsule. In the crystal structure, the external ethyl acetate solvate molecules are embedded within the lower rim alkyl legs at the base of each of the macrocycles, and the nanocapsules are arranged in hexagonal closest packing.

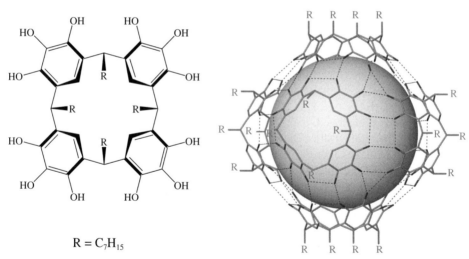

R = C$_7$H$_{15}$

Fig. 20.4.38.
Assembly of six *C*-heptylpyrogallol[4]arene molecules by intra- and intermolecular hydrogen bonds to form a globular supermolecule with a host cavity of volume ∼ 1.2 nm^3. H atoms are omitted for clarity, and hydrogen bonds are represented by dotted lines. From G. V. C. Cave, J. Antesberger, L. J. Barbour, R. M. McKinley and J. L. Atwood, *Angew. Chem. Int. Ed.* **43**, 5263-6 (2004).

With reference to the unique architecture of this hydrogen-bonded hexameric capsule **I** [Fig. 20.4.39(a)], it was noted that the pyrogallol[4]arene building block has the potential of serving as a multidentate ligand through deprotonation of some of the upper rim phenolic groups. As envisaged, treatment of

(a) 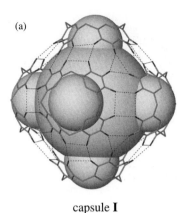 (b)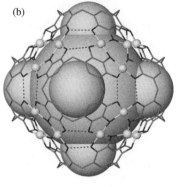

capsule **I**   capsule **II**

Fig. 20.4.39.
Hydrogen-bonded supramolecular capsule **I** compared with coordination capsule **II**. The external aliphatic groups are omitted for clarity. Note that a pair of intermolecular O–H···O hydrogen bonds is replaced by four square-planar Cu–O coordination bonds in the metal-ion insertion process that generates the isostructural inorganic analog. From R. M. McKinley, G. V. C. Cave and J. L. Atwood, *Proc. Nat. Acad. Sci.* **102**, 5944-8 (2005).

C-propan-3-ol pyrogallol[4]arene with four equivalents of $Cu(NO_3)_2 \cdot 3H_2O$ in a mixture of acetone and water yielded a large neutral coordination capsule **II** $[Cu_{24}(H_2O)_x(C_{40}H_{40}O_{16})_6 (acetone)_n]$ where $x \geq 24$ and $n = 1$–6. Single-crystal X-ray analysis established that retro-insertion of 24 Cu(II) metal centers into the hexameric framework results in substitution of 48 of the 72 phenolic protons, leaving the remaining 24 intact for intramolecular hydrogen bonding (O···O 0.2400–0.2488 nm). As shown in Fig. 20.4.39(b), the large coordination capsule **II** may be viewed as an octahedron with the six 16-membered macrocyclic rings located at its corners, and each of its eight faces is capped by a planar cyclic $[Cu_3O_3]$ unit of dimensions Cu–O 0.1911 to 0.1980 nm, O–Cu–O 85.67° to 98.23°, and Cu–O–Cu 140.96° to 144.78°.

Definitive location of all guest molecules inside the cavity is somewhat ambiguous owing to inexact stoichiometry and disorder. However, the (+)-MALDI mass spectra of **II** indicate that each individual capsule encloses different mixtures of water and acetone. In particular, two peaks implicated the presence of 24 entrapped water molecules that occupy axial coordination sites orientated toward the center of the cavity.

The pair of supramolecular capsules **I** and **II** represents a landmark in the construction of large coordination cages using multicomponent ligands. This elegant blueprint approach is facilitated by the robustness of the hydrogen-bonded assembly **I**, which serves as a template for metal-ion insertion at specific sites with conservation of structural integrity. Notably, capsule size is virtually unchanged as the center-to-corner distance of the four phenolic O atoms belonging to the eight-membered intermolecular hydrogen-bonded ring (0.1883–0.1976 nm) in **I** closely matches the Cu–O bond distance of 0.1913–0.1978 nm in **II**.

20.4.8 Reticular design and synthesis of porous metal–organic frameworks

The design and synthesis of metal–organic frameworks (MOFs) has yielded a large number of solids that possess useful gas and liquid adsorption properties. In particular, highly porous structures constructures constructed from discrete

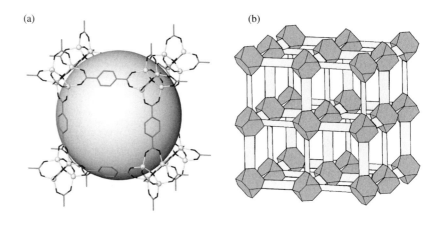

Fig. 20.4.40.
Crystal structure of MOF-5. (a) Zn_4O tetrahedra joined by benzenedicarboxylate linkers. H atoms are omitted for clarity. (b) The topology of the framework (primitive cubic net) shown as an assembly of $(Zn_4O)O_{12}$ clusters (represented as truncated tetrahedra) and *p*-phenylene ($-C_6H_4-$) links (represented by rods). From O. M. Yaghi, M. O'Keeffe, N. W. Ockwig, H. K. Chae, M. Eddaoudi, and J. Kim, *Nature* **423**, 705-14 (2003).

metal–carboxylate clusters and organic links have been demonstrated to be amenable to systematic variation in pore size and functionality.

Consider the structure of the discrete tetranuclear $Zn_4O(CH_3CO_2)_6$ molecule, which is isostructural with $Be_4O(CH_3CO_2)_6$, the structure of which is shown in Fig. 9.5.2. Replacement of each acetate ligand by one half of a linear dicarboxylate produces a molecular entity that can conceivably be interconnected to identical entities to generate an infinite coordination network. An illustrative example is $Zn_4O(BDC)_6$, referred to as MOF-5, which is prepared from Zn(II) and benzene-1,4-dicarboxylic acid (H_2BDC) under solvothermal conditions. In the crystal structure, a $Zn_4O(CO_2)_6$ fragment comprising four fused ZnO_4 tetrahedra sharing a common vertex and six carboxylate C atoms constitute a "secondary building unit" (SBU). Connection of such octahedral SBUs by mutually perpendicular *p*-phenylene ($-C_6H_4-$) links leads to an infinite primitive cubic network, as shown in Fig. 20.4.40. Alternatively, the smaller Zn_4O fragment (an oxo-centered Zn_4 tetrahedron) can be regarded as the SBU and the corresponding organic linker is the whole benzene-1,4-dicarboxylate dianion.

The resulting MOF-5 structure has exceptional stability and porosity as both the SBU and organic link are relatively large and inherently rigid. This reticular [which means 'having the form of a (usually periodic) net'] design strategy, based on the concept of discrete SBUs of different shapes (triangles, squares, tetrahedra, octahedra, etc.) considered as "joints" and organic links considered as "struts", has been applied to the synthesis and utilization of a vast number of MOF structures exhibiting varying geometries and network topologies. Based on the $Zn_4O(CO_2)_6$ SBU in the prototype MOF-5 (also designated as IRMOF-1), a family of isoreticular and isostructural cubic frameworks with diverse pore sizes and functionalities has been constructed, including IRMOF-6, IRMOF-8, IRMOF-11, and IRMOF-16, which are illustrated in Fig. 20.4.41.

An example of a porous framework that is isoreticular, but not isostructural, with MOF-5 is MOF-177, which incorporates the extended organic linker 1,3,5-benzenetribenzoate (BTB). The framework of crystalline MOF-177, $Zn_4O(BTB)_2 \cdot (DEF)_{15}(H_2O)_3$, where DEF = diethyl formamide, has an ordered structure with an estimated surface area of $4,500 \text{ m}^2\text{ g}^{-1}$, which greatly exceeds those of zeolite Y ($904 \text{ m}^2\text{g}^{-1}$) and carbon ($2,030 \text{ m}^2\text{ g}^{-1}$).

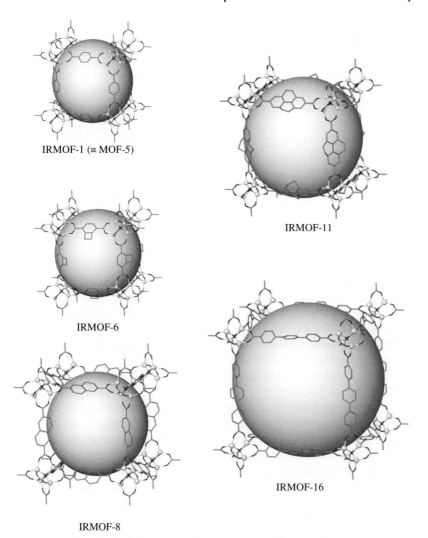

Fig. 20.4.41.
Comparison of cubic fragments in the respective three-dimensional extended structures of IRMOF-1 (≡ MOF-5), IRMOF-6, IRMOF-8, IRMOF-11, and IRMOF-16. From M. Eddaoudi, J. Kim, N. Rosi, D. Vodak, J. Wachter, M. O'Keeffe and O. M. Yaghi, *Science* **295**, 469-72 (2004).

As shown in Fig. 20.4.42, the underlying topology of MOF-177 is a (6,3)-net with the center of the octahedral $Zn_4O(CO_2)_6$ cluster as the six-connected node and the center of the BTB unit as the three-connected node. Its exceptionally large pores are capable of accommodating polycyclic organic guest molecules such as bromobenzene, 1-bromonaphthalene, 2-bromonaphthalene, 9-bromoanthracene, C60, and the polycyclic dyes Astrazon Orange R and Nile Red.

Furthermore, the ability to prepare these kinds of MOFs in high yield and with adjustable pore size, shape, and functionality has led to their exploration as gas storage materials. Thermal gravimetric and gas sorption experiments have shown that IRMOF-6, bearing a fused hydrophobic unit C_2H_4 in its organic link, has the optimal pore aperture and rigidity requisite for maximum uptake of methane. Activation of the porous framework was achieved by exchanging the included guest molecules with chloroform, which was then removed by

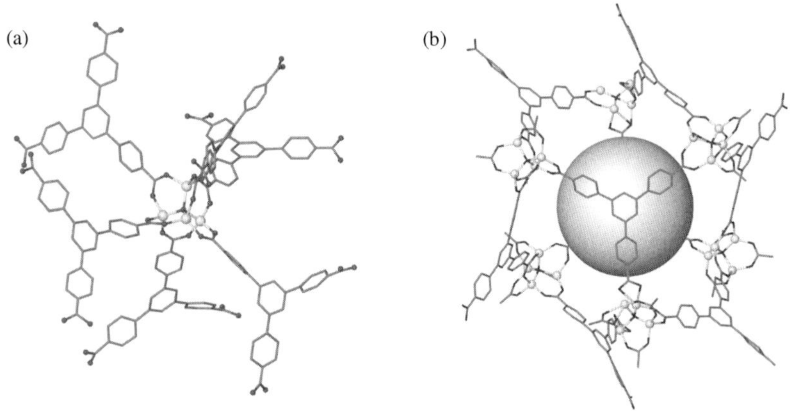

Fig. 20.4.42.
Crystal structure of MOF-177. (a) A central Zn$_4$O unit coordinated by six BTB ligands. (b) The structure viewed down [001]. From H. K. Chae, D. Y. Siberio-Pérez, J. Kim, Y. B. Go, M. Eddaoudi, A. J. Matzger, M. O'Keeffe and O. M. Yaghi, *Nature* **427**, 523-7 (2004).

gradual heating to 800°C under an inert atmosphere. The evacuated framework has a stability range of 100-400°C, and the methane sorption isotherm measured in the range 0-40 atm at room temperature has an uptake of 240 cm^3(STP)/g [155 cm^3(STP)/cm^3] at 298 K and 36 atm. On a volume-to-volume basis, the amount of methane sorbed by IRMOF-6 at 36 atm amounts to 70% of that stored in compressed methane cylinders at ∼205 atm. As compared to IRMOF-6, IRMOF-1 under the same conditions has a smaller methane uptake of 135 cm^3(STP)/g.

The isoreticular MOFs based on the Zn$_4$O(CO$_2$)$_6$ SBU also possesses favorable sorption properties for the storage of molecular hydrogen. At 77 K, microgravimetric sorption measurements gave H$_2$ (mg/g) values of 13.2, 15.0, 16.0, and 12.5 for IRMOF-1, IRMOF-8, IRMOF-11, and MOF-177, respectively. At the highest pressures attained in the measurements, the maximum uptake values for these frameworks are 5.0, 6.9, 9.3, and 7.1 molecules of H$_2$ per Zn$_4$OL$_x$ unit where L stands for a linear dicarboxylate.

MOF-177 has been demonstrated to act like a super sponge in capturing vast quantities of carbon dioxide at room temperature. At moderate pressure (about 35 bar), its voluminous pores result in a gravimetric CO$_2$ uptake capacity of 33.5 mmol/g, which far exceeds those of the benchmark adsorbents zeolite 13X (7.4 mmol/g at 32 bar) and activated carbon MAXSORB (25 mmol/g at 35 bar). In terms of volume capacity, a container filled with MOF-177 can hold about twice the amount of CO$_2$ versus the benchmark materials, and 9 times the amount of CO$_2$ stored in an empty container under the same conditions of temperature and pressure.

The previous structural and sorption studies of MOFs are all based on the discrete Zn$_4$O(CO$_2$)$_6$ SBU. Recent development has demonstrated that rod-shaped metal–carboxylate SBUs can also give rise to a variety of stable solid-state architectures and permanent porosity. Three illustrative examples are presented below.

In the crystal structure of Zn$_3$(OH)$_2$(BPDC)$_2$·(DEF)$_4$(H$_2$O)$_2$ where BPDC = 4,4'-biphenyldicarboxylate (MOF-69A), there are tetrahedral and octahedral Zn(II) centers coordinated by four and two carboxylate groups, respectively, in the *syn,syn* mode, with each μ_3-hydroxide ion bridging three metal centers [Fig. 20.4.43(a)]. The infinite Zn–O–C rods are aligned in parallel fashion and laterally connected to give a three-dimensional network [Fig. 20.4.43(b)], forming rhombic channels of edge 1.22 nm and 1.66 nm along the longer diagonal, into which the DMF and water guest molecules are fitted.

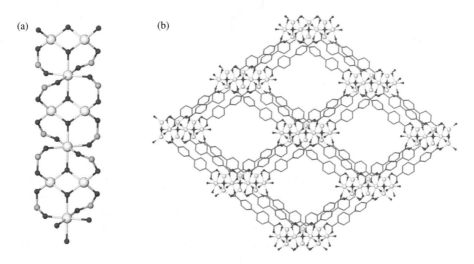

Fig. 20.4.43.
Channel structure of MOF-69A: (a) ball-and-stick representation of inorganic SBU; (b) SBUs connected by biphenyl links. The DEF and water molecules have been omitted for clarity. From N. L. Rosi, J. Kim, M. Eddaoudi, B. Chen, M. O'Keeffe and O. M. Yaghi, *J. Am. Chem. Soc.* **127**, 1504–18 (2005).

The Mn–O–C rods in MOF-73, $Mn_3(BDC)_3 \cdot (DEF)_2$, are constructed from a pair of linked six-coordinate Mn(II) centers [Fig. 20.4.44(a)]. One metal center is bound by two carboxylate groups acting in the *syn,syn* mode, one in the bidentate chelating mode, and a fourth one in the *syn,anti* mode. The other metal center has four carboxylates bound in the *syn,syn* mode and two in the *syn,anti* mode. Each rod is built of corner-linked and edge-linked

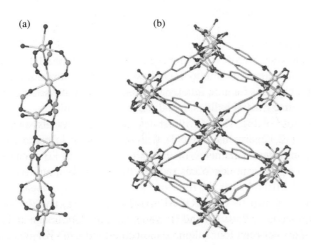

Fig. 20.4.44.
Channel structure of MOF-73: (a) ball-and-stick representation of inorganic SBU; (b) SBUs connected by *p*-phenylene links. The DEF molecules have been omitted for clarity. From N. L. Rosi, J. Kim, M. Eddaoudi, B. Chen, M. O'Keeffe and O. M. Yaghi, *J. Am. Chem. Soc.* **127**, 1504–18 (2005).

MnO$_6$ octahedra, and is connected to four neighboring rods by *p*-phenylene links. The resulting three-dimensional host framework [Fig. 20.4.44(b)] has rhombic channels of dimensions 1.12 × 0.59 nm filled by the DEF guest molecules.

The structure of MOF-75, Tb(TDC)·(NO$_3$)(DMF)$_2$ where TDC = 2,5-thiophenedicarboxylate, contains eight-coordinate Tb(III) bound by four carboxylate groups all acting in the *syn,syn* mode, one bidentate nitrate ligand, and two terminal DMF ligands [Fig. 20.4.45(a)]. The Tb–O–C rod orientated in the *a* direction consists of linked TbO$_8$ bisdisphenoids with the carboxyl carbon atoms forming a twisted ladder. Lateral linkage of rods in the *b* and *c* directions by the thiophene units generate rhombic channels measuring 0.97 × 0.67 nm, as illustrated in Fig. 20.4.45(b), which accommodate the DMF guest molecules and nitrate ions.

20.4.9 One-pot synthesis of nanocontainer molecule

Dynamic covalent chemistry has been used in an atom-efficient self-assembly process to achieve a nearly quantitative one-pot synthesis of a nanoscale molecular container with an inner cavity of approximately 1.7 nm^3.

In a thermodynamically driven, trifluoroacetic acid catalyzed reaction in chloroform, six cavitands **1** and twelve ethylenediamine linkers condense to generate an octahedral nanocontainer **2**, as shown in Fig. 20.4.46. Each cavitand has 4 formyl groups on its rim, and these react with the 24 amino groups of the linkers to form 24 imine bonds. After reduction of all the imine bonds with NaBH$_4$, the hexameric nanocontainer can be isolated via reversed-phase HPLC as the trifluoroacetate salt **2**·24CF$_3$COOH in 63% yield based on **1**. Elemental analysis of the white solid corresponds to the stoichiometric formula **2**·24CF$_3$COOH·9H$_2$O. The simplified ^{1}H and ^{13}C NMR spectra of **2** and **2**·24CF$_3$COOH are consistent with their octahedral symmetry.

If the same reaction is carried out with either 1,3-diaminopropane or 1,4-diaminobutane in place of ethylenediamine, the product is an octaimino hemicarcerand composed of two face-to-face cavitands connected by four diamino bridging units.

20.4.10 Filled carbon nanotubes

Much research has been devoted to the insertion of different kinds of crystalline and non crystalline material into the hollow interior of carbon nanotubes. The encapsulated species include fullerenes, clusters, one-dimensional (1D) metal nanowires, binary metal halides, metal oxides, and organic molecules.

The left side of Fig. 20.4.47 illustrates the van der Waals surfaces of the (10,10) and (12,12) armchair SWNTs with diameters of 1.36 and 1.63 nm and corresponding internal diameters of approximately 1.0 and 1.3 nm, respectively, as specified by the van der Waals radii of the sp^2 carbon atoms forming the walls. The (10,10) tube has the right size to accommodate a linear array of C$_{60}$ molecules, as shown by the HRTEM images marked (a) and (b) on the right side of Fig. 20.4.47. Parts (c) and (d) show the HRTEM image and modeling of an interface between four fullerene molecules and a 1D FeI$_2$ crystal.

Supramolecular Structural Chemistry

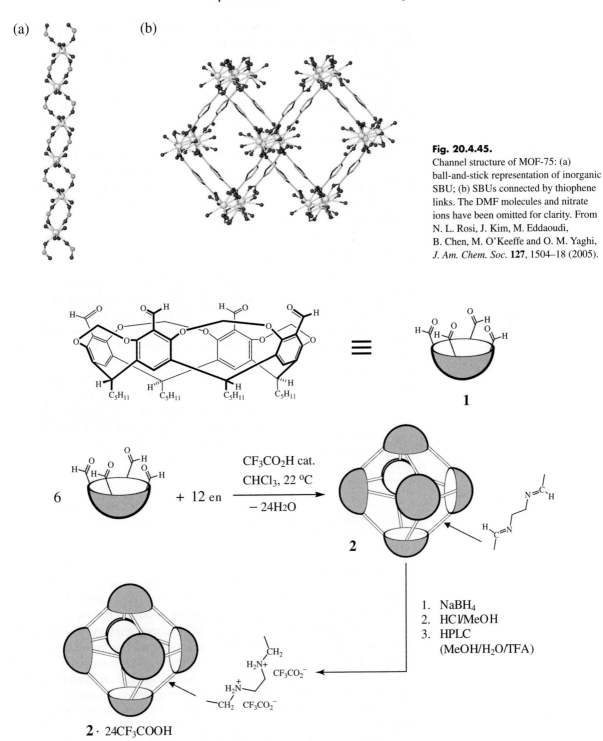

Fig. 20.4.45.
Channel structure of MOF-75: (a) ball-and-stick representation of inorganic SBU; (b) SBUs connected by thiophene links. The DMF molecules and nitrate ions have been omitted for clarity. From N. L. Rosi, J. Kim, M. Eddaoudi, B. Chen, M. O'Keeffe and O. M. Yaghi, *J. Am. Chem. Soc.* **127**, 1504–18 (2005).

Fig. 20.4.46.
Reaction scheme showing the thermodynamically controlled condensation of tetraformylcavitand **1** with ethylenediamine to form an octahedral nanocontainer **2**, which undergoes reduction to yield the trifluoroacetate salt **2·24CF$_3$COOH**.

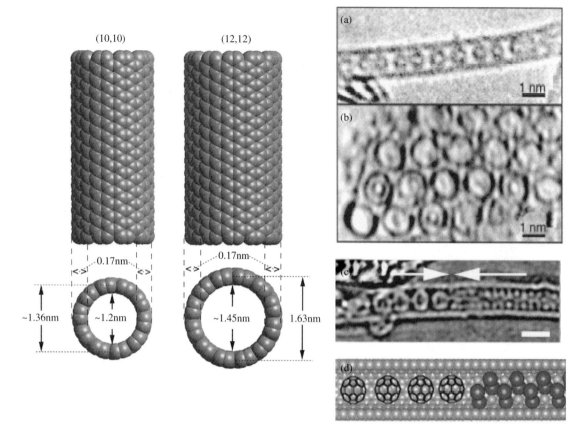

Fig. 20.4.47.
Left: Schematic representations of the van der Waals surfaces of (10,10) and (12,12) armchair SWNTs. Right: (a) HRTEM image showing a (10,10) SWNT filled with C_{60} molecules. (b) Second image from the same specimen as in (a), showing a cross-sectional view of an ordered bundle of SWNTs, some of which are filled with fullerene molecules. (c) HRTEM image (scale bar = 1.5 nm) of a filled SWNT showing an interface between four C_{60} molecules (a possible fifth molecule is obscured at the left) and a 1D FeI_2 crystal. (d) Schematic structural representation of (c) (Fe atoms = small spheres; I atoms = large spheres). Courtesy of Professor M. L. H. Green.

The formation of an ordered 2 × 2 KI crystal column within a (10,10) SWNT with $D \sim 1.4$ nm is shown in Fig. 20.4.48. The structural model is illustrated in (a), and a cross-sectional view is shown in (b). The coordination numbers of the cation and anion are changed from 6:6 in bulk KI to 4:4 in the 2 × 2 column. Each dark spot in the HRTEM image (c) represents an overlapping I–K or K–I arrangement viewed in projection, which matches the simulated image (d). The spacing between spots along the SWNT is ~0.35 nm, corresponding to the {200} spacing in bulk KI, whereas across the SWNT capillary the spacing increases to ~0.4 nm, representing a ~17% tetragonal expansion.

The incorporation of a 3 × 3 KI crystal in a wider SWNT ($D \sim 1.6$ nm) is depicted in Fig. 20.4.49. In this case, three different coordination types (6:6, 5:5, and 4:4) are exhibited by atoms forming the central, face, and corner ···I–K–I–K··· rows of the 3 × 3 column, respectively, along the tube axis. The <110> projection shows the K^+ and I^- sublattices as pure element columns,

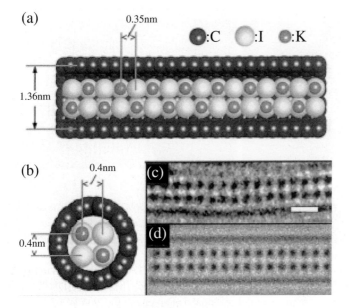

Fig. 20.4.48.
A 2 × 2 KI crystal column filling a (10,10) SWNT. (a) Cutaway structural representation of composite model used in the simulation calculations. (b) End-on view of the model, showing an increased lattice spacing of 0.4 nm across the capillary in two directions (assuming a symmetrical distortion). (c) HRTEM image. (d) Simulated Image. Courtesy of Professor M. L. H. Green.

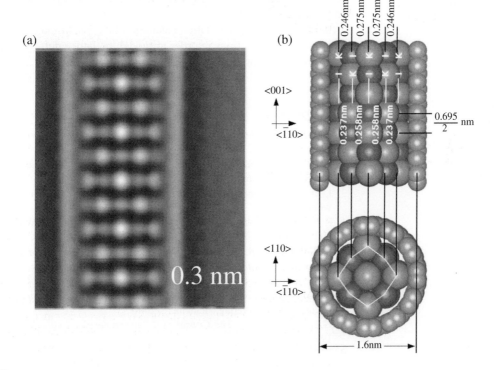

Fig. 20.4.49.
(a) Reconstructed HRTEM image (averaged along the tube axis) of the <110> projection of a 3 × 3 KI crystal in a $D \sim 1.6$ nm diameter SWNT. Note that the contrast in this image is reversed so that regions of high electron density appear bright and low electron density appear dark. (b) Structural model derived from (a). Courtesy of Professor M. L. H. Green.

Fig. 20.4.50.
(a) An isolated DWNT filled with KI. (b) Reconstructed image of the KI@DWNT composite; the marked area was structurally analyzed in detail. (c) Crystal model showing the distortions imposed and the three subsections considered for its construction. (d) Structural model of the marked section in (b). (e) Reconstructed versus simulated image of the three sections. Courtesy of Professor M. L. H. Green.

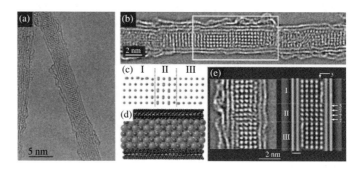

which are distinguishable by their scattering powers. The iodine atoms located along <110> all show a slight inward displacement relative to their positions in bulk KI, whereas the K atoms located along the same cell diagonal exhibit a small expansion.

It was found that KI can also be used to fill a DWNT, as shown in Fig. 20.4.50. The HRTEM image (a) and reconstructed image (b) of the KIDWNT composite was analysed using a model composed of an inner (11, 22) SWNT ($D \sim 2.23$ nm) and an outer (18, 26) SWNT ($D \sim 3.04$ nm). The measured averaged spacings between the atomic columns gave values of 0.37 nm across the DWNT axis and 0.36 nm along it, which agree well with the {200} d-spacing of rock-salt KI (0.352 nm). In the simulation calculation, the encapsulated fragment was divided into three sections in order to model the lattice defects such as plane shear and plane rotation.

References

1. G. A. Jeffrey, *An Introduction to Hydrogen Bonding*, Oxford University Press, New York, 1997.
2. G. R. Desiraju and T. Steiner, *The Weak Hydrogen Bond in Structural Chemistry and Biology*, Oxford University Press, New York, 1999.
3. J.-M. Lehn, *Supramolecular Chemistry: Concepts and Perspectives*, VCH, Weinheim, 1995.
4. P. J. Cragg, *A Practical Guide to Supramolecular Chemistry*, Wiley, Chichester, 2005.
5. J. W. Steed, D. R. Turner and K. J. Wallice, *Core Concepts in Supramolecular Chemistry and Nanochemistry*, Wiley, Chichester, 2007.
6. J. W. Sneed and J. L. Atwood, *Supramolecular Chemistry*, Wiley, Chichester, 2000.
7. K. Ariga and T. Kunitake, *Supramolecular Chemistry—Fundamentals and Applications*, Springer-Verlag, Heidelberg, 2006.
8. G. R. Desiraju (ed.), *The Crystal as a Supramolecular Entity*, Wiley, Chichester, 1996.
9. G. R. Desiraju (ed.), *Crystal Design: Structure and Function*, Wiley, Chichester, 2003.
10. A. Bianchi, K. Bowman-James and E. Garcia-España (eds.), *Supramolecular Chemistry of Anions*, Wiley–VCH, New York, 1997.
11. J.-P. Sauvage (ed.), *Transition Metals in Supramolecular Chemistry*, Wiley, Chichester, 1999.

12. M. Fujita (ed.), *Molecular Self-assembly: Organic versus Inorganic Approaches* (Structure and Bonding, vol. 96), Springer, Berlin, 2000.
13. F. Toda and R. Bishop (eds.), *Separations and Reactions in Organic Supramoilecular Chemistry*, Wiley, Chichester, 2004.
14. N. Yui (ed.), *Supramolecular Design for Biological Applications*, CRC Press, Boca Raton, FL, 2003.
15. I. Haiduc and F. T. Edelmann (eds.), *Supramolecular Organometallic Chemistry*, Wiley–VCH, Weinheim, 1999.
16. W. Jones and C. N. R. Rao (eds.), *Supramolecular Organization and Materials Design*, Cambridge University Press, Cambridge, 2002.
17. D. Wöhrle and A. D. Pomogailo, *Metal Complexes and Metals in Macromolecules*, Wiley–VCH, Weinheim, 2003.
18. E. R. T. Tiekink and J. J. Vittal (eds.), *Frontiers in Crystal Engineering*, Wiley, Chichester, 2006.
19. J. L. Atwood and J. W. Steed (eds.), *Encyclopedia of Supramolecular Chemistry*, Marcel-Dekker, New York, 2004.
20. D. N. Chin, J. A. Zerkowski, J. C. MacDonald and G. M. Whitesides, Strategies for the design and assembly of hydrogen-bonded aggregates in the solid state, in J. K. Whitesell (ed.), *Organised Molecular Assemblies in the Solid State*, Wiley, Chichester, 1999.
21. G. A. Ozin and A. C. Arsenault, *Nanochemistry: A Chemistry Approach to Nanomaterials*, RSC Publishing, Cambridge, 2005.
22. Q. Li and T. C. W. Mak, Novel inclusion compounds with urea/thiourea/selenourea-anion host lattices, in M. Hargittai and I. Hargittai (eds.), *Advances in Molecular Structure Research*, vol. 4, pp. 151–225, Stamford, Connecticut, 1998.
23. M. I. Bruce and P. J. Low, Transition metal complexes containing all-carbon ligands. *Adv. Organomet. Chem.* **50**, 179–444 (2004).
24. V. W. W. Yam and E. C. C. Cheng, Silver organometallics, in D. M. P. Mingos and R. H. Crabtree (eds.), *Comprehensive Organometallic Chemistry III*, vol. 2 (K. Meyer, ed.), pp. 197–249, Elsevier, Oxford, 2007.
25. T. C. W. Mak, X.-L. Zhao, Q.-M. Wang and G.-C. Guo, Synthesis and structural characterization of silver(I) double and multiple salts containing the acetylenediide dianion. *Coord. Chem. Rev.* **251**, 2311–33 (2007).
26. L. Zhao and T.C.W. Mak, Multinuclear silver-ethynide supramolecular synthons for the construction of coordination networks. *Chem. Asian J.* **2**, 456–467 (2007).
27. L. Zhao, X.-L. Zhao and T. C. W. Mak, Assembly of infinite silver(I) columns, chains, and bridged aggregates with supramolecular synthon bearing substituted phenylethynides, *Chem. Eur. J.* **13**, 5927–36 (2007).
28. J. Lagona, P. Mukhopadhyay, S. Charkrabarti and L. Issacs, The cucurbit[*n*]uril family. *Angew. Chem. Int. Ed.* **44**, 4844–70 (2005).
29. N. W. Ockwig, O. Delgardo-Friedrichs, M. O'Keeffe and O. M. Yaghi, Reticular chemistry: occurrence and taxonomy of nets and grammar for the design of frameworks. *Acc. Chem. Res.* **38**, 176–82 (2005).
30. X. Liu, Y. Liu, G. Li and R. Warmuth, One-pot, 18-component synthesis of an octahedral nanocontainer molecule. *Angew. Chem. Int. Ed.* **45**, 901–4 (2006).
31. J. Sloan, A. I. Kirkland, J. L. Hutchison and M. L. H. Green, Integral atomic layer architectures of 1D crystals inserted into single walled carbon nanotubes. *Chem. Commun.*, 1319–32 (2002).
32. J. Sloan, A. I. Kirkland, J. L. Hutchison and M. L.H. Green, Aspects of crystal growth within carbon nanotubes. *C. R. Physique* **4**, 1063–74 (2003).

33. P. M. F. J. Costa, S. Friedrichs, J. Sloan and M. L. H. Green, Imaging lattice defects and distortions in alkali-metal iodides encapsulated within double-walled carbon nanotubes. *Chem. Mater.* **17**, 3122–9 (2005).
34. P. M. F. J. Costa, J. Sloan and M. L. H. Green, Structural studies on single- and double-walled carbon nanotubes filled with ionic and covalent compounds. *Ciencia e Tecnologia dos Materiais* **18**, no. 3–4, 78–82 (2006).

Index

ab initio methods, 142
acentric crystal classes, 304
acetylenediide, 530
acetylenediide, alkaline-earth, 530
acetylenediide, copper(I) complex, 532
acetylenediide, ternary metal, 530
acetylenide, 530
acid hydrates, 626
active space, 146
adamantane, disordered cubic form, 357, 361
adamantane, ordered tetragonal form, 358, 361
adamantane-1,3,5,7-tetracarboxylic acid, 741
adamantane-like cage compounds, 358, 362
adamantane-like molecular skeleton, 174
agostic bond, 402, 425
agostic-bond complexes, 428
alkali metal complexes, 436
alkali metal oxides, 433
alkali metal suboxides, 433
alkalides, 447
alkaline-earth metallocenes, 455
alkaline-earth oxides, 367
allophanate, 781
alnicos, 392
alums, $M^I M^{III}(XO_4)_2 \cdot 12H_2O$, 353, 356
ammonia, 578
amorphous carbon, 506
angular correlation, 47
angular momentum quantum number, 31, 55
angular momentum, orbital, 55
angular momentum, spin, 55
angular momentum, total, 56
angular wavefunction, 30, 31, 33
anionic carbonyl cluster, 718
antibonding effect, 83
antibonding molecular orbital, 85, 89
argentophilic interaction, 785, 792, 795, 796, 797
argentophilicity (argentophilic attraction), 724
asymmetric molecule, 170
asymmetric unit, 319, 323
atomic orbital, 8
atomic orbitals, 31
atomic orbitals, energies of, 55
atomic units, 42
aurophilicity (or aurophilic attraction), 721
axial ratios, 301

azide ion, coordination modes, 562
azide ion, N_3^-, 562
azides, 562

band gap, 130
bands d–d, 271, 292
band theory, 128
barium titanate, $BaTiO_3$, 388
basic beryllium acetate, $Be_4O(CH_3COO)_6$, 336, 338
basis set, 142
benzene, supercrowded, 510
benzene-1,4-dicarboxylic acid, 800
benzenetribenzoate, 1,3,5-, 800
beryl, $Be_3Al_2[Si_6O_{18}]$, 351, 353
betaine, 787
binaphthyl, 1,1'-, 345
binaphthyl, 1,1'-, 347
biphenyl, 333, 336, 337
bismuth-bismuth double bond, 606
bismuthine, tetrameric, 605
bismuthonium ylide, 604
body-centered cubic packing (bcp), 381
Bohr radius, 7, 34
boiling points of metallic elements, 133
Boltzmann weighting factor, 135
bond order, 94
bond valence, 703
bond valence of metal-metal bond, 711
bond-angle relationships for T_d molecules, 175
bonding effect, 111
bonding molecular orbital, 84
boranes, 470
boranes, macropolyhedral, 479
boranes, metalla, 483
boranes, topological description, 471
boranes, *arachno*, 474
boranes, *hypercloso*, 483
boranes, *hypho*, 473, 475
boranes, *closo*, 477
boranes, *nido*, 474
borate, dihydrogen, 487
borates, structural principles, 489
borates, structural units, 487
borax, 489
borazine, 468
boric acid, 486

borides, metal, 464
borides, non-metal, 467
borides, rare-earth metal, 466
Born-Landé equation, 124
Born-Mayer equation, 124
boron carbides, 467
boron halides, 469
boron nitrides, 468
boron, α-R12, 461
boron, α-rhombohedral, 350, 352
boron, β-R105, 463
boron, β-R105 electronic structure, 481
boron, B_{12} icosahedron, 461
Borromean link, 753
Bravais lattices, 306, 309
butadiene, cyclization, 113
butadiene, equilibrium bond length, 113
butadiynediide, 792

cadmium iodide, CdI_2, 377
calcium carbide (Form I), CaC_2, 346, 348
calcium carbide, CaC_2, 369
calcium carbide, modifications, 530
calcium nitridoberyllate, $Ca[Be_2N_2]$, 346, 348
calix[4]arene bismelamine, 744
calomel, Hg_2Cl_2, 346, 348
carbide complex, terminal, 528
carbide-centered carbonyl clusters, 528
carbon atom, naked, 527
carbon black (soot), 506
carbon bridgehead, inverted bond configuration, 525
carbon fibers, 506
carbon nanotube, double-walled (DWNT), 807
carbon nanotube, filled, 804
carbon nanotube, multi-walled (MWNT), 508
carbon nanotube, single-walled (SWNT), 507, 804
carbon onions, 506
carbon, activated, 506
carbon, bond lengths, 520
carbon, coordination numbers, 520
carbon, covalent bond types, 517
carbon, hybridization schemes, 518
carbon-carbon bonds, abnormally long, 524
carbon-carbon single bonds, abnormally short, 526
carbonyl stretching modes in metal complexes, 246
carborane, actina, 485
carboranes, 470
carboranes, metalla, 483
carboxylic acid dimer synthon, 740
carcerand, 753
catenane, 753
cavitand, 804

centrifugal distortion constant, 159
cesium chloride, CsCl, 384
cesium iron fluoride, $Cs_3Fe_2F_9$, 352, 354
character tables, 180, 183
charge transfer transitions, 271
charge transfer, L $\rightarrow$ M, 291
charge transfer, M $\rightarrow$ L, 291
chelating β-diketiminate ligand, 494
cisplatin, 340, 342
clathrate hydrate, type I, $6G_L \cdot 2G_S \cdot 46H_2O$, 361, 365
clathrate hydrate, type II, $8G_L \cdot 16G_S \cdot 136H_2O$, 360, 363
clathrate hydrates, 625
cluster complexes, of Ge, Sn and Pb, 551
cluster compounds, 703
coal, 506
coke, 506
color center, 20
color center, or F-center, 368
complete basis set (CBS) methods, 151
complexes containing naked carbon atom, 527
composite methods, 151
concerted reactions, 113
condensed phase, 139
conduction band, 129
configuration energy, 67
configuration interaction (CI), 145
conglomerate, 338, 340
conrotatory process, 114
conservation of orbital symmetry, 113
coordinates of equipoints, 313, 317
coordination compounds of Group 2 elements, 451
coordination polymer, 757, 776
correlation energy, 145
corundum, α-Al_2O_3, 379
coulomb integral, 52
coupled cluster (CC) method, 146
coupled cluster singles and doubles (CCSD) method, 146
coupling, L–S, 56
coupling, j–j, 62
covalent radii, 99, 109
crinkled tape, 744
croconate, 783
crystal classes (crystallographic point groups), 301
crystal engineering, 737
crystal field theory (CFT), 261
crystal form, 300
crystal radii of ions, 121
crystal system, 307, 310
cubane, 349, 351
cubane-like molecular skeleton, 174
cubic closest packing (ccp), 364
cubic groups, 177

Index

cubic molecular box, 763
cucurbituril, 772
cyanuric acid, 744
cyaphide, 599
cyclic conjugated polyenes, 221
cycloheptatrienyl ligand, 520
cyclopentadienyl complexes, of Ge, Sn and Pb, 549

de Broglie wavelength, 4
degenerate states, 19
degree of interpenetration, 769
degrees of freedom, 236
deltate, $C_3O_3^{2-}$, 524
density functional theory (DFT), 142
density of states function, 129
depolarization ratio, 238
depolarized vibrational band, 239
determinantal wavefunction, 48
Dewar benzene derivative, $C_{18}H_{14}$, 344, 346
diamond, cubic, 500
diamond, hexagonal, 500
diamondoid network, 768
diatomic molecules, 91
diazene (or diimide), 578
dibismuthene, 605
diethylbarbituric acid, 5,5-, 744
diffuse functions, 144
dihydrogen bond, X–H···H–E, 413
dihydrogen borate, 780
diiodides of Tm, Dy and Nd, 700
dinitrogen complex of samarium, 700
dinitrogen complexes, 564
dinitrogen, bonding to transition metals, 568
dinitrogen, oordination modes, 566
dinuclear complexes, 707
dioxygen carriers, 618
dioxygen metal complexes, 616
dioxygen species, 610
direct product, 185
disilene, 539
disilenes, 537
dispersion energy, 136
disrotatory process, 114
dissymmetric molecule, 171
disulfide, coordination modes, 632
dodecahedron, pentagonal, 177
dodecahedron, triangulated, 177
donor-acceptor complexes, of Ge, Sn and Pb, 554
double cage, 787
double group, 281
double zeta basis set, 143
dual nature of matter, 4

effective ionic radii, 122
effective nuclear charge, 54, 67

eigenfunction, 9
eigenvalue, 9
eighteen-electron rule, 288
electrical conductivity, 128
electrides, 447
electron affinity, 64, 66
electron cloud, 8
electron correlation, 47, 144
electronegativities, 133
electronegativity, 64, 67
electronic configuration, 48, 64
electronic configuration, equivalent, 60
electronic configuration, ground, 72
electronic spectra of some metal complexes, 274
electronic spectra of square planar complexes, 289
electronic transition, 53
electronic wavefunction, 6
electrostatic interaction, 135
enantiomorphous pairs, 314, 318
energy level diagram for octahedral complex, 268
energy level diagram for square planar complex, 291
energy states, 129
enthalpies of atomization of metallic elements, 133
equipoint (equivalent position), 318, 322
equivalent position, special, 319, 322
equivalent positions, assignment of atoms and groups to, 333, 336
equivalent positions, general, 318, 321
exchange integral, 52
excited-state chemistry, 113
exodentate multitopic ligand, 760
expectation value, 9

face-centered cubic (fcc), 364
Fermi energy, 129
ferric wheel, 753
f orbitals, cubic set, 296
f orbitals, general set, 295
f orbitals, nodal characteristics, 296
f orbitals, shapes, 295
ferrocene, 336, 337
first-order crystal field interactions, 270
fluorite, CaF_2, 370
form symbol, 301
fractional coordinates, 313, 317
free-electron model, 16
Friedel's law, 321, 323
frontier orbital theory, 113
frozen core approximation, 148
fullerene adducts, 513
fullerene coordination compounds, 513

fullerene oligomers and polymers, 515
fullerene supramolecular adducts, 515
fullerene-C_{50}, 505
fullerene-C_{60}, 502
fullerene-C_{70}, 505
fullerene-C_{78}, geometric isomers, 505
fullerenes, 502
fullerenes, as π-ligands, 513
fullerenes, endohedral, 516
fullerenes, hetero, 516
fullerenic compounds, 511
fulleride salts, 514

Gaussian functions, 143
Gaussian-n (Gn) methods, 151
glide plane, axial, 309, 313
glide plane, diagonal, 309, 313
glide plane, diamond, 309, 313
glide plane, double, 310, 313
gold clusters, 721
gold(I)-thallium(I) interaction, 729
gold-xenon complexes, 678
graph-set descriptor, 743
graphical symbols for symmetry elements, 310, 313
graphite, hexagonal, 501
graphite, rhombohedral, 502
Grignard reagents, 454
ground-state chemistry, 113
guanidinium sulfonate, 744
guanidinium-carbonate, 749
guanidinium-trimesate, 749
guidance, N''-cyano-N, N-diisopropyl, 341, 344

Hückel molecular orbital theory, 110
Hückel theory for cyclic conjugated polyenes, 221, 228
halite, NaCl, 366
halogen anions, homopolyatomic, 654
halogen oxides, binary, 662
halogen oxides, ternary, 664
halogen oxoacids, 666
halogen, charge-transfer complex, 660
halogens, 654
Hamiltonian operator, 9
helicate, 754
hemicarcerand, 753, 804
hemoglobin, 618
Hermann-Mauguin notation, 301
heteronuclear diatomic molecules, 96
hexachlorocyclophosphazene, $(PNCl_3)_2$, 344, 346
hexaferrocenylbenzene, 510
hexagonal closest packing (hcp), 375

hexamethylbenzene, 340, 341
hexamethylenetetramine N-oxide, 359, 362
hexamethylenetetramine, $(CH_2)_6N_4$, 355, 358
hexamethylenetetramine, quaternized derivatives, 359, 363
hexanuclear clusters, 715
hexanuclear metal string cationic complexes, 727
hexatriene, cyclization, 114
high spin complexes, 264
high-nuclearity clusters, 717
homonuclear diatomic molecules, 92, 94
Hund's rule, 60
hybrid orbitals, 100
hybrid orbitals, construction of, 232
hybridization schemes, 99, 232
hybridization schemes involving d orbitals, 234
hydrazine, 578
hydride complex, covalent metal, 417
hydride complex, high coordinate, 420
hydride complex, interstitial metal, 419
hydride complex, lanthanide, 421
hydride ion, 400
hydride ligand, five coordinate, 421
hydride ligand, four coordinate, 421
hydrido cluster complex, rhodium, 421
hydrogen bond, 401
hydrogen bond, geometry, 403
hydrogen bond, inverse, 415
hydrogen bond, non-conventional, 411
hydrogen bond, symmetrical, 406
hydrogen bond, X–H···H–M, 414
hydrogen bond, X–H···M, 412
hydrogen bond, X–H···π, 411
hydrogen bonding, generalized rules, 405
hydrogen bonding, universality and importance, 409
hydrogen bonds in organometallic compounds, 408
hydrogen bridged bond, 401
hydrogen carbonate dimer, 745
hydrogen molecular ion, 79, 82
hydrogen peroxide, 614
hydrogen, isotopes, 399
hydrogen, polymeric, 401
hydrogenic orbitals, 39
hydronium, 626
hydronium (or hydroxonium), 400
hydroxylamine, 578
hyperconjugation, 523

ices, high-pressure, 621
icosahedral groups, 177
identity element, 170
improper rotation axis, 169

Index

induction interaction, 136
infinite square tube, 761
infrared (IR) activity, 237
inorganic supermolecule, 757
interfacial angles, 300
interhalogen compounds, 657
interhalogen ions, 659
intermetalloid cluster, 554
intermetalloid clusters of As and Bi, 606
intermolecular energy, 137
intermolecular interaction, 138, 733
interstitial heteroatoms, 718
intrinsic semiconductors, 130
inverse crown compounds, 457
inverse crown ether, 456
inversion axis, n-fold, 301
inversion center, 169
iodine compounds, polyvalent organic, 668
ionic bond, 121
ionic liquid, 126
ionic radii, 121
ionic size, 121
ionization energy, 64
iron(II) phthalocyanine, 341, 344
iron-sulfur cluster $[Fe_4S_4(SC_6H_5)_4]^{2-}$, 344, 345
irreducible representation, 180
iso-bond valence, 721
iso-structural clusters, 721

Kapustinskii equation, 125
Keesom energy, 135
kinetic energy operator, 12
Koch cluster, 368
krypton compounds, 679

lanthanide complexes, cyclopentadienyl, 694
lanthanide complexes, structures and properties, 694
lanthanide compounds, magnetic properties, 687
lanthanide contraction, 682
lanthanide(II) reduction chemistry, 699
lanthanides, 682
lanthanides, oxidation states, 684
lanthanides, term symbols, 685
lanthanum square antiprism, 763
Laplacian operator, 30
Laporte's rule, 188
lattice (or space lattice), 305, 307
lattice energy, 124, 126
Laue groups (Laue classes), 325, 329
lead icosahedron, $[Pt@Pb_{12}]^{2-}$, 348, 350
Lennard-Jones 12-6 potential, 138
linear catenation, 494
linear combination of atomic orbitals, 77

linear combinations of ligand orbitals, AL_4 molecule with T_d symmetry, 228
linear homocatenated hexanuclear indium compound, 494
linear tape, 744
linear triatomic molecules, 99
lithium π complexes, 443
lithium alkoxides and aryloxides, 437
lithium amides, 438
lithium halide complexes, 439
lithium nitride, 435
lithium phosphide complexes, 441
lithium silicide complexes, 440
local spin density approximation (LSDA), 147
localized orbitals, 101
low spin complexes, 264
low-valent oxides and nitrides, 452

Madelung constant, 124
magnetic materials, 391
magnetic quantum number, 42
magnets, permanent, 393
many-electron atoms, 53
melamine, 744
melting points of chlorides salts, 127
melting points of metallic elements, 134
metaboric acid, 487
metal complex, X–H σ-bond coordination, 424
metal hydride, binary, 416
metal string molecules, 724
metal–metal bonds, 119
metal-based infinite chains and networks, 729
metal-metal bonds, 133
metal-organic framework (MOF), 799
metal-oxo complexes, 616
metallacarboranes with 14 and 15 vertices, 486
metallic bonds, 134
metallic radii, 132
metalloid cluster, 494
metalloid cluster, high-nuclearity, 553
methyllithium, $(CH_3Li)_4$, 357, 360
microstate, 58
Miller indices, 301
minimal basis set, 142
mixed gold-silver clusters, 724
mixed-valent silver(I,II) compound, 790
molecular Borromean link, 756
molecular capsule, 798
molecular container, 804
molecular hydrogen, coordinate bond, 402
molecular hydrogen, coordination compound, 422
molecular metal, 120
molecular necklace, 773
molecular orbital, 8
molecular orbital theory, 85

molecular orbital theory for octahedral complexes, 283
molecular orbital theory for square planar complexes, 289
molecular orbital theory, applications, 213
molecular orbital theory, essentials, 84
molecular panel, 765
molecular point group, 170
molecular polyhedra, 762
molecular recognition, 734
molecular skeleton, bond valence, 472
molecular term symbols, 189
molecular trefoil knot, 755
molecular vibrations, applications, 236
molecular wires, inorganic, 729
molten salt, 126
multi-configuration self-consistent field, 145
multiple bonding, between heavier Group 14 elements, 557
multiplicity, 318, 322
myoglobin, 618

nano-sized tubular section, 761
nanocapsule, 797, 798
nanocontainer, 804
nanoscale dodecahedron, 766
naphthacene radical cation, 520
naphthalene, 335, 337, 341, 343
nickel asenide, NiAs, 376
nickel wheel, 760
niobium monoxide, NbO, 368
nitric oxide, 573
nitride ion, N^{3-}, 561
nitrogen, catenation of, 564
noble gas complexes, organometallic, 676
nitrogen oxides, 569
nitrosyl complexes, 574
nitrosyl halides, 573
nitrosyl salts, 573
noble gas atom, as donor ligand, 676
noble gas compounds, structural chemistry, 670
node, 758
non-benzenoid aromatic compound, 781
non-benzenoid aromatic compounds, 511
nonbonding molecular orbital, 84
normal modes, 236
normal vibration, 236
normalization constant, 8
normalization integral, 80

octahedral complexes, π bonding, 285
octahedral complexes, σ bonding, 283
octathio[8]circulene, $C_{16}S_8$, 511
oligo(α-pyridyl)amido ligand, 725

orbital energy level diagram, 97
order of the group, 184
organoaluminum compounds, 490
organoantimony compounds, 607
organobismuth compounds, 607
organolithium compounds, 443
organometallic compounds of Group 2 elements, 454
organometallic compounds, Group 13, 491
organometallic compounds, Group 13 containing M–M bonds, 492
organometallic compounds, of heavier Group 14 elements, 549
organouranium(IV) complex, dinuclear, 524
organoxenon compounds, 677
Orgel diagrams, 268
orthogonality, 8
orthonormality, 8
overlap integral, 80
oxide superconductors, 389
oxides of cyclic poly-sulfur, 637
oxides of sulfur, 634
oxides, of Ge, Sn and Pb, 546
oxo ligand, 616
oxo-acids of nitrogen, 575
oxo-centered Zn_4 tetrahedron, 800
oxoacids of sulfur, 637
oxocarbon dianions, 781
oxoniobates, 369
oxonium, 626
oxygen, crystalline phases, 612
oxygen, singlet, 611
oxygenyl, 613
ozone, 614

P_n group, bond valence, 586
P_n groups, in metal complexes, 581
paramagnetic species, 95
particle in a one-dimensional box, 13
particle in a ring, 21
particle in a three-dimensional box, 17
particle in a triangle, 23
Pauli Exclusion Principle, 42
pentanitrogen cation, N_5^+, 564
pentanuclear metal clusters, 715
perovskite, $CaTiO_3$, 385
peroxide, 610
perturbation methods, 146
phenylenediethynides, 794
phosphaalkenes, 596
phosphaalkynes, coordination modes, 597
phosphazanes, 591
phosphazenes, cyclic, 594
phosphazenes (phosphonitrilic compounds), 593
phosphazenes, generalizations, 596

Index

phosphinidenes (phosphanylidenes), 596
phosphorus bonding types, 586
phosphorus, allotropic forms, 579
phosphorus, orthorhombic black, 344, 346
phosphorus, stereochemistry and bonding, 587
phosphorus–nitrogen compounds, 590
phosphorus-carbon compounds, 596
phosphorus-carbon compounds, π-coordination complexes, 600
physical properties of non-centrosymmetric materials, 304
Planck's constant, 4
point group, 170
point group identification, flow chart, 178
point group identification, projection diagrams, 178
polarization functions, 143
polyatomic anions, of Ge, Sn and Pb, 547
polyiodides, 654
polyphosphide anions, 581
polypyridylamine, 725
polyrotaxane, 753, 773
polyselenides, 648
polysulfide, coordination compounds, 632
polysulfides, 631
polytellurides, 648
potassium octachlorodirhenate dihydrate, $K_2[Re_2Cl_8]\cdot 2H_2O$, 341, 344
potential energy curves, 82
potential energy operator, 12
principal quantum number, 34
probability, 5, 39
probability density function, 6, 35
proper rotation, 167
pyrogallol[4]arene, 797

quadratic configuration interaction (QCI) method, 147
quadruple bond, 708
quantum mechanical operator, 15
quintuple bond, 712

racemic crystal, 338, 340
radial correlation, 47
radial equation, 42
radial function, 30, 33, 34
radial probability distribution function, 39
Raman (R) activity, 238
rare-earth cations, coordination geometries, 691
rare-earth metals, 682
rare-earth metals, crystalline forms, 683
rare-earth metals, halides, 689
rare-earth metals, organometallic compounds, 694

rare-earth metals, oxides, 688
rational indices, 301
reductive cyclotrimerization of CO, 524
relativistic effects, 64, 71, 72, 75, 133
resonance integral, 88
reticular design, 799
rhenium trioxide, ReO_3, 390
rhodizonate, 781
rhombic dodecahedron, 768
rhombohedral lattice, obverse setting, 306, 309
rhombohedral lattice, reverse setting, 306, 309
rings and clusters of Group 15 elements, 605
rock salt, NaCl, 366
rosette, 744
rosette ribbon, 745
rotation axis, 167
rotation-reflectionaxis, 169
rotaxane, 753
rule of mutual exclusion, 238
Russell-Saunders coupling, 56
Russell-Saunders terms, 56
rutile, TiO_2, 380

samarocene, decamethyl, 700
sandwich compound, metal monolayer, 520
Schönflies notaion, 301
Schrödinger equation, 18
Schrödinger equation for hydrogen atom, 29
screw axis, 309, 312
second-order crystal field interactions, 270
secondary building unit (SBU), 800
secular determinant, 90
secular equations, 90
selection rules in spectroscopy, 187
selenium allotropes, 644
selenium polyatomic cations, 644
selenium polymeric cations, 646
selenium stereochemistry, 649
self-consistent field (SCF) method, 54
semiconductors n-type, 130
semiconductors p-type, 130
semi-empirical methods, 142
separation constants, 30
sigma complexes, generalizations, 428
sigma-bond complexes, 428
silanes, 534
silatranes, 536
silicates, 540
silicates, general structural rules, 543
silicides, 534
silicon bridgehead, inverted tetrahedral bond configuration, 554
silicon dumbbell, unsubstituted, 554
silicon, octahedral hexa-coordinate, 536
silicon, stereochemistry, 535
silicon, structural chemistry, 533

silver acetylenediide, 531, 785
silver acetylide, 531, 785
silver carbide, 531
silver ethynide complexes, 792
silver iodide, AgI, 383
silver(I) arylethynide complexes, 796
silver(I) polyhedron, 785
silver-ethynide supramolecular synthons, 796
silyl anion radical, 539
silyl radical, 539
site of symmetry, 333
Slater-type orbital (STO), 143
sodium and potassium, π complexes, 445
sodium coordination complexes, 441
sodium thallide, 495
space group, polar, 376
space groups, 308, 312
space groups, commonly occurring, 319, 323
space groups, from systematic absences, 332
space groups, occurrence in crystals, 338, 340
space groups, symmetry diagrams, 317, 321
space-group diagrams, 316, 320
space-group symbol, Hermann-Mauguin nomenclature, 317, 321
space-group symbol, Schönflies notation, 317, 321
space-group symbols, 310, 316
space-group symmetry, application in crystal structure determination, 339, 341
spacer ligand, 758
spectrochemical series, 264
spectroscopic terms, 56, 60
spherical harmonics, 31
spin function, 48
spin multiplicity, 57
spin variable, 49
spin-orbit (L–S) coupling, 273
spin-orbit interaction, 62
spin-orbit interaction, first-order, 273
spin-orbit interaction, second-order, 273
spinel, inverse, 374
spinel, $MgAl_2O_4$, 359, 363, 373
split valence basis set, 143
splitting pattern of d orbitals in square planar complex, 266
splitting pattern of d orbitals in tetragonally distorted octahedral complex, 265
splitting pattern of f orbitals in tetrahedral crystal field, 298
splitting pattern of d orbitals in octahedral complex, 262
splitting pattern of d orbitals in tetrahedral complex, 262
splitting pattern of f orbitals in octahedral crystal field, 298
stacking, π–π, 796
standing wave, 10

stereochemistry of As, Sb and Bi, 602
stereographic projection, 303
steric strain, 522
strong crystal field, 279
strong field ligands, 266
structure-property correlation, 118
subnitrides, 453
subvalent halides, of heavier Group 14 elements, 544
sucrose, 340, 343
sulfide, coordination modes, 631
sulfur allotropes, 627
sulfur oxides, 635
sulfur, oxoacids, 637
sulfur, polyatomic cations, 630
sulfur-nitrogen compounds, 641
super inverse crown ether, 457
super-adamantoid cage, 765
supermolecule, 733
superoxide, 610
supramolecular capsule, 799
supramolecular chemistry, 733
supramolecular rosette, 745
supramolecular self-assembly, 735
supramolecular synthon, 737
symmetries and activities of the normal modes, 236
symmetry elements, 167
symmetry operations, 167
symmetry plane, 168
synthon X · · · X trimer, 738
systemic absences, 328, 329

Tanabe–Sugano diagrams, 274
tellurium polyatomic cations, 644
tellurium polymeric cations, 646
tellurium stereochemistry, 649
term symbols, 56
tetrahedral iron(II) host–guest complex, 762
tetranuclear clusters, 714
tetraphenylporphinato)iron(II),*meso*-, 347, 350
thallium naked anion clusters, 496
thermochemical radii, 126
thioallophanate, 3-, 781
tin metalloid cluster, high-nuclearity, 554
torsion angle, 315, 317
total wavefunctions of atomic orbitals with $n = 1$–4, 35
totally symmetric representation, 181
transition-metal clusters, 703
transitions d–d, 271
trefoil knot, 753
triazine, -*s*, 350, 349, 351
trimethylaluminum, Al_2Me_6, 343, 344

Index

trinuclear clusters, 713
triple zeta basis set, 143

Uncertainty Principle, 16
unit cell, 306–308
unit cell transformation, 307, 310
unstable inorganic/organic anions, stabilization of, 780
unsupported In-In single bonds, 494
uranocene, 341, 343, 519
urea, 345, 347
urea, non-stoichiometric inclusion compounds, 350, 352
urea/thiourea complexes, 780

valence band, 129, 130
valence bond theory, 86, 100
valence tautomer, 782
van der Waals forces, 135
van der Waals interaction, 135
van der Waals interactions, 135
van der Waals radii, 139
variational method, 46, 86
vector sum, 57
vertical detachment energies (VDEs), 155
vibrational spectra of benzene, 254
vibrational spectra of five-atom molecules, 244

vibrational spectra of linear molecules, 252
vibronic interaction in transition metal complexes, 294
virial theorem, 13

Wade's rules, 474
water, 620
water, liquid, 623
wave equation, 6
weak crystal field, 279
weak field ligands, 266
Weizmann-n (Wn) methods, 152
Woodward–Hoffmann rules, 113
Wyckoff notation, 318, 322

xenon difluoride, XeF_2, 346, 348
xenon fluorides, bonding description, 672
xenon fluorides, complexes of, 674
xenon, stereochemistry, 671

zeolites, 542
zinc sulfide, cubic form, zinc blende, sphalerite, 371
zinc sulfide, hexagonal form, wurtzite, 376
Zintl phases, 496
zircon, ZrO_2, 371